T0136553

Martin Etuge (1966–). The most outstanding plant collector of the Bakossi Mts. Brought up in northern Bakossi, at Mejelet, he was recruited there in 1986 by Duncan Thomas who trained him as a plant collector. Now field biologist and herbarium manager of Conservation and Research for Endangered Species (CRES) at Nyasoso, he has been a leading player in Royal Botanic Gardens, Kew–National Herbarium of Cameroon field expeditions for many years. *Kupea martinetugei, Dorstenia poinsettifolia* var. *etugeana* and *Psychotria martinetugei* are named in his honour. The portrait above was taken at Kupe Village in Dec. 1999 (Martin Cheek).

The Plants of Kupe, Mwanenguba and the Bakossi Mountains,
Cameroon

– A Conservation Checklist –

with introductory chapters on the physical environment, vegetation, endemics,
invasives, phytogeography and refugia, ethnobotany, bryophytes, the macrofungi,
the vertebrate fauna, the protected areas system, sacred groves,
and IUCN Red Data species

Martin Cheek, Benedict John Pollard,

Iain Darbyshire, Jean-Michel Onana and Chris Wild

Royal Botanic Gardens, Kew
National Herbarium of Cameroon

Published by the Royal Botanic Gardens, Kew

First published in 2004 by
Royal Botanic Gardens, Kew
Richmond, Surrey, TW9 3AB, UK
www.kew.org

ISBN 1-84246-074-9

Typeset by Lydia Elstone
Formatted by Kate Hardwick and Benedict John Pollard
Checked by Beth Lucas
Figures sized and adjusted by Matthew Sankey

Cover design by Jeff Eden, Media Resources,
Information Services Department,
Royal Botanic Gardens, Kew.

Printed in the UK by Cromwell Press

For information or to purchase all Kew titles please visit
www.kewbooks.com or email publishing@kew.org

Front cover: the critically endangered *Kupea martinetugei* Cheek & S.Williams, type species of a new African tribe of Triuridaceae. This species is only known from two sites, both in forest above Kupe village, on the southern slopes of Mt Kupe. Photograph of *Cheek* 10225, taken at Kupe village, Dec. 1999.

Rear cover: spot satellite image of Kupe, Mwanenguba and the Bakossi Mts, with checklist boundary shown in yellow and specimen collection sites in red. Note the large areas of intact vegetation still unsurveyed for plants in the western part of our checklist area. Produced by Justin Moat and Susana Baena from an image presented by Conservation and Research for Endangered Species (CRES), Zoological Society of San Diego, U.S.A.

CONTENTS

LIST OF FIGURES

Frontispiece: portrait of Martin Etuge

PREFACE

We can only be successful in preserving the rich variety of plant species if we know them. This book is a very valuable resource not only to the Bakossi people but to all those who share the same common goal of preserving the rich natural heritage of mankind.

Plants have always, and will continue to, play an important role in Bakossi tradition, culture and lifestyle. They are a source of income, the main component of our food, a vital resource for construction, medicine and recreational purposes.

It is a matter of pride for the Bakossi people, that this book reveals that our forests, rocks, crater lakes and rivers have such a wide variety of plants. It is also a matter of principle that this rich natural resource be used in a way that takes into account the tradition, needs and the benefits of the Bakossi people.

I would like to thank and encourage the authors of this book and look forward to its widespread dissemination.

Nhon ' Mbwog Alexander Ngome NTOKO
Paramount Chief of Bakossi

ACKNOWLEDGEMENTS

Firstly we thank the late Paramount Chief of Bakossi, Ntoko and his successor Alexander Ntoko for their interest and support of our plant survey work in Bakossi. In particular, the late Paramount Chief never failed to invite each of our survey teams to his palace at Nyasoso in order to discuss our progress and plans. His successor, based in Switzerland, has continued his father's interests and we are very grateful to him for the preface to this book. We also thank the people, elders and chiefs of the different villages at which we have had the honour to be welcomed, hosted and supported. Kupe Village, Kodmin, Nyasoso, Tape Etube, Ngomboaku, Enyandong, Edib, Nyale, Nyandong, Messaka and Nsoung have all hospitably provided us with accommodation, domestic helpers and guides, and shared with us their routes through the forest, and the traditional Bakossi names and uses of the plants that they helped us to study. Epie Ngome, Daniel Ajang, Max Ebong, Chief Abweh, and Daniel Lyonga were amongst the foremost of these ethnobotanical informants.

Chris and Liz Bowden, founders of the original BirdLife International Mt Kupe Forest Project (MKFP) at Nyasoso, are thanked for encouraging Martin Cheek to begin an exhaustive survey at Mt Kupe in 1993 and for helping us when we began in early 1995. Their staff, particularly Ebong Asabe and Pascale Ngome gave us every assistance. Chris Wild and Gerrit Vossebelt, co-managers of the resurrected MKFP under WWF-Cameroon, continued to give us support in the field from 1998 onwards, as did their staff particularly Pascale Ngome and Ewang. By 2001 Chris Wild had set up a new, research-focused organization, CRES (Conservation and Research for Endangered Species, formerly the Center for Reproduction of Endangered Species), at Nyasoso and he and his staff, particularly Bethan Morgan, were able to continue supporting our fieldwork. CRES provided the satellite image that is the basis of our back cover and Chris is credited as co-author of the book on the basis of being first author on many of the important introductory chapters that follow.

The Darwin Initiative grant to R.B.G., Kew has been the main single source of funds to our 'Conservation of the Plant Diversity of Western Cameroon' Project, from which this volume is one of the main outputs. We are extremely grateful for this assistance, without which the fieldwork, specimen identifications and species descriptions which form the basis of this book would not have been completed. A substantial part of the publication costs of this book were met by this grant. In particular we thank Valerie Richardson, Sarah Collins and Carrie Calhoun for their administration of this grant at DEFRA, as well as Des Bennett, financial accountant at R.B.G., Kew, for assistance in managing the finances at Kew. The Edinburgh Centre for Tropical Forestry, contracted by DEFRA to monitor the grant, are thanked for their constructive reviews of our project throughout its life.

The Earthwatch Institute, Europe have been major supporters of R.B.G., Kew's fieldwork in Cameroon since 1993. While the Darwin grant paid for National Herbarium field expenses 1999–2003, Earthwatch has provided all the other fieldwork funding in Bakossi, apart from an R.B.G., Kew Overseas Fieldwork Committee grant in 2003. We are grateful to all the Earthwatch African fellows and volunteers that gave their time to help us in the forests of Bakossi. From head office in Oxford, Sally Moyes was wonderful in co-ordinating and communicating information on volunteers, logistics, funding and publicity. We greatly regret her retirement to her garden and bees in Eynsham. Julian Laird, Gill Barker, Lucy Beresford-Stooke and Pamela Mackney all helped greatly in recruiting volunteers with a wide variety of backgrounds for us in Europe, as has Tania Taranovski in the U.S.A. We are particularly grateful to Robert Barrington for his successful fundraising and for joining our first expedition to the Bakossi area, at Mt Kupe in 1995, and to Tania for joining the most recent, in western Bakossi, in 2003. We also thank Robert Llewellyn Smith and Robin Reeve-Johnson of the Earthwatch African Fellows Programme for providing us with many botanists and other scientists from all over Africa to work with us at Kupe and Bakossi, and joining us at Enyandong in 2001.

Earthwatch funds allowed the employment of staff in Cameroon. We thank camp managers Madeleine Groves and Karen Sidwell (1995), Pascale Ngome and Henry Ekwoge (1996), Theophilus Ngwene (1998), Martin Mbong and Hoffmann Ngolle Ngolle (1999 to 2003).

Martin Etuge (see frontispiece) has been our head of protocol, kindly making arrangements in advance of our stays with the chiefs and elders of the villages where we have been hosted, and, for example, making arrangements for libation ceremonies on our behalves. His skills as a specimen collector are of the first order: many of the rarest Bakossi species were first discovered by him. All in all, he has made a huge contribution to advancing the botany of Bakossi. We also thank his assistants, Freddy Epie and Edmondo Njume, for their help with collecting plants and processing specimens, and Peter Akume, Dansala Njire Japhet Wain and Godlove as our LandRover drivers and mechanics.

We thank all the local staff who helped with cleaning, cooking, carrying wood and water, and washing clothes during our stay in Bakossi, and as our guides into the forests.

The Global Environment Facility part-funded our 1996 expedition by support for the participation of the National Herbarium's involvement, through the Cameroon Biodiversity Conservation and Management Programme.

At the National Herbarium of Cameroon, Yaoundé, we thank the present leader, Gaston Achoundong, the past leader, Benoit Satabié, Elvire Biye, researcher, and technicians Jean-Paul Ghogue, Fulbert Tadjouteu Victor Nana, Nana Felicité and Bartholomie Tchiengue, all of whom have joined us in fieldwork at Bakossi. Boniface Tadadjeu, Laurent Kemeuzeu (drivers) and Jorobabel Moussa (general assistant) are also thanked.

We thank the staff of the Missouri Botanical Gardens, U.S.A., in particular Jim Miller, for the gift of about 11,000 ex TROPICOS electronic specimen records mostly from the Cameroon area (see Database chapter). Although some are represented by duplicates at Kew, many are not and for these we have put faith in the determinations of Missouri staff botanist Roy Gereau, and guest researchers at MBG, such as Ake Assi. Amongst these 11,000 collections, those of Duncan Thomas and associates in 1986 and 1987 who were then in MBG employ have been of very great importance in making this checklist as complete and representative as it is. This is because Thomas *et al.* collected in several parts of the checklist area which we have not yet visited, notably the lowland forest on the Kumba–Mamfe road and at e.g. Etam and Ngombombeng. They also made important collections from submontane forest habitat W of Bangem including the first collections of some of the endemic Bakossi taxa. These MBG-Thomas collections probably account uniquely for 10–20% of the taxa in this book, and help broaden the range of about the same percentage of other taxa, perhaps more.

Louis Zapfack (and his graduate students Sonwa Denis and Placide Simo) of the University of Yaoundé I are thanked for their assistance with fieldwork. Zapfack's urbane cordiality has always been a great asset to us in the field. Many volunteers have benefited from his leadership of field teams, including Ben Pollard, on his first expedition in Cameroon, nearly ten years ago. Professor Bonaventure Sonké and his graduate student Charlemagne are also thanked for their contributions in the field.

In Douala we thank Constance, Genevieve Fauré and her successor Valérie de Tailly, of the British Consulate and the Brothers of La Procure Générale des missions Catholiques, for help and hospitality.

Craig Hilton-Taylor, IUCN officer for Red Data, based at the World Conservation Monitoring Centre in Cambridge, U.K., is thanked for reviewing the Red Data assessments, and providing us with the IUCN RedListing criteria tables for incorporation in the Red Data taxa chapter.

At the Royal Botanic Gardens, Kew we thank Lydia Elstone for help in typesetting so much of this book, and Kate Hardwick and Laura Pleasants for assistance in that department. The Publications Committee agreed to contribute towards the publication costs and arranged review of the manuscript. John Harris and Ruth Linklater are thanked for guidance on producing the pdf file from which this book was digitally produced, and for arranging scanning of images and design work. Jeff Eden, Christine Beard and Matthew Sankey are thanked for their various contributions in the design of the cover, and arrangement of the colour plates and the figures. Beth Lucas has been most constructive and helpful in the course of preparing this volume for publication, and given generously of her time, saving a considerable amount of ours in the process. Ian Turner is thanked for his patient review of the manuscript and the valuable comments he made on its content. Mike Gilbert helped enormously in writing a macro to simplify construction of the index. William Milliken is thanked for reviewing the ethnobotany chapter.

At Kew we also thank Simon Owens, Keeper of the Herbarium, who has long supported our work in Cameroon, as has Daniela Zappi, Assistant Keeper for Regional teams, who has also championed our work, and, moreover joined our expedition to Nyandong in 2003. Eimear Nic Lughadha and Peter Crane have also been consistently supportive of our efforts in Cameroon. We also thank Mike Lock and Di Bridson for their support and guidance in their roles as the previous two Assistant Keepers under whom this project was run. Justin Moat is thanked for preparing the Satellite image on the back cover of this volume. Tivvy Harvey is thanked for her editorial contributions and improvements, and for her considerable time spent entering data, particularly towards the Leguminosae account. Karen Sidwell, Suzanne White, Julian Stratton and Penelope Doyle are thanked for their efforts in data entry and data management.

Determinations of the specimens gathered in the course of our fieldwork were made by ourselves and by contributions from the following botanists, often world experts in their fields. We sincerely thank them all, and their institutions, for their work, often done without reservation, towards producing the names which are used in this book. Herbaria aconyms are provided and follow Holmgren *et al.* (1991), except for KUPE, which refers to the reference set of specimens held at Nyasoso on Mt Kupe.

G.Achoundong (YA) (Violaceae); H.Atkins (E) (Zingiberaceae); P.C.Boyce (formerly K) (Araceae); the late J.J.Bos (formerly WAG) (Dracaenaceae); F.J.Breteler (WAG) (Anacardiaceae & Dichapetalaceae); B.L.Burtt (E) (Gesneriaceae); D.Champluvier (BR) (*Brachystephanus*, Acanthaceae); C.C.Davis (H) (Malpighiaceae); J.J.F.E. de Wilde (WAG) (Begoniaceae); M.Etuge (KUPE) (Sapindaceae); R.B.Faden (US) (Commelinaceae); F.Gonzalez (Aristolochiaceae); D.J.Harris (E) (Irvingiaceae & Zingiberaceae); C.C.H.Jongkind (WAG) (Combretaceae & Connaraceae); Z.Kaplan (Institute of Botany, Průhonice, Czech Republic) (Potamogetonaceae); H.Kennedy (UBC) (Marantaceae); A.J.M.Leeuwenberg (WAG) (Apocynaceae & Loganiaceae); K.Å.Lye (NLH) (Cyperaceae); G.Mathieu (Gent, Belgium) (Piperaceae); G.Mwachala (EA) (Dracaenaceae); I.Nordal (O) (Anthericaceae, Amaryllidaceae); M.Parsons (Florida State University) (Aristolochiaceae); V.Plana (E) (Begoniaceae); A.D.Poulsen (E) (Anthericaceae); Sebsebe Demissew (ETH) (Asparagaceae & *Maytenus*, Celastraceae); B.Sonké (University of Yaoundé I, Cameroon) (*Aulacocalyx, Oxyanthus & Rothmannia,* Rubiaceae); M.S.M.Sosef (WAG) (Begoniaceae & Ochnaceae); R.D.Stone (University of California, Berkeley, USA) (*Memecylon & Warneckea*, Melastomataceae); T.C.H.Sunderland (African Rattan Programme, Limbe Botanic Garden, Cameroon) (Palmae); J.J.Symoens (Gent, Belgium) (Potamogetonaceae); M.Thulin (UPS) (Campanulaceae).

At R.B.G., Kew we thank K.B.Vollesen (Acanthaceae); C.C.Townsend (Amaranthaceae and Mosses); G.Gosline (Annonaceae & Celastraceae, Ebenaceae, Icacinaceae, Olacaceae, Opiliaceae); D.G.Frodin (Araliaceae); D.J.Goyder (Asclepiadaceae); S.Bidgood (Bignoniaceae); D.Zappi (Cactaceae); Y.B.Harvey (Campanulaceae & Gentianaceae, Myrsinaceae, Sapotaceae); G.T.Prance & C.A.Sothers (Chrysobalanaceae); H.J.Beentje (Compositae); P.Hoffmann (*Phyllanthus*, Euphorbiaceae); P.Bhandol (Gesneriaceae); M.V.Norup (Lauraceae); B.A.Mackinder (Leguminosae); R.M.Polhill (Loranthaceae & Viscaceae); E.Woodgyer (Melastomataceae); N.A.Brummitt (Monimiaceae); T.M.A.Utteridge (Myrsinaceae: *Maesa*); P.S.Green (Oleaceae); S.E.Dawson & D.Bridson (Rubiaceae); T.M.Heller (Rutaceae); C.M.Wilmot-Dear (Urticaceae); P.Wilkin (Dioscoreaceae); S.M.Phillips (Eriocaulaceae); T.A.Cope & P.Doyle (Gramineae); P.J.Cribb (Orchidaceae); W.J.Baker & J.Dransfield (Palmae); P.J.Edwards (Pteridophyta) and P.R.Roberts (Fungi) for their contributions.

We are most grateful to Stuart Cable of R.B.G., Kew, for persuading the first author to make a reconnaissance trip to Mt Kupe in 1993, for co-leading and also leading expeditions there in 1995 and 1996, and for making identifications of resulting specimens in 1996–1997. For the Mount Cameroon Project, in 1997 and 1998, he originated the species database, from which many of the plant names, both accepted names and synonyms, and references in this volume are derived.

Last, but not least, we all thank George Gosline who first visited Cameroon as an Earthwatch volunteer on the 1995 expedition to Mt Kupe, and joined most of the subsequent expeditions to Kupe and Bakossiland, eventually co-leading the 2001 expedition to Enyandong. George has been instrumental in the considerable development and enhancement of the western Cameroon specimen and species databases. He also enabled us to print off data labels whilst still in the field, a real time-saver. He has since developed a system by which data could be exported from the Access database, using XML (Extended Markup Language) and XSLT (Extended Stylesheet Language) software, into its presently used Microsoft Word format.

BJP concurs with the above acknowledgements, and is grateful for the generosity of Halldóra Blair. Funding for his participation on the 10-week 1998 expedition to Kupe, Mwanenguba and Bakossi Mts was generously provided by The Merlin Trust and The Really Useful Group, for which he thanks Valerie Finnis and Brigadier A.B.D.Gurdon respectively. He also thanks everyone who attended the fundraising evening for his participation in that expedition. Alice Sanders of Olympus Optical Co. (UK) Ltd. kindly loaned photographic equipment. Many of the resulting images are being published in this volume for the first time. Syd Hill and Vernon Hales of Elstree School are thanked for their support of his 2001 expedition to Bakossiland. Thanks also to Philipp Schwalber.

NEW NAMES

The following plant names are published in this volume for the first time:

ACANTHACEAE

TILIACEAE

FOREWORD

Martin Cheek

Herbarium, Royal Botanic Gardens, Kew, Surrey, TW9 3AE, U.K.

Kupe, Mwanenguba and Bakossi Mts lie toward the southern end of the Cameroon Highlands that run from Bioko and Mt Cameroon in the South, northwards to Tchabal Mbabo and then continue eastwards as the Adamaoua area (see Fig. 1).

This is the third in a series of 'plant conservation checklists'; its predecessors dealt with the plants of Mt Cameroon (Cable & Cheek 1998) and of Mt Oku and the Ijim Ridge (Cheek, Onana & Pollard 2000). In contrast to these, the greater part of our current area, the Bakossi Mts, has been mostly unexplored by botanists. Indeed, inspection of Fig. 1 shows no mention of the Bakossi Mts, although it is the standard map for discussion of the biology of the Cameroon Highlands. This is despite the fact that, with the adjoining Rumpi Hills, they constitute what is possibly the largest intact pristine block of submontane forest, 800–1900(–2000) m alt., in Africa.

It is this submontane forest, together with the adjoining lowland forest, that contributes the bulk of the remarkably high number of taxa endemic to our checklist area (82), those threatened with extinction (232), and the overall total (2412), emphasising the extraordinary biodiversity of the Bakossi tribal area.

A major purpose of this book is to enable identification of the taxa within the checklist area. It is our presumption that it is likely that anyone wishing to identify plants from this part of Cameroon should have access to the Flora of West Tropical Africa, and perhaps the Flore du Cameroun. With these floras at hand, they should be able to determine most specimens at least to genus. We have included short descriptions for each published taxon to assist with identification to the species level and below. Although not necessarily diagnostic, these descriptions should prove invaluable to the field researcher. In particular, this book will help with identification of those taxa threatened with extinction, which must be the highest priorities for conservation. For each of these taxa, details to aid their monitoring and management are given in a separate Red Data chapter, which includes 16 pages of colour plates of a large number of those species. There are also nine line-drawings of threatened taxa. It is hoped that this information will aid the long-term survival of these threatened taxa. It is noted here that perhaps the scope of this publication is more than that of a true checklist, but we maintain the title 'A Conservation Checklist', as this is part of an onging series of conservation products bearing that same description.

A further aim of the book has been to record traditional Bakossi names and uses of the indigenous plants, since, with increased urbanization and westernization, there is a danger that this information will be lost.

The extent to which the species of this forest have lain unknown to science is illustrated by the University of Aberdeen undergraduate survey of 1993. In the course of a month's gatherings between Nyasoso and the summit of Mt Kupe, 98 specimens were made, of which ten were new to science, only two of which have so far been published. The route mentioned is the most visited by botanists in the whole checklist area. In 1998, in a previously unsurveyed area of the Bakossi Mts, and with more experience, I gathered about 14 novelties per hundred specimens.

Although our knowledge of the plants of Kupe–Bakossi has grown enormously, it is still very far from being complete. 123 species were listed from Mt Kupe (Stone 1993) which we increase twentyfold here, albeit from a larger area. However, great tracts of forest still remain unexplored (see Vegetation chapter) and are likely to yield hundreds of species new to the Bakossi area when they are researched.

More work is also needed to complete full scientific publication of many of the new species listed in this book. Of the 82 species that appear to be endemic to the checklist area, eight were published previous to our survey work which began in 1995, and 13 since then. While several of the remaining 61 endemic species are in various states of publication, many lack sufficient representation of either material or flowering/fruiting stages, for them to be published; more specimens are needed.

We hope that this book will encourage more people to visit this beautiful part of Cameroon, and to identify the plants that they meet, whether for professional (e.g. agricultural, silvicultural or conservation) or for touristic purposes.

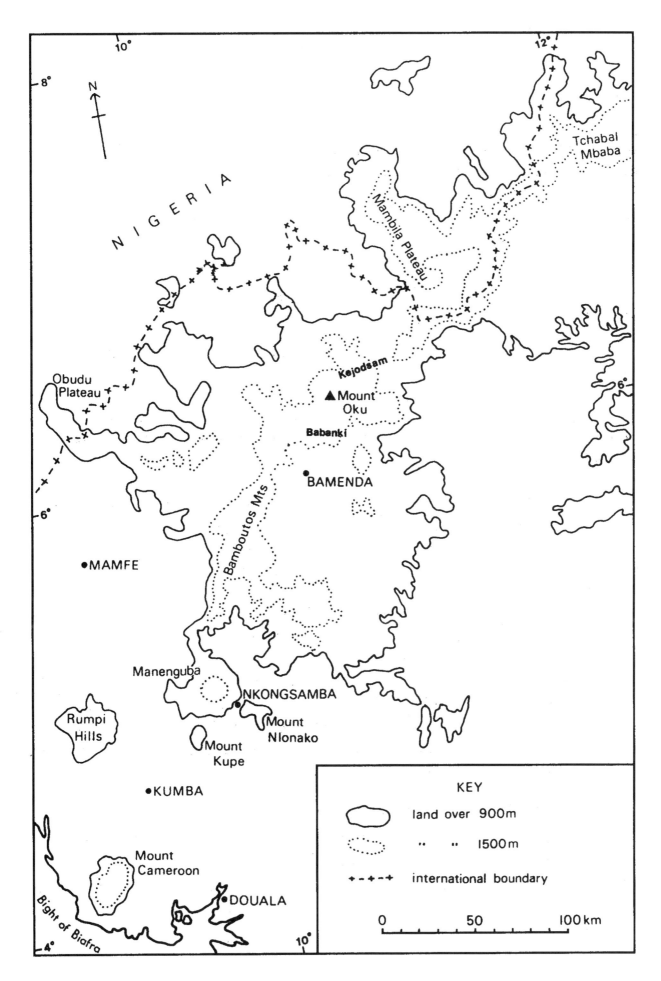

FIG. 1. THE CAMEROON HIGHLANDS.
Reproduced with the permission of BirdLife International from Stuart (1986).

7

THE CHECKLIST AREA

Martin Cheek

Herbarium, Royal Botanic Gardens, Kew, Surrey, U.K., TW9 3AE

The Kupe, Mwanenguba and Bakossi Mts area, for the purpose of this checklist, includes Mt Kupe, the Bakossi Mts and the lowlands that surround them, and the upper part of the Mwanenguba Massif (see Fig. 2). The easternmost boundary is longitude 9°55'E, just excluding Nkongsamba and Mt Nlonako. The boundary continues south-eastwards, following the road from Nkongsamba to Loum, along the Dibombe river, from the village of Ngwa, and so passes through, or near, Eboné, Manengole, Manjo, Mantem I, Kola, Lala, Nloe until the junction with the Loum-Kumba road is reached. We then take this road as the Southern Boundary through Ebonji and Etam, but include also the southern extensions of the Mungo River Forest Reserve and the Bakossi Forest Reserve. Near Kumba, where the Loum road has its junction with the Kumba-Mamfe road, our boundary moves northwards, following the Mamfe road, through Ikiliwindi, Baduma, Kurume, Ndoi Bakundu, Konye, Dikome-Bafaw, Supe, Kumbe, Wone, Babensi, Ekita and Manyemen until, before Nguti is reached, latitude 5°14'N. This latitude is taken as the northernmost part of the boundary, eastwards as far as the Mbu river. The Mbu river then forms the boundary which moves southeastwards to Enyandong and Bangem. At Bangem the boundary again moves due eastward, following latitude 5°5'N until finally it meets the easternmost boundary, longitude 9°55'E.

This boundary encompasses an area of approximately 2390 km²; it was chosen because:

1. It more or less agrees with the Bakossi tribal area.
2. It includes Kupe, Mwanenguba and the Bakossi Mts.
3. The fact that roads are used as the main boundaries helps give precision on deciding whether a specimen is 'in or out' of the checklist area.
4. It encompasses a fairly 'natural unit', the boundaries mostly comprising valley bottoms that are postulated as partial barriers to the movement of submontane species (see chapter on Phytogeography and refugia).

However, the species listed in this checklist almost all come from a much smaller area within this boundary (see rear cover). These are the W, N and S sides of Mt Kupe (the most intensively collected part of the whole) and from locations along the Jide valley and along a W–E line linking Nyandong in western Bakossi, with Nyale, Mwanzum, Kodmin in the heart of the Bakossi Mts, with Mwambong in the Jide Valley. Apart from some sampling in the Mwanenguba caldera, at Enyandong in northern Bakossi, and the Loum, Bakossi and Mungo forest reserves in the south, large areas have not been visited by us. It is fortunate that some of these gaps have been at least partly filled by the work of Duncan Thomas and his associates in 1986–1987, collecting along the western boundary, and near Bangem. On a smaller scale, about thirty other collectors have contributed a few specimens from the checklist area.

Within our checklist area, only four protected areas were gazetted, until recently. These were all production forest reserves:

Bakossi F.R.
Loum F.R.
Manehas F.R.
Mungo River F.R.

The new protected areas, still in the process of being created, are discussed in the chapter entitled 'The Protected Areas System'.

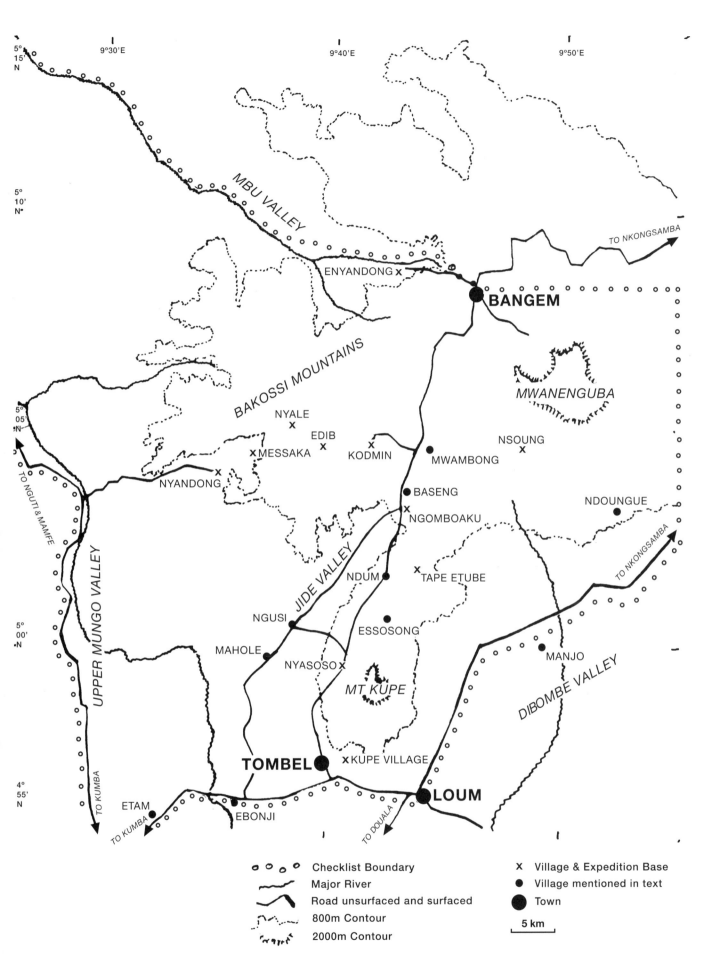

FIG. 2. THE CHECKLIST AREA.
Map of Kupe, Mwanenguba & the Bakossi Mts.

9

THE EVOLUTION OF THIS CHECKLIST

Martin Cheek & Benedict John Pollard

Herbarium, Royal Botanic Gardens, Kew, Surrey, TW9 3AE, U.K.

This conservation checklist was planned as one of the key products in the first proposal made in 1997, to the Darwin Initiative of the UK government's Department of Environment, Food and Rural Affairs (DEFRA) for a grant towards what later became the 'Conservation of the Plant Diversity of Western Cameroon' project. By this time, Stuart Cable and Martin Cheek had already led several expeditions to Mt Kupe in 1995 and 1996, sponsored by Earthwatch. In 1998, with the prospect of the Darwin Grant looking favourable, Martin Cheek was able to extend this survey work to the neighbouring and much larger and more poorly known area of the Bakossi Mts. This was with the encouragement of Chris Wild, the assistance of the National Herbarium of Cameroon (sponsored by GEF-Cameroon) and the continued support of Earthwatch volunteers. Expeditions continued in 1999, 2001 and 2003, all with Darwin Initiative and Earthwatch support. The rationale and methodology used, from targeting areas to collecting, processing and databasing specimens, obtaining the necessary permits and organising specimen export to Kew, has already been documented at length in Cheek & Cable (1997). Historic specimens were databased at Kew, and additional records sourced from FWTA and Flore du Cameroun. The ordering (1 to 5) of authors on the front cover recognises and reflects the respective levels of their contributions towards this publication.

Ben Pollard was one of the volunteers who joined the expeditions to the Bakossi Mts in 1998, following on from his participation on the 1996 expedition to Mt Oku (see Cheek *et al.* 2000). When eventually, in September 1999, the Darwin grant was awarded, he was the successful candidate for the post of Western Cameroon Darwin Initiative Officer at R.B.G., Kew. Taking up his post in early 2000, he was to be the main dynamo of the Conservation of the Plant Diversity of Western Cameroon project for two and a half years, organizing the bundles of specimens, databasing and barcoding new determinations, as well as writing accounts for labiates, menisperms, and many other smaller families. Notably, before he had concluded his work in 2002, he had already produced a draft of the Monocotyledons for this checklist. To do this he worked closely with specialists Phil Cribb and Kåre Lye to ensure completion of the accounts for the large families Orchidaceae (187 taxa) and Cyperaceae (46 taxa). He was also initially co-leader, and then leader, of the 2001 expeditions to northern Bakossi. He was instrumental in helping devise the current format of this checklist, modified from that used by Cheek *et al.* (2000).

Jean-Michel Onana, who was with us in the Bakossi Mts in 1998 and 2003 as leader of the National Herbarium field team, made two visits to R.B.G., Kew, funded by the Darwin grant. During these visits he was able to authoritatively name all the material of the major groups Compositae and Pteridophytes with the assistance of Kew specialists Henk Beentje and Peter Edwards respectively. In addition he was able to produce the account for the Burseraceae, his own specialism.

In January of 2003, Iain Darbyshire joined the core staff of the Wet Tropics of Africa team at R.B.G., Kew and ably carried on where Ben Pollard had left off the previous August, in addition to his curatorial duties. He co-led with Jean-Michel Onana and MC our most recent expedition to Bakossi, at Nyandong, in April 2003. He was heavily involved in producing family accounts, most notably the Acanthaceae, with Kaj Vollesen.

Chris Wild, currently the project manager of San Diego Zoo's Center for the Reproduction of Endangered Species at Nyasoso, Mt Kupe, has been involved with taxonomic and ecological research in the Bakossi tribal area for over a decade, working predominantly on snakes and other reptiles. He has also been involved in managing research programmes into a wide variety of other faunal groups, including drills, prosimians (and other primates), birds, fish, crocodilians, and amphibians, and is currently undertaking a study of Bakossi lore. He has consistently championed the botanical research reported in this volume, and been an instrumental part in promoting and developing strategies fundamental to our surveying and inventory programme.

George Gosline joined our first expedition to Bakossiland as a volunteer in early 1995. After taking up residence at R.B.G., Kew later that year, he joined every other expedition to the area, with the exception of that in 2003, becoming co-leader. In 1996 he instigated, supervised and led, as he has done ever since, the procedure of databasing our specimens in the field. In 1998 he reworked the specimen database, then in BRAHMS, into its current, more accessible and user-friendly, ACCESS format (see the specimen database chapter). Since then he has been its overall manager and developer. It was he that manipulated and loaded the c. 11650 TROPICOS specimen records sent to Kew by Missouri Botanical Garden, which solely account for about 10–20% of the taxon records in this book. Finally he has identified the material and described the taxa for several fair-sized families in this book, notably Annonaceae, Celastraceae, Ebenaceae, Icacinaceae, Olacaceae and Opiliaceae.

THE R.B.G., KEW WESTERN CAMEROON SPECIMEN DATABASE

George Gosline

Herbarium, Royal Botanic Gardens, Kew, Surrey, TW9 3AE, U.K.

The data for the Kupe, Mwanenguba, Bakossi Mts checklist are held on a database that contains records of over 64000 specimens, mostly collected in the western Cameroon region (SW and NW Provinces). The database has been accumulated from a number of sources, starting in 1993. The primary sources are, with approximate total numbers of specimens:

- 3250 specimen citations from the Flora of West Tropical Africa (2nd edition).
- 18350 specimens collected by the Mount Cameroon Project, or by Kew expeditions to Mt Cameroon.
- 19500 specimens collected by Earthwatch expeditions (1993–) under the direction of Dr Martin Cheek.
- 11650 specimen data records from the Missouri Botanic Garden (MO) TROPICOS database.
- 11500 historical specimens in the Kew herbarium (data entry sponsored by the Darwin Initiative).

We consider that about 35000 of these specimens have reliable determinations, either by family specialists, or based on verification at Kew; many await identification.

Of the total number of specimens on the database, just under 14000 originate within the area of this checklist, and approximately 9000 of these, with good determinations, are the basis of this checklist. Some of the determinations made at MO have not been accepted for this checklist, as we have not seen the specimens, and in several plant families identifications have been made by non-specialists.

The actual computer database was established at R.B.G., Kew for the Mount Cameroon project, with funding from the UK Government's Overseas Development Agency (ODA), now known as the Department for International Development (DFID). This was using BRAHMS software (Botanical Research and Herbarium Management System), developed by Denis Filer at Oxford (see www.brahms.co.uk). The data, using the same file structure, were imported into a Microsoft Access database in 1998.

The checklist that comprises the major part of this document is generated from the database, using a rather baroque process that reads the data using Microsoft ADO and formats them into an Extensible Markup Language (XML) document, using a program written in the Python scripting language. The XML is then transformed in HyperText Markup Language (HTML) documents, using Extensible Stylesheet Language Transformations (XSLTs). The HTML is then read into Microsoft Word and saved as a Word document that makes extensive use of styles, allowing easy changes to the formatting. The complexity of this system is largely due to the author's interest in exploring XML technologies and tools. The BRAHMS system, for example, is able to directly generate various checklist formats directly from its database.

The database has successfully served the needs of the various projects and expeditions in the Cameroon Botanical Surveys and Inventories Programme for over a decade. The data will eventually be incorporated into the Kew Herbarium Catalogue, HerbCat. The success of the database can be attributed to a number of factors, including:

- Persistent attention to quality of data. Databases are particularly subject to entropy as data are incorporated.
- Standards for the use of data fields.
- Care in the addition of data from outside sources.
- The solid design of the file structure of the BRAHMS system.
- The flexibility of a good relational database system when new data fields or structures are required.
- Data entry in the field. There is no doubt that the use of portable computers in the field and supervised data entry have made possible the impressive numbers of specimens collected and processed by the Earthwatch teams.
- Not treating the database as a prototype, but limiting the scope to the immediate needs of the projects using it.

STATISTICAL SUMMARY

Iain Darbyshire

Herbarium, Royal Botanic Gardens Kew, Surrey, TW9 3AE, U.K.

INTRODUCTION

This chapter provides a statistical breakdown of the vascular plant diversity recorded within the checklist area, reviewing in turn total diversity and diversity at the family and generic level. Comparison to other botanical checklists from the Cameroon Highlands is made.

TOTAL VASCULAR PLANT DIVERSITY RECORDED

Group	Taxa with accepted name	Species with accepted name	Undescribed new taxa	Taxa of uncertain identification	Total taxa
Dicotyledonae	1592	1550	65	147	1804
Monocotyledonae	440	425	5	19	464
Pinopsida	3	3	0	0	3
Gnetopsida	1	1	0	0	1
Lycopsida	10	9	0	0	10
Filicopsida	121	119	1	8	130
TOTAL	2167	2107	71	174	2412

The above table provides a breakdown of the total vascular plant diversity recorded in the checklist area. Note that Bryophytes are not included (see separate chapter: Bryophytes & Water Capture). In addition to the 2412 taxa recorded, a further 35 entities are listed in the checklist but are here omitted as they represent incomplete material which could not be fully determined (e.g. *Dissotis rotundifolia* vel *D. prostrata*) and thus may not represent taxa different to those otherwise recorded.

The discrepancy between 'taxa with accepted name' and 'species with accepted name' is due to the inclusion of infraspecific taxa within the former when more than one infraspecific taxon occurs for a single species; this was recorded on 56 occasions. 'Taxa of uncertain identification' represents either potentially new species for which more material is required to be certain, or taxa named tentatively, usually due to lack of sufficient material or to naming difficulties in groups where species delimitation is difficult.

Within these figures, 18 dicotyledon taxa and four monocotyledon taxa are cultivated as ornamentals, and 33 dicotyledons, nine monocotyledons and two gymnosperms are cultivated for their uses; further information on these is listed in the chapter on ethnobotany. These taxa are often now established in anthropic vegetation types, often closely associated with human settlement. A further 45 dicotyledons and two monocotyledons are alien to Cameroon but are fully naturalised in a variety of habitats. Of the remainder, approximately 40 taxa are of doubtful origin, either being pantropical or cosmopolitan weeds for which their geographical origin is unknown, or being taxa native to Cameroon but also often cultivated, their status in the checklist area therefore being unclear.

A comparison of total species numbers with other areas in the Cameroon Highlands is presented below, being expanded from those presented in Cable & Cheek (1998) and Cheek *et al.* (2000).

Checklist site	Area (km^2)	No. of taxa	Taxa / km^2
Bioko	2018	842	0.42
Korup Project Area	2510	1693	0.67
Mt Cameroon Area	2700	2435	0.90
Kupe-Mwanenguba-Bakossi	2390	2412	1.01
Mt Oku & Ijim Ridge Area	1550	920	0.59

These data indicate that the current checklist area contains a very high level of plant diversity, comparable with that of Mt Cameroon, though with a significantly higher count per km². Such high diversity is not unexpected in light of these two areas being recognised as locations for Pleistocene forest microrefugia, a theory discussed further in the chapter on Phytogeography. Several reasons are postulated for the higher total species diversity at Mt Cameroon. First, the Mt Cameroon checklist (Cable & Cheek 1998) covered a larger area (see table above), secondly it covers a range of habitats not found within the current checklist area, including littoral forest, mangroves and extensive montane grassland, all of which yield taxa not encountered in Kupe-Bakossi. Thirdly, the Mt Cameroon area is better explored botanically than the current checklist area. Further exploration of some of the, as yet unexplored, parts of the current checklist area may yield many extra taxa, particularly the species-rich lowland forests. Finally, several groups of taxa are believed to be under-represented in the current checklist, either because their favoured habitats have not been visited or because they have been under-collected, as no family specialist has visited the region. Of most note is the Gramineae family, with further exploration of the montane grasslands for grass taxa likely to add many more species to the current total of 60 species, compared to 110 on Mt Cameroon and 95 on Mt Oku.

Postscript: Since writing the above, two redundant *Asplenium* species names have been removed from the checklist.

ANALYSIS OF PLANT FAMILIES

The ten most species-rich families in the checklist area are listed in the table below.

Family	Taxa with accepted name	Species with accepted name	Undescribed new taxa	Taxa of uncertain identification	Total taxa	% of total taxa in checklist
Rubiaceae	213	205	39	27	279	11.6
Orchidaceae	183	175	1	0	184	7.6
Leguminosae	126	123	1	10	137	5.7
Euphorbiaceae	99	94	2	12	113	4.7
Acanthaceae	79	79	2	0	81	3.4
Compositae	61	61	0	2	63	2.6
Gramineae	59	59	0	1	60	2.5
Apocynaceae	45	45	1	6	52	2.2
Labiatae	50	49	1	0	51	2.1
Moraceae	46	42	0	4	50	2.1
TOTAL	**961**	**932**	**47**	**62**	**1070**	-
% of total taxa in checklist	44.39	44.23	66.2	35.06	44.41	-

The predominance of Rubiaceae reflects the fact that the submontane forests, rich in Rubiaceae taxa, represent both the largest and most widely collected vegetation type in the checklist area. The top eight families agree with those from Mt Cameroon, a region similarly dominated by wet forest types, with the exception that Gramineae is recorded as the fourth species-rich family on Mt Cameroon. Orchidaceae, Leguminosae, Euphorbiaceae, Acanthaceae and Apocynaceae are all important components of wet forest in the Guineo-Congolian vegetation, thus their high representation here is again due to the predominance of this habitat. The lower number of Gramineae species in the current checklist area is due in part to the smaller area of montane grassland here, though, as discussed previously, is also an artefact of the apparent under-collection of specimens from this family. The only other difference to Mt Cameroon is the presence of Labiatae as the ninth most diverse family here, a position occupied by Annonaceae on Mt Cameroon; this is an artefact of the inclusion of the genera *Clerodendrum*, *Premna* and *Vitex* within Labiatae in the current checklist, which were treated under Verbenaceae in the Mt Cameroon publication. Cyperaceae, the most diverse family recorded in the Mt Oku checklist area, falls just outside the top ten here; this is indicative of the much smaller area of swamp and damp grassland investigated in the current checklist area, these being the habitats which contribute the most Cyperaceae taxa. It is again possible that this family is under-represented here due to the limited surveys of sedge taxa from such habitats at sites such as Mwanenguba.

ANALYSIS OF PLANT GENERA

A total of 943 genera were recorded, one of which is undescribed (Acanthaceae *genus nov.*), with a further two possible genera (*Sapindaceae Indet.* and *Family Indet.*). Of these, 729 are Dicotyledons, 157 are Monocotyledons, two belong to Pinopsida, one to Gnetopsida, three to Lycopsida and 51 to Filicopsida.

The ten most genera-rich families in the checklist area are presented in the table below.

Family	Genera	% of total genera
Leguminosae	81	8.6
Rubiaceae	73	7.7
Orchidaceae	47	5.0
Euphorbiaceae	44	4.7
Acanthaceae	35	3.7
Gramineae	34	3.6
Compositae	29	3.1
Apocynaceae	19	2.0
Annonaceae	17	1.8
Labiatae	16	1.7

The fact that more genera are recorded in the Leguminosae than in the Rubiaceae, despite the latter family contributing more than twice as many taxa than the former, points to the low species diversity of many of the genera of Legumes in our area; not one of the 81 genera recorded occur within the ten most species-rich genera, presented in the table below, compared to two Rubiaceae genera.

Genus	Family	Taxa with accepted name	Species with accepted name	Unde-scribed new taxa	Taxa of uncertain identif-ication	Total taxa	% of taxa in family	% of taxa in checklist
Psychotria	Rubiaceae	26	24	15	10	51	18.4	2.1
Bulbophyllum	Orchidaceae	42	35	0	0	42	22.8	1.7
Ficus	Moraceae	27	26	0	4	31	62.0	1.3
Polystachya	Orchidaceae	30	29	0	0	30	16.3	1.3
Asplenium	Aspleniaceae	24	24	0	5	29	100.0	1.2
Begonia	Begoniaceae	27	26	0	1	28	100.0	1.2
Cyperus	Cyperaceae	24	22	1	0	25	54.5	1.0
Cola	Sterculiaceae	18	18	3	1	22	61.1	0.9
Pavetta	Rubiaceae	13	12	2	3	18	6.5	0.8
Salacia	Celastraceae	16	12	1	1	18	52.9	0.8
TOTAL		**247**	**228**	**23**	**25**	**294**	-	**12.2**

The current checklist area is remarkably similar to that of Mt Cameroon in terms of the most diverse genera, sharing nine of the top ten taxa. Only *Diospyros* is absent from the current list, being the eighth most diverse genus on Mt Cameroon; it is replaced in Kupe-Bakossi by *Pavetta*, although Mt Cameroon records one more species of this genus than the current area. The genus *Psychotria* is particularly diverse in Kupe-Bakossi, with many new, undescribed species. At Mt Cameroon, only 37 *Psychotria* taxa were recorded, it being only the third most diverse genus there. Kupe-Bakossi shares only four of the top ten genera with the Mt Oku checklist area, with the most species-rich genera in the latter area being those common in montane grassland and swamps, such as *Cyperus* (Cyperaceae), *Vernonia*, *Helichrysum* (both Compositae), *Hyparrhenia* and *Pennisetum* (both Gramineae). These genera are poorly represented in the current checklist area due both to the paucity of montane grassland and swamp areas and to the under-collection of these habitats.

ANALYSIS OF PHYTOGEOGRAPHICAL DIVISIONS

Distributional and chorological data were listed for a total of 2154 taxa. Of the 258 taxa not analysed for distribution, the majority (140) are ferns and fern allies, the remainder being taxa of uncertain identity for which a distribution could not therefore be applied. The numerical and percentage breakdown of chorological data is presented in the table below.

Chorology	Number of taxa	Percentage of total taxa analysed
Guineo-Congolian	461	21.4
Lower Guinea	287	13.3
Lower Guinea & Congolian	232	10.8
Afromontane	197	9.1
Upper & Lower Guinea	192	8.9
Tropical Africa	162	7.5
Pantropical	127	5.9
Western Cameroon Uplands	122	5.7
Narrow Endemic	94	4.4
Cameroon Endemic	76	3.5
Palaeotropical	63	2.9
Tropical & subtropical Africa	45	2.1
Tropical Africa & Madagascar	20	0.9
Other	76	3.5

The 'Other' category includes taxa with an unusual distribution, and thus a rare chorology, such as 'Amphi-Atlantic', 'Tropical Africa and Indian Ocean Islands' and 'Cosmopolitan'. It also includes taxa introduced but not widely naturalised in Cameroon, which cannot therefore be classified as Pantropical, Palaeotropical or Amphi-Atlantic.

The most striking aspect of the chorological data is the significant number and percentage of endemic or near-endemic taxa. The 4.4% of taxa entirely restricted, or almost entirely, to the checklist area is a conservative figure as it does not include uncertain taxa, several of which are likely to be endemic once more complete collections are made. 'Narrow Endemic' does not necessarily mean endemic to the checklist area. If the Western Cameroon Uplands centre of endemism is onsidered as a whole, it contributes over 10% of the total taxa. Further discussion on the high level of endemism in the checklist area is presented within the chapter on Phytogeography and Refugia.

Of the limited comparison possible with the Mt Cameroon data, from which chorological divisions have been modified, it is notable that a much lower percentage (21% as compared to 43% at Mt Cameroon) are Guineo-Congolian taxa. This halving of the proportion of widespread W African species further points towards the unusually high number of endemic or locally distributed taxa within the Kupe-Bakossi region.

NEW SPECIES DESCRIBED FROM THE CHECKLIST AREA

In the course of our work on the flora of the Kupe, Mwanenguba and Bakossi Mountains checklist, the following 27 new taxa, listed in the table below in chronological order by year of publication, have been described:

Taxon name	Family	Author(s) of publication	Year
Coffea montekupensis	(Rubiaceae)	Stoffelen *et al.*	1997
Diospyros kupensis	(Ebenaceae)	Gosline & Cheek	
Belonophora ongensis	(Rubiaceae)	Cheek & Dawson	2000
Aphelariopsis kupemontis	(Mycota)	Roberts	
Plectranthus cataractarum	(Labiatae)	Pollard & Paton	2001
Psychotria martinetugei	(Rubiaceae)	Cheek & Csiba	2002
Cola metallica	(Sterculiaceae)	Cheek	
Impatiens frithii	(Balsaminaceae)	Cheek & Csiba	
Coffea bakossii	(Rubiaceae)	Cheek & Bridson	

Taxon name	Family	Author(s) of publication	Year
Rhaptopetalum geophylax	(Scytopetalaceae)	Cheek *et al.*	
Angraecum sanfordii	(Orchidaceae)	Cribb & Pollard	
Polystachya kupensis	(Orchidaceae)	Cribb & Pollard	
Stelechantha arcuata	(Rubiaceae)	Dawson	
Newtonia duncanthomasii	(Leguminosae-Mimosoideae)	Mackinder & Cheek	2003
Dorstenia poinsettiifolia var. *etugeana*	(Moraceae)	Pollard *et al.*	
Phyllanthus caesiifolius	(Euphorbiaceae)	Hoffmann & Cheek	
Phyllanthus nyale	(Euphorbiaceae)	Hoffmann & Cheek	
Uvariopsis submontana	(Annonaceae)	Kenfack *et al.*	
Manniella cypripedioides	(Orchidaceae)	Salazar *et al.*	
Kupea martinetugei	(Triuridaceae)	Cheek *et al.*	
Ledermanniella onanae	(Podostemaceae)	Cheek	
Rinorea fausteana	(Violaceae)	Achoundong & Cheek	2004
Rinorea thomasii	(Violaceae)	Achoundong & Cheek	
Laccosperma korupensis	(Palmae)	Sunderland	
Justicia leucoxiphus	(Acanthaceae)	Vollesen *et al.*	
Peucedanum kupense	(Umbelliferae)	Darbyshire & Cheek	
Bulbophyllum kupense	(Orchidaceae)	Cribb & Pollard	

At least ten other taxa are currently in the course of publication, with many more in the early stages of description. It is hoped that, through analysis of the material currently available and collection of further material to complete gaps in our knowledge, the many undescribed species listed in this checklist will be described, and their conservation status assessed, in the near future.

THE PHYSICAL ENVIRONMENT

Chris Wild

Zoological Society of San Diego's Conservation and Research on Endangered Species
CRES Cameroon Office, B.P. 3055 Messa, Yaoundé, Cameroon

INTRODUCTION

The Cameroon Highlands are an extensive mountain chain sufficient in height and extent for the development of a montane climate (Tye 1986a). The Highlands are aligned along a N-E to S-W axis of a major fault in the earth's crust that separates the W Africa and Congo tectonic plates. The resulting volcanic chain along this fault extends offshore to the island peaks of Bioko, Principé, São Tomé and Annobon.

Mounts Kupe (2064 m), Mwanenguba (2411 m) and Bakossi (1895 m), the first inland peaks after Mt Cameroon, straddle the SW and Littoral Provinces of Cameroon covering an area of 3000 km^2 traditionally known as Bakossiland. These mountains form the first inland peaks at the southerly extent of the extensive *'Dorsale Camerounaise'*, -the main 'backbone' of the Cameroon Highlands. The Bakossi Mountains are a larger elevated land mass, albeit of lesser height than Kupe or Mwanenguba. The Bakossi Mts comprise two highland areas separated by the Mbwe Valley: the Mwenzekong Mts at 1760 m in the north, and the Mwendolengo Mts at 1895 m to the south. The Edib Hills, from 500–1500 m extend from the Mwendolengo Mts further southwards down to the lowlands at the confluence of the Mungo and Jide Rivers. This mountain complex is topographically linked by inter-montane ridges of above 1000 m elevation between Kupe and Mwanenguba, 1400 m between Mwanenguba and Mwendolengo, and 1200 m between Mwanenguba and Mwenzekong.

The Kupe, Mwanenguba and Bakossi forests are among the wettest in Africa. This mountain complex has provided a fertile environment for speciation and endemism acting as ancient Quaternary refugia for forest communities (Gartshore 1986; Maley 1987; Pook & Wild 1997; Cheek *et al.* 1999; Bowden 2001). Due to relatively high levels of rainfall and orographic uplift of warm, moist monsoon air from the coast, these mountains experience higher levels of precipitation than the surrounding lowlands. In such a wet environment frequently characterized by mists enveloping the topography at the vegetation level, these tropical montane cloud forests (TMCFs) are noted for their abundance of clear, fast-flowing streams and associated rheophile communities.

GEOLOGY

The Kupe-Mwanenguba mountain complex was formed by a combination of faulting and volcanic activity during two separate geological periods. Two extinct volcanoes occur in the area: Mwanenguba and Edib. These mountains, together with those of Bakossi, originate from Cameroon's second eruptive phase of post-Cretaceous volcanic activity that formed a white inferior geological series dating to the end of Neogene (Sieffermann 1973).

Mt Mwanenguba is a substantial Hawaiian-type extinct volcano. The 4 km wide Ebwoge Caldera contains two crater lakes that are a major feature for which the mountain is well known. The caldera rim, which reaches 300 m in height has numerous rocky outcrops of trachite and phonolite, and was formed by down-faulting and subsidence of the crater floor which is formed of basalt (Buckle 1978). To the SE of the caldera are the Elengoum peaks formed of rhyolite that reach up to 2411 m (Gèze 1943). Situated to the NE of the mountain is the Mbo Plain.

Mt Kupe is a steep-sided massive block horst formed by geological instability and uplifting along a major fault that formed during Cameroon's third eruptive phase in the Quaternary. This mountain is much younger than that of Mwanenguba or the Bakossi Mts Evidence of recent volcanic activity is seen in several small surface lava flows on the middle slopes. Smaller volcanic cones of recent Quaternary origin occur, many of which cluster below the southern slopes between Tombel and Loum. The largest of these cones in the area is Mt Amelo (1310 m), a few kilometres to the N of Kupe (IGN 1958).

Inselbergs of granite and syenite are a prominent feature of the Bakossi landscape, and are found in both western Bakossi near Nyale, Mwendolengo, and on the south-western slopes of Mt Kupe above the town of Tombel.

SOILS

The soils of the area are highly variable and linked to differences in volcanic geological origins between the older Neogene white inferior series and the recent Quaternary black superior series. Variations in soil characteristics are also related to elevation (both slope and land mass elevation), and their associated topoclimates.

The highly fertile lower eastern slopes of Mt Kupe have been studied in detail by Sieffermann (1973). For the most part these appear similar to the soil profiles seen on the western middle and lower slopes. They comprise of two different types that are difficult to distinguish: andosoils and brown eutrophic soils, situated on black basaltic rocks. They are both generally characterised by being well-drained and porous, with rich organic topsoils, highly porous subsoils, and high mineral infusion readily available for plant uptake. Topsoils are slightly acidic pH 5, subsoils pH 6.5 (Sieffermann 1973). In addition, a number of small edaphic enclaves can be seen from satellite imagery on Kupe's upper south-eastern slopes, and near the summits which support grassland rather than forest.

The soils on Kupe's lower slopes are very fertile, and are of high agricultural interest both to commercial plantations and village-based subsistence and cash-crop agriculture. The agricultural value varies according to soil depth and altitude, and is the principal reason for the attraction of many immigrant farmers from the W and NW Provinces in search of new farmland.

In contrast, the soils of the Edib volcano are poor with little nutrient cycling of the leaf litter, and low organic content of the topsoils which rest upon degraded metamorphic schists (Wild *pers. obs.*).

HYDROLOGY

The Bakossiland mountain complex gives rise to the sources of three major rivers. The source of the Cross River rises from the Ebwoge Caldera of Mt Mwanenguba at 1800 m, flowing NW towards Mamfe. The tributary, the Mbwe, bisects the Bakossi Mts, separating the Mwendolengo from the Mwenzekong Mountains of upper Bakossi, passing the villages of Enyandong and Babubok.

The Mungo River rises in the W of the Bakossi near the village of Boka, flowing southwards to the coast W of Douala. The Mungo also gains surface drainage from the Rumpi Hills to the W (Olivry 1986).

In the E, Mts Kupe and Mwanenguba provide the northwestern watershed of the Wouri River in Littoral Province, where the Dibombe tributary flows S and eastwards towards Douala (Olivry 1986).

The region supports an abundance of small streams, cascading waterfalls and plunge pools. The highest waterfalls reach in excess of 150 m along the western Bakossi escarpment in the Mwendolengo Mountains.

There are six crater lakes in Bakossiland. The famous twin lakes of Mwanenguba Ebwoge Caldera, male and female according to tradition, and comes from the name 'Mwanenguba', which means 'man and wife' (Buckle 1978); Lake Edib (1150 m), an upland 'bog-eyed' crater lake; and the craters of Lake Beme (450 m) in Bakossi, and the two recently discovered Lakes Mwandon (1200 m) on the Kupe-Mwanenguba intermontane ridge E of Mwambong village.

Thermal springs also occur including a thermomineral source at Melong (Marechal 1976). A number of spring-fed pools and surface seepages of mineral rich ground water exist, such as Nye-Mekang Spring near Mwambong. An area of submontane *Raphia* swamp, a rare habitat type in the Cameroon Highlands, occurs in the Mbumbe Hills E of Ngomboaku on the Kupe-Mwanenguba intermontane ridge.

CLIMATE

In the tropics, temperatures are high over most of the year, and contrasts in humidity play the greatest role in weather changes. As the tropical maritime air of the Atlantic nears the equator it acquires increasing moisture until it is transformed into the warm, moist, highly unstable air of the Intertropical Convergence Zone (ITCZ) (Buckle 1996). In November the ITCZ moves south bringing with it the dry, moisture-absorbing winds from the Sahara. These dusty, north-easterly winds, known as the Harmattan, are concurrent with the region's dry season which occurs from November through to March. The St Helena anti-cyclone moves the ITCZ N in April when low pressure forms heavy thunderstorms. The warm moist south-westerly winds begin around March when a large build-up of cumiliform clouds over the windward (southern) aspect of the mountains and hills are often seen in late afternoon. The monsoon precipitation increases in intensity in the following months, peaking in August, with regular and heavy cyclonic rainfall occurring until October.

Thunderstorms are a major feature of the climate of western Cameroon, with a mean annual frequency of 180 thunderstorms per annum in the area of Kupe-Mwanenguba (Griffiths 1972), the highest figures known for Africa. Mt Cameroon on the coast experiences around 140 thunderstorms annually. These storms often appear in series across the Cameroon Highlands, known as line squalls (Ojo 1977). Approximately 80% of the annual rainfall in Kupe-Mwanenguba occurs in the April-October wet season. Regular cloud cover and frequent mists result in low insolation, with sunshine reduced to one or two hours daily during August. In most years a week or two of bright sunshine without rainfall occurs during the month of July, a result of the influence of the four-season equatorial climate of Central Africa found S of Latitude 6° N. Although rainfall typically occurs in every month of the year throughout the area, there is a marked climatic contrast between the wet and dry seasons. Humidity is also exceptionally high, with the minimum recorded relative humidity of 81% during the influence of the Harmattan at Tombel (Ejedepang-Koge 1986).

RAINFALL

In addition to seasonal cyclonic rains, orographic uplift of the warm, moist, monsoon from the S rises over the mountains where it cools adiabatically. When it cools to below the dewpoint (RH=100%) the water vapour condenses and forms cloud. Precipitation follows when the water droplets coalesce to form larger and larger drops with the majority of rain falling on the south-westerly (windward) aspect of the mountains. Consequently, there are much higher levels of rain falling in the mountains than in the surrounding lowlands. The maximum levels of rainfall generally coincide with the mean cloud base height at that location.

The ORSTOM hydrological map of Cameroon (1972) shows these mountains to be enclosed by a 4 m isohyet. Mean annual rainfall is, however, highly variable within Bakossiland ranging from 3 m per annum at Loum on the south-eastern base of Mt Kupe, to 6–7 m on the south-western slopes of Mt Kupe (Sieffermann 1973). Local rainfall data taken from a range of periods are available from Christian Missionary data from Bangem (W aspect Mwanenguba, N-E Bakossi), Nyasoso (W aspect Kupe), Cameroon Development Corporation (CDC) data from Essossong (N aspect Kupe) (Ejedepang-Koge 1986), and for Nkongsamba, Loum and Tombel (Suchel 1972). Mean monthly rainfall data for six sites in the area are given in a table at the end of this chapter.

TEMPERATURE

The relatively high variability of daily cycles, contrasting with the low variability of annual cycles, are the major features of a tropical mountain climate (Sarmiento 1986). The environmental lapse rate of decreasing temperature with increasing altitude of about 0.6°C per 100 m rise in elevation (0.5°C in saturated air within cloud) is the predominant variable which determines the range of extremes encountered in these mountains. This is slowed slightly above the dewpoint to about 0.5°C per 100 m as saturated air gives off latent heat as it condenses, and this extra heat reduces the temperature change to a minimum. Being only a few degrees N of the equator, N-S aspects probably have little significant variation in levels of insolation received were it not for the interference of the cloud patterns at the vegetation level on windward (south-westerly) aspects. Additional significant differences can be expected to occur between E and W aspects (Sarmiento 1986) since easterly slopes typically receive greater insolation than western slopes because the latter aspect is covered by cloud by after midday, in contrast with east-facing slopes which receive early morning sunshine. Sites of easterly aspect receive greater direct insolation and are therefore drier, having higher maxima and lower minima (Sarmiento 1996). The S and W facing slopes are subject to a greater influence of rainfall and moisture than those of the N and E.

Mean monthly temperature data available for sites in the southern (Tombel) and northern (Bangem) parts of the Kupe-Mwanenguba-Bakossi area are given in a table at the end of this chapter (missionary data cited by Ejedepang-Koge 1986). These data show minor season fluctuations in monthly means of 23.4°–25.8°C at 458 m elevation and 20.6°–25.5°C at 1120 m elevation. The highest mean recorded temperature was 25.5°C from eastern base of Mt Kupe at 250 m (Sieffermann 1973).

BIOPHYSICAL PROCESSES

Altitude, through its effect on levels of temperature and precipitation over relatively short distances, is responsible for the sharp climatic and ecological gradients found on mountains (Adams *et al.* 1996). The lower temperatures affect the montane forest fauna both directly and indirectly, by affecting the composition and form of the vegetation (Tye 1986b). Species diversity, vegetation cover and productivity, and the growth forms of plants in general, decrease with rise in elevation (Gamachu, 1990).

The mid and upper elevations of the Kupe, Mwanenguba and Bakossi Mts are affected by biophysical processes which give rise to the formation of tropical montane cloud forest (TMCF *sensu* Hamilton *et al.* 1995). TMCF is a distinct forest type limited by the altitudinal amplitude of stable, and regular soaking mists at the vegetation level. It may be formed by reduced insolation resulting from frequent cloud cover at the vegetation level, also known as 'occult' precipitation, which depresses the daytime temperatures below the normal lapse rate and reduces visibility to as little as 20 m, often obscuring the canopy from the ground. Enveloping cloud influences the atmospheric interaction through reduced solar radiation and vapour deficit, canopy wetting, and general suppression of evapo-transpiration. The net precipitation (throughfall) is significantly enhanced (beyond rainfall contribution) through direct canopy interception of cloud water and low water use by the vegetation. In comparison with lower altitude tropical moist forest, the stand characteristics generally include reduced tree stature and increased stem density (Hamilton *et al.* 1995).

A number of species among both the flora and fauna are reliable indicators of TMCF in the Cameroon Highlands. A typical feature is a concurrent and marked increase in the abundance of epiphytic mosses. Lichens, more tolerant of desiccation, begin to replace the epiphytic mosses at higher elevations. There is an abundance of epiphytes with mosses draping from branches, and numerous orchids, filmy ferns and *Impatiens* species unique to the local cloud forest environment. Frequent stands of the tree fern *Cyathea manniana* are also typical of the cloud forests in these mountains. Faunal TMCF indicators are best represented among the poikilothermic lower vertebrates, in particular some of the moisture-loving anuran amphibians that are adapted to this environment.

Mt Mwanenguba has a well-developed TMCF environment that extends to the Ebwoge caldera at 1800 m, becoming very unstable above that altitude, the effects of which are seen in a transition of epiphytic mosses to dessication resistant lichens on the caldera rim at around 2100 m. The southern slopes of the mountain are still well forested down to around 1200 m at Mwanenguba village, which approximately coincides with the typical cloud base height associated with the mountain.

The windward sub- and pre-montane forests of the Edib Hills along the Elambah Ridge and around the crater lake experience persistent cloud cover at much lower elevations, from 800 to 1500 m. The TMCF environment, caused by mass elevation known as the 'Massenerhebung effect' (Grubb 1971), is manifested in the fog-bound vegetation, the low tree height, and gleying of soils. This results in poor nutrient cycling (Grubb 1971).

Mt Kupe has a more unstable cloud forest environment compared with that of Mwanenguba or the Bakossi Mts due to its topographic isolation and relatively smaller land mass, and this is reflected in the physiognomy of the vegetation. The typical altitude of the mountain cloud base is 1300 m, the lower limit of which is highly unstable, occasionally extending to below 1000 m. TMCF is therefore not fully developed over this mountain and there is no corresponding reduction in woody climbers, and stands of larger lowland trees occur up to 1600 m. Above 1850 m cloud is less important in maintaining an epiphytic community where lichens predominate.

Tropical montane cloud forests are among the rarest and most important of the world's rainforests for biodiversity and endemism (Aldridge *et al.* 1997; Bubb *et al.* 2004). The Kupe, Mwanenguba and Bakossi Mts represent a major association of TMCF in Central and W Africa with a distinct montane biota of international importance. Moreover, the role of TMCF in providing favourable moist environments and refugia for forest communities during dry glacial maxima in the northern hemisphere may be an important factor in explaining the high levels of diversity and endemism observed in these mountains.

REFERENCES & BIBLIOGRAPHY

Adams, W.M., Goudie, A.S. & Orme, A.R. (1996). The Physical Geography of Africa. Oxford University Press, Oxford

Aldridge, M., Billington, C., Edwards, M. & Laidlaw, R. (1997). Tropical Montane Cloud Forests: an urgent priority for conservation. WCMC Biodiversity Bulletin No. 2. The World Conservation Monitoring Centre, Cambridge.

Barry, R.G. (1978). Diurnal effects of topoclimate on an equatorial mountain. 14th Int. Tagung fur Alpine Meteorologie **72**: 1–8.

Bowden, C.G.R. (2001). The birds of Mt Kupe, southwest Cameroon. Malimbus **23**: 13–44.

Bubb, P., May, I., Miles, L. & Sayer, J. (2004). Cloud Forest Agenda. UNEP-WCMC Biodiversity Series No. 20. Cambridge.

Buckle, C. (1978). Landforms in Africa. Longman, London.

Buckle, C. (1996). Weather and Climate in Africa. Longman, Harlow.

Cheek, M., Gosline, G. & Etuge, M. (1999). Botanical summary of Kupe-Manenguba-Bakossi Mountains. Preliminary report to the WWF Mt Kupe Forest Project, Nyasoso.

Coe, M.J. (1967). Microclimate and animal life in the equatorial mountains. Zool. Africana **4**: 101–128.

Ejedepang-Koge, S.N. (1971). Southern Bakossi–A Geographic Study. Thesis, Faculty of Arts, University of Yaoundé.

Ejedepang-Koge, S.N. (1986). Tradition of a People Bakossi. ARC Publications, Yaoundé.

Ewane, D., Komesue, A. & Zachee, N.N. (2002). Geography of the Mwanenguba Region. CERUT's Extension Series, Limbe.

Furon, R. (1963). Geology of Africa. Oliver & Boyd, Edinburgh.

Gamachu, D. (1990). Some patterns of altitudinal variation of climatic elements in the mountainous regions of Ethiopia. *In*: Messerli, B. & Hurni, H. (*eds*). African Mountains: Problems and Perspectives. African Mountains Association, Walsworth Press, Missouri.

Gartlan, S. (1989). La Conservation des Ecosystèmes forestiers du Cameroun. IUCN, Gland, Switzerland and Cambridge, UK.

Gartshore, M.E. (1986). The status of the montane herpetofauna of the Cameroon Highlands. *In*: Stuart, S.N. (*ed.*). Conservation of Cameroon Montane Forests, International Council for Bird Preservation, Cambridge: 205–238.

Genieux, M. (1958). Climatologie du Cameroun. *In*: Atlas du Cameroun. ORSTOM, Yaoundé.

Gèze, B. (1943). Géographie physique et géologie du Cameroun occidentale. - Mém. Mus. Natn. d'Hist. Nat. **17**: 1–271.

Goudie, A.S. (1996). Climate: Past and Present. *In*: Adams, W.M., Goudie, A.S. & Orme, A.R. (*eds*). The Physical Geography of Africa. Oxford University Press, Oxford: 34–59.

Griffiths, J.F. (*ed.*) (1972). Climates of Africa. Amsterdam.

Grubb, P.J. (1971). Interpretation of the 'Massenerhebung Effect' on Tropical Mountains. Nature **229**: 44–45.

Hamilton, A.C. (1981). The Quaternary history of African forests: its relevance to conservation. African Journal of Ecology: 1–6.

Hamilton, A.C., Taylor, D. & Vogel, J.C. (1986). Early forest clearance and environmental degradation in south-west Uganda. Nature. **320**: 164–7.

Hamilton, L., Juvik, J.O. & Scatena, F. (1995). The Puerto Rico Tropical Cloud Forest Symposium: Introduction and Workshop Synthesis. *In*: tropical montane cloud forests, Ecological Studies **110**. Springer-Verlag, New York: 1–23.

HMSO (1983). Tables of temperature, relative humidity, precipitation and sunshine for the world. Part **IV**: Africa, the Atlantic Ocean South of 35° and the Indian Ocean. Met. O. 856d Her Majesty's Stationery Office, London.

IGN (1958). Douala, Carte de l'Afrique Centrale au 1:200,000, NB-32-IV. Institut Geographique National, Paris.

Kadomura, H. (1982). Summary and conclusions. *In* Kadomura, H. (*ed.*). Geomorphology and Environmental Changes in the Forests and Savanna of Cameroon. Hokkaido University, Japan: 90–100.

Kugbe, C.A. (*ed.*) (1975). Proceedings on the conference on the geology of Nigeria. Ile-Ife, University of Ife, Nigeria.

Laclavère. G. (1980). Atlas of the United Republic of Cameroon. - Editions Paul Lechevalier, Paris.

Maley, J. (1987). Fragmentation de la forêt dense humide africaine et extension des biotopes montagnards au Quaternary récent: nouvelles données polliniques et chronogiques. Implications paléoclimatiques et biogéographiques. Paleoecology of Africa **18**: 307–334.

Marechal, A. (1976). Géologie et géochimie des sources thermominerales du Cameroun. ORSTOM, Paris.

Moisel, M. (1912). Karte von Kamerun 1: 300, 000, 1st ed. - Dietrich Reimer (Ernst Vohsen), Berlin.

Neba, A.S. (1982). Modern geography of the United Republic of Cameroon. Hamilton, New York.

Nieuwolt, S. (1974). The influence of aspect and elevation on daily rainfall: some examples from Tanzania. *In*: Agroclimatology of the Highlands of Eastern Africa. Proceedings of the Technical Conference, Nairobi, October 1–5, 1973. WMO no. **389**. Geneva: FAO-UNESCO-WMO.

Ojo, O. (1977). The climates of West Africa. Heinemann, London.

Olivry, J.C. (1986). Fleuves et rivières du Cameroun. Monographies Hydrologiques, ORSTOM **9**: 1–733.

Pook, C.E. and Wild, C.J. (1997). The Phylogeny of the *Chamaeleo (Trioceros) cristatus* species-group from Cameroon inferred from direct sequencing of the mitochondrial 12S ribosomal RNA gene: Evolutionary and paleobiogeographic implications. *In*: Böhme, W., Bischoff, W. & Ziegler, T. (*eds*). Herpetologia Bonnensis, Bonn: Societas Europaea Herpetologica : 297–306.

Sarmiento, G. (1986). Ecological Features of Climate in High Tropical Mountains. *In* Vuilleumier, F. and Monasterio, M. (*eds*). High Altitude Tropical Biogeography. Oxford University Press, Oxford and New York: 11–45.

Sieffermann, G. (1973). Les sols de quelques régions volcaniques du Cameroun. Mémoires ORSTOM No. **66**, Office de la Recherche Scientifique et Technique Outre-Mer. Paris.

Stadtmuller, T. (1987). Cloud Forests in the Humid Tropics. A bibliographic review. United Nations University, Tokyo, and CATIE, Turrualba, Costa Rica.

Suchel, J.B. (1972). La répartition des pluies et les régimes pluviométriques du Cameroun. Travaux et Documents de Géographie Tropicale, CEGET-CNRS (Centre d'Etude et de Géographie Tropicale, Centre National de la Recherche Scientifique). **5**: 1–287.

Sumengole, A. (1998). Unsustainable anthropic activities and environmental degradation in afromontane forests: the case of the Mwanenguba montane forest, Cameroon. Unpublished dissertation, University of Yaoundé I.

Taylor, D. (1996). Mountains. *In* Adams, W.M., Goudie, A.S. & Orme, A.R. (*eds*). The Physical Geography of Africa. Oxford University Press, Oxford: 287–306.

Tye, H. (1986a). Geology and Landforms in the Highlands of Western Cameroon. *In*: Stuart, S.N. (*ed.*). Conservation of Cameroon Montane Forests. Report of the ICBP Cameroon Montane Forest Survey, International Council for Bird Preservation, Cambridge: 15–17.

Tye, H. (1986b). The Climate of the Highlands of Western Cameroon. *In*: Stuart, S.N. (*ed.*), Conservation of Cameroon Montane Forests. Report of the ICBP Cameroon Montane Forest Survey, International Council for Bird Preservation, Cambridge: 18–19.

USBGN (1962). Cameroon: official standard names approved by the US Board on Geographic Names, Office of Geography, Department of the Interior, Washington, D.C. (Gazetteer, **60**), 255 pp.

WCMC (1997). A global directory of tropical montane cloud forests (Draft). Aldridge, M., Billington, C., Edwards, M. & Laidlaw, R. (*eds*). The World Conservation Monitoring Centre, Cambridge.

Wuetrich, B. (1993). Forests in the clouds face a stormy future. Science News **144**: 23.

Mean rainfall data (mm) for six sites, cited in Ejedepang-Koge (1986).

Location	Altitude (m)	Period	Jan	Feb	Mar	Apr	May	Jun	Jul	Aug	Sep	Oct	Nov	Dec	Total (mm)
Loum	224	33 years													2992
Nkongsamba	882	34 years	16	53	151	199	226	261	431	482	476	345	103	19	2762
Tombel	498	24 years	53	121	248	240	267	507	489	480	554	474	160	65	3657
Nyasoso	829	5 years	21	107	230	182	298	402	822	720	645	453	77	45	4045
Essossong	1050	-	48	87	223	253	270	383	416	453	459	478	139	44	3253
Bangem	1120	2 years	27	15	25	159	273	296	513	515	383	231	84	22	2768

Temperature data (°C) for Tombel and Bangem, cited in Ejedepang-Koge (1986).

Location	Altitude (m)	Jan	Feb	Mar	Apr	May	Jun	Jul	Aug	Sep	Oct	Nov	Dec	Mean
Tombel	498	24.2	24.6	23.4	25.5	25.8	24.7	23.5	-	23.8	23.8	24.3	24.7	24.8
Bangem	1120	22.8	22.7	25.5	22.2	23.0	21.7	20.6	-	21.1	21.7	23.3	23.9	23.3

VEGETATION

Martin Cheek

Herbarium, Royal Botanic Gardens, Kew, Surrey, TW9 3AE, U.K.

INTRODUCTION

In this chapter, the eleven vegetation types represented in Fig. 3 (mapping units 1–11) are characterized: the percentage of the checklist area that they occupy is estimated, studies previous to those by R.B.G., Kew–National Herbarium of Cameroon expeditions (1995–2003) are reviewed, sampling, by way of specimen collecting levels is gauged, the physiognomy and species composition of the vegetation is examined, endemic taxa and threats to the vegetation type are listed, and phytogeographical links are assessed.

MAPPING UNITS

1 **LOWLAND EVERGREEN FOREST**
150–800 m altitude

2 **SEMI-DECIDUOUS FOREST**
c. 300 m altitude

3 **FRESHWATER SWAMP AND LOWLAND RIVERSIDE VEGETATION**
c. 150 m altitude

4 **SUBMONTANE FOREST**
800–1900(–2000) m altitude

5 **SUBMONTANE 'GRASSLAND' (SAVANNA THICKET)**
1000–1600 m altitude

6 **ROCK FACES OF CLIFFS AND INSELBERGS; LITHOPHYTIC COMMUNITIES**
600–1500 m altitude

7 **RHEOPHYTIC VEGETATION OF UPLAND STREAMS AND RIVERS**
400–1450 m altitude

8 **CRATER LAKES**
1150–1950 m altitude

9 **MONTANE FOREST & FOREST-GRASSLAND EDGE**
1900–2000(–2050) m altitude

10 **MONTANE GRASSLAND**
1900–2000(–2050) m altitude

11 **ANTHROPIC**
150–2000 m altitude

The table (on page 26) gives an overview of the vegetation types, each of which is then treated in more detail in the ensuing pages. Note that some Red Data taxa occur in more than one vegetation type. For more information on the Red Data species of each vegetation type see the Red Data chapter.

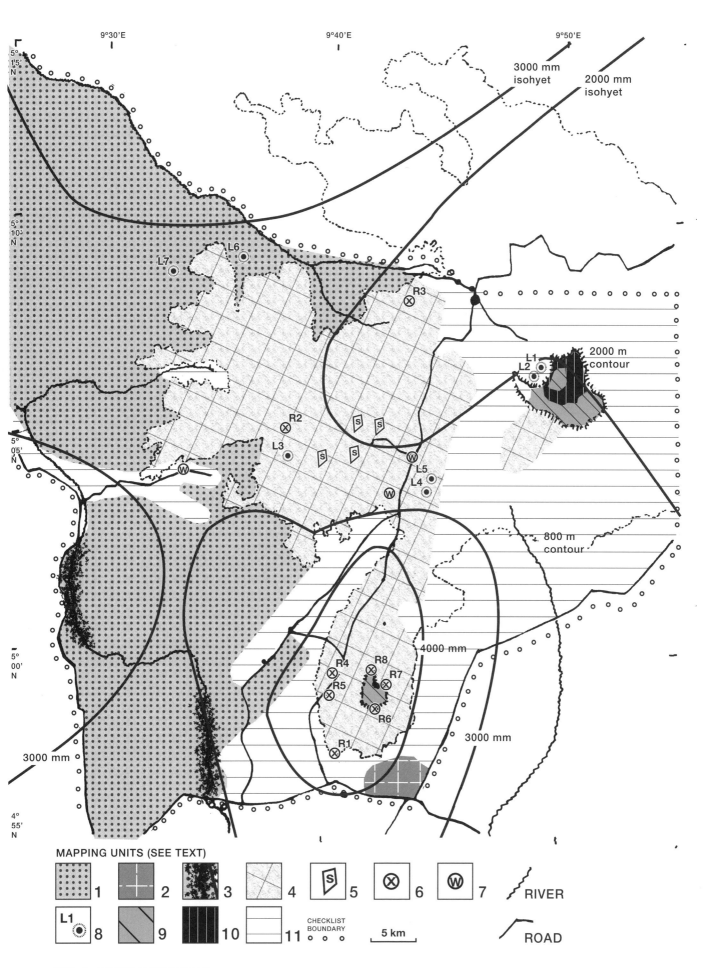

MAPPING UNITS (SEE TEXT)

1   2   3   4   5   6   7   RIVER

8   9   10   11   CHECKLIST BOUNDARY   5 km   ROAD

FIG. 3. VEGETATION MAP.
Showing Kupe, Mwanenguba & the Bakossi Mts. Isohyets (rainfall) are taken from Courade (1972).

25

Mapping Unit	Vegetation Type	Alt. range (m)	Approx. % check list area	Previous studies	Sampling	N° strict endemics	N° Red Data Taxa	Est. total N° taxa	Main Phytogeographic links
				0=nil, 1=low, 2=moderate, 3=high					
1	Lowland evergreen forest	150–800	30–40	0	1	8	77	> 1000	Mt Cameroon foothills
2	Semi-deciduous forest	c. 300	3	1	1	2?	6	150	Cameroon & W Africa
3	Freshwater swamp and lowland riverside vegetation	150–800	1	0	1	0	3	20	SW Cameroon
4	Submontane forest	(see below)	25–30			68	140**	> 1000	Mt Etinde, Rumpi Hills, Mt Nlonako, Bamenda Highlands & Crystal Mts (Gabon)
i	Block 1: Mwanenguba	800–1900 (–2000)		0	1	0			
ii	Block 2: Mt Kupe & the submontane bridge	800–1900 (–2000)		2	3	30*			
iii	Block 3: Bakossi Mts	800–1780		0	1–2	15*			
5	Submontane 'grassland' (savanna-thicket)	1000–1600	< 0.5	0	2	0	0	100	Mwanenguba montane grassland
6	Rock faces of cliffs and inselbergs; lithophytic communities	600–1500	< 0.5	0	2	0	6	40	W Africa
7	Rheophytic vegetation of upland streams and rivers	400–1450	< 0.5	0	2	1	8	15	SW Province
8	Crater lakes	1150–1950	< 0.5	0	2	0	4	70	Cameroon Highlands & E Africa?
9	Montane forest & forest-grassland edge	1900–2000 (–2050)	3	-	-	-	17	-	Cameroon Highlands & E Africa
	A. Mwanenguba		-	0	1	1	14	150	
	B. Mt Kupe		-	2	2–3	1	10	100	
10	Montane grassland	1900–2000 (–2050)	1–2	-	-	-	6	-	Cameroon Highlands & E Africa
	A. Mwanenguba		-	1	1	0	5	70	
	B. Mt Kupe		-	0	2–3	2	2	10	
11	Anthropic	150–2000	25–30	0	2	1	3	150	Pantropical

*A further 23 taxa are endemic to blocks 2 & 3 combined; ** Includes 39 taxa occurring in combinations of the 3 blocks.

The first quantitative studies of vegetational composition in the Kupe, Mwanenguba and Bakossi area appear to be those conducted by undergraduates of the Aberdeen University expeditions of the early 1990s in Lane (1993a&b, 1994a&b, 1995) and Stutter (1994). These were confined to the submontane forest (as classified here) of Mt Kupe above the village of Nyasoso. Prior to this, observations on the vegetation of the same area were published by Letouzey (1968a & 1968b) and Thomas (1986). All the foregoing are reviewed below under mapping unit 5, submontane forest, block 2 (Mt Kupe).

The results of an early German study of Mwanenguba grassland are not available (referred to in Letouzey 1968b, but reference not located). In any case, the study is believed to be sketchy. The specimens on which it is based were destroyed at Berlin in 1943 (Thorbecke, 1911: see mapping unit 10). Specimens collected by Ledermann in 1908–09 at Loum and at Ndoungue were later identified and used to discuss the vegetation at these places by Engler (1925). However, since canopy trees were generally not collected, Engler's work is not very informative regarding characterization of the vegetation.

Sampling, by way of specimen collecting locations, is shown on the rear cover of this book where collection sites are superimposed on a spot satellite image of the area (courtesy CRES). The concentration of collections from Mt Kupe is very clear, as is the fact that large areas of western Bakossi in particular are, botanically speaking, unknown.

Various sorts of pristine and near pristine evergreen forest still dominate the checklist area, although on the eastern side, near Nkongsamba and the Nkongsamba–Loum road, and also in the valley bottoms, anthropic vegetation is extensive.

The diversity of plant species (2412 species in 2390 km^2) ranks amongst the highest in tropical Africa and the 82 strictly endemic species (endemic to our checklist area) is an extremely high number. Most of these are confined to the submontane forest–taken together with the Rumpi Hills to the west, one of the most extensive intact areas in tropical Africa. The main threats to the natural vegetation types are clearance for firewood and timber, followed by conversion to agriculture. Illegal logging is thought to be the single most important threat. Phytogeographical links vary extremely widely, from those of the crater lakes (where links may be with Europe, NE or S-central Africa) to those of inselbergs and rock faces where most specialist species of this habitat type are restricted to W Africa (heliophiles) or immediately adjoining areas of the Cameroon Highlands (ombrophiles).

The vegetation map was made using a basemap obtained from the Mt Kupe Forest Project in 1995. On to this was superimposed isohyets from Courade (1974). The 800 and 2000 m contours are used as boundaries between lowland, submontane and montane vegetation types. The extent of natural vegetation cover is largely taken from the 1:200,000 'Map of Cameroon' series, specifically the Buea-Douala NB-32-IV and the Mamfe NB-22-x sheets, dated 1975, modified by personal observations on the ground, and by data from a recent landsat image, courtesy of Susana Baena. Letouzey's Notice de la carte phytogéographique du Cameroun (1985) was not considered until after this chapter was first drafted. That part of his map covering our checklist area compares tolerably well to the vegetation map presented here. His monumental work was based on scrutiny of aerial photographs of Cameroon, supplemented by extensive ground-truthing and extrapolation. His itinerary shows that he visited only the perimeter of our checklist area, apart from a day trip to Mt Kupe. He never visited the Bakossi Mts. Discrepancies between his representation of vegetation and that presented here are noted where they occur, under the mapping units below. The well known vegetation map of Africa published by White (1983) is at too coarse a scale to have any relevance here.

MAPPING UNIT 1: LOWLAND EVERGREEN FOREST; 150–800 m alt.

This unit accounts for the single most important vegetation type in terms of area (30–40% of the total). It is effectively restricted to the western and (less extensively) southern part of the checklist area although a tongue extends eastwards along the Mbu valley and remnants extend north-eastwards along the Jide trough in the S centre. This unit once occurred on the eastern edge of the checklist area where it has been replaced by anthropic vegetation (mapping unit 11).

Sampling of this vegetation type has been poor, so that although it is almost certainly the most species-diverse vegetation type in the checklist area, this is not reflected in the checklist. Our 'Earthwatch' expeditions have only sampled the Mungo River F.R. to the S in the vicinity of the bridge on the Kumba–Loum road, and the Bakossi F.R. in the vicinity of Mahole-Bulutu, spending about two to three days at each.

We have also sampled some of the remnants along the Jide valley, particularly near Kupe village and Ngomboaku, both near the upper end of the altitudinal range. In the W we have sampled near Nyandong, but these samples only arrived at Kew shortly before completion of the checklist, so have mostly not been analysed or included.

Fortunately, general collecting by Duncan Thomas and associates at sites along the Kumba–Mamfe road have done a great deal to fill this sampling gap, as have other earlier, predominantly road-based collectors. Despite this, large tracts in the SW and NW of the checklist area remain almost completely unknown botanically, and further survey work there is certain to reveal additional species.

No previous studies of this vegetation type in Bakossi appear to have been published, but Letouzey's treatment (1985), based largely on extrapolation from other areas, demands comment and is addressed below under altitudinal variation.

VARIATIONS IN SPECIES COMPOSITION

Geographical variation. Given the large and dissected nature of the lowland evergreen forest area, it is likely that there is geographical variation in species composition within it. Factors in this might be differences in rainfall and soils. Rainfall patterns in the checklist area are incompletely known, but appear to vary from 2 m in parts of the Mbu valley in the NW, to above 4 m in the Bakossi F.R. towards Nyasoso (see Fig. 3). In the SE corner of the checklist area semi-deciduous forest occurs. Studies of adjacent blocks of evergreen forest would yield an objective list of species of drier type evergreen forest.

At the moment, however, sampling is partially complete, so that no strong patterns of variation have yet emerged. Having said that, one can observe that well known and striking species such as *Desbordesia glaucescens* (Irvingiaceae) and *Baillonella toxisperma* (Sapotaceae) have only been recorded at Nyandong and (the second) also at Ikiliwindi, but are unknown at the moderately well-collected Mungo River F.R. and Bakossi trough areas.

Altitudinal variation within the lowland altitudal range represented in Bakossi (c. 150–800 m alt.), while also less-than-ideally sampled, does seem to show a pattern. Species lists resulting from dense sampling at the upper part of the range (c. 700–800 m alt.) in the Nyasoso and Kupe Village area can be compared with that derived from the moderate sampling in the Mungo River F.R. (150–250 m alt.). Many of the species present at the upper part of the range at the former sites have never been recorded at the Mungo River F.R. An example is *Diospyros kupensis* (Ebenaceae), conspicuous and identifiable even when vegetative. It is extensively recorded at 700–1000 m at several sites in Bakossi, and at Nyandong as low as 430 m alt, but has never been recorded in Mungo River F.R. despite intensive searching there by Gosline, Etuge and myself. Conversely, other species, such as *Quassia silvestris* (Simaroubaceae) and *Thomandersia laurifolia* (Acanthaceae) have only been recorded at altitudinal ranges of c. 150–350 m. Letouzey (1985) subdivides this mapping unit in our area into two units, based on altitude, the 500 m contour dividing them. Below this contour he defines the forest as being mapping unit 228: 'Atlantic Biafran forest with Caesalpiniaceae'. Above, occupying the 500–800 m band, he recognizes mapping unit 203: 'Atlantic forest with Caesalpiniaceae rare, N.W. type'. The division between these two types is based largely on his observation that caesalpinioid leguminous trees are more diverse, and more likely to be gregarious, at lower altitudes. This tallies with what I have seen, but I am not sure that it is a sufficient basis for formal subdivision, even given the evidence of the examples that I have cited above. Ironically, *Loesenera talbotii,* the most commonly recorded caesalpinoid legume in our area, and also the main gregarious caesalpinoid in Bakossi, is commonest in the 500–800 m band (see also monotypic stands below).

MONOTYPIC STANDS

In contradiction to the usual observation that lowland tropical evergreen forest is extremely diverse at all levels is the phenomenon of the monotypic stand, where one species dominates the canopy as is common in temperate zones. Leguminosae–Caesalpinioideae are the main source of this phenomenon in tropical Africa, and *Gilbertiodendron dewevrei* is the most widespread and most commonly involved of these species in the lower Guinea-Congolian area. However, in Bakossi the main species forming monotypic stands is *Loesenera talbotii*, the most collected legume of the checklist area (18 collections). This species is apparently not previously recorded as being gregarious. Nevertheless, I have documented in detail, with the assistance of Daniel Ajang, one site, above Kupe village where along one path it dominates the canopy over a distance of about 500 m, with over 20 trees being involved. At other sites in Bakossi it also appears gregarious, although no detailed studies have been conducted at these places. Outside of Bakossi it has only rarely been collected, and from no great distance away, both E and W.

Strychnos staudtii (Loganiaceae) is another contender for forming monotypic stands, since it is extremely common (21 collections) and occurs in almost every area of this mapping unit that has been sampled. However this possibility needs to be evaluated further by mapping the occurrence of individuals.

CHARACTERIZATION OF LOWLAND EVERGREEN FOREST

Emergents include *Ceiba pentandra, Desbordesia glaucescens, Baillonella toxisperma.*

Common canopy trees include *Allanblackia floribunda, Strombosia grandiflora, Tapura africana, Octoknema affinis, Panda oleosa, Oxystigma gilbertii, Berlinia bracteosa, Hylodendron gabonensis, Klainedoxa gabonensis, Diospyros viridicans, D. suaveolens, D. mannii, Ricinodendron heudelotii, Irvingia gabonensis, Vitex grandifolia, Tetrapleura tetraptera, Strychnos staudtii, Loesenera talbotii, Scottellia klaineana, Quassia silvestris, Microcos coriacea* and *Eriocoelum macrocarpum.*

Pioneer trees are *Pycnanthus angolensis, Musanga cecropioides, Ochroma lagopus, Macaranga* spp.*, Zanthoxylum gilletii, Trema orientalis, Maesopsis eminii.*

Common understorey shrubs and trees include *Massularia acuminata, Heckeldora staudtii, Cola cauliflora, Heisteria parvifolia, Chazaliella sciadephora, Diospyros conocarpa, D. preussii, Erythrococca anomala, Psychotria latistipula, P. martinetugei, P. leptophylla, Antidesma laciniatum, A. vogelianum, Maesobotrya* spp. *Crotonogyne preussii, Discoclaoxylon hexandrum, Oncoba glauca, Garcinia mannii, Lasianthera africana, Desmostachys tenuifolius, Anthonotha macrophylla, Angylocalyx oligophyllus, Penianthus longifolius, Glossocalyx brevipes, Olax latifolia, Microdesmis cf. puberula, Carpolobia alba, Aulacocalyx caudata, Belonophora coriacea, Ixora guineensis, Pavetta rigida, Tricalysia gossweileri, T. atherura, Octolepis casearia* and *Rinorea oblongifolia.*

Lianas include *Lavigeria macrocarpa, Dichapetalum* spp.*, Strychnos* spp.*, Salacia* spp.*, Triclisia dictyophylla, Jateorhiza macrantha, Cissampelos owariensis, Syntriandrium preussii, Calochone acuminata, Atractyogyne gabonica, Tetracera* spp.*, Hugonia* spp.*, Tarenna fusco-flava, Microcos barombiensis.*

Common herbs include *Stenandrium guineense, Psychotria globosa, Hymenocoleus subipecacuanha, H. rotundifolius, Culcasia dinklagei, C. striolata, Nephthytis poissonii, Marantochloa leucantha, Chlorophytum sparsiflorum, Anchomanes difformis, Commelina capitata, Palisota barteri, Pollia condensata, Costus englerianus, Cyanastrum cordifolium, Guaduella* spp.*, Dorstenia ciliata, D. elliptica, Piper umbellatum.*

Conservation priority species, owing to the poor sampling of lowland evergreen forest, are very incompletely worked out for this vegetation type. However, 77 Red Data taxa have been recorded (listed in the Red Data chapter), making it second in importance only to mapping unit 4. In addition, the two most important species for conservation in the Kupe, Mwanenguba, Bakossi Mt area are known from lowland evergreen forest. *Ossiculum aurantiacum* and Acanthaceae gen. nov. (*Cheek* 10152, *Onana* 987) qualify as 'most important species for conservation' in the checklist area because they are each so distinct as to be accorded their own genus, because they are extremely rare (one and two collections respectively) and because they are known from nowhere else in the world. *Onana* 987 was made within an area containing semi-deciduous forest, suggesting that the taxon prefers a drier type of lowland evergreen forest. Other strictly endemic species are *Psydrax bridsoniana*, known only from the Bakossi trough (two sites), *Cyperus microcristatus* (Kupe village, southern slopes of Mt Kupe) and *Memecylon* sp. nov. 2 (Nyandong valley, five collections). Near endemic species in this mapping unit are *Stelechantha arcuata* in the upper Mungo Valley and southern Bakossi trough, also being for example at Mt Cameroon, and *Rhaptopetalum* sp. nov., near Nyasoso, also occurring S of the checklist area, as is the case with *Psilanthus* sp. nov. (upper Mungo valley and Kumba), *Diospyros kupensis* (Kupe village area, Mbu and Nyandong valleys, also sometimes being found in submontane altitudes), and finally, from near the upper end of the altitudinal range of this mapping unit, *Afrothismia pachyantha* and *Kupea martinetugei*, both only known with certainty to exist on the lower slopes of Mt Kupe.

Threats to lowland evergreen forest are severe. Owing to the highly fertile soils of the Jide trough, Nyandong area, and the Dibombe valley, lowland forest in these areas has been replaced by agriculture almost totally. Elsewhere in our area forest occurs on less fertile soils. Nonetheless clearance has occurred along the Kumba–Mamfe road and is likely to intensify as this road is being greatly upgraded.

Selective logging has occurred over large tracts of this forest, most notably in the production forest reserves of the Mungo River and Bakossi in the S-centre of our checklist area. Where executed at low density and followed by lack of disturbance, selective logging does not necessarily constitute a conservation threat. However forest species are threatened where logging density is high, extraction practices poor, and where agricultural expansion follows logging, as occurs in parts of the Mungo River F.R., for example.

PHYTOGEOGRAPHY

The affinities of our lowland evergreen forest vegetation seem strongest with that of Mt Cameroon to the W in terms of overall assemblage of species (see Cable & Cheek 1998). However, very few other areas of lowland forest in SW Province, or indeed in Cameroon, have been documented, in terms of their detailed species composition. One that has is the southern part of the Korup National Park (Thomas 1993). This has several vegetational features not yet known in our checklist area, including areas of forest dominated by *Oubangia alata*, and monotypic groves of three

Caesalpinioid taxa, two of which are unknown in our area. The two units that Letouzey recognizes as occurring in this vegetation type (his mapping units 228 and 203) occur from Korup in the west, to the Mamfe basin in the N and eastwards to the Sanaga.

MAPPING UNIT 2: SEMI-DECIDUOUS FOREST; c. 300 m alt.

For the purpose of this checklist, this forest type is restricted to the *Triplochiton–Celtis* forest found in the SE corner of our area, in the Loum Forest Reserve (c. 3% of the checklist area). These elements accord strongly with Letouzey's definition of semi-deciduous forest in Cameroon (e.g. Letouzey 1968b: 185). *Triplochiton*, and a concentration of *Celtis*, are known from nowhere else in the checklist area. *Triplochiton*, in particular, is unmistakable due to its unusual, digitately-lobed leaves, and is unlikely to be overlooked wherever it occurs. The *Celtis* species are *C. adolfi-friderici* and *C. philippensis*. Now almost surrounded by anthropic vegetation, this forest type may once have been more extensive and is likely to have been continuous with the semi-deciduous forest of Mt Cameroon (see Cable & Cheek 1998). Semi-deciduous forest is not characterized by these two genera alone. Both Letouzey (op. cit.), and Hall & Swaine in classifying forest types of Ghana, list numerous other tree species that commonly occur in this forest type. Those that are known in Bakossi are as follows (arranged as per the checklist order):

	Milicia regia
Lannea welwitschii	*Trilepisium madagascariense*
Terminalia superba	*Sterculia oblonga*
Cylicodiscus gabunensis	*Sterculia rhinopetala*
Piptadeniastrum africanum	*Celtis gomphophylla*
Khaya ivorensis	*Celtis zenkeri*

However, these taxa are not confined to the Loum F.R., but are scattered along the Loum–Kumba and Kumba–Mamfe roads, as well as occurring in the Bakossi trough from Tombel to Nyasoso, in the Nyandong valley, with some recorded from the Mbu valley at Enyandong. The presence of these taxa, if not taken as an indication of well-developed semi-deciduous forest itself, at least suggests strongly a drier type of evergreen forest in the areas in which they occur. The presence of semi-deciduous forest is unexpected in an area mostly receiving 2–4 m or more of rain per annum; 2 m is usually considered sufficient to support evergreen forest. The explanation may be the free-draining nature of the volcanic derived substrate. The significance of semi-deciduous forest for phytogeography and refugia is treated under the chapter of that name.

Letouzey (1985) does not differentiate Loum F.R. as distinct from the other lowland forest at the periphery of our checklist area, treating it all as lowland evergreen with semi-deciduous elements. This may be because he did not visit this site, as far as is known.

Within the semi-deciduous forest at Loum F.R., evergreen forest also occurs along water courses, in the same way that evergreen gallery forest occurs in savanna area. At Loum, this evergreen forest is characterized by such species as *Phyllobotryum spathulatum* and *Annonidium mannii*.

Semi-deciduous forest of the *Triplochiton-Celtis* type is widespread in Cameroon. Large areas occur S of Yaoundé and extend into the E of Cameroon at the boundary with Congo-Brazzaville. Semi-deciduous forest is also widespread in W Africa, extending in a band westwards from Guinea-Conakry into Congo-Kinshasha, N of the evergreen forest of the Congo basin, reaching S Sudan and Uganda. Both *Triplochiton scleroxylon* and several of the *Celtis* species mentioned extend through much of this range. However, there are variations in species composition within this range.

Sampling of semi-deciduous forest has been poor, mostly restricted to one-day visits by our expeditions in 1995 and 1998, when based at Kupe village and Nyasoso. The only other specimens made at Loum (though not necessarily present day Loum F.R.) are those by Ledermann in 1908, most of which were destroyed in 1943.

Conservation priority species at Loum F.R. number seven. They are mostly widespread timber species of international trade e.g. *Khaya ivorensis* and *Entandophragma* spp. However three species described from Ledermann's Loum collections, *Thyrosalacia racemosa*, *Memecylon griseo-violaceum* and *Beilschmiedia crassipes* have not been located since. Although poorly understood (no specimens or illustration available and only sketchy descriptions) they may represent very rare, narrowly endemic species, hopefully still surviving at Loum but possibly now extinct. The critically endangered *Microberlinia bisulcata* is also recorded from Loum F.R., its only known location in Bakossi. One of the two specimens of the new genus of Acanthaceae referred to in mapping unit 1 is also located from Loum F.R.

Threats: Loum F.R. has suffered damage from clearance of forest and its replacement by cacao farms in some areas, and by wholesale replacement of much of the understorey of the surviving forest by cocoyam (*Colocasia* & *Xanthosoma* spp.) cultivation. It is not clear whether these threats are increasing or not.

MAPPING UNIT 3: FRESHWATER SWAMP FOREST AND LOWLAND RIVERSIDE VEGETATION; 150–800 m alt.

Perhaps 1% of the checklist area falls under this mapping unit, which should be a priority for more intensive investigation in future. Freshwater swamp forest areas are likely to be found in Bakossi but have neither been located nor inventoried to date apart from a small patch at Ngomboaku (see below). The high incidence of highly porous volcanic soils at low altitudes probably diminishes the incidence of impounded streams and so swamps. However enquiries should be made in lowland western Bakossi and the Mungo River F.R. S of the Kumba–Loum road for the existence and whereabouts of inundated forest.

Many of the tree species usually found in freshwater swamp forest also occur along lowland river banks. *Spondianthus preussii, Protomegabaria stapfiana, Ficus vogelioides, Hallea stipulosa, Homalium africanum* and *Crateranthus talbotii* are examples found along the lower reaches of the Mungo and Jide Rivers. All these taxa but the last are fairly widespread along rivers and in freshwater swamp in the forest areas of Cameroon. However, most of the riverbank species along the Mungo River section explored by us in detail, over two days, by boat (upstream of the Kumba-Loum bridge) were typical of lowland forest and are not noted for their affinities with rivers or swamps.

Lowland riverside forest, Mungo River, Loum-Kumba road, 26 Nov. 1999, 150 m alt., *Gosline* 219 *et seq.*

Brillantaisia lancifolia
Biophytum talbotii
Strophanthus thollonii
Hypselodelphys poggeana
Spondianthus preussii
Homalium africanum
Oxystigma gilbertii (*Cheek* 10238)

All the above taxa are noted for occurring near water, and usually near lowland rivers. The *Brillantaisia* is a rheophyte usually restricted to upland streams, an unexpected link with mapping unit 7. *Biophytum talbotii* is also restricted to habitats with flowing water, but usually occurs on the banks of slow flowing, lowland rivers, so is not treated in mapping unit 7. Large lowland rivers commonly have few, if any, obligate rheophyte taxa. The combination of slow-moving, highly turbid water and unstable earth banks is anathema to most obligate rheophytes which require clear, turbulent water on rock substrates such as are found in the higher reaches of the rivers in the checklist area.

A community of annual herbs grows on the Mungo river sandbanks when they are exposed in the dry season, including *Hydrolea glabra,* the only representative of its family in the checklist area. This community requires more scrutiny.

The Mungo River downstream of the Kumba–Loum bridge should be investigated.

Raphia swamp forest, Ngomboaku, 1999. Mbume Forest to E of village, 13 Dec. 1999, alt. 700 m. *Cheek* 10348 *et seq.*

Raphia cf. *africana*
Rungia buettneri
Floscopa mannii
Nelsonia smithii
Crateranthus talbotii

Raphia swamps are common throughout southern Cameroon wherever sluggish small rivers occur. Various species of *Raphia* are involved in such swamps, but usually just one species of *Raphia* per swamp. *Raphia* swamp canopies are often open, allowing high light levels to reach the ground level. The swamp mentioned above was wet underfoot even well into the dry season. The inventory at Mbume was not exhaustive. Two of the species listed are threatened, *Crateranthus talbotii* (VU) and *Floscopa mannii* (EN).

Letouzey (1985 4: 128) notes a wetland community along the Mungo as it winds to the W at a point just E of the Kumba–Mamfe road. He characterizes it as 'Prairies marécageuse', his mapping unit 244. This vegetation type is periodically inundated, eventually being dominated by *Pennisetum purpureum*. He states that 'being unaware of this site until recently, we have not yet inspected or surveyed it'.

None of the species endemic to the checklist area are known from this mapping unit, but four threatened species are known (see Red Data chapter). Threats to this mapping unit are unknown, but a high water-table in an area of heavy rainfall makes such areas a low priority for agriculture. *Hallea stipulosa* is a timber tree in international commerce.

MAPPING UNIT 4: SUBMONTANE (LOWER MONTANE, CLOUD, MOSS OR ELFIN) FOREST; 800–1900(–2000) m alt.

The Bakossi submontane forest is one of the largest pristine blocks of such forest in Tropical Africa. It forms about 25–30% of our checklist area, but a much higher proportion of the proposed new protected area system (see that chapter). This mapping unit now exists in three disjunct blocks (see vegetation map). Since they differ in their area, sampling, threats and endemic taxa, these subjects are treated below on a block-by-block basis after examining the vegetation type as a whole. Mainly within the submontane altitudinal band, several non-forest vegetation types can be recognized, namely savanna-thicket, inselbergs and rock faces, rheophytic communities, and crater lakes. These are treated subsequently, as mapping units 5–8 respectively.

The altitudinal boundaries of this mapping unit do not show an abrupt change in species composition on the ground and are somewhat arbitrary. A more complete analysis of our data on the altitudinal frequency of taxa on e.g. Mt Kupe, would be of value. Some submontane taxa can occur below 800 m altitude, and many lowland forest species can occur up to 1000 m altitude or more (e.g. *Oncoba dentata* 150–1700 m alt.). Nonetheless most submontane taxa do not occur much below 800 m in our area. The altitude at which the montane-submontane transition occurs varies between 1900–2000 m, being closer to the second figure on Mt Kupe and to the first figure at Mwanenguba. The transition to montane vegetation is more abrupt than the transition to lowland forest. Letouzey (1985) uses the same altitudinal boundaries for submontane vegetation.

Previous vegetational studies on submontane forest in Kupe-Mwanenguba-Bakossi Mts have been more or less confined to the W side of Mt Kupe, the second of the blocks. There have been differences in the altitudinal classification of vegetation adopted. On the basis of a rapid ascent to the summit in 1954, Letouzey (1968a) noted *Podocarpus milanjianus* between 1600–2000 m; remarkable, very old, *Santiria trimera* 1000–2000 m; numerous large Guttiferae of the genera *Allanblackia, Pentadesma, Symphonia, Garcinia*, and also *Cola*, between 1000–2000 m; with *Nuxia congesta* representing a montane stage at about 2000 m.

The next observations on vegetation were those by Thomas (1986). This work is out of print and for knowledge of it I rely on the comments of Lane (1993b, 1994a & b, 1995). According to Lane, Thomas's classification recognizes forest up to 1600 m as lowland, and above that as submontane: radically different to the classification adopted here. The early Aberdeen University reports adopt the Thomas classification (Stone 1993, Lane 1993b, Cable 1993). The later ones however, adopt a different approach, e.g. Lane (1994a & 1994b) which is close to that used here:

Lane's classification of vegetation on Mt Kupe (Lane 1994b).

Secondary Forest	< 1100 m	
Lowland Forest	< 800 m	(where it may exist)
Lower Transitional Zone	700–900 m	(where it may exist)
Submontane forest	800–1900 m	
Upper Transitional Zone	1800–2060 m	

Forest plot data

Stutter (1994) set up two permanent sample plots, one on a ridge at 1200 m, the other in the saucer at the summit area at 2000 m alt. Lane (1994) gives a useful summary of the data from these plots:

A half hectare permanent sample plot is situated at 1200 m on the Shrike Trail above Nyasoso. However it must be borne in mind that the plot exists on a ridge top and as such the results taken from Stutter (1994) may not necessarily truly reflect the situation elsewhere at this elevation.

Trees > 10 cm diameter at breast height (dbh) in 0.5 ha:-

Number of trees	338
Mean dbh	20.49 cm
Mean height	12 m
Number of buttressed trees	36
Mean dbh buttressed trees	63.81 cm
Number of species	47
Number of families	18

5 most common families:-

Sapotaceae	(2 spp., 15.2% of trees)
Myrtaceae	(1 sp., 14.9% of trees)
Burseraceae	(4 spp., 14.0% of trees)
Olacaceae	(3 spp., 10.7% of trees)
Euphorbiaceae	(5 spp., 9.8% of trees)

5 most common species:-

Eugenia zenkeri	(Myrtaceae, 14.8%)
Englerophytum stelecantha	(Sapotaceae, 13.9%)
Santiria trimera	(Burseraceae, 10.1%)
Drypetes leonensis	(Euphorbiaceae, 6.25%)
Strombosia scheffleri	(Olacaceae, 6.25%)

Trees between 5 –10 cm dbh in 0.1 ha:-

Number of trees	46	
Mean dbh	6.76 cm	
Mean height	6.5 m	
Number of species	20	
Number of families	15	
Most common trees	*Penianthus longifolius*	(Menispermaceae)
	Eugenia zenkeri	(Myrtaceae)
	Englerophytum stelecantha	(Sapotaceae)

[Note: *Englerophytum stelecantha* is restricted to ridges and rocky areas with thin soils on Mt Kupe; '*Eugenia zenkeri*' is almost certainly our *Eugenia sp. 1* and '*Drypetes leonensis*' is *Drypetes sp. A*, the last two taxa appearing more or less restricted to the submontane forest of Kupe–Bakossi Mts.]

A further half hectare permanent sample plot is located at 2000 m:

Trees > 10 cm dbh:-

Number of trees	397
Mean dbh	22.76 cm
Mean height	12.36 m
Number of buttress trees	4
Mean dbh buttress trees	87.94 cm
Number of species	26
Number of families	14

5 most common families:-

Guttiferae	(2 spp., 58.1% of trees)
Meliaceae	(2 spp., 14.5% of trees)
Annonaceae	(1 sp., 8.5% of trees)
Rubiaceae	(7 spp., 4.0% of trees)
Araliaceae	(2 spp., 3.0% of trees)

5 most common species:-

Garcinia smeathmannii	(57.8% of trees)
Carapa grandifolia	(13.7% of trees)
Xylopia staudtii	(8.5% of trees)
Zanthoxylum gilletii	(2.5% of trees)
Scheffleria mannii	(2.2 % of trees)

Trees between 5 cm–10 cm dbh in 0.1 ha:-

Number of trees	100
Mean dbh	7.13 cm
Mean height	8 m
Number of species	10
Number of families	7
Most common species	*Garcinia smeathmannii* (60.6%, Guttiferae)
	Rubiaceae indet. (18.2%)
	Pavetta kupensis (12.1%, Rubiaceae)

Trees < 5 cm dbh in 0.02 ha:-

Number of trees	75
Mean height	3 m
Number of species	10
Number of families	3
Most common species	*Psychotria peduncularis* (41.9%), Rubiaceae
	Psychotria sp. (29.7%, Rubiaceae)
	Garcinia smeathmannii (16.2%, Guttiferae)

[Note: '*Xylopia staudtii*' is *Xylopia africana*; '*Zanthoxylum gilletii*' is likely to be *Z. leprieurii*; Rubiaceae indet. remains indet. since no voucher has been located.]

Vascular epiphytes: quantitative data

Lane (1993a&b) studied the epiphytes on 11 trees of several species between 1000–1150 m alt. finding a total of 63 epiphytic species, an average of 18 species per tree, with 26 pteridophytes and 31 orchid taxa. He also investigated the distribution of epiphytes according to zones on the phorophyte. Pteridophytes (26 of the 63 spp.) were most diverse and denser at the inner zones, whereas orchids (31 of the 63 spp.) were more diverse and dense on the outermost branches. The other epiphytic taxa encountered were *Begonia*, *Peperomia*, *Utricularia* (two taxa each), *Remusatia* and *Rhipsalis*.

In contrast, 5 trees at 1650 m alt. produced only 17 epiphytic species, on average six species per tree, with 10 pteridophytes and two orchids, three *Peperomia* and two *Begonia*.

Tchiengue's doctoral thesis

Barthelemie Tchiengue's doctoral thesis at the University Yaoundé I, under the supervision of Dr. Bernard Nkongmeneck, is believed to be very close to finalization at the time of writing. This work focuses on the forest ecology of Mt Kupe and seems certain to transcend all previous studies on this subject when it is published.

Identification problems

In common with so many quantitative ecological studies, lack, or total absence, of voucher specimens has hampered the identification or confirmation of identification of the species referred to in most of such studies on Mt Kupe, with the exception of Lane's epiphyte study (Lane 1993a&b). Later specimen-based survey work has brought to light the improbability of some of the identifications published in these ecological studies and led to the alternative suggestions featured as 'notes' in the paragraphs on forest plot data above.

CHARACTERIZATION AND SUBDIVISION OF SUBMONTANE FOREST

Common and characteristic tree species of submontane forest in the Kupe, Mwanenguba and Bakossi Mts are *Xylopia africana*, *Oncoba lophocarpa*, *Garcinia smeathmannii*, *Dasylepis racemosa*, *Oncoba ovalis*, *Quassia sanguinea* (*Hannoa ferruginea*), shrubs such as *Psychotria peduncularis* var. *hypsophila*, *P. gabonica*, *Pauridiantha paucinervis* and *Deinbollia sp. 1*, and herbs such as *Sanicula elata*. These taxa occur throughout the 800–2000 m altitudinal range and are almost entirely restricted to this range. Some species of lowland forest also penetrate into this mapping unit, generally declining in frequency with altitude and so are less significant in upper than in lower submontane forest. Mass-flowering herbs are recorded from submontane forest, but are discussed under mapping unit 10, from which they are better known.

[Submontane forest can be subdivided into two altitudinal types–lower and upper. Again, no abrupt boundary occurs on the ground and although two different sets of species characterize the two subdivisions, many others occur throughout the entire submontane range and occur nowhere else. Some taxa, such as the extremely conspicuous *Psychotria camptopus*, with 3–4 m long dangling peduncles, blur the distinction between upper and lower montane, having a range of 1100–1700 m.]

Lower submontane forest (800–1400 m alt.)

Characteristic and common tree taxa of this altitudinal range are: *Allanblackia gabonensis, Pentadesma grandifolia, Garcinia lucida, G. conrauana* (the 'numerous large Guttiferae' of Letouzey 1968a); *Pseudagrostistachys africana* (700–1450 m – possibly restricted to ridges); *Santiria trimera* (200–1400 m); *Rhaptopetalum geophylax* (900–1500 m), *Chionanthus africanus, C. mannii* subsp. *congesta* and *C. mildbraedii*. Pioneer trees are *Cylicomorpha solmsii* (450–1450 m), *Polyscias fulva, Macaranga occidentalis* (600–1450 m) and *Cyathea manniana* (1200 m). Understorey treelets and shrubs are: *Sorindeia grandifolia* (930–1175 m), *Myrianthus preussii* subsp. *preussii* (520–1400 m), *Coffea montekupensis* (700–1500 m).

Lianas include *Uvaria heterotricha* (850–1600 m), *Pararistolochia ceropegioides* (600–1200 m), *Jasminum bakeri* (c. 1200 m), *J. dichotomum* (1200–1300 m) and *J. preussii* (870–1400 m).

Terrestrial herbs include: *Polyspatha paniculata* (700–1450 m), *Cardamine trichocarpa* (900–1550 m), *Graptophyllum glandulosum* ((200–) 800–1750 m) and *Buforrestia mannii* ((350–) 700–1200 m). Of the twenty-eight taxa of *Begonia* in our checklist area, all but seven are restricted to forest above 650 m, and seventeen only occur above 700 m alt. Eleven of the twelve *Impatiens* taxa in our checklist area are more or less restricted to this altitudinal range (two taxa descend to 350–400 m).

Epiphytes in lower submontane forest

The vascular epiphytic flora of Kupe, Mwanenguba and Bakossi Mts is concentrated in the lower submontane forest and must contend as the most diverse epiphytic flora in Africa. As documented by Lane (1993a&b, 1994b: see above), pteridophytes and orchids dominate the epiphytic community in terms of diversity of species and biomass. Most of the 187 orchid taxa that we have documented for our checklist area are epiphytes and most of these are restricted to submontane forest, including *Bulbophyllum*, which, with forty-two taxa is one of the most diverse genera in our checklist. Apart from pteridophytes and orchids, other families and genera contributing to the epiphytic flora are *Begonia* (thirteen hemi-epiphytic or canopy epiphytic taxa), Melastomataceae (*Calvoa hirsuta, Dicellandra barteri, Medinella mannii, M. mirabilis* and *Preussiella kamerunensis*), *Impatiens* (three epiphytic taxa: the Cameroon highlands having higher diversity of epiphytic taxa documented than anywhere else in Africa–Cheek & Csiba 2002), Lentibulariaceae (*Utricularia mannii, U. striatula*), Rubiaceae (*Chassalia petitiana, Mussaenda sp. nov.* and, a facultative epiphyte, *Hymenodictyon biafranum*), Cyperaceae (*Bulbostylis densa* var. *cameroonensis*, facultative, montane), Costaceae (*Costus letestui*), Cactaceae (*Rhipsalis baccifera*, predominantly lowland), Araceae (*Remusatia vivipara*), Loganiaceae (*Anthocleista scandens*, upper submontane), Urticaceae (*Procris crenata*), Acanthaceae (*Dischistocalyx grandifolius*, facultative, lowland forest), Moraceae (*Dorstenia astyanactis*–Pollard *et al.*, 2003), Piperaceae (*Peperomia*, seven species).

In addition, the species of *Schlefflera*, and many *Ficus* begin life as epiphytic shrubs, only later becoming terrestrial trees through the 'strangler' life form.

Many of the 135 species of ferns and fern-allies listed in this book are epiphytes, and are most numerous in submontane forest. Twelve species of filmy fern are known from our area. Compared with vascular plants, this group is undercollected in our checklist area.

Bryophytes (mosses and liverworts) are particularly conspicuous in submontane forest (also known as 'moss forest' in some parts of the globe), and have great ecological importance. The epiphytic species contribute significantly to the biomass of the canopy. Some of the epiphytic species are believed to be important in cloud stripping, i.e. absorbing water directly from clouds and so adding to the water received as rainfall. These species also provide an important substrate for many of the vascular epiphytes. Bryophytes are treated in a separate chapter.

Saprophytes-achlorophyllous mycotrophs

These shade-loving herbs can form small communities of up to six species in a 50 × 50 m plot, as we have discovered when conducting such sampling exercises. Two of such species assemblages have already been documented at length in Cheek & Williams (1999). However, some saprophyte taxa such as *Epipogium roseum*, always occur in isolation from other saprophytes. Most of the saprophytes recorded in the checklist area are from lower submontane forest at the boundary with lowland forest, the 750–1000 m altitudinal band harbouring most taxa, although species also occur below this band, and at altitudes up to 1300 m. Mt Kupe and the Bakossi Mts harbour twelve species and is currently the most diverse area of its size known for this ecological group in Africa, and possibly the world. In fact no other African country has as many saprophyte species as does our checklist area (Cheek & Williams 1999). The saprophyte species recorded thus far in Bakossi are as follows:

Sebaea oligantha		(Gentianaceae)
Afrothismia pachyantha	(E)	(Burmanniaceae)
Afrothismia saingei	(E)	
Afrothismia winkleri		
Burmannia densiflora		
Burmannia hexaptera		
Gymnosiphon longistylus		
Gymnosiphon sp. A		
Auxopus macranthus		(Orchidaceae)
Epipogium roseum		
Kupea martinetugei	(E)	(Triuridaceae)
Sciaphila ledermannii		

Species known only from Bakossi are marked E (endemic) above. *Kupea martinetugei* is notable for being the type not only of a new genus, but of a new tribe in its family (Cheek & Williams 2003, Cheek *et al.* 2004). Twelve species of 'saprophyte' were also known from Mt Cameroon (Cheek & Ndam 1996), but at least three of the taxa known from that location have not been seen for nearly a century (*Afrothismia pachyantha, A. winkleri* and *Oxygyne triandra:* Cable & Cheek 1998), despite repeated searches over the last ten years, and are most probably extinct at Mt Cameroon. For this reason we treat *A. pachyantha* as now being endemic to Bakossiland.

The explanation for the super-diversity of saprophyte species seen within Bakossi is probably due to it being part of a postulated microrefugium (see Phytogeography chapter) and to its possessing what is believed to be the largest pristine tracts of lowland-submontane evergreen forest within this microrefugium.

Upper submontane forest (1400–1900 m)

This forest type appears much less diverse than lower submontane forest in both tree diversity and epiphyte diversity, both overall and per unit area (see Stutter 1994, Lane 1993a&b, 1994b). The numbers of species that are restricted to this forest type are relatively few in number and generally not very conspicuous, in contrast to the situation described above for lower submontane forest. The bulk of the individual stems and basal area are made up of species which range throughout the whole submontane range e.g. *Garcinia smeathmannii* (see Stutter 1994, Lane 1994b above). The predominance of the last species in Stutter's plot is, as Lane (1994b) suggests, not entirely representative of the upper submontane forest as a whole. In large parts of the shallow saucer that forms the top of Mt Kupe, *Zenkerella citrina* replaces it as the dominant canopy species. Although we record this species with a range of 300–2000 m, below 1500 m it becomes infrequent and in practice it is a good indicator of upper submontane forest. Similarly, *Podocarpus milanjianus* descends to 900 m according to our records, but below 1700–1900 m it is rare. Even at altitudes where it is commonest, its distribution is patchy. At the upper end of the altitudinal range of this zone, montane species such as *Carapa grandifolia, Schefflera mannii* and *Nuxia congesta* occur with increasing frequency, causing Lane (1994b) to recognize a transition zone between these two vegetation types.

One of the few tree species restricted to upper submontane forest is *Tricalysia* sp. B *aff. ferorum*. Other shrub–small tree taxa more or less restricted to this forest type are *Maytenus buchananii* (1400–2050 m), *Maesa rufescens* (1500–2050 m) and *Anthocleista scandens* (1300–2000 m); climbers are *Salacia erecta* var. *erecta* (1350–1900 m) and *Tragia benthamii* (1300–1970 m).

Kupe and Bakossi Mts: endemic and near-endemic taxa of submontane forest.

Species restricted to Mt Kupe and the tongue to the N are listed separately under Block 2, below, and are not included here. Similarly, species restricted to the Bakossi Mts are listed under Block 3, below, and are not included here. The submontane forest of Mwanenguba is almost unknown scientifically, and accordingly it is possible to list only a single apparently endemic taxon from Block 1. Endemic taxa are defined as being known only from the area under which they are listed, and near endemic taxa as also occurring at one other site, such as the Rumpi Hills.

Begonia duncan-thomasii (also Rumpi Hills)
Salacia elegans var. aff. *inurbana*
Diospyros kupensis (also in lowland forest)
Drypetes sp. A.
Eugenia sp. 1
Chassalia aff. *laikomensis* sp. B
Chassalia aff. *laikomensis* sp. C
Coffea bakossii
Coffea montekupensis
Ixora sp. A
Lasianthus sp. A (including Mwanenguba)
Mussaenda sp. nov.

Pavetta owariensis subsp. *satabiei* (also Rumpi Hills)
Pentaloncha sp. nov.
Psychothria sp. aff. *camerunensis*
Psychothria sp. A aff. *gabonica*
Psychothria sp. B aff. *gabonica*
Sericanthe sp. A
Tricalysia sp. F
Allophylus sp. 1 (also Banyang Mbo)
Rhaptopetalum geophylax
Cola sp. nov. '*etugei*'
Dracaena cf. *phanerophlebia*

The phytogeographical links of the submontane vegetation of Kupe, Mwanenguba and the Bakossi Mts not surprisingly appear closest to those of the adjoining submontane areas. These are, to the W, the Rumpi Hills, to the E, Mt Nlonako and to the S, Mt Cameroon and more particularly Mt Etinde. The flora of the first two is incompletely known and for this reason it is difficult to gauge the extent of their linkage. Mt Nlonako shares with Mt Kupe *Ardisia koupensis* (absent from e.g. the Bakossi Mts). Several rare and threatened *Begonia* species, e.g. *B. adpressa, B. pelargoniiflora, B. pseudoviola* and *B. schaeferi* are also shared by Kupe, Mwanenguba and Bakossi Mts with Nlonako and, principally the Bamboutos Mts to the N.

The Rumpi Hills share two of the 23 taxa that otherwise occur only in both Mt Kupe and the Bakossi Mts, and five of the taxa that otherwise only occur in the Bakossi Mts.

When botanical surveys of the Rumpi Hills and Mt Nlonako are more complete, it is likely that increased linkages with our checklist area will be revealed.

Links with Mt Etinde

To the S, the flora and vegetation of the Mt Cameroon area are relatively well-studied (e.g. Cable & Cheek, 1998). The 'closed canopy' submontane forest of Mt Etinde (1760 m alt), the geologically older subpeak of Mt Cameroon, has been recognized as being quite separate from the 'discontinuous canopy' submontane forest of the more geologically recent main massif of Mt Cameroon (Cable & Cheek 1998, Thomas & Cheek 1992). Mt Etinde shares several striking phytogeographic links with Mt Kupe, in particular the *Garcinia smeathmannii* dominated forest in both their summit areas, the existence of Sapotaceae, particularly *Englerophytum stelechantha,* and of *Pseudogrostistachys africana* on ridges, all features absent from the main massif of Mt Cameroon. *Psychotria* sp. B aff. *gabonica* and *Pavetta brachycalyx* are species only known from submontane Mt Kupe and the Bakossi Mts, and Mt Etinde. *Cyperus rheophytorum, Impatiens frithii* and *Plectranthus cataractarum* are only known from both the Bakossi Mts and Mt Etinde, while *Manniella cyprepedioides* also occurs on Bioko. *Dovyalis sp. nov.* is only known from Mts Etinde and Kupe.

Links with submontane Upper Guinea

Gladiolus aequinoctialis (also Sierra Leone), *Dorstenia astyanactis* (also Mt Tonkui, Nimba area), *Brachystephanus nimbae* (also Nimba Mts) and *Croton aubrevilleii* (also Ghana and Ivory Coast, up to 850 m alt.) are all submontane species shared by the Kupe, Mwanenguba and Bakossi Mts, and in most cases some other parts of the Cameroon Highlands, with the mountains of Upper Guinea, in particular the Nimba Mts.

Links with the Crystal Mts, Gabon

Uvaria heterotricha, Sarcophrynium villosum, Pauridiantha venusta, Costus letestui, Barteria solida, Allanblackia gabonensis and *Calochone acuminata* all link Kupe, Mwanenguba, and the Bakossi Mts with the Crystal Mts of northern Gabon. In the case of the *Sarcophrynium*, the taxon is known from nowhere else but these two sites. In the case of *Allanblackia*, the taxon is also known from other geologically older (i.e. not Mt Cameroon) parts of the Cameroon highlands. Several of these submontane taxa (*Costus letestui, Pauridiantha venusta, Allanblackia gabonensis*) also occur on small submontane 'stepping stones' located in S Province, Cameroon e.g. at Bolobo (*P. venusta*), or Sangmelina and Ebolowa (*Allanblackia gabonenesis*).

THE SUBMONTANE FOREST BLOCKS OF KUPE, MWANENGUBA AND THE BAKOSSI MOUNTAINS

Block 1. Mwanenguba

Now the smallest of the three blocks, this remnant of a much larger expanse of forest appears from the 1:200,000 IGN map to have outlying islands in the anthropic vegetation near Nkongsamba. Letouzey's vegetation map of Cameroon shows 9–10 bands of submontane forest below the main block, radiating down watercourses on the S and E side of the massif from c. 1500 to c. 800 m alt. This block lies on the southern and southeastern flanks of the Mwanenguba caldera. Sampled by us only recently at Nsoung (three days in April 2003, expedition led by Darbyshire & Etuge), the resultant specimens have only become available recently and so have not been analysed fully.

This block was first sampled in 1908 at Ndoungue (c. 850 m alt.) by Ledermann, when it was pristine forest. Now, according to the IGN 1:200,000 map, the forest at Ndoungue and the vicinity has been cleared. Of the c. 200 specimens collected there by Ledermann, almost all have been destroyed. While the identifications of most of these are not available, those of some are. These were type specimens; see the paragraph on endemic species below.

In Nov. 1968 Sanford collected in the 1200–1800 m altitudinal range making a considerable number of specimens, but mostly of orchids (his specialism). The locality for most of these specimens is given as 'Massif Manengouba, near Nkongsamba, above upper farm of 'La Pastorale' 5°N, 9°58' E'. However, it is possible to piece together from his collecting notes that the area he worked in was a mosaic of grassland and forest. The grassland had a similar composition to that of the montane grassland recorded in the caldera of Mwanenguba (see mapping unit 10) while the composition of the forest trees is unknown, since he did not collect any of these.

It is not clear whether this grassland is secondary or natural, nor is it clear how it relates to the submontane grassland of the Bakossi Mts (mapping unit 5). Letouzey (1985) treats it as anthropic, his mapping unit 128: see anthropic vegetation below (mapping unit 11).

Leeuwenberg and Villiers also collected in the Mwanenguba submontane forest in the 1970s and 80s, but only a small part of the specimens of the first and only one or two of the second are available at the time of writing.

Block 1 is by far the most poorly sampled of the three submontane blocks and is highest priority for re-investigation. Believed formerly to have been linked with Block 2, it may nevertheless have unique features. Characterization of the submontane vegetation of Block 1 is almost impossible given the lack of data. More surveying in this area is needed.

Only a single **endemic taxon** is known for this block: *Beilschmiedia ndongensis*, based on material collected by Ledermann in 1908. It has not been recorded since, and may be extinct, given the habitat destruction at this site. However, although this taxon is accepted in the Flore du Cameroun account of the Lauraceae, species identification in this genus remains very difficult and it may be that material of the species concerned has been recollected but not identified. Given the diversity and endemicity seen in the other two submontane blocks, it is likely that, with further botanical exploration work, more endemics will come to light in future from the Mwanenguba block, so long as the forest continues to exist.

Threats to Block 1 are poorly known but are thought to be a) grassland dry season fires set in the adjoining montane grassland and anthropic fallow vegetation; b) clearance for new agricultural land, especially from the direction of the major town of Nkongsamba in the east.

Block 2. Mt Kupe and the submontane bridge

Second in size of the blocks is that comprising Mt Kupe and a tongue extending N along the submontane bridge (± continuous at 1000 m contour *fide* Wild) towards Mwanenguba. It was probably formerly contiguous with the first block. Letouzey (1985) shows this block to be fragmented into three sub-blocks due to conversion of forest into anthropic vegetation in the area immediately to the N of Mt Kupe. However, his map was based mainly on aerial observations and it is possible that he has misinterpreted as anthropic vegetation the band of submontane thicket that exists in this area (see mapping unit 5). The second block has been well sampled by us (1995–2000) at Mt Kupe (above Nyasoso, Kupe Village and Tape Etube).

Nonetheless, even in these relatively well-explored places, new and exciting discoveries of rare and undescribed plants are likely to be made in future. This is illustrated by the collection by Moses Sainge of a spectacularly peculiar new species to science of *Afrothismia, A. saingei*, above Mbulle on Mt Kupe in 2002 (Franke 2004), despite numerous surveys by our expeditions in previous years and despite our proficiency in locating saprophytes of this sort! Most of our surveys of this unit on Mt Kupe have been made above Nyasoso or Kupe village. The Sainge collection suggests that employing other starting points for surveys on Mt Kupe could lead to discoveries of other taxa hitherto unknown on the mountain.

Most of the taxonomic and ecological surveys made in the checklist area prior to our efforts in the late 1990s were conducted above Nyasoso (see 'previous vegetational studies' above). The eastern side of Mt Kupe is almost unknown. The northern tongue was fairly well sampled by us in 1999 at Ngomboaku and Mwambong, but is unknown N of Mwambong.

Endemic & near-endemic taxa of Block 2

The taxa listed below are unknown from the Bakossi Mts (a separate list for taxa endemic and near-endemic to both Mt Kupe and the Bakossi Mts is given above). Some of the taxa listed occur just below the 800 m contour. These are marked 'b', for being atypical in their lower altitudinal limt.

Brachystephanus kupeensis
Justicia sp. 1 (Ngomboaku only)
Friesodielsia enghiana var. *nov.*
Monanthotaxis sp. nov.
Uvariopsis submontana (also at Yabassi)
Drypetes sp. B
Tragia sp. B
Memecylon sp. 1
Ardisia koupensis (also at Nlonako)
Eugenia sp. 2
Calycosiphonia sp. A
Ixora sp. B
Psychotria sp. aff. *dorotheae*
Psychotria sp. aff. *foliosa* (Kupe & Ngomboaku)
Psychotria sp. A aff. *leptophylla*
Psychotria sp. B aff. *leptophylla*
Psychotria sp. A aff. *subobliqua*
Rytigynia sp. B

Rytigynia sp. C
Tricalysia sp. B aff. *ferorum* (also at Mwanenguba and Bali Ngemba F.R.)
Tricalysia sp. G
Vepris sp. 1
Pancovia sp. 1 (Kupe & Ngomboaku)
Microcos sp. A
Afrothismia pachyantha (considered extinct at Mt Cameroon, its type locality) **b**
Afrothismia saingei
Costus sp. A
Cyperus tenuiculmis subsp. *mutica*
Dracaena sp. aff. *phrynioides*
Bulbophyllum jaapii
Bulbophyllum kupense
Polystachya kupensis
Kupea martinetugei (report from Mt Cameroon not yet confirmed) **b**

Threats to Block 2 have been heaviest from the E, evidenced by the more extensive forest clearance there, reaching higher contours, than that on the western side. It is not clear to what extent, if any, forest clearance continues. Satellite monitoring is needed. On the western side, above Nyasoso (c. 850 m alt.) the main threats to the forest seen in mid 1995 (*pers. obs.*) were small-scale removal of timbers for local construction, removal of firewood and cultivation of cocoyams in one flat patch, at c. 1100 m alt. The forest boundary is now believed to be stable (Wild *pers. comm.* 2003). Above Kupe village (c. 500 m alt.), to the S of Mt Kupe, agriculture ascends to c. 850 m alt., mainly *Coffea canephora*. Observation on the ground over five years (1995–2000) shows the forest boundary here also to have been fairly stable.

Block 3. Bakossi Mts

The largest block, in the W of the checklist area, is in the Bakossi Mts, twice the size, or more, of blocks 1 and 2 combined. This block has been sampled by us along an E–W line from Mwambong to Kodmin, Mwanzum, Nyale and Nyandong, with excursions N and S from Kodmin (1998). It has also been sampled in the N from Enyandong, both by us in 2001 and by Thomas and associates in 1986. S of the E–W line referred to, and N of it, are vast tracts which are botanically unknown and still require surveying.

The character of the submontane vegetation of the Bakossi Mts is similar to that of Mt Kupe. There are, however, significant differences. While 23 species are listed above as endemic or near endemic to both areas, Mt Kupe is host to 32 other endemic or near endemic taxa that appear not to occur in the Bakossi Mts (listed above under Block 2). Some of these (e.g. *Ardisia koupensis*) are relatively frequent in the forest understorey of Kupe. Similarly the Bakossi Mts are host to 24 endemic or near endemic taxa that do not occur on Mt Kupe (listed below). Most of these endemics are so rare or inconspicuous as to be easily overlooked, but a few are both common and conspicuous (e.g. *Justicia leucoxiphus* and *Impatiens letouzeyi*), emphasising the distinctness of the submontane forest of the Bakossi Mts.

A further disparity between blocks 2 and 3 is the absence on Mt Kupe of one of the dominant submontane ridge trees of the Bakossi Mts, *Newtonia duncanthomasii* (also in the Rumpi Hills). A possible explanation for this disparity is discussed in the chapter on Phytogeography and Refugia.

Endemic and near-endemic taxa of Block 3

Impatiens letouzeyi

Impatiens frithii (also at Mt Etinde)

Begonia prismatocarpa subsp. *delobata* (also in the
    Rumpi Hills and at Nta-ali)

Salacia sp. nov. A

Phyllanthus nyale

Amphiblemma sp. 1 (see also mapping unit 6)

Ledermanniella letouzeyi (also in the Rumpi Hills) see
    also mapping unit 7

Ledermanniella onanae (see also mapping unit 7)

Bertiera sp. nov. (also in the Rumpi Hills)

Keetia bakossii

Pavetta muiriana (also at Banyang Mbo)

P. rubentifolia

Psychotria sp. aff. *fuscescens*

Rothmannia ebamutensis (also at Banyang Mbo)

Sericanthe sp. B

Tricalysia sp. C aff. *gossweileri*

Tricalysia sp. D cf.. *oligoneura*

Tricalysia sp. H

Rinorea fausteana (also in the Rumpi Hills & at
    Banyang Mbo)

Gymnosiphon sp. A

Hypolytrum pseudomapanioides

Hypolytrum subcompositus

The barrier between the Bakossi Mts and Mt Kupe appears not to be solely the Bakossi Trough (the Jide valley that runs from Tombel to Bangem). Several of the Bakossi Mts species cross the Trough and are abundant at Ngomboaku in the tongue of forest that extends northwards from Mt Kupe towards Mwanenguba. These species do not occur on Mt Kupe itself, so a second barrier must exist somewhere between Mt Kupe and Ngomboaku. This barrier may be the band of thicket immediately N of Mt Kupe (see mapping unit 5).

Endemic taxa of the Bakossi Mts & Ngomboaku (but absent from Mt Kupe)

Justicia leucoxiphus

Talbotiella bakossii (also Rumpi Hills)

Octoknema sp. nov.

Phyllanthus caesiifolius

Threats to the submontane forest of the Bakossi Mts seem most pronounced at the lower altitudes where illegal commercial logging has occurred on the southern and western slopes (Etuge *pers. comm.*) as recently as 2003. It is to be hoped that the proposed National Park status for this area will help protect the area. Development associated with higher-altitude villages, e.g. small-scale reservoir conservation, may pose difficulties for some narrowly distributed taxa, but with careful planning there, potential conflicts between development and conservation can be avoided. On the whole, populations of people in the higher altitude villages appear to be declining, with the possibility of some villages, such as Mejelet, being abandoned (Etuge *pers. comm.*). This phenomenon appears due to the lower soil fertility when compared with lower altitude villages and poorer and so more uncertain and more expensive communications. However, some higher-altitude communities, notably that of Kodmin, sustain themselves by transhumance. For several months of the year most of the population move to Mekom, at low-altitude in the fertile Nyandong valley, in order to farm the cash-crop cocoa.

MAPPING UNIT 5: SUBMONTANE 'GRASSLAND' (SAVANNA-THICKET); 1000–1600 m alt.

A remarkable feature of the Bakossi Mts is the occurrence of patches of 'grassland' in the forest. It was Chris Wild in early 1998, who first alerted us to the existence of this phenomenon and requested an opinion on their origin.

GIS techniques have not yet been employed to map or quantify this vegetation type. A Landsat image examined with the help of Susana Baena proved to be obscured by cloud in the areas where we have studied this vegetation on the ground. To the N and E of Nyandong, Letouzey (1985) shows four patches of what he has interpreted from aerial photography as anthropic vegetation but which may prove to be the same as that discussed below if inspected on the ground. The following notes are based on what we recorded in the vicinity of Kodmin and Edib in 1998. The total area occupied by mapping unit 5 is miniscule as a proportion of the total checklist: probably less than 0.5% of the total. However the physiognomy and assemblage of species involved is distinct and worthy of note.

Mapping unit 5 is only known as a subunit of Block 3 of Mapping Unit 4. That is, it is absent from the submontane forest of both Mwanenguba (poorly explored) and Mt Kupe (relatively well explored). However, immediately to the N of Mt Kupe, in the vicinity of Tape Etube, lies a very large expanse of *Aframomum-Hypselodelphys* thicket that Letouzey has interpreted as anthropic vegetation. The character of this vegetation seems entirely different from the grassland of the Bakossi Mts that is characterized below. I have not seen this area myself, since I was absent from the expedition there in January 1995 led by Cable and Lane. An apparently similar vegetation type exists on Mt Cameroon (Cable & Cheek 1998: xvi) where it is characterized as mapping unit 7. Whether this vegetation is entirely natural, or derived from land-slips on unstable slopes, or from elephant damage, or from man-made disturbance, perhaps some form of abandoned agriculture as has been suggested on Mt Cameroon, is unknown. More investigation is needed.

Six areas of 'grassland' were documented by us in 1998 in the Bakossi Mts. None of these was thought to be more than 100 ha and the smallest was only c. 25 × 25 m. The first, and largest, area encountered lies at c. 1450 m alt., almost immediately W of Kodmin, after traversing the stream that marks the boundary, and its associated riverine forest, taking the road (footpath) for Mwanzum and Edib, the dichotomy for which later occurs within the grassland. Taking the Mwanzum fork, at length a thickly forested ridge is encountered, through which the footpath cuts, emerging into another area of 'grassland' on the other side, also many hectares in extent. Continuing along the path to Mwanzum, and branching to the N for Mt Nloh takes one through another forest area to a third patch of grassland, also several hectares in extent, occupying much of the eastern flank of Mt Nloh, on a slope of about 45° from the horizontal. Returning to Kodmin, and taking the path for Edib, a fourth patch also of several hectares, is located about 2 hours walk distant, before entering the village of Edib. Finally, between Kodmin and Nzimbeng, to the N, a small patch of only about 25 × 25 m occurs W of the path, also on a slope of 45° or more, on rocky ground.

The first area mentioned, W of Kodmin, was investigated on 15 Feb. 1998 (drier season). The height of the canopy was 3 m, comprising *Pteridium aquilinum* (c. 35% of surface area) and *Dissotis brazzae* (c. 10%). The understorey, reaching 1 m from the ground was dominated by *Panicum hochstetteri* (c. 60%). Other species accounted for about 5% of surface area, prominent among these being *Virectaria major*, young *Clausena anisata*, *Hypericum roeperianum* and *Platostoma rotundifolium*. A pit dug at this site showed fine, red-brown soil to a depth of 60 cm, below which stones c. 2 cm diameter appeared. The area surveyed was 25 × 25 m, our standard plot size. More investigations of this sort would be rewarding since considerable variation in species composition was noted within and between areas. Composition is also likely to vary with season.

Enquiries at Kodmin produced no enlightenment as to the origin or purpose of these grassland areas. However, they are normally burnt each year, facilitating travel through them. 'Grassland' in submontane altitudes in SW Province, Cameroon, is unknown, although at montane altitudes, 1900–2000 m, grassland occurs naturally at Mt Cameroon, and elsewhere in the Cameroon Highlands.

The term 'grassland' is a misnomer for Europeans, since grasses in the sense of *Gramineae* are a relatively minor, and often inconspicuous part of the vegetation. In Pidgin English however, 'grass' means any herb (in contrast to a shrub or tree: 'stick') and so is appropriate. Physiognomically the vegetation consists of dense annual stems 1–3 m high, with few climbers, and a few shrubs or trees, usually in clumps. In terms of species diversity, *Compositae* are dominant.

Species surveys in 'grassland' patches gave the following results:
Grassland patches 0.5–3 km E of Kodmin, c. 1450 m alt. All are herbs, unless otherwise stated.
17 Jan. 1998 (*Cheek 8872–8907*)

Agrocharis melantha

Dissotis thollonii var. *elliotii*

Tetrorchidium didymostemon (tree)

Platostoma rotundifolium

Vernonia glabra var. *hillii*

Gomphocarpus semilunatus

Ipomoea tenuirostris

Triumfetta rhomboidea

Helichrysum odoratissimum

Setaria longiseta

Euphorbia schimperiana var. *pubescens*

Pouzolzia parasitica

Dalbergia saxatilis (shrub)

24 Jan., 7 & 12 Feb. 1998 (*Cheek 9077 et seq.*)

Wahlenbergia perrottetii

Viola abyssinica

Urena lobata

Cyperus fluminalis

Pennisetum polystachion subsp. *polystachyion*

Helichrysum foetidum

Virectaria major var. *major*

Solanecio mannii

Lactuca inermis

Elephantopus mollis

12 & 14 Nov. 1998 (*Cheek* 9571–9633)

Lantana ukambensis (shrub)
Marsdenia latifolia
Harungana madagascariensis
Gomphocarpus physocarpus
Lobelia rubescens
Premna angolensis (shrub)
Neonotonia wightii subsp. *pseudojavanica*

Dalbergia lactea (small tree)
Virectaria major var. *major*
Vernonia blumeoides
Phyllanthus nummulariifolius
Spermacoce intricans
Adenostemma mauritianum
Conyza attenuata

Grassland patch E side of Loh Mt, c. 1600 m alt.

23 Jan. 1998 (*Cheek* 9075 *et seq.*). Taxa additional to those already recorded.
Dissotis brazzae
Desmodium setigerum

Grassland patch E side of Kodmin Mt, c. 1500 m alt.

20 Jan. 1998 (*Cheek* 8977 *et seq.*). Taxa additional to those already recorded.
Kalanchoe crenata
Chromolaena odorata
Rubus rosifolius
Cyperus cyperoides subsp. *cyperoides*

Grassland patch Kodmin–Nzimbeng, c. 1450 m alt.

14 Nov. 1998 (*Cheek* 9208 *et seq.*). Taxa additional to those already recorded.
Vernonia blumeoides
Pteridium aquilinum

All but the W Kodmin grassland patch are very incompletely sampled, and that at Edib not at all. This reflects the survey system used: specimens were usually collected only when considered novelties in the checklist area, or when there was a specific need to voucher their existence at a site.

Most of the species recorded from this vegetation type also occur in the montane grassland and savanna at Mwanenguba but are unknown e.g. on Mt Kupe e.g. *Agrocharis melanantha, Vernonia glabra* var. *hillii, V. blumeoides, Helichrysum odoratissimum, Lactuca inermis, Ipomoea tenuirostris, Euphorbia schimperi* var. *pubescens, Dalbergia lactea, Platostoma rotundifolium, Dissotis brazzae, D. thollonii* var. *elliotii, Lobelia rubescens*. Some other of these species are well known as widespread and fire-tolerant: *Hypericum roeperianum, Pteridium aquilinum*.

Several taxa are post-cultivation pioneers, common in farmbush: *Neonotonia wightii* subsp. *pseudojavanica, Chromolaena odorata, Triumfetta rhomboidea, Urena lobata, Elephantopus mollis*.

A few of the species listed fit neither of those categories, but occur in other open areas at this altitudinal range within our checklist area: *Dalbergia saxatilis, Pouzolzia parasitica* (Cocoa farm weed, and on *Sphagnum* at Lake Edib) & *Virectaria major* (Montane grassland/forest edge of both Mt Kupe and Mwanenguba).

There are very few taxa which, within our checklist area, are known only from this vegetation type. Those that are, e.g. *Lantana ukambensis, Marsdenia latifolia, Gomphocarpus physocarpus, Solanecio mannii, Helichrysum foetidum, Cryptotaenia africana, Wahlenbergia perrottetii*, are not particularly rare, and are widespread outside the area. They may yet be found in Mwanenguba grassland. No strictly endemic or near endemic taxa are known from this vegetation type.

From the above it can be concluded that the submontane 'grassland' of the Bakossi Mts is of relatively recent origin given the lack of endemic or otherwise rare taxa. It has greatest affinities with the Mwanenguba montane grassland, whence most of its species appear derived. The post-cultivation pioneer taxa are consistent with previous agricultural activity. The deep soil recorded at the large grassland area W of Kodmin is not consistent with shallow-soil-induced grassland. Therefore there seems no reason why, apart from the usual setting of annual fires, most of these areas should not revert to forest eventually. Those sites on rock outcrops and with thin soils, e.g. that at Kodmin–Nzimbeng, are probably the only ones that are not man-made and which will persist in the absence of human influence. They might also have provided the stock of species that colonized the presumably later-created larger patches.

These larger areas were probably originally cleared of forest to provide land for agriculture. Research by Chris Wild (pers. comm. 1998) has revealed that Bakossi cattle once grazed these lands. This domesticated small-statured breed is now on the edge of extinction. A few are believed to survive at Dikome Balue (Wild pers. comm.). 'Living fences' of *Dracaena arborea* were deployed to keep the cattle out of plots of food crops immediately around the village. These fences, now enormous (10–20 m height) in size, are still much in evidence today at Kodmin, and at other submontane Bakossi villages.

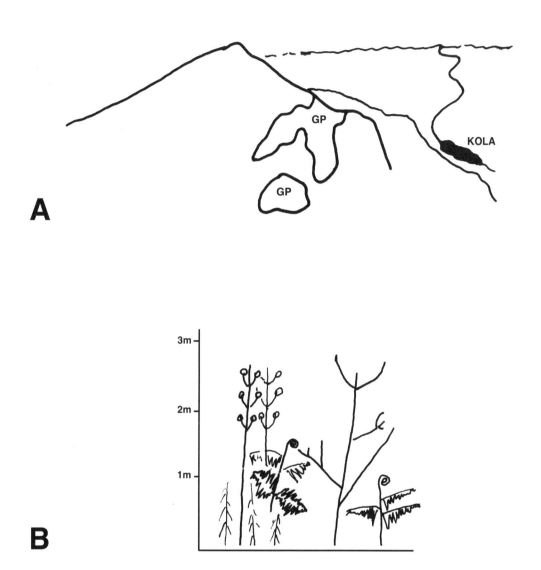

Fig. 4A (above). MONTANE GRASSLAND.
Sketch view from N of highest of three peaks (peak 1) showing grassland patches (marked GP), Feb. 1995, Mt Kupe.
Fig. 4B (below). SUBMONTANE 'GRASSLAND'.
Profile of patch Feb. 1998, W of Kodmin, Bakossi Mts.

MAPPING UNIT 6: ROCK FACES OF CLIFFS AND INSELBERGS: LITHOPHYTIC COMMUNITIES; 600–1500 m alt.

The vertical faces, several hundreds of metres high, of the uplifted syenitic-granitic inselbergs 'Rocks' scattered through Bakossi remain poorly inventoried in the absence of rope-equipped rock climbers. Those inselbergs that we have researched are (in chronological order), Kupe Rock near Kupe village, behind Tombel; Nyale Rock at Nyale village between Nyandong and Kodmin and finally Bime Rock, SE of Enyandong. Undoubtedly others occur. We have glimpsed large rock faces on the S, N and E faces of the summit area of Mt Kupe but these have not been studied botanically by us. Letouzey (1985), on the basis of aerial photography, shows two arcs of 'saxicole' communities corresponding with the SW and NE part of the rim of the Mwanenguba caldera, only two and not three summit outcrops on Mt Kupe, and in the Bakossi Mts 22 rock outcrops varying greatly in area, the largest extending in an arc c. 5 km long. Both Mwanenguba and Bakossi Mts outcrops need field inspection and surveying, as do those unvisited on Mt Kupe.

Smaller rock surfaces, only several metres in diameter, can also support specialized lithophytic communities. That at Ellas-Pong rock above Mbulle is detailed below. Several other small rock faces have been located by searching our specimen database for specialist lithophyte species (*Nodonema lineatum* and *Begonia duncan-thomasii*). One of these rock faces occurs W of Bangem (Thomas & McLeod collections), several above Nyasoso (Wheatley, and Thomas collections) and one in the vicinity of Kodmin at Abwoh (Biye, and Etuge collections). While these small rock faces have been localized by this means, they have not been fully surveyed for lithophytes. With the exception of Etuge's Kodmin Abwoh rock face, only one or two rock face specimen numbers were made by each collector, inadequate to characterize their plant communities in any detail.

Large rock surfaces are easily seen from many kilometres away. They may be so prominent as features as to be well known to villagers far beyond this distance. I first heard of Nyale Rock when at Kodmin, many hours distant, for example. Smaller rock surfaces can be easily spotted in the almost sheer valley walls of the Jide or Mbu troughs. They are most conspicuous, as pale grey, almost white areas in a carpet of dark green forest, in the dry season.

However, rock faces can be entirely concealed by the forest canopy and are then only to be located by enquiring of helpful villagers, or by accident. Without guides, locating all but the largest rock faces in forest areas is difficult.

Rock faces, although often locally conspicuous (and occasionally visually dominating some parts of Bakossi) must occupy less than 0.5% of our checklist area.

Threats to rock face communities are absent, apart from a) rock-falls, which could eliminate species at a site and b) forest clearance, which could remove the shade demanded by some lithophytes. Quarrying of stone from rock faces in Bakossi is unknown.

Kupe Rock is the best studied rock face. Facing SW it is capped by forest and embedded in the side of Mt Kupe. During the wet season much of the rock face is green with the annual growth of a carpet of herbs, but within a few days of the onset of the dry season in October-November, these shrivel to brown.

We have accessed this site only from the base. The herbs of the bottom-most few metres of the rock face, shaded by forest and receiving water that drips down from perennial seepages above, remain largely undesiccated in the dry season, and form a separate assemblage from those above in full sun. Collecting specimens from the exposed part of the rock involved climbing up piles of boulders and scree and using telescopic poles to bring down specimens from above. Only a minute portion of the plant community on the exposed part of the rock could be reached. The following list of taxa is the result of that exploration. Better sampling of the last community could be obtained by abseiling from the top of the rock face after climbing the footpath that ascends through the forest at the right hand side.

Kupe Rock (R1 of vegetation map) 1100–1400 m alt., 4°47' 24" N, 9°41' 24" E

7 Nov. 1995 Cheek 7688 *et seq.*

Coleochloa abyssinica var. *abyssinica*
Gladiolus aequinoctialis
Impatiens burtoni
Streptocarpus elongatus
Scadoxus cinnabarinus
Selaginella sp.
Panicum ?hochstetteri
Mikania chenopodifolia
Dissotis tubulosa
Virectaria multiflora
Vigna gracilis var. *multiflora*

Trichopteryx elegantula
Bidens mannii
Oplismenus hirtellus
Crassocephalum bauchiense
Spermacoce intricans
Brillantaisia vogeliana
Oleandra distenta
Pennisetum laxior
Achyranthes sp.
Nodonema lineatum
Ipomoea alba

Taxa seen but not collected (either sterile or beyond reach) were: *Begonia duncan-thomasii* (field det., therefore unreliable), *Cercestis kamerunensis*, two *Culcasia* spp., a large sedge and numerous *Selaginella*.

Nyale Rock is a classical inselberg, being an isolated rock pinnacle. Arising from the forest at an altitude of 950–1000 m, its height is hard to estimate. Due to its sheer walls, we were not able to ascend beyond the tree-shaded part of the rock so that the extensive exposed rock community of the top half remains unknown to science. Proficient rock climbers are needed to make more botanical progress here. Further collecting in the shaded portion of the rock is also likely to reveal scientific novelties since although we circumnavigated the rock in the course of a day, our survey of rock plants was not exhaustive.

Nyale Rock (R2 of vegetation map) Base 950 m alt., 4°50' N, 9° 37' E

17 Nov. 1998 Cheek 9664 *et seq.*

Begonia laporteifolia
Amphiblemma sp. 1
Utricularia andongensis
Blotiella sp.

Selaginella vogelii
Selaginella kraussiana
Pennisetum laxior
Impatiens kamerunensis subsp. *kamerunensis*

Bime Rock is also a free-standing rock pinnacle, but is forest capped and can be climbed with effort to the top. Many of the collections made there, by Pollard & Etuge, remain unidentified, having reached K only recently.

Bime (Mwendolengoe) Rock (R3 of vegetation map), 1760 m alt., 5°6' 24" N, 9°42' 59" E

12/13 Nov. 2001 *Etuge* 4514r *et seq.* & 23 Nov. 2001 *Etuge* 4434r *et seq*; *Pollard* 901 *et seq.*

Asplenium sp.
Nodonema lineatum
Mesanthemum jaegeri
Plectranthus decumbens

Ellasa-Pong Rock (Mbulle) R5 of vegetation map, 850 m alt., 4°48' 9" N, 9°41' E

30 Oct. 1998 *Cheek* 9510 *et seq.*

Spermacoce intricans
Amorphophallus preussii
Selaginella sp.
Impatiens kamerunensis subsp. *kamerunensis* (epilithic or terrestrial)

Ellasa-Pong Rock has a rock surface only 3–4 m high and perhaps 8 m across. Set in the forested side of the Jide-Bakossi trough, it is completely shaded. Its entire surface is covered by mosses and *Selaginella* with a few herbaceous species growing in this.

Rock face at Ndile waterfall (Baseng) W of vegetation map, SW of 4' L4, 750 m alt., 4°55' 7" N, 9°46' 43" E

16 Dec. 1999 *Mackinder* 331 *et seq.*, 'Rocky outcrop at edge of waterfall, saturated by spray'.

Amphiblemma sp. 1
Impatiens kamerunensis subsp. *kamerunensis*
Acanthonema strigosum

Elatostema welwitschii
Brillantaisia lancifolia

The last species of the list is also a rheophyte (see mapping unit 7) and points to the overlap between lithophytes and rheophytes. The last are mostly rock specific, although always associated with temporarily or permanently free-flowing water.

Other epilithic species are restricted, not to large or small rock faces but to free-standing boulders, e.g. *Argostemma africanum* (rocks in streams) or *Epithema tenue* (boulders in forest); other taxa are limited to vertical mineral substrate, but as often on earth banks as rocks: *Acanthonema strigosum*, *Antrophyum mannianum*.

Rock at Abwoh (Kodmin) 1650 m alt., 4°57' N, 9°43' E

20 Jan. 1998 *Etuge* 4024 *et seq.*, 'growing on a wet rock'.

Begonia duncan-thomasii
Adenostemma caffrum
Brillantaisia owarensis

Chlorophytum sparsiflorum
Coelachne africanum
Pilea sublucens

Rock W Bangem 1550 m alt., 5°05' N, 9°42' E

3 Jan. 1986 *Thomas & McLeod* 5294–5295
Begonia duncan-thomasii
Nodonema lineatum

Large granite boulders, Shrike Trail, Nyasoso (R4 of vegetation map) 1525 m alt., 4°47' N, 9° 43' E

9 July 1992 *Wheatley* 445
Nodonema lineatum

Note: rock faces/inselbergs R6–8 of the vegetation map are the three summit peaks of Mt Kupe: the sloping tops, the summit grasslands, are documented under mapping unit 10. Their vertical rock faces are botanically unexplored.

Rock face (or lithophyte) species can be divided into two ecological guilds: shade-demanders and sun-demanders, and both of these can be more or less subdivided further into a) obligate lithophytes, mostly restricted to rock habitats, and b) facultative lithophytes which appear in another, or more, habitat(s), often more commonly than on rocks. In most cases these facultative lithophytes appear 'opportunistic'. They have a wide ecological tolerance that can include embracing those difficulties of a rock face that are evidently anathema to most species–a vertical substrate, extremely thin or absent soil layer, and in the case of sun-exposed rock faces, complete desiccation in the drier months.

Red data species account for a high proportion of the obligate and near-obligate lithophytes, particularly those that are shade demanding. Threatened species are indicated as such in the lists below using NT (near threatened), CR (critically endangered), EN (endangered) or VU (vulnerable).

Nodonema lineatum (VU) a close relative of the african violet, *Saintpaulia*. About half the known localities of this monotypic genus are in the Kupe, Mwanenguba and Bakossi Mts. Its range extends to Nta Ali near Mamfe, and the Obudu Plateau of Nigeria.

Begonia duncan-thomasii (VU) is known only from rock faces in our checklist area and one other site in the Rumpi Hills. Usually it occurs with *Nodonema lineatum*.

Begonia laporteifolia (NT). Kupe-Bakossi accounts for about a third of all known localities of this taxon.

Begonia schaeferi (VU) is recorded from Bare, SE of Mwanenguba at a site not revisited in nearly a century. The other nine sites, all believed to be rock faces, occur northwards to the Bamenda Highlands.

Impatiens kamerunensis subsp. *kamerunensis* occurs as far W as Ghana. It is the only obligate lithophyte in Cameroon.

Amphiblemma sp. 1 is restricted to the Bakossi Mts and known from only three sites: when published as new to science it will rate threatened status.

SHADE-DEMANDING FACULTATIVE LITHOPHYTES

(i)   near-obligate species

These taxa have such a strong preference for shaded rock faces that they are only known from this habitat in our checklist area. Elsewhere they have also been recorded from other habitats.

Utricularia andongensis is widespread but rare in W Africa, and is usually confined to inselbergs.

Amorphophallus preussii (VU) is restricted to western Cameroon and is poorly known. Not hitherto recorded as a lithophyte.

Amorphophallus staudtii (NT) is restricted to Cameroon and Rio Muni, is poorly known and perhaps for that reason has not been previously recorded as a lithophyte.

Selaginella vogelii and *S. kraussiana*, as well as various Hymenophyllaceae, also known from shaded rock faces, have not been assessed for habitat specificity or threatened status.

(ii) 'opportunistic' species

Impatiens burtoni, Scadoxus cinnabarinus, Oplismenus hirtellus, Brillantaisia vogeliana, B. owariensis, Spermacoce intricans, Oleandra distenta, Asplenium sp., *Elatostema welwitschii, Adenostemma caffrum, Chlorophytum sparsiflorum, Pilea sublucens*, are all taxa abundant in other habitats in the checklist area but which have been recorded, usually just once, on a rock face. All are usually terrestrial, apart from *Oleandra* and *Asplenium* which are epiphytes. Epiphytes are often occasional opportunistic lithophytes. Given the similarities in their niches, this is to be expected. Ferns and orchids are the main vascular epiphytes in our checklist area–see Mapping Unit 4. Only a few species of these groups, e.g. *Liparis nervosa* var. *nervosa*, are nearly as often epilithic as epiphytic.

SUN-DEMANDING OBLIGATE LITHOPHYTES

Coleochloa abyssinica is a sedge recorded from N Nigeria, Ethiopia, Sudan, Uganda and Tanzania in 'crevices of rock pavement and outcrops' (FWTA).

Trichopteryx elegantula 'slopes of hills, between rocks' (FWTA), an afromontane taxon.

Pennisetum (Beckeriopsis) laxior is recorded from inselbergs from Ghana to Sudan (FWTA).

Mesanthemum jaegeri (NT) is known only from five sites in Sierra Leone, one each in Ivory Coast, Nigeria and newly recorded here for Cameroon (see checklist part).

Amongst the heliophile obligate lithophytes that might well be found if the exposed rock surfaces of our checklist area are more fully surveyed are the shrubby, fire-resistant Cyperaceae *Afrotrilepis* (poikilohydric and poikilochlorophyllous) and *Microdracoides*. In addition, several taxa of Scrophulariaceae and *Cyanotis* are known as inselberg specialists in W Africa, and may also be found (see Porembski & Barthlott 2000).

SUN-DEMANDING FACULTATIVE LITHOPHYTES
(i) Species of montane grassland on thin soils.

Gladiolus aequinoctialis is recorded in FWTA from Sierra Leone, Cameroon and Bioko as 'epilithic' or in 'stony grassland'.

Panicum hochstetteri occurs widely in montane Tropical Africa from Sierra Leone to Kenya, 'in forest and in grassland above the forest' (FWTA).

Bidens mannii (VU) known only from montane Cameroon and adjoining Nigeria, usually in stony grassland (FWTA).

Crassocephalum bauchiense (VU) known only from montane Cameroon and adjoining Nigeria, usually in stony grassland (FWTA).

(ii) Near obligate species

Virectaria multiflora 'wet places and rock outcrops', Mali to Cabinda (FWTA), known only from rock faces in our checklist area.

(iii) 'Opportunistic' species

Mikania chenopodifolia, Dissotis tubulosa, Vigna gracilis var. *multiflora, Achyranthes* sp., *Ipomoea alba*.

These species are more usually known from forest, roadsides, or as weeds of cultivation rather than from rock-face habitats.

The phytogeographic links of the lithophytes of the Kupe, Mwanenguba and Bakossi Mts depend upon their ecological guild. The shade-demanding obligate species are generally restricted to submontane (-montane) locations in the southern Cameroon Highlands. For a high proportion of these taxa, our checklist area is either a major or the main part of their range.

The sun-demanding obligate or near-obligate species of our checklist area are much more widespread, usually being spread westwards across W Africa to Sierra Leone (e.g. *Mesanthemum jaegeri* and *Gladiolus aequinoctialis*), and/or eastwards to Sudan or southwards to Cabinda.

The most detailed review of inselberg vegetation in W Africa is presented in Porembski and Barthlott (2000), and notes on 'Saxicolous' communities are also given by Letouzey (1968b). However, only a few of the sun-demanding taxa (*Coleochloa abyssinica*, *Virectaria multiflora* and *Gladiolus aequinoctialis*) are mentioned and the shade-demanding species appear not to have been covered. No previous studies of lithophytes in western Cameroon have been located.

MAPPING UNIT 7: RHEOPHYTIC VEGETATION OF UPLAND STREAMS AND RIVERS; 400–1450 m alt.

Rheophytes are plants adapted to fast-flowing water. The fifteen species discussed here are obligate rheophytes–being restricted entirely to such habitats. Waterfalls and rapids have the most highly developed rheophytic communities, and several of these occur in our checklist area. In total, the rheophytic vegetation studied in our checklist area probably only amounts to about one hectare, comprising half-a-dozen sites.

At all these river bed sites the vegetation is sparse, occupying only 2–10% of the surface area as a whole, the surface being of bare rock, usually black basalt.

The only previous record of a rheophyte in our area was made by Ledermann in 1908, the year in which he collected the material in Cameroon of the type of the most species-diverse genus of African Podostemaceae: *Ledermanniella*. *Ledermanniella thalloidea* was collected by him at Ndoungue (*Ledermann* 6328a). We have not relocated the exact site. Other sites for Podostemaceous rheophytes nearby, but just out of our checklist area are at Bare and Bakaka (records from Flore du Cameroun). Presumably these sites are on the upper reaches of the Dibombe River.

Rheophytes of 'Esense' streams on Mt Kupe

The submontane slopes of Mt Kupe (above c. 800 m), which we studied in detail in 1995–97, have no permanent streams, but seasonal ones, known as 'Esense' in Bakossi. Most months of the year they contain no surface water, and appear as shady, steep, mostly bare, boulder-lined gullies. Obligate rheophytes here are *Anubias barteri*, *Triplophyllum securidiforme* var. *securidiforme* and *Bolbitis fluviatilis*.

At Nyandong in March 2003 we discovered numerous colonies at c. 400 m alt. of *Ledermaniella thalloidea*, growing in rapids with *Tristicha trifaria*. Other rheophytes were *Virectaria angustifolia*, *Achyranthes talbotii* and *Brillantaisia lancifolia*. Other associated specimens, which may yield other obligate rheophytes, have arrived at Kew only recently and have not been identified yet. Site details are as follows:

Rapids on river Ndipesungkale at Nyandong, 400 m alt., 4°58'12" N, 9°34'48" E.

27 March 2003, *Cheek* 11459 *et seq.* Marked 'W' on vegetation map.
These specimens have not yet been identified at Kew

Enyandong in northern Bakossi in Oct. 2001

Numerous rapids were found in the two rivers that drain this trough (also referred to as the Mbu valley) but water levels were so high, and the flow so fast, that study was not possible.

The Jide River

The most productive river for rheophyte sites has been the upper reaches (above c. 1000 m) of the Jide which drains the Bakossi trough. Four sites along this river are noteworthy for rheophytes, including the most species-diverse of all, with nine obligately rheophytic taxa. The account of this site, near Mwambong, is taken from text for a book 'Rheophytes of Africa', now in preparation.

The Jide River at Mwambong (see Fig. 5)

The Jide River waterfall site is located in lowland/submontane forest at c. 1000 m alt. in the eastern Bakossi mountains of Western Cameroon (Grid Ref. 4°59' N, 9° 43' E, marked W on vegetation map). The waterfall has a drop of about 20 m, and falls over a cliff of vertical basalt pillars. I visited this site in February (peak dry season) and November (end of wet season) 1998, studying the area at the top of the falls. Here the river bed, of solid basalt, is about 7–10 m wide, and so large areas are shaded by trees of the adjoining forest, although other areas are in full sun. This site is notable for a newly described species of *Ledermanniella*, *L. onanae* (*Onana* 558, *Cheek* 9120) (see Cheek 2003a) and for providing the second-only collection (*Cheek* 9119) of the *Plactycerium*-leaved podostemad *Ledermanniella letouzeyi*. The first collection of this last species was made in the Rumpi Hills. A third podostemad, *Macropodiella pellucida* (*Cheek* 9118) has been identified at the same site. The generally ubiquitous rheophyte *Tristicha trifaria* is absent at the study site, though present below the falls, further downriver. Other obligate rheophyte species present at the top of the falls are *Virectaria angustifolia* (*Cheek* 9707), *Brillantaisia lancifolia* (*Cheek* 9701), *Anubias barteri* (no voucher collected), *Cyperus rheophytorum* (*Cheek* 9702) and *Menisorus pauciflorus* (*Cheek* 9704). A Gramineae, *Pennisetum monostigma* (*Cheek* 9705) also grows as a rheophyte in this community.

The rheophytes at the Jide site show distinct zonation, apparently due to ecological preferences for different light intensity levels or for duration of submergence. Indeed, at this site, the three species of podostemad present do not intermingle and instead show strong niche separation. *Ledermanniella onanae* is the most conspicuous podostemad, lining the top of the river channels abundantly above the falls where full sunlight falls on some minor rapids. It appears as a dense dark green band in full-flower in February, with perhaps only 10% of fertile branches still to open flowers, but was only just becoming exposed by the falling water level in November. There was no sign of post-fruiting die-back in February, so this may be a perennial species. *Macropodiella pellucida* lacks leaves entirely and consists merely of a flat, thallus-like root from which arise the minute flowers. It is less conspicuous and grows in a stratum below the *Ledermanniella onanae*. In February, the uppermost plants of *Macropodiella pellucida* were only just being exposed, and the first few flowers were open. Most plants however, were still submerged and detectable only as a dark green film over the rock surface. In November, when the water was perhaps 30 cm deeper and the current stronger, it was difficult to be sure that this species was present at all.

Ledermanniella letouzeyi is a large and eye-catching podostemad. It was however, inconspicuously present, being found only in the almost inaccessible vertical chute which houses the main flow of the waterfall. This area of the fall is shaded by trees from above. This species seems to require highly oxygenated water or constant spray and to be a shade-bearer. Collecting this podostemad was hazardous, incurring the risk of falling down the chute. The population of this species seems to grow the length of the chute. Jon Riley kindly collected plants from its base, by swimming across the pool at the base of the fall. He reported plants going up the chute as far as he could see, at least for several metres. Plants collected in February had already fruited and the flat, pendent, leathery, dichotomously branched, dark green leaves, up to 10 cm long and 1.5 cm wide, of several plants at the top of the falls had become detached by decay at their base, leaving only the rubbery thalloid root and short thick erect (c. 4 × 0.5 cm) stem (presumably the perennial parts of the plant) to which the fruits were attached. Plants collected in November were in flower.

Between the area occupied by the first two species, and that occupied by the second, was a quiet, shaded stretch of river c. 10 m long that appeared to lack any podostemads. It may be that the three species present are all so specialized that none could occupy this area, due to the low light levels or because of insufficiently oxygenated water. The non-podostemad rheophytes at the Jide site also show some habitat partitioning, into light demanders and shade bearers. In full sunlight for example, was *Virectaria angustifolia*, an obligate rheophyte widespread in W Africa. It fastens itself tightly to the substrate, not by its roots penetrating deeply into rock fissures, but by forming a mat of roots closely appressed to the rock surface and which grip irregularities in its surface. When pulled away from the substrate, the root mass appears to form a mould cast of the rock on which it was growing. Also in full sun, but deeply rooted into rock crevices were *Cyperus rheophytorum* and *Pennisetum monostigma*, the last of which is a facultative rheophyte. Both were completely exposed in both November and February, but would have been submerged under fast-flowing water throughout the wet season. Above the rocky main river flood bed, at the bank, occurred *Brillantaisia lancifolia*, which seems less tightly adapted to the rheophytic habit than the foregoing species for, although it has stenophyllous leaves, its roots do not seem so tightly anchored as those of other rheophytes. However, within its range, in the Gulf of Guinea, it is only known from the courses of seasonal or mountain streams and small rivers (*pers. obs.*), with just one record from a lowland river (see mapping unit 3) and so appears to be an obligate rheophyte.

In the partial shade of trees, though also rooted into rock crevices in the rock flood bed of the river were the rheophytic fern *Menisorus pauciflorus* and the aroid *Anubias barteri*.

Ndip River (Jide tributary) at ford on footpath from Kodmin to Nzimbeng, alt. 1150 m, 5° N, 9°43' E

14 Feb. 1998, *Cheek* 9196.
Ledermanniella onanae

No other rheophytes recorded at these rapids. This is the type and only other locality for the species (Cheek 2003).

Ndip River, falls at Ela Ndip Emechang near Kodmin, alt. 1450 m, 5°00' N, 9°41' E

21 Jan. 1998, *Satabie* 1109 and 16 Nov. 1998, *Pollard* 207.
Plectranthus cataractarum

This is the only known site in our checklist area (Pollard & Paton 2001). No other rheophyte was located at this spot.

Jide River at Baseng, Ndile falls near dispensary; alt. 750 m, 4°55' N, 9°44'42" E, Marked W on vegetation map

16 Dec. 1999, *Cheek* 10409 *et seq.*

At this point the Jide has a series of at least three falls in close proximity. The only point investigated was the plunge pool accessible from the medical dispensary.

The obligate rheophytes recorded were as follows:

Brillantaisia lancifolia	*Ledermanniella letouzeyi*
Achyranthes talbotii	*Tristicha trifaria*
Anubias barteri	*Menisorus pauciflorus*
Virectaria angustifolia	*Triplophyllum securidiforme* var. *nana*

Sampling of rheophyte sites in our checklist area is far from complete. The highest priority for investigation are the rivers of the Mbu valley which is part of the Cross river catchment area, otherwise unsampled by us for rheophytes. In addition, the upper clear water reaches of the Mungo River and its tributaries in western Bakossi are also still uninvestigated apart from the river at Nyandong. Here I heard of large falls in the area, but had insufficient time to investigate. Finally, the Jide river is still incompletely surveyed for rheophytes, and may yet yield more discoveries. The ideal time for such surveys is probably Dec.–Feb. inclusive when water levels are dropping, promoting flowering and fruiting of rheophytes, but by which time they have not yet dried up completely.

The phytogeography of the rheophyte communities in Bakossi is as follows. One species, *Tristicha trifaria*, is pantropical, resembling the predominant pattern seen in crater-lake species. The remaining species are limited to Guineo-Congolian Africa, and most are restricted to Lower Guinea (e.g. *Brillantaisia lancifolia*), while *Plectranthus cataractarum* and *Cyperus rheophytorum* are restricted to a few other localities in the western Cameroon uplands (namely at Mt Cameroon and Bioko), the remaining four Podostemaceae are restricted to SW Province Cameroon, of which three are nearly endemic to our checklist area (known from only one or two other sites outside Bakossi) and one is strictly endemic.

Links between this vegetation type and other aquatic vegetation types, such as mapping unit 8 (crater lakes) are absent, or in the case of mapping unit 3 (lowland swamp and riverine), restricted to one species: *Brillantaisia lancifolia*. Presumably aquatic species of the last two vegetation types would be uprooted and swept away in this fast-flowing environment. Most rheophytes require rock substrate in addition to clear fast-flowing water. As a consequence, there is some geographical overlap with rock-free communities, especially the shade-demanding assemblages (see mapping unit 6). This can be seen at the Ndile falls of Baseng, where, only a few metres above the flood zone in which *Brillantaisia lancifolia* (obligate rheophyte) occurs, *Amphiblemma* sp. 1 (obligate lithophyte) can be found. Shade-loving lithophytes that require high humidity are often to be found near waterfalls. Species such as *Plectranthus cataractarum*, included as a rheophyte here on the basis of its habitat, has relatively poorly developed rheophytic morphological traits (e.g. relatively weakly anchored and stream-lined) and may have originated relatively recently from a lithophytic ancestor. Several other species of *Plectranthus* genus show lithophytic, but not rheophytic tendencies, e.g. *P. decumbens* and *P. epilithicus*, also of this checklist.

Only one endemic rheophyte is known from our checklist area, *Ledermanniella onanae*. However *Ledermanniella letouzeyi* is only known from one other collection outside of the checklist area, and *L. thalloidea*, *Macropodiella pellucida*, *Cyperus rheophytorum*, *Plectranthus cataractarum*, *Brillantaisia lancifolia* and *Achyranthes talbotii* are all known from less than ten (and usually many fewer) gatherings outside our checklist area and all are threatened or near threatened with extinction.

Threats to rheophyte communities are thought to be mainly from a) increased turbidity of water due to sediment run-off after logging, and b) pollution of water courses with e.g. sewage, or surfactants, leading to increased algal growth. Algal growth was conspicuous at the Nyandong site, but the Jide sites appeared in good health. Logging has been a threat in the checklist area but appears now to have diminished.

Rheophytes are plants adapted to fast-flowing water. The fifteen species discussed here are obligate rheophytes–being restricted entirely to such habitats. Waterfalls and rapids have the most highly developed rheophytic communities, and several of these occur in our checklist area. In total, the rheophytic vegetation studied in our checklist area probably only amounts to about one hectare, comprising half-a-dozen sites.

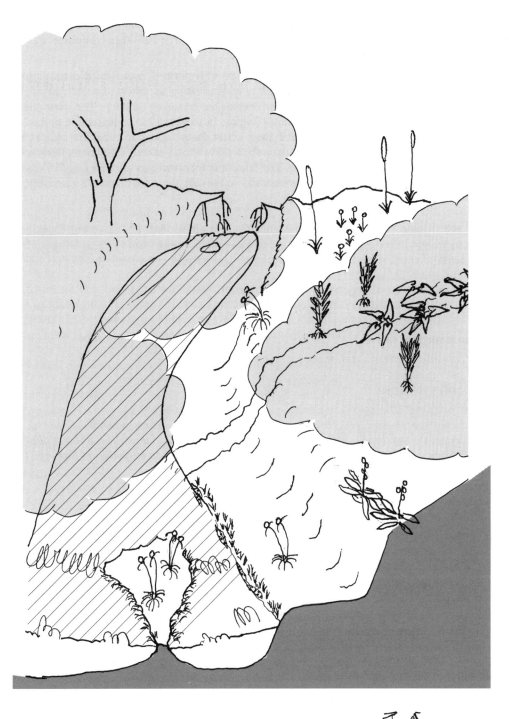

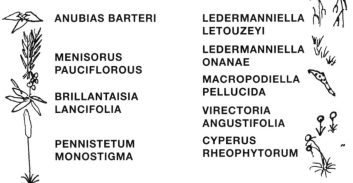

ANUBIAS BARTERI

MENISORUS
PAUCIFLOROUS

BRILLANTAISIA
LANCIFOLIA

PENNISTETUM
MONOSTIGMA

LEDERMANNIELLA
LETOUZEYI

LEDERMANNIELLA
ONANAE

MACROPODIELLA
PELLUCIDA

VIRECTORIA
ANGUSTIFOLIA

CYPERUS
RHEOPHYTORUM

Fig. 5. A RHEOPHYTE COMMUNITY.
The Jide River waterfall at Mwambong, Feb. 1998.

MAPPING UNIT 8: CRATER LAKES; 1150–1900 m alt.

Crater lakes occur throughout the Cameroon Highlands. They were formed in the third main period of volcanic activity that built these mountains. During the Quaternary, magma rising up volcanic vents was of high viscosity, causing blockages in the subterranean vents and so explosions, causing craters. The best known of these internationally is Lake Nyos in the Bamenda Highlands, which, in August 1986, released a cloud of gas with fatal consequences. Bioko and Mt Cameroon also have their crater lakes, and our checklist area is no exception. Examination of the 1:200,000 Buea-Douala IGN map shows five crater lakes. Two more were revealed by the researches of Martin Etuge in 1998, at Mwambong. The lakes and their margins are less than 0.5% of the checklist area. Three of the seven lakes have not been botanically surveyed; the remainder are relatively well studied. Aquatics were obtained by swimming.

The twin ('male' and 'female') lakes of the Mwanenguba Caldera are the best known in our checklist area and the only ones known by people living outside the area, receiving sporadic visits from tourists. Of these lakes, only one has been surveyed, the other, close by, having such precipitous sides that it is not easily reached.

Mwanenguba 'female' lake, L1 of vegetation map, 1950 m alt., 5°02' N 9°49' E

Cheek 7245 *et seq.* 5 Feb. 1995 & 9404 *et seq.* 28 Oct. 1998.

Submerged, bottom rooting, species

Myriophyllum spicatum forms mats c. 2 m wide, c. 5 m from shore.

Marginal, only partly submerged

Salix ledermannii
Osmunda regalis
Schoenoplectus corymbosus var. *brachyceras*
Polygonum cf. *pulchrum*
P. salicifolium
Hydrocotyle sibthorpioides

Micractis bojeri
Echinochloa crus-pavonis
Oldenlandia lancifolia var. *lancifolia*
Rhynchospora corymbosa (*Thomas* 3131, Feb. 1984)

Mwanenguba 'male' lake, L2 of vegetation map. Adjacent to 'female' lake.

Unsurveyed. Reputed to be 120 m deep (Tye 1986) c. 0.2 km diameter (1:200,000 IGN).

Lake Edib, L3 of vegetation map, 1250 m alt., 4°57.6' N, 9°39.2' E

11 Feb. 1998 (*Cheek* 9129–9168).

This crater-lake is exceptional in Cameroon for its surface being partly occluded by a floating mattress of *Sphagnum planifolium*, on which a distinct zonation of vascular plant species occurs as described below. It has the most complex ecosystem of all the crater lakes in Bakossi, and, almost certainly, in Cameroon and west-central Africa. It is reputed to have an undescribed but endemic species of leach, and fish are absent (Wild *pers. comm.*). We found the leach to be highly evident when sampling the aquatic vegetation. Jumping on the surface of the vegetation induces in it a heaving motion.

Three survey visits were made, all in 1998. The first was led by Etuge (24 Jan. 1998) aimed at reconnaissance and making general collections; the second was led by Onana and Cheek (11 Feb. 1998) and aimed at doing a vegetational transect; the third was led by Gosline and Etuge, with support by Pollard (20 Nov. 1998), aimed at reaching the hitherto uncollected S side of the lake, but was not successful in this endeavour due to time constraints.

Access to the lake is from the N side where the crater is steep but scaleable. The access path is a short branch from the main Edib village to Nyandong footpath. The S side of the crater wall is cliff-like, and possibly 200 m high. The transect was made along a perpendicular line from the access path to the water's edge. At this place the floating marginal layer is about 70 m across. Its width appears greater on the southern side, but at one point on the W side open water almost reaches the crater edge, making access to the S side difficult. Access in a clockwise direction is obstructed by the outlet stream in the NE part of the lake. The open water appears about 100–180 m wide, from the N bank to the S. Eight more-or-less sharply defined zones could be recognized. All species mentioned are vouchered under the Cheek series unless otherwise indicated. Proceeding from the open water to the crater edge, there were as follows (see Fig. 6):

A. Open water with a 2 m wide marginal band of *Nymphaea lotus*, amongst which *Utricularia* cf. *stellaris* was abundant. Other submerged aquatics might be found by exploring the open water.

B. A belt c. 1.25 m wide, 0.7 m high, of *Elaeocharis variegata*.

C. A belt c. 1 m wide, 1 m high, of *Cyperus pectinatus*.

D. A belt c. 4.75 m wide, 1.5 m high, the substrate raised above adjoining zones by several cm (presumably by *Pneumatopteris* material), and appreciably more solid underfoot, of:

Pneumatopteris afra	85%
Cyperus pectinatus	10%
Oplismenus hirtellus	3%
Dead stems	2%

Unfortunately, no voucher was taken of the *Pneumatopteris* and the *Oplismenus* was sterile, rendering the identifications uncertain.

E. A belt c. 7.8 m wide dominated by *Sphagnum* moss, the second dominant reaching c. 60 cm tall.

Sphagnum planifolium	75%
Cyperus pectinatus	10%
Oplismenus hirtellus	6%
Others	9%

Belt E contains a great diversity of other species restricted to wetlands:

Utricularia appendiculata	*Satyrium crassicaule*
Drosera madagascariensis	*Habenaria weileriana*
Laurembergia tetrandra subsp. *brachypoda*	*Ludwigia abyssinica*
Elaeocharis variegata	*Senecio ruwenzoriensis* (not usually a wetland
Lycopodiella cernua (*Pollard* 241)	species)
Pseudolycopodiella affine (*Gosline* 193)	

An unexpected feature is the presence of species of climbers growing prostrate on the moss, and of otherwise obligate epiphytes, growing directly in the moss. Not all these taxa are confined to zone E, some occurring in bands F–H.

Climbers
Urera trinervis
Sacosperma paniculatum
Artabotrys stenopetala
Pouzolzia parasitica
Secamone racemosus

Epiphytes
Impatiens letouzeyi
Begonia ampla
Begonia poculifera
Anthocleista microphylla

F. About 8.8 m wide, 1.4 m high, densely vegetated.

Pneumatopteris afra	50%
'*Imperata cylindrica*'	20%
Oplismenus hirtellus	20%
Dead matter	10%

Only two other species were recorded in this zone, a sterile papilionoid legume (*Cheek* 9153) and the first shrub encountered, a *Hymenodictyon floribundum*, 4 m tall. The '*Imperata cylindrica*' is a form name for a sterile grass, of which the identity is unconfirmed.

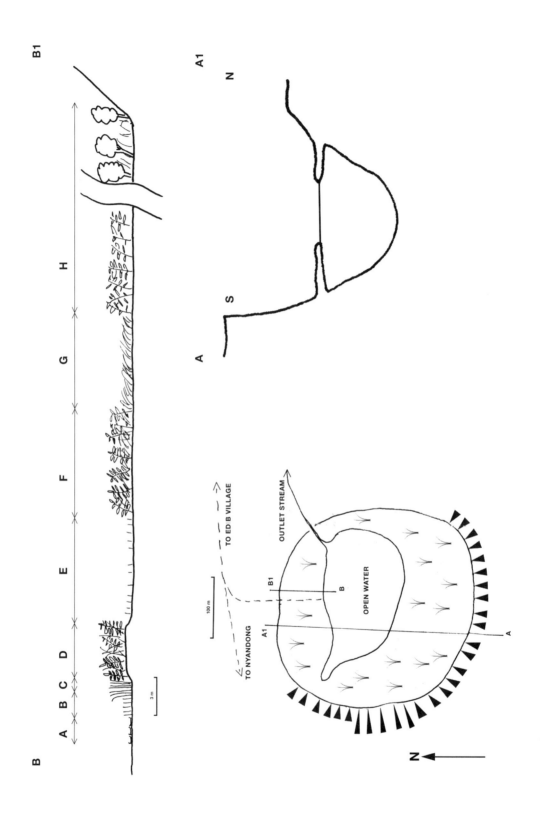

Fig. 6. LAKE EDIB.
Left: plan of lake; right: profile of lake, section A-A1; top: transect B-B1, showing vegetation zonation.

G. Excavations in this zone revealed the water table to be 15 cm below the surface. About 7.4 m wide, the plant layer about 1 m high, the species recorded in this zone were:

Oplismenus hirtellus	50%		Dead matter	10%
Fuirena umbellata	10%		*Cyclosorus striatus*	7%
Scleria vogelii	10%		*'Imperata cylindrica'*	7%

Other species present were
    Pouzolzia parasitica (*Onana* 591)
    Ludwigia abyssinica
    Emilia coccinea
    Guyonia ciliata

H. About 30 m wide, the herbaceous vegetation about 2 m tall. Shrubs, trees and woody climbers increasing in density towards the crater edge, where presumably the organic substrate is thicker and the water table lower. The first tree encountered was 13 m from the boundary with zone G: *Ficus preussii*, 6 m tall, with *Asplenium cancellatum* epiphytic upon it. Nearer the crater wall the main shrub was *Alchornea cordifolia* (a common swamp forest species elsewhere) growing c. 2 m tall and accounting for c. 10% of cover, with the occasional *Cyathea manniana*, *Shirakiopsis elliptica*, *Securidaca welwitschii*, *Landolphia buchananii*, *Bertiera racemosa* var. *racemosa*, *Dalbergia* sp. and *Heisteria parvifolia*.

The main herbs were as follows:

Cyclosorus striatus	40%		Dead matter 5%
Pneumatopteris afra	40%		
Oplismenus hirtellus	15%		
Scleria vogelii	10%		
Pouzolzia parasitica	5%		
'Imperata cylindrica'	5%		

Other species present were: *Brillantaisia* sp., *Hypselodelphys* sp., *Tristemma littorale* subsp. *biafranum*, *Hypoestes aristata*.

The zones listed above are not uniform in width around the periphery of the lake. Although only a segment of the circumference was explored, at some points, for example, zone E was found to dwindle to nil.

A different set of species were found in the outlet stream:

Polygonum salicifolium	*Lemna paucicostata*
Utricularia gibba	*Nymphaea* cf. *maculata*
Potamogeton nodosus	

Other species recorded from the lake margin, but occurring outside the transect and not usually located in a zone by their collector were:

Polygonum setosulum	*Cynorkis anacamptoides*
Oldenlandia lancifolia (stream edge)	*Eulophia horsfallii* (Zone F)
Floscopa glomerata subsp. *lelyi*	*Habenaria bracteosa*
Rhynchospora corymbosa	*Habenaria procera*
Schoenoplectus mucronatus	*Athyrium ammifolium*

Mwambong Lakes

Lake 1. L4 of vegetation map, 1150 m alt., 4°58' N, 9°56' E
Lake depth 1–2 m, length 50 m, width 25 m, situated in a shallow crater.

26 Oct. 1998 (*Cheek* 9358 *et. seq.*).

Pneumatopteris afra	Dominant herb at water's edge, forming sward 80 cm high; common in swamps.
Acalypha manniana	Herb at water's edge.
Gouania sp.	Occasional at edge.
Triumfetta cordifolia	Occasional at edge.
Ottelia ulvifolia	The only submerged species; covering the muddy lake bottom; wetland specialist.
Panicum sp. aff. *hochstetteri*	At edge.

Oldenlandia lancifolia var. scabridula	Water's edge; specialised wetland species.
Cyperus distans subsp. longibracteatus var. longibracteatus	Growing in the inlet stream; wetland species.

The lake is situated in submontane forest, with a broad herbaceous margin. Trees include *Musanga, Polyscias, Cyathea manniana, Turraeanthus africanus* and *Macaranga*. Herbs at the rim of the crater include *Rubus, Aframomum, Pennisetum* and *Hypselodelphys*: all secondary taxa suggestive of, in the absence of anthropic action, trampling by elephants.

Lake 2. L5 of vegetation map, 1200 m alt., 4°58' N, 9°46' E
Lake depth c. 2 m.

26 Oct. 1998. *Cheek* 9374–9375

Cyperus sp.	Leaves 2 m long, of which the top 15 cm emersed. Not identified further since sterile when collected.
Oldenlandia lancifolia	The only submerged aquatic, forming a blanket over the lake bottom, the top of the blanket at least 0.5 m below the water surface.

Situated in submontane forest, without an herbaceous margin: no specialist aquatics were found among the marginal species.

Lake Beme. L6 of vegetation map, c. 500 m alt., 5°10' N, 9°38' E
Unsurveyed. c. 0.75 km diameter (data from 1: 200,000 IGN). Contains numerous species of endemic cichlid fish: the greatest concentration in Africa (*fide* Wild).

The flora of the four lakes enumerated contributes numerous species otherwise unknown from the checklist area. It also adds seven flowering plant families: Droseraceae, Haloragaceae, Nymphaeaceae, Salicaceae, Hydrocharitaceae, Lemnaceae and Potamogetonaceae. Four of these families are only known from Lake Edib which also contributes the bulk of the species in this vegetation type. Although several of the species are threatened, these are few, generally forest species that have established on the *Sphagnum* blanket at Lake Edib, e.g. *Impatiens letouzeyi*. No species are endemic to this vegetation type. This reflects the fact that most aquatic-wetland species are widespread. *Osmunda regalis* and *Myriophyllum spicatum* also occur in Britain, for example.

Presumably all of the aquatic specialist species arrived from other lakes or wetland area. Many of these species do not have wind-dispersed seeds and would appear to be reliant on migrant birds, particularly wetland specialists, as vectors for dispersal. It would be useful to have information on which bird species are associated with the crater lakes discussed above, and whether any of these are migratory or not. It would also be useful to employ genetic fingerprinting techniques in a study of one or more species widespread in crater lakes, perhaps *Ottelia ulvifolia* or *Myriophyllum spicatum*, to determine whether a single introduction has arrived in the Cameroon Highlands and has been dispersed locally from lake to lake, or whether multiple introductions have been made. These techniques might also pinpoint the geographical source or sources of these species: for the higher altitude lakes, Europe, the E African Mts and southern Africa are all possibilities.

From the evidence presented above, only one species, *Oldenlandia lancifolia*, is recorded from all four lakes, suggesting that inter-lake dispersal within our checklist area is almost non-existent. *Oldenlandia lancifolia* is apparently not always a lake-specialist (we have two records of it in farmbush) although it appears so at all the lakes studied, either forming underwater mats or at least being partly submerged when reproductive. It is characterized in FWTA as occurring in 'moist places', which might include river margins and swampy areas. Consequently its existence at all four lakes is not proof of inter-lake dispersal.

There are several convincing explanations for the lack of shared species between the four lakes:

1. Differences in altitudinal range may make it difficult for species occurring at the higher altitude lakes (Mwanenguba c. 1950 m alt.) to survive in the lower altitude lakes (e.g. Mwambong c. 1200 m alt.). However this does not account for the disparity between the Mwambong lakes and that at Edib, which are at similar altitudes.

2. The small size of the Mwambong Lakes (one is only c. 50 × 25 m and 1–2 m deep) may render them difficult to locate by bird vectors, and also more vulnerable to small-scale events, such as elephant trampling.

3. The shallowness of the Mwambong Lakes might render them markedly seasonal, possibly drying up entirely in the dry season in some years, if not every year, making them less hospitable in the long term to any wetland species that might arrive.

4. Differences in water chemistry. The presence of *Sphagnum* at Lake Edib, which is the basis for a diverse ecosystem, is likely to be based upon a nutrient-poor water supply, presumably absent (as is *Sphagnum*) from e.g. Lake Mwanenguba. Sampling of the water chemistry of each of the lakes is desirable.

5. Inadequate sampling. Most of the area of Lake Edib, and most of that of the female lake of Mwanenguba, has not yet been surveyed, and may yet yield extra taxa, conceivably containing shared species. Inflatable boats are needed to help redress this deficiency. Efforts should also be made to obtain authorization and to reach and survey Lakes Epen, Beme and the male lake of Mwanenguba. The first two of these three, being of low altitude, are likely to yield additional species to Bakossi.

Threats to lake vegetation are absent, apart from the female lake of Mwanenguba, which is used as a cattle watering hole. This use risks eutrophication through contamination by deposits of cattle urine and faeces, abundantly evident at the most accessible slopes (*pers. obs.* 1995, 1998, 1999). The proposed introduction of exotic fish to one lake to develop a fishery is not likely to adversely affect the vascular plant species. It would, however, be an extremely grave threat to the algal and invertebrate communities,.

MAPPING UNIT 9: MONTANE FOREST & FOREST-GRASSLAND EDGE; (1900–)2000–2050 m alt.

Montane forest forms c. 3% of the vegetation of our checklist area. It is confined to two areas: the Mwanenguba caldera in the NE, and the summit area of Mt Kupe towards the SE. The last is just high enough to capture montane species although these are mixed with a high proportion of submontane taxa. The Bakossi Mts are generally too low to support true montane forest. The highest point, Mt Loh was recorded by us as only c. 1780 m alt. We recorded *Syzygium staudtii* and *Carapa grandiflora* there in 1998, but no other montane tree species.

Although Thorbecke, a geologist, surveyed Mwanenguba vegetation in 1907–1908 (Thorbecke 1911: 298–300) he remarked only upon its physiognomy and the preponderance of easily recognized families in the grasslands, such as Compositae. No montane forest species were mentioned by name, and he mentions having collected specimens only in grassland vegetation.

Letouzey (1968b: 340–341), reviewing the vegetation of Cameroon's mountains gives scanty detail on that of Mwanenguba montane forest, based on a quick visit, noting only that *Podocarpus* is present, and that the vegetation is strongly degraded, lower altitudes being under coffee, higher altitudes under pasture.

Regarding Mt Kupe, Letouzey (1968b: 341) points out that its vegetation is also poorly known, but notes *Podocarpus, Nuxia*, and the shrubs *Erica mannii, Hypericum revolutum* and *Adenocarpus mannii* at the 'prairie' on the summit.

Our sampling of montane forest at Mwanenguba has been restricted to three or four day-trips to the patches around the crater lakes (c. 1900 m) departing from either Nyasoso or Enyandong. These crater lakes lie near the centre of a c. 4 km wide grassy caldera. We have not yet reached the far larger montane forest areas on the SE rim and upper flank of the caldera which appear (1:200,000 map and satellite imagery) to ascend from 2000 m to the highest part of both Mwanenguba and our checklist area at 2390 m alt. (2411 m on the 1:200,000 IGN map). It is not clear whether these forests, so much further from Bangem, the main starting point and lacking easy vehicular access, have ever been botanically surveyed or not. Therefore much may remain to be discovered.

Sampling of montane forest at Mt Kupe has been more complete, if only because it is smaller in area. Once the summit itself has been achieved, however, navigation under the usual cloud cap is difficult, and in the rainy season uncomfortable. The top of the mountain is roughly saucer-shaped, c. 1900 m alt., with a stream. From the edge of the saucer rise three peaks, the tallest of which is 2050 m alt. (2064 m according to Letouzey, 1985). These three can be explored comfortably in the course of a day from a camp on the summit as we did in Oct./Nov. 1995.

Day trips to the summit from Nyasoso (800 m alt.) and Kupe Village (500 m alt.) were made by our expeditions in early 1995 and late 1996 but left little time for survey at montane elevations. In addition to our own specimen records from the summit, we also have those from the ascents in the 1980s by Thomas, and in 1992 by the University of Aberdeen expeditions (Stone 1993).

The physiognomy of Mt Kupe's montane forest is worth noting. The three summits themselves, and the ridges linking them, are clothed in a dense, short-statured forest (trees mostly 4–8 m, rarely to 20 m tall) that descends into scrub. This I attribute to the windswept nature of much of the peak areas. Large *Schefflera,* so characteristic of the montane forest of Mt Cameroon and Mt Oku, are more or less restricted to the more sheltered saucer-like summit area at 1900 m, where they occur mixed with submontane taxa such as *Zenkerella. Zenkerella* and the equally submontane *Xylopia africana* grow right to the summit peaks and the usually submontane *Garcinia smeathmannii* is dominant in some areas, e.g. ridges, such as at the Aberdeen University forest plot.

From Bioko in the Gulf of Guinea, to Tchabal Mbabo inland, the species composition of montane forest of the Cameroon line is remarkably uniform from one location to the other. About 20% of the tree species are more or less restricted (marked **x** in the table below) to this line of mountains. The remainder are a subset of those present in the E Africa mountains and can be classed as 'afromontane'. The diversity of species is much lower than that of submontane forest. As remarked upon for Kilum-Ijim, only ten species form 90% or more of the canopy, (Cheek *et al.* 2000).

The demarcation of submontane forest from montane forest is discussed under the first mentioned vegetation type.

The table below shows that most montane forest species of our checklist area also occur in other mountains and mountain ranges to the N (Mt Oku) and S (Mt Cameroon) in the Cameroon Highlands. However, two phytogeographical patterns are worth noting.

1. *Striking absences*

Kupe-Mwanenguba appears to lack records of several species that occur in mountains to the N and S. Neither *Myrsine neurophylla* (formerly *Rapanaea melanophloeos*), among the top ten commonest tree species of montane forest in the Cameroon line, nor *Gnidia glauca* (formerly *Lasiosiphon*), a very common fire-resistant pioneer, nor *Brucea antidysenterica*, a fairly infrequent small forest tree, nor *Discopodium penninervium* a large understorey shrub, nor *Psydrax dunlapii*, an understorey tree, have been found in our checklist area. Some, indeed all, of these might yet be found in the main Mwanenguba forest block which is almost completely unsurveyed. This block might also yield taxa such as *Carapa, Olea capensis, Cassipourea malosana* and *Maytenus undata* which are known from both Mt Kupe and the Bamenda Highlands, but are so far unknown from Mwanenguba.

It is quite possible that the only patch that we sampled (at the crater lakes) was too small to support these taxa, all of which (except the *Carapa*) are infrequent even in the forest in which they occur. The crater lake tree assemblage itself was strongly reminiscent of montane forest seen on Mt Cameroon and Mt Oku (Cheek *pers. obs.*).

The montane forest of Mt Kupe is unusual in several respects. The absence of *Clausena anisata, Maesa lanceolata* and *Pavetta hookeriana*, all otherwise frequent in montane forest of the Cameroon Highlands, including Mwanenguba, is noteworthy. Given the fairly good sampling of the summit area, it seems unlikely that such easily collected and generally common woody species would be overlooked. The absence of *Maesa lanceolata* is underscored by four collections of the generally rarer and less widespread *M. rufescens* (both taxa occur at Mwanenguba) and the absence of *Pavetta hookeriana* by its replacement by *Pavetta kupensis*, known only from the summit of Mt Kupe. Other species absent from Mt Kupe are *Ilex mitis, Morella arborea* and *Cassine aethiopica*. The first and last are infrequent where they do occur, e.g. Mt Oku, and may have been overlooked. The *Morella* is also infrequent and often occurs as isolated trees in grassland outside the forest, a scarce habitat on Mt Kupe. All six of these absent species have small red/orange, juicy berries so are presumably bird-dispersed. Since distance from nearby Mwanenguba or Mt Cameroon would present no impediment to these seed vectors, it may be that Mt Kupe's nearly permanent and complete cloud cover causes birds to pass it by.

Despite these absences the montane forest of Mt Kupe does contain species not known elsewhere in the checklist area. Apart from the endemic and threatened species listed later, these include: *Leptoderris fasciculata, Batesanthus purpureus* (also submontane), *Ficus oreodryadum, Zanthoxylum leprieurii* and *Aframomum zambesiacum*.

Table below: montane forest (above c. 2000 m alt.) at Mwanenguba and Mt Kupe. Species composition, present (Y), absent (N). Taxa restricted to the Cameroon Highlands marked **x**. Herbs have been omitted from this list due to considerations of space and time.

Canopy Trees	Mt Oku	Mt Mwanenguba	Mt Kupe	Mt Cameroon
Schefflera abyssinica	Y	Y	Y	Y
Schefflera mannii **x**	Y	Y	Y	Y
Syzygium staudtii **x**	Y	Y	Y	Y
Carapa grandiflora	Y	N	Y	N
Ixora foliosa **x**	Y	Y	Y	Y
Olea capensis	Y	N	Y	Y
Prunus africana	Y	Y	Y	Y
Clausena anisata	Y	Y	N	Y
Cassipourea malosana	Y	N	Y	N
Podocarpus milanjianus	Y	Y	Y	N
Bersama abyssinica	Y	Y	Y	Y
Ilex mitis	Y	Y	N	Y
Fire resistant/forest edge trees				
Agarista salicifolia	Y	Y	Y	Y
Nuxia congesta	Y	Y	Y	Y
Morella arborea **x**	Y	Y	N	Y
Maesa lanceolata (pioneer)	Y	Y	N	Y
Maesa rufescens (pioneer)	Y	Y	Y	Y
Neoboutonia glabrescens (pioneer)	Y	Y	N	N
Croton macrostachyus (pioneer)	Y	Y?	N	N
Understorey Shrubs				
Xymalos monospora	Y	Y	Y	Y
Pavetta hookeriana **x**	Y	Y	N	Y
Pavetta kupensis **x**	N	Y	Y	N
Pittosporum viridiflorum 'mannii' **x**	Y	Y	Y	Y
Cassine aethiopica	Y	Y	N	N
Maytenus undata	Y	N	Y	N
Rytigynia sp. A (*R. neglecta*) **x**	Y	Y	Y	Y
Psydrax dunlapii **x**	Y	N	N	Y
Rhamnus prinoides	Y	Y	N	N
Maytenus buchananii	Y	Y	Y	Y
Discopodium penninervium	Y	N	N	Y
Ardisia staudtii	Y	Y	Y	Y
Lianas				
Embelia schimperi	Y	N	N	N
Embelia mildbraedii **x**	N	N	Y	Y
Clematis simense	Y	Y	Y	Y
Clematis hirsuta	Y	Y	N	N
Jasminum dichotomum	Y	Y	N	N
Jasminum preussii	N	Y	Y	Y
Stephania abyssinica	Y	Y	Y	Y
Dalbergia oligophylla **x**	N	Y	Y	Y

The Mwanenguba-Kupe montane barrier

A second phytogeographic pattern that can be detected in montane forest species is a barrier to Cameroon highland species occurring between Mwanenguba and Mt Kupe. Table 7 documents seven woody species that are widespread in the Bamenda Highlands and most occur in the mountains further N still, but have their southern outpost at Mwanenguba, being unknown on Mts Kupe, Cameroon, or on Bioko. These species are: *Rhamnus prinoides, Embelia schimperi, Clematis hirsuta, Jasminum dichotomum, Croton macrostachyus, Neoboutonia glabrescens* and *Salix ledermannii*.

The nature of this barrier can only be conjectured. In the case of the *Salix*, lack of water bodies on the southern mountains is the probable cause. For the remainder, the wetter climate of the southern mountains might be at least part of the explanation. Dispersal difficulties may contribute to the matter, but it is worth noting that the *Salix* and *Clematis* have wind-dispersed seeds, so that dispersal difficulties do not explain their distribution.

Threatened montane forest taxa common to both Mwanenguba and Mt Kupe are those listed in the table above as restricted to, but widespread in, the Cameroon Highlands (*Pavetta hookeriana* and *Morella arborea* being absent from Mt Kupe, however) or in montane Africa (*Prunus africana*). In addition *Dalbergia oligophylla*, a liana, is an endangered species confined to both areas, and to Mt Cameroon. *Polystachya albescens* var. *manengouba*, an epiphyte, is known only from Mwanenguba, as is *Cyperus atrorubidus* var. nov.

Mt Kupe montane forest boasts its endemic *Pavetta kupense*, but other endemic taxa may yet come to light. *Clerodendrum inaequipetiolatum* is unknown from elsewhere in the Cameroon Highlands but is prominent and fairly common at the summit of Mt Kupe and descends to 1500 m alt. The glabrous stems of the Mt Kupe plants separate it from other populations and may warrant taxonomic recognition. Other upper submontane (c. 1500–2000 m alt.) potentially threatened species that reach to the summit area are *Eugenia* sp. 1, *Deinbollia* sp. 1 and *Tricalysia* sp. B aff. *ferorum*. *Embelia mildbraedii* and *Cassipourea malosana* are other montane threatened or near-threatened taxa on Mt Kupe.

The **threat** to montane forest on Mwanenguba appears to be that recorded by Letouzey: pasturing animals. The Mboroboro people who graze cattle and horses on the grasslands set dry season fires which eat into the forest edge. A GIS study is needed to quantify the extent to which this has and is occurring, and an on-the-ground survey is needed to assess the threats to the main montane forest block, which I have not visited. By contrast, the montane forest of Mt Kupe appears pristine: cultivation and grazing do not impinge upon it and no immediate man-made threats are apparent.

Synchronous mass-flowering of understorey monocarpic herbs

Elsewhere I have mentioned and discussed the above phenomenon at Mt Cameroon (Cable & Cheek 1998: xvii) and Mt Oku (Cheek *et al.*, 2000: 21). While all four species involved (*Mimulopsis solmsii, Brachystephanus giganteus*–formerly *Oreacanthus mannii, Acanthopale decempedalis* and *Plectranthus insignis*) have been recorded from our checklist area, they have not been recorded flowering together either at Mwanenguba (*Acanthopale* absent) or at Mt Kupe (*Mimulopsis* and *Plectranthus insignis* absent). The only evidence for multi-species synchronicity in this group in our area is the fact that in Dec. 1971 Leeuwenberg collected the *Mimulopsis* and the *Brachystephanus* in flower together at Mwanenguba. The second was recorded by him as locally abundant there.

As noted above, the Mwanenguba montane forest block is almost unexplored, so it is still not impossible that the four species assemblage might occur there, flowering in synchrony with each other at intervals probably of seven years.

While this four-species assemblage has been recorded as montane at Mt Oku and Mt Cameroon, in our checklist area the species concerned are predominantly submontane. Only *Brachystephanus giganteus* is restricted to truly montane altitudes (Mt Kupe and Mwanenguba, one collection each). *Mimulopsis solmsii*, known from 2000 m at Mwanenguba is also recorded at 1400 m (Kodmin) and 1000 m (Mwambong), *Plectranthus insignis* at 1300 m (Mwanenguba–Nsoung) and *Acanthopale* at 1000–1500 m alt.

Although numerous collections (nine) have been made from our checklist area (except Mwanenguba) of *Acanthopale*, none have been linked with flowering of other of the taxa, and in only one instance is there even any indication of mass-flowering. *Ensermu* 3557 (8 Nov. 1995) records the species as 'locally very common in an area of about 10 sq. metres, gregarious and mass-flowering'.

There does seem some evidence of within-species synchronicity for *Acanthopale* in our area. However the Bakossi Mts–Mwambong–Ngomboaku collections (collected in flower 1998–9) all appear to be on a different time clock from those on Mt Kupe (1994–95). This ties in with the phytogeographic observations made under mapping unit 4.

Since significant collecting work was done on Mt Kupe in 1993, 1996, 1998 and 1999, and *Acanthopale* was not then recorded in flower, the evidence suggests that it does indeed only flower at intervals of several years in our checklist area. It would be useful to pin down exactly what the interval is. *Plectranthus insignis* has only been collected once, near Mwanenguba, in our checklist area (*Darbyshire* 68, Mar. 2003). This population was at the end of its flowering cycle, so only scraps remained, making assessment of the area of this population difficult (Darbyshire *pers. comm.*).

Mass-flowering in *Brachystephanus kupeensis* and *Plectranthus* sp. nov.

The low evidence of reporting of mass-flowering (only 2 of the 15 collections mentioned above) does not negate its occurrence. In collecting a specimen it is normal to focus on recording features likely to be lost in preservation, at the expense of populational features. The notes of my own collection (made in 1993) of the submontane *Brachstephanus kupeensis*, endemic to Mt Kupe, fail to mention that it was mass-flowering over an area of at least 50 × 50 m. There is little doubt that this species only flowers over long intervals since its site, the 'nature trail' at Nyasoso, is easily accessible and is very well visited by botanists. Apart from the two collections made in 1993/94 it has not been seen before or since in flower. I myself have visited this site most years subsequent to 1993 and have never seen it in flower since. Finally, another mass-flowering, monocarpic species is *Plectranthus* sp. nov. (*Cheek* 1137, 20 March 2003, Massaka-Nyale).

Montane forest-grassland edge

The boundary between forest and grassland forms a distinct community which contains numerous species not found elsewhere, some of which are of conservation concern. Presumably the forest is too dark and the grassland too exposed to wind-damage or trampling to support these species. However, the demarcation between this community and grassland is less distinct than that with forest. Since grassland-forest is primarily forest dependent, it is classified here under mapping unit 9 rather than 10. Its area is difficult to estimate, but is probably very much less than 0.5% of the checklist area. The width of the band that comprises this community can vary from 2 m to 10 m or more. Fire-resistant trees that often occur at this boundary are indicated in Table 6. Fire-resistant shrubs characteristic of this community are as follows:
Hypericum revolutum
Hypericum roeperianum
Adenocarpus mannii

Mwanenguba

Sampling of this community by our expeditions has been restricted to the edge of the forest patch by the crater lakes in the centre of the Caldera. The main area of montane forest at Mwanenguba has not been sampled so its forest-grassland edge is likely to provide new taxa for our checklist area. The list below is representative, but not exhaustive.

Mwanenguba crater lakes 1900 m alt., grassland-forest transition. Feb. 1995 *Cheek* 7237 *et seq.*

Cynoglossum coeruleum subsp. *johnstonii*
Satureja punctata
Helichrysum forskahlii
Wahlenbergia silenoides
Alchemilla cryptantha
Laggera crispata
Asparagus warneckei
Dissotis thollonii var. *elliotii*
Vernonia biafrae
Bidens mannii
Commelina africana var. *africana*
Rumex abyssinicus
Trema orientalis
Hyparrhenia umbrosa
Pavonia urens var. *urens*
Crotalaria incana var. *incana*
Geranium arabicum subsp. *arabicum*

Veronica abyssinica
Pycnostachys eminii
Platostoma rotundifolium
Oldenlandia lancifolia
Desmodium salicifolium
Tephrosia interrupta
Pentas schimperiana subsp. *occidentalis*
Vernonia glabra var. *hillii*
Plectranthus kamerunensis
Crassocephalum vitellinum
Erica mannii
Amaranthus spinosus
Kosteletzkya adoensis
Phyllanthus nummariifolius
Triumfetta rhomboidea
Triumfetta annua
Justicia unyorensis

Due to the presence of scattered, fire-resistant trees and dense herbs 0.3–1 m tall, grassland-forest edge resembles a temperate woodland in physiognomy.

The domination of this community by Compositae, Labiatae and small, shrubby papilionoid legumes is evident. Such taxa as *Rumex, Satureja, Cynoglossum, Kosteletzkya,* Geraniaceae and *Erica* are only to be found within this community and nowhere else within the checklist area.

Several of the species listed are weedy, often associated with cultivation and may owe their presence here to disturbance and soil nutrient enhancement by cattle. These are *Amaranthus, Phyllanthus* and *Triumfetta* spp.

Most of these species also occur on other mountains along the Cameroon Highlands such as Mt Oku (see Cheek *et al.* 2000: 19) and Mt Cameroon (Cable & Cheek 1998). Indeed, the appearance of this community at Mwanenguba is strongly reminiscent of the same community at those places.

Threatened or near-threatened species in this community are:
Bidens mannii
Vernonia blumeoides.
Vernonia calvoana var. *calvoana*
Pentas ledermannii

Threats to this community need further research. Cattle trampling and grassland fires give cause for concern but it is likely that fires from lightning (perhaps less frequent or less intense than man-made fires) might be necessary to maintain the diversity of this community. Fire and grazing exclusion plots are needed to resolve this matter.

Mt Kupe

The summit areas are well-sampled on a piecemeal basis but the forest-grassland community appears poorly developed. The same three fire-resistant shrubs occur, as on Mwanenguba, namely *Hypericum revolutum, H. roeperianum* and *Adenocarpus mannii*.

Mt Kupe, c. 2000 m alt., grassland-forest edge. List from collections by Cable (e.g. Sept. 1992, June 1996) Cheek (Feb. 1995, Oct.–Nov. 1995), and Sebsebe (Oct.–Nov. 1995); species marked X are not present on Mwanenguba or elsewhere within the checklist area.

Laggera crispata
Vernonia holstii **x**
Erica mannii
Amphiblemma mildbraedii
Pentas schimperiana var. *occidentalis*
Pentas ledermannii

Virectaria major var. *major*
Carex echinochloë **x**
Cyperus mannii **x**
Isachne mauritiana **x**
Pteridium aquilinum

Numerous common species present in this community at Mwanenguba and other montane sites in the Cameroon Highlands are absent on Mt Kupe, reinforcing the idea of a strong isolation barrier which was suggested by similar absences in both the montane forest community (see above) and the montane grassland area (mapping unit 10). Most notable by their absence are:

Cynoglossum spp.
Lactuca glandulifera
L. inermis
Geranium arabicum subsp. *arabicum*
Platostoma rotundifolium
Pycnostachys eminii
P. meyeri

Satureja spp.
Kosteletzkya adoensis
Pavonia urens
Rumex abyssinica
R. nepalensis
Alchemilla spp.
Veronica abyssinica.

Lower incidence of fires, grazing and trampling relative to other montane sites and/or the small size of the three montane grassland areas on which this community borders (q.v. mapping unit 10) may also be factors that have dictated the absence of these taxa.

The only threatened species known thus far within the community is *Pentas ledermannii.*

MAPPING UNIT 10: MONTANE GRASSLAND; c. 1900–2050 m alt.

It occurs at two very distinct sites, Mwanenguba and Mt Kupe.

Mwanenguba

The grassland at Mwanenguba accounts for almost the entire surface area of this vegetation type in the checklist area, so minute are the patches at Mt Kupe. Our sampling, however, has been confined to the interior of the 3 km diameter caldera (c. 1900 m alt.). The greater part of the montane grassland of Mwanenguba lies on the SE shoulder of the caldera, which contains all the elevations above 2000 m (and the highest point, c. 2400 m). These remain unvisited by us, and appear to have been barely touched upon by earlier collectors to Mwanenguba, such as Letouzey, Sanford, de Gironcourt, Nicklés, Leeuwenberg and Villiers. However our knowledge of their itineraries and specimens is incomplete. Thomas, who collected at Mwanenguba in 1984 seems to have confined himself, as we did, to the area around the crater lakes, where he made an impressively comprehensive series of gatherings. It is unfortunate that the collections of Thorbecke from Mwanenguba, amongst the earliest specimens made in our checklist area (1907/08) do not figure in this checklist.

Thorbecke records having collected specimens of grassland taxa at Mwanenguba and that these were identified by Engler and used by him in a descriptive memoir to accompany a vegetation map in Passarge's publication 'Kamerun' which I have not been able to obtain (Thorbecke 1911). Most specimens collected in the German colonial era were unicates and were destroyed in Berlin in 1943. For these reasons Thorbecke's specimens and Engler's identifications of them, do not figure in this book.

Letouzey (1985) classifies the vegetation of the caldera floor as his mapping unit 112 ('paturage montagne a *Sporobolus africanus* 1600–2800m, avec galeries forêt'). He shows this as extending to the shoulders of the caldera and girdling the main montane forest block. However this map was largely the result of observation by aerial photography, and ground-truthing probably did not occur.

In brief, over 90% of the montane grassland in the checklist area remains botanically unexplored, and if this is rectified, is likely to add a substantial number of extra taxa to the area.

The grasslands of the Mwanenguba caldera are extensive, yet, probably due to trampling by cattle and annual burning by their Mboroboro herders, lack some of the diversity seen in the same habitat on Mt Cameroon.

Sporobolus africanus is the dominant species, with clumps scattered, bare black soil between. In the Bamenda Highlands, Hawkins & Brunt (1965) have documented the anthropic succession from montane forest, through *Hyparrhenia* grassland by forest clearance and fires, to *Sporobolus africanus* grassland, specifically by cattle trampling. Most of the species listed occur thinly scattered through the grassland, typically only 3–4 species occurring per m². Some annual taxa seem particularly well-adapted to the inter-*Sporobolus* tussock niche, these being *Radiola linoides, Wahlenbergia ramosissima* and *Antherotoma naudinii*.

The walls of steep banks have a denser vegetation and higher diversity of species than the open, flat grassland, perhaps due to reduced incidence of trampling by livestock.

The list below is representative, not exhaustive. Most of the species listed below are annuals that die as the dry season begins. However, several perennial species with underground rootstocks that can survive the aridity of the dry season, and fires, were noted and are marked 'p' in the list below.

Mwanenguba. Montane grassland inside the Caldera, N of the female lake, c. 1900 m alt.

Oct. 1998 *Cheek* 9416 *et seq.*

	Antherotoma naudinii			*Linum volkensii*
	Sporobolus africanus			*Radiola linoides*
p	*Nephrolepis undulata* var. *undulata*			*Wahlenbergia ramosissima*
	Swertia eminii		p	*Cyanotis barbata*
	Swertia mannii			*Impatiens burtoni*
	Sebaea brachyphylla		p	*Sonchus angustissimus*
	Spermacoce pusilla		p	*Lactuca inermis*
	Polygala albida subsp. *stanleyana*			*Trifolium usambarense*
	Uebelinia abyssinica			*Indigofera mimosoides* var. *mimosoides*
	Galium simense			*Crotalaria subcapitata* subsp. *oreadum*
	Antopetitia abyssinica			*Kotschya strigosa*

Neonotonia wightii	*Polygonum nepalense*
Desmodium uncinatum	*Tephrosia paniculata*
Desmodium repandum	p *Habenaria mannii*
Crotalaria incana subsp. *purpurascens*	p *H. peristyloides*
Polygala myriantha	p *H. cornuta*
Galinsoga parviflora	*Dichrocephala integrifolia* subsp. *integrifolia*
Plantago palmata	*Conyza pyrrhopappa*
Cerastium octandrum	

The appearance and diversity of montane grassland at Mwanenguba caldera is most similar, not to the geographically-close grassland at Mt Kupe or Mt Cameroon, but to that of Mt Oku (see Cheek *et al.* 2000: 18, and compare species listed there with those above).

Many of the distinctive elements of the Mt Cameroon grassland flora, such as *Hesperantha, Romulea, Thesium* and the diversity of terrestrial orchids have not been recorded from Mwanenguba, although if the higher altitudes of Mwanenguba are explored, some of these elements may be discovered especially if *Sporobolus*-free grassland is found. Above all it is the *Sporobolus africanus*-dominated grassland that unites Mt Oku with Mwanenguba rather than with Mt Cameroon, where this grassland type is unknown. Many of these species are widespread in montane Africa, as the epithets suggest.

Several of the species listed above are ruderal and are suggestive of pasture fertilized by cattle excrement, *Cerastium octandrum, Uebelinia abyssinica* and *Polygonum nepalense* being prime examples.

Threatened species are absent from the list above, apart from *Wahlenbergia ramosissima* (VU). This is to be expected, since at Mt Oku, Red Data species in montane grassland 'are largely confined to a subset of three grassland types where *Sporobolus africanus* is absent or rare'. So far, only *Sporobolus* grassland has been recorded from Mwanenguba although there is still much scope for discovering other types.

Threats to the caldera grassland are low. It is hard to see how it might be degraded further, although conceivably, heavier stocking with livestock and increased incidence of fires might reduce species diversity.

Mt Kupe

Montane grassland at the three peaks. Dates and number series as for mapping unit 9. 1950–2050 m alt.

Panicum hochstetteri	*Peucedanum kupense*
Loudetiopsis trigemina	*Lobelia hartlaubii*
Gladiolus aequinoctialis	*Plectranthus decumbens*
Monocymbium ceresiiforme	*Dissotis* sp. 1

Three patches of montane grassland are known on Mt Kupe, corresponding with each of the three peaks. Each patch is up to 100 m wide, and angled at about 45° from the horizontal near the peaks, the angle increasing to 90° as the distance from the peaks increases. In fact, these grassland patches appear to be the caps of inselbergs embedded in the mountain, with one face exposed. For this reason they are marked as inselbergs on the vegetation map, as R6–R8. Sampling of these grassland patches has occurred at the same time and by the same people as are documented for the montane forest under mapping unit 9. However, the steeper slopes have not been explored, for safety reasons. Abseiling equipment might facilitate such exploration in future.

Letouzey (1985) did not map the occurrence of grassland on Mt Kupe, although he visted the summit area previously. This may be a recording error, or he may have treated the patches as 'groupement saxicole divers', which he does record.

Thus far, only two species, *Panicum hochstetteri* and *Gladiolus aequinoctialis*, are common to these summit grasslands and the inselberg vegetation characterized earlier as mapping unit 6. Neither of these two taxa are obligate lithophytes. The low species overlap may be explained by the incompleteness of the survey of inselberg vegetation (see there), the altitudinal discrepancy (no sun-exposed rock faces have been surveyed at altitudes above 1000 m) or the differences in species due to slope (c. 45° from the horizontal at the summits, verses 90° on the faces of the inselbergs). If these grassland areas are induced by shallow soils over inselberg rock (pits need to be dug to confirm this) this might help account for the great differences in diversity, species composition and appearance with the montane grassland of Mwanenguba, where soils appear to be much deeper.

There are no species in common between the two lists for Mwanenguba and Mt Kupe. The dominant grass at Mt Kupe is not *Sporobolus africanus*, but *Panicum hochstetteri*. This forms dense blankets c. 60 cm deep, presumably suppressing most annual herbs and so reducing the species diversity. While evidence of fire (charred stems of woody plants) was found at 'Peak 2' there is no evidence that these are set so frequently as at Mwanenguba, nor indeed of them being set by man.

Two endemic species that have been recorded from these patches, *Peucedanum kupense* (VU) and *Dissotis* sp. 1 (more material needed to describe). Apart from these, the *Gladiolus* and *Plectranthus* are rated as near-threatened. Although species poor, this community has one of the highest proportions of threatened and near- threatened species of all those in the checklist area.

Man-made **threats** appear absent. However lightning-induced fires or land-slides may be natural threats.

So far as I am aware, the Mt Kupe summit grasslands have no parallel with any other grassland community.

MAPPING UNIT 11: ANTHROPIC VEGETATION; 200–1900 m alt.

Within the checklist area about 25–30% of the vegetation is anthropic, formed by the intervention of man. Most of this area occurs E of Mt Kupe and Mwanenguba, on the W side of the Loum–Nkongsamba road, and is excluded from the proposed protected areas. This road corresponds with the valley of the Dibombe river which is probably a fault, or trough, corresponding with the better known Bakossi trough W of Mt Kupe in which the Jide river flows. Both these valleys have numerous cinder cones along their length, indication of recent volcanic activity, and so high soil fertility. For this reason cultivation of crops is intensive.

Letouzey (1985) recognizes five types of anthropic vegetation in our checklist area:

A. Mapping unit 113. Montane *Sporobolus africanus* pasture (treated here under mapping unit 10).
B. Mapping unit 119. Facies de degradation ultime des forêts submontagnarde. This is applied to almost all anthropic vegetation in the 800-2000 m interval.
C. Mapping unit 128. Prairies et jacheres submontagnardes, paturées et habitués entre 1200–1800m, avec galleries forestières nettes. This mapping unit is shown as almost encircling the Mwanenguba caldera.
D. Mapping unit 234. Facies degradée, forêt toujour vert. Applied to anthropic areas formerly occupied by lowland evergreen forest.
E. Mapping unit 236. Facies de degradation ultime des forêts semi-caducifolie et forêt atlantique toujours verts. Applied as above, but to mixed semi-deciduous and evergreen forest.

Dessert bananas (*Musa*, originally Asian) grown for export to Europe are cultivated by commercial companies along the Loum–Nkongsamba and Loum–Tombel roads (see colour plate 2C). S of the Loum–Kumba road are large plantations of rubber (*Hevea brasiliensis*).

The two main cash crops in Bakossi are small-holder 'farms' of cacao (*Theobroma cacao*, S American) and coffee (*Coffea canephora*, tropical African). These are also cultivated in the valleys, and on fertile soils up to about 500 m and 700 m alt. respectively. These crops, especially the first, are cultivated along the Kumba–Mamfe road to some extent, and very intensively along the valley bottom at Nyandong, and at Enyandong in the E–W Mbu valley, also a trough. Generally, apart from the area E of Mt Kupe, it is only the valley-trough floors that are cultivated, the steep sides being much less fertile and so clothed in more or less pristine forest. These farms, after establishment, are maintained solely by use of the machete. Machines and herbicides are not used. The coffee bushes, especially if left untended as occurs when world prices are low, accumulate large quantities of epiphytes that must arise as seed from the adjacent forest canopy. In collecting fertile material from these coffee bushes, and from cultivated fruit trees such as mango (*Mangifera indica*, Indian) two new and threatened species of orchid have been located, so anthropic vegetation is not necessarily without some value for conservation. Shade trees, which accompany these crops in other parts of the world, and which make these farms more or less useful to biodiversity conservation, are either sparse or absent.

The home gardens and 'chop farms' (food crops) of lowland Bakossi villages, have a great diversity of crops. Cocoyams (*Xanthosoma*, S American and *Colocasia*, Asian), are probably the main starch staple cultivated, but Garri (also known as manihot, cassava or bobolo) (*Manihot esculenta*, S American) and plantains (*Musa* sp.) are also cultivated, together with some maize (*Zea mays*, S American), the latter apparently not faring as well as in areas with a longer dry season and less cloud cover. A diversity of yams (*Dioscorea*, mainly African in origin) are grown. Fruit trees include guava (*Psidium guayava*, S American), rose apple (*Syzygium malaccense*, SE Asian), orange (*Citrus sinensis*, E Asian), plum (*Dacryodes edulis*, native), pawpaw (*Carica papaya*, S American) and dessert bananas. These are generally cultivated adjacent to homes, but are sometimes found in chop farms, which can be a kilometre or more distant from a settlement. In addition to fruit, a variety of green vegetables such as water leaf (*Talinum triangulare*), bitter leaf (*Vernonia amygdalina*), *Corchorus olitorius*, *Solanum nigrum* and *Telfairia occidentalis* are cultivated. These are all indigenous African species.

While the above are grown for local consumption, some market gardening of vegetables and spices occurs on the eastern side near the Loum–Nkongsamba road which, with fast connections to the city of Douala to the SE and to the densely populated W Province to the N, provides opportunities for sales in those places. In addition to those crops mentioned, such species as *Zingiber officinale* (ginger, SE Asian), onions (*Allium cepa*) and garlic (*Allium sativum*, both Mediterranean) are also cultivated for this purpose.

Weeds

Anthropic vegetation is composed not just of cultivated plants, but of the weeds that occur as a result of cultivation. In intact natural vegetation, these weeds of cultivation are often rare or absent. Although our surveys have included species in farm bush and have picked up many cultivated species that linger in farm bush and can appear as if wild, I have only conducted one brief survey of weeds on a farm. The results are as follows:

Cocoa (*Theobroma cacao*) farm weeds collected at the cocoa farm of the Chief of Bintulu near the Mungo River Forest Reserve at 200 m alt. on 29 Nov. 1999, 4°47' 30° N, 9°36' 12" E. Specimen numbers consecutive in the series *Cheek* 10172–10189. This work was conducted as part of plot NA23, one of a series set up by the Mount Kupe Forest Project with EU funding to examine the environmental impact of upgrading the Ebonji–Ngusi road.

The 17 species listed were collected in an area of c. 2 × 2 m. Most are annual or short-lived perennial herbs, forming a dense sward 0.5–1 m high, completely covering the ground.

Pneumatopteris afra
Brillantaisia vogeliana
Elytraria marginata
Aneilema beniniense
Pouzolzia guineensis
Oxalis barrelieri
Vernonia stellulifera
Laportea aestuans
Oplismenus hirtellus

Synedrella nodiflora
Microsorum punctatum
Melanthera scandens
Acalypha arvensis var. *arvensis*
Stachytarpheta cayennensis
Asystasia gangetica
Sida garckeana
Sida rhombifolia var. α

Some of the weeds listed above also occur as natives of naturally disturbed areas of forest (e.g. due to tree falls) in our checklist area, e.g. *Brillantaisia vogeliana*, *Vernonia stellulifera*, *Oplismenus hirtellus*, *Aneilema* beniniense and Elytraria *marginata*. Others of these species appear native, but very rarely occur in naturally disturbed areas, being more usual as roadside weeds: *Synedrella nodiflora*, *Melanthera scandens*. A further group are not native, but introduced, probably by accident, from other countries or continents: *Oxalis barrelieri*, *Acalypha arvensis*, *Stachytarpheta cayennensis* and *Asystasia gangetica*. Those are not found in naturally disturbed areas of forest and are discussed in more detail in the invasives, alien and weedy plants chapter.

The higher altitude parts of our checklist, e.g. above 1000 m, lack the plantation and cash crops of the lower altitudes because temperatures are too cold to support them, and because, away from the fertile volcanic soils of the valley bottoms, substrates are far less suitable for crops. 'Robusta' coffee is sometimes grown as high as 1400 m alt., but much less commonly than at lower altitudes, presumably being less productive there.

One plantation crop that was established at higher altitudes was tea (*Camellia sinensis*, Chinese). A plantation was established at the beginning of the 20th century, but has been derelict for many years. The tea bushes at Essosong (c. 1000 m alt.), unchecked, have grown into substantial trees.

A list of all cultivated plant species encountered at one montane village is as follows:

Cultivated plants at Edib village, alt. 1170 m, recorded Thursday, 12 Feb. 1998.

Food crops

Brassica oleracea
Manihot esculenta
Ocimum gratissimum subsp. *gratisiimum* var.
    gratissimum
Persea americana
Coffea canephora

Colocasia esculenta
Xanthosoma sagittifolia
Saccharum officinale
Musa sp.
Elaeis guineensis

Ornamental / medicinal plants

Amaranthus sp.

Asclepias curassavica

Trichosanthes cucumerina subsp. *anguina*

Ricinus communis

Senna septemtrionalis

Rosa sp.

Cestrum nocturnum

Lantana camara

Canna indica

Hedychium sp.

As a result of the near absence of cash crops, higher altitude Bakossi anthropic vegetation may appear much as it has done for centuries. It has a distinctive naturalistic appearance formed by a small number of native species with distinctive architectures that are employed for very different purposes. This vegetation is best seen along the Tombel–Bangem road, once the 1000 m contour is approached. It is also apparent in the Kodmin–Nzimbeng area. Valleys appear covered with a mono-species woodland of the flat-topped, deciduous *Albizia gummifera* which is left in farmed areas since it is believed to encourage soil fertility; scientifically credible due to its probable nitrogen-fixing abilities. Even in apparently uncultivated areas, such woodland can be taken as indicating fallow, or 'farmbush'. Interspersed with the *Albizia* are lines of *Dracaena arborea*: the tallest (c. 20 m) and broadest leaved of all *Dracaena*, forming dense dark green leaf-rosettes 2 m in diameter and trunks with such well-developed *Rhizophora*-like roots that it is difficult to approach them. These seem to be the remnants of old stock hedges, used to exclude livestock, particularly cattle, from food crop areas. They are no longer used today, but persist (see mapping unit 5). *Phoenix reclinata*, unknown at lower altitudes in the checklist area, appears in the vicinity of villages in submontane Bakossi. Just W of Kodmin a substantial grove of well-tended, multi-stemmed plants can be found, none more than 3–4 m tall. These are tapped for palm-wine which is exported to other villages. Elsewhere *Phoenix reclinata*, untended can produce single stems 10–15 m high. These three elements together form the distinctive physiognomy of submontane Bakossi anthropic vegetation.

Strictly speaking, the montane *Sporobolus africanus* grasslands of the Mwanenguba caldera are also anthropic, formed by repeated burning and trampling. See mapping unit 10, montane grassland for more information. It is likely that at one stage, this caldera was forested.

A more comprehensive list of the cultivated species in our checklist area can be found in the ethnobotany chapter.

ENDEMIC VASCULAR PLANT TAXA

Benedict John Pollard & Martin Cheek

Herbarium, Royal Botanic Gardens, Kew, Surrey, TW9 3AE, U.K.

In the course of providing determinations for the many herbarium specimens collected from our checklist area, 82 vascular plant taxa (listed below in the order they appear in the checklist, the left column before the right column) have been identified as being endemic to Kupe, Mwanenguba and the Bakossi Mts. Eight of these were described and published before our surveys began in 1995, and 13 have been published since. There are also 13 taxa which we propose as new species but we do not count them here as being endemic as we cannot be certain that they actually do represent undescribed species. Often they are lacking generative stages, either flowers or fruits and so more material is required, or the taxonomy is not clear, e.g. *Scaphopetalum* sp. 1 of Kupe-Bakossi checklist (Sterculiaceae), where the genus needs revision. These 13 are listed after the endemics in a table entitled 'Putative Endemics'. The other 69 known endemics are all thought to be new to science and await formal description.

Several other species, such as *Thyrosalacia racemosa* (Celastraceae) are near-endemics, known to have occurred within our checklist area and at one or two localities just outside, in this case at one location just outside the E boundary, and at one location not far to the SE. In the checklist these may well be listed under our 'Narrow Endemic' chorological grouping, but this does not necessarily mean they are endemic to our checklist area. Another example of this is *Justicia leucoxiphus* (Acanthaceae), also a species described since 1995, known by 10 collections from inside our checklist boundaries but also by one collection originating beyond our W boundary, in the Rumpi Hills. Again, this is a 'Narrow Endemic', but is not endemic to our checklist area.

Endemic Vascular Plant Taxa
Dicotyledonae

Brachystephanus kupeensis
Justicia sp. 1
Genus nov.
Friesodielsia enghiana var. nov.
Monanthotaxis sp. nov.
Piptostigma sp. 1
Impatiens letouzeyi
Salacia elegans var. aff. *inurbana*
Salacia sp. nov. A
Diospyros kupensis
Drypetes sp. A
Drypetes sp. B
Phyllanthus caesiifolius
Phyllanthus nyale
Tragia sp. B
Beilschmiedia crassipes
Beilschmiedia ndongensis
Amphiblemma sp. 1
Diisotis sp. 1
Memecylon griseo-violaceum
Memecylon Sect. *Mouririoides* sp. nov. 1
Memecylon Sect. *Mouriroides* sp. nov. 2
Ardisia sp. cf. *etindensis*
Ardisia sp. near *staudtii*
Eugenia sp. 1
Eugenia sp. 2
Rhabdophyllum affine subsp. nov.
Octoknema sp. nov.
Ledermanniella onanae
Atroxima sp. aff. *afzeliana*
Calycosiphonia sp. A
Chassalia aff. *laikomensis* sp. B
Chassalia aff. *laikomensis* sp. C

Coffea bakossii
Coffea montekupensis
Ixora sp. A
Ixora sp. B
Keetia bakossii
Keetia sp. nov. aff. *ripae*
Lasianthus sp. 1
Mussaenda sp. nov.
Pavetta kupensis
Pavetta rubentifolia
Pentaloncha sp. nov.
Psychotria sp. aff. *camerunensis*
Psychotria sp. aff. *dorotheae*
Psychotria sp. A aff. *foliosa*
Psychotria sp. A aff. *fuscescens*
Psychotria sp. B aff. *fuscescens*
Psychotria sp. A aff. *gabonica*
Psychotria sp. aff. *latistipula*
Psychotria sp. A aff. *leptophylla*
Psychotria sp. B aff. *leptophylla*
Psychotria sp. aff. *subobliqua*
Psydrax bridsoniana
Rytigynia sp. B
Rytigynia sp. C
Sericanthe sp. A
Sericanthe sp. B
Vepris sp. 1
Allophylus cf. *schweinfurthii*
Pancovia sp. 1
Rhaptopetalum geophylax
Cola sp. nov. *'etugei'*
Microcos sp. A
Peucedanum kupense

Monocotyledonae

Afrothismia pachyantha
Afrothismia saingei
Gymnosiphon sp. A
Costus sp. A
Cyperus atrorubidus var. nov.
Cyperus microcristatus
Cyperus tenuiculmis subsp. *mutica*
Hypolytrum subcompositus
Hypolytrum pseudomapanioides

Dracaena cf. *phanerophlebia*
Dracaena sp.aff. *phrynioides*
Bulbophyllum jaapii
Bulbophyllum kupense
Ossiculum aurantiacum
Polystachya albescens subsp. *manengouba*
Polystachya kupensis
Stolzia sp. nov.
Kupea martinetugei

Putative Endemics

Ficus sp. 1
Ficus sp. 2
Psychotria sp. aff. *leptophylla*
Psychotria sp. 1
Psychotria sp. 3
Psychotria sp. 4
Psychotria sp. 5

Psychotria sp. 9
Rytigynia sp. D
Sabicea sp. A
Leptonychia sp. 1 of Kupe-Bakossi checklist
Leptonychia sp. 2 of Kupe-Bakossi checklist
Scaphopetalum sp. 1 of Kupe-Bakossi checklist

INVASIVE, ALIEN & WEEDY PLANTS

Martin Cheek

Herbarium, Royal Botanic Gardens, Kew, Surrey, TW9 3AE, U.K.

The terms used below are not entirely mutually exclusive. The definitions are intended to draw distinctions between the commonly used terms for the ecologically similar species treated below. Many of the introductions are believed to have spread from what was called the Victoria Botanic Garden, now the Limbe Botanic Garden, henceforth 'Limbe'.

Invasive plants are usually non-native species that invade, usually natural, habitat. They are of concern because they can displace native species, and so threaten them.

Aliens are species that are not native but which have become naturalized. They are not necessarily invasive or weedy but are of interest because they may become invasive. Also termed 'adventives' (plants recently introduced), 'exotics', many being 'garden-escapes', that were originally deliberately introduced for amenity horticulture, as ornamental flowering plants, while others were introduced as potential crop or fodder plants, and others inadvertently, presumably as seeds in the soil of intentionally introduced plants. Those listed below are species that have not (yet) become either invasive, or, in the sense of invading cultivated or fallow land, weedy.

Weeds are any plants growing in what a person considers the 'wrong' place. This usually refers to plants growing in cultivated situations which have not been planted but are spontaneous, and may threaten the yield of, e.g., crop plants by competing for water, space, nutrient or light resources. Details of weed species occurring in a cacao farm are given in the vegetation chapter of this book. Many weeds are pioneers in natural, disturbed situations, natives that have transferred to cultivated habitats. Other weeds are aliens.

In was not the main intention of our plant survey at Kupe, Mwanenguba and Bakossi Mts to document invasive, alien or weedy species. Indeed, it is likely that many common weedy species have been omitted from this checklist. However, lately there has been much international interest in this subject, primarily because of the problems that invasive species can pose to natural habitats, and also, less recently, because of interest in the effect that weed species can have on the yields of crops. In both cases it is considered important to record the presence of invasive species and of weeds in order to document their spread, and also to record the existence of alien species, since these can become invasive or weedy in future years. For this reason, new records for Cameroon or Africa are documented in detail below, with reference to the standard regional weed references for W Africa (Ivens *et al.* (1978), Okezie *et al.* (1987) and Terry (1983)) and by reference to FWTA.

Invasive species

Two invasive alien species were documented in lowland forest habitats, although neither has been found in pristine forest. Both appear to be newly recorded here as invasive species for Cameroon and possibly for Africa. No established invasive species are known. Indeed some species, invasive and problematic in other parts of the world are native and relatively uncommon in Bakossi, e.g. *Maesopsis eminii* (Rhamnaceae), very troublesome in some E African coastal forests. *Lantana camara* (Verbenaceae) an aggressive and noxious invader of natural habitat in S Africa, is fairly common as an ornamental in villages in Bakossi, but appears never to be invasive in natural habitat.

Ochroma lagopus (Balsaminaceae) is a pioneer tree which competes with *Musanga* (Cecropiaceae) in forest gaps and disturbed areas of forest. A fast-growing tree with wood of specialist applications (the Balsa wood of commerce), it is a native of S America. It was probably introduced to Cameroon through Limbe, in the German period (1892–1914) and persists there still. In the early twentieth century German plantation owners introduced many foreign commercial crops into Bakossi (cacao and coffee still being very evident today). *Ochroma* may have been introduced as an experiment that was found not to produce significant commercial returns. In Bakossi, *Ochroma* is especially evident when in fruit along the Tombel–Nyasoso road, at which time the brown fruit fibres are so numerous that they form a mist-like blanket over the road (*pers. obs.*). This species also occurs in the Mungo River Forest Reserve (Cheek 10151) where seedlings have been seen (*pers. obs.*), so it is clearly regenerating, and not merely persisting having been abandoned. The earliest collection from Bakossi is from the N (*Etuge* 464, Jan. 1987, Mejelet).

Alternanthera brasiliana (Amaranthaceae). This S American herb is newly recorded here for Cameroon, and possibly for Africa as a whole, since it is not listed in Lebrun & Stork (1991–1997), nor are any African specimens represented at the Kew Herbarium prior to our research. The first Bakossi record was identified by Cliff Townsend at Kew, based on *Etuge* 1776, March 1996, collected at the far eastern side of the checklist area, at Kolla, where it was noted 'used for hedges', possibly the original reason for its introduction. Subsequently it was recorded by Iain Darbyshire in 2003 at Nyandong village in western Bakossi (*Darbyshire* 4, 16 March 2003); this latter specimen was collected from 'open forest with coffee plantation, 420 m alt.'. Darbyshire (*pers. comm.*) reports it as being fairly common here and along the roadsides from Nyandong towards Konye, and confirms that he found it in disturbed forest. This species was known in Nyandong as a good goat fodder (*pers. obs.*) and so its further spread may be brought about both by goats, and by goat owners. Several species of Amaranthaceae are forest dwellers, withstanding shade, so this species may have the qualities needed to invade forest, where its blanket-like growth could suppress rarer, native species and so be a cause for conservation concern. Monitoring is recommended.

Cecropia peltata (Cecropiaceae) is likely to be found in Bakossi in the future. It was introduced to Cameroon through Limbe, whence it is slowly spreading outwards through the forest, competing on equal terms with the native forest gap pioneer *Musanga cecropioides* (Cecropiaceae).

Non-invasive alien taxa

Cuphea carthagenenis (Lythraceae) is newly recorded here as naturalized in Africa. It is not listed in Lebrun & Stork (1991–1997), nor in Keay (1954–1958), nor Hepper (1963–1972). It is extensively naturalized along motorable roads in the Bakossi area (*pers. obs.*) and has also been found along an old motorable road at Bu, near Wum, NW Province, Cameroon (*pers. obs.*). It is not treated here as an invasive because it has not been found in natural habitats such as forest, and is not treated as a weed because it has not been found on farms. Its origin may be as an ornamental exotic, since other species of the genus are planted in the area for this purpose. However this particular species is not particularly ornamental and may just as well have been inadvertently introduced with soil on plants from S America, possibly through Limbe. The earliest collection known of the taxon from Bakossi, and from Africa, appears to be *Etuge* 1640 (Nyasoso, Jan. 1996), since when three other collections have been made and numerous other sight observations throughout roadsides in Bakossi noted.

Cuphea is a S American genus of about 260 species (Mabberley 1997). *Cuphea carthagenensis,* formerly *C. balsamona* Cham. & Schlecht., has a wide ecological and geographical amplitude in its native habitat, being recorded from 2000 m alt. in Colombia, 50–100 m alt. in Panama, cut-over Cerrado (dry forest) in Brazil and swales and lake edges in Guyana (observations made from specimens in the Kew Herbarium). Unfortunately, no revision of this large genus is available, so it is not entirely clear as to the extent of the natural range of the species in the New World. The determinations of specimens at Kew by the noted Lythraceae specialist T.B. Cavalcanti have been relied upon for this account.

There are nineteenth century records of *Cuphea carthagenensis* from Rio de Janeiro, Guatemala, Martinique, Paraguay and Guyana suggesting that it is either naturally widespread in the New World or was spread by Europeans relatively early. In the USA it occurs from New Hampshire to Kansas and Louisiana under the name *C. petiolata*, and is also known as 'blue waxweed' or 'clammy cuphea'. The earliest records I have seen from various Pacific Islands are: Hawaii, dated 1895, Fiji 1947, New Caledonia 1960, and Vanuatu 1983. In SE Asia it was recorded in Java in 1950. One record from Java (*Forman* 136) specifies the Botanic Garden at Chibodas in 1956, suggesting that this may have been the port of entry to SE Asia. By 1975 it had been recorded in a Malaysian forest reserve and in 1979 from W Sumatra. It is clear from these records that this species has been spreading westwards, through the Pacific to SE Asia, and is now set to become a pantropical alien by occupying Africa. More research is needed to determine what the nature of the problems are, if any, that it causes elsewhere, and to learn from this experience to assist with preparation of a strategy for dealing with it in Africa.

Thunbergia alata (Acanthaceae), a native of E and S Africa, is an ornamental species which, like *Cuphea carthagenensis*, is often naturalized along roadsides but appears neither weedy nor invasive. According to Hepper (1963): 'frequently cultivated and often naturalized in warmer parts of the world'.

Crop aliens

The following crop species are thought to have reached Bakossi by the same route as *Ochroma* (see above) and are not, or very little, used today. They differ in that although they appear to be naturalized, they are not spreading, merely persisting. Observations suggest that fruits of the *Bixa* are used locally, as are, on a very limited scale, those of the

Artocarpus. This factor may favour the persistence of these species. The numerous other crop species actively cultivated near houses and on farms and occasionally persisting in secondary forest are not listed here due to lack of time and space. Many of these can be found in the Ethnobotany chapter.

Bixa orellana (Bixaceae):	occurs near dwellings on a very small and erratic scale, but also persists in farmbush.
Manihot glaziovii (Euphorbiaceae):	was formerly cultivated as 'Ceara rubber', is neotropical in origin, and rarely persists in Bakossi.
Artocarpus altilis (Moraceae):	is intermittent in secondary forest and farmbush near villages, never occurring in pristine forest.
Camellia sinensis (Theaceae):	is known only from Essosong where tea plants from a plantation, abandoned decades ago, are now trees.

Weedy aliens

Impatiens balsamina (Balsaminaceae):	a native of SE Asia, often grown as an ornamental, is recorded rarely as an escape into farmbush, but is long known to have been widely naturalized in Africa.
Cleome spinosa (Capparaceae):	is a neotropical ornamental that is sometimes naturalized in farmbush.
Drymaria villosa (Caryophyllaceae): subsp. *villosa*	a native of the neotropics, is fairly common as a weed.
Chromolaena odorata (Compositae):	is a neotropical species that is the major farm weed in our area. Among other neotropical Compositae (Asteraceae), that are weeds in our area are *Elephantopus mollis* and *Galinsoga spp.*
Ipomoea alba & *I. indica* (Convolvulaceae):	natives of S America and Asia respectively, both sometimes occur in farmbush, probably having been introduced as ornamentals.
Acalypha arvensis var. *arvensis* (Euphorbiaceae):	a native of Brazil, it has recently become a common lowland weed in Bakossi and in Cameroon generally, but is not recorded in either FWTA, nor any of the standard regional weed references (see below). The earliest Cameroonian specimen record is dated 1970. It was previously confused with the native, submontane, *A. manniana*.
Euphorbia heterophylla (Euphorbiaceae):	a native of the neotropics, it has long been fairly common as a weed in Tropical Africa.
Clerodendrum chinense (Labiatae)	a native of Asia, sometimes used as an ornamental, it is sporadically naturalized as a weed in farmbush in our area, although it is not listed in the standard regional weed references given below.
Senna spp., e.g. *S. hirsuta* (Leg.-Caes.):	occur as weeds of cultivation and farmbush. Our taxa are mainly neotropical in origin.
Mimosa invisa & *M. pudica* (Leg.-Mim.):	are both neotropical taxa that have long been naturalized in Tropical Africa.

The following papilionoid legumes derive from the neotropics and have long been naturalized in W Tropical Africa: *Calopogonium mucunoides, Centrosema pubescens, Crotalaria incana* subsp. *incana* and *Desmodium uncinatum*. Those originating from Tropical Asia are: *Crotalaria retusa* var. *retusa* and *Pueraria phaseoloides*.

Ludwigia decurrens (Onagraceae):	is a native of the neotropics but has long been widespread along roadsides in W Africa.
Oxalis barrelieri (Oxalidaceae):	is the most commonly collected species of the Oxalidaceae in Kupe-Bakossi. It is not recorded in any of the standard weed references below, nor in FWTA (FWTA 1:159 (1954)) and although there are numerous collections in Africa, (e.g. at Herb. K), none date before the early 1960s. Elsewhere in Guineo-Congolian Africa it appears in Ivory Coast and Congo (Kinshasha) in the 1970s (Herb. K). It can be inferred from this that *O. barrelieri* may have been introduced to Guineo-Congolian Africa in the 1950s, possibly entering through Cameroon (earliest collections known to the author). Its origin may have been SE Asia, e.g. the Malay Peninsula, where collections date back to the early 20th century. Despite the fact that this species can only have been found in the farmbush of Kupe-Bakossi for 50 years or less, it has already acquired a Bakossi name (Korkoremba) and is known to be useful for treating dysentery (see Ethnobotany chapter). Whilst this use may have been adopted from that used for the native *O. corniculata*, 'Korkoremba' is not applied to any other species of Oxalidaceae in Bakossi, so far as we know.
Oxalis corymbosa (Oxalidaceae):	native of the neotropics, it has long been naturalized in W Africa, and is a rare weed of farmbush in Bakossi. S America, It is cultivated around the world as an ornamental, was known as naturalised in 1954 from only two records in the FWTA area (FWTA 1:159 (1954)), both towns having botanic gardens (Aburi and Buea) so presumably being garden escapes. Our recent record of a naturalised population in Kupe-Bakossi suggests that it has now spread in the wild far beyond Buea, despite being known not to set seed (it produces basal bulbils).
Passiflora foetida (Passifloraceae):	also a native of the neotropics, it has long been a widespread roadside plant in Tropical Africa.
Peperomia bangroana (Piperaceae):	long known as *P. rotundifolia*, it has been an insignificant epiphytic weed of cacao and coffee plantations in Cameroon, spreading also onto native *Musanga*. Its earliest record on cacao in Bakossi is from 1950.
Polygonum nepalense (Polygonaceae):	a species of C and S Africa to Asia, not previously recorded from Cameroon or mainland W Africa (FWTA and regional weed references cited below) although known from Bioko. It now appears to be spreading as a submontane and montane weed in Cameroon and appears to be on the brink of invading natural grassland, although possibly only in nitrate-rich situations. This species is a cause for conservation concern since it may pose a threat to native grassland species by its spread. Monitoring is advised.
Rubus rosifolius (Rosaceae):	introduced from Asia, presumably for its edible fruit, it has long become widely naturalized in the Cameroon Highlands, extending southwards to Mt Cameroon.
Stemodia verticillata (Scrophulariaceae):	is neotropical in origin, but is now widely naturalized in W Africa, although occurring only occasionally.
Browallia americana (Solanaceae):	of which *B. viscosa* Humb., Bonpl. & Kunth is a synonym, is known from several records at three locations in Bakossi, of which the earliest is *Bruneau* 1075 (Jan. 1995, Nyasoso). Although it occurs along tracks, it can also occur in crop fields (*pers. obs.*). A native of the neotropics, its vivid

blue and white flowers are ornamental, and its presence in Cameroon, is probably due, as recorded on the earliest specimen known from W Central Africa (Buea, Cameroon, Jan. 1967, 3,000', collector indecipherable) to its being 'an escape from cultivation'. Specimens of the species at Kew record it as a cultivated plant in Zambia, Chile, India, Fiji and Malaysia, with the earliest records going back to the nineteenth century, being European: France (1829) and Kew (1895). It is not recorded in FWTA, nor in any of the standard regional weed references cited below.

Physalis angulata *P. lagascae* & *P. peruviana* (Solanaceae):	are all neotropical species established in farmbush or as farm weeds in Bakossi, all are long naturalized in W Africa.
Schwenckia americana (Solanaceae):	is also a neotropical derived weed, long known in W Africa though rare in Bakossi.
Eryngium foetidum (Umbelliferae):	is another neotropical species that is a weed of farms and roadsides in W Africa.
Pennisetum clandestinum (Gramineae):	is a native of E Africa, introduced to the montane grasslands of the Cameroon Highlands during the period of British administration in order to improve the quality of fodder for stock. It appears to have spread spontaneously at Mwanenguba.

REFERENCES

Hepper, F.N., (*ed.*) (1963–1972). Flora of West Tropical Africa. Volumes 2–3. Second edition. Crown Agents, London, U.K.

Ivens, G.W., Moody, K., Egunjobi, J.K. (1978). West African Weeds, Ibadan, Oxford University Press.

Keay, R.W.J. (*ed.*) (1954–1958). Flora of West Tropical Africa: Volume 1. Second edition. Crown Agents, London, U.K.

Lebrun, J.-P. & Stork, A.L. (1991–1997). Énumération des Plantes à Fleurs d'Afrique Tropicale. 4 volumes. Conservatoire et Jardin botaniques de la Ville de Genève, Geneva, Switzerland.

Mabberley, D.J. (1997). The Plant Book. Second edition. Cambridge University Press, U.K.

Okezie Akobundu, I. & Agyakwa, C.W. (1987). A Handbook of West African Weeds, International Institute of Tropical Agriculture, Ibadan.

Terry, P.J. (1983). Some Common Crop Weeds of West Africa and their control. USAID, Dakar.

PHYTOGEOGRAPHY & REFUGIA

Martin Cheek

Herbarium, Royal Botanic Gardens, Kew, Surrey, TW9 3AE, U.K.

In this chapter the phytogeographic affinities of each of the 11 vegetation types recognized are evaluated. These vary widely, for example the assemblage of species confined to shaded areas of rocks and inselbergs (mapping unit 6) being largely restricted to adjoining sites in the Cameroon Highlands with similar habitats, whilst at the other extreme, crater lake vegetation (mapping unit 8) seems to have affinities outside of W Africa, possibly with NE Africa, S-central Africa or Europe.

However the most important single vegetation type of the checklist area, in terms of containing the highest proportion of the species, the largest number of Red Data species and the highest number of strictly endemic species is the submontane forest (mapping unit 4). This vegetation type, in terms of numbers of shaded near-endemic species, unexpectedly seems to have links as strong or stronger with the Crystal Mts of Gabon (often apparently via 'stepping stones' in between) as with the very much closer Mt Etinde–Mt Cameroon. Only 60 km separates the peaks of Mt Etinde and Mt Kupe. Scrutinizing the species composition of Mt Etinde–Mt Cameroon and our checklist area shows some big differences at the species level. The group in which the differences are most marked are the terrestrial *Begonia* species of Sects. *Loasibegonia* and *Scutibegonia*. Six taxa of this group occur on Mt Etinde–Cameroon (Cable & Cheek 1998) and ten at the Kupe–Mwanenguba–Bakossi Mts. The overlap between the two areas comprises only two taxa. Added significance to this disparity is given by the fact this group are refuge indicators. They are shade-demanders and cannot survive outside the understorey of evergreen forest. They are self-incompatible and their small seeds lack wings, oil bodies or fleshy structures. They appear to have no mechanism of seed dispersal; their fruits are held close to the ground, and release their seeds by slowly rotting. It is believed that these factors classify these taxa as indicators of Pleistocene Refugia (Sosef 1994a). During the Pleistocene, when ice-sheets covered much of what is now the northern temperate zone, planetary temperatures as a whole dropped, and so did sea-levels and, in the tropical belt, rainfall. In Africa, evergreen forest disappeared, apart from in a few 'refuge' areas (e.g. Maley 1997, Maley & Brenac 1998). Initially two main refugia, Lower Guinea and the Albertine Rift, and two auxiliary refugia, in Upper Guinea and central Tanzania, were proposed. Since then additional refugia have been postulated, and the Lower Guinea refuge has been subdivided into several micro-refugia, one of which has been in the general neighbourhood of Mt Cameroon (e.g. Maley 1997, Maley & Brenac 1998), renowned as a centre of plant diversity (Cheek *et al.* 1994).

Finally, most species of *Begonia* that have been investigated are self-compatible, limiting the capacity of individual plants to establish new populations. It is believed that 'refuge' begonias remain in the former refuge sites because they lack the ability to colonise new sites for the above reasons.

In his taxonomic revision of the refuge begonias, Sosef (1994a) included an exhaustive analysis of their geography. He recognized 25 areas of endemism, defined by geographic clusters of taxa and then analysed these using cladistic methodology to produce 'areagrams' (see Fig. 7).

Our checklist area appears to fall in Sosef's (1994a) 'W Cameroon Mts' area, quite separate from, although geographically adjacent to, Mt Cameroon. Fifteen taxa occur in this area of endemism, of which eight are endemic to the area (extracted from Sosef 1994a, Table 13.1 and Fig. 13.6), giving it both the highest concentration of taxa and by far the highest concentration of endemics of all the areas of endemism. Thus, if refuge begonias are an accurate index of location and importance of refugia, the 'W Cameroon Mts' area (excluding Mt Cameroon) is the most important!

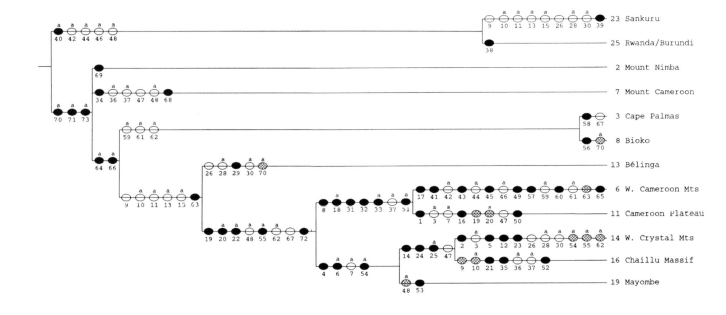

Fig. 7. GENERALIZED AREAGRAM, of 'refuge Begonia' areas of endemism, showing their relationships. Ellipses represent taxa: open = present due to dispersal; hatched = extinction; solid = apomorphy (taxon appears). Reproduced with permission from Sosef (1994a).

10 of these 15 'W Cameroon Mts' taxa occur in our checklist area. Of the remaining 5, two occur within 3–10 km of unexplored parts of our checklist area and so are likely to be found within it, and the remaining three are more distant, one occurring just over the border in Nigeria. From scrutiny of Sosef's records of specimen localities for the 'W Cameroon Mts' species, it is evident that Kupe, Mwanenguba and the Bakossi Mts have a higher number of refuge taxa than any other comparable place in this area of endemism. This exceptionally high concentration of refuge indicator taxa, it can be argued, is correlated to the extremely high number of endemic and threatened taxa in the checklist area. Of course, not all narrowly endemic or threatened taxa are indicative of refugia. Recently differentiated taxa can also be narrowly endemic (neo-endemics) and can also be threatened. *Peucedanum kupense* is an excellent example of a threatened putative neoendemic. It is an annual or biennial known only from the summit grasslands of Mt Kupe. *Peucedanum* is a large genus, with several species in Cameroon, among which two share so many similarities with *P. kupense* that it is possible to speculate that it may be of hybrid origin, and then to have undergone subsequent genetic drift (Darbyshire & Cheek 2004). There are few other such clear examples of putative neoendemics in our checklist area. Refuge endemics are more likely to be palaeoendemics, which appear more ancient in origin, often relictual in distribution, and in more slowly evolving groups.

Examples of palaeoendemics found in our checklist area occur in several groups. Among the saprophytes (achlorophyllous mycotrophs), many palaeoendemic refuge indicators can be found (Cheek & Ndam 1996). Mt Kupe and the Bakossi Mts have the greatest concentration of living saprophytes in Africa, just surpassing that of Mt Cameroon which included two extinct taxa in this group. While some saprophytes have wind-dispersed seeds and are widespread, probably for this reason (e.g. *Epipogium roseum*), most have no means of dispersal, and are shade-dependent, the same set of characteristics that give the 'refuge Begonias' their refuge indicator status (Cheek & Ndam 1996). Saprophytes at Mt Kupe with these characteristics include the three species of *Afrothismia* (two endemic) and *Kupea martinetugei*. Numerous Rubiaceae have been proposed as refuge indicators (Robbrecht 1996) although here the argument is slightly circular in as much as taxa have been nominated on the basis of their occurrence in previously recognized Refuges. Most Rubiaceae have fleshy, potentially bird-dispersed fruit and so have the capacity for fairly rapid spread, it would seem. However, Robbrecht has pointed out that what evidence there is available, points to the family being mostly self-incompatible (Robbrecht 1996). Therefore, a single plant, as might result from a 'dispersal event', is not likely to be able to reproduce itself, and so to be unviable with regard to establishing a new population. Amongst the taxa in our checklist nominated by Robbrecht as refuge indicators are the monotypic genus *Petitocodon*, *Calochone acuminata*, *Pentaloncha* sp. nov. (the other two species are confined to other microrefuges in the Gulf of Guinea), and several *Tricalysia* for which genus our checklist area has so many potentially new taxa. Other notable palaeoendemics in our checklist area, in the sense that, by definition, they have no close relatives, are the monotypic genera *Ossiculum* and Acanthaceae *gen. nov.*

Semi-deciduous forest as a barrier with Mt Cameroon

If it is accepted that our checklist area forms part of a microrefugium, here termed the W Cameroon Mts refugium to follow Sosef's term for this 'area of endemism', then what has divided it geographically from Mt Cameroon? On Sosef's areagram (Fig. 7) Mt Cameroon is far removed from the W Cameroon Mts, being paired with Bioko. W Cameroon Mts are sister to the Cameroon Plateau (the Yaoundé – Sangmelina area, towards Gabon) and these two areas are sister to a clade that includes the W Crystal Mts, the Chaillu Massif and Mayombe 'areas of endemism'. The most credible barrier between Mt Cameroon and Kupe – Bakossi Mts, is *Triplochiton* – Ulmaceae semi-deciduous forest, of which a band is present to the north of the first (see vegetation map in Cable & Cheek 1998) and in the SE corner (the Loum F.R.) of the second (see Fig. 3). It is perfectly possible that these two are or were continuous. Unfortunately surveys have not been conducted to verify this in the intervening, now largely anthropic vegetation in between the two. Letouzey (1968b) refers to semi-deciduous forest remnants occurring along the Nkongsamba – Loum and Loum – Kumba roads. In his 1985 vegetation map the Kupe – Bakossi Mts are shown sitting in a cup or arc of evergreen forest with semi-deciduous elements. This ties in and amplifies our own records of *Celtis* and other semi-deciduous forest taxa along the Loum – Kumba and Kumba – Mamfe roads (see mapping units 1 and 2).

During the last ice-age, c. 12,000 years BP when conditions were drier as well as colder, this evergreen-with-semi-deciduous forest would probably have become fully developed semi-deciduous forest, if not deciduous forest, savanna, or grassland. Most evergreen forest shade-demanding species would not have been able to survive the high light levels reaching the ground's surface, and would have been eliminated, except perhaps in gallery forest.

Due to the colder conditions, altitudinal vegetation bands would have dropped, or become more compressed. Given the current lapse rate of 0.6°C per 100 m alt., and a 4°C drop in the Pleistocene, submontane forest can be postulated to have had its boundary with lowland forest at only 100–200 m alt. Without the phytogeographic 'dry barrier' represented today by remnants of semi-deciduous forest, today's submontane species of Mt Etinde–Mt Cameroon and of Kupe, Mwanenguba and Bakossi Mts would have been able to intermingle.

It is conceivable that the Mungo, which drains the southern part of the Bakossi Mts, may have provided a corridor to the eastern flank of Mt Cameroon for evergreen forest shade-demanders.

Barriers between West, East, North and within the Kupe, Mwanenguba and Bakossi Mts
The East & West 'dry barriers'

The semi-deciduous forest that appears to have restricted the movement of plant species between Kupe, Mwanenguba and the Bakossi Mts and Mt Cameroon, had northward extensions up the Jide Valley (the Bakossi trough along the Tombel–Bangem road), the upper Mungo Valley (Kumba–Mamfe road) and the Dibombe Valley (the Loum–Nkongsamba road).

These northward extensions of the dry barrier would have restricted migration of submontane species between E and W and vice versa, it can be argued. As one moves northwards up the valleys, their floors either become higher (e.g. rising above 1000 m in the Jide) or the gap between the submontane areas on each side dwindles to almost nil, e.g. at Supe on the Kumba–Mamfe road where highland areas of the Bakossi Mts to the E and the Rumpi Hills on the W come very close together, giving the impression of a gorge, potentially allowing submontane species to traverse the barrier, especially in the Pleistocene. Therefore these barriers are much more permeable than that to the S, separating Mt Cameroon. This might account for the large number of endemics (19) common to both submontane Bakossi Mts and Mt Kupe, but occurring nowhere else, as far as is known. At the same time it explains the fact that 26 taxa are endemic to Mt Kupe, and are unknown in the Bakossi Mts, while 16 taxa are endemic to the Bakossi Mts and do not occur on Mt Kupe, nor anywhere else. The anomalous situation of the tongue-like submontane northern extension from Mt Kupe, which lacks the 26 endemic taxa of Mt Kupe, despite being continuous with it at the submontane altitude may best be explained by the *Aframomum*–Marantaceae thicket (see mapping unit 5) just north of Mt Kupe, also acting as an important barrier.

On the basis of available evidence, this northern tongue has closer links with the Bakossi Mts (uniquely sharing four endemic taxa) than with Mt Kupe (uniquely sharing only two endemic taxa).

The incompleteness of the dry barrier west of the Bakossi Mts is shown by the fact that the Bakossi Mts and (to the west) the Rumpi Hills share five endemic taxa, and a further two if Barombi Mbo, near Kumba, is treated as part of the Rumpi Hills. Two of the taxa nearly endemic to both Kupe and Bakossi also occur in the Rumpi Hills. These figures are likely to increase as the Rumpi Hills become better surveyed.

While the Rumpi Hills are moderately well surveyed, thanks to the efforts of Réné Letouzey and Duncan Thomas, areas to the east are much less well known so it is more difficult to address the connections with Kupe, Mwanenguba and the Bakossi Mts and the impact of the putative 'dry barrier' extension along the Loum – Nkongsamba route. Mt Nlonako, just SE near Nkongsamba is almost a twin of Mt Kupe geologically and we do know that the two share at least one species uniquely, *Ardisia koupensis*. A full survey there is needed to find how many other species that appear today to be strictly endemic to Mt Kupe are actually located there also. Further east in the Yabassi area lie low hills extending to the Sanaga which include some submontane areas. These areas are very poorly surveyed. However, *Uvariopsis submontana*, otherwise only known from Mt Kupe, has a record there. Furthermore *Loesenera talbotii* is also recorded from Yabassi, but is otherwise barely known outside our checklist area although very common within.

Barriers and connections to the North

Due N of Mwanenguba lies the Mbo plain, a lowland swampy basin (of which the largest town is Santcho). Drier areas here have savanna species which otherwise occur only to the N so this area is likely to have been an effective barrier to the interchange of submontane species with the north in the Pleistocene, as it is now. The link between the submontane areas in our checklist and the extensive Bamenda Highlands to the north (including the Bamboutos Mts) is an arc of submontane areas to the west of the Mbo Plain, which begin with the Ntale Mts (also known as the Mwenzekong Mts) just west of Mwanenguba, and separated from the northern Bakossi Mts only by the Mbu valley (also known as the Mbwe valley). This valley runs E–W and may also have presented a barrier to the interchange of submontane species between the Bakossi Mts and highland areas to the N. To the N of the Ntale Mts, separated by another E–W lowland valley, lies another submontane area, the Ekomane Mts. These were early explored by both Ledermann and Conrau (and then referred to as Kongoa-Gebirge), but only fragmentarily and never since. Two other small submontane areas further north then link with the southern Bamboutos of the Bamenda Highlands.

It is these submontane areas, and their outliers, such as the Obudu Plateau in Nigeria, and adjoining lowland areas, that appear to comprise the 'W Cameroon Mts' area of endemism, or refuge.

The distributions of Sosef's Refuge begonias within this area can be interpreted by reference to these submontane blocks and the barriers referred to above. *Begonia pseudoviola*, for example, is known from Nlonako, the Mt Kupe northern submontane 'tongue' and Mwanenguba, reaching the Bamenda Highlands presumably via the western arc submontane areas referred to above. It does not occur on Mt Kupe or in the Bakossi Mts proper, nor in the Rumpi Hills, so one can speculate that it was prevented from reaching these respectively by the *Aframomum* thicket just to the N of Mt Kupe, the Jide, Mbu and Upper Mungo valleys. *Begonia prismatocarpa* subsp. *delobata* on the other hand, is only known from the Rumpi Hills (and Nta Ali just to the N), and the northern Bakossi Mts. Its absence from the Bamenda Highlands, Mwanenguba, Mt Kupe and Nlonako can be attributed to it having failed to cross the Mbu and Jide valleys. *Begonia duncan-thomasii* is only known from the Rumpi Hills, Bakossi Mts, and Mt Kupe. This distribution is explained by it failing to cross the Mbu and Dibombe valleys.

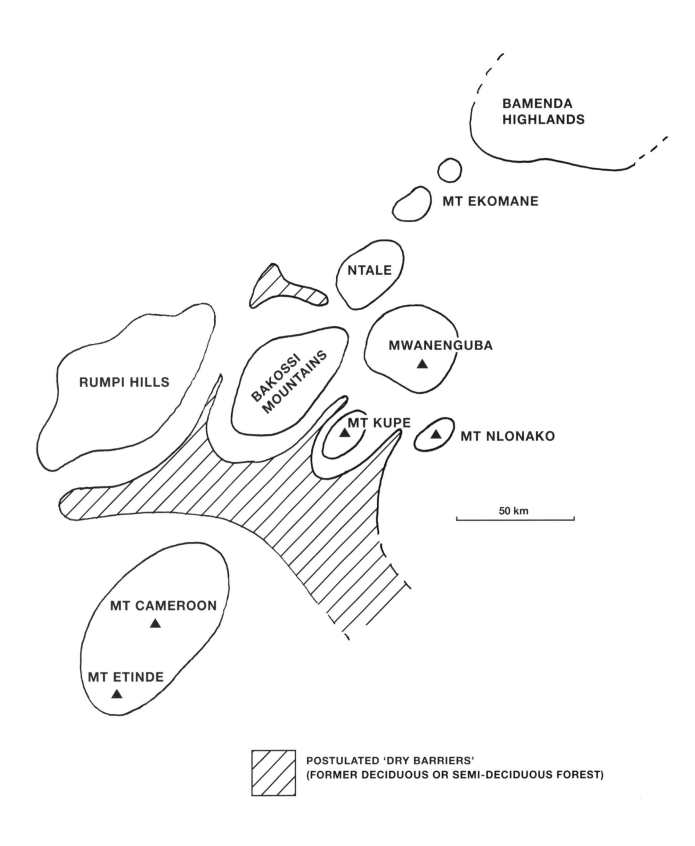

BAMENDA
HIGHLANDS

MT EKOMANE

NTALE

MWANENGUBA
▲

RUMPI HILLS

BAKOSSI
MOUNTAINS

MT KUPE
▲

MT NLONAKO
▲

50 km

MT CAMEROON
▲

MT ETINDE
▲

POSTULATED 'DRY BARRIERS'
(FORMER DECIDUOUS OR SEMI-DECIDUOUS FOREST)

FIG. 8. SKETCH MAP OF POSTULATED PLEISTOCENE BARRIERS between submontane areas of the southern Cameroon Highlands. The enclosed areas are: unshaded = submontane areas (above 800m); shaded = postulated dry barriers (former deciduous or semi-deciduous forest).

ETHNOBOTANY

Benedict John Pollard

Herbarium, Royal Botanic Gardens Kew, Surrey, TW9 3AE, U.K.

There are two main groups of plants in our checklist area that are significant to the local people: crop plants (mainly introduced) and mostly-native wild species. Expansion of cropland and non-sustainable harvesting of wild species contribute threats to natural vegetation.

Many of the plants listed in our checklist are of great importance to the local people and are inextricably linked to local customs and beliefs.

In compiling this checklist of plants from Kupe, Mwanenguba and the Bakossi Mountains, it became apparent that a number of our plant specimen collectors (particularly Cheek and Etuge) invested considerable time investigating the ethnobotanical knowledge held by the local people. Although this was never an absolute focus of our inventory programme, a great deal of information regarding local plant names and uses has been recorded with the agreed collaboration of local informants. Most of these records involve wild species, but where records are available for introduced species, such as mango and breadfruit, these have also been incorporated.

The local names have been incorporated in the checklist, near the end of each taxon account, to provide a clear link between the scientific and indigenous knowledge and classification systems. They are all Bakossi names, excepting a few in Pidgin English and occasionally local English names. We have decided to exclude from the checklist the finer details of local medicinal plant uses, as the intellectual property rights to this information belong with the Bakossi people, and is not to be disseminated without their express permission. The data regarding plant uses that were reported to us and recorded during our fieldwork are summarised with reference to standardised terms proposed by Cook (1995). This work provides a system whereby uses of plants can be described using standardised descriptors and terms and attached to taxonomic data sets. These descriptors operate on three levels, the third level being most detailed. We have restricted our use of these descriptors to Levels 1 and 2, with a further general descriptor for all uses except medicines, (only Level 1 is used for medicinal information), to provide a simple overview of different plant uses without disclosing the most sensitive information. Chief Abweh, Daniel Ajang, Max Ebong, Epie Ngome and Daniel Lyonga were amongst the foremost of the Bakossi ethnobotanical informants who provided us with the traditional names and uses of the plants that they helped us to study. Additional information on cultivated species is provided after the lists of wild species.

It is quite likely that there are many plants of potential economic value in our checklist area, some of which may be endemic. If these plants are to be investigated for medicinal properties, horticultural value or any other useful traits, it is essential that the correct developmental frameworks and strategies are in place. It is hoped that these issues can be addressed and perhaps discussions with stakeholders initiated so that permission to develop a more thorough and complete ethnobotanical survey may in future be granted. The data gathered to date may be valuable as a starting point for a more focused ethnobotanical survey. We intend to develop this important aspect of our study in future fieldwork, if feasible.

Wild plants are used in a multitude of ways in Bakossiland, and are listed below according to LEVEL 1 states. The dashed lines indicate separate plant groups: Dicotyledonae are listed first, followed after the dashed lines by Monocotyledonae, and thirdly Gymnospermae and Pteridophyta.

Caveat: be advised that the indigenous traditional knowledge contained in this volume has been given in good faith as a contribution to the common good and the furtherance of mutual understanding and the preservation of all life on our planet. It is not be used for personal or commercial gain and must be treated with respect and used only for the purpose for which it was gifted. Anyone who reads this volume assumes the moral and ethical obligations implied by this statement.

FOOD

Dicliptera laxata	(Acanthaceae)
Nelsonia canescens	
Amaranthus hybridus	(Amaranthaceae)
subsp. *cruentus*	
Celosia isertii	

Celosia trigyna	
Mangifera indica	(Anacardiaceae)
Spondias mombin	
Trichoscypha acuminata	
Landolphia buchananii	(Apocynaceae)

Orthopichonia cirrhosa	
Impatiens burtoni	(Balsaminaceae)
Impatiens macroptera	
Impatiens mannii	
Begonia adpressa	(Begoniaceae)
Begonia longipetiolata	
Canarium schweinfurthii	(Burseraceae)
Dacryodes edulis	
Lobelia rubescens	(Campanulaceae)
Buchholzia coriacea	(Capparaceae)
Buchholzia tholloniana	
Carica papaya	(Caricaceae)
Myrianthus preussii	(Cecropiaceae)
subsp. *preussii*	
Salacia debilis	(Celastraceae)
Salacia staudtiana	
var. *staudtiana*	
Emilia coccinea	(Compositae)
Vernonia amygdalina	
Vernonia hymenolepis	
Ipomoea alba	(Convolvulaceae)
Ipomoea batatas	
Brassica oleracea	(Cruciferae)
Citrullus lanatus	(Cucurbitaceae)
Cucumeropsis mannii	
Telfairia occidentalis	
Trichosanthes cucumerina	
subsp. *anguina*	
Manihot esculenta	(Euphorbiaceae)
Plukenetia conophora	
Ricinodendron heudelotii	
subsp. *africanum*	
var. *africanum*	
Garcinia kola	(Guttiferae)
Garcinia lucida	
Afrostyrax lepidophyllus	(Huaceae)
Lavigeria macrocarpa	(Icacinaceae)
Irvingia gabonensis	(Irvingiaceae)
Persea americana	(Lauraceae)
Tetrapleura tetraptera	(Leguminosae: Mim.)
Arachis hypogaea	(Leguminosae: Pap.)
Kotschya strigosa	
Phaseolus vulgaris	
Vigna adenantha	
Tristemma littorale	(Melastomataceae)
subsp. *biafranum*	
Tristemma mauritianum	
Bersama abyssinica	(Melianthaceae)
Artocarpus altilis	(Moraceae)
Treculia africana	
subsp. *africana*	
var. *africana*	
Psidium guajava	(Myrtaceae)
Syzygium malaccense	
Panda oleosa	(Pandaceae)
Sesamum radiatum	(Pedaliaceae)
Hilleria latifolia	(Phytolaccaceae)
Piper capense	(Piperaceae)
Piper guineense	
Piper umbellatum	

Talinum triangulare	(Portulacaceae)
Coffea canephora	(Rubiaceae)
Spermacoce intricans	
Citrus × *aurantium*	(Rutaceae)
Scoparia dulcis	(Scrophulariaceae)
Solanum americanum	(Solanaceae)
Solanum anguivi	
Solanum betaceum	
Solanum lycopersicum	
Solanum nigrum	
Solanum terminale	
Cola argentea	(Sterculiaceae)
Cola ficifolia	
Cola lateritia	
var. *lateritia*	
Cola lepidota	
Cola nitida	
Theobroma cacao	
Camellia japonica	(Theaceae)
Corchorus olitorius	(Tiliaceae)
Centella asiatica	(Umbelliferae)

Colocasia esculenta	(Araceae)
Xanthosoma sagittifolium	
Ananas comosus	(Bromeliaceae)
Dioscorea spp.	(Dioscoreaceae)
Saccharum officinarum	(Gramineae)
Zea mays	
Musa cvs	(Musaceae)
Elaeis guineensis	(Palmae)
Aframomum flavum	(Zingiberaceae)

Gnetum africanum	(Gnetaceae)
Lunathyrium boryanum	(Woodsiaceae)

FOOD ADDITIVES

Cleistopholis staudtii	(Annonaceae)
Xylopia aethiopica	
Landolphia incerta	(Apocynaceae)
Vernonia doniana	(Compositae)
Phyllanthus muellerianus	(Euphorbiaceae)
Ricinodendron heudelotii	
subsp. *africanum*	
var. *africanum*	
Tragia benthamii	
Garcinia kola	(Guttiferae)
Afrostyrax lepidophyllus	(Huaceae)
Cinnamomum verum	(Lauraceae)
Tetrapleura tetraptera	(Leguminosae: Mim.)
Glycine max	(Leguminosae: Pap.)
Xymalos monospora	(Monimiaceae)
Piper capense	(Piperaceae)
Piper guineense	
Piper umbellatum	
Rubus rosifolius	(Rosaceae)
Zanthoxylum gilletii	(Rutaceae)
Rhaptopetalum geophylax	(Scytopetalaceae)
Capsicum annuum	(Solanaceae)
Capsicum frutescens	

Laportea ovalifolia	(Urticaceae)

Aframomum melegueta	(Zingiberaceae)

ANIMAL FOOD

Landolphia landolphioides	(Apocynaceae)
Salacia staudtiana	(Celastraceae)
var. *staudtiana*	
Galinsoga quadriradiata	(Compositae)
Calopogonium mucunoides	(Leguminosae: Pap.)
Penianthus longifolius	(Menispermaceae)
Cuviera longiflora	(Rubiaceae)
Psychotria gabonica	
Virectaria major	
var. *major*	
Quassia silvestris	(Simaroubaceae)
Cola cf. *verticillata*	(Sterculiaceae)
Triumfetta cordifolia	(Tiliaceae)
var. *cordifolia*	

BEE PLANTS

Cylicomorpha solmsii	(Caricaceae)
Plectranthus glandulosus	(Labiatae)

MATERIALS

Sclerochiton preussii	(Acanthaceae)
Annickia chlorantha	(Annonaceae)
Xylopia africana	
Alstonia boonei	(Apocynaceae)
Rauvolfia vomitoria	
Ochroma lagopus	(Bombacaceae)
Cordia aurantiaca	(Boraginaceae)
Canarium schweinfurthii	(Burseraceae)
Musanga cecropioides	(Cecropiaceae)
Terminalia superba	(Combretaceae)
Kalanchoë crenata	(Crassulaceae)
Lagenaria siceraria	(Cucurbitaceae)
Diospyros pseudomespilus	(Ebenaceae)
subsp. *pseudomespilus*	
Diospyros zenkeri	
Euphorbia schimperiana	(Euphorbiaceae)
var. *pubescens*	
Hevea brasiliensis	
Macaranga occidentalis	
Manihot glaziovii	
Ricinodendron heudelotii	
subsp. *africanum*	
var. *africanum*	
Dasylepis racemosa	(Flacourtiaceae)
Garcinia kola	(Guttiferae)
Garcinia mannii	
Hypericum roeperianum	
Isodon ramosissimus	(Labiatae)
Vitex grandifolia	
Loesenera talbotii	(Leguminosae: Caesalp.)
Albizia adianthifolia	(Leguminosae: Mim.)
Pterocarpus soyauxii	(Leguminosae: Pap.)

Tapinanthus ogoensis	(Loranthaceae)
Hibiscus surattensis	(Malvaceae)
Sida sp. aff. *rhombifolia*	
Carapa procera	(Meliaceae)
Entandophragma angolense	
Guarea cf. *cedrata*	
Guarea mayombensis	
Lovoa trichiliodes	
Trichilia prieuriana	
subsp. *vermoesenii*	
Turraeanthus africanus	
Penianthus longifolius	(Menispermaceae)
Ficus ardisioides	(Moraceae)
subsp. *camptoneura*	
Ficus exasperata	
Milicia excelsa	
Coelocaryon botryoides	(Myristicaceae)
Coelocaryon preussii	
Pycnanthus angolensis	
Ardisia koupensis	(Myrsinaceae)
Eucalyptus spp.	(Myrtaceae)
Campylospermum laxiflorum	(Ochnaceae)
Diogoa zenkeri	(Olacaceae)
Strombosia grandiflora	
Carpolobia alba	(Polygalaceae)
Polygonum setosulum	(Polygonaceae)
Rumex nepalensis	
Maesopsis eminii	(Rhamnaceae)
Nauclea diderrichii	(Rubiaceae)
Nauclea vanderguchtii	
Pauridiantha venusta	
Rothmannia hispida	
Tricalysia gossweileri	
Zanthoxylum gilletii	(Rutaceae)
Zanthoxylum rubescens	
Quassia sanguinea	(Simaroubaceae)
Quassia silvestris	
Cola cauliflora	(Sterculiaceae)
Cola lateritia	
var. *lateritia*	
Triumfetta cordifolia	(Tiliaceae)
var. *cordifolia*	
Triumfetta rhomboidea	
Pouzolzia parasitica	(Urticaceae)
Urera trinervis	

Gloriosa superba	(Colchicaceae)
Carex echinochloë	(Cyperaceae)
Dracaena arborea	(Dracaenaceae)
Dracaena fragrans	
Bambusa vulgaris	(Gramineae)
Hypselodelphys scandens	(Marantaceae)
Marantochloa leucantha	
Thaumatococcus daniellii	
Eremospatha cuspidata	(Palmae)
Eremospatha macrocarpa	
Laccosperma opacum	
Oncocalamus tuleyi	
Raphia cf. *africana*	
Raphia hookeri	

Raphia regalis

Cupressus lusitanica	(Cupressaceae)
Cyathea camerooniana	(Cyatheaceae)
var. *zenkeri*	

FUELS

Alangium chinense	(Alangiaceae)
Schefflera barteri	(Araliaceae)
Canarium schweinfurthii	(Burseraceae)
Bridelia macrantha	(Euphorbiaceae)
Harungana madagascariensis	(Guttiferae)
Piptadeniastrum africanum	(Leguminosae: Mim.)
Maesa rufescens	(Myrsinaceae)
Eucalyptus spp.	(Myrtaceae)

SOCIAL USES

Marsdenia latifolia	(Asclepiadaceae)
Nuxia congesta	(Buddlejaceae)
Cannabis sativa	(Cannabaceae)
Acmella caulirhiza	(Compositae)
Laggera crispata	
Solanecio biafrae	
Vernonia doniana	
Rourea thomsonii	(Connaraceae)
Diospyros zenkeri	(Ebenaceae)
Cyrtogonone argentea	(Euphorbiaceae)
Neoboutonia mannii	
var. *glabrescens*	
Phyllanthus muellerianus	
Garcinia lucida	(Guttiferae)
Isodon ramosissimus	(Labiatae)
Pavonia urens	(Malvaceae)
var. *glabrescens*	
Dorstenia barteri	(Moraceae)
var. *barteri*	
Ardisia koupensis	(Myrsinaceae)
Peperomia fernandopoiana	(Piperaceae)
Piper umbellatum	
Oldenlandia lancifolia	(Rubiaceae)
var. *scabridula*	
Oxyanthus unilocularis	
Psychotria peduncularis	
var. *hypsophila*	
Nicotiana tabacum	(Solanaceae)
Cola anomala	(Sterculiaceae)
Urera trinervis	(Urticaceae)

Aneilema umbrosum	(Commelinaceae)
subsp. *umbrosum*	
Commelina benghalensis	
var. *hirsuta*	
Commelina capitata	
Murdannia simplex	
Paspalum conjugatum	(Gramineae)
Elaeis guineensis	(Palmae)
Phoenix reclinata	
var. *reclinata*	

Raphia cf. *africana*
Raphia hookeri

Cupressus lusitanica	(Cupressaceae)

VERTEBRATE POISONS

Emilia coccinea	(Compositae)
Massularia acuminata	(Rubiaceae)

MEDICINES

Dischistocalyx grandifolius	(Acanthaceae)
Eremomastax speciosa	
Hypoestes aristata	
Justicia tenella	
Nelsonia canescens	
Celosia isertii	(Amaranthaceae)
Mangifera indica	(Anacardiaceae)
Annickia chlorantha	(Annonaceae)
Artabotrys aurantiacus	
Cleistopholis staudtii	
Rauvolfia caffra	(Apocynaceae)
Rauvolfia vomitoria	
Voacanga africana	
Impatiens letouzeyi	(Balsaminaceae)
Impatiens mackeyana	
subsp. *mackeyana*	
Impatiens macroptera	
Kigelia africana	(Bignoniaceae)
Newbouldia laevis	
Carica papaya	(Caricaceae)
Acmella caulirhiza	(Compositae)
Adenostemma mauritianum	
Ageratum conyzoides	
subsp. *conyzoides*	
Bidens pilosa	
Crassocephalum montuosum	
Emilia coccinea	
Galinsoga quadriradiata	
Melanthera scandens	
Microglossa pyrifolia	
Synedrella nodiflora	
Vernonia amygdalina	
Vernonia stellulifera	
Jollydora duparquetiana	(Connaraceae)
Kalanchoë crenata	(Crassulaceae)
Zehneria minutiflora	(Cucurbitaceae)
Zehneria scabra	
Dichapetalum sp. 1	(Dichapetalaceae)
Acalypha arvensis	(Euphorbiaceae)
var. *arvensis*	
Alchornea cordifolia	
Croton gratissimus	
Drypetes aframensis	
Euphorbia hirta	
Ricinus communis	
Tragia benthamii	
Geranium arabicum	(Geraniaceae)
subsp. *arabicum*	

Garcinia kola (Guttiferae)
Garcinia lucida
Garcinia mannii
Garcinia smeathmannii
Irvingia gabonensis (Irvingiaceae)
Clerodendrum silvanum (Labiatae)
   var. *buchholzii*
Ocimum gratissimum
   subsp. *gratissimum*
   var. *gratissimum*
Plectranthus monostachyus
   subsp. *monostachyus*
Plectranthus sp. *nov.*
Desmodium adscendens (Leguminosae: Pap.)
   var. *adscendens*
Desmodium repandum
Physostigma venenosum
Pterocarpus soyauxii
Trifolium baccarinii
Vigna vexillata
Sida rhombifolia (Malvaceae)
   var. *β*
Sida sp. aff. *rhombifolia*
Dichaetanthera africana (Melastomataceae)
Tristemma mauritianum
Ficus chlamydocarpa (Moraceae)
   subsp. *chlamydocarpa*
Coelocaryon botryoides (Myristicaceae)
Syzygium staudtii (Myrtaceae)
Campylospermum laxiflorum (Ochnaceae)
Ludwigia erecta (Onagraceae)
Oxalis barrelieri (Oxalidaceae)
Oxalis corniculata
Hilleria latifolia (Phytolaccaceae)
Piper capense (Piperaceae)
Piper umbellatum
Plumbago zeylanica (Plumbaginaceae)
Gouania longispicata (Rhamnaceae)
Calochone acuminata (Rubiaceae)
Euclinia longiflora
Oldenlandia lancifolia
   var. *lancifolia*
Pauridiantha venusta
Spermacoce intricans
Tricalysia gossweileri
Zanthoxylum rubescens (Rutaceae)
Allophyllus sp. *1* (Sapindaceae)
Cardiospermum halicacabum
Synsepalum msolo (Sapotaceae)
Brucea guineensis (Simaroubaceae)
Physalis angulata (Solanaceae)
Physalis peruviana
Solanum torvum
Melochia melissifolia (Sterculiaceae)

   var. *mollis*
Eryngium foetidum (Umbelliferae)
Laportea ovalifolia (Urticaceae)
Pilea rivularis
Pilea sublucens

Gloriosa superba (Colchicaceae)
Aneilema umbrosum (Commelinaceae)
   subsp. *umbrosum*
Commelina benghalensis
   var. *hirsuta*
Murdannia simplex
Costus letestui (Costaceae)
Cyperus distans (Cyperaceae)
   subsp. *distans*
Cyperus flavescens
   subsp. *flavescens*
Cyperus pinguis
Cyperus renschii
   var. *renschii*
Echinochloa colona (Gramineae)
Eleusine indica
Zeuxine heterosepala (Orchidaceae)
Aframomum melegueta (Zingiberaceae)
Aframomum subsericeum
   subsp. *glaucophyllum*

ENVIRONMENTAL USES

Thunbergia vogeliana (Acanthaceae)
Alternanthera brasiliana (Amaranthaceae)
Alternanthera nodiflora
Adenostemma mauritianum (Compositae)
Ipomoea indica (Convolvulaceae)
Alchornea floribunda (Euphorbiaceae)
Codiaeum variegatum
   cv. *variegatum*
Oncoba dentata (Flacourtiaceae)
Garcinia kola (Guttiferae)
Irvingia gabonensis (Irvingiaceae)
Calliandra houstoniana (Leguminosae: Mim.)
   var. *callothyrsus*
Leucaena leucocephala
   subsp. *leucocephala*
Pueraria phaseoloides (Leguminosae: Pap.)
Duranta erecta (Verbenaceae)

Canna indica (Cannaceae)
Gloriosa superba (Colchicaceae)
Commelina benghalensis (Commelinaceae)
   var. *hirsuta*
Dracaena arborea (Dracaenaceae)
Dracaena fragrans

CROP PLANTS
(Table compiled with assistance from Iain Darbyshire)

There are a number of crop plants that are utilised for food, in agroforestry, for timber or for other inherent properties such as their medicinal powers. Below is a list of all those species reported to us to be grown as crops in Bakossiland.

FAMILY	SCIENTIFIC NAME	ENGLISH NAME	CATEGORY OF USE
DICOTYLEDONAE			
Amaranthaceae	**Amaranthus hybridus** subsp. **cruentus**	-	FOOD
Anacardiaceae	**Mangifera indica**	Mango	FOOD, MEDICINES
	Spondias mombin	-	FOOD
Bombacaceae	**Ochroma lagopus**	Balsa	MATERIALS
Burseraceae	**Canarium schweinfurthii**	Bush Plum	FOOD
	Dacryodes edulis	Plum	FOOD
Cannabaceae	**Cannabis sativa**	Cannabis	SOCIAL USES
Caricaceae	**Carica papaya**	Papaya & Pawpaw	FOOD, MEDICINES
Compositae	**Vernonia amygdalina**	Bitter Leaf	FOOD
	Vernonia hymenolepis	Sweet Bitter Leaf & Bitter Leaf	FOOD
Convolvulaceae	**Ipomoea batatas**	Sweet Potato	FOOD
Cruciferae	**Brassica oleracea**	Cabbage	FOOD
Cucurbitaceae	**Citrullus lanatus**	Water-melon & Egusi	FOOD
	Cucumeropsis mannii	Egussi	FOOD
	Lagenaria siceraria	Bottle Gourd	MATERIALS
	Telfairia occidentalis	-	FOOD
	Trichosanthes cucumerina subsp. **anguina**	Snake Gourd	FOOD
Euphorbiaceae	**Croton gratissimus**	-	MEDICINES
	Hevea brasiliensis	Rubber	MATERIALS
	Manihot esculenta	Cassava	FOOD
	Manihot glaziovii	Ceara Rubber	MATERIALS
	Ricinus communis	Castor Oil Plant	MEDICINES
Guttiferae	**Garcinia kola**	-	FOOD, FOOD ADDITIVES, MATERIALS, MEDICINES, ENVIRONMENTAL USES
	Garcinia lucida	-	FOOD, FOOD ADDITIVES, SOCIAL USES, MEDICINES
Labiatae	**Ocimum gratissimum** subsp. **gratissimum** var. **gratissimum**	Tea Bush	FOOD ADDITIVES, MEDICINES
Lauraceae	**Cinnamomum verum**	Cinnamon	FOOD ADDITIVES
	Persea americana	Avocado	FOOD
Leguminosae: Mimosoideae	**Calliandra houstoniana** var. **calothyrsus**	-	ENVIRONMENTAL USES
	Leucaena leucocephala subsp. **leucocephala**	-	ENVIRONMENTAL USES
Leguminosae: Papilionoideae	**Arachis hypogaea**	Groundnut & Peanut	FOOD
	Glycine max	Soya bean	FOOD ADDITIVES
	Phaseolus vulgaris	Bean	FOOD
	Pueraria phaseoloides	-	ENVIRONMENTAL

			USES
	Vigna adenantha	Koki Beans	FOOD
Malvaceae	*Abelmoschus esculentus*	Okra	FOOD
Moraceae	*Artocarpus altilis*	Breadfruit	FOOD
Myrtaceae	*Eucalyptus spp.*	Eucalyptus & Gum Tree	MATERIALS, FUELS
	Psidium guajava	Guava	FOOD
	Syzygium malaccense	Bush Apple & Apple	FOOD
Pedaliaceae	*Sesamum radiatum*	Sesame	FOOD
Portulacaceae	*Talinum triangulare*	Water Leaf	FOOD
Rubiaceae	*Coffea canephora*	Coffee	FOOD
Rutaceae	*Citrus × aurantium*	Orange	FOOD
Solanaceae	*Capsicum annuum*	Capsicum	FOOD ADDITIVES
	Capsicum frutescens	Chilli Pepper, Hot Pepper & Fire	FOOD ADDITIVES
	Nicotiana tabacum	Tobacco	SOCIAL USES
	Solanum betaceum	Tree Tomato	FOOD
	Solanum lycopersicum	Tomato	FOOD
	Solanum nigrum	Huckleberry	FOOD
Sterculiaceae	*Cola nitida*	Cola nut	SOCIAL USES
	Theobroma cacao	Cacao & Cocoa	FOOD
Theaceae	*Camellia japonica*	Tea	FOOD
Tiliaceae	*Corchorus olitorius*	-	FOOD

MONOCOTYLEDONAE

Araceae	*Colocasia esculenta*	Cocoyam	FOOD
	Xanthosoma sagittifolium	Cocoyam	FOOD
Bromeliaceae	*Ananas comosus*	Pineapple	FOOD
Dracaenaceae	*Dracaena arborea*	-	ENVIRONMENTAL USES
	Dracaena fragrans	-	SOCIAL USES, ENVIRONMENTAL USES
Gramineae	*Bambusa vulgaris*	Bamboo	MATERIALS
	Saccharum officinarum	Sugar Cane	FOOD
	Zea mays	Maize	FOOD
Musaceae	*Musa* cvs.	Banana, Plantain	FOOD
Palmae	*Elaeis guineensis*	Oil Palm	FOOD ADDITIVES
	Raphia hookeri	Raphia Palm	FOOD, MATERIALS
Zingiberaceae	*Aframomum melegueta*	Alligator Pepper	FOOD ADDITIVES, MEDICINES

PINOPSIDA

Cupressaceae	*Cupressus lusitanica*	Cypress	MATERIALS

DATA SUMMARY

TOTAL NUMBER OF TAXA IN CHECKLIST	2412
NUMBER OF TAXA WITH REPORTED USES	297
USES RECORDED	374

MEDICINAL PLANT SYSTEMATICS AND VALUE IN BAKOSSI

Martin Etuge[1]

[1] Zoological Society of San Diego's Conservation and Research of Endangered Species, CRES Cameroon Office, B.P. 3055 Messa, Yaoundé, Cameroon

The information provided in the tables below is based on the personal observations of the author as a born and bred, son-of-the-soil Bakossi man. The lists are arranged alphabetically by scientific name, if the latter is known. In most cases these observations are not recorded on herbarium specimens, and are here published separately from those data recorded in the chapter on ethnobotany, and have not been added to plant uses detailed in the checklist. Some of the non-native species, cultivated in medicinal gardens, and not being found naturalized or in farmbush or farms, are not otherwise recorded in the checklist, e.g. *Helianthus annuus* and *Aloe vera*. Be advised that the indigenous traditional knowledge contained in this chapter has been given in good faith as a contribution to the common good and the furtherance of mutual understanding and the preservation of all life on our planet and is not be used for personal or commercial gain and must be treated with respect and used only for the purpose for which it was gifted. Anyone who reads this volume assumes the moral and ethical obligations implied by this statement.

SCIENTIFIC NAME	LOCAL NAME	COMMON NAME	MEDICINAL ATTRIBUTES
Acmella caulirhiza	Medmekube		Typhoid, boils, toothache etc.
Aloe vera	Achang de-chiog	Indian aloe	Poisons, snake bites, skin infection
Asystasia gangetica	Esume choug		Emetic, vomiting
Ageratum conyzoides	Esumesioh	Goats weed	Stomach problems, headache
Acanthus montanus	Mecholechine	Pears testle	Abdominal pains, boils and abscesses
Begonia sp.			Night poison
Bidens sp.	Kodekode	Black jack	Febrifuge, periodic fever
Biophytum sp. with *Kalanchoe*	Chual	Life plant	Cough
Carica papaya	Pawpe	Pawpaw	Sore throat; durative alternative
Centella asiatica	Ehiog		Appendicitis etc.
Commelina benghalensis	Nkoleke	Never die	Ringworms, typhoid, blood clotting
Cymbopogon citratus	Mehang metea	Fever grass	Febrifuge, expectorant, fever, cough, tea
Cynodon dactylon	Semesm	Bahama grass	Dizziness
Cynodon dactylon	Nzezong	Bahama grass	Hypertension, rib pains
Desmodium sp.	Pee-mbodeh	Clover	Dysentery, piles
Dichrocephala integrifolia	Esysio-mboug		Eye problems, conjunctivitis
Dorstenia sp.	Eseh-emuseh	Manpower	Aphrodisiac, sexual stimulant
Elaeis guineensis	Nlom-ndee	Young palm leaves	Syphilis and gonorrhoea
Emilia coccinea	Mballe etuebweh	Emilia	Gastric pains, ear-ache, pratual bath
Eremomastax speciosa	Ehentane		Blood tonic, anaemia
Eryngium foetidum	Chiangweh, or Esume-nyagusse		Abscesses and boils
Eulophia horsfallii	Akwo Ikwog	Swamp orchid	Bleeding piles
Euphorbia hirta	Mekide ambae		Gastric pains, eye drop
Helianthus annuus		Sun flower	Piles
Hibiscus rosa-sinensis		Hibiscus	Diarrhoea, dysentery
Ipomoea sp.	Sope-ahente		Cathartic
Kigelia africana	Esounzoung		Breast problems, promote sexual emotion
Leucaena leucocephala	Esum-achad		Ease conception
Laportea ovalifolia	Tuleboug	Burning grass	Palpitation and ear problems
Microglossa sp.	Ekeble		Enema for babies, stomach pains or problems
Momordica foetida	Ndume		Abortion, stomach pains
Musa sp.	Nyiake	Banana	Altertive, gastric

Musanga cecropiodes	Ekumbe	Umbrella tree	Body pains, cough
Ocimum gratissimum	Messeeb	Masepo	Expectorant, convulsion, febrifuge
Oxalis corniculata	Benbene	Wood sorrel	Poison?
Pennisetum sp.	Akaka		Convulsion
Phoenix reclinata	Menack	Dead palm	Stomachic
Piper umbellatum	Mebonmbong		Poison, kidney, haemorrhage
Plectranthus decurrens	Etangloh		Enema for pregnant women, ease delivery
Rauvolfia vomitoria	Nhimpaah, abude, Nhim-mbaah		Febrifuge/typhoid fever, eyes, anthelmintic
Ricinus communis	Mbangdiobe	Castor oil shrub	Purgative
Saccharum officinarum	Nkogeh	Sugar cane	Dysentery
Scoparia dulcis	Nzumbe		Asthma, diarrhoea
Selaginella vogelii	Abude		Kidney problems
Senna alata	Nkumenkalleh	Ringworm bush	Expels worms, eyes fever, fast delivery
Sida javanensis	Good milk, weed or esum-eyame		Ease delivery, liver problem etc.
Solanum torvum	Ndede Nzuog		Snake bite, antibiotic
Solanum sp.	Etame tame		Appetite, sore throat
Spathodea campanulata	Echib		Crooked eye problem
Stellaria media	Echim-ekede	Chick weed	Kidney, liver, piles, skin disease etc.
Zea mays	Mesonge mengwen	Corn silk	Urine, bladder

Scientific name unknown	Akwo-kwogg		Piles, bleeding
Scientific name unknown	Apaah akuomeh		Stomach
Scientific name unknown	Sabaasaad	Lime fruit	Cough, fever etc.
Scientific name unknown	Mierang	Monkey sugar cane	Clears voices

BRYOPHYTES & WATER CAPTURE

Sue Williams[1], Cliff Townsend[2], Bob Magill[3], Martin Cheek[2]

[1] 7 Middleton Hill Rd., Rowe, Massachusetts, MA 01367, U.S.A.
[2] Herbarium, Royal Botanic Gardens, Kew, Surrey, TW9 3AE, U.K.
[3] Herbarium, Missouri Botanical Garden, P.O. Box 299, Saint Louis, Missouri, MO 63166-0299, U.S.A.

INTRODUCTION

Bryophytes - mosses and liverworts - are easily overlooked, and often neglected, even by those people conducting intensive general botanical surveys. However, bryophytes are an important group of plants - in fact they are the second largest group of green land plants and have important roles in water storage, nutrient uptake from rain, and ecological interactions - both in providing a humid, nutrient-rich seedbed for many vascular plants as well as a habitat for animals. They are also useful as bio-indicators of air and water quality, and other contaminants due to the fact that they absorb both water and nutrients almost exclusively through their surfaces. There is no filtering by soil, roots or waxy cuticles (Frahm 2003). Bryophytes capture and store water, helping to regulate water levels and preventing erosion. They are also important in absorbing essential nutrients from precipitation and leachates from the canopy as well as from dust and gases which would otherwise be lost due to leaching during heavy rainfalls (Bates 2000). These nutrients build up as phytomass and eventually are returned to the ecosystem when the bryophyte masses fall to the ground and decompose. Other plant checklists produced in this series have omitted mention of these non-vascular plants entirely. Yet, bryophyte checklists are as important as vascular plant checklists. Due to their sensitivity to disturbance, they can be useful tools in conservation biology. Bryophytes can show trends earlier than those of vascular plants. Without checklists, frequency and distribution of bryophytes are unknown meaning that rare or endangered species are also unknown preventing timely protection of threatened habitats. It is now known that bryophytes do make a contribution to the plant diversity of an area and can provide an ecological function which is of great potential value to those people living in the 20 or so villages that surround Mt Kupe. Authors of moss names and bryophyte families are not given here.

PROVIDING WATER IN THE DRY SEASON

Although the Bakossi area attracts a high annual rainfall (see Climate chapter), the drier season (November to April inclusive) heralds such a drop in the levels of streams that descend Mt Kupe as to give concern to the communities that depend upon them for drinking, bathing and washing (Chris Bowden *pers. comm.*). Over recent decades this difficulty has intensified and has been attributed to destruction of the forest above the villages. However, the mechanism by which forest cover can 'catch' water, the extent of the contribution it can make, and the involvement of bryophytes, particularly mosses, is generally poorly known.

Natural forest cover at altitudes of above c. 800 m, that is, at elevations that coincide with cloud (hence 'cloud forest'), can 'strip' water directly from these clouds by condensing cloud-water droplets on leaf surfaces (Hamilton *et al.* 1995). This source of water reaching the ground is generally referred to in the literature as horizontal precipitation (HP) to distinguish it from rain, i.e. vertical precipitation (VP). The contribution of HP to total precipitation is most marked in the dry season. One study that shows this dramatically was conducted in Hawaii, the tropical state of the U.S.A. During a hundred-day period in the dry season, the study area of forest received 336 mm of rainfall (measured by collecting in a forest clearing) while the total precipitation reaching the ground through the forest canopy ('throughfall' composed of HP and VP) amounted to 606 mm. Thus the presence of natural forest nearly doubled the total amount of water reaching the ground (Hamilton *et al.* 1995), this being potentially of considerable significance in affecting water-levels of forest-fed streams and equally so to the dependent local communities. The extent of the contribution by cloud forest to 'catching water' varies greatly according to aspect, elevation and other factors that are incompletely understood. Relatively few studies have been conducted and our efforts to locate satisfactory ones assessing the issue in African 'cloud forest' were unfruitful. Nyasoso in Bakossi would make an excellent base for such studies in future.

Mosses and liverworts, particularly those that are epiphytic on tree branches, are considered extremely important for cloud-stripping, that is, generating HP. This is because their structure is highly conducive to capturing cloud droplets, due to their high surface area to mass ratio, and their thin (usually only one to two cells thick) and acutely apexed leaves. Furthermore at altitudes above 800 m, liverworts, and especially mosses, can form a large part of the tree canopy, enveloping twigs of trees and shrubs and forming dangling skeins a metre or more long that are thought to be particularly effective at increasing water capture. *Pilotrichella*, *Orthostichella* and *Neckeropsis* are the main pendent mosses on Mt Kupe. No satisfactory studies conducted in Africa could be located, but work in Colombia reported by Bruijnzeel and Proctor (1995) showed that bryophyte layer thickness on tree twigs, and total bryophyte biomass, changed dramatically with altitude. Below 1000 m altitude at that study site, biomass and layer thickness was negligible. However, at 1000 m the level shot up significantly, then remained ± constant to 2000 m, where it trebled and remained ± constant up to the highest elevation at 3500 m.

THE INVENTORY OF THE BRYOPHYTE TAXA OF BAKOSSI

During 1995 Sue Williams, American bryophyte and saprophyte specialist, joined our expeditions as an Earthwatch volunteer on Mt Kupe from February to March and in November, making about 50 bryophyte collections under her own series of collecting numbers which were later identified by various specialists credited in the list below. Particularly notable amongst these specialists is Bob Magill (mosses). Below are presented determinations of her specimens in the order in which they were collected, arranged by her eight collection localities.

During the period 1996–1999, another eleven specimens, comprising nine taxa were collected by R.B.G. Kew staff and associates and identified by Kew's Cliff Townsend, whose identification list of these includes global distributions and general notes. Only two of these nine taxa were already recorded in Sue Williams's work and the records for these two have been integrated, in square brackets, into the Williams list, along with several records that fit one or other of the eight Williams localities. Three other localities not represented by Williams specimens are added.

BRYOPHYTES OF KUPE, MWANENGUBA AND BAKOSSI MTS

The 63 bryophyte specimens, representing the 59 taxa that are known from this area are ordered below by locality, following Sue Williams's treatment. All specimen numbers are those of Williams unless otherwise indicated.

Location 1.   **Top of Mt Kupe, 2000m, montane forest, 7 Feb. 1995.**

(all *Williams* s.n., det. by R.E.Magill)

Pilotrichella ampullacea	– hanging from branches (common)
Trachypodopsis serrulata	– epiphyte
Brachymenium revolutum	– epiphyte
Campylopus hensii	– dead log
Campylopus pilifer	– exposed rock
Porthamnium stipitatum	– epiphyte
Neckeropsis lepineana	– epiphyte

Cable 2933   *Macromitrium levatum*, Earthwatch peak 1, epiphyte, 1950 m. (det. Cliff Townsend)
Widespread in Africa and recently synonymised with the asiatic *M. sulcatum* (Hook.) Brid., a step which needs further testing but may well be correct.

Cheek 7344a   *Orthostichella turgidellacea*, pendent on *Psychotria* sp. aff. *peduncularis*, 1980 m, 7 Feb. 1995 (det. Cliff Townsend)
Only known from Cameroon. P. Dusén, its original collector, observed (Dusén 1895): 'not found below…500 m on the west side of the Cameroons Mountains. In the neighbourhood of Bomana, 670 m above the sea, I very often saw this moss, quite covering the tree branches in very dense tufts. On the west side of the Little Cameroons' Peak, I found it at a height of 800 m, growing comparatively thinly at the ends of the branches of trees'.

Location 2.   **West side of Mt Kupe on Main Trail to top from Nyasoso Village, lower submontane forest, 11 Feb. 1995.** (All dets R.E. Magill unless otherwise noted)

Williams 179	*Bryum sp.*	– rock in path; 850m
Williams 180	*Racopilum tomentosum*	– rock in path; 850 m
Williams 181	*Phyllodon truncatus*	– rock in path; 850 m; (det. Brian O'Shea)
Williams 182	*Bryum sp.*	– on rock; 850 m
Williams 184	*Mastigophora sp.*	– large fallen tree branch, secondary forest, 850 m,

		(det. Alan Whittemore)
Williams 185	*Lejeunea sp.*	– large fallen tree branch, 850 m (det. Alan Whittemore)
Williams 187	*Orthostichidium involutifolium*	– large fallen tree branch; 850 m
Williams 188	*Entodon lacunosus*	– trunk of *Cylicomorpha* ~1.5 m up; 850 m
		(det. Liz Kungu)
Williams 189	*Thuidium sp.*	– trunk of *Cylicomorpha*
Williams 191	*Vesicularia sp.*	– growing on old log in forest
Williams 192	*Radulina borbonica*	– end of log in forest (det. Brian O'Shea)
Williams 193	*Fissidens sp.*	– disturbed soil of path
Williams 195	*Taxilejeunea sp.*	– base of large tree; 1000 m (det. Alan Whittemore)
Williams 196	*Mittenothamnium sp.*	– on liana; 1100 m
Williams 197	*Porotrichum sp.*	– buttress of large tree from base to ~2m high; 1100m
Williams 198	*Pyrrhobryum spiniforme*	– rotten log; 1200 m

Location 3. **West side of Mt Kupe on Mt Kupe Forest Project Nature Trail (850 m), lowland-lower submontane forest, 11 March 1995.**

Williams 200	*Porella sp.*	– large rotten log (det. Alan Whittemore)
Williams 201	*Cyathodium sp.*	– shaded rock (det. Alan Whittemore)
Williams 202	*Vesicularia sp.*	– rock next to brook
Williams 205	*Dumortiera sp.*	– bare soil of steep bank near brook (det. Alan Whittemore)
Williams 206	*Vesicularia sp.*	– base of small dead tree
Williams 208	*Leucomium strumosum*	– rotten legume stump
Williams 210	*Hypnum sp.*	– dead log
Williams 212	*Thuidium tenuissimum*	– rock near path
Williams 215	*Fissidens sp.*	– disturbed soil of path
Williams 216	*Riccardia sp.*	– rotten stump (det. Alan Whittemore)

Etuge 1360A   *Orthostichella turgidellacea*, Bush-Shrike trail, 1190 m, 26 Oct. 1995, (det. Cliff Townsend), see note under **location 1**.

Location 4. **South side of Mt Kupe on a southeast running ridge North of Kupe village, 7 Nov. 1995.**

| *Williams 225* | *Radulina borbonica* | – forest near top of steep ridge growing at the base of a small tree ~ 0.5 m up; 1000 m (det. Brian O'Shea) |

Location 5. **South side of Mt Kupe, main trail to summit from Kupe village, 8 Nov. 1995.**

| *Williams 231* | *Philonotis sp.* | – forest, vertical bank of shaded stream |

Location 6. **South side of Mt Kupe, main trail to summit starting from school compound, 10 Nov. 1995.**

Williams 237	*Anthoceros sp.*	– disturbed soil at path edge; 750 m (det. Alan Whittemore)
Williams 239	*Vesicularia sp.*	– base of stump; 750 m
Williams 240	*Acroporium megasporum*	– fallen tree; 850 m
Williams 241	*Calymperes lonchiphyllum*	– underside of large boulder in middle of now-dry stream (det. Brian O'Shea)

Cheek 7823A   *Orthostichidium involutifolium*, Kupe Village, 4°47′N 9°42.1′E, main trail, 1260 m, 15 Nov. 1995 (det. Cliff Townsend)

Zapfack 724   *Pyrrhobryum spiniforme*   – epiphyte, 1300 m.
Cable 2578   *Pyrrhobryum spiniforme*, epiphyte, 1300 m., 23 May 1996, on a rotting tree trunk in primary forest, 1960 m. (both det. Cliff Townsend)
A widespread species in the tropical regions of both the Old and New Worlds, extending northwards to Japan. Surprisingly, not previously recorded from Cameroon.

Location 7. **South side of Mt Kupe on trail to 'Kupe Rock' to 'saddle area' from west side of Kupe village, Lower submontane forest, 14 Nov. 1995.**

Williams 242	*Porothamnium sp.*	– on liana; 900 m
Williams 243	*Isopterygium sp.*	– buttress of large forest tree; 900 m
Williams 244	*Callicostella sp.*	– root base of large tree near ground; 900 m

Saddle area	**(1100m):**	
Williams 246	*Pilotrichella ampullacea*	– fallen twig
Williams 247	*Orthostichidium involutifolium*	– fallen twig
Williams 248	*Mastigophora diclados*	– trunk of large fallen tree (det. Alan Whittemore)
Williams 248a	*Herbertus grossevittatus*	– mixed with the *Mastigophora diclados*; (det. Nick Hodgetts, who said it was the only specimen of this group yet collected in W Africa).
Williams 249	*Wijkia sp.*	– large rotten log in relatively open area
Williams 250	*Campylopus sp.*	– trunk of large fallen tree
Williams 253	*Bazzania sp.*	– trunk of large fallen tree (det. Alan Whittemore)
Williams 254	*Syrrhopodon lamprocarpus*	– base of large broken tree (det. Brian O'Shea)
Williams 256	*Pyrrhobryum spiniforme*	– rotten log
Williams 257	*Symphyogyna sp.*	– tree trunk ~ 1 m up (det. Alan Whittemore)

Location 8.   **Top of Kupe Rock, alt. 850 m, 29 May 1996.**

Ryan 392   *Campylopus chevalieri*   – epiphytic on tree; 850 m, (det. Cliff Townsend)
A species confined to the W African countries around the Gulf of Guinea.

Ryan 392A   *Brachymenium nepalense*   – (det. Cliff Townsend)
Mixed in with *Campylopus chevalieri*, 29 May 1996. A widespread species in S,W & C Africa and Asia.

Location 9.   **Lake Edib, component of *Sphagnum* moss mat floating on the south side of this volcanic lake, alt. 1200 m, 20 Nov. 1998.**

Gosline 192   *Sphagnum planifolium*   – (det. Cliff Townsend)
The most widely distributed African species of *Sphagnum*

Location 10.   **Mwanenguba, 5°02′N 9°49′E, grazed lakeside grassland by crater lake, alt. 2000 m, 28 Oct. 1998.**

Pollard 163   *Campylopus filiformis* J.-P.Frahm & C.C.Towns.
A previously unknown species of which the description has been sent for publication.

Location 11.   **Ngomboaku, 4°55.57′N 9°44.17′E, Abang road then right to forest, undisturbed evergreen lowland forest, alt. 850 m, 11 Dec. 1999.**

Cheek 10318   *Lepidopilidium devexum*   – (det. Cliff Townsend)
Described from Mt Cameroon, since found in Bioko, Rio Muni and Tanzania.

Location 12.   **Nyandong, W Bakossi, alt. 300 m, 27 March 2003.**

Cheek 11476   *Bryum alpinum vel aff.*   – (det. Cliff Townsend)
Differs from normal mature *Bryum alpinum* in its thin-walled cells and narrowly elliptic leaf outline. These characters may be due to the aquatic or semi-aquatic habitat. The species is a widespread one (including in Africa) and is recorded from Cameroon.

NEW DISCOVERIES

The most exciting discovery of our embryonic bryophyte collecting programme was that made by Ben Pollard, whose specimen *Pollard* 163 has been found by Cliff Townsend and J.-P. Frahm, to represent a species new to science, currently in press in Journal of Bryology as *Campylopus filiformis* J.-P.Frahm & C.C.Towns. Apart from this, Townsend also identified the first records from Cameroon of *Pyrrhobryum spiniforme* (*Cable* 2578 and *Zapfack* 724).

REFERENCES

Bates, J.W. (2000). Mineral Nutrition, Substrate Ecology and Pollution: 248–311. *In*: Shaw & Goffinet (*eds*). Bryophyte Biology. Cambridge University Press.

Bruijnzeel, L.A. & Proctor, J. (1995). Hydrology and Biogeochemistry of TMCF: what do we really know? *In*: Hamilton, L.S., Juvnik, J.O. & Scatena, F.N. Tropical Montane Cloud Forests. Springer-Verlag.

Dusén, P. (1895). New and some little-known mosses from the west coast of Africa. Kongl. Svenska Vetenskaps Akademiens Handligar 28(2): 32.

Frahm, J.-P. (2003). Manual of Tropical Bryology 23.

Hamilton, L.S., Juvnik, J.O. & Scatena, F.N. (1995). Tropical Montane Cloud Forests. Springer-Verlag.

A1

A3

A2

B2

B1   B3

1 mm

Fig. 9. BRYOPHYTES
A (above). *Syrrhopodon lamprocarpus*
B (below). *Pyrrhobryum spiniforme*
Scale bar = 1mm. Both drawn by Sue Williams.

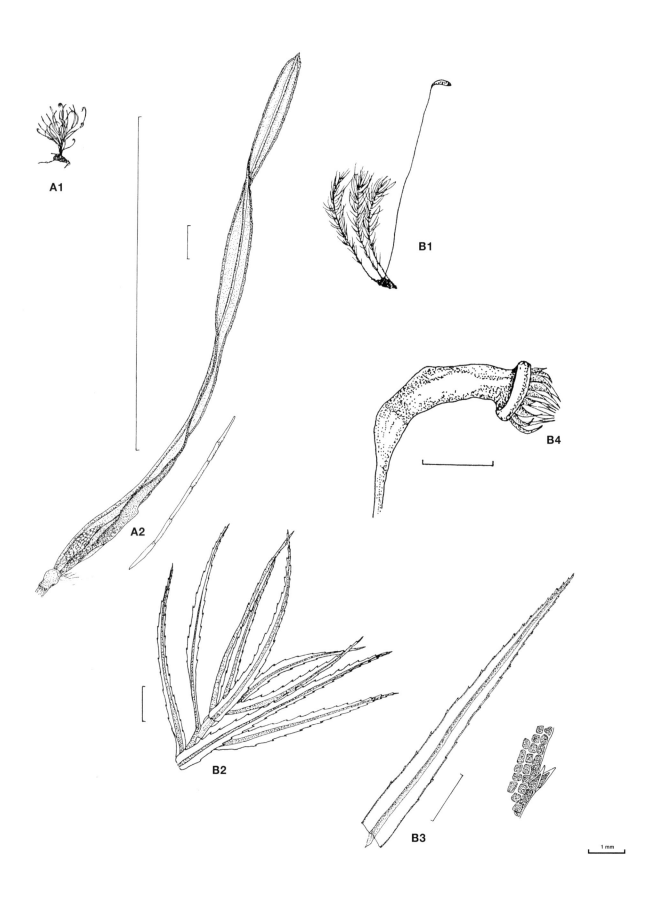

Fig. 10. BRYOPHYTES
A (above). *Syrrhopodon lonchiphyllum*
B (below). *Herbertus grossevittatus*
Scale bar = 1mm. Both drawn by Sue Williams.

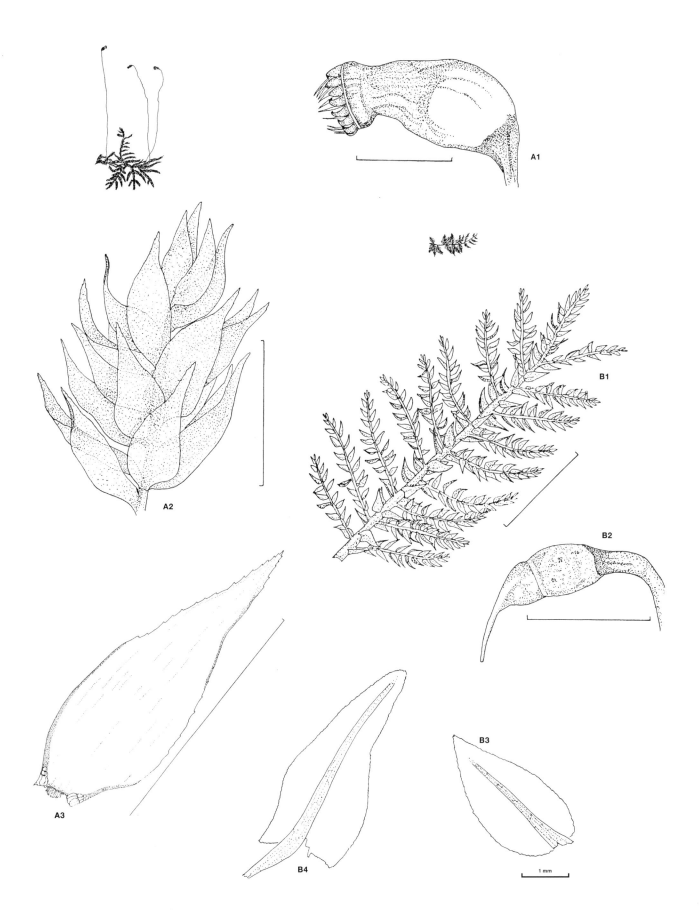

Fig. 11. BRYOPHYTES
A (above). *Wijkia* sp.
B (below). *Thuidium tenuissimum*
Scale bar = 1mm. Both drawn by Sue Williams.

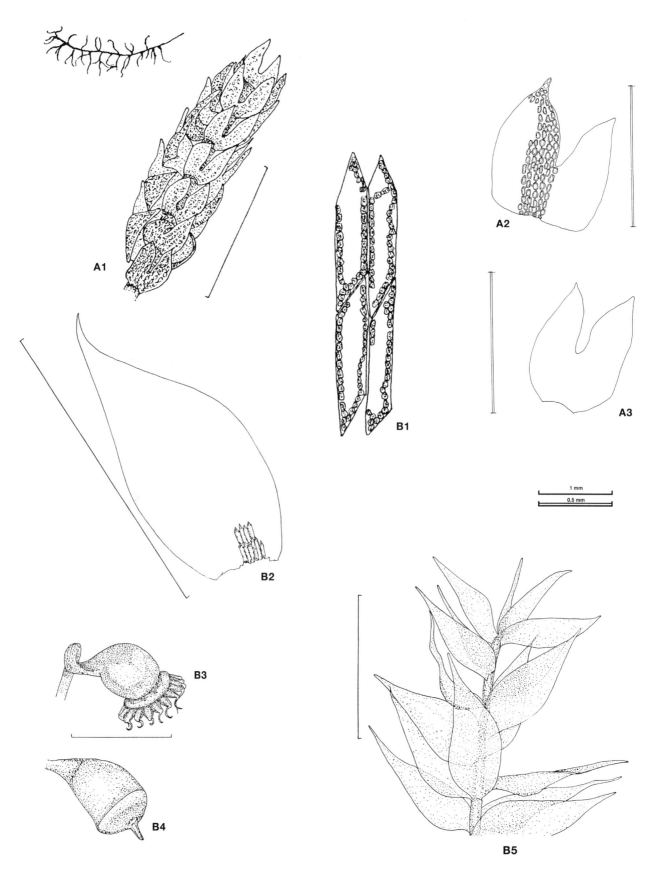

Fig. 12. BRYOPHYTES.
A (above). *Mastigophora diclados*
B (below). *Vesicularia* sp.
Scale bar = 1mm. Both drawn by Sue Williams.

NOTES ON THE MACROFUNGI (*BASIDIOMYCOTA*) OF MT KUPE

Peter R. Roberts

Mycology Section, Royal Botanic Gardens, Kew, Surrey, TW9 3AE, U.K.

INTRODUCTION

No systematic survey of the fungi of the Mt Kupe area has yet been undertaken, but a few specimens of the larger *Basidiomycota*, particularly bracket fungi, have been collected by interested botanists and examined at Kew. Together, these amount to some 35 collections representing 20 species, a small initial sample from a mycota which might be expected to comprise over 3000 species if fully researched.

The extent to which the fungi of tropical Africa are little known can be gauged from the fact that, even from such a small and effectively random sample, one new species has been described. This is the heterobasidiomycete *Aphelariopsis kupemontis*, a branching, coral-like species with 'auricularioid' (tubular, septate) basidia (see Fig. 13). No other members of the genus are known from Africa, its closest relatives having been described from Borneo and from South America (Roberts 2000). Two other heterobasidiomycetes recorded from Mt Kupe are the pantropical: *Auricularia delicata*, a gelatinous, ear-like species, which is sometimes collected and eaten in Africa (Rammeloo & Walleyn 1993), and the white, gelatinous, seaweed-like *Tremella fuciformis*, which is also edible and cultivated for food in Asia.

Eleven species of poroid or bracket fungi have been recorded from Mt Kupe, most of them pantropical or even cosmopolitan in distribution. Three large, conspicuous species, *Ganoderma australe*, *Laetiporus sulphureus* and *Rigidoporus ulmarius*, are, for example, commonly found in Surrey, England as well as Cameroon. Descriptions of these can be found in the standard text by Ryvarden & Johansen (1980).

Related to the poroid fungi are two species of *Lentinus*, a genus of rather tough, gilled, mushroom-like fungi growing on wood. Both species are common in tropical Africa and said to be edible (Pegler 1983). The true agarics are represented by two small, litter-rotting species, *Marasmiellus subcinereus* and *Marasmius haediniformis*. The gill-less *Podoscypha involuta*, a tawny-brown, trumpet-shaped fungus found in swarms on logs (Reid 1965), is the only stipitate stereoid fungus recorded to date.

Specimens were collected between 1993 and 1996 by S. Cable, M. Cheek, M. Etuge, and P. Lane. Collections were determined by D.N. Pegler, P. Roberts, and L. Ryvarden.

SPECIES LIST

Agaricoid
Lentinus brunneofloccosus Pegler
Lentinus sajor-caju (Fr.) Fr.

Marasmiellus subcinereus (Berk. & Broome) Singer
Marasmius haediniformis Singer

Auricularioid
Auricularia delicata (Fr.) Henn.

Clavarioid
Aphelariopsis kupemontis P.Roberts

Poroid
Antrodiella versicutis (Berk. & M.A. Curtis) Gilb. & Ryvarden
Ganoderma australe (Fr.) Pat.
Hexagonia tenuis (Hook.) Fr.
Laetiporus sulphureus (Fr.) Murr.
Lignosus sacer (Fr.) Ryvarden
Microporus affinis (Blume & Nees) Kuntze

Microporus incomptus (Fr.) Kuntze
Polyporus grammocephalus Berk.
Rigidoporus microporus (Fr.) Overeem.
Rigidoporus ulmarius (Fr.) Imazeki
Trametes africana Ryvarden
Trametes pubescens (Fr.) Pilát

Stereoid

Podoscypha involuta (Klotzsch) Imazeki

Tremelloid

Tremella fuciformis Berk.

REFERENCES

Pegler, D.N. (1983). The genus *Lentinus*. Kew: Royal Botanic Gardens.

Rammeloo, J. & Walleyn, R. (1993). The edible fungi of Africa south of the Sahara. Meise: Nat. Bot. Gard., Belgium.

Reid, D.A. (1965). A monograph of the stipitate stereoid fungi. Beih. Nova Hedwigia **18**: 1–382 + 48 pl.

Roberts, P. (2000). *Aphelariopsis kupemontis*: a new auricularioid species from Cameroon. Persoonia **17**: 491–493.

Ryvarden, L. & Johansen, I. (1980). A preliminary polypore flora of East Africa. Oslo: Fungiflora.

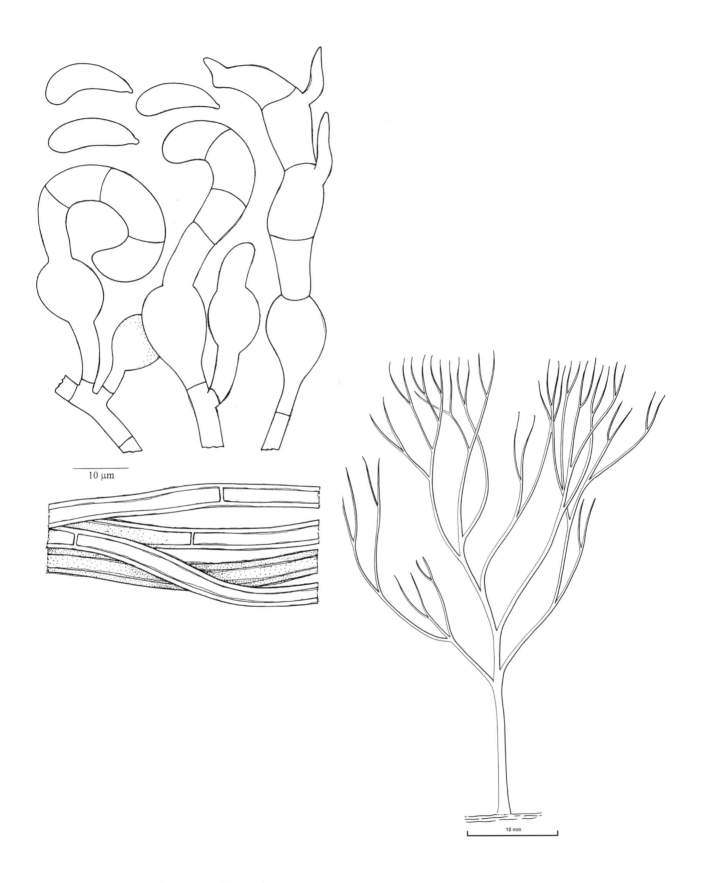

10 μm

10 mm

FIG. 13. *Aphelariopsis kupemontis* (Fungus)
Drawn by Peter Roberts.

THE VERTEBRATE FAUNA

Chris Wild[1], Bethan Morgan[1] and Roger Fotso[2]

[1] Zoological Society of San Diego's Conservation and Research of Endangered Species, CRES Cameroon Office, B.P. 3055 Messa, Yaoundé, Cameroon
[2] Cameroon Biodiversity Programme, Wildlife Conservation Society, BP 3055 Messa, Yaoundé, Cameroon

INTRODUCTION

A fauna is the totality of different animal species found in one place, be that anything in scale from a field to a continent (Hughes 1983). Here we are concerned with the vertebrate fauna occurring within an area of approximately 3000 km^2 covering Bakossiland in south-western Cameroon, and including the Kupe, Mwanenguba and Bakossi Mountains, and the upper reaches of three river systems, the Cross, Mungo and Wouri.

Regional patterns of diversity and endemism in West and Central Africa are evident in a range of faunal groups (Booth 1958; Moreau 1966; Hall & Moreau 1970; Grubb 1978, 1982, 1990 & 2001; Hughes 1983; Prigogine 1988; Colyn *et al.* 1991; Maley 1996a, 1996b & 1996c; Fjeldså & Lovett 1997; Poynton 1999). The isolation of the Congo Forest Block from both the Guinea Forest Block and the East Coast forests is ancient, probably dating back hundreds of thousands of years in the case of the former, and to millions in the latter (Rasmussen 1993). The Atlantic coastal forests of south-eastern Nigeria, Cameroon, Equatorial Guinea and Gabon are a major evolutionary centre of radiation and endemism where many taxa are thought have evolved *in situ* within the coastal forest Pleistocene refuge of Biafra (Kingdon 1990).

The Kupe, Mwanenguba and Bakossi forests have provided a fertile environment for speciation and endemism acting as ancient Quaternary refugia for forest communities (Gartshore 1986; Maley 1987; Maley & Brenac 1997; Pook & Wild 1997). Topographic diversity, steep ecotones and habitat heterogeneity have given rise to parapatric and allopatric species along altitudinal gradients, forming complex ecological communities and zonation of taxa. Faunal assemblages are characterized by high levels of endemism, especially among the amphibians, with notable radiations among crater lake fish, montane lizard genera, and an important complement of montane endemic bird species. The Atlantic coastal lowland forest faunas are also typical of the region, and very well represented in all vertebrate classes. Endemism among the mammals is relatively low at the local level, but several Cross-Sanaga endemic species occur among the primates. The Kupe, Mwanenguba and Bakossi fauna shows closest resemblance with those in the neighbouring Rumpi Hills and Mt Nlonako, and to a lesser extent Mt Cameroon.

Historically, Bakossiland has received attention from zoologists since the late 19th century, when M. Conradt set up a station on the northern aspect of Mt Kupe to develop a tea plantation at Essossong in 1898. It was from here that Moisel conducted geographic surveys as far as Lake Edib (Moisel 1912) and collections of snakes were included in Cameroon's first snake species checklist (Sternfeld 1907). William Serle made the first serious attempt at systematic bird collections from Kupe and Mwanenguba in the 1940s, when he discovered the Mount Kupe bush shrike *(Malaconotus kupensis)* and the white-throated mountain babbler *(Kupeornis gilberti)* in the forest above Nyasoso (Serle 1950, 1951, 1954 & 1965; Bowden 2001). Other notable zoological collections include that of Perret and Mertens (1957) who amassed herpetological material from Nkongsamba, Eisentraut who mist-netted for birds and bats in the 1960s (Eisentraut 1968 & 1973), and Amiet, who made an important collection of amphibians (Amiet 1975) and killi-fish (Amiet 1987). Joger made valuable herpetological collections from Mt Kupe (Joger 1982). Stucki-Stirn (1979) made zoological collections from across the former anglophone West Cameroon in the 1950s and 60s, but the extensive collection (c. 6000 snake specimens) upon which his work was based has since been lost.

Birdlife International (then known as the International Council for Bird Preservation - ICBP) carried out a survey of the birds of the west Cameroon mountain chain which included visits to Kupe and Mwanenguba in 1984 (Stuart 1986). It was on this expedition that the first assessment of the status of the montane amphibian fauna of the Kupe and Mwanenguba Mountains was compiled (Gartshore 1986). The results of the ICBP survey eventually led to the establishment of the first international conservation NGO project in the area in 1991, the Mount Kupe Forest Project (MKFP). Subsequent long-term inventories following the facilitation of visiting scientists and research associates of the MKFP have enabled sufficient time for taxonomic investigation, allowing a more thorough picture of the fauna, including records and discoveries of many rare and endangered species not typically seen in rapid assessments. This is the first time, for example, that long-term bird-ringing data have been acquired for the area (Bowden 2001). Rigorous ecological data has also been published by Hofer *et al.*, (1999 & 2000) who established altitudinal parameters and the organisation of submontane assemblages of the herpetofauna on Mt Kupe. Surprisingly, the mammalian fauna was among the last of the faunal groups to be explored (King 1994).

In this chapter we summarise information on the vertebrate fauna of Bakossiland collated by the Mount Kupe Forest Project to date. Sources include the published literature, museum records, unpublished reports and field records from MKFP staff, visiting zoologists, students and birdwatchers, maintained by the MKFP since 1991. All species records have been critically examined and only those thought to be reliable are included here.

THE FAUNA

Fish

Three distinct riverine ichthyofaunas occur in the Mungo, Mbwe (Cross) and Dibombe (Wouri) Rivers of the region, with 41, 132, and 69 fish species respectively (Teugels & Guegan 1994). The Mungo River fauna shares faunal affinities with the Meme River to the south, and with the Cross at the headwater tributaries of these two rivers where they interdigitate northwest of Bakossiland (Trewavas 1974). The Cross is the richest (and largest) of the three rivers, and Reid (1991) lists 90 species from the upper reaches in South West Province.

Of the five volcanic crater lakes found in Bakossiland, Lake Beme (Bermin) is the only lake known to be occupied by naturally occurring fish populations. Lake Beme is the richest freshwater site for fish species worldwide (relative to the 60 ha surface area), with an intact micro-flock of *Tilapia* comprising nine endemic cichlid species: *Tilapia bakossiorum, T. bemini, T. bythobates, T. flava, T. gutturosa, T. imbriferna, T. spongotroktis, T. snyderae,* and *T. thysi.* This explosive radiation is unique among tilapine assemblages in that the Beme species are a biparental custodial substrate-spawning community. The endemic species include phytoplankton and sponge-eating representatives. Two other non-cichlid species also occur in the lake, *Barbus abionensis* and *Fundulopanchax mirabilis* (Thys van den Audenaerde 1972; Stiassny *et al.* 1992; Schliewen *et al.* 1994).

In addition, small seasonal forest pools and streams throughout the Bakossi lowlands support at least five species of killifish (*Aphyosemion* spp.), including *A. cinnamomeum*, which is endemic to the Bakossi and adjacent Rumpi Hills, and *A. celiae,* which is endemic to the premontane forests around Nkongsamba in Littoral Province (Amiet 1987).

Amphibians

A total of 113 species of amphibian occur and the area is notable for the most diverse assemblage of anuran species in Africa, ranking among the highest anywhere for both species diversity and endemism. Over 50% of the anuran fauna is endemic to the Cameroon Highlands and Atlantic coastal forest of the country. Strict endemics to the Kupe, Mwanenguba and Bakossi Mountains include: *Cardioglossa trifasciata* and *Leptodactylodon erythrogaster* which are only known from the submontane forest of Mwanenguba (Amiet 1975), and the recently discovered guitar frog *Leptodactylodon wildi* from the Edib Hills in Bakossi (Amiet & Dowsett-Lemaire 2000). At least ten species (six of which are as yet undescribed) within the genera *Afrixalus, Arthroleptis, Cardioglossa, Leptodactylodon* and *Phrynodon* are considered to be strictly endemic to these mountains (Amiet *pers. comm.*).

The cloud forests and submontane slopes and hills of Bakossiland are important for a diversity of regional endemics unparalleled in Africa. Notable examples abundant in these mountains include the torrent frog (*Conraua robusta*) and hairy frog (*Trichobatrachus robustus*) which are dependant on the clear, fast flowing mid-altitude streams and associated hill forest. Other restricted-range species for which these forests are important include: *Astylosternus batesi, A. diadematus, A.* cf. *montanus, Bufo tuberosus, Cardioglossa elegans, C. gracilis, C. mertensi, C. venusta, Leptodactylodon bicolor, L. ornatus, L. mertensi, Petropedetes cameronensis, P. newtoni, P. parkeri, P. perreti, Phrynobatrachus cricogaster, P. werneri, Werneria bambutensis, W. preussi, W. tandyi,* and *Wolterstorffina parvipalmata* (Amiet 1975; Gartshore 1986; Dowsett-Lemaire & Dowsett 1998 & 1999; Hofer *et al.,* 1999 & 2000; Schmitz *et al.,* 2000).

The world's largest frog, the goliath (*Conraua goliath*) is present up to 700 m in the eastern tributaries of Kupe and Mwanenguba which drain to the Dibombe River in Littoral Province. The record from Bakossi (Dowsett-Lemaire & Dowsett 2000) is erroneous, nor does it occur elsewhere in the Cross or Mungo river systems in South West Province. A variety of tree frogs and tree toads occur. These are represented in the genera *Acanthixalus, Afrixalus, Alexteroon, Chiromantis, Hyperolius, Leptopelis, Nectophryne, Opisthothylax,* and *Wolterstorffina.* Leaf litter anurans are dominated in the primary forests by the genera *Arthroleptis* and *Bufo.* The large Cameroon browed toad (*Bufo superciliaris*) is found throughout these forests, breeding in the lowland rivers, and the rare *Nyctibates corrugatus* was discovered in the Edib Hills in 1998. *Xenopus amieti* occurs in Lake Mwandon on the Kupe-

Mwanenguba intermontane ridge. Most of the naturally occurring species are forest-dependant with few exceptions such as *Kassina decorata* in the open crater calderas of Edib, Mwanenguba and Mwandon. In addition, two lowland species of legless caecilian occur, *Geotrypetes seraphini* and *Cynisca manni*.

Reptiles

A verified total of 105 reptile species occur within the area, comprising 35 lizard, 62 snake, 5 chelonian and 3 crocodilian species. Four genera of lizard, *Chamaeleo, Lacertaspis, Leptosiaphos* and *Panaspis,* have radiated in the montane environment in these mountains.

Mwanenguba and Bakossi host the richest assemblage of chameleon species on the African mainland with nine species, including five regional endemics: the Cameroon dwarf chameleon (*Chamaeleo camerunensis*), the Cameroon mountain chameleon (*C. montium*), Pfeffer's chameleon (*C. pfefferi*), Perret's chameleon (*C. wiedersheimi perreti*) and the four-horned chameleon (*C. q. quadricornis*) (Klaver & Böhme 1992; Wild 1993). Two of these are restricted to these mountains *C. q. quadricornis* (also present on Mt Nlonako) and *C. wiedersheimi perreti* (unique to Mwanenguba and Bakossi). These assemblages reach their highest density in the Mwanenguba forests where seven species of *Chamaeleo* occur, although only three species are truly syntopic at any one altitudinal zone. The western pygmy chameleon (*Rhampholeon spectrum*) is euryzonal across altitudinal gradients commonly occurring in sympatry within all the *Chamaeleo* zonal assemblages (Wild 1994).

Endemic montane and hill forest species of skink of the closely related genera *Lacertaspis, Leptosiaphos* and *Panaspis* present include *Lacertaspis chriswildi* and *L. lepesmei, Leptosiaphos amieti, L. pauliani* and *L. vigintiserierum,* and *Panaspis gemmiventris*. Regional endemic lowland species include *Lacertaspis reichenowi, L. rohdei* and *Panaspis breviceps*. In addition, one rare gecko species has a montane distribution in these mountains and for which few records exist elsewhere is the Koehler's tree gecko (*Cnemaspis koehleri*), which is present above 1100 m on Kupe, Mwanenguba and Bakossi (Hofer *et al.* 1999 & 2000).

African lowland forest lacertid lizards are represented by few museum records and are rarely encountered in the field. Observation and capture of these fast heliophilic lizards is also difficult, records of which are very rare and worthy of note. The African Forest Lizard (*Adolphus africana*), which exhibits a montane distribution in Cameroon has been confirmed from Kupe, Bakossi and Mwanenguba; the Western Green Lizard (*Gastropholis echinata*) has been collected on the Kupe-Mwanenguba intermontane ridge and from Nyasoso village; Guenther's Flying Lizard (*Holaspis guentheri*) has been collected on Kupe; and Ford's Lizard (*Poromera fordi*) has been recorded from Kupe and Bakossi riparian habitats. Other lowland lizards include Conrau's green day gecko (*Lygodactylus conraui*) in the herbaceous understorey of the Bakossi lowland primary forest.

A total of 62 snake species representing 36 genera occur within 3000 km² area, the highest recorded species richness relative to area on the African continent, although similar species densities probably occur elsewhere in Cameroon's coastal forests. The fauna represents 72% of the 88 recorded forest snake species in Cameroon. Although unlike the East African Highlands, mass elevation has not been a factor in snake speciation and endemism in the region, and no strictly montane snakes are known from Cameroon, a number of rare lowland species occur in the montane forests and grasslands of Kupe, Mwanenguba, Bakossi that are otherwise rare as lowland species in the country. The following species are most frequently encountered and considered well-adapted to the montane environment and are therefore euryzonal: *Bothrolycus ater, Buhoma depressiceps, Dipsadoboa unicolor* in montane forest above 1300 m, and *D. viridis* in the lower elevation submontane forest 900–1500 m. The egg-eating snake, *Dasypeltis scabra,* is locally abundant in the submontane savanna of the Edib Hills and Kodmin around 1400 m. Common venomous snakes include the Jameson's mamba (*Dendroaspis jamesoni*), forest cobra (*Naja melanoleuca*), gaboon viper (*Bitis gabonica*) and rhinoceros viper (*Bitis rhinoceros*). The rare Gold's tree cobra (*Pseudohaje goldii*) has been collected from both Kupe and Bakossi, and a new species of tree viper (*Atheris subocularis*) was recently described from the western limit of the Bakossi, and may be considered the only snake endemic to the South West Province (Lawson 2001).

Three terrestrial and two aquatic chelonian species have been recorded in the area, and all three African crocodilian species are present. The dwarf crocodile (*Osteolaemus tetraspis*) is the most widespread and common crocodile occurring in forest ponds and lakes, marshes, and smaller slow-moving rivers with sand or silt substrate. It has been recorded from multiple locations over Bakossiland, including atypical habitats such as mid-altitude submontane vegetation of Lake Edib, 1150 m (Wild 2000), and from a rocky stream above Nyasoso at 880 m on Mt Kupe. The slender-snouted (*Crocodylus cataphractus*) and Nile crocodile (*Crocodylus niloticus*) are very rare, with records of both from the Mungo River in the southern Bakossi Mountains and the Mbwe (Mbu) tributary of the Cross River.

Birds

A total of 330 bird species have been recorded so far on Mt Kupe, including 17 of the 29 bird species of the Cameroon mountains endemic bird area, and also 30 of the 44 bird species typical of the Afrotropical highlands biome that occur in Cameroon (Dowsett-Lemaire & Dowsett 2000; Fotso *et al.* 2001). Until very recently the Mount Kupe bush-shrike (*Malaconotus kupeensis*) was thought to be endemic to the mountain, with only about 7 pairs located in 1990 between 950–1450 m. Green breasted bush-shrike (*Malaconotus gladiator*) is not uncommon above 1500 m, with preference to canopy clearings. Monteiro's bush-shrike *Malaconotus monteiri* has been recorded (Andrews 1994). Bates's weaver (*Ploceus batesi*) is only of marginal occurrence; here, as elsewhere, this species seems to prefer secondary forest. White-naped pigeon (*Columba albinucha*) and Zenker's honeyguide (*Melignomon zenkeri*) are mainly observed in secondary forest 1000–1100 m. These endemic species occur together with at least 25 other montane species (those having lower altitudinal limits), and the high diversity of the well-represented lowland forest communities are the main reasons for the high number of species, despite the lack of any wetland or other major habitat types in the area (Bowden 2001).

A total of 270 bird species are now known to occur on Mt Mwanenguba, including 19 of the 29 restricted range species confined to the Cameroon mountains endemic bird area (Stattersfield *et al.* 1998) and also 36 of the 44 bird species typical of the Afrotropical highlands biome that occur in Cameroon (Dowsett-Lemaire & Dowsett 2000; Fotso *et al.*, 2001). The montane forest swift (*Schoutedenapus myoptilus*) was discovered for the first time in Cameroon here in 1999 (Dowsett-Lemaire & Dowsett 1999). The crater grasslands are home to cape grass owl (*Tyto capensis*) (Mwanenguba is one of the only three definite Cameroon localities), where it is not rare.

A total of 334 bird species have been recorded so far in the Bakossi mountains (Anye Ndeh & Mbah Bian 2003), including 18 of the 29 restricted ranged species confined to the Cameroon mountain endemic bird area (Stattersfield *et al.* 1998), and also 34 of the 44 bird species typical of the Afrotropical highlands biome (Dowsett-Lemaire & Dowsett 2000; Fotso *et al.* 2001). With Mt Cameroon and Mt Oku, Bakossi is one of the three most important sites for the conservation of local endemics along the Cameroon chain, given that its area of upland forest is much greater than those on nearby Mt Kupe and Mwanenguba (which are however equally rich in montane bird species). The most endangered of these, the Mount Kupe bush-shrike (*Malaconotus kupeensis*), is not uncommon in the Edib Hills in primary forest at 1000–1250 m. Bakossi is also particularly important for its population of white-throated mountain babbler (*Kupeornis gilberti*), abundant throughout the mid-altitude forest. Bannerman's weaver (*Ploceus bannermani*) reaches its western limit at Kodmin, the only part of Bakossi with suitable habitat (forest clearing for this ecotone species). Green breasted bush-shrike (*Malaconotus gladiator*) is common in the hills around Kodmin and Edib at 1100–1450 m, also taking advantage of exposed canopy at forest edges (Dowsett-Lemaire & Dowsett 1999 & 2000).

Mammals

A total of 110 mammal species have been recorded, including 16 primate species, 13 carnivores, 12 ungulates and 3 pangolin species. The existence of forest elephants (*Loxodonta africana cyclotis*) in the area must date back to Pliocene period following the discovery of a fully mineralized molar tooth on the southern slopes of Mt Mwanenguba in 1992. Elephants were hunted out from the Kupe and Mwanenguba area by 1966 and 1984 respectively, but still occur in low numbers in parts of southern Bakossi along the Mungo River. It is uncertain, however, whether elephants still migrate between Bakossi and the Banyang Mbo Wildlife Sanctuary to the north where they occur in greater numbers.

Ten simian primate species occur in Bakossiland. The chimpanzee (*Pan troglodytes*) occurs locally in Bakossi, is stable on Mt Kupe, and is probably extinct from Mwanenguba. The drill (*Mandrillus leucophaeus*) occurs in large groups on Kupe with recent sightings in excess of 300 individuals there, in smaller fragmentary groups in Bakossi, and is still present in low numbers on Mwanenguba (Wild *et al.* 2004). Preuss's red colobus (*Piliocolobus preussi*) is known from the Bakossi near the Mungo River where it is very rare (King 1994; Faucher 1999) and the red-capped mangabey (*Cercocebus torquatus*) occurs locally in low numbers in the lowlands of Kupe and Bakossi. Five guenon species occur, including two species endemic to the region north of Sanaga River, the red-eared guenon (*Cercopithecus erythrotis camerunensis*), and Preuss's guenon (*C. preussi*), both of which are known from the Bakossi, Mwanenguba and the forest east of Ngomboaku on the Kupe-Mwanenguba intermontane ridge, but the latter is extinct or absent from Mt Kupe. The ubiquitous mona (*C. mona*), putty-nosed (*C. nictitans*) and crowned guenon (*C. pogonias*) also exist throughout the forests of Bakossiland, with the latter species being relatively uncommon. A sixth guenon species, the tantalus (*C. tantalus*), occurs as an invasive farmbush species and may not be considered as part of the naturally occurring fauna of the area (Kavanagh 1978).

At least six prosimian primates occur. The Potto (*Perodicticus potto*) occurs at altitudes up to 1600 m throughout the area, and is considered common. Pallid needle-clawed galago (*Euoticus pallidus*), Allen's galago (*Galago alleni*), Thomas's dwarf galago (*Galagoides thomasi*) and Demidoff's dwarf galago (*G. demidoff*) are all common and widespread, including in degraded forest and farmbush. The Angwantibo has been recorded from Bakossiland on just three occasions in the 1990s. A specimen from Nyasoso captured in 1998 was later conserved as a skin and skeleton. Whilst it is generally assumed that *Arctocebus* occurring north of the Sanaga are assigned to *A. calabarensis*, and to the south *A. aureus* occurs, the Nyasoso specimen has been assigned to the latter southerly species on the basis of pelage colour and length as seen in life. A new genus of lorisiform primate, *Pseudopotto martini* Schwarz (1996), may exist on Mt Kupe. The genus, never before observed in the wild, may occur on the mountain's lower slopes after several sightings conforming to the species description were made in 1998. There is, however, disagreement as to the validity of this taxon (Sarmiento 1998), and this problem may not be satisfactorily resolved until another specimen is obtained. What is certain, however, is that both long- and short-tailed lorisiform primates occur in the submontane forest of Mt Kupe.

Thirteen confirmed carnivore species have been recorded. Golden cat (*Felis aurata*) is known from a few records from Bakossi and Kupe. Leopards (*Panthera pardus*) still occur in Bakossi near Edib, Nyale and Bangone. The last individual from Kupe was shot by a hunter in Mbulle village in 1956. West African linsang *(Poiana leightoni)* is known from a skin in northern Bakossi, servaline genet (*Genetta servalina*), and blotched genet (*G. tigrina*) occur throughout the primary forests. The spot-necked otter (*Lutra maculicollis*) occurs in the Jide River. Both the African civet (*Civettictis civetta*) and African palm civet (*Nandinia binotata*) are widespread. The flat-headed cusimanse (*Crossarchus platycephalus*) and four species of mongoose have also been recorded: marsh *(Atilax paludinosus),* slender *(Herpestes sanguinea),* long-snouted (*Herpestes naso*) and the black-legged dog mongoose *(Bdeogale nigripes*).

Twelve ungulate species are known, including forest buffalo (*Syncerus caffer nanus*) from the Kupe-Mwanenguba ridge where it is now extinct, and from the Mungo River where few may remain. The red river hog (*Potamochoerus porcus*) is still widespread over much of the area. The sitatunga (*Tragelaphus spekei*) is known from Bakossi where seven records come from above 1300 m altitude in the Edib Hills. Bushbuck (*T. scriptus*) are widespread but in low numbers. The water chevrotain (*Hyemoschus aquaticus*) has been recorded from the Mungo River, and seven duiker species occur including yellow-backed (*Cephalophus silvicultor*), black-fronted (*C. nigrifrons*), and Ogilby's (*C. ogilbyi*).

The area supports three species of pangolin, including the rare giant pangolin (*Smutsia gigantea*). Western tree hyraxes (*Dendrohyrax dorsalis*) are generally widespread, probably in low densities, whereas red-headed rock hyraxes (*Procavia ruficeps*) are restricted to grassy exposed habitats such as the Mwanenguba caldera and the Lake Edib caldera. Twelve insectivore species have been recorded, including the Mwanenguba shrew *Crocidura manengubae* (Hutterer 1982; Dowsett-Lemaire & Dowsett 1998). The giant otter-shrew (*Potamogale velox*) is widespread throughout these mountains favouring the small fast-flowing streams. It has also been recorded from the lakeshore vegetation in the Mwanenguba caldera.

Of the remaining mammalian fauna, thirty-two species of rodent have been recorded, including Lady Burton's rope squirrel (*Funisciurus isabella*), Cooper's mountain squirrel (*Paraxerus cooperi),* and the long-eared flying mouse (*Idiurus macrotis).* Three anomalures occur, the Lord Derby's (*Anomalurus derbianus),* Beecroft's (*A. beecrofti*) and Lesser (*A. pusillus).* At least 21 bat species have been collected, although this group has also been poorly studied in Bakossiland (Dowsett-Lemaire & Dowsett 1999). The chiropteran fauna is likely to be well represented in this area given the diversity of habitats associated with topographic and physiognomic diversity of the vegetation.

SCIENTIFIC PRIORITIES FOR THE CONSERVATION OF THE FAUNA

Archaeological evidence from the region suggests that the Bakossi forests have been inhabited and influenced by Man since the Neolithic (Mercader & Marti 1999). The process of population extinction of various megafauna species probably started in the 19th Century following the expansion of the Bamileke and related tribal kingdoms, invasions of mounted Bororo warriors from the north, followed by the agricultural development of the Jide and Dibombe valleys by the German colonial administration shortly after the turn of the century. By the mid-20th Century the use of dane guns and, later, shotguns became widespread, and hunting proliferated by the 1970s. In recent decades hunting has, as elsewhere in the forest zone of Africa, become a major threat to wildlife and the ecological stability of the rain forest (Wild *et al.* 2004).

The conservation status of the megafauna is the most dramatically affected of faunal groups. Although no single species has yet been shown to have become extinct from Bakossi, some populations of the larger species have been locally severely reduced in size or are on the verge of extinction in this forest. A number of megafauna species have already been lost during the latter part of the 20th Century from Mwanenguba (elephant, leopard, chimpanzee, giant pangolin) and Kupe (elephant, leopard, Preuss's guenon, giant pangolin, sitatunga). The future of elephant in Bakossi is now in doubt. Given the important role of forest elephants as seed dispersers in tropical forests (e.g. Lieberman *et al.* 1987) and the fact that in some forests elephants have been proposed as the main dispersal agents for up to 30% of tree species (Alexandre 1978), botanists may have some concern as to their imminent extinction from Bakossiland (also see Hawthorne & Parren 2000).

Solutions to over-hunting and the hunting of protected endangered species of large mammals must be found to stem the loss of a number of important mammal populations, in particular for simian primates, ungulates and elephant. Protection of large mammals in Cameroon's Protected Areas has proven difficult to achieve, and questions must be asked as to the appropriateness of management models and practises. Experience has shown in Kupe and Bakossi that this may best be achieved by enforcement based on partnerships between government, communities and NGOs, supported by diverse land-use management options within a Protected Areas Support Zone. Capacity building and empowerment of government field staff and community-based organisations are integral to the development of such partnerships, with the attachment of Protected Areas legal counsel to ensure that protection measures are enforced. The formation of such a complex management system has not been developed to date, and the risk of degradation of faunal diversity is imminent under current trends. Despite this, certain initiatives by Bakossi traditional authorities have contributed significantly to reduction in hunting locally, particularly on Mt Kupe (Wild *et al.* 2004).

Encroachment of cloud forests and associated submontane hill forests are also of similar concern, because of the importance of middle altitude forests to submontane birds and the stream-loving rheophile amphibians for which they are so important. Bakossi is not only one of the most extensive wet submontane cloud forests in Cameroon, it is also one of the last remaining forests in Cameroon that maintains a continuous altitudinal ecotone from montane forest to a lowland riverine habitat. It is in the middle altitudinal forests, now lacking in so much of the Cameroon Highlands, that the richest assemblages of endemic forest-dependant amphibians, reptiles and birds occur. There is an increasing risk that if the montane forests become too isolated from adjacent lowland communities, then faunal dispersal by complementing lowland faunal elements, in particular the frugivores, seed dispersers and pollinators, may become damaging to the montane forest ecological stability. Little is known of the role of the lowland mammals and birds in the maintenance of montane forest communities. The lack of premontane lowland forest around Mts Kupe and Mwanenguba is, therefore, a cause for consideration in the planning of adjacent land-use and potential corridors between these sites and the neighbouring Bakossi, which has a more extensive lowland forest community.

The introduction of *Tilapia* fish to one of the two crater lakes of Mwanenguba by missionaries and Peace Corps volunteers is cause for concern. The ecological impact of this introduction is as yet unknown, but similar cases have caused the extinction of aquatic amphibians, notably *Xenopus* spp., in East African crater lakes (Tinsley & Kobel 1996). Similar fish introductions have also caused extinction among endemic lake cichlids elsewhere in Africa, protection from which is a conservation interest. Deliberate poisoning of rivers and lakes with agricultural pesticides for the purposes of fishing is also a serious matter of environmental concern, in particular in the still-water crater lakes where heavy metal and other toxins may not readily dissipate under natural conditions.

In addition, there is unknown potential for damaging effects from atmospheric pollutants from pesticides aerially applied in adjacent banana plantations to the south, especially since the location of the plantations lie in the direction of prevailing monsoon winds, which subsequently precipitate out on the mountains due to orographic uplift and occult precipitation associated with cloud forest. Amphibians are among the most sensitive indicators to such pollutants (Bubb *et al.* 2004), and with the lack of any monitoring system for this group in Central Africa, future research options for developing amphibian monitoring should be considered, especially on Mt Kupe where the anuran cloud forest fauna is so well known.

Knowledge of museum material should be fundamental to any inventory of tropical biodiversity because museum collections record the natural history of the area collected. However, museum material does not provide a means of determining how complete an inventory is, nor a measure of the relative abundance of component species. The latter is influenced by the intent, activity times, etc. of the collector as well as heterogeneities of time and space of the collection. As Gaston (1996) points out, comparisons of the richness of different areas often remain insecure, because there is no basis for demonstrating the relative completeness of different inventories. If a rich and varied sample is obtained then it may be possible to estimate absolute faunal diversity, although few data exist as to determine what sample sizes or search efforts are required for respective groups.

The development of practical, cost-effective and reliable biological monitoring systems of selected groups is an important component of forest management. However, such monitoring systems can only be of value if data quality can be verified, and few groups readily lend themselves to this problem. Moreover, the life-histories of so much of the fauna are poorly known, and drawing conclusions from the presence or absence of species can be ambiguous. Even with careful selection and design of monitoring systems, the problems of skilled staff and finance limit their sustainability. Among the vertebrates, further survey work on fish and amphibians may offer informative results, together with a continuation of bird-ringing, but the latter is labour intensive. Monitoring of bushmeat off-take has shown to be a useful indicator of large mammal abundance in many studies, and further work on this issue is desirable.

Much information from the area remains unpublished, while certain groups are still lacking sufficient data (e.g. Chiroptera), and the problems of relative sampling effort in mountainous tropical forests mean that comparative values for species diversity, and the organisation of ecological communities, are difficult to evaluate. The extensive sampling and surveys carried out on Mt Kupe have provided a basis to estimate near-absolute values of most terrestrial vertebrate orders there, with the exception of bats and rodents, for which we only have a fragmentary picture of the dominant species. However, more new species of lower vertebrate have been discovered and described from this area than at any other period during the history of scientific investigation spanning more than 100 years. New species will likely continue to emerge from the area, as well as from more exhaustive reviews of cosmopolitan pan-African species complexes.

REFERENCES & BIBLIOGRAPHY

Alexandre, D.Y. (1978). Le role disseminateur des éléphants en forêt de Taï, Côte d'Ivoire. Revue Ecologie (La Terre et la Vie) **32**: 47–72.

Amiet, J.-L. (1975). Ecologie et distribution des amphibiens anoures de la region de Nkongsamba (Cameroun). Annales de la Faculté de Sciences, Yaoundé **20**: 33–107.

Amiet, J.-L. (1987). Faune du Cameroun: Le genre *Aphyosemion* Myers. Editions Sciences Nat., Compiègne.

Amiet, J.-L. & Dowsett-Lemaire, F. (2000). Un nouveau *Leptodactylodon* de la Dorsale camerounaise (Amphibia, Reptilia). Alytes **18**: 1–14.

Andrews, S.M. (1994). Rediscovery of Monteiro's Bush Shrike in Cameroon. Bull. Afr. Bird Club **1**: 24–25.

Anye Ndeh, D. & Mbah Bian, R. (2003). Ornithological surveys of Nkwende Hills, Bakossi Mounts, Ejagham and Mawne Forest reserves, Rumpi Hills, and UFA (11-001 and 11-002). A report to the WCS Cameroon Biodiversity Programme.

Booth, A.H. (1958). The zoogeography of West African primates: a review. Bull. I.F.A.N. **20**: 587–622.

Bowden, C.G.R. (2001). The birds of Mt Kupe, southwest Cameroon. Malimbus **23**: 13–44.

Bowden, C.G.R. & Andrews, M. (1994). Mount Kupe and its birds. Bull ABC **1**: 1–13.

Bubb, P., May, I., Miles, L. & Sayer, J. (2004). Cloud Forest Agenda. UNEP-WCMC Biodiversity Series No. **20**. Cambridge.

Colyn, M., Gautier-Hion, A. & Verheyen, W. (1991). A Reappraisal of Paleoenvironmental History in Central Africa: Evidence for a Major Fluvial Refuge in the Zaire Basin. Journal of Biogeography **18**: 403–407.

Dowsett-Lemaire, F. & Dowsett R.J. (1998). Zoological survey of small mammals, birds and frogs in the Bakossi and Kupe Mts. Report for WWF-Cameroon.

Dowsett-Lemaire, F & Dowsett, R.J. (1999). Survey of birds and amphibians on Mt Manengouba, Mt Nlonako, North Bakossi and around Kupe in 1998–99. Report for WWF-Cameroon.

Dowsett-Lemaire, F & Dowsett, R.J. (2000). Further biological surveys of Manengouba and central Bakossi in March 2000 and an evaluation of the conservation importance of Manengouba, Bakossi, Kupe and Nlonako Mts, with special reference to birds. Report for WWF-Cameroon.

Eisentraut M. (1968). Beitrag zur Säugetierfauna von Kamerun. Bonn. zool. Beitr. **19**: 1–14.

Eisentraut, M. (1973). Die Wirbeltierfauna von Fernando Poo und WestKamerun. Bonn. zool. Monogr. **3**.

Faucher, I. (1999). Preliminary zoological survey of the Bakossi Mountains with special reference to large and medium-size mammals and birds. Unpublished report to WWF Cameroon.

Fjeldså, J. & Lovett, J.C. (1997). Geographical patterns of old and young species in African forest biota: The significance of specific montane areas as evolutionary centres. Biodiversity and Conservation **6**: 325–346.

Fotso, R.C., Dowsett-Lemaire, F., Dowsett, R.J., Cameroon Ornithological Club, Scholte, P., Languy, M. & Bowden, C. (2001). Cameroon. *In:* Fishpool, L.D.C. & Evans M.I. (*eds*), Important Bird Areas in Africa and Associated Islands: Priority Sites for Conservation. Pisces Publications and Birdlife International (Birdlife Conservation Series No. 11), Newbury and Cambridge, UK: 133–159.

Gartshore, M.E. (1986). The status of the montane herpetofauna of the Cameroon Highlands. *In:* Stuart, S.N. (*ed.*), Conservation of Cameroon Montane ForestsCambridge: International Council for Bird Preservation: 205–238.

Gaston, K.J. (1996). Biodiversity: A Biology of Numbers and Difference. Blackwell Science Ltd., Oxford.

Grubb, P. (1978). Patterns of speciation in African mammals. Bull. Carnegie Mus. Natur. Hist. **6**: 152–167.

Grubb, P. (1982). Refuges and Dispersal in the Speciation of African Mammals. *In:* Prance, G.T. (*ed.*). Biological Diversification in the Tropics. Columbia University Press, New York: 553–557.

Grubb, P. (1990). Primate geography in the Afro-tropical forest biome. *In:* Peters, G. & Hutterer, R. (*eds*). Vertebrates in the tropics, Museum Alexander Koenig, Bonn: 187–214.

Grubb, P. (2001). Forest endemism and biogeography. *In:* Weber, W., White, L.J.T., Vedder, A. & Naughton-Treves, L. (*eds*). African RainForest Ecology and Conservation. Yale University Press, New Haven, USA and London, UK: 88–100.

Hall, B.P. & Moreau, R.E. (1970). An Atlas of Speciation in African Passerine Birds. British Museum (Natural History), London.

Hamilton, A.C. (1981). The Quaternary history of African forests: its relevance to conservation. African Journal of Ecology: 1–6.

Hamilton, A.C. (1989). Guenon evolution and forest history. *In:* Gautier-Hion, A., Bourlière, F., Gautier, J.-P. & Kingdon, J. A primate radiation: evolutionary biology of the African guenons. Cambridge University Press, Cambridge: 13–34.

Hawthorne, W.D. & Parren, M.P.E. (2000). How important are forest elephants to the survival of woody plant species in Upper Guinean forests? Journal of Tropical Ecology **16**: 133–150.

Hofer, U., Bersier, L.-F. & Borcard, D. (1999). Spatial organization of a herpetofauna on an elevational gradient revealed by null model tests. Ecology **80**: 976–988.

Hofer, U., Bersier, L.-F. & Borcard, D. (2000). Ecotones and gradient as determinants of herpetofaunal community structure in the primary forest of Mount Kupe, Cameroon. Journal of Tropical Ecology **16**: 517–533.

Hughes, B. (1983). African snake faunas. Bonn. zool. Beitr. **24**: 311–355.

Hutterer R. (1982). *Crocidura manengubae* n. sp. (Mammalia: Soricidae), eine neue Spitzmaus aus Kamerun. Bonn. zool. Beitr. **32**: 241–248.

Joger, U. (1982). Zur Herpetofaunistik Kameruns (II). Bonner Zoologische Beiträge **33**: 313–339.

Kavanagh, M. (1978). Monkeys new home in the forest. New Scientist **77**: 515–517.

King, S. (1994). Utilisation of Wildlife in Bakossiland, Cameroon - with particular reference to Primates. Traffic Bulletin **14**: 63–73.

Kingdon, J. (1990). Island Africa: The Evolution of Africa's Rare Animals and Plants. Collins, London.

Klaver, C. & Böhme, W. (1992). The species of the *Chamaeleo cristatus* species-group from Cameroon and adjacent countries, West Africa. Bonn. zool. Beitr. **43**: 433–476.

Lawson, D., Noonan, B.P. & Ustach, P.C. (2001). *Atheris subocularis* (Serpentes: Viperidae) Revisited: Molecular and Morphological Evidence for the Resurrection of an Enigmatic Taxon. Copeia **3**: 737–744.

Lieberman, D., Lieberman, M. & Martin, C. (1987). Notes on seeds in elephant dung from Bia National Park, Ghana. Biotropica **19**: 365–369.

Maley, J. (1987). Fragmentation de la foret dense humide africaine et extension des biotopes montagnards du Quaternaire recent: nouvelles donnees polliniques et chronologiques. Implications palaeoclimatiques et biogeographiques. Palaeoecol. Afr. **18**: 307–334.

Maley, J. (1989). Late Quaternary climatic changes in the African rain forest: the question of forest refuges and the major role of sea surface temperature variations. *In:* Leinen, M. & Sarnthein, M. (*eds*). Palaeoclimatology and Palaeometeorology. Modern and past patterns of global atmospheric transport. NATO Adv. Sci. Inst., Ser. C, Math. Phys. Sci. **282**: 585–616.

Maley, J. (1996a). Le cadre paleoenvironmental des refuges forestiers africains: quelques données et hypotheses. *In:* Van der Maesen, L.J.G. *et al.* (*eds*). The Biodiversity of African Plants. Kluwer: 519–535.

Maley, J. (1996b). The African rain forest - main characteristics of changes in vegetation and climate from the Upper Cretaceous to the Quaternary. Proc. Royal. Soc. Edinburgh. **104B**: 31–73.

Maley, J. (1996c). Les fluctuations majeures de la forêt dense humide africaine au cours des vingt derniers millénaires. *In:* Hladik, M. *et al.* (*eds*). L'alimentation en forêt tropicale: Interactions bioculturelles et applications au développement. UNESCO, Parthenon Publ., Paris: 55–76.

Maley, J. & Brenac, P. (1997). Vegetation dynamics, palaeoenvironments and climatic changes in the forests of western Cameroon during the last 28,000 years B.P. Rev. Paleobotany and Palynology **99**: 157–187.

Mercader, J. & Marti, R. (1999). Archaeology in the tropical forest of Banyang-Mbo, Southwest Cameroon. Nyame Akuma **52**:17–24.

Moisel, M. (1912). Karte von Kamerun 1: 300,000, 1st ed. Dietrich Reimer (Ernst Vohsen), Berlin.

Moreau, R.E. (1969). Climatic changes and the distribution of forest vertebrates in West Africa. J. Zool. Lond. **158**: 39–61.

Perret, J.-L. & Mertens, R. (1957). Étude d'une collection herpétologique faite du Cameroun de 1952 à 1953. Bulletin de l'Institute Français d'Afrique Noire **19**: 548–601.

Pook, C.E. & Wild, C. (1997). The Phylogeny of the *Chamaeleo (Trioceros) cristatus* species-group from Cameroon inferred from direct sequencing of the mitochondrial 12S ribosomal RNA gene: Evolutionary and paleobiogeographic implications. *In:* Böhme, W., Bischoff, W. & Ziegler, T. (*eds*). Herpetologia Bonnensis. Bonn: Societas Europaea Herpetologica: 297–306.

Poynton, J. (1999). Investigating biogeographical patterns: Small steps between the obvious and the obscure. J. herpetol. Assoc. Africa **43**: 1–5.

Prigogine, A. (1988). Speciation patterns of birds in the Central African forest refugia and their relationship with other refugia. Acta XIX Congr. Int. Orn.: 2537–46.

Rasmussen, J.B. (1993). A taxonomic review of the *Dipsadoboa unicolor* complex, including a phylogenetic analysis of the genus (Serpentes, Dipsadidae, Boiginae). Zool. Mus. Univ. Cop. **19**: 129–196.

Reid, G.McG. (1991). Threatened rainforest cichlids of Lower Guinea, West Africa. A case for conservation. Ann. Mus. R. Afr. Centr, Sci. Zool. **262**: 109–119

Sarmiento, E. (1998). The validity of '*Pseudopotto martini*'. African Primates **3**: 44–45.

Schliewen, U.K., Tautz, D. & Paabo, S. (1994). Sympatric speciation suggested by monophyly of crater lake cichlids. Nature **368**: 629–632.

Schmitz, A., Euskirchen, O. & Bohme, W. (2000). Zur Herpetofauna einer montanen Regenwaldregion in SW-Kamerun (Mt Kupe und Bakossi Bergland). Herpetofauna **22**: 16–27.

Schwartz, J.H. (1996). *Pseudopotto martini*: A new genus and species of extant Lorisiform primate. Anthropological Papers of the American Museum of Natural History **78**: 14.

Serle W. (1950). A contribution to the ornithology of the British Cameroons. Ibis **92**: 343–376, 602–638.

Serle W. (1951). A new species of shrike and a new race of apalis from West Africa. Bull. Brit. Orn. Club **71**: 41–43.

Serle W. (1954). A second contribution to the ornithology of the British Cameroons. Ibis **96**: 47–80.

Serle W. (1965). A third contribution to the ornithology of the British Cameroons. Ibis **107**: 60–94, 230–246.

Stattersfield, A.J.M., Crosby, M.J., Long, A.J. & Wege, D.C. (1998). Endemic Bird Areas of the World: Priorities for biodiversity conservation. BirdLife International Series No. **7**. Cambridge, UK.

Sternfeld, R. (1907). Die Schlangenfauna von Kamerun. Mit einer Bestimmungstabelle. Mitt. zool. Mus. Berl. **3**: 397–432.

Stiassny, M.L.J., Schliewen, U.K. & Dominey, W.J. (1992). A new species flock of cichlid fishes from Lake Bermin, Cameroon with a description of eight new species of *Tilapia* (Labroidei/ Cichilidae). Ichthyol. Explor. Freshwaters **3**: 311–346.

Stuart, S.N. (*ed.*) (1986). Conservation of Cameroon Montane Forests. International Council for Bird Preservation, Cambridge.

Stucki-Stirn, M.C. (1979). Snake Report 721: A comparative study of the herpetological fauna of the former West Cameroon/Africa. Herpeto-Verlag, Teuffenthal, Switzerland.

Teugels, G.G. & Guegan, J-F. (1994). Diversité biologique des poissons d'eaux douces de la Basse-Guinée et de l'Afrique Centrale. *In:* Teugels, G.G., Guegan, J.-F. & Albaret, J.-J. (*eds*). Biological Diversity of African Fresh and Brackish Water Fishes. Ann. Mus. R. Afr. Centr., Zool. **275**: 67–85.

Teugels, G.G., Reid, G.M. & King, R.P. (1992). The fishes of the Cross River Basin: taxonomy, ecology and conservation. Ann. Mus. R. Afr. Centr, Sci. Zool. **266**: 1–132.

Tinsley, R.C. & Kobel, H.R. (*eds*) (1996). The Biology of *Xenopus*. Clarendon Press, Oxford.

Thys van den Audenaerde, D.F.E. (1972). Description of a small new *Tilapia* (Pisces, Cichilidae) from West Cameroon. Rev. Zool. Bot. Afr. **85**: 93–98.

Trewavas, E. (1974). The freshwater fishes of rivers Mungo and Meme and Lakes Kotto, Mboandong and Soden, West Cameroon. Bulletin of the British Museum (Natural History) Zoology, London **26**: 331–419.

Wild, C.J. (1993). Notes on the rediscovery and congeneric associations of the Pfeffer's chameleon *Chamaeleo pfefferi* Tornier, 1900 (Sauria: Chamaeleonidae) with a brief description of the hitherto unknown female of the species. British Herpetological Society Bulletin **45**: 25–32.

Wild, C.J. (1994). Ecology of the western pygmy chameleon *Rhampholeon spectrum* Buchholz 1874 (Sauria: Chamaeleonidae). British Herpetological Society Bulletin **49**: 29–35.

Wild, C.J. (2000). Report on the Status of Crocodilians in the Cameroon Forest Zone. Pages in Crocodiles: Proceedings of the 15th working meeting of the Crocodile Specialist Group of the Species Survival Commission of IUCN, Varadero, Cuba, 17–20th January 2000. World Conservation Union, Gland, Switzerland.

Wild, C.J., Morgan, B.J. & Dixson, A.F. (2004). Historical trends and current status of the drill (*Mandrillus leucophaeus*) in Bakossiland, Cameroon. International Journal of Primatology (in press).

THE PROTECTED AREAS SYSTEM

Chris Wild[1], Atanga Ekobo[2], Bernard Fosso[3] and Alexander Ntoko[4]

[1] Zoological Society of San Diego's Conservation and Research of Endangered Species, CRES Cameroon Office, B.P. 3055 Messa, Yaoundé, Cameroon
[2] WWF Central Africa Regional Programme Office, BP 6776, Yaoundé, Cameroon, Central Africa
[3] Department of Wildlife & Protected Areas, Ministry of Environment & Forests, Yaoundé, Cameroon
[4] Nhon 'Mbwog Alexander Ngome Ntoko, Paramount Chief of Bakossi, International Telecommunication Union (ITU), Place des Nations, CH-1211 Geneva 20, Switzerland

INTRODUCTION

The Government of Cameroon's Department of Wildlife & Protected Areas (MINEF-DFAP) Action Plan 2003–2007 is a five-year vision to classify 30% of the national territory in the permanent forest domain (MINEF 2003). An important contribution to the implementation of the Government's Land Zone Plan for Cameroon is the proposed Protected Areas system in Bakossiland (Wild and Ekobo, 2003; Wild *et al.* 2003, Tanyi-Mbianyor 2003). These proposals include one new National Park, four Integral Ecological Reserves and twelve Sacred Groves. In this chapter we outline the background and current processes that will lead to the formation of a new Protected Areas system in Bakossiland.

TRADITIONAL AUTHORITY AND COMMUNITY-BASED ACTIONS IN CONSERVATION

Traditional forms of conservation were practised by the Bakossi People before the colonial administration in the form of tree planti
ng in the north-east of the Bakossi Mountains to protect sacred areas and burial grounds (Ejedepang-Koge 1971; Wild *et al.* 2003). Chieftaincy and community responses of the Bakossi against external interventions of timber companies date back to the 1940s (Ejedepang-Koge 1971). The Bakossi were not only the first to set aside and protect forests, they were also the first to oppose large scale timber logging in their tribal homeland.

More recently, traditional support for conservation in the area has been remarkably strong, with over a decade of important actions such as the Bakossi Paramount Chief's traditional hunting ban on Mt Kupe in 1994, which came with widespread chieftaincy and community support. The Bakossi Chiefs proposal to maintain Mt Kupe under the strongest legislative protection available (as an Integral Ecological Reserve) was another important example (Chungong 2002). The formation of the Mt Kupe Hunters Association in 2004 is the most recent attempt to organise the communities against hunting infringements in the area. Elsewhere in Bakossi, local hunting issues have been addressed by other Bakossi communities in the Edib Hills, and in 2002 the villages of Edib and Kodmin independently declared traditional hunting bans in their forests.

The appointment of the District Head and first Paramount Chief of Bakossi, Fritz Ntoko Epie, by the early British Colonial Administration effectively established a tribal focal point in Nyasoso village on the western slopes of Mt Kupe. This initiative was based on the chieftaincy reforms started by the German administration in 1891, and later developed by the British colonial administration, and by the Government of Cameroon following independence in 1962. The focal point of the Bakossi Paramount Chief and the Royal Palace in Nyasoso in local matters relating to conservation, rural development and local governance continues to be of influence in communities and government to the present day.

FOREST EXPLOITATION

A number of activities have threatened the forest and wildlife of the Bakossi forests. Over the past ten years environmental degradation has included: increasing commercial timber exploitation; agricultural and pastoral encroachment; commercial bushmeat hunting; subsistence hunting; agricultural pesticide poisoning of lakes and rivers with heavy metal toxins, and commercial exploitation of *Prunus africana* by pharmaceutical companies (Sumengole 1998; CERUT 1999). Of particular concern has been the excessive hunting of the megafauna throughout the area, and the recent increase in commercial timber exploitation in southern Bakossi.

Timber has been commercially exploited from Bakossiland since the 1940s when the SAFA Timber Company from the then French Cameroons was exploiting timber in the Tombel area. In the 1950s the Swiss Lumber Company based at Etam exploited the southern Bakossi near the Mungo River. Other companies also arrived and the Cameroon Industrial Forest Company exploited other parts by 1970, including the plateau areas of Mt Kupe up to 1050 m above Nyasoso. A number of concessions were granted in the Bakossi forests in the 1980s but were never capitalised upon (Ejedepang-Koge 1971; MINEF archives, Bangem).

In the late 1990s PMF Wood exploited a 'Vente de coupe' (2500 ha) in western Bakossi in the Mulongo area, logging beyond permitted limits (Global Witness 2003). In 2001, the company Complexe Helena Bois solicited the Dutch subsidiary Compagnie Industrielle des Bois Exotiques Camerounais (CIBEC) to exploit forest under the premise of Community Forest. The latter affair provoked social conflict and a concerted response from the local communities lead by the Bakossi community-based organisation EKADA (Manga 2003; Greenpeace International 2003; Konstant 2003). In July 2003 CIBEC were reported as bankrupt as a possible result of the Provincial injunction placed on the logging company to stop activities, and their timber depot in Douala was closed. Further efforts to log the Bakossi forests under the auspices of 'Community Forestry' by Helena Bois and associates in late 2003 fell into further operational difficulties and the companies withdrew from the forest.

PAST CONSERVATION INTERVENTIONS

The first legally recognised state Forest Reserves in Bakossiland were established by the colonial administration in response to the clearance of lower lying areas for cash-crop production which began in 1910. Throughout the 20th Century plantations of tropical cash crops of rubber, cocoa, oil palm, tea, coffee, pepper and banana were developed (Ejedepang-Koge 1975). The first cocoa farm in Tombel was cleared for planting in 1920, and this opened up accelerated deforestation for cash crop farming on the southern pre-montane flank of Mt Kupe, and to the east along the Jide Valley. These developments justified conservation measures and the creation of the Forest Reserves of Bakossi, Mungo, Manehas and Loum by the British colonial administration of the former West Cameroons.

In October 1991 Birdlife International initiated the first international NGO assisted management project in Bakossiland aimed at conserving the forest biodiversity of Mt Kupe (Bowden & Bowden 1993). The Mt Kupe Forest Project (MKFP) was funded by the UK Government JFS and the European Union. Despite some progress in conservation education and community micro-projects, the project closed in 1995 due to a lack of funds. Between 1995 and 1997 the MKFP remained closed until October 1997 when management responsibility was transferred to WWF Cameroon and a new management team arrived. The MKFP was co-managed by the Worldwide Fund for Nature (WWF) and MINEF under the World Bank's Global Environment Facility (GEF) programme with matching funds from the UK government Department for International Development's Joint Funding Scheme (DFID-JFS) and WWF-UK.

In April 1998 a participatory planning workshop was held that developed the first management log-frame for Mt Kupe (Vabi 1998). In 2000 WWF initiated a farm-forest boundary demarcation process validated by the Kupe Chiefs. WWF facilitated the demarcation of the farm-forest boundary of the Mt Kupe Forest with participation from villages adjacent to the mountain (Ngane et al. 2001). Funds for the project ended in June 2001 when the management responsibility for Mt Kupe was transferred to the WWF Coastal Forests Programme which maintained the site office in Nyasoso and completed the boundary demarcation process in May 2002.

On Mt Mwanenguba, the Cameroon environmental NGO, the Centre for the Environment and Rural Transformation (CERUT), assisted by funds from IUCN Netherlands and ICCO, Zeist initiated the Mwanenguba Mountain Forest Project (MMFP) in 1995. CERUT have been engaged with local government and communities on Mt Mwanenguba with a number of community micro-projects, environmental awareness campaigns, road development, rapid forest surveys and participatory mapping since 1995 (CERUT 1995, 1997 & 1999).

Conservation policies favoured by international NGOs in the early 1990s focused mainly on conservation education programmes, community micro-projects and the development of an ecotourism structure aimed primarily at birdwatchers. Throughout the 1990s the international NGOs concerned with the management of Mt Kupe developed the MKFP with a view to legalizing a Community Forest on Mt Kupe (Nkouette 1995; Sosler 1999). The strategy to create a Community Forest on Mt Kupe was, however, a concept that did not support traditional values, or provide legislative solutions to illegal encroachment by non-indigènes.

Despite management interventions by international conservation agencies for more than ten years, no legislative basis has been effected in law to gazette Protected Area status for any of the sites in Bakossiland, including biologically important sites such as Mt Kupe. This sacred mountain had always been regarded as a taboo forest by the Bakossi traditional authorities, and the strategy of international NGOs to develop community utilization and tourism was something that weakened the same indigenous belief systems that had conserved the forest within its pristine state for so long. Moreover, no *in situ* capacity had been developed to validate a community-based management forum adapted to integrate tradition and modern decision-making processes in the management of the forest.

An alternative option was proposed by WWF and the Zoological Society of San Diego's Conservation and Research of Endangered Species (CRES) to nominate Kupe as an Integral Ecological Reserve at a stakeholder workshop presided by MINEF in Loum in 2002. Integral Ecological Reserves represent the strictest non-intervention status under Cameroon law affording higher protection than that of a National Park. This new option was favoured over that of the former proposed Community Forest, and was unanimously adopted by the Chiefs, Councillors and Elites representing all 16 villages surrounding the mountain (Chungong 2002). The subsequent removal of illegal immigrant farms and settlements on the eastern slopes of Kupe was celebrated by the traditional Bakossi authorities involved in the project. Their responses also indicated a renewed wave of support for the role of conservation projects in their rural affairs.

Without formal Government recognition and protection of the Bakossi forests it has not only been difficult to ensure that conservation objectives could be achieved, but it has also been difficult to secure a continuity of donor funds to such a poorly defined and planned environment. The vulnerability of the absence of Protected Areas status in the Bakossi forests was seriously exposed by illegal logging and lack of effective governance when the timber company CIBEC exploited the area in 2002–03. In an attempt to rectify this situation, CRES began a partnership in 2002 with the WWF Coastal Forests Programme with a view to developing a revised conservation strategy for the area. The illegal logging was addressed by the Bakossi community-based organization EKADA and the logging was stopped through legal challenges and community actions.

With financial support from Conservation International's Global Conservation Fund, CRES and WWF jointly submitted a proposal to the Government to gazette Kupe, Mwanenguba, and the Bakossi Mountains in 2003. The proposal was the first to be supported by the Ministry of Environment and Forests with a Ministerial declaration of the government's intention to gazette the Bakossi National Park (Tanyi-Mbianyor 2003). This was followed by a second initiative to gazette twelve Sacred Groves of National importance and two World Heritage Sites in December 2003 (Wild *et al.* 2003) a part of which is published in this volume (see Chapter Sacred Groves).

THE NEW PROTECTED AREAS SYSTEM

The Kupe, Mwanenguba and Bakossi Mts have been selected by the Government of Cameroon as part of its new Protected Areas and Critical Sites Network (MINEF 2003). In order to fulfil conservation objectives, and reflect the diverse needs of stakeholders, a system of Protected Areas with differing management options was conceived. These forests are a complex and partially fragmented landscape. The most important areas for biodiversity conservation have been identified based on extensive scientific field research over the past ten years, much of which is presented in the present volume for the first time, including references cited herein. The proposed status and area (ha) of these sites are as follows (Wild & Ekobo 2003).

Site	Proposed Status	Size (ha)
Bakossi Mountains	National Park	76,551
Mount Mwanenguba	Integral Ecological Reserve	5,252
Mount Kupe	Integral Ecological Reserve	4,676
Lake Edib	Integral Ecological Reserve	80
Lake Beme	Integral Ecological Reserve	64

In addition, two of the above sites, Mt Kupe and Mt Mwanenguba, have been recommended to the Ministry of Environment and Forests and the Ministry of Culture as proposed World Heritage Sites on the basis of cultural values to numerous traditional societies other than the Bakossi, and the cultural landscape of the South West and Littoral Provinces (Wild *et al.* 2003); the recommendation also being supported on a dualistic basis of their respective biological importance.

Additional nominations for Lake Edib and Lake Beme as potential RAMSAR sites are also in preparation in the Department of Wildlife and Protected Areas on the basis of the unique floristic community of Lake Edib, and the unique fish community of Lake Beme (the richest site in the world for endemic fish species, relative to surface area).

FOREST ZONING

In support of the Protected Areas formation, a stratified and complex Support Zone is planned that reflects the diversity of stakeholders' interests and needs. These are intended to support traditional and economic development needs of stakeholders and communities, and provide a compromise between stakeholders with the minimum adverse impact on the biological communities and processes the proposed Protected Areas are designed to conserve: nine zones adjacent to the Protected Areas that are currently planned include:

1. Forest Corridor Zones
2. Montane Forest Restoration Zones
3. Sacred Groves
4. Village Enclavement Zones
5. Community Forests
6. Community Hunting Zones
7. Rural Development Zone
8. Tourism Development Zone
9. Sites of Special Scientific Interest

In October 2003 the World Commission on Protected Areas identified protection and management of Sacred Groves as an important component of National Parks. Among the WCPA recommendations, it was stated that Protected Areas should be managed in full compliance with the rights of indigenous peoples. In the Protected Areas formation process in Bakossiland Sacred Groves comprise part of the Support Zone originally proposed by Wild & Ekobo (2003). Certain Sacred Groves of national importance have already been proposed to the Ministry of Territorial Administration in Kupe-Mwanenguba Division (Wild *et al.* 2003), and are currently under gazettement with the support of Bakossi traditional authorities. The sites are important in their national functions concerning the tribal origins, problem solving, and the ancestral heritage of the Bakossi People for which the forests are so important. These include:

1. Ancestral Burial Ground at Mwekan
2. Mwanenguba Lakes and Elangum
3. Mount Kupe Forest
4. Mwendolengoe
5. Mekog
6. Lake Beme
7. Lake Edib
8. Kuku Falls
9. Abonkume
10. Dion d'Eseh
11. Ekwel'Mbwe
12. Njengele at Mwamenam
13. Njengele at Ngusi
14. Ndibenyok at Mahole

PARTNERSHIPS IN THE DEVELOPMENT OF A MANAGEMENT STRUCTURE

The new Protected Areas system for Bakossiland will be Cameroon's next test case in governance and partnerships aiming to conserve rainforest biodiversity and develop rural areas. The integrity of a new Protected Areas system will depend on the understanding and commitment of partners from diverse cultural, political and economic backgrounds. The validity of management plans and the effectiveness of their implementation must therefore be based on a participatory process that is sufficiently adaptable to incorporate ancient cultural beliefs and practices within a modern political and development framework. If *in situ* solutions to conflicts in natural resource-use, such as illegal logging or bushmeat exploitation are to be found, then culturally-based mechanisms for their monitoring, regulation and enforcement must be developed. The initial formulation of a management structure that supports socio-cultural coping strategies within a practical legal framework is essential to implement at the earliest stage of the Protected Areas formation.

Although many of the environmental threats and social administrative problems common to the Bakossi are typical of those found elsewhere in the country, there is increasing potential to provide working solutions to these problems with indigenous NGOs in the area. Over forty self-established community-based organisations have been formed in Bakossi. They vary widely in their capacities and levels of organisation. Among the most notable of these are: the Ekambode Development Association (EKADA), which represents 24 Bakossi villages; the Western Bakossi Development Association (WBDA) representing 11 Bakossi villages; and the Sumediang Women's Group in Northern Bakossi. Both EKADA and WBDA have recently solicited international agencies including CRES and WWF for support in managing the Bakossi forests.

Part of the new Protected Areas management structure will include the Bakossi Traditional Council under the patronage of the Paramount Chief of Bakossi, Nhon Mbwog A. N. Ntoko. The Council, currently in the early stages of conception, will represent traditional and development interests of the Bakossi People as the principal stakeholders in the Bakossi forests. It is envisioned that the Council will assume formal responsibility for the management of Sacred Groves, as well as being instrumental in the management of the National Park, Integral Ecological Reserves and other components of the forest Support Zone. A large resource base already exists with the community-based organisations in Bakossiland, and the Bakossi Traditional Council may also act as an umbrella NGO with a capacity to implement management components and micro-projects within the Protected Areas system.

In collaboration with the Government of Cameroon, CRES and WWF are assisting in the establishment of a new management structure with Bakossi traditional authorities, community-based organizations and other stakeholders for the new Protected Areas system in Bakossiland. Forthcoming legislation is expected to secure the Bakossi Mountains as a National Park, along with the establishment of Integral Ecological Reserves and Support Zone.

ACKNOWLEDGEMENTS

Thanks are given to the Government of Cameroon, particularly the Department of Wildlife & Protected Areas (DFAP-MINEF), for their constructive collaboration in Cameroon. For financial support of current activities supporting CRES partnerships with government and communities in the Protected Areas formation we are grateful to the Zoological Society of San Diego, Conservation International Global Conservation Fund, the Offield Family Foundation, and WWF Cameroon. Thanks to Peter Mbu, Oscar Tchuisseu, Filip Verbelen, Illanga Itoua, William Konstant, Alan Dixson, Bernard Ivo Ekinde-Sone, Chief Anthony Komesue, Theophilus Ngwene, Benjamin Ngane, Ekwoge Abwe and Nkwelle Jacob for their efforts in the formation of the new Protected Areas. We thank all our various collaborators including the Late Paramount Chief of Bakossi, Nhon Mbwog R. M. Ntoko.

REFERENCES

Bowden, C.G.R. & Bowden, E.M. (1993). The Conservation of Mount Kupe, Cameroon. Proc. VIII Pan-Afr.Orn.Congr. 231–235.

CERUT (1995). National Workshop for the Conservation of the Mwanenguba Mountain Forest. Bangem 20–23 November, 1995. Centre for the Environment and Rural Transformation, Limbe.

CERUT (1997). Mwanenguba Mountain Forest Conservation Project. From year 1998–2001, Plan of Action Proposal. Centre for the Environment and Rural Transformation, Limbe.

CERUT (1999). The State of *Prunus africana* in the Mwanenguba Mountain Forest. Centre for the Environment and Rural Transformation, Limbe.

Chungong, A. (2002). Chiefs Support Mount Kupe Forest Reserve. Cameroon Tribune, No. 7703 / 3992, October 15, 2002.

Ejedepang-Koge, S.N. (1971). Forest and Man in Bakossi. Unpubl. dissertation, University of Yaoundé.

Ejedepang-Koge, S.N. (1975). Tradition and Change in Peasant Activities: a study of the indigenous people's search for cash in the South West Province of Cameroon. Dept. of Private Education, Ministry of National Education, Yaoundé.

Global Witness (2003). Rapport de l'Observateur Independent No. 065. Mission conjointe UCC – Observateur Independent. Inspection de Vente de Coupe (VC) 11-06-13, departement du Koupe-Manenguba. Province du Sud-Ouest. 23rd May, 2003.

Greenpeace International (2003). Forest Crime File. Corporate Crimes. Chainsaw Criminal CIBEC. April 2003.

Konstant, W (2003). The EKADA Gate. Zoonooz, Zoological Society of San Diego, October 2003: 18–22.

Manga, P. (2003). Chiefs Protest Acquisition of Community Forest. The Post, No. 0440, p. 6, January 21, 2003, Buea (Cameroon).

MINEF (2003). Plan D'Action / Action Plan 2003–2007. PSFE Composante, Department of Wildlife & Protected Areas, Ministry of Environment & Forestry, Yaoundé, January 2003.

Ngane, B.K., Nzie, S.N., Ntoko, V.J. & Ngwene, T.N. (2001). Farm-Forest Boundary Demarcation Report. Mt Kupe Forest Project, WWF Cameroon.

Nkouette, J.M. (1995). Les populations locales et le developpement des forets communautaires dans la region du Mont Koupe: implication de la nouvelle loi forestiere. Memoire, Institut National de Developpment Rural, Universite de Dschang.

Sosler, A. (1999). Common Ground: The Mount Kupe Forest Project and the Community. Report to the Mount Kupe Forest Project, WWF Cameroon.

Sumengole, A. (1998). Unsustainable anthropic activities and environmental degradation in afro-montane forests: the case of the Mwanenguba montane forest, Cameroon. Unpublished dissertation, University of Yaoundé I.

Tanyi-Mbianyor, C.O. (2003). Declaring of Public Interest the so called Bakossi National Park in Kupe Manenguba Division, South West Province. Public Notice Order 0969, Ministry of Environment and Forestry, Yaoundé.

Vabi, M. (1998). Report of the Participatory Planning Workshop for the Mount Kupe Forest Project, Nyasoso, 21–22 April 1998. Ministry of Environment & Forestry / Global Environment Facility / WWF Cameroon, Yaoundé.

Wild, C., Abwe, E.E. & Dixson, B. (2003). Sacred Groves of National Importance in Bakossiland, Cameroon. Report to the Ministry of Territorial Administration, The Government of the Republic of Cameroon, Yaoundé.

Wild, C. & Ekobo, A. (2003). Boundary Descriptions & Recommendations for the Gazettement of Seven New Protected Areas in the South West & Littoral Provinces. Report prepared for the Department of Wildlife and Protected Areas, Ministry of Environment and Forestry, The Government of the Republic of Cameroon. Yaoundé.

SACRED GROVES

Chris Wild

Zoological Society of San Diego's Conservation and Research of Endangered Species, CRES Cameroon Office, B.P. 3055 Messa, Yaoundé, Cameroon

INTRODUCTION

Sacred groves are a fundamental component of African traditional societies (Wenger, 1977). They have, however, been afforded little recognition and poor legal protection in Protected Areas management in Africa to date. The recent growth in literature, and attention to international legislation concerning indigenous peoples rights and Protected Areas, suggest that political trends may be changing in their favour (e.g. Nawa 1990; Cox & Elmquist 1991; Carmichael *et al.* 1997; Pimbert & Pretty 1997; Ramakrishnan *et al.* 1998; Anaya 2000; Beltran 2000; Butt & Price 2000; Gulliford 2000; Posey 2000; Van Gemerden 2000; Greene 2002; Kamden 2002; Lynch & Maggio 2002; Alden Wiley 2003; Rakoto Ramiarantsoa 2003). Gombya-Ssemmbajjwe (1995) described the traditional roles in the management of sacred forests in Uganda. Amoako-Atta (1998) initiated one of the first West African studies of a sacred grove within a community management context in Ghana. Sigu *et al.* (undated) also described rule enforcement and human impact on sacred groves in Kenya, whilst numerous other indigenous initiatives across African 'ethnoforests' are currently in progress (M. Sheridan, *pers. comm.*).

The international community has shown recent indications towards greater recognition of indigenous people's rights and sacred forests. In the September 2003 World Parks Congress, the World Commission on Protected Areas identified protection and management of sacred groves as a specific output of the congress with the planned production of preliminary guidelines for the management of sacred natural sites. Among the recommendations of the Congress, was a policy statement that Protected Areas should be managed in full compliance with the rights of indigenous peoples. Some African states, however, may be more reticent to adopt such policies in the interests of ethnic groups, and this underlines the importance for tribal institutions to be better organized for constructive dialogue with governments.

This chapter is based on an earlier report to the Government of Cameroon that nominated sacred sites of national importance to the Bakossi People of south-western Cameroon for official gazettement (Wild *et al.* 2003). The sacred groves comprise one of the nine zones supporting the Protected Areas formation in Bakossiland outlined by Wild & Ekobo (2003). The proposed Bakossi National Park declared by the Minister of Environment and Forestry (Public Notice Order No. 0969 of 08 August 2003), and the proposed Integral Ecological Reserves of Mts Kupe and Mwanenguba, and crater lakes of Beme and Edib (MINEF 2003; Wild & Ekobo 2003), have important implications for stakeholders concerned with their management. The Bakossi traditional authorities, the Chiefs and secret societies, are not only concerned with stakeholder decision-making processes within a Government-led forum of Protected Areas management, but also in ensuring the continuity and development of their own indigenous institutions and traditions.

Sacred groves are referred to herein on the basis of their cultural profiles relative to the Bakossi People and Cameroon. Consideration has also been given to sacred sites associated with outstanding natural features because of multiple stakeholder interests in such places within the proposed Forest Zoning Plan. The Bakossi cult of graves is also associated with many sacred precincts and groves of trees. The rites and rituals concerning the tribal assemblies and secret societies' use of sacred groves are strictly guarded by those with access to such information. Due to the sensitive cultural interests associated with sacred groves, the scope of this report is limited to information considered appropriate for public dissemination. Information gathered has, therefore, been treated as 'privileged knowledge', and comments presented here based on traditional oral sources are discretionary and restricted to non-sensitive information, or to those that are already published or widely known in Bakossi folklore. In addition, during the course of this investigation a number of prehistoric sites of related archaeological interest came to light, and brief mention of these is given in the appropriate sections.

The history of the coastal forests of South West Cameroon has been given by Ardener (1956, 1996), Austen & Derrick (1999) and Wilcox (2002). Numerous reports and specialized studies concerning the anthropology and culture of the Bakossi and related tribes are cited herein. These include publications and manuscripts by indigenous authors: e.g. Ejengele (1966); Kome (1970); Ejedepang-Koge (1971, 1976, 1986); Ajanoh (1988); Ebah (1988); Ngole (1988); Etouke Etouke (1992); Ekollo (1997), and Nyah (1998); early British and French colonial intelligence reports: e.g. Carr (1922, 1923); Rutherford (1923); Smith & Arthur (1931); Vaux (1933); Hailey (1951); Kouosseu

(1989); missionary reports and theological discussions on indigenous tradition and Christianity: e.g. Staub (1936); Bureau (1968); Monyaise (1975); Balz (1976, 1977, 1984, 1995a, 1995b); Dah (1986); Gardi (1994), and Ngome (1994). For the purpose of the present report, however, documentation compiled by missionaries must be considered biased since the missionary aim was the dissolution of indigenous institutions (Hallden 1968; Northern 1984; Miller 2002) and socio-economic control (Beaver & Danker 2002) and, moreover, such missionary reports are typically personal in perspective.

Bakossiland

Bakossiland is the tribal territory of the Bakossi People. The area is located in a mountainous landscape of the forest interior of the Atlantic coast running 90 km north to south, and 60 km east to west at the widest points. Mounts Kupe (2064 m), Mwanenguba (2411 m) and Bakossi (1895 m) form the physical centre of the ancestral homeland occupied by the Bakossi with an area of 3000 km^2. These mountains are the first inland peaks after Mt Cameroon which straddle the South West and Littoral Provinces of Cameroon. The region has a dramatic landscape dominated by volcanoes and mountains, caves, inselbergs, cinder cones, lava flows, high waterfalls and plunge-pools, calderas and crater lakes. The mountains receive high levels of rainfall and support a tropical montane cloud forest with frequent mists at the vegetation level. For the most part the area is covered in dense rainforest except some of the fertile lowland valleys and the slopes of Mt Mwanenguba which have been given over to agriculture. The source of the Cross River flows from the Ebwoge Caldera of Mt Mwanenguba, while the source of the Mungo River flows from the Bakossi Mountains. Mts Kupe and Mwanenguba also represent the north-western watershed of the Wouri River which reaches the coast at Douala.

The Bakossi People

The Bakossi are the largest tribe comprising some 70,000 people within the Mbo linguistic group (Hedinger 1987). They have a mystical origin as the descendents of the founding ancestor and tribal hero, Ngoe, (translated means 'leopard') who is buried in a planted sacred grove at Mwekan in the Bakossi Mts. The tribes that developed from the sons of Ngoe include: the Bakundu, Balue, Bafaw, Bakem, Bongkeng, Balondo, Basossi, Mbo, Abo, Balong, Miamillo, Mwamenam, Bakaka, Minie, Manehas, Baneka, and Bakaka inhabiting lands neighbouring the Kupe-Mwanenguba in South West and Littoral Provinces. These areas include the Rumpi Hills (Bakundu and Balue), Mt Nlonako (Balondo, Bakem, Baneka and Bakaka), Mbo Plain (Mbo), and Wouri River coastal lowlands north of Douala (Abo). The foundation and distribution of these ethnic groups in the densely forested interior of coastal south west Cameroon were probably formed early in the mid-19th Century, a period of extensive tribal warfare, displacement, and migration. They are interrelated in their social and political institutions and in their art forms.

The Traditional Role of Sacred Groves in Bakossi Society

The secret societies of the coastal forest region share many cultural similarities, including the so-called '*Leopard Society*', men associated with leadership, nobility, power and tribal status (Ruel 1969). These societies were responsible for instructing young males in the norms of social and moral behaviour by inculcating the ethics of cooperation, civic order, hygiene, sexual mores, and the tactics of village defence. Simultaneously, the societies also enforced the norms, acting as an institutionalised social regulatory body (Northern 1984). In Ejedepang-Koge's words (1976), they acted like a 'social security system' responsible for the redistribution of wealth. The instrumental belief centred around the concept of the continuing influence of the dead, the ancestors, upon the living. Ancestors as spirit spectres controlled the welfare of the living, who depended on their goodwill for health, fertility, good harvest, luck in hunting, in short, their material well-being. The secret societies that formed ancestral cult lodges sought this goodwill by sacrificing to the ancestors, especially under critical circumstances (Northern 1984).

In every system of religious beliefs, places in which the rites are performed must be in evidence. All over Bakossiland one sees isolated, protected sacred and secret forests or groves (Ejedepang-Koge 1971). Sacred groves may be natural or planted. The sites of many sacred groves are recognisable from the conspicuous peregun trees (*Dracaena arborea*) which were planted to protect the sites. The Bakossi secret juju societies were the first to designate and manage these as reserved forests. These groves have been set-aside for a long time as places of meeting and ritual. Forests have been protected for tribal, clanal, village or family ancestors or gods to inhabit. Sacred groves are both respected and feared, trees are left alone to protect them as groves, thus maintaining their mystery. There are many different types of sacred forest: ancient meeting places, or groves of the ancestors, ancient meeting places of the secret societies, 'shrine-bushes' where communities made sacrifices, planted groves that signified social boundaries, taboo forests, as well as sacred natural features such as boulders, inselbergs, waterfalls, thermal springs and crater lakes. Isolated buma trees are also conspicuous in the farmland around villages, and many of these are thought to be inhabited by ancestral spirits who protect the village.

As Ejedepang-Koge (1986) noted, the abundance and widespread distribution of groves reflects the need of the people for an elaborate structure for maintaining peace and harmony amongst them. Many sacred groves also served

118

an important function in society by bringing people within and among villages together for feasting and dancing ceremonies. Animals were sacrificed in the name of ancestors, and this facilitated an essential social exchange and intercourse among the people. Bakossi was traditionally ordered without force. Peace was maintained, morality kept high, with a seriousness of purpose for the service of society were all maintained, thanks to Bakossi institutions. Whenever a traditional society such as this suddenly abandons its long-lived institutions without adequate safeguards by the forces of disintegration and without any adequate substitution, there is bound to result a certain maladjustment (Ejedepang-Koge 1976 & 1986).

THE SACRED GROVES

Sacred Groves of National Importance to all Bakossi and Cameroon

The following three sites in Bakossiland are considered of National importance to all Bakossi: the Mbo linguistic Group, and Cameroon (Ejedapang-Koge 1986; Ngome 1994; Balz 1995; Wild *et al.* 2003). They are distinguished from other sacred groves due to their central and fundamental significance to Bakossi mythology, folklore and tradition.

Ancestral Burial Ground at Mwekan

Located in a sacred grove situated between the villages of Mwekan, Eloum and Mwaku. The shrine, which contains remnant antiquities, is marked by peregun trees located at 9°45'56" E, 5°02'34" N situated off the Mwaku-Mwambong road, and the Mwaku-Mwekan footpath.

The ancestral burial ground of Ngoe and Sumediang, founding ancestors of the Bakossi, Minie and Mbo tribes. The area of Mwekan is also the beginning point of migration of Ngoe's sons. His sons are the names of different tribal groups across Bakossi, and Mwekan is the epicentre whence they spread. The shrine at Mwekan is *'the remaining centre from which wisdom, law and peace radiates to those who, though having their own areas of settlement never severed their links to the cradle'* (Balz; 1995). In the early part of the 20th Century many Bakossi families used to be able to trace their ancestry to Ngoe through his sons who founded different tribes and clans (Carr 1922). Certain family representatives of the above villages attending the shrine must be peace-loving and humble. Aspiring Bakossi individuals, in particular politicians, are blessed here. Formerly attended by the Nninong clan who left following the foundation of Ekwel'Mbwe. Of major cultural importance to the Bakossi People, and numerous related tribes in the region that are thought to number more than a quarter of a million people in total, who can relate their heritage back to one common ancestor and location. An important cultural site of the Bakossi, Mbo linguistic group, and Cameroon.

Access to the shrine is prohibited without the consent of the secret society (Ahon Society) members of Mwekan, Mwaku and Eloum. Individuals are not permitted to visit the grove under any circumstances. The site is currently under restoration by the Nninong clan, including rebuilding of the traditional roundhouse that formerly housed antiquities which were carved from wood.

Mwanenguba Lakes and Elangum

The Mwanenguba Lakes are situated within the Ebwoge caldera of Mt Mwanenguba at an altitude of 1915 m. Lake Sumediang – the so-called 'female lake', which is blue in colour is larger than Lake Ngoe – the so-called 'male lake', which is green in colour. Elangum is a rocky peak projecting from the peaks above the caldera. The maximum diameter of the Ebwoge caldera is 4100 m with an area of 926 ha.

The name 'Mwanenguba' is derived from Bakossi tradition and means 'man and wife', represented by Ngoe and Sumediang, the founding ancestors of the Bakossi, Minie and Mbo tribes. It is widely quoted in Bakossi folklore and mythology that Ngoe mystically appeared from the summit peak called Elangum. On this mountain, Ngoe the hunter made use of the animals and water-supply to live. During his time here he encountered a woman named Sumediang, fishing at the female lake, whom he married. The lakes are important because they are the meeting place of the common ancestors, making them culturally and historically significant. It is claimed that it is from Mwanenguba that the ancestral spirits observe those entering and leaving Bakossiland. Power emanates mysteriously from the lakes during times of stress and danger so as to enter a chosen son of the land to deliver the people. Water from the male lake is sacred and only accessible to the invisibles. According to Ngome (1994), Kupe assures health, wealth and prosperity to the people, while Mwanenguba maintains the ever exhaustive link between the people and their 'god'. The rites and rituals concerning Mwanenguba are performed by clans from the so-called 'Four Eyes' clan sanctuaries of Abonkume, Ekwel'Mbwe, Njengele at Mwamenam and Dion d'Eseh. Representatives from these areas come together at the male lake once per year to see how the coming year will look. The information is then disseminated to the Bakossi People through the clan sanctuaries and local sacred groves. A famous and important cultural site of Cameroon.

The Mwabi, Nninong and Mwamenam clans continue to conduct annual sacrifices at the lakes on behalf of the Bakossi People. The Mbo People are reported also to make sacrifices at the eastern foot of the mountain. The caldera plain is inhabited by Bororo cattle herdsmen as it is an important holding station for fattening cattle sold to the Bakossi and others in the coastal south. Rudimentary permanent features have been put in place, including a small school for the Bororo children. In 2003 the Bangem Rural Council built concrete steps along the path between the two lakes, and two traditional Bakossi roundhouses on the shore of the female lake. These are unlikely to remain usable without the constant use of fire inside the buildings which preserves the roofing thatch. The initiative has been considered by some traditional authorities as a challenge to the sanctity of the sites. The female lake can be entered by visitors; access to the male lake is, however, strictly prohibited in Bakossi tradition. Visitors to the lakes must report to the police station at Bangem in advance where a tourist fee is payable.

Mount Kupe Forest

The proposed Mt Kupe Integral Ecological Reserve encompasses most of the remaining montane and submontane rainforest of the mountain. The total area of the Reserve, demarcated by adjacent communities, is 4,676 ha; he summit is 2064 m.

Mt Kupe is considered the most mystical and dangerous of Bakossi sacred places. The cloud forest above the villages that surround the mountain is the world of many of the Bakossi ancestors who were sent there by juju elders of the secret societies. The three summits have been described as places where different groups of ancestors may choose to meet, as well as the basin near the summits. There are many other sacred areas on the mountain where secret societies meet in order to communicate with the ancestors. Wealth and poverty are both believed to come from Kupe. Intelligence, diseases and mosquitoes are among the mystical bundles given to secret society members by the ancestral spirits. In addition, Kupe is reputed to be the home of wizards, witches and malevolent forest spirits which take the form of animals and trees, and are often associated with natural features such as boulders, caves, waterfalls and ravines. Although of tremendous mystical and powerful significance, Kupe is only so because of its keeping by the secret societies who regard the mountain differently to other Bakossi sacred places. Kupe is known to certain elders as '*The Forest of the Dead*' where the souls of men in bondage are sold and sent to work for the ancestors. It is the Ahon Society in particular, with their dreaded 'esam' precincts (two-doored roundhouses formerly decorated with human skulls) that are responsible for keeping the traditions of Kupe. Steeped in myth throughout the country, Kupe is home to many gods of the tribe, a taboo forest that until recently only initiated individuals and elite groups were permitted to enter, and only at specific times. Even in the main cities of Douala and Yaoundé, and among the numerous other tribes such as the Fang, Bamileke, as well as the Igbo of Nigeria, Kupe is reputed to be the most feared place on earth. Some say '*Mount Kupe - God no live there!*'. It is interesting to note therefore, that Kupe's reputation still prevails given that for over 100 years the village of Nyasoso, the regional geographic stronghold of the Basel Mission - Presbyterian Church in Cameroon, and village of highest elevation on the mountain, has concentrated its anti-traditionalist efforts. This fact is indicative of the robust nature by which Kupe has been and continues to play a role in the traditional world of the Bakossi, and others. A famous and important cultural site of Cameroon.

The proposed status of the site as an 'Integral Ecological Reserve' (IER) was accepted by the sixteen Bakossi chiefs representing the villages surrounding the forest at a workshop in Loum in 2002 (Chungong 2002). The boundary demarcation has been completed through a participatory process initiated by the Senior Divisional Officer Kupe-Mwanenguba and H.R.H. the Late Paramount Chief of Bakossi in November 2000. The proposed status of the mountain as an IER, the most exclusive of Cameroon's protected areas due to its strict non-intervention policy, was considered favourable by the Bakossi Chiefs after an earlier proposal to designate Mt Kupe as a Community Forest. Conservation strategies by international NGOs were geared towards nominating Kupe's forest as a Community Forest during the 1990s, a concept that did not support traditional values or provide legislative solutions to illegal encroachment by non-indigenes. This was because the Kupe had always been regarded as a 'taboo' forest, as opposed to a forest exploited and used by the community at large. Moreover, Bakossi traditional authorities were encouraged by the removal of non-indigenous settlements and farms encroaching the eastern aspect of the mountain. This action being taken by Government after the traditional acceptance of the mountain's proposed IER status. Up until the early 1990s access to the mountain was given by the traditional authorities of villages surrounding the mountain. Since the initiation of conservation NGO activities on the Kupe forest in 1991, ecotourists have tended to frequent Nyasoso where a system of fees is run on behalf of the communities by the Mt Kupe Forest Project.

Sacred Groves Associated With Outstanding Natural Features

The mountains, lakes and other natural features are of important cultural significance since they are held to be the home of the ancestral spirits and the source of all wealth. Specific areas of interest include: inselbergs, caves, waterfalls, plunge-pools, valleys, platforms, cinder cones, plains, swamps and crater lakes. The Jide River in particular is regularly punctuated with falls and plunge-pools. Although mostly of local importance they are referred to here because of their relevance to the Protected Areas formation and interest to non-indigenous stakeholders. Among these are two crater lakes, Beme and Edib, which are listed as proposed Integral Ecological Reserves based on their outstanding biological values, but are also considered sacred in their own right, being strongly associated with ancestor cults in a similar way to the lakes of Mwanenguba. As with Mt Kupe, the proposed status of 'Integral Ecological Reserve' for Lakes Beme and Edib is thought to favour traditional interests as it prohibits unauthorised access and damaging activities at the designated sites.

Lake Beme

Lake Beme is located in the northwest Bakossi Mountains at an altitude of 450 m. A small outlet stream on the north edge of the lake drains northwards into the Mbwe (Mbu) River, a tributary of the Cross River. The lake is centred on: 9°38'11" E, 5°09'29" N. The area of internal caldera of the crater is 64 ha. Perimeter boundary of the proposed Integral Ecological Reserve follows the highest point of the caldera rim to protect the watershed, landscape and ecological processes pertaining to the lake. Regarded by the Beme and Babubok villages to be the home of their ancestors. Permission to visit the lake is required from the Beme Village traditional authorities. Fishing and swimming are practised there by the local communities.

Lake Edib

Lake Edib is located in the spectacular caldera of the Edib Volcano of the Edib Hills in the Bakossi Mountains at an altitude of 1150 m. The caldera is centred on: 9°39'15" E, 4°57'36" N. The caldera rim of this extinct volcanic crater reaches 1280 m in elevation and has an area of 80 ha. Perimeter boundary of the Proposed Integral Ecological Reserve follows the highest point of the caldera rim to protect the watershed, landscape and ecological processes pertaining to the Lake. Regarded by Edib village to be the home of their ancestors. Permission is required to visit the lake from the Edib Village traditional authorities. Swimming is forbidden.

Mwendolengoe

Mwendolengoe is a mountain inselberg prominent on the horizon of the Bakossi landscape near Mwendolengoe and Mwaku villages, southwest of Bangem. A resting place for spirits from Kupe.

Mwekog (also known by botanists as Nyale Rock)

Mwekog is another inselberg of traditional significance near the village of Nyale in western Bakossi. A resting place for spirits from Kupe.

Kuku Falls

Numerous waterfalls and plunge-pools are to be found across the Bakossi landscape. The highest of these are to be found along the western Bakossi escarpment stretching from the Elambah Ridge south of Lake Edib, north-west to Mahusom and Nyale, including the Kuku Falls which are about 100 m in height. A remote location in pristine condition that is considered to be a meeting place for the ancestors. The highest falls along the Jide River are the Elephant Falls (Ela-Nziog) near Bekume where an elephant is said to have fallen down the waterfall. Another Jide site, Ela-Nlonde, is a sacred pool on a small tributary, the Nlonde, between Ndisse and Bekume. This site was formerly owned by the Bafaw Tribe. Many other freshwater sites also exist, such as the Mbeng falls in Western Bakossi, the falls at the western rim of the Mwanenguba caldera; the 'natural bridge' at 1900 m elevation on Mt Kupe; Ebonji plunge-pool, Ndibe e Ndele falls at Baseng on the River Jide; Lake Mwandon and Nye Mekang Spring at Mwambong; Mwandib falls at Ndom; Mami Water at Njombe, Buba Falls on the Mungo River, and the Ekomtolo Falls in the Bakaka Forest.

Sacred Groves as Clan Sanctuaries

At least nine other shrines have existed in the interests of all the Bakossi People. Although their significance and functions have been severely challenged by missionaries (e.g. Balz 1995b), they nonetheless represent the infrastructure of a complex organisation of secret societies that transcend clan interests alone, and are responsible for the continuity of Bakossi tradition. It is within this context therefore, that they are considered of relevance to the national interest. Of the sites listed below, five are still in use by secret societies, including the so-called 'Four Eyes of Mwanenguba' (Abonkume, Dion d'Eseh, Ekwel'Mbwe and Njengele at Mwakumel), while four others seem to have been destroyed or abandoned by local communities. Most are relatively small in size, below 1 ha. Access is strictly controlled by the secret societies.

Abonkume

Sacred grove located at Mwabi village on Mt Mwanenguba. A meeting place for tribal matters concerning the environment: calamities, famine and danger. Serious clan problems are judged and solved here. Aspiring young men are blessed with wine following divination. Meetings are held every five days with permanent seats for family representatives. A grove also reported to play a part in the annual sacrifices at the Mwanenguba lakes in the Ebwoge caldera by the Nhia clan (Ngome 1994). One of the so-called 'Four Eyes of Mwanenguba'.

Ahid de Nkang

A small hill betwenn Mwambong II (Mwesok) and Mwambong III (Nkang) Village. Resting place for spirits from Kupe. A geographically central meeting place for all senior Bakossi representatives. Related in function to Dion d'Eseh. Colonial masters' resthouse built on the area of the shrine is seen as a challenge to traditional authority.

Asongwa

A sacred grove located on the traditional boundary between Ngombombeng and Kack Villages. Reported to have been abandoned in recent years. A second site by the same name was formerly located on the traditional boundary between Mwambong and Ngomboaku. This sacred grove was destroyed by road construction after rituals were performed to 'close' the shrine.

Dion d'Eseh

Sacred Grove located near Mwambong between the villages of Manta I and Manta II. National function for problem solving. Frequently used by all Bakossi Ahon Society members. Meetings are held every five days. Permanent seats are maintained for family/village representatives of the clan. New seats can be given to aspiring and humble young men. One of the so-called 'Four Eyes of Mwanenguba'.

Ekwel'Mbwe

This house shrine is located in Nkack village on the Kupe-Mwanenguba intermontane ridge, south of Bangem. A sacred precinct and clan sanctuary of the Nninong clan. Formerly all-Bakossi meeting place for internal tribal affairs and traditional judicial court. Decisions taken here may not be challenged within Bakossiland. A grove also reported to play a part in the annual sacrifices at the Mwanenguba lakes in the Ebwoge caldera (Ngome 1994) and one of the so-called 'Four Eyes of Mwanenguba'. Frequently used by Ahon Juju Society members of Nninong Clan to date, often with neighbouring and other Bakossi clans. Traditional stools are still maintained at the site for higher ranking members, including those of external clans.

M'Mbom M'Mwan'Ngoe

Located in Mwambong Village, a geographically central meeting place for all senior Bakossi clan/society representatives. Destroyed due to the construction of a Government school directly on the area of the shrine.

Ngwanekode

A sacred grove located on the traditional boundary between Bekume and Mekedmbeng villages. Matters concerning the Asomengoe Clan were formerly resolved here, where information from Mwanenguba and Kupe on the coming year were discussed and measures taken to fortify the clan. Reported to have been abandoned in recent years.

Njengele at Mwamenam

Located Mwamenam and Mwamekeng villages on the southern slopes of Mt Mwanenguba. Traditional house sanctuary surrounded by Peregun trees. Regularly used by the Mwamenam clan. Held in very high status by the secret societies which are regulated by decisions made there. A branch from the great tree at Njengele at Ngusi (below) was carried here to connect the two places. Similar in function to Abonkume and Dion d'Eseh. One of the so-called 'Four Eyes of Mwanenguba'.

Njengele at Ngusi

Located near the village of Ngusi in the Jide River valley between the Edib Hills and Mt Kupe. A major meeting place for the Asomengoe subclans of Mwetuk, Mwetan, Mbwogmut, Mwasundem and Mwanyo. An ancient Bakossi traditional court. Used also in matters concerning war as recently as 1997 in the preparation of over 50 Bakossi soldiers (including non-Asomengoe individuals) before being sent by the government to fight in the Bakassi border conflict with Nigeria. Perhaps the most important of the clan sanctuaries with respect to the spiritual keeping of Mt Kupe. Superior in status to Asongwa and Ngwanekode.

Ndibenyok at Mahole

The stream called Ndibenyok near Mahole village is a tributary of the Mungo River. This is among the most dangerous of Bakossi shrines. Formerly there was a three-door roundhouse to host its meetings. The site's function is to send away unwanted evils and problems and said to be the last resort for problem solving in Bakossi. A person who had broken the social norm and could not find peace with the Society at Njengele was sent to Ndibenyok for mortal punishment, to be carried away by the Mungo. The pool in the stream was also a site where missionaries encouraged the Bakossi to throw away sacred objects such as masks and statues in the early 20th Century.

Other Sacred Groves of Historical Importance

Numerous other sacred sites of local importance to clans, villages and secret societies are scattered across the Bakossi landscape. Of particular importance are the other burial sites of Ngoe's sons, grandsons and the graves of Chiefs. Originally all Bakossi villages had a least one grove maintained for the whole community, as well as local sites often associated with natural features with mystical powers linked to the otherworld of the ancestors.

Prehistoric Sites of Archaeological Interest

The prehistory of Bakossiland has been surveyed recently by Mercader and Marti (1999) who discovered Middle Stone Age, Late Stone Age and Iron Age sites at both open air and rock shelter locations in northern Bakossiland around the Mwenzekong Mountains. Twenty archaeological locales suggest that archaeological resources are abundant and portray a deep historical and environmental record, as yet mostly unknown. Cave rock-shelters are of particular importance to the study of African prehistory (Brain, 1981). A number of such caves on Mt Kupe, and in western Bakossi near Mekog, Nyale Village, may yield further important information on the prehistory of the area. Other sites in the Edib Hills of uncertain age, including man-made rock formations, megaliths and forest enclaves that have not regenerated into forest, are also known to exist but have not been formally studied or recorded to date.

DISCUSSION

Sacred groves represent an important institutional dimension to traditional societies. As Colchester (1993) observed: *'Ecological balance will not be restored in Equatorial Africa without the resurgence of long-submerged traditions of equality and accountability which require a long and slow process of rebuilding community institutions and controls'*. The historical decline of sacred groves has had a detrimental impact on the state of Bakossi organisation and culture. This trend is likely to continue without the legal protection and influence of the Bakossi indigenous institutions in the management of sacred groves.

Today some of the ancient secret societies of Bakossi that predate the colonial administration and the tribal chieftaincy system are still active: they are well organised, and continue to exert prevalent executive and judicial roles within the Bakossi communities. However, with the weakening and elimination of secret societies throughout Cameroon and Africa, fundamental changes in the regulation of social norms have also occurred. As Iyam (1995) found among the Biase People of the Cross River, chieftaincy disputes suggest that homogeneity no longer guarantees decision by consensus. Many of these changes result from local and foreign influences that have left in place only skeletal images of once-effective indigenous practises and institutions. At the same time, these cultural vestiges have been accepted as indigenous both by those who practice them and by development agencies. Where no vestiges are left, old and forgotten practices and behaviours are resurrected to contest present-day realities.

The Bakossi still revere a number of sacred groves in a national context. The landscape, weather and environment are all inextricably linked to the Bakossi ancestor-cult and belief system. As part of the rich cultural and biological heritage of Bakossiland, the sacred groves comprise the environmental aspect that is fundamental to the cultural framework of Bakossi Society. However, many sacred groves have been lost in Bakossiland in recent decades. These include groves of national or clanal function such as Asongwa and Ngwanekode, while an unknown number of clanal, village and family groves have been lost to village development, forest exploitation and abolishment by missionaries. Extant sacred groves are threatened by similar activities that could be averted under formal government recognition of the sites.

Three sites in Bakossiland, the Ancestral Burial Ground at Mwekan, the Mwanenguba lakes and Mt Kupe, are of the profound significance in Bakossi history, folklore and tradition. Moreover, in light of their regional importance in the cultural landscape of nineteen tribes in the South West and Littoral Provinces, and popular influence beyond, they must also rank highly in Cameroon's rich national heritage.

The sacred groves associated with outstanding natural features are prominent in the imagination of Bakossi People as well as to visitors to the area. In addition to Mt Kupe, the Mwanenguba lakes and Elangum, five other proposed sites associated with lakes, falls and inselbergs may also be of interest to tourists and the public. These include the inselbergs of Mwendolengoe and Mwekog, the crater lakes of Beme and Edib, and the Kuku Falls. Official recognition and protection of these sites as sacred groves may help avoid conflicts of interest, and maintain them within acceptable levels of intrusion or disturbance.

The four extant clan sanctuaries of national importance: Abonkume, Dion d'Eseh, Ekwel Mbwe, and Njengele at Mwamenam, are of fundamental importance to the continuity and organization of secret societies that regulate the chieftaincy systems and certain internal tribal affairs. They represent the so-called 'Four Eyes of Mwanenguba' responsible for carrying out sacrifices to Ngoe on behalf of the Bakossi People. Njengele at Ngusi is of particular importance to the spiritual keeping of Mt Kupe, while Ndibenyok is most important for problem solving for all the Bakossi. These precincts are surrounded in secrecy, and although they do not concern the general public directly, their status and protection are of importance to Bakossi traditional society.

The future of many sacred groves in Bakossiland is threatened, and along with them the infrastructure of Bakossi culture and organisation. Agricultural expansion of plantations has consumed many such places along with the development of village amenities. Commercial timber exploitation has destroyed some sacred groves, as well as certain magic trees in the forest known as 'Kingsticks' that were inhabited by the ancestral spirits. In addition, evangelisation and the campaigns of the Church against traditional ancestor-cults continues to marginalize and weaken the indigenous institutions that maintain the groves, a process described by Unchendo (1976) as *ancestorcide*. In some cases sacred groves have been abandoned by communities under pressure by missionaries to *'forsook the little gods'* (Dah 1996). The increase in eco-tourism, exploration and field research in recent years may also have contributed to the degradation of indigenous belief systems associated with sacred places such as the Mt Kupe forest.

Sacred groves have direct relevance to the formation of the proposed Protected Areas and their management options, as well as to the continuity of Bakossi institutions formerly so important in shaping society. They represent a significant component of the proposed Forest Zoning Plan to support the Protected Areas formation. The development of community coping strategies and regulation of internal affairs pertaining to the self-regulation of forest resource-use, are all integral to building the capacity of communities as essential partners in Protected Areas management. Government recognition by official gazettement of sacred groves of national importance is therefore considered to be a vital strategic option towards supporting traditional mechanisms for an enhanced indigenous stakeholder capacity in natural resource management, and an advantage in preserving Bakossi culture.

ACKNOWLEDGEMENTS

This chapter is based on an earlier report to the Government of Cameroon with field assistance from Ekwoge Enang Abwe and Barnaby Dixson. Thanks are given to all collaborators and Bakossi Traditional Authorities consulted for this work, including the Late Paramount Chief of Bakossi, Nhon Mbwog R. M. Ntoko, and his successor, Nhon Mbwog A. N. Ntoko.

REFERENCES & BIBLIOGRAPHY

Ajanoh, S. (1988). 'The Choice of Muekan Was Not Fortuitous'. Eshotan (Bulletin of the Nninong Cultural and Development Association). 1: 28-29 Yaoundé.

Alden Wiley, L. (2003). Community Roles in Protected Areas Management in Africa. *In:* Jaireth, H. & Smyth, D. (*eds*). Innovative Governance: Indigenous Peoples, Local Communities and Protected Areas. Ane Books, New Delhi, Chennal, Kolkata: 29–59.

Alobwede Epie, C.E. (1982). The Language of Traditional Medicine in Bakossi. Thesis for the award of Doctorat d'Etat, University of Yaoundé.

Amoako-Atta, B. (1998). Preservation of Sacred Groves in Ghana: Esukawkaw Forest Reserve and its Anweam Sacred Grove. Working Paper No. 26, South-South Co-operation Programme on Environmentally Sound Socio-Economic Development in the Humid Tropics, UNESCO, Paris.

Anaya, S.J. (2000). Indigenous Peoples and International Law. Oxford University Press.

Ardener, E. (1956). Coastal Bantu of the Cameroons. International African Institute, London.

Ardener, E. (1996). Kingdom on Mt Cameroon: studies in the History of the Cameroon Coast 1500–1970. Berghan Books, Oxford.

Austen, R.A. & Derrick, J. (1999). Middlemen of the Cameroons Rivers: The Duala and their Hinterland, c. 1600–1960. African Studies Series 96, Cambridge University Press, Cambridge.

Balz, H. (1976). Was it wrong what they did? English translation of the reports by H. Ntungwa & E. Keller on the abolition of jujus in Nyasoso in 1934. Cyclostyled, Theological College, Nyasoso (Cameroon).

Balz, H. (1977). Secret Societies and the Church in Cameroon. Cyclostyled, Theological College, Nyasoso (Cameroon).

Balz, H. (1984). Where the Faith has to Live: Studies in Bakossi Society and Religion. Part I: Living Together. The Basel Mission, Basel (Switzerland).

Balz, H. (1995a). Where the Faith has to Live: Studies in Bakossi Society and Religion: The Living, the Dead and God. Part II, 1: Chapters I and II. Dietrich Reimer Verlag, Berlin.

Balz, H. (1995b). Where the Faith has to Live: Studies in Bakossi Society and Religion: The Living, the Dead and God. Part II, 2: Chapters III and IV. Dietrich Reimer Verlag, Berlin.

Barbier, J.C., Champaud, J. & Gendreau, F. (1983). Migrations et developpement: la region du Moungo au Cameroun. ORSTOM, Paris.

Beaver, R.P. & Danker, W.J. (2002). Profit for the Lord: Economic Activities in Moravian Missions and the Basel Mission Trading Company. Wipf & Stock Publishers.

Beltran, J. (2000). Indigenous and Traditional Peoples and Protected Areas: Principles, Guidelines and Case Studies. Island Press.

Brain, C.K (1981). The Hunters or the Hunted? An introduction to African cave taphonomy. University of Chicago Press, Chicago.

Bureau, R. (1968). Influence de la Christianisation sur les institutions traditionelles des ethnies cotieres du Cameroun. *In:* Baetta, G.G. (ed.). Christianity in Tropical Africa. London: 165–181.

Butt, N. & Price, M.F. (2000). Mountain People, Forests, and Trees: Strategies for Balancing Local Management and Outside Interests. The Mountain Institute, Harrisonburg.

Butt-Thompson, F.W. (2003). West African Secret Societies. Kessinger Publishing Co.

Carmichael, D.L., Hubert, J. Reeves, B. & Schanche, A. (1997) (*eds*). Sacred Sites, Sacred Places. Routledge, Taylor and Francis Books Ltd.

Carr, F.B. (1922). Assessment Report from 1922 on the Bakossi, Nninong and Elung Tribes of the Kumba Division. Cameroon National Archives, Buea, File no. 748/22, 172Ae 3a.

Carr, F.B. (1923). Assessment Report on the Tribal Area of Bassossi. Cameroon National Archives, Buea.

Chungong, A. (2002). Chiefs Support Mount Kupe Forest Reserve. Cameroon Tribune, No. **7703 / 3992**, October 15, 2002.

Colchester, M. (1993). Slave and Enclave: Towards a Political Ecology of Equatorial Africa. The Ecologist **23(5)**: 166–173.

Cox, P.A. & Elmquist, T. (1991). Indigenous control of tropical rain-forest reserves: an alternative strategy for conservation. Ambio **20**: 317–321.

Dah, J.N. (1996). The Gospel in Bakossi 1896–1996. Pforzheim/Hohenwart, Germany.

Ebah, L.E. (1988). 'Feasts and Ceremonies at Nninong'. Eshotan (Bulletin of the Nninong Cultural and Development Association). **1**: 64–75. Yaoundé.

Ejedepang-Koge, S.N. (1971). Forest and Man in Bakossi. Unpublished dissertation, University of Yaoundé.

Ejedepang-Koge, S.N. (1976). Tradition and Christianity in the Bakossi Society. Lecture delivered in the Presbyterian Theological College, Nyasoso, January 6th, 1976 (unpublished ms).

Ejedepang-Koge, S.N. (1986). Tradition of a People Bakossi. ARC Publications, Yaoundé.

Ejengele, H. (1966). La puissance de l'évangile face au paganisme Minieh des Mbo. Thèse de licence, Faculté de Théologie Protestante, Paris.

Ekollo, S. (1997). L'évolution des chefferies traditionelles dans la région du Mungo: le cas de la chefferie superieur Bakaka 1945–1960. Mémoire DIPES, ENS, Yaoundé.

Etouke Etouke, J.-L. (1992). Histoire des Peuples Mbos Ngo (Mboos). Cameroun dans les l'histoire ancienne de l'Afrique, Tome I. Le Sociologie Africain, Yaoundé.

Fisiy, C.F. (1995). Chieftaincy in the modern state: an institution at the crossroads of democratic change. Paideuma **41**: 49–62.

Gardi, B. (1994). Kunst in Kamerun. Begleitheft zur Ausstellung, Basel.

Gombya-Ssemmbajjwe, J.W.S. (1995). Sacred Forests as a Traditional Management Arrangement in Modern Ganda Society. Uganda Journal **42**: 32–44.

Greene, S.E. (2002). Sacred Sites and the Colonial Encounter: A History of Meaning and Memory in Ghana. Indiana University Press.

Guiffo, J.-P. (1990). Nkongsamba: mon beau village. Editions de l'Essoah, Yaoundé.

Gulliford, A. (2000). Sacred Objects and Sacred Places: Preserving Tribal Traditions. University Press of Colorado.

Hailey, Lord. (1951). Native Administration in British African Territories. H.M.S.O., London.

Hallden, E. (1968). The Culture Policy of the Basel Mission in the Cameroons, 1886–1905. Uppsala University Press, Uppsala.

Hedinger, R. (1987). The Manenguba Languages (Bantu A.15, Mbo cluster) of Cameroon. School of Oriental and African Studies, University of London.

Infield, M. (1989). Hunters stake a claim in the forest. New Scientist. Nov: 52–55.

IUCN (in press). Preliminary Guidelines for the Management of Sacred Natural Sites.

Iyam, D.U. (1995). The Broken Hoe: Cultural Reconfiguration in Biase Southeast Nigeria. University of Chicago Press, Chicago and London.

Kamdem, E. (2002). Management et interculturalité en Afrique: Experience Camerounaise. Les Presses de L'Université Laval, Paris.

Kome, A.E. (1970). 'The importance of Ndieh'. Bakossi Cultural Review 1970: 32–33. Mutengene.

Kougnia, H. (1922). Les Mbos entre deux frontieres. Résistance à la division. Mémoire DIPEN, Yaoundé.

Kouosseu, J. (1989). Populations autochtones, populations allogènes et administration coloniale dans le Mungo sous administration française, 1916–1960. Mémoire Maitrise, Université de Yaoundé.

Lynch, O.J. & Maggio, G.F. (2002). Mountain Laws and Peoples: Moving Towards Sustainable Development and Recognition of Community-Based Property Rights. The Mountain Institute, Harrisonburg.

Manga, P. (2003). Chiefs Protest Acquisition of Community Forest. The Post, No. **0440**, p. 6. January 21st, Buea (Cameroon).

Manon, G.C. (1977). Evolution et tendancies de la chefferie supérieure Bakaka. Mémoire licence en droit public, Université de Yaoundé.

Mekobe, S.D. (1977). La résistance traditionelle face au droit foncier camerounais: expérience des Mbos dans le Mungo. Mémoire licence droit privé, Université de Yaoundé.

Mercader, J. & Marti, R. (1999). Archaeology in the tropical forest of Banyang-Mbo, Southwest Cameroon. Nyame Akuma **52**:17–24.

Miller, J. (2002). Missionary Zeal and Institutional Control: Organizational Contradictions in the Basel Mission on the Gold Coast 1828–1917. Studies in the History of Christian Missions, Routledge Curzon.

MINEF (2003). Plan D'Action / Action Plan 2003–2007. PSFE Composante, Department of Wildlife & Protected Areas, Ministry of Environment & Forestry, Yaoundé.

Monyaise, D.P.S. (1975). Bogosi Kupe. Schaik Uitgewers, Netherlands.

Nawa, N. (1990). The role of traditional authority in the conservation of natural resources in the Western Province of Zambia, 1878–1989. MA thesis, University of Zambia, Lusaka.

Ngole, P.N. (1988). 'Groves and Secret Societies'. Eshotan (Bulletin of the Nninong Cultural and Development Association). **1**: 52–64. Yaoundé.

Ngome, E. (1994). The Concept of the God-Man in Traditional Bakossi Thought, Socialization and Change. Master's Thesis, Faculty of Protestant Theology, Yaoundé.

Northern, T. (1984). The Art of Cameroon. Smithsonian Institution, Washington.

Nyah, M. (1998). Bare-Bakem. Des origins a nos jours. Essai d'étude historique. Mémoire DIPES II, Yaoundé.

Pimbert, M. & Pretty, J.N. (1997). Parks, People and Professionals: Putting 'Participation' into Protected Areas Management. *In*: Ghimire, K.B. & Pimbert, M.P. (*eds*), Social Change and Conservation: Environmental Politics and Impacts of National Parks and Protected Areas. Earthscan, London: 297–330.

Posey, D.A. (2000). Cultural and Spritual Values of Biodiversity. ITDG Publishing.

Rakoto Ramiarantsoa, H. (2003). Pensée zéro, pensée unique, La <robes des ancetres> ignorée. *In* : Rodary, E., Castallanet, C. & Rossi, G. (*eds*). Conservation de la nature et developpement : L'integration impossible? Gret-Kartala, Paris: 105–120.

Ramakrishnan, P.S., Saxena, K.G. & Chandrashekara, U.M. (1998). Conserving the Sacred for Biodiversity Management. Science Publishers Inc.

Ruel, M. (1969). Leopards and Leaders: Constitutional Politics among a Cross River People. Tavistock, London.

Rusillon, J. (1968). Le rôle social d'une société secrète d'Afrique Equatoriale. Genève-Afrique VII **2**: 51–59.

Rutherford, J.W.C. (1923). An Assessment Report on the British Section of the Mbo Tribal Area in the Mamfe Division in the Cameroons Province. Cameroon National Archives, Buea.

Sigu, G.O., Omenda, T.O., Ongugo, P.O. & Opiyo, A. (undated). Sacred Groves Institutions, Rule Enforcement and Impact on Forest Condition: The case of Ramogi Hill Forest Reserve, Kenya. Kenya Forestry Institute, Nairobi (unpublished draft).

Smith, I.S. & Arthur, O.R. (1931). An Assessment Report on the Bakossi, Elong and Ninong Clans of Kumba Division. Cameroon National Archives, Buea.

Staub, E. (1936). 'Beware of False Gods' – The rise and fall of Paramount Chief Ntoko Epie and the success of the gospel in Bakossi-Cameroon. Basel, Switzerland.

Studstill, D.J. (1970). L'arbre ancestral. *In:* Calame-Griaule, G. (*ed.*). Le thème de l'arbre dans les contes africains II. Bibliothèque de la Selaf. **20**: 119–137.

Unchendo, V.C. (1976). Ancestorcide! Are the African Ancestors Dead? *In:* Newell, W.H. (*ed.*). Ancestors. The Hague: 283–296.

Van Gemerden, B.S. (2000). The Social Dimension of Rainforest Management in Cameroon: Issues for Co-management. Tropenbos Foundation.

Vaux, H. (1933). Intelligence report on the Bakossi Clan. Cameroon National Archives, Buea, Oct. 1932, 27pp.

Wenger, S. (1977). The timeless mind of the sacred: Its new manifestation in the Osun Groves. Institute of African Studies, University of Ibadan, Ibadan.

Wilcox, R.G. (2002). Commercial transactions and cultural interactions from the Niger Delta to Douala and Beyond. African Arts, Spring 2002: 42–55.

Wild, C. & Ekobo, A. (2003). Boundary Descriptions & Recommendations for the Gazettement of Seven New Protected Areas in the South West & Littoral Provinces. Report to the Department of Wildlife and Protected Areas, Ministry of Environment and Forestry, The Government of the Republic of Cameroon, Yaoundé.

Wild, C., Abwe, E.E. & Dixson, B. (2003). Sacred Groves of National Importance in Bakossiland, Cameroon. Report to The Ministry of Culture, The Ministry of Environment and Forestry, The Ministry of Territorial Administration, The Government of the Republic of Cameroon, Yaoundé.

RED DATA TAXA OF KUPE, MWANENGUBA AND BAKOSSI MTS

Martin Cheek

Herbarium, Royal Botanic Gardens, Kew, Surrey, TW9 3AE, U.K.

Departing from previous conservation checklists in this series (Cable & Cheek 1998; Cheek *et al.* 2000), **all** flowering plant taxa recorded in the checklist with an accepted name have been assessed on a family by family basis for their level of threat, i.e. as threatened (CR–Critically Endangered, EN–Endangered or VU–Vulnerable), Near Threatened (NT), Least Concern (LC) or Data Deficient (DD) following the IUCN guidelines (IUCN 2001). By implication, Gymnosperms, Pteridophytes and taxa without an accepted name were not assessed. The table below outlines the numbers of taxa per IUCN category:

Category	CR	EN	VU	NT	LC	DD
No. of taxa	33	52	147	296	1497	1

Of the 2412 taxa in the checklist, 2026 (c. 84%) have been assessed using the IUCN guidelines, and 528 of those (c. 26%) were found to be threatened or Near Threatened. The main part of this chapter consists of taxon treatments, giving detailed information on the **232** threatened species known to be present in the checklist area. This is undoubtedly the greatest concentration of Red Data plant taxa known in Cameroon, and probably in West and Central Africa (see Phytogeography and Refugia). Most of these taxa were assessed for the first time in the process of writing this book. All of the treatments have been reviewed by Craig Hilton-Taylor, the IUCN Red Data officer. All have been accepted apart from a) some of the assessments of infra-specific taxa since IUCN rules do not allow these to be accepted unless the 'mother' species has also been assessed; b) one assessment which was queried by an orchid reviewer. IUCN rules also do not allow acceptance of taxa unless they are either published or on the brink of publication. Consequently, an estimated 70 new species to science that are not yet at this point, most of which are only known from Bakossi, have not been assessed and are not mentioned in this chapter.

This, the introductory part of the chapter, details the methodology used in making the assessments, followed by a series of lists in which Red Data taxa are detailed by vegetation type.

ASSESSMENTS–METHODOLOGY

Taxa of Least Concern

In the first place all taxa that were found to be fairly widespread e.g. Extent of Occurrence greater than 20,000 km^2 and/or from 20 or more localities were listed as LC (IUCN (2001) does not indicate a number of localities for NT or LC, so we chose 20 as our cut-off point here for LC). These facts were established principally using the Flora of West Tropical Africa (FWTA) as an indication of range and number of collections sites, supplemented by other published sources, such as Flore du Cameroun, or by research into specimens at K, in cases of doubt. Taxa which, by these measures, are widespread and common do not qualify as threatened under Criterion B of IUCN (2001), the main criterion used here for assessing threatened species. Under Criterion B only taxa with an Extent of Occurrence less than 20,000 km^2 and/or occurring at no more than 10 localities are considered threatened. Under Criterion A, widespread and common taxa may, in contrast, still be assessed as threatened, but only if their habitat, or some other indicator of their population size has been, or is projected to be, reduced by at least 30% in the space of three generations, so long as this does not exceed 100 years. Most of the taxa assessed in this work are restricted to evergreen forest. In assessing widespread and common taxa that extend into the Congo basin forests, there has been no difficulty in disbarring them as threatened under Criterion A since these forests have seen little reduction in coverage. Logging in this area is projected to increase in future but is still considered unlikely, at this stage, to surpass the limit above. However for widespread and common taxa assessed in this work that extend their range westwards, into West Africa, specifically Upper Guinea, it has been contended (Hawthorne *pers. comm.*) that over the last century more than 30% of the forest habitat in this area has probably been lost due to logging, followed by agriculture, almost any species of long-lived perennial, such as forest tree species, can be assessed under IUCN (2001) as Vulnerable under Criterion A. Hawthorne, in IUCN (2003, downloaded on 08 November 2004) has assessed several such taxa thus, almost always those of widespread economic importance, such as *Lophira alata* (local name: Azobé; uses: timber). Since the exploitation of these taxa for economic purposes has not usually been executed on a sustainable basis, this has brought additional threats to such taxa. Although these are, in the main, included and accepted in the taxon treatments that follow, other taxa in this checklist that have similar ranges and are also common have been assessed here as Least Concern. Admittedly, these do not, in the main, have the same levels of economic importance and associated threats as those listed by Hawthorne.

However, it is possible that in future these taxa might be reassessed as threatened if more precise past and present distributional data become available for them and these can be compared with accurate maps showing loss of forest habitat. Recently the ECOSYN project, based at Wageningen University, has compiled and made available, sets of maps that address the last issue for the Upper Guinea area.

Not all widespread taxa are common. *Neoschumannia kamerunensis* although known from Ivory Coast, Cameroon and CAR, is known from only five sites, and so is threatened. In contrast, some common taxa are localized, and not at all widespread. *Coffea montekupensis*, known only from Mt Kupe and the Bakossi Mts, is fairly common within submontane forest in this area, and seems set to be well protected in future, and is treated as Near Threatened (NT). *Cyphostemma mannii* is only known from Cameroon and Bioko, but is treated by us as being of Least Concern (LC) as it is common within its range.

Threatened and Near Threatened Taxa

Those taxa in the checklist that were not assessed as LC, were then checked for level of threat using IUCN (2001). A summary of the criteria used is provided in the eight tables that follow 'Using Criterion D' below. Most of these taxa were at least fairly narrowly endemic (restricted) in their distributions, e.g. endemic to Cameroon, or to SW Province, Cameroon and Nigeria, or to Cameroon and Gabon. Many of the submontane or montane taxa in this group were either restricted to the Cameroon Highlands or less commonly showed disjunct distributions, also occurring at Mt Nimba in Upper Guinea or the Crystal Mts of Gabon (see Vegetation and Phytogeography chapters). Except in very rare cases, such as *Kupea martinetugei*, Criterion C, which demands a knowledge of the number of individuals of a species, was not used, since these data were not available. Criterion E, which depends on quantitative analysis to calculate the probability of extinction over time, was also not used. Those taxa which did not qualify as threatened under the criteria above, but which were only known from, e.g. 11–20 localities, were treated as Near Threatened (NT). Frequently, supporting data on the distribution of these taxa occur in the checklist under the notes section of the relevant taxon treatment. NT taxa are not treated further in this Red Data chapter. The IUCN Category 'Data Deficient' (DD) is only used once, for *Memecylon griseo-violaceum* (Melastomataceae): see explanation under the notes section for this taxon in the checklist.

1) **Using Criterion B**

About two-thirds of the assessments were made using Criterion B (usually B2ab), since the nature of the data available to us for our taxa lends itself to this criterion. Knowledge of the populations and distributions of most tropical plant species is mainly dependent on the existence of herbarium specimens. This is because there are so many taxa, most of which are poorly known and have never been illustrated. For this reason observations based only on sight-records are particularly unreliable and so undesirable in plant surveys of diverse tropical forest. Exceptions are a family or genus specialist working with a monograph at hand, or a proficient tree spotter identifying timber tree species. In contrast surveys of birds and primates are not specimen dependent, since species diversity in these groups is comparatively low, and comprehensive, well-illustrated identification guides are available.

For the purpose of Criterion B we have taken herbarium specimens to represent 'locations'. Deciding whether two specimens from one general area represent one or two locations is open to interpretation, unless they are from the same individual plant. Generally in the case of several specimens labelled as being from one town e.g. 'Bipinde', or one forest reserve e.g. 'Bafut Ngemba F.R.', these have been interpreted as one 'location'. Where a protected area has been divided into several geographical subunits, as at Mt Cameroon (see Cable & Cheek 1998), and it is known that, say six specimens of a taxon occurring at Mt Cameroon fall into two such subunits, then this is treated as two locations. 'Area of occupancy' (AOO) and 'extent of occurrence' (EOO) have been measured by extrapolating from the number of locations at which a species is known. The grid cell size used for calculating area of occupancy has depended on the taxon concerned–figures of 1 or 2 km^2 have often been employed. Information on declines in habitat quality in Criterion B has been obtained from personal observations, sometimes supplemented by local observers, such as Chris Wild.

ii) **Using Criterion A**

Most of the remaining assessments, about one third of the total, were made using Criterion A. The lack of knowledge in many cases of the degree of loss of habitat or population for a taxon across its whole range meant that this criterion was not used more frequently. Taxa with ranges extending to Nigeria and or South Cameroon, or beyond were difficult to assess under Criterion A because estimates of loss in these areas were either imprecise or unavailable. In contrast, montane and submontane taxa of the Cameroon Highlands that extend to the Bamenda Highlands were relatively easy to assess under Criterion A thanks to a recent GIS study of forest loss over eight years in one part of the Bamenda Highlands (Moat in Cheek *et al.* 2000).

Thes data have been used to estimate that during the last century, over 30% of the forest habitat of montane species (above 2000 m alt.) has been lost in the Cameroon Highlands, so qualifying species with this range as vulnerable. Examples of such species are *Morella arborea* and *Schefflera mannii*, formerly treated as unthreatened (e.g. Cheek *et al.* 2000) because they are secure on Mt Cameroon.

iii) **Using Criterion D**

Assessments using Criterion D generally depend on a knowledge of the global numbers of mature individuals (D1). This knowledge is not available for plant taxa in Bakossi except in a few cases such as *Keetia bakossii* (CR D1). However VU D2 is a special case which allows threatened status for taxa on the basis simply of a low number of sites (five or less) or a restricted area of occupancy (20 km^2 or less), in the absence of other factors, such as declines or direct threats, see e.g. *Peucedanum kupense*. VU D2 has therefore been used in a substantial minority of cases in this Red Data chapter.

THE 2001 IUCN RED LIST CRITERIA (VER. 3.1) FOR CRITICALLY ENDANGERED, ENDANGERED AND VULNERABLE

A: Reduction in population size

A: Reduction in population size	Critically Endangered (CR)	Endangered (EN)	Vulnerable (VU)
A1: An observed, estimated, inferred or suspected population size reduction of …			
	≥ 90%	≥ 70%	≥ 50%
over the last 10 years or 3 generations whichever is the longer, where the causes of the reduction are: clearly reversible AND understood AND ceased, based on (and specifying) any of the following: (a) direct observation (b) an index of abundance appropriate for the taxon (c) a decline in area of occupancy, extent of occurrence and/or quality of habitat (d) actual or potential levels of exploitation (e) the effects of introduced taxa, hybridisation, pathogens, pollutants, competitors or parasites.			

A: Reduction in population size	Critically Endangered (CR)	Endangered (EN)	Vulnerable (VU)
A2: An observed, estimated, inferred or suspected population size reduction of …			
	≥ 80%	≥ 50%	≥ 30%
over the last 10 years or three generations, whichever is the longer, where the reduction or its causes may not have ceased OR may not be understood OR may not be reversible, based on (and specifying) any of the following: (a) direct observation (b) an index of abundance appropriate for the taxon (c) a decline in area of occupancy, extent of occurrence and/or quality of habitat (d) actual or potential levels of exploitation (e) the effects of introduced taxa, hybridisation, pathogens, pollutants, competitors or parasites.			

A: Reduction in population size	Critically Endangered (CR)	Endangered (EN)	Vulnerable (VU)
A3: A population size reduction of ….			
	$\geq$ 80%	$\geq$ 50%	$\geq$ 30%

projected or suspected to be met within the next 10 years or 3 generations whichever is the longer (up to a maximum of 100 years), based on (and specifying) any of the following:

(b) an index of abundance appropriate for the taxon
(c) a decline in area of occupancy, extent of occurrence and/or quality of habitat
(d) actual or potential levels of exploitation
(e) the effects of introduced taxa, hybridisation, pathogens, pollutants, competitors or parasites.

A: Reduction in population size	Critically Endangered (CR)	Endangered (EN)	Vulnerable (VU)
A 4: An observed, estimated, inferred, projected or suspected population size reduction of ..			
	$\geq$ 80%	$\geq$ 50%	$\geq$ 30%

over any period of.10 years or 3 generations whichever is longer (up to a maximum of 100 years), where the time period includes both the past and the future, and where the decline or its causes may not have ceased OR may not be understood OR may not be reversible, based on (and specifying) any of the following:

(a) direct observation
(b) an index of abundance appropriate for the taxon
(c) a decline in area of occupancy, extent of occurrence and/or quality of habitat
(d) actual or potential levels of exploitation
(e) the effects of introduced taxa, hybridisation, pathogens, pollutants, competitors or parasites.

B: Geographic range

B: Geographic range ...	Critically Endangered (CR)	Endangered (EN)	Vulnerable (VU)
... in the form of either B1 (extent of occurrence) OR B2 (area of occupancy) OR both:			
B1: Extent of occurrence estimated to be (km$^{2)}$, and estimates indicating any two of a-c:	< 100	< 5000	< 20000
B2: Area of occupancy estimated to be (km$^{2)}$, and estimates indicating any two of a-c:	< 10	< 500	< 2000
a: Severely fragmented or known to exist at	only 1 location	$\leq$5 locations	$\leq$10 locations

b. Continuing decline, observed, inferred or projected, in any of the following:
(i)   extent of occurrence
(ii)   area of occupancy
(iii)   area, extent and/or quality of habitat
(iv)   number of locations or subpopulations
(v)   number of mature individuals.

c. Extreme fluctuations in any of the following:
(i)   extent of occurrence
(ii)   area of occupancy
(iii)   number of locations or subpopulations
(iv)   number of mature individuals.

C: Population size

C: Population size ...	Critically Endangered (CR)	Endangered (EN)	Vulnerable (VU)
... estimated to number fewer than (mature individuals) and either	< 250	< 2500	< 10000
C1. An estimated continuing decline of at least	25%	20%	10%
in (years)	3	5	10
or (generations)	1	2	3
whichever is longer (up to a maximum of 100 years in the future) OR			
C2: A continuing decline, observed, projected, or inferred, in numbers of mature individuals AND at least one of the following: (a-b)			
a) Population structure in the form of one of :			
(i) no subpopulation estimated to contain more than (mature individuals), OR	50	250	1000
(ii) at least (%) of mature individuals are in one subpopulation	90%	95%	All (100%)
(b) Extreme fluctuations in number of mature individuals.			

D: Population size

D1: Population size	Critically Endangered (CR)	Endangered (EN)	Vulnerable (VU)
estimated to number fewer than (mature individuals)	< 50	< 250	< 1000

VU D2. Population with a very restricted area of occupancy (typically less than 20km^2) or number of locations (typically 5 or less) such that it is prone to the effects of human activities or stochastic events within a very short time period in an uncertain future, and is thus capable of becoming Critically Endangered or even Extinct in a very short time period

E: Quantitative analysis

E. Quantitative analysis	Critically Endangered (CR)	Endangered (EN)	Vulnerable (VU)
showing the probability of extinction in the wild is at least	50%	20%	10%
within (years)	10	20	100
or (generations)	3	5	-
whichever is the longer (up to a maximum of 100 years).			

Changes in IUCN criteria

It is to be hoped that there will be a moratorium on the almost annual changes in IUCN criteria which have been made in recent years. These frequent changes have made the work of assessors more difficult and have also reduced comparability of assessments made in different years. Assessments made in earlier Cameroonian checklists (Cable & Cheek 1998, Cheek *et al.* 2000) under IUCN criteria (1994) have been updated according to IUCN criteria (2001) where the taxa occur in the present checklist.

Red Data species by vegetation type

The Red Data species of Kupe, Mwanenguba and the Bakossi Mts are presented below under each vegetation type in which they occur. The vegetation types follow the classification used in the vegetation chapter. An individual species may occur in more than one vegetation type, e.g. *Stelechantha arcuata* is known from two specimens, one at Mile 15, Kumba–Mamfe road (c. 200 m alt.), the other near Kupe Village (c. 950 m alt.), so is listed below under both mapping unit 1 (lowland evergreen forest, 150–800 m alt.) and mapping unit 4, Block 2 (submontane forest 800–1900/2000 m alt.). In both cases the name is listed with a suffix to indicate that it occurs also at a higher (+) or lower (-) vegetation type respectively. Species known only from the boundary, or near the boundary of a vegetation type are noted with the altitude in brackets, e.g. *Pavetta rubentifolia* (800 m). Species known only from the submontane tongue of forest extending northwards from Mt Kupe are listed with their abbreviated location in brackets, e.g. *Oxyanthus montanus* (Ngombom.) indicating that this taxon is only known from Ngombombeng, whereas (Ngom.) indicates Ngomboaku, a little to the north, (Mwamb.) indicates Mwambong. Species are listed in the same order as in the main checklist, i.e. alphabetically by family (not indicated in the lists) and within family first by genus, then by species.

By far the most important vegetation type for threatened taxa is the submontane forest (140 Red Data taxa), with the lowland evergreen forest also contributing a very large number (77 Red Data taxa). The remaining vegetation types each also contribute a significant number of Red Data taxa, except submontane savanna-thicket which has none. The occurrence of a threatened taxon in anthropic vegetation (*Cyperus tenuiculmis* subsp. *mutica*) should not be taken as evidence for this vegetation being particularly important for conservation. Rather it is likely that this taxon occurs naturally in some other vegetation type–perhaps Esense gullies–and has strayed out into the roadside ditch in which it was collected. Two other Red Data species, *Polystachya kupense* and *Bulbophyllum kupense*, are also only known from anthropic vegetation (on a coffee bush and a mango tree, respectively) and have obviously arrived there by seed from the adjacent submontane forest and so are listed both there and under anthropic vegetation.

Total numbers of Red Data taxa per vegetation type are presented in the first table of the vegetation chapter.

MAPPING UNIT 1: Lowland evergreen forest; 150–800 m alt.

Dicliptera silvestris	VU	*Calochone acuminata* +	VU
Whitfieldia preussii	VU	*Chazaliella obanensis* +	VU
Cleistopholis staudtii +	VU	*Coffea bakossii* -	EN
Uvariodendron giganteum (& mu4)	VU	*Hymenocoleus glaber* +	VU
Pararistolochia goldieana	VU	*Mitriostigma barteri*	EN
Neoschumannia kamerunensis	CR	*Nauclea diderrichii*	VU
Salacia lehmbachii var. *pes-ranulae*	VU	*Pauridiantha divaricata*	VU
Diospyros kupensis +	VU	*Pauridiantha venusta* +	VU
Crotonogyne impedita	CR	*Pavetta brachycalyx* +	EN
Crotonogyne strigosa	VU	*Pseudosabicea batesii* +	VU
Crotonogyne zenkeri	VU	*Pseudosabicea medusula*	VU
Drypetes molunduana +	VU	*P. pedicellata*	VU
Drypetes preussii	VU	*Psychotria camerunensis*	VU
Drypetes staudtii	VU	*P. densinervia* +	VU
Thecacoris annobonae	EN	*P. minimicalyx*	CR
Garcinia kola	VU	*P. podocarpa*	VU
Pyrenacantha cordicula	EN	*Psydrax bridsoniana* (800–850m)	EN
Vitex lehmbachii +	VU	*Rutidea nigerica*	VU
Vitex yaundensis	EN	*Stelechantha arcuata* +	CR
Napoleonaea egertonii +	VU	*Tarenna baconoides* var. *baconoides*	VU
Eurypetalum unijugum	VU	*Tricalysia talbotii*	VU
Gossweilerodendron joveri	VU	*Oricia lecomteana* +	VU
Loesenera talbotii +	VU	*Deinbollia insignis*	VU
Angylocalyx talbotii	VU	*Placodiscus caudatus*	EN
Hugonia macrophylla	VU	*P. opacus* +	VU
Hugonia micans	VU	*Baillonella toxisperma*	VU
Strychnos elaeocarpa	VU	*Cola metallica*	CR
Strychnos staudtii +	VU	*C. praeacuta*	CR
Medusandra richardsiana	VU	*Dicranolepis polygaloides*	VU
Amphiblemma amoenum	EN	(to 870 m)	
Entandrophragma angolense +	VU	*Rinorea thomasii*	VU
Entandrophragma cylindricum	VU	*Afrothismia pachyantha*	CR
Entandrophragma utile	VU	*Afrothismia winkleri* +	CR
Dorstenia prorepens +	VU	*Aneilema silvaticum* +	VU
Ardisia koupensis +	EN	*Cyperus microcristatus*	CR
Eugenia fernandopoana	VU	*Marantochloa mildbraedii*	EN
Campylospermum letouzeyi	VU	*Ossiculum aurantiacum*	CR
Lophira alata	VU	*Kupea martinetugei* (c. 700 m)	CR
Belonophora ongensis	CR		

MAPPING UNIT 2: Semi-deciduous forest; c. 300 m alt.

Thyrosalacia racemosa	CR	*Microberlinia bisulcata*	CR
Hamilcoa zenkeri	VU	*Khaya ivorensis*	VU
Afzelia pachyloba	VU	*Tricalysia atherura* (& mu4)	VU

MAPPING UNIT 3: Freshwater swamp and lowland riverside vegetation; 150–800 m alt.

Crateranthus talbotii	VU
Hallea stipulosa	VU
Floscopa mannii	EN

MAPPING UNIT 4: Submontane forest; 800–1900(–2000) m alt.

Submontane Kupe, Mwanenguba, Bakossi Mts

Magnistipula conrauana	VU
Oncoba lophocarpa	VU
Diaphananthe polydactyla	VU

Submontane Mwanenguba & Bakossi

Angraecopsis tridens	VU

Submontane Mwanenguba & Mt Kupe

Isolona zenkeri	VU
Bidens mannii	VU
Peperomia kamerunana +	EN
Polystachya cooperi	EN

Submontane Bakossi & Mt Kupe

Acanthopale decempedalis	VU	*Lovoa trichilioides*	VU
Staurogyne bicolor	VU	*Triclisia macrophylla*	CR
Cyathula fernando-poensis	VU	*Dorstenia astyanactis*	EN
Xylopia africana	VU	*Ficus chlamydocarpa*	VU
Begonia adpressa	VU	subsp. *chlamydocarpa*	
Begonia bonus-henricus	VU	*Calochone acuminata*	VU
Begonia furfuracea	VU	*Coffea bakossii* -	EN
Begonia preussii	VU	*Hymenocoleus glaber* -	VU
Dielsantha galeopsoides	EN	*Pavetta owariensis*	EN
Cylicomorpha solmsii	VU	var. *satabiei*	
Mikaniopsis maitlandii	VU	*Vepris trifoliolata*	VU
Diospyros kupensis -	VU	*Quassia sanguinea*	VU
Pseudagrostistachys africana	VU	*Mapania ferruginea*	VU
Allanblackia gabonensis	VU	*Bulbophyllum bifarium*	VU
Napoleonaea egertonii -	VU	*Bulbophyllum nigericum*	VU
Loeesnera talbotii -	VU	*Bulbophyllum scaberulum*	VU
Anthocleista microphylla	VU	var. *fuerstenbergianum*	

Submontane Mt Mwanenguba (Block 1)

Begonia schaeferi	VU
Polystachya farinosa	EN
Polystachya geniculata	EN

Submontane Mt Kupe (Block 2)

Asystasia glandulifera	VU	*Tiliacora lehmbachii*	EN
Brachystephanus kupeensis	CR	*Dorstenia prorepens* -	VU
Brachystephanus longiflorus	VU	*Ardisia koupensis*	EN
Brachystephanus nimbae	VU	*Chassalia petitiana*	VU
Justicia camerunensis	VU	*Chazaliella obanensis* -	VU
Sclerochiton preussii	EN	*Cuviera talbotii* (& Ngom.) -	VU
Uvariodendron giganteum (& mu1)	VU	*Ixora foliosa* +	VU
Uvariopsis submontana	CR	*Oxyanthus montanus* (& Ngombom.)	VU
Uvariopsis vanderystii	EN	*Pauridiantha venusta* -	VU
Pararistolochia ceropegioides		*Pavetta brachycalyx* -	EN
(& Mwamb.) -	VU	*Psychotria densinervia* -	VU
Salacia mamba -	VU	*Stelechantha arcuata* -	CR
Dactyladenia johnstonei	CR	*Oricia lecomteana* -	VU
Crassocephalum bauchiense (& mu6)	VU	*Allophylus bullatus*	VU
Mikaniopsis vitalba	VU	*Deinbollia maxima* (900 m)	VU
Momordica enneaphylla (& Ngom.)	VU	*Afrothismia saingei*	CR
Dichapetalum altescandens	VU	*Afrothismia winkleri* -	CR
Croton aubrevillei	VU	*Aneilema silvaticum* (& Ngom.) -	VU
Drypetes molunduana -	VU	*Dracaena viridiflora* +	VU
Homalium hypoplasium	EN	*Angraecum pyriforme*	VU
Clerodendrum anomalum	EN	*Angraecum sanfordii*	EN
Vitex lehmbachii -	VU	*Bulbophyllum comatum*	EN
Anthocleista scandens	VU	*Bulbophyllum jaapii*	CR
Strychnos staudtii -	VU	*Bulbophyllum kupense*	CR
Tapinanthus preussii	VU	*Bulbophyllum pandanetorum*	EN
Memecylon dasyanthum	VU	*Disperis mildbraedii*	VU
Entandophragma angolense -	VU	*Genyorchis micropetala*	EN

Habenaria thomana	VU	*Polystachya cooperi*	EN
Polystachya bicalcarata	VU	*Polystachya kupensis*	CR
(& Mwamb.)		*Raphia regalis* (& Ngom., 850 m)	VU

Submontane Bakossi Mts (Block 3)

Afrofittonia silvestris	VU	*Calycosiphonia macrochlamys*	VU
Justicia leucoxiphus (& Ngomb.)	VU	*Keetia bakossii*	CR
Justicia orbicularis	VU	*Pavetta muiriana*	EN
Pseuderantherum dispermum	VU	*Pavetta rubentifolia* (800 m)	CR
Piptostigma calophyllum (& Ngomb.)	VU	*Pseudosabicea batesii* -	VU
Schefflera hierniana	EN	*Psychotria lanceifolia*	VU
Impatiens frithii	VU	*Rothmannia ebamutensis*	EN
Impatiens letouzeyi	EN	*Sabicea xanthotricha*	EN
Begonia pelargoniiflora	CR	*Trichostachys interrupta* (1000 m)	VU
Begonia prismatocarpa subsp.		*Placodiscus opacus* -	VU
delobata	VU	*Rinorea fausteana*	EN
Begonia pseudoviola (Ngomb. only)	VU	*Rhaphidophora pusilla*	VU
Drypetes magnistipula	EN	*Hypolytrum pseudomapanioides*	EN
Phyllanthus caesiifolius (& Ngomb.)	CR	*Hypolytrum subcompositus*	CR
Phyllanthus nyale	CR	*Scleria afroreflexa*	EN
Triclisia lanceolata	EN	*Sarcophrynium villosum*	EN
Dorstenia poinsettiifolia var.	EN	*Bulbophyllum teretifolium* (800 m)	CR
etugeana		*Habenaria batesii*	EN
Ficus tremula subsp. *kimuenzensis*	VU	*Habenaria weileriana*	VU
Cassipourea acuminata	EN	*Manniella cypripedioides*	VU

MAPPING UNIT 5: Submontane 'grassland' (savanna-thicket); 1000–1600 m alt.

No threatened species known.

MAPPING UNIT 6: Rock faces of cliffs and inselbergs; lithophytic communities; 600–1500 m alt.

Begonia duncan-thomasii	VU	*Nodonema lineatum*	VU
Bidens mannii	VU	*Amorphophallus preussii*	VU
Crassocephalum bauchiense (& mu4)	VU		

MAPPING UNIT 7: Rheophytic vegetation of upland streams and rivers ; 400–1450 m alt.

Brillantaisia lancifolia	VU	*Ledermanniella onanae*	EN
Achyranthes talbotii	VU	*Ledermanniella thalloidea*	VU
Plectranthus cataractarum	VU	*Macropodiella pellucida*	EN
Ledermanniella letouzeyi	EN	*Cyperus rheophytorum*	VU

MAPPING UNIT 8: Crater lakes; 1150–1950 m alt.

Secamone racemosa	VU
Impatiens letouzeyi	VU
Anthocleista microphylla	VU
Habenaria weileriana	VU

MAPPING UNIT 9: Montane forest and forest-grassland edge; 1900–2000(–2050) m alt.

Mwanenguba & Mt Kupe

Brachystephanus giganteus	VU
Peperomia kamerunana	EN
Ixora foliosa	VU

Mwanenguba

Impatiens sakerana	VU	*Disperis nitida*	EN
Morella arborea	VU	*Polystachya albescens*	CR
Chassalia laikomensis	CR	subsp. *manengouba*	

Mt Kupe

Pavetta kupensis	CR
Palisota preussiana	VU
Dracaena viridiflora	VU

Mwanenguba, Mt Kupe & Bakossi Mts

Schefflera mannii	VU
Begonia oxyanthera -	VU

Montane forest–grassland edge; 1900/2000 m to 2050 m alt.

Mwanenguba

Vernonia calvoana var. *calvoana*	VU

Mwanenguba & Mt Kupe

Dalbergia oligophylla (& mu10) -	EN
Pentas ledermannii	VU

Mt Kupe

No threatened species known.

<u>**MAPPING UNIT 10**</u>: **Montane grassland; 1900–2000(–2050) m alt.**

Mwanenguba

Crotalaria ledermannii	VU
Rhabdotosperma ledermannii	VU
Habenaria nigrescens	VU
Wahlenbergia ramosissima subsp.	
ramosissima	VU

Mwanenguba & Mt Kupe

Bulbostylis densa	
var. *cameroonensis*	VU

Mt Kupe

Peucedanum kupense	VU

<u>**MAPPING UNIT 11**</u>: **Anthropic vegetation; 150–2000 m alt.**

Cyperus tenuiculmis subsp. mutica	CR
Polystachya kupense	CR
Bulbophyllum kupense	CR

TAXON RED DATA TREATMENTS

Martin Cheek, Iain Darbyshire & Benedict John Pollard

Herbarium, Royal Botanic Gardens, Kew, Surrey, TW9 3AE, U.K.

DICOTYLEDONAE

ACANTHACEAE
(I.Darbyshire)

Acanthopale decempedalis C.B.Clarke
VU B2b(iii)c(iv)
Range: Nigeria, Cross River State (1 coll.); Bioko (5 coll.); Cameroon, SW Province: Kupe-Bakossi (9 coll.), Mt Cameroon (9 coll.), Bamenda Highlands (Mt Oku: 1, elsewhere: 3 coll.).
This taxon forms gregarious communities together with *Mimulopsis solmsii*, *Brachystephanus giganteus* (both Acanthaceae) and *Plectranthus insignis* (Labiatae), flowering together en masse on 7–9 year cycles followed by die-off and subsequent regeneration (Cheek *et al.* 2000). *M. solmsii* is a widespread Afromontane taxon but the remainder are restricted to the W Cameroon uplands phytochorion. Kupe-Bakossi appears the most important area for *A. decempedalis*, being recorded at 6 locations here. Following mass-flowering years, the dead woody stems of these taxa provide a source of firewood for local populations in the Bamenda Highlands.
Habitat: closed canopy submontane and montane forest understorey, occasionally in disturbed forest; (700–)1000–2300m alt.
Threats: much of the montane forest in SW Province remains relatively undisturbed, but clearance for agriculture is widespread in the Bamenda Highlands, threatening these populations. Furthermore, the cyclical mass-flowering habit of this species results in large fluctuations in mature populations, making it susceptible to short-term stochastic change, for example local fire events or landslides which could adversely affect seedling populations.
Management suggestions: current population sizes in terms of both number of individuals and area of occupancy should be assessed as this is not clear from the specimen data. A better understanding of the flowering cycle of this community should be gained, including population fluctuations between cycles. Bamenda Highlands populations outside protected areas may provide valuable information on how immature populations respond to increased anthropogenic pressures.

Afrofittonia silvestris Lindau
VU A2c
Range: Nigeria, Cross River State (5 coll.); Bioko (1 coll.); Cameroon, SW Province: Bakossi (2 coll.), Korup (7 coll.), Mt Cameroon (14 coll.), Banyang-Mbo (1 coll.).
Predominantly a lowland forest taxon, it is recorded as 'locally abundant' in Korup and widespread on Mt Cameroon (Cable & Cheek 1998, where it was first assessed as VU), but appears rare in Bakossi and absent from Mt Kupe.
Habitat: lowland closed-canopy forest, often growing in deep shade on forest floor; 0–500m alt. Rarely in mid-elevation forest; 900–1400m alt. (Bakossi).
Threats: continued forest clearance. However, higher-altitude sites discovered in Bakossi are less threatened.
Management suggestions: the high altitude sites in Bakossi may prove important for the conservation of this species and so should be assessed more fully; atypically for Acanthaceae taxa, this species is readily identifiable when sterile, aiding population assessments.

Asystasia glandulifera Lindau
VU B2ab(iii)
Range: Cameroon, SW Province: Kupe-Bakossi (6 coll.), E Province: Bertoua and Betare Oya (2 coll.), W Province: Bafoussam-Foumbot (1 coll.), unlocated (2 coll.).
Previously known from five scattered locations, inventory work in Kupe-Bakossi has considerably increased the known range, with three additional sites.
Habitat: mid-elevation to submontane forest, preferring forest margins and clearances and lingering in farmbush; 800–1550m alt.
Threats: intensification of land use at lower elevations, particularly the conversion of lower forest margins to agriculture; this is a significant threat around larger habitations such as Nyasoso in Bakossi, where this taxon has been collected four times, and around Bafoussam, a very large town in W Province around which almost all natural habitat has long since been anthropogenically modified.

Management suggestions: monitoring of the populations above Nyasoso should provide further information on the habitat requirements of this species, particularly the extent to which it tolerates anthropogenic disturbance. Active protection of threatened sites may be required.

Brachystephanus giganteus Champl. ined.
VU B2b(iii)c(iv)
Range: Bioko (2 coll.); W Cameroon: Kupe-Bakossi (2 coll.), Rumpi Hills (1 coll.), Mt Cameroon (13 coll.), Bamenda Highlands (Mt Oku: 3 coll., Bafut Ngemba: 1 coll.).
A species of the gregarious mass-flowering herbaceous community of upland forest, it is more restricted in altitudinal range than *Acanthopale decempedalis*. It appears locally common on Mt Cameroon but is scarce elsewhere, being restricted to Mwanenguba and on Mt Kupe in the current checklist area. Both are pre-1980 collections, and this species was not recorded during extensive inventory work at these sites in the 1990s, perhaps because these surveys did not coincide with mass-flowering years.
Previously named *Oreacanthus mannii* Benth., this taxon has been reassigned to *Brachystephanus* by D. Champluvier during her monograph of this genus, currently in preparation.
Habitat: closed canopy montane forest; (1300–)2000–2600m alt.
Threats: as for *Acanthopale decempedalis*.
Management suggestions: as for *Acanthopale decempedalis*. In addition, a survey of this taxon should be carried out during the next mass-flowering year on Mt Kupe and Mwanenguba.

Brachystephanus kupeensis Champl. ined.
CR B1ab(iii)
Range: Cameroon, SW Province: Mt Kupe, above Nyasoso (2 coll.).
Unknown until the inventory work of the 1990s, this taxon has been collected at two sites within 1km of each other from the lower slopes of Mt Kupe near Nyasoso; no record of the abundance of this taxon at these sites has been made.
Habitat: submontane forest and secondary regrowth forest; 900–1100m alt.
Threats: the continued expansion of Nyasoso town has resulted in increased agricultural activity on Mt Kupe up to approx. 1000m alt., with further encroachment inevitable in the near future.
Management suggestions: it is imperative that the collection sites are revisited to assess current population levels; care should be taken to separate this taxon from similar *Brachystephanus* species. A better understanding of its habitat requirements should be gained, in particular the extent to which it tolerates disturbance. Notifying the Nyasoso population of its existence and rarity, perhaps through issuing a conservation poster on this species to local schools and community halls, may help in the protection of these sites. Other potential sites on the mountain should be surveyed for new populations.

Brachystephanus longiflorus Lindau
VU B2ab(iii)
Range: Nigeria: Ogoja (1 coll.); Bioko (3 coll.); Cameroon, SW Province: Mt Kupe (2 coll.), Mt Cameroon (2 coll.).
The paucity of collections from Cameroon, despite extensive survey work, and its absence from the Bakossi Mountains, the Rumpi Hills and the Bamenda Highlands, indicate that this species is highly restricted in range.
Habitat: submontane and montane forest undergrowth; (1000–)1300–1550m alt.
Threats: forest clearance threatens the Nigerian population.
Management suggestions: so long as forest remains largely undisturbed above 1000m, no intervention is required in Bioko or Cameroon; confirmation of its continued existence in Nigeria is required.

Brachystephanus nimbae Heine
VU B2ab(iii)
Range: Guinea (Conakry) and Liberia: Nimba Mts (4 coll.); Ivory Coast: Mt Tonkoui (1 coll.); Ghana (3 coll. at 2 locations); Cameroon, SW Province: Mt Kupe (3 coll.).
First described from the Nimba Mountains in Upper Guinea, this species has a highly disjunct distribution on isolated mountain ranges across the Guinean phytochorion. The three collections from above Nyasoso on Mt Kupe represent a major increase in the extent of occurrence for this species; it has strangely not been found in other montane areas of Cameroon despite extensive collection in seemingly suitable sites.
Habitat: mid-elevation forest, often in wet rocky areas; 900–1100m alt.
Threats: forest clearance for agriculture at elevations less than 1000m represents a significant threat to the Cameroon population. The montane forests of Ghana and Ivory Coast are similarly threatened. Extensive forest remains on the Nimba Mountains.

Management suggestions: survey of other likely sites in W Africa, e.g. Mt Nlonako in W Cameroon. Survey of current known populations may provide a greater understanding of its specific habitat requirements and so where it is most likely to be found in future.

Brillantaisia lancifolia Lindau
VU B2ab(iii)

Range: Nigeria, Cross River State (2 coll.); Cameroon, SW Province: Bakossi (3 coll.), Mt Cameroon (2 coll.); Rio Muni (1 coll.); Gabon (2 coll.).

This taxon is apparently obligately rheophytic. It is easily recognised from other species in the genus by its narrowly elliptic leaves. Although having a large extent of occurrence, this species is rare across its range, being known from only 10 locations, though recorded as 'locally abundant' at both Ndile waterfall, Ngomboaku (*Mackinder* 335) and on the Onge River (*Watts* 930).

Habitat: rocky river margins in low altitude to submontane forest; 0–1350m alt.

Threats: forest clearance in upstream sites may result in loss of populations through increased run-off causing flooding or silting of its favoured habitat. This is a potential threat at, for example, the Mungo River in Bakossi, as the level of upriver activity is high. Increased use of rivers by humans, for example providing water for livestock, may cause excessive disturbance.

Management suggestions: monitoring of populations where upstream disturbance is occurring to assess the sensitivity of this species, for example along the Mungo River. Prevention of forest clearance in catchments where this species is recorded.

Dicliptera silvestris Lindau
VU A2c

Range: C.A.R.: St. Floris N.P., Yalinga, Boukoko (1 coll. each); Cameroon, SW Province: Nguila near Nguti (1 coll.), Mt Kupe (1 coll.), E Province: Dengdeng to Gojoum (1 coll.).

This taxon appears rare throughout its range, with only 1 collection made during the extensive inventory work at Mt Kupe. It has not yet been recorded in the Bakossi Mountains, despite being recorded near Nguti to the west. Its apparent absence from Central Province is also unusual in view of the populations near Dengdeng and in C.A.R. to the east.

Habitat: open low altitude forest; 0–750m alt.

Threats: as a low altitude species, it is vulnerable to habitat loss through conversion to plantations or through logging activity.

Management suggestions: survey work in potential sites in central and eastern Cameroon may reveal further populations of this easily overlooked species; care should be taken to separate it from other similar *Dicliptera* species, such as *D. umbellata* and *D. laxata*. The habitat requirements of this species require greater understanding; at Mt Kupe it appears tolerant of some disturbance.

Justicia camerunensis (Heine) I.Darbysh. *comb. nov.*
VU A2c

Range: Nigeria: Gembu Dist. (1 coll.); Cameroon, SW Province: Mt Kupe (8 coll.), Mt Cameroon (1 coll.), Kumba (1 coll.), NW Province: Bafut-Wum (1 coll.), Central Province: nr. Yaoundé (2 coll.), S Province: Obang (1 coll.), E Province (2 coll.), unlocated (2 coll.).

In Cameroon this taxon has a patchy distribution, being absent from several seemingly suitable areas including the Bakossi Mountains and much of Mt Cameroon.

Habitat: understorey of closed canopy mid-elevation forest; (600–)900–1200m alt.

Threats: throughout its range, it is threatened by extensive forest clearance. The Northwest Province population is highly threatened as closed-canopy forest is scarce there today; it was noted in the collection of 1975 that the forest patch in which it was found was 'now in exploitation by local people with handsaws' (*Leeuwenberg* 8648); this site is therefore likely to be lost. At Mt Kupe, it was recorded most often above Nyasoso, at around 1000m alt., where agricultural encroachment is causing significant losses of forest. Sites around Yaoundé are also likely to have been lost.

Management suggestions: protection of existing mid-elevation forest where possible.

Justicia leucoxiphus Vollesen, Cheek & Ghogue
EN B1ab(iii)

Range: Cameroon, SW Province: Bakossi Mountains and Rumpi Hills (10 coll.).

First collected in the 1980s, this species is now well known from the Bakossi Mountains following extensive survey work in the 1990s. It is recorded from 5 separate locations; Kodmin, Mejelet, Mwambong-Jide, Ngomboaku and Mbu Bolomi (Rumpi Hills), having an extent of occurrence of less than 5,000km². It is locally not uncommon,

Nhon Mbwog Richard Mambo NTOKO, Late Paramount Chief of Bakossi, champion of forest conservation in Bakossi. Born 23 April 1924, Died 23 June 2001.

Plate 1

Mt Kupe, submontane forest, view of summit area from Nyasoso, Oct. 1998. BJP

Summit grassland of Mt Kupe; second peak, view towards Dibombe Valley, Nov. 1995. MC

Cindercone (foreground with banana plantation), from Loum – Kumba road, Oct. 1998. BJP

Submontane savanna thicket, 1600m alt., Mt Loh, Jan. 1998. MC

Plate 2

Mwambong Crater Lake. Note botanist at left-middle of picture. Mwambong, Oct. 1998. BJP

Mwanenguba female crater lake in dry season, grassland with montane forest, 2 Feb. 1995. MC

Mwanenguba caldera: montane grassland with *Sporobolus africana*. BJP

Kupe Rock in dry season, inselberg with forest cap, Jan. 1995. MC

Plate 3

A. Bime Rock, view from Mwendolengo, Nov. 2001. BJP
B. View of Mbu valley from top of Bime Rock, looking toward Enyandong, Nov. 2001. BJP
C. Inselberg: Nyale Rock, Nov. 1998. MC
D. Lake Edib, with Dr. Satabie (former head of YA) in the middle foreground, Feb. 1998. MC

Plate 4

A. *Staurogyne bicolor* (*Acanthaceae*), Ngomboaku, Dec. 1999, *Cheek* 10363. MC
B. *Justicia leucoxiphus* (*Acanthaceae*), Kodmin, Nov. 1998, sine coll. MC
C. *Achyranthes talbotii* (*Amaranthaceae*), Baseng, Dec. 1999, *Cheek* 10401. MC
D. *Piptostigma calophyllum* (*Annonaceae*), Edib – Kodmin, Feb. 1998, *Cheek* 9177. MC
E. *Uvariopsis submontana* (*Annonaceae*), Kupe Village, Jan. 1995, *Cheek* 7131. MC
F. *Xylopia africana* (*Annonaceae*), Kodmin, Feb. 1998, *Cheek* 9192. MC

Plate 5

A. *Pararistolochia ceropegioides* (*Aristolochiaceae*), Kupe Village, Jan. 1995, *Cheek* 7002. MC

B. *Neoschumannia kamerunensis* (*Asclepiadaceae*), Kupe Village, Nov. 1995, *Etuge* 1443, MC

C. *Secamone racemosa* (*Asclepiadaceae*), Lake Edib, Feb. 1998, *Cheek* 9162. MC

D. *Impatiens frithii* (*Balsaminaceae*), Kodmin, Oct. 1998, *Cheek* 9526. BJP

E. *Impatiens letouzeyi* (*Balsaminaceae*), Kodmin, Nov. 1998, *sine coll.* BJP

F. *Begonia adpressa* (*Begoniaceae*), Kodmin, Oct. 1998, *sine coll.*, photo. det. Sosef. BJP

Plate 6

A. *Begonia pseudoviola* (*Begoniaceae*), Ngomboaku, Dec. 1999, *Cheek* 10319. MC
B. *Dielsantha galeopsoides* (*Campanulaceae*), Manehas F.R., Oct. 1998, *Etuge* 4355. MC
C. *Cylicomorpha solmsii* (*Caricaceae*), Nyasoso, Feb. 1995, *Cheek* 7299. MC
D. *Magnistipula conrauana* (*Chrysobalanaceae*), Kupe Village, Nov. 1995, *Cheek* 7844. MC
E. *Phyllanthus caesiifolius* (*Euphorbiaceae*), Kodmin – Mwanzum, Nov. 1998, *Cheek* 9636. MC
F. *Dovyalis* sp. nov. (*Flacourtiaceae*), Kupe Village, Jan. 1995, *Cheek* 7049. MC

Plate 7

A. *Oncoba lophocarpa* (*Flacourtiaceae*): flagelliflory, Kupe Rock, Jan. 1995, *Cheek* 7188. MC
B. *Oncoba ovalis* (*Flacourtiaceae*), Kodmin, Nov. 1996, *Cheek* 8911. MC
C. *Nodonema lineatum* (*Gesneriaceae*), Bime Rock, Nov. 2001, *Etuge* 4520r. BJP
D. *Allanblackia gabonensis* (*Guttiferae*), Kodmin, Feb. 1998, *Cheek* 9092. MC
E. *Garcinia kola* (*Guttiferae*), Kupe Village, Jan. 1995, *Cheek* 6089. MC
F. *Plectranthus cataractarum* (*Labiatae*), Kodmin, Nov. 1998, *Pollard* 207. BJP

Plate 8

A. *Napoleonaea egertonii* (*Lecythidaceae*), Manehas F.R., Oct. 1998, *Pollard* 125. BJP
B. *Loesenera talbotii* (*Leguminosae-Caesalpinioideae*), Kupe Village, Nov. 1995, *Cheek* 7721. MC
C. *Zenkerella citrina* (*Leguminosae-Caesalpinioideae*), Mt Kupe, Nov. 1995, *Cheek* 7610. MC
D. *Newtonia duncanthomasii* (*Leguminosae-Mimosoideae*), Nyale, Nov. 1998, *Cheek* 9691. MC
E. *Millettia macrophylla* (*Leguminosae-Papilionoideae*), Loum F.R., Oct. 1998, *Cheek* 9297. MC
F. *Dissotis* sp. 1 (*Melastomataceae*), Mt Kupe summit, Oct. 1995, *Sidwell* 428. MC

Plate 9

A. *Memecylon* sp. nov. 2 (*Melastomataceae*), Nyandong, March 2003, *Stone* 2449. MC
B. *Tiliacora lehmbachii* (*Menispermaceae*), Kupe Village, Nov. 1995, *Cheek* 7809. MC
C. *Dorstenia astyanactis* (*Moraceae*), Ngomboaku, Dec. 1999, *Cheek* 10340. MC
D. *Dorstenia poinsettiifolia* var. *etugeana* (*Moraceae*), Nyale, Nov. 1998, *Etuge* 4462. MC
E. *Eugenia* sp. 1 (*Myrtaceae*), Kupe Village, Nov. 1995, *Cheek* 7868. MC
F. *Octoknema* sp. nov. (*Olacaceae*), Ngomboaku, Dec. 1999, *Cheek* 10321. MC

Plate 10

A. *Ledermanniella letouzeyi* (*Podostemaceae*), Mwambong, Nov. 1998, *Cheek* 9706. MC

B. *Ledermanniella onanae* (*Podostemaceae*), Mwambong, Nov. 1998, *Cheek* 9703. MC

C. *Macropodiella pellucida* (*Podostemaceae*), Mwambong, Feb. 1998, *Cheek* 9118. MC

D. *Bertiera* sp. nov. (*Rubiaceae*), Edib – Kodmin, Feb. 1998, *Cheek* 9178. MC

E. *Calochone acuminata* (*Rubiaceae*), Nyasoso, Jan. 1995, *Cheek* 7290b. MC

F. *Calycosiphonia* sp. A. (*Rubiaceae*), Kupe Village, Nov. 1995, *Cheek* 7882. MC

Plate 11

A. *Chassalia aff. laikomensis* sp. B (*Rubiaceae*), Kodmin, Nov. 1998, *Cheek* 9592. MC

B. *Coffea bakossii* (*Rubiaceae*), Ngomboaku, Dec. 1999, *Gosline* 260. MC

C. *Coffea montekupensis* (*Rubiaceae*), Kodmin – Mwanzum, Nov. 1998, *Cheek* 9686. MC

D. *Hymenocoleus glaber* (*Rubiaceae*), Kodmin, Jan. 1998, *Cheek* 8962. MC

E. *Ixora* sp. A. (*Rubiaceae*), Kupe Village, Nov. 1995, *Cheek* 7786. MC

F. *Pentaloncha sp. nov.* (*Rubiaceae*), Kupe Rock, Jan. 1995, *Cheek* 7183. MC

Plate 12

A. *Pentas ledermannii* (*Rubiaceae*), Kupe summit, Oct. 1995, *Cheek* 7573. MC
B. *Psychotria podocarpa* (*Rubiaceae*), Nyale, Nov. 1998, *Cheek* 9644. MC
C. *Trichostachys interrupta* (*Rubiaceae*), Nyale, Nov. 1998, *Etuge* 4461. MC
D. *Allophylus* sp. 1 (*Sapindaceae*), Nyasoso, Nov. 1998, *Gosline* 137. MC
E. *Deinbollia* sp. 1 (*Sapindaceae*), Kodmin, Nov. 1998, *Pollard* 234. BJP
F. *Pancovia* sp. 1 (*Sapindaceae*), Kupe Village, Jan. 1995, *Cheek* 6045. MC

Plate 13

A. *Placodiscus* sp. 1 (*Sapindaceae*), Kodmin – Edib, Feb. 1998, *Cheek* 9176. MC

B. *Rhaptopetalum geophylax* (*Scytopetalaceae*), Kodmin – Edib, Feb. 1998, *Cheek* 9175. MC

C. *Quassia sanguinea* (*Simaroubaceae*), Kodmin, Jan. 1998, *Cheek* 8873. MC

D. *Cola sp. nov.* 'etugei' (*Sterculiaceae*), Nyasoso, Oct. 1998, *Cheek* 9518. MC

E. *Cola sp. nov.* 'kodminensis' (*Sterculiaceae*), Mwanzum – Kodmin, Nov. 1998, *Cheek* 9682. MC

F. *Rinorea thomasii* (*Violaceae*), Baseng, Dec. 1999, *Cheek* 10413. MC

Plate 14

A. *Amorphophallus staudtii* (*Araceae*), Kodmin, Jan. 1998, *Cheek* 9061. MC
B. *Afrothismia pachyantha* (*Burmanniaceae*), Kupe Village, Nov. 1995, *Williams s.n.* MC
C. *Afrothismia winkleri* (*Burmanniaceae*), Kupe Village, May 1996, *Cheek* 8354. MC
D. *Gymnosiphon* sp. A (*Burmanniaceae*), Kodmin – Mwanzum, Jan. 1998, Cheek 9635. MC
E. *Costus* sp. A (*Costaceae*), Kupe Village, Jan. 1995, *Cheek* 7111. MC
F. *Bulbostylis densa* var. *cameroonensis* (*Cyperaceae*), Mt Kupe summit, Oct. 1995, *Cheek* 7575. MC

Plate 15

A. *Bulbophyllum bifarium* (*Orchidaceae*), Kodmin, Nov. 1998, *Pollard* 214. BJP

B. *Diaphananthe polydactyla* (*Orchidaceae*), Kodmin, Nov. 1998, *Pollard* 172. BJP

C. *Manniella cypripedioides* (*Orchidaceae*), Mwambong, Oct. 1998, *sine coll.* MC

D. *Ossiculum aurantiacum* (*Orchidaceae*), Mungo R.F.R. (photo ex cult. WAG), Dec. 1980, *Beentje* 1460A. van der Laan

E. *Raphia regalis* (*Palmae*), Ngomboaku, Dec. 1999, *Cheek* 10310. MC

F. *Kupea martinetugei* (*Triuridaceae*) male inflorescence, Kupe Village, Dec. 1999, *Cheek* 10225. MC

Plate 16

though its range does not extend to Mt Kupe or Mwanenguba in the east. The assessment above is taken from that by Cheek in the protologue (Vollesen *et al.* 2004).

Habitat: closed canopy forest understorey, rarely secondary forest; 900–1500m alt.

Threats: an inferred decline in habitat quality due to illegal logging activity which begun within its area of distribution in 2002, but which has recently reduced following efforts from local conservation groups and local communities.

Management suggestions: as a striking and easily recognisable species, *J. leucoxiphos* could contribute to the promotion of community-based forest protection. Conservation posters on the rarity and uniqueness of this species could be distributed to schools and community centres, encouraging local communities to protect areas where it is found. Further inventory work in the Bakossi Mountains and Rumpi Hills is likely to uncover new locations.

Justicia orbicularis (Lindau) V.A.W.Graham
VU A2c

Range: Nigeria: Cross River State (3 coll.); Bioko (1 coll.); Cameroon, SW Province: Bakossi (3 coll.), Kumba (1 coll.), Central Province: nr. Yaoundé (2 coll.), S Province: Ebolowa-Amban (1 coll.), unlocated (1 coll.).

A highly distinctive species known from just 12 locations. Populations appear disjunct, being absent in several seemingly suitable areas. In Bakossi, it is only known in the west, though further exploration of the lowlands of southern Bakossi may reveal further populations. It is also absent from the lowlands around Mt Cameroon, and has not been recorded in extensively collected forest areas around Bipinde and Kribi in S Province, Cameroon.

Habitat: lowland forest, often along streams; 0–700m alt.

Threats: throughout its range, deforestation has been extensive in the lowlands, and this continues outside protected areas.

Management suggestions: as the species is readily identifiable, population census data should be possible in protected areas, and an assessment of its habitat requirements made. Further inventory work in suitable lowland sites, such as the Loum Forest Reserve, may provide additional sites.

Pseuderanthemum dispersum Milne-Redh.
VU B2ab(iii)

Range: Nigeria: Cross River State (3 coll.); Cameroon, SW Province: Bakossi (1 coll.), Mt Cameroon (1 coll.), Babenga nr. Limbe (1 coll.), S Province: Bipinde (1 coll.).

This robust herb or shrub appears rare throughout its range, and was recorded only once during the extensive inventory work in SW Province, during the 1980s and 1990s: a collection by D.W.Thomas at Bangem in north Bakossi. It has never been recorded in central or southern Bakossi or Mt Kupe and has not recently been found on Mt Cameroon.

Habitat: undergrowth of low- to mid-elevation forest; 0–1000m alt.

Threats: a continued decline in habitat is inferred by clearance of extensive areas of lowland and mid-elevation forest throughout its range for plantation agriculture and timber.

Management suggestions: attempts to rediscover this species on Mt Cameroon should be made. The collection from Bangem should be verified, as the specimen was not available at Kew.

Sclerochiton preussii (Lindau) C.B.Clarke
EN B1ab(v)

Range: Nigeria: Obudu Plateau (1 coll.); Cameroon, SW Province: Mt Cameroon (2 coll.), Mt Kupe (3 coll.).

First collected 'near Buea' in Nov. 1891, it was rediscovered on Mt Cameroon 101 years later in the saddle of Mt Etinde-Mt Cameroon. It was also found in the 1960s at Obudu, Nigeria. The Mt Kupe collections from the 1990s, from above Kupe Village and Nyasoso, represent a significant expansion in the area of occupancy, but it has not been recorded from elsewhere in Bakossi. The assessment of EN, first made in Cable & Cheek (1998), is maintained here.

Habitat: an understorey shrub of closed-canopy submontane and montane forest; 1100–1400m alt.

Threats: stems cut for trap 'springing sticks' on Mt Kupe (Cheek, *pers. obs.*); this is likely to have an impact upon the population of mature individuals at the two Kupe sites.

Management suggestions: as stated in Cable & Cheek (1998), an evaluation of the threat posed by exploitation for trapping is advised, and local trappers should be encouraged to use alternative, more common taxa.

Staurogyne bicolor (Mildbr.) Champl.
VU B2ab(iii)

Range: Cameroon, SW Province: Kupe-Bakossi (11 coll.), Mbu-Bakundu (1 coll.), S Province: 15km north of Kribi (1 coll.), Bipinde-Ebolowa (1 coll.).

Previously known from only three disjunct locations, the botanical inventory work of the 1980s–90s identified the Kupe-Bakossi region as highly important for this species, with six new locations from Bangem in the north to Nyale in the west and Mt Kupe in the east. It has not so far been collected on the slopes of Mt Cameroon or in Nigeria or Bioko.

Habitat: an understorey herb of mid-elevation to submontane forest, often near water; 700–1400m alt.

Threats: as this species extends to forest below 1000m, it is threatened by widespread deforestation.

Management suggestions: protection of the remnant mid-elevation forest in Bakossi appears imperative for the survival of this species. Attempts should also be made to rediscover the disjunct populations in S Province.

Whitfieldia preussii (Lindau) C.B.Clarke
VU B2ab(iii)

Range: Bioko (1 coll.); Cameroon, SW Province: Bakossi (1 coll.), Matoh (1 coll.), Kumba (1 coll.), Barombi (1 coll.), S Province: Bipinde (3 coll.).

This taxon appears rare throughout its range, although at Malabo-Luba, Bioko in 1989 it was recorded as 'frecuente en colonias reducidas'. In Bakossi, it is known only from the Mungo River Forest Reserve. The Bipinde collections are from the early 1900s; no subsequent records are known from S Province.

Habitat: an understorey shrub of lowland forest, rarely secondary scrub; 0–350m alt.

Threats: forest clearance throughout its range. However, at Kumba this species was collected from a road verge, suggesting that it can tolerate some disturbance.

Management suggestions: further data are needed on the range and habitat requirements of this species, particularly the extent to which it can tolerate disturbance. The continued existence at Bipinde should be confirmed during future botanical inventory work at this site.

AMARANTHACEAE
(I.Darbyshire)

Achyranthes talbotii Hutch. & Dalziel
VU B2ab(iii)

Range: Nigeria: Cross River State (2 coll.); Cameroon, SW Province: Mt Cameroon (8 coll.), S Bakundu F.R. (1 coll.), Bakossi (2 coll.), Banyang-Mbo (1 coll.), NW Province: Bu (1 coll.), Littoral Province: nr. Nkongjok (1 coll.), nr. Yingui (1 coll.).

This taxon was treated in Cable and Cheek (1998) as endangered, being then recorded from only three sites, the type collection from Oban District, Nigeria and the Mt Cameroon collections on the Onge and Joke rivers. However, with the recent collections from Baseng in the Jide Valley, Nyandong in western Bakossi, the Banyang-Mbo Wildlife Sanctuary and near Bu, NW Province, together with the records of this taxon from two rivers in Littoral Province previously omitted, it is now known from 10 locations and is therefore downgraded to VU.

Habitat: fast-flowing rivers, growing half submerged, rooting among boulders; often around waterfalls; 0–750m alt.

Threats: the proposed conversion of lowland forest around Mt Cameroon to plantation is likely to threaten the plant communities of the rivers that drain the area, such as the Onge, an important site for this taxon, through flooding and excessive silting of their habitat. Illegal logging for timber in lowland Bakossi is likely to have a similar impact.

Management suggestions: botanical surveys of other suitable sites in SW and Littoral Provinces should reveal further populations of this taxon; monitoring of the riparian plant communities in response to increased upstream run-off should be carried out.

Cyathula fernando-poensis Suess. & Friedrich
VU D2

Range: Bioko (3 coll.); Cameroon, SW Province: Mt Kupe (1 coll.), Bakossi Mts (1 coll.).

This taxon was previously known from only three collections made in 1951 at El Pico, Bioko by A.S. Boughey, with no subsequent collections from the island. The collection on Mt Kupe above Nyasoso (*Lane* 242) represented the first record for continental Africa; this was followed by discovery of a second site on the trail from Kodmin to Mwanzum (Muahunzum) in north Bakossi. There are currently no records of abundance of this species at its three known localities. The paucity of collections of this taxon is perhaps due in part to the inconspicuous nature of this species, although it is clearly rare throughout its range.

Habitat: understorey of closed-canopy submontane and montane forest; 1330–2100m alt.

Threats: the forests of Bioko and west Cameroon over 1000m alt. are currently under only minimal anthropogenic pressure. However, the low number of locations, and their apparent isolation, render this species vulnerable to stochastic events such as localised landslide or overgrazing by native mammals. It is possible that following the exhaustion of lowland forest resources, there will be increased human encroachment into the montane forests in future, unless protected by law.

Management suggestions: a survey of the current known populations should be undertaken, including its rediscovery on Bioko. This species should be sought during any future botanical inventory work in additional montane sites such as the highest parts of the Rumpi Hills.

ANACARDIACEAE
(M.Cheek)

Two species of this family, both occurring in Bakossi, were previously assessed as CR on Mt Cameroon, namely *Trichoscypha bijuga* Engl. and *T. camerunensis* Engl. (Cable & Cheek 1998). Further, *Sorindeia mildbraedii* was listed as VU by WCMC (1997, cited in IUCN 2003, downloaded 8th November 2004). Both genera have at last been revised, as has been desperately needed for many years, by F.J.Breteler (see notes under this family in the checklist). As a result of the broader species concepts he has adopted (numbers of previously accepted species having been reduced to synonymy), while the first maintains its name by virtue of nomenclatural priority, the second is now regarded as a synonym of *T. laxiflora* Engl. and the third a synonym of *S. winkleri*. Further, the geographic ranges of all three taxa have been expanded due to the change in species concept (see checklist part), so these taxa are downgraded here, the first and third to NT, the second to LC.

ANNONACEAE
(M.Cheek)

Cleistopholis staudtii Engl. & Diels
VU B2ab(iii)
Range: W Nigeria; C.A.R.; Gabon (all 1 coll.); Cameroon: Korup (1 coll.), Kumba (2 coll.), Douala-Edea, Kribi (all 1 coll.), Bipinde (several coll.), Bakossi (3 coll.).
This 30m tree was first described from Kumba. Closely related to *C. patens*, it is widespread but localised. Its absence from the intensively collected Mt Cameroon checklist area (Cable & Cheek 1998) shows that this observation is not merely an artefact of under-collection. Through some parts of its range (Nigeria, C.A.R., Kumba), forest loss has been extensive since the last collections were made and the species may no longer occur at those places.
Habitat: lowland evergreen forest; 500–950m alt.
Threats: clearance of forest for timber, followed by agriculture.
Management suggestions: *Cleistopholis staudtii* is probably most secure at Korup National Park but its protection in other parts of its range is advisable in order to conserve a good tranche of its genetic diversity. The tree or trees at Kupe Village in Bakossi should be examined to see if protection there is supportable.

Isolona zenkeri Engl.
VU B2ab(iii)
Range: Cameroon: Kupe-Bakossi (3 coll. at 3 sites), Bipinde (several coll.), Kribi (1 coll.); Gabon (4 coll. from 2 sites).
Described from Zenker's early 20th Century material from Bipinde (numerous collections), the only other record we have for South Province is *Bos* 4866, 12 km from Kribi (1969). In neighbouring Gabon two sites are known: Haut Ngoungé (*Le Testu* 5117, 1928) and Libreville (*Klaine* 2583, 1922). The three new sites for the taxon in Kupe-Bakossi extend the range of the species considerably. The absence of *I. zenkeri* from Mt Cameroon is remarkable (Cable & Cheek 1998). In all, only seven sites are known.
Habitat: evergreen forest; 0–1200m alt.
Threats: clearance for timber, agriculture and (e.g. at Libreville), urban expansion.
Management suggestions: a survey to revisit the known sites, rediscover the species and assess the likelihood of protecting those subpopulations that do survive is required. Kupe-Bakossi may offer the best hope for the survival of this treelet, assuming that recent conservation proposals are enacted; it has the highest density and most recently discovered sites that are included in a probable protected area for forest.

Piptostigma calophyllum Mildbr. & Diels
VU B2ab(iii)
Range: Cameroon: Bakossi (3 coll. at 3 sites), Edea, Ebolowa, Sangmelima (1 coll. each); Gabon (2 coll.).
Spectacular for its large and beautiful leaves, this small cauliflorous tree is not easily overlooked. It was first described in 1915, based on *Mildbraed* 5791 from Ebolowa.
Habitat: pristine submontane evergreen forest; 900–1320m alt.
Threats: forest clearance for logging and agriculture.

Management suggestions: the three known sites in Bakossi constitute over a third of the global total. Education, perhaps through a conservation poster, might facilitate protection of this charming tree here. None of these three sites seem under immediate threat, although this cannot be said of those elsewhere.

Uvariodendron giganteum (Engl.) R.E.Fr.

VU B2ab(iii)

Range: Cameroon: Mt Cameroon, Yaoundé (1 coll. each), Bakossi (2 coll.); Gabon (3 coll. from 2 sites).

First collected in the early 20th Century from Mt Cameroon (*Lehmbach* 230) and Yaoundé (*Zenker & Staudt* specimen), this rare species has not been refound at earlier locations in recent years, despite, for example, intensive surveys in the 1990s at the first. Forest destruction at the second site has been particularly extensive in recent decades following the expansion of Yaoundé. Apart from the two sites in Gabon (1960s collections) this gigantic-leaved tree is otherwise only known at present from two sites in Bakossi.

Habitat: lowland and submontane evergreen forest; 200–1600m alt.

Threats: forest clearance.

Management suggestions: while both sites in Bakossi (Mungo River F.R. and Mt Kupe above Nyasoso) are officially protected, surveys of tree(s) at both sites would facilitate future monitoring. Public education might reduce harvesting of this species for firewood.

Uvariopsis submontana Kenfack, Gosline & Gereau

EN B2ab(iii)

Range: Cameroon: Mt Kupe (12 coll.), Yabassi-Yingui (1 coll. *fide* Kenfack).

In January, the gnarled trunks of this medium-sized canopy tree are largely covered in big pink, fleshy, velvety-hairy flowers in the males. The 12 collections on Mt Kupe were made at three sites. The site near Kupe village occurs very close to the upper level of cultivation and may be adversely affected if this level is elevated further. The absence of this species from the Bakossi Mts is notable.

Habitat: submontane evergreen forest, 840–1200m alt.

Threats: clearance of forest for agriculture, especially at the lower parts (below 1000m) of its altitudinal range.

Management suggestions: public education of the importance of this tree is advisable to assure its continued existence at its stronghold on Mt Kupe; a poster explaining this should be produced for local use.

Uvariopsis vanderystii Robyns & Ghesq.

EN B2ab(iii)

Range: Cameroon: Takamanda (1 coll.), Mt Kupe (2 coll.); Gabon (1 coll.); Congo (Kinshasa) (1 coll.).

Flowers in this small tree are only produced at ground level and radiate from the trunk on pedicels 4–5cm long. Although moderately widespread, it is both patchy and rare: despite intensive inventories at Korup and Mt Cameroon it has not been recorded there. We have no data on threats (or its continued existence) at sites in Gabon and Congo, but the prospects of continued forest quality at Takamanda have not been enhanced by construction of a new road there; the site at Mt Kupe, above Kupe village, is within 100m of cultivated land and has been earmarked for agricultural expansion. However, the small plot containing this *Uvariopsis* has been rented for conservation purposes by us from the farmer and monitored since 1995 and has not since deteriorated.

Habitat: lowland and submontane evergreen forest; 800–1000m alt.

Threats: forest clearance for agriculture.

Management suggestions: at Mt Kupe continued monitoring and rental payments are needed to ensure the continued survival of this species. The Takamanda site is partly protected as a wildlife reserve but the tree(s) of *U. vanderstyii* should be refound and monitored. Consideration should be given to educating the inhabitants of the area as to the importance of this plant. At the other locations, trees should be refound (if still existing), censused and considered for protection by the appropriate authorities.

Xylopia africana (Benth.) Oliv.

VU A2c

Range: São Tomé (1 coll.); SE Nigeria: Obudu (1 coll.); Cameroon: Mt Cameroon (4 coll.), Bakossi (10 coll.), Bali Ngemba (numerous coll.).

Of all the forests that we have surveyed, *Xylopia africana* has been found in greatest density at the Bali Ngemba Forest Reserve in North West Province. This, now the largest remnant of the forest that cloaked the Bamenda Highlands at the 1300–1900m range, is only 1000ha in extent and shrinking fast due to illegal clearance for farming. Presumably this species was once common in the Bamenda Highlands where it is now all but extinct. While there are no figures for rates of forest loss in the Bamenda Highlands as a whole, in one area which has been studied, the Kilum-Ijim area, forest loss of 25% over 8 years in the 1980s–1990s has been recorded (Moat in Cheek *et al.* 2000). Past forest loss in the Bamenda Highlands is therefore the main basis for the threat to *Xylopia africana*. On Mt Cameroon it appears rare, being found only twice in the surveys of 1992–1994. Elsewhere in the mountains of the

Cameroon line it is also known from the extension into Nigeria: the Obudu Plateau where it is also threatened due to forest clearance, if indeed, it is still extant there. It is also known from São Tomé in the Gulf of Guinea. Strangely, it is not known from Bioko, the Rumpi Hills or the Bamboutos Mts. Mt Kupe and the Bakossi Mts probably now support the largest single subpopulation of *Xylopia africana*.

Habitat: submontane and lower montane forest; 800–2000m alt.

Threats: clearance of forest for timber and agricultural land.

Management suggestions: if proposals to protect forest above 1000m alt. in much of Bakossi are enacted and respected, this subpopulation seems secure.

ARALIACEAE
(M.Cheek)

Schefflera hierniana Harms
VU B2ab(iii)

Range: Bioko; Cameroon: Belo to Lake Oku, Rumpi Hills (1 coll. each), Mt Cameroon, Bakossi Mts (2 coll. each).

A strangling epiphytic shrub of cloud forest recently resurrected by David Frodin from synonymy in the much commoner *S. barteri*, this rare species is known from only six sites.

Habitat: evergreen forest; 900–2100m alt.

Threats: forest clearance for agriculture and wood. It has not recently been seen at the Belo to Lake Oku site despite intensive surveys in the late 1990s (Cheek *et al.* 2000) and since forest has almost disappeared between these two locations, it is probably extinct there. Habitat degradation has been steady on some parts of Mt Cameroon at these altitudes and the species was not refound there in intensive surveys of the early 1990s.

Management suggestions: the best hope for the survival of this species may be the Bakossi Mts at the Kodmin and Nzee Mbeng sites since pressure here is relatively low and since the species was seen here most recently (1998). Continued surveys of the unexplored parts of the Bakossi Mts might well reveal new sites. Further surveys at its other sites, if focused on this species, might yet reveal that it survives so long as significant areas of forest exist.

Schefflera mannii (Hook.f.) Harms
VU A2c

Range: Annobon (1 coll.); São Tomé (1 coll.); Bioko (10 coll. at 4 sites); SE Nigeria: Obudu (1 coll.); Cameroon: Mt Cameroon (7 coll.), Mt Oku (7 coll.), Kupe and Mwanenguba Mts (6 coll. at 4 sites), Bamboutos and Bafut Ngemba (each 1 coll.).

One of the very few montane (above 2000m alt.) trees that are endemic to the Cameroon uplands (another is *Morella arborea*), this evergreen canopy tree begins life as an epiphytic shrub.

Habitat: montane forest; (1400–)2000–2400m alt.

Threats: forest clearance for agriculture and wood has reduced the habitat of this species by an estimated 30% or more over its whole range due principally to loss in the Bamenda Highlands, which, having the largest area above 2000m in the Cameroon uplands, was probably once the stronghold for this species. Between 1987 and 1995, 25% of forest was lost in one area of the Bamenda Highlands (Moat in Cheek *et al.* 2000). Extensive losses of habitat have also occurred at Mwanenguba, Obudu, Bamboutos and Bafut Ngemba.

Management suggestions: the status of *S. mannii* on the Gulf of Guinea islands requires elucidation, perhaps by surveying. At Mts Kupe, Cameroon and Oku (Kilum-Ijim) the species seems currently secure.

ARISTOLOCHIACEAE
(M.Cheek)

Pararistolochia ceropegioides (S.Moore) Hutch. & Dalziel
VU A3c

Range: Cameroon (6 pre-1978 coll.), Mt Kupe (19 coll., 3 sites); Gabon (3 pre-1978 coll.).

This canopy climber is easily spotted when fertile since the orange, bijoux, flowers are produced in clusters at eye-level from the figure-of-eight-shaped stems. It is widespread but fairly rare in semi-deciduous and submontane forest in the forest belt of Cameroon and northern Gabon. In Bakossi, although common (19 coll.), that part of its population occurring below 1000m alt. is under threat of forest clearance, while there has been extensive forest loss at some of its known locations elsewhere, e.g. at Nlonako and Yaoundé. Overall, a projection of 30% habitat loss for this species over the next three generations seems conservative. Data on collections cited under range for this species and *P. goldieana* are taken from Poncy (1978).

Threats: forest clearance for timber and agriculture.

Habitat: semi-deciduous and submontane forest; 600–1200m alt.

Management suggestions: that part of the subpopulation in Bakossi that occurs above the 1000m contour has a fairly good chance of long term survival.

Pararistolochia goldieana (Hook.f.) Hutch. & Dalziel

VU A2c

Range: Sierra Leone (2 coll.); Nigeria (6 coll.); Bioko (1 coll.); Cameroon (2 coll. pre-1978, 3 coll. post-1978).

This climber almost certainly has the largest flowers (c. 30 × 30 × 30 cm) of any African species of flowering plant. It is the African equivalent of the SE Asian *Rafflesia*, matching it in habitat, colour, size and scent of the flowers. Although widespread, it is rare. Despite intensive surveys over several years at Mt Cameroon, only a single individual was found (Cable & Cheek 1998). Forest loss has been extensive throughout Sierra Leone and Nigeria and is continuing; it is unlikely that this plant survives at its Calabar or Lagos localities, for example. The life cycle of this species is unknown, but it appears to be a perennial, producing annual shoots. It is estimated that a generation might last for 10 years.

Habitat: lowland evergreen forest, often seen in disturbed areas; 0–400m alt.

Threats: clearance of forest for timber and agriculture.

Management suggestions: a plant of this species in semi-natural forest at Limbe Botanic Garden in Cameroon offers excellent possibilities for research into the environmental and propagation requirements of *P. goldieana*. The subpopulation at Mahole (Bakossi F.R.) offers a good chance of conserving this magnificent species in the wild, possibly by a poster education scheme. Flowering plants offer the potential to attract tourists.

ASCLEPIADACEAE
(M.Cheek)

Neoschumannia kamerunensis Schltr.

CR B2ab(iii)

Range: Cameroon: Mt Cameroon (1 pre-1988 coll.), Mt Kupe; Ivory Coast (1 coll.); C.A.R. (1 coll.).

The following text is modified from the assessment in Cable & Cheek (1998). Discovered at Man O' War bay near Mabeta-Moliwe in 1899 by Schlechter, it has not yet been recollected there. One specimen was located in the 1960s at Mt Tonkui in Ivory Coast (probably extirpated there according to data from Gautier *pers. comm.*), but it was not until March 1995 that an Earthwatch-sponsored expedition at Mt Cameroon rediscovered this extremely rare vine above Likombe, in farmbush. Subsequently, one of the original discoverers, Etuge (the other being Meve, an Asclepiad specialist), found a population of seven plants at Mt Kupe in forest at the edge of a farm. Since Oct. 1995 the area at Mt Kupe has been rented from the local farmer in order to protect the plant. Another possible site was later found by Etuge in nearby northern Bakossi but the specimen has still to be verified. Finally, David Harris discovered this species in the Dzanga-Sangha reserve in southern C.A.R. in 1996. The assessment of critically endangered is maintained for this taxon, on the basis of its geographic range being severely fragmented, its area of occupancy being less than 10km^2 (given three probable extant sites, an area of occupancy of 3km^2 is calculated), and continued decline of habitat quality being projected.

Habitat: lowland to submontane forest, withstanding and perhaps benefiting from some disturbance; 0–1000m alt.

Threats: although known to withstand some disturbance, presumably being able to regenerate after cutting, there is no doubt that intensive agriculture and tree clearance would destroy this rare species.

Management suggestions: the now well-known plant above Likombe should be protected by agreement with the local landowner. A search for more individuals in the same area was carried out without success in 1995. This species may be a candidate for propagation at Limbe Botanic Garden for reintroduction to a protected, managed area or areas. The site at Mt Kupe remains the primary hope for the survival of this species in the wild. Regular monitoring by Martin Etuge and the author over the last eight years has shown the population at this site to be stable.

Secamone racemosa (Benth.) Klack.

VU A2c

Range: Bioko; Cameroon: Mt Cameroon (where possibly now extinct), Bakossi Mts (1 site); Congo (Kinshasa); Rwanda; Burundi; Uganda.

This twiner, which was assessed as VU by Cheek (in Cable & Cheek 1998) has since been transferred to the genus *Secamone*. The assessment has been modified to take account of recent changes in IUCN criteria. No new data has come to light since 1998.

Habitat: forest and lake edge; 1200–2400m alt.

Threats: forest clearance for agricultural purposes, e.g. on Mt Cameroon.

Management suggestions: a census of the species should be conducted at Lake Edib, the only known extant population in continental W Africa.

146

BALSAMINACEAE
(M.Cheek)

Impatiens frithii Cheek
EN B2ab(iii)
Range: Cameroon: Bakossi Mts, Kodmin area (4 coll.) and Mt Etinde (4 coll.).
First described in 2002, the slender, hairpin-shaped flowers of this epiphytic species are flame-red. They are produced in the wet season from stems often only 15cm long which scramble along mossy branches of low trees and shrubs. Assessed as vulnerable (Cheek & Csiba 2002), *Impatiens frithii* is here reassessed as endangered given an estimated area of occupancy of 10km², being known from only two broad, but very fragmented (100km interval), sites and the threats mentioned below, referred to in detail in Cheek & Csiba (2002).
Habitat: moss forest; (800–)1100–1700m alt.
Threats: planned plantation expansion (Mt Etinde) and reservoir scheme (Bakossi Mts).
Management suggestions: see *Impatiens letouzeyi*.

Impatiens letouzeyi Grey-Wilson
EN B2ab(iii)
Range: Cameroon: Bakossi Mts (5 coll. from 4 sites).
This robust, epiphytic herb has the largest flowers of all Cameroonian, and perhaps African, *Impatiens*. Described in 1981 from a single specimen in western Bakossi collected in the 1970s (*Letouzey* 15353), it was rediscovered by an Earthwatch team led by George Gosline in 1998 and subsequently found to be fairly common in the Kodmin-Edib area. At the end of the wet season, plants are readily spotted by their fallen flowers below their phorophytes. This plant has not been found on Mt Kupe despite searches in all months over several years.
Habitat: epiphytic in crown of trees 4–6m from ground, in shrubs, over streams or terrestrial, or on lakeside *Sphagnum* blanket; 1200–1350m alt.
Threats: the planned reservoir scheme near Kodmin may threaten part of the population of this species.
Management suggestions: a detailed study of this species at Kodmin-Edib would provide more precise data on population density and demography, as well as assisting in the placement of future development schemes. Botanical surveys in other parts of the Bakossi Mts, or Rumpi Hills, might yet discover more sites which would reduce the threat assessment of this taxon.

Impatiens sakerana Hook.f.
VU A2c
Range: Bioko; Cameroon: Mt Cameroon (6 coll.), Mwanenguba (2 coll.), Fossimondi, Bamboutos Mts (1 coll.), Bamenda Highlands (1 coll.), Bali Ngemba F.R. (3 coll.) to Mt Oku (12 coll.).
This often robust, locally common terrestrial herb has the highest altitudinal range of all Cameroonian *Impatiens*. Secure on Mt Cameroon, and probably Bioko, forest at Mwanenguba, the Bamenda Highlands and Bamboutos Mts has been under pressure from grassland fires set by graziers and for clearance for agriculture; forest in the Kilum-Ijim area outside the protected area having seen a reduction in cover of c. 25% in eight years of the 1980s and 1990s (Moat in Cheek *et al.* 2000). Assuming a generation time of ten years, it is estimated that about 30% habitat loss may have occurred in the last 30 years.
Habitat: understorey of montane forest; 2000m alt.
Threats: see above.
Management suggestions: enforcement of protected area boundaries. Demographic studies are needed to elucidate generation time and ecological requirements of this taxon.

BEGONIACEAE
(M.Cheek)

Begonia adpressa Sosef
VU B2ab(iii)
Range: Cameroon: Kupe-Bakossi, Rumpi Hills, Mt Nlonako, S Bamboutos (8 pre-1995 coll.).
Described by Sosef in 1992, when only eight collections were known (Sosef 1994b: 150), all of which were from W of Bangem in in the northern Bakossi Mts apart from three collections immediately to the north in the Bamilike area, one to the west, in the Rumpi Hills, and one immediately to the east, at Mt Nlonako. To these we can add six further collections from three sites made since 1995, one from Kupe village, one from Nyasoso, the others all from Kodmin. This constitutes nine locations and an area of occupancy of 14km² calculated at 1km² per collection. The habitat quality at the Bakossi sites is projected to remain stable if proposed conservation methods come to pass, but at all

other sites except the Rumpi Hills natural forest at this taxon's altitude range are under severe pressure, or have been eliminated.

Habitat: terrestrial and on rocks in forest; 1000–1750m alt.

Threats: forest clearance for timber, followed by agriculture.

Management suggestions: the future for *Begonia adpressa* seems bleak outside of the Rumpi Hills and Bakossi Mts. However, at these last sites it seems secure if proposed conservation measures are respected.

Begonia bonus-henricus J.J.de Wilde

VU A3c

Range: Cameroon: Rumpi Hills, Ebolowa (1 coll. each), Mt Kupe and Bakossi Mts (17 coll.).

First described in 1980, this species is unique in sect. *Squamibegonia* Warb. (characterised by peltate scales, indehiscent globose berries, inflorescences enclosed in sheathing bracts) in having long (c. 1m) slender (1.5mm diam.) pendulous or rooting stems. It was then known from just five collections at four sites: c 15km SE of Ebolowa, Rumpi Hills and, in Bakossi, Ngomboaku and Ngussi (coll. from last two sites not seen at K). This geographical range is very similar to that of *Newtonia duncan-thomasii*, although the *Begonia* also occurs on Mt Kupe and has a slightly lower altitudinal range. We have added fifteen more collections in Bakossi to those cited by de Wilde & Arends (1980). Excellent illustrations, notes and a description are given by those authors.

Habitat: hemiepiphytes in forest; 660–1300m alt.

Threats: clearance of forest for timber and agriculture at lower altitudes (below 1000m alt.) seems unfortunately inevitable in the Bakossi area.

Management suggestions: assuming that proposed conservation plans in the Bakossi area are enacted, this locally common species seems fairly secure there above the 1000m contour and no other conservation actions seems necessary.

Begonia duncan-thomasii Sosef

VU B2ab(iii)

Range: Cameroon: Kupe-Bakossi and Rumpi Hills.

Described by Sosef in 1992 and known only from four pre-1995 collections (Sosef 1994b: 156), all but one of which are in Kupe-Bakossi, the exception being *Satabie* 259 from the Rumpi Hills; the remainder were collected by Duncan Thomas. These three sites are at the summit and on the west side of Mt Kupe, and west of Bangem. To these in 1998, we have added only two more collections from a fifth site, near Kodmin at Abo'h (*Biye* 31 and *Etuge* 4025).

Habitat: *B. duncan-thomasii* appears to be restricted to shady, damp, vertical rock faces in forest, sometimes with the rare *Nodonema lineatum*; 1550–2000m alt.

Threats: see *Begonia schaeferi*.

Management suggestions: the five sites of this species merit revisiting and regular monitoring for numbers of individuals and threats.

Begonia furfuracea Hook.f.

VU B2ab(iii)

Range: Bioko (4 coll., 2 sites); Cameroon: Bamenda, Kondo (1 coll.), Mt Kupe and the Bakossi Mts (8 coll.).

This epiphytic species of *Begonia* sect. *Tetraphila* was recently revised by de Wilde (2002). He cites evidence that the subpopulation on Bioko may have been affected by cultivation of cocoa, coffee and bananas, for which most of the original vegetation below 700m was cleared. *Begonia furfuracea* was first collected there by Mann in 1860, recollected many decades later by Mildbraed and then relatively recently (c. 1960s) by Sanford and Carvalho. The Cameroonian subpopulation first came to light in the 1970s at Kondo, 40km northwest of Bamenda (*Letouzey* 14186) an area where most forest has now been extirpated. The species probably now has its stronghold at Kupe-Bakossi; de Wilde cites three collections from the area (de Wilde 2002: 89). To these can be added a further five (see checklist).

Habitat: evergreen forest, 1000–1300m alt.

Threats: assuming that proposed conservation plans in Bakossi are enacted this species seems secure there, its stronghold. Some habitat destruction occurs at its known sites, but levels are low.

Management suggestions: yearly monitoring of plants to assess population trends seems advisable. An attempt should be made to rediscover and assess the status of the Bioko subpopulation.

Begonia oxyanthera Warb.

VU A2c

Range: Nigeria (1 coll.); Bioko (6 coll. from 2 sites); Cameroon: Mt Cameroon (4 coll.), Mt Kupe and the Bakossi Mts (5 coll.), Rumpi Hills (1 coll.), Bamboutos Mts (2 coll.), Mwanenguba (1 coll.), Bamenda Highlands (17 coll.).

The range data above are taken from the account of the species in the recently published revision of *Begonia* sect. *Tetraphila* (de Wilde 2002). In previous Red Data assessments (Cable & Cheek 1998 and Cheek *et al.* 2000) this taxon was listed as LR nt. Here it is reassessed as vulnerable because of habitat destruction in what appears to be its main subpopulation in the Bamenda Highlands from which most of the c. 30 specimens listed by de Wilde derive. Moat (in Cheek *et al.* 2000) records forest loss in the Kilum-Ijim protected area itself. Since the 'generation' duration of this taxon might easily be five years, it is estimated that habitat loss for the species over its area of occupancy as a whole is likely to have been over 30% in the last 15 years. The increasing frequency of occurrence of this species moving northwards (Mt Cameroon and Bioko only c. 2 sites each; Bamenda Highlands numerous sites and collections) is perhaps a reflection of a longer dry season requirement.

Habitat: submontane and montane forest; 1200–2200(–2400)m alt.

Threats: forest clearance for wood and agriculture (mainly in the Bamenda Highlands).

Management suggestions: subpopulations in SW Province, and probably also in Bioko, are fairly secure, lacking threats. In NW Province the substantial subpopulation in the well protected Kilum-Ijim site (Mt Oku and the Ijim Ridge) may be the only locality where the taxon will survive, unfortunately, thus protection there is important.

Begonia pelargoniiflora J.J.de Wilde & J.C.Arends
CR B2ab(iii)

Range: Bioko (2 coll.); Cameroon: Bakossi Mts and Mt Nlonako (1 coll. each).

First published in 1992, an updated account was included in de Wilde's revision of *Begonia* sect. *Tetraphila* (2002: 183). *B. pelargoniiflora* shows a similar disjunct distribution to that of *B. furfuracea*, that is, Bioko, northwards to the Bakossi Mts but omitting the very-well-collected Mt Cameroon. However, whereas de Wilde records eight collections for *B. furfuracea*, there are only four known for *B. pelargoniiflora*, two from Bioko and one each from the Bakossi Mts and the adjoining Mt Nlonako. The fact that it has not been recollected in the last 10 years despite intensive botanical inventory work e.g. in Bakossi, illustrates that this species really is very rare. By comparison, 30 specimens of the related and similar *B. longipetiolata* were made between 1995–2003 in Kupe-Bakossi. On the basis of an area of occupancy of $1km^2$ per site, the total area of occupancy is calculated as just $4km^2$ for this taxon. Its range is considered severely fragmented on the basis of its absence from Mt Cameroon. If the latter is disallowed, its rating will be lowered to endangered.

Habitat: forest; 1000–1500m alt.

Threats: forest clearance for wood and agriculture, e.g. ongoing at Mt Nlonako, the type locality.

Management suggestions: an effort should be made to rediscover this taxon and to assess the size of the subpopulation and its demography in detail. In view of its rarity this species might be suitable for propagation and reintroduction in suitable localities.

Begonia preussii Warb.
VU A3c

Range: Nigeria (3 coll.); Bioko (1 coll.); Cameroon: Edea-Kumba (15 coll.).

This taxon was previously assessed as EN (Cable & Cheek 1998) but is here downrated to VU in the light of the taxonomic revision by de Wilde (2002, source of the range data above) showing it to be more common and widespread in Cameroon than was previously thought. In addition, the discovery of eight further specimens at five sites in Kupe-Bakossi confirms the foregoing. Nonetheless, due to the low altitudinal range of this taxon, habitat loss of about 30% over the next 15 years (estimated as equating to three generations) can be postulated.

Habitat: evergreen forest; 850–1000m alt.

Threats: clearance of forest for wood and agricultural land is a major threat throughout its range and probably accounts for the lack of collections from Bioko in the last century (where forest was largely cleared below 1000m alt.).

Management suggestions: this taxon is most likely to survive in lowland forest reserves such as the Bakossi F.R. and Mokoko F.R., but only if their boundaries are respected. Resources need to be found to achieve this.

Begonia prismatocarpa Hook. subsp. *delobata* Sosef
VU B2ab(iii)

Range: Cameroon: Nta-ali (1 coll.), Rumpi Hills (2 coll.), Kupe-Bakossi (2 pre-1995 coll.).

First recognized in 1994 (Sosef 1994: 179), this subspecies was first collected in October 1946 in the northwest part of Kupe-Bakossi (*Dundas* in FHI 15329). It seems restricted to lowland and submontane sites bordering the upper reaches of the Mungo valley, extending eastwards down the Mbu valley (*Thomas* 5351). The fact that in our inventory of Bakossi plants in the late 1990s only 1 in c. 9,000 specimens (*Cheek* 9645) belong to this taxon indicates its rarity, and also its restriction to western Bakossi, which we have sampled sparingly.

Habitat: terrestrial on soil, or on rocks, or epiphytic on the base of tree trunks, in forest; 400–1100m alt.

Threats: forest clearance for timber, followed by agriculture. Being restricted to forest at lower altitudes, this rare plant is especially vulnerable to illegal logging, which is currently so prevalent throughout its range (*pers. obs.*).

Management suggestions: the low altitudinal range of this taxon leaves it outside of protected areas within its range. Therefore public education as to the advisability of conserving this taxon is recommended. The site at Nyale Rock is remarkable for the lack of pressure and the possibilities of long-term conservation.

Begonia pseudoviola Gilg
VU B2ab(iii)
Range: Cameroon: S Bamboutos Mts (6 coll.), Kupe-Bakossi (3 coll.), Mt Nlonako (3 coll.).
Described by Gilg in 1904 from *Conrau* 10 collected in Nov. 1898 between Banti and Babesong at 600–700m alt., this beautiful species is easily recognized by its diminutive size and its purple-black leaf-blades which bear long, white, patent-curved hairs on their upper surfaces. The nine collections recorded by Sosef (1994b: 182) are immediately to the northeast (Nkongsamba-Mt Nlonako) or north (S and W Bamboutos Mts) of Kupe-Bakossi. Until we worked at Ngomboaku in 1999, recording three sites for the species in a fairly small area east of the village, *B. pseudoviola* was entirely unknown from Kupe-Bakossi. Owing to its relatively low altitudinal range, *B. pseudoviola* is extremely vulnerable to forest clearance (see discussion at beginning of this chapter). For the purposes of this assessment we rate this species as VU under criterion B, being known from ten locations and having an area of occupancy of 12km^2 (see *B. adpressa*). The prognosis for habitat destruction for this taxon is high. This taxon might be better assessed under criterion A, but lack of data on the state of sites west of the Bamboutos Mts makes this difficult to apply.
Habitat: wet mid-elevation forest; 600–850m alt.
Threats: see above.
Management suggestions: revisting the Ngomboaku site to assess the population size and potential threats should be made; active protection should be sought at this location.

Begonia schaeferi Engl.
VU D2
Range: Nigeria: Obudu Plateau (1 coll.); Cameroon: Mwanenguba, Kongoa Mts, Mt Nlonako (each 1 coll.), Bamboutos Mts (3 coll.), Bamenda Highlands (3 coll.).
First collected in Nov. 1900 at Bare, Mwanenguba, by Schaefer and described in 1921, the species has not since been seen in the Kupe-Bakossi area and may be extinct there. Personal observations of *B. schaeferi* in the Bamenda Highlands shows that this species is demanding in its habitat requirements and likely to occur at extremely few sites within a given area (Cheek *et al.* 2000). The assessment of this species as VU D2 in Cheek *et al.* (2000) is maintained here: no new data are available on the taxon. The information presented below is mainly taken from that work.
Habitat: on rocks and vertical rock faces in moist to comparatively dry places in primary submontane to montane forest, the latter sometimes with trees not taller than 6–12m; c. 1500–2300m alt. (Sosef 1994).
Threats: while cliff faces generally are unlikely to be disturbed, clearance of adjoining forest for fuel and agriculture could endanger this species by removing the shade necessary for its survival.
Management suggestions: during surveys of cliff spaces (in forest) this species should be looked for and, if located, the number of plants and locality recorded.

CAMPANULACEAE
(M.Cheek)

Dielsantha galeopsoides (Engl. & Diels) E.Wimm.
EN A3c
Range: SE Nigeria; Bioko (both 1 coll.); Cameroon, SW Province: Barombi Mbo (1 pre-1963 coll.), Kupe-Bakossi (3 coll. in 1990s).
This sprawling monotypic herb resembles a *Lobelia*, to which it is closely related, in flower. It differs in the peculiar, longitudinally-dehiscent fruit and shade-demanding ecology. Discovered in the late 19th century at Barombi near Kumba, it appears not to have been collected at any of its three previously known locations for over sixty years. Despite intensive surveys in the 1990s this taxon remains unknown in the Mt Cameroon area. In Bakossi *Dielsantha galeopsoides* is widespread but rare. Known from only three sites, it is endangered by farm expansion in two of these: Nyasoso Nature Trail and Manehas F.R. At Nyale it is probably fairly secure owing to the remote nature of Nyale and the low level of human habitation there. A population reduction of two-thirds over the next ten years of what is the world's only known extant sites for *Dielsantha* is thus projected.
Habitat: lowland evergreen forest; 650–1000m alt.
Threats: forest clearance for agriculture: suitable habitat at Bioko was destroyed for cocoa plantations, habitat at Lake Barombi Mbo was reported destroyed in 2002 (Aaron Davis. *pers. comm.*).

Management suggestions: Bakossi is the most important area for this genus according to available data, indeed it is only here that plants are known to survive. Research in population density, demography and recruitment would best be conducted at Nyasoso since the plant exists on the very edge of the town in the 'Nature Trail' above the school. In the long term, the Nyale site offers best scope for protection.

Wahlenbergia ramosissima (Hemsl.) Thulin subsp. *ramosissima*
VU B2ab(iii)
Range: Cameroon: Mt Cameroon (4 coll.), Mwanenguba, Chappal Waddi, Mambilla Plateau (1 coll. each), Bamenda Highlands (2 coll.).
This small (4–15cm tall) erect, blue-flowered annual herb is found between grass tussocks and is known from only nine specimens at five mountain sites along the Cameroon line. It has previously been assessed (Cable & Cheek 1998, Cheek *et al.* 2000) as VU. This rating is maintained here.
Habitat: between *Sporobolus* grass tussocks and on shady banks; 1500–2600m alt.
Threats: unknown, but possibly trampling by cattle during the wet (growing) season.
Management suggestions: assessing the size of the subpopulations, ideally when the plants are in flower and most conspicuous, is advisable. Monitoring variation in numbers of individuals present from year to year would also help better assess the threat to this taxon.

CARICACEAE
(M.Cheek)

Cylicomorpha solmsii (Urb.) Urb.
VU B2ab(iii)
Range: Cameroon: Mt Cameroon (Buea-Bulifambo 2 coll.), Kumba (Barombi, 3 coll.), Kupe-Bakossi (5 sites), Wum (2 coll.), Mujung (1 coll.), Yaoundé (1 *obs.*, M.Cheek), 80km SSW Yokodouma (1 coll.).
Known from nine submontane forest sites, this dramatic dioecious tree, one of only two *Caricaceae* on the African continent, is unmistakable even when sterile due to the broad, soft, spiny trunks which yield white exudate, and the digitately lobed leaves. It appears to be a pioneer, but is always very local, and usually gregarious.
Habitat: submontane forest; (500–)1000–1400m.
Threats: forest clearance for agriculture and wood. It may now be extinct on Mt Cameroon and at Barombi, Kumba since no records have been found for either of these sites in nearly 100 years.
Management suggestions: the demography of this taxon is not documented and requires study. Nyasoso at Mt Kupe would be a good base for such research, and as a centre for conservation of the species since a sizeable population exists on the main trail to the summit at 1100m alt. Good populations also exist at Nkom-Wum and in the Bakossi Mts, near Kodmin. Extensive traditions attached to these trees are likely to help protect them in Bakossi.

CELASTRACEAE
(I.Darbyshire & B.J.Pollard)

Salacia lehmbachii Loes. var. *pes-ranulae* N.Hallé
VU B2ab(iii)
Range: Nigeria, Cross River State: Oban (1 coll.); Cameroon, SW Province: Bakossi Forest Reserve (1 coll.), 15km S of Akwaya (1 coll.), S Province: S of Ebolowa (3 coll.).
First described in 1986, all previous collections of this taxon were made prior to 1980, when lowland forest south of Ebolowa appeared to be the centre of its distribution. These forests have experienced significant reductions following expansion of lowland plantation agriculture. The discovery of this taxon in the protected Bakossi Forest Reserve is of significance to its future conservation. *Salacia lehmbachii sensu lato* is distributed from Sierra Leone to Tanzania; the complex comprises seven varieties of which var. *pes-ranulae* is both the most localised in distribution and the most distinct morphologically.
Habitat: dense lowland forest understorey; 450–700m alt.
Threats: continued loss of lowland forest throughout the species range, particularly in Cross River State, Nigeria and the forests of South Province, Cameroon. Illegal encroachment of agriculture into the Bakossi Forest Reserve, facilitated by European Community road building, threatens this population.
Management suggestions: heightened protection of remaining forest at the Bakossi Forest Reserve may help to preserve this taxon. Further botanical inventory work in lowland sites, such as the neighbouring Loum Forest Reserve, may reveal further populations; care should be taken to accurately identify all future collections of *S. lehmbachii* to variety level.

Salacia mamba N.Hallé

VU B2ab(iii)

Range: Cameroon, SW Province: Mt Kupe (3 coll.), Littoral Province: Bakaka (1 coll.), Nkongsamba-Loum (2 coll.), Central Province: Otélé, nr. Yaoundé (1 coll.), S Province: 10km SW of Amban (1 coll.); Gabon: nr. Makokou (3 coll.); Congo (Brazzaville): Mayombe (1 coll.), Komono to Mbila (1 coll.).

First described in 1986, this taxon is now known from collections at 10 sites. Although relatively widespread, it is clearly rare throughout its range and absent from many seemingly suitable sites. The discovery of this species on the western slopes of Mt Kupe extends its western range limit; it is also known from the lowlands adjacent to this mountain to the east, although the expansion of settlements along the Loum-Nkongsamba road may have caused local population losses since its collection there in the 1970s.

Habitat: dense riverine, lowland and mid-elevation forest; 500–1000m alt.

Threats: continued loss of lowland forest in Cameroon and Congo (Brazzaville); the Gabonese sites are likely to have remained relatively undisturbed.

Management suggestions: protection of existing lowland forest sites. Care should be taken to separate this taxon from other lowland *Salacia* species in future botanical inventory work in Lower Guinea.

Thyrosalacia racemosa (Loes. ex Harms) N.Hallé

CR A2ac

Range: Cameroon, Littoral Province: Yabassi (1 coll.), Babong (1 coll.), Lom (= Loum) (2 coll.).

This species is only known by 4 Ledermann collections from 1908 and 1909 in the lowland forests around the Yabassi-Loum area, at and beyond the SE corner of our checklist area.

Habitat: lowland forest, c. 200m.

Threats: as a climber, this species is particularly sensitive to understorey clearance, and from personal observation at the Loum F.R., an estimate of 80% reduction of population size over three generations is made here, based on the continuing development of extensive cocoa and banana small-holdings within the reserve. The Babong subpopulation is likely to have disappeared due to urban development, and probably only the Yabassi sub-population has survived the last 100 years.

Management suggestions: protection of existing lowland forest sites. Surveys in the Yabassi area would be most welcome, and this species made one of the targets for re-collection and population size monitoring.

Postscript: the assessment above will be scrutinized by IUCN for inclusion in the 2005 Red list.

CHRYSOBALANACEAE
(M.Cheek)

Dactyladenia johnstonei (Hoyle) Prance & F.White

CR A3c

Range: Nigeria: Obudu (1 coll.); Cameroon: Bamenda Highlands (5 coll.), Mt Kupe (1 coll.).

This rare, hispid tree was first collected in the Bamenda Highlands in 1931 (*Johnstone* 74/31). Since that time four other collections from scattered locations in the Bamenda Highlands have been made, from near Bum, Wum, Nkambe and Fonfuka, as well as one record from adjacent Nigeria. Our record from Mt Kupe (*Cheek* 10158) is the first from SW Province. Given the threats below and an estimated generation time of 30 years, with projected forest loss at these sites of at least 80% over 90 years, this taxon is here assessed as CR under criterion A3.

Habitat: forested valley slopes; 950–1600m alt.

Threats: felling for timber and clearance for agricultural land: these problems are exacerbated in NW Province by a shortage of both commodities for a dense human population. Forest at this altitude is particularly under threat and may well no longer survive at the localities where this taxon was previously collected. A 25% reduction in forest area over eight years was recorded by Moat (in Cheek *et al.* 2000) near Mt Oku using GIS techniques. A similar situation is recorded in Cross River State, Nigeria. Forest at Mt Kupe is less threatened, at least above 1000m alt. However the Kupe material is recorded at 950m and is thus vulnerable.

Management suggestions: efforts should be made to rediscover this taxon in NW Province and to protect it. Individuals may yet survive in royal, sacred or village forests and might be protected there. It appears to be unknown at Bali Ngemba F.R. the only extant protected area covering the altitudinal range of this species in the Bamenda Highlands. At Mt Kupe the one individual known should be rediscovered and an effort made to find others. Education of the local population is suggested regarding the importance of this species for conservation.

Magnistipula conrauana Engl.

EN A3c

Range: Cameroon: Bamboutos Mts (5 coll.), Mwanenguba (1 coll.), Kupe-Bakossi (4 coll. at 4 sites).

Letouzey and White (Fl. Cameroun 20, 1978) record the collections for this taxon as above. During the late 1990s we found the taxon also to occur, rarely although fairly widespread, at Mt Kupe and the Bakossi Mts (four sites). It is estimated here that this canopy tree will become extinct at Bamboutos and Mwanenguba (given the threats indicated below) in the next 100 years but has a good possibility of surviving in the newly discovered southern half of its range.

Habitat: submontane forest, 1000–1500m alt.

Threats: the Bamboutos Mts are densely populated and intensively cultivated, for such crops as *Coffea arabica*. What few fragments of forest remain are under great pressure to supply wood and for further agricultural land. The rate of forest area loss of 25% in eight years for the Oku area cited elsewhere in this chapter is probably matched in the Bamboutos area.

Management suggestions: Letouzey & White (1978 *loc. cit.*) mention this plant occurring in hedges in Bamboutos. Encouragement of this practice, and also evaluating sacred forests there for the existence of this species, might be the means to survival in the northern part of the range. In the southern part of its range *M. conrauana* seems likely to survive if existing conservation plans go ahead.

COMPOSITAE
(M.Cheek)

Bidens mannii T.G.J.Rayner
VU B2ab(iii)

Range: Cameroon: Mt Cameroon (10 coll.), Mwanenguba, Kupe Rock, Bamboutos, Lake Aweng, Santa Mt (1 coll. each), Mt Oku (2 coll.).

This floriferous, robust herb looks like a garden escape, and is certainly worthy of horticultural attention. It is abundant, common and secure on old lava flows on Mt Cameroon, but elsewhere it appears rare and its chances of survival precarious (see threats below). Seven sites are known.

Habitat: grassland, near forest, often at well-drained sites; (1000–)2000m alt.

Threats: as for *Vernonia calvoana* var. *calvoana*; in addition, cultivation of habitat for crops e.g. at Lake Aweng, Santa and Bamboutos, is a threat.

Management suggestions: as for *Crassocephalum bauchiense*.

Crassocephalum bauchiense (Hutch.) Milne-Redh.
VU B2ab(iii)

Range: Nigeria: Bauchi Plateau, Jos Plateau, Pawpaw Mt, Amedzefe (1 coll. each); Bioko (3 coll.); Cameroon: Bakossi (3 coll. at 2 sites), Mt Cameroon (11 coll.), Mt Oku (6 coll.).

This erect blue-flowered herb is known from nine montane sites in Lower Guinea. If new sites are discovered it may be downgraded to NT, especially in view of the fact that little data are available regarding loss of its habitat.

Habitat: open woodland, savanna, forest edge; 900–2000m alt.

Threats: clearance of trees for agriculture and wood.

Management suggestions: the effect of forest loss on this, and other, rare, sun-loving Compositae, many of which share the same range of habitats, is poorly understood. Research is needed to redress this. Limited 'habitat destruction' at intervals of years, in the form of fires or limited long-fallow small-holder agriculture or small-scale tree felling may be helpful to the survival of such taxa by keeping their habitat open and allowing establishment of probably short-lived perennials such as these.

Postscript: range of this taxon now believed to extend to Congo (Kinshasa) and Uganda (Beentje, *pers. comm.*) which may diminish the threat to this taxon when reassessed.

Mikaniopsis maitlandii C.D.Adams
VU B2ab(iii)

Range: Nigeria: Chappal Waddi (2 coll.); Bioko (1 coll.); Cameroon: Mt Cameroon (5 coll.), Bakossi-Kupe (2 coll. at 2 sites).

This climber, often found in gaps in cloud forest cover was previously assessed (Cable & Cheek 1998: li) as LR nt. Here it is reassessed as VU due to only 5 sites being known with the threats indicated below being prevalent. On Mt Cameroon it was not detected during the intensive surveys of the early 1990s, so is at least very rare there.

Habitat: montane forest gaps, sometimes with *Mimulopsis solmsii*; 1000–2200m alt.

Threats: forest clearance for agriculture and wood, particularly likely at the lower part of its altitudinal range at sites such as Chappal Waddi and Mt Cameroon (plantation expansion to the 1000m contour is a major threat).

Management suggestions: this rarely-seen species is perhaps most easily refound and studied at the sites where it has most recently been recorded, both in Bakossi. Data on demography and population size would better inform management decisions. In the meantime, education of the public by means of e.g. conservation posters might help reduce pressure on subpopulations that survive.

Mikaniopsis vitalba (S.Moore) Milne-Redh.
VU B2ab(iii)
Range: Cameroon: Nyasoso, Mt Kupe (1 coll.), Djang, 40 km W of Bertoua (1 coll.), Yaoundé (2 sites); Uganda (2 sites); Gabon; Angola; Congo (Kinshasa) (1 coll. each).
This rare climber, although very widespread, is known only from nine sites scattered over five countries.
Habitat: lowland gallery, swamp, submontane forest or savanna; 700–1600m alt.
Threats: forest clearance for agriculture and wood, and (e.g. Yaoundé area) urban expansion.
Management suggestions: more material is needed of this taxon from Nyasoso (*Etuge* 2177) to confirm that it is indeed *M. vitalba*. Although in *Mikaniopsis* it is best placed here, it differs from the specimens at other sites in having leaves rounded (not cordate) at the base and more than twice as long as broad, and elongated, not congested, racemes, and in growing at 1600m, not 700–750m alt. Site surveys are required to investigate subpopulations and to gather data on demography and recruitment. This taxon should be looked for in ongoing surveys at the proposed National Park at Mefou, which might form a secure base for the taxon. The Nyasoso site is probably secure.

Vernonia calvoana (Hook.f.) Hook.f. subsp. *calvoana* var. *calvoana*
VU D2
Range: Cameroon: Mt Cameroon (14 coll.), Mwanenguba (1 coll.) and Mt Oku (2 coll.).
This large herb, which can grow to the size of a small tree, 12m tall, appears relatively common on Mt Cameroon, perhaps due to the absence of livestock grazing there. A total of three mountain sites are known.
Habitat: forest grassland boundary; 1900–2300m alt.
Threats: poorly known, but grazing and the associated annual firing of grassland are considered likely to be detrimental to this species.
Management suggestions: as for *Crassocephalum bauchiense*.

CUCURBITACEAE
(I.Darbyshire)

Momordica enneaphylla Cogn.
VU B2ab(iii)
Range: Cameroon, SW Province: Kupe-Bakossi (3 coll., 2 sites), Central Province: 20km NW of Eséka (1 coll.), S Province: Bipinde (1 coll.); Gabon (1 coll.); Congo (Kinshasa), Forestier Central: Yangambi (3 coll.), Watsi, Boende (1 coll.), Yabwesa, Isangi (1 coll.).
This little-known liana was first collected in Gabon, being described in 1888. Collections from Congo (Kinshasa) are all from the central Congo Basin. In Cameroon, the discovery of 2 sites in the Kupe-Bakossi area, on Mt Kupe above Kupe Village and at Ngomboaku in the Bakossi Mts, extends this species' range northwestwards. This is a highly distinctive member of the Cucurbitaceae family, being the only bipinnate species of west Africa, with 3 ternate leaflets. *Sidwell* 443 appears to be the first collection of the currently undescribed female flowers of this species; this specimen demonstrates how this taxon could be easily overlooked, as it is flowering prior to the development of leaves. A collection of a single *Momordica* flower on a trailing leafless stem at Nyandong, western Bakossi (c. 500m alt.) by M.Etuge in March 2003 may prove to be this species on further investigation.
Habitat: primary forest, including riverine and swamp forest; occasionally in secondary forest growth; 450–1070m alt.
Threats: on Mt Kupe, this species is recorded in forest down to 900m, which is below the lower limit of effective protection on this mountain, thus this population is threatened by forest clearance.
Management suggestions: further investigation of the *Etuge* collection at Nyandong, Bakossi may reveal a new site for this taxon. If the low levels of forest clearance continue in Gabon and Congo (Kinshasa), this taxon should remain extant there, though its rediscovery in these two countries is advisable in light of the age of the known collections.

EBENACEAE
(M.Cheek & I.Darbyshire)

Diospyros kupensis Gosline
VU B2ab(iii)

Range: Cameroon, SW Province: Mt Kupe (8 coll.), Nyale, central Bakossi (2 coll.), Nyandong, west Bakossi (1 coll.), Enyandong, north Bakossi (1 coll.).

First collected in 1995, this species is restricted to Mt Kupe and the Bakossi Mountains. It remains fairly numerous on the slopes of Mt Kupe above Kupe Village; a colony of 30–40 individuals was recorded in one location here in the mid 1990s. Elsewhere it is rare, though is easily overlooked when sterile. In 2003 it was recorded for the first time in west Bakossi, referring to a single sterile specimen adjacent to the track from Nyandong to Maseka, extending the taxon's altitudinal range down to 460m. It was also recorded for a third time at the base of Nyale Rock (Etuge *pers. obs.*). This species appears most closely related to *D. conocarpa* Gürke & K.Schum, but it lacks the modifications for ant association of this taxon. *D. kupensis* displays an interesting form of vegetative reproduction; in many specimens the main trunk of the monopodial growth is angled at approximately 45°, with side shoots growing vertically and producing adventitious roots, which eventually produces several vertical stems from originally monopodial growth (Gosline & Cheek 1998, Kew Bull. 53: 463).

Habitat: understorey of undisturbed mid-elevation to submontane forest on well-drained slopes; 460–1300m alt.

Threats: several populations have been recorded on Mt Kupe below 1000m alt., which is below the limit of effective protection through restricted land use at this site; these populations are therefore threatened by agricultural encroachment. The Nyandong site is threatened by disturbance along the well-used track to Maseka.

Management suggestions: further botanical inventory work will likely reveal further locations for this taxon in the Bakossi Mountains. Active conservation of the larger populations on Mt Kupe may be required; informing the community at Kupe Village and neighbouring settlements of the rarity and uniqueness of this species has begun with a conservation poster specific to *D. kupensis*, with the intention that this will promote community-led protection.

EUPHORBIACEAE
(M.Cheek)

Croton aubrevillei J.Léonard
VU A2c; B2ab(iii)

Range: Ivory Coast; Ghana; Cameroon.

This rare tree was assessed by Hawthorne in 1997 as VU A1c; B1 + 2c, but listed only from Ivory Coast and Ghana (IUCN 2003, downloaded on November 8th 2004). It is maintained here as VU. The specimen cited in the checklist extends the range of this taxon considerably, the Cameroon specimen being the first from Lower Guinea. Accordingly, the extent of occurrence of this taxon now exceeds 20,000km^2. Furthermore, since 1997, criterion B has been modified. The assessment for this taxon is adjusted accordingly. Given continued forest loss in Ivory Coast since 1997 and the precariousness of the taxon in Cameroon, future re-evaluation is likely to rate this taxon as endangered.

Habitat: lowland evergreen forest; to 850m alt.

Threats: clearance of forest for agriculture.

Management suggestions: the location for the only known Cameroonian specimen, Nyasoso, 850m alt., indicates that this species might already be lost to forest clearance. A search should be made to rediscover the tree, and a survey conducted to discover if any other trees of the species occur in the area, and whether regeneration is proceeding or not. Artificial propagation and public education might be suitable options to ensure the survival of this taxon.

Crotonogyne impedita Prain
CR A2c

Range: Cameroon: Bakossi-Kumba (3 sites), Mt Kala (1 site).

A poorly known rainforest shrub, *Crotonogyne impedita* appears to be known from only four collections at as many sites; three of these are on the western and southern boundary of the Bakossi checklist area in SW Province. Two of these provided the original material (*Büsgen* 163 from Kumba and *Ledermann* 6397 from Loum) at the beginning of the 20th century, but there are no modern collections from either site. In the 1980s, D.W.Thomas, specialist in Cameroonian Euphorbiaceae, made the other two collections known, at Konye in western Bakossi (*Thomas* 5167) and Mt Kala (*Thomas* 3473).

Habitat: understorey of semi-deciduous and evergreen forest; to 400m alt.

Threats: most of the understorey of Loum F.R. has already been lost to agriculture; much of the forest that survived at Kumba itself, e.g. Barombi Mbo, has been lost in recent decades. Recent reports are that the forest at Mt Kala has been extensively cut-over. Konye lies on the Kumba-Mamfe road that is in the process of being upgraded so increased deforestation along this route is predictable.

Management suggestions: rediscovering this rare species, not seen in 20 years, is desirable. It is advised that the starting point be the sites of the D.W.Thomas collections. *In situ* protection would be ideal, but if unrealistic, cultivation in a botanic garden, such as Limbe, may be warranted, with a view to reintroductions to secure sites in future.

Crotonogyne strigosa Prain
VU A3c; B2ab(iii)

Range: SE Nigeria: Calabar-Oban (3 coll.); SW Cameroon: Mamfe to Mundemba (5 coll.).

This forest shrub, unusual in its red exudate, is unique in its genus due to its long, shaggy indumentum. It is only known from eight specimens, all of which occur west of the Cameroon mountain line. Its rarity is highlighted by its absence from Mt Cameroon (Cable & Cheek 1998). The note of this species being 'also in French Cameroons' (Keay 1963: 400) appears erroneous.

Habitat: lowland evergreen forest; c. 100m alt.

Threats: logging of forest followed by agriculture, amplified by ongoing, massive upgrading of the Kumba-Mamfe road through the heart of its range.

Management suggestions: subpopulations of this shrub in Korup N.P. are probably secure, and the site at Etam, Mungo River F.R. is due to have its protection status upgraded soon. Surveying the numbers, regeneration and demography of the subpopulation at Etam is suggested.

Crotonogyne zenkeri Pax
VU A2c; B2ab(iii)

Range: coastal Cameroon: W Bakossi, Kribi, Batanga, Nyong at Makak, Bipinde (1 coll. each); Gabon: Libreville (1 coll.).

This forest shrub appears restricted to wet forest close to the Atlantic coast. Despite intensive recent inventories it is unknown from the Mt Cameroon area (Cable & Cheek 1998). Only seven collections from six sites are known. Four of these sites are threatened.

Habitat: wet lowland evergreen forest; to 200m alt.

Threats: forest clearance followed by agriculture in the Kumba-Mamfe area (see notes under *Crotonogyne strigosa*), forest clearance for touristic development (Kribi-Batanga area, *pers. obs.* 2000) and for urban expansion in the Libreville area, Gabon (*pers. obs.* 2002).

Management suggestions: the western Bakossi location for this taxon should be surveyed in order to rediscover this plant and evaluate the size and demography of its subpopulation, if extant. Surveys of the same nature are advised at other sites. This data could form the basis of a plan to protect the species.

Drypetes magnistipula (Pax) Hutch.
EN B2ab(iii)

Range: Cameroon: Bakossi Mts (2 sites), Bipinde (2 coll.); Gabon: Crystal Mts (1 coll.).

This 4m treelet is named for its large persistent ovate stipules, probably the largest of any African *Drypetes*, measuring 3.5 × 1.5cm.

Habitat: lowland evergreen forest; 800–1000m alt.

Threats: clearance of forest for timber and agriculture is ongoing in both host countries but, so far as is known, is not intense at any of the four sites known for the species.

Management suggestions: although distinctive, *D. magnistipula* is a poorly known species. Since neither the Fl. Cameroun Euphorbiaceae account, nor that for Fl. Gabon, are available, no recent literature is available on the taxon. It is entirely possible that more collections, representing further sites, exist in herbaria other than that of Kew, such as BR, MO, P and WAG. These should be checked. Demographic and other populational data desirable in assessing the conservation requirements of this species could be obtained if the sites of *Thomas* 5313 and *Doumenge* 519 are found intact.

Drypetes molunduana Pax & K.Hoffm.
VU A2c

Range: SE Nigeria: Omo, Okomu, Oban; Cameroon: Kumba, Maguka, S Bakundu, Etinde, Bakossi (3 sites), Nkolbisson, Sangmelima.

A small forest tree, this *Drypetes* has oblong-ligulate, persistent, ridged stipules, most unusual in the genus. Although fairly widespread and well-collected, most of its known sites have suffered forest clearance over the last three decades.

Habitat: lowland evergreen forest; to 1200m alt.

Threats: clearance of forest for wood, agriculture and urban expansion (Kumba and Nkolbisson) is recorded at 8 of the 11 known sites and it has been extensive at many of these.

Management suggestions: enforcement of existing protected areas would ensure the survival at several sites, although it is uncertain whether any forest survives at the Nigerian localities. Demographic and other populational data for the taxon could be obtained by research at Mahole near the Bakossi Forest Reserve.

Drypetes preussii (Pax) Hutch.

VU B2ab(iii)

Range: SE Nigeria (1 coll.); Cameroon: Mt Cameroon (6 coll.), Kumba, Kumba-Mamfe (1 coll. each), Bipindi (6 coll.); Gabon: Koumemayong (1 coll.).

Assessed by WCMC in 1997 as VU B1 + 2c (IUCN 2003, downloaded on November 8th 2004), *Drypetes preussii* was there treated as being restricted to Cross River and adjoining forests in Cameroon. Inspection at the Kew Herbarium shows that its range extends along the coast to Gabon. Accordingly its extent of occurrence now exceeds 20,000km^2, so it is re-evaluated here accordingly to the new model for Criterion B (IUCN 2001). It is considered that eight sites are known. Threats to lowland forest in the Mt Cameroon area are documented in Cable & Cheek (1998) and in western Bakossi under *Crotonogyne strigosa*.

Habitat: lowland evergreen forest; 200m alt.

Threats: see above.

Management suggestions: as for *Crotonogyne zenkeri*.

Drypetes staudtii (Pax) Hutch.

VU B2ab(iii)

Range: SE Nigeria: Omo, Ikom, Calabar; Cameroon: Mt Cameroon (2 sites), Wum (1 coll.), Bakossi (2 sites), Lolodorf.

This canopy tree has very large, glossy, leathery, finely-veined oblong leaves resembling those of *D. magnistipula* but lacking persistent stipules. Its geographic distribution is patchy. Although known from only nine sites there are indications that it is locally fairly common. At Omo there are four collections, and at the Mokoko River F.R. (Mt Cameroon) there are eight. Meanwhile, adjacent forest reserves such as Onge or Bakundu have no records of the taxon. Were better data on local threats available throughout the range of this taxon, it would be better assessed under criterion A and then would be likely to rate EN or CR. Extensive losses of forest areas have occurred in Nigeria and are ongoing at Wum (*pers. obs.*) The forest at Mokoko has also been under great pressures for clearance.

Habitat: wet lowland evergreen forest, tending to sand substrates; 400m alt.

Threats: clearance of forest for timber and expansion of agriculture, both large scale commercial and small-holder.

Management suggestions: investigation of unpublished forest plot data from e.g. Mokoko might give data useful for developing a plan for the management of this species. In the absence of this the Nyandong or Mokoko River sites in Bakossi could be surveyed for this data. The latter looks to be a secure site for the taxon if correct conservation plans are enacted.

Hamilcoa zenkeri (Pax) Prain

VU B2ab(iii)

Range: Cameroon: Bipindi (5 coll.), Longii (2 coll.), Eseka, Barombi-Kang (2 coll.), S Bakundu F.R. (3 coll.), Loum F.R. (2 coll.).

An understorey shrub, vegetatively resembling a *Cola* but with slightly serrate leaves, this monotypic genus is only known from six sites, at several of which it is known from more than one collection and so may well be gregarious, even locally fairly common. Its absence from all Mt Cameroon forests except for the drier S. Bakundu F.R., and its existence in Bakossi only at the drier, semi-deciduous forest of Loum F.R., both suggest that the taxon might avoid the wetter evergreen forests.

Habitat: evergreen and semi-deciduous forest; to 400m alt.

Threats: much of the understorey of the Loum F.R. has been replaced by small-holder plots of e.g. coco-yam (*Colocasia*) and cacao (*Theobroma*). A large part of S Bakundu F.R. is converted to yam cultivation and the specimens from Longii and Barombi-kang both record its presence in logged or degraded forest suggesting that it can tolerate some disturbance, including logging. No data is available on the other sites. In future this taxon might be better assessed under Criterion A if data are available.

Management suggestions: reducing incursions into government forest reserves would protect this species in SW Province. This might be assisted by a public education programme. The sites in Littoral and S Province should be assessed for the existence of *Hamilcoa*, and of threats, in particular conversion of forest to other uses.

Phyllanthus caesiifolius Petra Hoffm. & Cheek

CR C2a(i); D1

Range: Cameroon: Bakossi Mts (2 coll.).

A monopodial, stoloniferous forest shrublet, *Phyllanthus caesiifolius* is immediately recognized even when not fertile by its bicolored leaves; the upper surface is dark green in the upper half and whitish blue in the bottom half. The paper in which it was published carries the conservation assessment (Hoffmann & Cheek 2003), reproduced below in modified form. Neither of the two known subpopulations (c. 15km apart) contain more than 50 individuals. It appears to be vegetatively apomictic.

Habitat: undisturbed submontane forest with *Gymnosiphon* sp. nov., *Xylopia africana*, *Oncoba ovalis*, *Dielsantha galeopsoides*, *Pentaloncha* sp. nov.; 1000–1275m alt.

Threats: forest clearance for agriculture.

Management suggestions: continued surveys are advised in the altitudinal range indicated in the hope of discovering further sites for the taxon. Production of a conservation poster for use in Bakossi might help locate more sites and educate people as to the rarity of this species.

Phyllanthus nyale Petra Hoffm. & Cheek

CR C2a(i); D1

Range: Cameroon: Bakossi Mts (2 sites).

This monopodial forest shrublet is distinguished from the other six species of *Phyllanthus* in Bakossi by its winged, not terete, branchlets. It is named for the type locality, Nyale Rock, a dramatic inselberg located east of Nyandong on the footpath to Kodmin. Neither of the two known subpopulations contain more than 50 individuals. *Phyllanthus nyale* was published with *P. caesiifolius* in a paper that included the conservation assessment above (Hoffmann & Cheek 2003).

Habitat: submontane forest; 1000m alt.

Threats: forest clearance for agriculture.

Management suggestions: as for *Phyllanthus caesiifolius*.

Pseudagrostistachys africana (Müll.Arg.) Pax & K.Hoffm. subsp. *africana*

VU A2c; B2ab(iii)

Range: Ghana; São Tomé; Bioko; SE Nigeria: Obudu Plateau (1 site); Cameroon: Mt Etinde; Mt Kupe and Bakossi Mts (4 sites), Bali Ngemba F.R.

Listed as VU A1c, B1 + 2c by Hawthorne in 1997 (IUCN 2003, downloaded on November 8th 2004), this monotypic genus, a tree restricted to submontane forest (apart from at one lowland site in Ghana), probably now has its largest subpopulation in Bakossi, where it is fairly secure. It is reassessed here on the basis of more extensive disturbance data in Cameroon and according to the modified IUCN criteria of 2001. In the field it is readily recognizable by the large Irvingiaceae-like sheathing apical stipule and the long-scalariform tertiary leaf venation.

Habitat: submontane, or rarely lowland, forest; 500–1500m alt.

Threats: forest clearance for wood and agriculture (Obudu Plateau and Bali Ngemba F.R.).

Management suggestions: the status of this taxon in São Tomé and Bioko needs more investigation. Bali Ngemba F.R. represents the most easily accessible and dense population of the taxon, followed by Mt Kupe; these are the more promising sites for demographic studies of the taxon.

Thecacoris annobonae Pax & K.Hoffm.

EN A2c

Range: Annobon; Cameroon, SW Province: Korup, Mungo River F.R., Ngusi, Kumba-S Bakundu and Kumba-Mamfe.

Judging by the number of collections (nine), this species was most abundant in the Kumba-S. Bakunda area. It has not been seen in S Bakundu in decades, although a single collection was made nearby in the Mokoko F.R. (Cable & Cheek 1998) in the 1990s. From Kumba its range extends northwest along the valley towards Mamfe (3 collections), eastwards towards Tombel, then up the Jide trough to Ngusi (1 collection each).

Habitat: lowland evergreen forest; to 400m alt.

Threats: clearance of forest for wood, agriculture and urban expansion; large parts of S Bakundu have been cleared for yam plantations (Cable & Cheek 1998) and the Kumba-Mamfe road is being massively upgraded with concomitant expansion of agriculture likely. 50% loss of the population of *T. annobonae* is estimated to have occurred in recent decades and this loss is continuing.

Management suggestions: the type, and modern material of *T. annobonae* from Annobon, need to be examined in more detail to establish whether it is the same entity as the Cameroonian material (see FWTA 1: 372, 1963). A survey in Annobon is needed to investigate survival of the species there. Although it has not been found in surveys at Bakossi in the late 1900s, lowland forest was not comprehensively investigated and it may well survive there. Once rediscovered, a plan for the management of *T. annobonae* can be made.

FLACOURTIACEAE
(M.Cheek)

Homalium hypolasium Mildbr.
EN B2ab(iii)

Range: Rio Muni; Cameroon: Yaoundé-Deng Deng and Mt Kupe (1 coll. each).

Described in the early 20th century from two collections (*Tessmann* 485 collected in 1908 from Rio Muni and *Mildbraed* 8508 collected in 1914 NE of Yaoundé), there appears to have been no subsequent collection of *H. hypolasium* until 1985 (*Thomas* 5100 at Mt Kupe).

Habitat: lowland and submontane evergreen forest, to 1100m alt.

Threats: forest clearance for agriculture and timber at Mt Kupe and probably (site visits needed to confirm) at the other two sites.

Management suggestions: Mt Kupe offers the most likely opportunity of refinding this species. It is advised that an effort be made to do this, and if successful, that further individuals be sought and censured in the area, together with levels of regeneration. If refound, conservation education, perhaps in the form of a poster, is suggested to benefit the local population. Surveys at the other two sites should be conducted if possible.

Oncoba lophocarpa Oliv.
VU A2c

Range: Cameroon: Mt Cameroon (c. 6 coll.), Kupe-Bakossi (13 coll. at 6 sites), Bamenda Highlands: Bali Ngemba. This canopy tree of submontane forest is notable for the rope-like inflorescences that can bear fried-egg-like flowers several metres from the trunk, where they emerge from the leaf litter like parasites. Previously this species was listed as LR nt (Cable & Cheek 1998) on the basis that it was restricted to Mt Cameroon where its habitat was considered unthreatened. Independently it was assessed as VU D2 under the name *Caloncoba lophocarpa* on the basis of it occurring at Bangem, Mamfe, Bakossi Mts and Mt Cameroon (WCMC 1997, based on forms by Peguy, T: (IUCN 2003, downloaded on November 8th 2004)). We now know that its range and number of sites are larger than previously thought (see above) and that it no longer qualifies as VU D2 but VU A2c since forest losses in the Bamenda Highlands part of this extended range equate to over 30% habitat loss over the last 100 years over the species range as a whole.

Habitat: submontane evergreen forest; (400–)800–1950m alt.

Threats: forest clearance for wood and agriculture. This is most significant at the lower parts of the altitudinal range (below 1000m) at all three areas. In the Bamenda Highlands the tree is probably now restricted to the Bali Ngemba Forest Reserve. These extensive highlands were probably once home to the main subpopulation of the species given the density of the species at the Bali Ngemba remnant. Elsewhere in the Bamenda Highlands, 25% loss of forest in the eight years between 1987–1995 is recorded by Moat in Cheek *et al.* (2000). With 13 collections at six sites, Kupe-Bakossi appears to be the stronghold for this species, where it is relatively secure and unthreatened.

Management suggestions: a population census, focussing upon the known sites at Kupe-Bakossi and Bali Ngemba, should be carried out and specific threats recorded. A poster campaign focussing on this striking species may serve to promote forest conservation to the local communities.

GESNERIACEAE
(M.Cheek)

Nodonema lineatum B.L.Burtt
VU B2ab(iii)

Range: Nigeria: Ogoja, Obudu Cattle Range and Boshi Extension F.R. (5 coll.); Cameroon: Nta Ali (1 coll., 1 obs.), Kupe-Bakossi Mts (5 sites).

First collected by Jane Medler in 1973 in Nigeria, and described in 1981, this monotypic genus resembles an African violet (*Saintpaulia*). Its habitat requirements are highly specific, but occur at several of the 'rocks' (inselbergs) in Kupe-Bakossi, notably Bime Rock, Kupe Rock and two smaller rocks above Nyasoso.

Habitat: perennially wet, mossy, vertical granite in forest shade; 800–1800m alt.

Threats: removal of shade due to forest clearance is the most likely threat. The rock-face habitat itself is not likely to be mined but plants are vulnerable to rock-falls. Much forest clearance in Ogoja has occurred in recent decades.

Management suggestions: data on the population size and demography of this species would inform better conservation management of this taxon. Such studies would best be carried out from Kupe village or Nyasoso both of which have subpopulations nearby.

GUTTIFERAE
(M.Cheek)

Allanblackia gabonensis (Pellegr.) Bamps
VU A2c

Range: Cameroon: Kupe-Bakossi (11 coll. at 5 sites), Bali Ngemba, Bamenda Highlands (several coll.), Mt Bana, Batcham, Yaoundé, Ebolowa, Sangmelima (1 coll. each); Gabon: Moubighou, Moucongo and Tcyengue (1 coll. each).

This tree is conspicuous for carpeting the forest floor with its pale lemon-coloured fallen flowers (usually red in the Bali Ngemba subpopulation), each about 6cm across. The largest part of its domain was probably once the Bamenda Highlands, where submontane forest is now confined to a few small parcels, the largest of which is at Bali Ngemba. Forest in these highlands is still being lost: 25% of forest in one area disappeared in an eight year period ending in 1995 (Moat in Cheek *et al.* 2000). Overall, more than 30% of the habitat of this species has probably disappeared over the last 100 years. Mt Kupe and the Bakossi Mts are now probably the stronghold for *A. gabonensis*. Elsewhere the species occurs on several of the small hills dotted through the forest belt in South and Central Province, finally extending into the Crystal Mts of Gabon.

Habitat: submontane forest; 700–1500m alt.

Threats: forest clearance for agriculture and wood particularly in the Bamenda Highlands and Yaoundé area.

Management suggestions: this species is fairly secure in the upper part of its altitudinal range in Kupe-Bakossi, but enforcement of the forest reserve boundary is needed if it is to survive at Bali Ngemba. Surveys should be conducted on the forested hills from which it has been collected elsewhere in Cameroon and also in the Crystal Mts, to determine whether it survives there and whether it can be protected at any of these sites.

Garcinia kola Heckel
VU A2cd

Range: Sierra Leone, Liberia, Ivory Coast, Ghana, Benin, Cameroon, Gabon, Congo (Kinshasa).

The assessment above is given by Hawthorne (IUCN 2003, downloaded on November 8th 2004), on the basis that it 'is probably the most important source of chewsticks. Over-exploitation has caused population declines. Seedlings are uncommon and slow-growing'. That assessment is apparently based only on its occurrence in Ghana and Congo (Kinshasa) since these are the only two countries cited under distribution. The range data cited above is taken from FWTA. In SW Province the main use for the species is not as a chew-stick (for which the most important species is *Garcinia mannii*), but for the comestible-medicinal seeds ('Bitter cola'), probably harvested sustainably from the fallen fruits. The seeds are marketed extensively by vendors all over at least the southern part of Cameroon. The species is at least occasionally cultivated (*pers. obs.*). On the basis of these observations in Cameroon, and given the large range of the species, *G. kola* would not otherwise be assessed as threatened. However, it is perfectly possible that general habitat loss and felling for dental hygiene in the western part of its range is sufficient to justify Hawthorne's rating, so this is maintained here.

ICACINACEAE
(I.Darbyshire)

Pyrenacantha cordicula Villiers
EN B2ab(iii)

Range: Ivory Coast: 60km N of Sassandra, Davo River Gorge (1 coll.); Ghana: Draw River Forest Reserve (1 coll.); Cameroon, SW Province: Ngombombeng near Nyasoso (1 coll.); Equatorial Guinea: location unknown.

Despite the wide range of this taxon, it appears extremely scarce throughout. It was first described by Engler under the invalid name *Chlamydocarya tessmannii* Engl., and was recorded as being found in Cameroon, though no specimens were cited. J.F.Villiers published the taxon, originally under the illegitimate name *Pyrenacantha cordata* Villiers, following the collection of a specimen with male flowers in the Ivory Coast in 1959 (*Leeuwenberg* 2084), from where it has not subsequently been recollected. Villiers also recorded the species as occurring in Cameroon and Equatorial Guinea, but cited no specimens. The Ghana collection, made in 1974, contains male flowers. Female flowers and fruits were unknown until the collection by M.Etuge and D.W.Thomas (*Etuge* 28) at Ngombombeng, Cameroon in 1987. Data are here derived from Fl. Cameroun.

Habitat: an understorey climber in dense humid forest or secondary forest; c. 750m alt.

Threats: widespread and continued loss of lowland and mid-elevation forest in the Ivory Coast is likely to threaten any extant populations of this species here. The site at Ngombombeng in Cameroon lies along the route from Nyasoso to Ngomboaku, a somewhat populous area with resultant widespread loss of forest below 1000m alt., the limit of effective forest protection on the adjacent Mt Kupe.

Management suggestions: more data are required on the distribution of this species in Cameroon and Equatorial Guinea, including previous collecting locations. The forest around Ngombombeng and adjacent forest areas should be surveyed to try to rediscover this taxon here. Formal description of the fruits of this taxon, from the *Etuge* 28 collection, should be made to aid field botanists in identification of this taxon in future.

LABIATAE
(B.J.Pollard)

Clerodendrum anomalum Letouzey
EN B2ab(iii)
Range: Cameroon: Kupe-Bakossi (1 coll.), Ngambé (1 coll.); Gabon (3 coll. at 3 sites).
C. anomalum was described in 1974, based on *Leeuwenberg* 9540 (type), and also collected from near the edge of our eastern boundary in 1972 (*Letouzey* 11085), 100km E of Douala, this being the only other Cameroonian subpopulation apart from at Kupe Village. Three Gabon collections were made by *Le Testu* (6096, 7594 & 8162), between 1926 and 1930.
Habitat: degraded lowland to submontane forest formations, overgrown edges to forest paths; c. 300–1000m alt.
Threats: degradation of forest may actually favour this species, but wholescale clearance for urbanisation or plantation establishment is a serious threat.
Management suggestions: attempt to relocate this species, and to introduce it into cultivation.

Plectranthus cataractarum B.J.Pollard
VU A2c; B2ab(ii,iv,v); C2a(i); D2
Range: Bioko: Ureka-Moca (3 coll.); Cameroon, SW Province: Bakossi (3 coll.), Mt Cameroon (1 coll.), Etinde (3 coll.).
This species was previously assessed in Pollard & Paton (2001); the details presented here follow that publication. Despite the wide extent of occurrence of this species, its area of occupancy is small, perhaps no more than 5–10km^2. Due to its extreme habitat specifity and the likely extinction of the Njonji subpopulation on Mt Etinde due to clearance of forest for plantation agriculture here, it is assessed as vulnerable.
Habitat: spray zone of waterfalls on wet rocks in or around fast flowing water within lowland to submontane forest, rarely non-epilithic on edge of watercourses; 300–1450m alt.
Threats: see above; in addition, changes in run-off regime following deforestation within the stream catchments may result in habitat changes which are unfavourable to this species, such as increased silting along watercourses.
Management suggestions: the proposed protection of the Bakossi sites should help to preserve this species. Confirmation of its continued existence on Mt Etinde or otherwise should be made.

Vitex lehmbachii Gürke
EN A2c
Range: Cameroon: Buea (4 coll.), Mt Etinde (1 coll.), Barike-Manya, Kumba (1 coll.), Kupe-Bakossi (3 coll. at 3 sites), Bamenda Highlands (1 coll.).
First collected at Buea, as *Lehmbach* 11, in April 1897, this species has been collected at ± regular intervals since, but often in areas which have undergone considerable decline in habitat quality.
Habitat: mid-elevation evergreen forest; 150–1330m alt.
Threats: only the 2 collections from Mwambong and Ngomboaku, in Bakossiland, are likely to have been unaffected by logging activities, and clearance of forest for agriculture. At all the other known locations, much clearance has occurred in the distant and recent past, and is likely to continue in the future. I estimate that the population size will have reduced by ≥50% over the last three generations, which I here estimate to be between c. 50 and 100 years.
Management suggestions: the two Bakossi sites should be revisited and seed collected, if available, to help introduce this species into cultivation. Searches around Buea could also be instigated, for the same purpose.

Vitex yaundensis Gürke
CR A4c; B1ab(i,ii,iii,iv,v) + 2ab(i,ii,iii,iv,v)
Range: Cameroon: Mungo River Forest Reserve (1 coll.) and Yaoundé (1 coll.).
This tree was only known from the type collection until *Cheek* 10139, from the Mungo River F.R., in 1999. It was first collected in jungle by the station at Cameroon's capital city, Yaoundé: *Zenker* 1412, fl., June 1897, at 800m altitude.
Habitat: low- to mid-elevation evergreen forest; 150–800m alt.
Threats: there can be little doubt that since Zenker's 1897 collection the Yaoundé location will have suffered enormously from the ever-growing population size, and resulting expansion of urbanisation. At the Mungo River

location, we know of illegal logging activity and clearance of habitat for agricultural purposes, which suggests that this taxon could be on the verge of extinction. Although known from two locations, they are severely fragmented, and indeed *V. yaundensis* could now be said to likely occur at just one location (Bakossiland).

Management suggestions: visits to the 2 known locations should be instigated to assess the chances of survival of this species. Inventory work should be continued in other uncollected areas of suitable habitat, between and around the two locations. If found, seed of this species should be distributed for germination at suitable sites, both *in situ* and *ex situ*.

LECYTHIDACEAE
(I.Darbyshire & M.Cheek)

Crateranthus talbotii Baker f.
VU A2c
Range: SE Nigeria, Cameroon.

The assessment VU B1+2ac is given in IUCN (2003, downloaded on November 8th 2004), which cites WCMC (1997) as its source together with recent (1996) information from Klaus Schmitt on its status in Nigeria: 'occurring in an area extending from south-east Nigeria into Cameroon. The largest, if not the only, remaining population occurs in the Oban Division of the Cross River National Park in Nigeria and the contiguous Korup National Park in Cameroon'. However, we now know that the species extends into the Rumpi Hills (*Mambo & Thomas* 6, 1986!) and into the Bakossi Mts (5 coll. at 3 sites), so that 13 sites are known (Darbyshire *obs.*). Accordingly, the taxon is no longer threatened under criterion B and so is reassessed here under criterion A, it being suspected that over 30% of its habitat has been lost in the last 100 years, largely in Nigeria, from the statement of Schmitt.

Habitat: lowland freshwater swamp forest (where it is gregarious, *pers. obs.*) and submontane forest near streams; 200–1000m alt.

Threats: forests outside protected areas have largely been logged and cleared for commercial crops and subsistence farming (IUCN 2003, downloaded on November 8th 2004).

Management suggestions: continued or improved protection of the Oban, Korup and proposed Bakossi National Parks is advised.

Napoleonaea egertonii Baker f.
VU B2ab(iii)
Range: Nigeria: Cross River State, Oban (3 coll., 2 sites), Atolo to Mamfe (1 coll.); Cameroon, SW Province: Kupe-Bakossi (3 coll., 3 sites), Takamanda Forest Reserve (1 coll.), Korup N.P. (1 coll.); Gabon: Mahounda (1 coll.), Ikembélé (1 coll.).

This striking forest tree was known from very few sites prior to the plant inventory work in western Cameroon beginning in the 1980s. Discoveries of this species at Takamanda and Korup are important as it is relatively well protected at these sites; however, it is not common at Korup, only 1–2 trees having been found (Cheek, *pers. obs.*), and its abundance at Takamanda is unknown. At Kupe Village and the adjacent Manehas Forest Reserve, the species is again uncommon, one plant being found at each location. However, several specimens were observed within close proximity to Nyandong in W Bakossi.

Habitat: low- to mid-elevation evergreen forest, often occurring on rocky slopes; 250–1000m alt.

Threats: the Nigerian sites are likely to have been either lost or under severe threat from widespread logging of lowland forest here. The two sites at Mt Kupe are below the 1000m alt. lower limit of effective forest protection and thus vulnerable to agricultural encroachment. At Nyandong, several trees were recorded close to the village and adjacent to tracks; these are highly vulnerable to future expansion of the village and road improvement.

Management suggestions: a survey of the number of trees of this species at Korup and Takamanda should be carried out to determine its abundance at these sites, as they are the best protected, thus offering the best opportunity for conservation of this species. Informing local communities, most notably at Nyandong where this species appears most common and in closest proximity to human settlement, of the scarcity of this species may help to promote community-led conservation, particularly as it is such a striking and easily recognisable taxon.

LEGUMINOSAE-CAESALPINIOIDEAE
(M.Cheek)

Afzelia pachyloba Harms
VU A1d
Range: Nigeria to Congo (Kinshasa).

Assessed thus by the African Regional Workshop (1997) on the basis of being a rainforest species heavily exploited for its commercial timber, with relatively few seed trees remaining through its range (IUCN 2003, downloaded on November 8th 2004). This assessment is maintained here.

Eurypetalum unijugum Harms
VU B2ab(iii)
Range: Cameroon: Baduma, S Bakundu, Lake Ejagham, Campo (each 1 coll.), Bipinde (3 coll.).
Not known from either adjoining Nigeria (FWTA) or Gabon (Fl. Gabon), this tree, though widespread in Cameroon, does seem endemic there. The range data above are taken from material at K and Fl. Cameroun.
Habitat: lowland evergreen forest; 0–400m alt.
Threats: the rarity of this species makes it vulnerable to forest clearance. At S Bakundu, yam farming has threatened the forest (Cable & Cheek 1998) while at Campo, extensive illegal commercial logging was observed in Feb. 1998 (*pers. obs.*). Threats at the other sites are unknown.
Management suggestions: both the Baduma and Ejagham collections record the species as locally abundant and at the second site profuse regeneration was recorded (*Letouzey* 14612, 13529). This may be a gregarious, grove forming species, which reduces the likelihood of extinction at any one site due to loss of a single individual. Surveys are advised at all known sites to assess the state and size of subpopulations and to investigate how well these sites are protected. This information could form the basis of a species management plan.

Gossweilerodendron joveri Normand ex Aubrév.
VU B2ab(iii)
Range: Cameroon: Nolbewoa, Kong, W Ngoulemakong, N Kumba, Yaoundé-Edea, Yaoundé-Mbalamayo (each 1 coll.); Angola, Rio Muni: Sendye, Okuamkos (each 1 coll.); Gabon: Goualé (1 coll.), Oveng (3 coll.).
The range data above are taken from the revision of *Prioria* by Breteler (1999) where the species is treated as *Prioria*, a genus not accepted ubiquitously.
Habitat: lowland evergreen forest; c. 0–500m alt.
Threats: forest clearance due to agriculture and logging.
Management suggestions: a survey to refind and establish subpopulation sizes and demography would inform a management plan for the species.

Loesenera talbotii Baker f.
VU A3c; B2ab(iii)
Range: SE Nigeria: Oban and Calabar-Mamfe (1 coll. each); SW Cameroon: Yingui-Yabassi, WSW Mamfe (1 coll. each), Mt Kupe-Bakossi (numerous sites and coll.).
This gregarious tree is only known from four sites outside Kupe-Bakossi, within which it is rather common across its altitudinal range. If this taxon is to be treated under Criterion B above, interpretation of what defines a site is critical to its status as VU. It could be argued that the Kupe-Bakossi area houses as many as 10 sites if the usual site definition adopted in this work is maintained. This taxon is also considered VU under Criterion A given at least an estimated 30% loss of habitat, particularly in the lower part of its altitudinal range, in the next 100 years.
Habitat: lowland and submontane evergreen forest; 180–1000m alt.
Threats: forest clearance for agriculture and wood, e.g. at Kupe village.
Management suggestions: research into the demography of this gregarious tree would inform development of a management plan. Since *Loesenera talbotii* is so frequent in Kupe-Bakossi, it is advisable that conservation efforts are focused here. This tree could form the basis of a public education campaign due to its conspicuousness in Bakossi.

Microberlinia bisulcata A.Chev.
CR A1c; 2c
Range: SW Cameroon.
The assessment above, made in Cable & Cheek (1998) is maintained here. Although the number of known sites has increased, that of Loum F.R. being added in this assessment, this is also threatened by forest destruction.

LEGUMINOSAE-PAPILIONOIDEAE
(M.Cheek)

Angylocalyx talbotii Baker f.
VU A3c; B2ab(iii)
Range: SE Nigeria: Oban, (3 coll.); SW Cameroon: Mt Cameroon (2 coll. at 2 sites), Korup (several coll.), Bambuko F.R. (1 coll.), Bakossi, Mungo River F.R. (1 site).

A cauliflorous treelet or shrub with beautiful, mottled flowers borne on the trunk, succeeded by long, large, brittle-succulent cylindrical fruit, this species is very similar to the closely related and much more common *A. oligophyllus* (see checklist section for differences). Only six sites are known.

Habitat: lowland evergreen forest; 100–200m alt.

Threats: forest clearance for logging and agriculture. Small-holder agriculture was found to be eating into the part of the Mungo River F.R. where we saw this species most recently. Market gardening at the Bambuko F.R. has destroyed much habitat there and forest around Mt Cameroon is also under threat (Cable & Cheek 1998).

Management suggestions: a revision of *Angylocalyx* taxa would clarify the distinctiveness and characteristics of this tree. As matters stand it is threatened with extinction at all its known sites, except at Korup N.P. Enforcement of existing reserve boundaries would secure the protection of *A. talbotii*.

Crotalaria ledermannii Baker f.
VU D2

Range: Nigeria: Mambilla Plateau (1 coll.); Cameroon: Mwanenguba Mts (1 coll.), Bamenda Highlands (7 coll.). This herb was assessed as above in Cheek *et al.* (2000). No new data have come to light since then, apart from the recent collection at Bali Ngemba F.R., so the original conservation assessment is maintained here.

Dalbergia oligophylla Baker ex Hutch. & Dalziel
EN A2c; A3c

Range: Cameroon: Mt Cameroon (4 coll.), Bafut Ngemba (1 coll.), Bali Ngemba (1 coll.), Mt Kupe (6 coll.) and Mwanenguba (1 coll.).

Known only from six sites, this shrub-liana dries a distinctive black, helping to distinguish it from similar species from Sierra Leone and Gabon, with which it has been confused.

Habitat: submontane and montane forest edge; (900–)1500–2000m alt.

Threats: forest clearance for agriculture and wood: the subpopulations at Bafut Ngemba, Obudu Plateau and Mwanenguba have quite possibly been lost due to this already. The Bamenda Highlands may once have been the main range of this species (judging by its existence at Bali Ngemba and Bafut Ngemba) but forest loss here has been as high as 25% between 1987–1995 in one area studied (Moat in Cheek *et al.* 2000). Overall, over 50% habitat loss is postulated over the last 100 years, and at least 50% of that which remains could be lost in the next century.

Management suggestions: unless Bali Ngemba F.R. is protected from incursion more vigorously, the only hope for the survival of this species is the summit area of Mt Kupe and the upper tree line of Mt Cameroon both of which are reasonably secure from threat.

Millettia macrophylla Benth
VU A1c; B1 + 2c

Range: SE Nigeria, Bioko and Cameroon.

The assessment above was made by WCMC (1997) on the basis that it is threatened by heavy logging and clearance for agriculture (IUCN 2003, downloaded on November 8th 2004). Only eight collections are held at K from Cameroon and the species occurs only in threatened lowland forest, thus the assessment seems justified and is maintained here.

Habitat: lowland evergreen forest; c. 400m alt.

Threats: see above.

Management suggestions: the best hope for the survival of this species is in well-protected lowland forest reserves, such as Korup. Future plans to better protect the Mungo River F.R. should also help secure the survival of this tree.

LINACEAE
(M.Cheek)

Hugonia macrophylla Oliv.
VU A2c; B2ab(iii)

Range: S Nigeria (4 coll. at 3 sites); Cameroon: Mt Cameroon (2 coll. at 1 site), W Bakossi (1 site); Gabon (4 coll. at 4 sites).

This liana, with nine known sites, has a very patchy distribution. This is possibly due in part to under-collecting, but even if some more sites are discovered, it is likely that, overall, its area of occupancy will be reduced by over 30% in the next 50 years. Range data are taken from FWTA, Fl. Cameroun and Fl. Gabon.

Habitat: lowland evergreen forest; to c. 650m alt.

Threats: clearance of forest for agriculture (Mt Cameroon, W Bakossi), wood (e.g. Nigeria), urban expansion (Sibange Farm, Gabon) or a combination of these.

Management suggestions: a survey of known sites is needed to find where the taxon is extant, and to assess the best means to conserve it.

Hugonia micans Engl.
VU A2c; B2ab(iii)

Range: Cameroon: W Bakossi, Kribi-Lolodorf, Nsambi (1 coll. each); Gabon (11 coll. at 5 sites).

With just eight known sites, this liana is clearly rare, and vulnerable under criterion B. The extent of habitat loss (estimated as over 30% of area of occupancy in the last 50 years), principally in Gabon, also qualify it as VU under criterion A. Range data are taken from FWTA, Fl. Cameroun and Fl. Gabon.

Habitat: lowland evergreen forest; c. 500m alt.

Threats: urban expansion in the Libreville area has probably led to the removal of forest at what was the stronghold for this species. (6 collections); elsewhere forest clearance for agriculture and wood has been and continues to be a threat to this taxon.

Management suggestions: see under *H. macrophylla*.

LOGANIACEAE
(M.Cheek)

Anthocleista microphylla Wernham
VU A3c; B2ab(iii)

Range: Ghana (Atewa Hills, 1 coll.); SE Nigeria (2 coll.); Bioko (2 coll.); Principé (1 coll.); São Tomé (7 coll. at 4 sites); Cameroon: Kupe-Bakossi (6 coll. at 2 sites).

This strangling shrub of lower submontane forest has its stronghold in São Tomé with four sites and many more collections, the most recent in 1993. Elsewhere it is very rare, and is known from a total of only 10 sites. Range data are partly taken from the revision by Leeuwenberg (1961). Loss of its habitat of over 30% in the next 50 years is predicted. Limited habitat disturbance appears beneficial for *A. microphylla*; several collections occur in areas of regenerating (secondary) forest.

Habitat: disturbed submontane forest; 800–1200m alt.

Threats: forest clearance for agriculture and wood; it may already have been lost from both Nigerian sites (Oban and Obudu) since significant forest loss has occurred there in recent decades. Although it is fairly common at Kupe village, it occurs there at such low altitudes (c. 800m) that it is vulnerable to agricultural expansion.

Management suggestions: more data are needed on the survival of this species at its various sites and the extent to which it is already protected. São Tomé should be the priority for a survey.

Anthocleista scandens Hook.f.
VU A2c

Range: Bioko: Clarence Peak, Moca; SE Nigeria: Obudu (1 site); São Tomé (5 sites); Cameroon: Bafut Ngemba (2 coll.), Gepka, Nkambe (1 coll.), Mt Etinde (3 coll.), Mt Kupe (9 coll. at 3 sites), Mt Oku (2 sites).

It is estimated here that, over the last 100 years, over 30% of the habitat of this species, mostly in the Bamenda Highlands and Bamboutos, has been lost due to forest clearance (see threats below). This strangling epiphytic shrub or small tree is spectacular in flower.

Habitat: submontane forest; 1200–2000m alt.

Threats: forest clearance for agriculture and wood, particularly in the Bamenda Highlands, where forest loss has been running at 25% over eight years at one sample area (Moat in Cheek *et al.* 2000).

Management suggestions: forest losses in recent decades suggest that *A. scandens* may no longer survive at Bafut Ngemba, Gepka or Obudu. Conservation efforts are probably best focused where known sites are most concentrated, namely São Tomé (5 sites, but protection levels unknown) and Mt Kupe (3 sites, protection levels currently high). At Mt Oku this taxon occurs at the lower altitudinal boundary of protection and is likely to become extinct; one of the two sites there occurs outside this boundary. The population at Mt Etinde occurs at the peak and is inaccessible to all but trekkers. Unless the summit area is cleared for touristic purposes, *A. scandens* seems secure here.

Strychnos elaeocarpa Gilg ex Leeuwenb.
VU B2ab(iii)

Range: Cameroon: Edea-Kribi 58–65 km, Kumba-Mamfe mile 15, Loum-Kumba at Mungo Bridge, Likumba-Tiko, Mt Cameroon (Mabeta-Moliwe and Onge).

Known from only ten collections at six sites (data from Leeuwenberg 1969, and Cable & Cheek 1998), this is a rare lowland species of restricted distribution.

Habitat: lowland evergreen forest; 50–200m alt.

Threats: clearance of forest for agriculture, both small-holder (Kumba-Mamfe) and large-scale plantation (Likomba-Tiko, where forest is now absent, and both Onge and Mabeta-Moliwe, scheduled for CDC plantation expansion); also building-sand extraction at Mungo Bridge.

Management suggestions: as is usual with lianas, this taxon is probably often overlooked. More intensive surveys in the Mungo River and Bakossi Forest Reserves might locate subpopulations which, if those boundaries are respected, are reasonably secure. This aside, no known site for the species appears to be within an officially protected area.

Strychnos staudtii Gilg
VU A2c
Range: Cameroon: Mt Cameroon (7 coll. at 3 sites), Kumba area (3 coll.), Kupe-Bakossi (21 coll. at 8 sites); Gabon (2 coll. at 2 sites).

This lowland to submontane tree is unusual in a genus predominantly consisting of lianas. Although the most commonly collected and widespread *Strychnos* in Kupe-Bakossi, it is vulnerable there to forest clearance because its low altitudinal range puts several of its sites outside of protected areas. It is also very heavily threatened in the lowlands around Mt Cameroon (Cable & Cheek 1998), as it is in the Kumba area. Although threats in the Gabon area are unknown, even assuming it is secure here, an overall loss of habitat for *S. staudtii* of over 30% in the next 100 years can be confidently predicted.

Habitat: evergreen forest; 200–1200m alt.

Threats: clearance of lowland forest for commercial plantation expansion (Mt Cameroon area) and small-holder agriculture (elsewhere). Threats in Gabon are not known.

Management suggestions: *Strychnos staudtii* is fairly secure at Mt Kupe on the lower slopes; such is the density of individuals there that currently no action is needed to ensure its survival.

LORANTHACEAE
(M.Cheek & B.J.Pollard)

Tapinanthus preussii (Engl.) Tiegh.
VU A2c + 3c
Range: SE Nigeria (1 site); coastal Cameroon (7 sites); Gabon (2 sites); Cabinda (1 site).

This parasitic shrub is known from only 11 sites in lowland forest (Polhill & Wiens 1998). It appears to have a very patchy distribution, not being recorded from some extensive areas which have been well surveyed, such as Mt Cameroon (Cable & Cheek 1998). The low altitude at which it occurs makes it vulnerable to forest clearance throughout its range. This taxon is conspicuous for its large flowers, possibly the longest in *Tapinanthus*.

Habitat: lowland evergreen forest; to c. 900m alt.

Threats: clearance of forest for agriculture and wood.

Management suggestions: efforts should be made to rediscover the species at its known sites and to obtain protection, perhaps by public education, e.g. through a poster campaign. All Loranthaceae are perceived of as nuisance plants in Cameroon since some taxa spread to crop trees.

MEDUSANDRACEAE
(M.Cheek)

Medusandra richardsiana Brenan
VU A3c
Range: Cameroon: SW Province.

The analysis of specimens of this tree by Brenan in 1952 showed it not only to be a new species to science, but a new genus and family. It behaves as a pioneer (D.W.Thomas, *pers. comm.*) so some forest disturbance may favour it, although forest clearance is a threat. It is here estimated that over 30% of its habitat will be destroyed in the next 100 years. It is almost confined to SW Province at the locations indicated below.

Habitat: lowland evergreen forest; 200–600m alt.

Threats: clearance of forest for expansion of plantations (Mt Cameroon, eastern slopes at Bimbia), increased small-holder agriculture and logging along the upgraded Kumba-Mamfe road, clearance of forest for yam farming in S Bakundu.

Management suggestions: increased and more effective policing of existing areas gazetted as protected, particularly the S Bakundu F.R., would secure the future of this species.

MELASTOMATACEAE
(M.Cheek)

Amphiblemma amoenum Jacq.-Fél.
EN B2ab(iii)
Range: Cameroon, SW Province: northern Rumpi Hills to Nta Ali, western Bakossi Mts.
Known only from four *Letouzey* collections, this prostrate herb was first described in 1976. It is particularly distinct within *Amphiblemma*.
Habitat: lowland to submontane forest, sometimes on rocks near streams; 850m alt.
Threats: clearance of forest for agriculture and wood: only one site (Nta Ali) is protected by a reserve, as far as is known.
Management suggestions: the known sites should be revisited and surveyed for the species to assess whether they are included in protected areas or not. Public education might help to reduce pressure on the surviving plants.

Memecylon dasyanthum Gilg & Ledermann ex Engl.
VU A2c
Range: Cameroon: Mt Kupe (1 coll. at 1 site) to Bamenda Highlands and Bamoun (numerous coll. at several sites).
This tree of upper altitude forest appears often to have been confused (e.g. in Cable & Cheek 1998) with the lower altitude *M. myrianthum*. In the Bamenda Highlands it is the commonest Memecylonoid in e.g. Bali Ngemba but this forest is diminishing rapidly. There is no doubt that over 30% of its habitat will be lost in the next century. It is likely that if better data were available for forest loss in the Bamoun area, the assessment would be EN.
Habitat: upper submontane forest; 1550–1900m alt.
Threats: clearance of forest for agriculture and wood, especially in the Bamenda Highlands where between 1987–1995 25% of forest cover was lost in one sample area (Moat in Cheek *et al.* 2000). The forest of the Bamoun highlands are also afflicted by forest loss, particularly due to dry-season fires (Ndam *pers. comm.*).
Management suggestions: although the species is reasonably secure at Mt Kupe, this population seems to be an outlier. The Bali Ngemba F.R. in NW Province has the densest subpopulation of *M. dasyanthum* known, and should be the priority for protection of the species. Enforcement of existing boundaries would probably be adequate to secure the future of the species there. More data are needed on the survival of forest habitat in the Bamoun area and the continued existence of this species there.

MELIACEAE
(M.Cheek)

Entandrophragma angolense (Welw.) C.DC.
Entandrophragma cylindricum (Sprague) Sprague
Entandrophragma utile (Dowe & Sprague) Sprague
Khaya ivorensis A. Chev.
Lovoa trichilioides Harms
VU A1cd
The above are all internationally traded timber species of the mahogany family which were listed as VU by Hawthorne (Hawthorne 1997 in www.redlist.org), and (IUCN 2003, downloaded on November 8th 2004) using the 1994 criteria of IUCN. They all have a wide range in Africa and, were they reassessed in this book, without reference to their use as timber, they would probably be downlisted. Hawthorne variously cites over-exploitation, poor levels of regeneration, fire damage, and slow growth to support his assessments of these species.

MENISPERMACEAE
(B.J.Pollard)

Platytinospora buchholzii (Engl.) Diels var. *macrophylla* Diels
CR A2c + 3c
Range: Nigeria (1 coll.); Cameroon: Mt Kupe (2 coll.), Douala-Edea Reserve (1 coll.), Bipindi (1 coll.), Yangamok II (1 coll.).
Zenker 3014a was cited by Diels (Engler 1910) ollected at Bipindi, whence *Zenker* 4008, st., 1911 was also collected. From 1969, three collections were made within 10 years of each other: *Letouzey* 9610, fr., 26 Nov., at Yangafok II, 25 km ENE from Bafia, the fruiting Nigerian collection of *Wit & Daramola* in FHI 64887, from the Ago-Owu Forest Reserve, West State, 28 Dec. 1971, and *McKey* 260, st., 26 May 1979, fallow fields by the village of Cité-Lac Tissongo, Douala-Edea Reserve. Two further collections were made from Bakossi, along Walter's trail

on Mt Kupe itself in October 1995 (*Cheek* 7535, st., 1000m) and June 1996 (*Cable* 2898b, st., 1000m). Additional material is desirable to enhance our knowledge of this monospecific genus.

Habitat: lowland and submontane forest; 0–1000m alt.

Threats: forest clearance, especially at lower altitudes, for agriculture. Even if the trees are left standing, but the undergrowth is cleared, vines can all be cut at ground level, as in the Loum F.R. understorey.

Management suggestions: the location of the subpopulation on Mt Kupe should be revisited, and an assessment of the number of individuals and regeneration made.

Tiliacora lehmbachii Engl.
EN A2c + 3c; B2ab(iii)

Range: Cameroon: Mt Cameroon (2 coll.), Mt Kupe (1 coll.); Congo (Kinshasa) (1 coll.).

Engler described *Tiliacora lehmbachii* in 1899 based on *Lehmbach* 905 (type), fl., Nov. 1897 collected at Buea, 1000m, the same approximate location as the second known collection, *Maitland* 313, fr., Jan. 1929. A Congo (Kinshasa) collection, *Germain* 404, fl., 29 July 1943, was made in 'Forestier Central, Yangambi, île Tutuku' at 470m. The only other collection is *Cheek* 7809, fl., 15 Nov. 1995 at 960m near Kupe Village.

Habitat: lowland and submontane forest; 470–1000m alt.

Threats: forest clearance, especially at lower altitudes, for agriculture. Even if the trees are left standing, but the undergrowth is cleared, vines can all be cut at ground level, as in the Loum F.R. understorey.

Management suggestions: the subpopulation near Kupe village should be revisited, and an assessment of the number of individuals and regeneration made. There has been much vegetation clearance around Buea, so Mt Kupe offers the best opportunity for relocating this species in Cameroon.

Triclisia lanceolata Troupin
EN A2c + 3c; B2ab(iii)

Range: Cameroon: Bakossi Mts (1 coll.), M'balmayo (1 coll.); Congo (Kinshasa) (3 coll.).

Troupin described *Triclisia lanceolata* in 1949, from two Gillett specimens collected in the 'District du Bas-Congo, environs de Kisantu' citing *Gillett* 3395, from 1903 as the type and *Gillett* s.n., from 1909. A third congolian collection, *Compère* 1239, was made on 21 Jan. 1960 (Léopoldsville Province, Territ Thysville). *W.J.J.O de Wilde* 1900 was the first Cameroon collection, made on 12 Feb. 1964, about 5km S of M'balmayo, S of Yaoundé, on the border of the Nyong river. It then remained uncollected until *Cheek* 9169, on 2 Nov. 1998 at Edib village in the heart of Bakossiland, Cameroon. This collection was of fallen fruits only, which match well those on all the other specimens held at Kew. A 1946 collection from SW Nigeria, *Jones & Onochie* in FHI 17554, during the vegetational reconnaissance to Omo and Shasha Forest Reserves, is annotated (at K) as *sp. nr. lanceolata*. It differs in having larger, membranaceous (not coriaceous) leaves, and is not included in this assessment. The ♂ flowers are unknown, a fact less surprising when one considers that these vines are probably undercollected. This could be due to the difficulty of collecting fertile specimens from high up in the canopy. Although widespread, it appears to be rare within its range, and the subpopulations are severely fragmented.

Habitat: lowland and submontane forest; 0–1500m alt.

Threats: forest clearance, especially at lower altitudes, for agriculture. Even if the trees are left standing, but the undergrowth is cleared, vines can all be cut at ground level, as in the Loum F.R. understorey.

Management suggestions: the subpopulation near Edib village should be revisited, and an assessment of the number of individuals and regeneration made.

Triclisia macrophylla Oliv.
CR A2c + 3c

Range: Sierra Leone (1 coll.); Bioko (1 coll.); Cameroon: Mt Cameroon (1 coll.), Mt Kupe (1 coll.), Bakossi Mts (1 coll.).

Daniel Oliver described *Triclisia macrophylla* in 1868, based on Mann's collection of Jan. 1860 from Bioko. It was subsequently collected on the African mainland at Mt Cameroon in Nov. 1891, *Jungner* 144 (UPS), and soon after by the Swedish botanist Afzelius in Sierra Leone (1896, *Afzelius* s.n., UPS). It remained uncollected for 100 years, before *Cable* 2739, fl., 30 May 1996, 1000m alt., near Kupe Village. As part of a series of 25m × 25m plots conducted during botanical surveys in Cameroon, a final collection (st.) was made in the Bakossi Mts, in 1998, as *Plot Voucher* B96, from Kodmin at 1500m. Although widespread, it is rare within its range, and the subpopulations are severely fragmented.

Habitat: lowland and submontane forest; 0–1500m alt.

Threats: continued clearance of low altitude forest for plantations, agriculture and urban expansion. Troupin (1962: 93) describes its habitat as 'fôrets denses humides sempervirentes de basse altitude', which would suggest below 1000m. It is likely that all of these collections were from at or below 1000m, and on Mt Kupe the level of protection is above this altitude. The proposed Bakossi Mts National Park should provide a level of protection for the Kodmin subpopulation, but it is quite probable that the other subpopulations have already disappeared.

Management suggestions: the subpopulations near Kupe Village and Kodmin should be revisited, and an assessment of the number of individuals made.

MORACEAE
(M.Cheek & B.J.Pollard)

Dorstenia astyanactis Aké Assi
EN B2ab(iii)
Range: Ivory Coast (Mt Tonkui, several coll.); Cameroon, SW Province: Bakossi Mts (Nyale Rock, Ngomboaku), Mt Kupe (Nyasoso - Nature Trail), NW Province: Nkom-Wum F.R., below Bu village.
To Martin Etuge goes the distinction of discovering all four of the Cameroonian sites for this species, probably the most bizarre, and certainly the only epiphytic, *Dorstenia* in Africa. First discovered in Ivory Coast by Aké Assi where it is only known from Mt Tonkui, *D. astyanactis*, published in 1997, is there either extinct or on the verge of extinction due to forest loss (Laurent Gautier *pers. comm.*). In Cameroon it is on the edge of extirpation by agricultural expansion at all its sites save Nyale. It is to be hoped that this expansion will cease. The assessment above was first published in Pollard *et al.* (2003).
Habitat: submontane forest; 600–1000m alt.
Threats: forest clearance for agriculture and wood.
Management suggestions: conservation education as to the significance of this species may help the survival of this species at Nyasoso, Nyale and at Bu.

Dorstenia poinsettiifolia Engl. var. *etugeana* B.J.Pollard
EN B2ab(iii)
Range: Cameroon, SW Province: Rumpi Hills (1 site), Bakossi Mts (Nyale Rock, Nyale-Kodmin and Nyandong).
Four sites are known for this very distinct variety of *Dorstenia*. The assessment above was first published in Pollard *et al.* (2003).
Habitat: lowland and submontane forest; 450–1000m alt.
Threats: forest loss due to expansion of agriculture.
Management suggestions: annual monitoring of sites to detect deleterious changes, and further surveys of unexplored areas to detect potential new sites for the taxon.

Dorstenia prorepens Engl.
VU B2ab(iii)
Range: SE Nigeria: Okumu F.R., Oban; Bioko; Cameroon: Buea, Mt Kupe (Kupe village, Nyasoso: 1 coll. each).
Eight sites, all within a small range, are known for this creeping to ascending herb.
Habitat: lowland and submontane forest; 400–1550m alt.
Threats: forest loss due to agricultural expansion and wood excavation, particularly at Bambuko (Cable & Cheek 1998), and due to urban expansion at Kumba.
Management suggestions: surveys of subpopulations at each known site to detect whether the taxon survives at these, and if so, to gather quantitative data on individuals, regeneration and threats. This data could inform a plan to guide management of the survival of this species.

Ficus chlamydocarpa Mildbr. & Burret subsp. *chlamydocarpa*
VU A2c; B2ab(iii)
Range: Bioko (1 site); Cameroon: Mt Cameroon (3 coll. at Buea), Mt Kupe, Bakossi Mts, Bamboutos-Dschang (2 coll.), Bafang-Fotouni, Mt Oku-Lumeto, Nkambe-Tabenken, Nkambe-Mben.
This large tree is only known from ten sites. The range data above are taken from FWTA, Fl. Cameroun, Cable & Cheek (1998), Cheek *et al.* (2000) and the Kew herbarium. Records for the taxon in FWTA for Principé and São Tomé appear unsubstantiated.
Habitat: submontane and montane forest; 1300–2300m alt.
Threats: forest clearance for agriculture and wood; at Mt Cameroon it was collected above Buea and recorded as 'frequent' in 1929–1931, but was not refound in the intensive survey of Mt Cameroon in the early 1990s. At Bafang and Dschang almost all forest has been lost in recent decades. 25% forest loss in one area of the Bamenda Highlands 1987–1995 (Moat in Cheek *et al.* 2000) suggest that more than 30% of the habitat of this species over its whole range has been lost in the last century.
Management suggestions: a concerted effort should be made to find this tree in the Buea area, and at its other sites and to assess the size of its subpopulations, regeneration and threats. Surveys of sacred and royal forests should pay particular attention to *Ficus* since *Ficus* are often revered, and such places may yield new sites for this taxon.

Ficus tremula Warb. subsp. *kimuenzensis* (Warb.) C.C.Berg

VU B2ab(iii)

Range: SE Nigeria: Calabar (1 coll.); Cameroon: Bakossi (Kodmin, 2 coll.), Kribi, Kribi-Bipinde, Barundi-Muramuya; Congo (Kinshasa): Kinkosi, Kimuenza; Angola (1 coll. each).

This very distinctive strangler is only known from the seven sites above. Although widespread, it seems extremely local. It has not been recorded in recent exhaustive surveys at either Mt Cameroon or Mt Oku (Cable & Cheek 1998; Cheek *et al.* 2000).

Habitat: lowland (less usually submontane) evergreen forest; 0–2000m alt.

Threats: forest clearance for urbanization (Calabar) or touristic development (Kribi) or more generally for agriculture and wood.

Management suggestions: surveys to rediscover this species at its known sites are advised. Subpopulational data including regeneration levels and threats should be gathered, where the plant can be refound. This could form the basis for future monitoring and development of a management plan for the species. In the short term, public education at Kodmin regarding the conservation significance of this taxon is desirable.

MYRICACEAE
(M.Cheek)

Morella arborea (Hutch.) Cheek
(syn. *Myrica arborea* Hutch.)

VU A3c

Range: Bioko (Moka area, 4 coll.); Cameroon: Mt Cameroon (two sites, 6 coll.), Mwanenguba (6 coll.), Bamboutos-Djuttitsa, Bamenda Highlands, Djottin, massif Mbam, Ijim Ridge (1 coll.).

Morella arborea is one of the few tree species restricted to the Cameroon Highlands. Within this area it has a wide range, but is generally rather rare.

Habitat: submontane and montane forest edge, or isolated in grassland; 1300–2400m alt.

Threats: clearance of montane forest for agriculture (e.g. by fire for both pastoral and tilled land) and wood, particularly firewood. It is here estimated that over 30% of the habitat of *Morella arborea* will be lost over the next 100 years, predominantly in the Mwanenguba-Bamenda Highlands part of its range. At Mt Cameroon, and probably also at Bioko, pressures on forest at higher altitudes is low.

Management suggestions: research is needed on the ecology and demography of this species. *Morella arborea* appears to have some fire-resistant qualities (*pers. obs.*) but more exact data are required. Limited grassland fires may equally favour the species as threaten it.

MYRSINACEAE
(I.Darbyshire)

Ardisia koupensis Taton

EN B1ab(iii)

Range: Cameroon, SW Province: Mt Kupe (19 coll. from 3 sites), Mt Nlonako (1 coll.).

This taxon was first described by Taton in 1979 from single collections from the western slopes of Mt Kupe, above Mbule town, and from the western slopes of Mt Nlonako, Kupe's sister peak to the northeast, both collected in 1976 (*Letouzey* 14669, 14466). Further inventory work on Mt Kupe has added 17 more collections from 2 additional sites on the western slopes, being particularly prominent above Kupe Village. However, this species has not been found in the adjacent Bakossi Mountains.

Habitat: mid-altitude to submontane closed canopy forest understorey, occasionally secondary forest; 650–1250m alt.

Threats: the lower elevation sites for this taxon on Mt Kupe are significantly below the lower limit of effective forest conservation on the mountain and are therefore vulnerable to continuing habitat loss, particularly above the expanding town of Nyasoso. At 650m alt. near Kupe village it was recorded in forest very close to farmland; such sites are particularly at risk. The local custom of using branches of this species to drive away flies from corpses may also lead to losses of individual plants within the vicinity of settlements.

Management suggestions: further inventory work at other submontane locations on Mt Kupe and Nlonako will likely reveal further sites for this taxon; it may then be downgraded to VU under IUCN criteria. Continued protection of submontane forest above 1000m on Mt Kupe should ensure its future survival.

MYRTACEAE
(M.Cheek)

Eugenia fernandopoana Engl. & Brehmer
VU B2ab(iii)
Range: Bioko (2 coll.); Corisco Is.; C.A.R.; Cameroon: Kribi, Longii Beach, Bangem (1 coll. each).
Known from only six sites, this forest shrub appears very rare within its range. It is not certain that any of these sites are effectively protected.
Habitat: lowland evergreen forest; 0–800m alt.
Threats: forest clearance for agriculture (e.g. Bangem and Bioko) and for tourism development (Kribi); the threats at the sites are Corisco Isl. and C.A.R. are unknown.
Management suggestions: a revision of African *Eugenia* is desirable. The identity of the Bangem specimen should be checked. Site visits to establish whether the species survives at each of its known locations is suggested. These could be combined with a demographic study.

OCHNACEAE
(I.Darbyshire)

Campylospermum letouzeyi Farron
VU A2c; B2ab(iii)
Range: Cameroon, SW Province: W Bakossi (2 coll.), Mt Cameroon (4 coll.), Littoral Province: nr. Melong (2 coll.), W Province: Bangati (1 coll.).
First described from lowland forest near Bouda, Melong, Littoral Province in 1969, this taxon has been recorded at 8 further sites, with the botanical inventory work of the 1980s and 1990s identifying SW Province as the stronghold of this species. At Mt Cameroon, it has been collected from Munyenge, Mundongo, Onge and the S Bakundu Forest Reserve. In the Kupe-Bakossi area the taxon appears rare, being recorded only twice, from Kurume and Mekom in western Bakossi; however, the paucity of collections here may be in part due to the limited inventory work carried out at lower altitudes.
Habitat: lowland forest especially in secondary regrowth, appearing to favour only light shade; 0–300m alt.
Threats: conversion of large areas of lowland forest in western Cameroon to permanent agricultural land, particularly intensive plantation agriculture, have severely reduced this species' habitat. Proposals for further conversion of forest to plantation around Mt Cameroon will result in further losses.
Management suggestions: further botanical surveys of lowland forest, for example in S Bakossi and Kumba District may reveal further sites for this species; care should be taken to separate it from other large-leaved *Campylospermum* spp. More data on its ecology are required; if it does require early secondary regrowth, areas of such vegetation should be maintained through active management in lowland protected areas, such as the Mungo-Bakossi Forest Reserves.

Lophira alata Banks ex Gaertn.f.
VU A1cd
Range: Guinea (Conakry) to Congo (Kinshasa).
This tree was assessed as above by the African Regional Workshop of 1997, as cited in IUCN (2003; www.redlist.org) on the basis of large scale destruction of wet evergreen forest throughout its range, over-exploitation as a timber source, slow rates of growth and poor regeneration levels in less than optimum conditions. Its timber, Azobe, rich in silica, is resistant to marine-borer and favoured for use in jetties. In Cameroon it is common, widespread and regenerates easily in many areas and does not appear threatened; however, this is evidently not the case in other parts of its range, so this assessment is retained here.
Habitat: lowland evergreen forest; 400–1000(–1300)m alt.
Threats: see above.
Management suggestions: improved protection and management (e.g. felling regimes) of existing forest reserves.

PIPERACEAE
(M.Cheek)

Peperomia kamerunana C.DC.
EN B2ab(iii)
Range: Bioko (2 coll.); Cameroon: Mt Cameroon (5 coll., 2 or more sites), Mt Kupe (2 coll.), Mwanenguba (1 coll.)

In 1998 (Cable & Cheek 1998), I assessed this species as LR nt. At that time it was only known from Bioko and Mt Cameroon, and I assumed, from the fact that it was known from two locations on the mountain, that it was widespread there, but overlooked. There were, and are, no known threats at the altitudinal range at which it occurs, there or on Bioko. The range of the species has now been extended to Mt Kupe, where it is also secure, and to Mwanenguba where it is threatened (see below). Owing to imprecise data on the seven specimens known from Bioko and Mt Cameroon, it is known with certainty from only three sites at these places. If the new specimens are added, the species is known from a total of ten collections at five sites. While the species is locally plentiful (*fide* Brenan 9371), it must be extremely local, given that it has not been collected again on Mt Cameroon in over 50 years, despite intensive surveys there in the 1990s, including a large and expert botanical survey team based at or very near Brenan's site for a week in 1992. Although two specimens were recently discovered (1992) at Mt Kupe, this is in comparison to 24 specimens of *P. fernandopoiana* from the same area.

Habitat: submontane and montane forest; 1400–2300m alt.

Threats: forest reduction at Mwanenguba for firewood and due to man-made fires set by pastoralists.

Management suggestions: although threatened at Mwanenguba, it appears secure at the other known sites. Nonetheless, it should be rediscovered and monitored at these. Quantitative data on each of the subpopulations would be valuable for developing a management plan for the taxon.

PODOSTEMACEAE
(M.Cheek)

Ledermanniella letouzeyi C.Cusset

EN B2ab(iii)

Range: Cameroon, SW Province: Rumpi Hills (1 coll.), Bakossi Mts (2 sites).

Only known from the above three sites.

Habitat: vertical rock faces in waterfalls in submontane forest; 750–1350m alt.

Threats: stochastic changes due to small area of occupancy (probably less than 0.2ha) and low number of sites (see above); future logging activities are likely to contaminate its aquatic environment due to surface run-off after logging.

Management suggestions: the Rumpi Hills site should be assessed. Efforts should be made to monitor all subpopulations for numbers of individuals annually. *Podostemaceae* are vulnerable to increased turbidity of water which reduces the ability of seedlings to establish.

Ledermanniella onanae Cheek

EN B2ab(iii)

Range: Cameroon: Bakossi Mts (2 sites).

This rheophyte resembling a *Lycopodium*, is known from only two sites. The assessment above was published in Cheek (2003a).

Habitat: perennial waterfalls and rapids in submontane forest; 1000–1200m alt.

Threats: as for *L. letouzeyi*.

Management suggestions: efforts should be made to monitor the subpopulations for numbers of individuals annually.

Ledermanniella thalloidea (Engl.) C.Cusset

VU B2ab(iii)

Range: Cameroon: Bakossi area (Nyandong, Nkongsamba, Ndoungué), Sanaga-Nachtigal, Sanaga 10km N Edea, Bipindi.

This species is known only from six sites, but probably often overlooked. It was first discovered at Ndoungué.

Habitat: rapids in large open rivers in lowland evergreen forest; c. 400m alt.

Threats: a probable future decline in habitat quality is predicted due to contamination of its aquatic environment from surface run-off after logging.

Management suggestions: a survey of the known sites for this species would provide data on whether the species survives at these, and could obtain quantitative subpopulation data needed for a management plan. Further surveys in the Bakossi-Rumpi Hills area are likely to yield new sites for the species. Unless in flower, the species is easily overlooked.

Macropodiella pellucida (Engl.) C.Cusset

EN B2ab(iii)

Range: Cameroon: Baré near Nkongsamba, Mana at Mundemba, Jide at Mwambong.

Only known from the above three sites, this species is very easily overlooked since the leaves are highly reduced and stems are absent.

Habitat: as *Ledermanniella onanae.*

Threats: as for *L. letouzeyi.*

Management suggestions: monitoring the subpopulations for numbers of individuals annually is advised.

RHIZOPHORACEAE
(I.Darbyshire)

Cassipourea acuminata Liben
EN B2ab(iii)

Range: Cameroon, SW Province: Menyum (1 coll.); Gabon: Miledi (1 coll.); Congo (Kinshasa): environs of Mobanga, Bas-Katanga, (1 coll.), Yangambi (5 coll.).

This species, described in 1986 by Liben, was previously known from 3 disjunct locations in Gabon and Congo (Kinshasa). The collections from Yangambi in the forests of central Congo (Kinshasa), seemingly the most significant site for this taxon, were all made in 1939. The collection from Menyum in W Bakossi is the first for Cameroon and greatly extends the species' extent of occurrence. However, it appears to be absent from large areas of apparently suitable forest stands within its range.

Habitat: lowland to mid-altitude rainforest, particularly along watercourses, including periodically flooded forest; 450–1000m alt.

Threats: low level anthropogenic disturbance at the Cameroon site may threaten this taxon here; stochastic events such as severe river flooding at the Congo (Kinshasa) sites may result in loss of local populations.

Management suggestions: rediscovery of the populations at all the listed locations is important, as the majority refer to old collections. Future surveys of *Cassipourea* taxa in the Lower Guinea & Congolian forests may reveal further populations of this species; care should be taken to separate it from similar species such as *C. malosana.*

RUBIACEAE
(M.Cheek)

Belonophora ongensis S.E.Dawson & Cheek
CR A1c + 2c

Range: Cameroon, SW Province: known only from foothills northwest of Mt Cameroon (5 coll.), the disjunct northern part of Korup N.P. (1 coll.) and Kupe-Bakossi (1 coll.).

An earlier assessment of this species (Cheek & Dawson 2000, listed as Cheek & Cable 1999 in IUCN 2003, downloaded on November 8th 2004) is maintained, although the range has been expanded by the collection of another specimen. This monopodial treelet is known from a single site in the Kupe-Bakossi area, in the Mungo River Forest Reserve, just north of Bulutu on the Ebonji-Ngusi road ascending the ridge to the east (*Cheek* 10193, fr., 30 Nov. 1999). At this site two plants only were found, growing with *Hylodendron, Guarea cedrata, Guibourtia demeuseii, Pachystela* sp. and *Garcinia mannii.* The forest at this site was threatened by upgrading of the Ebonji-Ngusi road, then in progress under EU sponsorship (*pers. obs.* 1999).

Habitat: lowland evergreen forest; 200–500m alt.

Threats: forest clearance for agriculture. In the western foothills of Mt Cameroon, the type locality, *Belonophora ongensis* is threatened by expansion of plantations.

Management suggestions: the site at Bulutu should be revisited. If this forest has not been cleared, as is feared, a survey of the subpopulation of *Belonophora ongensis* should be conducted as a baseline for future monitoring and an education programme initiated concerning the species. Conservation efforts for the species should be concentrated at the Onge Forest (Mt Cameroon) since the densest population is found there.

Calochone acuminata Keay
VU A3c

Range: SW Cameroon (7 pre-1995 coll.); Gabon: Crystal Mts (7 coll.); Cabinda (1 coll.).

This spectacular species was first collected in 1951 in the Mungo valley between the Rumpi Hills and Bakossi Mts and published as a new genus by Keay (1958: 30). A further six collections were made over the next 45 years in western Cameroon, all along the roads that mark the western and eastern boundaries of the Kupe-Bakossi area, apart from one collection in the Rumpi Hills (*Letouzey* 14542) and one at Mt Etinde (*Watts* 567). Seven collections are cited in Fl. Gabon (17: 216, 1970), all from the Crystal Mts. Between 1995 and 1999 11 collections were made in the Kupe-Bakossi area where it appears relatively frequent and is used medicinally. In conservation terms, Kupe-Bakossi appears to be the main stronghold for the species. When in flower (Jan. and Feb.) the plants are so spectacular that they are easily detected, so long as they have not reached the forest canopy. It is postulated that over 30% of the habitat of this species will be lost over the next 50 years.

Habitat: submontane evergreen forest, (250–)800–1000(–1250)m alt.

Threats: forest clearance for agriculture.

Management suggestions: monitoring of the population of *Calochone acuminata* is recommended. The survival of this species might be enhanced by targeted conservation education in Bakossi.

Calycosiphonia macrochlamys (K.Schum.) Robbr.

VU B2ab(iii)

Range: Ghana; Bioko; Gabon (1 coll. each); Congo (Kinshasa) (2 sites: Kama-Lemuna and Bankaie); Cameroon: Lolodorf, Mapanja and Edib (1 coll. each).

Although spread throughout the Guineo-Congolian forest, this tree, so conspicuous in flower, is known from only nine sites. The range data above are taken from K and Robbrecht's revision (1981: 377).

Habitat: lowland, or more usually submontane, forest; to 1200m alt.

Threats: forest clearance for agriculture and wood; given that trees are so thinly spread, even a small-scale event such as a single tree-fall could eliminate a subpopulation. Forest at Mapanja is under pressure from agricultural expansion (*pers. obs.*). Apart from the Mapanja and Edib collections, the species has not been recorded in the last 50 years.

Management suggestions: efforts should be made to rediscover the trees at the known sites and assess their subpopulations, regeneration (if any) and possibilities for protection. Conservation education regarding the species at the village level may assist survival.

Chassalia laikomensis Cheek

CR A2c

Range: Nigeria: Mambilla Plateau (1 coll.); Cameroon: Mwanenguba (1 coll.), Bamenda Highlands (several sites).

The assessment above was made in Cheek & Csiba (2000), listed as having been assessed by Cheek & Pollard (2000) in IUCN (2003; www.redlist.org). That assessment is maintained here. A new location to the south, that of Mwanenguba is added here. Mwanenguba has seen similar forest loss to that in the Bamenda Highlands.

Habitat: montane evergreen forest; 1650–2000(–2400)m alt.

Threats: about 95% of the original forest cover of the Bamenda Highlands has been lost to e.g. agriculture (Cheek *et al.* 2000, Cheek & Csiba 2000) and there have been similar losses at Mambilla and Mwanenguba.

Management suggestions: more information is needed on the numbers of individuals at the known sites and levels of regeneration. Enforcement of existing protected area boundaries would help protect a significant proportion of the surviving plants.

Chassalia petitiana F.Piesschaert

VU B2ab(iii)

Range: Bioko (1 site); Cameroon: Mt Kupe (2 sites); Gabon: Crystal Mts (2 sites); Congo (Kinshasa): Kivu (2 sites).

Known from only seven sites, this rare epiphytic shrub has a highly-fragmented population. Although given a name as early as 1922 (*Psychotria epiphytica* Mildbr., based on the Bioko material), it was not legitimised as a species, or fully described, until many decades had elapsed (Piesschaert *et al.* 1999, from which most of the range data above are taken). This is the only known published fully epiphytic African Rubiaceae.

Habitat: submontane forest; 750–1200m alt.

Threats: forest clearance for agriculture and wood at Mt Kupe, especially at the lowest altitudes.

Management suggestions: more data is needed on the survival of this species at its known sites, and on the extent of the subpopulations. At Mt Kupe the survival of this species might be enhanced by targeted conservation education.

Chazaliella obanensis (Wernham) Petit & Verdc.

VU A3c

Range: Nigeria (1 coll.); Cameroon: lower slopes of Mt Cameroon and Mt Kupe.

This shrub has the largest leaves in its genus. First published as a *Psychotria* in 1913, it was only known from a single specimen, from Oban, Nigeria, until the 1990s when botanical surveys on the lower slopes of Mt Cameroon found it to be fairly common in the Onge forest and also present at Bimbia; here it is highly threatened by expansion of plantations. *Chazaliella obanensis* is also fairly common on the lower slopes of the western side of Mt Kupe, indeed it is the commonest member of the genus in Bakossi, with over 20 specimens known. However most of these specimens occur below 1000m alt. and so are particularly vulnerable to expansion from agriculture. It is estimated here that 30–50% of the global habitat of this species is likely to be lost in the next ten years.

Habitat: lowland evergreen rainforest; rainfall exceeding 3m *per annum*; 250–1250m alt.

Threats: clearance of forest for expansion of agriculture.

Management suggestions: efforts should be made to protect a portion of this species' habitat so that it does not become extinct as a result of the apparently unrelenting expansion of agriculture in the lowlands of SW Cameroon.

Coffea bakossii Cheek & Bridson
EN B2ab(iii)
Range: Cameroon: Mt Kupe (1 site), Bakossi Mts (2 sites).

This rare tree was assessed in the species protologue (Cheek *et al.* 2002) as VU, but is reassessed here as EN since it is only known from three sites at such low altitudes that it is vulnerable to agricultural pressure. The other endemic *Coffea, C. montekupensis* Stoffelen, with 30 collections, is much commoner than *C. bakossii*, for which only four are know.

Habitat: evergreen lowland to submontane forest; 700–900m alt.

Threats: forest clearance for small-holder agriculture, and wood.

Management suggestions: a targeted conservation education campaign, with a conservation poster, might well help the survival of this species in those parts of its range outside of the proposed protected areas in Bakossi.

Cuviera talbotii (Wernham) Verdc.
VU B2ab(iii)
Range: SE Nigeria: Oban (2 coll.); Cameroon, SW Province: Korup (1 coll.) and Mt Kupe (13 coll. at 4 sites).

This shrub or small tree appears to be narrowly distributed (e.g. absent from Mt Cameroon) and rare, except at Mt Kupe where it is abundant at several sites on the lower slopes.

Habitat: submontane evergreen forest; 840–1300m alt.

Threats: forest clearance (at lower altitudes) for wood and agricultural land.

Management suggestions: a taxonomic reassessment of this species with *C. leiochlamys* is needed. Mt Kupe, with two-thirds of the known sites for *C. talbotii*, is the key area for the conservation of this species. So long as the boundary of the proposed Mt Kupe protected area is respected, this species seems secure, although some pressure can be expected in the lower part of the altitudinal range.

Hallea stipulosa (DC.) Leroy
VU A1cd
Range: Gambia to Congo (Kinshasa).

This widespread timber tree was given the assessment above by WCMC (1997) cited in IUCN (2003; www.redlist.org), on the basis that in many places it suffers from over-exploitation, e.g. clear-cutting. This has not been observed by us in Cameroon, but evidently occurs elsewhere.

Habitat: lowland evergreen forest, often in swamps where it may be dominant; 150m alt.

Threats: as above.

Management suggestions: none.

Hymenocoleus glaber Robbr.
VU B2ab(iii)
Range: Cameroon (3 pre-1997 coll.).

Described in 1977, this species was first collected at Limbe (*Schlechter* 12366). The two other localities recorded by its author, Robbrecht, were 8km west of Masok (*Leeuwenberg* 5416) and Mambe Forest near Boga (*Letouzey* 12279). A further collection was made near Limbe in 1993 (*Cable* 260 cited in Cable & Cheek 1998: 108). Numerous extra collections were made at three new sites in Kupe-Bakossi in 1995–present. This species is assessed as VU, being known only from six sites. Each colony appears to occupy only 1–2m^2.

Habitat: lowland evergreen and submontane forest; 750–1400m alt.

Threats: clearance of forest for agricultural expansion and wood, particularly in the lower part of its range in the Mt Cameroon area (Cable & Cheek 1998) and also below 1000m in Kupe-Bakossi.

Management suggestions: since Kupe-Bakossi has the densest number of sites and collections for the species, it seems to offer the best possibility of survival for the species. It seems likely that several areas in which the species occurs will be protected in future, though not those at lower altitudes.

Ixora foliosa Hiern
VU A2c + 3c
Range: Nigeria: Chappal Waddi and Chappal Hendu; Cameroon: Mt Cameroon (numerous coll.), Mt Kupe, Mwanenguba, Bamenda Highlands (numerous collections from many sites).

About half the area where this characteristic tree of wet montane forest occurred was in the Bamenda Highlands. They are now destitute of natural forest except for a very few exceptions. It is estimated that over 30% of the habitat of this tree has been lost over the last century and that over 30% of that remaining will be lost in the next century.

Habitat: montane forest.

Threats: forest clearance for agriculture and wood, especially in the Bamenda Highlands, once probably the main area for this species. In one study area of these highlands, 25% of forest was lost between 1987-1995 (Moat in Cheek *et al.* 2000).

Management suggestions: implementation and policing of protected area boundaries would ensure the survival of this species in most of its range.

Keetia bakossii Cheek ined.

CR D1

Range: Cameroon: Bakossi Mts, Kodmin area (3 coll.).

Only three collections are known of this striking lianescent shrub. All appear to be within an area of c. 1km^2 at, or near which, a new reservoir has been proposed, which may threaten the species. The data in this assessment are taken from Cheek (in press).

Habitat: submontane evergreen forest; 1400–1500m alt.

Threats: see above.

Management suggestions: if reservoir construction goes ahead, a survey to refind the plants is advised. If any are in danger of destruction, their relocation to another site should be investigated.

Mitriostigma barteri Hook.f.

EN A3c

Range: Bioko (2 coll.); SW Cameroon (9 coll.).

This small shrub, discovered on Bioko in the 1860s, is probably extinct there, not having been collected there since. Most of the forest on Bioko below 1000m alt. was cleared for cacao production. On the lower slopes of Mt Cameroon, three collections were made at the Bimbia-Bonadikombo forest (then Mabeta-Moliwe) in the 1992 botanical survey there, and three near Upper Boando, below Mt Etinde, in the 1993 survey (Cable & Cheek 1998). A single collection is known from Korup National Park, near Baro village, another at Bolo Forest in the Mungo valley north of Kumba in western Bakossi (both in the 1980s) and, going eastwards, one specimen between Bafang and Yabassi (*Letouzey* 11168 in 1972). I have visited all but the last localities and am acutely aware that all but that at Korup lack a good chance of long-term security from clearance in the next ten years. On the basis that a population reduction of 50% is suspected in the next ten years, *Mitriostigma barteri* is here assessed as EN using the criteria of IUCN (2001).

Habitat: undisturbed lowland evergreen forest; 0–450m alt.

Threats: clearance for agricultural expansion, particularly plantations.

Management suggestions: it is proposed that the sites identified above are surveyed to assess the size of their subpopulations of this species, and that measures be taken to protect at least the most significant sites.

Nauclea diderrichii (De Wild. & T.Durand) Merrill

VU A1cd

Range: Sierra Leone to Congo (Kinshasa).

This widespread timber tree was given the assessment above by an African Regional Workshop in 1997 (cited in IUCN 2003, downloaded on November 8th 2004), on the basis that it is heavily exploited for its timber used in general construction work. Regeneration is good in large canopy gaps but the species is out-competed by other pioneers after clear-felling.

Habitat: lowland evergreen forest; to 800m alt.

Threats: see above.

Management suggestions: good management of lowland production forest reserves should ensure the survival of this species.

Oxyanthus montanus Sonké

VU B2ab(iii)

Range: Bioko (2 coll. at 2 sites); Cameroon: Mt Cameroon (6 coll. at 3 or more sites), Bakossi Mts (1 coll.).

This taxon, first formally published in 1994, was initially collected at Mt Cameroon in the 1930s, near Buea at Musake camp; a second location there, given as 'Nyanga camp', is now lost. In all, six specimens are known from Mt Cameroon, which appears to be the stronghold for the species. There it appears confined to the belt of forest between Etinde and Buea; this belt is steadily being reduced by agriculture along its lower boundary. A total of six sites are known for the species.

Habitat: submontane evergreen forest; 1400–1600m alt.

Threats: forest clearance for agriculture, particularly at Mt Cameroon (see above).

Management suggestions: a survey is advised to rediscover the species at its known sites, to quantify the subpopulations and assess regeneration. The forested eastern slope of Mt Cameroon should be given priority.

Pauridiantha divaricata (K.Schum.) Bremek.

VU B2ab(iii)

Range: Cameroon, SW Province: Kumba-S Bakundu area (6 coll. at 4 sites: Barombi Mbo, Banga, Bombe, Etam), S Province: Bipinde, Mbanga, Nyong and Lolodorf.

A nondescript understorey shrub of lowland evergreen forest, *P. divaricata* has two disjunct areas of distribution. When more data on habitat loss are available throughout its range, this taxon is likely to be reassessed as either EN or CR under criterion A.

Habitat: lowland evergreen forest; c. 150m alt.

Threats: clearance of forest for agriculture; Barombi Mbo forest has been largely cleared (Aaron Davis, *pers. comm.* 2003) and a large area of S. Bakundu has been cleared for yam farming (Cable & Cheek 1998).

Management suggestions: the status of *P. divaricata* in the Kumba-S Bakundu area needs reassessment in the wake of the depredations of recent decades. A survey in S Province to rediscover its subpopulations and assess the possibilities of their protection is also advisable.

Pauridiantha venusta N.Hallé

VU B2ab(iii)

Range: Cameroon, S Province: Bolobo (1 coll.), SW Province: Mt Cameroon (11 coll. at 3 sites), Mt Kupe-Mungo River F.R. (8 coll. at 4 sites); Gabon: Crystal Mts (1 coll.).

Known from only nine sites, this understorey shrub is distinctive for its pink- or red-drying leaves. Disjunct between the Crystal Mts of Gabon, where it was first discovered in 1966, and Mt Cameroon and Kupe-Bakossi, a single specimen is known in the interlude, from Bolobo in S Province.

Habitat: lowland evergreen forest; 0–950m alt.

Threats: clearance of forest for agriculture and wood, particularly in the Mt Cameroon area (Cable & Cheek, 1998). If reassessed using criterion A, this species might well rate as EN.

Management suggestions: due to the proposed enhanced protection of forest at Mt Kupe, Mungo River F.R. and the Bakossi F.R., this taxon has a reasonable chance of survival at these sites. At Kupe village and at Ngomboaku, education for the public in the protection of this species may assist its conservation.

Pavetta brachycalyx Hiern

EN B2ab(iii)

Range: Cameroon: Mt Cameroon (9 coll. from 2 sites) and Kupe-Bakossi (3 coll. from 2 sites).

Once thought restricted to Mt Cameroon, the range of this nondescript species is here extended to Bakossi (see checklist section).

Habitat: lowland and submontane forest; 300–1500m alt.

Threats: clearance of forest for agriculture especially in the lower part of the altitudinal range.

Management suggestions: this species is reasonably secure where it occurs above c. 1000m alt. in both the Bakossi and Etinde parts of its range. Elsewhere it is threatened. Populations outside of protected areas could be assisted in their survival by a poster campaign.

Pavetta hookeriana Hiern var. *hookeriana*

VU A2c + 3c

Range: Cameroon: Mt Cameroon (numerous coll.), Mwanenguba (c. 3 coll.), Bamenda Highlands (numerous coll. at several sites).

It is estimated that over 30% of the habitat of this species has been lost in the last century and that over 30% of that remaining will be lost in the next century.

Habitat: montane forest; 1900–2000(–2400)m alt.

Threats: secure from threat at Mt Cameroon, *P. hookeriana* is threatened by forest clearance for agriculture and wood throughout the extensive Bamenda Highlands, probably once the main area for the species. Study of one large area in the highlands between 1987–1995 showed that 25% of the surviving forest was lost (Moat in Cheek *et al.* 2000).

Management suggestions: improved policing of existing forest reserve boundaries would prevent extinction of this species in the Bamenda Highlands, where its survival is precarious, except at Kilum-Ijum. At Mwanenguba and Bamboutos Mts (presence inferred) it may not survive much longer. The species is most secure at Mt Cameroon, where the narrowly endemic variety *pubinervia* also occurs.

Pavetta kupensis S.Manning

CR B2ab(iii)

Range: Cameroon, SW Province: Mt Kupe (6 coll.) and western Bakossi (1 coll.).

This shrub, which bears some resemblance to *P. owariensis* var. *satabiei*, is more or less confined to the summit of Mt Kupe.

Habitat: montane forest; c. 2000m alt. (*Thomas* 5197 at c. 400m alt.).

Threats: the subpopulations on the summits of Mt Kupe are vulnerable to stochastic events, such as landslides.

Management suggestions: the geographically and altitudinally anomalous *Thomas* 5197 (MO) should be studied to confirm or reject its placement in this taxon; the latter is most likely. A study to quantify the demography of the population at Mt Kupe would help develop a plan to manage the species.

Pavetta muiriana S.Manning

EN B2ab(iii)

Range: Cameroon: Bakossi Mts (4 coll. at 2 sites) and Barombi Mbo (1 coll. at 1 site).

This very distinctive shrub (see checklist), known from only three sites, is almost entirely restricted to the Bakossi Mts

Habitat: lowland and submontane forest; 400–1780m alt.

Threats: clearance of forest for agriculture, wood and public events. Aaron Davis (*pers. comm.* 2003) reports that much, perhaps most, of the crater forest of Barombi Mbo has been cleared.

Management suggestions: Kodmin, in the heart of the Bakossi Mts, with three collections, has the densest subpopulation of *Pavetta muiriana* and is the logical site from which to conduct a demographic survey that will also quantify individuals of this taxon, to provide data for a management plan.

Pavetta owariensis P.Beauv. var. *satabiei* S.Manning

EN B2ab(iii)

Range: Cameroon, SW Province: Rumpi Hills (1 coll. at 1 site), Bakossi Mts and Mt Kupe (5 coll. at 3 sites).

Known from only four sites, this shrub differs from other varieties in the absence of domatia, the large (10mm diameter) fruit with persistent calyces and tall (to 5m) stature.

Habitat: submontane forest; 800–1400m alt.

Threats: forest clearance for agriculture and wood, especially at the lower part of its altitudinal range at the densely populated Nyasoso.

Management suggestions: this taxon should prove reasonably easy to refind, given the number of recent collections. A survey to do this should also quantify the subpopulations and assess regeneration levels and demography. A poster campaign would aid public education on the importance of protecting this shrub.

Pavetta rubentifolia S.Manning

CR B2ab(iii)

Range: Cameroon, SW Province: Bakossi Mts (1 coll. at 1 site).

Known only from the type collection (*Thomas & McLeod* 5343, west of Bangem, 1986).

Habitat: submontane forest; 800–1600m alt.

Threats: forest clearance for agriculture and wood.

Management suggestions: a survey should be made to refind this distinctive species (leaves maroon when live), quantify the subpopulation and develop a plan to protect it.

Pentas ledermannii K.Krause

VU B2ab(iii)

Range: SE Nigeria: Obudu (1 coll.); Cameroon: Mt Oku area (3 coll.), Bamboutos (1 coll.), Mt Kupe-Mwanenguba (2 sites), Santa, Lake Aweng (1 coll. each).

At Mt Kupe known from four collections on the grassy summits (*Cable* 108, 1992; *Cheek* 7573, 1995, *Sebsebe* 5086, 1995), also known from Mwanenguba (*Leeuwenberg* 9970, 1972). Mt Kupe is the southernmost point for *Pentas ledermannii*; it does not appear to extend to Mt Cameroon. Mt Oku and the Ijim Ridge appear to be the northern extreme of its range.

Habitat: montane forest-grassland edge; 1000–2060m alt.

Threats: forest clearance is a major threat in the Bamenda Highlands, with 25% loss in one area between 1987–1995 (Moat in Cheek *et al.* 2000). This rate of loss probably continues and probably also extends to the 1800–2000m altitudinal range in Obudu, Bamboutos and Mwanenguba. Frequent human-set fires in grasslands in these same areas probably also adversely affect the grassland-forest interface as a habitat for this taxon, although occasional natural fires may aid its regeneration.

Management suggestions: research to resolve these uncertainties would aid in management planning for the conservation of this and other species in this habitat in the Cameroon uplands. For the moment, only the summit area of Mt Kupe lacks the two threats outlined above, and so is alone in offering a secure base for *Pentas ledermannii*.

Pseudosabicea batesii (Wernham) N.Hallé

VU B2ab(iii)

Range: Cameroon: Mokoko (7. coll.), Nyandong, Menyum, Bipinde; Gabon: Abanga, Moumba, Libreville, Kinguele, Mfoa (1 coll. each).

Known from only nine sites, this climber appears thinly scattered throughout its range, being absent from some very well-collected areas, such as Douala-Edea and Mt Cameroon (apart from Mokoko, where unusually it appears common).

Habitat: lowland evergreen forest; up to 1000m alt.

Threats: forest clearance for agriculture and wood (Mokoko F.R.) or due to urbanization (Libreville).

Management suggestions: this bizarre *Rubiaceae* (appearing to have alternate leaves) is, by the number of collections, particularly common at Mokoko F.R. in the western foothills of Mt Cameroon. This would therefore be the logical site to concentrate on the protection of this species, were it not for the threat of its deforestation for plantation expansion (Cable & Cheek 1998). Ideally, a survey of all the known sites to quantify subpopulations, regeneration and vulnerability, is suggested.

Pseudosabicea medusula (K.Schum.) N.Hallé

VU B2ab(iii)

Range: Cameroon: Mokoko (3 coll.), Mamfe (2 coll.), Baduma, Dja, Kribi-Lolodorf, Douala, Bipinde, Ebolowa-Lolodorf, Ebolowa-Nkondo, Lake Tissongo (2 coll.).

Known from 10 sites widely scattered in the evergreen coastal forest belt of Cameroon, the preponderance of collections are from the south-eastern part of the range.

Habitat: lowland evergreen forest; 0–200m alt.

Threats: forest clearance for agriculture or wood (e.g. Mokoko; Cable & Cheek 1998).

Management suggestions: this creeper seems most common (6 coll.) in Littoral/South Province (Douala-Edea-Ebolowa-Lolodorf-Bipindi) and conservation efforts are presumably best focused here.

Pseudosabicea pedicellata (Wernham) N.Hallé

VU B2ab(iii)

Range: SE Nigeria: Afi River, Orem, Okwangwo, Boshi, Oban; Cameroon, SW Province: Korup-Fabe Rd, Kumba-Mamfe Rd (1 coll. each).

A scandent shrub apparently more common in SE Nigeria than adjoining Cameroon, this species grows in a particularly threatened habitat. Only seven sites are known.

Habitat: lowland evergreen forest; c. 400m alt.

Threats: clearance of forest for agriculture and wood; this has been particularly prevalent in Nigeria in recent decades, although several collection sites of this species are from protected areas. However, those in Cameroon appear unprotected.

Management suggestions: the status of this plant in Nigeria needs surveying, in particular since historic collections are concentrated there. Quantitative evaluation of the size of subpopulations, regeneration and threats would help in the formulation of a management plan.

Psychotria camerunensis Petit

VU B2ab(iii)

Range: Cameroon: Kumba, Yaoundé and Bipinde areas.

This taxon was described in 1966 from eight specimens centred around three towns in the evergreen forest area of Cameroon: Kumba, Yaoundé and Bipinde. Since then only two more specimens have been found, both from Bakossi, of which one has been verified (*Etuge & Thomas* 192, Tombel). Several other specimens have been attributed to this taxon since, but they lack the diagnostic densely brown pubescent stems and obovate leaves that dry brown. Forest clearance around Yaoundé and Tombel has been extensive and is continuing. Since there are also less than 10 locations, and an area of occupancy less than 2,000km², *P. camerunensis* is here assessed as VU under criterion B of IUCN (2001). More intensive research may show that it merits a higher category of threat.

Habitat: evergreen forest understorey; 150–1700m alt.

Threats: clearance of forest for urban expansion (Yaoundé area) and cultivation of crops (Tombel and Yaoundé).

Management suggestions: an attempt should be made to rediscover this species in the wild in the Bakossi area since it has not been seen since 1986 despite intensive collection there since 1995. It should also be looked for in the Bipinde area, and threats there assessed, since currently these are unknown.

Psychotria densinervia (K.Krause) Verdc.

EN B2ab(iii)

Range: Cameroon: Bipinde and Bakossi.

This 6m tree is remarkable for the pendulous, cord-like peduncles several metres long which bear fist-sized, globular inflorescences. Only the much commoner *P. camptopus* is likely to be confused with it (differences are recorded in the checklist). First described from a single specimen from the NW slopes of the Mimfia Mts of Bipinde (*Zenker* 4683) it was not recollected until the 1950s, in Mungo Ndaw, western Bakossi (*Dundas* in FHI 15324). A further specimen was found in 1996 near Kupe village at the base of Kupe Rock (*Ryan* 225). Lowland forest in these areas is generally under threat of clearance although the site at Kupe Rock seems reasonably secure at present. In summary, only three sites are known, at each of which no more than a single tree was recorded.

Habitat: lowland evergreen forest; up to 1000m alt.

Threats: clearance for agriculture.

Management suggestions: it is recommended that attempts be made to rediscover this rare species at Bakossi and to educate local villagers as to the importance of its conservation. See also the comments made for *P. minimicalyx*. It is assumed that neither of the Bakossi sites can be formally protected due to their low altitude.

Psychotria lanceifolia K.Schum.

VU D2

Range: Cameroon: Bipinde-Kribi and Bakossi.

This taxon was described from several Zenker collections (6 at K) from Bipinde in S Province, made between 1904 and 1912 and otherwise unlocated. It was not re-recorded until Bos (WAG) made at least two collections in 1968 and 1969 at km 19 and 20 from Kribi on the Bipinde road. The two specimens listed from two sites in Bakossi (*Manning* 481, *Etuge* 414, MO) have not been seen by us. This is a very distinct, even ornamental, species: 'small tree about 2.5m high, slender main stem and gracefully arching, very dark green branches, leaves thinly leathery, smooth, glossy' (*Bos* 3061). We assess *P. lanceifolia* as VU on the basis that it is known only from three sites, occupying only a few m^2 at each site, and is affected by the threats given below.

Habitat: stream-sides, usually in lowland evergreen forest; 0–1600m alt.

Threats: clearance of forest for agriculture is suspected, but needs confirmation.

Management suggestions: the identity of the MO specimens cited needs checking, although this is a very distinct species, not likely to be confused with any other. The collection sites should be resurveyed in order to (a) determine whether the taxon is still extant, (b) decide whether it is indeed rheophytic (restricted to watercourses), and (c) determine whether it is threatened by human activities or other factors.

Psychotria minimicalyx K.Schum.

CR A3c

Range: Cameroon, SW Province: Kumba and Kumba-Mamfe areas.

Described in 1899 on the basis of a single specimen (*Staudt* 606) from Kumba, it was rediscovered nearby at Banga, S Bakunda F.R. in 1948 (*Brenan* 9269). Subsequently it has been found at two sites along the Mungo valley running from Kumba towards Mamfe (*Thomas & Nemba* 5185, Konye, 1985 and *Mambo & Thomas* 64, Mbu, 10km W of Wone, 1986). It has been confused with *P. bifaria* on account of its similar leaf-shape and size but is distinguished by the terete, evenly densely pubescent stem (not 4-angular, with 2 lines of hairs). This shrublet is assessed under criterion A as CR due to the likelihood of 80% of its habitat being destroyed in the next 30 years. A generation period of ten years is estimated.

Habitat: lowland evergreen forest; c. 300m alt.

Threats: clearance of its habitat for agriculture, such as cacao plantations or food crops, is likely along the Mungo valley owing to the prevalence of fertile soils.

Management suggestions: the lowland forest habitat of *P. minimicalyx* does not lend itself to formal protection in a gazetted area because it grows in fertile soils close to centres of population and lines of communication. Assuming that a new site cannot be found in one of the areas already mooted for protection, it is recommended that an attempt be made to rediscover plants at one of the known sites and to educate the population as to the desirability of protecting it, possibly by using posters or rental agreements.

Psychotria podocarpa Petit

VU B2ab(iii)

Range: Nigeria (1 coll.); Cameroon, SW Province (7 coll.).

This shrub is highly attractive in fruit and has great potential as a pot plant. The profuse glossy black, pendulous fruits are held on contrasting bright red, carrot-shaped pedicels. When first described in 1964, it was known from only a single Nigerian collection but in the early 1990s it was found at Bomana and Njonji, both at the foot of Mt Cameroon, and since then three sites have been found at Bakossi, namely Ngomboaku, Nyale and Konye. The first two of these seem secure sites for this species (*pers. obs.*). A total of six sites are known; where site observations have been made, it is known to occupy only 1–2m^2. Threats are given below.

Habitat: lowland evergreen forest; 300–800m alt.

Threats: clearance for agriculture especially in the Mt Cameroon area, where the planned expansion of plantations is likely to destroy the subpopulations listed above.

Management suggestions: a survey of the subpopulations present at each of the sites is desirable. At the Njonji site, hundreds of individuals were recorded in a small area, but this has not been noted at other sites. Although distinctive in fruit, in flower it may be confused with the very common *P. latistipula* but can be separated by the pubescent stem and inflorescence.

Psydrax bridsoniana Cheek & Sonké in press
EN B2ab(iii)
Range: Cameroon, SW Province: Jide valley.
Known from only two sites, this rare tree is in great danger of extinction due to agricultural expansion. The assessment above is taken from Cheek & Sonké (Kew Bull., in press).
Habitat: lowland to submontane evergreen forest; 800–1100m alt.
Threats: clearance for agriculture and wood. Both sites, near Kupe village and Nyasoso, are on the edge of cultivated land. Further clearance for agriculture could result in extinction of this species.
Management suggestions: a survey to rediscover the species should be conducted urgently. The Bakossi public should be informed of the uniqueness of these rare trees, and to help ensure their survival. New protected areas in Bakossi are likely to help secure the future of this tree species.

Rothmannia ebamutensis Sonké
EN B2ab(iii)
Range: Cameroon, SW Province: known only from the Bakossi Mts including the montane part of the Banyang Mbo Wildlife Sanctuary.
This shrub or small tree was discovered to be a new species by Prof. Bonaventure Sonké, one of Cameroon's leading botanists (Sonké 2000), on the basis of eight specimens collected by him from at least six plants in the neighbourhood of the village of Ebamut. All these specimens derive from the mountainous eastern part of the Banyang Mbo Wildlife Sanctuary, to the north of Ebamut. In September 2001, while determining specimens at the Kew Herbarium of the genera *Oxyanthus*, *Rothmannia* and *Aulacocalyx* for this checklist, Prof. Sonké discovered a further specimen of this species, *Biye* 50, collected in 1998 from the heart of the Bakossi Mts, near Kodmin. *Rothmannia ebamutensis* is assessed here as EN due to its estimated area of occupancy of 5km^2, and its being known from only two sites, in an area where there is decline in habitat quality, albeit at a low level.
Habitat: submontane forest; 1100–1500m alt.
Threats: forest clearance even on a local scale, of a fraction of a hectare, if coinciding with one of the sites of this species, could reduce the size of, or destroy, a subpopulation. Placement of a proposed new reservoir near Kodmin should be made with caution!
Management suggestions: Dr. Sonké's material has already been used by the management team of Banyang Mbo to promote public awareness and pride in this very local and spectacular species. This could be emulated in the Kodmin area. Survey teams in explored parts of Bakossi should look out for new sites for this plant.

Rutidea nigerica Bridson
VU B2ab(iii)
Range: Benin; Nigeria: Lagos, Egbada, Benin-Iyek, Akpaka, Ijebu, Idranre; Cameroon, SW Province: Nyasoso, Baduma, Kumba.
This liana is known from only ten sites, at some of which (Lagos, Kumba) it may very well already be extinct due to the threats discussed below.
Habitat: lowland forest; 250–800m alt.
Threats: forest clearance for agriculture and wood; it is very likely that its habitat has been lost at Lagos and Kumba where there has been extensive forest clearance in recent decades as there has been in much of Nigeria.
Management suggestions: surveys should be made to attempt to rediscover this species at its known sites, and to evaluate the size of subpopulations, regeneration, local threats and possibilities for conservation. When these data are available it may well be possible and necessary to re-evaluate this taxon under criterion A, at a higher level of threat.

Sabicea xanthotricha Wernham
EN B2ab(iii)
Range: SE Nigeria: Oban; SW Cameroon: Mt Cameroon (Etinde and Mokoko), Bakossi Mts (Kodmin).
A pithy shrub in a genus consisting mostly of lianas, this species is currently only known from four sites.
Habitat: lowland and submontane evergreen forest; 0–1400m alt.
Threats: forest clearance for agriculture and wood, particularly at Oban and at Mokoko F.R.

Management suggestions: a survey to rediscover plants of this taxon at the known sites and gather the usual data on each of the subpopulations is recommended. The site at Kodmin, assuming that it is not affected by a proposed reservoir scheme, is particularly secure, being remote from human habitation.

Stelechantha arcuata S.E.Dawson
CR A2c + 3c + 4c

Range: Cameroon: Mt Cameroon at Bimbia-Bonadikombo and at Mokoko, Kupe-Bakossi in western Bakossi and at Kupe Village, S Province at N'Kolandom (1 coll. each)

This blue-flowered, cauliflorous treelet, even within its range, seems very rare given the few collections known despite intensive collecting over recent years in several of the forests where it occurs, e.g. at Mt Cameroon. The assessment above is listed in IUCN (2003; www.redlist.org) under the citation Dawson (2003); the species was published with an assessment in 2002 (Dawson 2002).

Habitat: lowland evergreen forest; 200–950m alt.

Threats: forest clearance for agriculture, especially plantations in the Mt Cameroon area (Cable & Cheek 1998) and in Bakossi, small-holder agriculture.

Management suggestions: individual trees of this taxon merit special protection given their rarity. A survey to relocate plants at their known sites should assess what measures are most suitable at each site. At Kupe village for example, renting the forest area concerned together with a poster campaign might be the best means of supporting the survival of the subpopulation there.

Tarenna baconioides Wernham var. *baconioides*
VU B2ab(iii)

Range: SE Nigeria: Oban, Ikom, Iyila-Ibere, Akamkpa; Cameroon: Mt Cameroon (Onge), W Bakossi, Baduma. Six sites are known for this nondescript climber.

Habitat: lowland evergreen forest; 0–400m alt.

Threats: forest clearance for agriculture and wood, particularly at Onge (see Cable & Cheek 1998) and at the Nigerian sites. Forest clearance in Nigeria has been extensive in recent decades.

Management suggestions: quantitative data on habitat loss for this taxon would probably result in it being given an elevated threat level under criterion A. These data, and qualitative data on the subpopulations, including regeneration levels and local threats, are needed to inform a management plan for the taxon, and should be obtained by visits to the known sites.

Tricalysia atherura N.Hallé
VU B2ab(iii)

Range: Cameroon: Kupe village, Loum, Ngoase, Etinde, Mt Kala, Mt Fébé, Nteigne; Gabon: Belinga, Mt Babiel. Known from nine sites, this understorey shrub, although first described correctly from the last site, in Gabon, earlier received a name and description based on material from Loum in Cameroon. The sites above are taken from Robbrecht (1987: 76), apart from those of Etinde (Cable & Cheek 1998) and Kupe village (newly recorded here).

Habitat: lowland forest often near the boundary with semi-deciduous forest (*pers. obs.*); 200–1250m alt.

Threats: clearance or part-clearance of forest for agriculture; much of the understorey of the Loum F.R. has been cleared and replaced by crops; other parts have been cleared of natural forest and replaced by cocoa (*pers. obs.*); Mt Kala has had its forest damaged in recent years and Mt Fébé has had prestige buildings erected upon it; the lower slopes of Mt Etinde (where one plant was recorded at 400–600m alt.) is earmarked for plantation expansion.

Management suggestions: a survey to refind this species at each of its sites, gather quantitative data on subpopulation size, recruitment and protection levels would inform development of a management plan for the species. In the absence of this, conservation efforts should be concentrated at Kupe village, which adjoins the Loum area. At the former, 15 collections were made in recent years (compared with one at Etinde, despite similar sampling effort) suggesting Kupe village to have a very high density of the species. The new Mt Kupe protected area should help secure the future of the species.

Tricalysia talbotii (Wernham) Keay
VU B2ab(iii)

Range: SE Nigeria: Oban, Awi-Akampka, Boli-Bateriko, Iyamoyong F.R.; SW Cameroon: Inokum-Mbenyan, Ebone-Yabassi, Ndikinimeki, Minso and western Bakossi (Kumba-Mamfe: 2 sites).

Known from only the ten sites listed above. All but the last two are taken from Robbrecht (1979: 353).

Habitat: lowland forest; c. 250–350m alt.

Threats: clearance of forest for agriculture and wood.

Management suggestions: a survey to rediscover this taxon, quantify subpopulations, demography, regeneration and threats, is advisable. Given far-reaching forest loss in Nigeria and anticipated increased forest loss along the upgraded Kumba-Mamfe Road, the prospects for this species do not seem good. Reassessment under criterion A (requiring better threat data than now available) is likely to result in EN or CR ratings.

Trichostachys interrupta K.Schum.
VU B2ab(iii) ·

Range: SE Nigeria (2 coll.); Cameroon: Korup N.P., Lake Barombi Mbo (1 coll. each), Bakossi Mts (2 coll. at 2 sites).

Published in 1903 on the basis of *Preuss* 466 from Barombi near Kumba, two additional localities were later found in Nigeria (Oban, *Talbot* 1045 and Kwa Falls, *Brenan* 9238). Rediscovered in Cameroon in 1983 (*Thomas* 2291, south Korup N.P.), the only other locations for this subshrub are in the Bakossi Mts below Nyale Rock (*Etuge* 4176, Feb. 1998) and between Nyale and Kodmin (*Etuge* 4461, Nov. 1998).

Habitat: understorey of lowland evergreen forest; 400–1000m alt.

Threats: clearance of forest for agriculture has occurred extensively in Nigeria, and at Barombi Mbo, where the species may no longer occur.

Management suggestions: the Nyale-Kodmin area has least pressure of all known sites and offers the best hope for the survival of this herb. Surveys to determine whether the species survives at the other sites are suggested. If so, an assessment of the best means of protecting them, if feasible, should be made.

RUTACEAE
(M.Cheek)

Oricia lecomteana Pierre
VU B2ab(iii)

Range: SE Nigeria, Cameroon & Gabon.

This monopodial tree, though fairly widespread and conspicuous, is rare (less than 10 sites are known) and threatened, as outlined below.

Habitat: evergreen forest; to 1000m alt.

Threats: clearance of lowland forest for agriculture and wood, particularly in the Mt Cameroon area (Cable & Cheek 1998) and in Nigeria.

Management suggestions: more effective policing of protected areas is advised, together with annual monitoring of individuals at known sites.

Vepris trifoliolata (Engl.) Mziray
(syn. *Oricia trifoliolata* (Engl.) Verdoorn)
VU B1 + 2c

Range: Cameroon: Korup, Mt Cameroon, Mt Kupe and the Bakossi Mts.

This species, under its synonym, was assessed as above by WCMC (1997) and is listed thus in IUCN (2003; www.redlist.org). The assessment was based on data from Peguy provided in 1997. Two collections were made at Limbe before 1931 (Fl. Cameroun 30, 1963) but the taxon was not refound in the Mt Cameroon area in the intensive surveys of the 1990s (Cable & Cheek 1998).

Habitat: lowland and submontane evergreen forest; 0–1600m alt.

Threats: habitat decline through logging and the expansion of areas under cultivation (Peguy data).

Management suggestions: a revision of central African *Vepris* species would enable development of a more effective identification guide than is now available. Confusion with other species of *Vepris* is possible currently. Mt Kupe and the Bakossi Mts may now be the stronghold for this species. Data on numbers of individuals per site and demography should be gathered to develop a management plan for the species.

SAPINDACEAE
(M.Cheek)

Allophylus bullatus Radlk.
VU A2c

Range: SE Nigeria, Cameroon, Principé & São Tomé (Cameroon line mountains).

This understorey tree of upper submontane to montane forest, while secure on Mt Cameroon and Mt Kupe, has lost large tracts of its habitat in recent decades in the Bamenda Highlands. Over 30% of its overall habitat is estimated to have been lost in the last 100 years.

Habitat: upper submontane and montane forest; 1600–2400m alt.

Threats: clearance of forest for agriculture and wood, particularly in the Bamenda Highlands of Cameroon, once probably the main area for *A. bullatus*. Study of one area here (Moat in Cheek *et al.* 2000) showed that 25% of forest was lost between 1987–1995.

Management suggestions: improved policing of the existing protected areas would secure the future of this species.

Deinbollia insignis Hook.f.

VU B2ab(iii)

Range: Nigeria (Sapoba F.R., 4 coll.); Bioko (2 coll.); Cameroon, SW Province: Bambuko F.R. (2 coll.) and Mt Kupe (2 coll., two sites).

This large treelet, with leaves to 1m long, is only known from six sites, all of which are threatened with, or have suffered, forest clearance. It is suspected that, when better data are available, this species may prove to be CR.

Habitat: lowland forest; 400m alt.

Threats: extremely vulnerable due to clearance of lowland forest for agriculture. It may well be extinct on Bioko due to extensive forest clearance there for cacao plantations in the late 19th and 20th centuries. It may also be extinct in Nigeria due to extensive forest loss there in the late 20th century. Forest loss at Bambuko is documented in Cable & Cheek (1998). It is notable that the species was not found elsewhere around Mt Cameroon during the intensive surveys of the early 1990s. At Mt Kupe it is VU due to it occurring at low altitude, placing it outside of the proposed new protected area.

Management suggestions: surveys should be made to rediscover this species at its known sites. It seems possible that it may only survive at Mt Kupe (where the most recent collections are recorded). It is therefore advised that efforts to protect the tree might be centred here. Individuals need to be rediscovered, demographic data obtained and protection secured with the help of local communities.

Deinbollia maxima Gilg

VU B2ab(iii)

Range: Sierra Leone (type coll.); S Nigeria (1 coll.); Cameroon: Mt Cameroon (7 coll. at 3 sites), Mt Kupe (1 coll.), Bipinde (1 site); Gabon (numerous coll. at 3 sites, including Libreville).

This tree or shrub only just qualifies as VU under criterion B since 10 sites are known.

Habitat: lowland to submontane forest; c. 900m alt.

Threats: forest clearance for timber and agriculture, particularly in Nigeria, Mt Cameroon and the Libreville area. Threats in Sierra Leone are unknown.

Management suggestions: the best hope for the survival of this species is probably in Gabon, where numerous collections have been made and pressure on forest is relatively low. However, attempts should be made to rediscover the species at all its known sites and to gather demographic data for a management plan.

Placodiscus caudatus Pierre ex Radlk.

EN B2ab(iii)

Range: Cameroon, SW Province: Korup (1 coll.), Bakossi Mts (confirmation of identity needed); C.A.R. (1 coll.); Gabon (2 coll., 1 at Libreville).

Placodiscus caudatus is a cauliflorous tree known with some certainty only from four sites.

Habitat: lowland evergreen forest; c. 300m alt.

Threats: forest clearance for logging, followed by agriculture and urbanization (Libreville site).

Management suggestions: a revision of the species of *Placodiscus*, many of which are poorly known, would result in a better understanding of their geographic ranges and identification. From the available evidence Gabon, with two known sites, most warrants concentration of resources in conserving this poorly known species. A survey is advised to rediscover the plant there, and gather demographic data to aid formulation of a management plan. Currently, the species is only known to be protected at its Korup site.

Placodiscus opacus Radlk.

VU B2ab(iii)

Range: Cameroon, SW Province: Bakossi Mts (2 sites); C.A.R.: Boukoko; Rio Muni (1 coll.); Gabon (3 sites).

Known from seven sites, *Placodiscus opacus* shares many similarities with its congener, *P. caudatus* (see checklist for diagnostic characters).

Habitat: lowland and submontane evergreen forest; 200–1000m alt.

Threats: see *Placodiscus caudatus*.

Management suggestions: as for *Placodiscus caudatus*.

SAPOTACEAE
(I.Darbyshire)

Baillonella toxisperma Pierre
VU A1cd
Range: E Nigeria, Cameroon, Gabon, W Congo (Kinshasa) and Angola: Cabinda.
This species was assessed as vulnerable by L.White (1997) and is listed so by IUCN (2003; www.redlist.org), on the basis that the species is overexploited for its timber and is therefore declining in large parts of its range, that maturation rates are slow (90–100 years), and that regeneration is limited by occurring only under a closed canopy. *B. toxisperma* is recorded as the second most important exported timber species in Gabon. We maintain this assessment here.
Habitat: closed canopy lowland primary rainforest and old secondary forest, 0–500m alt.
Threats: see above.
Management suggestions: minimum exploitable bole diameters have been set in several countries (White in IUCN 2003, downloaded on November 8th 2004); these should be enforced where possible. Better protection and prevention of illegal logging in protected areas may ensure the future survival of this species.

SCROPHULARIACEAE
(M.Cheek)

Rhabdotosperma ledermannii (Murb.) Hartl.
VU A2c
Range: Nigeria: Chappal Wadi; western Cameroon: Bamenda Highlands (Ndu, Bambili Lakes, Bamenda-Mba Kokeka, Bambui and Laikom) and Mwanenguba.
This 1m or more tall, yellow-flowered herb is difficult to overlook. It overlaps in range with *R. densifolia*, which is also restricted to the Cameroon mountain line, but is more common and less threatened. Only eight sites are known.
Habitat: grassland near edge of forest; c. 2000m alt.
Threats: loss of montane forest due to agriculture is thought to be the main concern for this species. 25% of forest cover was lost in a sample area of the Bamenda Highlands between 1987–1995 (Moat in Cheek *et al.* 2000). This equates to about a 30% loss in area of occupancy for this species over 10 years.
Management suggestions: conservation efforts should be concentrated in the Bamenda Highlands where the species seems to be most abundant (six collection sites). The ecological relationship of *R. ledermannii* with forest edge needs further confirmation. Research is also needed on its demography, (for example is it a biennial?) and requirements for establishment in the wild. It is possible that the two species of *Rhabdotosperma* in our area are conspecific; further investigation is required to establish their relationship.

SIMAROUBACEAE
(M.Cheek)

Quassia sanguinea Cheek & Jongkind ined.
VU A2c
Range: Nigeria: Obudu Plateau; Cameroon: SW, NW & W Provinces.
Restricted to the submontane forests of the Cameroon Highlands, from Mt Cameroon in the S to the Bamenda Highlands in the N, with an extension into Nigeria at Obudu, this small tree is distinctive in its red rhacheae (axes of the leaves).
Habitat: submontane forest; 800–1750m alt.
Threats: forest clearance for wood, followed by agriculture, particularly in the northern part of its range, the Bamboutos Mts and the Bamenda Highlands. In the latter, a remote sensing study over eight years (1987–1995) by Moat (in Cheek *et al.* 2000) of one area showed 25% loss of forest. By extrapolation, it is here estimated that over 30% of its overall population has been lost due to habitat destruction over the last three generations, or sixty years (estimating one generation at twenty years).
Management suggestions: although forest loss in W and NW Provinces has seriously reduced the population of *Q. sanguinea* in those areas, it seems relatively secure at Mt Cameroon and Kupe-Bakossi in SW Province. So long as these areas remain protected, no further action is needed to ensure the survival of the species. However, data on generation duration and other aspects of demography, together with data on densities, are desirable.

STERCULIACEAE
(M.Cheek)

Cola metallica Cheek
CR A2c
Range: Cameroon: SW Province.
This shrub was assessed as CR A1c + 2c in Cheek (2002). That assessment is maintained here in modified form since use of A1 using IUCN (2001) guidelines now implies that causes of population reduction have ceased.
Habitat: lowland evergreen forest; c. 50–700m alt.
Threats: all known sites for this species are believed to be on land threatened by clearance for plantations, other forms of agriculture, or logging.
Management suggestions: improved policing of existing protected areas would help to secure the survival of this species.

Cola praeacuta Brenan & Keay
CR A2c
Range: SW Cameroon: Mt Cameroon and Mt Kupe.
This treelet was originally assessed as CR A1c + 2c in Cable & Cheek (1998). That assessment is maintained here in modified form (see *Cola metallica*, above).
Habitat: lowland evergreen forest; 0–400m alt.
Threats: clearance of forest for timber, followed by agriculture.
Management suggestions: improved policing of existing protected areas and reinforcement of the level of official protected status, would help secure the future of the species. The main site for this species is the Onge forest in the western foothills of Mt Cameroon. It is advised that resources for the conservation of this species be focused there.

THYMELAEACEAE
(M.Cheek)

Dicranolepis polygaloides Gilg ex H.H.W.Pearson
VU B2ab(iii)
Range: Cameroon: Mundame-Etam, Bakossi F.R., Nyasoso, Bambuko F.R., Essam (Nanga-Eboko), Bipinde.
Known only from the six sites above, this very distinctive shrub is the smallest by far of all Cameroonian *Dicranolepis*.
Habitat: lowland evergreen forest; to c. 900m alt.
Threats: forest clearance due to logging, followed by agriculture. Although the species is centred in SW Province, all the known sites here have been affected by habitat destruction to some extent or another. Threats to the last two sites (outside of SW Province) are unknown.
Management suggestions: revisiting all known sites for the species is advised in order to attempt to rediscover it and to assess levels and nature of threats, if ongoing. Data on numbers of individuals per population and demography are also required to inform a management plan for the species.

UMBELLIFERAE
(I.Darbyshire & M.Cheek)

Peucedanum kupense I.Darbysh. & Cheek
VU D2
Range: Cameroon, SW Province: Mt Kupe (2 coll.).
The area of occupancy of this taxon is less than 1–2km², this being our estimate for the grassland areas comprising the two (of three) peaks of Mt Kupe from which it has been collected. Since these grassy areas are very precipitous below the peaks, we have not fully explored them and so can only roughly estimate the size of the populations of *P. kupense*. It is likely that the number of mature individuals is only a few hundred, but could fall in the range 50 to several thousand. We here assess this taxon as VU due to the very restricted area of occupancy and number of locations such that it is vulnerable to stochastic change (see below). Subsequent attempts to relocate this species in 2003 to 2004 have been unsuccessful to date (M. Etuge, *pers. comm.*). It is possible that this conservation assessment will be down-rated in future if the plant is found in other adjacent montane grassland areas, such as the poorly known Mt Nlonako to the northeast. However, investigations of montane grassland in the Bakossi Mts, the Mwanenguba Mts and Mt Cameroon have so far not revealed populations of this plant.
Habitat: montane grassland; 1900–2000m alt.

Threats: stochastic change such as lightning-induced fires or landslides. The species is not endangered by man at present, being found far above the highest limits (c. 1100m alt.) of agriculture on the mountain.

Management suggestions: rediscover and investigate more fully the populations of this taxon in the grassy areas on the summit of Mt Kupe, possibly using ropes and harnesses. Surveys of other grassy areas on the mountain, such as Kupe Rock, might reveal further sub-populations. Ideally, surveys should be carried out annually over several years so that fluctuations can be recorded and their significance assessed. Surveys are best conducted between October and March, the dry season, when the plant is flowering and fruiting and so is conspicuous and easily identifiable.

VIOLACEAE
(M.Cheek & I.Darbyshire)

Rinorea fausteana Achoundong

EN B1ab(iii)

Range: Cameroon, SW Province: Banyang Mbo Wildlife Sanctuary (1 coll.), Rumpi Hills (2 coll.), Bakossi (3 coll. at 3 sites, 1 further unlocated specimen).

This species occupies an area of occupancy less than 5000km^2 and occurs at five sites, with a continuing decline being anticipated in the quality of habitat in this area, and is thus assessed as EN (see Achoundong & Cheek 2004). Despite being a submontane species, it is not known from Mt Kupe; it seems that the Jide valley forms a barrier to its expansion eastwards from the Bakossi Mts. It is also not known from Mt Cameroon; here a band of semi-deciduous lowland forest likely forms a barrier to expansion southwards.

Habitat: dense submontane evergreen forest; 1100–1500m alt.

Threats: in Bakossi its habitat is threatened by continued sporadic illegal logging operations.

Management suggestions: formal protection of the Bakossi Mts Forests may help preserve this species; further populations should be sought for in the currently-protected Banyang Mbo wildlife sanctuary.

Rinorea thomasii Achoundong

VU A3c

Range: Cameroon, SW Province: Mt Cameroon (8 coll. at 3 sites), Korup National Park (6 coll.), Bakossi Mts (1 coll.)

This species is assessed as VU on account of a suspected decline in the extent of occurrence of 30%, and its quality of habitat, over the next three generations (estimated as c. 45 years) (see Achoundong & Cheek 2004).

Habitat: lowland evergreen forest, generally near rivers; 0–800m alt.

Threats: of the 3 areas in which the taxon is known to occur, it seems secure at only the Korup National Park, where it is afforded a high level of protection. The surviving forest in the Jide valley of the Bakossi Mts has mostly already been extirpated by agricultural activities, particularly prevalent there owing to the high fertility of the soil; this is likely to continue and the species is already rare here, as evidenced by the single collection, made from the bottom of a steep-sided gorge. Its habitat at Mt Cameroon is vulnerable to clearance for timber extraction and plantation agriculture; already much of the lower altitude forest has been converted to rubber, banana and oil palm plantations.

Management suggestions: census of the populations at the most protected areas mentioned above should provide better data on which to formulate a conservation strategy.

MONOCOTYLEDONAE

ARACEAE
(I.Darbyshire)

Amorphophallus preussii (Engl.) N.E.Br.

VU B2ab(iii)

Range: Cameroon, SW Province: Mt Kupe (1 site, 3 coll.), Bakossi (1 coll. at Bangem, recorded but not collected elsewhere), Mt Cameroon (1 site, 2 coll.), Littoral Province: Mt Nlonako (1 coll.), W Province: Bangou nr. Bangwa (1 coll.).

This taxon was originally described in 1892 based upon 2 specimens from Buea, Mt Cameroon, from where it has not subsequently been collected and may be extinct. However, it has since been recorded in the 1960s at Bangou and in the 1970s on the western slopes of Mt Nlonako. Inventory work in the 1990s revealed this species to be locally abundant above Nyasoso on the western slopes of Mt Kupe; it has also been collected at Bangem and recorded flowering elsewhere in the Bakossi Mts (Cheek, *pers. obs.*). This species is sometimes locally gregarious, as recorded on Mt Kupe and on Mt Nlonako, where Letouzey recorded it as 'abundant'. *A. preussii* is not easily confused with other African species of the genus when in flower; it more closely resembles some Asian species (Hetterscheid & Ittenbach 1996: 113). In SW Province, Cameroon, it has been recorded in flower in December and January. The specimen from Mt Nlonako has more lanceolate lobing to the leaves and was determined by Ittenbach as '*Amorphophallus preussii* (Engl.) N.E.Br. *var.*'.

In Cable & Cheek (1998), this taxon was listed under EN A1c + 2c; however, it is here reduced to VU due to the discovery of further sites for the taxon. The use of criterion B is preferred here as the taxon's habitat is not uniformly threatened (see below).

Habitat: understorey of montane forest, particularly on rocky slopes; 800–1600m alt.

Threats: populations at the lower altitudinal limit of its range (c. 800m) on Mt Kupe are threatened by agricultural encroachment linked to the expansion of Nyasoso town. The Mt Cameroon population is feared extinct following habitat loss above Buea.

Management suggestions: rocky areas and cliff-faces around Buea should be searched during the flowering period in order to rediscover the species on Mt Cameroon. Further analysis of the subpopulation on Mt Nlonako should be made to determine whether this is a distinct subspecies. A fuller assessment of its status in the Bakossi Mts should be carried out.

Rhaphidophora pusilla N.E.Br.

VU D2

Range: Cameroon, SW Province: Bakossi Mts (1 coll.); Gabon: Crystal Mts (1 coll.).

This taxon was previously known only from the type collection made in 1862 in the Crystal Mts, Gabon, by Gustav Mann. It has not subsequently been collected there. The discovery of this taxon in the Bakossi Mts near Kodmin therefore greatly increased the known extent of occurrence of this taxon. It is, however, classified as vulnerable here on the basis that it is known from only 2 locations and is therefore threatened by stochastic events, such as local fires or landslides, or future anthropogenic pressures. This taxon was, until recently, considered synonymous with *R. africana* N.E.Br., but has been resurrected by P.Boyce (*pers. comm.*). It is readily separated from the latter taxon, being a slender (not robust) climber and having solitary (not clustered) and erect (not pendent) inflorescences.

Habitat: submontane forest; c. 1500m alt.

Threats: the locations of the two known collections remain largely unthreatened, although the site near Kodmin was adjacent to the well-used track from Kodmin to Muahunzum (and eventually to Nyandong), thus it may be threatened by future increased use, and thus expansion, of this route.

Management suggestions: rediscovery of the two populations, and an assessment of their size and any local threats, is imperative. Further taxonomic work on the *Rhaphidophora* complex in west Africa may reveal further populations until now treated as *R. africana*.

BURMANNIACEAE

(M.Cheek)

Afrothismia pachyantha Schltr.

CR A2c; B1 + 2ab(iii); C2a(i,ii); D1

Range: Cameroon: Mt Cameroon (almost certainly extinct), Mt Kupe (1 site).

This species was first assessed as CR in 1998 (Cable & Cheek 1998: lxv: CR A1c + 2c; B1 + 2abcde; C2b; D). Since then no new sites have been located, and further searching on Mt Cameroon has failed to rediscover the species there (Franke *et al.* in press), where it is almost certainly extinct. Changes to the IUCN criteria in recent years, (IUCN 2003), have resulted in a modification of the assessment. Less than 50 individuals are known at the only site, and numbers appear to have declined between 1995 and 2003.

Habitat: lowland evergreen forest; c. 700m alt.

Threats: forest clearance for agriculture, stochastic events.

Management suggestions: continued monitoring and rental of the single known site for this species is advised.

Afrothismia saingei T.Franke

CR B1 + 2ab(iii); C2a(i,ii); D1

Range: Cameroon, SW Province: Mt Kupe (1 site).

Known only from the type, *Sainge* 1053 (7 Oct. 2002, Mbulle, 970m), this species is one of the most bizarre members of its genus.

Habitat: lowland to submontane evergreen forest; 970m alt.

Threats: forest clearance for timber of agriculture; a single stochastic event could destroy the single known site. Forest habitat at this altitude in the Mbulle area has declined in quality, especially towards Nyasoso. It is inferred that less than 50 individuals are known, as is usual for *Afrothismia* sites.

Management suggestions: this species should be rediscovered and protected by educating people at Mbulle of the presence of this species, known only in the forest above their village.

Afrothismia winkleri (Engl.) Schltr.
CR B1ab(iii)
Range: SE Nigeria, Cameroon, Uganda.
Known from only seven sites throughout its range, this species is probably extinct at Mt Cameroon, the type locality, since it has not been seen there in about 100 years, despite considerable searching. The species was assessed as CR A1c + 2c in Cable & Cheek (1998), but since then the IUCN criteria have changed and new sites have been discovered, at Korup and at Banyang Mbo, both in SW Province, Cameroon; accordingly a new assessment is made here, based on a severely fragmented distribution. The Nigerian material has been suggested as belonging to a different, new, unpublished species.
Habitat: lowland evergreen forest; 0–700m alt.
Threats: forest clearance for timber and agriculture. The extent of occurrence is less than 100km^2 (at each site only 1–3m^2 are occupied), severely fragmented and declining in quality.
Management suggestions: continued monitoring and rental of the site at Mt Kupe, and protection of the new sites at Banyang Mbo and Korup.

COMMELINACEAE
(M.Cheek)

Aneilema silvaticum Brenan
VU B2ab(iii)
Range: Nigeria, Cameroon and Congo (Kinshasa).
Known at K from three collections in Nigeria (*Meikle* 637, 1949; *J.D.Kennedy* 2674 & 1758, respectively, 1935 and 1931, all from Sapoba), three in Congo-Kinshasa (*Lemaire* 379, Mobwassa; *Seret* 69, Uele; *Tilquin* 183, Kinshasa) and two in Cameroon (*Zenker* 1110, Bipinde, 1896 and *Mbatchou* 470, Mt Cameroon, 1992).
In our area it was first collected between Kupe village and the Loum Forest Reserve (*Etuge* 2778, 15 July 1996) and more recently at Ngomboaku (*Ghogue* 493, 15 Dec. 1999). The second specimen was recorded as being from farm fallow, so *A. silvaticum* may be favoured by limited forest disturbance. It is much rarer (only 2 collections in our area) than *A. beniniense* (10 collections) with which it is most likely to be confused. Its apparent rarity may partly be attributed to its small and inconspicuous habit.
Habitat: lowland forest; 300–880m alt.
Threats: clearance of lowland forest for timber and/or agriculture is known to be a cause for concern in Nigeria generally, Mt Cameroon, Kinshasa and the Loum Forest Reserve (*pers. obs.*) with only seven sites known, the area of occupancy of this species is far below the 2000km^2 threshold of criterion B. Where the author has seen this species, it only occupied a site of about 4–5m^2.
Management suggestions: better policing and higher protection of existing forest reserves would help secure the future of this species. Further intensive survey work is likely to yield additional sites.

Floscopa mannii C.B.Clarke
EN B2ab(iii)
Range: Nigeria, Cameroon, Gabon & Congo (Kinshasa).
This taxon was previously known (K) from one collection in Nigeria (*Talbot* 756, Oban 1911), one in Gabon (*Mann* 1867, Corisco Bay, 1862) and one in Cameroon (*Leeuwenberg* 6800, 3 km NE Lomié, 1965). In our area it is known only from a single site to the east of Ngomboaku, in Mbumbe forest (*Cheek* 10349, 13 Dec. 1999). Here it was gregarious, and thriving in a swampy valley-bottom (c. 700m alt.) where the forest had been partly cleared, apart from *Raphia*. This species is probably often overlooked due to its small size and unspectacular nature. However, only four collections are known.
Habitat: lowland swamp forest; c. 700m alt.
Threats: clearance of lowland forest for timber and for agriculture. With an area of occupancy estimated at less than 500km^2 at only four sites, *F. mannii* is particularly susceptible to stochastic events.
Management suggestions: it is not clear that any of the sites for this species are currently gazetted as protected: this should be rectified if possible, and if governmental protection cannot be secured, efforts should be made to invoke protection from local communities. Ngomboaku, where the species was seen recently at a site that is easily revisited, might be a suitable focus for conservation efforts of *F. mannii*.

Palisota preussiana K.Schum. ex C.B.Clarke
VU D2
Range: Bioko (1 coll.); Cameroon, SW Province: Mt Cameroon (7 coll.) and Mt Kupe.

Known from only a single collection in our area: *Cable* 3383, collected on Max's trail, Mt Kupe, at 2000m alt. on 26 June 1996 when it was recorded as bearing immature fruit. Only two duplicates were collected, suggesting that only two individuals were seen. Distinguished from all other *Palisota* species with aerial stems in that the stems are slender (c. 6mm diameter when dried) and that the leaves are not in pseudowhorls, but evenly scattered along the stem. *P. preussiana* is also unusual in the single terminal inflorescence. The short petioles usually recorded in this species are longer than usual (3–7cm) in our specimen.

Habitat: submontane and montane forest; 1200–2000m alt.

Threats: clearance of forest for agriculture (e.g. on Mt Cameroon, cocoyam farms).

Management suggestions: efforts to protect this species should be centred at the eastern slopes of Mt Cameroon where it appears to be most common. More material on the Mt Kupe subpopulation is required to determine whether *Cable* 3383 is either aberrant, or represents a distinct taxonomic entity.

CYPERACEAE
(I.Darbyshire, B.J.Pollard & M.Cheek)

Bulbostylis densa (Wall.) Hand.-Mazz. var. *cameroonensis* Hooper
VU D2
Range: Cameroon, SW Province: Mt Cameroon (1 site, 2 coll.), Mt Kupe (1 coll.), Littoral Province: Mt Mwanenguba (1 coll.), NW Province: Bali Ngemba F.R. (1 coll.).

Until the 1990s, this variety was known only from the type collection (*Mann* 1360b) on Mt Cameroon, believed to be from the Mann's Spring area. It was then rediscovered here in 1992 (*Thomas* 9407), and has subsequently been recorded on three other montane grassland sites, at Mt Kupe, Mwanenguba and the Bali Ngemba Forest Reserve. It has not been recorded elsewhere in the Bakossi Mts. Although the extent of occurrence is now known to be much greater than when first assessed in Cable & Cheek (1998), the four known sites are highly isolated and limited in size, thus they are still vulnerable to local stochastic change, such as lava flow on Mt Cameroon or natural fires, thus the vulnerable status is still valid.

Care must be taken not to confuse this plant with the sympatric typical variety.

Habitat: montane grassland; 1800–3000m alt.

Threats: see above.

Management suggestions: more data on the size of each population is required to assess further its vulnerability to local stochastic events. Further botanical inventory work on montane grassland sites in e.g. the Bamenda Highlands, or Mt Nlonako in Littoral Province, may reveal further populations, in which case the conservation status would be downgraded.

Cyperus microcristatus Lye in press
CR B2ab(ii,iii)
Range: Cameroon, SW Province: Mt Kupe (1 coll.).

Despite the extensive botanical inventory work on Mt Kupe, this taxon is currently known from only a single collection (*Patterson* 11) collected in 1995 near Kupe Village, and is thus likely to be very rare within its extremely limited distribution (< 100km^2). This species most closely resembles the sympatric *C. densicaespitosus* Mattf. & Kük., differing in the more perennial habit, and the smaller spikelets and glumes which are more prominently winged.

Habitat: roadsides in agricultural area, 500m alt.

Threats: intensification of small-scale agriculture on Mt Kupe below 1000m alt. is resulting in an inferred decline in habitat quality within this taxon's distribution. The only known site is by a road, thus it is vulnerable to disturbance, as dirt roads are often annually 'improved' following rain damage.

Management suggestions: a survey of the previous collection site and other suitable locations on the lower slopes of Mt Kupe is imperative in order to rediscover and assess the population size of this taxon. The extent to which it can tolerate disturbance should be studied further, as in its one known location, it was found in an area of relatively high disturbance.

Cyperus rheophytorum Lye in press
VU B2ab(iii)
Range: Cameroon, SW Province: Mt Cameroon, Njonji (4 coll.), Mungo River W of Mbu (1 coll.), Bakossi Mts (4 sites, 4 coll.).

This species was first collected from the Mungo River in 1986 (*Manning* 542), and has subsequently been collected along the River Jide at Mwambong (*Onana* 585) and near Nyale (*Cheek* 10405), at a stream near Kodmin (*Etuge* 4063), and at Ndile waterfall, Baseng (*Cheek* 10405). It is locally gregarious, but as it is known from less than 10 locations it is considered VU.

Habitat: on rocks and stones in or beside streams and rivers, usually submerged during rains; 500–1350m alt.

Threats: illegal logging activity in catchment areas in the Bakossi Mts is likely to result in increased run-off and thus higher river levels with large sediment loads which may wash away existing populations of this taxon or destroy its habitat through excessive silting.

Management suggestions: this species should be sought-for in all future botanical work along rivers within its range. Monitoring of populations in catchments affected by land-use change, such as the Mungo River, should be carried out to assess this taxon's sensitivity to increased run-off or silting.

Cyperus tenuiculmis Boeck. subsp. *mutica* B.J.Pollard & Lye in prep.
CR B2ab(iii)
Range: Cameroon, SW Province: Nyasoso, Mt Kupe (1 coll.).
This new subspecies has been collected only once, as recently as 1998 (*Pollard* 113), when it was found in a ditch by the Tombel to Bangem road in Nyasoso town, where it was recorded as common. However, due to the lack of other records of this taxon from this area despite extensive botanical inventory work, and the significant threat to the known location (see below), it is considered CR.

Habitat: damp roadside ditch; 880m alt.

Threats: the Tombel to Bangem road is a well-used, unpaved route and so is prone to degradation during periods of heavy rain. Annual repairs are therefore made, including redigging of the drainage ditches. The only known population of this taxon is therefore severely threatened, if not already lost.

Management suggestions: the site should be revisited in order to determine if the population is extant; care should be taken to distinguish it from other *Cyperus* spp.; publication of the formal description is therefore important.

Hypolytrum pseudomapanioides D.A.Simpson & Lye in press
EN B2ab(iii)
Range: Cameroon, SW Province: Bakossi Mts (2 coll.).
This taxon is known from only 2 collections made in 1998, one 0.5km from Kodmin towards Ndip (*Cheek* 8919) and the second on the path from Kodmin to Muawhojom (*Ghogue* 33). It has not been recorded from similar habitat on the adjacent Mt Kupe or elsewhere in the Cameroonian highlands. It is considered EN due to the highly restrictred area of occupancy ($<500km^2$) and the perceived decline in habitat quality (see below). Recent molecular studies (Simpson *et al.* 2003) have cast doubt over the separation of the genus *Hypolytrum* from *Mapania*; future more-detailed work may support their amalgamation into a single genus.

Habitat: montane forest undergrowth; 1470–1500m alt.

Threats: localised encroachment of small-scale agriculture along the paths from which the 2 collections were made threaten these populations, though this is limited at present.

Management suggestions: further investigation of sedge taxa in the montane forest of the Bakossi Mts may reveal further populations which are less threatened by agricultural encroachment. Care should be taken to separate this species from the superficially similar *H. subcompositus* Lye & D.A.Simpson in prep.

Hypolytrum subcompositus Lye & D.A.Simpson in press
CR B2ab(iii)
Range: Cameroon, SW Province: Kodmin, Bakossi (1 coll.).
Known only from the type collection from Kodmin in the Bakossi Mts, collected in January 1998 (*Etuge* 4007), this taxon has a highly restricted area of occupancy ($<10km^2$) and is clearly very rare within its range, as it has not subsequently been found in Bakossi despite extensive survey of its montane forest habitat.

Habitat: montane forest undergrowth; 1500m alt.

Threats: due to its highly restricted area of occupancy, this species is highly threatened by local disturbance, such as small-scale agricultural encroachment which is widespread in the Bakossi Mts

Management suggestions: rediscovery, and a subsequent census, of the known population is a priority; subsequent surveys of sedge taxa in the Bakossi Mts may reveal further sites.

Mapania ferruginea Ridl.
VU B2ab(iii)
Range: São Tomé: Contador (1 coll.), Macambrá, Vanhulst (1 coll.), Casa del Pico (2 coll.), Callario (1 coll.), unlocated (1 coll.); Principé: W of Pico (1 coll.); Cameroon, SW Province: Bakossi (1 coll.), Mt Kupe (2 coll.).
The collections above Kupe Village on Mt Kupe (*Cable* 3773 and *Etuge* 2853) in July 1996 represented the first records of this species on continental Africa. It has subsequently been found once in the Bakossi Mts at Kodmin. It

was previously known only from São Tomé & Principé where it was recorded as 'a rare species … confined to fairly high altitudes' (Simpson 1992). Care should be taken to separate this taxon from the similar *M. soyauxii* (Boeck.) H.Pfeiffer which is recorded at lower altitudes in Cameroon; notes on the separation of these taxa can be found in Simpson (1992).

Habitat: montane forest undergrowth; 700–2000m alt.

Threats: the subpopulation at the lower altitudinal range on Mt Kupe (1100m) is threatened by future encroachment of agriculture and resultant forest clearance. Montane forest on São Tomé is also threatened by small-scale human encroachment.

Management suggestions: enforced protection of the montane forest habitat on Mt Kupe. A survey of the São Tomé populations, this species' stronghold, should be carried out and important areas protected where necessary.

Scleria afroreflexa Lye in press
EN B2ab(iii)
Range: Cameroon, SW Province: Bakossi Mts (1 coll.), NW Province: Boyo (1 coll.), Bali Ngemba F.R. (2 coll.).
This conservation assessment was originally made by B.J.Pollard in Lye & Pollard (in press, Nordic Journal of Botany), but here we add the additional site of the Bali Ngemba Forest Reserve from where this taxon was collected twice in 2001. It was first collected in 1999 in two disjunct locations 170km apart in western Cameroon: between Laikom and Fundong in Boyo Division, NW Province, and near Kodmin in the Bakossi Mts, SW Province. It was recorded as locally common at the latter site. Further subpopulations are likely to exist in locations that have not yet been included in botanical inventories, given the large areas of grassland at altutudes of around 1500m in the Bamenda Highlands, Bamboutos and, to a lesser extent, in SW Province (Pollard, *pers. obs.*).

Habitat: montane grassland and grassland patches in submontane forest; 1450–1550m alt.

Threats: deliberate burning of the montane grasslands in NW Province may result in subpopulation losses or may lead to long-term habitat changes which do not favour this taxon. The site near Kodmin appears relatively unthreatened though may be lost to forest encroachment if human disturbance remains low.

Management suggestions: studies of the subpopulations in the Northwest Highlands could be made to better understand the ecology of this taxon, including its tolerance of fire and human disturbance. This species should also be searched-for in suitable habitat elsewhere in western Cameroon; discovery of further subpopulations would lead to a downgrading of its conservation status.

DRACAENACEAE
(I.Darbyshire)

Dracaena viridiflora Engl. & K.Krause
VU B2ab(iii)
Range: Nigeria, Cross River State: Oban (1 coll.), Obudu Plateau (1 coll.); Cameroon, SW Province: Mt Kupe (2 sites, 2 coll.), Douala to Likomba (1 coll.), S Province: Bipinde (1 coll.); Rio Muni (1 coll.).
Despite the relatively large altitudinal range and extent of occurrence of this species, it is rare throughout, being known from only 7 locations, those in Nigeria, Rio Muni and S Province, Cameroon, from early 20th century collections. In addition to the locations cited, this taxon is also possibly known in Cameroon from Mevou southeast of Ebolowa (S Province) and from 5km NW of Yaoundé (Central Province), although determination of these collections is to be confirmed. *D. viridiflora* is very close to *D. mildbraedii* K.Krause, an uncommon taxon of Rio Muni with which *D. viridiflora* was treated as synonymous before the taxonomic revision by Bos (1984). However, it is still possible that *D. viridiflora* is under-recorded due to confusion with this sympatric species.

Habitat: closed-canopy lowland to submontane and montane forest; c. 200–2000m alt.

Threats: the sites in Nigeria and the lowland site(s) in S Province, Cameroon, are under threat from forest clearance for agriculture and logging. The Mt Kupe sites are, however, currently unthreatened, being within the altitudinal range of effective protection on this mountain.

Management suggestions: continued protection of the submontane and montane forest on Mt Kupe; a survey of the population of this taxon here will help determine how important a site this is for conservation of this taxon.

MARANTACEAE
(B.J.Pollard)

Marantochloa mildbraedii Loes. ex Koechlin
EN B2ab(iii)
Range: Cameroon: Mungo River F.R. (1 coll.); Gabon (2 coll.).

Mildbraed 3856a, from Moloundou, Gabon is the type collection, with *Jeffrey* 158, 11 August 1957, 600m, Mitzic, being the other collection from Gabon, and Martin Etuge's 1998 specimen from the Mungo River Forest Reserve, extends its range into Cameroon.

Habitat: lowland forest; 200–600m alt.

Threats: there has been illegal tree-felling in and around the Mungo River F.R., and it is certain that the habitat quality there is in decline.

Management suggestions: a search of the Cameroonian locality should be made to assess the status of this taxon, and its habitat preferences. Jeffrey's notes indicate this taxon was found growing in young secondary forest, which may indicate a preference for disturbance, and so allow us to focus future searches on this habitat type.

Sarcophrynium villosum (Benth.) K.Schum.

EN B2ab(iii)

Range: Cameroon: Bakossi Mts (5 coll. at 3 sites), Mt Nlonako (1 coll.); Gabon (2 coll.).

This taxon was known only from *Mann* 1032 (type), collected July 1861 from the Gaboon River, until *Letouzey* 14465, 17 March 1976, Mt Nlonako, just outside our checklist area. A second Gabon specimen from ± 1020m, is *Louis et al.* 980, from Massif du Chaillu, Songou Mountain, 28 Nov. 1983, followed by the first Bakossi specimen, *Etuge* 170, in 1986. Four further Bakossi collections have been made since, bringing the total number of known locations to 5.

Habitat: submontane forest; 1000–1500m alt.

Threats: the Bakossi subpopulations should be well protected now, at altitudes of above 1000m, but the Nlonako site is not afforded the same level of protection and there is an increasing level of small-scale agriculture on the mountain.

Management suggestions: monitor populations within the Bakossi Mts.

ORCHIDACEAE
(B.J.Pollard & I.Darbyshire)

Angraecopsis tridens (Lindl.) Schltr.

VU A2c

Range: Bioko (4 coll.); Cameroon: Mt Cameroon (3 coll.), Mwanenguba (1 coll.), Bamboutos Mts (1 coll.), Bakossi Mts: Kodmin (2 coll.) and Enyandong (1 coll.).

This taxon was originally assessed in Cable & Cheek (1998). The three additional collections from the Bakossi Mts are within the expected range of this species, and alter the original assignation of EN to VU, as threats throughout its range are still estimated to be at 1998 levels, except at the new localities which fall within the Bakossi Forest proposed National Park, and so we here reduce our estimate of population size reduction to c. 30% over the last 10 years.

Habitat: montane forest; 1200–1500m alt.

Threats: conversion of forest to small-scale agriculture or plantations.

Management suggestions: this species should be introduced into cultivation, commencing at the CRES site at Nyasoso.

Angraecum pyriforme Summerh.

VU B2ab(iii)

Range: Ivory Coast (3 coll.); Nigeria (3 coll.); Cameroon (1 coll.).

Described from *Talbot* 888 (type), Oban, Nigeria in 1911, two more Nigerian collections near Sapele (*Wright* 132 and 154), were made in July 1953 & 14 June 1954 respectively. Three Ivory Coast collections were made in the 1960s: *Aké Assi* 6028, Tai forest, 21 Oct. 1961, and 9002, near Dahiri, 27 June 1966, then *Breteler* 5232, border of Bandama, 10 July 1968. The only other collection is that of Sharon Balding at Nyasoso in Aug. 1993. This represents a new record for Cameroon, this taxon not having been included in the Fl. Cameroun accounts (1998–2001).

Habitat: an epiphyte in forest; c. 1450m alt.

Threats: clearance of forest for conversion to agricultural small-holdings or plantations, particularly threatening the Nigerian and Ivory Coast sites where clearance has been widespread in recent decades, although the Tai Forest is thought to remain intact.

Management suggestions: further orchid surveys are to be encouraged around Nyasoso to attempt to relocate this species and, if successful, introduce it to the CRES orchid garden there.

Angraecum sanfordii P.J.Cribb & B.J.Pollard

EN B2ab(iii)

Range: Cameroon, SW Province: Mt Cameroon (1 coll.), Mt Kupe (3 sites).

First collected by W.W.Sanford above Buea on Mt Cameroon in 1968 (*Sanford* 550/65), this species was not formally described until 2002 (Cribb & Pollard 2002: 653). It has since been collected at 1100–2000m on Mt Kupe at 3 different locations, 2 above Nyasoso and 1 on the summit of Kupe Rock. The Buea site was at 800m alt.; clearance of forest for plantation agriculture and timber has been widespread at such altitudes in the vicinity of Buea since Sanford's collection was made (Cheek, *pers. comm.*), thus this species may now be extinct on Mt Cameroon and so restricted to Mt Kupe.

Habitat: an epiphyte in submontane and montane forest; 800–2000m alt.

Threats: see above for threats to the Mt Cameroon site; on Mt Kupe, much of the forest above 1000m is protected, thus disturbance is limited, though the site at 1100m above Nyasoso is threatened by encroachment of agriculture following the continued expansion of this town.

Management suggestions: efforts to rediscover the population on the eastern slopes of Mt Cameroon should be made. A survey of its population on Mt Kupe, particularly at lower elevations, would allow a better assessment of its status on the mountain, seemingly the stronghold of this species.

Bulbophyllum bifarium Hook.f.

VU B2ab(iii)

Range: Cameroon endemic: Bakossi Mts (3 coll. at 1 site), Mt Kupe (2 coll. at 2 sites), Mt Cameroon (3 coll. at 3 sites), Douala to Bimbia (1 coll.), Mfongu near Bagangu (1 coll.), Bana-Bateha near Fibé (1 coll.), Nkokom Massif near Ndom (1 coll.).

First described from a collection by Gustav Mann on Mt Cameroon (*Mann* 2121), this species is now known from 10 sites; the specimen cited in FWTA from Ivory Coast is now now referred to *B. bidenticulatum* J.J.Verm. subsp. *bidenticulatum* (Vermeulen 1987: 167). On Mt Kupe it was first recorded in April 1899 at Nyasoso (*Schlechter* 12896), and relocated there almost 90 years later (*Thomas* 5062, Nov. 1985), but was not recorded there during the intensive surveys of the 1990s. The Bakossi Mts appear to be a stronghold for this species, with 3 collections made in the Kodmin area in 1998.

Habitat: an epiphyte in submontane and montane forest; 800–1800m alt.

Threats: the lower altitude sites for this taxon, particularly that on the Douala to Bimbia road and at Nyasoso on Mt Kupe, where it was recorded at c. 800m, are threatened by forest clearance for plantation agriculture at the former and for small-holder farming at the latter; these sites are likely to be already lost. The higher altitude sites on Mt Cameroon, Mt Kupe and at Kodmin are, however, less threatened at present.

Management suggestions: conservation of this taxon should focus upon the Kodmin area of the Bakossi Mts where it appears to be most gregarious and where threats to its habitat are currently limited. A survey of its abundance should be carried out in November, its known flowering period there.

Bulbophyllum comatum Lindl. var. *comatum*

EN B2ab(iii)

Range: Nigeria: Cross River State (1 coll.); Bioko (2 coll.); Cameroon: Kupe Village-Loum Forest Reserve (1 coll.).

This variety, first collected on Bioko, *Mann* 642, is restricted to the west Cameroon uplands. It is seemingly absent from large areas of suitable altitudinal range and habitat within its extent of occurrence, for example Mt Kupe and Mt Cameroon. *B. comatum* var. *comatum* is separated from the Guineo-Congolian var. *inflatum* (Rolfe) J.J.Verm. by the denser and longer hairs on the ovary and by the more uniformly long-white-haired abaxial side of the sepals (Vermeulen 1987: 86).

Habitat: mid-elevation to submontane forest; (300–)600–1300m alt.

Threats: forest clearance is widespread in lower elevation sites throughout this taxon's range, for example on Bioko much of the forest below 1000m alt. has been converted to plantation and in the Loum Forest Reserve, small-holder farms continue to encroach into the forest despite its protected status.

Management suggestions: enforcement of the Loum Forest Reserve boundary may protect this species there. This species should also be brought into cultivation at the CRES orchid garden as part of an *ex situ* conservation programme.

Bulbophyllum jaapii Szlach. & Olszweski

VU D2

Range: Cameroon: Mt Kupe (1 coll.).

This species was discovered in November 1985 at the summit of Mt Kupe by D.W.Thomas & H.L.MacLeod (*Thomas* 5049), where it was recorded as a 'common branch epiphyte' in forest and scrub. However, intensive botanical surveys on this mountain during the 1990s, including the summit forests and scrub, have revealed no further collections, indicating that this species is extremely limited in its distribution. It is perhaps restricted to the east-facing slopes near the summit, which were less well-collected during the 1990s.

Habitat: epiphyte of montane forest and scrub; 1800m alt.

Threats: this species is found well beyond the highest altitude of destructive anthropogenic activity on the mountain and is within the zone of protection, thus its habitat receives little disturbance. However, its extremely limited distribution renders it vulnerable to stochastic events, particularly localised fire or mudslides, which could cause small-scale loss of forest cover and thus loss of habitat. Such an event would threaten the single known population of this species.

Management suggestions: the summit of Mt Kupe should be visited in Nov. (recorded flowering time) to assess the only known location for this taxon. Information on numbers of individuals, and estimates of population size may then allow us to use Criteria A, C & D1 in future, and prepare a management plan, to include both *in situ* and *ex situ* conservation measures.

Bulbophyllum kupense P.J.Cribb & B.J.Pollard
CR B1ab(iii) + 2ab(iii).

Range: Cameroon: Mt Kupe (1 coll.).

The following conservation assessment is taken largely from Cribb & Pollard (2004). *B. kupense* is known only from an individual epiphyte on one 5m tall mango tree, in a private garden, where one individual plant of this taxon continues to grow. The type is a small portion of that individual. It is probable that this species also occurs in the Mt Kupe forest, which exists only 100m away from the type locality.

Habitat: phorophyte: *Mangifera indica* L. (cultivated); an epiphyte, 830m alt.

Threats: loss of the only known location due, for example, to the cutting down of the host tree or due to disturbance during harvesting of the mangos is possible. If this taxon also occurs in the surrounding forest above Nyasoso, it is threatened by the continuing decline in habitat area and quality due to clearance of trees for firewood and small-scale agriculture.

Management suggestions: a certain measure of protection has been afforded this species, as this garden area has been fenced off to demarcate the garden's boundary (Pollard, *pers. obs.* 2002). A portion of the individual plant was removed from the garden and is now growing as part of the living orchid collection at the CRES headquarters in Nyasoso. A propagation plan is currently being developed by the authors, in collaboration with Martin Etuge, manager of the CRES project garden.

Bulbophyllum nigericum Summerh.
VU A2c; B2ab(iii)

Range: Nigeria: (5 coll., at 5 sites); Cameroon: Mt Kupe (1 coll.), Bakossi Mts (1 coll.), Bali Ngemba (1 coll.), unlocated (2 coll.).

This species was described from *King* 124 collected in October 1958 from Plateau Province, Nigeria, from where King made 3 additional collections. A further record, from the Mambilla Plateau in November 1993 (*Sporrier* 18) remains unconfirmed, the specimen being labelled '*Bulbophyllum nigericum ?*'. The specimen cited in FWTA from the Ivory Coast is now referred to *B. bidenticulatum* J.J.Verm. subsp. *bidenticulatum* (Vermeulen 1987: 167). It was first collected in Cameroon on the southern side of Mt Kupe in (?)1970, *Letouzey* 408, but has not since been recorded there. Two additional specimens are recorded from western Cameroon by Vermeulen (1987: 92), but no site locations are listed. Recent intensive surveys in Cameroon have revealed only one additional site, at Enyandong in the Bakossi Mts (*Salazar* 6322, Oct. 2001), and from a plot voucher specimen collected at Bali Ngemba in 2002.

Habitat: an epilith or epiphyte in submontane and montane forest; c. 800–2050m alt.

Threats: the Nigerian sites are threatened by continued extensive clearance of forest up to high elevations; one or more of these subpopulations are likely to have been lost. The plant at Enyandong was found growing in largely-cleared forest, on a tree in the village. It is therefore likely to occur in the surrounding forest, some of which is being cleared for small-holder farming, thus threatening this subpopulation. In all, a loss of over 30% of the population is estimated over the past 3 generations, which we here estimate to be 10 years, much of this loss being irreversible.

Management suggestions: as this species is found within the village of Enyandong, on a tree in front of the house of the Chief of the village, this is an ideal location for promoting community-based conservation. Local residents here could be encouraged to search for this species in the surrounding forest, perhaps using a species conservation poster as an aid to identification, and to promote protection of any locations where it is found.

Bulbophyllum pandanetorum Summerh.
EN B2ab(iii)

Range: Cameroon: Mt Kupe (1 coll.); Gabon: Ngounyé (2 coll., at 2 sites).

This taxon was previously known from two collections made in September 1925 in the Ngounyé River area of Gabon, the type from near Kembélé (*LeTestu* 5527), and the second from along the River Dévèla (*LeTestu* 5547). Both were recorded growing on *Pandanus* trees, hence the specific epithet. The discovery of this species growing in submontane forest on Mt Kupe in June 1996 (*Zapfack* 607) greatly extends the extent of occurrence of this species, but it is clearly scarce, not having been recollected in Gabon and not having been found elsewhere in western or southern Cameroon, despite extensive botanical survey work there.

Habitat: lowland or submontane forest; c. 200–950m alt.

Threats: the location above Nyasoso on Mt Kupe is below the lower altitudinal limit of effective protection on the mountain and thus vulnerable to habitat loss through encroachment of farms; threats to the Gabon sites are unknown.

Management suggestions: attempts should be made to rediscover this species in its previous collection localities, and to bring this species into cultivation for *ex situ* conservation. Status of habitat at the Gabon locations needs to be researched.

Bulbophyllum scaberulum (Rolfe) Bolus var. *fuerstenbergianum* (De Wild.) J.J.Verm.

VU B2ab(iii)

Range: Nigeria, Ogoja Province: Ikwette (2 coll.), Mambilla Plateau (1 coll.); Bioko (1 coll.); Cameroon, SW Province: Mt Kupe (1 coll.), Bakossi Mts (1 coll.), NW Province: Ijim Ridge (1 coll.), unlocated (1 coll.); Congo (K): Haut Katanga (1 coll.).

This taxon has a large extent of occurrence but highly limited area of occupancy, being known from disjunct populations in the west Cameroon uplands and in the Congo Basin. It was described from Bioko, *Furstenberg* s.n., but has not since been re-recorded there. It was first noted as occurring in Cameroon by Vermeulen (1987: 118), but no site details were recorded. It has since been collected once in NW Province, along the Ijim Ridge (*DeMarco* 19, fl., Dec. 1999), once at Kodmin in the Bakossi Mts (*Pollard* 252, Nov. 1998), and once near Kupe Village on Mt Kupe (*Gosline* 253, Dec. 1999).

Habitat: an epiphyte in lowland to montane forest; c. 500–1500m alt.

Threats: the Nigerian populations are threatened by continued, widespread, forest clearance for agriculture. Similarly, the site near Kupe Village, at 800m elevation, is below the lower limit of effective protection on Mt Kupe and thus vulnerable to future agricultural encroachment.

Management suggestions: in light of the significant threat to several of the known locations of this species, it should be a high priority to bring this taxon into cultivation and to begin an *ex situ* conservation programme, with eventual reintroduction into protected areas.

Bulbophyllum teretifolium Schltr.

CR B1ab(iii) + 2ab(iii)

Range: Cameroon, SW Province: Mt Cameroon, Bibundi (1 coll.), (?)Bakossi: Bangem (2 coll.).

This species is known (for certain) only from the type collection (*Schlechter* 12362), deposited at the Berlin herbarium where it was destroyed during World War II. It was recorded as common at Bibundi by Schlechter in April 1899 (Vermeulen 1987: 160), but has not since been collected there. The material from near Bangem, northern Bakossi, was collected and determined by D.W.Thomas in Jan. 1986 (*Thomas* 5281 & 5950), and deposited at Missouri (MO). These specimens refer to the same plant, which was found growing on a mango tree in the village of Nkud, then taken into cultivation in Kumba, flowering in March. They have not yet been seen at Kew and so their determinations are to be confirmed. It is possible that the Bangem material in fact refers to *B. kupense*, which closely resembles the description in the protologue of *B. teretifolium*, but differs in having dorsiventrally compressed leaves (not terete), a pendent habit (not erect to erect-patent), and is known from fertile material (not sterile). It is possible that Schlechter erroneously described *B. teretifolium* as erect to erect-patent based on the herbarium sheet, but as he had observed it as being common, his interpretation of the habit was likely based on personal familiarity with its mode of growth. This difference, combined with that of leaf morphology, lead to the conclusion that *B. kupense* and *B. teretifolium* are indeed two different entities (Cribb & Pollard 2004) in press). The former taxon must be ruled out before the Bangem material can be confirmed as a second site for *B. teretifolium*, thus this conservation assessment is based solely on the material from Bibundi.

Habitat: an epiphyte in forest; 800m alt.

Threats: the fact that this species has not been recollected in the Mt Cameroon area despite extensive botanical inventory work there in the past century, suggests that this species may have been lost here; suitable habitat in the type locality is threatened by forest clearance for agriculture.

Management suggestions: confirmation of the identity of the Bangem material is important; it should therefore be compared to the forthcoming protologue of *B. kupense*. Efforts to rediscover this species at the type locality should also be made.

Postscript: Dr Zapfack, one of the IUCN assessors (Orchid Red List Authority) has rejected the above assessment, stating that this taxon is very widespread in semi-deciduous rainforest, and that he has collections from Nguti, Ebolowa and Akolinga. The author of this account (BJP) disagrees with that position (see Cribb & Pollard 2004), but for the moment we are leaving the assessment as it stands, until we are able to discuss the statement with him.

Diaphananthe polydactyla (Kraenzl.) Summerh.

VU B2ab(iii)

Range: Cameroon, SW Province: Kupe Bakossi (7 coll., from 6 sites), NW Province: Tatum (1 coll.), Bafut Ngemba F.R. (1 coll), Baba II (1 coll.).

This taxon was described from a single collection from Tatum (Kufum) 30km ENE of Mt Oku in the Bamenda Highlands (*Ledermann* 5716A). It has since been collected twice in that region, in 2001 and 2002. It was recorded at Mwanenguba in November 1968 (*Sanford* 5478 & 5486). Inventory work in the Kupe-Bakossi region in the 1980s and 1990s revealed five further locations for this taxon: Bangem, Ehumseh-Mejelet, Enyandong, Kodmin and on Mt Kupe, above Nyasoso. Although considered rare within its range, this is probably a function of collection intensity, as it has appeared time and again in recent surveys of suitable habitat (Pollard, *pers. obs.*).

Habitat: an epiphyte of submontane to montane forest and woodland; c. 1400–2000m alt.

Threats: clearance of montane forest is widespread in the Bamenda Highlands, and at Bangem and Mwanenguba to the south. Botanical inventory work in the 1990s at Mwanenguba failed to provide new records, indicating that this population may no longer be extant.

Management suggestions: as Kupe-Bakossi now appears the stronghold of this taxon, a survey of the epiphytic orchids of the submontane and montane zone should be carried out to better assess the abundance of this, and other rare taxa. Maintenance of the protection of the montane forests in this area should ensure the survival of *D. polydactyla*.

Disperis mildbraedii Schltr. ex Summerh.

VU B2ab(iii)

Range: Nigeria (2 coll. at 2 sites); Bioko (1 coll.); W Cameroon (3 coll., at 3 sites).

This species was described from *Mildbraed* 6312, collected at Sta Isabel, Bioko, 16 Aug. 1911, 1100–1400m. *Ujor in FHI* 29965 was the second collection, made on 15 Aug. 1951 in the Bamenda Highlands: Banja in Bafut Ngemba F.R. to Lake Bambili. Two Nigerian collections were made in 1973: *Hall* 2950, 18 June, N slopes of Babanke (Obudu Plateau), and *Chapman* 85 *(in FHI* 70882*)*, 16 Aug., 1800m, Mambilla Plateau. *Cable* 117 & 3378, both from Mt Kupe, are the only other known collections.

Habitat: forest; c. 1100–1800m.

Threats: continued clearance of forest on Bioko, in the Bamenda Highlands and in neighbouring parts of Nigeria.

Management suggestions: attempt to relocate this species and introduce it into cultivation.

Disperis nitida Summerh.

EN B1ab(iii) + 2ab(iii)

Range: W Cameroon: Mwanenguba (1 coll.), Bamenda Highlands (5 coll. from 4 sites).

This taxon was assessed by Cheek *et al.* (2000). Here details of the assessment are modified slightly to include the relevant subcriteria, and sub-subcriteria.

Habitat: montane forest on lower branches or leaning trunks of trees in densely canopied areas, rarely terrestrial; 1800–2800m alt.

Threats: continued clearance of remaining montane forest at Mwanenguba and in the Bamenda Highlands outside of protected areas, together with possible future encroachment into protected areas, severely threaten this species.

Management suggestions: see Cheek *et al.* (2000); management is best focused upon the Kilum-Ijim population which is best protected and at which the species appears from our surveys to be at its most numerous.

Genyorchis micropetala (Lindl.) Schltr.

EN B2ab(iii)

Range: Bioko (3 coll., at 3 sites); Cameroon (3 coll., at 2 sites).

Described from Bioko, based on *Mann* 644, Dec. 1860, 1330m, with two further Bioko collections: *Melville* 437A, 2 Sept. 1959, 1600m, from Moka, and *Sanford* 4267, 4 Jan. 1967, 1800m, Pico Boca. A collection from Congo (K), *Burn* s.n., is a doubtful record, and determined as *G. ? micropetala*, and is not included in this assessment. *Sanford* 5256 & 5274, both made on 27 Oct. 1968, Batouri district, near Dimako, and *Letouzey* 477, from Mt Kupe represent the only other known collections.

Habitat: forest; c. 1330–1800m.

Threats: small-scale forest clearance of montane forest on Bioko and in Cameroon may threaten the subpopulations. Though this is currently limited within this species' altitudinal range, expansion of agriculture into higher altitudes is likely to occur in the near future due to growing food demands, thus threatening unprotected sites.

Management suggestions: attempt to relocate this species in its Bioko and Cameroon locations, and introduce it into cultivation. Further investigation of the Congo (K) material is needed in order to confirm its identity; the conservation status would be reassessed as VU under criterion B, if it is shown to belong to this species.

Habenaria batesii la Croix
EN B2ab(iii)
Range: Cameroon, SW Province: Bakossi Mts (1 coll.), S Province: Efulen (1 coll.).
This taxon was described in 1993 from a single collection from Efulen, near Ebolowa in S Province, Cameroon, made in Oct. 1895 (*Bates* 453). It has not been recorded at this site since, and remained the only record of this species until *Pollard* 731, collected at Enyandong in Bakossiland, in Oct. 2001. This taxon was originally published in error under the later homonym *Habenaria praetermissa* la Croix (la Croix 1993: 371) but was corrected in 1996 (la Croix: 364).
Habitat: a terrestrial herb of dense lowland and submontane forest; c. 200–900m alt.
Threats: Efulen is located on the main route from Lolodorf to Ebolowa, two sizable towns in S Province, and is therefore likely to have experienced significant habitat disturbance, due to both population and agricultural expansion; this site may well now be lost as the species was last recorded here over a century ago (if this is confirmed, this taxon will warrant CR status under IUCN criterion B). The Bakossi site is threatened by encroachment of small-holder farming into the submontane forest habitat.
Management suggestions: attempts to rediscover the species at its two collection locations should be made (preferably in Oct., its recorded flowering time). This species may be suitable for a species conservation poster to raise awareness within the local communities as to the unique flora of their forests.

Habenaria nigrescens Summerh.
VU A2c; B2ab(iii)
Range: Nigeria: Obudu Plateau (1 coll.); Cameroon: Bafut Ngemba F.R. (4 coll., 1 site), Mt Neshele, 10km ESE Bamenda (1 coll.), Banyo, Mayo Tankou (1 coll.), Mwanenguba (1 coll.).
The previous stronghold of this highly localised taxon was previously the Bafut Ngemba Forest Reserve in NW Province, Cameroon, from where the type specimen was collected (*Daramola in FHI* 41568, July 1959), and from where 3 other collections were made between 1959 and 1975. Outside of the Bamenda Highlands, it is known from a single collection in 1973 from the Obudu Plateau of Nigeria (*Hall* 2952), and from a 1971 collection from Mwanenguba, Littoral Province, (*Leeuwenberg* 8473). An unconfirmed record from 1967, made near Ngaoundéré, Adamawa Province, is recorded at a considerably lower altitude (1100m) and in drier climatic conditions to the other known collections, the specimen being labelled as '*Habenaria* sp. aff. *nigrescens*' and likely refers to a separate entity (*Meurillon in CNAD* 844).
Habitat: montane grassland; c. 1700–2300m alt.
Threats: a significant reduction in both the quality and area of natural habitat has been recorded in the Bafut Ngemba F.R. (Pollard, *pers. obs.* 2002), and it is likely that this will have had deleterious effects on all plant taxa, irrespective of habitat preferences. Similarly, high pressures from human activities occur in the montane sites at Obudu, Nigeria, and Mwanenguba, Cameroon.
Management suggestions: a survey of this taxon should be made at the Bafut Ngemba Forest Reserve and protective measures should be put in place here if it remains extant.

Habenaria thomana Rchb.f
VU B2ab(iii)
Range: Bioko (1 coll.); São Tomé (4 coll., at 4 sites); Cameroon (9 coll., at 4 sites).
This taxon was described based on *Mann* s.n., on São Tomé, from where it was believed to be endemic and known from a further three collections: two in the late nineteenth century (*Moller* 95 (?) and *Quintas* 96) and in August 1997 at Calvario (*Stévart* 39). In 1972 it was collected for the first time in Cameroon, along the Kumba-Douala road (*Leeuwenberg* 10617). It is now also known from above Bwassa village on Mt Cameroon, and from 7 collections made in the 1990s from 2 locations on Mt Kupe, and from Bioko.
Habitat: a terrestrial herb of lowland to submontane forest, rarely in secondary forest; 50–1330m alt.
Threats: clearance of forest throughout its range, especially below 1000m.
Management suggestions: management should focus upon its two strongholds of São Tomé and Mt Kupe. A survey should be carried out at the former in order to assess its current status, as 3 of the 4 known collections were made in the nineteenth century. At the latter, material from the more-threatened lower altitude sites should be considered for cultivation and propagation at the CRES headquarters living orchid collection.

Manniella cypripedioides Salazar, T.Franke, Zapfack & Beenken
EN B2ab(iii)
Range: Bioko (1 coll.); Cameroon, SW Province: Banyang Mbo Wildlife Sanctuary (1 coll.), Mt Cameroon-Mt Etinde (1 coll.), Bakossi Mts (2 coll. from 2 sites).
This taxon is currently known from only 5 locations, all within the west Cameroon uplands chain. It was first described in 2002 (Salazar *et al.*: 240), from which all the above location data were recorded, with the exception of an additional collection made in November 2001 at Bime Rock, Mwendolengo, Bakossi (*Etuge* 4515r). In Salazar *et*

al. (2002), this taxon was given a conservation rating of DD (data deficient), on the basis that the status of the first Bakossi site (Kodmin) and the Bioko population could not be assessed. However, we here modify this assessment on the basis of criterion B. *M. cypripedioides* is known from only 5 locations. Due to the presence of a threat to the Kodmin population (a reservoir scheme has been proposed near the village) this qualifies it as EN under criterion B. Its habitat is believed to be largely unthreatened at some of the other locations, these being: Bioko (where forest at c. 1300m alt. remains largely intact), Banyang Mbo (which is a well-protected wildlife sanctuary), and Mwendolengo, (where disturbance above c. 1000m alt. is minimal). At the Bioko location, this species was recorded as 'frequent in dense shade' (*Sanford* 4326), thus this species may be gregarious within its specialised habitat.

Habitat: a terrestrial herb along streams in primary submontane forest; 950–1350m alt.

Threats: the proposal to expand plantations to the 1000m contour around Mt Etinde-Mt Cameroon may well adversely affect the subpopulation there; placement of the reservoir at Kodmin may threaten the population there.

Management suggestions: attempts should be made to discover this species within its preferred habitat in other parts of the western Cameroon uplands (its recorded flowering period is Oct. to Dec.).

Ossiculum aurantiacum P.J.Cribb & Laan
CR B2ab(iii)

Range: known only from the type collection (*Beentje* 1460A), 13km from Kumba on the road to Loum, Mungo River Forest Reserve, W Cameroon.

This monospecific genus has not been seen since it was first discovered in 1980, despite several thousand botanical collections since being made within 50 miles of the type location.

Habitat: an epiphyte in lowland evergreen rainforest; 200m alt.

Threats: recent road construction along the border of the Mungo River Forest Reserve, aided by European Union funding, provides greater accessibility into this species' only known location and may result in intensification of forest clearance outside of the reserve boundary and encroachment of small-scale cash-crop agriculture into the reserve, if not carefully managed.

Management suggestions: urgent action is required to address the issues of timber extraction from the Mungo River area, and to protect the existing forest reserves from illegal encroachment of agricultural activity. Intensive orchid surveying should be carried out for the CRES orchid programme at Mungo and other nearby lowland areas, such as the adjacent Loum Forest Reserve.

Polystachya albescens Ridl. subsp. *manengouba* Sanford
CR B1ab(iii) + 2ab(iii)

Range: Cameroon: Mwanenguba (1 coll.).

Known only from the type collection, *Sanford* 5557, Nov. 1968, this subspecies has never been recollected from Mwanenguba despite botanical inventory work there in both the 1970s by Leeuwenberg and in the 1990s by Cheek and others. *P. albescens* Ridl. *sensu lato* comprises six subspecies, ranging from Nigeria to Tanzania and south to Zimbabwe. The similar *P. albescens* subsp. *albescens* is recorded at significantly lower altitudes than subsp. *manengouba*. Whilst subsp. *manengouba* is the scarcest, subsp. *angustifolia* (Summerh.) Summerh. is also restricted to Cameroon, being known from only 4 collections on Mt Cameroon and not extending to the Kupe-Bakossi Mts.

Habitat: montane woodland; 2000m alt.

Threats: high levels of disturbance and montane woodland clearance have occurred in recent decades at Mwanenguba.

Management suggestions: rediscovery of this taxon and introduction into cultivation in order to develop a controlled propagation plan. Restoration of forest habitat on Mwanenguba is desirable and could include re-introduction of threatened orchids after a propagation programme.

Polystachya bicalcarata Kraenzl.
VU A2c + 3c; B2ab(iii)

Range: Bioko (1 doubtful coll.), Cameroon, SW Province: Buea (3 coll.), Mt Cameroon unlocated (1 coll.), Mt Etinde (1 coll.), Mt Kupe (2 coll. at 2 sites), W Province: Bamboutos Mts (1 coll.), NW Province: Mt Oku (1 coll.).

Assessed in both Cable & Cheek (1998) and Cheek *et al.* (2000) as EN A1c + 2c, the assessment is here altered in order to take into account the submontane sites on Mt Kupe, where this taxon's forest habitat remains largely intact, thus reducing the estimated percentage of habitat loss from >50% to >30%. This taxon is here also assessed under criterion B, being known from only 8 locations, one of which remains doubtful, with an area of occupancy of less than 500km^2.

Habitat: an epiphyte of submontane and montane forest; 950–2000m alt.

Threats: forest clearance for agriculture and firewood, particularly in the Bamboutos Mts and above Buea on Mt Cameroon.

Management suggestions: confirmation of this species' presence on Bioko is required. Surveys of this taxon at each of the collection localities should be made in order to assess its abundance. Continued protection of the submontane forest on Mt Kupe should ensure this species' future survival.

Postscript: Dr Zapfack, one of the IUCN assessors (Orchid Red List Authority) has added a new locality for this taxon: Banyang Mbo. We have not yet had the opportunity to verify his identification..

Polystachya cooperi Summerh.
EN B2ab(iii)
Range: Nigeria: Mambilla Plateau (1 coll.), Obudu Plateau (2 coll. at 2 sites); Cameroon: Mwanenguba (1 coll.), Mt Kupe (1 coll.).

This species was first collected in 1961 on the Obudu Plateau (*Cooper* 2A) and was subsequently cultivated at Kew. Further collections were made at the Obudu Cattle Ranch in Aug. 1973 (*Lowe* 2662) and on the Mambilla Plateau at Ngel Nyaki Forest Reserve in July 1976 (*Chapman* 207), where it was recorded as a 'common epiphytic orchid' at 5000ft alt. It was first collected in Cameroon at Mwanenguba in Nov. 1968 (*Sanford* 5464), but has not since been rediscovered at that site. The only recent collection is from Mt Kupe in Aug. 1993 (*Balding & Sivell* 43).

Habitat: submontane and montane forest and woodland; 1400–1700m alt.

Threats: extensive and continued forest clearance at high altitudes in E Nigeria and at Mwanenguba threaten any remaining populations at these sites. The submontane location on Mt Kupe, however, has experienced little disturbance to date.

Management suggestions: as Mt Kupe appears to offer the best chance for survival of this taxon, attempts to rediscover and census the population here should be made. Continued protection of the forest above 1000m alt. here should ensure this species' survival, though it is clearly rare, having been collected only once during the extensive surveys on this mountain.

Polystachya farinosa Kraenzl.
EN B2ab(iii)
Range: São Tomé; Cameroon: Mwanenguba and Mt Cameroon.

This species is known from only two collections in Cameroon, the first from above Buea on Mt Cameroon, collected in the nineteenth century (*Preuss* 1064), and the second from Mwanenguba, collected in Nov. 1968 (*Sanford* 5463). It has also been recorded from São Tomé, though data are sparse on its distribution there. *P. farinosa* was considered in FWTA to be conspecific with *P. bifida* Lindl., but was reinstated as a separate entity by Szlachetko & Olszewski (Fl. Cameroun 35: 536 (2001)). The latter species is more widely distributed in the Lower Guinea & Congolian phytochorion.

Habitat: an epiphyte of dense montane forest; 1450–1650m alt.

Threats: continued clearance of montane forest at Mwanenguba threatens this population which may no longer be extant, as this species was not rediscovered there during botanical survey work in the 1970s and 1990s.

Management suggestions: more data on the São Tomé population are required. Attempts to rediscover this species at both Mwanenguba and Mt Cameroon should be encouraged.

Polystachya geniculata Summerh.
EN B2ab(iii)
Range: Cameroon: SW Province: Mamfe (2 coll.), Banyang-Mamfe (1 coll.), Littoral Province: Mwanenguba (1 coll.).

First described from material collected in May 1947 growing on rock-slabs near Mamfe (*Gregory* 124), subsequent material from Banyang-Mamfe (*Eyeku in FHI* 22304) and Mamfe (*Gregory* 323) are likely to refer to the same location, as these specimens were again collected from 'outcrops of granite' and 'rock-slabs with running water in the wet season' respectively. The single collection from Mwanenguba was made in June 1948 (*Gregory* 302), where this species was found in marshy grassland.

Habitat: an epilith on seasonally-wet rocks, or terrestrial in marshy grassland; (?)200–1850m alt.

Threats: the Mwanenguba population is threatened by habitat deterioration from farming at high elevations. The sites near Mamfe, although not directly threatened by humans, may deteriorate through, for example, changes in run-off regime following human disturbance in the proximity of these rocky sites.

Management suggestions: attempt to relocate the area in which the collections were made near Mamfe; efforts should then be made to relocate this species there as it is a more likely location for its future survival than the heavily disturbed Mwanenguba Massif.

Polystachya kupensis P.J.Cribb & B.J.Pollard
CR B2ab(iii)
Range: Cameroon: Mt Kupe (1 coll.).

The following conservation assessment is taken largely from Cribb & Pollard (2002: 636–638). This taxon is known from only a single specimen, above Kupe Village on the path to Kupe Rock, made in May 1996 (*Cable* 2521). It has not been rediscovered on the mountain despite attempts from Orchidaceae specialists during several months of collecting there, nor has it been found in the now well-collected Bakossi Mountains, to the west. It is therefore likely to be very rare within its extremely limited distribution.

Habitat: an epiphyte in submontane forest, growing on *Coffea* sp.; 1050m alt.

Threats: the only known locality for this species is in an area of forest being encroached upon by plots of *Coffea canephora* Pierre ex A.Froehner, and may thus be lost in the near future. However, the type specimen was collected from a *Coffea* tree (possibly a wild species) so this taxon may be able to withstand such a change in habitat in the short term. However, well-tended farms are 'cleaned' of epiphytes, thus *P. kupensis* is unlikely to survive in the long-term in coffee plantations.

Management suggestions: revisiting the type location and rediscovering this species are a priority. Once this has been achieved, further assessment of its ecological requirements, including its potential for survival in plantation agriculture, can be assessed. Ultimately, enforced forest protection or cultivation followed by propagation may be the only means of conserving this taxon.

PALMAE
(I.Darbyshire)

Raphia regalis Becc.
VU B2ab(iii)

Range: Nigeria, Cross River State: Oban (1 coll.), Equi Issu Hills (1 coll.); Cameroon, SW Province: Bakossi (1 coll., 1 additional sight record), E Province: Djouo, nr. Somalono (1 coll.); Gabon: Nyanga (1 coll.); Congo (Brazzaville): 'Nschaggebod' (1 coll.); Cabinda: Nkanda Mbaku (1 coll.)

First described from a 1910 collection from the Oban area, Nigeria, this species was thought extinct in that country until a conscious effort to rediscover it in the 1970s by Otedoh proved successful, it being recorded in large numbers in the Equi Issu hills near the Cameroonian border. In Cameroon it appears restricted to the Bakossi Mountains, where it has been collected at Ngomboaku and recorded at Nyandong (Darbyshire & Etuge *pers. obs.* 2003). It is also known from Gabon and Angola (Cabinda). This highly distinctive palm, with no aerial trunk and with leaves rating amongst the largest in the plant kingdom, is likely under-recorded due to difficulties in collecting specimens of it, and in the fact that it has received limited taxonomic attention, being treated only briefly in Otedoh's 1982 revision of the genus *Raphia*, with no specimen citations.

The fronds of *R. regalis* are used as building material and for mat-weaving in Nigeria and Cameroon, but use in the latter country is reported as limited (*Cheek* 10310). In Nigeria, it is also tapped for palm wine, but this is not done in Cameroon, where *R. hookeri* G.Mann & H.Wendl. is favoured.

Habitat: mid-elevation wet forest, often on ridges, where it may be gregarious; 500–850m alt.

Threats: forest clearance for timber and for agricultural expansion in Nigeria and Bakossi threatens this species. Selective felling for use in building, mainly in Nigeria, provides a further threat.

Management suggestions: a full survey of the populations of this taxon should be carried out throughout its range; this should not prove difficult as the species is so distinctive. This will likely lead to the discovery of more sites, thus the conservation assessment may be downgraded in the future. Local populations should be encouraged to use the commoner *R. hookeri* in preference to *R. regalis* in building and palm wine production.

TRIURIDACEAE
(M.Cheek)

Kupea martinetugei Cheek & S.Williams
CR C1

Range: Cameroon, SW Province: Mt Kupe (2 sites).

This species was assessed in Cheek *et al.* (2003) as CR because less than 250 individuals are known (at only two sites), these being estimated to be likely to decline by more than 25% in the next generation (estimated at c. 50 years).

Habitat: lowland to submontane evergreen forest; c. 700m alt.

Threats: see above.

Management suggestions: continued monitoring of the two known sites, both at Mt Kupe, is advised. Verification of a possible new site at Onge Forest (Franke *et al.* in press) is needed.

READ THIS FIRST! EXPLANATORY NOTES FOR THE CHECKLIST

Iain Darbyshire

Herbarium, Royal Botanic Gardens, Kew, Surrey, TW9 3AE, U.K.

Before using this checklist, the following explanatory notes to the conventions and format used should be read.

The checklist is compiled in an alphabetical arrangement: species within genera, genera within families and families within the groups Dicotyledonae, Monocotyledonae, Pinopsida, Gnetopsida, Lycopsida (fern allies) and Filicopsida (true ferns), following Kubitzki, in Mabberley (1997: 771–781). Indeterminate taxa are placed at the end of the relevant section, for example 'Acanthaceae genus indet.' is located at the end of the Acanthaceae account.

Identifications and descriptions of the species were carried out on a family-by-family basis by both family specialists and general taxonomists; these authors are credited at the head of each account. As a general rule, if two authors are listed, the primary author is responsible for the determinations and the second author for the compiling of the account, including writing of descriptions, distributional data and conservation assessments. Exceptions to this rule do occur, for example G.Mathieu is listed as the second author on the Piperaceae account in recognition of his assistance in delimiting two difficult *Peperomia* taxa from our region, the remaining determinations and all descriptions having been carried out by M.Cheek. Where someone is responsible for only one or two genera within a larger family, this is noted in the author accreditation.

As the incomplete Flore du Cameroun (1963–present) is a particularly relevant source of information on the plants of the checklist area, a reference to the volume and year of publication is listed at the head of each family account where available.

The families and genera accepted here follow Brummitt (1992), with recent updates on the R.B.G., Kew Vascular Plant Families and Genera database, with the following exceptions:

- in Ochnaceae we follow C.Farron in recognising *Campylospermum* Tiegh., *Rhabdophyllum* Tiegh. and *Idertia* Farron in place of *Gomphia* Schreb. (the latter not occurring in our checklist area).
- in Cyperaceae we follow K.Å.Lye in sinking *Pycreus* Pal. and *Kyllinga* Rottb. into *Cyperus* L.
- in Peridiscaceae we follow V.Savolainen (*pers. comm.*) in placing the genus *Soyauxia* Oliv. within this family; it is placed within Medusandraceae by Brummitt (1992).

ANGIOSPERMAE & GYMNOSPERMAE

Each Angiosperm and Gymnosperm taxon within the family account is treated in the following manner:

Taxon name

The species name adopted follows the most recent taxonomic work available. The bibliographic works of Lebrun & Stork (1991–1997, & 2003) are used as a guide throughout, though some discrepancies do occur, for example we accept the genus *Heterosamara* Kuntze (Polygalaceae) in accordance with Brummitt (1992) and thus recognise the species *Heterosamara cabrae* (Chodat) Paiva, which is listed by Lebrun & Stork (2003: 192) as *Polygala cabrae* Chodat. Standard forms for author citations follow Brummitt & Powell (1992).

Species names not validly published at the time of publication of the checklist are noted as 'in press', 'in prep.' or 'ined.' depending upon the extent to which the publication process has advanced, the former indicating that the protologue has been accepted for, but is awaiting publication, the second that the protologue is at least part written and edited but is not yet in its final form for publication, and the latter that the species concept is firmly established but formal descriptions await.

Not all names listed are straightforward binomials with authorities. A generic name followed by '*sp.*' generally indicates that the material was inadequate to name to species, for example *Manilkara sp.* (Sapotaceae). Use of '*sp. 1*', '*sp. 2*' etc. generally indicate unmatched specimens which may be new to science or may prove to be variants of a currently accepted species; these taxa usually require additional material in order to confirm identity, for example *Psychotria sp. 1* (Rubiaceae), or new taxa for which sufficient material is available, but are awaiting formal description, for example *Deinbollia sp. 1* (Sapindaceae). Unless otherwise stated, these provisional names are applicable only to the current checklist, thus '*sp. 1*' indicates '*sp. 1 of the Kupe-Bakossi checklist*'. The use of '*sp.*

nov.' is a firm statement that the taxon is new to science but awaiting formal description; sufficient material may or may not be available for this process. A generic name followed by '*cf.*': indicates that the specimens cited should be compared with the associated specific epithet, for example *Wheatley* 415 (*Ardisia cf. buesgenii*), should be compared with *Ardisia buesgenii* (Gilg & Schellenb.) Taton. This is an indication of doubt (sometimes due to poor material), suggesting that the taxon is close to (but may differ from) the described taxon. The terms '*aff.*', indicating that the taxon has affinity to the subsequent specific epithet, and '*vel. aff.*', indicating that the specimen refers to the taxon listed or a closely allied entity, are applied in a similar fashion. These uncertainties are generally explained in the taxon's 'Notes' section (see below).

Taxon reference

The majority of species referred to within the checklist are found in the 2nd edition of Flora of West Tropical Africa (FWTA: Keay 1954–58; Hepper 1963–72), the standard regional flora. Only species names which do not occur in FWTA are given a reference here; if no reference is cited the taxon name is currently accepted and occurs in FWTA. The references listed are not necessarily the place of first publication of the name; rather, we have tried where possible to use widely accessible publications which provide useful information on that taxon, such as a description and/or distribution and habitat data. The reference is recorded immediately below the taxon name. In the case of scientific journals, we list the journal name, volume and page numbers and date of publication, with recording of volume part number where it aids in access to the publication. In the case of books, we list the surname of the author(s), the book title, the page number for the taxon in question and the year of publication. Journal and book titles are often abbreviated in the interest of economy of space. Several notable publications are:

- Fl. Cameroun: Flore du Cameroun (1963–). Muséum National d'Histoire Naturelle, Paris, France & MINREST, Yaoundé, Cameroon.
- Fl. Congo, Rw. & Bur.:   Flore du Congo, du Rwanda et du Burundi (1948–). Jardin Botanique National de Belgique, Brussels, Belgium, which has also been variously called Flore du Congo Belge, et du Ruanda-Urundi, and Flore d'Afrique Centrale (Zaïre-Rwanda-Burundi). One abbreviation for all these works is applied here for clarity.
- Fl. Gabon: Flore du Gabon (1961–). Muséum National d'Histoire Naturelle, Paris, France.
- FTEA: The Flora of Tropical East Africa (1952–). Crown Agents, London & A.A.Balkema, Lisse, Netherlands.

Synonyms

In some instances, names used in FWTA have been superseded and are thus reduced to synonymy; these are listed below the accepted name, with the prefix 'Syn.:'. Names listed in synonymy in FWTA are not recorded here. Other important synonyms are, however, recorded; for example, the members of the genus *Polygonum* recorded in this checklist are often assigned to *Persicaria* Mill. but, in agreement with Lebrun & Stork (2003), are maintained within the former genus here, thus the equivalent taxon names within *Persicaria* are listed in synonymy.

Taxon description

The short descriptions provided for each taxon are based primarily upon the material cited in order to provide the most accurate representation for field botanists working within the checklist area. However, where necessary, they are supplemented by extracted details from the descriptions in FWTA, Fl. Cameroun and the cited taxonomic works. The descriptions are not exhaustive or necessarily diagnostic; rather, they aim to list the key characters to enable field identification of live or dried material, thus microscopic or complex characters are referred to only when they are essential for identification. Where two or more taxa closely resemble one another, a comparative description may be used, by for example stating 'Tree resembling *Uapaca guineensis*, but ...'; such comparisons are only made to other taxa occurring within the checklist area.

Several abbreviations are used in the descriptions, most notably d.b.h. (referring to 'diameter at breast height', being a standard measure of the diameter of a tree trunk), the use of 'c.' as an abbreviation for 'approximately', '±' meaning 'more or less', and (n.v.) for '*non vide*' to indicate that a specimen has not been seen by the author. In addition, in the Menispermaceae account, where male and female flowering parts are repeatedly referred to, the symbols '♂' and '♀' are applied for 'male' and 'female' respectively.

Habitat

The habitat, recorded at the end of the description, is derived mainly from the field notes of the cited specimens and therefore does not necessarily reflect the entire range of habitats for that taxon; rather, those in which it has been recorded within the checklist area. Habitat information is taken from published sources only where field data is not available, for example, where the only specimens recorded were not available to us, but are cited in Fl. Cameroun. Altitudinal ranges, listed together with habitat, are generated directly from the database of specimens from the checklist area, and thus do not necessarily reflect the entire altitudinal range known for the taxon. For example, the recorded altitude for *Afrofittonia silvestris* Lindau in the checklist area is 1400m, but this taxon is more commonly associated with lowland rainforest in the rest of its range. Where no altitudinal data was recorded with the specimens, it is omitted.

Distribution

For the sake of brevity, country ranges are generally recorded for each taxon rather than listing each separate country, for example 'Sierra Leone to Uganda' is taken to include all or most of the intervening countries. Only where taxa are recorded from only two or three, rarely more, countries within a wide area of occurrence are the individual countries listed, for example 'Sierra Leone, Cameroon & Uganda'. For more widespread taxa, a more general distribution such as 'Tropical Africa' or 'Pantropical' is recorded. Where a species is alien to the checklist area, its place of origin is noted, together with its current distribution. Several country abbreviations are used:

- Guinea (B): Guinea (Bissau), or former Portuguese Guinea.
- Guinea (C): Guinea (Conakry) - the Republic of Guinea, or former French Guinea.
- CAR: Central African Republic
- Congo (B): Congo (Brazzaville)–the Republic of Congo, or former French Congo.
- Congo (K): Congo (Kinshasa) - the Democratic Republic of Congo, or former Zaïre.

Abbreviations for parts of the country are also used (N: north, S: south, E: east, W: west, C: central, and also combinations of these, e.g. SW: southwest, WC: west-central) where appropriate. Where appropriate, Equatorial Guinea is divided into Bioko, Annobon (both islands) and Rio Muni (mainland), and the Angolan enclave of Cabinda north of the Congo River is recorded separately from Angola itself (south of the river).

In addition to country range, a chorology, largely based upon the phytochoria of White (1983) but with modifications to reflect localised centres of endemism in W Africa, is recorded in brackets. The main phytochoria used are:

- **Upper Guinea**: broadly the humid zone following the Guinean coast from Senegal to Ghana.
- **Lower Guinea**: separated from Upper Guinea by the 'Dahomey Gap', an area of drier savanna-type vegetation, that reaches the Atlantic coast. Lower Guinea represents the humid zone from Nigeria to Gabon, including Rio Muni, Cabinda, and the wetter parts of western Congo (Brazzaville).
- **Congolian**: the basin of the River Congo and its tributaries, from eastern Congo (Brazzaville) and southern CAR through Congo (Kinshasa) and to Uganda, Zambia and Angola.
- **Afromontane**: a series of vegetation types restricted to montane regions, principally over 2000m alt.
- **W(estern) Cameroon Uplands**: a subdivision of the Afromontane phytochorion, used for taxa restricted to the mountain chain running from the Gulf of Guinea islands (Annobon, São Tomé, Principé and Bioko) to western Cameroon and southeast Nigeria.
- **Cameroon endemic**: for those taxa restricted to Cameroon, a subdivision of Lower Guinea. Taxa endemic to montane western Cameroon are however recorded under W Cameroon Uplands unless they are endemic to the checklist area, when they are listed as a Narrow Endemic (see below).
- **Narrow endemic**: for those taxa restricted or very nearly restricted to the checklist area, again a subdivision of Lower Guinea or W Cameroon Uplands depending upon the taxon's altitudinal range.

These phytochoria are variously combined where appropriate, for example 'Guineo-Congolian (montane)' refers to an Afromontane taxa restricted to the mountains of the Upper and Lower Guinea and Congolian phytochoria. Taxa with ranges largely confined to the Guineo-Congolian phytochorion, but with small outlying populations in wet forest in, for example, west Tanzania or northern Zimbabwe, are here recorded as Guineo-Congolian rather than Tropical African, as the latter would provide a more misleading representation of the taxon's true phytogeography. A range of other chorologies are used for more widespread species, including 'Tropical Africa', 'Tropical & southern Africa', 'Tropical Africa & Madagascar', 'Palaeotropics' (taxa from tropical Africa and Asia or some other Old World region), 'Amphi-Atlantic' (taxa from tropical Africa and S America), 'Pantropical' and 'Cosmopolitan'. If these terms are used in the distribution, no separate phytochorion is listed.

For some taxa, such as those native to one area of the tropics but widely cultivated elsewhere, the chorology is difficult to define and is thus omitted. Both distribution and chorology are omitted for taxa where there is uncertainty over its identification.

Orthography of Bakossi place names

There is significant variation in the orthography of Bakossi place names. For example, Manengouba, Muanengouba, Muanenguba all refer to the same place and have been widely used on maps and in the literature. This place is referred to in this checklist as Mwanenguba, following the advice of Chris Wild, who over many years in Bakossi has sought to establish the single most acceptable typography for the majority of Bakossi place names. Throughout this book we have followed his guidance on place names. The Bakossi place names adopted and their locations can be viewed at a glance by reference to Fig. 2. Having adopted Chris Wild's spellings of certain Bakossi place names, where they differed from those recorded by us in the field, many herbarium specimens held at YA and K, and any duplicates distributed already, or in the future, to the herbaria at EA, ETH, MA, MO, P, SCA, SRGH and WAG will subsequently have the old variant spellings on specimen data labels.

Accepted orthography in this volume	Previously recorded orthographic variants
Edib, Lake Edib	Edip, Lake Edip
Jide River, Jide Valley	Chide River, Chide Valley
Mwambong	Muambong
Mwandon, Lake Mwandon	Muambong, Lake Muambong
Mwanenguba (& Lake)	Muanenguba, Muanengouba, Manengouba (& Lake)
Mwendolengo(e)	Muandelengoh, Mwandelengoh
Ngomboaku	Ngomboku, Ngombuku

Conservation assessment

The level of threat of future extinction on a global basis is assessed for each taxon that has been fully identified and has a published name, or for which publication is imminent, under the guidelines of the IUCN (2001). Under the heading 'IUCN:', each taxon is accredited one of the following Red List categories:

     CR: Critically Endangered
     EN: Endangered
     VU: Vulnerable
     NT: Near-threatened
     LC: Least Concern
     DD: Data Deficient

Those taxa listed as CR, EN and VU are treated in full within the chapter on Red Data species, where the criteria for assessment are recorded. Those listed as LC or NT are not treated further in this publication, but it is recommended that further investigation of the threats to those taxa recorded as NT are made. Undescribed taxa, or those with an uncertain determination, are not assessed. In addition, we do not consider it appropriate to assess taxa from the poorly-known genera *Beilschmiedia* (Lauraceae), *Aframomum* and *Renealmia* (both Zingiberaceae) for which species delimitation is currently poorly understood and thus for which conservation assessments would be somewhat meaningless; these genera should be revisited once full taxonomic revisions have been completed.

Specimen citations

Specimens from the checklist area are recorded below the distribution and conservation assessment, with the following information:

Location: these are listed in alphabetical order, and refer to either the nearest village or town to the exact collection location (e.g. 'Kupe Village', 'Konye', 'Nyasoso', 'Tombel'), a general geographical location (e.g. 'Mwanenguba', 'Mt Kupe') or a protected area ('Loum F.R. (Forest Reserve)', 'Mungo River F.R.'). These are meant as a rough indicator as to the actual location, which can be provided by reference to the cited specimen.

Collector and number: within each location, collectors are ordered alphabetically and, together with the unique collection number, are underlined. Only the principal collector is listed here, thus for example, collections listed under 'Balding' were originally recorded as 'Balding & Sivell'; this alteration has been made for the sake of brevity. Specimens listed under 'Plot', with the prefix 'B' refer to collections made within a 25 × 25m plot in representative forest in the vicinity of Kodmin in January 1998 by M.Cheek *et al.*

Phenology: this information is derived directly from the Cameroon specimens database at R.B.G., Kew and is thus dependent upon recording of such information at the time of collection or at the point of data entry onto the database; if this was not done no phenological information is listed. Collections are recorded as flowering (fl.), fruiting (fr.) or sterile (st.) where applicable.

Date: the month and year of collection is recorded for each specimen; within each collector from each location, collections are ordered chronologically. In some instances, particularly where records were derived from Fl.

Cameroun accounts, the month and year of collection are unknown; in such cases we have stated 'date unknown', or have listed a range of possible collection dates.

The specimens cited are of herbarium material derived from a variety of sources, mostly from the many expeditions run jointly between Earthwatch, Herbier National Camerounais and R.B.G., Kew since 1992, but also from those historic collections accessioned at Kew, Missouri and Yaoundé. In addition, a very few confident sight records of uncollected taxa are also included, most of which concern cultivated plants.

Notes

Items recorded in the notes field at the end of the taxon account include:

- Taxonomic notes, for example in the case of new or uncertain taxa, how they differ from closely related species.
- Notes on the source of specimen data, for example for those specimens recorded in Fl. Cameroun, or for specimens not seen by the author(s) of the family account.

Occasionally, notes are placed at either the head of a family account (for example, in Labiatae, Verbenaceae and Zingiberaceae) or at the head of a genus (for example, *Napoleonaea* in Lecythidaceae and *Habenaria* in Orchidaceae) which explain particular points of taxonomy, for example, where the species within a genus are poorly delimited, or where we have used broad generic or species concepts.

Ethnobotanical information

Local names and uses are listed for each taxon where appropriate; these are derived largely from local residents and field assistants and are reproduced here with their consent. Local names are in the Bakossi language unless otherwise stated; English names are listed only where they are widely used in Cameroon. For each local name or use listed, a source is attributed, usually by reference to the collector and number of the specimen from which the information was derived, including the source of the information where recorded, for example, the local name listed for *Merremia umbellata* subsp. *umbellata* (Convolvulaceae), 'Mendi Mendiba' is attributed to 'Epie Ngome in Pollard 105', indicating that Epie Ngome was the local informant who applied that local name to the specimen collected under *Pollard* 105. When a specimen is cited, all the preceding information was derived from that specimen. This might include both the local name and the use(s). This is particularly important to note with regard to taxa with multiple names and uses. In circumstances where the ethnobotanical information was provided verbally with no attached specimen, or where it is known by the author(s) of the family account, the terms '*fide*', '*pers. comm.*' (personal communication) or '*pers. obs.*' (personal observation) are used to attribute the information to a source.

The layout of the information on local uses follows the convention of Cook (1995), listing level 1 state categories of use in capital letters, followed by the level 2 state categories, then the specific use. In order to comply with guidelines set by the Convention on Biological Diversity (2002), the detailed uses of medicinal plants are not listed here; only the level 1 state 'MEDICINES' is recorded. Negative ethnobotanical information, for example if a plant has no local name and/or no known local use, is also recorded.

PTERIDOPHYTA

As the primary focus of this checklist is on the Spermatophyta, for ferns and fern allies only the taxon name, together with a reference and synonyms where appropriate, collection details including location and altitude, and ethnobotanical notes are recorded. No descriptions, distributional data or IUCN assessment are made for these taxa.

PLANT REFERENCES & BIBLIOGRAPHY

The references listed below concern those cited in the introductory chapters on plants, excluding the Red Data chapter, which has references inserted under each taxon account, and the chapters on the physical environment, sacred groves, the macrofungi, fauna, birds, and the protected area system, which have references at the end of each individual chapter.

Achoundong, G. & Cheek, M. (2004). Two new species of *Rinorea* (*Violaceae*) from western Cameroon. Kew Bull. **58**: 957–964.

Breteler, F.J. (1999). A revision of *Prioria*, including *Gossweilerodendron, Kingiodendron, Oxystigma,* and *Pterygopodium (Leguminosae-Caesalpinioideae-Detarieae)* with emphasis on Africa. Wag. Agric. Univ. Pap. **99(3)**.

Bos, J.J. (1984). *Dracaena* in west Africa. Wag. Agric. Univ. Papers. **84**: 126pp.

Brummitt, R.K. & Powell, C.E., (eds) (1992). Authors of Plant Names. Royal Botanic Gardens, Kew, U.K.

Brummitt, R.K. (1992). Vascular Plant Families and Genera. Royal Botanic Gardens, Kew, U.K.

Cable, S. (1993). A vegetative tree Key for Mt Kupe, Cameroon. B.Sc. thesis, Tropical Environmental Science, Aberdeen University (cited in Lane 1994a).

Cable, S. & Cheek, M. (comps, eds.) (1998). The Plants of Mt Cameroon, a Conservation Checklist. Royal Botanic Gardens, Kew. lxxix + 198pp.

Cheek, M. (1995). *Dischistocalyx* T. Anderson ex Benth.: terrestrial herbs, climbers, then epiphytes! Acanthus **6**: 3–4.

Cheek, M. (1997). Ghosts and ghost hunters. Kew Magazine, Autumn 1997: 16–19.

Cheek, M. (2001). Cameroon's Rainforest Checklists. Earthwatch Fellowship Programme Newsletter **5**: 1.

Cheek, M. (2002). Three new species of *Cola* (*Sterculiaceae*) from western Cameroon. Kew Bull. **57**: 402–415.

Cheek, M. (2003a). A new species of *Ledermanniella (Podostemaceae)* from western Cameroon. Kew Bull. **58**: 733–737.

Cheek, M. (2003b). Cameroon: 50 new species published. Kew Scientist **24**: 4.

Cheek, M. (2004a). A new species of *Afrothismia* (*Burmanniaceae*) from Kenya. Kew Bull. **58**: 951–955.

Cheek, M. (2004b). Kupeaeae, a new tribe of *Triuridaceae* from Africa. Kew Bull. **58**: 939–949.

Cheek, M. (in press). A new species of *Keetia* (*Rubiaceae–Vanguerieae*) from western Cameroon. Kew Bull.

Cheek, M. & Bridson, D. (2002). A new species of *Coffea* (*Rubiaceae*) from western Cameroon. Kew Bull. **57**: 675–680.

Cheek, M. & Cable, S. (1997). Plant inventory for conservation management: the Kew-Earthwatch programme in Western Cameroon, 1993–96. In: Doolan, S. (ed.) African Rainforest and the Conservation of Biodiversity. Proceedings of the Limbe Conference. Earthwatch Europe, Oxford: 29–38.

Cheek, M. & Csiba, L. (2000) A new species and a new combination in *Chassalia* (*Rubiaceae*) from western Cameroon. Kew Bull. **55**: 883–888.

Cheek, M. & Csiba, L. (2002). A new epiphytic species of *Impatiens (Balsaminaceae)* from western Cameroon. Kew Bull. **57**: 669–674.

Cheek, M. & Csiba, L. (2002). A revision of the *Psychotria chalconeura* complex (*Rubiaceae*) in Guineo-Congolian Africa. Kew Bull. **57**: 375–387.

Cheek, M., Csiba, L. & Bridson, D. (2002). A new species of *Coffea* (*Rubiaceae*) from western Cameroon. Kew Bull. **57**: 675–680

Cheek, M. & Dawson, S.E. (2000). A synoptic revision of *Belonophora* (*Rubiaceae*). Kew Bull. **55**: 65–80.

Cheek, M. & Etuge, M. (2000). Bakossi tea: *Dicliptera laxa* C.B.Cl. Acanthus **7**: http://www.wits.ac.za/museums/herbarium/acanthus7.htm

Cheek, M., Gosline, G. & Csiba, L. (2002). A new species of *Rhaptopetalum* (*Scytopetalaceae*) from western Cameroon. Kew Bull. **57**: 661–667.

Cheek, M. & N. Ndam (1996). Saprophytic Flowering Plants of Mount Cameroon. In: van der Maesen, van der Burgt & van Medenbach de Rooy (eds). The Biodiversity of African Plants. Proceedings XIV AETFAT Congress. Kluwer: 612–617.

Cheek, M., Onana, J.-M. & Pollard, B.J. (comps, eds.) (2000). The Plants of Mount Oku and the Ijim Ridge, Cameroon, a Conservation Checklist. Royal Botanic Gardens, Kew. iv + 211 pp.

Cheek, M., Thomas, D.W., Besong, J.B., Gartlan, S. & Hepper, F.N. (1994). Mount Cameroon, Cameroon. In: Davies, S.D., Heywood, V.H. & Hamilton, A.C. (eds). Centres of Plant Diversity: A Guide and Strategy for their Conservation. WWF/IUCN. Cambridge: 163–166.

Cheek, M. & Williams, S. (2000). A review of African saprophytic flowering plants. *In*: Timberlake, J.R. & Kativu, S. (*eds*). African plants: biodiversity, taxonomy and uses. Proceedings of the XV AETFAT congress, Harare, Zimbabwe. Royal Botanic Gardens, Kew: 39–49.

Cheek, M., Williams S. & Etuge M. (2003). *Kupea martinetugei,* a new genus and species of *Triuridaceae* from western Cameroon. Kew Bull. **58**: 225–228.

Convention on Biological Diversity (2002). Convention on Biological Diversity: Text and Annexes. United Nations Environment Programme, Montreal, Canada.

Cook, F.E.M. (1995). Economic Botany Data Collection Standard. Royal Botanic Gardens, Kew, U.K.

Courade, G. (1974). Commentaire des Cartes. Atlas Regional, Ouest I. ORSTOM, Yaoundé.

Cribb, P.J. (1978). A revision of *Stolzia* (*Orchidaceae*). Kew Bull. **33(1)**: 79–89.

Cribb, P.J. & Pollard, B.J. (2002). New orchid discoveries in western Cameroon. Kew Bull. **57**: 653–659.

Cribb, P.J. & Pollard, B.J. (2004). *Bulbophyllum kupense* P.J. Cribb & B.J. Pollard, an unusual new orchid from western Cameroon. Kew Bull. **59**: 137–139.

Darbyshire, I. & Cheek, M. (2004). A new species of *Peucedanum* L. (*Umbelliferae*) from Mt Kupe, western Cameroon. Kew Bull. **59**: 133–136.

Dawson, S.E. (2002). A new species of *Stelechantha* Bremek. (*Rubiaceae, Urophylleae*) from Cameroon. Kew Bull. **57**: 397–402.

de Wilde, J.J.F.E. (2002). Studies in *Begoniaceae* VII. Wageningen University Papers. Backhuys, Leiden.

de Wilde, J.J.F.E. & Arends, J.C. (1980). *Begonia* section *Squamibegonia* Warb.: a taxonomic revision. Miscellaneous Papers Landbouwhogeschool. Wageningen. **19**: 377–421.

Engler, A. (1910). Pflanzenr. **46**: 168.

Engler, A. (1925). *In*: Engler, A. & Drude, O. Die Vegetation der Erde 9, Die Pflanzenwelt Afrikas 5, 1. Engelmann, Leipzig. 341 pp.

Franke, T. (2004). *Afrothismia saingei* (*Burmanniaceae*: *Thismiae*) a new myco-heterotrophic plant from Cameroon. Syst. Geogr. Pl. **74**: 27–33.

Franke, T., Sainge, M.N. & Agerer, R. (in press). A new species of *Afrothismia* (*Burmanniaceae*; Tribe: *Thismieae*) from the western foothills of Mt Cameroon. Blumea.

Gosline, G. & Cheek, M. (1998). A new species of *Diospyros* (*Ebenaceae*) from Southwest Cameroon. Kew Bull. **53**: 461–465.

Hall, J.B. & Swaine, M.D. (1981). Distribution and Ecology of Vascular Plants in a Tropical Rain Forest Vegetation: Forest Vegetation in Ghana. Junk, The Hague.

Hawkins, P. & Brunt, M. (1965). The Soils and Ecology of West Cameroon. 2 vols. FAO, Rome. 516 pp., numerous plates and maps.

Hepper, F.N., (*ed.*) (1963–1972). Flora of West Tropical Africa. Volumes 2–3. Second edition. Crown Agents, London, U.K.

Hetterscheid, W. & Ittenbach, S. (1996). Everything you always wanted to know about *Amorphophallus*, but were afraid to stick your nose into! Aroideana **19**: 7–131.

Hoffmann, P. & Cheek, M. (2003). Two new species of *Phyllanthus* (*Euphorbiaceae*) from southwest Cameroon. Kew Bull. **58**: 437–446.

Holmgren, P.K., Holmgren N.H. & Barnett, L.C. (1990). Index Herbariorum. 8th ed. New York Botanical Garden. 693 pp.

IUCN (1994). IUCN Red List Categories. IUCN, Gland, Switzerland.

IUCN (2001). IUCN Red List Categories and Criteria: Version 3.1. IUCN Species Survival Commission, IUCN, Gland, Switzerland & Cambridge, U.K.

IUCN (2003). 2003 IUCN Red List of Threatened Species. <www.redlist.org>.

Keay, R.W.J. (*ed.*) (1954–1958). Flora of West Tropical Africa: Volume 1. Second edition. Crown Agents, London, U.K.

Keay, R.W.J. (1958) *Randia* and *Gardenia* in West Africa. Bull. Jard. Bot. Nat. Belg. **28**: 15–72.

la Croix, I.F. (1993). Two new species of *Habenaria* sect. *Podandria* (*Orchidaceae*) from West Africa. Kew Bull. **48**: 369–373.

la Croix, I.F. (1996). A new name in *Habenaria* (*Orchidaceae*). Kew Bull. **51**: 364.

Lane, P. (1993a). Vascular Epiphyte Distribution in the Montane Forests of Mount Kupe, Cameroon. BSc. thesis (Plant Science-Ecology), Aberdeen Univ., 50pp. Cyclostyled.

Lane, P. (1993b). Vascular Epiphyte Distribution in the Forests of Mount Kupe, Cameroon. *In*: Stone, R. (*ed.*). Aberdeen Cameroon Montane Forest project 1992. Cyclostyled: 23–28.

Lane, P. (1994a). Botanical Studies on Mt Kupe. Report for Mount Kupe Forest Project, 7pp. Cyclostyled.

Lane, P. (1994b). The Vegetation of Mt Kupe, Cameroon. Report for Mount Kupe Forest Project, 8pp. Cyclostyled.

Lane, P. (1995). Vegetation Monitoring for the Submontane Forest on Mt Kupe, Cameroon. 35 pp. Bound.

Lebrun, J.-P. & Stork, A.L. (1991–1997). Énumération des Plantes à Fleurs d'Afrique Tropicale. 4 volumes. Conservatoire et Jardin botaniques de la Ville de Genève, Geneva, Switzerland.

Lebrun, J.-P. & Stork, A.L. (2003). Tropical African Flowering Plants: Volume 1. *Annonaceae-Balanitaceae*. Conservatoire et Jardin botaniques de la Ville de Genève, Geneva, Switzerland.

Leeuwenberg, A.J.M. (1961). The *Loganiaceae* of Africa: I. *Anthocleista*. Acta Bot. Neerl. **10**: 1–53.

Leeuwenberg, A.J.M. (1969). The *Loganiaceae* of Africa VIII. *Strychnos* III. Belmontia **10**: 1–316.

Letouzey, R. (1968a). Cameroun, *In*: Hedberg & Hedberg (*eds.*) Conservation of Vegetation in Africa South of the Sahara. Acta Phytogeographica Suecica **54**. Uppsala: 115–121.

Letouzey, R. (1968b). Étude Phytogéographique du Cameroun. Encyclopédie Biologique LXIX. Éditions Paul Lechevalier, Paris.

Letouzey, R. (1985). Notice de la carte phytogéographique du Cameroun au 1:500 000. IRAD, Yaoundé, Cameroon.

Liben, L. (1986). Deux *Cassipourea* (*Rhizophoraceae*) nouveaux d'Afrique centrale. Bull. Jard. Bot. Nat. Belg. **56**: 139–144.

Mabberley, D.J. (1997). The Plant Book. Second edition. Cambridge University Press, U.K.

Mackinder, B. & Cheek, M. (2003). A new species of *Newtonia* (*Leguminosae-Mimosoideae*) from Cameroon. Kew Bull. **58**: 447–452.

Maley, J. (1997). Middle to late Holocene changes in Tropical Africa and other continents: Palaeomonsoon and sea surface temperature variations. Palaeoenvironnments & Palynology (CNRS/ISEM & ORSTOM). NATO ASI series, Vol. **149**: 611–640

Maley, J. & Brenac, P. (1998). Vegetation dynamics, palaeoenvironments and climatic changes in the forests of western Cameroon during the last 28,000 years B.P. Review of Paleobotany and Palynology. **99**: 157–187.

Otedoh, M.O. (1982). A revision of the genus *Raphia* Beauv. (*Palmae*). J. Niger. Inst. Oil Palm Res. **6(22)**: 145–189.

Peguy, T., Edwards, I., Cheek, M., Ndam, N. & Acworth, J. (2000). Mount Cameroon cloud forest. *In*: Timberlake, J.R. & Kativu, S. (*eds*). African plants: biodiversity, taxonomy and uses: proceedings of the XV AETFAT congress, Harare, Zimbabwe. Royal Botanic Gardens, Kew: 263–277.

Piesschaert, F., Jansen, S., Huysmans, S., Smets, E. & Robbrecht, E. (1999). *Chassalia petitiana* (*Rubiaceae-Psychotrieae*), an overlooked epiphytic species hidden in the African canopy. Syst. Bot. **24**: 315–322.

Polhill, R.M. & Wiens, D. (1998). Mistletoes of Africa. Royal Botanic Gardens, Kew.

Pollard, B.J. & Paton, A.J. (2001). A new rheophytic species of *Plectranthus* L'Hér. (*Labiatae*) from the Gulf of Guinea. Kew Bull. **56**: 975–982.

Pollard, B.J., Cheek, M. & Bygrave, P. (2003). New *Dorstenia* (*Moraceae*) discoveries in western Cameroon. Kew Bull. **58**: 185–193.

Poncy (1978). Le Genre *Pararistolochia* (*Aristolochiaceae*) d'Afrique Tropicale. Adansonia **17(4)**: 465–494.

Porembski, S. & Barthlott, W. (*eds*) (2000). Inselbergs: biotic diversity of isolated rock outcrops in tropical and temperate regions. Springer.

Robbrecht, E. (1977). The tropical African genus *Hymenocoleus* (*Rubiaceae: Psychotrieae*): additions. Bull. Jard. Bot. Nat. Belg. **47**: 3–29.

Robbrecht, E. (1979). The African genus *Tricalysia* A. Rich. (*Rubiaceae - Coffeeae*): 1. A revision of the species of subgenus *Empogona*. Bull. Jard. Bot. Nat. Belg. **49**: 239–360.

Robbrecht, E. (1981). Studies in tropical African *Rubiaceae*: 2. Bull. Jard. Bot. Nat. Belg. **51**: 359–378.

Robbrecht, E. (1987). The African genus *Tricalysia* A.Rich. (*Rubiaceae*): 4. A revision of the species of section *Tricalysia* and section *Rosea*. Bull. Jard. Bot. Nat. Belg. **57**: 39–208.

Robbrecht, E. (1996). Geography of African *Rubiaceae* with reference to glacial rain forest refuges. *In*: van der Maesen, van der Burgt & van Medenbach de Rooy (*eds*). The Biodiversity of African Plants. Proceedings XIV AETFAT Congress. Kluwer, Netherlands: 564–581.

Roberts, P. (2000). *Aphelariopsis kupemontis*: a new Auricularioid species from Cameroon. Persoonia **17**: 491–493.

Salazar, G.A., Franke, T., Zapfack, L. & Beenken, L. (2002). A new species of *Manniella* (*Orchidaceae, Cranijideae*) from western tropical Africa, with notes on protandry in the genus. Lindleyana **17**: 239–246.

Sigvaldason, G.E. (1989). Conclusions and recommendations. International conference on Lake Nyos disaster, Yaoundé, Cameroon, 16–20 March, 1987. Journal of Volcanology and Geothermal Research. **39**: 97–109.

Simpson, D.A. (1992). *Mapania*: a revision of the genus. Royal Botanic Gardens Kew, UK. 189pp.

Simpson, D.A., Furness, C.A., Hodkinson, T.R., Muasaya, A.M. & Chase, M.W. (2003). Phylogenetic relationships in *Cyperaceae* subfamily *Mapanioideae* inferred from pollen and platid DNA sequence data. Amer. J. Bot. **90**: 1071–1087.

Sonké, B. (1999). *Oxyanthus* (*Rubiaceae*) en Afrique Centrale. Opera Botanica Belgica **8**: 106pp.

Sonké, B. (2000). Une nouvelle espèce de *Rothmannia* (*Rubiaceae, Gardenieae*) de Banyang Mbo, Cameroon. Syst. Geogr. Pl. **70**: 149–153.

Sosef, M.S.M. (1994a). Refuge begonias. Taxonomy, phylogeny and historical biogeography of *Begonia* sect. *Loasibegonia* and sect. *Scutibegonia* in relation to glacial rain forest refuges in Africa. Wageningen Agric. Univ. Papers **94–1**: 1–306.

Sosef, M.S.M. (1994b). Studies in *Begoniaceae* V. Wageningen University Papers. Backhuys, Leiden.

Sosef, M.S.M. (1996). Begonias and African rain forest refuges: general aspects and recent progress. *In*: van der Maesen, van der Burgt & van Medenbach de Rooy (*eds*). The Biodiversity of African Plants. Proceedings XIV AETFAT Congress. Kluwer, Netherlands: 602–611.

Stoffelen, P., Cheek, M., Bridson, D. & Robbrecht, E. (1997). A new species of *Coffea* (*Rubiaceae*) and notes on Mount Kupe (Cameroon). Kew Bull. **52**: 989–994.

Stone, R. (1993). Aberdeen Cameroon Montane Forest Project 1992. Expedition Report. 49 + 16pp. Cyclostyled.

Stutter, O.W. (1994). A study of Forest Characterisation and Altitudinal Zonation on Mt Kupe, Cameroon.8 + 84 pp. Cyclostyled.

Taton, A. (1979). Contribution à l'étude du genre *Ardisia* Sw. (*Myrsinaceae*) en Afrique tropicale. Bull. Jard. Bot. Nat. Belg. **49**: 81–120.

Thomas, D.W. (1993a). Korup Project Plant List (all species). Revised July 1993. Cyclostyled, 55 pp.

Thomas, D.W. (1993b). Provisional vascular plant species list for Mt Kupe. Cyclostyled.

Thomas, D.W. (1986). Vegetation in the Montane Forests of Cameroon. *In*: Stuart, S.N. (*ed.*) Conservation of Cameroon Montane Forests. International Council of Bird Preservation. Cambridge: 20–27.

Thomas, D. & Cheek, M. (1992). Vegetation and plant species on the south side of Mount Cameroon in the proposed Etinde reserve. Royal Botanic Gardens, Kew. Report to Government of Cameroon from ODA. 42 pp.

Thorbecke, F. (1911). Aus dem Schutzgebiete Kamerun. Das Manenguba-Hochland. Mitteilungen au den Deutschen Schutzgebieten **24 (heft 5)**: 279–310.

Troupin, G. (1962). Monographie des *Menispermaceae* africaines. Acad. Roy. Sci. Outre-Mer, Classe Sci. Nat. Méd. Mémoires in 8°, **N.S. 13/2**, 312 pp.

Tye, H. (1986). The Climate of the Highlands of Western Cameroon. *In*: Stuart, S.N. (*ed.*), Conservation of Cameroon Montane Forests. Report of the ICBP Cameroon Montane Forest Survey, International Council for Bird Preservation, Cambridge: 18–19.

Vermeulen, J.J. (1987). A taxonomic revision of the continental African *Bulbophyllinae*. Orchid Monographs **2**. 300 pp. 101 figs. 11 plates.

Vollesen, K., Cheek, M. & Ghogue, J.-P. (2004). *Justicia leucoxiphus* (*Acanthaceae*), a spectacular new species from Cameroon. Kew Bull. **59**: 129–131.

White, F. (1983). The Vegetation of Africa: a descriptive memoir to accompany the Unesco / AETFAT / UNSO vegetation map of Africa. Unesco, Switzerland.

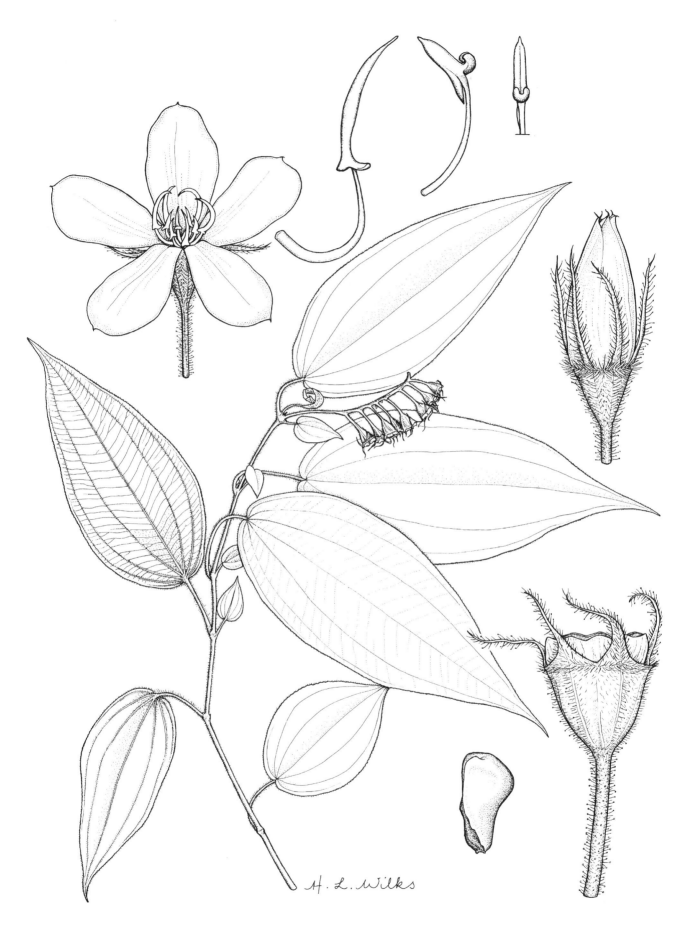

Fig. 14. *Amphiblemma* sp. 1 (Melastomataceae)
Drawn by Hazel Wilks.

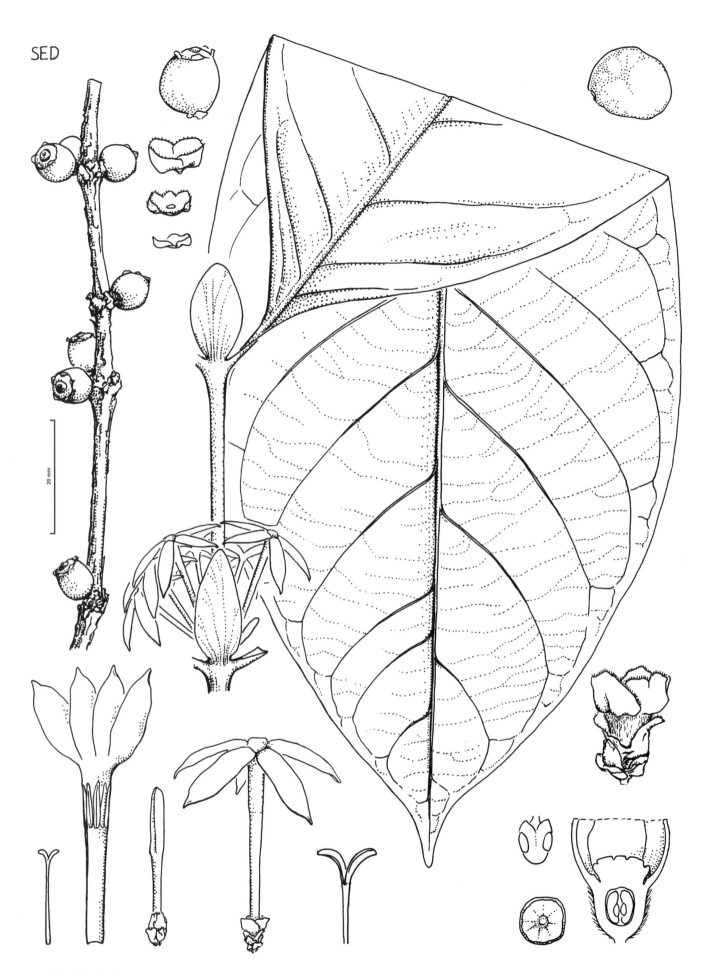

SED

Fig. 15. *Belonophora ongensis* (Rubiaceae)
Drawn by Sally Dawson.

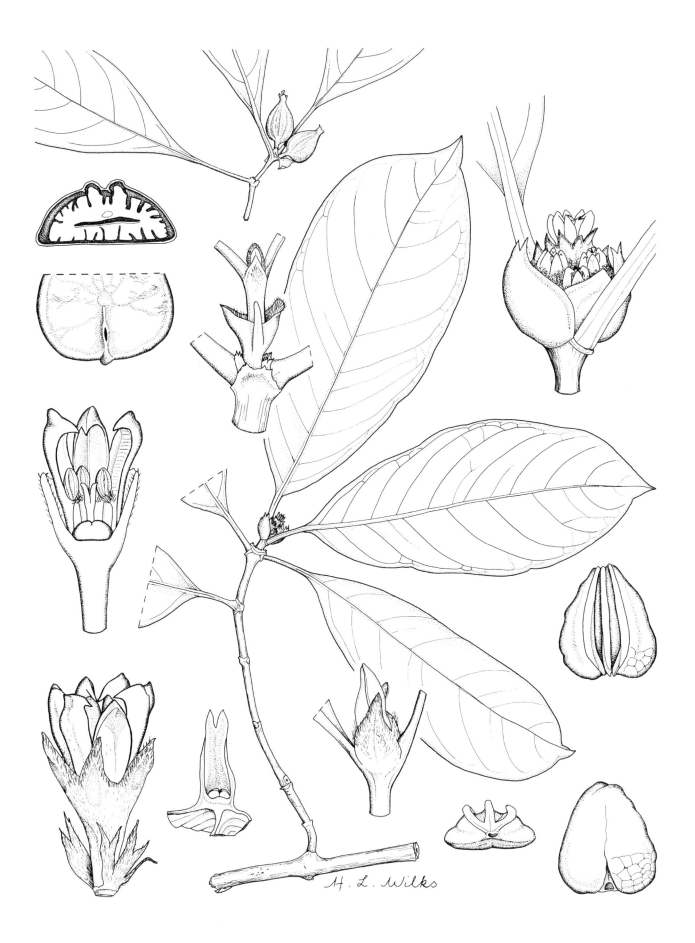

Fig. 16. *Psychotria* sp. A. aff. *gabonica* (Rubiaceae)
Drawn by Hazel Wilks.

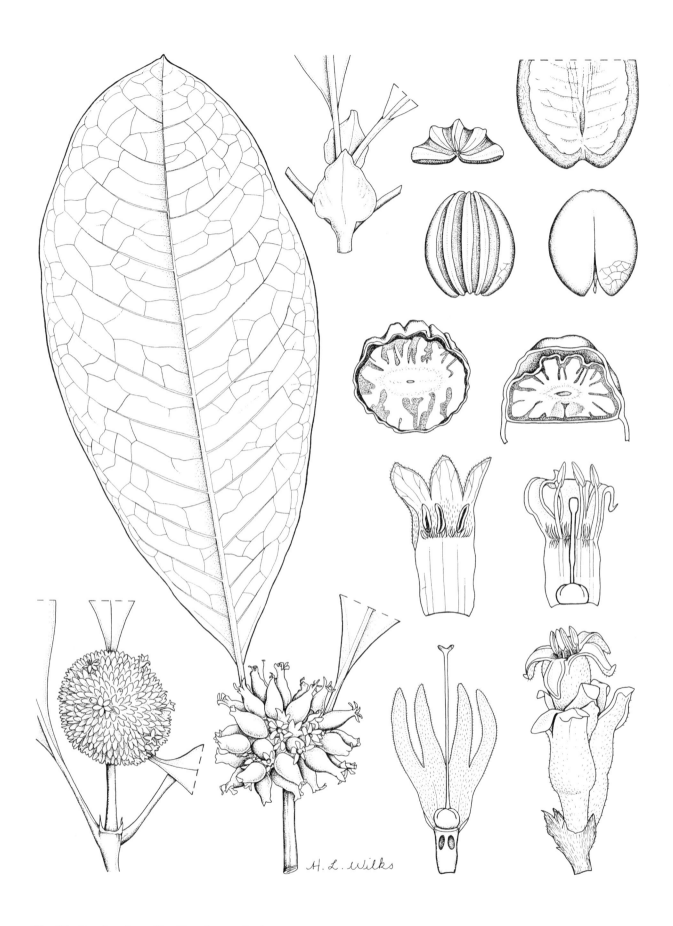

Fig. 17. *Psychotria* sp. B. aff. *gabonica* (Rubiaceae)
Drawn by Hazel Wilks.

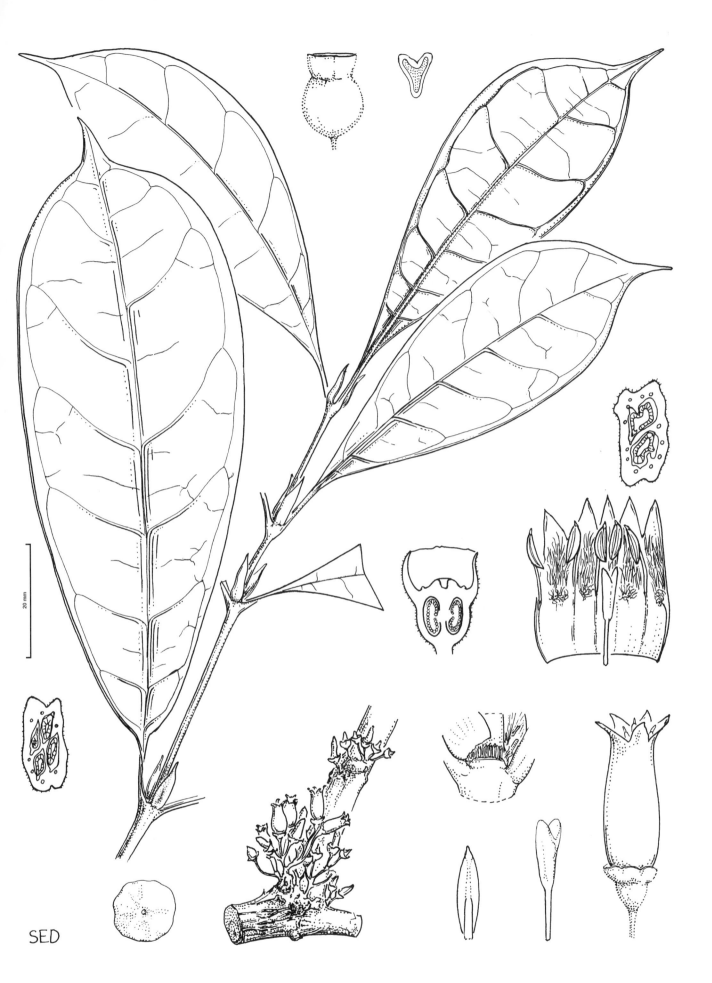

SED

Fig. 18. *Stelechantha arcuata* (Rubiaceae)
Drawn by Sally Dawson.

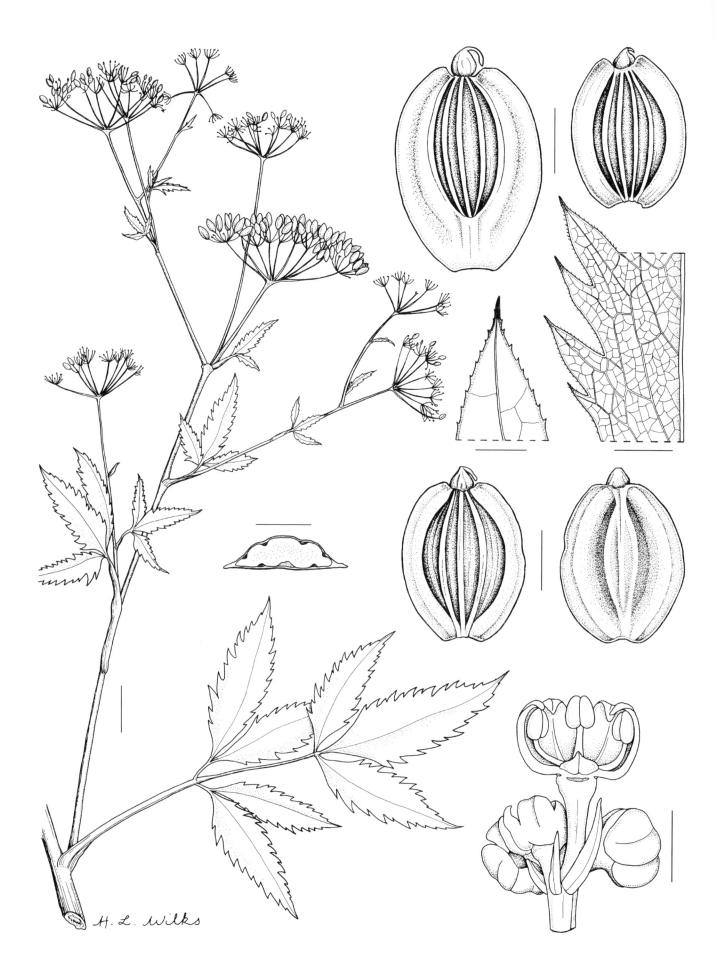

Fig. 19. *Peucedanum kupense* (Umbelliferae)
Scale bars 2cm (habit), 1mm (bottom right); 2mm (fruit). Drawn by Hazel Wilks.

Fig. 20. *Cyperus rheophytorum* (Cyperaceae)
Drawn by Hazel Wilks.

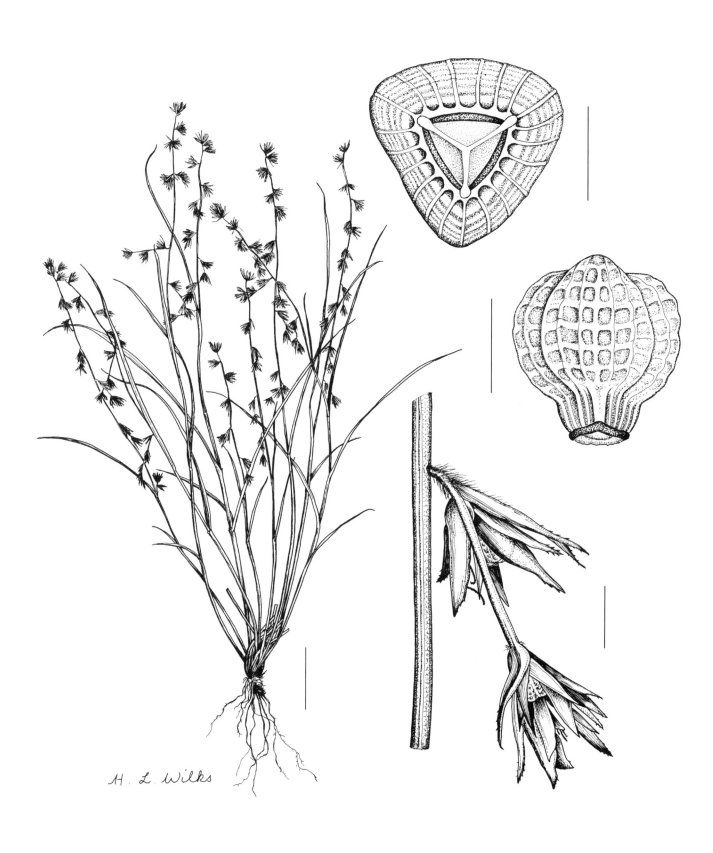

Fig. 21. *Scleria afroreflexa* (Cyperaceae)
Scale bars 2cm (habit, left); 2mm (bottom right); 0.5 mm (top right). Drawn by Hazel Wilks.

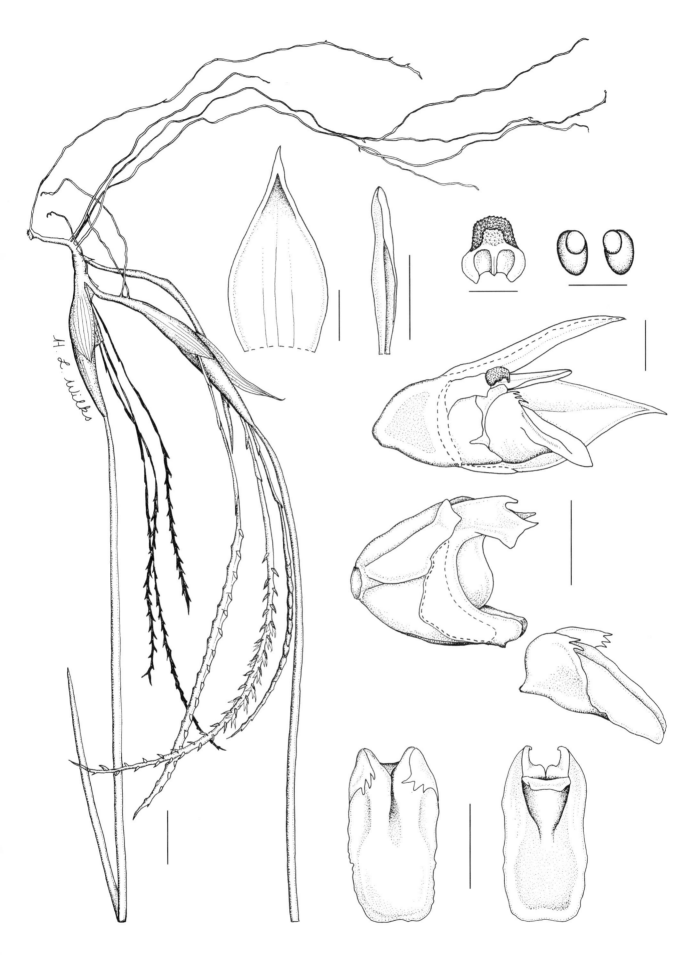

Fig. 22. *Bulbophyllum kupense* (Orchidaceae)
Scale bars 2cm (habit, left); 0.5cm (pollinia); 1mm (other flower parts). Drawn by Hazel Wilks.

ANGIOSPERMAE

DICOTYLEDONAE

ACANTHACEAE

K.B. Vollesen & I. Darbyshire;
Brachystephanus by D. Champluvier
& I. Darbyshire

Acanthopale decempedalis C.B.Clarke

Mass-flowering herb, 0.5–1m; stems becoming woody at base, glabrous; leaves broadly elliptic, 15–19 × 7–9cm, pairs often unequal, base long-attenuate, apex acuminate, both surfaces pilose; petiole winged; racemes axillary or terminal to 8cm, 3–4-flowered; bracts oblanceolate, 1.8cm, pilose; buds pubescent; calyx lobes lanceolate, 0.9cm; corolla pendulous, funnel-shaped c. 3cm long, white with purple streaking; stamens 4. Forest; 970–1550m.
Distr.: SE Nigeria, Bioko & W Cameroon. (W Cameroon Uplands).
IUCN: VU
Edib: Pollard 238 11/1998; **Kodmin:** Cheek 10264 fr., 12/1999; **Kupe Village:** Cable 771 1/1995; **Ensermu 3557** fl., fr., 11/1995; **Muahunzum:** Etuge 4445 11/1998; **Mwambong:** Cheek 9382 10/1998; **Ngomboaku:** Ghogue 438 fl., 12/1999; **Nyasoso:** Lane 171 fl., 11/1994; 221 fl., 11/1994.

Acanthus montanus (Nees) T.Anderson

Erect herb to 1.5m; stems becoming woody at base, hirsute at nodes and apices; leaves subcoriaceous, oblanceolate, 25–30 × 6.5–13.5cm, ± pinnatisect, spinosely toothed, upper surface variegated, scabrid below; petiole 8mm; inflorescence a dense terminal raceme to 14cm; bracts and calyx lobes ovate, spinose, glabrescent, 2.5cm; corolla lobes form a lip below androecium, 3.5 × 3cm, partially deflexed, white with purple streaking; stamens 4; anthers densely white-haired on one side, hairs to 3mm. Forest and secondary growth; 150–1000m.
Distr.: Benin to CAR, Angola & NW Zambia. (Guineo-Congolian).
IUCN: LC
Ekona Mombo: Etuge 411 fl., fr., 12/1986; **Enyandong:** Cheek 10956 10/2001; Etuge 4481r 11/2001; **Kodmin:** Cheek 8957 1/1998; 8971 1/1998; **Konye:** Thomas 5163 fl., 11/1985; **Kupe Village:** Etuge 1493 11/1995; Lane 263 fl., 11/1994; 380 fl., 1/1995; **Mungo River:** Gosline 222 fl., 11/1999; **Ndum:** Cable 887 fl., 1/1995; **Nyandong:** Cheek 11298 3/2003; **Nyasoso:** Cable 647 fl., 12/1993; Cheek 5667 fl., 12/1993; Etuge 1526 fl., 1/1996; Sidwell 341 fl., 10/1995.
Local names: Echinengwa (Lane 263); Mesommechum (Cheek 11298).

Afrofittonia silvestris Lindau

Creeping herb, rooting at nodes, internodes c. 5.5cm; stems finely pubescent; leaves broadly elliptic, 3.2 × 2.5cm, base and apex rounded, margin subentire, surfaces with numerous raphides, upper surface often variegated, dark green with paler patches around the principal veins, pale green below; petiole to 2.1cm, pubescent; spikes axillary, to 7cm long; peduncle to 1.5cm; bracts obovate, 1 × 0.7cm, pubescent,

purple-veined; calyx included; corolla bilabiate, c. 1.2cm, white with purple spots in throat; stamens 4. Forest; 1400m.
Distr.: SE Nigeria, Bioko & SW Cameroon. (Lower Guinea).
IUCN: VU
Enyandong: Cheek 10935 10/2001; **Kodmin:** Plot B285 1/1998.

Ascotheca paucinervia (T.Anderson ex C.B.Clarke) Heine

Fl. Gabon 13: 210 (1966).
Syn.: *Rungia paucinervia* (T.Anderson ex C.B.Clarke) Heine
Decumbent herb to 70cm; stems dark red, tomentose; leaves ovate, 10–16 × 4.5–7cm, base obtuse to rounded, apex acuminate, margin subentire, raphides numerous on laminae; petiole to 8.5cm, tomentose; inflorescence a 4-faceted terminal raceme, c. 4.5cm; peduncle 7.5cm; bracts obovate, 7 × 5mm, shallowly emarginate but with central apiculum, ciliate, pale green with deep red margins; calyx enclosed within bracts; corolla tube exserted, white; stamens 2. Forest, disturbed forest; 700–1000m.
Distr.: SE Nigeria to Gabon. (Lower Guinea).
IUCN: NT
Mbule: Etuge 4399 10/1998; **Ngomboaku:** Cheek 10365 fl., 12/1999; Mackinder 296 12/1999.

Asystasia decipiens Heine

Herb to 1(–2)m; stems robust, glabrous; leaves elliptic, 10–15 × 3.5–6.5cm, base asymmetric attenuate, apex acuminate, margin subcrenate, glabrous; petiole to 3cm, glabrescent; racemes axillary or terminal, to 7.5cm long; flowers held on one side; peduncle minutely puberulent; bracts and bracteoles linear, to 0.3cm; calyx lobes lanceolate, 0.3cm; corolla 5-lobed, 0.8cm long, yellow-green; fruit 1.9cm long, glabrescent. Forest; 850–1400m.
Distr.: Sierra Leone to NW Tanzania. (Guineo-Congolian).
IUCN: LC
Enyandong: Ghogue 1203 fl., 10/2001; **Nyasoso:** Balding 34 fl., fr., 8/1993; Cable 639 fl., 12/1993; Cheek 5649 fl., fr., 12/1993; 7462 10/1995; Lane 148 fl., 6/1994; 160 fl., 10/1994; Thomas 5034 11/1985; Wheatley 431 fl., 7/1992.
Note: according to Ensermu Kelbessa (*ined.*), *A. decipiens* is a synonym of *A. leptostachya*; we maintain them as separate taxa here.

Asystasia gangetica (L.) T.Anderson

Straggling herb to 1m; stems sparsely pubescent; leaves ovate, c. 7 × 3cm, base attenuate then abruptly rounded, apex acuminate, margin subentire, laminae sparsely pubescent or glabrous; petiole to 1.5cm; inflorescence an axillary or terminal spike to 14cm, 6–11-flowered; bracts lanceolate, c. 1mm, ciliate; calyx lobes subulate, c. 4mm, pubescent; corolla tube 1.3 × 0.8cm, puberulent, white with purple markings at base of throat; stamens 4; ovary and base of style pubescent; fruits 2 × 0.3cm, puberulent, green. Farmbush, forest edges; 200–1000m.
Distr.: widespread in palaeotropics; introduced to neotropics.
IUCN: LC
Kupe Village: Ensermu 3515 11/1995; **Loum F.R.:** Etuge 4282 fl., 10/1998; **Mwanenguba:** Leeuwenberg 8538 fl., 9/1971; 9569 fl., 4/1972; **Mungo River F.R.:** Cheek 10187 fl., fr., 11/1999; **Nlonako:** Etuge 1773 fl., fr., 3/1996; **Tombel:** Etuge 367 fl., fr., 11/1986.

Asystasia glandulifera Lindau

Straggling herb or climber to 2m; stems robust, finely pubescent towards apex; leaves ovate, 8.3–10 × 2.8–4cm, base acute-obtuse, apex acuminate, margin subcrenate,

upper surface sparsely pilose; petiole to 0.8cm; racemes terminal, to 4cm long, dense, glandular-hairy throughout; bracts ovate, 0.8 × 0.25cm; calyx lanceolate, 0.6cm long; corolla funnel-shaped, 2 × 1.3cm, white with purple streaking; pedicels to 1.5cm; fruit 3cm long, glandular-hairy. Farmbush, secondary forest; 850–1550m.
Distr.: Cameroon Endemic.
IUCN: VU
Essossong: Cable 928 fl., 2/1995; **Kupe Village:** Ensermu 3597 11/1995; **Nyasoso:** Cable 1227 fl., 2/1995; Cheek 7489 10/1995; Lane 216 fl., 11/1994; 512 fl., fr., 2/1995.

Asystasia leptostachya Lindau
Bot. Jahrb. 57: 22 (1920).
Rhizomatous herb to 2m; stems robust, becoming woody at base, glabrescent; leaves elliptic 9–12.5 × 3–5cm, base acute, asymmetric, apex acuminate, margin subcrenulate, glabrous; petiole 1.5cm; inflorescence a raceme or sparsely-branched panicle to 4cm, solitary or clustered, 9–15-flowered, puberulent throughout; pedicels 1–2mm; bracts triangular, 1mm; calyx lobes subulate, 3mm; corolla tube pale green with cream throat spotted red; stamens 4; fruits 2 × 0.3cm, green. Forest; 770–1430m.
Distr.: Cameroon. (W Cameroon Uplands).
IUCN: NT
Kodmin: Barlow 9 fl., 1/1998; Cheek 8921 fl., 1/1998; Etuge 4069 fl., fr., 1/1998; Ghogue 34 fl., 1/1998; **Kupe Village:** Ensermu 3511 11/1995; 3540 11/1995; Etuge 2695 fl., 7/1996; **Ngomboaku:** Ghogue 442 fl., fr., 12/1999; **Nsoung:** Darbyshire 36 3/2003; **Nyasoso:** Cable 3572 fl., fr., 7/1996; Sidwell 327 10/1995.

Asystasia lindaviana Hutch. & Dalziel
Fl. Gabon 13: 138 (1966).
Syn.: *Filetia africana* Lindau
Herb to 1m; stems glabrescent; leaves ovate-elliptic, 14–15.5 × 4.5–7cm, base rounded, apex acuminate, margin subentire, upper surface often variegated, dark green with pale green-white patches, raphides numerous; petiole to 2.5cm; racemes terminal, to 25cm long, narrow, occasionally branched at base; flowers numerous; bracts and bracteoles linear, to 0.3cm; calyx lobes subulate, 0.35cm long, glandular-hairy; corolla bilabiate, 1.3cm long, tube narrow, reddish or white, with pink spotting on lips. Forest; 200m.
Distr.: Cameroon & Gabon. (Lower Guinea).
IUCN: NT
Ikiliwindi: Nemba 554 6/1987.
Note: specimen not at Kew, the description here is based upon other W Cameroon material.

Asystasia macrophylla (T.Anderson) Lindau
Shrub to 3m; leaves oblanceolate, 27–37 × 10–12cm, base long-cuneate, apex caudate, margin subcrenulate, glabrous; petiole 2.5cm; inflorescence a terminal thyrse to 24cm, 6–10 flowers per cluster; bracts triangular, 2mm, pubescent; calyx lobes subulate, 3.5mm, pubescent; corolla tube 3.5 × 1cm with an elongate throat, white, drying red-black; stamens 4, style to 3cm, persistent. Forest; 250–1100m.
Distr.: SE Nigeria & Bioko to Gabon. (Lower Guinea).
IUCN: NT
Baduma: Nemba 674 11/1987; **Kurume:** Thomas 5459 1/1986; **Mejelet-Ehumseh:** Etuge 477 1/1987; **Mungo River F.R.:** Onana 936 fl., 11/1999.

Asystasia vogeliana Benth.
Herb or subshrub to 2.5m; stems robust, finely pubescent along the furrows; leaves elliptic, 9.5–17.5 × 4–6cm, base attenuate, apex long-acuminate, margin shallowly crenulate,

glabrous to finely pilose; petiole 2cm, pubescent; terminal racemes solitary or paired, all parts puberulent, 7–10-flowered; bracts lanceolate, 2–3mm; calyx lobes linear, to 6mm; corolla to 7cm, with a throat 5 × 0.3cm, opening to rounded lobes, white to violet or blue; stamens 4, style to 4.5cm, persistent. Forest and secondary growth; 250–1400m.
Distr.: Sierra Leone to Congo (K), Sudan & NW Tanzania. (Guineo-Congolian).
IUCN: LC
Baduma: Nemba 673 11/1987; **Bangem:** Thomas 5382 1/1986; **Ekona Mombo:** Etuge 409 fl., 12/1986; **Kodmin:** Etuge 4016 fl., 1/1998; Onana 583a fl., 2/1998; 601a fl., 2/1998; **Kupe Village:** Cable 768 fl., 1/1995; Ensermu 3509 fl., 11/1995; 3543 11/1995; Lane 280 fl., 1/1995; 373 fl., 1/1995; **Ndum:** Lane 483 fl., 2/1995; **Ngomboaku:** Etuge 4655 fl., 12/1999; Ghogue 465 fl., 12/1999; Gosline 259 fl., 12/1999; **Nyasoso:** Cable 643 fl., 12/1993; Cheek 5628 fl., 12/1993; 7479 fl., 10/1995; Lane 170 fl., 11/1994; Sidwell 325 fl., 10/1995; 342 fl., 10/1995; Thomas 5037 11/1985.

Barleria brownii S.Moore
Climber or subshrub; stems robust, sparsely pubescent towards apex; leaves ovate, c. 8.5 × 4.5cm, base rounded, apex acuminate, margin subcrenulate, sparsely pilose on both surfaces; petiole to 1.3cm, upper leaves subsessile; inflorescence a dense terminal cluster, interrupted by pubescent elliptic leaves to 1.1 × 0.8cm; calyx lobes 4, narrowly elliptic, 0.7 × 0.25cm, 1 broader; corolla trumpet-shaped, 3.5 × 2.5cm, blue; stamens 2, long-exserted; fruit ellipsoid, 1.2cm long, glabrous, drying black. Forest; 400–1000m.
Distr.: Ghana to Angola & Uganda. (Guineo-Congolian).
IUCN: LC
Kupe Village: Etuge 1480 11/1995; **Nyasoso:** Cheek 7294 fr., 2/1995.

Barleria opaca (Vahl) Nees
Straggling subshrub or climber to 2m; stems pubescent; leaves ovate-elliptic, 7.5–10.3 × 3–3.6cm, base acute-cuneate, apex acuminate, margin subentire, pilose on both surfaces, densest along the lower veins, raphides numerous on the upper surface; petiole to 1cm, pubescent; inflorescence a dense terminal cluster to 5.5cm with interrupting pilose ovate leaves, 1.9 × 0.8cm; calyx lobes lanceolate, 1.1cm, pubescent, 1 lobe forked at the apex; corolla trumpet-shaped, 3 × 2.7cm, pale blue; stamens short-exserted. Secondary forest, streamsides; 250m.
Distr.: Sierra Leone to Cameroon. (Upper & Lower Guinea).
IUCN: LC
Ikiliwindi: Nemba 676 12/1987.
Note: specimen not at Kew; the description based upon other W Cameroon material.

Brachystephanus giganteus Champl. ined.
Syn.: *Oreacanthus mannii* Benth.
Mass-flowering herb or subshrub to 2m; stems robust, puberulent towards apex; leaves elliptic, 14–25.5 × 6–10.3cm, base cuneate-attenuate, apex acuminate, acumen to 1.5cm, margin subcrenulate, surfaces sparsely pubescent, principally along veins; inflorescence a many-branched panicle 18–35 × 8–12cm, glandular-hairy throughout, hairs to 2mm; bracts and bracteoles linear, to 5mm; calyx lobes subulate, 1.5–2cm; corolla bilabiate, glabrous, tube short, c. 1mm, white, lips blue to purple, upper lip ovate, c. 5 × 3mm, lower lip quadrangular, c. 5 × 5mm; stamens 2, exserted; fruits linear, c. 1.2cm long, glabrous. Forest and grassland; 2050–2100m.
Distr.: Cameroon & Bioko. (W Cameroon Uplands).

IUCN: VU
Mt Kupe: Letouzey 2128 fr., 8/1959; **Mwanenguba:** Leeuwenberg 8925 fl., fr., 12/1971.

Brachystephanus jaundensis Lindau
Kew Bull. 51: 760 (1996).
Straggling herb with erect stems to 80cm; stems ± pubescent; leaves ovate-elliptic, 8.5–17.5 × 3.7–6.7cm, base attenuate, apex acuminate, margin subentire, glabrescent, raphides numerous on both surfaces; petiole to 3.5cm, pubescent; racemes terminal, to 15cm long, many-flowered, glandular-hairy; bracts linear, to 1cm long; calyx lobes linear, to 1(–1.5)cm long; corolla puberulent, 2.4(–3.5)cm with a long narrow tube, apex bilabiate, violet; stamens exserted; fruit 1.4cm long, glabrous. Forest; 1000–1360m.
Distr.: Bioko, Cameroon & Gabon. (Lower Guinea).
IUCN: NT
Kodmin: Cheek 8984 1/1998; **Kupe Village:** Cheek 7195 fr., 1/1995; **Nyasoso:** Cable 3493 7/1996.

Brachystephanus kupeensis Champl. ined.
Sprawling herb, 0.3–2m; stems glabrescent; leaves elliptic, c. 12 × 4.2cm, base attenuate, apex acuminate, margin subcrenulate, glabrous, raphides numerous particularly on the under surface; petiole to 1.6cm; inflorescence a many-flowered panicle, 18 × 5cm, glandular-hairy throughout, hairs to 2.5mm; bracts and bracteoles linear, to 0.35cm long; calyx lobes subulate, 0.5cm long; corolla bilabiate, 1.6cm long, glabrous, purple, lower lip 3-lobed, upper somewhat rolled backwards; stamens shortly exserted. Primary and secondary forest; 900–1100m.
Distr.: Mt Kupe, Cameroon. (Narrow Endemic).
IUCN: CR
Nyasoso: Cheek 5673 fl., 12/1993; Lane 186 fl., 11/1994.

Brachystephanus longiflorus Lindau
Herb to 1.5m; stems tomentose towards apex; leaves elliptic, 12–17.5 × 5.5–8.2cm, base cuneate then abruptly rounded, apex acuminate, margin subcrenulate, sparsely pubescent on the upper surface and margins and veins of the lower surface; petiole to 5.5cm; spikes terminal, 5(–11)cm, sessile, dense; bracts elliptic, 1.7 × 0.6cm, apex caudate, ciliate; calyx lobes linear, 1.5(–3)cm, ciliate; corolla bilabiate, 3cm long, deep pink; fruit 0.7cm long, glabrous. Forest; 1360–1540m.
Distr.: SE Nigeria, Bioko & SW Cameroon. (W Cameroon Uplands).
IUCN: VU
Kupe Village: Cheek 7206 fl., fr., 1/1995; **Nyasoso:** Wheatley 444 fl., 7/1992.

Brachystephanus mannii C.B.Clarke
Fl. Gabon 13: 241 (1966).
Syn.: *Oreacanthus cristalensis* Champl. & Figueiredo
Subshrub, 60–100cm high; stem with two opposite rows of hairs in the upper part, glabrous beneath; leaves elliptic, 6.5–18 × 3.5–7cm, base cuneate, apex acuminate, glabrous above, pubescent to puberulous and often red-violet beneath; petiole 1.5–4cm; inflorescence a dense thyrsoid panicle 9–19 × 2–4cm, axis retrorsely pubescent; bracts lanceolate, 6–7 × 1–2mm; calyx lobes 8–13 × 0.7–1mm, both purple-rose; corolla bilabiate, c. 1.4cm long, glabrous, tube white, upper lip rolled backwards, oblong, tapering into a subacute apex, white-blue, lower lip tridentate, blue or violet; fruit c. 1.1cm long; seeds minutely tuberculate. Forest, often along streams; 850m.
Distr.: Bioko, Cameroon, Gabon. (Lower Guinea).
IUCN: NT
Ngomboaku: Cheek 10287 fl., fr., 12/1999.

Brachystephanus nimbae Heine
Kew Bull. 51: 761 (1996).
Subshrub, 0.8–1.2m high; stems slender, glabrous or with two opposite puberulent rows; leaves elliptic, 2.2–14 × 1.2–5.3cm, base cuneate to attenuate, apex obtuse to acuminate, glabrous above, glabrescent or puberulent on the nerves beneath; petiole 0.3–2.5cm; inflorescence spiciform, 2.5–9 × 0.5cm, axis puberulent, 2–7 flowers per node; bracts triangular-lanceolate, 5–9 × 1mm glabrous or glandular-hairy; calyx lanceolate, 7–10 × 0.5mm, lobes free to the base, usually glabrous; corolla bilabiate, puberulent, white or pale mauve to violet with purple streaks, minutely puberulent, tube 1.25–1.9cm long, upper lip oblong, shortly ciliate, lower lip tridentate; fruit 1.1cm long; seeds minutely tuberculate, c. 1.5mm diameter. Forest, often in wet rocky areas; 900–1050m.
Distr.: Guinea (C) to Cameroon. (Upper & Lower Guinea).
IUCN: VU
Nyasoso: Cable 18 fl., 7/1992; Cheek 9281 10/1998; Lane 35 fl., 8/1992.

Brillantaisia lamium (Nees) Benth.
Erect herb to 1.5m; stems tomentose towards apices, hairs to 4mm; leaves ovate, 8.5–11 × 5–6cm, base cordate or truncate, apex long-acuminate, margin subcrenulate, laminae pilose; petiole to 6.5cm, densely tomentose; inflorescence a lax panicle, glandular-hairy throughout; bracts ovate, 9 × 8mm, apiculate; pedicels to 1.3cm; calyx lobes linear, 0.8cm; corolla bilabiate, 2.5cm long, purple to blue; stamens 2; fruit linear, 3–3.5cm long, glabrous, brown-green. Farmbush, forest margins; 700–1700m.
Distr.: Guinea (C) to Angola & W Tanzania. (Guineo-Congolian).
IUCN: LC
Bangem: Manning 398 fl., 10/1986; **Basengi Village:** Sidwell 418 fl., fr., 10/1995; **Edib:** Etuge 4497 fl., fr., 11/1998; **Kupe Village:** Ensermu 3517 11/1995.

Brillantaisia lancifolia Lindau
Scrambling herb; stems glabrous except when young, then puberulent; leaves narrowly elliptic 7.5–10 × 2–3cm, base attenuate, apex long-acuminate, margin serrulate, laminae sparsely pilose, raphides numerous on both surfaces; petiole 1–2cm, pubescent; terminal panicles lax, 11 × 6cm, densely glandular-pubescent; bracts linear, 0.3cm; calyx lobes linear, uneven in length, to 0.8cm; corolla bilabiate, 3cm long, purple with a white throat, yellow spotted; fruit linear c. 2cm, glabrous. Rocky river margins in forest; 150–1350m.
Distr.: SE Nigeria to Gabon. (Lower Guinea).
IUCN: VU
Mungo River: Gosline 219 fl., 11/1999; **Ngomboaku:** Mackinder 335 12/1999; **Nyale:** Cheek 9701 11/1998.

Brillantaisia madagascariensis T.Anderson ex Lindau
Herb to 1m; similar to *B. lamium*, but leaves larger (20–24 × 8.2–9cm), base cuneate, merging into winged petiole, margin more clearly crenulate, raphides numerous on both leaf surfaces; inflorescence less branched and denser, to 14.5cm long; bracts obovate, ciliate, 1.2 × 0.8cm, linear calyx lobes longer, to 2cm; corolla white or blue; fruits minutely puberulent. Forest, secondary growth; 850m.
Distr.: Guinea to Congo (K), E Africa & Madagascar. (Tropical Africa & Madagascar).
IUCN: LC
Nyasoso: Thomas 5045 11/1985.

Note: specimen not at Kew; description based upon other Cameroon material.

Brillantaisia owariensis P.Beauv.

Syn.: *Brillantaisia nitens* Lindau
Erect herb to 2.5m; stems robust, glabrous; leaves broadly ovate, 18–21 × 9–11cm, base acute to truncate with a cuneate extension grading into the winged petiole, apex acuminate, margin serrate, laminae sparsely pilose, raphides numerous on upper surface; petiole to 4cm, winged towards apex, tomentose, hairs to 2mm; panicles lax, to 20cm long, many-flowered, glandular-hairy throughout; bracts caducous; bracteoles linear to ovate, c. 7mm long; calyx lobes linear, uneven, longest 1.2cm; corolla bilabiate, c. 3cm, blue-violet with white throat; fruit linear, 2.2cm long, pubescent. Forest, forest-grassland transition; 700–1650m.
Distr.: Nigeria to S Sudan & W Tanzania. (Lower Guinea & Congolian).
IUCN: LC
Basengi Village: Sidwell 423 10/1995; **Essossong:** Leeuwenberg 9244 fl., fr., 1/1972; 9245 fl., fr., 1/1972; **Kodmin:** Cheek 8916 1/1998; Etuge 3960 fl., fr., 1/1998; 4027 fl., fr., 1/1998; Ghogue 5 fl., fr., 1/1998; **Kupe Village:** Etuge 1437 11/1995; Sidwell 434 fl., 11/1995; 436 11/1995; 437 fl., 11/1995; **Mejelet-Ehumseh:** Etuge 451 fl., fr., 1/1987; **Mt Kupe:** Thomas 5106 12/1985; **Mwambong:** Onana 556a fl., fr., 2/1998; **Mwanenguba:** Leeuwenberg 8554 fl., 10/1971; Sanford 5437 fl., fr., 11/1968; **Nyasoso:** Cable 1244 fl., 2/1995; 638 fl., 12/1993; 704 fl., 1/1995; Etuge 1617 fl., 1/1996; Sidwell 331 10/1995; 352 fl., fr., 10/1995.

Brillantaisia vogeliana (Nees) Benth.

Erect herb to 1.2m; similar to *B. owariensis*, but leaf base truncate to subcordate, raphides abundant on both leaf surfaces; petiole to 9cm, less clearly winged, cuneate extension of leaf base may be absent; bracts usually persistent, ovate, to 2.2 × 1cm, becoming narrower towards the apex; bracteoles oblanceolate, 0.5cm long; calyx lobes linear, only 1cm long. Forest, forest margins; 50–1550m.
Distr.: Ghana to Congo (K), Sudan to Kenya. (Tropical Africa).
IUCN: LC
Baduma: Nemba 622 9/1987; **Kodmin:** Satabie 1111 fl., 1/1998; **Kupe Village:** Cable 3776 fl., fr., 7/1996; 3877 fr., 7/1996; Cheek 7703 11/1995; Ensermu 3516 fr., 11/1995; 3549 fl., fr., 11/1995; Etuge 2626 fl., 7/1996; Sidwell 438 fl., 11/1995; **Mwanenguba:** Leeuwenberg 8419 fl., fr., 9/1971; **Mungo River F.R.:** Cheek 10173 fl., 11/1999; Leeuwenberg 10628 fl., fr., 11/1972; **Nyasoso:** Cable 62 fl., 8/1992; Cheek 7492 10/1995; Lane 225 fl., 11/1994; Sidwell 303 10/1995; 304 fl., 10/1995; 350 fl., fr., 10/1995.

Chlamydocardia buettneri Lindau

Herb, 0.3–1.5m; stems robust, hairs run in a line on each side; leaves subsessile, elliptic-ovate, 10.7–12.5 × 3–4cm, base attenuate, apex long-acuminate, margin subentire, raphides numerous on both surfaces; inflorescence spikes terminal, 6(–10)cm, glandular-hairy; bracts variable, usually spathulate, 1 × 0.3–0.7cm, apex shortly acuminate or tri-lobed; calyx linear, 0.8cm; corolla bilabiate, 1cm long, white-green; stamens 2. Forest, forest margins; 870–1000m.
Distr.: Ivory Coast to Gabon. (Upper & Lower Guinea).
IUCN: LC
Kupe Village: Cable 841 fl., 1/1995; Cheek 7162 fr., 1/1995; 7655 11/1995; **Nyasoso:** Cable 1216 2/1995.

Dicliptera laxata C.B.Clarke

Suberect herb to 0.5m; stems glabrous; leaves ovate to elliptic, 6.5–11 × 2.5–5cm, base acute, apex acuminate, margin subcrenulate, upper surface sparsely pilose, raphides present on under surface; petiole to 2.5cm, puberulent; flowers in axillary umbels, 2 per axis; peduncle to 1cm; pedicels to 0.4cm, puberulent; paired leafy obovate glabrous bracteoles 0.8 × 0.4cm enclose the flowers; corolla partially exserted, white. Farmbush, cultivation, forest margins; 1400m.
Distr.: Cameroon, Bioko, Congo (K) & E Africa. (Tropical Africa).
IUCN: LC
Kodmin: Cheek 8884 fl., 1/1998; Etuge 3985 fl., 1/1998.
Local names: Nzuh (Cheek 8884); Bakossi Tea (Etuge 3985). **Uses:** FOOD – leaves – boil leaves in water, colour changes to reddish, add sugar and milk, and drink your Bakossi tea any time of the day (Etuge 3985), which is taken to strengthen blood (Cheek 8884).

Dicliptera obanensis S.Moore

Herb to 40cm; stems decumbent, rooting at lower nodes, glabrescent; leaves similar to *D. laxata*; inflorescence a few-flowered terminal raceme to 4.5cm; peduncle puberulent; flowers enclosed within paired leafy bracteoles, obovate, 1.1 × 0.6cm, minutely puberulent; calyx included; corolla 0.7cm long, purple; pedicels 0.4cm; fruit 0.6cm long, puberulent. Forest, secondary growth; 700–1300m.
Distr.: Sierra Leone to Cameroon. (Upper & Lower Guinea).
IUCN: LC
Edib: Etuge 4493 11/1998; **Kupe Village:** Lane 319 fr., 1/1995; **Nyasoso:** Cable 655 fl., 12/1993.

Dicliptera silvestris Lindau

Wiss. Ergebn. Deutsch. Zentr.-Afr. Exped. 1907–8, 2: 302 (1911).
Creeping or scrambling herb; similar to *D. laxata*, but bracteoles densely pilose on margins and midrib with hairs to 1mm; corolla pink-purple with a mottled throat. Secondary forest; 750m.
Distr.: Cameroon & CAR. (Lower Guinea).
IUCN: VU
Kupe Village: Cheek 7108 1/1995.
Note: the similar *D. umbellata* (Vahl) Juss., distinguishable by the subsessile, apiculate bracteoles, is recorded near Kumba and therefore likely in the checklist area.

Dischistocalyx grandifolius C.B.Clarke

Fl. Gabon 13: 25 (1966).
Herb, shrub or epiphyte to 2m; stems glabrous; leaves elliptic-oblong, (12–)25–27 × 6.5–8cm, pairs unequal, base acute, asymmetric, apex acuminate, margin crenulate, lamina red below, raphides numerous on both surfaces; petiole 2.5cm, glabrous; inflorescence a dense terminal raceme to 10cm; bracts elliptic, c. 3cm long, base acute, apex acuminate; calyx lobes lanceolate, 1.8–3 × 0.2–0.4cm with numerous raphides; corolla funnel-shaped, 5–7.5 × 2–3.5cm long, lobes emarginate, purple; stamens 4; fruit 2.8 × 0.4cm, glabrous, brown. Forest including stream margins; 900–1450m.
Distr.: Cameroon & Gabon. (Lower Guinea).
IUCN: NT
Edib: Cheek 9121 fl., 2/1998; **Kodmin:** Cheek 8954 fl., 1/1998; Etuge 4123 fl., 2/1998; Gosline 52 fl., 1/1998; **Mbule:** Cable 3347 fl., 6/1996; **Mwambong:** Gosline 126 fl., 10/1998; Onana 535 fl., 2/1998; **Nyandong:** Tchiengue 1736 3/2003; **Nyasoso:** Cable 3324 fl., 6/1996; 3506 fl.,

7/1996; Etuge 2061 fl., fr., 6/1996; 2363 fl., 6/1996; Lane 38 fl., 8/1992; Zapfack 707 fl., 6/1996.
Local name: N'zo Kunte. **Uses:** MEDICINES (Gosline 52).

Elytraria marginata Vahl

Herb to 75cm, erect or subprostrate; stems pubescent; leaves in rosettes towards apices, obovate or oblanceolate, 8–13 × 3–4cm, base cuneate, apex acute or acuminate, margin subentire, upper surface sparsely pilose; petiole to 1cm; inflorescence a dense terminal spike to 13cm; bracts rigid, ovate, 0.5cm with spinose apiculum 0.2cm and ciliate margins; corolla exserted, c. 0.5cm, white; stamens 2; fruits c. 0.5cm long, glabrous, green. Forest, farmbush; 200–1000m.
Distr.: Guinea (C) to Uganda. (Guineo-Congolian).
IUCN: LC
Kodmin: Plot B246 fr., 1/1998; **Kupe Village:** Cable 2454 fr., 5/1996; Ensermu 3528 fl., 11/1995; Kenfack 272 fr., 7/1996; Lane 277 fl., 1/1995; **Mungo River F.R.:** Cheek 10174 fl., 11/1999; **Nyale:** Etuge 4209 fl., 2/1998; **Nyandong:** Cheek 11338 3/2003; **Nyasoso:** Lane 213 fl., 11/1994.

Eremomastax speciosa (Hochst.) Cufod.

Fl. Gabon 13: 30 (1966).
Syn.: *Eremomastax polysperma* (Benth.) Dandy
Syn.: *Clerodendrum eupatorioides* Baker
Herb to 3m; stems robust, tomentose towards apex, hairs to 1.5mm; leaves ovate, 5–7 × 3–5cm, base truncate to subcordate, apex acuminate, margin crenate, surfaces pubescent along veins; petiole to 5.7cm; inflorescence a many-flowered panicle; bracts ovate, 1 × 0.5cm, finely pilose; calyx lobes linear, c. 1.5cm, tomentose; corolla 3–5cm long with neck 3cm, pale blue-lavender, 5 lobes all below the androecium; stamens 4, anthers blue; fruit 1.7 × 0.3cm, pubescent. Forest, farmbush; 400–1450m.
Distr.: W Africa to Uganda & NW Tanzania. (Guineo-Congolian).
IUCN: LC
Edib: Etuge 4093 fl., 1/1998; **Enyandong:** Cheek 10982 10/2001; **Kodmin:** Cheek 9042 1/1998; Gosline 159 11/1998; **Kupe Village:** Ensermu 3569 fl., 11/1995; 3595 11/1995; Etuge 1479 11/1995; Lane 376 fl., 1/1995; **Mejelet-Ehumseh:** Etuge 462 fl., 1/1987; **Mwanenguba:** Leeuwenberg 8548 fl., 10/1971.
Uses: MEDICINES (Etuge 1479).

Graptophyllum glandulosum Turrill

Herb or shrub to 2.5m; stems robust, woody at base, glabrescent or finely pubescent; leaves elliptic, 14–19 × 6.5–9cm, base acute, apex acuminate or caudate, acumen to 2.5cm, margin subcrenule, raphides numerous on upper surface; petiole c. 2cm; inflorescence a terminal raceme to 5.5cm, glandular-pubescent throughout; bracts lanceolate, to 4mm long; calyx lobes subulate, c. 4mm; corolla 3 × 0.7cm, bilabiate, orange-red; stamens 2, orange; pedicels c. 0.7cm. Forest; 200–1750m.
Distr.: SE Nigeria & Cameroon. (W Cameroon Uplands).
IUCN: NT
Bangem: Manning 451 fl., 10/1986; 489 fl., 10/1986; **Ehumseh - Mejelet:** Etuge 339 fl., 10/1986; **Enyandong:** Cheek 10909 10/2001; **Kodmin:** Barlow 1 fl., 1/1998; Cheek 8866 fl., 1/1998; Etuge 4410 fl., 11/1998; **Kupe Village:** Cable 2475 fl., 5/1996; 742 fl., 1/1995; Cheek 6082 fl., 1/1995; 7052 fl., 1/1995; Ensermu 3513 11/1995; Ryan 217 fl., 5/1996; **Mbule:** Cable 3341 fl., 6/1996; **Mile 12 Mamfe Road:** Nemba 311 fl., 10/1986; **Mt Kupe:** Leeuwenberg 8815 fl., 12/1971; **Mungo River F.R.:** Onana 939 fl.,

11/1999; **Ngomboaku:** Etuge 4683 fl., 12/1999; Ghogue 464 fl., 12/1999; **Nyandong:** Cheek 11334 3/2003; Darbyshire 24 3/2003; **Nyasoso:** Cable 3565 fl., 7/1996; 3609 fl., 7/1996; 642 fl., 12/1993; Etuge 2552 fl., 7/1996; Lane 248 fl., 11/1994.

Hypoestes aristata (Vahl) Soland. ex Roem. & Schult.

Erect herb to 2m; stems pubescent, hairs white; leaves ovate, 11.5–18.5 × 4.5–8cm, base acute, apex gradually acuminate, margin subentire, veins, margins and upper surface ± pilose, raphides present on upper surface; petiole 1.5–5cm; flowers in dense axillary clusters; bracts ovate, to 1.3cm long, with apiculum 0.5cm, flexible; calyx included; corolla bilabiate, c. 2cm long, pink with white base; stamens 2, 2.5cm, long-exserted; fruit 1.7cm long, glabrous. Secondary forest, farmbush; 700–1250m.
Distr.: Nigeria to E & S Africa. (Tropical Africa).
IUCN: LC
Basengi Village: Sidwell 421 fl., 10/1995; **Lake Edib:** Cheek 9168 fl., 2/1998; **Ndum:** Williams 145 1/1995; **Ngomboaku:** Etuge 4654 fl., 12/1999; **Nyasoso:** Cheek 5643 fl., 12/1993; 7900 11/1995.
Local name: Isangando (Language: Bakweri) (Lyonga Daniel in Cheek 7900). **Uses:** MEDICINES (Cheek 7900).

Hypoestes forskaolii (Vahl) R.Br.

Fl. Gabon 13: 229 (1966).
Herb to 0.5m; stems puberulent; leaves ovate-elliptic, 13–18 × 5–6.5cm, base acute, apex acuminate, margin subcrenule, sparsely puberulent on lower surface; petiole to 7cm; inflorescence a spike to 5cm; bracts 0.6 × 0.1cm, apex acute, pubescent, overlapping; corolla bilabiate, c. 1cm long, white with pink spotting; stamens 2, short-exserted. Farmbush; 800–850m.
Distr.: widespread in tropical Africa.
IUCN: LC
Bangem: Thomas 5355 1/1986; **Nyasoso:** Etuge 1523 fl., 12/1995; Thomas 3056 fl., fr., 2/1984.

Hypoestes rosea P.Beauv.

Erect herb 1(–1.5)m; stems pubescent towards apex; leaves elliptic, 11–15.5 × 3.2–6.5cm, base attenuate, apex acuminate, margin subentire, raphides present on upper surface; inflorescence a dense axillary or terminal spike, occasionally shortly-branched, numerous in the upper section of the plant, puberulent throughout; flowers surrounded by several bracts / bracteoles, outer ones lanceolate, to 1.2cm long, apex caudate, ciliate; calyx included; corolla bilabiate, c. 2cm long, lower lip 3-lobed, deflexed, upper lip ovate, pink-purple throughout; stamens 2, exserted, filaments purple. Secondary forest, farmbush; 250m.
Distr.: Nigeria to Congo (K) & Burundi. (Lower Guinea & Congolian).
IUCN: LC
Konye: Nemba 440 1/1987; **Kurume:** Thomas 5461 fl., fr., 1/1986.

Hypoestes triflora (Forssk.) Roem. & Schult.

Straggling herb to 40cm high; stems fine, pubescent towards apex; leaves ovate, 3–4 × 2–2.5cm, pairs uneven, base rounded, apex acuminate, margin crenate, surfaces pilose; petiole to 1.5cm, pilose; inflorescence terminal, cymose, 2–3-flowered; bracteoles 2 per flower, oblanceolate, 0.9 × 0.3cm, pubescent; calyx lobes lanceolate, 0.6cm; corolla bilabiate, c. 1.2cm long, white-mauve with purple spotting on lower lip; stamens 2, exserted. Forest; 1450–2150m.
Distr.: tropical Africa. (Afromontane).

ACANTHACEAE

IUCN: LC
Kodmin: Cheek 8943 fl., 1/1998; Etuge 4047 fl., 1/1998;
Mwanenguba: Leeuwenberg 8922 fl., 12/1971; Sanford 5523 fl., 11/1968.

Isoglossa glandulifera Lindau

Erect herb to 0.6(–3)m; stems glabrous; leaves thin, ovate, 4–5 × 3.5–4cm, base rounded to subcordate, apex acuminate, margin subcrenulate, raphides numerous; petiole to 3.7cm; inflorescence a lax terminal panicle to 12.5cm long; peduncles, pedicels and calyx glandular-hairy; bracts linear, to 3mm; calyx lobes lanceolate, 2mm; corolla tube 2.5cm long, lower-lip 3-lobed, pink with purple spots; stamens 2, included; fruit 1.4cm long, sparsely glandular-hairy. Forest; 1450–1550m.
Distr.: SE Nigeria, Bioko & W Cameroon. (W Cameroon Uplands).
IUCN: NT
Kodmin: Cheek 9582 fl., 11/1998; **Nyasoso:** Lane 238 fl., 11/1994.
Local name: Mbeneh (Max Ebong in Cheek 9582).

Justicia buchholzii (Lindau) I.Darbysh. *comb. nov.*

Basionym: *Duvernoia buchholzii* Lindau, Bot. Jahrb. 20: 43 (1894).
Type: Cameroon, Buchholz s.n. (B, holotype; destroyed).
Syn.: *Adhatoda buchholzii* (Lindau) S.Moore
Herbaceous climber; stems glabrous except when immature; leaves obovate, 27 × 10cm, base acute, apex abruptly acuminate, margin subentire to crenulate, glabrous; petiole 4.5cm; panicle terminal, primary peduncle to 21cm; flowers paired at 2–4cm intervals; bracts suborbicular c. 3 × 2cm, base subcordate, apex acuminate, glabrous; calyx lobes lanceolate, to 0.8cm; corolla bilabiate, 1.7cm, white; stamens 2. Forest; 200–420m.
Distr.: S Nigeria to Congo (B). (Lower Guinea).
IUCN: NT
Baduma: Nemba 668 11/1987; **Mile 12 Mamfe Road:** Nemba 307 fl., 10/1986; **Mungo River F.R.:** Onana 943 11/1999.
Note: *Adhatoda* in entirety is now considered synonymous with *Justicia* (see e.g. Enum. Pl. Af. Trop. 4: 484 (1997)).

Justicia camerunensis (Heine) I.Darbysh. *comb. nov.*

Basionym: *Adhatoda camerunensis* Heine, Kew Bull. 16: 165 (1962).
Type: Cameroon, Keay in FHI 37391 (K, holotype; FHI, isotype).
Shrub to 2.5m; stems, petioles, lower leaf surfaces and inflorescences tomentose; hairs brown, persistent; leaves ovate to obovate, to 36 × 15.5cm, base acute to cuneate, apex acuminate, margin subentire to crenulate; petiole to 6cm; racemes axillary, dense, 2.5–5.5cm; bracts and calyx lobes lanceolate, c. 1cm long, green-purple; corolla bilabiate, 1.5cm, white with purple markings; stamens 2. Forest; 900–1190m.
Distr.: E Nigeria & Cameroon. (W Cameroon Uplands).
IUCN: VU
Mt Kupe: Cheek 10157 fl., 11/1999; Leeuwenberg 8826 fl., 12/1971; Schoenenberger 3 11/1995; **Nyasoso:** Cable 1133 fr., 2/1995; Cheek 7544 10/1995; Etuge 1349 10/1995; Lane 246 fl., 11/1994; Sebsebe 5020 10/1995.

Justicia extensa T.Anderson

Herb or shrub to 2m; stems woody at base, densely pubescent towards apices; leaves elliptic, 17–20 × 8–9cm, base acute,

apex acuminate, margin subcrenulate, glabrescent; petiole 3.5–5.5cm; inflorescence a many-flowered terminal panicle to 21cm long; peduncles pubescent; bracts ovate, 2.5 × 1.5cm, base cordate, apex acute; calyx lobes triangular, 0.4cm, puberulent, drying blackish; corolla bilabiate, c. 1cm long, white-green with purple markings inside lower lip; fruit 1.8cm long, glabrescent, drying blackish. Forest; 0–1100m.
Distr.: Guinea (C) to Congo (K) & CAR. (Guineo-Congolian).
IUCN: LC
Kupe Village: Cheek 7125 fr., 1/1995; 7181 fr., 1/1995; Ensermu 3536 11/1995; Etuge 2696 7/1996; Schoenenberger 28 fl., 11/1995; 44 fl., 11/1995; **Mwambong:** Onana 552 fr., 2/1998; **Mungo River F.R.:** Onana 930 fl., fr., 11/1999; **Nyale:** Cheek 9659 fl., fr., 11/1998; **Nyasoso:** Cable 1220 2/1995; 680 fl., 12/1993; Cheek 7452 10/1995; Lane 247 fl., 11/1994; Sebsebe 5032 10/1995; Sidwell 326 fl., fr., 10/1995; Thomas 5032 fl., 11/1985.

Justicia insularis T.Anderson

Erect or straggling herb to 1m; stems pubescent, hairs to 1.5mm; leaves ovate, 7–8 × 2.5–3.3cm, base acute, apex acute to shortly acuminate, margin subentire, surfaces pilose; petiole to 3.3cm; inflorescence a subsessile axillary cluster, pubescent throughout; bracts elliptic, 1.2 × 0.3cm; calyx lobes lanceolate, 0.2–0.3cm; corolla bilabiate, 1.3cm long, lower lip pink-purple with white spots, upper lip white; stamens 2, exserted; fruit 0.8cm long, glabrescent. Forest, farmbush; 700–1500m.
Distr.: Nigeria to Congo (K). (Lower Guinea & Congolian).
IUCN: LC
Kodmin: Etuge 4048 1/1998; Ghogue 36 fl., 1/1998; Gosline 91 fl., 2/1998; **Kupe Village:** Cheek 6093 fl., 1/1995; Ensermu 3518 fl., 1/1995; Etuge 2617 7/1996; Lane 314 fl., 1/1995; **Ndum:** Williams 134 fl., 1/1995; **Ngomboaku:** Etuge 4658 12/1999; **Nyasoso:** Cable 1225 fl., 2/1995; Lane 179 fl., 11/1994; Sidwell 322 10/1995; 343 fl., 10/1995.

Justicia ladanoides Lam.

Symb. Bot. Ups. 29: 88 (1989).
Erect herb to 1.5m; resembling *J. insularis*, but corolla deeper pink-purple; style exceeding length of stamens; stigmas uneven (style shorter than stamens and stigmas even in *J. insularis*). Riverine forest; 770m.
Distr.: Senegal to Congo (K), Uganda, Ethiopia & Kenya. (Guineo-Congolian).
IUCN: LC
Mt Kupe: Schoenenberger 22 fl., 11/1995.

Justicia laxa T.Anderson

Herb or shrub to 1.5m; similar to *J. extensa*, but stems glabrescent; leaves glabrous; petiole to only 2.5cm; bracts inconspicuous, linear, c. 0.5cm; calyx lobes lanceolate, 3mm, drying green, not black. Forest; 350–1000m.
Distr.: SE Nigeria & Bioko to Congo (K). (Lower Guinea & Congolian).
IUCN: LC
Bakossi F.R.: Etuge 4286 fl., fr., 10/1998; **Mungo River F.R.:** Cheek 10220 fl., 11/1999; **Ngomboaku:** Mackinder 306 fl., 12/1999; **Nyasoso:** Lane 214 fl., 11/1994.

Justicia leucoxiphus Vollesen, Cheek & Ghogue

Kew Bull. 59: 129 (2004).
Herb to 2m; stems robust at base, pubescent towards apices; leaves elliptic, 15–20 × 6–9cm, pairs occasionally unequal, base acute to obtuse, apex acuminate, margin subentire, midrib on under-surface puberulent, raphides ± numerous; petiole 2(–4.5)cm, puberulent; inflorescence a terminal spike,

flattened in one plane; 2 distinct sets of ovate bracts each c. 1.3 × 2cm, base obtuse to rounded, apex acute, ciliate, the outer set held at c. 50 degrees to the peduncle, green-white, the inner set upright and held close to the peduncle, white; calyx lobes lanceolate, c. 0.9cm, included within the bracts; corolla bilabiate, only the lips emerging from the bracts, lower lip c. 0.7cm wide, shallowly 3-lobed, deflexed, pink-purple, upper lip ovate, 0.3cm wide, white with pink margin; stamens white. Forest; 950–1500m.

Distr.: Bakossi & Rumpi Hills. (Narrow Endemic).
IUCN: EN

Enyandong: <u>Gosline 340</u> 11/2001; **Kodmin:** <u>Cheek 8867</u> fr., 1/1998; <u>Etuge 3998</u> fl., 1/1998; <u>4411</u> fl., 11/1998; **Mejelet-Ehumseh:** <u>Etuge 314</u> fl., 10/1986; **Mwambong:** <u>Cheek 10276</u> 12/1999; <u>Onana 544a</u> fr., 2/1998; **Ngomboaku:** <u>Cheek 10393</u> fl., 12/1999; <u>Ghogue 439</u> fl., 12/1999; <u>Gosline 264</u> fl., 12/1999.

Justicia orbicularis (Lindau) V.A.W.Graham
Kew Bull. 43: 587 (1988).
Syn.: *Adhatoda orbicularis* (Lindau) C.B.Clarke
Herb or shrub to 1.5m; stems densely tawny-puberulent; leaves elliptic, 10–18 × 5.5–11cm, base attenuate, apex acute, margin crenulate, lower surface puberulent, particularly along veins; petiole to 4cm, puberulent; inflorescence a 4-faceted spike to 18cm long; bracts suborbicular, 1.3cm diameter., nervose, puberulent, overlapping; corolla exserted, bilabiate, white-yellow with purple-brown streaking. Forest including stream margins; 300–700m.
Distr.: SE Nigeria, Bioko & Cameroon. (Lower Guinea).
IUCN: VU
Konye: <u>Thomas 5192</u> fl., 11/1985; **Nyale:** <u>Etuge 4171</u> 2/1998; **Nyandong:** <u>Darbyshire 113</u> 3/2003.

Justicia paxiana Lindau
Bot. Jahrb. 20: 63. (1894).
Syn.: *Rungia paxiana* (Lindau) C.B.Clarke
Herb or subshrub to 1.5m; stems glabrescent; leaves elliptic, 9.5–15 × 3.5–5cm, pairs unequal, base attenuate, apex acuminate, margin subentire, veins and margins sparsely pilose, raphides numerous on both surfaces; petiole to 4.2cm, puberulent; inflorescence an axillary spike to 5.5cm; bracts obovate, 0.8 × 0.5cm, apex rounded to emarginate, ciliate, green with a 2mm hyaline white margin; calyx included; corolla bilabiate, c. 0.7cm long, pink; stamens 2. Forest, montane grassland; 1330–1400m.
Distr.: Guinea (C) to Uganda & NW Tanzania. (Guineo-Congolian).
IUCN: LC
Kodmin: <u>Biye 47</u> fl., 1/1998; <u>Cheek 10262</u> 12/1999; <u>Etuge 4425</u> fl., 11/1998.

Justicia preussii (Lindau) C.B.Clarke
Herb or slender shrub to 3m; stems robust, pubescent towards apex; leaves ovate-elliptic, 11–12 × 4–4.7cm, base acute to obtuse, apex shortly acuminate, margin subcrenulate, pubescent along midrib, sparsely so on lower surface and margins; petiole 0.7cm, pubescent; inflorescence a dense terminal raceme to 15cm, pubescent throughout; bracts oblanceolate, to 0.6cm long; calyx lobes lanceolate, 0.7cm long; corolla bilabiate, 4.8cm long, flattened dorsi-ventrally, lower lip 2cm broad, white to blue, darkest on lower lip, with dark blue veins. Forest and farmbush; 1000–1400m.
Distr.: Cameroon & Bioko. (W Cameroon Uplands).
IUCN: NT
Edib village: <u>Cheek 9170</u> fl., 2/1998; **Kodmin:** <u>Cheek 10273</u> fl., 12/1999; <u>Etuge 4040</u> fl., 1/1998; **Nyasoso:** <u>Groves 93</u> fl., 2/1995.

Note: species has a very restricted distribution and is rare over much of this but appears numerous on parts of Mt Cameroon.

Justicia striata (Klotzsch) Bullock subsp. *occidentalis* J.K.Morton
Symb. Bot. Ups. 29: 110 (1989).
Suberect herb to 50cm; stems puberulent; leaves elliptic, 4–4.5 × 1.8–2.5cm, base acute, apex barely acuminate, obtuse, margin subcrenulate, finely pilose on both surfaces, raphides numerous on upper surface; petiole 1–1.5cm, pubescent; flowers in subsessile axillary clusters of 2–3, puberulent throughout; bracts oblanceolate, 0.8 × 0.2cm; calyx lobes lanceolate, 0.4cm; corolla bilabiate, 0.7–1.2cm long, lower lip trilobed, white with purple markings. Farmbush, villages; 1700m.
Distr.: Ghana, Cameroon & CAR. (Upper and Lower Guinea).
IUCN: LC
Bangem: <u>Manning 399</u> fl., 10/1986.

Justicia tenella (Nees) T.Anderson
Creeping or subprostrate herb to 20cm; stems pubescent, hairs white; leaves ovate, c. 2 × 1.3cm, base obtuse, apex acute to obtuse, margin subentire, glabrescent; petiole c.1.3cm; flowering spikes 1.8cm long, axillary in clusters of 2–3, 4-faceted; peduncle to 1.7cm; bracts suborbicular, 2.5mm diameter., pale green, overlapping; corolla bilabiate, lower lip purple with white margin; fruit c. 3mm long, glabrous. Farmbush, grassland and forest margins; 650–1500m.
Distr.: tropical Africa & Madagascar.
IUCN: LC
Kodmin: <u>Cheek 8970</u> fl., 1/1998; <u>Ghogue 75</u> fl., 2/1998; **Ngomboaku:** <u>Cheek 10284</u> fl., 12/1999; <u>Ghogue 482</u> 12/1999; **Nyasoso:** <u>Etuge 1605</u> 1/1996; <u>Sebsebe 5006</u> 10/1995; <u>Thomas 5030</u> 11/1985.
Uses: MEDICINES (Etuge 1605).

Justicia tristis (Nees) T.Anderson
Bot. J. Linn. Soc. 7: 38 (1864).
Syn.: *Adhatoda tristis* Nees
Herb or shrub to 2m; stems robust, glabrous; leaves obovate, 24–30 × 9.5–12.5cm, base acute-cuneate, apex shortly acuminate, margin subentire, midrib and secondary veins ± pubescent; flowering spikes to 28cm, narrow (c. 0.8cm), many-flowered, sparsely puberulent, drying black; bracts triangular, 0.4cm; calyx lobes lanceolate, to 0.6cm; corolla bilabiate, c. 1cm long, white-green, purple-tinged; fruit 1.7cm long, densely puberulent, style persistent. Forest; 0–1000m.
Distr.: SE Nigeria, Bioko & Cameroon. (Lower Guinea).
IUCN: NT
Basengi Village: <u>Sidwell 422</u> 10/1995; **Kupe Village:** <u>Cheek 7157</u> fl., 1/1995; <u>7651</u> 11/1995; <u>Ensermu 3546</u> 11/1995; <u>Schoenenberger 43</u> 11/1995; **Mungo River F.R.:** <u>Leeuwenberg 10603</u> fl., 11/1972; <u>Onana 933</u> 11/1999; **Ngomboaku:** <u>Cheek 10299</u> 12/1999.

Justicia unyorensis S.Moore
Journ. Bot. 49: 308 (1911).
Straggling herb to 60cm; stems pubescent; leaves ovate, 2.6–5.5 × 0.6–1.9cm, base attenuate, apex acuminate or acute, margin subentire, surfaces pilose; petiole to 1.3cm, pilose; inflorescence a subsessile axillary cluster, 2–3-flowered, pubescent throughout; bracts linear, to 0.9cm; calyx lobes lanceolate, c. 2.5mm; corolla bilabiate, 0.7cm long, pink with

purple and white markings on lower lip; fruit 0.6cm long, glabrous. Montane grassland; 2000m.
Distr.: Nigeria to Tanzania. (Afromontane).
IUCN: LC
Mwaneneguba: Cheek 9458 fl., fr., 10/1998.

Justicia sp. 1
Suberect or decumbent herb to 15cm; stems puberulent; leaves elliptic, 4.5–5 × 1.8–2.8cm, base acute, apex short-acuminate, margin subcrenulate, raphides numerous on both surfaces; petiole to 0.6cm, puberulent; flowers axillary, solitary or in cymes of 2–5; pedicel length variable, 0.2-1.7cm; inflorescence puberulent throughout; bracteoles lanceolate, c. 1.5mm; calyx lobes subulate, 0.4–0.5cm long; corolla bilabiate, c. 0.5cm long, white, lower lip 3-lobed, upper lip short, hooded. Swamp with partially cleared forest; 700m.
Distr.: Bakossi. (Narrow Endemic).
Ngomboaku: Cheek 10364 12/1999.
Note: *sp. nov.*, more material required.

Lankesteria barteri Hook.f.
Erect herb to 2.5m; stems robust, pubescent; leaves obovate-oblanceolate, 14–17 × 4.3–6.7cm, base cuneate, apex shortly acuminate, margin shallowly crenulate, upper surface ± pilose, raphides numerous; petiole to 2cm; inflorescence spikes terminal, 5(–7)cm, pubescent throughout; bracts elliptic-obovate, 1.3 × 0.5cm, apex acuminate-apiculate, ciliate; calyx lobes subulate, c. 1cm long, emerging from between the bracts; corolla tube to 2cm, narrow, lobes spreading, white with yellow throat. Forest; 250–1100m.
Distr.: Nigeria to Congo (K). (Lower Guinea & Congolian).
IUCN: NT
Ikiliwindi: Nemba 675 12/1987; **Konye:** Thomas 5150 fl., 11/1985; **Mejelet-Ehumseh:** Etuge 472 fl., 1/1987.

Lankesteria brevior C.B.Clarke
Erect herb to 1.2m; stems robust, densely pubescent; leaves obovate, 14.5–17 × 5.8–7cm, base cuneate, apex acuminate, acumen prominent, margin shallowly crenate, raphides numerous on upper surface; petiole 1.8cm, puberulent; inflorescence a terminal spike to 14cm; peduncle 0.3cm; bracts ovate, tube 2 × 1cm, pilose, green-purple, overlapping; calyx lobes to 0.5cm, included within bracts; corolla 1.5cm long, neck narrow, white with yellow throat. Forest; 300m.
Distr.: Sierra Leone to Cameroon. (Upper & Lower Guinea).
IUCN: LC
Ekona Mombo: Etuge 423 fl., 12/1986; **Mungo River F.R.:** Cheek 10120 11/1999; **Tombel:** Etuge 4681 fl., 12/1999.

Lankesteria elegans (P.Beauv.) T.Anderson
Erect herb to 2.5cm; similar to *L. brevior*, but bracts glabrous except for ciliate margin; corolla 3–3.7cm long, orange. Forest; 200–1100m.
Distr.: Sierra Leone to Congo (K), Uganda, Sudan & SW Ethiopia. (Guineo-Congolian).
IUCN: LC
Mejelet-Ehumseh: Etuge 466 fl., 1/1987; **Southern Bakossi F.R.:** Thomas 5558 fl., 2/1986.

Mendoncia gilgiana (Lindau) Benoist
Slender climber to 4m; stems twining, pubescent; leaves ovate-elliptic, 6.8–9.5 × 2.7–4cm, base rounded, apex acuminate-apiculate, margin subcrenulate, young leaves pubescent, becoming glabrescent; petiole 1.8–4.5cm, pubescent; inflorescence an axillary fascicle, 1–3 flowers per node; pedicel to 2cm, pubescent; flowers enclosed within paired ovate bracteoles 1.6 × 1cm, apex acuminate, densely

pubescent, margins pink; corolla tube exserted, 1.7cm long, white; stamens 4; fruit a green-black ellipsoid drupe, c. 1cm long, enclosed within the bracts. Forest, farmbush; 800–1150m.
Distr.: Cameroon to Congo (K), S Sudan to NW Tanzania. (Lower Guinea & Congolian).
IUCN: LC
Bangem: Thomas 5280 1/1986; **Kupe Village:** Cheek 7092 fr., 1/1995; Schoenenberger 51 fl., 11/1995; **Mt Kupe:** Schoenenberger 1 fr., 11/1995; **Mwambong:** Cheek 9366 10/1998; **Nyasoso:** Cable 2882 6/1996; Cheek 5660 fr., 12/1993; 7487 10/1995; Etuge 2515 7/1996; Sidwell 384 10/1995.

Mendoncia phytocrenoides (Gilg) Benoist
Benoist, Not. Syst. 2: 143 (1944).
Robust woody climber; stems twining, densely pubescent; leaves ovate, 10–14 × 6.5–8.5cm, base rounded to subcordate, apex obtuse to shortly acuminate, apiculate, margin subentire, veins on under surface puberulent; petiole 2.5cm, pubescent; inflorescence a 3–4-flowered axillary fascicle; pedicels to 3cm, pubescent; paired bracteoles ovate-deltoid, 3 × 2cm, densely puberulent, pale green with prominent reddish nerves; corolla white-yellow; fruits ellipsoid drupes c. 2cm long, enclosed within the bracts, puberulent, green-black; style persistent. Forest, bushland; 700–1400m.
Distr.: S Nigeria to Gabon. (Lower Guinea).
IUCN: NT
Kodmin: Plot B126 1/1998; **Kupe Village:** Cable 3884 7/1996; Etuge 2769 7/1996.

Mimulopsis solmsii Schweinf.
Mass-flowering herb to 1(–4)m; stems 4-angled, glabrescent; leaves ovate, 9–9.5 × 5–6cm, base shallowly cordate, apex acuminate, margin dentate, glabrescent; petiole to 5.3cm, puberulent; inflorescence a terminal cyme; peduncles to 1.8cm, pubescent; bracts elliptic, 1.3 × 0.3cm; bracteoles linear, pubescent; calyx lobes linear, to 1.3cm, glandular pubescent; corolla tube funnel-shaped, to 2.5cm long, pale pink. Forest, forest edge; 1050–2100m.
Distr.: Bioko & Cameroon to E Africa. (Afromontane).
IUCN: LC
Kodmin: Etuge 4413 fl., 11/1998; Plot B4 1/1998; B279 10/1998; **Mwambong:** Etuge 4325 fl., 10/1998; **Mwaneneguba:** Leeuwenberg 8926 fl., 12/1971.

Monechma depauperatum (T.Anderson) C.B.Clarke
Erect herb to 2m; stems woody towards base, densely pubescent; leaves oblong-lanceolate, 12.5–15 × 3.8–5cm, base acute to obtuse, apex long-acuminate, margin subcrenulate, surfaces pilose; petiole c. 2cm, pilose; inflorescence a terminal or axillary spike to 10.5cm; bracts dense, ovate, 1.4 × 0.4cm, acuminate, overlapping, green with pale margins; calyx included, lanceolate, to 0.6cm long; corolla bilabiate, c. 1cm long, white; stamens 2; fruit c. 1.1cm long, puberulent at apex. Forest margins, wet grassland; 900–1500m.
Distr.: Senegal to Angola. (Guineo-Congolian).
IUCN: LC
Enyandong: Etuge 4446r 11/2001; **Kodmin:** Etuge 4018 fl., fr., 1/1998; **Mwambong:** Mackinder 151 1/1998; **Mwaneneguba:** Etuge 4386 fl., 10/1998.

Nelsonia canescens (Lam.) Spreng.
Creeping or straggling herb; stems decumbent, densely villose; leaves ovate, 2.9–3.5 × 1.9–2.8cm, base truncate to

obtuse, apex acute, margin subcrenulate, surfaces densely pubescent, particularly when young; petiole to 1.8cm, villose; inflorescence a cylindrical terminal spike to 4cm long, densely white-hairy throughout; bracts ovate, 0.5 × 0.2cm, acuminate, overlapping; calyx included; corolla pink-purple, c. 2mm, upper 3 lobes rounded to emarginate, not reflexed. Farmbush, disturbed environments; 500–1450m.
Distr.: palaeotropical, introduced in tropical America, the Pacific & Australia.
IUCN: LC
Kodmin: Cheek 9022 fl., 1/1998; Etuge 4017 fl., 1/1998; **Kupe Village:** Cheek 7914a 11/1995; **Nyasoso:** Etuge 1600 fl., 1/1996.
Local name: Abwi Cah (Cheek 9022). **Uses:** FOOD – inflorescences (Cheek 9022); MEDICINES (Etuge 1600).

Nelsonia smithii Oersted

Kew Bull. 33: 401 (1979).
Creeping herb; similar to *N. canescens*, but less densely hairy throughout, hairs on leaves sparse and stout, clearly multicellular; petiole longer, to 3cm; corolla tube white with pink lobes, not uniformly pink-purple, upper 3 lobes acute and reflexed, not rounded and erect. Swamps, riverine habitats; 250–1200m.
Distr.: tropical Africa.
IUCN: LC
Kupe Village: Lane 342 fl., 1/1995; **Mwanenguba:** Leeuwenberg 8597 fl., 10/1971; **Mungo River F.R.:** Cheek 9356 10/1998; **Ngomboaku:** Cheek 10350 fl., 12/1999; Ghogue 446 fl., 12/1999.
Note: recognised by Enayet Hossain as a variety of *N. canescens* (Willdenowia 14: 403, 1985).

Phaulopsis angolana S.Moore

Symb. Bot. Ups. 31: 99 (1995).
Syn.: *Phaulopsis silvestris* (Lindau) Lindau
Erect herb to 50cm; stems puberulent; leaves ovate-rhombic, 5.5–6 × 2.5–3cm, pairs unequal, base attenuate, apex shortly acuminate, margin subcrenulate, midrib pubescent below, raphides numerous on upper surface; petiole c. 1.5cm; inflorescence a dense ellipsoid axillary spike to 3.5cm; bracts broadly ovate, 0.7cm long, apex acuminate, densely ciliate, green turning purplish, overlapping; calyx lobes subulate, 0.35cm, pilose; corolla bilabiate, 0.5cm long, white; fruits to 6mm, glabrous. Forest edge, farmbush; 150–1450m.
Distr.: Sierra Leone to Angola & E Africa. (Tropical Africa).
IUCN: LC
Kodmin: Cheek 9049 fl., 1/1998; **Kupe Village:** Ensermu 3556 11/1995; **Loum F.R.:** Leeuwenberg 10349 fl., 9/1972; **Mt Kupe:** Schoenenberger 58 fl., 11/1995; **Mwambong:** Cheek 9367 fl., 10/1998; **Ngomboaku:** Etuge 4664 fl., 12/1999; **Nyasoso:** Cheek 5629 fl., 12/1993.
Note: *P. ciliata* (Wild.) Hepper is recorded in Kumba District and therefore likely in the checklist area.
Local name: Mbi-Mbura = Goat Breast. **Uses:** use unknown (Epie Ngome in Cheek 9367).

Phaulopsis imbricata (Forssk.) Sweet

Similar to *P. angolana*, but inflorescence to only 1.7cm long, subspherical not ellipsoid; bracts suborbicular, to 1cm diameter. Forest margins, secondary vegetation; 300–1300m.
Distr.: tropical Africa.
IUCN: LC
Loum F.R.: Onana 993 fl., 12/1999; **Mwanenguba:** Leeuwenberg 8983 fl., 12/1971.
Note: FWTA indicates that this species has larger leaves, to 8cm, but in our specimens the maximum leaf length recorded is 5.5cm.

Pseuderanthemum dispermum Milne-Redh.

Herb or shrub, 1(–3)m; stems robust, becoming woody towards base, pubescent towards apex; leaves elliptic, 9.3–11.5 × 4–4.5cm, base attenuate, apex acuminate, margin subcrenulate, upper surface and lower midrib sparsely pilose; petiole to 1.5cm; inflorescence a narrow terminal spike to 23cm; peduncle pubescent; flowers paired; bracts caducous; calyx lobes linear, to 0.7cm; corolla tube to 3.2cm, lobes spreading, 0.6cm wide, brick red with darker centre; fruit 1.4cm long, glabrescent, 2-seeded. Forest; 1000m.
Distr.: SE Nigeria & Cameroon. (W Cameroon Uplands).
IUCN: VU
Bangem: Thomas 5269 1/1986.
Note: specimen not seen at Kew; description based on other W Cameroon material.

Pseuderanthemum ludovicianum (Büttner) Lindau

Erect herb or shrub to 2m; stems robust, glabrous; leaves elliptic, 13.5–28 × 4.2–15cm, base acute, apex shortly acuminate, margin subcrenulate, raphides numerous on both surfaces; petiole to 2.2cm; inflorescence a terminal thyrse to 14cm, with peduncle to 9cm; flower-clusters closely spaced; bracts inconspicuous; calyx lobes lanceolate, c. 3mm long; corolla tube 3.5 × 0.15cm, bilabiate, upper lip 2-lobed, reflexed, lower lip 3-lobed, white speckled purple, puberulent; stamens 2, exserted; fruit to 3cm long. Forest; 300–1150m.
Distr.: Liberia to Cameroon & S Sudan to NW Tanzania. (Guineo-Congolian).
IUCN: LC
Kodmin: Cheek 8865 fl., 1/1998; Plot B131 fl., 1/1998; **Konye:** Thomas 5164 11/1985; **Kupe Village:** Cable 723 1/1995; 765 fl., 1/1995; Cheek 6061 fl., 1/1995; Ensermu 3553 fl., fr., 11/1995; Lane 281 fl., 1/1995; Ryan 277 fl., 5/1996; **Ndum:** Cable 877 fl., 1/1995; 918 fl., 2/1995; **Ngomboaku:** Cheek 10342 fl., 12/1999; **Ngusi:** Schlechter 12905 fl., 1/1900; **Nyasoso:** Cable 665 fl., 12/1993; Cheek 6007 fl., 1/1995; Lane 211 fl., 11/1994; Sidwell 383 fl., 10/1995; Thomas 3050 fl., 2/1984; 5039 11/1985.

Pseuderanthemum tunicatum (Afzel.) Milne-Redh.

Herb to 30cm; stems glabrous; leaves oblanceolate, 15–17 × 5.5–6cm, base cuneate, apex shortly acuminate, margin irregularly crenulate, raphides numerous; petiole to 2cm; inflorescence a slender (sub)terminal thyrse to 23.5cm long; peduncle to 9cm; clusters widely spaced to 5cm; calyx lobes linear-lanceolate, to 0.35cm; corolla 5-lobed, bluish white; fruit to 1.8cm, glabrous. Forest; 850–970m.
Distr.: W Africa to Congo (K). (Guineo-Congolian).
IUCN: LC
Kupe Village: Cable 760 fl., 1/1995; Ryan 305 fl., fr., 5/1996; **Mombo:** Olorunfemi in FHI 30574 fl., fr., 5/1951; **Nyasoso:** Cable 860 fl., 1/1995.

Rhinacanthus virens (Nees) Milne-Redh.

Herb or shrub to 1.25m; stems puberulent; leaves ovate-elliptic, 5.5–17 × 2.5–6cm, base acute, apex acuminate, margin subentire, glabrous or sparsely pilose; petiole to 3.3cm; inflorescence a lax, few-flowered panicle to 27cm long, usually shorter; peduncles fine; flowers subsessile; bracts linear-lanceolate; calyx lobes lanceolate, 4mm, glandular-puberulent; corolla to 1.7cm, tubular with short lobes, white; fruit 1.1cm long, puberulent. Forest; 800m.
Distr.: Guinea (C) to Congo (K). (Guineo-Congolian).
IUCN: LC

Bangem: Thomas 5369 1/1986.
Note: specimen not held at Kew; description based upon other W Cameroon material.

Rungia buettneri Lindau

Bot. Jahrb. 20: 46 (1894).
Erect herb to 30cm; stems slender, puberulent; leaves ovate, 6.5–8 × 2.5–3cm, base obtuse, apex acuminate, margin subcrenate, veins and margins sparsely pubescent, raphides numerous on upper surface; petiole to 2cm, puberulent; spikes terminal or axillary, to 2.7cm; bracts ovate, 0.8 × 0.35cm, apex acuminate, puberulent towards base, hyaline margin < 1mm; calyx included; corolla bilabiate, white; stamens 2. Disturbed areas; 700–1200m.
Distr.: Cameroon to Congo (K), Uganda & Sudan. (Lower Guinea & Congolian).
IUCN: LC
Mwanenguba: Leeuwenberg 8849 fl., 12/1971; **Ngomboaku:** Cheek 10348 fl., 12/1999.

Sanchezia nobilis Hook.f.

Bot. Mag. 92: t. 5594 (1866).
Robust herb or subshrub to 3m; stems quadrangular, glabrous; leaves oblong-obovate, 15–20 × 4.5–5.5cm, apex acuminate, base attenuate into short winged petiole, margin shallowly dentate, midrib and secondary nerves in cultivars prominent, yellow, surfaces glabrous, raphides numerous; inflorescence an interrupted spike 10–20cm long; flowers in clusters of 3, each cluster subtended by a red ovate bract c. 2 × 1.5cm, glabrous; corolla yellow-orange, 3–3.5cm long; stamens and style long-exserted; fruit to 2.5cm long, glabrous. Cultivated; 420–950m.
Distr.: S America, widely cultivated elsewhere. (Pantropical).
IUCN: LC
Mwanenguba: Leeuwenberg 9860 fl., fr., 5/1972; **Nyandong:** Cheek 11299 fl., 3/2003.

Schaueria populifolia C.B.Clarke

Herb to 1m; stems glabrescent; leaves ovate, 11.5–12.5 × 5.8–7.5cm, base cordate, apex shortly acuminate, margin subentire, raphides numerous on both surfaces; petiole to 3.7cm; inflorescence a dense terminal spike to 8cm long; bracts linear, 1.2 × 0.1cm, minutely glandular-hairy towards apex, raphides numerous; calyx subulate, 0.6cm long; corolla bilabiate, 1.7cm long, pink-purple; fruit 1.6cm long, glabrous. Forest; 150–900m.
Distr.: SE Nigeria, Bioko & Cameroon. (Lower Guinea).
IUCN: NT
Kupe Village: Cheek 7654 11/1995; 7771 11/1995; Ensermu 3533 11/1995; **Mungo River F.R.:** Cheek 10153 fl., 11/1999; **Ngomboaku:** Mackinder 292 fl., 12/1999.
Note: to be reassigned to *Justicia* by D.Champluvier.

Sclerochiton preussii (Lindau) C.B.Clarke

Scrambling shrub to 2m tall; stems glabrous; leaves elliptic, 6.5–8.3 × 14.5–23cm, base attenuate, apex shortly acuminate, margin subentire, lamina glabrous, papery; petiole to 1.8cm; inflorescence a short terminal spike of 2–3 flowers; bracts and calyx lobes stiff, broadly lanceolate, to 4.5cm long, striate, brown-green, shiny; corolla fused into a 5-lobed deflexed lip, orange-pink with red markings; stamens 4. Forest; 1200–1400m.
Distr.: SE Nigeria & SW Cameroon. (W Cameroon Uplands).
IUCN: EN
Kupe Village: Cheek 7846 11/1995; **Nyasoso:** Etuge 2477 fl., 6/1996; Wheatley 427 fl., 7/1992.

Uses: MATERIALS – traps/snares – stems cut for trap 'springing sticks' (fide Cheek).

Staurogyne bicolor (Mildbr.) Champl.

Bull. Jard. Bot. Nat. Belg. 61: 103 (1991).
Stoloniferous herb to 20cm tall; stems succulent, tomentose, smelling of almonds when crushed; leaves elliptic, 4.4–5.7 × 8.7–9.3cm, base and apex obtuse(-acute), margin subcrenulate, veins on under-surface tomentose; petiole to 3cm, tomentose; inflorescence a globose terminal spike to 3cm diameter.; bracts imbricate, elliptic, ciliate, green or red-tinged; calyx included within bracts; corolla bilabiate, to 2cm long, upper lip 3-lobed, purple, lower lip 2-lobed, white, throat with purple striations; stamens 4. Forest, often near water; 700–1370m.
Distr.: Cameroon. (Lower Guinea).
IUCN: VU
Bangem: Thomas 5383 1/1986; **Kodmin:** Biye 49 fl., 1/1998; Cheek 8982 1/1998; Gosline 67 fl., 1/1998; **Kupe Village:** Cable 3828 7/1996; Cheek 7678 11/1995; **Ngomboaku:** Cheek 10363 fl., 12/1999; Ghogue 437 fl., 12/1999; **Nyale:** Etuge 4210 fl., 2/1998; 4459 fl., 11/1998; **Nyasoso:** Cable 654 fl., 12/1993.

Stenandrium guineense (Nees) Vollesen

Kew Bull. 47: 182 (1992).
Syn.: *Crossandra guineensis* Nees
Syn.: *Stenandriopsis guineensis* (Nees) Benoist
Stoloniferous herb to 50cm; stems tomentose; leaves in rosettes, elliptic, 9–14.5 × 4.5–5.7cm, base abruptly rounded, apex obtuse, margin subcrenulate; lamina pale green or purplish below, dark green above, variegated when young, veins pubescent on lower surface; petiole to 2.5cm, tomentose; inflorescence a narrow terminal spike to 13cm long; bracts oblanceolate-rhombic, 1.7 × 0.6cm, apex acute, uppermost third with a regularly toothed margin, striate, pink-tinged; calyx subulate, 0.7cm long; corolla with a linear tube to 1.8cm, lobes spreading, pale purple with darker markings; fruit 1.1cm long, glabrous. Forest; 350–1450m.
Distr.: Guinea (C) to Angola (Cabinda) & Uganda to Sudan. (Guineo-Congolian).
IUCN: LC
Kodmin: Cheek 8965 1/1998; **Kupe Village:** Cheek 6078 fr., 1/1995; 7826 11/1995; 7827 11/1995; Etuge 2625 fl., 7/1996; 2874 fl., 7/1996; **Manehas F.R.:** Cheek 9403 fl., 10/1998; **Mungo River F.R.:** Onana 974 12/1999; Tadjouteu 317 12/1999; **Nyasoso:** Cable 3287 6/1996; 3296 6/1996; 3494 fl., 7/1996; 3550 fl., 7/1996; 81 fl., 8/1992; 82 fl., 8/1992; Etuge 2544 7/1996; Lane 63 fl., 8/1992; Sebsebe 5031 10/1995.

Stenandrium talbotii (S.Moore) Vollesen

Kew Bull. 47: 186 (1992).
Syn.: *Crossandra talbotii* S.Moore
Syn.: *Stenandriopsis talbotii* (S.Moore) Heine
Similar to *S. guineense*, but leaves obovate, 9–9.5 × 6–6.5cm, upper surface sparsely pubescent; bracts of inflorescence oblong, 2.2 × 0.5cm, apex truncate, margin toothed in upper two thirds but irregular and concentrated towards the apex; calyx lobes lanceolate, to 1cm long. Forest; 1050m.
Distr.: Nigeria & Bioko to Gabon. (Lower Guinea).
IUCN: NT
Nyasoso: Balding 16 fl., 7/1993.

Thomandersia laurifolia (Benth.) Baill.

Shrub to 5m; stems glabrous; leaves coriaceous, elliptic, 10–16.5 × 5.5–6cm, pairs uneven, base acute or obtuse, apex acuminate, acumen to 1.5cm, margin entire, glabrous; petiole

to 4.5cm; inflorescence a slender, many-flowered raceme, 10–30cm; bracts < 1mm; calyx lobes triangular, c. 1mm; corolla 0.8–1.2cm long, bilabiate with a prominent 3-lobed lower lip c. 0.5cm long, white to yellow-green; stamens 4; fruit ovoid, c. 1.5cm long. Secondary forest; 150–350m.
Distr.: Cameroon to Congo (K). (Lower Guinea & Congolian).
IUCN: LC
Bakolle Bakossi: Etuge 145 5/1986; **Bolo:** Nemba 4 4/1986; **Etam:** Etuge 83 3/1986; **Ikiliwindi:** Etuge 503 3/1987.

Thunbergia alata Boj. ex Sims
Herbaceous climber; stems twining, pubescent; leaves sagittate, c. 5 × 2.2cm, apex acute, margin irregularly undulate, pubescent; petiole to 2.5cm, winged, pubescent; flowers solitary, axillary; pedicels to 4.8cm, pubescent; subtending bracts paired, ovate, 1.5 × 0.7cm, subcordate, pubescent; calyx included; corolla trumpet-shaped, 4cm long, lobes rounded, 1.5cm broad, orange-yellow with darker throat; stamens 4; fruit base orbicular, 0.9cm diameter., linear above, 1.2 × 0.3cm, puberulent. Farmbush; 600–800m.
Distr.: native to E & S Africa; cultivated and naturalised elsewhere. (Pantropical).
IUCN: LC
Nyasoso: Etuge 1647 1/1996; Pollard 577 fl., 9/2001.

Thunbergia fasciculata Lindau
Herbaceous climber to 8m; stems twining, puberulent at nodes; leaves ovate, 8–9.5 × 5.5–7cm, base cordate, apex acuminate, margin irregularly undulate with apiculate vein apices, puberulent on both surfaces; petiole to 9cm; axillary inflorescence 2–3-flowered; peduncle to 12.5cm, puberulent; sessile bracts paired, leafy, cordiform, 5.5 × 4cm, margin dentate; bracteoles elliptic, 2.5cm long, apex mucronate; corolla trumpet-shaped, to 5cm long, purple with yellow throat, hidden by bracts until fully out. Forest; 1200m.
Distr.: Nigeria to Congo (K), Uganda, Sudan & SW Ethiopia. (Guineo-Congolian).
IUCN: LC
Mwambong: Cheek 9369 fl., 10/1998.

Thunbergia vogeliana Benth.
Shrub to 3m; stems glabrous; leaves elliptic, 17–20 × 7–8.5cm, base acute, apex acuminate, acumen to 2cm, margin shallowly dentate, glabrous; petiole c.1cm; flowers solitary, axillary; pedicel to 3cm; paired bracts ovate, 2.5 × 1.5cm, glabrous; calyx lobes lanceolate, to 1.2cm long; corolla trumpet-shaped, 7cm long, rounded lobes 3cm wide, purple with white, yellow-spotted, throat; fruit to 3cm long, swollen at base. Forest, forest margins, sometimes cultivated; 350–1600m.
Distr.: Ghana to Cameroon, Uganda & W Tanzania. (Guineo-Congolian).
IUCN: LC
Kodmin: Cheek 10274 fl., 12/1999; **Kupe Village:** Cable 2636 fl., 5/1996; Cheek 7773 11/1995; 8380 5/1996; Etuge 2045 fl., 5/1996; 2614 fl., 7/1996; Ryan 268 fl., 5/1996; 367 fl., 5/1996; **Mungo River F.R.:** Onana 965 11/1999; **Nyasoso:** Cable 2834 6/1996; 41 fl., 8/1992; Cheek 7459 10/1995; Etuge 1339 10/1995; 2114 fl., 6/1996; 2369 fl., 6/1996; Sidwell 320 fl., 10/1995; Sunderland 1513 fl., 7/1992; 1529 fl., 7/1992; Wheatley 424 fl., 7/1992.
Uses: ENVIRONMENTAL USES – ornamentals (Etuge 2045).

Whitfieldia elongata (P.Beauv.) De Wild. & T.Durand
Woody climber or shrub to 3m; stems glabrous; leaves elliptic-ovate, 12–13 × 4.5–5.5cm, base shortly attenuate, apex acuminate, margin subcrenulate, glabrous; petiole to 2cm; inflorescence a lax terminal raceme, occasionally branched, to 17.5 × 13cm; bracts narrowly ovate, to 1.4cm long; bracteoles elliptic, 1.5 × 0.7cm, green-white; calyx lobes linear, 2.3 × 0.35cm, glandular-puberulent, cream; corolla tube linear, 3.3 × 0.15cm, lobes elliptic, 2.3cm long, white. Forest, forest margins; 800–1200m.
Distr.: Nigeria to Angola, Sudan, coastal Kenya, Tanzania & Zanzibar. (Tropical Africa).
IUCN: LC
Mt Kupe: Thomas 5105 12/1985; **Nyasoso:** Cheek 7910 fl., 11/1995.

Whitfieldia preussii (Lindau) C.B.Clarke
Shrub to 1m; stems glabrous; leaves elliptic, 14.5–21 × 6.3–8cm, base attenuate, apex acuminate, margin subentire, glabrous, raphides present on upper surface; petiole to 4.5cm; inflorescence a lax terminal raceme, rarely branched; peduncle to 23cm; bracts linear, 3mm; bracteoles ovate, 0.7 × 0.35cm; calyx lobes oblong, 1 × 0.4cm, apex rounded; corolla bilabiate, 2.3cm long, cream-yellow with brown markings; stamens 4. Forest; 200m.
Distr.: Cameroon & Bioko. (Lower Guinea).
IUCN: VU
Mungo River F.R.: Etuge 4321 10/1998.

Genus nov.
Herb to 80cm; stems with 2 lines of pubescence; leaves elliptic, c. 14.5 × 6.5cm, base attenuate, apex acuminate, margin subcrenate, raphides numerous on both surfaces; inflorescence a terminal spike to 7cm long; bracts overlapping, elliptic, c. 1 × 0.7cm, ciliate; calyx included within bracts; corolla 4-lobed, c. 0.7cm long, yellow-green. Forest; 150–200m.
Distr.: W Cameroon. (Narrow Endemic).
Loum F.R.: Onana 987 12/1999; **Mungo River F.R.:** Cheek 10152 11/1999.
Note: fruiting material required.

ALANGIACEAE

M. Cheek

Fl. Cameroun 10 (1970).

Alangium chinense (Lour.) Harms
Tree 5–18m, glabrescent; leaves alternate, ovate, c. 10 × 6cm, acuminate, obliquely rounded, entire, digitately 5-nerved, scalariform; inflorescences cymose, axillary, c. 20-flowered, 4cm; peduncle 1.5cm; flowers orange, 1.1cm; fruit ellipsoid, 1cm, fleshy. Forest margins and farmbush; 900–2000m.
Distr.: palaeotropical, excluding upper Guinea.
IUCN: LC
Kodmin: Cheek 9596 11/1998; **Mwanenguba:** Leeuwenberg 9959 fl., 6/1972; **Nyasoso:** Cable 2813 fr., 6/1996; Elad 117 2/1995; Etuge 1812 fl., 3/1996; 2375 fr., 6/1996.
Local name: Achee. **Uses:** FUELS – fuelwood (Max Ebong in Cheek 9596).

AMARANTHACEAE

C.C. Townsend & I. Darbyshire

Fl. Cameroun 17 (1974).

Achyranthes aspera L. var. *aspera*

Branching herb to 1m tall; stems sparsely pubescent or glabrescent; leaves opposite, elliptic-ovate, 10–19 × 4–9.4cm, apex shortly acuminate, base attenuate, margin subentire, sparsely pilose, more so in young leaves; petiole 1–1.5cm; inflorescence a narrow terminal spike, 7–36cm long; flowers solitary, subsessile, dense in upper section, more widely spaced below; peduncle pubescent; flowers becoming deflexed in fruit; bracteoles 0.4–0.5cm, spinulose with membranous wings to half the bracteole length with attenuate apex; tepals lanceolate, 0.6–0.7cm, green; stamens and ovary purple-red. Farmbush, forest edges; 800–1000m.
Distr.: pantropical.
IUCN: LC
Kupe Village: Cheek 7151 fr., 1/1995; **Ndum:** Williams 136 fl., 1/1995; **Nyasoso:** Cable 703 fl., 1/1995; Cheek 6023 fr., 1/1995.

Achyranthes aspera L. var. *pubescens* (Moq.) C.C.Towns.

F.T.E.A. Amaranthaceae: 102 (1985).
Herb to 1m, resembling var. *aspera*, but leaves elliptic and more prominently acuminate, plant more densely pubescent throughout, particularly in young shoots and leaves. Farmbush, forest edges; 900–2000m.
Distr.: pantropical.
IUCN: LC
Mwanenguba: Cheek 9411 10/1998; **Nyasoso:** Cheek 5636 fl., 12/1993.

Achyranthes aspera L. var. *sicula* L.

Fl. Cameroun 17: 32 (1974).
Herb resembling var. *aspera*, but generally less robust; leaves ovate, c. 5.5 × 2.7cm, apex acuminate, base acute or shortly attenuate; spikes shorter, max. 10cm in our specimens, with smaller flowers; tepals 3–4mm; bracteole wings truncate at apex. Farmbush; 650–1450m.
Distr.: pantropical.
IUCN: LC
Essossong: Leeuwenberg 9254 fl., 1/1972; **Kodmin:** Cheek 9020 1/1998; Ghogue 68 1/1998; **Nyasoso:** Etuge 1597 1/1996.
Local name: Kore Koreh. **Uses:** none (Cheek 9020).

Achyranthes talbotii Hutch. & Dalziel

Erect rheophytic herb to 50cm; stems woody at base, glabrous, purple-red; leaves linear, 3.2–4.5 × 0.7–0.9cm, apex acute, base cuneate, margin irregularly undulate, sparsely puberulous; inflorescence similar to other *Achyranthes* spp., but spikes generally 5–8cm; flowers widely-spaced in lower section, tepals c. 3mm. Rocks in or by streams; 400–750m.
Distr.: SE Nigeria & SW Cameroon. (Lower Guinea).
IUCN: VU
Baseng: Cheek 10401 12/1999; **Nyandong:** Cheek 11466 3/2003.

Alternanthera brasiliana (L.) Kuntze

Kuntze: Rev. Gen.: 537 (1891).
Scrambling or erect herb to 1.5m; stems red-brown, white-pubescent, swollen at nodes; leaves opposite, widely-spaced (internodes 5.5–14cm), ovate, 3.3–5.7 × 0.8–1.4cm, apex and base acute, margin subentire, red-green, pubescent; petiole to 0.5cm; inflorescence axillary on lateral shoots, a dense white cylindrical spike c. 1.5 × 1cm; bracts ovate, 3.5 × 1.5mm, pubescent; peduncle 1–7cm, white-pubescent. Farmbush, villages; 420m.
Distr.: C & S America; introduced to Cameroon.
IUCN: LC
Nlonako: Etuge 1776 3/1996; **Nyandong:** Darbyshire 4 fl., 3/2003.
Note: Etuge 1776 represents the first collection of this species for Cameroon and possibly for W Africa.
Uses: ENVIRONMENTAL USES – boundaries – used for hedging (Etuge 1776).

Alternanthera nodiflora R.Br.

Decumbent herb; rootstock woody; stems angled, sparsely pubescent; leaves opposite, oblanceolate to elliptic, c. 4.5 × 0.8cm, apex acute, base attenuate-cuneate, margin subentire, glabrescent, subsessile; inflorescence a sessile axillary cluster, 0.5–0.7 × 0.5cm, white; tepals ovate, acuminate, c. 0.4 × 0.15cm. Villages; 470m.
Distr.: tropical Africa, India, Malaysia, Australia. (Palaeotropics).
IUCN: LC
Nyandong: Darbyshire 2 fl., 3/2003.
Uses: ENVIRONMENTAL USES – boundaries – used for low hedging in gardens (Darbyshire 2).

Amaranthus hybridus L. subsp. *cruentus* (L.) Thell.

Erect herb to 1.5(–2)m; stems furrowed, glabrous; leaves alternate, ovate-trullate, 11.5–18 × 4.2–10.2cm, apex acute to rounded, base asymmetrical, cuneate or acute, margin irregularly undulate, glabrescent; petiole 4–9.5cm; spikes axillary or terminal, the latter many-branched, dense, c. 18 × 8cm, cream-yellow to green; peduncles pubescent; bracteoles ovate, caudate, c. 1.5–2mm long. Farmbush, cultivation; 500–650m.
Distr.: widespread in the tropics & S Europe.
IUCN: LC
Kupe Village: Cable 2696 5/1996; **Ngusi:** Etuge 1554 1/1996; **Nlonako:** Etuge 1786 3/1996.
Local name: known in local markets as 'green vegetables'.
Uses: FOOD – leaves – common all over and in markets; slice leaves, then prepare as any other vegetable (Etuge 1554, 1786).

Amaranthus spinosus L.

Erect herb to 1m; stems succulent, red-green, glabrescent, paired axillary spines 1–1.2cm; leaves alternate, ovate, c. 3 × 1.3cm, apex acuminate-mucronate, base acute, margin irregularly undulate; terminal inflorescence a several-branched spike, dense, to 9cm long, axillary inflorescences spicate or globose, the latter 0.5cm in diameter; bracts and tepals mucronate, red-green. Villages, farmbush; 1500–2000m.
Distr.: widespread in the tropics & subtropics.
IUCN: LC
Bangem: Pollard 844 11/2001; **Mwanenguba:** Cheek 9410 10/1998; Sanford 5438 fl., fr., 11/1968.

Celosia argentea L.

Robust herb to 1.5m; stems ridged; leaves alternate, elliptic-lanceolate, to 13.5 × 4cm, apex acuminate, base attenuate, glabrous; petiole to 3cm; spikes terminal, conical or elongate, to 10(–15) × 1.8cm; flowers dense; tepals lanceolate, to 8mm long, pink-purple, turning silvery. Cultivated in gardens; 950m.

Distr.: widespread in the tropics, possibly originating in Africa.
IUCN: LC
Mwanenguba: Leeuwenberg 9887 fl., 5/1972.

Celosia isertii C.C.Towns.
F.T.E.A. Amaranthaceae: 15 (1985).
Climbing or scrambling herb to 2m; stems glabrous; leaves drying green-black, alternate, ovate, 6.8–9.3 × 3–5cm, apex acuminate-apiculate, base attenuate then abruptly truncate, margin irregularly undulate, glabrous or scabridulous, veins occasionally purplish below; petiole c. 2.5cm; inflorescence spikes axillary or terminal, occasionally branched, 2.5–10cm; peduncle to 3.5cm, glabrescent; bracteoles and tepals ovate-elliptic, white, the latter 3–4mm long; style bifid. Forest, forest edge; 700–1350m.
Distr.: tropical African highlands. (Afromontane).
IUCN: LC
Kodmin: Ghogue 73 2/1998; **Kupe Village:** Cable 3887 7/1996; Cheek 6040 fl., 1/1995; Lane 282 fl., 1/1995; **Mwambong:** Onana 543 2/1998; **Ndum:** Williams 133 fl., 1/1995; **Ngomboaku:** Cheek 10286 12/1999; Ghogue 478 12/1999; **Nyasoso:** Cable 2809 6/1996; Cheek 7483 fl., 10/1995; Etuge 2578 7/1996.
Local name: Mogbe (Bakweri Language). **Uses:** FOOD – unspecified parts; MEDICINES (Lyonga Daniel in Cheek 7483).

Celosia leptostachya Benth.
Suberect branching herb to 40cm; stems furrowed, glabrous; leaves alternate, ovate-deltate, 3.4–5.6 × 2.3–3.8cm, apex acute, base attenuate then abruptly truncate, margin irregularly undulate, glabrous; petiole c. 2cm; spikes terminal, 8–23cm; peduncle fine, glabrous; flowers in laxly arranged glomerules, c. 6-flowered, green-brown when dry; bracteoles and tepals ovate, c. 2.5mm long; style bifid. Forest; 800m.
Distr.: Nigeria to Congo (K). (Lower Guinea & Congolian).
IUCN: LC
Ngomboaku: Cheek 10354 12/1999.

Celosia pseudovirgata Schinz
Fl. Cameroun 17: 13 (1974).
Syn.: *Celosia bonnivairii sensu* Keay in FWTA 1: 147
Suberect or procumbent herb to 1m; resembling *C. isertii*, but leaves 6.8–14.5 × 3.2–7.2cm, less truncate at the base; bracteoles and tepals brown; tepals ovate, c. 2mm. Forest, forest edges; 700–1550m.
Distr.: Nigeria to Congo (K). (Lower Guinea & Congolian).
IUCN: LC
Kodmin: Cheek 8985 1/1998; 9054 1/1998; Etuge 4011 1/1998; **Kupe Village:** Cable 3769 7/1996; 776 fl., 1/1995; Etuge 2826 7/1996; Ryan 211 5/1996; **Muahunzum:** Etuge 4441 11/1998; **Ngomboaku:** Cheek 10366 12/1999; **Nyasoso:** Cheek 5679 fl., 12/1993; Etuge 1317 fl., 10/1995; Lane 173 fl., 11/1994; 241 fl., 11/1994.

Celosia trigyna L.
Syn.: *Celosia laxa* Schum. & Thonn.
Suberect herb to 1.5m; stems fine, furrowed, glabrous; leaves alternate, ovate, 5–7 × 3–3.8cm, apex acuminate, base shortly attenuate then rounded, margin subentire, glabrous; petiole to 3.5cm; inflorescence a lax terminal panicle or spike, 3.5–15cm; branching to 3.2cm, glomerules widely-spaced, 3–5-flowered; bracteoles and tepals ovate, white, the latter c. 2mm long; style generally trifid. Forest, forest edges; 650m.
Distr.: widespread in (sub)tropical Africa.
IUCN: LC

Ngusi: Etuge 1553 1/1996.
Uses: FOOD – unspecified parts – used locally as a vegetable (Etuge 1553).

Cyathula fernando-poensis Suess. & Friedrich
Herb to 30cm; stems shallowly furrowed, pubescent towards apex, swollen above nodes, red-green; leaves opposite, elliptic, 5.2–6.5 × 2.8–3.4cm, apex acuminate, base attenuate, margin subentire, pubescent particularly on upper surface; petiole 0.5–1cm; flowering spikes axillary or terminal, subsessile or pedunculate, to 3cm, puberulent; flowers in dense glomerules; tepals ovate, green; infertile flowers contain several hooked bristles to 3mm. Forest; 1330–1550m.
Distr.: Bioko & SW Cameroon. (W Cameroon Uplands).
IUCN: VU
Kodmin: Cheek 10269 12/1999; **Nyasoso:** Lane 242 fl., 11/1994.
Note: Lane 242 represents the first record of this species for continental Africa.

Cyathula prostrata (L.) Blume var. *pedicellata* (C.B.Clarke) Cavaco
Fl. Cameroun 17: 46 (1974).
Syn.: *Cyathula pedicellata* C.B.Clarke
Scrambling or decumbent herb to 70cm high; stems furrowed, ± pilose; leaves opposite, elliptic-rhombic, 2.8–4.5 × 1.6–2.5cm, apex acute to shortly acuminate, base attenuate, margin subentire, sparsely pilose; petiole 0.3–0.8cm; racemes terminal, 4–13cm; peduncle puberulent, 2–3-flowered cymules on peduncles to 1.5mm; bracts and tepals ovate, to 1.5cm long, green; hooked bristles of infertile flowers numerous, radiating, 1.5mm. Farmbush, forest; 650–1500m.
Distr.: Sierra Leone to Congo (K) & Tanzania. (Guineo-Congolian).
IUCN: LC
Kodmin: Cheek 9199 2/1998; Ghogue 10 1/1998; **Nyale:** Etuge 4191 2/1998; **Nyasoso:** Cable 705 fl., 1/1995; Etuge 1593 1/1996; Lane 178 fl., 11/1994.

Cyathula prostrata (L.) Blume var. *prostrata*
Scrambling herb to 1m, resembling var. *pedicellata*, but cymules subsessile; hooked bristles of infertile flowers less radiating; stem and leaves more densely pilose throughout, hairs to 1mm; leaves to 6 × 3cm. Farmbush; 650–1000m.
Distr.: pantropical.
IUCN: LC
Mbule: Leeuwenberg 9306 fl., 1/1972; **Nyasoso:** Etuge 1596 1/1996.

Cyathula uncinulata (Schrad.) Schinz
Suberect or climbing herb to 2m; stems robust, densely pilose, c. 3mm diameter; leaves opposite, ovate, 7.7–9 × 4.6–6.3cm, apex shortly acuminate, base rounded, margin irregular to subentire, densely appressed-pubescent; petiole 0.7–1.7cm, pilose; inflorescence terminal on lateral shoots, globose, c. 2cm diameter, yellow-brown with dense flexible hooked bristles to 0.7cm; bracts lanceolate, c. 5mm. Forest edges, grazed areas; 2000m.
Distr.: widespread in tropical Africa & Madagascar.
IUCN: LC
Mwanenguba: Cheek 9413 10/1998; Pollard 856 11/2001.

ANACARDIACEAE

F.J. Breteler & M. Cheek

Note: most of the taxa in this account are based on the determinations of Breteler on material at K, WAG & YA. Breteler is in the process of revising the main genera in Africa, e.g. Breteler (2003), 'The African genus *Sorindeia...*' Adansonia (sér. 3) 25: 93–113; Breteler (2004, in press), ' The genus *Trichoscypha...*' Adansonia (sér. 3). In the course of these revisions a wider species concept has been adopted; this results in the two taxa assessed as CR in Cable & Cheek (1998) being downgraded here to NT. One of the taxa is reduced to synonymy. Two taxa (*Lannea welwitschii* & *Sorindeia juglandifolia*) are based on determinations by Gereau of specimens at MO, not seen by us. The species of *Trichoscypha* are distinguished from those of *Sorindeia* by styles and stigmas 3 (not 1); flowers 4(–5)-merous (not 5).

Antrocaryon klaineanum Pierre

Tree 15m, no exudate; leaves pinnate, 30–40cm, 8-jugate; leaflets elliptic c. 6 × 2.5cm, subacuminate, obtuse, nerves 5 pairs; petiolule 0.5cm; petiole 13cm; inflorescence spike-like, 18cm, supra-axillary; flowers numerous, 1–2mm; fruit top-shaped, c. 2 × 3cm, fleshy, endocarp with 5 valves on upper surface. Forest; 600–800m.
Distr.: S Nigeria to Gabon. (Lower Guinea).
IUCN: NT
Bakossi F.R.: Cheek 9331 10/1998; **Manehas F.R.:** Gosline 128 10/1998; **Ngusi:** Etuge 45 4/1986.

Lannea welwitschii (Hiern) Engl.

Tree 20–25m; leaves 30–40cm, 2–3-jugate; leaflets leathery, drying dark brown below, ovate, c. 10 × 5cm, acuminate, obtuse, nerves 8 pairs, raised; petiolule 1cm; petiole 15cm; inflorescence terminal or from upper 2–3 axils, 7–20cm; branches 2–7cm long; flowers white, 2mm; fruit discoid, 8mm, foveolate, fleshy. Forest; 100–200m.
Distr.: Ivory Coast to Congo (K). (Guineo-Congolian).
IUCN: LC
Etam: Thomas 7164 7/1987; **Ikiliwindi:** Nemba 532 6/1987; **Nyandong:** Ghogue 1453 fr., 3/2003.

Mangifera indica L.

Tree c. 15m; slash with translucent mango-scented exudate; leaves simple, long-elliptic-oblong, c. 24 × 7cm subacuminate, obtuse; petiole c. 5cm; inflorescence a terminal erect, diffuse panicle or thyrse, c. 30 × 20cm; flowers white, 4mm; fruit fleshy, 1-seeded, oblique, c. 9 × 5 × 4cm. Farmbush, cultivated; 650m.
Distr.: native of India. (Pantropical).
IUCN: LC
Ngusi: Etuge 1540 1/1996.
Note: persisting in secondary forest.
Local names: Mangole; Mango. **Uses:** FOOD – infructescences – an important cash crop; MEDICINES (Etuge 1540).

Pseudospondias microcarpa (A.Rich.) Engl. var. *longifolia* (Engl.) Keay

Tree 8–15m; broken stems scented of mango; leaves pinnate, c. 40cm, 5-jugate; leaflets shining below ovate-elliptic, largest at leaf apex up to 12 × 4.5cm, acuminate, obtuse to rounded; petiolule 0.5cm; petiole 8cm; inflorescence c. 20 × 12cm, highly diffuse; flowers 3mm, greenish yellow, petals 5, stamens 5, ovaries 3; fruit ellipsoid, 2–5cm, 1-seeded, red. Forest near streams; 350–1000m.
Distr.: S Nigeria & Cameroon. (Lower Guinea).

IUCN: NT
Mwambong: Cheek 9109 2/1998; **Ngomboaku:** Ghogue 454 12/1999; **Nyandong:** Thomas 7008 5/1987.

Pseudospondias microcarpa (A.Rich.) Engl. var. *microcarpa*

Tree 5–30m, resembling var. *longifolia*; leaves often only 3–4-jugate, oblique, to 20 × 8cm; distinguished by leaflets 4–12 (not 15–17); flowers ≥4mm (not 3); stamens 8 (not 6). Farmbush and secondary forest; 560–1200m.
Distr.: Senegal to Malawi. (Tropical Africa).
IUCN: LC
Kupe Village: Cable 2536 5/1996; 709 fr., 1/1995; Cheek 8407 5/1996; Etuge 1903 5/1996; 1994 5/1996; Kenfack 210 7/1996; Lane 268 fl., 1/1995; **Mwambong:** Etuge 4145 2/1998; Mackinder 153 1/1998; **Ndum:** Cable 908 fr., 2/1995; **Ngomboaku:** Ghogue 486 12/1999; **Nyasoso:** Cable 1170 fr., 2/1995; 3216 6/1996; 3635 7/1996; Etuge 1514 12/1995; Sebsebe 5034 10/1995; **Nzee Mbeng:** Gosline 92 2/1998.
Local names: Monkey Plum; Eke (*fide* Cheek).

Sorindeia africana (Engl.) van der Veken

Bull. Jard. Bot. Nat. Belg. 24: 242 (1959).
Syn.: *Sorindeia gilletii* De Wild.
Syn.: *Sorindeia nitidula* Engl.
Tree 5–7m; exudate white; leaves 1–3-jugate, the c. 35cm, apical leaflets drying brown, leathery, larger, 21 × 8cm, acuminate, unequally obtuse; lateral nerves c. 9 pairs, matt, margin revolute, undulate; petiolule 0.5cm; petiole 8cm; inflorescence terminal, c. 35 × 35cm; flowers white, 2mm; fruits ovoid 1.5cm. Forest; 200–1400m.
Distr.: Nigeria to Angola. (Lower Guinea & Congolian).
IUCN: LC
Ikiliwindi: Nemba 521 6/1987; **Mejelet-Ehumseh:** Etuge 164 fr., 6/1986; **Mile 15, Kumba-Mamfe Road:** Nemba 146 7/1986; **Mwanenguba:** Leeuwenberg 9575 fl., 4/1972.

Sorindeia grandifolia Engl.

Syn.: *Sorindeia warneckei* Engl.
Syn.: *Sorindeia zenkeri* Engl.
Stout tree, often leaning 1.5(–4)m; leaves matt dark green above and below, 30–60cm, 1–3-jugate, oblong-elliptic, to 25(–44) × 10(–14)cm, shortly acuminate, obliquely rounded, lateral nerves 8–10 pairs, yellow below; petiolule 0.5–1.5cm; petiole 7–15cm; inflorescence cauliflorous, pendulous, 4–30cm, 1–8-branched; pink in bud; flowers 2–3mm; fruit ellipsoid, 2cm. Forest; 300–1175m.
Distr.: Guinea (C) to S Sudan & São Tomé, N of Equator. (Guineo-Congolian).
IUCN: LC
Bolo: Nemba 55 3/1986; **Kupe Village:** Cheek 6050 1/1995; 6088 fl., 1/1995; 7137 fl., 1/1995; Etuge 2893 7/1996; Kenfack 268 7/1996; Lane 271 fl., 1/1995; 404 fr., 1/1995; Ryan 314 5/1996; **Nyasoso:** Cable 1118 fl., 2/1995; Wheatley 397 fr., 7/1992.

Sorindeia juglandifolia (A.Rich.) Planchon ex Oliver

Syn.: *Sorindeia adolfi-friederici* Engl. & Brehmer
Syn.: *Sorindeia claessensii* De Wild.
Tree 5–10m; leaves c. 30cm, 4-jugate; leaflets oblong to narrowly elliptic, 12cm, acuminate, truncate-decurrent, lateral nerves 10–12 pairs; petiole 4cm; inflorescence terminal, erect, c. 25 × 25cm; branches sparse, glabrous; flowers pink, 2mm; fruit yellow, ellipsoid 0.8cm. Forest; 300m.
Distr.: Guinea (C) to Burundi. (Lower Guinea & Congolian).

IUCN: LC
Bolo: <u>Nemba 5</u> 4/1986; **Kumba-Baduma:** Nemba & Thomas 149 (MO), cited by Breteler.

Sorindeia winkleri Engl.
Bot. Jahrb. 43: 413 (1909).
Syn.: *Sorindeia mildbraedii* Engl. & Brehmer
Tree to 15m; leaves 4-jugate; leaflets oblong, 6–12 × 2.5–5.2cm, acuminate, obtuse, lateral nerves 10–12 pairs; inflorescence terminal, to 50cm, densely rusty-puberulous; petals pale green; fruit ellipsoid, orange or red, 2.5cm. Forest; 350m.
Distr.: SE Nigeria, Cameroon, Gabon to Rwanda. (Lower Guinea & Congolian).
IUCN: NT
Mekom: <u>Nemba 24</u> 4/1986.
Note: description from FWTA, name change from Breteler dets on K material. About 30 specimens are known.

Spondias mombin L.
Tree 15m; leaves pinnate, c. 30cm, 7-jugate; leaflets papery, dark green, elliptic-oblong, c. 7 × 2.5cm, acutely acuminate, obtuse, serrate-sinuate, c. 20 pairs of lateral nerves; petiolule 0.2cm; petiole c. 7cm; inflorescence terminal; fruit c. 6cm diameter, fleshy, orange, watery, acidic and mango-like in taste, endocarp c. 3cm diameter, including large spines. Farmbush, farms, cultivated for fruit; 350–1000m.
Distr.: Senegal to Congo (K) & tropical America. (Guineo-Congolian; South America).
IUCN: LC
Bakolle Bakossi: <u>Etuge 139</u> fr., 5/1986; **Kupe Village:** <u>Etuge 1901</u> 5/1996.

Trichoscypha acuminata Engl.
Tree 15m; leaves 1.2m, c. 15-jugate; middle leaflets leathery oblong-lanceolate, c. 20–25 × 5–10cm, acuminate, rounded, c. 15 nerve pairs; petiole 15cm; inflorescence cauliflorous, 15–30 × 6cm; flowers 2–3mm, pink; fruit red, 3cm. Forest; 200–900m.
Distr.: S Nigeria to Congo (K). (Lower Guinea & Congolian).
IUCN: LC
Enyandong: <u>Etuge 4390r</u> 10/2001; **Etam:** <u>Dundas 15218</u> fl., 5/1946.
Local name: marketed as 'bush bon-bon'. **Uses:** FOOD – fruits edible, harvested from the wild (*fide* Cheek).

Trichoscypha bijuga Engl.
Syn.: *Trichoscypha preussii* Engl.
Tree 10m; white exudate drying black; leaves c. 70cm, 5-jugate; leaflets even in size, narrowly oblong-elliptic, to c. 23 × 9cm, acumen c. 2cm, oblique, base obtuse, lateral nerves c. 16 pairs; petiolule c. 1cm; petiole c. 14cm; inflorescence terminal, very dense-flowered, c. 17 × 12cm, c. 5-branched, densely red-pubescent; flowers 3mm, cream. Forest; 220–750m.
Distr.: Liberia to Congo (K). (Guineo-Congolian).
IUCN: NT
Bakolle Bakossi: <u>Etuge 259</u> 9/1986; **Konye:** <u>Thomas 5158</u> 11/1985; **Kupe Village:** <u>Lane 426</u> fl., 1/1995; **Mile 15, Kumba-Mamfe Road:** <u>Nemba 201</u> 9/1986.

Trichoscypha ? bijuga Engl.
Tree 3–4m, resembling *T. bijuga*, but inflorescence cauliflorous not terminal, c. 15cm, pendulous, inconspicuously puberulent. Forest; 880m.
Kupe Village: <u>Cheek 7066</u> 1/1995.

Note: more material required to establish that the distinguishing characteristics are consistent; possibly a new species to science.

Trichoscypha sp., probably bijuga Engl.
Tree 3–8m, resembling *T. bijuga*, but inflorescence lax, diffuse, pendent. Forest; 900–960m.
Kupe Village: <u>Cable 730</u> fl., 1/1995; <u>845</u> fl., 1/1995.
Note: more material required to establish if conspecific with *T. bijuga*.

Trichoscypha laxiflora Engl.
Bot. Jahrb. 15: 110 (1892).
Syn.: *Trichoscypha camerunensis* Engl.
Syn.: *Trichoscypha talbotii* Baker.f.
Shrub 2.5m, with ± rusty-pilose branchlets and inflorescence, longer hairs reflexed; leaves 1–4-jugate; leaflets oblong-elliptic, 8–25 × 2.5–14cm, acuminate, lateral nerves 6–13 pairs; inflorescence terminal, much-branched; petals 1–2mm; disk glabrous. Forest; 350m.
Distr.: SE Nigeria to Congo (B). (Lower Guinea).
IUCN: NT
Bakolle Bakossi: <u>Etuge 154</u> 5/1986.
Note: approx. 30 specimens known.

Trichoscypha lucens Oliv.
F.T.A. 1: 444 (1868).
Treelet 2.5m; leaves 60cm, 11-jugate; leaflets oblong-elliptic, to 20 × 6cm, acuminate, obtuse, to 10-nerved; petiole 9cm; inflorescence terminal, glabrescent, c. 50 × 25cm; branches numerous, patent, to 13cm; flowers 2–3mm, white; stamens bright orange. Forest; 1190m.
Distr.: Ivory Coast to Mozambique. (Tropical Africa).
IUCN: LC
Nyasoso: <u>Sidwell 415</u> 10/1995.

Trichoscypha mannii Hook.f.
Syn.: *Trichoscypha longipetala* Baker f.
Tree or shrub to 15m; leaves to 1m, 17-jugate; leaflets papery, ovate-elliptic to oblong-lanceolate, 12–20 × 3–8cm, mostly with long hairs; inflorescence shorter than leaves, hispid to softly hairy; fruits beaked, 3 × 1.5cm, hairy. Forest; 300m.
Distr.: SE Nigeria to Congo (B). (Lower Guinea).
IUCN: NT
Bolo: <u>Nemba 7</u> 4/1986.

Trichoscypha oddonii De Wild.
Ann. Mus. Congo (Sér. 5) 1: 282 (1906).
Syn.: *Trichoscypha abut* Engl. & Brehm.
Tree to 15m, resembling *T. acuminata*, differing in the leaves 1–1.8m long; inflorescence with numerous foliaceous elliptic bracts, c. 4 × 4cm, clustered at base. Forest; 2000m.
Distr.: Cameroon to Cabinda. (Lower Guinea).
IUCN: NT
Mwanenguba: <u>Leeuwenberg 10010</u> (WAG), cited by Breteler (in press).

Trichoscypha patens (Oliv.) Engl.
Tree 15–20m; leaves 20–30cm, 1–3-jugate; leaflets leathery, drying black, elliptic c. 10 × 6cm, acuminate, obtuse-acute; petiolule 0.7cm; petiole 5–10cm; inflorescence glabrous, terminal c. 30 × 30cm; branches c. 6, to 15cm; flowers 2mm, yellow; fruit ellipsoid, 3cm. Forest; 300–600m.
Distr.: Cameroon to Congo (K). (Lower Guinea).
IUCN: LC
Baduma: <u>Nemba 118</u> fr., 6/1986; **Ngombombeng:** <u>Etuge 125</u> 5/1986.

ANISOPHYLLEACEAE

Trichoscypha rubicunda Lecomte
Bull. Soc. Bot. France 52: 656 (1906).
Treelet 1.5–4(–8)cm; white exudate drying black; leaves c.
40cm; 3-jugate; leaflets papery leathery, dull olive-green
below, oblong or oblong-lanceolate, c. 16 × 7cm, acumen
ligulate, c. 1 × 0.2cm, base obtuse, lateral nerves c. 8 pairs;
petiolule 0.5cm; petiole 5–12cm; inflorescence terminal;
25cm; branches 3–4cm apart, to 15cm long; bearing partial
inflorescences 1.5cm long, interrupted; pedicel 5mm; corolla
pink 2mm; receptacle red. Forest; 875–1200m.
Distr.: Cameroon & Gabon. (Lower Guinea).
IUCN: NT
Kupe Village: Cheek 7028 fl., 1/1995; **Ndum:** Lane 471 fl.,
2/1995; **Nyasoso:** Cable 1213 fl., 2/1995; Elad 135 fl.,
2/1995; Lane 527 fl., 2/1995.
Note: only 16 specimens are known (Breteler, in press).

Trichoscypha sp. nov. 1
Tree to 16m; slash red-brownish with white exudate; leaves
80cm, 7-jugate; petiole terete, 17cm, brown puberulent;
leaflets oblong, to 20 × 8cm, including acumen 2.5 × 0.3cm,
obtuse-acute, dull white below, lateral nerves 14 pairs;
petiolule 1cm. Forest; 1200m.
Distr.: Cameroon & Rio Muni. (Lower Guinea).
Edib: Etuge 4504 11/1998.
Note: fertile material required by Breteler in order to name;
only 3 specimens are known (Breteler, in press).

ANISOPHYLLEACEAE

M. Cheek

Anisophyllea polyneura Floret
Adansonia (sér. 4) 8: 377 (1986).
Tree 40cm d.b.h.; bole sinuous; bark dark red, hard, 1cm
thick; leaves alternate, elliptic, 15 × 10cm, apex acuminate-
mucronate, base rounded-decurrent, nerves palmate, 5,
tertiary nerves subscalariform; fruit ovoid, 4 × 3cm. Forest.
Distr.: SE Nigeria to Congo (K). (Lower Guinea &
Congolian).
IUCN: LC
Manyemen: Letouzey 13923 fr., 6/1974.

Anisophyllea cf. polyneura Floret
Tree sapling 2m; branches whorled; leaves uniform,
alternate, obliquely lanceolate, 21 × 7cm, including 3cm
acumen, base obtuse, nerves 4 pairs, the upper 2 inserted on
the midrib up to 3.5cm from the petiole; scalariform; petiole
0.5cm. Forest; 850m.
Ngomboaku: Cheek 10313 12/1999.
Note: possibly the juvenile stage of *A. polyneura*. More
collections needed from Ngomboaku to elucidate.

ANNONACEAE

G. Gosline

Annickia chlorantha (Oliv.) Setten & P.J.Maas
Taxon 39: 676 (1990).
Syn.: *Enantia chlorantha* Oliv.
Tree to 30m; wood and sap bright yellow; leaves elliptic-
oblong, 5–22 × 2–7cm, 7–14 pairs lateral nerves, abaxially
appressed-pubescent; flowers solitary, extra-axillary; petals

3, opposite sepals, lanceolate, densely appressed-pubescent,
triangular in cross section, 2–3 × 0.6–1.2cm; fruit with
numerous ovoid monocarps, 2.5–3 × 1cm; stipes 2–3cm;
pedicel 1.5–2cm. Forest; 300–710m.
Distr.: Nigeria to Congo (K). (Lower Guinea & Congolian).
IUCN: LC
Bolo: Nemba 102 fr., 5/1986; **Kupe Village:** Etuge 2049
5/1996; **Mungo River F.R.:** Cheek 9354 10/1998; **Nyale:**
Cheek 9654 11/1998; Etuge 4235 2/1998.
Local name: Ebor. **Uses:** MATERIALS – wood – used for
carpentry (Marcel Njume in Cheek 9654); MEDICINES
(Etuge 4235 & Marcel Njume in Cheek 9654).

Anonidium mannii (Oliv.) Engl. & Diels var. mannii
Tree to 30m; leaves elliptic-obovate to obovate-oblong, 20–
45 × 7–18cm, sharply acuminate, acumen 2–3cm long, 10–20
pairs lateral nerves; inflorescence pendant from trunk or bare
branches, cymes to 2–3m long; flowers hermaphrodite or
male; petals 6, greenish yellow, elliptic to obovate, 2.5–5 ×
1.5–4cm; fruits syncarpous, large cylindric-ovoid, 25–50 ×
10–30cm, to 10 kilos, surface reticulate-areolate. Forest;
250–1000m.
Distr.: Ghana to Congo (K). (Guineo-Congolian).
IUCN: LC
Loum F.R.: Cheek 10240 12/1999; **Nyasoso:** Cable 2801 fr.,
6/1996; Cheek 6009 fl., 1/1995; 7896 11/1995; Sunderland
1494 fr., 7/1992.

Artabotrys aurantiacus Engl. & Diels
Notizbl. Bot. Gart. Berlin 2: 300 (1899).
Woody climber; leaves coriaceous, elliptic-oblong, 7.5–10 ×
2.5–6cm; inflorescence a recurved hook, c. 2cm long, few-
flowered; pedicels 7–10mm long; petals 15–30 × 4–9mm,
tomentose, white to yellow-orange; monocarps red at
maturity, few, ellipsoid, both extremities rounded, 15–30 ×
7–13mm. Forest; 1200–1520m.
Distr.: Cameroon to Congo (K). (Lower Guinea &
Congolian).
IUCN: LC
Kodmin: Cheek 9063 1/1998; **Nyasoso:** Elad 132 2/1995;
Nzee Mbeng: Gosline 99 fl., 2/1998.
Uses: MEDICINES (Gosline 99).

Artabotrys congolensis De Wild. & T.Durand
Fl. Congo, Rw. & Bur. 2: 317 (1951).
Syn.: *Artabotrys rhopalocarpus* Le Thomas
Woody climber to canopy; leaves elliptic, 15–26 × 6–11cm,
8–11 widely-spaced lateral nerves; inflorescence a recurved
hook, 1.7–2.4cm long, with 1–3 flowers; petals 6, tomentose,
outer broadly triangular 17–25 × 8–14mm, inner elliptic-
lanceolate, 11–17 × 5–7mm; fruits unique, 15–20 monocarps,
sessile but club-shaped, c. 6 × 2.5cm, each containing 2
discoid seeds. Forest; 700–1150m.
Distr.: Cameroon to Congo (K). (Lower Guinea &
Congolian).
IUCN: LC
Kupe Village: Cable 3894 7/1996; Gosline 241 11/1999;
Nyasoso: Cable 2851 6/1996.
Note: formerly *Artobotrys rhopalocarpus* LeThomas, but put
in synonymy by LeThomas without having seen fruits of *A.
congolensis*.

Artabotrys stenopetalus Engl. & Diels
Liana or scandent shrub; leaf papery, ovate-oblong, 5.5–17 ×
2.5–5cm, with 9–11 pairs lateral nerves; inflorescence a
recurved hook 1–2cm long, with few to many flowers;
flowers yellow (?); pedicel c. 0.5cm; outer petals linear, 6–15

× 1.5–4mm, densely appressed pubescent; inner petals subequal with flat section; monocarps sessile, ellipsoid, c. 2 × 1cm. Forest; 1000–1450m.
Distr.: Ghana to Congo (B). (Upper & Lower Guinea).
IUCN: LC
Kodmin: Gosline 149 fr., 11/1998; **Lake Edib:** Cheek 9143 fl., fr., 2/1998; **Mwambong:** Onana 523 2/1998.

Cleistopholis glauca Pierre ex Engl. & Diels
Fl. Congo, Rw. & Bur. 2: 300 (1951).
Tree to 35m, resembling *C. patens* and *C. staudtii*, but inflorescence subumbellate, pedunculate (2–7mm), with 2–3 hooded bracteoles; monocarps ellipsoid, subsessile, smooth. Forest; 100m.
Distr.: Cameroon to Congo (K). (Lower Guinea & Congolian).
IUCN: LC
Mungo River F.R.: Thomas 2561 11/1983.
Note: specimen cited not seen.

Cleistopholis patens (Benth.) Engl. & Diels
Tree to 30m; petioles 3–12mm; blades oblong to somewhat falcate, 4.5–25 × 2.5–6cm, with 10–25 pairs lateral nerves, shiny above; flowers 2–8 in fascicles; pedicels 0.8–2.5cm long; outer petals erect, green, obovate-oblong, 5–12 × 2–4mm, apex rounded; inner petals ovate, folded over androecium; fruiting pedicels 1.5–3cm long; monocarps 4–9, subglobose, tuberculate; stipes c. 7mm long. Forest; 150m.
Distr.: Sierra Leone to Uganda. (Guineo-Congolian).
IUCN: LC
Mungo River F.R.: Cheek 10229 fl., 12/1999.

Cleistopholis staudtii Engl. & Diels
Tree to 30m; petioles 10–15mm; blades elliptic-oblong, 9–17 × 3–6.5cm, both faces matt, with 10–12 pairs lateral nerves; flowers axillary in 2–3-flowered fascicles; outer petals linear, 15–20 × 2–3mm;, inner petals ovate, 2–2.5 × 3–3.5mm; fruits as *C. patens*. Forest; 500–950m.
Distr.: Nigeria, Cameroon, Gabon. (Lower Guinea).
IUCN: VU
Kupe Village: Cheek 7915a fl., 11/1995; Etuge 2884 7/1996; Gosline 240 fl., 11/1999.
Local name: Etchatan. **Uses:** FOOD ADDITIVES – infructescences; MEDICINES (Ngolle Abraham in Cheek 7915a).

Duguetia staudtii (Engl. & Diels) L.W.Chatrou
Syn.: *Pachypodanthium staudtii* Engl. & Diels
Tree to 30m; trunk straight, 60–70cm diameter; leaves linear-oblong, 15–32 × 3–5(–8)cm, apex attenuate, no acumen; petiole 2–5mm, with ridges carrying onto twig in young leaves; blade with stellate hairs below, with 15–20 pairs lateral nerves; inflorescence extra-axillary, 3–5-flowered; pedicels 1–2cm; petals white, oblong, 2.5–4 × 1–1.5cm; fruits globular, 4–6cm diameter, consisting of many mericarps with pyramidal apices compressed together, easily separating. Forest; 1050m.
Distr.: Sierra Leone to Congo (K). (Guineo-Congolian).
IUCN: LC
Enyandong: Gosline 367 11/2001.

Friesodielsia enghiana (Diels) Verdc.
Fl. Gabon 16: 240 (1969).
Syn.: *Oxymitra obanensis* (Baker f.) Sprague & Hutch.
Woody climber to 30m or understorey shrub 1–2m when young; shoots densely rusty-hirsute; leaves oblong-lanceolate, 8–34 × 3.5–8cm, acuminate, inferior of leaf strikingly blue-grey, with 11–20 pairs lateral nerves; nerves

and lamina hirsute; inflorescence extra-axillary in 2–3-flowered fascicles; outer petals papery, elliptic-ovate, 15–25 × 12–20mm, rufous-tomentose; inner petals coriaceous, smaller, glabrescent; fruit pedicel 2–3cm long; monocarps ellipsoid or moniliform, rufous-hirsute. Forest; 420m.
Distr.: Sierra Leone to Congo (K). (Guineo-Congolian).
IUCN: LC
Mungo River F.R.: Onana 947 11/1999.

Friesodielsia enghiana (Diels) Verdc. var. ***nov.***
Woody climber, understorey shrub when young; differing from the eponymous variety in being entirely glabrous. Forest understorey; 760–1050m.
Distr.: Bakossi mountains. (Narrow Endemic).
Kupe Village: Cable 2523 5/1996; 3683 7/1996; **Mt Kupe:** Cheek 10154 fl., 11/1999; **Ngomboaku:** Ghogue 500 fl., 12/1999.

Greenwayodendron suaveolens (Engl. & Diels) Verdc.
Adansonia (sér. 2) 9: 90 (1969).
Syn.: *Polyalthia suaveolens* Engl. & Diels
Tree to 10–25m; bark smooth, pale to dark grey; leaves oblong-elliptic, 4–12 × 3–5cm, apex acute to long-acuminate, with 5–13 lateral nerves; flowers polygamous, solitary or in 2–8-flowered fascicles; petals subequal, linear-oblong or lanceolate, 10–28 × 1.5–2.5mm, finely pubescent; fruiting pedicels 0.3–2.2cm long; monocarps c. 8–10, globose, 1.2–1.8 × 0.9–1.5cm, 1–2-seeded; stipes 5–8mm long; seeds compressed-hemiglobose, c. 1cm diameter, strongly wrinkled, with a circumferential groove. Forest; 200–1780m.
Distr.: Nigeria to Uganda. (Lower Guinea & Congolian).
IUCN: LC
Enyandong: Etuge 4452r 11/2001; **Ikiliwindi:** Nemba 564 6/1987; **Kodmin:** Cheek 9088 fl., fr., 1/1998; Plot B70 1/1998; **Kupe Village:** Elad 95 1/1995; **Mile 12 Mamfe Road:** Nemba 291 10/1986; **Ngombombeng:** Etuge 30 fr., 4/1986.
Note: LeThomas does not recognize Verdcourt's separation of *Greenwayodendron* from *Polyalthia*.

Hexalobus crispiflorus A.Rich.
Tree to 30m; trunk deeply channeled; bark brown, fissured; leaves elliptic to ovate-lanceolate, 7.5–20cm × 3–8cm, with 12–17 pairs lateral nerves; flowers solitary, or 2–3 in axils; sepals 3, 1–2 × 0.5–1cm; petals pale yellow, united at base, with crispate borders, linear 2.5–8 × 0.5–1.5cm, anthers numerous; carpels 6–10; fruits on a very thick pedicel, 1–2cm long; mericarps oblong, 8–9 × 4–5cm. Forest; 450m.
Distr.: Guinea (C) to Congo (K) & Sudan. (Guineo-Congolian).
IUCN: LC
Nyale: Cheek 9716 fl., 11/1998.

Isolona zenkeri Engl.
Engl. & Prantl, Pflanzenf. Nachtr. 1: 161 (1897).
Tree to 5m; leaves elliptic-oblong to obovate-oblong, 11–25 × 3.5–9cm, with 10–13 pairs lateral nerves; flowers pendant, axillary or ramiflorous, bright yellow; sepals small, to 6mm long; petals joined at base in a tube to 1/4 the length of the flower, lobes linear-lanceolate, 10–22mm long, linear, narrowing to a point; fruit syncarpous, ovoid-oblong, 4–8 × 2.5–3.5cm. Forest; 1200m.
Distr.: Cameroon & Gabon. (Lower Guinea).
IUCN: VU
Kupe Village: Etuge 2000 fl., 5/1996; **Mwanenguba:** Leeuwenberg 9550 fl., 4/1972; **Nyasoso:** Etuge 1794 fl., 3/1996.

Monanthotaxis cauliflora (Chipp) Verdc.

Adansonia (sér. 2) 8: 242 (1968).

Syn.: *Popowia cauliflora* Chipp

Climber; branchlets ferruginous puberulent; leaf elliptic-oblong to obovate-oblong, 9–20 × 3–8cm, with 11–19 lateral nerves, lower face tomentose; flowers unisexual; male flowers axillary, in few-flowered cymes; female inflorescence cauliflorous or on old wood, a dense multiflorous raceme, 7–15cm long, the whole tomentose; female flowers puberulent, three outer petals 6 × 5mm, inner very much reduced and difficult to see; carpels numerous; infructescence a dense raceme; monocarps moniliform, of 1–3–6 ellipsoid articles, c. 1 × 0.6cm, the whole puberulent. Forest; 720–1750m.

Distr.: Nigeria to Congo (K). (Lower Guinea).

IUCN: LC

Bangem: Thomas 5274 1/1986; **Kodmin:** Cheek 9067 fr., 1/1998; 9202 fr., 2/1998; Etuge 4122 fr., 2/1998; Gosline 175 fr., 11/1998; **Kupe Village:** Cheek 7711 fl., 11/1995; **Muahunzum:** Etuge 4442 fr., 11/1998.

Monanthotaxis filamentosa (Diels) Verdc.

Kew Bull. 25: 31 (1971).

Syn.: *Popowia filamentosa* Diels

Woody climber; shoot golden-brown long-hirsute; leaves oblong or oblong-elliptic, 11–19 × 4.5–7.5cm, shortly acuminate, metallic grey beneath; flowers unisexual in axillary few-flowered lax racemes; rachis to 13cm; sepals more than half as long as petals; petals to 15mm long, outer pale brown, inner pink. Forest; 1100–1700m.

Distr.: Nigeria to Congo (K) and CAR. (Lower Guinea & Congolian).

IUCN: LC

Kupe Village: Cable 2575 5/1996; **Nyasoso:** Cable 1185 fl., 2/1995; Etuge 1729 fl., 2/1996; 2377 fl., 6/1996.

Monanthotaxis foliosa (Engl. & Diels) Verdc.

Kew Bull. 25: 21 (1971).

Syn.: *Enneastemon foliosus* (Engl. & Diels) Robyns & Ghesq.

Woody climber, densely ferruginous tomentose; leaves elliptic-obovate, 5.5–13.5 × 3–7cm, underside densely appressed-pubescent, with 8–11 pairs lateral nerves; inflorescence in fascicles of 2–5 cymes, of 3–5 flowers each, 8–18mm long; flowers globular, 3–3.5mm diameter, bisexual; stamen filaments pediform. Forest; 800–1700m.

Distr.: Ghana to Cameroon & Congo (K). (Guineo-Congolian).

IUCN: LC

Kupe Village: Cheek 7794 fr., 11/1995; 7880 fr., 11/1995; Gosline 254 12/1999; **Nyasoso:** Cable 1169 fl., 2/1995; Etuge 1748 2/1996.

Monanthotaxis laurentii (De Wild.) Verdc.

Kew Bull. 25: 26 (1971).

Syn.: *Popowia congensis* (Engl. & Diels) Engl. & Diels

Climber to 10m; leaves obovate, to 15 × 5cm, acumen 1.5 × 1cm, with 6–11 lateral nerves; inflorescence solitary, extra-axillary from new shoots; pedicel 4–18mm, one bract towards top of pedicel; flowers bisexual, buds cream-green, c. 1cm diameter; infructescence stipes 1–2cm long; monocarps with 1–3(–5) strongly cylindrical green segments, c. 2 × 0.4cm, terminated by a sharp point, 4mm long. Forest; 850m.

Distr.: Sierra Leone to Congo (K). (Guineo-Congolian).

IUCN: LC

Ngomboaku: Etuge 4676 fl., fr., 12/1999.

Monanthotaxis sp. nov.

Liana; stems medium brown, appressed-pubescent; leaves obovate, 8–14 × 3.5–6cm, with c. 8 lateral nerves, dark green below, pubescent below; flowers unknown; stipe 2cm; c. 7 monocarps; peduncles 2cm; fruits dark brown moniliform, with 1 or more cylindrical segments 3 × 0.8cm, terminal segment apiculate. Forest; 1000m.

Distr.: Mt Kupe. (Narrow Endemic).

Nyasoso: Cable 3526 7/1996.

Note: one specimen from Nyasoso with unique fruits, more material required.

Monodora brevipes Benth.

Tree to 13m; leaves obovate to oblong, 15–30 × 5–20cm; flowers pendant; sepals round, 1cm long; outer petals to 4.5cm long; inner petals ovate, with conspicuous crimson blotches; fruit ribbed, globose, 7.5cm diameter. Forest; 200–1300m.

Distr.: Guinea (C) to Cameroon. (Upper & Lower Guinea).

IUCN: LC

Edib: Etuge 4488 fr., 11/1998; **Ikiliwindi:** Nemba 556 6/1987; **Nyasoso:** Lane 532 fr., 2/1995.

Monodora myristica (Gaertn.) Dunal

Tree to 30m; leaves elliptic to obovate, 9–50 × 2–20cm, with 10–21 pairs lateral nerves; flowers pendant; pedicel 5–25cm; outer petals ovate-lanceolate, 4–9 × 2.5–3.5cm, with crispate margins; inner petals 2.5–4.5 × 2–3cm, apex obtuse; fruits with a peduncle to 25cm long, 1.5cm diameter, body subspherical, 10–20cm diameter. Forest; 150–1200m.

Distr.: Sierra Leone to Kenya. (Tropical Africa).

IUCN: LC

Enyandong: Gosline 342 11/2001; Pollard 797 fr., 10/2001; **Kupe Village:** Cable 2544 5/1996; Etuge 1497 fl., 11/1995; 2017 fr., 5/1996; **Mungo River F.R.:** Cheek 10228 fl., 12/1999; **Ngomboaku:** Cheek 10357 fr., 12/1999; Ghogue 494 fr., 12/1999; **Ngusi:** Etuge 1576 fl., 1/1996; **Nyandong:** Darbyshire 22 3/2003; Ghogue 1487 3/2003; **Nyasoso:** Cable 1215 fl., 2/1995; 2789 6/1996; 3285 fr., 6/1996; 3507 7/1996; 653 fl., 12/1993; Cheek 7519 fr., 10/1995; Etuge 2516 fl., 7/1996; Zapfack 672 6/1996.

Local name: not known (Gosline 342).

Monodora tenuifolia Benth.

Tree to 20m; leaves papyraceous, 6–20 × 2.5–7.5cm, with 9–14 pairs lateral nerves; flowers solitary in axils, pendant; outer petals ovate-lanceolate, 4–7 × 2–3cm, margins undulate; inner petals cochleariform, 1.5–3.5cm long, with 2 oblong, pubescent appendages in the middle from margins; fruiting pedicel 3.5–7cm long; body spherical and smooth, 4–8cm diameter. Forest; 750m.

Distr.: Guinea (C) to Angola. (Upper & Lower Guinea).

IUCN: LC

Manehas F.R.: Cheek 9399 10/1998.

Local name: Et-chat-eh. **Uses:** unknown (Epie Ngome in Cheek 9399).

Piptostigma calophyllum Mildbr. & Diels

Bot. Jahrb. 53: 443 (1915).

Tree to 12m; leaves subsessile, very large, obovate, 60–80 × 20–32cm, densely pubescent below, with 30–70 pairs lateral nerves; inflorescence a cauliflorous very contracted cyme; flowers with sepals and outer petals creating an apparent 6-lobed calyx; inner petals yellow-ochre, linear-lanceolate, 5–6cm long, densely tomentose. Forest; 900–1320m.

Distr.: Cameroon & Gabon. (Lower Guinea).

IUCN: VU

Edib village: Cheek 9177 fl., 2/1998; **Kodmin:** Gosline 198 fl., 11/1998; **Ngomboaku:** Gosline 256 fl., 12/1999.

Piptostigma fasciculatum (De Wild.) Boutique
Fl. Congo, Rw. & Bur. 2: 306 (1951).
Tree to 20m; leaves glabrous, obovate-oblong, 10–35 × 5–11.5cm, with 14–21 pairs lateral nerves; flowers 1–3 in axils, with fine pedicels 1.5cm long; inner petals linear, ribbon-like, 4.5–10 × 0.5cm; fruits with a thick pedicel 1.5–2cm long, ellipsoid-oblong, rounded at each end, 4–10.5 × 4–4.5cm, yellow-orange when ripe. Forest; 950–1200m.
Distr.: Ivory Coast to Congo (K). (Guineo-Congolian).
IUCN: LC
Kupe Village: Etuge 1844 fr., 3/1996; Gosline 234 11/1999.

Piptostigma sp. 1
Tree to 20m, with straight bole; stem and young leaves with long appressed hairs; flowers solitary; brachyblasts absent. Forest; 350–1100m.
Distr.: Bakossi mountains. (Narrow Endemic).
Kupe Village: Cable 787 fl., 1/1995; Cheek 7849 11/1995; 8328 fr., 5/1996; Etuge 2698 7/1996; **Mungo River F.R.:** Cheek 10197 11/1999; **Nyasoso:** Etuge 2420 fr., 6/1996.

Polyceratocarpus parviflorus (Baker f.) Ghesq.
Tree to 12m; leaves obovate-oblong, 9–23 × 3–10cm, acuminate, underside of leaf glaucous with few hairs, with 9–11 pairs lateral nerves; flowers axillary (?), solitary, male or hermaphrodite, ovoid, c. 1cm diameter; stamens forming a cone; 2–4 carpels when present; fruits with cylindrical mericarps to 4 × 1cm, contracted around seeds in dried specimens. Forest; 950–1000m.
Distr.: Guinea (C) to Gabon. (Upper & Lower Guinea).
IUCN: LC
Kupe Village: Cable 2721 5/1996; Elad 86 fr., 1/1995; Gosline 236 fl., 11/1999; 237 fr., 11/1999.
Note: some fruits are borne on the trunk in our specimens; this is in contrast to the flora description (axillary).

Uvaria angolensis Welw. ex Oliv.
F.T.E.A. Annonaceae: 15 (1971).
Syn.: *Uvaria angolensis* Welw. ex Oliv. subsp. *guineensis* Keay
Liana or shrub to 6m; glabrescent with simple and stellate hairs; leaves elliptic oblong 4–13(–17) × 3–6(–7)cm, shortly acuminate, base rounded to subcordate, with 6–12 pairs lateral nerves, very prominent beneath; flowers yellow, solitary or a few extra-axillary; pedicel 2–6mm; sepals joined in bud; petals subequal, 12–20 × 8–11mm; fruiting pedicel 7–10mm; mericarps c. 20, cylindric 1.5–4 × 0.5–2cm, apiculate, reddish tomentose, contracted around seeds; stipes sublateral, 8–12mm long. Forest; 750m.
Distr.: Sierra Leone to Uganda & Zimbabwe. (Tropical Africa).
IUCN: LC
Baseng: Cheek 10415 fr., 12/1999.

Uvaria baumannii Engl. & Diels
Liana or shrub to 10m; leaves lanceolate-oblong, 5–16(–9) × 2–6cm, underside densely covered with 2 layers of stellate hairs, with 17–25 pairs lateral nerves; flowers solitary, extra-axillary or terminal; sepals joined in bud; petals 12–20mm long, outer oblong, inner ovate; fruiting pedicel 2–2.5cm; receptacle subglobose, 15–30mm diameter; mericarps numerous, subglobose, 6–10mm diameter, verruculose; stipes lateral, 3–4cm long. Forest; 1450m.
Distr.: Sierra Leone to Congo (K). (Guineo-Congolian).
IUCN: LC

Kodmin: Cheek 8950 1/1998; Gosline 147 fl., 11/1998.

Uvaria heterotricha Pellegr.
Bull. Soc. Bot. France 96: 173 (1950).
Liana to 10m; leaves elliptic-oblong 7–16 × 3–6cm, bottom of leaf sparsely covered with two types of hairs, very small simple or stellate, and long appressed hairs (visible with microscope), with 10–13 pairs lateral nerves; flowers large, terminal; calyx joined in bud; petals yellow with red spot at base; external petals 15–25 × 12–22mm; inner larger 20–35 × 10–22mm; fruits with numerous globular mericarps mounted distinctly laterally on long stipes. Forest; 850–1600m.
Distr.: Cameroon & Gabon. (Lower Guinea).
IUCN: NT
Essossong: Leeuwenberg 9249 fl., 1/1972; **Kodmin:** Cheek 8863 fl., 1/1998; **Kupe Village:** Cable 2500 fr., 5/1996; 2674 fr., 5/1996; 3809 fr., 7/1996; Etuge 2659 fr., 7/1996; Kenfack 297 fr., 7/1996; Ryan 355 fr., 5/1996; **Mbule:** Cable 3344 fr., 6/1996; **Mwambong:** Gosline 83 fl., 2/1998; **Nyasoso:** Cable 3459 fr., 7/1996; 3558 fr., 7/1996; Cheek 7457 fr., 10/1995; Etuge 1819 fl., 3/1996; 2092 fr., 6/1996.
Note: this species was previously known only from 4 sites in Gabon and 1 in Cameroon in the Rumpi Hills; it is the commonest *Uvaria* in the checklist area. The fruits are previously undescribed.

Uvariodendron calophyllum R.E.Fr.
Tree to 20m; leaves oblanceolate, rounded base and apex with abrupt acumen c. 2cm long, 40–60 × 12–20cm; petioles and lower surface of leaves densely tomentose; cauliflorous; flowers nearly sessile in clusters of 2–3; bracts, calyx, outer petals all velvety-brown; petals crimson inside, with broad yellow margins; flowers c. 4cm diameter; fruiting carpels velvety-brown. Forest understorey; 300m.
Distr.: Ghana, Nigeria, Cameroon. (Upper & Lower Guinea).
IUCN: NT
Bolo Moboka: Nemba 64 fl., 3/1986; **Mungo River F.R.:** Olorunfemi in FHI 30561 fr., 5/1951.
Note: *U. calophyllum* and *U. giganteum* are similar, except for the indumentum. There may be some confusion.

Uvariodendron connivens (Benth.) R.E.Fr.
Tree to 13m; leaves elongate-obovate, to 40 × 12cm, abruptly acuminate, glabrous, with c. 20 pairs lateral veins; flowers axillary on young shoots, solitary; pedicel c. 1.5cm; sepals green, c. 1cm; petals purple, ovate, c. 3 × 2cm, tomentellous; mericarps glabrous, ellipsoid, c. 4 × 3cm. Forest understorey; 300–1200m.
Distr.: SE Nigeria, Bioko & Cameroon. (Lower Guinea).
IUCN: NT
Bakolle Bakossi: Etuge 156 fr., 5/1986; **Bolo:** Nemba 56 3/1986; **Kupe Village:** Gosline 235 11/1999; **Mungo River F.R.:** Gosline 209 fr., 11/1999; **Ndum:** Lane 501 fl., 2/1995; **Nyasoso:** Etuge 2390 6/1996; 2396 6/1996.

Uvariodendron giganteum (Engl.) R.E.Fr.
Tree to 10m; cauliflorous; leaves oblanceolate, rounded base and apex with abrupt acumen c. 2cm long, 40–72 × 12–23cm, with 28–30 pairs lateral nerves; flowers cauliflorous, solitary, to 8cm diameter; bracts, calyx and petals tomentose externally; inner petals yellow, purple within; fruits sessile to 20cm diameter, mericarps numerous, obovoid-oblong, 2.5–7 × 1.5–2.5cm; seeds 14–18, biseriate. Forest understorey; 200–1600m.
Distr.: Cameroon & Gabon. (Lower Guinea).
IUCN: VU
Mungo River F.R.: Cheek 9337 10/1998; **Nyasoso:** Sunderland 1536 fr., 7/1992.

APOCYNACEAE

Uvariodendron mirabile R.E.Fr.
Tree to 10m; leaves glabrous, 15–30 × 3–9cm, lateral nerves 15–20, prominent beneath; cauliflorous; calyx lobes short triangular leaving petals exposed; petals 2.5–3.5 × 2–2.5cm, ovate, pubescent outside; mericarps cylindrical with transverse grooves, grey-pubescent, c. 4 × 2cm; stipes 1cm long. Forest understorey; 1200m.
Distr.: Ivory Coast to SW Cameroon. (Upper & Lower Guinea).
IUCN: LC
Nyasoso: Lane 142 fr., 6/1994.

Uvariopsis submontana Kenfack, Gosline & Gereau
Novon 13: 444 (2003).
Tree to 20m; leaves simple, obovate with acute apex; flowers densely cauliflorous on first 2–4m of bole, in fascicles of up to 500; monoecious, with female flowers below the male; 4 petals; 2 sepals; flowers pyramidal in aspect, scarlet, the calyx to half the length of the petals; pedicels 4–5cm long; fruit irregular ovoid, smooth, orange, 4 × 2cm; peduncle red. Forest; 840–1200m.
Distr.: Cameroon. (W Cameroon Uplands).
IUCN: EN
Kupe Village: Cable 2736 5/1996; Cheek 7034 fl., 1/1995; 7131 fl., 1/1995; 7795 11/1995; Elad 69 fl., 1/1995; Gosline 233 11/1999; **Ndum:** Lane 490 fl., 2/1995; **Nyasoso:** Cable 1221 fl., 2/1995; Etuge 2562 fr., 7/1996; Lane 131 fr., 6/1994; 534 fl., 2/1995; Sebsebe 5035 10/1995.
Note: fairly common on Mt Kupe and also known from a collection from Yabassi to the SE. Spectacular in flower, with flowers covering the lower part of the bole.
Local name: Michile (Cheek 7131).

Uvariopsis vanderystii Robyns & Ghesq.
Ann. Soc. Sc. Brux. (Sér. B) 53: 64 (1933).
Treelet to c. 4m; trunk somewhat swollen at base; leaves obovate, 20–30 × 7–12cm, glabrous; cauliflorous; monoecious, with female flowers below the male; 4 petals; 2 sepals; female flowers on very long thin pedicels, to 30cm, some trailing on ground; buds globular with petals united; petals extremely thick and woody, ovate, c. 1.5 × 1.2cm, joined at base; male flower pedicels straight, 4–5cm long, smaller than female; fruits unknown. Forest understorey; 800–950m.
Distr.: Cameroon, Gabon, Congo (K). (Lower Guinea & Congolian).
IUCN: EN
Kupe Village: Cheek 7032 fl., 1/1995; Gosline 232 11/1999; 244 fl., 12/1999.
Note: this species was originally described from two specimens from the Congo and Gabon. We now know of these two trees above Kupe village and another collection from the Takamanda Forest Reserve.

Xylopia aethiopica (Dunal) A.Rich.
Tree to 30m; lightly buttressed to 1m; leaves elliptic, 7–20 × 3–9cm, ± long-acuminate, with 7–8 pairs lateral nerves, not conspicuous; flowers axillary, solitary or 2–6-fasciculate, needle-like, 4–5cm long; pedicel thin, 0.5–1cm long; mericarps to 30, subsessile, long-cylindrical, 5–6 × 0.5cm, dehiscing to reveal bright-red interior and black seeds. Forest and farms; 350m.
Distr.: Senegal to Sudan & Mozambique. (Tropical Africa).
IUCN: LC
Bakolle Bakossi: Etuge 148 fr., 5/1986.

Uses: FOOD ADDITIVES – commonly used as a pepper spice (*fide* Cheek); MATERIALS – wood used as timber (*fide* Etuge).

Xylopia africana (Benth.) Oliv.
Tree to 12m; leaves obovate-elliptic, 9–16 × 4–8cm, strongly reticulate, nerves on underside dark red; flowers yellow-orange, subglobose, c. 1.5cm diameter, remaining closed except for a small apical opening; fruits with pedicel c. 3cm; monocarps c. 10, cylindrical, 3–12 × 1cm, pointed, constricted between seeds; stipe c. 2cm. Montane forest; 840–2050m.
Distr.: SE Nigeria, SW Cameroon & São Tomé. (W Cameroon Uplands).
IUCN: VU
Kodmin: Cheek 9192 fl., 2/1998; Gosline 75 fl., 1/1998; Plot B73 1/1998; B86 1/1998; **Kupe Village:** Cable 3814 7/1996; Elad 90 1/1995; Etuge 1441 11/1995; **Mt Kupe:** Cheek 7605 11/1995; **Nyasoso:** Cable 2870 6/1996; Etuge 1740 fl., 2/1996.
Local names: Nkii (Etuge 1740 & Cheek 9192); N'ki (Gosline 75). **Uses:** MATERIALS – wood used as timber (Etuge 1740); wood used in building houses, the bark for ropes (for carrying baskets), poles for sickles in harvesting cocoa pods (Cheek 9192); very good stick for plucking cacao; used for roof of traditional houses (Gosline 75).

Xylopia quintasii Engl. & Diels
Tree to 30m; leaves obovate to obovate-oblong, 5.5–13 × 2.5–6cm, apex rounded emarginate or slighty acuminate, bases cuneate, with 7–8 pairs lateral nerves; flowers clustered, axillary; inner and outer petals subequal, 10–20 × 2.5–3mm; inner slightly thickened, with an enlarged rectangular base to 1/4 length of petal; fruits with pedicel c. 1cm long; mericarps cylindrical, moniliform, 3–5 × 0.6–0.8cm; stipes 0.5–1.0cm. Forest; 200–600m.
Distr.: Sierra Leone to CAR, Congo (K). (Guineo-Congolian).
IUCN: LC
Mile 12 Mamfe Road: Nemba 293 10/1986; **Ngusi:** Etuge 56 fr., 4/1986.

APOCYNACEAE
M. Cheek & A.J.M. Leeuwenberg

Alafia lucida Stapf
Liana, glabrous; leaves elliptic, 9 × 5cm, obtuse, lateral nerves 4–5 pairs; petiole 5mm; panicles terminal, dense, c. 3 × 3cm; corolla twisting to right, yellow with red centre, 8mm; fruit bifid; follicles c. 26 × 1cm, glabrous; seed-body 1.8cm, hairs 2cm. Forest; 300–1400m.
Distr.: Guinea (C) to Tanzania. (Guineo-Congolian).
IUCN: LC
Bolo Moboka: Nemba 238 9/1986; **Kodmin:** Etuge 3983 1/1998; Nwaga 10 fl., fr., 1/1998; **Nyale:** Cheek 9692 11/1998.
Note: several other species, e.g. *A. multiflora*, are likely to be found in lowland Bakossi.

Alstonia boonei De Wild.
Tree 10–40m, glabrous; branches and leaves whorled; leaves obovate to oblanceolate, 7–25 × 3–11cm, subsessile, lateral nerves 30–55 pairs; panicles terminal; corolla white, 9–15mm including lobes 3–6mm, twisting to right; follicles paired, 50 × 0.5cm, villous. Forest; 750m.

Distr.: Senegal to Tanzania. (Tropical Africa).
IUCN: LC
Mt Kupe: <u>Schoenenberger 60</u> 11/1995.
Local names: Sobia or Wild Rubber. **Uses:** MATERIALS – timber (*fide* Etuge).

Baissea axillaris (Benth.) Hua
Liana, densely brown-pubescent; internodes c. 1.5cm; leaves ovate-oblong, c. 4.5 × 2.5cm, apex acute, base cordate or truncate, with 4–5 pairs lateral nerves; petiole 2mm; panicles axillary, 1–2cm, 2–6-flowered; corolla twisting to right, 5mm, white; fruit not seen. Forest; 1000m.
Distr.: Senegal to Congo (K). (Guineo-Congolian).
IUCN: LC
Nyasoso: <u>Cheek 7533</u> 10/1995.

Baissea baillonii Hua
Bull. Jard. Bot. Nat. Belg. 64: 98 (1995).
Syn.: *Baissea breviloba* Stapf
Liana, resembling *B. axillaris*, but leaves often ovate or deltoid, acuminate, acumen 1–8mm; sepals narrowly ovate (not triangular). Forest; 400m.
Distr.: Guinea (C) to Congo (K). (Guineo-Congolian).
IUCN: LC
Eboné: <u>Leeuwenberg 9209</u> 1/1972.

Baissea campanulata (K.Schum.) De Kruif
Bull. Jard. Bot. Nat. Belg. 64: 102 (1995).
Syn.: *Baissea subsessilis* (Benth.) Stapf ex Hutch. & Dalziel
Liana, minutely puberulent; leaves oblong to 10 × 4.5cm, acumen acute, 1cm, base acute to truncate, lateral nerves 3–4 pairs, domatia densely hairy; petiole 2mm; panicles 2cm, axillary and on leafless stem apices, 2–10-flowered; corolla campanulate, white, c. 1 × 1cm, twisting to right; fruit not seen. Forest; 1200m.
Distr.: Ivory Coast to Congo (K). (Guineo-Congolian).
IUCN: LC
Nzee Mbeng: <u>Gosline 104</u> 2/1998.

Baissea cf. leonensis Benth.
Liana, glabrous; resembling *B. campanulata*; internodes c. 3cm; leaves elliptic, c. 6 × 2.5cm, acumen 1cm, base obtuse, lateral nerves c. 15–20 pairs, domatia absent; petiole 5mm; flowers and fruit not seen. Forest; 900m.
Distr.: Senegal to Uganda. (Guineo-Congolian).
Kupe Village: <u>Kenfack 295</u> 7/1996.

Baissea multiflora A.DC.
Syn.: *Baissea laxiflora* Stapf
Liana, finely puberulent; internodes c. 2cm; leaves elliptic, c. 7–8 × 2–3cm, acumen 0.5cm, base obtuse, lateral nerves 3–4 pairs; petiole c. 3mm; panicles axillary, 7cm; pedicels to 8mm; corolla 3mm, white, twisting to right; fruit not seen. Forest; 300m.
Distr.: Senegal to Congo (K). (Guineo-Congolian).
IUCN: LC
Southern Bakossi F.R.: <u>Latilo in FHI 31190</u> fl., 3/1953.
Note: *B. subrufa* (*B. mortehanii*) and other spp. of the genus are also likely to be found in lowland Bakossi.

Baissea subrufa Stapf
Bull. Jard. Bot. Nat. Belg. 64: 143 (1995).
Liana, resembling *B. campanulata*, but leaf base rounded-obtuse, domatia crescent-shaped; petiole 1cm; panicles 4cm, c. 20-flowered; corolla 4mm, including tube 2mm. Forest; 800m.
Distr.: Cameroon to Congo (K). (Lower Guinea & Congolian).

IUCN: LC
Kupe Village: <u>Lane 273</u> fl., 1/1995.

Baissea sp. 1
Liana, resembling *B. subrufa* (and perhaps conspecific, but fertile material needed), but lacking domatia entirely, lateral nerves 5 pairs; petioles 3–5mm. Forest; 1400m.
Kodmin: <u>Plot B185</u> 1/1998.

Callichilia bequaertii De Wild.
Meded. Land. Wag. 78(7): 12 (1978).
Syn.: *Callichilia macrocalyx* Schellenb.
Shrub 2m; leaves elliptic or obovate, to 26 × 11cm, subacuminate, lateral nerves 13; petiole 2cm; axillary panicles c. 10cm, pendulous, 2–4-flowered; calyx 3cm, divided to base; corolla white, 12cm, including tube 6cm. Forest; 720m.
Distr.: Nigeria to Congo (K). (Lower Guinea & Congolian).
IUCN: LC
Nyale: <u>Etuge 4170</u> 2/1998.

Clitandra cymulosa Benth.
Liana, closely resembling *Landolphia*, glabrous; leaves papery, drying blackish green, elliptic, c. 9 × 5cm, acumen 0.5cm, base rounded, lateral nerves 7 pairs; flowers axillary and terminal, subsessile; fruit globose, c. 3 × 2cm. Forest; 400m.
Distr.: Guinea (C) to Tanzania. (Guineo-Congolian).
IUCN: LC
Mombo: <u>Olorunfemi in FHI 30560</u> fr., 5/1951; **Southern Bakossi F.R.:** <u>Gartlan 34</u> 9/1968.

Dictyophleba cf. lucida Pierre
Liana; stems and petioles puberulent; leaves obovate, 23 × 9cm, acumen acute, 1cm, base subauriculate, lateral nerves 9 pairs; petiole 1.5cm. Forest; 1400m.
Distr.: Guinea (C) to Cameroon. (Upper & Lower Guinea).
Kodmin: <u>Plot B270</u> 1/1998.
Note: fertile material needed to confirm identification.

Funtumia africana (Benth.) Stapf
Tree resembling *F. elastica*, but leaf domatia hairy, without pits; sepals 1.5–4mm; corolla lobes usually longer than tube; fruit fusiform, to 32 × 1cm, acute. Forest; 250m.
Distr.: Senegal to Mozambique. (Tropical Africa).
IUCN: LC
Konye: <u>Thomas 5926</u> fl., 3/1986.
Note: reputed to provide poor quality rubber.

Funtumia elastica (Preuss) Stapf
Tree to 15m, glabrous; stem drying black, smooth; leaves elliptic to 19 × 7cm, acumen cuneate, 1.5cm, base acute-decurrent, lateral nerves c. 10 pairs, domatial pits to 1.5mm, margin undulate; fascicles axillary, 2cm; sepals 3–5mm; corolla white, c. 9mm, twisted to right, lobes shorter than tube; follicles paired, patent, c. 17 × 2cm, obtuse, angled; seeds hairy. Forest; 420–700m.
Distr.: Guinea (C) to Uganda. (Guineo-Congolian).
IUCN: LC
Kupe Village: <u>Cheek 7019</u> fl., fr., 1/1995; <u>7920a</u> fl., 11/1995; **Loum F.R.:** <u>Cheek 9295</u> 10/1998.
Note: provides excellent quality rubber according to the scientific literature.
Local name: Cotton Tree (not 'black cotton') (Cheek 7920a).

Holarrhena floribunda T.Durand & Schinz var. *floribunda*

Tree or shrub, 1–15m, glabrous; leaves elliptic, thinly papery, c. 12 × 4cm, subacuminate, obtuse-decurrent, lateral nerves c. 10 pairs, tertiary nerves reticulate; petiole 1cm; panicles axillary, 4 × 6cm; corolla white, 1.2cm, twisted to right, lobes rounded; fruit bifid, linear, each 30 × 0.5cm; seeds plumed. Deciduous forest; 220–300m.
Distr.: Mali to Congo (B). (Upper & Lower Guinea).
IUCN: LC
Bakolle Bakossi: Etuge 288 9/1986; **Bolo Moboka:** Nemba 256 9/1986.

Hunteria umbellata (K.Schum.) Hallier f.

Syn.: *Hunteria eburnea* Pichon
Syn.: *Hunteria mayumbensis* Pichon
Tree 7–12m, glabrous; leaves oblong-elliptic, to 20 × 8cm, acumen cuneate, 1cm, base obtuse, lateral nerves 10–20 pairs; petiole 1–2cm; panicle terminal, dense, 1.5cm on peduncle 5cm; corolla white, 8mm, twisting to right; fruit paired, globose, 4cm, beaked, warted, orange. Forest; 200–500m.
Distr.: Ghana to Cameroon. (Upper & Lower Guinea).
IUCN: LC
Mile 15, Kumba-Mamfe Road: Nemba 617 7/1987; **Mombo:** Olorunfemi in FHI 30579 5/1951; **Wone:** Nemba 582 7/1987.
Note: *Isònema buchholzii* and *Pleioceras zenkeri* are examples of broadly similar species known from Kumba and so likely to be found in lowland Bakossi.

Landolphia bruneelii (De Wild.) Pichon

Wag. Agric. Univ. Papers 92(2): 25 (1992).
Liana, glabrous; resembling *L. buchananii*, but leaves lanceolate to elliptic, to 11.5 × 5cm, acumen 1cm, lateral nerves c. 15 pairs, lower surface black-spotted; petiole 5mm; flowers axillary, simple, sessile, white, 4cm. Forest; 400m.
Distr.: Cameroon to Congo (K). (Lower Guinea & Congolian).
IUCN: NT
Eboné: Bamps 1520 date unknown.
Note: only 17 specimens documented of this taxon.

Landolphia buchananii (Hallier f.) Stapf

Wag. Agric. Univ. Papers 92(2): 27 (1992).
Liana, glabrous, lenticellate; leaves pale matt-brown, drying green, midrib yellow, papery, elliptic-oblong, c. 8 × 3cm, subacuminate, obtuse, lateral nerves c. 15 pairs, conspicuously tertiary-reticulate; petiole 5mm; panicle terminal, c. 1cm; corolla white, 13mm, twisting to left; fruit globose, 4cm, green with large white circles. Forest; 1000–1600m.
Distr.: SE Nigeria to S Mozambique. (Tropical Africa).
IUCN: LC
Kodmin: Barlow 4 1/1998; Cheek 9569 fr., 11/1998; Etuge 4060 1/1998; Onana 598 2/1998; **Lake Edib:** Cheek 9159 2/1998; **Nyasoso:** Etuge 1825 3/1996; **Nzee Mbeng:** Gosline 96 2/1998.
Local name: Nhoum. **Uses:** FOOD – fruits edible (Edmondo Njume in Cheek 9569).

Landolphia congolensis (Stapf) Pichon

Syn.: *Cyclocotyla congolensis* Stapf
Liana, glabrous; leaves thinly coriaceous, slightly grey-green above, green and often black-dotted below, elliptic, c. 11 × 3.5cm, acumen ligulate, 1cm, cuneate, lateral nerves 4–7 pairs below; petiole 0.5–1cm; flowers single, axillary; corolla white, 3cm; fruit narrowly turbinate, 6 × 2.5cm. Forest; 1000m.
Distr.: Cameroon to Congo (K). (Lower Guinea & Congolian).
IUCN: LC
Ndibise: Etuge 527 6/1987.

Landolphia dulcis (Sabine) Pichon

Syn.: *Landolphia dulcis* (Sabine) Pichon var. *barteri* (Stapf) Pichon
Liana, pilose when young; leaves papery to leathery, ovate to obovate, 4–22 × 1.8–11cm, acumen up to 2.5cm, base cuneate to auriculate, glandular-dotted below, lateral nerves 4–7 pairs; petiole 2–17mm; inflorescence axillary, 1–3 per axil, 1–7-flowered, 1.5 × 1.5cm or more; peduncle 0–5.5mm; calyx lobes to 4 × 2mm; corolla tube 7–20 × 0.9–2mm, tomentose to glabrous, lobes 5–22 × 1.5–4mm; fruit 2–10 × 2–6cm. Forest; 250m.
Distr.: Senegal to Gabon. (Upper & Lower Guinea).
IUCN: LC
Mile 15, Kumba-Mamfe Road: Nemba 214 9/1986.
Note: specimen cited not seen.

Landolphia cf. glabra (Pierre ex Stapf) Pichon

Liana with tendrils, glabrous; leaves lanceolate, 11 × 3.5cm, including 2cm spatulate acumen, obtuse to rounded, lateral nerves 10 pairs, straight, patent; petiole 0.5cm. Forest; 1400m.
Distr.: Nigeria to Congo (K). (Lower Guinea & Congolian).
Kodmin: Plot B202 1/1998.
Note: fertile material needed, may equally be *L. ligustrifolia*.

Landolphia incerta (K.Schum.) Persoon

Wag. Agric. Univ. Papers 92(2): 92 (1992).
Syn.: *Aphanostylis mannii* (Stapf) Pierre
Liana, glabrous; leaves ovate to lanceolate, 4.5–9 × 2.5–3.5cm, acumen spatulate, base rounded-obtuse, lateral nerves c. 15 pairs; petiole 3mm; infructescence axillary; fruit subellipsoid to c. 4 × 2cm, sometimes waisted, evenly speckled with white lenticels. Forest; 1000–1450m.
Distr.: Guinea (B) to Congo (K). (Guineo-Congolian).
IUCN: LC
Kodmin: Cheek 9619 11/1998; **Mwambong:** Cheek 9113 2/1998.
Local name: Nkab. **Uses:** FOOD ADDITIVES – exudates – milk used in palm wine (Cheek 9619).

Landolphia jumellei (Jumelle) Pichon

Wag. Agric. Univ. Papers 92(2): 98 (1992).
Liana, brown-hirsute; leaves drying brown, obovate or elliptic, to 15 × 7cm, acumen acute, cordate, lateral nerves 8–9 pairs; petiole 0.5cm; fascicles axillary, 1cm; corolla 1cm, white, twisted to left. Forest; 250m.
Distr.: Cameroon & Gabon. (Lower Guinea).
IUCN: LC
Baduma: Nemba 229 9/1986; **Mungo River:** Daramola in FHI 9815 fl., 4/1959.

Landolphia landolphioides (Hallier f.) A.Chev.

Liana, resembling the related *L. buchananii*, but leaves to 19 × 8cm, lateral nerves c. 10 pairs; petiole 1cm; panicle terminal or terminal and axillary, c. 3cm; corolla 24mm; fruit globose, to 8cm (dried), white spots 1mm diameter, orange (fresh) to 10cm; rind leathery, 1cm thick. Forest; 1000–1300m.
Distr.: Nigeria to Uganda. (Lower Guinea & Congolian).
IUCN: LC

Mt Kupe: Nemba 906 3/1988; **Mwambong:** Mackinder 172 1/1998; **Ndum:** Cable 878 fr., 1/1995; Lane 457 fr., 1/1995; **Nyasoso:** Cheek 7291b fr., 2/1995; Etuge 1756 3/1996.
Uses: ANIMAL FOOD – eaten by monkeys (Cheek 7291b).

Landolphia owariensis P.Beauv.
Liana, puberulent when young; stems drying black, with minute lenticels; leaves leathery, drying glossy grey-green above, brown below, elliptic-oblong, c. 12 × 5cm, acumen to 1cm, obtuse, lateral nerves 12–15 pairs; petiole 0.5–1cm; panicles terminal, 2 × 2cm, dense; peduncle 2–5cm; corolla 9mm, pubescent, yellow; fruit globose, to 4cm; rind hard, fragile, like eggshell. Forest, *Aframomum* thicket; 200–1300m.
Distr.: Guinea (C) to Uganda. (Guineo-Congolian).
IUCN: LC
Ikiliwindi: Nemba 555 6/1987; **Mwambong:** Onana 541 2/1998; **Mwanenguba:** Leeuwenberg 8337 fr., 12/1971; **Ndum:** Cable 902 fl., 1/1995; **Nyasoso:** Cheek 7512 10/1995; 7517 10/1995; **Wone:** Nemba 590 7/1987.
Note: several other spp. of *Landolphia*, and other allied lianescent Apocynaceae, such as *Saba comorensis* (*S. florida* of FWTA), *Vahadenia laurentii* and species of *Dictyophleba* and *Ancyclobotrys* are very likely to be found in lowland Bakossi in future.

Landolphia robustior (K.Schum.) Persoon
Wag. Agric. Univ. Papers 92(2): 172 (1992).
Syn.: *Anthoclitandra robustior* (Schumann) Pichon
Liana, glabrous; leaves drying brown-black, narrowly elliptic, c. 16 × 5cm, acumen 0.5cm, base obtuse, lateral nerves c. 8 pairs; petiole 1cm; fascicles axillary, dense, 1cm; corolla 1cm, white; fruit ellipsoid, 4 × 1.5cm, smooth. Forest; 250–760m.
Distr.: Nigeria to Angola. (Guineo-Congolian).
IUCN: LC
Ikiliwindi: Nemba 112 fr., 6/1986; **Ngombombeng:** Etuge 40 4/1986.

Landolphia cf. violacea (K.Schum. ex Hallier f.) Pichon
Liana with tendrils, glabrous; leaves elliptic to obovate, 13 × 6cm, acumen 0–5cm, base acute, abruptly subauriculate, lateral nerves 7–8 pairs, brochidodromous; petiole 0.5cm. Forest; 900m.
Distr.: SE Nigeria to Congo (K). (Lower Guinea & Congolian).
Nyasoso: Cable 2946a 6/1996.
Note: identification tentative.

Motandra guineensis (Thonn.) A.DC.
Liana, puberulent; leaves papery, elliptic-oblong or oblanceolate, to 12 × 3.5cm, acumen 1–2cm, base rounded, nerves c. 4 pairs; petiole 0.5cm; panicles terminal; rachis to 12cm; corolla twisting to right, white, 5mm; fruit bifid or entire, fleshy, narrowly ovoid, c. 8 × 1.5cm, apex pointed, densely hairy, winged. Forest; 300m.
Distr.: Guinea (C) to Uganda. (Guineo-Congolian).
IUCN: LC
Bolo: Thomas 5902 fl., 3/1986.

Oncinotis glabrata (Baill.) Stapf ex Hiern
Liana, glabrous; leaves papery, elliptic-oblong, 12 × 4.5cm, acumen 1cm, lower surface glossy, lateral nerves 6–8 pairs, brochidodromous; petiole 1cm; panicles axillary; rachis to 8cm, many-flowered, puberulent; corolla yellow, twisting to right, 5mm; stamens exserted; fruit bifid; follicles each c. 15 × 1.5cm; seeds winged. Forest; 900m.

Distr.: Guinea (C) to Uganda. (Guineo-Congolian).
IUCN: LC
Ngomboaku: Gosline 258 12/1999.

Orthopichonia cirrhosa (Radlk.) H.Huber
Kew Bull. 15: 437 (1962).
Syn.: *Orthopichonia nigeriana* (Pichon) H.Huber
Liana, glabrous; raised lenticels dense, concolorous; leaves leathery, grey-green, lanceolate, c. 10 × 4cm, acumen ligulate, 1cm, base acute, lateral nerves c. 40 pairs, straight; petiole 1.5–2cm; panicles contracted, 2cm, axillary and terminal; corolla white, 1.5cm; fruit globose, 2cm, rugose, orange, edible. Forest; 850–1050m.
Distr.: SE Nigeria to Congo (K). (Lower Guinea & Congolian).
IUCN: LC
Kupe Village: Cable 2729 5/1996; Cheek 7805 fr., 11/1995; Etuge 1498 11/1995; Ryan 207 5/1996; **Mbule:** Letouzey 14675 fr., 4/1976; **Mt Kupe:** Schoenenberger 46 fr., 11/1995.
Local names: Mbide or Nhum. **Uses:** FOOD – infructescences (Marcus in Cheek 7805).

Picralima nitida (Stapf) T.Durand & H.Durand
Tree or shrub 3–20m, glabrous; resembling *Hunteria umbellata*, but corolla 3–4cm; fruit follicles to 20cm. Forest; 250–800m.
Distr.: Ivory Coast to Uganda. (Guineo-Congolian).
IUCN: LC
Ikiliwindi: Etuge 499 3/1987; **Kupe Village:** Etuge 1978 5/1996; Kenfack 227 7/1996; **Nyandong:** Cheek 11356 3/2003.

Pleiocarpa bicarpellata Stapf
Shrub 8m, resembling *P. rostrata*, but leaves c. 10 × 4cm, acumen 1.5cm, ligulate, lateral nerves 19–25 pairs; inflorescence 10–15-flowered; anthers up to 1.3mm from throat; fruit with 2 carpels. Forest; 450–910m.
Distr.: Nigeria to Kenya. (Tropical Africa).
IUCN: LC
Ekona Mombo: Etuge 419 12/1986; **Nyandong:** Ghogue 1436 fr., 3/2003; **Nyasoso:** Zapfack 667 6/1996.

Pleiocarpa cf. mutica Benth.
Shrub resembling *P. rostrata*, but lateral nerves 9–12 pairs; inflorescence 10–15-flowered; corolla less than 2mm wide (pressed); anthers 0.2–1.3mm from corolla throat; fruits 1-seeded. Forest; 870–1200m.
Distr.: Sierra Leone to Gabon. (Upper & Lower Guinea).
Kupe Village: Cable 2452 5/1996; 740 fl., 1/1995; Etuge 2640 7/1996.
Note: material cited mostly lacks flowers and is identified tentatively on nerve number. May prove to be aberrant *P. rostrata*.

Pleiocarpa rostrata Benth.
Wag. Agric. Univ. Papers 96(1): 152 (1996).
Syn.: *Pleiocarpa talbotii* Wernham
Shrub 1–3m, glabrous; leaves opposite, lanceolate or elliptic, c. 18 × 8cm, acumen 1cm, base obtuse-decurrent, lateral nerves up to 9 pairs; petiole 1cm; fascicles 1–4-flowered; corolla white, tube 2–2.5cm, 3mm wide (pressed); anthers inserted 3–8mm from throat, lobes twisting to left; fruit apocarpous; follicles 4–5, berries, ovoid, 3cm, rounded, stipitate or not, 1–5-seeded. Forest; 150–1400m.
Distr.: SE Nigeria to Gabon. (Lower Guinea).
IUCN: LC

Bangem: <u>Thomas 5339</u> 1/1986; **Etam:** <u>Etuge 207</u> 7/1986; **Kupe Village:** <u>Cable 2496</u> 5/1996; <u>2573</u> 5/1996; <u>2610</u> 5/1996; <u>2659</u> 5/1996; <u>3716</u> 7/1996; <u>716</u> fl., 1/1995; <u>788</u> fl., fr., 1/1995; <u>804</u> fl., 1/1995; <u>Cheek 6027</u> fl., fr., 1/1995; <u>7650</u> 11/1995; <u>8370</u> 5/1996; <u>Ensermu 3554</u> 11/1995; <u>Etuge 1697</u> 2/1996; <u>2848</u> 7/1996; <u>Kenfack 304</u> 7/1996; **Mt Kupe:** <u>Schoenenberger 65</u> fl., 11/1995; **Nyale:** <u>Etuge 4179</u> 2/1998.
Note: rostrate, ornamented fruits are not known in Bakossi.

Rauvolfia caffra Sond.
Syn.: *Rauvolfia macrophylla* Stapf
Tree 15–30m; trunk smooth, grey; leaves in whorls, 4, oblanceolate, to c. 30 × 9cm, subacuminate, decurrent-sessile; petiole to 2cm, lateral nerves c. 30 pairs; inflorescence as *R. vomitoria*; fruit with carpels completely united. Farmbush; 350–1300m.
Distr.: tropical and subtropical Africa.
IUCN: LC
Kupe Village: <u>Etuge 1712</u> 2/1996; <u>2036</u> 5/1996; <u>2858</u> 7/1996; <u>Groves 11</u> fl., fr., 1/1995; **Manehas F.R.:** <u>Cheek 9395</u> 10/1998; **Mwanenguba:** <u>Leeuwenberg 9259</u> fl., 1/1972; **Ngusi:** <u>Etuge 1537</u> 1/1996; **Nyasoso:** <u>Thomas 3066</u> 2/1984.
Local name: Elome. **Uses:** MEDICINES (Cheek 9395).

Rauvolfia mannii Stapf
Shrub 1–3m, glabrous; leaves opposite or in whorls of 3, elliptic, c. 9(–15) × 3(–7)cm, acumen 1cm, cuneate-decurrent, lateral nerves 10 pairs; petiole 1cm; panicle 1.5cm, 3–10-flowered; corolla 8mm, white with 5 red spots, rarely pink; berry bilobed, rarely entire. Forest; 200–1500m.
Distr.: tropical Africa.
IUCN: LC
Baduma: <u>Nemba 576</u> 7/1987; **Bangem:** <u>Thomas 5367</u> 1/1986; <u>5916</u> 3/1986; **Bolo Moboka:** <u>Nemba 248</u> 9/1986; **Kodmin:** <u>Barlow 11</u> 1/1998; <u>Cheek 10265</u> 12/1999; <u>8918</u> 1/1998; <u>Etuge 4418</u> 11/1998; <u>Ghogue 43</u> fl., 1/1998; <u>Onana 569</u> 2/1998; **Konye:** <u>Thomas 5200</u> 11/1985; **Mekom:** <u>Thomas 5397</u> fl., fr., 1/1986; **Mungo River F.R.:** <u>Cheek 10115</u> 11/1999.

Rauvolfia vomitoria Afzel.
Shrub or tree 2–8m; stems greyish white; leaves in whorls of 3, membranous-papery, elliptic c. 20 × 7cm, subacuminate, cuneate, lateral nerves 12 pairs; petiole 2cm; panicles puberulent, 10cm; corolla white, 8mm; fruit with 2 ovoid berries 8mm. Forest, savanna; 700–1750m.
Distr.: Senegal to Uganda. (Guineo-Congolian).
IUCN: LC
Ehumseh - Mejelet: <u>Etuge 315</u> 10/1986; **Essossong:** <u>Leeuwenberg 9227</u> fl., 1/1972; **Kodmin:** <u>Biye 38</u> 1/1998; <u>Cheek 8885</u> 1/1998; <u>Plot B262</u> 1/1998; <u>Satabie 1115</u> fl., 1/1998; **Konye:** <u>Tadjouteu 545</u> 3/2003; **Kupe Village:** <u>Cable 2419</u> 5/1996; <u>Cheek 7020</u> fr., 1/1995; <u>Etuge 1374</u> 11/1995; <u>2621</u> 7/1996; <u>Lane 334</u> fr., 1/1995; <u>367</u> fl., 1/1995; **Mwanenguba:** <u>Leeuwenberg 9577</u> fl., fr., 4/1972; **Nyasoso:** <u>Cable 2821</u> 6/1996; <u>3474</u> 7/1996.
Local name: Ngimpah. **Uses:** MATERIALS – wood used for cutlass handles; MEDICINES (Cheek 8885).

Strophanthus bullenianus Mast.
Liana; exudate white or clear; stems red, glabrous to hispid; leaves ovate to obovate, 3–16 × 1.3–7.5cm, acumen slender, 1–2cm, cuneate to subcordate, undulate, translucent dots, lateral nerves 7–13 pairs; petiole 2–7mm; inflorescence 1–10-flowered; peduncle 0–5cm; bracts 0.5–1cm; pedicels 0.5–2cm; calyx 5–12mm; corolla cream, turning red, 7–15mm,

upper, wider part of tube 2.5–3.5mm, lobes ovate, reflexed, with tails 1–4cm; anthers exserted. Forest; 400m.
Distr.: Cameroon to Congo (K). (Lower Guinea).
IUCN: LC
Eboné: <u>Leeuwenberg 8646</u> 11/1975.

Strophanthus hispidus DC.
Shrub or liana, deciduous, hispid; resembling *S. bullenianus*; inflorescence 1–70-flowered; calyx 1.3–3.5cm; upper, wider part of corolla tube 5.5–12mm; anthers included. Forest; 400m.
Distr.: Senegal to Tanzania. (Guineo-Congolian).
IUCN: LC
Mungo River: <u>Olorunfemi in FHI 30529</u> fl., 4/1951.

Strophanthus preussii Engl. & Pax
Shrub to 5m, or liana to 12m; stems glabrous; resembling *S. bullenianus*, but 1–24-flowered; bracts 0.4–2cm; calyx 0.4–2.5cm; corolla 1.2–2.6cm; corolla lobes over 12cm. Forest; 400m.
Distr.: Guinea (C) to Tanzania. (Guineo-Congolian).
IUCN: LC
Banyu: <u>Olorunfemi in FHI 30699</u> fl., 7/1951.

Strophanthus thollonii Franch.
Liana to 20m, glabrous; resembling *S. bullenianus*, but inflorescence 1–2-flowered; bracts 0.1–1.2cm; pedicels 0.2–1cm; calyx 1–2.6cm; corolla tube 2.4–3.8cm, upper, wider part 11–22mm wide; corona lobes 1–2.7cm; corolla lobes ovate, 2–4 × 1–2cm, patent to slightly reflexed, tails absent. Riverbanks in forest; 150–300m.
Distr.: SE Nigeria to Gabon. (Lower Guinea).
IUCN: LC
Mekom: <u>Thomas 5392</u> fl., 1/1986; **Mungo River:** <u>Gosline 223</u> fl., 11/1999.
Note: amongst other spp. of the genus likely to be found in lowland Bakossi are S. gratus and S. sarmentosus.

Tabernaemontana brachyantha Stapf
Tree 8–15m, glabrous; leaves elliptic, c. 30 × 17cm, acumen 0.7cm, obtuse, lateral nerves c. 12 pairs; petiole 1cm; panicle c. 20 × 16cm; peduncle 7–23cm; corolla white, tube 1.2–1.8cm; fruit bicarpellate; carpels divided to base, each depressed, subglobose or transversely oblong, 4–5 × 5.5–8 × 4–6cm, sutured; wall 5–15mm thick, containing a mass of ellipsoid seeds, 12mm. Forest; 200–1100m.
Distr.: SE Nigeria to Congo (K). (Lower Guinea).
IUCN: LC
Ikiliwindi: <u>Nemba 457</u> 1/1987; **Kupe Village:** <u>Cable 715</u> fl., 1/1995; <u>Cheek 7055</u> fl., 1/1995; <u>Etuge 1620</u> 1/1996; <u>Lane 266</u> fl., 1/1995; **Ndum:** <u>Cable 883</u> fl., 1/1995.
Note: remarkable for the short corolla tube.

Tabernaemontana crassa Benth.
Tree 6–20m, resembling *T. brachyantha* and perhaps not reliably distinguishable in fruit; peduncle 3.5–11cm; corolla tube 3.5–6cm; carpels obliquely subglobose, 5–12 × 4–11 × 4–10cm; wall 9–40mm. Forest; 760–1260m.
Distr.: Sierra Leone to Congo (K). (Guineo-Congolian).
IUCN: LC
Kupe Village: <u>Cable 3692</u> 7/1996; <u>Cheek 7831</u> 11/1995; <u>Ryan 249</u> 5/1996; **Nyasoso:** <u>Cable 2802</u> 6/1996; <u>3490</u> 7/1996; <u>Etuge 2430</u> 6/1996.
Note: our material is all in fruit; confusion with *T. brachyantha* is therefore possible.

Tabernaemontana eglandulosa Stapf

Small, night-flowering liana; stem 1–3cm diameter; anisophyllous or not; leaves subelliptic, 4–22 × 1.5–10cm, base and apex variable, lateral nerves 4–10 pairs; petioles 4–50mm; intrapetiolar stipules; inflorescence congested, to 14 × 10cm, 3–12-flowered, pedunculate; sepals 1.5–7mm; corolla tube 2.5–6cm; carpels subglobose, to 8 × 7 × 5cm, yellow, with two blunt angles. Forest; 700m.
Distr.: Benin to Congo (K). (Lower Guinea & Congolian).
IUCN: LC
Nyandong: Thomas 6699 2/1987; **Southern Bakossi F.R.:** Binuyo in FHI 35160 fl., 1/1956.
Note: specimens cited not seen.

Tabernaemontana inconspicua Stapf

Kew Bull.: 120 (1894).
Syn.: *Pterotaberna inconspicua* (Stapf) Stapf
Shrub or tree, glabrous; stems white; leaves membranous, drying brown, elliptic, c. 12 × 5.5cm, acumen 1cm, cuneate-decurrent, lateral nerves 9–10 pairs; petiole 1–1.5cm; inflorescence 4cm; peduncle 2cm, c. 5-flowered; corolla tube 1cm; fruit carpels separate, white, ovoid, 3cm, rostrate, diverging; calyx persistent. Forest; 350m.
Distr.: Cameroon to Congo (K). (Lower Guinea & Congolian).
IUCN: LC
Bakolle Bakossi: Etuge 155 fr., 5/1986.

Tabernaemontana pachysiphon Stapf

Tree 8m, resembling *T. crassa*, but fruit globose, c. 20cm diameter (fresh), 12cm (dried); fruit wall as thick as seed cavity is wide; flowers not seen. Forest; 450m.
Distr.: Ghana to Malawi. (Tropical Africa).
IUCN: LC
Bakossi F.R.: Cheek 9313 10/1998.
Note: other species likely to be found in lowland Bakossi are *T. contorta* and possibly *T. psorocarpa*.

Tabernaemontana penduliflora K.Schum.

Shrub 3–5m, resembling *T. brachyantha*, but leaves c.18 × 8cm, acumen 1cm, lateral nerves 10–12 pairs; petiole 1cm; fascicles 3–10-flowered; corolla tube 1.2cm; fruit not seen. Forest; 520m.
Distr.: Nigeria to Congo (K). (Lower Guinea & Congolian).
IUCN: LC
Eboné: Leeuwenberg 9370 11/1972.

Tabernaemontana ventricosa Hochst. ex A.DC.

Tree (6–)10–20m, broadly resmbling *T. brachyantha*; leaves papery, drying pale green, oblong-elliptic to 21 × 9cm, acumen 0.5cm, obtuse-rounded decurrent, lateral nerves 12 pairs, drying yellow; peduncle 4–6cm, stout, 10–20-flowered; sepals 5mm; corolla not seen; ripe fruit globose, 8cm, green with rough surface. Forest; 1100–1600m.
Distr.: Cameroon to Kenya. (Afromontane).
IUCN: LC
Kodmin: Etuge 4013 1/1998; Plot B67 1/1998; **Kupe Village:** Cable 3790 7/1996; Ryan 347 5/1996; **Nyasoso:** Elad 138 fl., 2/1995; 143 fl., 2/1995; Etuge 2393 6/1996.

Tabernaemontana sp. 1

Tree 25m; leaves elliptic-oblong, c. 23 × 8cm, subacuminate, cuneate-decurrent, c. 9 pairs lateral nerves; petiole 2cm; fruit globose, 4cm (dry), 5–8cm (fresh). Forest; 420m.
Mungo River F.R.: Onana 942 11/1999.
Note: not matched with any other species, but better and more ample material needed to resolve identification.

Voacanga africana Stapf

Tree 7–12m; stems greyish white; leaves obovate, c. 22 × 10cm, rounded, decurrent-sessile, lateral nerves c. 15 pairs; inflorescence erect, 6–15cm, c. 20-flowered; corolla white, tube 8mm, lobes patent, 20mm; fruit pendulous; follicles 2, each 4cm, green, mottled white, splitting to show black seeds in orange mass; wall 1cm thick. Forest; 500–1200m.
Distr.: Senegal to Tanzania. (Guineo-Congolian).
IUCN: LC
Kupe Village: Etuge 1830 3/1996; 1912 5/1996; **Manehas F.R.:** Cheek 9394 10/1998; **Ngombombeng:** Etuge 27 4/1986; **Nyasoso:** Etuge 2507 7/1996; Thomas 3046 2/1984; **Wone:** Nemba 593 7/1987.
Local names: Elongeh-longeh (Etuge 1912); El-il-long (Epie Ngome in Cheek 9394). **Uses:** MEDICINES (Etuge 1830 & 1912) & (Epie Ngome in Cheek 9394).

Voacanga bracteata Stapf

Syn.: *Voacanga diplochlamys* K.Schum.
Shrub 2–4m; leaves elliptic, c. 9 × 3cm, acuminate, acute, lateral nerves 7 pairs; petiole 1cm; inflorescence pendulous; peduncle 2cm, 2–5-flowered; calyx less than 12mm; corolla brown, tube 1.7cm, lobes 1cm. Forest; 710m.
Distr.: Sierra Leone to Congo (K). (Guineo-Congolian).
IUCN: LC
Enyandong: Cheek 10940 10/2001; **Nyale:** Etuge 4177 2/1998.

Voacanga psilocalyx Pierre ex Stapf

Syn.: *Voacanga bracteata* Stapf var. *zenkeri* (Stapf) H.Huber
Shrub closely resembling and perhaps conspecific with *V. bracteata*, differing principally in the calyx more than 12mm (FWTA). Forest; 300–1000m.
Distr.: S Nigeria to Gabon. (Lower Guinea).
IUCN: NT
Bangem: Thomas 5910 3/1986; **Konye:** Thomas 5159 11/1985.
Note: specimens cited det. by Leeuwenberg; fairly rare, only 17 specimens at K.

Voacanga sp. 1

Tree 6(–9)m, glabrous; stems greyish white, white pustular lenticellate, puberulent when young; leaves oblong or oblanceolate, to 24 × 7cm, acumen 0.5cm, cuneate, lateral nerves to 14 pairs; petiole 2.5cm; flowers not seen; fruiting peduncle 8–12cm, 1-fruited, pendulous; fruit with carpels united, only slightly bilobed before dehiscence (when to 4 × 5cm), drying black, finely ridged, puberulent, dehiscing along a single terminal suture, when 4.5 × 7 × 4cm. Forest; 950–1400m.
Distr.: Kupe-Bakossi & Bali-Ngemba. (W Cameroon Uplands).
Kodmin: Gosline 179 11/1998; 69 1/1998; **Kupe Village:** Etuge 2711 7/1996; 2861 7/1996; **Mbule:** Cable 3365 6/1996.
Note: probably a new species; flowers required.

AQUIFOLIACEAE

M. Cheek

Fl. Cameroun 19 (1975).

Ilex mitis (L.) Radlk. var. *mitis*

Tree c. 5m; leaves alternate, variable, elliptic, to 10 × 4cm, acute with mucron, base acute, margin entire or slightly serrate, lateral nerves 8–15; petiole 1.5cm; fascicles 1–3-flowered; peduncle 1cm; flowers white, 2mm; fruit ellipsoid, to 6mm, red with apical button; calyx persistent, 5-lobed. Forest; 2000m.
Distr.: Sierra Leone to S Africa. (Afromontane).
IUCN: LC
Mwanenguba: Thomas 3119 fr., 2/1984.

ARALIACEAE

D.G. Frodin & M. Cheek

Fl. Cameroun 10 (1970).

Cussonia bancoensis Aubrév. & Pellegr.
Tree 30m, glabrous; resembling *Schefflera barteri*, but leaflets drying black, ovate, 27 × 14cm, acumen 0.5cm, sessile, lateral nerves 10 pairs, subserrate; petiole 80–90cm; peduncle 30–40 × 2cm; umbels of 7–9 spikes, each 15–30cm, aroid-like; fruits sessile, 5mm, white, turning red. Forest; 970m.
Distr.: Liberia to Cameroon. (Upper & Lower Guinea).
IUCN: NT
Nyasoso: Cheek 7530 10/1995.
Note: the only fertile specimen known from Cameroon; leaves match Nigerian material well. Rare in most of its range.

Polyscias fulva (Hiern) Harms
Pioneer tree 35m, felty-puberulent; leaves alternate, 40cm, imparipinnate, 6-jugate, elliptic-oblong, 4.5–8 × 2.5–4cm, subacuminate, rounded, densely pale brown-felty below; umbels of panicles, each 30–50cm; partial peduncles racemose, to 5cm; fruits 5mm, purple. Forest; 960–990m.
Distr.: Guinea (C) to Kenya. (Tropical Africa).
IUCN: LC
Kupe Village: Cheek 7818 11/1995; **Ndum:** Elad 96 fr., 2/1995; **Nyasoso:** Etuge 3189 9/1996.

Schefflera abyssinica (Hochst. ex. A.Rich.) Harms
Tree resembling *S. barteri*, but deciduous; leaflets ovate, 15 × 9cm, gradually acuminate, shallowly cordate, serrate, lateral nerves 10 pairs; petiolules 7cm. Forest; 1300–1800m.
Distr.: Cameroon, Uganda, Ethiopia. (Afromontane).
IUCN: LC
Mwanenguba: Etuge 4377 10/1998; **Nsoung:** Darbyshire 79 3/2003.

Schefflera barteri (Seem.) Harms
Tree 4–20m, initially an epiphyte, then a strangler on e.g. *Albizia* (Elad 109) or *Pycnanthus* (Cheek 7529); leaves c. 7-digitately compound, 35cm; leaflets elliptic, 7–13 × 7cm, acumen 0.5cm, obtuse-decurrent, lateral nerves c. 6 pairs; petiolules to 3cm; stipule intrapetiolar; umbels of 15 spikes 30cm; partial peduncles numerous, 1cm; pedicels 5mm; flowers white, 2mm; stamens 6, erect; fruit 5mm, red. Forest; 200–1450m.
Distr.: Guinea (C) to Congo (K). (Guineo-Congolian).
IUCN: LC
Kodmin: Cheek 9576 fr., 11/1998; Plot B217 1/1998; **Kupe Village:** Cable 3656 7/1996; 3893 7/1996; Cheek 8372 5/1996; **Mejelet-Ehumseh:** Etuge 179 6/1986; **Mekom:** Nemba 18 4/1986; **Mile 12 Mamfe Road:** Nemba 298

10/1986; **Ndum:** Elad 109 2/1995; **Ngomboaku:** Cheek 10360 12/1999; **Nyasoso:** Cable 2812 6/1996; 3245 6/1996; Cheek 7529 10/1995; Wheatley 403 fl., fr., 7/1992; Zapfack 710 6/1996.
Note: material in E Africa now referred to separate taxa, e.g. *S. stuhlmannii*.
Local names: Ebenokoup (Max Ebong in Cheek 9576); Ebin (Cheek 10360). **Uses:** FUELS – fuelwood (Max Ebong in Cheek 9576).

Schefflera barteri × *mannii ?*
Tree to 15m, intermediate between *S. barteri* and *S. mannii* in morphology and altitudinal range: leaflets oblong, lateral nerves 9 pairs; pedicels 2mm. Forest; 1700m.
Nyasoso: Etuge 1751 2/1996.

Schefflera hierniana Harms
Epiphytic, climbing shrub 4–8m; resembling *S. barteri*, but with acumen slender, 1–2cm, persistent fleshy-leathery bracts at base of inflorescence, most conspicuous in immature inflorescences; inflorescence congested, densely brown-scurfy-pubescent; partial peduncles 2–3mm; pedicels 2mm. Forest; 1200–1300m.
Distr.: Cameroon & Bioko. (W Cameroon Uplands).
IUCN: EN
Kodmin: Etuge 3974 1/1998; **Nzee Mbeng:** Gosline 98 2/1998.
Note: taxon resurrected by Frodin from synonymy with *S. barteri*; much rarer.

Schefflera mannii (Hook.f.) Harms
Tree resembling *S. barteri*, but leaves 18 × 9cm, subacuminate, lateral nerves 15–20 pairs; petiolules 5cm; flowers sessile. Forest; 1400–2000m.
Distr.: Nigeria, Bioko, Cameroon. (W Cameroon Uplands).
IUCN: VU
Kodmin: Ghogue 46 1/1998; Plot B64 1/1998; **Mt Kupe:** Sebsebe 5092 11/1995; **Mwanenguba:** Thomas 3137 2/1984; **Nyasoso:** Elad 125 fr., 2/1995; Etuge 1684 1/1996.

ARISTOLOCHIACEAE

M. Parsons, F. Gonzalez & M. Cheek

Pararistolochia ceropegioides (S.Moore) Hutch. & Dalziel
Adansonia (sér. 2) 17: 482 (1978).
Climber 5–15m; stem 8-shaped in section, 1–2cm diameter, flexible; leaves leathery, ovate, c. 13 × 8cm, 3-nerved, acuminate, rounded-truncate; inflorescence cauliflorous, 1–2m from ground, 3–8-flowered; flowers 5–7cm; perianth tube pouched at base, pale brown outside; tepals 3, equal, 1–2cm, slightly splayed, throat orange; fruit fleshy, ellipsoid, c. 5–6 × 3–4cm, 5-ridged, base and apex truncate; pedicel 3–4cm. Forest; 600–1200m.
Distr.: Cameroon & Gabon. (Lower Guinea).
IUCN: VU
Kupe Village: Cable 721 fl., 1/1995; 843 fr., 1/1995; Cheek 7002 fr., 1/1995; Etuge 2923 7/1996; Groves 18 fl., 1/1995; Parsons 1 11/1999; 2 11/99; 3 11/99; 4 12/99; **Mwambong:** Parsons 7 12/1999; **Nyasoso:** Cable 1135 fl., fr., 2/1995; 3316 6/1996; Cheek 7318 2/1995; 9522 10/1998; Etuge 2100 6/1996; 2555 7/1996; Sebsebe 5048 10/1995; Sidwell 377 10/1995.

Note: *P. preussii* of Kumba, of which the type is destroyed, appears to be continuous with this taxon insofar as perianth length, a key character in Poncy (1978).

Pararistolochia goldieana (Hook.f.) Hutch. & Dalziel
Climber resembling *P. ceropegioides*, but stems annual, terete; leaves membranous, ovate, c. 30 × 30cm, acuminate, deeply cordate; flowers at ground level, to 30 × 20 × 20cm, red and white striped; fruit cylindrical, c. 30 × 3cm. Forest; 300–400m.
Distr.: Sierra Leone to Cameroon. (Upper & Lower Guinea).
IUCN: VU
Kupe Village: Etuge 2033 5/1996; **Mahole:** Parsons 14 12/1999.

Pararistolochia promissa (Mast.) Keay
Syn.: *Pararistolochia talbotii* (S.Moore) Keay
Syn.: *Pararistolochia tenuicauda* (S.Moore) Keay
Climber resembling *P. ceropegioides*, but perianth tube 8cm, white, veined purple, lobes dangling, linear-spatulate, each c. 18 × 0.2cm, smelling of stale urine. Forest; 350–1400m.
Distr.: Ghana to Congo (K). (Guineo-Congolian).
IUCN: LC
Kodmin: Plot B216 fl., 1/1998; **Mt Kupe:** Nemba 913 3/1988; **Mwambong:** Parsons 5 12/1999; 10 12/1999; **Nyasoso:** Cheek 7328 fl., fr., 2/1995.

Pararistolochia triactina (Hook.f.) Hutch. & Dalziel
Climber resembling *P. ceropegioides*, but leaves shallowly 3-lobed, deeply cordate-auriculate, 12–15 × 9–12cm; perianth tube 9cm; fruit 25 × 3–4cm, 6-angled. Forest; 1000–1150m.
Distr.: Benin to Uganda. (Guineo-Congolian).
IUCN: LC
Kodmin: Cheek 9206 2/1998; **Mwambong:** Parsons 11 12/99; 6 12/1999.
Note: *P. flos-avis* and *P. zenkeri* are amongst the other species likely to be found in lower altitude forest of the checklist area, since they are known nearby.

Pararistolochia cf. zenkeri (Engl.) Hutch. & Dalziel
Climber, resembling *P. ceropegioides*, but flowers larger, often red, perianth lobes triangular, 2–2.5 × 1.2–2cm, reflexed; leaves oblong-elliptic, 7–11 × 5–6.5cm; fruit 15 × 3cm, 6-grooved with rounded ridges. Forest; 1000m.
Distr.: SE Nigeria & Cameroon. (Lower Guinea).
Nyasoso: Sebsebe 4999 10/1995.
Note: the specimen cited has fruit only; flowers are needed to confirm identification.

ASCLEPIADACEAE
D.J. Goyder

Anisopus efulensis (N.E.Br.) Goyder
Kew Bull. 49: 743 (1994).
Syn.: *Anisopus mannii* N.E.Br. *sensu* Bullock in part, excl. type.
Slender twiner; latex white; leaves opposite, to c. 15 × 9cm, obovate, glabrous; inflorescences axillary, single or in opposite pairs, with flowers in subsessile clusters; corolla lobes 1.5–3(–10)mm long, velvety-pubescent; corona lobes

subglobose with an upward- or inward-pointing tooth; follicles usually paired, divergent. Evergreen forest; 200–1700m.
Distr.: Ghana to N Zambia. (Guineo-Congolian).
IUCN: LC
Kupe Village: Etuge 2847 7/1996; **Nyasoso:** Etuge 1732 2/1996; 2195 6/1996; **Southern Bakossi F.R.:** Thomas 5559 fl., 2/1986.

Asclepias curassavica L.
Erect herb to c. 1m; latex white; leaves opposite, 7–18 × 1–3cm, narrowly elliptic, acute, ± glabrous; inflorescences extra-axillary at upper nodes, umbelliform, with erect flowers; corolla bright red, lobes reflexed, c. 7.5mm long, glabrous; corona lobes arising high up the staminal column, orange, fleshy, cucullate and with conspicuous teeth; follicles usually occuring singly, narrowly fusiform, erect. Cultivated; 1170m.
Distr.: a pantropical weed, native to S America.
IUCN: LC
Edib village: Cheek (sight record) 2/1998.

Batesanthus parviflorus Norman
J. Bot. 67, suppl. 2: 91 (1929).
Climber resembling *B. purpureus*, but corolla much smaller, lobes to 7mm. Forest edge; 1900m.
Distr.: Sierra Leone to Cameroon. (Upper & Lower Guinea).
IUCN: NT
Mwanenguba: Polhill 5190 fl., 3/1984.
Note: taxon not recognised by FWTA or Lebrun & Stork but maintained by Venter (BLFU) who is revising the genus. Species needs delimitation before a full conservation assessment. Two specimens at K only.

Batesanthus purpureus N.E.Br.
Woody twiner; latex white; leaves opposite, to c. 16 × 10cm, oblong, glabrous, with interpetiolar stipular fringe; flowers in lax axillary panicles; corolla purple, rotate, lobes to c. 18 × 11mm, rounded; corona indistinct, annular; follicles stout, erect, parallel. Forest; 1000–1950m.
Distr.: Cameroon. (W Cameroon Uplands).
IUCN: VU
Kupe Village: Ryan 429 5/1996; 430 5/1996; **Nyasoso:** Cable 2915 6/1996.

Ceropegia talbotii S.Moore
Slender twiner; latex clear; leaves opposite, to c. 5 × 3cm, ovate, subglabrous; inflorescences extra-axillary; flowers with long slender calyx lobes; corolla white, blotched purple, with slender curved tube 2–4cm long, expanding suddenly at the mouth, lobes c. 1cm long, united at tip. Forest margins; 960–1000m.
Distr.: Sierra Leone to Cameroon. (Upper & Lower Guinea).
IUCN: LC
Enyandong: Etuge 4492r 11/2001; **Essossong:** Cheek 9387 10/1998.

Cynanchum adalinae (K.Schum.) K.Schum. subsp. *adalinae*
Ann. Missouri Bot. Gard. 83: 291 (1996).
Slender twiner; latex white; leaves opposite, to c. 7.5 × 5cm, ovate with cordate base, glabrous; inflorescences extra-axillary, subsessile, with sweetly-scented flowers; corolla rotate, creamish green, lobes 3.5–4mm long; corona white, 2.5–3mm long, obscuring the gynostegium. Margins of secondary forest; 660–1100m.
Distr.: Ivory Coast to Congo (K). (Guineo-Congolian).
IUCN: LC

Kupe Village: Cheek 7038 fl., 1/1995; 7095 1/1995; 8327 5/1996; Etuge 1908 5/1996; Kenfack 231 7/1996; Ryan 338 5/1996; Zapfack 910 7/1996; **Mt Kupe:** Nemba 911 fl., 3/1988; **Ndum:** Cable 886 1/1995; **Ngomboaku:** Etuge 4669 12/1999; **Nyasoso:** Cable 2775 6/1996; 863 fl., 1/1995; Elad 120 fl., 2/1995; Groves 102 2/1995; Sidwell 306 10/1995.

Gomphocarpus physocarpus E.Mey.
Kew Bull. 56: 788 (2001).
Shrubby herb to 2.5m; latex white; leaves opposite, to c. 9 × 1.5cm, narrowly oblong to lanceolate; umbels extra-axillary, with white flowers; corolla reflexed, lobes 5–8mm long; corona lobes with tooth on upper margin; follicles inflated, globose, covered with filiform processes. Seasonally wet pastures and flood plains, disturbed areas; 1450m.
Distr.: S Africa to Mozambique, naturalised in other tropical and subtropical regions of the world. (Pantropical).
IUCN: LC
Kodmin: Cheek 9584 11/1998.
Local name: Echim eh nyoh = 'Snake' (Cheek 9584).

Gomphocarpus semilunatus A.Rich.
Kew Bull. 56: 791 (2001).
Stout shrubby herb to 2.5m; latex white; leaves opposite or whorled, to 12 × 1.5cm, linear or linear-lanceolate; umbels extra-axillary, with pink or white flowers; corolla ± rotate, lobes 7–9mm long; corona lobes pink or purple, upper margins denticulate; follicles inflated, ovoid-subglobose, covered with filiform processes. Seasonally flooded alluvial grasslands and waste places, frequently in large numbers; 1975m.
Distr.: Nigeria to Ethiopia, E Africa, Zambia, Angola. (Tropical Africa).
IUCN: LC
Kodmin: Cheek 8870 1/1998; **Mwanenguba:** Cheek 7248 fl., fr., 2/1995.

Marsdenia latifolia (Benth.) K.Schum.
Omlor, Gen. Rev. Marsdenieae (Asclepiadaceae): 75 (1998).
Syn.: *Gongronema latifolium* Benth.
Slender to robust woody scrambler; latex white; leaves opposite, to c. 12 × 7cm, ovate to ovate-oblong, pubescent; inflorescences extra-axillary with cream or yellow-green flowers scattered along the axes; corolla lobes c. 2–2.5mm long; corona lobes fleshy and with an apical tongue; follicles single, subcylindrical. Margins of evergreen forest or scrub; 1450m.
Distr.: widespread in C & W Africa, to Zambia, Angola. (Tropical Africa).
IUCN: LC
Kodmin: Cheek 9572 11/1998.
Local name: Nkum mwankume. **Uses:** SOCIAL USES – 'Religious' uses – traditional use for countering juju (Max Ebong in Cheek 9572).

Neoschumannia kamerunensis Schltr.
Slender twiner; latex clear; leaves opposite, to c. 12 × 6.5cm, elliptic or oblong, glabrous; inflorescences extra-axillary, with racemes of green or dark purple flowers; corolla reflexed, the lobes 7–18mm long; corona in three series along the gynostegium. Evergreen forest; 800m.
Distr.: Ivory Coast, Cameroon & CAR. (Upper & Lower Guinea).
IUCN: CR
Kupe Village: Etuge 1443 11/1995; 1990 5/1996; **Mejelet-Ehumseh:** Etuge 2933 8/1996.

Pentarrhinum abyssinicum Decne. subsp. *angolense* (N.E.Br.) Liede & Nicholas
Kew Bull. 47: 482 (1992).
Slender twiner; latex white; leaves opposite, to c. 7 × 5cm, broadly ovate with a cordate base, pubescent; inflorescences extra-axillary on long peduncles, with 5–15 greenish yellow flowers; corolla lobes 6–7mm long; corona lobes 2–3mm long, white, slipper-shaped. Forest; 1800m.
Distr.: scattered across tropical Africa.
IUCN: LC
Kodmin: Ghogue 56 fr., 1/1998.

Pergularia daemia (Forssk.) Chiov.
Robust twiner; latex white; leaves opposite, to c. 11 × 11cm, broadly ovate with cordate base, pubescent; inflorescences extra-axillary; flowers white, cream or green; corolla tube 2–4mm long, lobes 5–13mm long, spreading; staminal corona lobes conspicuous, with basal and apical projections; follicles paired, smooth or echinate. Dry bushland; 260–1450m.
Distr.: widespread across tropical Africa, Arabia, Indian subcontinent. (Palaeotropics).
IUCN: LC
Baduma: Nemba 628 9/1987; **Kodmin:** Cheek 9608 fl., 11/1998; **Ndum:** Williams 138 1/1995; **Ngusi:** Etuge 1571 1/1996; **Nyasoso:** Cable 2824 6/1996; Etuge 1757 3/1996; Letouzey 14713 fl., 4/1976; Thomas 5042 11/1985; Zapfack 631 6/1996.

Secamone africana (Oliv.) Bullock
Woody climber; latex white; leaves opposite, to c. 8 × 3cm, lanceolate to ovate, glabrous; inflorescences terminal or extra-axillary, rusty-pubescent; flowers congested; corolla lobes 2–4mm long; follicles 5–10cm long, widely divergent. Forest margins; 1400m.
Distr.: Cameroon & N Angola to Kenya & Tanzania. (Tropical Africa).
IUCN: LC
Mejelet-Ehumseh: Etuge 175 fl., 6/1986.

Secamone afzelii (Schultes) K.Schum.
Slender woody climber; latex white; leaves opposite, to c. 7 × 3cm, oblong-elliptic, glabrous; inflorescences terminal or extra-axillary, lax, well-branched with minute yellow or orange flowers; corolla lobes c. 1mm long; follicles slender, paired, 5–7cm long. Forest; 300–700m.
Distr.: widespread across C & W Africa. (Tropical Africa).
IUCN: LC
Etam: Etuge 210 fl., 8/1986; **Kupe Village:** Etuge 2749 7/1996; **Mungo River F.R.:** Cheek 9335 10/1998.

Secamone letouzeyana (H.Huber) Klack.
Adansonia (sér. 3) 23: 332 (2001).
Syn.: *Toxocarpus letouzeyanus* H.Huber
Slender woody twiner; twiner with reddish hairs, latex white; leaves opposite, to c. 9 × 3cm, narrowly ovate to elliptic, pubescent below; inflorescences extra-axillary, lax, branched, with white, cream or greenish flowers; corolla lobes narrowly oblong, twisted, 3.5–5mm long. Marshy or disturbed areas in forest; 1300m.
Distr.: Cameroon to Congo (K). (Lower Guinea & Congolian).
IUCN: NT
Kodmin: Satabie 1121 fl., 2/1998.
Note: rare throughout its range, known from only 11 previous collections.

Secamone racemosa (Benth.) Klack.

Adansonia (sér. 3) 23: 322 (2001).
Syn.: *Rhynchostigma racemosum* Benth.
Syn.: *Toxocarpus racemosus* (Benth.) N.E.Br.
Slender woody twiner; twiner with reddish hairs, latex white; leaves opposite, to c. 12 × 5cm, usually elliptic, mostly glabrous; inflorescences opposite and axillary, lax, with white, yellow, green or pink flowers; corolla lobes oblong to obovate, not twisted, 3–4mm long; follicles slender, paired, 18–30cm long. Marshy or riverine areas in montane forest; 1250m.
Distr.: Bioko, Cameroon, Congo (K) to Burundi. (Afromontane).
IUCN: VU
Edib: Etuge 4097 1/1998; **Lake Edib:** Cheek 9162 2/1998.
Note: disjunct populations: W Cameroon uplands and Albertine Rift; slight morphological separation. Two previous collections in Cameroon.

Tacazzea apiculata Oliv.

Robust liana with reddish stems, latex white; leaves opposite, 6–10(–20) x 3–7(12)cm, broadly ovate to elliptic, tomentose beneath, interpetiolar ridge with reddish colleters; inflorescences axillary, paired, with lax panicles of greenish or reddish flowers; corolla rotate, lobes 5–7mm long; corona lobes filiform above; follicles paired, divergent, 3–8.5cm long. Forest margins and along freshwater lakes and rivers; 1975–2000m.
Distr.: widely distributed throughout tropical & subtropical Africa.
IUCN: LC
Bangem: Leeuwenberg 9533 fl., 3/1972; **Mwanenguba:** Cheek 7268 fl., 2/1995.

Tacazzea pedicellata K.Schum.

Ann. Missouri Bot. Gard. 88: 566 (2001).
Syn.: *Zacateza pedicellata* (K.Schum.) Bullock
Woody liana; stems reddish, latex white; leaves opposite, 10–20 × 3–5.5cm, elliptic or oblong, with conspicuous, closely parallel secondary venation; inflorescences opposite and axillary; pedicels c. 3–4cm long, very slender; flowers green or pink, c. 6mm long. Lowland forest, frequently along rivers; 1300m.
Distr.: Nigeria to Congo (K). (Lower Guinea & Congolian).
IUCN: NT
Edib: Etuge 4478 11/1998.
Note: fairly rare in Cameroon & Gabon, but 10 specimens in Congo.

Telosma africana (N.E.Br.) N.E.Br.

Woody climber; latex clear, often sparse; leaves opposite, to c. 10 × 5cm, oblong to ovate; inflorescences extra-axillary, subsessile; flowers green, sometimes tinged with red, with long hairs in throat; corolla with narrow tube 5–8mm long, lobes 5–8mm long, twisted. Scrub, forest margins; 200m.
Distr.: widespread in tropical & subtropical Africa.
IUCN: LC
Southern Bakossi F.R.: Binuyo in FHI 35556 fr., 2/1956.

Tylophora conspicua N.E.Br.

Slender woody twiner; latex cloudy; leaves opposite, to c. 20 × 13cm, ovate-oblong, glabrous above; inflorescences extra-axillary, zig-zag, with white to dark red flowers at the angles; corolla rotate, fleshy, 13–18mm across; corona lobes fleshy; follicles mostly single, to 10cm long. Evergreen or deciduous forest, often near streams; 400–550m.
Distr.: widely distributed across tropical Africa.
IUCN: LC

Kupe Village: Etuge 1447 11/1995; 2029 5/1996.

Tylophora oblonga N.E.Br.

Twining climber; leaves obtusely cuneate, truncate or cordate, 10–15cm long; flowers greenish yellow, 7mm diameter; corona of 5 laterally elongated, contiguous fleshy tubercles. Primary or secondary forest, disturbed areas; 1260–1900m.
Distr.: Liberia to Gabon. (Upper & Lower Guinea).
IUCN: LC
Kupe Village: Cheek 7832 11/1995; **Mt Kupe:** Etuge 1366 11/1995.

BALANOPHORACEAE

M. Cheek

Fl. Cameroun 33 (1991).

Thonningia sanguinea Vahl

Underground root parasite, leafless, achlorophyllose; flower-like inflorescences in forest floor, pink or red, c. 5cm diameter; outer bracts petal-like, stiff, 2cm; flowers minute, numerous, yellow, resembling stamens. Forest floor; 840–1200m.
Distr.: Sierra Leone to Zambia. (Guineo-Congolian).
IUCN: LC
Kupe Village: Cheek 7132 fl., 1/1995; 7736 11/1995; Ryan 381 5/1996; **Ndum:** Williams 146 1/1995; **Nyasoso:** Sebsebe 5040 10/1995; Sidwell 355 10/1995; 401 10/1995; Sunderland 1482 fl., 7/1992.

BALSAMINACEAE

M. Cheek

Fl. Cameroun 22 (1981).

Impatiens balsamina L.

Terrestrial herb to c. 1m; leaves in whorls, narrowly-elliptic, to c. 8 × 2cm; pedicel red, to c. 1.5cm; flowers salmon-pink; spur about 1.5cm; fruits hairy. Farmbush; 760m.
Distr.: native to India and SE Asia, widely naturalised in Africa.
IUCN: LC
Enyandong: Pollard 782 fl., fr., 10/2001.
Local name: Elah-Epenbel. **Uses:** MATERIALS – cosmetics – grind the leaves on a stone and add a few drops of kerosene; plaster the mixture on your nails, after 5 minutes remove it and it will have left a stain that lasts up to one month, or until the nail grows out (Lucas Ebako in Pollard 782).

Impatiens burtoni Hook.f.

Terrestrial herb c. 30cm; leaves alternate, lanceolate, c. 8 × 5cm; single-flowered; pedicel c. 3cm; flower white, c. 3 × 1cm, face concave; sepals densely long-hairy; spur about as long as petals. Farmbush; 760–2000m.
Distr.: Cameroon to Burundi. (Lower Guinea & Congolian).
IUCN: LC
Enyandong: Pollard 773 fl., 10/2001; **Kodmin:** Biye 17 fl., 1/1998; Ghogue 44 fl., 1/1998; **Kupe Village:** Cable 2668 fl., 5/1996; 3672 fl., 7/1996; **Mwanenguba:** Cheek 9460 fl., 10/1998; **Nlonako:** Etuge 1788 fl., 3/1996; **Nyasoso:** Cable 6 fl., 7/1992; Etuge 1302 fl., fr., 10/1995; 2549 fl., fr., 7/1996;

BALSAMINACEAE

Lane 105 fl., 6/1994; Thomas 5035 fl., fr., 11/1985; Wheatley 381 fl., 7/1992.
Note: common in farmbush, notable for having large white-hairy flowers.
Uses: FOOD – fruits edible (*fide* Pollard).

Impatiens filicornu Hook.f.
Grey-Wilson, *Impatiens* of Africa: 95 (1980).
Terrestrial or epilithic, rarely epiphytic, herb c. 30cm; leaves drying dark brown above, alternate, lanceolate, c. 9 × 4cm; peduncle c. 11cm; flowers fascicled at apex; bracts ovate, 3–4mm; pedicel 1.5cm; flowers pink to white with purple markings, face flat, c. 1.5cm diameter, spur c. 15mm. Forest; 800–1800m.
Distr.: Bioko, Cameroon to Cabinda. (Lower Guinea).
IUCN: LC
Bangem: Thomas 5325 1/1986; Kodmin: Cheek 9095 fl., 2/1998; Kupe Village: Etuge 2800 fl., 7/1996; Mt Kupe: Sidwell 433 fl., 11/1995; Nyasoso: Cable 74 fl., 8/1992; Wheatley 446 fl., 7/1992.

Impatiens frithii Cheek
Kew Bull. 57: 669 (2002).
Epiphytic herb 15–20cm, glabrous; leaves alternate, ovate, c. 7–13 × 3–5.5cm; inflorescence 1–4-flowered; peduncle 4.5–6cm; pedicel c. 2cm; flowers orange-red, mouth concave, c. 2.5 × 1cm; lateral sepals c. 5mm, entire; spur recurved at apex. Forest; 1300–1400m.
Distr.: Cameroon. (W Cameroon Uplands).
IUCN: VU
Kodmin: Cheek 9526 fl., 10/1998; Etuge 4420 fl., 11/1998; Gosline 57 fl., 1/1998; Plot B83 1/1998.
Note: only known from the Bakossi Mts and Mt Etinde.

Impatiens hians Hook.f. var. *hians*
Terrestrial herb 30–100cm, glabrous; leaves alternate, ovate, c. 12 × 6cm; peduncle c. 4cm; flowers 2–5 on a short rachis; pedicels 3cm; flower green, face concave, 5 × 2.5cm; spur red. Forest; 760–1700m.
Distr.: Bioko to W Congo (K). (Lower Guinea).
IUCN: LC
Edib: Pollard 237 fl., 11/1998; Kodmin: Muasya 2020 fl., 12/1999; Kupe Village: Cable 2614 fl., 5/1996; 3676 fl., 7/1996; Etuge 2854 fl., fr., 7/1996; Mwambong: Cheek 9370 fl., 10/1998; Nyasoso: Balding 75 fl., 7/1993; Cable 115 fl., 9/1992; Etuge 1728 fl., 2/1996; 2480 fl., 6/1996.
Note: distinct in the gaping mouth.

Impatiens kamerunensis Warb. subsp. *kamerunensis*
Grey-Wilson, *Impatiens* of Africa: 164 (1980).
Terrestrial or epilithic herb c. 30cm resembling *I. filicornu*, but leaves opposite, drying green above; flowers pink, spread along the upper part of the peduncle; bracts filiform. Wet rocks by streams in forest; 750–1350m.
Distr.: Ghana to Cameroon. (Upper & Lower Guinea).
IUCN: LC
Kodmin: Gosline 58 fl., 1/1998; 68 fl., 1/1998; Kupe Village: Cable 3673 fl., 7/1996; 3744 fl., 7/1996; Etuge 2655 fl., 7/1996; Mbule: Leeuwenberg 9301 fl., fr., 1/1972; Ngomboaku: Mackinder 332 fl., 12/1999; Nyale: Cheek 9671 fl., 11/1998; Nyasoso: Cable 3202 fl., 6/1996; 3294 fl., 6/1996; 3432 fl., 7/1996; 36 fl., 8/1992; Etuge 2161 fl., 6/1996; Wheatley 478 fl., 7/1992.

Impatiens kamerunensis Warb. subsp. *obanensis* (Keay) Grey-Wilson
Grey-Wilson, *Impatiens* of Africa: 94 (1980).
Syn.: *Impatiens obanensis* Keay
Terrestrial or epilithic herb 15–30(–120)cm; resembling *I. kamerunensis* subsp. *kamerunensis*, but leaves alternate. Wet rocks by streams in forest; 800–1200m.
Distr.: Liberia to Cameroon. (Upper & Lower Guinea).
IUCN: LC
Mt Kupe: Schoenenberger 24 fl., fr., 11/1995; Thomas 5101 fl., fr., 12/1985; Nyasoso: Balding 96 fl., 8/1993; Cheek 7448 fl., 10/1995.

Impatiens letouzeyi Grey-Wilson
Fl. Cameroun 22: 25 (1981).
Epiphytic herb 50–150cm; leaves alternate, oblong-elliptic, 14 × 5cm; peduncles 0.4cm, 1–4-flowered; pedicels 4cm; flowers pink, face concave, c. 6 × 4cm; lateral spur 1cm, dentate; spur c. 6mm, tightly coiled. Forest; 1200–1350m.
Distr.: Cameroon. (Narrow endemic).
IUCN: EN
Edib: Etuge 4092 fl., 1/1998; Kodmin: Etuge 4086 fl., 1/1998; Gosline 59 fl., 1/1998; Lake Edib: Satabie 1123 fl., 2/1998; Mwambong: Cheek 9371 fl., 10/1998.
Note: known only from the Bakossi Mts.
Local name: Fume. Uses: MEDICINES (Gosline 59).

Impatiens mackeyana Hook.f. subsp. *mackeyana*
Grey-Wilson, *Impatiens* of Africa: 221 (1980).
Terrestrial herb 30–40cm; leaves alternate, ovate, c. 11 × 7cm; peduncle < 0.5cm, usually single-flowered; pedicels 2.5cm; flower purple-pink, face concave, 4.5 × 3cm, glabrous; lateral sepals 6mm, dentate on one edge; spur 1cm, recurved. Forest; 1210–2000m.
Distr.: Nigeria to Gabon. (Lower Guinea).
IUCN: NT
Mwambong: Pollard 121 fl., 10/1998; Mwanenguba: Cheek 9456 fl., 10/1998.
Note: rare, only 6 collections known in Cameroon (1982).
Local name: Mbotlore. Uses: MEDICINES (Epie Ngome in Pollard 121).

Impatiens macroptera Hook.f.
Terrestrial herb 0.3–1m, glabrous; leaves alternate, lanceolate, c. 14 × 6.5cm; peduncle c. 15cm, with numerous flowers in fascicle at apex; flowers purple-blue, face concave, c. 4.5 × 2cm; lateral sepals 6mm, entire; spur 7mm, recurved. Open areas in forest; 420–1300m.
Distr.: Nigeria to Gabon. (Lower Guinea).
IUCN: NT
Bangem: Thomas 5288 1/1986; Kupe Village: Cable 2412 fl., 5/1996; Ensermu 3552 fl., 11/1995; Etuge 1716 fl., 2/1996; 1952 fl., 5/1996; Lane 309 fl., 1/1995; Schoenenberger 55 fl., 11/1995; Mbule: Etuge 4393 fl., fr., 10/1998; Ngomboaku: Mackinder 300 fl., 12/1999; Nyale: Etuge 4186 fl., 2/1998; Nyandong: Cheek 11304 3/2003; Darbyshire 106 3/2003; Nyasoso: Cable 20 fl., 7/1992; 3255 fl., 6/1996; 3295 fl., 6/1996; 3509 fl., 7/1996; Etuge 2373 fl., 6/1996; Lane 36 fl., 8/1992; 45 fl., 8/1992; Sunderland 1498 fl., 7/1992; Wheatley 383 fl., 7/1992; Zapfack 697 fl., 6/1996.
Local name: Ehpol (Etuge 1716). Uses: FOOD – leaves (Etuge 1716 & 2373); unspecified parts – eat as vegetable (Cheek 11304); MEDICINES (Cheek 11304).

Impatiens mannii Hook.f.
Syn.: *Impatiens deistelii* Gilg
Terrestrial herb 0.5–1m, glabrous; leaves alternate, ovate, c. 7 × 5cm; single-flowered; pedicel c. 1cm; flower white, the lower sepal with conspicuous transverse purple bars, face

concave, 1.5–2 × 0.6cm, glabrous; spur shorter than petals. Open areas in forest; 800–2000m.
Distr.: Bioko & Cameroon to Congo (K). (Lower Guinea & Congolian).
IUCN: LC
Bangem: Thomas 5289 1/1986; **Ehumseh:** Manning 394 10/1986; **Kodmin:** Biye 16 fl., 1/1998; Cheek 8989 fl., 1/1998; **Kupe Village:** Cable 2543 fl., 5/1996; 3654 fl., 7/1996; Cheek 8411 fl., 5/1996; Etuge 2691 fl., 7/1996; Kenfack 222 fl., 7/1996; **Nyasoso:** Cable 3397 fl., 6/1996; Etuge 1824 fl., 3/1996; 2190 fl., 6/1996.
Uses: FOOD – leaves (Etuge 1824).

Impatiens niamniamensis Gilg
Terrestrial, epiphytic or epilithic herb, 0.3–1m, glabrous; leaves alternate, lanceolate, c. 20 × 8cm; flowers in axillary fascicles of 2–4; peduncle sessile; pedicels c. 4cm; flowers red, with white hood, face concave, c. 3 × 1.5cm; lateral sepals 2mm; spur recurved, erect. Forest; 350–1400m.
Distr.: Cameroon to NW Tanzania. (Lower Guinea & Congolian).
IUCN: LC
Bakolle Bakossi: Etuge 137 5/1986; **Enyandong:** Gosline 350 11/2001; **Kodmin:** Etuge 3989 fl., 1/1998; **Kupe Village:** Cable 2450 fl., 5/1996; 2664 fl., 5/1996; 741 fl., 1/1995; Cheek 6031 fl., 1/1995; Etuge 1415 fl., 11/1995; Groves 19 fl., 1/1995; Kenfack 253 fl., 7/1996; Lane 440 fl., 1/1995; Ryan 306 fl., 5/1996; Zapfack 908 fl., 7/1996; **Mekom:** Nemba 16 4/1986; **Mwambong:** Etuge 4334 fl., 10/1998; **Nyandong:** Tadjouteu 539 3/2003; **Nyasoso:** Cable 1212 fl., 2/1995; 2768 fl., 6/1996; 2897 fl., 6/1996; 3317 fl., 6/1996; 3431 fl., 7/1996; 37 fl., 8/1992; Etuge 1286 fl., 10/1995; 2075 fl., 6/1996; 2168 fl., 6/1996; Lane 43 fl., 8/1992; Sidwell 336 fl., 10/1995; Wheatley 391 fl., 7/1992; Zapfack 700 fl., 6/1996.
Note: our material may be better named as *I. bicolor* Hook.f. (Eberhard Fischer *pers. comm.*).

Impatiens sakerana Hook.f.
Terrestrial herb 0.3–1m; leaves in whorls of 3–4 or opposite, ovate-elliptic, c. 7.5–12 × 3.5–4cm, apex acuminate, margin crenate-toothed, pale brown-hairy below; racemes short, 3–4-flowered; peduncle c. 4cm; flowers 1.5–3cm long, lip red, gradually narrowing into short spur, greenish red, swollen at tip. Forest patches; 2000m.
Distr.: Bioko & Cameroon. (W Cameroon Uplands).
IUCN: VU
Mwanenguba: Leeuwenberg 9957 fr., 6/1972; Thomas 3143 fl., 2/1984.

BASELLACEAE

M. Cheek

Basella alba L.
Climbing herb to 4m, glabrous; stem twining; leaves alternate, ovate, c. 6 × 4.5cm, gradually acuminate, base truncate to cordate; petiole 3cm; stipules absent; inflorescence axillary, 8cm, several-flowered; flowers white and purple, scarcely opening, 4mm. Lake edges, riverbanks; 1250m.
Distr.: Sierra Leone to Kenya. (Tropical Africa).
IUCN: LC
Enyandong: Cheek 10998 10/2001; **Lake Edib:** Cheek 9147 fl., 2/1998.

BEGONIACEAE

M.S.M. Sosef, V. Plana, M. Cheek & J.J. de Wilde

Begonia adpressa Sosef
Studies in Begoniaceae 5. Wag. Agric. Univ. Papers 94(1): 149 (1994).
Terrestrial herb 20cm; blade ovate, c. 12 × 7 cm, acuminate, entire, nerves adpressed-hairy; petiole c. 12cm; peduncle 15cm, 3–4-flowered; perianth 1.5cm, yellow. Forest; 800–1400m.
Distr.: Cameroon. (W Cameroon Uplands).
IUCN: VU
Bangem: Thomas 5329 1/1986; **Kodmin:** Barlow 12 1/1998; Biye 7 1/1998; Gosline 41 1/1998; Plot B100 1/1998; **Kupe Village:** Cheek 7224 fl., 1/1995; **Nyasoso:** Cable 1159 fl., 2/1995.
Local name: M'diamaso. **Uses:** FOOD – unspecified parts – eaten (Gosline 41).

Begonia sp. aff. *adpressa* Sosef
Resembling *B. adpressa*, but with few or no hairs on the sutures between the locules (Sosef *pers. obs.*). Forest; 1330m.
Kodmin: Cheek 10267 12/1999.

Begonia ampla Hook.f.
Epiphytic herb, 1–5m from ground; stems 1cm diameter. or more, scales fimbriate, silvery; blades pale green, suborbicular, 22cm, acuminate, cordate; petiole c. 17cm; inflorescence concealed in cup-like bracts 3–7cm wide; perianth 1.5cm, white striped pink; fruit berry-like, globose, indehiscent, c. 1cm diameter. Forest; 660–1500m.
Distr.: Cameroon to Uganda. (Afromontane).
IUCN: LC
Kodmin: Etuge 3996 1/1998; Gosline 54 1/1998; **Kupe Village:** Biye 52 11/1999; Cable 2435 5/1996; 3661 7/1996; 773 fl., 1/1995; Cheek 7029 fl., 1/1995; 8341 5/1996; Ensermu 3565 11/1995; Etuge 1381 11/1995; 2921 7/1996; Groves 17 fl., 1/1995; Plana 85 11/1999; Ryan 244 5/1996; Zapfack 907 7/1996; **Lake Edib:** Cheek 9166 2/1998; **Muahunzum:** Plana 89 11/1999; **Mungo Ndaw:** Dundas in FHI 15327 10/1946; **Ngomboaku:** Mackinder 312 12/1999; **Nyandong:** Ghogue 1470 fl., 3/2003; **Nyasoso:** Balding 67 fl., 8/1993; Cable 2781 6/1996; 3269 6/1996; 3461 7/1996; 3508 7/1996; 3549 7/1996; 3602 7/1996; 46 fl., 8/1992; Etuge 2138 6/1996; 2143 6/1996; 2403 6/1996; Sidwell 351 10/1995; Wheatley 473 fl., fr., 7/1992; Zapfack 698 6/1996.

Begonia bonus-henricus J.J.de Wilde
Misc. Papers Landbouwhogesch. Wag. 19: 409 (1980).
Epiphyte 3–5(–30)m from ground, leaves narrowly elliptic or lanceolate, c. 8.5 × 2.75cm, acuminate, obtuse, subentire, densely grey-peltate-scaley on both surfaces; inflorescence with 1 female and 1–3 male flowers, concealed in sheathing bracts; perianth 1.4cm, white, lower lobe with red streaks; fruit globose, indehiscent, 1cm diameter. Forest; 660–1300m.
Distr.: Cameroon Endemic.
IUCN: VU
Kodmin: Plana 92 11/1999; **Kupe Village:** Cable 2432 5/1996; 3668 7/1996; Etuge 1998 5/1996; Plana 80 11/1999; Ryan 281 5/1996; 331 5/1996; **Nyasoso:** Cable 2840 6/1996; 3291 6/1996; 3420 7/1996; 3547 7/1996; Etuge 2356 6/1996; Lane 134 fl., 6/1994; Wheatley 462 fl., 7/1992; Zapfack 706 6/1996.
Local name: Ntung Nkim (Plana 92).

Begonia cilio-bracteata Warb.

Syn.: *Begonia hookeriana* Gilg ex Engl.

Terrestrial herb c. 25cm; aerial stem lacking or up to 5cm; sparsely hirsute and with minute glandular hairs; blades rarely peltate, elliptic or obovate, 5–21 × 3–27cm, obtuse, base cordate, entire to crenate; petiole 3–12cm, hairs white, pendent, wavy; peduncle 1.5–6cm; bracts clustered at apex; perianth 0.6–1.2cm, white, veined red. Forest; 250m.

Distr: SE Nigeria & Cameroon. (Lower Guinea).

IUCN: LC

Mungo Ndor: Dundas in FHI 15325 10/1946.

Begonia duncan-thomasii Sosef

Studies in Begoniaceae 5. Wag. Agric. Univ. Papers 94(1): 156 (1994).

Epilithic herb, to 25cm; lacking aerial stems; blade red-green, obliquely ovate, 13 × 7.5cm, acute, densely denticulate, bubble-bullate; petiole 15cm; peduncle 7cm; 3-flowered; perianth 1cm, yellow. Cliffs; 900–1750m.

Distr.: Cameroon. (W Cameroon Uplands).

IUCN: VU

Kodmin: Biye 31 1/1998; Etuge 4025 1/1998; **Nyasoso:** Cable 3209 6/1996; Thomas 5492 fl., 2/1986.

Begonia eminii Warb.

Terrestrial or epiphytic herb, resembling *B. fusialata* and *B. mannii*, but leaves oblong-elliptic, c. 11 × 5cm, base cordate, subequal; female inflorescence 3-flowered; flowers pink; fruit terete in transverse section, red, dehiscing by slits or valves. Forest; 250–1450m.

Distr.: Cameroon to Uganda. (Lower Guinea & Congolian).

IUCN: LC

Kodmin: Cheek 9581 11/1998; **Kupe Village:** Cable 2541 5/1996; Kenfack 310 7/1996; Mackinder 281 12/1999; Plana 99 12/1999; Schoenenberger 42 fl., fr., 11/1995; **Mt Kupe:** Plana 76 11/1999; Schoenenberger 75 fl., 11/1995; **Mwambong:** Onana 545 2/1998; **Ndum:** Groves 41 fr., 2/1995; **Nyasoso:** Cable 12 fl., 7/1992; 2779 6/1996; 3263 6/1996; 3528 7/1996; Etuge 2149 6/1996; 2453 6/1996; Sidwell 313 10/1995; Sunderland 1496 fl., fr., 7/1992; Wheatley 412 fl., 7/1992; 472 fl., fr., 7/1992; Zapfack 686 6/1996; **Tombel:** Thorold CD10 fl., 3/1953.

Note: *B. eminii* is easily confused with other species of *Begonia* sect. *Tetraphila* as defined by J.J.F.E. de Wilde: Studies in Begoniaceae 7. Wag. Agric. Univ. Papers 01(2). Those in our area are: *B. furfuracea, B. longipetiolata, B. preussii, B. mannii, B. fusialata, B. polygonoides, B. oxyanthera* and *B pelargoniiflora*.

Begonia furfuracea Hook.f.

Epiphyte, resembling *B. mannii*; leaves oblong, c. 13 × 3.5cm, acuminate, rounded; nerves 3–4 pairs, dentate, lower surface densely pale brown scaly-hairy. Forest; 1000–1330m.

Distr.: Bioko & Cameroon. (W Cameroon Uplands).

IUCN: VU

Bangem: Thomas 5889 3/1986; **Kodmin:** Plana 93 11/1999; 97 11/1999; **Kupe Village:** Cable 2649 5/1996; **Mbule:** Cable 3354 6/1996; **Nyasoso:** Cable 1233 fl., 2/1995; Wheatley 419 fl., 7/1992.

Begonia fusialata Warb. var. *fusialata*

Studies in Begoniaceae 7. Wag. Univ. Papers 01(2): 97 (2002).

Terrestrial or epiphytic, often scandent, to 3m; resembling *B. eminii*, but leaves highly oblique at base; female inflorescence 2-flowered; fruit 3-angled, white. Forest; 300–2000m.

Distr.: Guinea (C) to Congo (K). (Guineo-Congolian).

IUCN: LC

Kupe Village: Cable 2526 5/1996; 2539 5/1996; 3681 7/1996; Cheek 7000 fr., 1/1995; Etuge 2772 7/1996; Kenfack 309 7/1996; Lane 372 fl., fr., 1/1995; Plana 100 12/1999; 104 12/1999; Schoenenberger 76 fr., 11/1995; **Ngomboaku:** Etuge 4685 12/1999; **Nyasoso:** Cable 3394 6/1996.

Begonia laporteifolia Warb.

Syn.: *Begonia schlechteri* Gilg

Epilithic herb, 7–20 cm, lacking aerial stems; blade elliptic-oblong, c. 16 × 7cm, acute, crenate, brown-white below, vein-reticulum densely brown-hairy; petiole 7cm, brown-woolly; peduncle to 8cm, 3–4-flowered; perianth 1cm, white, veined red. Rocks in forest; 710–1200m.

Distr.: Cameroon. (W Cameroon Uplands).

IUCN: NT

Bangem: Thomas 5330 1/1986; **Muahunzum:** Cheek 9639 11/1998; **Nyale:** Cheek 9664 11/1998; Etuge 4231 2/1998.

Note: only 9 specimens known, apart from those cited here.

Begonia longipetiolata Gilg

Studies in Begoniaceae 4. Wag. Agric. Univ. Papers 91(6): 195 (1991).

Epiphyte, sometimes epilithic; stems c. 30cm, rooting along their length; leaves oblong, 10–20 × 4–9cm, acuminate, rounded-truncate, entire; petiole 3–10cm; female inflorescence 1-flowered, subsessile; male inflorescence 10–20-flowered, diffuse; peduncle c. 3cm; perianth c. 5mm, white or pink. Forest; 660–1400m.

Distr.: SE Nigeria to Congo (K). (Lower Guinea & Congolian).

IUCN: LC

Bangem: Thomas 5261a 1/1986; **Essossong:** Groves 42 fl., 2/1995; **Kodmin:** Biye 40 1/1998; Etuge 4119 2/1998; 4161 2/1998; Plot B286 1/1998; Satabie 1102 fl., 1/1998; **Kupe Village:** Cable 2446 5/1996; 3736 7/1996; 3763 7/1996; 761 fl., 1/1995; 814 fl., 1/1995; Cheek 8330 5/1996; Etuge 1696 2/1996; Groves 16 fl., 1/1995; Lane 355 fl., 1/1995; 366 fl., 1/1995; 441 fl., 1/1995; Plana 101 12/1999; Ryan 228 5/1996; 293 5/1996; Zapfack 905 7/1996; **Mt Kupe:** Plana 79 11/1999; **Mwambong:** Onana 529 2/1998; **Nyale:** Cheek 9681 11/1998; 9699 11/1998; **Nyandong:** Stone 2492 3/2003; **Nyasoso:** Cable 1210 fl., 2/1995; Etuge 2169 6/1996; Lane 524 fl., 2/1995; Zapfack 708 6/1996.

Uses: FOOD – leaves (Ryan 293); unspecified parts – eaten (Plana 79).

Begonia macrocarpa Warb.

Erect, unbranched terrestrial herb; stem c. 30cm; leaves clustered at stem apex, obliquely obovate or oblong, c. 10 × 4.5cm, acuminate, unequally obtuse, slightly serrate; petiole 0.5–3cm; stipules obovate, 1.2cm, serrate; inflorescence axillary, c. 3cm; flowers white, 1.5cm; fruit strongly 3-winged, 1.5 × 1.5cm. Forest; 710–1350m.

Distr.: Guinea (C) to Cameroon. (Upper & Lower Guinea).

IUCN: LC

Kodmin: Cheek 8991 1/1998; **Kupe Village:** Cable 851 fl., 1/1995; Cheek 7136 fl., fr., 1/1995; Etuge 2836 7/1996; Lane 283 fl., fr., 1/1995; Ryan 330 5/1996; **Menyum:** Thomas 7138 5/1987; **Nyale:** Etuge 4180 2/1998.

Note: closely related to, and easily confused with, *B. sessilifolia*.

Begonia mannii Hook.

Hemiepiphytic herb up to 8m from ground, resembling *B. eminii*, but androecium symmetrical, cone-shaped and stipules longer persistent, upper blade black-brown in dried leaves, contrasting with pale brown lower surface, thick (not concolorous, membranous); male and female inflorescences not on the same branch. Forest; 600m.

Distr.: Sierra Leone & Liberia, Nigeria to Gabon. (Upper & Lower Guinea).

IUCN: LC

Ngombombeng: Etuge 130 5/1986.

Note: also known in our area from Supe (Letouzey 14358, *fide* de Wilde 2002: 171).

Begonia oxyanthera Warb.

Syn.: *Begonia jussiaeicarpa* Warb.
Shrubby, scandent epiphyte; leaves elliptic, c. 10 × 4cm, acuminate, obtuse to rounded, entire or slightly dentate, lower margin and nerves red, lower surface glabrescent; petiole 1cm; inflorescence 1.5cm; perianth 0.8cm cream-red; anther top strongly acuminate to acute; fruit cylindric, 4 × 0.4cm, red. Forest; 800–2330m.

Distr.: Bioko & Cameroon. (W Cameroon Uplands).

IUCN: VU

Kodmin: Plana 94 11/1999; **Kupe Village:** Ryan 243 5/1996; **Mt Kupe:** Cheek 7592 11/1995; **Muahunzum:** Plana 91 11/1999; **Mwanenguba:** Sanford 5543 11/1968; **Nlonako:** Leeuwenberg 9747 fl., 4/1972; **Nyasoso:** Cable 2924 6/1996.

Begonia oxyloba Welw. ex Hook.f.

Terrestrial herb, 0.3–1m; stem ascending; leaves palmately 5–7-lobed to about half the blade radius, transversely unequally elliptic in outline, c. 19 × 23cm, base shallowly cordate; petiole c. 14cm; inflorescence axillary, 3–4-flowered, 3cm; flowers pink, c. 2cm; fruit ellipsoid, juicy, indehiscent, 2cm. Forest; 800–1600m.

Distr.: Guinea (C) to Tanzania & Madagascar. (Tropical Africa & Madagascar).

IUCN: LC

Kodmin: Cheek 8992 1/1998; **Kupe Village:** Ensermu 3525 11/1995; 3566 11/1995; Etuge 2804 7/1996; 2895 7/1996; Ryan 307 5/1996; **Nyasoso:** Balding 4 fl., 7/1993; Cable 116 fl., fr., 9/1992; 2939 6/1996; 3419 7/1996; 3505 7/1996; Etuge 1288 10/1995; 2568 7/1996; Sidwell 353 10/1995; Sunderland 1530 fl., 7/1992; Wheatley 474 fl., fr., 7/1992.

Begonia pelargoniiflora J.J.de Wilde & J.C.Arends

Studies in Begoniaceae 7. Wag. Agric. Univ. Papers 01(2): 183 (2002).
Hemiepiphyte resembling *B. longipetiolata*, but female inflorescence (4–)6–15-flowered (not 1(–3)-flowered), outer perianth segments 17–22mm (not less than 17mm), androecium subactinomorphic, filaments free to base (not strongly zygomorphic, filaments fused at base). Forest; 1000m.

Distr.: Bioko & Cameroon. (W Cameroon Uplands).

IUCN: CR

Menyum: Doumenge 491 5/1987.

Begonia poculifera Hook.f. var. *poculifera*

Epiphyte resembling *B. ampla*, but blade axis at right angles to petiole, dark green, veined red, obliquely ovate-lanceolate, 11 × 5.5cm, acuminate, truncate, entire. Forest; 870–2000m.

Distr.: SE Nigeria to Tanzania. (Afromontane).

IUCN: LC

Lake Edib: Pollard 243 11/1998; **Kodmin:** Cheek 9071 1/1998; Gosline 165 11/1998; **Kupe Village:** Cable 2616 5/1996; 2645 5/1996; Cheek 8395 5/1996; Ryan 240 5/1996; Zapfack 906 7/1996; **Mbule:** Cable 3367 6/1996; **Muahunzum:** Plana 90 11/1999; **Nyasoso:** Balding 47 fl., 8/1993; Cable 114 fl., 9/1992; 2923 6/1996; 3205 6/1996;

3395 6/1996; 75 fl., 8/1992; Etuge 2165 6/1996; 2192 6/1996; Lane 123 fl., 6/1994; 138 fl., 6/1994; 150 fl., 6/1994; 54 fl., 8/1992; Wheatley 418 fl., fr., 7/1992.

Begonia poculifera Hook.f. var. *teusziana* (J.Braun & K.Schum.) J.J.de Wilde

Misc. Papers Landbouwhogesch. Wag. 19: 404 (1980).
Epiphytic herb resembling var. *poculifera*, but leaf ratio 3:2 to 1:1 (not 2:1 to 3:1); petioles 2.5–10(–14)cm (not 1–4(–6) cm); leaf margin coarsely toothed to obscurely lobed (not usually entire). Forest; 800–1150m.

Distr.: Nigeria to Gabon. (Lower Guinea).

IUCN: NT

Kupe Village: Etuge 2664 7/1996; **Mbule:** Cable 3355 6/1996; **Mwanenguba:** Letouzey 14384 fl., 8/1975; **Nyasoso:** Cable 2833 6/1996; 3568 7/1996; Lane 32 fl., 8/1992; Zapfack 699 6/1996.

Begonia polygonoides Hook.f.

Shrubby canopy epiphyte; branches pendulous, 60–80cm, glabrous; leaves elliptic-oblong, 7–14 × 1.5–2cm, acute, base oblique; petiole 1mm; stipules persistent; inflorescence 3cm, 3-flowered; perianth 8mm, white. Forest; 400–1700m.

Distr.: Sierra Leone to Ghana, SE Nigeria to Congo (K). (Guineo-Congolian).

IUCN: LC

Kupe Village: Cable 2429 5/1996; 2430 5/1996; 3670 7/1996; Plana 102 12/1999; **Nyasoso:** Cheek 7497 10/1995; Etuge 1770 3/1996; Sunderland 1542 fl., 7/1992; Zapfack 626 6/1996; **Tombel:** Thorold CD11 fl., 3/1953.

Begonia potamophila Gilg

Studies in Begoniaceae 5. Wag. Agric. Univ. Papers 94(1): 171 (1994).
Terrestrial or epilithic herb to 40cm, lacking aerial stems; blade elliptic, c. 20 × 12cm, acumen 1cm; petiole hairs 4mm; peduncle 15cm, few-flowered; perianth 1cm, red outside, yellow within. Forest, on rocks; 700–1200m.

Distr.: Cameroon to Congo (K). (Lower Guinea).

IUCN: NT

Bakossi F.R.: Thomas 5331 fl., 1/1986; **Kupe Village:** Cable 759 fl., 1/1995; Cheek 6029 fl., 1/1995; Etuge 1398 11/1995; 1439 11/1995; Lane 315 fl., 1/1995; **Nyasoso:** Cheek 6022 fl., 1/1995.

Begonia preussii Warb.

Bot. Jahrb. 22: 36 (1895).

Syn.: *Begonia sessilanthera* Warb.
Epiphytic herb resembling *B. oxyanthera*, but occurring at lower altitudes; male inflorescence slender; peduncle 0.5–5cm; 7–15-flowered (not contracted, 0.1–0.8cm, less than 7-flowered); perianth pink or red (not white, red-edged); anthers 1.5mm, truncate (not 2.7–4.5mm, acute or acuminate). Forest; 850–1000m.

Distr.: Bioko, SE Nigeria & Cameroon. (W Cameroon Uplands).

IUCN: VU

Kodmin: Plana 96 11/1999; 98 11/1999; **Kupe Village:** Etuge 1999 5/1996; Zapfack 940 7/1996; **Mwambong:** Etuge 4331 10/1998; **Ndum:** Cable 939 fl., 2/1995; **Nyasoso:** Cable 17 fl., fr., 7/1992; 3282 6/1996.

Begonia prismatocarpa Hook. subsp. *delobata* Sosef

Studies in Begoniaceae 5. Wag. Agric. Univ. Papers 94(1): 179 (1994).

Terrestrial herb, 10cm, lacking aerial stems; blade ovate, 4 × 3 cm, acumen 1cm, cordate, dentate; petiole 2.5cm; peduncle 3cm, 3-flowered; perianth 0.7cm, yellow. Forest; 600–800m.
Distr.: Cameroon. (W Cameroon Uplands).
IUCN: VU
Bambe: <u>Dundas in FHI 15329</u> fl., fr., 10/1946; **Bangem:** <u>Thomas 5351</u> 1/1986; **Nyale:** <u>Cheek 9645</u> 11/1998.

Begonia pseudoviola Gilg
Terrestrial herb, 14cm, lacking aerial stems; blade transversely ovate, 6cm, acute, cordate, subserrate, upper surface usually purple-black, with long white patent hairs; petiole 10cm; peduncle 3.5cm; perianth 1cm, yellow. Forest; 760–850m.
Distr.: Cameroon. (W Cameroon Uplands).
IUCN: VU
Ngomboaku: <u>Cheek 10319</u> 12/1999; <u>Mackinder 285</u> 12/1999; <u>294</u> 12/1999.

Begonia quadrialata Warb. subsp. *quadrialata* var. *quadrialata*
Terrestrial herb, 20cm; lacking aerial stems; blade suborbicular, 9 × 7.5cm, obtuse, entire; petiole c. 10cm, hairy; peduncle c. 6cm; perianth 0.8cm, yellow. Forest; 400–1100m.
Distr.: Sierra Leone to Togo, SE Nigeria to Angola. (Upper & Lower Guinea).
IUCN: LC
Bakossi F.R.: <u>Etuge 4310</u> 10/1998; **Mejelet-Ehumseh:** <u>Etuge 494</u> 2/1987.

Begonia schaeferi Engl.
Studies in Begoniaceae 5. Wag. Agric. Univ. Papers 94(1): 203 (1994).
Terrestrial herb 10–22cm, aerial stems lacking or to 2cm, scattered with minute glandular hairs; blade peltate, elliptic or elliptic-ovate, 3.5–14 × 2.5–7.2cm, acuminate, denticulate, sparsely-hirsute; petiole 4–21cm, hairs patent, wavy; inflorescences either bisexual or male; peduncle simple or 3-branched; bracts elliptic, 4–8mm, dentate-ciliate, distichous; perianth 0.8–2.2cm, yellow. Vertical Rock-faces; c. 1900m.
Distr.: Cameroon & Nigeria. (W Cameroon Uplands).
IUCN: VU
Mwanenguba: <u>Schaefer 72</u> 11/1900.
Note: specimens at BR, B; not seen at K.

Begonia sciaphila Gilg ex Engl. var. *sciaphila*
Bull. Jard. Bot. Nat. Belg. 39: 94 (1969).
Herb resembling *B. macrocarpa* and *B. sessilifolia*, but stipules suborbicular, 1cm, bidentate. Forest; 1000m.
Distr.: Cameroon to Cabinda. (Lower Guinea).
IUCN: NT
Mbule: <u>Leeuwenberg 9308</u> fr., 1/1972.
Note: rare, only 7 specimens at K.

Begonia sessilifolia Hook.f.
Erect herb very similar to *B. macrocarpa*, but petiole less than 5mm; leaves unequally sided at base, relatively slender (2.5–5cm broad, not 4–7cm broad). Forest; 700–1250m.
Distr.: Bioko & Cameroon. (W Cameroon Uplands).
IUCN: NT
Kupe Village: <u>Cable 2438</u> 5/1996; <u>782a</u> fl., 1/1995; <u>Cheek 6062</u> fr., 1/1995; <u>Etuge 1385</u> 11/1995; **Nyasoso:** <u>Lane 189</u> fl., 11/1994.

Begonia staudtii Gilg
Studies in Begoniaceae 5. Wag. Agric. Univ. Papers 94(1): 210 (1994).

Syn.: *Begonia staudtii* Gilg var. *dispersipilosa* Irmsch. Resembling *B. potamophila*, but leaves minutely bullate, hairs on upper surface with broadened bases. Forest, often on rocks; 800–1300m.
Distr.: SE Nigeria & Cameroon. (Lower Guinea).
IUCN: NT
Kupe Village: <u>Etuge 1695</u> 2/1996; <u>1974</u> 5/1996; <u>2801</u> 7/1996; <u>2899</u> 7/1996; **Nyasoso:** <u>Cable 3299</u> 6/1996; <u>3495</u> 7/1996; <u>Etuge 2056</u> 6/1996; ·<u>2167</u> 6/1996.

BIGNONIACEAE
S. Bidgood & I. Darbyshire

Fl. Cameroun 27 (1984).

Kigelia africana (Lam.) Benth.
Tree to 15m, glabrous; leaves opposite, imparipinnate, to 50cm, 5–6-jugate; leaflets oblong-elliptic, 11–20 × 3.5–8cm, apex acuminate, base acute to rounded; petiole 8–25cm; petiolules 3–6mm; panicles pendent, lax, to 40cm; lateral branches to 7cm; peduncle to 20cm; pedicels 1.2cm; calyx cupular, 2.8 × 2.5cm, lobes triangular, 0.6–1cm; corolla campanulate, 6(–8)cm, dark red; fruit sausage-shaped, 20–50(–100)cm. Forest, farmbush; 250–1700m.
Distr.: throughout tropical Africa, widely cultivated elsewhere.
IUCN: LC
Bangem: <u>Manning 499</u> 10/1986; **Bolo-Meboka:** <u>Thomas 7151</u> 7/1987; **Kodmin:** <u>Cheek 8882</u> 1/1998; <u>Ghogue 76</u> fl., 2/1998; **Kupe Village:** <u>Etuge 2004</u> 5/1996; <u>2701</u> fl., 7/1996; <u>Kenfack 251</u> 7/1996; <u>Lane 295</u> fl., fr., 1/1995; **Ndum:** <u>Elad 111</u> fl., fr., 2/1995; <u>Groves 25</u> fl., 2/1995; **Nyasoso:** <u>Cable 2796</u> 6/1996; <u>Sunderland 1524</u> fl., fr., 7/1992.
Local names: Aso sorre (Cheek 8882); Sausage Tree (*fide* Pollard). **Uses:** MEDICINES (Cheek 8882).

Markhamia lutea (Benth.) K.Schum.
Tree to 20m, glabrous; stems lenticellate; pseudostipules orbicular, 2cm; leaves 4-jugate, to 30cm; leaflets oblong-elliptic, 11–14 × 4–5.5cm, acuminate, base asymmetric-acute; petiole to 9cm; petiolules c. 5mm; panicles to 8cm, few-branched, 6–10-flowered; calyx lepidote, c. 2cm, beaked in bud, split on one side; corolla 6cm, yellow with red lines in throat; fruit linear, to 50 × 1.5cm, valves single-nerved. Forest, farmbush; 400–1000m.
Distr.: Ghana to Congo (K), E Africa. (Tropical Africa).
IUCN: LC
Bakossi F.R.: <u>Etuge 4284</u> 10/1998; **Enyandong:** <u>Pollard 779</u> 10/2001; **Mwanenguba:** <u>Leeuwenberg 9017</u> fr., 12/1971; **Ndum:** <u>Groves 38</u> fr., 2/1995; **Nyasoso:** <u>Cheek 7454</u> 10/1995.
Local name: Mbibeng (Etuge 4284).

Newbouldia laevis (P.Beauv.) Seeman ex Bureau
Tree to 15m, glabrous; leaves to 30cm, 4-jugate; leaflets subsessile, coriaceous, oblong-elliptic, 12–15.5 × 5–6.5cm, acumen 1.5cm, base asymmetric-acute, margin serrate to subentire; petiole to 11cm; panicles racemose, to 8cm, dense; pedicels 7mm; calyx to 2cm, splitting on one side; corolla 5cm, pink to purple and white; fruit linear, to 34 × 2.3cm, valves 3-nerved; seeds biwinged, 1 × 4.5cm. Forest and cultivated; 300–850m.
Distr.: Senegal to Congo (K). (Guineo-Congolian).
IUCN: LC

Konye: Thomas 5180 11/1985; **Kupe Village:** Cheek 7825 11/1995; **Ngusi:** Etuge 1533 fl., fr., 1/1996; **Nyale:** Etuge 4178 2/1998; **Nyasoso:** Thomas 3076 fl., 2/1984.
Uses: MEDICINES (Etuge 1533).

Spathodea campanulata P.Beauv. subsp. *nilotica* (Seem.) Bidgood
Proc. XIII Plenary Meet. AETFAT, Zomba, Malawi (1991) 1: 330 (1994).
Tree 10m; stems lenticellate; leaves c. 30cm, 5–6-jugate; leaflets subsessile, oblong-elliptic, c. 12.5 × 4.5cm, acuminate, base asymmetric, rachis and lower surface velvety-hairy; racemes few-flowered; pedicels to 6cm; calyx recurved, 6–8cm, split on one side, velvety-hairy; corolla 10–12cm, scarlet with yellow margin; fruit linear, to 19.5 × 3cm; seeds winged, 2cm. Forest edge; 1520m.
Distr.: Nigeria to Ethiopia & Tanzania. (Tropical Africa).
IUCN: LC
Kodmin: Pollard 200 11/1998.
Local name: African Tulip Tree (*fide* Pollard).

Stereospermum acuminatissimum K.Schum.
Tree to 25m, glabrous; leaves to 35cm, 5(–6)-jugate; leaflets oblong-ovate, 14–19.5 × 5.5–7cm, acumen to 2cm, base asymmetric; petiole 9cm; petiolules to 1cm; panicles subcorymbose, 15 × 14cm, many-flowered; pedicels 0.5–1cm; calyx tubular, 1.5cm; corolla 5cm, pink; fruit linear, to 60cm; seeds bi-winged, 2.5–3cm. Forest, forest edge; 900m.
Distr.: Guinea (C) to Cameroon. (Upper & Lower Guinea).
IUCN: LC
Enyandong: Ghogue 1240 fl., fr., 10/2001; **Nyasoso:** Etuge 2455 fl., 6/1996:

Tecoma stans (L.) Juss. ex Kunth
Fl. Cameroun 27: 58 (1984).
Treelet or shrub 4m; stems lenticellate; leaves 8–12cm, 2–3-jugate; leaflets sessile, ovate-lanceolate, c. 7 × 1.7cm, acuminate, base acute-attenuate, margin serrate, puberulent below; petiole to 3.5cm; racemes to 2.5cm, 10–20-flowered; pedicel 6mm; calyx cupular, 5mm, teeth 1mm; corolla 5.5cm, yellow with red lines in throat; fruit linear, 20 × 0.6cm, lenticellate. Cultivated, becoming naturalised; 900m.
Distr.: native of S & C America, widely cultivated as an ornamental elsewhere. (Pantropical).
IUCN: LC
Mwanenguba: Leeuwenberg 9016 12/1971.

BIXACEAE

I. Darbyshire

Fl. Cameroun 19 (1975).

Bixa orellana L.
Small tree or shrub to 3m; stems; petioles and inflorescence branches rusty-scaled; leaves ovate, 11–15 × 5.5–8.5cm, apex acuminate, base truncate, margin subentire to obscurely undulate; lateral nerves 6–7 pairs; petiole to 5.5cm; panicles terminal, many-branched, 7 × 10cm; pedicels 0.5–1.3cm; sepals 5, rounded, 0.8cm, brown with a pale circular gland at base c. 1.5mm; petals 5, oblanceolate, to 2cm long, white-pink; stamens numerous, free; fruit ovoid, 1.5–4cm long, densely hispid; seeds covered in red pulp. Cultivated, persisting in forest edges; 600m.
Distr.: widely introduced in the tropics, originally from S America. (Pantropical).

IUCN: LC
Etam: Etuge 221 fl., fr., 8/1986; 234 fl., fr., 8/1986;
Nyasoso: Pollard (sight record) fl., 10/1998.

BOMBACACEAE

M. Cheek

Fl. Cameroun 19 (1975).

Note: *Bombax buonopozense* (large red flowers) can also be expected in Bakossi.

Ceiba pentandra (L.) Gaertn.
Deciduous spiny tree to 60m; triangular buttress to 2 × 2m; leaves alternate palmately compound; leaflets c. 7, normally elliptic-oblong, c. 17 × 4cm, sessile; inflorescence fasciculate on leafless wood; flowers 3cm, white; pedicels 4cm; fruit pendulous, c. 20cm long, valves falling to reveal white fibres. Forest and farmbush; 460m.
Distr.: tropical America & Africa. (Amphi-Atlantic).
IUCN: LC
Nyandong: Cheek 11412 3/2003; **Nyasoso:** Pollard (sight record) 10/1998.

Ochroma lagopus Sw.
Syn.: *Ochroma pyramidale* (Cav. ex Lam.) Urb.
Tree c. 15m; leaves alternate, orbicular in outline, c. 30cm diameter, palmately 5-lobed; softly-hairy on both sides; petiole 10cm; stipules orbicular-ovate, 1cm, persistent; flowers erect, 8cm, white. Farmbush, forest; 150–1100m.
Distr.: tropical America, introduced to tropical Africa. (Pantropical).
IUCN: LC
Mejelet-Ehumseh: Etuge 464 fl., 1/1987; **Mungo River F.R.:** Cheek 10151 11/1999.
Note: possibly introduced to Mt Kupe by German plantation owners. Spreading, but otherwise not yet widespread in Cameroon. The brown fruit fibres form 'carpets' below the trees. Common in spots along Bangem to Tombel road.
Local name: Balsa (of commerce) (*fide* Cheek).

BORAGINACEAE

M. Cheek

Cordia africana Lam.
Tree 30–40m; slash yellow, quickly oxidising brown-green; leaves orbicular, 11–17cm, obtuse-acute, base obtuse to shallowly cordate, serrate, densely whitish brown puberulent below apex; nerves 3–4 pairs; petiole 3–5cm; inflorescence terminal; calyx 0.6cm, ridged; corolla white, 2cm diameter. Farmbush; 920m.
Distr.: Guinea (C) to Kenya. (Tropical Africa).
IUCN: LC
Nyasoso: Cheek 7504 10/1995.
Note: *Cordia millenii* and *C. senegalensis* are also likely to occur in our area.

Cordia aurantiaca Baker
Shrub or tree 2.5–3.5m; slash pale brown, turning green, then black; leaves papery, drying dark brown above, mid-brown below, elliptic, c. 14 × 7cm, acutely acuminate, with apical

hair, obtuse-rounded, entire, puberulent on nerves; nerves 3–4 pairs; petiole 1–1.5cm; inflorescence 2–10cm; flowers clustered; calyx 1 × 0.8cm, deeply-ribbed, puberulent; petals yellow 1.5cm; stamens long-exserted; berry ovoid 1.5cm; calyx cup-like. Forest; 250–1000m.
Distr.: S Nigeria, Bioko & Cameroon. (Lower Guinea).
IUCN: NT
Bolo-Meboka: Thomas 7152 7/1987; **Ngusi:** Etuge 1558 1/1996; **Nyandong:** Tadjouteu 564 3/2003; **Nyasoso:** Cable 2786 6/1996; Groves 103 2/1995.
Local name: Achum. **Uses:** MATERIALS – timber (*fide* Etuge).

Cordia platythyrsa Baker

Tree 15–20m; leaves elliptic or ovate-elliptic, 9–15cm, subacuminate, obtuse to truncate, entire or undulate; nerves 4–5 pairs, densely grey-white-puberulent below; petiole 8–10 cm; inflorescence 15–20cm, diffuse; fruit 1–2cm. Forest; 900–1200m.
Distr.: Sierra Leone to Cameroon. (Upper & Lower Guinea).
IUCN: LC
Mombo: Olorunfemi in FHI 30554 fr., 5/1951; **Nyasoso:** Etuge 2417 6/1996; Sunderland 1505 fl., 7/1992.

Cynoglossum amplifolium A.DC. var. *subalpinum* (T.C.E.Fr.) Verdc.

F.T.E.A. Boraginaceae: 106 (1991).
Syn.: *Cynoglossum amplifolium* Hochst. ex A.Rich. fa. *macrocarpum* Brand
Erect herb to 1m, hispid; leaves ovate, over 5cm broad, acuminate, narrowed at base with short-winged petiole; flowers conspicuous, bluish or white, 3–4mm diameter. Farmbush; 1900–2000m.
Distr.: Guinea (C) to Kenya. (Tropical Africa).
IUCN: LC
Mwanenguba: Letouzey 14400 fl., 8/1975; Pollard 158 10/1998; Thomas 3157 fl., fr., 2/1984.

Cynoglossum coeruleum A.DC. subsp. *johnstonii* (Baker) Verdc. var. *mannii* (Baker & C.H.Wright) Verdc.

F.T.E.A. Boraginaceae: 110 (1991).
Syn.: *Cynoglossum lanceolatum* Forssk. subsp. *geometricum* (Baker & C.H.Wright) Brand
Herb 0.3–1.2m tall; cauline leaves elliptic to lanceolate, 2–22 × 0.8–5.5cm, sessile or short-petiolate; radical leaves long-petiolate; nutlets with glochidia only on meridian and top surface. Submontane forest, swamp forest, margins of cultivation; 900–2000m.
Distr.: Guinea (C) to Kenya. (Tropical Africa).
IUCN: LC
Kodmin: Biye 27 1/1998; **Mwanenguba:** Cheek 7238 fl., fr., 2/1995; Leeuwenberg 8447 fl., fr., 9/1971; Letouzey 14393 fl., 8/1975; Pollard 161 10/1998; **Nyasoso:** Cable 2868 6/1996.

Ehretia cymosa Thonn. var. *cymosa*

Tree 20–30m; leaves 2–3 per node, ovate, to 28 × 19cm, papery, drying black above, greenish black below, shortly acutely-acuminate, base rounded-truncate; nerves 4–6 pairs, subglabrous below; petiole 8–12cm; inflorescence 20–30cm, dense; flowers white, 3mm; calyx cup-like; berries violet, 5mm. Farmbush; 350–1000m.
Distr.: Sierra Leone to Uganda. (Guineo-Congolian).
IUCN: LC
Bakolle Bakossi: Etuge 140 5/1986; **Kupe Village:** Etuge 1948 5/1996; **Nyasoso:** Zapfack 638 6/1996.

Note: resembles *Cordia*, but styles once-divided (not twice); stamens included; flowers smaller.

BUDDLEJACEAE

M. Cheek

Nuxia congesta R.Br. ex Fresen.

Syn.: *Lachnopylis mannii* (Gilg) Hutch. & M.B.Moss
Tree 2–8m; bole white, fibrous; stems orange-brown; leaves in whorls of 3, dimorphic: type 1 ovate-elliptic, c. 4 × 2.5cm, apex rounded, base cuneate, serrate; petiole 0.2cm; type 2 c. 11 × 5.5cm, entire; petiole 2cm; inflorescence terminal, 7 × 10cm, dense, many-flowered; flowers white, c. 7mm. Forest, forest edge; 1500–1980m.
Distr.: Guinea (C) to South Africa. (Afromontane).
IUCN: LC
Kodmin: Cheek 8972 1/1998; **Mt Kupe:** Cheek 7571 10/1995; **Mwanenguba:** Etuge 4382 10/1998; **Nyasoso:** Cheek 7345 fl., 2/1995.
Uses: SOCIAL USES – smoking materials/drugs – snuff adulterants – spice for snuff made from the dried flowers (Cheek 7345).

BURSERACEAE

J.-M. Onana & I. Darbyshire

Canarium schweinfurthii Engl.

Tree to 40m; exudates copious, drying whitish; young branches and young leaves rusty-pubescent; leaves clustered towards branch apices, c. 30–60cm long, imparipinnate, 8–12-jugate; leaflets oblong-ovate, 5–20 × 3–5.5cm, base cordate, lateral nerves pubescent below; petiole flat on upper side towards base; panicles to 30cm long; pedicels c. 3mm; flowers cream-white, c. 1cm; fruits ellipsoid, c. 3.5 × 2cm, epicarp glabrous, purple-black, endocarp trigonous. Forest.
Distr.: Senegal to Sudan & Tanzania. (Guineo-Congolian).
IUCN: LC
Nyandong: Etuge (sight record) 3/2003.
Local names: Atoii, Bush Plum or Bush Candle. **Uses:** FOOD – infructescences – fruits edible; MATERIALS – timber; FUELS – tinder – latex used for lighting fires (fide Etuge).

Dacryodes edulis (G.Don) H.J.Lam

Tree to 20m; exudate brown; branchlets, petioles, inflorescence rachis and outside of flowers stellately brown-hairy; leaves imparipinnate, 4–7-jugate; leaflets oblong-ovate, 12.5–20.5 × 3.5–5.5cm, acumen to 1.8cm, base acute, oblique in lateral leaflets, undersurface sparsely stellate-hairy with long simple hairs along midrib; lateral nerves 11(–14) pairs; petiole 9cm; petiolules 0.5cm; panicles many-branched; partial peduncles to 14cm, grooved; flowers 0.4cm, cream; fruit ovoid-ellipsoid, 5.5 × 2.2cm. Forest, and planted near villages.
Distr.: S Nigeria to Angola. (Lower Guinea & Congolian).
IUCN: LC
Konye: Etuge (sight record) 3/2003; widely planted in Bakossi.
Local name: Plum (Pidgin English). **Uses:** FOOD – infructescences – fruits edible, deliberately planted for this purpose (*fide* Etuge).

Dacryodes klaineana (Pierre) H.J.Lam

Tree to 45m, smelling strongly of turpentine when cut, resembling *D. edulis*, but branchlets glabrescent; leaves 3(–4)-jugate; base of lateral leaflets (sub)symmetrical; lateral nerves 10–12 pairs, stellate hairs on lower surface inconspicuous or absent; petiolules to 2cm; inflorescence rusty-scaled, not stellately-hairy; fruit oblong-ellipsoid, 2–3 × 1.5–2cm. Forest; 200–1000m.

Distr.: Sierra Leone to Gabon. (Upper & Lower Guinea).
IUCN: LC
Ikiliwindi: <u>Nemba 558</u> 6/1987; **Karume:** <u>Nemba 67</u> 5/1986; **Kodmin:** <u>Onana 600</u> 2/1998; **Kupe Village:** <u>Elad 81</u> 1/1995; **Mungo River F.R.:** <u>Onana 953</u> 11/1999; **Nyasoso:** <u>Cable 3629</u> 7/1996.
Note: Nemba 587 from Wone, identified by Gereau (MO) as *D. letestui*, has not been observed by the authors and is here omitted as the determination is likely erroneous in view of the fact that this taxon is known only from Gabon.

Santiria trimera (Oliv.) Aubrév.

Tree to 20m, with flattened stilt roots, glabrous; strongly smelling of turpentine when cut; leaves imparipinnate, 2–3-jugate; leaflets oblong-elliptic, 10.5–16.5 × 4–6.5cm, acumen narrow, to 2.2cm long, base acute to obtuse; lateral nerves 7–10 pairs; petiole 9.5cm; petiolules 1cm; panicles axillary, to 9cm; fruit green-black, asymmetrically oblate, 1.2 × 2 × 1.5cm; style remains lateral; fruiting pedicel 0.8cm. Forest; 200–1400m.

Distr.: Sierra Leone to Congo (K). (Guineo-Congolian).
IUCN: LC
Bolo: <u>Nemba 51</u> 3/1986; **Ikiliwindi:** <u>Nemba 560</u> fr., 6/1987; **Karume:** <u>Nemba 65</u> fr., 5/1986; **Kodmin:** <u>Plot B45</u> 1/1998; **Kupe Village:** <u>Cable 2489</u> 5/1996; **Nyandong:** <u>Tchiengue 1711</u> 3/2003.

CACTACEAE

D. Zappi & M. Cheek

Rhipsalis baccifera (J.Miller) Stearn

Syn.: *Rhipsalis cassutha* Gaertn.
Epiphyte, pendulous, 45cm; stems terete, grey-green, succulent, glabrous; branching at intervals of c. 10cm; flower buds red, 1cm; petals white at anthesis; fruit globose 8mm, white, red-tinged. Forest; 650–1050m.

Distr.: tropical America & Africa. (Amphi-Atlantic).
IUCN: LC
Enyandong: <u>Cheek 10979</u> 10/2001; **Kodmin:** <u>Cheek 9175a</u> 2/1998; **Kupe Village:** <u>Zapfack 960</u> 7/1996; **Ndum:** <u>Cable 924</u> fr., 2/1995; **Ngusi:** <u>Etuge 1586</u> fr., 1/1996; **Nyasoso:** <u>Thorold 73</u> 11/1950.

CAMPANULACEAE

M. Thulin & Y.B. Harvey

Dielsantha galeopsoides (Engl. & Diels) E.Wimm.

Terrestrial herb, prostrate or scrambling up to 25cm tall; several succulent angled stems, often rooting below; leaves ovate to oblong-ovate, 30–70mm long, margins serrate; petioles to 30mm; inflorescence lax; sessile; hypanthium narrowly obconical, to 8mm long; calyx lobes linear, to 4mm long; bracts at base of hypanthium to 5mm long; corolla to 9mm long, white to pale purple; capsule to 20mm long, narrowly urceolate prior to splitting into 5 long papery 'lobes'; seeds numerous. Forest; 650–1000m.

Distr.: Cameroon Endemic.
IUCN: EN
Manehas F.R.: <u>Etuge 4355</u> fl., 10/1998; **Nyale:** <u>Etuge 4460</u> fl., 11/1998; **Nyasoso:** <u>Etuge 1592</u> 1/1996.

Lobelia hartlaubii Buchenau

Straggling or erect herb 15–35(–50)cm long; stems ribbed, subglabrous, often rooting at the lower nodes; leaves triangular, 15–35 × 10–30mm, margin coarsely dentate; petiole 5–20mm long, occasionally "winged"; racemes 2–4-flowered; pedicels c. 4mm long; bracts 3–8mm long; bracteoles c. 1.5mm long, near top of pedicel; hypanthium to 4mm long in flower, extending to 9mm in fruit; calyx lobes linear; corolla c. 6mm long; seeds 0.7–1mm long, numerous, pale or reddish brown. Grassland-forest edge, forest; 800–1900m.

Distr.: Nigeria to Uganda & Burundi, Madagascar. (Tropical Africa).
IUCN: LC
Bangem: <u>Thomas 5337</u> 1/1986; **Edib:** <u>Etuge 4502</u> 11/1998; **Kodmin:** <u>Ghogue 22</u> fl., 1/1998; <u>Gosline 171</u> fl., 11/1998; **Mt Kupe:** <u>Cheek 7583</u> 10/1995.

Lobelia molleri Henriq.

Scrambling herb to about 15cm high; stems rooting at lower nodes, with narrowly-winged edges; leaves up to 24–32 × 11–16mm, lanceolate to ovate, narrowing towards apex, margin shallowly crenate; flowers in leafy racemes; pedicels to 12mm long; bracts to 7mm long; hypanthium 2.5mm long (to 8mm in fruit); calyx lobes narrowly triangular; corolla to 5mm long, white, purple-tinged; seeds to 0.6mm long, brown. Upland forest, forest margins, streamsides; 1940m.

Distr.: São Tomé, Bioko, Cameroon to Malawi. (Tropical Africa).
IUCN: LC
Kupe Village: <u>Ryan 415</u> fl., 5/1996.

Lobelia rubescens De Wild.

Syn.: *Lobelia kamerunensis* Engl. ex Hutch. & Dalziel
Straggling herb to 45cm long; stems triangular with narrowly-winged edges, often rooting at nodes; leaves lanceolate, gradually narrowing upwards on stem, 13–30 × 3–10mm, margin serrate, subglabrous; petioles c. 5mm long; flowers in leafy racemes; pedicels 7–14mm long; bracts to 14mm long; hypanthium 5–6mm long (extending to 7mm in fruit); calyx lobes narrowly triangular; corolla to 10mm long, blue with yellow throat; seeds c. 0.5mm long, brown. Farmbush, grassland; 800–1450m.

Distr.: Nigeria to Tanzania. (Lower Guinea & Congolian).
IUCN: LC
Bangem: <u>Thomas 5344</u> 1/1986; **Kodmin:** <u>Cheek 8894</u> 1/1998; <u>9587</u> 11/1998.
Local name: no name (Cheek 8894); unknown (Max Ebong in Cheek 9587). **Uses:** FOOD – leaves – used in cooking (Cheek 8894).

Monopsis stellarioides (Presl) Urb. subsp. *schimperiana* (Urb.) Thulin

F.T.E.A. Lobeliaceae: 47 (1984).

Syn.: *Monopsis stellarioides* (Presl) Urb. var. *schimperiana* (Urb.) E.Wimm.
Annual or perennial decumbent herb; stems 5–60cm long, rooting at nodes; leaves subsessile; linear to elliptic, up to 35 × 9 mm; mucronate at the apex, margin incrassate; flowers

solitary; pedicels 8–35mm long; bracts at base of pedicel to 16mm long; hypanthium obconical; calyx lobes 2–6mm long; corolla 6.5–11.5mm long, dirty-yellow, dull-pink or brownish purple; capsule up to 10mm long; seeds broadly elliptic, c. 0.6mm long, dark brown. Upland grassland, moorland, forest margins, rocky places; 2000m.
Distr.: Bioko, Cameroon, Congo (K), Ethiopia to Malawi & Comoros. (Afromontane).
IUCN: LC
Mwanenguba: Pollard 863 fl., 11/2001.

Wahlenbergia perrottetii (A.DC.) Thulin
Symb. Bot. Ups. 21: 199 (1975).
Syn.: *Cephalostigma perrottetii* A.DC.
Annual herb to 45cm tall; stem hirsute at base, glabrous elsewhere; lobes narrowly elliptic, 23–26 × 6mm, margin undulate; inflorescence lax; pedicels to 15mm long; bracts 4.5mm long at base of pedicels; hypanthium obconical, c. 1.5mm long (to 4mm in fruit); calyx lobes to 1mm long; corolla 1.5–2mm long, blue, split almost to base; style blue; seeds c. 0.6mm long, elliptic. Grassland; 1330–1500m.
Distr.: tropical Africa & Madagascar, also in S America.
IUCN: LC
Kodmin: Cheek 10275 12/1999; 9077 fl., 1/1998.

Wahlenbergia ramosissima (Hemsl.) Thulin subsp. *ramosissima*
Symb. Bot. Ups. 21: 189 (1975).
Syn.: *Lightfootia ramosissima* (Hemsl.) E.Wimm. ex Hepper
Erect annual herb, c. 10cm high; stems laxly-pubescent; leaves sessile, basal leaves broadly ovate, 5 × 4mm; stem leaves narrowly ovate, up to 10 × 4mm, margin undulate; inflorescence lax; bracts at base of pedicels to 3mm long; pedicels 12–14mm long; hypanthium obovoid, to 1mm long, expanding to 2mm in fruit; calyx lobes to 0.5mm long; corolla blue to 1mm long, split almost to base in linear lobes; seeds elliptic, to 0.6mm long. Grassland; 2000m.
Distr.: Nigeria & Cameroon. (W Cameroon Uplands).
IUCN: VU
Mwanenguba: Cheek 9429 fl., 10/1998; Pollard 847 11/2001.
Note: the identification of the Pollard specimen needs to be confirmed.

Wahlenbergia silenoides Hochst. ex A.Rich.
Symb. Bot. Ups. 21: 78 (1975).
Syn.: *Wahlenbergia mannii* Vatke
Perennial herb, up to 60cm; stems hirsute at base; leaves lanceolate to narrowly ovate, 5–25 × 1–7mm, margin sometimes undulate; inflorescence few-flowered, lax; pedicels up to 60mm long; hypanthium narrowly obconical; calyx lobes 1.6–4mm long; corolla 4–6.5mm long, +/- white, deeply-split into narrowly elliptic lobes; capsule narrowly obconical, up to 17mm long; seeds 0.6–0.7mm long. Grassland, forest-grassland transition; 1975–2000m.
Distr.: Nigeria, Bioko, W Cameroon, E to Sudan, Ethiopia & E Africa. (Afromontane).
IUCN: LC
Mwanenguba: Cheek 7241 2/1995; Pollard 867 fl., fr., 11/2001.

CANNABACEAE

I. Darbyshire

Fl. Cameroun 19 (1975).

Cannabis sativa L.
Erect herb, 0.3–2(–4)m; stems ridged, puberulent; leaves opposite or alternate, palmately (3–)5–8(–11)-foliolate, leaflets sessile, lanceolate, 8(–15) × 0.8(–2)cm, margin serrate, scabridulous above and with minute gland dots; petiole 3–7.5cm, puberulent; stipules subulate, c. 4mm; male flowers in lax axillary and terminal panicles, white-green, puberulent; female inflorescence more compact; achene ellipsoid, 2.5–5 × 2–3.5mm, shiny, enveloped by an acuminate bracteole. Planted in villages, occasionally persisting in abandoned areas; 420m.
Distr.: cosmopolitan, originating from Asia.
IUCN: LC
Nyandong: Nguembou 629 st., 3/2003.
Note: presumed planted for medicinal purposes, apparently spontaneous.
Uses: SOCIAL USES – smoking materials/drugs – smoked as an intoxicant (*fide* Pollard via Freddie Epie).

CAPPARACEAE

M. Cheek

Fl. Cameroun 29 (1986).

Buchholzia coriacea Engl.
Tree 15m; leaves simple, obovate, 10–30 × 4–11cm; up to 10 pairs of nerves conspicuous on upper surface; inflorescence unbranched, c. 20cm; flowers green c. 1.5cm; petals absent; stamens numerous; fruit globose, c. 5cm, 1-seeded. Forest; 250–300m.
Distr.: Guinea (B) to Gabon. (Upper & Lower Guinea).
IUCN: LC
Baduma: Nemba 224 9/1986; 620 9/1987; **Enyandong:** Etuge 4497r 11/2001; **Konye:** Thomas 5147 11/1985.
Local name: Mbam (Etuge 4497). **Uses:** FOOD – seeds (Etuge 4497) & infructescences – fruit flesh edible, tastes like peppery winter squash (Thomas 5147); grind seeds, eat with *Cola* nuts (*fide* Etuge).

Buchholzia tholloniana Hua
Fl. Cameroun 29: 18 (1986).
Tree 8m; resembling *B. coriacea*, but leaves narrowly elliptic; nerve pairs 11–17, inconspicuous on upper surface; inflorescence often-branched. Forest; 600m.
Distr.: Cameroon to Congo (K). (Lower Guinea & Congolian).
IUCN: NT
Kupe Village: Cheek 7921a 11/1995.
Local name: Mbam. **Uses:** FOOD – infructescences; MEDICINES (Cheek 7921).

Cleome afrospina Iltis
Fl. Cameroun 29: 40 (1986).
Erect herb resembling *C. spinosa*, but secondary nerves 17–28 (not less than 17); capsules 9–12mm wide (not 4mm); fruiting gynophore as long as pedicel (not twice as long); petals 0.9–1.6cm (not to 2.3cm). Secondary forest; 710–800m.
Distr.: Cameroon to Congo (K). (Lower Guinea & Congolian).

IUCN: LC
Kupe Village: Cable 2530 5/1996; Kenfack 330 7/1996.
Note: easily confused with *C. spinosa*.

Cleome iberidella Welw. ex Oliv.
Fl. Cameroun 29: 61 (1986).
Syn.: *Cleome montana* A. Chev. ex Keay
Herb resembling *C. rutidosperma*, but leaves to 5cm; leaflets to 2.5cm; flowers numerous at stem apex; most reliably differing in the hairy fruits. Thicket; 1000m.
Distr.: Guinea (B) to Zambia. (Afromontane).
IUCN: LC
Mwanenguba: Leeuwenberg 9570 fl., fr., 4/1972.

Cleome rutidosperma DC.
Fl. Cameroun 29: 57 (1986).
Syn.: *Cleome ciliata* Schum. & Thonn.
Herb; branches prostrate, radiating from taproot, c. 30cm; leaves 2.5cm; leaflets 3, 1.5cm; flowers single, erect, 0.7cm; petals 4, 0.5mm, erect; fruit cylindrical, 4cm, glabrous. Roadside; 420–900m.
Distr.: pantropical, originating in W Africa.
IUCN: LC
Mwanenguba: Leeuwenberg 8528 fl., 9/1971; **Nlonako:** Etuge 1779 3/1996; **Nyandong:** Darbyshire 1 fl., 3/2003.

Cleome spinosa Jacq.
Erect herb 1–1.5m; leaves palmately 7-foliolate; leaflets to 10cm, with stipular spines; inflorescence dense, unbranched, to 20cm, densely clothed with erect ovate bracts 1.5cm; pedicels 2–3cm; petals pale pink, 2–3cm; fruit cylindric, 10cm. Cultivated, farmbush; 900m.
Distr.: pantropical, originating in Colombia.
IUCN: LC
Mwanenguba: Leeuwenberg 9018 fl., fr., 12/1971.

Euadenia alimensis Hua
Fl. Cameroun 29: 74 (1986).
Syn.: *Euadenia pulcherrima* Gilg & Benedict
Shrub to 2m; leaves 25cm; leaflets 3, obovate, 19cm; petiolule 2mm; inflorescence erect, unbranched; flowers green, upper 2 petals oblanceolate, 10 × 2.7cm, lower petals shorter than sepals; sepals 5, oblong, 2.2 × 0.4cm; fruit not seen. Forest; 990m.
Distr.: Cameroon to Uganda. (Lower Guinea & Congolian).
IUCN: NT
Kupe Village: Etuge 1491 11/1995.
Note: Kers in Fl. Cameroun (1986) states that this species has only been collected in Cameroon in 1901, 1908 and 1928, and is rare or has disappeared; also that the Cameroon material is particularly large-flowered and with ornamental potential. Treated as a synonym of *E. eminens* Hook.f. by Lebrun & Stork (1991).

Euadenia trifoliolata (Schum. & Thonn.) Oliv.
Small tree c. 5m; resembling *E. alimensis*, but upper 2 petals c. 6 × 1cm, lower petals larger and more slender than the sepals. Forest; 1500m.
Distr.: Ivory Coast to Cameroon. (Upper & Lower Guinea).
IUCN: NT
Kodmin: Onana 577 2/1998.

Ritchiea erecta Hook.f.
Syn.: *Ritchiea polypetala* Hook.f.
Syn.: *Ritchiea brachypoda* Gilg
Syn.: *Ritchiea oreophila* Gilg & Benedict
Shrub to 1m, unbranched; leaves 1–3-foliolate; leaflets lanceolate to obovate, 13–17 × 6–10cm; inflorescence corymbose, 30–40-flowered, erect; flowers c. 9cm wide; pedicel 1–2.5cm; sepals 4, ovate, c. 2cm; petals 8–12, white, 4–5 × 0.1–0.3cm; stamens 50–60; fruit ellipsoid 3–7 × 1.5–3cm, fleshy, indehiscent. Forest; 1200m.
Distr.: Nigeria to Gabon. (Lower Guinea).
IUCN: NT
Ndoungue: Ledermann 6332 11/1909.

Ritchiea simplicifolia Oliv. var. *caloneura* (Gilg) Kers
Fl. Cameroun 29: 102 (1986).
Syn.: *Ritchiea caloneura* Gilg
Shrub; leaves simple and (apex) trifoliolate on main axis; simple leaves elliptic c. 22 × 7cm, acuminate, cuneate; nerves c. 10 pairs; petiole 0.5cm; compound leaves with central leaflet c. 22cm; laterals c. 17cm; petiole 9cm; inflorescence terminal, 2–3-flowered; pedicels 0.5cm; flowers c. 5cm diameter. Forest; 350m.
Distr.: Cameroon Endemic.
IUCN: NT
Bakolle Bakossi: Etuge 159 5/1986.
Note: specimen not at K, determined by D.W.Thomas (MO). Taxon otherwise known only from S of the Sanaga (Fl. Cameroun 29). Description from Bipindi material at Kew. *R. simplicifolia* var. *simplicifolia* is also likely to be found in our area (known from S Bakundu F.R.).

CARICACEAE

M. Cheek

Carica papaya L.
Dioecious tree to 5m, unarmed, glabrous; leaves orbicular in outline, c. 30cm diameter, palmately-lobed, lobes c. 7, 15cm, with 1–2 lateral lobes, sinuate; male inflorescence c. 20-flowered; flowers c. 3cm long, white. Persisting in farmbush; 1300m.
Distr.: widely cultivated in the tropics, originating from S America. (Pantropical).
IUCN: LC
Kupe Village: Etuge 1701 fl., 2/1996.
Local names: Papaya, Pawpaw (widely used names). **Uses:** FOOD – infructescences – a cash crop (*fide* Pollard); MEDICINES (Etuge 1701).

Cylicomorpha solmsii (Urb.) Urb.
Dioecious tree resembling *Carica*, but to 30m; spines on leafy branches; leaf-lobes entire. Secondary forest; 450–1450m.
Distr.: Cameroon. (W Cameroon Uplands).
IUCN: VU
Enyandong: Cheek 10963 10/2001; **Kodmin:** Cheek 8938 1/1998; **Kupe Village:** Cable 2503 fl., 5/1996; Kenfack 258 fl., 7/1996; **Mbule:** Leeuwenberg 9307 fl., 1/1972; **Nyandong:** Cheek 11365 3/2003; Ghogue 1477 fl., 3/2003; **Nyasoso:** Cable 2759 fr., 6/1996; 696 fl., 1/1995; 862 1/1995; Cheek 7299 fl., 2/1995; Etuge 1525 fl., 1/1996; Thomas 3043 fl., 2/1984.
Local names: Echooa (Cheek 8938); Ejooa. **Uses:** BEE PLANTS – holes are made in the trunk to encourage bees to nest (Cheek 8938); MATERIALS/SOCIAL USES – the hollow trunks once used as prison cells (*fide* Cheek).

CARYOPHYLLACEAE

M. Cheek

Cerastium octandrum Hochst. ex A.Rich.
Sticky, straggling herb c. 10cm; leaves opposite, elliptic, c. 1.5 × 0.5cm, sessile; inflorescence dischasial, terminal; flowers white c. 0.5cm; fruit glossy-yellow, cylindrical, 1cm, apex toothed. Montane grassland; 2000m.
Distr.: Cameroon to Tanzania. (Afromontane).
IUCN: LC
Mwanenguba: Cheek 9450 fr., 10/1998.

Drymaria cordata (L.) Willd.
Straggling, slightly viscid herb; stems to 90cm; leaves broadly ovate, 1–2.5cm long; inflorescence a terminal cyme; flowers few, white. Montane or submontane grassland; 650–1550m.
Distr.: pantropical.
IUCN: LC
Kupe Village: Cheek 7123 fl., 1/1995; **Ngusi:** Etuge 1564 fl., 1/1996; **Nyasoso:** Cable 1245 fl., 2/1995; Lane 218 fl., 11/1994.

Drymaria villosa (Cham. & Schltdl.) Duke subsp. *palustris* (Cham. & Schltdl.) Duke
Ann. Missouri Bot. Gard. 48: 227 (1961).
Straggling herb, as subsp. *villosa*, but sepals oblong to orbicular; petals usually divided to near the base; capsule often twice as long as sepals. Grassland, villages; 800–1650m.
Distr.: neotropics, introduced elsewhere. (Pantropical).
IUCN: LC
Mejelet-Ehumseh: Manning 429 10/1986; **Nyasoso:** Thomas 5028 11/1985.
Note: specimens not seen, confirmation required.

Drymaria villosa Cham. & Schltdl. subsp. *villosa*
Ann. Missouri Bot. Gard. 48: 226 (1961).
Straggling, slightly viscid herb; stems to 90cm long, sparsely long-hairy; leaves broadly ovate, smaller than in *D. cordata*; inflorescence a terminal cyme; flowers few, white; sepals elliptic, acute; petals divided to c. half their length; capsule length rarely exceeding sepals. Montane grassland; 2000m.
Distr.: neotropics, established elsewhere. (Pantropical).
IUCN: LC
Mwanenguba: Cheek 9477 fl., 10/1998.

Stellaria mannii Hook.f.
Straggling, sometimes sticky herb to 60cm; leaves long-petiolate; inflorescence a terminal cyme; flowers few, white. Forest edge; 1200–1500m.
Distr.: pantropical.
IUCN: LC
Kodmin: Cheek 8959 fr., 1/1998; 9031 fl., fr., 1/1998; **Mt Kupe:** Thomas 5110 fl., 12/1985; **Nyasoso:** Lane 229 fl., 11/1994.

Uebelinia abyssinica Hochst.
Bull. Jard. Bot. Nat. Belg. 55: 435 (1985).
Syn.: *Uebelinia hispida* Pax
Prostrate herb; stems straggling, hispid, to 45cm long; leaves elliptic to orbicular, 7–15mm long; flowers axillary, white. Montane grassland; 1900–2000m.
Distr.: Cameroon to Ethiopia. (Afromontane).
IUCN: LC

Mwanenguba: Cheek 9423 fr., 10/1998; Leeuwenberg 8459 fl., 9/1971.

CECROPIACEAE

M. Cheek & I. Darbyshire

Fl. Cameroun 28 (1985).

Musanga cecropioides R.Br. ex Tedlie
Tree to 30m, with an umbrella-like crown; stipule sheaths 10–20cm, deep red outside, silky within, caducous; leaves 12–15-digitately divided almost to base, less so when immature, lobes oblanceolate, c. 25 × 6cm, felty grey-puberulent below, dark green above; lateral nerves numerous, prominent below; male flowers in numerous globose heads, 4mm, on a many-branched panicle; female inflorescence dense, ellipsoid, 2cm on a peduncle to 12cm; fruit ellipsoid, yellow-green. Farmbush, forest; 1050m.
Distr.: Guinea (C) to Uganda. (Guineo-Congolian).
IUCN: LC
Nyasoso: Lane 33 fl., fr., 8/1992.
Note: a common and gregarious forest pioneer.
Local names: Ekombe or Umbrella Tree. **Uses:** MATERIALS – local timber (*fide* Etuge).

Myrianthus arboreus P.Beauv.
Tree 7(–25)m; leaves alternate, palmately-compound; petiole robust, 30cm; leaflets 7, elliptic oblong, c. 30 × 10cm, acute or acuminate, serrate, concolorous; male inflorescence c. 20 × 15cm; fruit compound, ellipsoid, c. 15 × 10cm, yellow, juicy. Forest, farmbush; 220–990m.
Distr.: Guinea (C) to W Tanzania. (Guineo-Congolian).
IUCN: LC
Bakolle Bakossi: Etuge 253 9/1986; **Enyandong:** Ghogue 1268 10/2001; **Ndum:** Elad 100 fr., 2/1995; **Nyandong:** Ghogue 1537 fr., 3/2003.

Myrianthus preussii Engl. subsp. *preussii*
Treelet 2–4(–9)m, resembling *M. arboreus*, but petioles slender, 15cm; leaflets 5(–7), bright-white below, obovate, long-acuminate; petiolules c. 3cm; male inflorescence c. 6 × 4cm; fruit globose, c. 5cm diameter. Forest; 520–1400m.
Distr.: SE Nigeria & Cameroon. (Lower Guinea).
IUCN: NT
Eboné: Leeuwenberg 8723 fl., 11/1971; **Kodmin:** Cheek 9004 1/1998; Etuge 3968 1/1998; 4430 11/1998; Nwaga 7 1/1998; **Kupe Village:** Cable 3863 7/1996; Cheek 7180 fr., 1/1995; 7774 11/1995; 7843 11/1995; Etuge 2851 7/1996; Ryan 354 5/1996; **Manehas F.R.:** Etuge 4349 10/1998; **Nyandong:** Cheek 11344 3/2003.
Local name: 'Bush Pineapple' (Ryan 354) & (Etuge 3968).
Uses: FOOD – unspecified parts – edible (Ryan 354).

Myrianthus sp. 1
Treelet 3–5m, resembling *M. arboreus*, but leaves palmately 3–5-lobed by ¾ the radius, white below with brown-hairy nerves; petioles unequal in length on same stem. Forest; 1400m.
Distr.: Cameroon. (W Cameroon Uplands).
Kodmin: Plot B234 1/1998.
Note: probably a new species. More material desirable. Conspecific with e.g. Letouzey 14532 from Rumpi Hills.

CELASTRACEAE

G. Gosline
Maytenus by Sebsebe Demissew

Fl. Cameroun 19 (1975) & 32 (1990).

Apodostigma pallens N.Hallé var. *buchholzii*
Fl. Gabon 29: 197 (1986).
Woody climber to 15–25m; leaves ovate, 8–20 × 4–10cm, with linear acumen to 2mm; 3–6 pairs lateral nerves; inflorescence axillary cymes, drying yellow-brown to red-brown; buds 1mm diameter; flowers 1.5–3mm diameter; infructescence with 4–6 short segments, without branching in cymes; fruits 1–3 mericarps, oblong to 4 × 2cm. Forest; 1200m.
Distr.: Ivory Coast, Ghana, Cameroon, Gabon, Angola. (Upper & Lower Guinea).
IUCN: LC
Nyasoso: Cable 1141 fl., 2/1995.
Note: uncommon over much of its range, though specimens are often not identified to variety in this species.

Cassine aethiopica Thunb.
Shrub or small tree to 7m; branches somewhat pubescent; leaves variable, 4–12.5 × 1.5–5cm, toothed; 6–8 pairs lateral nerves; inflorescence a solitary axillary cyme; flowers tiny to 2mm, light yellow-green; pedicel to 3mm; sepal exterior pubescent; ripe fruits reddish, ovoid, to 1.5cm diameter, persistent calyx at base. Forest, savanna, shrubland; 2000–2410m.
Distr.: Cameroon, Ethiopia to SW Africa, Comoros, Seychelles. (Tropical Africa & Indian Ocean Islands).
IUCN: LC
Mwanenguba: Letouzey 14406 fl., fr., 8/1975; Thomas 3111 fl., 2/1984.

Hippocratea myriantha Oliv.
Woody climber to 15–45m; resinous threads; stems lenticellate; leaves ovate, obovate to deltoid, 4–15 × 3–8cm; 4–7 pairs lateral nerves, drying brown; inflorescence a large, white, puberulent cyme often in panicles, with a peduncle 3–8cm long; cyme to 8cm long; flowers very numerous, white, puberulous, 1.5–4mm diameter; fruits with flat, dehiscent, oblong-attenuate mericarps, 5–9 × 2–3cm. Forest; 1450m.
Distr.: Guinea (C) to Congo (K). (Guineo-Congolian).
IUCN: LC
Kodmin: Cheek 8924 fl., 1/1998; Gosline 146 11/1998.

Loeseneriella africana (Willd.) R.Wilczek ex N.Hallé var. *africana*
Fl. Cameroun 32: 212 (1990).
Woody climber 2–15m; stems with lenticels; leaves variable, elliptic, to 12 × 8cm, drying brown; inflorescence a short contracted cyme, 1–2.5cm; buds conical, 3–5 × 3–4mm; petals triangular; flowers hardly opening; mericarps flat, dehiscent, 4–8 × 2–4cm. Forest, riverbanks; 1000m.
Distr.: widespread in tropical Africa.
IUCN: LC
Nyasoso: Etuge 1518 fl., 12/1995.

Loeseneriella apocynoides (Welw. ex Oliv.) N.Hallé ex J.Raynal var. *apocynoides*
Fl. Gabon 29: 247 (1986).
Woody climber to 10–30m; young branches reddish brown puberulent; plant climbing partly via paired twigs, 2–3cm long; leaves c. 8 × 5cm; 5–6 secondary veins; inflorescence a dichotomous cyme; peduncles and bracts tomentose; flowers 3–8mm diameter, puberulent; fruiting mericarps elliptic, to 7 × 4cm. Forest, riverbanks; 1400m.
Distr.: Guinea (C) to Tanzania. (Guineo-Congolian).
IUCN: LC
Kodmin: Plot B8 1/1998; B29 1/1998; B37 1/1998.
Note: without flowering material it is impossible to determine the variety definitively.

Loeseneriella rowlandii (Loes.) N.Hallé
Fl. Gabon 29: 244 (1986).
Syn.: *Hippocratea rowlandii* Loes.
Woody climber to 10–15(–50)m; leaves elliptic, to 15 × 7cm; secondary nerves 4–5(–7) pairs; inflorescences of axillary cymes, 3–4cm long, and few-flowered; flower buds green, ovoid, to 5cm long; flowers with lanceolate petals 10–15mm in diameter; mericarps elliptic, to 6 × 3cm. Forest rocks, semi-deciduous forest in sunny spots; 150m.
Distr.: Sierra Leone to Congo (K). (Guineo-Congolian).
IUCN: LC
Mungo River F.R.: Cheek 10232 fl., 12/1999.

Loeseneriella yaundina (Loes.) N.Hallé ex R.Wilczek
Fl. Cameroun 32: 201 (1990).
Climber to 10m; leaves 7–16 × 3–8cm, sometimes crenulate; 6–8 pairs lateral nerves; petiole 6–12mm; inflorescence branching or cymes 3–10cm long; flowers greenish to orange, 10–12mm diameter; petals oblong, thinning at the margins, 5–7 × 2mm, glabrous; mericarps elliptic, 8–10 × 3–4cm. Gallery forest, marshy areas; 250–300m.
Distr.: Cameroon to Congo (K). (Lower Guinea & Congolian).
IUCN: NT
Kurume: Thomas 5465 fl., 1/1986; **Mekom:** Thomas 5396 fl., 1/1986.

Maytenus buchananii (Loes.) R.Wilczek
Fl. Cameroun 19: 28 (1975).
Small tree; older branches with spines, 2–3cm long; leaves elliptic, 3.5–6 × 1.3–2.5cm, toothed, base strongly attenuate; inflorescence an axillary cyme, 1–1.5cm long; sepals as long as petals; flower to 7.5mm long; fruit a tri-valved capsule 7–8mm long, each valve with 1–2 brown seeds. Forest edge, riverbanks; 1410–2060m.
Distr.: Ivory Coast to Mozambique. (Tropical Africa).
IUCN: LC
Mt Kupe: Etuge 3824 fl., 4/1997; Sebsebe 5084 fl., 11/1995; **Mwanenguba:** Letouzey 14408 8/1975; **Nyasoso:** Cable 106 fl., 9/1992; 3391 fl., 6/1996.

Maytenus gracilipes (Welw. ex Oliv.) Exell subsp. *gracilipes*
Symb. Bot. Ups. 25(2): 51 (1985).
Syn.: *Maytenus serrata* (Hochst. ex A.Rich.) R.Wilczek var. *gracilipes* (Welw. ex Oliv.) R.Wilczek
Tree or shrub to 5m; spiny; leaves 8–16 × 3–17cm, elliptic, base strongly attenuate, apex angled, shortly acuminate; lateral nerves 10–12 pairs; inflorescence a loose axillary cyme, 5–14cm long; flowers tiny, c. 2 × 2mm; fruit a tri-locular capsule, red-violet when fresh, 1.5–2 × 1cm; 2–4 seeds per locule. Forest; 870–1600m.
Distr.: Cameroon to Angola. (Lower Guinea).
IUCN: LC
Ndum: Cable 920 fl., 2/1995; **Nyasoso:** Cable 1248 fr., 2/1995; 3272 6/1996; Etuge 2182 6/1996; Gosline 133 11/1998; Groves 92 fl., fr., 2/1995; Zapfack 668 6/1996.

Maytenus undatus (Thunb.) Blakelock

Shrub or tree, to 8m; spines absent or very small; leaves elliptic to lanceolate, 5.5–10 × 2–5.5cm, base attenuate to rounded, toothed; 7–10 pairs lateral nerves; inflorescence axillary in a compact dense subsessile cymule; flowers c. 3 × 3mm; capsule reddish, 6–8mm long, containing 2–3 seeds; Forest edges, riverbanks; 1950–2000m.
Distr.: tropical Africa.
IUCN: LC
Nyasoso: Cable 2901 6/1996; Cheek 7343 fl., 2/1995; Etuge 1652 1/1996.

Pristimera preussii (Loes.) N.Hallé

Fl. Gabon 29: 214 (1986).
Syn.: *Hippocratea preussii* Loes.
Woody climber to 15(–60)m; leaves to 10 × 6cm, acuminate; 6–9 pairs of secondary nerves; inflorescence an axillary lax cyme, c. 10cm long; flowers creamy-brown, c. 5–6cm diameter; pedicels c. 1mm; petals clawed, denticulate; fruits, flat dehiscent mericarps, 6 × 4cm. Forest; 400–700m.
Distr.: Cameroon & Congo (K). (Lower Guinea & Congolian).
IUCN: NT
Kupe Village: Etuge 1928 fl., 5/1996; 2732 fl., 7/1996.

Reissantia indica (Willd.) N.Hallé var. *astericantha* N.Hallé

Adansonia (sér 2) 17: 408 (1978).
Woody climber to 10–30m; leaves drying green, somewhat ovate with rounded base, 12 × 6cm; 5–6 pairs secondary veins; inflorescence cymes axillary or terminal with 2–3 initial ramifications, 2–3cm long, without flowers; branching up to 8 times; flowers (1–)2–4mm diameter; petals linear, attenuate with rolled edges, green, drying black; fruits of 3 dehiscent mericarps, 4 × 1cm. Forest; 800m.
Distr.: Guinea (C) to Uganda. (Guineo-Congolian).
IUCN: LC
Nyasoso: Etuge 2500 fr., 7/1996.

Reissantia indica (Willd.) N.Hallé var. *loeseneriana* (Hutch. & M.B.Moss) N.Hallé

Fl. Gabon 29: 190 (1986).
Liana resembling var. *astericantha*, but leaves often drying brown, not green; flowers 1–2mm, white to yellowish, drying brown. Forest; 200–250m.
Distr.: Senegal to Angola & Ethiopia. (Tropical Africa).
IUCN: LC
Ikiliwindi: Nemba 528 6/1987; **Kurume:** Thomas 8214 4/1988.

Salacia alata De Wild. var. *superba* N.Hallé

Fl. Gabon 29: 120 (1986).
Liana or shrub to 5m; stems verruculose and distinctly sinuously 4-winged; leaves elliptic to obovate, to 26 × 10.5cm, apex short-acuminate, base rounded; lateral nerves 11–12 pairs; petiole 0.5cm; glomerules axillary, 3–4cm, several to many-flowered; fruit subglobose, 3cm diameter. Forest; 350–1000m.
Distr.: Nigeria, Cameroon & Gabon. (Lower Guinea).
IUCN: NT
Mungo River F.R.: Tadjouteu 316 12/1999; **Ngomboaku:** Ghogue 4440 12/1999.
Note: the similar *S. alata* var. *alata* can also be expected in our area; it differs in having straight (not sinuous) stem-wings.

Salacia debilis (G.Don) Walp.

Woody climber to 30m; no resinous threads; branches dark brown, lenticellate; branchlets with 4 fine ridges; leaves with distinct teeth, elliptic, 5–13 × 2–6cm, acuminate; 6–8 pairs lateral nerves; inflorescence of sessile axillary fascicles, with 6–8 flowers each; pedicel filiform, 6–21mm long; buds globular, 1.8mm diameter; flowers 3–5mm diameter, yellow to orange; fruits spherical, 1.3–3.2mm, becoming red; 2–3 seeds. Forest; 850–1700m.
Distr.: Guinea (B) to Congo·(K). (Guineo-Congolian).
IUCN: LC
Bangem: Thomas 5316 fl., 1/1986; **Kodmin:** Cheek 9187 fl., 2/1998; **Kupe Village:** Cable 3844 fr., 7/1996; Etuge 1891 fr., 5/1996; Ryan 270 5/1996; 426 fr., 5/1996; **Mbule:** Cable 3349 fr., 6/1996; 3374 fr., 6/1996; **Nyasoso:** Cable 3492 fr., 7/1996; Etuge 1742 fl., 2/1996; 2128 fr., 6/1996; 2564 fr., 7/1996; 2566 fr., 7/1996; Groves 80 fl., 2/1995; Wheatley 459 fr., 7/1992.
Uses: FOOD – unspecified parts (Ryan 270).

Salacia dusenii Loes.

Small tree or woody climber to 15m; no resinous threads; bark dark to light grey; petiole 6–10mm; leaves oblanceolate to ovate, 7–16 × 4–8cm, acumen to 2cm; lateral nerves 6–8 pairs; flowers sessile; pedicels 7–20mm, thickening slightly towards the top; fruit globular, to 3cm diameter. Forest; 1050–1600m.
Distr.: Nigeria to Congo (K). (Lower Guinea & Congolian).
IUCN: LC
Kupe Village: Cheek 7787 fl., fr., 11/1995; **Mt Kupe:** Cheek 7620 fr., 11/1995; **Nyasoso:** Cable 3399 fr., 6/1996; 3590 fl., 7/1996; 3617 fl., 7/1996; Etuge 1351 fr., 10/1995; 2126 fr., 6/1996; 2134 fr., 6/1996; 2476 fl., 6/1996.

Salacia elegans Welw. ex Oliv. var. *aff. inurbana* N.Hallé

Woody climber 6–40m; resinous threads; stems square, dark brown; leaves elliptic to 10 × 4cm with distinct teeth; c. 10 pairs lateral nerves, drying reddish brown; petiole 5–8mm; inflorescence axillary with a very short peduncle, 1–4mm long; pedicels 2–6mm; buds c. 2mm diameter; flower pale yellow 3–4.5mm diameter; fruits often in groups, small, globular, 1–1.5cm diameter, orange at maturity; 3 persistent styles; persistent calyx. Forest; 150–1650m.
Distr.: Bakossi Mountains. (Narrow Endemic).
Kodmin: Cheek 9205 fr., 2/1998; Etuge 4028 fr., 1/1998; **Mungo River F.R.:** Cheek 10234 fl., 12/1999; **Mwambong:** Cheek 9105 fr., 2/1998; Onana 528 fr., 2/1998; **Nyasoso:** Etuge 1767 fr., 3/1996.
Note: our specimens have larger leaves than var. *inurbana*.

Salacia erecta (G.Don) Walp. var. *dewildemaniana* (R.Wilczek) N.Hallé

Fl. Gabon 29: 115 (1986).
As for var. *erecta*, but floral pedicel densely verrucose-pustulate. Forest; 800–1400m.
Distr.: Cameroon, Congo (K). (Lower Guinea & Congolian).
IUCN: LC
Mejelet-Ehumseh: Etuge 168 fr., 6/1986.

Salacia erecta (G.Don) Walp. var. *erecta*

Climber to 15m, no resinous threads; branchlets dark brown smooth; petioles to 5mm, channel-margins undulate; blade elliptic, 4–15 × 1.5–5cm; 7–10 lateral nerves, margin toothed; flowers in sessile axillary fascicles, sometimes on thin shoots appearing like peduncles; pedicels 3–7mm; floral pedicel ± smooth; buds ovoid; flowers c. 5mm diameter, yellow; fruits globular, 1–3cm diameter, orange-red. Forest; 1350–1900m.

Distr.: Guinea (C) to Zambia. (Tropical Africa).
IUCN: LC
Mejelet-Ehumseh: Etuge 168 fr., 6/1986; **Mt Kupe:** Cheek 7565 fl., 10/1995; **Nyale:** Cheek 9695 11/1998.

Salacia gabunensis Loes.
Fl. Gabon 29: 93 (1986).
Woody climber to 15m; no resinous threads; twigs green; branchlets light brown; petiole 10–12mm; leaf-blade elliptic to 17 × 8cm; c. 8 lateral nerves, midrib prominent above; flowers sessile in axils; pedicels c. 5mm long; buds triangular with thickened calyx; fruits red, globular, c. 4cm diameter. Forest; 700–1190m.
Distr.: Cameroon to Congo (K). (Lower Guinea & Congolian).
IUCN: LC
Kupe Village: Cable 3864 fl., 7/1996; **Nyasoso:** Etuge 1352 fr., 10/1995; Sebsebe 5047 fr., 10/1995.

Salacia lehmbachii Loes. var. lehmbachii
Shrub to 2m; without resinous threads; stems quadrangular, slightly pustulose; leaves ovate, to 19 × 9cm; pedicel 5 × 10mm; c. 5 lateral nerves, tertiary veins perpendicular to midrib across leaf; inflorescence axillary, short-pedunculate; flowers 5–8mm diameter, red to orange; fruit long and acuminate or turbinate, to 7 × 4cm. Forest; 850–1550m.
Distr.: Cameroon to Congo (K). (Guineo-Congolian).
IUCN: LC
Kodmin: Etuge 4165 fl., 2/1998; **Ndum:** Lane 487 fl., 2/1995; **Nyasoso:** Cable 3539 7/1996; 3548 fl., 7/1996; 3587 fr., 7/1996; Cheek 7360 fl., 2/1995; Lane 150b fr., 10/1994; Sebsebe 5010 fr., 10/1995; Thomas 3055 2/1984.

Salacia lehmbachii Loes. var. pes-ranulae
N.Hallé
Fl. Gabon 29: 75 (1986).
Shrub to 2m; without resinous threads; stems quadrangular; leaves sessile, elongate-elliptic, to 20 × 7cm, linear acumen 2.5cm long; c. 6 pairs lateral nerves, tertiary veins perpendicular to midrib across leaf; inflorescence sometimes in midrib grooves, with alternate divisions starting 4–20mm below summit of rachis-peduncle; flowers 5–8mm diameter, red to orange; fruit long and acuminate. Forest; 450m.
Distr.: SE Nigeria & Cameroon. (Lower Guinea).
IUCN: VU
Bakossi F.R.: Cheek 9326 fl., fr., 10/1998.

Salacia loloensis Loes. var. loloensis
Fl. Gabon 29: 79 (1986).
Shrub 2–4m; no resinous threads; branchlets green; leaves oblong-elliptic, (10–)13–27 × 4–9cm, discolorous, often drying greenish brown below; petiole 6–10mm; 10–16 pairs secondary nerves, little-ascending; inflorescences axillary, sessile, 1–5-flowered; pedicels variable (4–)7–22mm; flower creamish yellow, pinkish yellow or pale green, c. 3.5mm diameter; fruits ovoid, or ± fusiform, 3.5–7 × 2.5–3.5cm, orange or bright red at maturity. Forest; 150m.
Distr.: Nigeria, Cameroon, Gabon, Congo. (Lower Guinea & Congolian).
IUCN: LC
Etam: Etuge 85 fl., 3/1986.

Salacia longipes (Oliv.) N.Hallé var.
camerunensis (Loes.) N.Hallé
Fl. Gabon 29: 66 (1986).
Syn.: *Salacia camerunensis* Loes.
Woody climber to 15m; no resinous threads; branchlets green; leaves 5–18 × 2–8cm, drying green, acumen linear, to

1.5cm; petiole 2–12mm; 6–12 pairs lateral nerves; inflorescence with a peduncle 4–5cm long, carrying a fascicle to 10mm broad; pedicels 9–14mm; buds 2–3mm diameter; flowers to 10mm diameter, orange to yellow; fruits variable, oblong, ellipsoid or turbinate, 3–6.5 × 2–4cm; 2–6 seeds. Forest; 700–900m.
Distr.: Liberia to Congo (K). (Guineo-Congolian).
IUCN: LC
Kupe Village: Cheek 7012 fl., fr., 1/1995; Etuge 2608 fl., fr., 7/1996; 2622 fr., 7/1996; Kenfack 285 fl., fr., 7/1996.

Salacia mamba N.Hallé
Fl. Gabon 29: 145 (1986).
Woody climber 6–20m; resinous threads; brown stems; leaves pale green, elongate-elliptic, 6–15 × 2–7cm; nerves drying whitish, 6–12 pairs lateral nerves; inflorescence axillary, miniscule, with a peduncle of 0.5(–1)mm carrying a glomerule with few flowers; flowers 1.5mm diam; pedicel 1.2mm long; flower cupuliform; fruits globose 2–3 × 3–4cm, with 1–3 seeds. Forest; 760–950m.
Distr.: Cameroon, Gabon, Congo (B). (Lower Guinea).
IUCN: VU
Kupe Village: Cable 3690 fl., 7/1996; Kenfack 290 fl., fr., 7/1996; Ryan 343 fl., 5/1996.

Salacia pallescens Oliv.
Fl. Cameroun 32: 60 (1990).
Shrub to 3m; stems lenticellate; branches brown, lenticellate; no resinous threads; leaves ellipsoid, 7–16 × 2.5–6cm, toothed, margins undulate; 8–12 pairs lateral nerves; inflorescence sessile fascicles of 3–8 flowers; pedicel 4–13mm; flowers 3–4mm, flower violet-black, 4–6mm diam; fruit globose, 1.5–3cm diameter, orange-red with violet markings; 1–3 seeds. Forest; 1600m.
Distr.: Guinea (C) to Zimbabwe. (Guineo-Congolian).
IUCN: LC
Nyasoso: Etuge 1820 fl., 3/1996.

Salacia pyriformioides Loes.
Woody climber to 7m; no resinous threads, twigs olive when dried; leaves large, elliptic, 8–25 × 5–12cm, acumen 5–15mm, olive-green when dried; lateral nerves 6–9 pairs, arcuate; inflorescence sessile on stems, few- to many-flowered; pedicels long, filiform, 9–40mm; flower yellow, 4.5mm diameter; fruits ellipsoid, 3–7 × 3–5cm; seeds numerous. Forest; 700–1600m.
Distr.: Nigeria to Congo (K). (Lower Guinea & Congolian).
IUCN: NT
Kupe Village: Cable 733b fl., 1/1995; Cheek 7004 fl., 1/1995; 7026 fr., 1/1995; **Ndum:** Lane 507 fl., fr., 2/1995; **Nyasoso:** Cable 1122 fl., 2/1995; 2938 fr., 6/1996; Elad 119 fl., 2/1995; Lane 528 fl., 2/1995.
Note: a widespread but rare species, known over much of its range only from early- to mid-twentieth century collections.

Salacia staudtiana Loes. var. cerasiocarpa
(R.Wilczek) N.Hallé
Fl. Gabon 29: 63 (1986).
Woody climber resembling var. *staudtiana*, but internodes 4-angled (not cylindrical); generally with persistent bud scales (not generally absent); fruit globular, 20mm or less across, without a neck (not 20mm or more, attenuated into a basal neck). Forest; 1200–1500m.
Distr.: Cameroon, CAR, Congo (K). (Lower Guinea & Congolian).
IUCN: LC
Mt Kupe: Thomas 5104 fl., 12/1985.

CHRYSOBALANACEAE

Salacia staudtiana Loes. var. *leonensis* Loes.
Fl. Gabon 29: 62 (1986).
Syn.: *Salacia caillei* A.Chev.
Woody climber resembling var. *staudtiana*, but pedicels
shorter, 3–9mm; fruits smaller. Forest; 420m.
Distr.: Guinea (C), Sierra Leone, Cameroon & Gabon.
(Upper & Lower Guinea).
IUCN: LC
Mungo River F.R.: Onana 960 fr., 11/1999.

Salacia staudtiana Loes. var. *staudtiana*
Fl. Gabon 29: 60 (1986).
Woody climber to 8m; no resinous threads; branches
cylindric; branchlets drying green, nodes somewhat flattened;
leaves elliptic, drying green, paler below, acuminate, 6.5–15
× 3–8.5cm, margins entire; 6–8 pairs arcing lateral nerves;
inflorescence axillary with a peduncle of 6–8mm, holding a
glomerule or several bracteate branches, c. 1mm long;
pedicel 10–30mm long; flowers yellow to yellow-green, 4–
5mm diameter; fruit orange globose, 2–3cm diameter,
attenuated into a neck c. 8mm long. Forest; 350–1200m.
Distr.: Cameroon & Gabon. (Lower Guinea).
IUCN: NT
Kupe Village: Cable 2483 5/1996; 2681 5/1996; 3718 fr.,
7/1996; Cheek 7009 fr., 1/1995; 7676 fr., 11/1995; Etuge
1838 fl., fr., 3/1996; 1996 fr., 5/1996; 2891 fl., fr., 7/1996;
Lane 292 fr., 1/1995; Ryan 269 fl., fr., 5/1996; 294 fl., fr.,
5/1996; **Mungo River F.R.:** Cheek 10206 fr., 11/1999;
Nyasoso: Cable 3561 fl., fr., 7/1996; 3564 fr., 7/1996; 3626
fr., 7/1996; Gosline 134 fr., 11/1998; Sebsebe 5028 fr.,
10/1995.
Note: widespread in Cameroon, rare in Gabon.
Local name: Epa'hed. **Uses:** FOOD – infructescences –
edible fruits, important in some way (Cheek 7676);
ANIMAL FOOD – unspecified parts – monkeys eat, but not
men (Cheek 7009).

Salacia sp. nov. *A*
Woody climber to 10m, with resinous threads; twigs
medium- to dark brown without lenticels; leaves ovate to
oblong; petiole to 1cm; 5–7 lateral nerves, acumen short;
flowers on a short peduncle to 5mm, always axillary; pedicel
to 5mm; flower cupuliform when open, dull-orange; fruit
unknown. Forest; 1000–1200m.
Distr.: Bakossi Mts. (Narrow Endemic).
Kodmin: Gosline 116 2/1998; **Mwambong:** Cheek 9117
2/1998; Onana 524 2/1998.
Note: no other West African *Salacia* has the combination of
resinous threads and short peduncle.

Salacighia letestuana (Pellegr.) Blakelock
Syn.: *Salacia letestuana* Pellegr.
Woody climber to 25m; stem to 10cm diameter; leaves
elliptic, to 18 × 9cm, often high in the canopy; inflorescence
on leafless branches to 60cm, carrying long, flowering spikes
to 140cm long; flowers sessile on glomerules; pedicels 6–
20mm, fragrant, 11–15mm diameter, white; fruit red-orange,
spherical, c. 3cm diameter, edible. Forest; 1000m.
Distr.: Ivory Coast to Congo (K). (Guineo-Congolian).
IUCN: LC
Nyandong: Darbyshire 109 fr., 3/2003; **Nyasoso:** Cheek
7539 fr., 10/1995.

Simirestis dewildemaniana N.Hallé
Adansonia (sér. 2) 3: 465 (1958).
Woody climber; leaves to 10 × 5cm, ours drying dark green
with lighter, reddish nerves; 5–6 pairs lateral nerves;
inflorescence an axillary dichotomous cyme divided up to 6

times, often in terminal panicles; flowers green, 4–6mm
diameter; fruits with flat, dehiscent mericarps to 7 × 3cm.
Forest; 1000m.
Distr.: Ivory Coast, Cameroon to Uganda & Sudan. (Guineo-
Congolian).
IUCN: LC
Nyasoso: Cable 3535 fl., fr., 7/1996.
Note: matching specimens at Kew were identified by Hallé
as *S. scheffleri* in 1982; he apparently put this species in
synonymy with *S. scheffleri* in 1981, and then restored it in
1984.

Thyrosalacia racemosa (Loes. ex Harms) N.Hallé
Fl. Gabon 29: 26 (1986).
Woody climber to c. 6m; branches mostly opposite; petiole
5–12mm; leaves 8–23 × 3–9.5cm; inflorescences terminal;
peduncle with 2 bracteoles; flowers lateral, borne singly, 15–
22mm diameter; disc simple, patelliform. Forest; 1000m.
Distr.: Cameroon. (Narrow endemic).
IUCN: CR
Loum: Ledermann 6420 fl., 12/1909; 6438 fl., 12/1909.
Note: this taxon is possibly conspecific with *T.*
pararacemosa N.Hallé, from Cameroon and Gabon, but
seems to differ in its simple, patelliform disc (not double,
with a lower, subcupuliform and upper, disciform layer);
specimens not seen, probably destroyed at B, but cited in Fl.
Cameroun 32: 18 (1990). The only other localities for this
taxon are at Yabassi and Babong, 25km S of Nkongsamba,
both being just outside our checklist area.

CHRYSOBALANACEAE
G.T. Prance & C.A. Sothers

Fl. Cameroun 20 (1978).

Dactyladenia johnstonei (Hoyle) Prance & F.White
Brittonia 31: 485 (1979).
Syn.: *Acioa johnstonei* Hoyle
Shrub or small tree; young branches hispid, hairs 2–3mm;
stipules triangular, c. 4mm with stipitate glands; petioles 5–
6mm, rugose; leaves ovate-elliptic, 6–12 × 2–5cm, acumen
blunt, base obtuse-rounded, scabrous beneath, soon
glabrescent; glands c. 5mm from margin and at base, midrib
sparsely hispid; lateral nerves c. 6 pairs; cymes terminal and
axillary, 1–2-branched, 10cm, rachis hispid and puberulous;
bracts and bracteoles deltate, glandular; pedicel 4–6mm long;
receptacle-tube 1–1.2cm, base gibbous; sepals c. 5mm; petals
obovate, white, 5–6mm; stamens c. 20, 2.5cm; fruit ovoid, c.
25 × 40 × 37mm; exocarp olive-brown tomentose. Forest;
950m.
Distr.: W Cameroon & Nigeria. (W Cameroon Uplands).
IUCN: CR
Mt Kupe: Cheek 10158 11/1999.

Magnistipula conrauana Engl.
Fl. Cameroun 20: 79 (1978).
Syn.: *Hirtella conrauana* (Engl.) A.Chev.
Shrub or tree to 12m; branches glabrous, lenticellate; stipules
foliaceous, ovate, base oblique, 3–4 × 2–3cm; leaves ovate to
oblong-elliptic, 22–25 × 10–12cm, base acute to rounded,
decurrent, glabrous; numerous glands towards base and under
acumen; lateral nerves 6–8 pairs; panicles spreading, terminal
and axillary, to 30cm, often subtended by foliaceous bracts,
rachis glabrous; normal bracts 3 × 2.5mm, triangular,

margins glandular; bracteoles 1mm, eglandular; pedicels 4–5mm, receptacle obliquely campanulate, curved, 6–7mm, glabrous outside, deflexed-villous within; sepals 4 × 2.5mm, ciliate, tomentellous within; petals white to pale violet, 2–4mm; stamens 7, 4–6mm long; staminodes c. 7, tooth-like; ovary glabrous; style curved, 3–5mm; fruit ovoid, c. 20 × 35 × 55mm. Forest; 1000–1500m.
Distr.: Cameroon Endemic.
IUCN: EN
Kodmin: <u>Onana 574</u> 2/1998; **Kupe Village:** <u>Cheek 7844</u> 11/1995; **Muahunzum:** <u>Etuge 4444</u> 11/1998; **Mwanenguba:** <u>Leeuwenberg 9572</u> fl., 4/1972; **Nsoung:** <u>Etuge 4973</u> st., 3/2003.

Magnistipula tessmannii (Engl.) Prance
Bol. Soc. Brot. (sér. 2) 40: 185 (1966).
Syn.: *Parinari tessmannii* Engl.
Tree to 40m; branches puberulent and appressed-hispid when young; stipules triangular, 8 × 1mm, caducous; petiole 1–1.2cm with 2–4 glands, puberulent and hispid; leaves elliptic-oblong or oblanceolate, 8–16(–30) × 3–11cm, abruptly acuminate, base acute, glabrous; several glands near base and midrib, midrib puberulent; lateral nerves 8–12 pairs; cymose panicles terminal and subterminal, 12–18cm, rachis puberulent; branches and pedicels tomentellous; bracts and bracteoles early caducous; pedicels 1–1.5mm; receptacle obliquely campanulate, subgibbous at base, 4–5mm, tomentellous outside, villous within; sepals triangular, 1.5–3mm, unequal, tomentellous; petals white, 3.5–4mm, ciliate; stamens 8–10, curved; staminodes 8–9, tooth-like; ovary villous; style curved, 3.5mm; fruit globose, c. 4.3 × 5 × 6cm; exocarp glabrous, lenticellate; endocarp thick, woody. Forest; 450–925m.
Distr.: Nigeria to Gabon & Cabinda. (Lower Guinea).
IUCN: NT
Enyandong: <u>Cheek 10921</u> 10/2001; **Nyandong:** <u>Cheek 11327</u> 3/2003; **Tombel:** <u>Letouzey 14653</u> fl., 4/1976.
Note: trees often gregarious.
Local name: Echi Be. **Uses:** FOOD – seeds – fallen fruits harvested by women for their edible seeds, which are used as groundnut substitutes (Cheek 10921).

Maranthes chrysophylla (Oliv.) Prance
Bull. Jard. Bot. Nat. Belg. 46: 295 (1976).
Syn.: *Parinari chrysophylla* Oliv.
Tree to 30m; leaves to 25 × 13.5cm, apex shortly acuminate-cuspidate, base obtuse; lower surface tomentellous, fulvous; pedicels 0.2–0.3cm; receptacle-tube and sepals both 0.4–0.6cm, fulvous-tomentellous outside. Forest; 800m.
Distr.: Liberia to Congo (K). (Guineo-Congolian).
IUCN: LC
Manehas F.R.: <u>Etuge 4344</u> 10/1998.
Note: specimen not determined to subsp.

Maranthes glabra (Oliv.) Prance
Bol. Soc. Brot. (sér. 2) 40: 184 (1966).
Syn.: *Parinari glabra* Oliv.
Tree 20(–40)m; leaves to 14 × 5.5cm, caudate-acuminate, base cuneate, lower surface glabrous or rarely sparsely white-arachnoid; pedicel c. 0.5cm; receptacle-tube c. 0.4 × 0.25cm; sepals c. 0.3cm; flower exterior glabrous except for whitish tomentellous 'protected' parts of sepals; drupe ellipsoid, c. 4.5 × 3.5cm; exocarp glabrescent. Forest; 280m.
Distr.: Sierra Leone to Congo (K). (Guineo-Congolian).
IUCN: LC
Karume: <u>Nemba 71</u> fl., fr., 5/1986.

COMBRETACEAE

I. Darbyshire & C.C.H. Jongkind

Fl. Cameroun 25 (1983).

Combretum bracteatum (M.A.Lawson) Engl. & Diels
Large liana or shrub to 3m; leaves ovate-oblong, 11.3–16.5 × 6–6.5cm, apex shortly acuminate, base subcordate, glabrous; lateral nerves 8–9 pairs; inflorescence a spike to 30(–40)cm, glandular-pubescent; bracts elliptic, 1.7–3cm long, yellow-green; upper receptacle tubular, curved, 2.8 × 0.9cm, yellow-orange; calyx lobes 0.4cm, triangular; petals elliptic, 0.7cm long, puberulent, yellow-pink; stamens extending to 0.8cm beyond calyx, red; samara 4-winged, 3.5cm diameter. Secondary forest; 1100m.
Distr.: Nigeria to Congo (K). (Guineo-Congolian).
IUCN: LC
Mejelet-Ehumseh: <u>Etuge 468</u> 1/1987; **Nlohe:** <u>Hédin 9</u> date unknown.

Combretum confertum (Benth.) M.A.Lawson
Liana to 6m or more; leaves as *C. bracteatum*, but to 20.5 × 10cm, base rounded; flowers in dense, capituliform axillary panicles to 5cm; peduncles puberulent; upper receptacle campanulate, to 0.5cm long, glabrescent; calyx lobes c. 1mm; petals ovate, 3mm, scarlet; stamens extend c. 1cm beyond calyx; fruit not seen. Riverine forest; 150m.
Distr.: Nigeria to Congo (K). (Lower Guinea & Congolian).
IUCN: NT
Mungo River F.R.: <u>Cheek 10227</u> 12/1999.

Combretum indicum (L.) R.A.DeFilipps
Contr. U.S. Natl. Herb. 45: 194 (2003)
Syn.: *Quisqualis indica* L.
Scandent shrub 4m or liana; leaves oblong-elliptic, c. 10 × 5cm, acuminate-apiculate, base rounded; lateral nerves 6–7 pairs, glabrous; petiole 0.5cm; inflorescence spikes terminal, 2–8cm; flowers clustered towards apex; upper receptacle 4–5.5cm, attenuate, puberulent, green; calyx lobes 1.5mm, triangular; petals ovate, 1–1.8cm, white outside, red within; stamens not extending beyond calyx; samara ellipsoid, 2.8 × 0.7cm, 4 wings subcoriaceous. Cultivated in villages, sometimes persisting along watercourses; 400m.
Distr.: cultivated widely in tropics as an ornamental, originally from Asia.
IUCN: LC
Nyandong: <u>Cheek 11477</u> 3/2003.

Combretum latialatum Engl. ex Engl. & Diels
Engl. Monogr. Afr. Pflanzenfam. Combretaceae: 86 (1899).
Syn.: *Quisqualis latialata* (Engl. ex Engl. & Diels) Exell
Liana; stems densely puberulent; leaves elliptic, 3.7–7.3 × 1.8–2.6cm, apex acuminate-apiculate, base obtuse, glabrous; midrib and 5–7 pairs of lateral nerves yellow below; petiole c. 1cm, puberulent; panicles with racemose partial-peduncles; flowers clustered towards apex; upper receptacle attenuate, 0.6–1cm, puberulent; calyx lobes c. 1mm; petals ovate, 4–5mm, yellow; stamens extend 1mm beyond calyx; samara 3.3cm diameter. Secondary forest; 200–250m.
Distr.: Nigeria to Angola. (Lower Guinea & Congolian).
IUCN: LC
Baduma: <u>Nemba 162</u> fl., 7/1986; **Ikiliwindi:** <u>Nemba 531</u> 6/1987.

Combretum paniculatum Vent.

Liana to 10m or more; lower part of petioles often becoming woody hooks; leaves elliptic, 9–11 × 5–6.5cm, apex shortly acuminate, base rounded, glabrescent; lateral nerves 6 pairs; petiole 1–2cm; panicles many-branched, to 30 × 12cm; flowers in dense spike-like glomerules 0.5–2cm on puberulous partial peduncles; upper receptacle c. 4mm long, puberulous; calyx minute-dentate; petals ovate, c. 2mm, red, glabrous; stamens extend 3–5mm beyond calyx, red; samara c. 2.7cm diameter. Forest edge; 850–1000m.
Distr.: Senegal to Mozambique. (Tropical Africa).
IUCN: LC
Kupe Village: Elad 67 fl., 1/1995; **Ndum:** Williams 137 fl., 1/1995; **Nyasoso:** Thomas 3053 fl., 2/1984.

Terminalia catappa L.

Tree 15–25m; stems puberulent towards apex; leaves crowded, obovate, 14–23 × 9–12.5cm, shortly acuminate, base cuneate; lateral nerves 9–11 pairs; petiole 0.7–1.8cm, felty-puberulent; inflorescence spikes axillary, to 18cm, puberulous; flowers 4mm, tomentose within; fruit sessile, fleshy, ellipsoid 5.5 × 3cm, 2 wings 2mm wide. Roadside, planted; 850m.
Distr.: native to India, introduced throughout the tropics.
IUCN: LC
Nyasoso: Darbyshire 174 3/2003.

Terminalia mantaly H.Perrier

Fl. Cameroun 25: 79 (1983).
Tree to 8m; stems spreading; branching in one plane; leaves obovate, 5–6.5 × 2.3–3.3cm, emarginate, base long-cuneate, margin subundulate; petiole glabrous, 4mm; spikes axillary, to 4cm; peduncle 4cm, puberulous; flowers 2mm, white-green; stamens 2.5mm; fruit ellipsoid, c. 1.8cm long, not winged. Villages, ornamental; 420m.
Distr.: originally from Madagascar, widely planted in the tropics.
IUCN: LC
Nyandong: Darbyshire 173 3/2003.

Terminalia superba Engl. & Diels

Tree 35(–50)m with flat crown; resembling *T. catappa*, but stems glabrous; leaves coriaceous, c. 12 × 6cm; lateral nerves 6–7 pairs, glabrous; petiole to 5cm with paired glands; spikes to 13.5cm; samara transversely oblong-elliptic, not fleshy, wings 2 × 2.7cm. Forest; 500m.
Distr.: Guinea (C) to W Congo (K). (Upper & Lower Guinea).
IUCN: LC
Nlog: Etuge 18 fr., 4/1986; **Nlohe:** Hedin 1611 date unknown; **Nyandong:** Cheek 11375 3/2003.
Local name: Nkom. **Uses:** MATERIALS – timber (*fide* Etuge).

COMPOSITAE

J.-M. Onana & H.J. Beentje

Acmella caulirhiza Delile

Fragm. Flor. Geobot. 36(1) Suppl.1: 228 (1991).
Syn.: *Spilanthes filicaulis* (Schum. & Thonn.) C.D.Adams
Syn.: *Spilanthes africana* DC.
Creeping herb; capitula conical with tiny yellow to orange flowers; ray florets present, tiny. Forest, streambanks; 650–1500m.

Distr.: tropical & subtropical Africa and Madagascar.
IUCN: LC
Kodmin: Cheek 8931 1/1998; Etuge 3986 1/1998; Ghogue 11 fl., 1/1998; **Kupe Village:** Cheek 8316 5/1996; Etuge 2726 fl., 7/1996; **Ngusi:** Etuge 1562 fr., 1/1996; **Nyasoso:** Cheek 7901 fl., 11/1995.
Local names: Derekube (Etuge 1562); Seke (Bakweri language, Lyonga Daniel in Cheek 7901); Me' Koobe = 'Eye of Fowl' (Cheek 8316). **Uses:** SOCIAL USES – smoking materials/drugs – drug adulterants – chew flower heads before drinking any alcohol; this reduces drunkenness (Etuge 1562); MEDICINES (Lyonga Daniel in Cheek 7901); (Cheek 8316); (Etuge 2726).

Adenostemma caffrum DC.

Erect or semi-procumbent herb to 1.5m, rooting at nodes on wet ground; stem ± fleshy; leaves opposite, ovate, 5–16 × 0.5–8cm; capitula 6–10mm across, with involucre in 1–2 series and many small white florets; pappus of (2–)3–4(–5) pegs. Wet ground; 950–1650m.
Distr.: widespread in tropical Africa.
IUCN: LC
Kodmin: Cheek 8990 1/1998; Etuge 4026 1/1998; **Ngomboaku:** Ghogue 510 12/1999.

Adenostemma mauritianum DC.

Erect annual herb c. 1m tall; cauline leaves ovate, 14 × 8cm, apex acute, base cuneate, margin serrate; petiole c. 6cm; capitula 5–10 per inflorescence, c. 5mm across, white, florets 2–3mm long; involucre of c. 10 linear-oblong bracts, c. 4mm long. Forest; 210–2000m.
Distr.: SE Nigeria to Zimbabwe & E Africa. (Afromontane).
IUCN: LC
Kodmin: Cheek 9618 fl., 11/1998; **Kupe Village:** Etuge 2719 7/1996; Kenfack 263 7/1996; **Mwanenguba:** Cheek 9475 10/1998; **Mungo River F.R.:** Etuge 4318 10/1998; **Nyasoso:** Cheek 5678 fl., 12/1993; Etuge 1595 1/1996; Lane 217 fl., 11/1994.
Local name: Nkab. **Uses:** MEDICINES; ENVIRONMENTAL USES – boundaries/barriers/supports – used to tie fences (Cheek 9618).

Ageratum conyzoides L. subsp. *conyzoides*

Annual herb 0.3–1.5m tall; ± pilose; cauline leaves ovate, 5 × 2.5cm, apex acute, base acute to truncate, margin serrate; petiole c. 1cm, white-pilose; capitula blue or whitish with small florets, c. 6mm diameter, numerous in dense terminal aggregations; involucre with bracts narrowly oblong, c. 4mm long, green, with two white lines, apex with 2–3 spines. Farmbush, weedy; 650–1450m.
Distr.: pantropical.
IUCN: LC
Enyandong: Pollard 774 fl., 10/2001; **Kodmin:** Cheek 9012 1/1998; **Kupe Village:** Cable 2462 5/1996; Etuge 2708 7/1996; **Ngomboaku:** Etuge 4670 12/1999; **Ngusi:** Etuge 1569 1/1996; **Nyasoso:** Cable 2784 6/1996; Cheek 5647 fl., 12/1993.
Local name: Mmwen. **Uses:** MEDICINES (Cheek 9012).

Aspilia africana (Pers.) C.D.Adams subsp. *africana*

Syn.: *Aspilia africana* (Pers.) C.D.Adams var. *minor* C.D.Adams
Shrub or herb to 2m tall, rarely scrambling; cauline leaves lanceolate, c. 7.5 × 2.8cm, apex acute, base rounded, margin serrate, upper and lower surfaces scabrid; capitula terminal, single or few, 2.5cm diameter, with rays, orange; ray florets c. 6–12 × 3mm, apex bilobed. Farmbush; 1200m.

Distr.: widespread in tropical Africa.
IUCN: LC
Mwanenguba: Leeuwenberg 8178 fl., 8/1971.

Bidens mannii T.G.J.Rayner
Kew Bull. 48: 459 (1993).
Syn.: *Coreopsis monticola* (Hook.f.) Oliv. & Hiern var. *monticola*
Syn.: *Coreopsis monticola* (Hook.f.) Oliv. & Hiern var. *pilosa* Hutch. & Dalziel
Herb to 2m tall; young stems and leaves reddish; cauline leaves deeply 3(–5)-lobed, to 6 × 4cm, the lobes deeply toothed to lobed; capitula pale yellow, c. 4.5cm diameter, rayed, rays elliptic, 2 × 1cm; involucre of 5 inner broadly triangular hairy bracts, c. 1cm long, and 5 outer narrowly oblong glabrous bracts, c. 2cm long. Grassland, forest-grassland transition; 1100m.
Distr.: Cameroon. (W Cameroon Uplands).
IUCN: VU
Kupe Village: Cheek 7697 fl., 11/1995; **Mwanenguba:** Cheek 7285 2/1995.

Bidens pilosa L.
Erect herb c. 1m tall; cauline leaves 3–5-lobed; lobes ovate or lanceolate, to 6 × 2cm, apex acute, base obtuse, margin serrate, glabrous; petiole 1–1.5cm long; capitulum c. 1.5cm diameter, rayed, rays white, oblong, c. 5 × 2mm; disc florets yellow; involucre with inner bracts narrowly elliptic, with scarious margins, outer much shorter, linear. Farmbush, grassland, forest edges; 650–2000m.
Distr.: pantropical.
IUCN: LC
Enyandong: Ghogue 1221 fl., fr., 10/2001; **Kodmin:** Cheek 9036 1/1998; **Mwanenguba:** Cheek 9455 10/1998; **Ngusi:** Etuge 1563 1/1996; **Nyasoso:** Cheek 5655 fl., 12/1993; Etuge 1337 10/1995.
Note: fruits stick on clothes.
Local name: Kure Kure Abah. **Uses:** MEDICINES (Cheek 9036).

Bothriocline schimperi Oliv. & Hiern ex Benth.
F.T.A. 3: 266 (1877).
Syn.: *Erlangea schimperi* (Oliv. & Hiern ex Benth.) S.Moore
Shrub to 1.5m, puberulent; leaves elliptic, to 15 × 6cm, serrate-dentate; lateral nerves 10–14 pairs, petiolate; capitula 5–8mm broad, tightly-clustered in broad cymes; peduncles 1–5mm; involucral bracts ovate, apices acute, erect, margins scarious; flowers white. Forest; 1300m.
Distr.: Sierra Leone, Nigeria, Cameroon, Ethiopia. (Afromontane).
IUCN: NT
Edib: Onana 588 2/1998; **Kodmin:** Biye 24 1/1998.
Note: not in E Africa; type of the genus.

Chromolaena odorata (L.) R.M.King & H.Robinson
Fragm. Flor. Geobot. 36(1) Suppl.1: 450 (1991).
Syn.: *Eupatorium odoratum* L.
Shrub or liana to 3.5m high; leaves opposite, ovate, 3–10 × 2–6cm, dentate; florets white or mauve, in capitula 10mm long, these arranged in corymbs; pappus of silky bristles. Roadsides, villages; 475–1650m.
Distr.: pantropical.
IUCN: LC
Kodmin: Cheek 8977 fl., 1/1998; Gosline 169 11/1998; **Kupe Village:** Lane 276 fl., 1/1995; **Ngusi:** Etuge 1557 1/1996; **Nyandong:** Darbyshire 3 fl., 3/2003.

Note: a major weed of farms and roadsides.
Local name: Ajacasala (Etuge 1557).

Conyza attenuata DC.
Fragm. Flor. Geobot. 36(1) Suppl.1: 89 (1991).
Syn.: *Conyza persicifolia* (Benth.) Oliv. & Hiern
Herb 0.4(–3)m tall; cauline leaves sessile, linear, 10 × 0.7cm, apex acute, base clasping the stem for half its diameter, margin distantly serrate-dentate; capitula 6–10mm wide, without rays, white, clustered in dense aggregations; involucre of c. 25 lanceolate bracts c. 4 × 1mm, margins scarious. Farmbush; 1450m.
Distr.: tropical Africa. (Afromontane).
IUCN: LC
Kodmin: Cheek 9606 fl., 11/1998; **Nyasoso:** Etuge 1846 4/1996.

Conyza bonariensis (L.) Cronquist
Fragm. Flor. Geobot. 36(1) Suppl.1: 59 (1991).
Syn.: *Conyza sumatrensis* (Retz.) E.Walker
Syn.: *Erigeron floribundus* (Kunth) Sch.Bip.
Syn.: *Erigeron bonariensis* L.
Annual weedy herb 40cm to 1.5m high, erect; leaves sessile and narrowly lanceolate, 2–15 × 0.2–3cm, entire or dentate; capitula 4–8mm in lax panicles, with tiny white florets with purple tips; pappus of white bristles. Grassland; 650–1450m.
Distr.: pantropical.
IUCN: LC
Kodmin: Cheek 9026 1/1998; **Mwanenguba:** Leeuwenberg 8984 fl., 12/1971; **Nyasoso:** Etuge 1594 1/1996.
Uses: locally useful, use unknown (Cheek 9026).

Conyza pyrrhopappa Sch.Bip. ex A.Rich.
F.T.A. 3: 318 (1877).
Syn.: *Microglossa angolensis* Oliv. & Hiern
Shrub or woody herb to 3m; leaves elliptic, 8 × 3cm, aromatic; capitula 4–7mm long, with tiny white or yellow flowers. Fallow, grassland; 2000–2160m.
Distr.: Nigeria to Sudan, S to Angola & Zambia. (Lower Guinea & Congolian).
IUCN: LC
Mwanenguba: Cheek 9491 10/1998; Sanford 5502 fl., 11/1968.

Conyza subscaposa O.Hoffm.
Subscapose herb; leaves obovate, to 20 × 5cm; capitula with tiny pale yellow flowers; involucre of 3–8mm long bracts, often tipped red. Grassland; 2000m.
Distr.: W Cameroon to Congo (K) & E Africa. (Afromontane).
IUCN: LC
Mwanenguba: Cheek 9474 10/1998; Sanford 5420 fl., 11/1968.

Crassocephalum bauchiense (Hutch.) Milne-Redh.
Herb 0.3–1m tall; stems with white crisped hairs; cauline leaves pinnately lobed, c. 7 × 3.5cm, lobes 5, divided almost to the base, margins denticulate; capitula 1cm diameter, without rays, blue or purple; involucre of c. 20 inner bracts, as long as capitulum, and c. 20 short, patent, linear, basal outer bracts 2–3mm long. Forest edge; 900–1200m.
Distr.: N Nigeria, Cameroon. (Lower Guinea (montane)).
IUCN: VU
Kupe Village: Cheek 7700 11/1995; **Nyasoso:** Cable 649 fl., 12/1993; Cheek 5650 fl., 12/1993.

Crassocephalum crepidioides (Benth.) S.Moore

Annual herb 25–120(–150)cm tall, rarely a short-lived perennial; leaves obovate to ovate, unlobed or 2–8-lobed, 5–26 × 2–10cm, serrate; capitula in dense to lax terminal corymbs, without rays; florets orange to red; style arm appendages distinct; pappus of bristles. Farmbush; 1300m.
Distr.: Guinea (C) to Mascarenes, naturalised in tropical Asia & Pacific islands. (Palaeotropics).
IUCN: LC
Kupe Village: Cheek 8387 5/1996.

Crassocephalum montuosum (S.Moore) Milne-Redh.

Herb 1.5m tall; cauline leaves elliptic in outline, 8.5 × 6cm, pinnately 5-lobed, the apical lobe largest, lower lobes decreasing in size, lobes dentate, glabrous; capitula 10–20, in aggregations, c. 0.5mm diameter, without rays, yellow or brownish yellow; involucre as in *C. bauchiense*. Forest, forest-grassland transition; 650–1450m.
Distr.: W Cameroon to Congo (K), to E Africa & Madagascar. (Afromontane).
IUCN: LC
Kodmin: Biye 18a fl., fr., 1/1998; Cheek 9056 fl., fr., 1/1998; **Kupe Village:** Etuge 1988 fl., fr., 5/1996; 2761 7/1996; 2764 fl., 7/1996; **Ngusi:** Etuge 1579 fl., 1/1996; **Nyasoso:** Cable 2782 fl., 6/1996; Cheek 5651 fl., 12/1993; 5654 fl., 12/1993; 7310 2/1995; 7473 fr., 10/1995.
Local names: Belekam (Cheek 7310); Ilee (Bakweri language, Lyonga Daniel in Cheek 7473). **Uses:** MEDICINES (Cheek 7310); (Lyonga Daniel in Cheek 7473).

Crassocephalum rubens (Juss. ex Jacq.) S.Moore

Syn.: *Crassocephalum sarcobasis* (DC.) S.Moore
Annual herb to 1m; capitula few, with many tiny purple or mauve flowers. Fallow, rocky slopes in grassland; 2000m.
Distr.: Cameroon, Sudan to S Africa, Madagascar, Yemen. (Afromontane).
IUCN: LC
Mwanenguba: Cheek 9467 10/1998; Leeuwenberg 10698 fl., 11/1972; **Nyasoso:** Etuge 1674 fr., 1/1996.

Crassocephalum vitellinum (Benth.) S.Moore

Sprawling perennial herb to 2m; leaves elliptic to obovate, 10 × 6cm, sometimes with lobed base; capitula with many tiny yellow or orange flowers. Forest edge; 900–2000m.
Distr.: SE Nigeria & W Cameroon to E Africa. (Afromontane).
IUCN: LC
Kupe Village: Cheek 7030 1/1995; **Mwanenguba:** Cheek 7284 fr., 2/1995; Cheek 9464 fl., 10/1998; Leeuwenberg 8142 fl., 8/1971; Sanford 5419 fl., 11/1968; **Nyasoso:** Etuge 2407 6/1996; 2441 fl., 6/1996; Lane 219 fl., 11/1994.

Dichrocephala integrifolia (L.f.) Kuntze subsp. *integrifolia*

Annual herb 10cm to 1m, erect or procumbent; leaves divided with a large terminal lobe and 1–3 pairs of smaller basal lobes, all dentate-serrate; florets white, in capitula 2–5mm long, in compound panicles; pappus absent or of few bristles. Forest, forest-grassland transition; 650–2000m.
Distr.: palaeotropical. (Montane).
IUCN: LC
Kodmin: Cheek 9028 1/1998; **Mwanenguba:** Cheek 9492 10/1998; **Ngusi:** Etuge 1565 1/1996; **Nyale:** Cheek 9642 fl., 11/1998; **Nyasoso:** Cable 1246 fl., 2/1995; Lane 240 fl., 11/1994.
Local name: Mbesug (Cheek 9028).

Elephantopus mollis Kunth

Perennial herb, 40–90cm; leaves oblanceolate, 6–16 × 2–5cm, the base clasping; flowers white, in capitula that are united into glomerules, these in compound cymes; pappus of 5–6 setae. Grassland, forest edge; 1300–1500m.
Distr.: tropical America, introduced in Africa & Asia.
IUCN: LC
Kodmin: Cheek 9079 1/1998; Ghogue 27 1/1998; Gosline 168 11/1998; **Mwanenguba:** Leeuwenberg 8980 fl., 12/1971.

Elephantopus scaber L. subsp. *plurisetus* (O.Hoffm.) Philipson var. *brevisetus* Philipson

Kew Bull. 43: 273 (1988).
Subscapose perennial herb 0.2–1.5m high, with basal rosette and few stem leaves; basal leaves elliptic, 6–60 × 1–10cm; florets white or mauve, the capitula in glomerules 1–3cm across, these arranged in cymes; pappus of 7–10 narrow scales. Lake edge; 1800m.
Distr.: Cameroon, Congo (K) to Kenya. (Guineo-Congolian).
IUCN: LC
Mwanenguba: Etuge 4389 10/1998.

Emilia coccinea (Sims) G.Don

Annual herb, 15–120cm high; leaves ± subsucculent, 1–20 × 0.4–6cm; capitula in terminal corymbs; disc florets bright-orange; pappus of white bristles. Forest, farmbush; 650–1450m.
Distr.: tropical Africa. (Afromontane).
IUCN: LC
Edib: Etuge 4089 1/1998; **Kodmin:** Cheek 9021 1/1998; **Kupe Village:** Cheek 7074 fl., 1/1995; 8313 5/1996; Ryan 377 5/1996; **Ngusi:** Etuge 1541 1/1996; **Nyasoso:** Cable 2762 6/1996; Etuge 2070 6/1996.
Local names: Etumbwe = 'Ear of a Dog' (Koge Lawrence in Cheek 8313); Mpale Emuseh (Cheek 9021). **Uses:** VERTEBRATE POISONS – unspecified vertebrates – for 'night poison', take one leaf with 7 seeds alligator pepper (Koge Lawrence in Cheek 8313); FOOD – leaves – young leaves eaten as vegetables; MEDICINES (Etuge 1541) & (Cheek 9021); many other uses (Cheek 9021).

Emilia lisowskiana C.Jeffrey

Kew Bull. 52: 208 (1997).
Annual herb similar to *E. coccinea*, but florets yellow and differing in some more technical characters. Farmland; 1200m.
Distr.: Sierra Leone to N Zambia. (Guineo-Congolian).
IUCN: LC
Mwanenguba: Leeuwenberg 8180 fl., 8/1971.
Note: specimen not seen.

Galinsoga parviflora Cav.

Annual herb 10–75(–100)cm high, erect, spreading or decumbent; leaves opposite, ovate, 1–11 × 0.5–7cm; 3-veined from base; capitula in axillary cymes; ray florets white, small; disc florets yellow; pappus absent or of 15–20 small scales. Montane grassland; 2000m.
Distr.: S America, introduced in W Cameroon. (Montane).
IUCN: LC
Mwanenguba: Cheek 9448 10/1998.

Galinsoga quadriradiata Ruiz. & Pav.

Fl. Masc. Compositae: 204 (1993).
Syn.: *Galinsoga ciliata* (Raf.) Blake
Annual herb to 30cm, scabridulous; capitula small, with white ray florets and yellow disc florets. Forest-grassland transition; 650–1450m.

Distr.: S America, introduced in W Cameroon. (Montane).
IUCN: LC
Kodmin: <u>Cheek 9027</u> 1/1998; **Kupe Village:** <u>Cheek 8315</u> 5/1996; **Ngusi:** <u>Etuge 1561</u> 1/1996.
Local names: Akaval (Koge Lawrence in Cheek 8315); Medemiheh (Cheek 9027). **Uses:** ANIMAL FOOD – unspecified parts – eaten by some birds (Cheek 9027); MEDICINES (Koge Lawrence in Cheek 8315).

Guizotia abyssinica (L.f.) Cass.
Syn.: *Guizotia scabra* (Vis.) Chiov.
Annual herb 1–2m high; leaves opposite, subconnate-perfoliate, 10–15 × 2–6cm, scabrid; capitula in terminal cymes; ray florets yellow; disc florets yellow; pappus absent. Grassland, lake edge; 1800m.
Distr.: Cameroon, Ethiopia to Malawi, India. (Palaeotropics (montane)).
IUCN: LC
Mwanenguba: <u>Etuge 4388</u> 10/1998.

Gynura scandens O.Hoffm.
Engl. Pflanzenw. Ost-Afr. C: 416 (1895).
Climbing herb, 1.5–12m long, slightly fleshy and with unpleasant smell; leaves ovate or triangular, 2–12 × 1–8cm; 3–5-veined from base; capitula in terminal corymbs, without rays; florets orange; pappus of white bristles. Farmbush, forest edge; 1000m.
Distr.: Cameroon to Malawi. (Tropical Africa).
IUCN: LC
Kupe Village: <u>Etuge 1947</u> 5/1996.

Gynura sp. near procumbens Merr.
Straggling herb to 4m, glabrous; stem square, hollow; leaves alternate, elliptic, to 8 × 4cm, acute, base unequally acute; lateral nerves 5 pairs, serrate-dentate; petiole 1cm; flowers axillary, from 20cm of stem; partial peduncles 11cm; peduncle 6cm; capitula 6, on stipes 2cm, 1.5 × 1cm, orange. Forest; 1000m.
Mwambong: <u>Etuge 4143</u> fr., 2/1998.
Note: possibly a new taxon, known only from this specimen; more material required.

Helichrysum cameroonense Hutch. & Dalziel
Herb to 1.2m, aromatic; capitula dense with pale yellow bracts and small yellow flowers. Grassland, forest-grassland transition; 2100m.
Distr.: W Cameroon. (W Cameroon Uplands).
IUCN: NT
Mwanenguba: <u>Leeuwenberg 8928</u> fl., 12/1971.
Note: detailed information on this rare species is given in Cheek *et al.* 2000: 59.

Helichrysum foetidum (L.) Moench
Annual or short-lived perennial, 0.3–1.5m high; stem usually unbranched; leaves aromatic, lanceolate, 2–10 × 0.5–2.5cm, glandular; capitula in leafy corymbs; involucre and florets yellow; pappus of small white or yellow bristles. Grassland; 1550m.
Distr.: Nigeria to Congo (K), Sudan to S Africa, Arabia & Spain. (Montane).
IUCN: LC
Kodmin: <u>Cheek 9182</u> 2/1998.

Helichrysum forskahlii (J.F.Gmel.) Hilliard & B.L.Burtt
Notes Roy. Bot. Gard. Edin. 38(1): 146 (1980).
Syn.: *Helichrysum cymosum* (L.) Less. subsp. *fruticosum* (Forssk.) Hedberg

Syn.: *Helichrysum cymosum sensu* C.D.Adams in FWTA 2: 264
Syn.: *Helichrysum helothamnus* Moeser var. *helothamnus*
Perennial herb or shrub, 0.2–1m high, densely leafy; leaves narrowly lanceolate, 0.4–3 × 0.1–1cm; capitula in clusters in corymbs; involucre and florets pale yellow; pappus of small white bristles. Forest-grassland transition; 1100–2000m.
Distr.: Nigeria to Zambia, Sudan to Tanzania & Yemen. (Afromontane).
IUCN: LC
Lake Edib: <u>Cheek 9161</u> 2/1998; **Mt Kupe:** <u>Thomas 3106</u> fl., 2/1984; **Mwanenguba:** <u>Cheek 7240</u> 2/1995; <u>Cheek 9497</u> 10/1998; <u>9498</u> 10/1998; <u>Leeuwenberg 8464</u> fl., 9/1971; <u>Sanford 5390a</u> fl., 11/1968.

Helichrysum globosum Sch.Bip. ex A.Rich.
Perennial herb 0.2–1m high; leaves elliptic, 4–20 × 1–8cm, stalked in rosette leaves, auriculate in stem leaves; capitula small, in clusters; involucre yellow or pink; florets yellow; pappus of small white bristles. Forest-grassland transition; 2000m.
Distr.: W Cameroon to Angola, Zimbabwe, Ethiopia & E Africa. (Afromontane).
IUCN: LC
Mwanenguba: <u>Cheek 9465</u> 10/1998; <u>Sanford 5423</u> fl., 11/1968.

Helichrysum odoratissimum (L.) Sweet
Perennial herb 0.3–1m high, sometimes a scrambler; leaves smelling of curry, lanceolate, 0.5–6 × 0.2–1.5cm, usually decurrent on the stem; capitula small, many, in dense terminal clusters; involucre and florets yellow; pappus of small white bristles. Forest-grassland transition; 1450–2100m.
Distr.: SE Nigeria to Angola, Sudan & Ethiopia. (Afromontane).
IUCN: LC
Kodmin: <u>Cheek 8904</u> 1/1998; <u>8905</u> 1/1998; <u>9567</u> fl., 11/1998; **Mwanenguba:** <u>Leeuwenberg 8929</u> fl., 12/1971; <u>Sanford 5407</u> fl., 11/1968; <u>Thomas 3146</u> fl., 2/1984.
Local name: Etoon-dat. **Uses:** not used (Cheek 8904 & 8905); unknown (Max Ebong in Cheek 9567).

Lactuca glandulifera Hook.f.
Syn.: *Lactuca glandulifera* Hook.f. var. *calva* R.E.Fr.
Perennial scrambling herb to 2m with white sap; capitula many, with small yellow or white ligular flowers. Forest and forest-grassland transition; 2000m.
Distr.: tropical Africa. (Afromontane).
IUCN: LC
Kodmin: <u>Cheek 8893</u> 1/1998; **Mwanenguba:** <u>Thomas 3155</u> fl., 2/1984.

Lactuca inermis Forssk.
Kew Bull. 39: 132 (1984).
Syn.: *Lactuca capensis* Thunb.
Perennial herb to 2m with white sap; capitula with small blue or mauve ligular flowers. Grassland; 1200–2000m.
Distr.: tropical & subtropical Africa. (Afromontane).
IUCN: LC
Kodmin: <u>Cheek 9096</u> 2/1998; **Mwanenguba:** <u>Cheek 9462</u> 10/1998; <u>Leeuwenberg 8498</u> fl., 9/1971; <u>Sanford 5446</u> fl., 11/1968.

Laggera crispata (Vahl) Hepper & J.R.I.Wood
Kew Bull. 38: 83 (1983).
Syn.: *Blumea crispata* (Vahl) Merxm. var. *crispata*
Syn.: *Laggera alata* (D.Don) Sch. Bip. ex Oliv. var. *alata*

Syn.: *Laggera pterodonta* (DC.) Sch.Bip. ex Oliv.
Annual or perennial herb to 2.4m; stems winged; aromatic;
capitula with many small pink to purple flowers. Forest;
1000–1980m.
Distr.: tropical Africa & Asia. (Palaeotropics).
IUCN: LC
Kodmin: Ghogue 62 fl., 1/1998; **Mwanenguba:** Cheek 7243
fr., 2/1995; **Nyasoso:** Cheek 7342 fr., 2/1995; Etuge 1754
3/1996.
Local names: Taku mohsee, 'Devil's snuff/tobacco'. **Uses:**
SOCIAL USES – used by herbalists (Cheek 7342).

Melanthera scandens (Schumach. & Thonn.)
Roberty
Perennial herb to 1m high, or scandent and then to 4m high;
leaves opposite, ovate, 2–13 × 1–7cm, scabrid, strongly 3-
veined from base; capitula solitary or in lax corymbs; ray
florets orange-yellow, disc florets yellow; pappus of bristles.
Roadsides, farmbush; 200–1450m.
Distr.: tropical Africa. (Afromontane).
IUCN: LC
Kodmin: Cheek 9032 1/1998; **Kupe Village:** Cheek 8319
5/1996; **Loum F.R.:** Biye 70 12/1999; Onana 992 12/1999;
Mungo River F.R.: Cheek 10184 11/1999; **Ndum:** Cable
879 fl., 1/1995; **Nyasoso:** Cable 2785 6/1996; Cheek 7468
fr., 10/1995; Etuge 2593 fl., fr., 7/1996; Zapfack 630 6/1996.
Local names: Bwassa (Bakweri language, Daniel Lyonga in
Cheek 7468); Warraday (Koge Lawrence in Cheek 8319).
Uses: MEDICINES (Daniel Lyonga in Cheek 7468); (Koge
Lawrence in Cheek 8319).

Micractis bojeri DC.
Gleditschia 18: 211 (1990).
Syn.: *Sigesbeckia abyssinica* (Sch.Bip.) Oliv. & Hiern
Annual herb; leaves opposite, connate; capitula small, with
few tiny yellow ray florets and yellow disc florets. Forest
edge; 2000m.
Distr.: Cameroon, Ethiopia to Malawi. (Afromontane).
IUCN: LC
Mwanenguba: Cheek 9407 10/1998.

Microglossa pyrifolia (Lam.) Kuntze
(Scandent) shrub to 3(–6)m; capitula many, tiny, with minute
cream to pale yellow flowers. Forest; 2000m.
Distr.: tropical Africa & Asia. (Palaeotropics).
IUCN: LC
Kodmin: Cheek 9191 fl., fr., 2/1998; **Mwanenguba:**
Thomas 3139 fl., 2/1984; **Nyasoso:** Etuge 1671 1/1996.
Local name: Apamueneh. **Uses:** MEDICINES (Etuge 1671).

Mikania chenopodifolia Willd.
Fragm. Flor. Geobot. 36(1) Suppl. 1: 460 (1991).
Syn.: *Mikania capensis* DC.
Shrub 5–9m; leaves opposite, ovate, 2–10 × 1–7cm,
subsagittate to hastate, glandular-punctate, with 3 main veins;
capitula in dense leafy corymbs, without rays; florets white;
pappus of bristles. Forest, thicket; 800–1800m.
Distr.: tropical Africa & Asia. (Palaeotropics).
IUCN: LC
Kodmin: Biye 22 fl., fr., 1/1998; Etuge 4429 fl., fr., 11/1998;
Ghogue 57 fl., 1/1998; Gosline 170 11/1998; **Kupe Village:**
Cable 2549 fl., fr., 5/1996; 3748 fl., 7/1996; Cheek 7047 fr.,
1/1995; 7691 11/1995; Ryan 382 5/1996; **Mwambong:**
Gosline 125 10/1998; **Ndum:** Cable 874 fl., 1/1995; Lane
473 fr., 2/1995; **Nyasoso:** Cheek 7541 10/1995; Etuge 1321
fl., 10/1995; Lane 224 fl., 11/1994.

Mikania microptera DC.
Fragm. Flor. Geobot. 36(1) Suppl. 1: 463 (1991).
Shrub to 5m; leaves opposite, ovate, 3–11 × 3–11cm, with
three main veins, glandular; capitula lax leafy corymbs; disc
florets white, without rays; pappus of bristles. Forest-
grassland transition; 1050m.
Distr.: tropical Africa & S America. (Montane).
IUCN: LC
Nyasoso: Cable 3611 fl., 7/1996.

Mikaniopsis maitlandii C.D.Adams
Shrubby climber with purplish stems; leaves fleshy, ovate
and cordate, to 8cm long; 5–7-veined from base; capitula in
dense clusters, without rays; florets yellow. Forest; 1000–
1400m.
Distr.: W Cameroon. (W Cameroon Uplands).
IUCN: VU
Kodmin: Etuge 4415 11/1998; **Kupe Village:** Etuge 2693
fr., 7/1996.

Mikaniopsis sp. near maitlandii C.D.Adams
Vine, glabrous, latex absent; leaves ovate in outline, 13 ×
10cm, acumen acute, 0.5cm, base broadly cordate, cleft 1cm
deep, subhastate with two short reflexed lobes 4cm from the
petiole, 5-nerved from the base; petiole twisting, forming
persistent hook, 6cm; racemes white-puberulent, 6cm; 5–9
capitula; capitula 7mm diameter, without rays. Forest edge;
800m.
Nyasoso: Etuge 2503 7/1996.
Note: possibly a new taxon; known only from this specimen;
more material required.

Mikaniopsis vitalba (S.Moore) Milne-Redh.
Exell, Suppl. Cat. Vasc. Pl. São Tomé: 57 (1956).
Climber to 8m, glabrous; stems terete, ridged; leaves ovate,
acumen 0.5cm, base oblique, entire, 3-nerved from base;
petioles 10cm; racemes 15cm; pedicels 1.5cm; capitula
without rays; florets white, 7mm. Forest edge; 1600m.
Distr.: Cameroon to Uganda. (Tropical Africa).
IUCN: VU
Nyasoso: Etuge 2177 6/1996.

Senecio ruwenzoriensis S.Moore
Perennial herb to 0.6m; leaves succulent, obovate, 10 × 5cm;
capitula many, small, with yellow ray florets and small
yellow disc florets. Grassland; 1250m.
Distr.: Nigeria to S Africa. (Afromontane).
IUCN: LC
Lake Edib: Cheek 9141 fr., 2/1998.

Solanecio biafrae (Oliv. & Hiern) C.Jeffrey
Kew Bull. 41: 922 (1986).
Syn.: *Crassocephalum biafrae* (Oliv. & Hiern) S.Moore
Scrambling subsucculent herb; leaves triangular-subhastate
or sagittate, 2–9 × 1–6cm; capitula many in loose terminal
corymbs, without rays; disc florets pale yellow; pappus of
bristles. Forest-grassland transition; 600–990m.
Distr.: Sierra Leone to Cameroon, Congo (K) & Uganda.
(Guineo-Congolian (montane)).
IUCN: LC
Kupe Village: Cheek 7919a 11/1995; **Ndum:** Williams 153
fr., 1/1995; **Nyandong:** Cheek 11456 3/2003.
Uses: SOCIAL USES – 'Religious' uses – ritual – leaves
added to the first water given to a newborn baby to drink
(Ngolle Abraham in Cheek 7919).

Solanecio mannii (Hook.f.) C.Jeffrey
Kew Bull. 41: 922 (1986).

Syn.: *Crassocephalum mannii* (Hook.f.) Milne-Redh.
Shrub or small tree to 4(–10)m; leaves subsucculent, obovate, to 50 × 15cm; capitula many, without rays, with small yellow disc florets. Forest; 1550m.
Distr.: Nigeria to Congo (K), Sudan to E Africa & Zimbabwe. (Afromontane).
IUCN: LC
Kodmin: Cheek 9183 2/1998.

Sonchus angustissimus Hook.f.
Perennial herb to 3m; leaves pinnatilobed; capitula with many yellow ligulate flowers. Grassland; 1170–2000m.
Distr.: Nigeria to Congo (K). (Lower Guinea & Congolian (montane)).
IUCN: LC
Edib village: Cheek 9173 2/1998; **Mwanenguba:** Cheek 9463 10/1998.

Synedrella nodiflora Gaertn.
Annual herb, erect, 15–120cm high; leaves opposite, ovate or elliptic, 2–12 × 1–5cm, scabridulous; capitula in dense sessile glomerules; ray florets small, yellow; disc florets yellow; pappus of 2–3 small awns. Forest, farmbush; 200–1000m.
Distr.: pantropical. (Montane).
IUCN: LC
Kupe Village: Cheek 8317 fr., 5/1996; **Mungo River F.R.:** Cheek 10182 11/1999; **Nyasoso:** Sidwell 390 10/1995.
Local name: Esu mish er. **Uses:** MEDICINES (Koge Lawrence in Cheek 8317).

Tithonia diversifolia (Hemsl.) A.Gray
Kirkia 6: 56 (1967).
Perennial herb to 5m; leaves large, ovate to obovate, 20 × 15cm, lobed; capitulum solitary, with swollen stalk and large yellow ray florets. Farms, farmbush; 1050m.
Distr.: originally from S America, widely naturalised in the tropics. (Pantropical).
IUCN: LC
Ndum: Cable 922 2/1995.

Vernonia amygdalina Delile
Shrub or small tree 0.5–10m high; leaves elliptic, 4–15 × 1–4cm, slightly hairy; capitula in large terminal corymbs; florets small, white, scented; pappus of small outer scales and longer inner bristles. Cultivated; 650m.
Distr.: Mali to Zimbabwe & Annobon. (Tropical Africa).
IUCN: LC
Ngusi: Etuge 1534 fr., 1/1996.
Local name: Ndu (*fide* Etuge), Bitter Leaf. **Uses:** FOOD – leaves – edible, used in the African dish 'Ndole', with rice or plantains; MEDICINES (Etuge 1534).

Vernonia biafrae Oliv. & Hiern
Syn.: *Vernonia tufnelliae* S.Moore
Scandent or straggling shrub 0.7–6m tall; leaves ovate, 1.3–7 × 0.8–5cm, entire to serrate; capitula in terminal and upper axillary corymbs; florets small, pale mauve or purplish, rarely white; pappus of small outer scales and larger inner bristles. Forest; 1000–1975m.
Distr.: widespread in tropical Africa. (Afromontane).
IUCN: LC
Mwambong: Onana 548 fl., 2/1998; **Mwanenguba:** Cheek 7289 fl., 2/1995; **Ndum:** Cable 895 fl., 1/1995.

Vernonia blumeoides Hook.f.
Small shrub to 1.2m; capitula many, with small mauve or purple flowers. Grassland; 1450–2000m.

Distr.: Nigeria & Cameroon. (Lower Guinea (montane)).
IUCN: NT
Kodmin: Cheek 9208 fr., 2/1998; 9585 fl., 11/1998; **Mwanenguba:** Cheek 9488 fl., 10/1998; Sanford 5381 fl., 11/1968.
Note: restricted to about 11 large sites, but the area of suitable habitat may be increasing.
Local name: Nzion Enkumbe. **Uses:** unknown (Max Ebong in Cheek 9585).

Vernonia calvoana (Hook.f.) Hook.f. subsp. *calvoana* var. *calvoana*
Kew Bull. 43: 237 (1988).
Woody herb to small tree 1–12m high; leaves elliptic, 6–34 × 2–9cm, serrate, thinly-pubescent; capitula in terminal corymbs; the involucre bracts with green appendages; florets small, white or purple; pappus of multiseriate bristles. Grassland; 2330m.
Distr.: Mt Cameroon, Mt Oku & Mwanenguba. (W Cameroon Uplands).
IUCN: VU
Mwanenguba: Sanford 5526 fl., 11/1968.

Vernonia doniana DC.
Syn.: *Vernonia conferta* Benth.
Tree 6–13m high; leaves elliptic to obovate, 10–90 × 8–26cm, glabrous above, tomentose beneath; capitula many, in a panicle to 1m long; florets small, white; pappus of small outer scales and longer inner bristles. Forest clearings, farmbush; 700–1550m.
Distr.: Guinea (C) to Bioko, Cameroon, Angola, Uganda & Sudan. (Guineo-Congolian).
IUCN: LC
Enyandong: Ghogue 1275 10/2001; **Kodmin:** Cheek 9194 2/1998; Gosline 115 fl., fr., 2/1998; **Kupe Village:** Cable 708 fl., 1/1995; Cheek 7021 fl., 1/1995; Kenfack 326 fr., 7/1996; **Mejelet-Ehumseh:** Etuge 467 fl., 1/1987.
Local name: N'boor. **Uses:** FOOD ADDITIVES – leaves – preservatives – used in preserving *Cola* nuts by northeners, especially Hausas; SOCIAL USES – used in trials of thieves: ten leaves stick together if a thief, but don't if not (Cheek 7021).

Vernonia frondosa Oliv. & Hiern
Shrub or small tree 4–6m high; leaves sessile, to 90 × 45cm, crowded near top of stem; capitula in clusters in a large panicle; florets small, purple. Forest clearings; 710m.
Distr.: Ivory Coast, Nigeria & Cameroon. (Upper & Lower Guinea).
IUCN: NT
Nyale: Etuge 4217 2/1998; **Nyandong:** Tchiengue 1747 fl., fr., 3/2003.
Note: known from only 13 sites, but possibly undercollected due to its very large leaves.

Vernonia glabra (Steetz) Vatke var. *hillii* (Hutch. & Dalziel) C.D.Adams
Perennial herb; stems 0.3–3m from a woody rootstock; leaves elliptic, 2–16 × 0.5–7cm, serrate; capitula many in dense corymbose cymes; florets small, blue; pappus of two rows of bristles, the outer smaller. Forest edge, thickets; 1975m.
Distr.: Cameroon & Kenya to S Africa. (Afromontane).
IUCN: LC
Kodmin: Cheek 8898 fr., 1/1998; **Mwanenguba:** Cheek 7281 fr., 2/1995.
Local name: Ndu Sol ate. **Uses:** not used (Cheek 8898).

CONNARACEAE

Vernonia holstii O.Hoffm.
Bot. Jahrb. 20: 220 (1894).
Woody herb or shrub to 4.5m; leaves ovate, 18 × 12cm; capitula many, small, with small white or pale mauve flowers. Margins and clearings in forest; 1980m.
Distr.: Cameroon to Mozambique. (Tropical Africa).
IUCN: LC
Nyasoso: Cable 1195 fl., 2/1995; Cheek 7341 fr., 2/1995.
Local name: Ndu Ngalle. **Uses:** resembles 'bitter leaf', but has no known uses (Cheek 7341).

Vernonia hymenolepis A.Rich.
Kew Bull. 43: 237 (1988).
Syn.: *Vernonia leucocalyx* O.Hoffm. var. *acuta* C.D.Adams
Syn.: *Vernonia leucocalyx* O.Hoffm. var. *leucocalyx*
Syn.: *Vernonia insignis* (Hook.f.) Oliv. & Hiern
Woody herb or shrub to 4m; leaves elliptic, 16 × 16cm, serrate; capitula large; bracts with pink appendages; small mauve or purple flowers. Forest; forest margins; sometimes cultivated; 650–1600m.
Distr.: Cameroon, Uganda, Sudan, Ethiopia & Kenya. (Afromontane).
IUCN: LC
Kupe Village: Cable 2527 fl., fr., 5/1996; Cheek 7148 fl., fr., 1/1995; 8322 fr., 5/1996; Etuge 1979 fl., fr., 5/1996;
Ngomboaku: Cheek 10285 12/1999; **Nyasoso:** Cheek 5631 fl., fr., 12/1993; Etuge 1604 1/1996; 2173 fl., fr., 6/1996.
Local names: 'Bitter Leaf', but not THE Bitter Leaf, but used as such here (Cheek 5631); 'Sweet Bitter Leaf' (Cheek 8322); 'Bitter Leaf' (Cheek 10285). **Uses:** FOOD – leaves – pounded as *V. amygdalina* to make 'Ndole' (Cheek 8322); extremely bitter leaves (Cheek 10285); slice, fry or put in sauce – African dish Ndole soup, always available in restaurants (Etuge 1604).

Vernonia ituriensis Muschl. var. *occidentalis* (C.D.Adams) C.Jeffrey
Kew Bull. 43: 231 (1988).
Syn.: *Vernonia glabra* (Steetz) Vatke var. *occidentalis* C.D.Adams
Woody herb 1–2m high; leaves elliptic, 3–25 × 1–9 cm, serrate; capitula in corymbose cymes, florets blue or mauve; pappus of small outer scales and longer inner bristles. Grassland; 1500m.
Distr.: Nigeria & Cameroon. (W Cameroon Uplands).
IUCN: NT
Mwanenguba: Sanford 5541 fl., fr., 11/1968.
Note: restricted to about 11 large sites, but habitat may be increasing.

Vernonia myriantha Hook.f.
Syn.: *Vernonia subuligera* O.Hoffm.
Shrub or small tree 1–6m; leaves elliptic or ovate, 11–50 × 2–23cm, serrate; capitula many in large terminal compound cymes; florets mauve or purple, fading to white; pappus of small outer scales and larger inner bristles. Forest; 1450m.
Distr.: Sierra Leone to Ethiopia & S Africa. (Afromontane).
IUCN: LC
Kodmin: Cheek 9565 fl., 11/1998.
Local name: Ngo-Pooma. **Uses:** unknown (Max Ebong in Cheek 9565).

Vernonia purpurea Sch.Bip.
Perennial herb to 2m; leaves elliptic, 10 × 3cm, dentate; capitula many, with small purple or mauve flowers. Montane grassland; 2000–2330m.
Distr.: Senegal to Mozambique. (Afromontane).
IUCN: LC

Mwanenguba: Cheek 9489 10/1998; Sanford 5529 fl., 11/1968.

Vernonia stellulifera (Benth.) C.Jeffrey
Fl. Zamb. 6(1): 142 (1992).
Syn.: *Triplotaxis stellulifera* (Benth.) Hutch.
Annual herb 0.1–1.2m high; leaves ovate or elliptic, 2–10 × 1–3cm, remotely crenate; capitula in corymbose cymes, the florets mauve to purple; pappus of a rim of scales with a few inner bristles. Forest and farmbush; 650–1530m.
Distr.: Guinea (C) to Zambia & Uganda. (Guineo-Congolian (montane)).
IUCN: LC
Enyandong: Cheek 10964 10/2001; **Kodmin:** Cheek 8896 1/1998; Ghogue 21 fl., fr., 1/1998; **Kupe Village:** Cheek 7147 fl., 1/1995; 8314 5/1996; Ryan 253 fl., fr., 5/1996; **Mungo River F.R.:** Cheek 10178 11/1999; **Ngusi:** Etuge 1577 fl., fr., 1/1996; **Nyasoso:** Cheek 5657 fl., 12/1993; Etuge 2592 fl., fr., 7/1996.
Local names: Akaval (Koge Lawrence in Cheek 8314); Seme-sim (Cheek 8896). **Uses:** MEDICINES (Etuge 2592); (Koge Lawrence in Cheek 8314); (Cheek 8896).

CONNARACEAE
C.C.H. Jongkind & I. Darbyshire

Agelaea paradoxa Gilg
Wag. Agric. Univ. Papers 89(6): 140 (1989).
Syn.: *Castanola paradoxa* (Gilg) Schellenb. ex Hutch. & Dalziel
Liana to 10(–40)m; branchlets puberulous; leaves trifoliolate, 13 × 10cm; leaflets ovate-elliptic, terminal leaflet c. 10 × 5.5cm, acumen to 1.4cm, base acute to obtuse, asymmetric in lateral leaflets, glabrous, mucous cells numerous on upper surface; petiole 7–8cm, puberulent; petiolules 0.4cm; panicles axillary, may be numerous on leafless branches, to 11cm, many-flowered; branches puberulent; pedicels 1mm; flowers 3mm, honey-scented; sepals ovate, 1–1.5mm, puberulent; petals oblanceolate, 3mm, white. Forest; 750–850m.
Distr.: Senegal to Congo (K). (Guineo-Congolian).
IUCN: LC
Baseng: Cheek 10417 fl., 12/1999; **Ngomboaku:** Cheek 10323 fl., 12/1999.

Agelaea pentagyna (Lam.) Baill.
Wag. Agric. Univ. Papers 89(6): 144 (1989).
Syn.: *Agelaea hirsuta* De Wild.
Syn.: *Agelaea dewevrei* De Wild. & T.Durand
Syn.: *Agelaea floccosa* Schellenb.
Syn.: *Agelaea grisea* Schellenb.
Syn.: *Agelaea obliqua* (P.Beauv.) Baill.
Syn.: *Agelaea preussii* Gilg
Syn.: *Agelaea pseudobliqua* Schellenb.
Liana 6(–25)m; branches furrowed, puberulent; leaves trifoliolate, 14 × 14cm; leaflets ovate, terminal leaflet 10 × 8cm, acumen 1.5cm, base rounded, finely tomentose on lower surface and veins; petiole 11.5cm; petiolule 3mm; panicles to 35cm, glabrous to tomentose; sepals 2.5–5mm, with fringing multi-cellular hairs; petals 3–5.5mm. Forest; 960m.
Distr.: tropical Africa & Madagascar.
IUCN: LC
Kupe Village: Cable 840 fl., 1/1995.

Cnestis corniculata Lam.

Syn.: *Cnestis aurantiaca* Gilg
Syn.: *Cnestis congolana* De Wild.
Syn.: *Cnestis grisea* Baker
Syn.: *Cnestis longiflora* Schellenb.
Syn.: *Cnestis sp. A sensu* Hepper in FWTA 1: 743
Liana to 10m; branchlets yellow-brown velutinous; leaves imparipinnate, to 45cm, 7–11-jugate; leaflets opposite to alternate, oblong 10.5 × 3.5cm, acumen 1.3cm, base asymmetric truncate or subcordate, veins beneath puberulent, basal leaflets much reduced, ovate; petiolules 1mm; racemes axillary or cauliflorous, to 6cm, single or in clusters of up to 5, rachis densely puberulent; flowers 0.3cm; sepals reflexed, puberulent; petals oblong-lanceolate, white; follicles 1–4, ellipsoid, 0.8 × 0.45cm (immature), elongating on maturity, densely long-hairy. Forest, thickets; 260–1030m.
Distr.: Senegal to E Tanzania. (Tropical Africa).
IUCN: LC
Baduma: Nemba 624 9/1987; Kupe Village: Groves 7 fr., 1/1995; Kenfack 260 7/1996; Ryan 264 5/1996; Nyasoso: Sebsebe 4988 10/1995.

Connarus congolanus Schellenb.

Wag. Agric. Univ. Papers 89(6): 247 (1989).
Liana; branches lenticellate, pubescent to glabrescent; leaves trifoliolate; leaflets coriaceous, ovate-elliptic, 5–25 × 3–10.5cm, apex acuminate, base cuneate to rounded, glabrous; petiole 2.5–17cm; petiolules c. 6mm; panicles to 27cm, brown-pubescent; sepals ovate or triangular, fleshy, c. 3mm, brown-pubescent outside; petals linear-elliptic, c. 7mm; follicles ovoid to obovoid, 5–7 × 2–3cm, obtuse to acute at apex, not stipitate. Forest, secondary forest; 250m.
Distr.: Liberia & SE Nigeria to Congo (K). (Guineo-Congolian).
IUCN: LC
Bolo-Meboka: Thomas 7154 7/1987.
Note: specimen not seen at K.

Connarus griffonianus Baill.

Liana; branchlets densely brown-pubescent; leaves 2–5-jugate; leaflets elliptic to obovate, 5–22 × 2–7cm, apex acuminate, base rounded to cuneate, densely brown-pubescent to glabrescent beneath; petiole 5–12cm; petiolules 4–7mm; panicles to 35cm, brown-pubescent; sepals spirally arranged, triangular to ovate, fleshy, 2–2.5cm, brown-pubescent outside; petals elliptic to obovate, c. 5.5 × 1.5mm, pilose outside; follicle obliquely pyriform, c. 2 × 1.5cm, oblique apex mucronate; stipe c. 3mm, soon glabrescent. Forest, secondary forest; 500m.
Distr.: Nigeria to Angola. (Lower Guinea & Congolian).
IUCN: LC
Nlog: Etuge 15 4/1986.
Note: specimen not seen at K.

Jollydora duparquetiana (Baill.) Pierre

Monopodial treelet 2(–8)m; leaves imparipinnate, to 46cm; leaflets 5, pairs subopposite, obovate, to 28 × 13cm, acumen 1.8cm, base acute-cuneate, glabrous; petiole c. 25cm; petiolules 1.2cm, swollen; racemes cauliflorous, 1–several, rachis to 2cm; sepals ovate to oblong, 2.5mm, puberulent outside; petals oblong, 5–9mm; fruit ellipsoid, 3 × 1.8cm, apiculate, stipitate, orange, smooth. Forest; 200–300m.
Distr.: E Nigeria to Congo (K). (Lower Guinea & Congolian).
IUCN: LC
Baduma: Nemba 117 fr., 6/1986; 573 7/1987; Bakolle Bakossi: Etuge 286 9/1986; Bakossi F.R.: Cheek 9312

10/1998; Ikiliwindi: Nemba 455 1/1987; Mile 15, Kumba-Mamfe Road: Nemba 206 9/1986.
Local name: Ajip-ambwe (meaning 'Bell of a dog'). Uses: MEDICINES (Epie Ngome in Cheek 9312).

Rourea cf. minor (Gaertn.) Alston

Liana to 2.5m; stems glabrous, cylindrical; leaves imparipinnate, to 21cm long, 3–4-jugate; leaflets alternate, elliptic, terminal leaflet 10 × 3.8cm, acuminate-apiculate, base acute, asymmetrical in lateral leaflets, glabrous; petiole 6.5cm; petiolules 1.5mm; panicles axillary, 5 × 2.5cm, puberulent; flowers 1–1.5mm in bud; follicles (1–)2–5, asymmetric-ellipsoid, 0.6cm long. Forest; 1400–1560m.
Kodmin: Onana 592 2/1998; Muahunzum: Etuge 4437 11/1998.
Note: differs from *R. minor* in the apiculate (not exapiculate) leaf-tips and the puberulent (not glabrous) inflorescences, and in usually having > 1 (not single) follicles in fruit. Possibly *sp. nov.*, more material needed for further identification.

Rourea myriantha Baill.

Wag. Agric. Univ. Papers 89(6): 342 (1989).
Syn.: *Paxia liberosepala* (Baker f.) Schellenb.
Syn.: *Paxia cinnabarina* Schellenb.
Liana to 10m, resembling *R. cf. minor*, but branchlets puberulent; leaves 2–3-jugate; leaflets ovate to oblong-obovate, apex subacuminate-apiculate, base rounded to acute; panicles to 15cm, puberulent to glabrescent; sepals ovate, 3–4mm, puberulent outside, red to white; petals lanceolate, 1–1.5cm; follicles usually 1 per flower, c. 3.5 × 1.7cm, rostrate, glabrous, often lenticellate. Forest; 1100m.
Distr.: E Nigeria to Angola. (Lower Guinea & Congolian).
IUCN: LC
Kupe Village: Cheek 7858 11/1995.

Rourea solanderi Baker

Wag. Agric. Univ. Papers 89(6): 355 (1989).
Syn.: *Spiropetalum heterophyllum* (Baker) Gilg
Syn.: *Spiropetalum solanderi* (Baker) Gilg
Liana to 6m, resembling *R. myriantha*, but branches furrowed; leaflets oblong-elliptic, terminal leaflet to 9 × 4.8cm; petiolules 3mm; panicles to 7cm, puberulous; sepals partly united, lobes 2.5 × 5mm, velutinous on both sides; petals 1–2cm; follicles 1(–3) per flower, 2 × 1cm, rostrate, velutinous brown; calyx lobes unevenly expanded to max. 7mm long. Forest; 350–1300m.
Distr.: Sierra Leone to Congo (K). (Guineo-Congolian).
IUCN: LC
Edib: Etuge 4099 fr., 1/1998; Mekom: Nemba 13 fr., 4/1986.

Rourea thomsonii (Baker) Jongkind

Wag. Agric. Univ. Papers 89(6): 359 (1989).
Syn.: *Jaundea pubescens* (Baker) Schellenb.
Syn.: *Jaundea pinnata* (P.Beauv.) Schellenb.
Liana to 20m, rarely a shrub; branches often lenticellate; leaves 2–4-jugate; leaflets elliptic to oblong, terminal leaflet 8.5–13.5 × 3.7–5.6cm, acuminate-apiculate, base cuneate to obtuse, often puberulent on under surface; petiole 8.5cm; petiolules to 5mm; panicles axillary or often numerous at ends of leafy branchlets, to 15cm, rachis puberulent; sepals 1.5mm, puberulent outside; petals oblong-lanceolate, c. 7mm, white; follicles 1 per flower, ovoid, 4 × 2.5cm, apex acute to acuminate, green, smooth. Wet tropical forest and semi-deciduous forest; 800–1450m.
Distr.: Guinea (B) to Mozambique. (Tropical Africa).

IUCN: LC
Kodmin: Cheek 9577 fl., 11/1998; **Kupe Village:** Cheek 7756 11/1995; **Ngomboaku:** Cheek 10307 12/1999; Gosline 269 fl., 12/1999; **Nyasoso:** Etuge 2131 fr., 6/1996; 2386 fr., 6/1996; 2484 fr., 7/1996.
Local name: Ekogeh-ehnbip. **Uses:** SOCIAL USES – unspecified social uses – in years when this plant flowers a lot there is much sickness in the village (Max Ebong in Cheek 9577).

CONVOLVULACEAE

M. Cheek

Calycobolus africanus (G.Don) Heine
Climber of forest canopy; fallen fruit with two of the calyx lobes greatly expanded, papery, unequal; larger ovate-orbicular c. 6 × 5cm; smaller ovate-triangular 2cm, cordate; nut 1cm. Forest; 700m.
Distr.: Sierra Leone to Congo (K). (Guineo-Congolian).
IUCN: LC
Kupe Village: Cheek 7007 fr., 1/1995; **Nyandong:** Cheek 11317 3/2003.
Note: *Neuropeltis acuminata* with similar winged fruits (but wings bract-derived) is also likely in our area.

Hewittia sublobata (L.f.) Kuntze
Climbing herb resembling *Ipomoea*, but pollen smooth; glabrous; leaves ovate, c. 10 × 7cm, subacuminate-mucronate, base obtuse-acute cordate; petiole 5cm; inflorescence c. 5-flowered, capitate; bracts lanceolate, 2 × 0.7cm; sepals leafy, ovate, c. 1.3cm; corolla 2.5cm, pale yellow, dark red at base; fruit globose. Farmbush; 1200–1400m.
Distr.: tropical Africa & Asia. (Palaeotropics).
IUCN: LC
Kodmin: Ghogue 72 fl., 2/1998; **Mwanenguba:** Leeuwenberg 9611 fl., 4/1972.

Ipomoea alba L.
Climbing herb, glabrous; leaves papery, ovate, c. 15 × 12cm, acumen slender, 1.5cm, cordate; petiole c. 17cm; inflorescence 1–5-flowered, racemose; peduncle as long as rachis; corolla white, opening at night, 8cm long, 10cm across; fruit ovoid, 3cm. Farmbush; 150–1190m.
Distr.: native to S America. (Pantropical).
IUCN: LC
Enyandong: Ghogue 1296 fl., 11/2001; **Kupe Village:** Cheek 7709 11/1995; **Mungo River:** Gosline 218 fl., 11/1999; **Ndum:** Cable 889 fl., 1/1995; Lane 455 fl., fr., 1/1995; Williams 144 fr., 1/1995; **Ngomboaku:** Ghogue 470 fl., 12/1999; **Nyasoso:** Sidwell 403 10/1995; Thomas 5047 11/1985.
Local name: Nkong or Sweet Potato. **Uses:** FOOD –edible fruits or tubers (*fide* Etuge).

Ipomoea batatas (L.) Lam.
Climber, tubers edible; stems to 15m, glabrous; leaves triangular-ovate, c. 10 × 8cm, long-acute, broadly cordate, or pedately 5-lobed, central lobe to 5cm; petiole c. 7cm; inflorescence cymose, 5–10-flowered, rachis to 3cm; peduncle 10cm; sepals lanceolate 0.6cm; corolla pale pink, 3cm. Farmbush, farms; 650–990m.
Distr.: native to S America. (Pantropical).
IUCN: LC

Bangem: Thomas 5249 1/1986; **Ndum:** Williams 152 fl., 1/1995; **Nyasoso:** Cable 1226 2/1995; Etuge 1607 1/1996.
Local name: Sweet Potato. **Uses:** FOOD – cultivated for root tubers (*fide* Cheek).

Ipomoea indica (Burm.) Merr.
Interpr. Rumph. Herb. Amboin.: 445 (1917).
Syn.: *Ipomoea acuminata* Baker
Syn.: *Ipomoea congesta* R.Br.
Climber, reflexed yellow-hairy; leaves ovate, 7–8 × 5–6cm, tapering acute, cordate, sometimes 1–2-lobed; petiole 2cm; inflorescence 3–4-flowered, subcapitate; peduncle 5.5cm; bracts 3–4, linear, 1.5cm; sepals narrowly lanceolate, 2cm; corolla pale pink, 7cm. Farmbush; 950–1000m.
Distr.: tropical Asia; planted elsewhere. (Pantropical).
IUCN: LC
Kupe Village: Cheek 10166 11/1999; **Nyasoso:** Etuge 1513 12/1995; Sunderland 1495 fl., 7/1992.
Uses: ENVIRONMENTAL USES – introduced as an ornamental but persisting and possibly spreading without help from man (*fide* Cheek).

Ipomoea involucrata P.Beauv.
Climber resembling other *Ipomoea* species, distinct in the entire, sheathing involucral bract, c. 1.5cm long, 7cm wide; flowers 5–15, capitate; corolla pink, 3–4cm. Farmbush; 425–1450m.
Distr.: Senegal to Kenya. (Tropical Africa).
IUCN: LC
Kodmin: Cheek 9039 1/1998; **Kupe Village:** Lane 324 fl., 1/1995; **Mwanenguba:** Leeuwenberg 8582 fl., 10/1971; **Ndum:** Williams 135 fl., 1/1995; **Nyandong:** Darbyshire 132 3/2003; **Nyasoso:** Cable 2783 6/1996; Etuge 1760 3/1996; Sidwell 319 10/1995.
Note: very common weed, identified by the sheathing involucral bract; Sidwell 319 is odd in the serrate-crenate leaf margins.

Ipomoea mauritiana Jacq.
Climber, glabrous; leaves to 20 × 16cm, subacuminate, truncate to shallowly cordate, entire or with 1–5 shallow or deep lateral lobes; petiole c. 12cm; inflorescence nearly capitate, rachis 3cm; bracts absent; peduncle 14cm; sepals ovate, chaffy, 0.7cm; corolla pink, 7cm. Farmbush; 250m.
Distr.: pantropical.
IUCN: LC
Mile 15, Kumba-Mamfe Road: Nemba 147 7/1986.
Note: specimen not seen at K.

Ipomoea tenuirostris Choisy
Slender climber; stems pilose; leaves lanceolate, c. 4–8 × 1.6–4.5cm, appressed-hairy; inflorescence subracemose, 3–12cm, rachis as long as peduncle, 3–10-flowered; calyx lobes c. 0.5cm; corolla 1.5cm, white, purple throat; fruit globose 0.5cm. Forest edge; 1000–2000m.
Distr.: Sierra Leone to Cameroon. (Upper & Lower Guinea).
IUCN: LC
Kodmin: Cheek 8891 1/1998; Etuge 4044 1/1998; Gosline 160 11/1998; **Mwanenguba:** Sanford 5545 fl., 11/1968; **Nyasoso:** Etuge 1771 3/1996.

Ipomoea cf. wightii (Wall.) Choisy
Climber to 8m, appressed white-puberulent; leaves broadly ovate, c. 16 × 14cm; inflorescence c. 10-flowered, loosely capitate; bracts numerous, triangular, 1.2 × 0.4cm; peduncle c. 15cm, apex white; sepals linear-lanceolate 1.5cm; corolla pale to dark pink, 6cm. Secondary forest, scrub; 760–1100m.

Ndum: Cable 875 fl., 1/1995; **Ngomboaku:** Cheek 10283 12/1999; Mackinder 299 12/1999.
Note: distinct from *I. wightii* in e.g. the concolorous leaves (not white below). Possibly *sp. nov.*

Lepistemon parviflorum Pilg. ex Büsgen
Climber, glabrous; leaves broadly ovate, c. 14 × 11cm; petiole 12cm; inflorescence capitate, dense, c. 4 × 4cm, c. 50-flowered; peduncle 1cm; bracts absent or inconspicuous; pedicels c. 1–5cm; flowers 0.8cm, 4–numerous. Secondary forest; 800m.
Distr.: Sierra Leone to Cameroon. (Upper & Lower Guinea).
IUCN: LC
Nyasoso: Biye 62 11/1999.

Merremia umbellata (L.) Hallier f. subsp. *umbellata*
Climber resembling *Ipomoea*, glabrous; leaves ovate c. 9 × 6cm, subacuminate, acutely cordate; inflorescence 5–15-flowered, loosely capitate; peduncle c. 20cm; bracts absent; pedicels 1.5cm; several corollas open at once, bright-yellow, c. 3cm. Roadside, farmbush; 150–420m.
Distr.: Gambia to Cameroon & tropical America. (Amphi-Atlantic).
IUCN: LC
Mungo River: Gosline 217 fl., 11/1999; **Mungo River F.R.:** Onana 985 12/1999; Pollard 105 10/1998; **Nyandong:** Cheek 11411 3/2003.
Note: extremely common along roadsides, flowering early in the morning.
Local name: Mendi Mendiba (Epie Ngome in Pollard 105).

CRASSULACEAE
M. Cheek

Crassula alsinoides (Hook.f.) Engl.
Fleshy straggling herb; stems with 2 lines of pale hairs; leaves ovate, 1–1.5 × 0.5–1cm, apex acute, base claw-like then short-attenuate; flowers terminal or axillary; pedicel fine, to 7mm; flowers 0.5cm; sepals and petals lanceolate, the latter white. Forest edge, open areas; 1650m.
Distr.: W Cameroon & Bioko to E & S Africa, Yemen. (Afromontane).
IUCN: LC
Mejelet-Ehumseh: Manning 435 10/1986.

Kalanchoe crenata (Andrews) Haw.
Syn.: *Kalanchoe laciniata sensu* Hepper in FWTA 1: 117
Erect, unbranched, succulent herb to 1m; leaves elliptic, c. 7 × 4.5cm, rounded, obtuse, crenate; petiole 1cm; inflorescence corymbose, 6cm across, c. 20-flowered; pedicels glandular-pubescent; flowers pink, yellow or red, 1.5cm long; petals 5. Farmbush; 650–1500m.
Distr.: Guinea (C) to Cameroon. (Upper & Lower Guinea).
IUCN: LC
Kodmin: Cheek 8976 fl., fr., 1/1998; **Mejelet-Ehumseh:** Etuge 496 2/1987; **Nyasoso:** Etuge 1608 fl., 1/1996.
Local name: Ehowal. **Uses:** MATERIALS – cleansers – warm leaves in fire to clean oily hands, as it acts like soap; MEDICINES (Etuge 1608).

Umbilicus botryoides Hochst. ex A.Rich.
Succulent herb 5cm, glabrous; leaves orbicular, 2cm. diameter, peltate, margin crenate; spikes erect 5cm; flowers pendulous, greenish white, 8mm. Rocky grassland; 2000m.

Distr.: Cameroon to Tanzania. (Afromontane).
IUCN: LC
Mwanenguba: Pollard 157 fl., fr., 10/1998.

CRUCIFERAE
M. Cheek

Fl. Cameroun 21 (1980).

Brassica oleracea L.
Robust erect herb to 1m or more; basal leaves petiolate, fleshy, glaucous, glabrous, lyrate, lateral lobes 1–5 pairs, terminal lobe large, rounded, to 50 × 30cm, margin undulate, crisped, nerves prominent, whitish; upper leaves becoming obovate, auriculate, finally linear and clasping; inflorescence a panicle; petals yellow or white, clasped, to 2cm; fruit linear, 5–10 × 0.5cm, beaked. Cultivated; 1170m.
Distr.: cosmopolitan.
IUCN: LC
Edib village: Cheek (sight record) 2/1998.
Local name: Cabbage. **Uses:** FOOD – leaves eaten (*fide* Pollard).

Cardamine trichocarpa Hochst. ex A.Rich.
Annual herb with aerial stem 10–30cm, glabrous; leaves pinnate, 6cm; leaflets 7, rounded, crenate-serrate; flowers lacking petals; sepals c. 2mm; fruit congested at apex, 1.5–3cm long. Forest edge; 900–1550m.
Distr.: Cameroon to Kenya, India. (Afromontane & India).
IUCN: LC
Ngomboaku: Ghogue 495 12/1999; **Nyasoso:** Lane 220 fr., 11/1994.

Rorippa madagascariensis (DC.) Hara
Fl. Cameroun 21: 19 (1980).
Syn.: *Rorippa humifusa* (Guill. & Perr.) Hiern
Annual herb, lacking aerial stem; leaves in basal rosette, numerous, bipinnatifid, c. 15cm; inflorescence prostrate, 5cm; sepals c. 1mm; petals smaller, white; fruit to 14mm. Villages; 700m.
Distr.: Senegal to Uganda, Madagascar. (Guineo-Congolian & Madagascar).
IUCN: LC
Nyale: Cheek 9640 fl., 11/1998.

CUCURBITACEAE
I. Darbyshire

Fl. Cameroun 6 (1967).

Cayaponia africana (Hook.f.) Exell
Slender climber; robust stems deeply-ridged, finely-pubescent; leaves trilobed, 7–9 × 5.5–8cm, base truncate to subcordate, lobe apices acute, margin dentate, scabridulous; fruits ellipsoid, c. 1.6 × 1.3cm, yellow-orange, with fine reticulate ornamentation, on short-branched or unbranched pubescent axillary peduncles 1–3cm; fruits may be present after leaves have died back. Farmland, disturbed forest; 780–1100m.
Distr.: Gambia to Cameroon. (Upper & Lower Guinea).
IUCN: LC
Enyandong: Cheek 10989 10/2001; **Ndum:** Cable 910 fr., 2/1995.

Citrullus lanatus (Thunb.) Mansf.

F.T.E.A. Cucurbitaceae: 46 (1967).

Syn.: *Colocynthis citrullus* (L.) Kuntze

Creeping herb; stems ridged, with soft white-yellow hairs to 1.5mm; tendrils bifid; leaves triangular in outline, 8–10 × 5.5–8cm, (bi-)pinnatisect, lobes obovate with rounded or abruptly acuminate apices, lobe margins irregularly undulate, surfaces pubescent, particularly along veins; petiole 3.5–7cm, pubescent; male and female flowers solitary, axillary, on finely-pubescent pedicels, 0.5–2.5cm; sepal lobes lanceolate, c. 3mm; petals c. 2.5cm, yellow; anthers highly-sinuous; fruit suborbicular-ellipsoid, smooth, green, fleshy with many pale or dark brown seeds. Cultivated; 420m.

Distr.: widely cultivated in the tropics. (Pantropical).

IUCN: LC

Nyandong: Darbyshire 172 fl., 3/2003.

Local names: Egusi, Water-melon. **Uses:** FOOD – seeds – in Bakossi it is usually cultivated for its seeds which are, together with *Cucumeropsis mannii* seeds, known as 'egusi' (Darbyshire 172).

Coccinia barteri (Hook.f.) Keay

Herbaceous climber; stems ridged, glabrous, angular; tendrils bifid; leaves variable, palmatisect, (3–)5 lobed to sublobed, 12–20 × 10–15cm, base ± deeply cordate, lobe apices acute or acuminate, margins ± serrate; dioecious, both sexes on racemes; peduncle 2–3cm, pubescent; flowers subsessile; males with 5–10 flowers; females 3–5; flowers c. 1.8 × 1.5cm; petals yellow; fruits ellipsoid or subglobose c. 4 × 2.5cm, green, drying blackish, smooth; seeds numerous, c. 4 × 2mm, white, smooth, with narrow rim. Farmbush, forest; 710–1600m.

Distr.: widespread in (sub)tropical Africa.

IUCN: LC

Kupe Village: Cable 2534 5/1996; Cheek 7886 11/1995; Etuge 1940 5/1996; 1941 5/1996; 2716 7/1996; Kenfack 329 7/1996; **Ndum:** Groves 28 fr., 2/1995; **Nyasoso:** Cable 3515 7/1996; Elad 141 fr., 2/1995; Etuge 1300 10/1995; 1815 3/1996.

Cucumeropsis mannii Naud.

Syn.: *Cucumeropsis edulis* (Hook.f.) Cogn.

Herbaceous climber; robust ridged stems densely pubescent, hairs to 1mm; leaves pentagonal, apices mucronate, base deeply cordate, margin broadly dentate; petioles 5–10cm with dense hairs to 1mm; monoecious, male flowers clustered at end of c. 2cm peduncle; flowers c. 0.5cm; calyx lobes lanceolate, partially reflexed; petals yellow; fruit ellipsoid, yellow-green, c. 6 × 3.5cm in our material, but can grow much larger, smooth, with a distinct conical base. Cultivated; 1000m.

Distr.: widely cultivated, originating in W Africa. (Tropical & subtropical Africa).

IUCN: LC

Nyasoso: Etuge 1755 3/1996.

Local name: Esake; this is the true 'Egusi' of W Africa, a term now used for the edible seeds of several Cucurbitaceae spp. **Uses:** FOOD – seeds edible (Etuge 1755).

Lagenaria breviflora Benth.

F.T.E.A. Cucurbitaceae: 49 (1967).

Syn.: *Adenopus ledermannii* Harms

Syn.: *Adenopus breviflorus* Benth.

Herbaceous climber; stems shallowly ridged, glabrous; leaves shallowly to deeply 3–5 lobed, c. 14 × 12cm, base subcordate or truncate with paired glands at the base of the lower lamina, apex obtuse, mucronate, margins broadly dentate, laminae scabrid, particularly above; petiole 3–5cm;

male flowers in a raceme, 4–6-flowered, clustered towards apex; peduncle 4–5cm, glabrous; pedicels c. 2cm; elongate calyx tube 5.5 × 0.8cm, swollen at base, lobes c. 3mm, acute; corolla white-pink, lobes 2.5 × 1cm, involute; anthers on a central column 2 × 0.5cm, included, fragrant; female inflorescence not observed; fruit ovoid, 7–10cm diameter. Secondary vegetation including farmland; 800m.

Distr.: Sierra Leone to Congo (K). (Guineo-Congolian).

IUCN: LC

Mt Kupe: Cheek 7905b fl., 11/1995.

Lagenaria siceraria (Molina) Standl.

Climber resembling *L. breviflora*, but leaves not or only slightly 3-lobed, laminae pubescent; petiole 5–20cm; male flowers solitary on a pedicel to over 20cm; calyx tube to 3cm long, campanulate; fruit variable, globose to bottle-shaped, 10–80cm long. Farmbush and cultivated; 420m.

Distr.: pantropical.

IUCN: LC

Nyandong: Darbyshire (sight record) 3/2003.

Local name: Bottle Gourd. **Uses:** MATERIALS – fruits used as containers for liquids; decorative ornaments (*fide* Cheek).

Momordica cf. angustisepala Harms

Herbaceous climber; stems thin, ridged, glabrous; leaves ovate c. 9 × 7.5cm, base deeply cordate, apex acuminate, margins entire, adaxial lamina scabrid, drying blackish; male inflorescence a c. 8-flowered umbel enclosed within a leafy, orbicular, sessile bract with entire margins and acuminate apex; peduncle c. 2cm long; pedicels c. 0.4cm; sepals lanceolate, c. 6mm long, enclosing corolla in bud; petals white; female inflorescence and fruits not observed. Forest; 400m.

Bakossi F.R.: Etuge 4304 10/1998.

Note: the inflorescence closely resembles *M. angustisepala*, but it otherwise differs in having unlobed (not trilobed), and entire (not serrate); leaves with glabrous (not pubescent), surfaces. Possibly *sp. nov.*

Momordica cf. cabraei (Cogn.) C.Jeffrey

Woody climber; principle stems woody 3.5mm, circular; herbaceous branches ridged, internodes 13–15cm; no leaves or flowers available in our material; pedicel 1cm; fruits solitary, suborbicular, 5.5 × 4.8cm, green-white, scabrous particularly towards the base; sepals persistent, lobes c. 1cm in length, reflexed; seeds elliptic with warted ornamentation. Farmbush; 870m.

Nyasoso: Cable 3247 6/1996.

Note: the fruits match those of *M. cabraei*, but further material, including leaves, required for confirmation.

Momordica charantia L.

Rank-smelling herbaceous climber; stems thin, ridged, tomentose; leaves palmatisect c. 7 × 6cm, 5–7 lobed with strongly sinuate margins, fine white hairs at vein margins on both surfaces; monoecious; flowers solitary; pedicel 7–12cm, sparsely-hairy, with sessile ovate entire bract, 1 × 0.7cm, approximately midway along; petals yellow, dark veined, 1–2cm; sepal lobes acute, c. 4mm; ovary fusiform, hispid; fruits c. 6.5 × 4cm, irregularly ridged, orange, 3-valved; seeds c. 1 × 0.6 × 0.3cm, coarsely warted. Disturbed vegetation, may be planted; 870–1100m.

Distr.: pantropical.

IUCN: LC

Ndum: Williams 141 1/1995; **Nyasoso:** Cable 11 7/1992; 2799 6/1996; 3231 6/1996; Cheek 5634 fl., 12/1993; 7524 10/1995; Sidwell 385 10/1995; Zapfack 642 6/1996.

Momordica cissoides Planch. ex Benth.

Herbaceous climber; stems thin, ridged, glabrous; leaves palmately compound, (3–)5-foliolate c. 8 × 5.5cm, slightly scabrid; leaflets elliptic, base acute, assymetric in lateral leaflets, apex mucronate, margins serrate; dioecious; male umbels and solitary female flowers both enclosed within toothed orbicular bract (2 × 2cm), with cordate base; peduncle 4–5cm; sepals lanceolate c. 3mm; petals elliptic, 2.5 × 1cm, white-yellow; fruits c. 3.5 × 2cm, ovate, orange, densely bristled. Forest and farmbush; 840–1500m.
Distr.: Guinea (C) to Angola, E Africa. (Tropical Africa).
IUCN: LC
Edib: Cheek 9124 2/1998; **Kodmin:** Cheek 8958 1/1998; Pollard 249 11/1998; Satabie 1112 fl., 1/1998; **Kupe Village:** Cheek 7116 1/1995; Etuge 1411 11/1995; 1939 5/1996; Kenfack 257 7/1996; Ryan 337 5/1996; Zapfack 909 7/1996; **Nyasoso:** Cable 1186 fl., 2/1995; 3589 7/1996; Cheek 7295 fr., 2/1995; Groves 76 2/1995; Sidwell 354 10/1995.

Momordica enneaphylla Cogn.

Fl. Cameroun 6: 156 (1967).
Woody or herbaceous climber to 15m; stems ridged, glabrous; leaves biternate, glabrous; leaflets ovate, base obtuse, assymetric in lateral leaflets, apex acuminate, mucronate, margin entire except for mucrons at vein apices; dioecious; male inflorescence a c. 4-flowered umbel enclosed within an orbicular sessile bract, c. 2 × 2cm; peduncle c. 2cm; female flowers solitary; pedicel c. 1cm; calyx reflexed; petals c. 1.5 × 1cm yellow; ovary ellipsoid, 2 × 0.3cm, ridged; fruits not observed. Forest, forest margins; 700–1070m.
Distr.: Cameroon, Gabon, Congo (K). (Lower Guinea & Congolian).
IUCN: VU
Kupe Village: Cable 2710 5/1996; Sidwell 443 11/1995; **Ngomboaku:** Cheek 10367 12/1999.

Momordica foetida Schum. & Thonn.

Syn.: *Momordica cordata* Cogn.
Herbaceous climber; stems deeply-ridged, finely-pubescent or glabrous; leaves ovate, 9–14 × 8–12cm, base deeply cordate, apex acuminate, margin shallowly-dentate, pubescent along veins on abaxial lamina; petioles c. 7cm; monoecious, male flowers in umbels; peduncle c. 10cm, finely-pubescent, may be kinked; pedicels 1–5cm, subtended by a leafy bract c. 4mm; sepals rounded at apex, pubescent, drying black; petals obovate 1.5 × 0.5cm; stamens 2 with sinuous anthers; female flowers solitary; pedicel c. 7cm; fruits densely hispid when immature, pale orange with sparser bristling when mature, ovoid to ellipsoid c. 5.5 × 3cm; seeds red. Secondary forest, farmbush; 1100–1550m.
Distr.: tropical & S Africa.
IUCN: LC
Kupe Village: Ensermu 3574 11/1995; **Mwambong:** Gosline 85 2/1998; **Nyasoso:** Cable 1241 fr., 2/1995; Lane 223 fl., fr., 11/1994.

Momordica gilgiana Cogn.

Herbaceous climber; stems ridged, glabrescent; tendrils simple; leaves trifoliolate, drying green-black, terminal leaflet ovate to obovate, 4.5–6.5 × 8.5–10cm, lateral leaflets asymmetric, margins with sparse prominent teeth to 1.5mm, upper surface finely pale punctate, lower surface glabrous except puberulent along midrib and major veins; petiole to 2cm; inflorescence umbellate; peduncle to 15cm, with prominent glands towards apex; pedicels 0.7–1.7cm, puberulent; calyx lobes ovate, 2–3mm, 2 with prominent glands at apex; petals obovate, to 2cm long; fruit obovoid, 6–

10(–15) × 3–7(–10)cm, highly tuberculate, glabrous, orange. Forest; 500m.
Distr.: Bioko, Cameroon & Gabon. (Lower Guinea).
IUCN: NT
Nyandong: Ghogue 1513 fr., 3/2003.
Note: currently known from only c. 10 sites.

Momordica multiflora Hook.f. var. multiflora

Fl. Cameroun 6: 176 (1967).
Herbaceous climber; stems thin, ridged, glabrous; leaves ovate, c. 9 × 7cm, base deeply-cordate, apex acuminate-mucronate, margin entire apart from mucrons at vein apices; petiole c. 5cm, pubescent; male inflorescence a raceme or panicle comprising numerous flowers, clustered towards the apex; peduncle 3–20cm; a bract c. 3mm sometimes present at the base of the lowest pedicel; pedicels c. 5mm, densely pubescent; petals white, elliptic c. 1cm long; sepals pubescent, lobes acute; anthers straight; female inflorescence and fruits not observed. Secondary forest, farmbush; 1000m.
Distr.: Ghana to Angola & Tanzania. (Guineo-Congolian).
IUCN: LC
Kupe Village: Etuge 2689 7/1996.

Momordica cf. obtusisepala Keraudren

Herbaceous climber; stems thin, ridged, finely-pubescent mainly around nodes; leaves 8–9 × 7–8.5cm, shallowly 5-lobed, base cordate, apex acuminate, margin broadly-dentate, scabrid on both surfaces; petiole pubescent, 5–6cm long; fruits solitary (on thin pedicels, 2–3cm long); c. 10 × 4cm, smooth, orange-red, ellipsoid with attenuate base and apex; sepals ± persistent, reflexed; seeds coarsely-ridged and warted, orbicular c. 0.7cm diameter. Bushland at forest margin; 500m.
Kupe Village: Etuge 1466 11/1995.
Note: the fruits of this specimen resemble those of other *Momordica* spp.; the leaves closely resemble those of *M. obtusisepala*, the only recorded *Momordica* sp. with lobing in this region, which is known only from the type collection in Bertoua, Cameroon and for which the fruit is undescribed.

Momordica sp. 1

Herbaceous climber, similar to *M. foetida*, but male umbels enclosed within a leafy orbicular sessile bract c. 2.5cm with entire margins and acuminate apex; peduncle c. 4.5cm (not c. 10cm); pedicels c. 0.5cm (not 1–5cm); petiole c. 2cm (not c. 7cm); sepal lobes acuminate (not rounded) and glabrous (not pubescent). Forest; 1120m.
Nyasoso: Sebsebe 5038 10/1995.
Note: more material required.

Oreosyce africana Hook.f.

Herbaceous climber to 3–4m; stems 1–2mm, ridged, setose, particularly when immature; leaves (5–)8–9 × (4–)6–7cm, shallowly 5-lobed, base cordate, apex mucronate, margins finely-dentate, surfaces densely setose, hairs white; petiole 4–10cm, setose; monoecious; flowers solitary or paired, axillary, both male and female flowers with pubescent yellow petals, elliptic, c. 1cm; calyx densely hairy with lanceolate lobes to 2mm; fruits ovoid, 1–2 × 1cm, densely hispid. Open disturbed ground, often at high elevations; 940m.
Distr.: Bioko & Cameroon, E & S Africa. (Afromontane).
IUCN: LC
Nyasoso: Cheek 7508 10/1995.
Note: the single specimen from Mt Kupe was recorded at a significantly lower altitude than at which this species is more commonly found (usually > 1500m).

Raphidiocystis mannii Hook.f.

Herbaceous climber; stems thin, ridged, glabrous or finely-pubescent; leaves (3–)5-lobed, c. 11 × 7cm, base deeply cordate, apex acuminate, margin shallowly undulate with small mucrons at apices of veins, upper surface scabrid; petioles c. 6.5cm, pubescent; dioecious; male flowers in branched or unbranched axillary clusters; pedicels 0.5–2cm; female flowers solitary or paired; sepals in both sexes strongly pinnatipartite, individual lobes lanceolate; petals 0.7 × 0.5cm (male), 2 × 1cm (female), yellow-orange with acute apex; fruits broadly ellipsoid, c. 4.5 × 3.5cm, densely orange-red hairy; calyx often persistent. Forest, forest margins; 300–1450m.

Distr.: Nigeria, Bioko & Cameroon. (Lower Guinea).
IUCN: NT
Kodmin: Cheek 9579 11/1998; **Kupe Village:** Cable 2548 5/1996; Cheek 7710 11/1995; Etuge 1911 5/1996; 2771 7/1996; Kenfack 224 7/1996; **Ngomboaku:** Ghogue 492 12/1999; **Nyasoso:** Cable 3284 6/1996; Etuge 1753 3/1996.
Note: locally common in Bakossi.
Local name: Ndun-ehmohsay. **Uses:** unknown (Max Ebong in Cheek 9579).

Telfairia occidentalis Hook.f.

Herbaceous climber; stems ridged, glabrous or pubescent; leaves palmately compound, 3(–5)-foliolate, central leaflet obovate with attenuate base; lateral leaflets assymetric, apices apiculate, margins shallowly-dentate; petiole c. 4cm; male inflorescence on racemes c. 20cm, c. 6-flowered; peduncle pubescent towards apex; pedicels c. 1.5cm, subtended by stalked crenate bract, c. 0.5cm; calyx lobes crenate, projected outwards, c. 0.5cm, pubescent; petals c.1.5 × 0.5cm; fruits large, ellipsoid, grey-green, ridged. Secondary vegetation, often cultivated; 840–1000m.

Distr.: tropical Africa.
IUCN: LC
Enyandong: Gosline 347 11/2001; **Kupe Village:** Cheek 7110 1/1995.
Local name: Ekongobong (*fide* Etuge). **Uses:** FOOD – cultivated for its edible leaves and, less commonly, its edible seeds (Cheek 7110 & *fide* Etuge).

Trichosanthes cucumerina L. subsp. *anguina* (L.) I.Grebenščikov

Verz. Landwirtsch. U. Gartn. Kulturpfl., Auf. 2: 928 (1986).
Herbaceous climber; stems ridged, sparsely pubescent; leaves ovate-orbicular in outline, 7–10 × 8–12cm, deeply 5(–7)-lobed, lobe apices rounded, leaf base cordate, margins irregularly sinuous-denticulate, surfaces sparsely pubescent; petiole 2–8cm, puberulent; male inflorescence a few-flowered raceme; peduncle c. 10cm, pubescent; pedicels 1–2cm; sepals triangular, 3mm; petals oblong, c. 1cm long, white, long-fimbriate; female flowers solitary on pedicels 0.5–5cm; fruit linear, to 1(–2)m long, c. 5cm diameter, green with paler streaking, turning orange. Cultivated; 1170m.

Distr.: palaeotropics, originating from Asia.
IUCN: LC
Edib village: Cheek, photographic record 2/1998.
Local names: Snake Gourd or Tomato. **Uses:** FOOD – infructescences – young fruits edible (*fide* Cheek).

Zehneria keayana R.Fern. & A.Fern.

Fl. Cameroun 6: 43 (1967).
Herbaceous climber to 4m; stems to 1mm, ridged, glabrous; leaves c. 6.5 × 4.5cm, ovate, base deeply cordate, apex acuminate, margin dentate, irregularly scabrid on upper surface; petioles 2–2.5cm; dioecious; male flowers on axillary racemes 4–5cm long, 1–2 per axis; pedicels c. 2mm, persistent; flowers c. 1mm, with white conical petals; calyx glabrous; female flowers and fruits not observed. Open forest near water; 1350m.

Distr.: Liberia, Ghana, Nigeria to Congo (K). (Guineo-Congolian).
IUCN: LC
Kodmin: Cheek 8980 1/1998.
Note: uncommon over most of its range.

Zehneria minutiflora (Cogn.) C.Jeffrey

Fl. Cameroun 6: 37 (1967).
Syn.: *Melothria minutiflora* Cogn. var. *hirtella* Cogn.
Syn.: *Melothria minutiflora* Cogn. var. *minutiflora*
Syn.: *Melothria minutiflora* Cogn. var. *parviflora* Cogn.
Herbaceous climber; stems c. 1mm, ridged, minutely pubescent; leaves c. 4.5 × 4cm shallowly 3-lobed, base cordate, apex acute or acuminate, margin shallowly-dentate, surfaces with white hairs to 1mm; petiole 2–5cm, pubescent; dioecious; male inflorescence not observed; female flowers solitary; pedicels 2.5–3cm long, thin; petals white (2–3mm), ovary ellipsoid c. 3mm long; fruit ellipsoid, c. 2 × 0.8cm, orange-red, smooth; seeds visible through thin epicarp. Disturbed vegetation including farmbush; 800–1450m.

Distr.: tropical & S Africa.
IUCN: LC
Kodmin: Cheek 9016 1/1998; Etuge 3987 1/1998; Satabie 1113 fl., 1/1998; **Mwambong:** Gosline 86 2/1998; **Ngomboaku:** Etuge 4671 12/1999.
Local names: Ndun or Ndum. **Uses:** MEDICINES (Cheek 9016).

Zehneria scabra (L.f.) Sond.

Fl. Cameroun 6: 44 (1967).
Syn.: *Melothria punctata* (Thunb.) Cogn.
Syn.: *Melothria mannii* Cogn.
Herbaceous climber or trailer; stems 1–2mm, ridged, scabrid around nodes; leaves 5–6(–9) × 3.5–4(–7.5)cm, ovate, base deeply-cordate, apex acute or acuminate, margin dentate, scabrous, particularly on upper surface, drying blackish; dioecious; male umbels c. 20-flowered; peduncle 1–3cm, finely-pubescent; pedicels 0.5cm; female umbels 5(–10)-flowered; peduncle 1–2mm; flowers 1–2mm; calyx pubescent, blackish; petals white-yellow; fruit orbicular, 0.5–1cm diameter, orange when mature, finely-pitted. Secondary vegetation, farmland; 600–2000m.

Distr.: tropical Africa & Asia. (Palaeotropics).
IUCN: LC
Kodmin: Cheek 9044 1/1998; 9566 fl., 11/1998; **Kupe Village:** Cheek 7146 fl., fr., 1/1995; Lane 316 fr., 1/1995; **Mwanenguba:** Cheek 9468 10/1998; Leeuwenberg 8448 fl., fr., 9/1971; **Ndum:** Groves 40 fr., 2/1995; **Nyasoso:** Etuge 1636 1/1996; Sidwell 406 10/1995.
Local names: Ndun or Ndum. **Uses:** many, including MEDICINES (Cheek 9566).

DICHAPETALACEAE

F.J. Breteler & M. Cheek

Fl. Cameroun 37 (2001).

Note: *Dichapetalum affine, D. minutiflorum, D. mombuttense, D. oblongum, D. tetrastachyum, D. umbellatum,* and *D. zenkeri* all occur on the edge of our checklist area and are likely to be found within it in future.

Most were collected at Bakaka Forest; km 11 Nkongsamba-Loum, by Leeuwenberg.

Dichapetalum altescandens Engl.
Fl. Cameroun 37: 12 (2001).
Climber; stems dark brown-black, puberulent; leaves elliptic to obovate, 8–12 × 3–4cm, acuminate, obtusely rounded or cuneate at base; 5–6 pairs of nerves, sparsely-hairy on nerves; petiole 0.2–0.4mm; stipules caducous; inflorescence axillary, c. 25-flowered, compact; peduncle 0.4–2cm; flowers 3–5mm; fruit globose, to 3-lobed, 1.5–4cm. Forest; 900m.
Distr.: Cameroon & Gabon. (Lower Guinea).
IUCN: VU
Mt Kupe: Thomas 5483 2/1986.

Dichapetalum angolense Chodat
Climber, c. 15m; stems and petioles shortly, softly, golden-hairy; leaves obovate or elliptic, c. 19 × 10cm, acute, rounded; nerves c. 9 pairs, raised; dark brown, softly-hairy below; petiole 1.5cm; stipules caducous; inflorescence axillary, c. 8 × 8cm; fruit ellipsoid or globose, 2cm; indumentum as stem. Forest; 900–1450m.
Distr.: Ivory Coast to Sudan, Uganda & Angola. (Guineo-Congolian).
IUCN: LC
Kodmin: Cheek 8941 1/1998; **Nyasoso:** Cheek 7536 10/1995; 9283 10/1998; Etuge 2380 6/1996.

Dichapetalum choristilum Engl.
Fl. Cameroun 37: 26 (2001).
Climber, 20m; stems and petioles sparingly scurfy-black hairy; leaves drying black above, elliptic, 7–11 × 4–4.5cm, acuminate, obtuse; lateral nerves 4–6 pairs; petiole 0.3–1cm; stipules filiform, 2mm, fairly caducous; inflorescence axillary, c. 1.5 × 3cm; flowers white, 7mm. Forest; 250–1400m.
Distr.: Liberia to Congo (K). (Guineo-Congolian).
IUCN: LC
Kurume: Thomas 5466 fl., 1/1986; **Ngomboaku:** Cheek 10383 12/1999; **Nyasoso:** Wheatley 463a fl., fr., 7/1992.

Dichapetalum gabonense Engl.
Fl. Cameroun 37: 43 (2001).
Syn.: *Dichapetalum nitidulum* Engl. & Ruhl.
Shrub 3m; stems glabrous, brown; leaves elliptic, c. 10 × 5cm, shortly acuminate, acute to rounded; nerves c. 5 pairs; petiole c. 1cm; stipules caducous; inflorescence axillary or petiolar, c. 2 × 3cm; flowers greenish white, 5mm. Forest; 200–250m.
Distr.: Nigeria to Congo (B) & Cabinda. (Lower Guinea).
IUCN: LC
Baduma: Nemba 226 9/1986; **Mungo River F.R.:** Biye 64 12/1999.

Dichapetalum heudelotii (Planch. ex Oliv.) Baill. var. *heudelotii*
Fl. Cameroun 37: 50 (2001).
Syn.: *Dichapetalum subauriculatum* (Oliv.) Engl.
Syn.: *Dichapetalum ferrugineum* Engl.
Syn.: *Dichapetalum johnstonii* Engl.
Climber, 12m; stem, petiole and midrib above shortly red-brown-hairy, midrib white-hairy below; leaves narrowly elliptic-oblong, c. 17 × 6cm, acute, rounded; nerves c. 9 pairs; stipules caducous; inflorescence axillary, dense, subsessile c. 0.6 × 0.6cm; flowers white, 5mm. Forest; 300–1500m.
Distr.: Guinea (C) to Congo (K). (Guineo-Congolian).
IUCN: LC

Bolo: Nemba 2 fl., 4/1986; **Kupe Village:** Cheek 8399 5/1996.

Dichapetalum heudelotii (Planch. ex Oliv.) Baill. var. *longitubulosum* (Engl.) Breteler
Fl. Cameroun 37: 54 (2001).
Syn.: *Dichapetalum scabrum* Engl.
Syn.: *Dichapetalum longitubulosum* Engl.
Climber, differing from var. *heudelotii* in the sepals 3.5–5mm (not 2–3mm), united for 2/5–2/3 their length (not united only at base); petals and stamens 4.5–9mm, united in tube 3.5mm long (not 3.5–6mm, free). Forest; 250–500m.
Distr.: SE Nigeria & SW Cameroon. (Lower Guinea).
IUCN: LC
Baduma: Nemba 187 8/1986; **Nlog:** Etuge 11 4/1986.
Note: specimen not at K, det. by Breteler at WAG.

Dichapetalum heudelotii (Planch. ex Oliv.) Baill. var. *ndongense* (Engl.) Breteler
Fl. Cameroun 37: 57 (2001).
Climber, differing from other varieties of this species most notably by the short petals and stamens, 2–3mm and 1.5–2.5mm (not both 3.5–6mm). Forest; 400m.
Distr.: Guinea (C) to Congo (K). (Guineo-Congolian).
IUCN: NT
Ndoungue: Ledermann 6296 fl., 9–11/1909.
Note: the cited specimen refers to the type of this taxon.

Dichapetalum aff. heudelotii (Planch. ex Oliv.) Baill.
Climber to 8m, resembling *D. heudelotii*, but stems glabrous, red-brown, with conspicuous, raised, white lenticels; leaves larger to 28 × 11cm acuminate, base auriculate, c. 4 pairs glands conspicuous on surface near lower midrib. Forest; 250–900m.
Mt Kupe: Cheek 10133 11/1999; **Nyasoso:** Etuge 2361 6/1996.
Note: identified as '*D. ? heudelotii*' by Breteler. More material, especially flowers, needed to elucidate.

Dichapetalum insigne Engl.
Fl. Cameroun 37: 57 (2001).
Shrub, at length climbing, 4–6m; stem white-hairy; leaves elliptic, c. 18 × 7cm, acuminate, asymmetrically rounded-obtuse; petiole 0.2–0.6cm; stipules persistent, triangular c. 15 × 4mm; inflorescence axillary; flowers white, 2mm; fruits cylindrical. Forest; 200m.
Distr.: Cameroon to Angola. (Lower Guinea).
IUCN: LC
Mt Kupe: Cheek 10170 11/1999.
Note: unusually in this genus, this species is often a shrub, as opposed to displaying the typical climbing habit of other species in the genus.

Dichapetalum madagascariense Poir. var. *madagascariense*
Fl. Cameroun 37: 70 (2001).
Syn.: *Dichapetalum floribundum* (Planch.) Engl.
Syn.: *Dichapetalum guineense* (DC.) Keay
Syn.: *Dichapetalum thomsonii* (Oliv.) Engl.
Shrub or climber, to 30m; stems hollow, minutely orange-puberulent; leaves glossy above and below, elliptic, c. 17 × 8cm, shortly acuminate, obtuse; nerves c. 5 pairs; petiole c. 1cm, drying black; stipules caducous; inflorescence axillary, diffuse, c. 8 × 13cm; flowers white, 4mm; fruits 1.5cm, warty. Forest; 720–1000m.

Distr.: Sierra Leone to Madagascar & Comores. (Tropical Africa & Madagascar).
IUCN: LC
Kupe Village: Etuge 2770 7/1996; Kenfack 293 7/1996; 321 7/1996; **Menyum:** Doumenge 587 5/1987.

Dichapetalum mundense Engl.

Fl. Cameroun 37: 82 (2001).
Climber 8m, glabrous; leaves glossy, drying black, elliptic to 15 × 5cm, acuminate, obtuse; nerves c. 5 pairs; petiole 0.5cm; stipules caducous; inflorescence axillary, gracile, c. 3 × 3cm; flowers white, c. 2mm. Forest; 350–1400m.
Distr.: Nigeria to Congo (K). (Lower Guinea & Congolian).
IUCN: NT
Kodmin: Etuge 4424 11/1998; **Mekom:** Nemba 20 fl., 4/1986.
Note: only 6 other specimens at 2 sites listed for Cameroon in Fl. Cameroun 37.

Dichapetalum pallidum (Oliv.) Engl.

Climber differing from other *Dichapetalum* species by the elliptic leaves, 14–16 × 4–5cm, long-acuminate, blades white below, apart from the brown nerves, margin undulate. Forest; 250–1750m.
Distr.: Guinea (C) to Congo (K). (Guineo-Congolian).
IUCN: LC
Kodmin: Cheek 8942 1/1998; Plot B28 1/1998; **Mile 15, Kumba-Mamfe Road:** Nemba 210 9/1986; **Mt Kupe:** Thomas 5482 2/1986.

Dichapetalum rudatisii Engl.

Climber or shrub; stems solid, densely red-brown powdery-puberulent; leaves elliptic to obovate, 9–16 × 3–9cm; acuminate, rounded-cuneate; nerves 5–8 pairs, glabrescent; petiole 3–5mm; stipules subulate, 1–3mm, caducous; inflorescence petiolar, subsessile, subumbellate, up to 30-flowered; flowers white, 6mm; fruit hairy. Forest; 250m.
Distr.: SE Nigeria to Gabon. (Lower Guinea).
IUCN: LC
Konye: Nemba 438 1/1987; **Manyemen:** Leeuwenberg 10328 fl., fr., 9/1972; **Mile 15, Kumba-Mamfe Road:** Nemba 208 9/1986.

Dichapetalum tomentosum Engl.

Shrub 2–3m, densely green-grey puberulent; leaves elliptic or obovate-elliptic, c. 7 × 3cm, broadly acute at base and apex; c. 4–5 pairs of nerves, densely puberulent, pale grey-green below; petiole 0.6cm; inflorescence axillary, 3cm; peduncle 2cm; flowers subsessile in clusters; fruit ellipsoid, warty, 2cm. Forest; 250–1150m.
Distr.: SE Nigeria to Congo (K). (Lower Guinea).
IUCN: LC
Mile 15, Kumba-Mamfe Road: Nemba 215 9/1986; **Nyasoso:** Cable 2853 6/1996; Zapfack 677 6/1996.

Dichapetalum unguiculatum Engl.

Fl. Cameroun 37: 114 (2001).
Climber resembling *D. tomentosum*, but leaves c. 12 × 4cm, shortly acuminate, dull grey-brown below. Forest; 750m.
Distr.: Cameroon to Congo (K). (Lower Guinea & Congolian).
IUCN: LC
Nyasoso: Etuge 2600 7/1996.

Dichapetalum sp. 1

Climber, densely dark brown pubescent; leaves oblanceolate, c. 23 × 10cm, apex rounded-mucronate, base cuneate; subscabrid below, four pairs of bright-red glands inserted near the base of the midrib on the lower surface, amongst dense red-brown pubescence; nerves c. 12 pairs. Forest; 1000m.
Distr.: Cameroon. (Lower Guinea).
Nyasoso: Cheek 9517 10/1998.
Note: not matching any other *Dichapetalum* in our area. Fertile material required to elucidate further.
Local name: Nyon Nyo. **Uses:** MEDICINES (Epie Ngome in Cheek 9517).

Tapura africana Oliv.

Tree, c. 15m, grey puberulent; leaves obovate to oblanceolate, drying glossy grey-black, c. 20 × 7cm, subacuminate, unequally cuneate; nerves c. 3 pairs; petiole 1–2cm; stipules caducous; inflorescence petiolar, fasciculate, c. 0.5cm diameter. Forest; 420–800m.
Distr.: Nigeria to Gabon. (Lower Guinea).
IUCN: NT
Bakossi F.R.: Cheek 9322 10/1998; **Eboné:** Leeuwenberg 9910 fr., 5/1972; **Kupe Village:** Cheek 7744 11/1995; Etuge 2663 7/1996; **Mungo River F.R.:** Cheek 9353 10/1998; Onana 954 11/1999.

DILLENIACEAE

M. Cheek

Tetracera cf. alnifolia Willd. var. *podotricha* (Gilg) Staner

Syn.: *Tetracera podotricha* Gilg
Liana; stem twisting, scabrid; leaves drying brown, elliptic, 14 × 8cm, apex rounded, mucronate, base rounded, upper surface smooth, lower scabrid; lateral nerves c. 10 pairs, quaternary nervation visible; petiole 1.5cm, rim long patent-hairy. Forest; 1150m.
Distr.: Nigeria to Congo (K). (Lower Guinea & Congolian).
Kodmin: Cheek 9200 2/1998.
Note: flowering material is needed.

Tetracera cf. stuhlmanniana Gilg

Syn.: *Tetracera potatoria* Afzel. ex G.Don
Liana resembling *T. cf. alnifolia*, but leaves drying green below, elliptic, c. 26 × 12cm, apex acute, base cuneate, margin serrate in distal ½ to ⅔, scabrid above and below; petiole 1.5cm; lateral nerves c. 17 pairs. Forest; 840m.
Distr.: Guinea (B) to Congo (K). (Guineo-Congolian).
Kupe Village: Cheek 7134 1/1995.

DROSERACEAE

M. Cheek

Drosera madagascariensis DC.

Carnivorous herb, c. 8cm tall; stem aerial, decumbent; leaves marcescent, obovate, c. 7 × 4mm, rounded, acute, upper surface with long-stalked glandular-hairs on margins, sessile hairs at centre; petiole 1–5cm; stipules ovate, 3mm, laciniate; inflorescence raceme-like, 15cm, rachis 2cm, c. 5-flowered; flowers pink, 5mm. *Sphagnum* bed; 1200–1250m.
Distr.: tropical & subtropical Africa, Madagascar.
IUCN: LC
Lake Edib: Cheek 9131 fr., 2/1998; Gosline 188 fl., fr., 11/1998.

EBENACEAE

G. Gosline

Fl. Cameroun 11 (1970).

Diospyros bipindensis Gürke

Shrub or tree 3–20m; bark very hard, shattering when struck; branchlets frequently with ellipsoid swellings c. 2–3cm diameter; leaves glabrous, oblong, 10–15(–26) × 4–6(–10)cm, acumen c. 2.5cm long, with 5–8 pairs lateral nerves, anastomosing some distance from leaf margin; petioles 5–10(–15)mm; flowers axillary, solitary or in small panicles of 2–3; male flower calyx c. 4mm, 4 lobes united; petals c. 2cm, united to 1/2; female flower calyx c. 10mm, 4 triangular lobes; corolla 18–20mm with 4 reflexed lobes; fruits with accrescent 4-lobed calyx 2.5cm long; fruit as long as calyx, 7–8 seeded. Forest; 150–950m.
Distr.: Cameroon to Uganda. (Lower Guinea & Congolian).
IUCN: LC
Bakossi F.R.: Etuge 92 fl., 3/1986; **Etam:** Etuge 220 fr., 8/1986; **Kupe Village:** Gosline 243 12/1999; **Mile 15, Kumba-Mamfe Road:** Nemba 126 fr., 7/1986; **Mungo River F.R.:** Gosline 120 10/1998; Onana 969 12/1999; **Nyale:** Etuge 4226 2/1998; **Nyasoso:** Gosline 131 10/1998.
Local name: Megire = 'The Bell of a Dog' (Gosline 120).

Diospyros canaliculata De Wild.

Tree to 20m, 0.6–1m d.b.h.; bole straight; bark black, almost smooth, brittle; slash black outside, pale pink inside, turning deep greenish yellow; leaves drying green-brown to brown-black, to 22 × 9cm, acuminate, acumen tip rounded, 8–11 lateral nerves, veins closely reticulate, raised on both surfaces; male flowers on short pedicels, mostly in clusters on main stem, a few axillary; females in clusters on stem; pedicel to 1cm; male calyx to 3mm; corolla 1cm with a narrow tube and four large lobes; female calyx 1cm, with 4 ridges alternating with short rounded lobes; long-stalked fruits in clusters on branches; fruiting calyx 4 × 2cm, concealing the fruit, 4 ridges distinct. Forest; 750m.
Distr.: Liberia to Congo (K). (Guineo-Congolian).
IUCN: LC
Kupe Village: Lane 434 fl., 1/1995.

Diospyros cinnabarina (Gürke) F.White

Syn.: *Diospyros simulans* F.White
Tree to 15m; bark smooth, black or dark grey, with superficial vertical striations; slash black outside, red and fibrous inside; leaves ovate-elliptic to lanceolate, 7–24 × 3–15cm, (sub)acuminate, with 5–7 pairs lateral nerves, lower surface distinctly pale grey; flowers in crowded cymes, 5–100-flowered, generally on old wood; male flowers 10 × 2mm, sharply conical; female flower 7.5mm with 3 lobes 1/3 length of the corolla; fruit globose to ovoid, red, 3–4 × 2–3cm, apex pointed, surface warty when dried, sometimes deeply grooved between the seeds. Forest; 710–1000m.
Distr.: E Nigeria to Congo (B). (Lower Guinea).
IUCN: NT
Kupe Village: Cheek 7165 fl., 1/1995; **Ndibise:** Etuge 523 fr., 6/1987; **Ngomboaku:** Gosline 255 12/1999; **Nyale:** Etuge 4224 2/1998; 4229 2/1998; 4473 11/1998.
Note: uncommon over most of its range.

Diospyros conocarpa Gürke & K.Schum.

Tree to 10m; trunk slender; bark black, with thin papery scales; slash black outside, pale yellow and fibrous inside; branchlets with bristle-like hairs; leaves harbouring ants; lamina papery, to 25 × 8cm, acuminate, base cordate, bristle-like hairs on midrib beneath, with 7–9 pairs lateral nerves; inflorescence axillary; male flowers in lax drooping 3–5-flowered racemes; female flowers solitary; calyx 1–2cm with 4 papery triangular lobes free almost to the base; corolla 2cm, tube 1cm, lobes 4, large; fruits to 3.5 × 1.5cm, ovoid-conical with few bristly hairs; fruiting calyx 1cm with 4 papery lobes. Forest; 200–500m.
Distr.: Nigeria to Congo (K). (Guineo-Congolian).
IUCN: LC
Bakossi F.R.: Gosline 119 10/1998; **Mile 15, Kumba-Mamfe Road:** Nemba 151 fr., 7/1986; 607 7/1987; **Mungo River F.R.:** Onana 948 11/1999; **Nlog:** Etuge 3 fl., 4/1986; **Nyandong:** Cheek 11352 3/2003.

Diospyros dendo Welw. ex Hiern

Tree to 15m; leaves oblong-lanceolate to elliptic, 10–15(–18) × 4–5(–7)cm, acuminate, often 2 glands near base of blade, with 5–6 pairs lateral nerves, arcing, lamina finely puberulent at base and apex; petiole 5–6mm; flowers axillary in contracted cymes of 5–20 male or 3–5 female flowers; males 5–6mm long; calyx pubescent; stamens numerous, exserted; females 8–10mm long; fruits ovoid, 1.5 × 1.2cm, surrounded by accrescent calyx, with 4–6 triangular lobes, 3 × 16cm, with undulate margins, either open in a star-shape or folded around fruit. Forest; 220–1500m.
Distr.: Nigeria to Congo (K). (Guineo-Congolian).
IUCN: LC
Baduma: Nemba 220 fr., 9/1986; **Bakolle Bakossi:** Etuge 260 fr., 9/1986; **Kodmin:** Etuge 3999 1/1998; Plot B60 1/1998; **Nyale:** Etuge 4474 11/1998.

Diospyros iturensis (Gürke) Letouzey & F.White

Fl. Cameroun 11: 93 (1970).
Tree to 20m; bark black with irregular fissures; slash black outside, reddish inside; leaves elliptic or lanceolate-elliptic, to 25 × 8cm, acumen tip rounded, with 7–9 pairs lateral nerves, deeply incised above, prominent below; flowers in 1–9-flowered clusters in the axils of twigs and older wood; calyx 2mm; corolla 7mm, with 3 short lobes; fruits yellow-orange, c. 2cm, globose, glabrous; fruiting calyx 0.5cm, irregularly lobed. Forest; 250m.
Distr.: Nigeria to Congo (K). (Lower Guinea & Congolian).
IUCN: LC
Ikiliwindi: Nemba 110 fr., 6/1986; **Mile 15, Kumba-Mamfe Road:** Nemba 143 fr., 7/1986.

Diospyros kamerunensis Gürke

Tree to 15m; stems with red-gold pubescence; leaves lanceolate, oblong or elliptic, 15–20 × 5–9cm, acumen to 2.5cm, acute, base rounded, young leaves densely red-gold pubescent, with 4–6(–8) pairs lateral nerves, tertiary nerves indistinct; petiole 1–2cm, pubescent; male flowers in axillary panicles of 3–6(–20) flowers, pubescent; calyx 8mm, deeply lobed; corolla 18–20mm; female flowers solitary or few in axils, larger than male; fruit on thick pedicel c. 5mm, subglobose, 5 × 4cm, red-orange, glabrescent. Forest; 250m.
Distr.: Liberia to Ghana, Cameroon, Gabon. (Upper & Lower Guinea).
IUCN: NT
Loum F.R.: Cheek 10258 12/1999.
Note: widespread but uncommon over its range.

Diospyros kupensis Gosline

Kew Bull. 53: 461 (1998).
Treelet 1.5–3(–5)m with verticillate branching, resembling *D. conocarpa*, but leaves not harbouring ants, base acute to obtuse, not cordate; branches and midrib appressed-tomentose, not hispid; male inflorescence a 1–5-flowered

glomerule, not racemose; calyx lobes ligulate, 2.5cm, completely concealing corolla which is divided to only ⅓ its length. Forest; 460–1300m.

Distr.: Kupe-Bakossi. (Narrow Endemic).

IUCN: VU

Enyandong: Etuge 4396r 10/2001; **Kupe Village:** Abu Juam EW95 81 fr., 11/1995; Cable 751 1/1995; Cheek 6064 fl., 1/1995; Etuge 1369 11/1995; 1723 fr., 2/1996; 2607 fl., 7/1996; 3800 fl., 4/1997; **Mt Kupe:** Gosline 36 fr., 1/1997; **Nyale:** Etuge 4220 2/1998; 4456 fl., 11/1998; **Nyandong:** Cheek 11409 3/2003.

Diospyros mannii Hiern

Tree to 30m; leaves elliptic, 10–12(–20) × 3–5(–7)cm, acumen acute, laminae glabrescent, with 6–8 pairs lateral nerves ascendant, nerves hirsute then glabrescent; petiole 6–8mm; male inflorescence a dense cyme of up to 30 flowers, on old wood, or axillary; female flowers in lax cymes of 2–3 flowers; male flowers c. 1.5cm long; calyx densely pubescent; female flowers to 2.5cm long, with pedicel 1cm, pubescent; fruits on robust pedicels 1–2(–3)cm long, subglobose 5(–9) × 4.5(–8)cm, orange, initially covered with irritant hairs; calyx lobed 3(–6)cm diameter. Forest; 250–280m.

Distr.: Sierra Leone to Congo (K). (Guineo-Congolian).

IUCN: LC

Baduma: Nemba 231 9/1986; **Ikiliwindi:** Nemba 113 6/1986; **Karume:** Nemba 70 5/1986.

Diospyros melocarpa F.White

Tree to 15m; leaves often drying red-brown, ovate-elliptic to lanceolate-elliptic, 9 × 5.5cm, acumen c. 1.5cm with parallel sides and rounded apex, with 3–5 pairs lateral nerves ascendent, tertiary nerves not visible; petiole c. 8mm; male flowers subsessile, axillary in groups of up to 20; calyx 2.5mm; corolla conical, 8mm, lobes 3, 1mm; fruits subglobose c. 3.5cm diameter, yellow, glabrous; calyx minimal. Forest; 200m.

Distr.: Nigeria to Congo (K). (Lower Guinea & Congolian).

IUCN: NT

Baduma: Nemba 567 7/1987.

Note: uncommon to rare over much of its range.

Diospyros monbuttensis Gürke

Tree to 10m with spreading crown; bole short, fluted; bark papery, purple-brown to yellow-brown, peeling off in papery scales; slash pale yellow; often with large spines on trunk and branches; leaves papery, oblanceolate, to 25 × 10cm, shortly acuminate, with 5–9 pairs lateral nerves; male clusters 3–5-flowered at the apex of flattened internodal peduncles; female flowers solitary; calyx c. 1cm, completely enclosing the corolla in bud, splitting open irregularly; fruits yellow or orange, about 2cm long, broadly ellipsoid or globose, surrounded in lower half by the cup-shaped calyx with papery irregular margin. Forest; 1000m.

Distr.: Ivory Coast to Congo (K). (Guineo-Congolian).

IUCN: LC

Mwambong: Etuge 4133 fl., fr., 2/1998.

Diospyros preussii Gürke

Tree to 9m; bark black with shallow vertical fissures; slash black outside, pale yellow inside; leaves oblanceolate or oblanceolate-elliptic, to 40 × 14cm, shortly acuminate, glabrous, with 6–9 pairs lateral veins, tertiary veins densely reticulate; flowers crowded in clusters of up to 30+ on branches and main stem; male calyx 2mm; corolla 14mm, lobes 4, 3mm; female calyx 4mm with 4 heart-shaped lobes;

fruit ellipsoid, 15 × 12mm, hidden within deep purple calyx with undulate margins. Forest, streambanks; 200–850m.

Distr.: SE Nigeria, Cameroon & Gabon. (Lower Guinea).

IUCN: LC

Kupe Village: Lane 279 1/1995; 396 fl., 1/1995; **Manehas F.R.:** Gosline 129 10/1998; **Mombo:** Olorunfemi in FHI 30575 5/1951; **Mungo River F.R.:** Cheek 10168 11/1999; **Nyasoso:** Gosline 130 10/1998.

Diospyros pseudomespilus Mildbr. subsp. pseudomespilus

Bull. Jard. Bot. Nat. Belg. 48: 312 (1978).

Syn.: *Diospyros undabunda* Hiern ex Greves

Tree 2–35m; bark dark grey to black, with longitudinal striations; slash brownish yellow inside; leaves (sub) oblanceolate, 18(–30) × 6(–10)cm, shortly acuminate, with 7–12 pairs lateral nerves; male flowers in cymes of 3–5, axillary or on old wood; females solitary or 2–3-flowered; calyx 8mm long; corolla 12–15mm; fruits subsessile, to 4cm diameter; calyx star-shaped, 2.5–4.5cm, with 4–5 lanceolate lobes, margins undulate. Forest; 1000m.

Distr.: Nigeria & Cameroon. (Lower Guinea).

IUCN: NT

Kupe Village: Etuge 1482 fr. 11/1995; **Nyale:** Etuge 4471 11/1998.

Local name: Ebony. **Uses:** MATERIALS – wood – timber and furniture (Etuge 1482).

Diospyros suaveolens Gürke

Tree 8–30m; bark scaly, deep purplish black; slash black outside, deep yellow turning orange inside; leaves lanceolate-elliptic, to 8(–14) × 2(–5)cm, acuminate, base rounded or subcordate, with c. 7 pairs of lateral nerves, distinctly hairy; flowers in lax many-flowered cymes on branches and main stems, never axillary; calyx 4mm, hairy; corolla 10mm, glabrous, 5-lobed; fruits broadly ovoid-conical, 35 × 22mm, densley covered with short brown hairs and longer, yellow irritant hairs; calyx 10mm with 4–5 tringular lobes as long as the tube. Forest; 150–300m.

Distr.: Nigeria, Cameroon & Gabon. (Lower Guinea).

IUCN: NT

Etam: Etuge 76 fl., 3/1986; **Kupe Village:** Etuge 2793 7/1996; **Loum F.R.:** Gosline 118 10/1998; **Mt Kupe:** Cheek 10169 11/1999; **Mungo River F.R.:** Gosline 122 10/1998.

Local name: Save savele = 'smaller and smaller' (Gosline 118).

Diospyros viridicans Hiern

Tree to 25m; branching sympodial; leaves elliptic, 17 × 11cm, shortly acuminate, with 5–8 pairs lateral nerves, tertiary nerves parallel, perpendicular to laterals, drying reddish brown to olive-brown; petioles 1.5–2cm, canaliculate; male flowers in loose cymes; calyx 2.5mm, 4-lobed; corolla 8mm; stamens 20–30, exserted; female flowers similar; fruits globose, 2–3cm diameter, black at maturity; calyx with reflexed lobes 4–6 × 3mm. Forest; 280–300m.

Distr.: Sierra Leone to Congo (K). (Guineo-Congolian).

IUCN: LC

Bolo: Thomas 5714 fl., 3/1986; **Karume:** Nemba 68 fl., 5/1986.

Diospyros zenkeri (Gürke) F.White

Tree to 20m; bole black and smooth; slash black outside, pink with a yellow inner edge inside; leaves drying black, elliptic to ovate, to 10 × 5cm, acumen to 2.5cm, with 4–5 pairs lateral veins looping far from edge; male flowers in clusters of 1–9 in leaf axils or on older twigs, c. 1cm long, cylindric; calyx with 3 short lobes; female flowers in clusters

of 1–3, shorter and wider than males; fruits globose, yellow turning black, 2.5cm, glabrous; calyx not accrescent. Forest; 200–350m.
Distr.: Nigeria to Congo (K). (Lower Guinea & Congolian).
IUCN: LC
Bolo: Nemba 9 4/1986; **Ikiliwindi:** Etuge 502 3/1987; **Loum F.R.:** Gosline 117 10/1998; **Mungo River F.R.:** Gosline 121 10/1998.
Local name: M'kira. **Uses:** MATERIALS/SOCIAL USES – used for hunting to mark the trail, give success (Gosline 117).

ERICACEAE

M. Cheek

Fl. Cameroun 11 (1970).

Agarista salicifolia G.Don
Gen. Syst. 3: 837 (1834).
Syn.: *Agauria salicifolia* (Comm. ex Lam.) Hook.f. ex Oliv.
Shrub or small tree, to 13m; leaves oblong-lanceolate, acute at ends; inflorescence a short axillary raceme; flowers pink. Forest edge; 1800–2000m.
Distr.: Cameroon to Madagascar. (Afromontane & Madagascar).
IUCN: LC
Mt Kupe: Cheek 7596 11/1995; **Mwanenguba:** Etuge 4383 10/1998; Thomas 3122 fl., 2/1984.

Erica mannii (Hook.f.) Beentje
Utafiti 3: 13 (1990).
Syn.: *Philippia mannii* (Hook.f.) Alm & Fries
Heath-like shrub to 4m; leaves ascending to erect, in whorls of 3; few purplish flowers at end of each branchlet. Forest edge; 1980–2100m.
Distr.: Cameroon to Kenya. (Afromontane).
IUCN: LC
Mt Kupe: Cheek 7593 11/1995; Sebsebe 5089 11/1995; **Mwanenguba:** Leeuwenberg 9962 fl., 6/1972; Sanford 5517 fl., 11/1968; Thomas 3160 fl., 2/1984.
Note: the absence of *E. tenuipilosa* from Bakossi is unexpected.

EUPHORBIACEAE

M. Cheek, *Phyllanthus* by P. Hoffmann

Acalypha arvensis Poepp. & Endl. var. *arvensis*
Nov. Gen. Sp. Pl. 3: 21 (1841).
Herb 0.15–1m, erect, white pubescent; leaves ovate, 3.5–6 × 1.5–2.8cm, acute, base cuneate, rarely rounded, margin shallowly (< 1mm) serrate-crenate; petiole 1–3cm; female inflorescence shortly cylindrical, 2–4 × 1–2cm, with c. 7 bract-awns, 5–7mm, spreading, white-pilose; male inflorescences pendulous, slender, red, 2–4cm, usually few and inconspicuous. Farmbush; 200–500m.
Distr.: Brazil, alien elsewhere. (Pantropical).
IUCN: LC
Baduma: Nemba 626 9/1987; **Kupe Village:** Cable 2707 fl., 5/1996; **Loum F.R.:** Pollard 95 fl., 10/1998; **Mungo River F.R.:** Cheek 10185 fl., 11/1999; **Nlonako:** Etuge 1784 fl., 3/1996.

Note: previously confused with *A. manniana*; a low altitude alien, recently established (not in FWTA). Earliest Cameroonian collection seen dated 1970.
Uses: MEDICINES (Etuge 1784); none known (Epie Ngome and George Ngole in Pollard 95).

Acalypha brachystachya Hornem.
Herb 0.45m, stinging, inconspicuously puberulent; leaf blade broadly ovate, 5 × 4cm, shortly acuminate, rounded to truncate; petiole 5cm; female inflorescences sessile, axillary, of c. 8 subsessile flowers each subtended by a deeply 3-lobed bract; bracts narrowly triangular, 4mm; male inflorescences inconspicuous. Farmbush.
Distr.: tropical Africa & Asia. (Palaeotropics).
IUCN: LC
Nlonako: Etuge 1789 fl., 3/1996.

Acalypha manniana Müll.Arg.
Shrub 0.5-1.5m, resembling *A. arvensis*, but occurring at higher altitudes; leaves cordate at base, deeply (2mm) serrate-crenate; petiole to 6cm; female inflorescence with bract awns appressed, not spreading, 3–4mm; male inflorescences often to 8cm, sometimes with a female flower at base and then, in absence of the rarer female inflorescence, resembling *A. neptunica*. Forest, forest edge; 800–1500m.
Distr.: Nigeria, Cameroon, Rwanda & Burundi. (Afromontane).
IUCN: NT
Bangem: Thomas 5315 1/1986; **Kodmin:** Cheek 8910 fl., 1/1998; Ghogue 3 fl., 1/1998; **Kupe Village:** Cable 3755 fl., 7/1996; **Mwambong:** Cheek 9359 fl., 10/1998; **Nyasoso:** Etuge 2367 fl., 6/1996.

Acalypha ornata Hochst. ex A.Rich.
Shrub 2.5m, resembling *A. manniana*, but leaves to 12 × 8cm, base truncate-rounded and shortly, abruptly acuminate; male inflorescences to 5cm; Farmbush; 1000m.
Distr.: tropical Africa.
IUCN: LC
Nyasoso: Cable 3478 fl., 7/1996.
Note: specimens with female inflorescences and/or fruit are needed to confirm this identification.

Acalypha racemosa Baill.
Herb 1–1.5m, puberulent; leaves ovate, c. 9 × 4.5cm, acuminate, truncate-rounded, coarsely serrate; petiole c. 12cm; male inflorescence axillary, pendulous c. 12cm; female inflorescence a diffuse terminal panicle 18 × 9cm; fruit 2mm, stalked, glandular-hairy. Farmbush; 850–1000m.
Distr.: tropical Africa & Asia. (Palaeotropics).
IUCN: LC
Nyasoso: Cable 3491 fl., fr., 7/1996; Cheek 7505 fr., 10/1995; Etuge 2514 fr., 7/1996.

Alchornea cordifolia (Schum. & Thonn.) Müll.Arg.
Shrub to 8m; leaves ovate, c. 17 × 13cm, acute-acuminate, shallowly cordate, margin serrate, crenate, large nectar craters in axils of midrib and lower lateral nerves; minutely pubescent below; petiole c. 12cm; male inflorescence axillary, pendant, yellow, c. 20cm, highly-branched; female unbranched; styles 2, red, to 2cm long; fruit 1cm diameter, green. Roadsides, farmbush; 650–800m.
Distr.: widespread in tropical Africa.
IUCN: LC
Kupe Village: Cable 2690 fr., 5/1996; Cheek 7018 fl., 1/1995; Etuge 1977 fr., 5/1996; Lane 332 fl., 1/1995.
Local name: Abun. **Uses:** MEDICINES (Cheek 7018).

Alchornea floribunda Müll.Arg.

Shrub, resembling *A. cordifolia*, but leaves often subclustered, oblanceolate, to 35 × 12cm, acumen short, base cuneate then abruptly truncate, nerves 10+ pairs; petiole 1cm; male inflorescence terminal, erect, red; female inflorescence with styles 3, 2cm, red. Farmbush, secondary forest; 200–1700m.

Distr.: Senegal to Uganda. (Guineo-Congolian).

IUCN: LC

Bolo: Nemba 104 5/1986; **Ehumseh:** Manning 464 10/1986; **Ikiliwindi:** Nemba 464 1/1987; **Kodmin:** Plot B122 10/1998; **Kupe Village:** Etuge 2021 fr., 5/1996; Lane 275 fl., 1/1995; **Nyandong:** Ghogue 1438 fl., 3/2003; **Nyasoso:** Cable 2753 fr., 6/1996; 2794 fl., 6/1996; 2858 fr., 6/1996; Etuge 1758 fr., 3/1996.

Uses: ENVIRONMENTAL USES – boundaries/supports – stems used for fencing (Etuge 1758).

Alchornea hirtella Benth. fa. *glabrata* (Müll. Arg.) Pax & K.Hoffm.

Shrub 4–5m, resembling *A. floribunda*, but minutely pubescent; leaves drying brown, not green, oblanceolate or obovate-elliptic, c. 14 × 5.5cm, nerves less than 10 pairs; petiole 0.3–1cm. Forest; 800–1500m.

Distr.: Sierra Leone to S Africa. (Tropical Africa).

IUCN: LC

Bangem: Thomas 5292 fr., 1/1986; **Kodmin:** Barlow 13 fl., fr., 1/1998; Cheek 8915 fl., 1/1998; Etuge 4427 fl., 11/1998.

Alchornea hirtella Benth. fa. *hirtella*

Shrub resembling fa. *glabrata*, but young shoots and petioles patent-pubescent, not minutely puberulent, midrib and main veins evenly to sparingly setose, not glabrous. Forest; 800m.

Distr.: Guinea (B) to Tanzania & Zambia. (Tropical Africa).

IUCN: LC

Bangem: Thomas 5291 1/1986.

Antidesma laciniatum Müll.Arg. var. *laciniatum*

Tree 8–10m, puberulent; leaves obovate-oblong, 15(–30) × 8(–15)cm, subacuminate, base abruptly rounded; petiole 0.5cm; stipules ovate, 3–6-lobed, the lobes lanceolate; infructescence non-interrupted, pendulous racemose, c. 15cm; fruits ellipsoid, 7mm, red, fleshy. Forest; 200–1275m.

Distr.: Ivory Coast, Nigeria to Congo (K). (Guineo-Congolian).

IUCN: LC

Ikiliwindi: Nemba 545 6/1987; **Kupe Village:** Etuge 2024 fr., 5/1996; 2657 fr., 7/1996; **Ndibise:** Etuge 533 fr., 6/1987; **Nyasoso:** Cable 1165 2/1995; Sunderland 1493 fr., 7/1992; Wheatley 414 fr., 7/1992.

Antidesma laciniatum Müll.Arg. var. *membranaceum* Müll.Arg.

Tree resembling var. *laciniatum*, but pubescent, especially on stems, petioles, midribs and stipules; stipule lobes 6–9, filiform. Forest; 1000–1200m.

Distr.: Guinea (C) to Congo (K). (Guineo-Congolian).

IUCN: LC

Kodmin: Etuge 4169 fr., 2/1998; **Kupe Village:** Cable 2601 fr., 5/1996; Ryan 267 fr., 5/1996; **Nyasoso:** Cable 2857 fr., 6/1996; 2892 fr., 6/1996; 3445 fr., 7/1996; Etuge 1809 fl., 3/1996; 2113 fr., 6/1996.

Antidesma venosum Tul.

Tree 6m, resembling *A. laciniatum*, but leaves obovate, c. 10 × 6.5cm, apex rounded or truncate, base obtuse; stipules entire; infructescence branched. Farms; 800–1300m.

Distr.: tropical Africa.

IUCN: LC

Bangem: Thomas 5247 fl., 1/1986; **Kodmin:** Cheek 9532 fr., 10/1998.

Note: left uncleared on farms (Cheek 9532).

Antidesma vogelianum Müll.Arg.

Tree or shrub 3–10m, pubescent; leaves elliptic, rarely slightly obovate, c. 15 × 5.5cm, acumen c. 2cm, base acute; petiole c. 5mm; stipule lanceolate, c. 6 × 2mm, entire; infructescence as *A. laciniatum*. Forest; 200–1400m.

Distr.: Nigeria to Tanzania. (Lower Guinea & Congolian).

IUCN: LC

Bolo: Nemba 63 fl., 3/1986; **Ikiliwindi:** Nemba 458 fl., 1/1987; 462 1/1987; **Konye:** Thomas 5145 fl., 11/1985; 5146 fr., 11/1985; **Kupe Village:** Cheek 6099 fl., 1/1995; Etuge 2802 fr., 7/1996; Lane 340 fl., 1/1995; **Kurume:** Thomas 5416 fr., 1/1986; **Mejelet-Ehumseh:** Etuge 177 6/1986; **Mungo River F.R.:** Olorunfemi in FHI 30604 fr., 5/1951; **Ngomboaku:** Cheek 10386 fl., fr., 12/1999; Mackinder 322 fl., 12/1999; **Nyasoso:** Etuge 2581 fr., 7/1996.

Bridelia atroviridis Müll.Arg.

Shrub 8–12m, inconspicuously puberulent; leaves distichous, elliptic-oblong, rarely oblanceolate-elliptic, drying black above, c. 12 × 5.5cm, acuminate, base rounded, ends of secondary nerves not joining to a marginal nerve; fruit axillary, fleshy, ellipsoid, 7mm. Semi-deciduous forest; 550m.

Distr.: Sierra Leone to Zimbabwe. (Tropical Africa).

IUCN: LC

Kupe Village: Zapfack 926 fr., 7/1996; **Loum F.R.:** Tadjouteu 312 fr., 11/1999.

Note: spiny stem reported in FWTA but not noted in our specimens.

Bridelia micrantha (Hochst.) Baill.

Shrub or tree, 14m; trunk spiny, glabrescent; leaves resembling *B. atroviridis*, but drying pale green below (bluish green when alive), the secondary nerves fawn, joining to form a marginal nerve; inflorescence axillary glomerules; flowers greenish white, 3mm diameter; fruit fleshy, ellipsoid, red, 7mm. Forest edge; 150–1500m.

Distr.: Senegal to S Africa. (Tropical Africa).

IUCN: LC

Etam: Etuge 94 3/1986; **Kodmin:** Cheek 8861 fl., 1/1998; Plot B259 10/1998; **Kupe Village:** Etuge 1394 fl., 11/1995; 1924 fl., 5/1996; 2738 fr., 7/1996; **Kurume:** Thomas 8210 4/1988; **Ngomboaku:** Cheek 10304 fl., 12/1999; **Nyasoso:** Lane 509 fr., 2/1995; Thomas 3062 fl., 2/1984.

Local anme: Awang. **Uses:** FUELS – fuelwood (Cheek 8861).

Bridelia stenocarpa Müll.Arg.

Tree 7m; resembling *B. micrantha*, but leaves drying dark brown below, secondary nerves c. 15 pairs, not c. 10 pairs; fruit blue. Forest; 250m.

Distr.: Senegal to Congo (K). (Guineo-Congolian).

IUCN: LC

Mile 15, Kumba-Mamfe Road: Nemba 170 7/1986.

Note: delimitation from *B. micrantha* needs more work (FWTA).

Codiaeum variegatum (L.) A.Juss cv. *luteomaculatum*

Shrub 1m, puberulent; leaves spotted yellow, spots to 1cm, oblanceolate or oblong-oblanceolate, c. 10 × 3cm, shortly acuminate, obtuse; petiole 1.5cm. Farms; 1050m.

Distr.: native to New Guinea, cultivated widely in tropics. (Pantropical).
IUCN: LC
Ngomboaku: Ghogue 491 12/1999.

Codiaeum variegatum (L.) A.Juss cv. *weismanii*
Shrub 5m, glabrous, no exudate; leaves leathery, narrowly oblong-ligulate, 30 × 4cm, midrib area and secondary nerves orange-red above, lower surface pink, apex subacuminate, base cuneate; petiole 4.5cm; inflorescence pinkish red, racemose, 25cm; pedicels 1.5cm; flowers white, 5mm. Farmbush, farms; 650m.
Distr.: native to New Guinea, cultivated widely in tropics. (Pantropical).
IUCN: LC
Ngusi: Etuge 1550 fl., 1/1996.
Note: persisting on abandoned farms.
Uses: ENVIRONMENTAL USES – boundaries – an ornamental boundary flower; use stems planted to share or divide boundary disputes in farmbushes (Etuge 1550).

Croton aubrevillei J.Léonard
Bull. Jard. Bot. Nat. Belg. 28: 113 (1958).
Syn.: *Croton sp. nr. mubango* Müll.Arg.
Tree 15m; stem and lower leaf surface densely covered in white, fimbriate, peltate scales; leaves drying dark brown above with minute stellate hairs, ovate, c. 8.5 × 5cm, acuminate, base rounded, basal glands prominent, stalked; petiole 4–8cm. Secondary forest; 850m.
Distr.: Ivory Coast to Ghana, Cameroon. (Upper & Lower Guinea).
IUCN: VU
Nyasoso: Thomas 3074 fr., 2/1984.

Croton gratissimus Burch.
Kew Bull. 45: 556 (1990).
Syn.: *Croton zambesicus* Müll.Arg.
Tree 5–6m, resembling *C. aubrevillei*, but leaves drying green and glabrous above, oblong-elliptic, 6–13 × 2–3.5cm, obtuse or acute, base cuneate, abruptly rounded, basal glands sessile, inconspicuous; petiole 1–3.5cm; stipule 0.4cm; inflorescence c. 2.5cm; fruit 1.4cm diameter. Cultivated; 700–1000m.
Distr.: widespread in tropical Africa.
IUCN: LC
Kupe Village: Cheek 7016 1/1995; **Ndibise:** Etuge 538 fr., 6/1987.
Note: a Subsahelian species of which the only other Cameroonian material at K is from near Maroua. The stock of the Bakossi material was presumably brought many 100s of kms south, from the north.
Local names: Ndumme (Etuge 538); Ndume (Daniel Epie in Cheek 7016). **Uses:** MEDICINES (Etuge 538); planted on farms for medicines (Daniel Epie in Cheek 7016)

Croton longiracemosus Hutch.
Tree 15–25m; stem and lower leaf blade sparsely stellate-lepidote hairy, almost glabrescent; leaves drying dark brown, elliptic c. 11 × 7cm, abruptly acuminate, obtuse, basal glands sessile, margin serrate; petiole 3–6cm; inflorescence 25cm; fruit 3cm diameter, golden stellate-hairy. Forest; 1000m.
Distr.: Cameroon, Gabon & Congo (K). (Lower Guinea & Congolian).
IUCN: NT
Nyasoso: Cable 2881 fr., 6/1996.
Note: rare throughout its range.

Croton macrostachyus Hochst. ex Del.
Tree 6m, resembling *C. aubrevillei*, but leaves densely stellate-hairy above, broadly ovate, c. 12 × 8.5cm, acute, deeply cordate, margin obscurely hooked, basal glands sessile; inflorescence 10cm; fruit 1.2cm diameter. Roadside, cultivated; 1900m.
Distr.: tropical Africa.
IUCN: LC
Mwanenguba: Etuge 4365 fr., 10/1998.

Crotonogyne impedita Prain
Shrub resembling *C. preussii*, but highly-branched; indumentum stellate and lepidote; leaves oblanceolate or elliptic, to 20 × 6cm, base cuneate, drying brown, glabrous below; petiole 1–3cm; stipules glabrous; male inflorescence with 1–3 branches; fruit with pedicel 3cm. Forest; 300m.
Distr.: Cameroon Endemic.
IUCN: CR
Konye: Thomas 5167 fl., 11/1985; 5170 fl., 11/1985; **Loum F.R.:** Ledermann 6397 12/1909.

Crotonogyne manniana Müll.Arg.
Shrub 2–8m, resembling *C. impedita*, but stem and inflorescence indumentum fimbriate-lepidote; leaves to 35 × 10cm, male inflorescence unbranched. Forest; 200–1000m.
Distr.: Ghana to Gabon. (Upper & Lower Guinea).
IUCN: NT
Bangem: Thomas 5919 fl., 3/1986; **Mile 15, Kumba-Mamfe Road:** Nemba 604 7/1987; **Nyale:** Etuge 4212 fl., 2/1998.
Note: listed LRnt (fide WCMC), Breteler (1997) at www.redlist.org.

Crotonogyne preussii Pax
Monopodial shrub 2m; leaves arranged in a litter-gathering funnel, narrowly oblanceolate, c. 60 × 10cm, acuminate, cordate, drying green, lower surface thinly scattered with entire peltate scales, denser on stem and inflorescence, base of leaf with scrotiform glands; petiole 0.5cm; stipules 1 × 0.5cm; inflorescence 1(–2) per stem, unbranched, 30cm; flowers minute, grey, in glomerules 0.5–1cm apart; fruit subspherical, 3-lobed, 1.5cm diameter, brown, sessile. Forest; 100–1000m.
Distr.: SE Nigeria, CAR & Cameroon. (Lower Guinea).
IUCN: NT
Bakossi F.R.: Cheek 9315 fl., fr., 10/1998; Etuge 4290 fl., fr., 10/1998; **Kupe Village:** Etuge 2744 fl., fr., 7/1996; Lane 408 fr., 1/1995; **Mungo River F.R.:** Cheek 10195 fl., fr., 11/1999; Onana 940 fl., fr., 11/1999; Thomas 2563 11/1983; 2563b 11/1983; **Nyasoso:** Balding 17 fr., 7/1993; Sunderland 1488 7/1992.

Crotonogyne strigosa Prain
Shrub 2–4m; red exudate; resembling *C. impedita*, but long, shaggy hairy, hairs c. 2mm; leaves oblanceolate, 18 × 5.5cm, including 2cm acumen, base cuneate-truncate; petiole 1cm; stipules narrowly triangular, 0.8cm; inflorescences c. 10cm; flowers 2–5 at apex, white, 1cm; fruit unknown. Forest; 150m.
Distr.: SE Nigeria & W Cameroon. (Lower Guinea).
IUCN: VU
Etam: Etuge 186 7/1986.
Note: occurring mainly W of the Cameroon highlands, 8 specimens at K.

EUPHORBIACEAE

Crotonogyne zenkeri Pax
Bot. Jahrb. 26: 327 (1899).
Shrub 3m, closely resembling *C. manniana*, but stem and inflorescence indumentum brown stellate-pubescent. Forest; 250m.
Distr.: Cameroon & Gabon. (Lower Guinea).
IUCN: VU
Mile 15, Kumba-Mamfe Road: Nemba 154 fl., fr., 7/1986.

Cyrtogonone argentea (Pax) Prain
Tree 10m, no exudate; bole brown-yellow; stems scattered with yellow-brown scales, scabrid; leaves dimorphic, juveniles large (c. 30 × 20cm), 3-lobed, the large central lobe with up to 3 irregular lobes on each side, upper surface glabrous, lower bright-white with overlapping scales; petiole 4–11cm, gland pair inconspicuous; mature leaves decreasingly lobed and smaller to obovate, 11 × 5cm, entire; inflorescence 20 × 5cm; fruit 3-lobed, densely yellow-brown-hairy, hard, 5 × 5cm. Evergreen forest; 200–450m.
Distr.: SE Nigeria to Gabon. (Lower Guinea).
IUCN: NT
Bakossi F.R.: Cheek 9314 10/1998; **Ikiliwindi:** Nemba 562 6/1987.
Local name: Mkama-lana = 3 fingers. **Uses:** SOCIAL USES – drugs – facilitants to drug effects – used to enhance strength of distilled spirit (Epie Ngome in Cheek 9314).

Discoclaoxylon hexandrum (Müll.Arg.) Pax & K.Hoffm.
F.T.E.A. Euphorbiaceae: 280 (1987).
Syn.: *Claoxylon hexandrum* Müll.Arg.
Pithy tree 3–5(–10)m, minutely puberulent; leaves elliptic, to 28 × 12cm, short acumen, acute, margin serrate; petiole c. 18cm; inflorescences numerous, one per axil, pendulous, spike-like, c. 12cm; flowers minute. Forest; 300–1700m.
Distr.: Liberia to Uganda. (Guinea-Congolian).
IUCN: LC
Bolo: Nemba 59 fl., 3/1986; **Essossong:** Groves 53 fl., 2/1995; **Kodmin:** Gosline 49 fl., 1/1998; **Kupe Village:** Cable 729 fl., 1/1995; Cheek 7072 fl., 1/1995; Elad 66 fl., 1/1995; Etuge 1452 fl., 11/1995; Lane 369 fl., 1/1995; Ryan 351 fl., 5/1996; **Mekom:** Thomas 5409 fl., 1/1986; **Molongo:** Nemba 780 2/1988; **Nsoung:** Darbyshire 69 3/2003; **Nyasoso:** Cable 1139 fl., 2/1995; Cheek 6017 fl., 1/1995; 7303 fl., 2/1995; Etuge 1738 fl., fr., 2/1996.
Local names: Chiensa (Gosline 49); Inchene (Cheek 7072).
Uses: none (Gosline 49); no uses known (Cheek 7072).

Discoglypremna caloneura (Pax) Prain
Dioecious tree c. 20m, glabrous; leaves ovate, c. 12.5 × 8.5cm, broadly acuminate, base rounded to obtuse, margin serrate, 3-nerved at base, with 1–2 pairs elliptic, brown, sunken glands at base, each 1mm; petiole 6cm; stipules caducous; male inflorescence a terminal panicle of spikes, c. 20cm long; flowers c. 3mm; fruit c. 8mm diameter, strongly 3-lobed, the valves falling to reveal red seeds. Farmbush, secondary forest; 250–400m.
Distr.: Guinea (C) to Congo (K). (Guineo-Congolian).
IUCN: LC
Baduma: Nemba 125 6/1986; 182 8/1986; **Ikiliwindi:** Etuge 516 3/1987; **Kupe Village:** Etuge 1930 5/1996; 2014 fr., 5/1996.

Drypetes aframensis Hutch.
Tree (8–)15–20m; stems glabrous yellow, shallowly grooved; leaves oblong-elliptic, 15–20 × 5–8cm, shortly acuminate, obliquely obtuse, nerves 6–8 pairs, midrib yellow below, blade yellowish green, tertiary nerves raised, conspicuous; petiole 0.5–1cm, yellow; stipules triangular 1mm; inflorescences axillary and below the leaves, brachyblasts c. 4mm diameter. Forest; 950–1150m.
Distr.: Ghana to Cameroon. (Upper & Lower Guinea).
IUCN: LC
Kupe Village: Cheek 7804 11/1995; **Mt Kupe:** Cheek 10162 11/1999; **Nyasoso:** Cable 2838 6/1996; 3597 7/1996.
Note: distinctive for its yellow-coloured entire leaves. New record for Cameroon; may prove distinct when flowers available.
Local name: not known. **Uses:** MEDICINES (Marcus in Cheek 7804).

Drypetes angustifolia Pax & K.Hoffm.
Engl. Pflanzenr. Euphorb. -Phyllanthoid.-Phyllanth. 261 (1922).
Tree 10m; stems sparsely brown-puberulous, shallowly - grooved; leaves elliptic, c. 10 × 3cm, acumen 2–3cm, base obliquely-acute, c. 15 serrate spines on each side; petiole 3mm; stipule caducous; inflorescence axillary; flowers single, sessile; female flower with sepals 4, 2mm, styles 2; fruit ellipsoid, 2cm, glabrous. Forest; 700m.
Distr.: Cameroon to Congo (K). (Lower Guinea & Congolian).
IUCN: LC
Nyandong: Thomas 7263 2/1987.

Drypetes cf. calvescens Pax & K.Hoffm.
F.T.E.A. Euphorbiaceae: 96 (1987).
Shrub 2m; stems puberulous, terete; leaves distichous, oblong-oblanceolate, 10 × 3.5cm, acuminate, unequally acute, nerves c. 6 pairs on each side, brochidodromous, midrib drying yellow below. Evergreen forest; 250m.
Distr.: Cameroon to Congo (K). (Lower Guinea & Congolian).
Loum F.R.: Cheek 10244 12/1999.
Note: fertile material required to confirm identity.

Drypetes chevalieri Beille
Shrub or small tree 2m; glabrous; stem triangular-ridged; leaves ovate, 14 × 5.5cm, acuminate, very unequally rounded, leathery, minutely spiny-serrate, nerves 5 pairs, looped 0.5cm from margin, sessile; stipules caducous; inflorescences axillary glomerules, 7mm diameter; flowers 2mm diameter. Forest; 300m.
Distr.: Sierra Leone to Cameroon. (Upper & Lower Guinea).
IUCN: LC
Bolo: Nemba 93 5/1986; **Bolo Moboka:** Nemba 239 9/1986.
Note: determination by D.Thomas, specimen not seen.

Drypetes ivorensis Hutch. & Dalziel
Tree 8m; stems terete, densely yellow-brown pubescent; leaves oblong, c. 17 × 7cm, acuminate, obliquely obtuse, margin with c. 20 shallow crenate-serrate teeth, nerves c. 7 pairs; petiole 3mm; stipules caducous; fruits not seen. Forest; 220m.
Distr.: Liberia to Ghana, Cameroon. (Upper & Lower Guinea).
IUCN: LC
Bakolle Bakossi: Etuge 284 fr., 9/1986.

Drypetes cf. klainei Pierre ex Pax
Tree 5m; stems dull white, grooved, brown puberulent; leaves drying brown below, elliptic, 11–18 × 4.8–7cm, slender acuminate, obliquely obtuse, nerves 5–6 pairs, brochidodromous, midrib hairy below, entire; inflorescences ramiflorous, sessile, few-flowered; female flowers with

sepals 4, puberulent; ovary pubescent; styles 2, separate at base; stigmas lobed. Forest; 1000m.
Distr.: Ivory Coast, Cameroon & Gabon. (Upper & Lower Guinea).
Ndibise: Etuge 530 fr., 6/1987.
Note: this specimen has much larger leaves than usual and the midrib is hairy. More material, ideally with male flowers is required; this may prove to be an undescribed taxon.

Drypetes leonensis Pax
Tree 20m; stems and pedicels patent-puberulent, angled; leaves elliptic, to 11 × 6cm, acumen 1cm, acute to obtuse, lateral nerves 4 pairs, glabrous; petiole 5mm; stipules caducous; fascicles axillary, few–30-flowered; pedicels 5mm, female flowers 3mm; male flowers with calyx lobes 4; stamens 4, inserted outside a glabrous disc; fruit globose, 1cm. Forest; 800m.
Distr.: Guinea (C) to Cameroon. (Upper & Lower Guinea).
IUCN: NT
Bangem: Thomas 5307 1/1986.
Note: specimen cited not seen at K.

Drypetes magnistipula (Pax) Hutch.
F.T.A. 6(1): 678 (1912).
Treelet 4m; stems 4-ridged, glabrescent; leaves oblong (-obovate), 30 × 11cm, acuminate, unequally rounded-truncate, margin subentire, nerves 8 pairs, looped, weakly linked near margin; petiole 1.5cm; stipules foliaceous, persistent, ovate-triangular, 3.5 × 1.5cm, acute, base cordate; inflorescence few-flowered, cauliflorous; flowers not seen. Forest; 800–1000m.
Distr.: Cameroon & Gabon. (Lower Guinea).
IUCN: EN
Bangem: Thomas 5313 1/1986; **Menyum:** Doumenge 519 5/1987.
Note: determination by D.Thomas, specimen not seen. Other collections are from S Cameroon.

Drypetes molunduana Pax & K.Hoffm.
Tree 5–8m; stem puberulent-glabrescent, slightly angled, black, streaked white; leaves drying dark brown, oblanceolate, to 28 × 9cm, acumen slender, unequally obtuse, midrib appressed-hairy, margin shallowly crenate; petiole 5mm; stipule oblong-triangular, 8 × 1.5mm, puberulent; inflorescence cauliflorous; flowers fasciculate; fruit ovoid-elliptic, c. 3 × 1.5cm. Evergreen forest; 400–1200m.
Distr.: S Nigeria & Cameroon. (Lower Guinea).
IUCN: VU
Kupe Village: Lane 415 fr., 1/1995; **Mahole:** Olorunfemi 30591 fr., 5/1951; **Nyasoso:** Sebsebe 5059 10/1995; Wheatley 471 fr., 7/1992.
Note: the long-persisting stipules and leaf colour are distinctive.

Drypetes paxii Hutch.
Tree, 40m; 5–6 buttresses; stem puberulent with a raised, glabrous pale brown ridge; leaves oblong-elliptic, c. 11 × 4.5cm, acumen 1.5cm, unequally obtuse, margin subentire, midrib as stem, nerves 6–10, looping 5mm from margin; petiole 5mm; stipules caducous; inflorescence axillary; flowers 3–6; fruit ellipsoid, 1cm. Forest; 250m.
Distr.: Nigeria to Congo (K). (Lower Guinea & Congolian).
IUCN: LC
Ikiliwindi: Etuge 509 3/1987.
Note: determination from D.Thomas, specimen not seen.

Drypetes preussii (Pax) Hutch.
Tree 10m; stems densely puberulent, terete or striate; leaves elliptic, to 17 × 6cm, acuminate, slightly unequally obtuse-rounded, serrate, nerves 7 pairs, strongly looped 1cm from margin; petiole 3mm; stipules caducous; inflorescence cauliflorous 7mm, white; pedicels 7mm; fruit ellipsoid, 3-lobed, 2cm, smooth; stigmas minute. Forest; 250m.
Distr.: SE Nigeria & Cameroon. (Lower Guinea).
IUCN: VU
Mile 12 Mamfe Road: Nemba 651 10/1987.
Note: determination by D.Thomas, specimen not seen.

Drypetes principum (Müll.Arg.) Hutch.
Tree 6–15m; stems brown-puberulent, shallowly furrowed; leaves resembling those of *Tapura africana*, drying blackish, shining below, oblanceolate, c. 15 × 6cm, acuminate, acute; petiole 8mm; stipules caducous; inflorescence fasciculate, axillary or on leafless branches; pedicels 3mm; flowers with sepals 5mm, pubescent; ovary pubescent, stigmas 2; fruit 15mm. Forest; 150–900m.
Distr.: Guinea (C) to Cameroon. (Upper & Lower Guinea).
IUCN: LC
Kupe Village: Lane 402 fr., 1/1995; **Manehas F.R.:** Etuge 4341 fl., 10/1998; **Mbule:** Etuge 4392 fl., 10/1998; **Mungo River F.R.:** Cheek 10231 fl., fr., 12/1999.

Drypetes staudtii (Pax) Hutch.
Tree or shrub, 15cm d.b.h..; stems strongly 4-angled, glossy purple-brown; leaves glossy, oblong, 24 × 9cm, obtuse, obliquely rounded, margin finely serrate-toothed, midrib red-brown, secondary nerves c. 12, quaternary nerves prominent; petiole 5mm; stipules caducous; inflorescence cauliflorous, fascisculate; male flowers 15mm diameter, yellow; stamens c. 30; pedicels 15mm. Evergreen forest; 420m.
Distr.: SE Nigeria to Cameroon. (Lower Guinea).
IUCN: VU
Mungo River F.R.: Onana 956 fl., 11/1999; **Nyandong:** Cheek 11314 3/2003.
Note: easily recognised by the glossy, finely-nerved, rounded to subcordate leaves and 4-angled purple-brown stem.

Drypetes sp. A
Tree 10–25m; stems sparsely and inconspicuously puberulent on one side; leaves drying pale grey-green above, pale green below, ovate-elliptic or elliptic-oblong, 10–15 × 4.5–6.5cm, acuminate, base obliquely obtuse, shortly decurrent, margin entire, nerves 5–6 pairs; sterile plot vouchers ovate, serrate, recurved; petiole 5mm; stipules oblong, caducous; inflorescence on trunk, fasciculate on burrs, sessile; pedicels c. 17mm; male flowers glabrous; sepals 5, 7mm; anthers 15; disc glabrous; female flowers with stigmas 3, obovate; fruit broadly ellipsoid, glabrous, c. 4cm diameter. Evergreen forest; 810–1400m.
Distr.: Cameroon. (Narrow Endemic).
Kodmin: Plot B42 1/1998; B62 1/1998; B63 1/1998; B104 1/1998; **Kupe Village:** Cable 2720 fr., 5/1996; Elad 84 fr., 1/1995; **Ngomboaku:** Cheek 10398 fl., 12/1999; Ghogue 458 fl., 12/1999; Muasya 2026 12/1999; **Nyasoso:** Cable 1137 fr., 2/1995; 3332 fr., 6/1996; Cheek 9514 10/1998; Etuge 2107 fr., 6/1996; 2567 fr., 7/1996.
Note: a new species awaiting description; most similar to *D. obanensis* and *D. afzelii*.
Local name: Ngele-Sale. **Uses:** related to a species which has edible fruit (Cheek 9514).

Drypetes sp. B
Tree 3–10m, puberulent; leaves matt, dark grey-green below, narrowly oblong, c.13–16 × 4–5cm, acuminate, acute, margin

spiny-serrate, nerves 6–8 pairs; petiole 3mm; inflorescence resembling *D. sp. A*, but pedicels c. 3mm; male flowers with sepals 5, 3–4mm; Evergreen forest; 1100–1200m.
Distr.: Cameroon. (Narrow Endemic).
Nyasoso: Cable 1150 fl., 2/1995; Etuge 2394 fr., 6/1996.
Note: probably a new species, but more material needed to describe.

Drypetes sp. C
Tree 4–5m, sparsely appressed-puberulent; stems grey-white, ridged; leaves drying black, oblong, to 20 × 7cm, acumen to 5cm, cuneate, lateral nerves 6 pairs, brochidodromous, looping over 1cm from margin; petiole 1cm; stipules caducous, small; fascicles axillary, up to 10-flowered; pedicels 4mm; male flowers 3–4mm diameter; sepals 4; stamens 4 around glabrous disc. Forest; 1100m.
Distr.: Cameroon. (W Cameroon Uplands).
Kupe Village: Cable 812 fl., 1/1995; **Nyasoso:** Lane 529 fl., 2/1995.
Note: probable *sp. nov.*, also at Bali Ngemba. More material and analysis needed. Keys closest to *D. leonensis* in FWTA.

Elaeophorbia drupifera (Thonn.) Stapf
Tree 15m, glabrous; stems c. 1cm thick, exudate white; stem spines paired; leaves oblanceolate-oblong, to 15 × 4cm, obtuse or rounded, base cuneate-decurrent, sessile, lateral nerves inconspicuous, 3 pairs. Forest; 710m.
Distr.: Ghana to Cameroon, Uganda. (Guineo-Congolian).
IUCN: LC
Nyale: Etuge 4227 2/1998.

Erythrococca anomala (Juss. ex Poir.) Prain
Shrub 1–3m tall, glabrous or pubescent; stem with stipular spines 1–4mm; leaves ovate or elliptic, to 3–8 × 1.5–3.5cm, long acuminate, obtuse, crenate; petiole 1cm; inflorescences many-flowered, subumbellate; peduncle c. 2mm; pedicels 4mm; flowers white, pendulous, 3mm diameter; fruit 2-lobed, 1cm wide, green, dehiscing to show red seeds. Evergreen forest; 200–1100m.
Distr.: Guinea (C) to Cameroon. (Upper & Lower Guinea).
IUCN: LC
Bangem: Thomas 5328 1/1986; **Ekona Mombo:** Etuge 418 fr., 12/1986; **Kupe Village:** Etuge 2760 fl., fr., 7/1996; **Mungo River:** Leeuwenberg 6835 fl., 10/1965; **Nyasoso:** Cable 3275 fr., 6/1996; 693 fr., 1/1995; Cheek 6008 fr., 1/1995; 9282 fl., 10/1998; Etuge 1295 fl., 10/1995; 1805 fl., fr., 3/1996; 2108 fr., 6/1996; Lane 198 fl., fr., 11/1994; 205 fr., 11/1994; Sebsebe 5002 fr., 10/1995; Sidwell 360 10/1995.
Note: unique in Cameroonian *Erythrococca* in the spiny stipules.

Erythrococca atrovirens (Pax) Prain
Ann. Bot. 25: 623 (1911).
Shrub or tree to 4.5m, sparingly to densely yellow-puberulent; spines absent; leaves drying green, ovate to elliptic-oblong, 8–13 × 3–7cm, long acuminate, cuneate to rounded, subentire; lateral nerves 4–6 pairs, pubescent below, densest along the nerves; petiole 1–1.5cm; male racemes 2–4cm, long-pedunculate, pubescent; pedicels capillary; female flowers with ovary bilobate; styles smooth; fruit green. Forest.
Distr.: Cameroon, Congo (K) to Tanzania. (Lower Guinea & Congolian).
IUCN: LC
Manyemen: Letouzey 13924 6/1975.
Note: specimen not seen, description ex FTEA.

Erythrococca hispida (Pax) Prain
Shrub or small tree resembling *E. membranacea*, but stems appressed-hairy, not pilose; infructescence racemose, not subumbellate; fruits hispid, not glabrous. Forest; 1100m.
Distr.: Cameroon. (W Cameroon Uplands).
IUCN: NT
Mejelet-Ehumseh: Etuge 448 fr., 1/1987.
Note: specimen not seen. Apparantly restricted to Mt Cameroon (22 records) apart from Etuge 448. Bamenda Highlands material differs in taxonomy.

Erythrococca membranacea (Müll.Arg.) Prain
Shrub 2–3(–5)m; unarmed, pilose, hairs yellowish brown, c. 2mm long; leaves oblong or elliptic, 8–12 × 3.5–5cm, long-acuminate, obtuse, margin coarsely and irregularly serrate; inflorescence 6–8-flowered, subumbellate, 3cm; peduncle 2cm; pedicels 5mm; flowers green, 3mm wide; fruit green, bilobed, 1.5cm wide; seeds red. Forest; 700–1800m.
Distr.: SE Nigeria & W Cameroon. (W Cameroon Uplands).
IUCN: NT
Bangem: Thomas 5326 1/1986; 5327 1/1986; **Kodmin:** Biye 10 fl., fr., 1/1998; Gosline 61 fr., 1/1998; Plot B132 10/1998; **Kupe Village:** Cheek 7830 fl., 11/1995; **Mt Kupe:** Thomas 5111 12/1985; **Ngomboaku:** Cheek 10368 fl., 12/1999; **Nyasoso:** Cable 3404 fr., 6/1996; 692 fr., 1/1995; 706 fr., 1/1995; Cheek 7334 fl., 2/1995; Elad 142 fr., 2/1995; Etuge 2119 fr., 6/1996; Lane 188 fl., 11/1994.
Local name: none known. **Uses:** none known (Cheek 7334).

Erythrococca cf. oleracea Prain
Ann. Bot. 25: 624 (1911).
Shrub or tree 4–7m tall, resembling *E. membranacea*, but glabrous; leaves ovate or elliptic, c. 9 × 6cm, acuminate, obtuse-truncate, finely, sparsely, serrate; petiole 1cm; infructescence racemose, pendulous, 8cm; peduncle flattened 4cm; fruits c. 6; pedicel 5mm; fruit white. Forest; 600–1000m.
Kupe Village: Cable 2545 fr., 5/1996; Etuge 1904 5/1996; **Ngombombeng:** Etuge 122 fr., 5/1986.
Note: differs from *E. oleracea* in the leaves and petioles ± glabrous (fide A. Radcliffe-Smith). Similar also to *E. africana*, but glabrous. Distinctive in the flattened peduncle; flowers needed. A revision of *Erythrococca* in Lower Guinea is needed before Red Data assessments can be sensibly made.

Euphorbia forskalii J.Gay
Webb & Berth., Phyt. Canar. 3: 240 (1847).
Syn.: *Euphorbia aegyptiaca* Boiss.
Prostrate herb; stems numerous, 15cm; white-pubescent on upper side, glabrous below; leaves alternate, obovate, 0.5 × 0.3cm, dentate, puberulent above and below; petiole 0.5mm; inflorescences sessile; flowers lacking perianth; ovary 1.5mm diameter, white-hairy. Roadside; 400m.
Distr.: Senegal to India, drier parts. (Palaeotropics).
IUCN: LC
Nlonako: Etuge 1775 fl., 3/1996.

Euphorbia heterophylla L.
Erect herb 60cm, glabrous; leaves alternate, elliptic, c. 7 × 3.5cm, subacuminate, acute; petiole 2cm; flowers lacking perianth; fruit 3-lobed, 5mm diameter. Farmbush; 1300m.
Distr.: originally American, now pantropical.
IUCN: LC
Kupe Village: Etuge 1703 fr., 2/1996.
Note: other weedy species likely to be found in our area (they are known in Kumba) are *E. thymifolia*, *E. prostrata* and *E. kamerunica*.

Euphorbia hirta L.

Herb 30cm; stems densely white-puberulent, with long yellow hairs mixed, several stems per plant; leaves opposite, oblong, c. 2 × 1cm; inflorescences dense axillary subumbels, c. 7mm diameter; peduncle 2mm; perianth absent; fruit 1mm diameter. Farmbush; 650–800m.
Distr.: pantropical.
IUCN: LC
Nlonako: Etuge 1785 fr., 3/1996; **Nyasoso:** Etuge 1598 fr., 1/1996; **Tombel:** Etuge 374 fl., 11/1986.
Uses: MEDICINES (Etuge 1785).

Euphorbia schimperiana Hochst. ex A.Rich. var. *pubescens* (N.E.Br.) S.Carter

Kew Bull. 40: 813 (1985).
Herb 1m, white exudate, glabrous apart from the puberulent stem apex; leaves alternate, narrowly oblanceolate-oblong, c. 6 × 1cm, acute, cuneate; petiole c. 0.5cm; inflorescence terminal, 4 branches from stem apex, each forking 2–4 times; flowers green, perianth absent; fruit 5mm diameter. Grassland, farmbush; 800–1500m.
Distr.: Cameroon, Ethiopia to Zimbabwe, Arabia. (Afromontane & Arabia).
IUCN: LC
Bangem: Thomas 5302 fr., 1/1986; **Kodmin:** Barlow 16 fr., 1/1998; Cheek 8903 fr., 1/1998; Pollard 251 fl., fr., 11/1998.
Uses: MATERIALS – snares/traps – used in trapping (Epie Ngome in Cheek 8903).

Grossera macrantha Pax

Tree 6–8m; bole fluted, greyish green; resembling *G. paniculata*, but stems, petioles and midribs puberulent; leaves drying grey below with pale brown nerves, elliptic-oblong, acumen 1cm, serrate, panicles to 12cm; flowers 5mm diameter. Forest; 350m.
Distr.: Cameroon and Congo (K). (Lower Guinea & Congolian).
IUCN: LC
Nyandong: Thomas 7014 5/1987.

Grossera multinervis J.Léonard

Bull. Jard. Bot. Nat. Belg. 25: 317 (1955).
Tree 15–20m tall; resembling *G. paniculata*, but leaves with 15–20 pairs of secondary nerves and with highly distinct marginal teeth; male receptacle pubescent; stamens 15–19. Forest; 1000m.
Distr.: Cameroon & Congo (K). (Lower Guinea & Congolian).
IUCN: LC
Menyum: Doumenge 560 5/1987.
Note: D.Thomas extended the range to Western Cameroon of this otherwise Congolian species in the 1980's. The specimen cited was determined by him and not seen by us.

Grossera paniculata Pax

Adansonia (sér. 2) 3: 70 (1963).
Tree 20m, glabrous; leaves drying brown, elliptic or obovate, c. 20 × 9cm, obtuse, acute, margin inconspicuously denticulate, nerves c. 10–13; inflorescence a terminal panicle, c. 26 × 26cm, lateral branches c. 10; female flowers pink and white c. 6mm diameter, petals 5; stamens 19–23. Forest; 930–1040m.
Distr.: Cameroon & Gabon. (Lower Guinea).
IUCN: LC
Kupe Village: Cable 3815 7/1996; **Ngusi:** Letouzey 14654 fl., 4/1976.

Hamilcoa zenkeri (Pax) Prain

Tree 15m or more; stems grey, glabrous; trunk cylindric; bark and slash whitish brown, with slow white exudate; leaves drying yellowish green below, papery, oblanceolate-oblong, 18 × 6cm, acumen 1cm, acute; secondary nerves acute, c. 6 pairs, raised, margin sparsely and acutely serrate, basal gland pair adaxial, inconspicuous; petiole variable, 1–4cm, swollen at base and apex; stipules caducous; inflorescence axillary racemes, 4cm; flowers 6–8, each c. 4mm. Semi-deciduous forest; 250m.
Distr.: Cameroon Endemic.
IUCN: VU
Loum F.R.: Cheek 10248 12/1999; 10251 12/1999.

Hevea brasiliensis (Kunth.) Müll.Arg.

Tree 20m, producing white exudate; leaves alternate, trifoliolate; leaflets elliptic to obovate, 8–20 × 2.5–8cm, entire, acute, lateral nerves c. 20 pairs, looped near margin; petiolules with basal gland; panicles below the leaves; capsule 3-lobed, 4.5cm; seeds 2.5cm, mottled.
Distr.: pantropical, originally Brazil.
IUCN: LC
Tombel: Cheek, sight record 11/1995.
Local name: rubber. **Uses:** MATERIALS – cultivated extensively in high rainfall lowland areas of Cameroon for the exudates produced when wounded (*fide* Cheek).

Klaineanthus gaboniae Pierre ex Prain

Tree 27m, 1.2m d.b.h.; no exudate; slash hard, white; stems densely puberulent; leaves drying dark brown, oblanceolate-oblong, 30 × 10.5cm, acuminate, acute, margin entire, c. 15 pairs secondary nerves; petiole swollen at base and apex, c. 4cm; stipules caducous; inflorescences single from the uppermost 3–4 leaves, each c. 12cm; branches up to 1cm; flowers white, 2mm. Evergreen forest.
Distr.: SE Nigeria to Congo (K). (Lower Guinea & Congolian).
IUCN: LC
Etam: Olorunfemi in FHI 30532 fl., 4/1951.

Macaranga cf. *hurifolia* Beille

Tree, puberulent; leaves triangular-ovate, 13 × 10cm, long-acuminate, margin coarsely dentate, teeth c. 12 per side, uneven, rounded at apex, nerves 4 per side, lower surface with dense, red, surface glands; petiole 7cm; stipule triangular, 2 × 2mm; fertile material not seen. Forest; 1400m.
Distr.: Sierra Leone to Cameroon. (Upper & Lower Guinea).
Kodmin: Plot B275 1/1998.
Note: our single specimen is puberulent, not pubescent as is normal in this species; nor is there any mention of spines in the collecting data.

Macaranga monandra Müll.Arg.

Tree 8–10m; trunk and stems (below leaves) spiny, rusty-stellate, glabrescent; leaves elliptic, c. 13 × 7cm, acuminate, base rounded or obtuse, margin dentate, c. 6 teeth per side; petiole 3–6cm; inflorescence pendulous, c. 3cm; bractlets laciniate; fruit subspherical, 4mm. Secondary forest, farmbush; 350–1150m.
Distr.: S Nigeria to Uganda. (Lower Guinea & Congolian).
IUCN: LC
Bakolle Bakossi: Etuge 138 fr., 5/1986; **Kodmin:** Cheek 9197 2/1998; **Nyasoso:** Thomas 3070 2/1984.

Macaranga occidentalis (Müll.Arg.) Müll.Arg.

Tree 3–25m; trunk often spiny, with red or clear exudate, glabrous; leaves suborbicular, 30–50cm diameter, shallowly 3–5-lobed, lobes acuminate, base cordate, lower surface often

bluish white below and with minute red glands; petiole c. 30cm; stipules 4 × 2cm; inflorescence axillary, pendulous, paniculate; bracts 5–10cm, deeply dentate-sinuate, densely pubescent; female inflorescences to 18cm; fruit 1–2-lobed 8mm. Forest edge; 600–1450m.
Distr.: SE Nigeria & Cameroon. (W Cameroon Uplands).
IUCN: NT
Kodmin: Cheek 9045 1/1998; **Kupe Village:** Cable 2499 fr., 5/1996; Cheek 7067 fl., 1/1995; 7745 11/1995; 8339 fl., 5/1996; Elad 91 1/1995; **Ndum:** Williams 142 fl., 1/1995; **Ngombombeng:** Etuge 118 5/1986; **Nsoung:** Etuge 4953 3/2003; **Nyasoso:** Cable 2811 fr., 6/1996; Thomas 3063 2/1984.
Note: the similar *M. schweinfurthii* is likely to occur in our area at low altitudes: male bracts up to 3mm long, glandular-puberulent, entire or denticulate; leaf sinuses ⅓ leaf length, female inflorescences to 8cm, usually a straggling shrub.
Local name: Nkeke. **Uses:** MATERIALS – leaves used for wrapping; stems used to make beds (Cheek 9045).

Macaranga spinosa Müll.Arg.
Tree 8m, spiny from trunk to leaf branches (spines c. 1cm), densely pilose-pubescent (hairs 1–2mm); leaves oblong, c. 9 × 5cm, acuminate, base truncate and abruptly cordate, margin sinuous-lobed, c. 4 shallow lobes on each side; petiole 3–5cm; fertile material not seen. Forest; 700m.
Distr.: Liberia to Angola. (Upper & Lower Guinea).
IUCN: LC
Nyale: Etuge 4244 2/1998.

Maesobotrya cf. barteri (Baill.) Hutch.
Tree 2–3m, pubescent; leaves elliptic, 19 × 7.5cm, long acuminate, cuneate-rounded, margin with hair tufts at nerve endings, midrib and secondary nerves hairy above and below; petiole 5–40mm; stipules 4 × 2mm, densely white-pubescent, caducous; inflorescence cauliflorous, 10cm; fruit ellipsoid, red, 6mm. Forest; 350–915m.
Enyandong: Etuge 4495r 11/2001; **Mungo River F.R.:** Cheek 10201 fr., 11/1999.
Note: the Cameroonian material of *Maesobotrya* needs delimitation at species level. More material of this taxon is also required.

Maesobotrya dusenii (Pax) Hutch.
Tree 8m, puberulous; leaves elliptic-oblong, c. 22 × 10cm, acuminate, base rounded, margin entire; petiole 8cm; stipules foliaceous, falcate, to 2 × 2cm; inflorescences cauliflorous, fasciculate racemes, 14cm. Forest; 150–350m.
Distr.: SE Nigeria to Gabon. (Lower Guinea).
IUCN: LC
Mekom: Nemba 11 4/1986; **Mombo-Bakossi:** Olorunfemi in FHI 30557 5/1951; **Mungo River F.R.:** Cheek 10146 11/1999.
Note: easily recognised by the foliaceous stipules.

Maesobotrya cf. staudtii (Pax) Hutch.
Tree 8m, pubescent; leaves elliptic, 22 × 11.5cm, subacuminate, rounded to obtuse, with subappressed hairs above and below on midrib and secondary nerves; petiole 8cm; stipules caducous; female inflorescences cauliflorous, in clusters, 4cm long. Forest; 700m.
Nyale: Etuge 4181 fr., 2/1998.
Note: see note to *M. cf. barteri*.

Maesobotrya sp. A
Tree 3m, brown-pubescent; leaves elliptic, 15 × 7cm, acuminate, base rounded, margin hairy, glabrous above, midrib and secondary nerves hairy below; petiole c. 5cm;

stipule obliquely obovate, c. 8 × 4cm, outer surface glabrous, margin pubescent, persistent; cauliflorous, racemes 15cm; fruit red, ellipsoid, 16mm, glabrous. Forest; 350m.
Bakolle Bakossi: Etuge 149 5/1986.
Note: see note to *M. cf. barteri*.

Mallotus oppositifolius (Geiseler) Müll.Arg.
Shrub 2–10m, puberulent; leaves opposite, blades ovate to c. 17 × 8cm, apex attenuate, base rounded, obscurely-toothed, lower surface with minute, red, surface glands; inflorescence axillary, erect, spike-like, c. 14cm; flowers 3mm diameter, yellow; stamens numerous; fruit 8mm diameter, glabrous. Forest edge; 500–1000m.
Distr.: tropical Africa & Madagascar.
IUCN: LC
Kupe Village: Etuge 1893 fl., fr., 5/1996; Lane 308 fl., 1/1995; **Loum F.R.:** Etuge 4277 fr., 10/1998; **Ngusi:** Etuge 1584 fl., 1/1996; **Nyasoso:** Cable 3520 fr., 7/1996.
Local name: Ayed (Etuge 4277).

Mallotus subulatus Müll.Arg.
Shrub or tree 1–5m, stellate-pubescent, resembling *M. oppositifolius*, but leaves not glandular below, acuminate, subcordate; stipules persistent, subulate; fruit bristly. Forest; 300m.
Distr.: Sierra Leone to Congo (K). (Guineo-Congolian).
IUCN: LC
Bolo: Nemba 42 3/1986.
Note: specimen not seen.

Manihot esculenta Crantz
Shrub 2–3m; leaf-blade c. 12cm, divided to within 1cm of the base, into 3 narrowly elliptic lobes, entire, white below; petiole c. 7cm; flowers white, c. 1cm diameter; fruit c. 1cm, winged. Introduced, farms and farmbush; 740–1300m.
Distr.: tropical America, now widely cultivated. (Pantropical).
IUCN: LC
Bazeng: Sebsebe 5061 fl., fr., 10/1995; **Edib village:** Cheek (sight record) 2/1998; **Kupe Village:** Cable 3848 fl., 7/1996; **Nyasoso:** Etuge 1614 fl., fr., 1/1996.
Local name: Cassava. **Uses:** FOOD – tubers edible. 1. grind, fry to make the type of food called 'Gari'. 2. boiled, then pounded in a malter to make the type of food called 'Fufu-casava'. 3. lastly mix ground cassava in a pot, boiled to make the type of meal called 'water-fufu cassava' or just chew the tubers raw, gives you starch. Cultivated all over Cameroon (Etuge 1614); tubers edible (Sebsebe 5061).

Manihot glaziovii Müll.Arg.
Shrub or small tree to 10m; producing white exudate, resembling *M. esculenta*, but leaves peltate, proportionally broader; flowers up to 2.5cm long; fruits without ridges. Introduced, farms and farmbush; 800m.
Distr.: tropical America, cultivated in palaeotropics. (Pantropical).
IUCN: LC
Nyasoso: Etuge 2498 fl., 7/1996.
Uses: MATERIALS - formerly cultivated as 'Ceara Rubber' (*fide* Cheek).

Mareya micrantha (Benth.) Müll.Arg. subsp. *micrantha*
Tree 4–8m resembling *Alchornea hirtella*; no exudate, puberulous; leaves drying grey-green, obovate or oblong-elliptic, to 11 × 4cm, subacuminate, acute, margin very inconspicuously serrate, secondary nerves 3–4 pairs, basal gland pair inconspicuous, adaxial; petiole c. 1cm, swollen at

base and apex; inflorescence spike-like, c. 13cm, axillary; flower c. 1mm. Forest; 220–300m.

Distr.: Guinea (C) to Congo (K). (Guineo-Congolian).

IUCN: LC

Baduma: <u>Nemba 119</u> fr., 6/1986; **Bakolle Bakossi:** <u>Etuge 264</u> fr., 9/1986; **Eboné:** <u>Bamps 1511</u> fl., fr., 12/1967; **Ekona Mombo:** <u>Etuge 424</u> fl., 12/1986.

Mareyopsis longifolia (Pax) Pax & K.Hoffm.

Shrub or tree 8–30m; no exudate; stems puberulent at apex; leaves drying brown, oblanceolate, 52 × 15cm, acuminate, obtuse then abruptly rounded, margin acutely serrate-glandular, the teeth not confluent with the brochidodromous secondary nerves; petiole variable, 2.7cm, swollen at base and apex; inflorescences spike-like, to 20cm, ramiflorous. Evergreen forest; 300–900m.

Distr.: S Nigeria to Congo (K). (Lower Guinea & Congolian).

IUCN: LC

Bolo: <u>Nemba 94</u> 5/1986; **Kupe Village:** <u>Etuge 2903</u> 7/1996; **Ngomboaku:** <u>Ghogue 468</u> 12/1999; **Nyale:** <u>Etuge 4216</u> 2/1998.

Margaritaria discoidea (Baill.) G.L.Webster var. *discoidea*

J. Arnold. Arb. 48: 311 (1967).

Syn.: *Phyllanthus discoideus* (Baill.) Müll.Arg.

Deciduous tree 10–25m, glabrous; leaves ovate–elliptic, 7–9 × 3.5cm, subacuminate, acute to obtuse, bluish white below, drying green, secondary nerves 10 pairs, arcuate, quaternary nerves reticulate, conspicuous; petiole 8mm; fruit 3–lobed, glossy, 7mm. Farmbush, forest edge; 660–1300m.

Distr.: widespread in tropical Africa.

IUCN: LC

Kupe Village: <u>Cheek 8337</u> fr., 5/1996; <u>Etuge 1709</u> fr., 2/1996; <u>2797</u> fr., 7/1996; **Nyasoso:** <u>Cable 3271</u> fr., 6/1996.

Margaritaria discoidea (Baill.) G.L.Webster var. *fagifolia* (Pax) Radcl.-Sm.

F.T.E.A. Euphorbiaceae: 66 (1987).

As var. *discoidea*, but petiole glabrous or sparingly puberulous (not puberulous or pubescent); leaves elliptic-lanceolate, usually acutely acuminate (not obovate-elliptic or elliptic to suborbicular-obovate, subacute, obtuse or rounded). Forest; 220–300m.

Distr.: Guinea (B) to S Africa. (Tropical Africa).

IUCN: LC

Bakolle Bakossi: <u>Etuge 261</u> 9/1986; **Bolo Moboka:** <u>Nemba 240</u> 9/1986.

Micrococca mercurialis (L.) Benth.

Herb, erect, 15–30cm; stems white-puberulent on one side; leaves elliptic to rhombic, 2.5–1.5cm, acuminate, base cuneate to rounded, margin serrate to lobed; petiole up to 1.5cm; inflorescence axillary, filiform, 3cm, interrupted; female flowers 1–2, sepals 3, ovate, free, 1.5mm, glands subpetaloid; fruit 6mm diameter. Forest; 1600m.

Distr.: tropical Africa & Asia. (Palaeotropics).

IUCN: LC

Nyasoso: <u>Etuge 1828</u> fr., 3/1996.

Neoboutonia mannii Benth. var. *glabrescens* (Prain) Pax & K.Hoffm.

Syn.: *Neoboutonia glabrescens* Prain

Tree 10–14m, stellate hairy; leaves transversely elliptic, c. 15 × 15cm, rounded, base shallowly cordate, margin entire, lower surface scurfy with white, bun-shaped hair-clusters on veinlets; petiole 7cm; stipules ovate, 3mm; inflorescence terminal, much-branched, 20 × 20cm; flowers minute, male calyx glabrous; fruit 1cm, 3-lobed. Secondary forest; 350–1300m.

Distr.: Liberia to Gabon. (Upper & Lower Guinea).

IUCN: LC

Kupe Village: <u>Etuge 1711</u> 2/1996; **Nyandong:** <u>Thomas 7003</u> 5/1987.

Note: the author maintains that there are two distinct entities within *N. mannii*. Specimens held at K have been labeled "*N. mannii* var. *glabrescens* (Prain) Pax & K.Hoffm.but it is unclear as to whether the combination was ever validly published. It is therefore likely that a new combination will have to be made for this taxon following further investigation.

Local name: Ebolog. **Uses:** SOCIAL USES – drugs – Duncan Thomas states that in Accra, Ghana, people smoke the roots as a drug (Etuge 1711).

Neoboutonia mannii Benth. var. *mannii*

Tree, to 23m resembling var. *glabrescens*, but leaf surface below with long, soft, spreading hairs as well as scurf; male calyx puberulent; leaves orbicular-cordate, 10–20cm long; flowers yellowish green; styles narrow with 2 linear lobes. Forest edge, farmbush; 350–1900m.

Distr.: S Nigeria to Cameroon. (Lower Guinea).

IUCN: NT

Kupe Village: <u>Etuge 1976</u> fl., fr., 5/1996; **Mbule:** <u>Letouzey 14681</u> fl., 4/1976; **Messaka:** <u>Thomas 7001</u> fl., 5/1987; <u>7002</u> fl., 5/1987; **Mwanenguba:** <u>Etuge 4366</u> fl., 10/1998; **Ngombombeng:** <u>Etuge 126</u> fl., 5/1986; **Nyasoso:** <u>Sunderland 1479</u> 7/1992; <u>1481</u> fl., 7/1992; <u>Zapfack 634</u> fl., fr., 6/1996.

Pentabrachion reticulatum Müll.Arg.

Tree or shrub to 8m, glabrous; leaves oblong, 24 × 10cm, abruptly acuminate, rounded or obtuse, entire, tertiary nerves conspicuously scalariform; petiole 0.7cm, stout; stipules caducous; inflorescences axillary, ramiflorous; flowers several from numerous woody burrs, 5mm diameter; fruiting pedicels 2.5cm; fruit 1.5cm. Forest; 150–400m.

Distr.: Cameroon & Gabon. (Lower Guinea).

IUCN: NT

Etam: <u>Etuge 189</u> 7/1986; **Mombo-Bakossi:** <u>Olorunfemi in FHI 30556</u> fr., 5/1951.

Phyllanthus sp. aff. boehmii Pax

Decumbent herb, 40cm, resembling *P. reticulatus*; stem red, glabrous; leaflets elliptic, 1 × 0.5cm; fruit not seen. Grassland; 2000m.

Distr.: Cameroon. (W Cameroon Uplands).

Mwanenguba: <u>Leeuwenberg 9535</u> fl., 3/1972.

Note: possibly a new species, but material incomplete - not matched in W-C Africa by Radcliffe-Smith or Hoffmann.

Phyllanthus caesiifolius Petra Hoffm. & Cheek

Kew Bull. 58: 439 (2003).

Monopodial shrublet 0.2–0.4m, glabrous; lateral branches horizontal, 8–16cm; leaves oblong to elliptic, 1–2.5 × 0.6–1cm, obtuse to rounded apex and base, upper surface with basal half bluish white, upper half dark green; staminate flowers pale green, c. 1.5 × 1.5mm; pedicels 4–5mm; pistillate flowers 1.5–2 × 2mm; pedicels 7–10mm. Submontane forest; 700–1275m.

Distr.: Bakossi Mts. (Narrow Endemic).

IUCN: CR

Kodmin: <u>Cheek 9636</u> fl., 11/1998; **Ngomboaku:** <u>Cheek 10376</u> fl., 12/1999.

Note: endemic to the Bakossi Mts, distinguished from other *Phyllanthus* by the bicoloured leaves.

Phyllanthus muellerianus (Kuntze) Exell

Climber 8m; stem spines 3-pronged, ultimate branches c. 15cm, glabrous; leaves ovate, 6 × 2.8cm, acuminate, truncate, bluish green below; petiole 3mm; inflorescences multi-flowered, c. 5cm, in fascicles, in axils of fallen-leaf scars; fruit red, berry, 2.5mm; pedicel 3mm. Farmbush; 800–1500m.

Distr.: tropical Africa.

IUCN: LC

Kodmin: Cheek 8887 1/1998; **Kupe Village:** Etuge 1727 fr., 2/1996; **Ndum:** Elad 113 fr., 2/1995; **Nyandong:** Cheek 11470 3/2003; Tchiengue 1713 3/2003; **Nyasoso:** Etuge 2490 fl., 7/1996.

Local names: Mahoum (Elad 113); Mahwoom (Cheek 8887). **Uses:** FOOD ADDITIVES – bark – used for flavouring palm wine (Cheek 8887); SOCIAL USES – smoking materials/drugs – facilitants to drug effects – used to put in palm wine to give more alcohol (Elad 113).

Phyllanthus nummulariifolius Poir.

F.T.E.A. Euphorbiaceae: 28 (1987).

Syn.: *Phyllanthus capillaris* Schumach. & Thonn.

Subshrub, straggling 0.6–2m, resembling *P. reticulatus*; stem red or brown; leaves obovate or elliptic to 1.5 × 0.8cm, mucronate, obtuse; fruit 3-carpellate, 2mm diameter; pedicel 8mm. Fallow and grassland; 1200–2000m.

Distr.: tropical Africa.

IUCN: LC

Kodmin: Cheek 8902 fl., 1/1998; 9633 fl., 11/1998; **Kupe Village:** Etuge 1843 fl., 3/1996; **Mwanenguba:** Cheek 9445 fr., 10/1998.

Local name: forgotten (Cheek 8902).

Phyllanthus nyale Petra Hoffm. & Cheek

Kew Bull. 58: 442 (2003).

Monopodial shrublet resembling *P. caesiifolius*, but leaves not bicoloured above; lateral stems strongly winged, the wings as wide as the main stem. Forest; 1000m.

Distr.: Bakossi Mts. (Narrow Endemic).

IUCN: CR

Nyale: Cheek 9679 fl., 11/1998; Etuge 4453 fl., 11/1998.

Note: endemic on Nyale Rock. Distinguished from other species of *Phyllanthus* in Bakossi by the winged lateral stems.

Phyllanthus odontadenius Müll.Arg.

Annual herb 0.3–0.6(–1)m; main stem drying white, ridged, glabrous; ultimate stems c. 18cm, 12-winged; leaves oblong, c. 1.5 × 0.6cm; petiole 2mm; flowers single, minute, pendulous, white; fruit 3-carpellate, c. 2mm diameter; pedicel 2mm. Farmbush; 700–1050m.

Distr.: Guinea (B) to Angola. (Guineo-Congolian).

IUCN: LC

Kupe Village: Etuge 2730 fl., 7/1996; Lane 307 fl., 1/1995; 341 fl., fr., 1/1995; **Nyasoso:** Lane 44 fr., 8/1992.

Phyllanthus sp. A

Herb 40cm, resembling *P. reticulatus*, glabrous; leaves elliptic, 1.5 × 0.8cm; fruit 3–4mm, 3 carpels; pedicels 5mm. Stream gully in submontane forest; 970m.

Kupe Village: Cable 779 fr., 1/1995.

Note: possibly a new species, but male and female flowers required to investigate further.

Plagiostyles africana (Müll.Arg.) Prain

Tree 10m; white exudate, glabrous; leaves drying green, narrowly elliptic or obovate elliptic, c. 14 × 4.5cm, acuminate, broadly acute, acutely serrate, nerves 5, brochidodromous, forking c. 8mm from margin; petiole 2cm, swollen at base and apex; inflorescence axillary, c. 3cm; flowers 3mm; fruit transversely ellipsoid, fleshy, red, c. 2 × 1.5cm, 1-seeded. Evergreen forest; 250–500m.

Distr.: SE Nigeria to Congo (K). (Lower Guinea & Congolian).

IUCN: LC

Ikiliwindi: Etuge 508 3/1987; **Konye:** Nemba 444 fl., 1/1987; **Kupe Village:** Etuge 1472 11/1995; **Nlog:** Etuge 22 fr., 4/1986; **Nyandong:** Cheek 11333 3/2003.

Plukenetia conophora Müll.Arg.

Flora 47: 530 (1864).

Syn.: *Tetracarpidium conophorum* (Müll.Arg.) Hutch. & Dalziel

Liana to 15m, glabrous; leaves ovate-elliptic, c. 10 × 5.5cm, acumen 0.5cm, base obtuse, crenate-serrate, 3-nerved at base; petiole 8cm; inflorescences axillary, paniculate, c. 6cm; flowers dull-white, 2mm; fruit capsule 4-lobed, 7cm diameter, winged and ridged. Secondary forest; 250–1000m.

Distr.: Sierra Leone to Congo (K). (Guineo-Congolian).

IUCN: LC

Konye: Nemba 434 1/1987; **Kupe Village:** Etuge 1968 fl., 5/1996; **Kurume:** Thomas 5419 1/1986; 5455 1/1986; 5456 1/1986; **Mwambong:** Cheek 9114 2/1998; **Ndibise:** Etuge 537 6/1987; **Nyandong:** Cheek 11306 3/2003.

Local names: Kat or Kay (Cheek 11306); Cashew Nut (*fide* Etuge). **Uses:** FOOD – seeds harvested extensively for sale as snacks; 'noisette' (Cheek 11306); seeds edible boiled (*fide* Etuge).

Protomegabaria stapfiana (Beille) Hutch.

Tree 10–15m, buttressed; stems glabrous, no exudate; leaves drying matt, papery, oblanceolate, 14–51 × 6–15cm, acute or obtuse, cuneate or obtuse, entire, midrib pubescent below; petiole to 7cm, swollen at base and apex; inflorescence ramiflorous or in lower leaf axils, 1–5 per node, spikes 3cm; flowers 3cm. River edge; 250–600m.

Distr.: Sierra Leone to Gabon. (Upper & Lower Guinea).

IUCN: LC

Bolo: Nemba 57 fl., 3/1986; **Konye:** Nemba 439 fl., 1/1987; **Ngombombeng:** Etuge 29 4/1986.

Pseudagrostistachys africana (Müll.Arg.) Pax & K.Hoffm. subsp. *africana*

Tree 5–20m tall; bark whitish with green spots; slash brown to red; leaves leathery, elliptic, 30 × 11–18cm, subacuminate, base rounded then slightly decurrent, with a pair of flat glands, ± serrate, nerves 21–24 pairs, venation scalariform; petiole 2.5–5cm; stipule single, 3cm, caducous, scar completely encircling stem. Forest; 710–1450m.

Distr.: Ghana, Bioko, São Tomé, Cameroon. (Upper & Lower Guinea).

IUCN: VU

Kodmin: Cheek 9602 11/1998; **Kupe Village:** Cable 3751 7/1996; Elad 92 fl., 1/1995; Ryan 224 5/1996; **Nyale:** Etuge 4238 2/1998; **Nyasoso:** Thomas 3045 2/1984.

Uses: unknown (Cheek 9602).

Pycnocoma cornuta Müll.Arg.

Sparsely-branched shrub, 1–1.5m; with terminal leaf rosettes; leaves oblanceolate, c. 40 × 13cm, acuminate, cuneate, decurrent, margin serrate-dentate; petiole 1cm; inflorescences spike-like, c. 16cm, erect in upper axils, rachis densely

pubescent; bracts orbicular, villous; female flowers c. 2cm long; ovary wings 8mm. Forest; 250m.
Distr.: Ghana to CAR. (Upper & Lower Guinea).
IUCN: NT
Ikiliwindi: Nemba 678 12/1987.

Pycnocoma macrophylla Benth.
Syn.: *Pycnocoma brachystachya* Pax
Shrub resembling *P. cornuta*, but rachis sparsely-pubescent to glabrescent; bracts ovate, glabrescent. Forest; 1000m.
Distr.: Ivory Coast to Congo (K). (Guineo-Congolian).
IUCN: LC
Nyale: Etuge 4448 fl., 11/1998.

Ricinodendron heudelotii (Baill.) Pierre ex Heckel subsp. *africanum* (Müll.Arg.) J.Léonard var. *africanum*
Bull. Jard. Bot. Nat. Belg. 31: 398 (1961).
Deciduous tree 30m; trunk grayish white, glabrous except in bud, brown-stellate-hairy; leaves alternate, palmately compound; seedlings lobed; leaflets 5, subelliptic, c. 24 × 10cm, acute at base and apex, denticulate; petiole c. 20cm, gland pair at apex; stipules foliaceous, 2 × 3cm, deeply dentate, caducous; inflorescence diffuse, c. 45 × 45cm, paniculate, with new leaves; flowers 2mm wide, greenish white; fruit subfleshy, indehiscent, 2-lobed, grey-green, c. 3 × 3cm; seed yellow, 0.8cm diameter. Old secondary forest; 400–850m.
Distr.: Guinea (B) to Tanzania. (Guineo-Congolian).
IUCN: LC
Kupe Village: Etuge 1938 fr., 5/1996; **Nlonako:** Etuge 1790 fl., 3/1996; **Nyasoso:** Etuge 2603 fr., 7/1996; Thomas 3059 fl., 2/1984.
Local names: Esange and Nja-sanga. **Uses:** FOOD – seeds – the edible seeds are ground and cooked in soup, particularly with fish; MATERIALS – wood – tools, musical instruments – the wood is used for planks, making fufu mortars and carving local guitars (Etuge 1790), timber (*fide* Etuge); FOOD ADDITIVES – seed used as spice (Etuge 2603).

Ricinus communis L.
Robust, subwoody, erect herb to 2m, glabrous; leaves alternate, digitately lobed, blades to 60cm, glandular-toothed, glaucous; inflorescence terminal, paniculate; female flowers above, 1cm, white; males below; capsule 2–3cm; seeds 8mm, mottled. Cultivated, rarely persisting in fallow; 1170m.
Distr.: pantropical, originally from NE & E Africa.
IUCN: LC
Edib village: Cheek (sight record) 2/1998.
Local name: castor oil plant. **Uses:** MEDICINES (*fide* Cheek).

Shirakiopsis elliptica (Hochst.) H.-J.Esser
Govaerts, Frodin & Radcliffe-Smith, World Checklist & Bibliogr. Euphorbiaceae 4: 1470 (2000).
Syn.: *Sapium ellipticum* (Krauss) Pax
Deciduous shrub or tree 3–25m; white exudate, glabrous; leaves drying black, leathery, elliptic, c. 9 × 4cm, obtuse, acute, finely serrate; petiole 3mm; inflorescence terminal spikes, 6cm; flowers numerous, 1mm, green; fruit globose or bilobed, fleshy, 8mm; styles 2, coiled. Forest or forest edge; 850–1400m.
Distr.: tropical & S Africa.
IUCN: LC
Bangem: Villiers 1335 fr., 5/1982; **Kodmin:** Plot B40 1/1998; **Lake Edib:** Cheek 9164 fl., 2/1998; **Mejelet-Ehumseh:** Etuge 174 fr., 6/1986; 492 2/1987; **Mwanenguba:** Leeuwenberg 10013 fr., 6/1972; 9268 fl.,

1/1972; **Mwambong:** Cheek 9104 fl., 2/1998; **Nyasoso:** Cable 2826 fr., 6/1996; Etuge 2074 fr., 6/1996; Thomas 3052 2/1984.

Sibangea similis (Hutch.) Radcl.-Sm.
Kew Bull. 32: 481 (1978).
Syn.: *Drypetes similis* Hutch.
Tree 8m; stems glabrous, terete, longitudinally wrinkled, greyish white; leaves oblong, to 32 × 9cm, apex narrowly acuminate, base obliquely acute, margin with c. 8 obscure spiny teeth, midrib drying black, blade dark brown; petiole 2cm; stipules woody; inflorescence ramiflorous, of single shortly-pedicellate flowers; flowers with sepals narrowly oblong, persisting in fruit. Evergreen forest; 200–300m.
Distr.: SE Nigeria & Cameroon. (Lower Guinea).
IUCN: NT
Ikiliwindi: Nemba 551 6/1987; **Mungo River F.R.:** Cheek 10122 11/1999.
Note: less than 20 specimens are known.

Spondianthus preussii Engl. var. *preussii*
Tree 10m, dense-crowned; leaves leathery, clustered in dense whorls of 5–7, clusters separated by internodes c. 15cm, sessile and long-petiolate leaves in each cluster; petiolate leaves with blades elliptic, c. 25 × 20cm, apex and base rounded; petiole 10cm; sessile leaves smaller, c. 10 × 6cm; infructescence terminal, c. 10cm, bearing numerous red globular fruits, 1.5cm diameter. River edge; 150m.
Distr.: Liberia to Bioko, Cameroon & Gabon. (Upper & Lower Guinea).
IUCN: LC
Mungo River: Gosline 228 fr., 11/1999; **Mungo River F.R.:** Cheek 10147 11/1999.

Tetrorchidium didymostemon (Baill.) Pax & K.Hoffm.
Syn.: *Tetrorchidium minus* (Prain) Pax & K.Hoffm.
Tree 5m, glabrous; stems zig-zag, producing dishwater-like exudate; leaves obovate or elliptic, c. 17 × 7cm, apex rounded, acumen abrupt, 1–2cm, base cuneate; petiole 1cm; stipules reduced to a line of finger-like glands; female flowers solitary, subterminal; ovary naked; stigmas 3, plate-like; male inflorescences yellow, to 4cm, leaf-opposed; fruit 3-angled, 7mm diameter. Secondary forest; 350–1500m.
Distr.: Guinea (C) to Tanzania. (Guinea-Congolian).
IUCN: LC
Bakolle Bakossi: Etuge 141 5/1986; **Edib village:** Cheek 9179 fl., 2/1998; **Kodmin:** Cheek 8879 fr., 1/1998; **Kupe Village:** Cable 2615 fr., 5/1996; Etuge 1971 fr., 5/1996; 2831 fr., 7/1996; Schoenenberger 48 fl., 11/1995; **Ngomboaku:** Etuge 4662 fl., 12/1999.

Thecacoris annobonae Pax & K.Hoffm.
Syn.: *Thecacoris cf. annobonae* Pax & K.Hoffm. of FWTA
Tree 3–10m; no exudate; stems glabrescent, grey-wrinkled; leaves drying dark green-brown, oblong to obovate, c. 20 × 9cm, acuminate, obtuse, entire, nerves c. 6 parts; petiole 0.5cm; stipule caducous; male inflorescence 1 per leaf axil, interrupted spikes c. 20cm; flowers yellow, 2mm; Forest; 300–600m.
Distr.: Cameroon & Annobon. (Lower Guinea).
IUCN: EN
Bolo: Nemba 99 fr., 5/1986; **Konye:** Nemba 14 fl., 4/1986; **Mungo River:** Olorunfemi in FHI 30531 4/1951; **Ngusi:** Etuge 48 fl., fr., 4/1986.

EUPHORBIACEAE

Thecacoris leptobotrya (Müll.Arg.) Brenan
Tree or shrub 2–10m; resembling *T. annobonae*, but leaves c. 14 × 6cm, nerves c. 8–10 pairs; male inflorescence 5–7cm. Forest; 350m.
Distr.: SE Nigeria to Congo (K). (Lower Guinea & Congolian).
IUCN: LC
Bakolle Bakossi: Etuge 153 5/1986.

Thecacoris stenopetala (Müll.Arg.) Müll.Arg.
Shrub 2–4m; leaves elliptic or obovate-elliptic, 5–18 × 2–9cm, acuminate or obtuse, cuneate to rounded, midrib setose below; male inflorescence 2–8cm, pubescent; petals greenish yellow, linear-oblanceolate, ciliate; capsule 3-lobed, 9mm diameter. Forest; 300m.
Distr.: Sierra Leone to Bioko, SW Cameroon & São Tomé. (Upper & Lower Guinea).
IUCN: LC
Konye: Thomas 5161 11/1985.

Tragia benthamii Baker
Climber 3m, stinging; stem long-hairy; leaves ovate, c. 6.5 × 3.5cm, acuminate, cordate, serrate, densely pubescent below; petiole 2cm; female calyx lobes 6, lacking foliaceous apex, deeply pectinate, the lobes setose. Montane scrub; 1300–1975m.
Distr.: tropical Africa.
IUCN: LC
Edib: Etuge 4494 fr., 11/1998; **Kodmin:** Cheek 8889 fl., 1/1998; 9605 fr., 11/1998; **Mwanenguba:** Cheek 7273 fr., 2/1995.
Local names: Tro bo ma pool (Cheek 8889); Toluh wombah (Cheek 9605). **Uses:** FOOD ADDITIVES – unspecified parts – used in cooking dog (Cheek 8889); MEDICINES (Cheek 9605).

Tragia preussii Pax
Climber, 6m; stem puberulent; leaves lanceolate, c. 14 × 6cm, acuminate-acute, base hastate, margin sinuate, glabrous; petiole c. 3–5cm; stipule triangular, 2mm; infloresence c. 6cm; calyx lobes as *T. benthamii*. Evergreen forest; 400–960m.
Distr.: S Nigeria to Congo (K). (Lower Guinea & Congolian).
IUCN: LC
Bakossi F.R.: Etuge 4308 fl., 10/1998; **Kupe Village:** Cheek 7812 fr., 11/1995.

Tragia sp. B
Climber 3m, stinging; stem sparsely-puberulent; leaves lanceolate, c. 9 × 3.5cm, long-acuminate, deeply cordate, margin serrate-sinuate, upper surface long-setose, lower inconspicuously puberulent; inflorescence axillary, 2–3cm; pedicel of basal female flower as long as the male part of the inflorescence; female calyx lobes 6, apex foliaceous, pectinate below, with 1–2 teeth on each side; Forest; 870–970m.
Distr.: Cameroon. (Narrow Endemic).
Nyasoso: Cable 3286 fr., 6/1996; Cheek 7520 fl., 10/1995.
Note: *Tragia sp. B* is distinct from *Tragia sp. A* of FWTA. Apparently a new species; the combination of long female pedicels and foliaceous calyx lobe apices is unique in W-C Africa.

Uapaca guineensis Müll.Arg.
Tree 4.5–18m, with conspicuous, arching stilt roots, glabrous; leaves obovate, c. 27 × 13cm, rounded, acute, margin slightly sinuous; petiole c. 3cm; stipules subulate, 4mm; inflorescences of single flowers from axils of fallen-leaf scars; fruits fleshy, globose, ellipsoid, 2cm diameter, warty, glabrous. Forest; 200–650m.
Distr.: Sierra Leone to Congo (K). (Guineo-Congolian).
IUCN: LC
Konye: Thomas 7015 /1987; **Kupe Village:** Lane 385 fr., 1/1995.

Uapaca staudtii Pax
Tree resembling *U. guineensis*, but stipules persistent, orbicular, foliaceous, c. 1 × 1cm; leaves oblanceolate. Evergreen forest; 200–710m.
Distr.: S Nigeria to Gabon. (Lower Guinea).
IUCN: NT
Bakole: Etuge 391 11/1986; **Mile 12 Mamfe Road:** Nemba 310 fr., 10/1986; **Nyale:** Etuge 4218 2/1998.

Uapaca togoensis Pax
Tree 12m, densely pubescent, with or without stilt roots; leaves leathery, scabrid, oblong-obovate, 11–25 × 5–15.5cm, rounded, base cuneate to rounded; petiole 1–8cm; stipules linear, caducous; male flowers 5mm diameter; calyx and fruit pubescent. Secondary forest; 520m.
Distr.: Mali to CAR. (Upper & Lower Guinea).
IUCN: LC
Eboné: Leeuwenberg 10222 7/1972.
Note: specimen cited not seen at K.

Uapaca vanhouttei De Wild.
Tree c. 40m, resembling *U. guineensis*, but stems, petioles and leaf midribs golden-pilose, the hairs c. 2mm long. Evergreen forest; 450–710m.
Distr.: S Nigeria to Congo (K). (Lower Guinea & Congolian).
IUCN: LC
Bakossi F.R.: Cheek 9329 fr., 10/1998; **Nyale:** Etuge 4234 2/1998; **Nyandong:** Ghogue 1488 3/2003.

Uapaca sp. A
Tree 9m resembling *U. guineensis*, but leaves 43 × 19cm, acumen 1cm; petiole 8cm; stipules ?foliaceous (poorly preserved). Evergreen forest; 900m.
Kupe Village: Etuge 2612 7/1996.
Note: this may be an aberrant specimen of *U. staudtii*. More and better material is needed to evaluate its status.

Vernicia montana Lour.
Syn.: *Aleurites montana* Wilson
Tree 12m, no exudate, glabrous; leaves broadly ovate, dimorphic, entire, c. 12 × 10cm, acuminate, cordate, basal glands connate on upper surface of petiole apex, or 3-lobed, c. 18 × 18cm, lobes c. 6cm deep, with a gland at the base of the sinuses; inflorescences terminal, paniculate, c. 20-flowered, 15cm long, in fascicles from flushing, often leafless stems; flowers c. 3cm wide; calyx 2-lobed, petals 5, white, anthers in 2 whorls of 5. *Cacao* farm, shade tree; 700m.
Distr.: SE Asia, planted in tropical Africa. (Palaeotropics).
IUCN: LC
Kupe Village: Cheek 7036 fl., 1/1995.

FLACOURTIACEAE

M. Cheek

Casearia barteri Mast.

Tree or shrub 5–12m; leaves distichous, elliptic-oblong, leathery, c. 20 × 8cm, subacuminate, rounded to obtuse, entire, nerves 8 pairs, glabrous; petiole 1cm; inflorescence axillary, fasciculate, 5–10-flowered; flowers 3mm, pale green; sepals 5; petals 10; stamens 10; fruit orange, ellipsoid, c. 4 × 3cm, 3-valved. Swamp forest; 200–300m.
Distr.: Sierra Leone to Congo (K). (Guineo-Congolian).
IUCN: LC
Bolo: Nemba 95 5/1986; **Ikiliwindi:** Nemba 561 6/1987.
Note: specimens not seen.

Dasylepis racemosa Oliv.

Tree (4–)6–15m; bark smooth, green-grey with brown plaques; leaves elliptic or oblong, c. 19 × 9cm, acuminate, obtuse, undulate-dentate; petiole 1cm; inflorescence axillary, spicate, 5–6cm; flowers 1–2cm diameter; sepals 4, pink-red, 0.7cm; petals c. 8, white; stigmas 3; fruit 3-valved, globose, 1.5–2cm diameter, thick-walled. Forest, *Aframomum* thicket; 860–1900m.
Distr.: SE Nigeria, Cameroon, Congo (K) & Uganda. (Lower Guinea & Congolian (montane)).
IUCN: LC
Kupe Village: Cable 809 fl., 1/1995; **Menyum:** Doumenge 449 5/1987; **Mt Kupe:** Cheek 7613 11/1995; **Ndum:** Elad 103 2/1995; **Nyasoso:** Cheek 7307 fl., 2/1995; Etuge 1529 1/1996; 1741 2/1996; 2467 6/1996; Groves 84 2/1995.
Local name: Mpun. **Uses:** MATERIALS – snares/traps – wood very strong, used for spring traps (Epie in Cheek 7307).

Dasylepis seretii De Wild.

Tree c. 15m, much resembling *D. racemosa*, but pedicels 2–3mm, not 6–10mm; leaves only 5–6cm wide, finely scalariform-reticulate. Forest; 600m.
Distr.: Cameroon & Congo (K). (Lower Guinea & Congolian).
IUCN: LC
Ngusi: Etuge 43 fr., 4/1986.
Note: *D. blackii* = *D. thomasii* Breteler, almost certainly occurs in our area: it is known from S Bakundu and Kumba.

Dovyalis zenkeri Gilg

Shrub 4–8m; stem spiny, puberulent; stems medium- or red-brown, with lenticels; leaves drying dark brown, elliptic-oblong, c. 10 × 4cm; inflorescences fasiculate; male flowers green, 6mm; sepals 6; fruit globose, densely hairy, 2cm; sepals 8. Forest; 700–1000m.
Distr.: Guinea (B) to Congo (K). (Guineo-Congolian).
IUCN: LC
Kupe Village: Cable 3883 7/1996; Etuge 1906 5/1996; 2720 7/1996; **Nyasoso:** Cable 3233 6/1996; 3439 7/1996.

Dovyalis sp. nov.

Shrub or tree, 3–7m; spiny, glabresent; stems whitish brown, lenticels conspicuous; leaves drying green, ovate or elliptic, 6–9 × 3.5cm, long-acuminate, rounded, margin undulate to sublobed, nerves 4–5 pairs, midrid glabrous; petiole 0.5cm; only female inflorescences known; flowers single; sepals 5; styles 5–7; fruit orange, fleshy, globose, 2cm, glabrous. Forest; 800–1300m.
Distr.: Cameroon. (W Cameroon Uplands).

Kupe Village: Cable 835 fr., 1/1995; Cheek 7049 fr., 1/1995; Etuge 1449 11/1995; 1713 2/1996; Gosline 250 12/1999; **Nyasoso:** Groves 85 fr., 2/1995; Lane 181 fr., 11/1994; Sebsebe 5000 10/1995.

Flacourtia indica (Burm.f.) Merr. *vel. aff.*

Dioecious tree or shrub 2–15m; leaves drying dark brown, oblong-elliptic, c. 13 × 5cm, acuminate, serrate; petiole 1.5cm; inflorescence 2cm, 3–5-flowered; flowers green, 2–3mm; sepals 4, white-hairy inside; styles 4–6, spreading, apically 2–3-lobed; fruit a 5–6-seeded, globose berry, 1–1.5cm. Forest; 750–870m.
Distr.: tropical Africa.
IUCN: LC
Kupe Village: Cable 2437 5/1996; 3807 7/1996.
Note: while it is clear that we have two taxa of *Flacourtia*, application of names is unclear.

Flacourtia vogelii Hook.f.

Shrub 1–2m; spines 1–2cm on leafy shoots; leaves drying green, ovate 6–12 × 4–5cm, acuminate, obtuse, nerves 4 pairs, serrate-crenate. Forest; 700–1400m.
Distr.: Sierra Leone to Cameroon & CAR. (Upper & Lower Guinea).
IUCN: LC
Kodmin: Plot B188 1/1998; **Kupe Village:** Cheek 6091 1/1995.
Note: fertile material needed to confirm identity.

Homalium africanum (Hook.f.) Benth.

Syn.: *Homalium sarcopetalum* Pierre
Syn.: *Homalium molle* Stapf
Tree 4.5–8m; leaves papery, drying grey-brown above, green below, long-elliptic to oblong, to 20 × 7cm, acuminate, rounded-obtuse, entire to obscurely coarsely serrate; lateral nerves 7 pairs, acute, glabrous; petiole 0.5cm; stipules falcate, to 3cm, persistent; inflorescence terminal, c. 20 × 15cm, with 5–15 branches; petals in fruit c. 2mm. Riverbank in forest; 150–800m.
Distr.: Sierra Leone to Congo (K). (Guineo-Congolian).
IUCN: LC
Kupe Village: Lane 405 fl., 1/1995; **Mwanenguba:** Leeuwenberg 9579 fl., 4/1972; **Mungo River:** Gosline 224 fl., 11/1999.

Homalium dewevrei DeWild. & T.Durand

Fl. Gabon 34: 72 (1995).
Tree 7–12m; leaves narrowly oblong-elliptic, to 18 × 6cm, acuminate, obtuse, serrate, veins 8–10 pairs; petiole to 2cm; inflorescence resembling *H. africanum*, but branching evenly along main axis; fruiting petals 6mm. Forest; 1100m.
Distr.: Cameroon to Congo (K). (Lower Guinea & Congolian).
IUCN: LC
Mejelet-Ehumseh: Etuge 491 2/1987.
Note: specimen not seen.

Homalium hypolasium Mildbr.

Notizbl. Bot. Gart. Berlin 8: 173 (1922).
Tree resembling *H. africanum*, but leaves elliptic, c. 13 × 8cm, apex emarginate, base rounded, margin finely serrulate; lateral nerves 12 pairs, lower surface densely velvety-yellowish brown tomentulose; stipules caducous. Forest; 1200m.
Distr.: Cameroon & Rio Muni. (Lower Guinea).
IUCN: EN
Mt Kupe: Thomas 5100 12/1985.
Note: specimen not seen.

FLACOURTIACEAE

Homalium letestui Pellegr.
Syn.: *Homalium skirlii* Gilg ex Engl.
Tree 20–30m; leaves leathery, oblong-elliptic, 25–11cm, shortly acuminate, cordate, crenate, glabrous; petiole 0.5cm; stipules caducous. Forest; 250–1000m.
Distr.: Guinea (C) to Gabon. (Upper & Lower Guinea).
IUCN: LC
Bolo: Nemba 49 fl., fr., 3/1986; **Ikiliwindi:** Etuge 515 3/1987; **Loum F.R.:** Onana 997 12/1999; **Mwambong:** Onana 559 2/1998.

Homalium longistylum Mast.
Bull. Jard. Bot. Nat. Belg. 43: 270 (1973).
Syn.: *Homalium macropterum* Gilg
Syn.: *Homalium aylmeri* Hutch. & Dalziel
Tree 10m; leaves leathery, elliptic, 9–13 × 6–8cm, shortly acuminate, obtuse-decurrent, serrate-undulate, glabrous; petiole 1cm; inflorescence axillary, unbranched, 10–15cm; fruiting petals c. 1.2cm. Forest; 200m.
Distr.: tropical Africa.
IUCN: LC
Ikiliwindi: Nemba 544 6/1987.
Note: specimen not seen.

Oncoba cf. bukobensis Gilg
Shrub 3m; leaves papery, elliptic, 15 × 9cm, rounded to shortly acuminate, base truncate to slightly cordate, serrate-spiny, glabrous, quaternary nerves prominent below; petiole to 6cm; infloresence axillary, below leaves, 4cm, c. 4-flowered; fruit globose, 1.5cm, including spines 0.5cm, orange, sparsely puberulent. Forest edge; 1450m.
Kodmin: Cheek 9629 fr., 11/1998.
Note: possibly a new species, most similar to *O. bukobensis* of Burundi. Known only from this specimen.

Oncoba dentata Oliv.
Fl. Gabon 34: 52 (1995).
Syn.: *Lindackeria dentata* (Oliv.) Gilg
Tree or shrub to 15m; leaves oblong-elliptic or ovate, 15–22 × 7–12cm, shortly acuminate, dentate to subentire; lateral nerves 6–10 pairs; petioles 3–16cm; inflorescence 5–10cm; flowers in fascicles of 1–5; sepals 3, 5mm; petals 6–10, 5–10mm; fruit orange, indehiscent, with long bristles, c. 2cm diameter. Forest; 150–1700m.
Distr.: Guinea (C) to Sudan. (Guineo-Congolian).
IUCN: LC
Bolo Moboka: Thomas 5717 fl., 3/1986; **Enyandong:** Ghogue 1300 fr., 11/2001; **Etam:** Etuge 80 3/1986; **Ikiliwindi:** Nemba 519 fr., 6/1987; **Kupe Village:** Cable 2469 5/1996; 3706 7/1996; Etuge 1894 5/1996; 2643 7/1996; Kenfack 238 7/1996; **Messaka:** Ghogue 1509 fl., 3/2003; 1516 3/2003; **Mombo:** Olorunfemi in FHI 30578 fr., 5/1951; **Nyandong:** Cheek 11315 3/2003; **Nyasoso:** Balding 21 fr., 7/1993; Cable 2879 6/1996; 3232 6/1996; 3511 7/1996; Etuge 12 fl., 4/1986; 1750 2/1996; 2502 7/1996; Sunderland 1474 fr., 7/1992; Zapfack 671 6/1996.
Uses: ENVIRONMENTAL USES – boundaries – sticks used for fencing (Etuge 1750).

Oncoba glauca (P.Beauv.) Planch.
Fl. Gabon 34: 44 (1995).
Syn.: *Caloncoba glauca* (P.Beauv.) Gilg
Tree c. 10m; leaves lanceolate to elliptic, 11–25 × 4–8cm, long-acuminate, entire; flowers to 10cm diameter; sepals 3, 2cm; petals c. 12–13, 3–6cm; stamens numerous; fruit orange, ovoid, pendulous, c. 6cm. Forest; 150–1200m.
Distr.: Nigeria to Gabon. (Lower Guinea).
IUCN: LC
Bolo: Thomas 5716 3/1986; **Bolo Moboka:** Nemba 235 fr., 9/1986; **Etam:** Etuge 77 3/1986; **Kupe Village:** Cheek 8331 5/1996; Etuge 1414 11/1995; 1986 5/1996; Kenfack 271 7/1996; Lane 297 fr., 1/1995; Ryan 248 5/1996; 428 5/1996; **Messaka:** Ghogue 1520 fr., 3/2003; **Mt Kupe:** Schoenenberger 49 fr., 11/1995; **Mungo River F.R.:** Cheek 10230 12/1999; **Ngomboaku:** Gosline 266 12/1999; **Nyandong:** Stone 2460 3/2003; **Nyasoso:** Cheek 7532 10/1995.

Oncoba lophocarpa Oliv.
F.T.A. 1: 117 (1868).
Syn.: *Caloncoba lophocarpa* (Oliv.) Gilg
Tree (6–)12–25m; leaves elliptic or elliptic-ovate, c. 14 × 7cm, long acuminate; inflorescences cauliflorous, c. 2cm long, or on long, whip-like, mostly leafless branches radiating 3–5m from the trunk on the ground; flowers 5–7cm diameter; sepals 3, 1.5cm; petals white, 8, c. 3cm; fruit ovoid, c. 5 × 4cm, strongly 8–12 winged. Forest; 800–1950m.
Distr.: Cameroon. (W Cameroon Uplands).
IUCN: VU
Bangem: Thomas 5314 1/1986; **Kodmin:** Etuge 3995 1/1998; Onana 570 2/1998; Pollard 193 11/1998; **Kupe Village:** Cable 2600 5/1996; 811 fl., 1/1995; Cheek 7188 fl., 1/1995; Ryan 256 5/1996; **Mt Kupe:** Leeuwenberg 8812 fl., 12/1971; Letouzey 14692 fl., 4/1976; Sebsebe 5078 11/1995; **Mwambong:** Mackinder 161 1/1998; **Nyasoso:** Cable 1238 fl., 2/1995.

Oncoba mannii Oliv.
Fl. Gabon 34: 56 (1995).
Syn.: *Camptostylus mannii* (Oliv.) Gilg
Tree (4–)8–15m; leaves leathery, elliptic, 14–17 × 7–9cm, subacuminate; inflorescence axillary; flowers 10–20 in sessile clusters of 2–4; pedicels 2cm; flowers 2.5–4cm diameter; sepals 3, 0.7cm; petals c. 12, 1.8cm; fruit ovoid, c. 3 × 2.5cm, green, speckled white. Forest; 200–1700m.
Distr.: SE Nigeria to Congo (K). (Lower Guinea & Congolian).
IUCN: LC
Bangem: Thomas 5285 1/1986; **Ehumseh:** Manning 459 10/1986; **Ikiliwindi:** Nemba 456 1/1987; **Kupe Village:** Cable 757 fl., 1/1995; Lane 403 fl., 1/1995; 429 fl., 1/1995; **Kurume:** Thomas 5420 fl., 1/1986; **Mombo:** Olorunfemi in FHI 30559 fr., 5/1951.

Oncoba ovalis Oliv.
F.T.A. 1: 118 (1868).
Syn.: *Camptostylus ovalis* (Oliv.) Chipp
Tree or shrub 3–8m; leaves elliptic, c. 13 × 5cm, acuminate; petiole 3cm; inflorescence axillary, 6cm, a 5–10-flowered raceme; flowers white, 1–2cm diameter; sepals 3, 5mm; petals 12, 1cm; fruit subcylindric, 4cm, orange, 6–8-winged; rostrum c. 0.5cm. Forest; 900–1750m.
Distr.: Cameroon. (W Cameroon Uplands).
IUCN: NT
Ehumseh - Mejelet: Etuge 331 10/1986; **Kodmin:** Cheek 8911 1/1998; 9595 fl., 11/1998; Etuge 3967 1/1998; Ghogue 29 1/1998; Onana 567 2/1998; **Kupe Village:** Cable 2606 5/1996; 3794 7/1996; Cheek 7200 fl., fr., 1/1995; 7758 11/1995; 7776 11/1995; Etuge 2647 7/1996; 2855 7/1996; **Mbule:** Cable 3342 6/1996; 3369 6/1996; **Mt Kupe:** Leeuwenberg 8804 fl., 12/1971; Schoenenberger 52 fr., 11/1995; **Mwambong:** Onana 539 2/1998; **Nyasoso:** Etuge 2374 6/1996; Sebsebe 5057 10/1995.
Note: restricted to Mt Cameroon (7 collections), Kupe-Bakossi (6 sites) and Nigeria (1 site), but habitat threatened only at lower part of altitudinal range.

Local name: Mukey. **Uses:** use unknown (Max Ebong in Cheek 9595).

Oncoba cf. spinosa L.

Spiny shrub, 3m; leaves leathery, elliptic, c. 7.5 × 4cm, obtuse at base and apex, margin serrate, glabrous; petiole 0.5cm; spines 0.5–3cm; Forest; 1975m.
Mwanenguba: Cheek 7256 2/1995.
Note: possibly *sp. nov.* of Bamenda Highlands; flowers and fruits required.

Ophiobotrys zenkeri Gilg

Tree 30–40m; leaves leathery, elliptic, c. 11 × 5cm, shortly-acuminate, acute, margin revolute, entire to slightly toothed, glabrous; petiole 0.5cm; inflorescence terminal, c. 20 × 12cm; partial peduncles 5–6, each with 3–5 basal branches, ultimate branches c. 20, 15cm, spicate; flowers 4mm wide; sepals and petals reflexed, adaxial appendages appressed to ovary. Forest; 200m.
Distr.: Ivory Coast to Gabon. (Upper & Lower Guinea).
IUCN: LC
Ikiliwindi: Nemba 557 6/1987.
Note: specimen not seen.

Phyllobotryon spathulatum Müll.Arg.

Fl. Gabon 34: 24 (1995).
Syn.: *Phyllobotryon soyauxianum* Baill.
Monopodial shrub 1m; leaves leathery, dark green, oblanceolate, c. 90 × 15cm, shortly acuminate, subsessile, entire or weakly-toothed; inflorescences sessile, from midrib of upper surface; flowers c. 0.7cm diameter, pink; fruits ovoid, purple, 1cm, ridged. Forest; 250m.
Distr.: Nigeria, Cameroon & Gabon. (Lower Guinea).
IUCN: NT
Loum F.R.: Cheek 10246 12/1999.

Scottellia klaineana Pierre var. *klaineana*

Fl. Gabon 34: 33 (1995).
Tree 15m or more; leaves leathery, elliptic, 15cm, acuminate, margin acutely and inconspicuously spiny-serrate in distal third, nerves brochidodromous; petiole 1.5cm. Forest; 150m.
Distr.: Sierra Leone to Gabon. (Upper & Lower Guinea).
IUCN: LC
Mungo River F.R.: Cheek 10140 11/1999.

GENTIANACEAE

Y.B. Harvey & I. Darbyshire

Sebaea brachyphylla Griseb.

Erect herb 20(–50)cm; stems glabrous, few-branched, dichotomous; leaves sessile, orbicular, 8mm; cymes terminal, dense, 5–20-flowered; pedicels c. 1mm; sepals oblong, 3mm, with a prominent central nerve; corolla 6mm, inflated at base, yellow; stamens exserted. Grassland; 1700–2000m.
Distr.: tropical Africa. (Afromontane).
IUCN: LC
Mwanenguba: Cheek 9420 10/1998; Letouzey 14390 fl., 8/1975.

Sebaea oligantha (Gilg) Schinz

Adansonia (sér. 2) 7: 216 (1967).
Erect, saprophytic, unbranched herb 1–4cm, glabrous, white throughout; leaves scale-like, 2mm; flowers few, terminal, white, 5-merous, 4–6mm; petals inflated at base. Forest floor; 1000m.

Distr.: Ivory Coast, Nigeria to Congo (K), Uganda, Sudan & Ethiopia. (Guineo-Congolian).
IUCN: NT
Bangem: Thomas 5386 1/1986.
Note: also known from Mt Kupe (Cheek, *pers. obs.*).

Swertia eminii Engl.

Enum. Pl. Afr. Trop. 4: 355 (1997).
Erect herb to 25cm, glabrous; lateral branches few, c. 3cm; leaves subsessile, elliptic-lanceolate, c. 13 × 4.5mm, strongly 3(–5) parallel-nerved; cymes terminal, 3–6-flowered; pedicels 6mm; sepals ovate-lanceolate, 4–5mm; petals 4–5, ovate, 5–7mm, white. Grassland; 1800–2000m.
Distr.: Cameroon, Congo (K), Uganda, Burundi, Sudan, Kenya. (Afromontane).
IUCN: LC
Mwanenguba: Cheek 9418 10/1998; Letouzey 14398 fl., 8/1975.

Swertia mannii Hook.f.

Erect herb to 15cm, as *S. eminii*, but lateral branches more numerous, to 8cm; leaves linear-lanceolate, to 1.1 × 0.2cm; cymes more lax, 6–8-flowered; pedicels to 1.8cm; sepals lanceolate, 3mm; petals 5, lanceolate, 5.5mm, white with a purple stripe externally. Grassland; 2000m.
Distr.: Guinea (C) to Cameroon. (Upper & Lower Guinea).
IUÇN: LC
Mwanenguba: Cheek 9419 10/1998.

GERANIACEAE

M. Cheek

Geranium arabicum Forssk. subsp. *arabicum*

Notes Roy. Bot. Gard. Edin. 42: 171 (1985).
Syn.: *Geranium simense* Hochst. ex. A.Rich.
Straggling herb 6–20cm; leaves orbicular in outline, 2–3cm diameter, deeply palmately-lobed, lobes 5, with 2–3 lateral lobes, sparingly pubescent; petiole 3cm; inflorescence 1–2-flowered; peduncle c. 8cm; flowers 1cm, pale pink (-white) with dark veins. Grassland-forest boundary, roadsides; 1300–2000m.
Distr.: Nigeria to Kenya. (Afromontane).
IUCN: LC
Kodmin: Cheek 9035 fl., 1/1998; Ghogue 69 fl., 1/1998; Gosline 109 fl., 2/1998; **Mwanenguba:** Cheek 7274 fl., fr., 2/1995; Cheek 9441 fl., 10/1998; Leeuwenberg 8415 fl., 9/1971; Pollard 131 fr., 10/1998; 852 11/2001; Thomas 3142 fl., 2/1984.
Note: G. mascatense, with a dark 'eye', is unknown in Bakossi.
Local name: Esagmekah. **Uses:** MEDICINES (Cheek 9035).

GESNERIACEAE

P. Bhandol & B.L. Burtt

Fl. Cameroun 27 (1984).

Acanthonema strigosum Hook.f.

Perennial herb, 5–10cm; stem short, unifoliate; leaf linear-oblong, 10–30cm long, base cordate, dark green upper surface, pubescent; inflorescence cymose; pedicels short; corolla funnel shaped, slightly curved, 2cm long, lobes

purple, tube white; stamens 4 (2 fused), apical tooth on filament. On rocks and banks in forest; 350–1000m.
Distr.: Bioko & Cameroon. (Lower Guinea).
IUCN: NT
Kupe Village: Etuge 1434 fl., 11/1995; **Muahunzum:** Schoenenberger 20 fl., 11/1995; **Mwambong:** Cheek 9377 10/1998; **Nyandong:** Thomas 7006 5/1987.

Epithema tenue C.B.Clarke
Succulent epilithic herb, 5–20cm; stem slender; petioles 2–2.5cm long; leaves mostly basal, 5–25 × 5–20cm, ovate, cordate, upper surface densely pubescent, lower surface glabrous and pale; peduncle 0.5–7cm; bracts 2, large, irregularly lobed; calyx 3mm, pilose, lanceolate, acute; corolla tubular 15mm, white, blue or purple; stamens 2; capsule globose, 2mm diameter. On rocks in forest; 850–1100m.
Distr.: Sierra Leone to Sudan & Uganda. (Guineo-Congolian).
IUCN: LC
Enyandong: Ghogue 1235 10/2001; **Nyasoso:** Balding 37 fl., 8/1993; Cable 644 fl., 12/1993; 871 1/1995; Sebsebe 4984 fl., 10/1995; Thomas 5040 11/1985; Wheatley 411 fl., 7/1992.

Nodonema lineatum B.L.Burtt
Fl. Cameroun 27: 22 (1984).
Rhizomatous, rosulate herb; petioles long; leaves cordate, 3–10 × 2.5–6.5cm, upper surface softly-pubescent, lower surface subglabrous; corolla tubular, lobed, white with purple guide stripes in tube; stamens short; fruit subglobose. Forest; 800–1800m.
Distr.: SE Nigeria and SW Cameroon. (Lower Guinea).
IUCN: VU
Bangem: Thomas 5276 1/1986; **Kupe Village:** Cheek 7707 fl., 11/1995; **Menyum:** Thomas 7123 5/1987; **Mt Kupe:** Thomas 5089 11/1985; **Mwendolengo:** Etuge 4520r 11/2001; **Nyasoso:** Wheatley 445 fl., 7/1992.

Streptocarpus elongatus Engl.
Erect fleshy herb to 1m; leaves ovate-elliptic, acuminate, rounded or subcordate at base, upper surface shortly-pubescent, lower surface subglabrous; cymes lax, pedunculate; corolla tubular, white, 1cm long; fruit bright-green, 5cm long, twisted, glabrous. Forest; 850–1850m.
Distr.: Sierra Leone & Cameroon. (Upper & Lower Guinea).
IUCN: LC
Enyandong: Tadjouteu 476 fl., fr., 11/2001; **Mwanenguba:** Sanford 5498 fl., 11/1968; **Nyasoso:** Cheek 5653 fl., 12/1993; Etuge 1326 fl., 10/1995; Lane 191 fl., 11/1994; Sebsebe 4993 fl., 10/1995; Thomas 5038 fl., 11/1985.

Streptocarpus nobilis C.B.Clarke
Epiphytic succulent herb to 1m; leaves ovate, 12 × 7cm, base rounded to subcordate, apex acute, serrate, shortly pubescent on both surfaces; inflorescence axillary; peduncles up to 15cm; bracts 4–8mm; corolla tubular, lobes deep purple; capsule 3–6cm long, shortly pubescent with some glandular hairs. Forest; 730m.
Distr.: Liberia to Cameroon, São Tomé & CAR. (Guineo-Congolian).
IUCN: LC
Mt Kupe: Schoenenberger 5 fl., 10/1995; **Nyasoso:** Pollard 86 fl., 10/1998.

GUTTIFERAE
M. Cheek

Allanblackia floribunda Oliv.
Tree 15–20m, 5-fluted, to 3m; leaves oblong-elliptic or ovate-elliptic, c. 15 × 4.5cm, resembling *A. gabonensis*, but glossy, midrib concolorous; flowers pinkish red; petal margins folding to give triangular apex; staminal phalanges with apical flange; anthers only visible on upper surface; central disc deeply-honeycombed; fruit cylindrical 25–50cm. Evergreen forest; 200–420m.
Distr.: Nigeria to Congo (K). (Lower Guinea & Congolian).
IUCN: LC
Mile 12 Mamfe Road: Nemba 313 fl., 10/1986; **Mungo River F.R.:** Onana 984 fl., 12/1999; **Nyale:** Cheek 9714 fl., 11/1998.

Allanblackia gabonensis (Pellegr.) Bamps
Bull. Jard. Bot. Nat. Belg. 39: 347 (1969).
Syn.: *Allanblackia sp.* of FWTA 1: 291
Tree 10–30(–45)m; leaves obovate, c. 12 × 5cm, acuminate, obtuse to rounded, lower surface matt; lateral nerves c. 15 pairs, resin canals inconspicuous, midrib pinkish red; petiole 1.5cm, with axillary cup; male inflorescence terminal, 3–15-flowered; flowers pale yellow or pink, 4.5cm diameter; petal apex rounded; staminal phalanges 5, with anthers on both upper and lower surface; central disc 5-lobed, slightly undulate; fruit ovoid c. 15cm. Forest; 710–1450m.
Distr.: Cameroon & Gabon. (Lower Guinea).
IUCN: VU
Kodmin: Cheek 9092 fl., 2/1998; **Kupe Village:** Cable 3738 fr., 7/1996; Elad 88 1/1995; Etuge 2632 fl., 7/1996; **Mbule:** Letouzey 14671 fl., 4/1976; **Nyale:** Etuge 4219 2/1998; **Nyasoso:** Cable 1151 fl., 2/1995; Etuge 1801 fl., 3/1996; 2125 fl., 6/1996; 2419 fl., 6/1996; Sebsebe 5037 fl., 10/1995.

Calophyllum inophyllum L.
Tree 10m, glabrous; exudate white; leaves leathery, oblong-elliptic, 15 × 8cm, rounded, base obtuse; lateral nerves c. 200 pairs; petiole 1cm; racemes axillary, 10–15cm, 10-flowered; pedicels 3cm; flowers 2cm; petals white; anthers numerous, free; fruit globose, 2cm. Planted; 150m.
Distr.: coast of Indian & Atlantic oceans, largely planted elsewhere.
IUCN: LC
Etam: Etuge 75 fl., fr., 3/1986.

Garcinia cf. afzelii Engl.
Tree 15m, with adventitious roots at base; leaves drying grey-green below, ovate-elliptic, c. 12 × 4.5cm, apex obtuse, acumen slender, subspatulate, 1.2cm, base obtuse, c. 30 pairs of secondary nerves, resin canals sinuous, faint; petiole 4mm; inflorescence not seen; flowers cream, 4mm diam; stamens as *G. cf. brevipedicellata*. Forest; 100m.
Distr.: Guinea (C) to Cameroon. (Upper & Lower Guinea).
Mungo River F.R.: Thomas 2564 fl., 11/1983.
Note: better material needed to confirm identity.

Garcinia cf. brevipedicellata (Baker f.) Hutch. & Dalz.
Tree 12–15m; leaves elliptic-oblong, c. 12 × 6cm obtuse, acumen slender, 1.5cm, base obtuse, resin canals black, conspicuous, at 45° from midrib; inflorescence axillary, subsessile, 3-flowered; flowers yellow, 1.8cm diameter; anthers in 4 reniform, stipitate masses; fruit globose, 3cm diameter. Forest; 1000m.

Distr.: S Nigeria & Cameroon. (Lower Guinea).
Nyasoso: Cable 3514 7/1996; Wheatley 457 fr., 7/1992.
Note: better material needed to confirm identity; description of flowers extra-Bakossi.

Garcinia conrauana Engl.

Tree (4–)15(–20)m; bole cylindrical; leaves drying pale brown below, elliptic, c. 14 × 10cm, apex rounded, acumen abrupt, 0.5cm, base obtuse, margin revolute, resin canals concolorous; petiole 2.5cm; inflorescence axillary, subsessile, 3-flowered; flowers white, 3cm diameter, staminal mass of four segments joining to form a dome, 1.5 × 1cm; fruit spherical, green, 12 × 12cm. Submontane forest; 1050–1500m.
Distr.: W Cameroon. (Lower Guinea).
IUCN: NT
Edib: Etuge 4100 fr., 1/1998; **Kodmin:** Cheek 10266 fl., 12/1999; Onana 571 fr., 2/1998; **Kupe Village:** Cheek 7766 11/1995; Ryan 233 fr., 5/1996; **Muahunzum:** Etuge 4435 fl., 11/1998.
Note: known from 15 localities, mostly submontane, mostly in coastal Cameroon.

Garcinia kola Heckel

Tree 5–12m; bole cylindrical; leaves elliptic, 13 × 6cm, acuminate, obtuse, resin canals black, parallel to nerves; petiole 1.5cm; inflorescence subumbellate, subsessile on short shoots, 10–15-flowered; flowers c. 1cm diameter; petals concave; staminal bundles 4; fruit orange, fleshy, 10cm diameter. Forest edge and cultivated; 700m.
Distr.: Sierra Leone to Congo (K). (Guineo-Congolian).
IUCN: VU
Kupe Village: Cheek 6089 fl., 1/1995; Etuge 1621 fl., 1/1996; **Ngomboaku:** Ghogue 509 fl., 12/1999.
Local names: Ne'eh (Cheek 6089), Bitter Kola (Etuge 1621), Nya (*fide* Etuge). **Uses:** MATERIALS – twigs are chewsticks (Cheek 6089); SOCIAL USES – drugs – stimulant – the mesocarp is edible and the seeds are very popular, eaten raw like kola-nuts; drug adulterants – bark is added to palm-wine as a flavouring; (Etuge 1621 & *fide* Etuge); MEDICINES (Etuge 1621); ENVIRONMENTAL USES – this species has potential for agroforestry and always grown from cuttings (Etuge 1621); seeds traded (Cheek 6089).

Garcinia lucida Vesque

Bull. Jard. Bot. Nat. Belg. 40: 284 (1970).
Tree (3–)9–15m; leaves elliptic, c. 35 × 16cm, acute, gradually acuminate, base obtuse, secondary nerves c. 16 pairs, conspicuous; petiole c. 2cm; male inflorescence terminal, branched, pendulous, c. 60cm; female axillary, subsessile; flowers white, 5mm diam; fruit ellipsoid, grey-green, 6 × 5cm, often rostrate. Forest; 700–1400m.
Distr.: Cameroon & Gabon. (Lower Guinea).
IUCN: NT
Bakossi F.R.: Thomas 5287 fl., 1/1986; **Edib:** Etuge 4110 fl., fr., 1/1998; **Enyandong:** Cheek 10920 10/2001; **Kupe Village:** Cable 2737 fl., 5/1996; 3830 fl., fr., 7/1996; 737 fl., 1/1995; Cheek 6085 fl., fr., 1/1995; 7674 fr., 11/1995; Etuge 1384 fl., fr., 11/1995; Lane 421 fr., 1/1995; 422 fl., 1/1995; 423 fl., 1/1995; Ryan 292 fl., 5/1996; 320 fr., 5/1996; **Mbule:** Cable 3350 fr., 6/1996; Leeuwenberg 8805 fl., 12/1971; **Mt Kupe:** Leeuwenberg 9269 fl., 1/1972; **Nyasoso:** Cable 1125 fl., 2/1995; 3560 fr., 7/1996; Etuge 2111 fr., 6/1996; 2573 fr., 7/1996; Gosline 139 fl., fr., 11/1998; Lane 96 fr., 6/1994.
Note: often gregarious. Known from about 15 sites.

Local names: Ngu (Cable 737; Etuge 1384 & 4110; Daniel Epie in Cheek 6085 & Cheek 10920); Ngoo (*fide* Etuge). **Uses:** FOOD – eat seeds raw (Ryan 320); SOCIAL USES – drugs – facilitants to drug effects – bark and fruits harvested for flavouring palm wine, and to make it strong; trees sometimes killed by bark harvesting (Cheek 6085); MEDICINES (Etuge 2111 & 4110).

Garcinia mannii Oliv.

Tree c. 30m; bole cylindric; leaves drying matt yellowish green below, elliptic-oblong, c. 13 × 6cm, obtuse to rounded, acumen slender, 2cm, base acute, c. 40 pairs secondary nerves; petiole 8mm; inflorescence subsessile, 3-flowered or terminal on spur shoots; flowers 1.8cm diameter; petals bright, glossy deep-yellow or red; androecium annular, surrounding ovary. Forest; 300–700m.
Distr.: Nigeria to Gabon. (Lower Guinea).
IUCN: NT
Bolo Moboka: Thomas 5895 fl., 3/1986; **Kupe Village:** Elad 87 1/1995; **Mungo River F.R.:** Cheek 10194 fl., 11/1999; Onana 955 11/1999; **Nyale:** Etuge 4994 fl., fr., 3/2003; **Nyandong:** Cheek 11363 3/2003.
Local name: Mbimbi (Etuge 1657), Mekak, Esé (*fide* Etuge). **Uses:** MATERIALS – the main *Garcinia* for chewsticks (Elad 87; Cheek *pers. obs.* & *fide* Etuge); MEDICINES (Etuge 1657).

Garcinia smeathmannii (Planch. & Triana) Oliv.

Syn.: *Garcinia polyantha* Oliv.
Tree 4–15m; leaves thickly leathery, drying pale brown below, narrowly oblong-elliptic, c. 20 × 8cm, obtuse, subacuminate, base obtuse, secondary nerves c. 20 pairs; petiole c. 1.5cm; inflorescence sessile, axillary, on leafy stems, umbellate-fascicled, 15–20-flowered; pedicels c. 15mm; flowers white, 1cm diameter; anthers with free filaments inserted on ligules, as long as petals; fruits globose, 2cm; stigmas 2; pedicels 4cm in fruit. Forest; 650–2000m.
Distr.: Guinea (B) to Zambia. (Guineo-Congolian).
IUCN: LC
Kodmin: Cheek 9084 1/1998; Ghogue 55 fl., 1/1998; Onana 572 fl., 2/1998; **Kupe Village:** Cable 2506 fr., 5/1996; Cheek 7199 fl., 1/1995; 7883 11/1995; Elad 94 fl., 1/1995; Lane 409 fr., 1/1995; Ryan 401 fr., 5/1996; **Mt Kupe:** Sebsebe 5069 10/1995; Thomas 5057 fr., 11/1985; 5480 fl., 2/1986; 5494 fr., 2/1986; **Nyasoso:** Cable 1190 fl., 2/1995; 3390 fr., 6/1996; Elad 127 fr., 2/1995; Etuge 1657 fl., 1/1996; 1688 fr., 1/1996.
Uses: MEDICINES (Etuge 1657).

Garcinia sp. aff. smeathmannii sp. 1

Tree 18–20m resembling *G. smeathmannii*, but leaves c. 30 × 13cm, pale grey below, secondary nerves c. 15 pairs; inflorescences 50–100-flowered, ramiflorous. Forest; 420–800m.
Kupe Village: Mackinder 280 fl., 12/1999; **Mungo River F.R.:** Onana 952 11/1999.
Note: possibly a new species.

Garcinia sp. aff. smeathmannii sp. 2

Tree 7–10m, resembling *G. smeathmannii*, but leaves obovate-elliptic or elliptic, c. 13 × 5cm, apex acute, acuminate, base acute; inflorescence 1–4-flowered. Forest; 950m.
Kupe Village: Gosline 245 fl., 12/1999; Ryan 290 fl., 5/1996.
Note: possibly a new species, matching Letouzey 13419 from Metchum, NW Province.

HALORAGACEAE

Harungana madagascariensis Lam. ex Poir.

Shrub or small tree 3–6m, glabrescent; leaves ovate c. 12 × 5.5cm, acuminate, base rounded, nerves c. 10 pairs; petiole 2cm; producing bright orange exudate when broken; inflorescence a dense terminal panicle, 7–15cm; flowers white, 2mm; petals hairy; berries orange, 3mm. Farmbush; 250–1750m.

Distr.: tropical Africa & Madagascar.

IUCN: LC

Bangem: Kenneth Tah 3 fl., 11/2001; **Ehumseh - Mejelet:** Etuge 316 fl., 10/1986; **Etam:** Etuge 241 fl., 8/1986; **Kodmin:** Cheek 9573 fl., 11/1998; Ghogue 15 fr., 1/1998; Gosline 111 fr., 2/1998; **Kupe Village:** Kenfack 325 fl., 7/1996; **Mile 15, Kumba-Mamfe Road:** Nemba 216 fl., 9/1986; **Mwambong:** Onana 550 fl., 2/1998; **Nsoung:** Darbyshire 59 3/2003; **Nyasoso:** Etuge 2449 fr., 6/1996.

Local name: Ngelle. **Uses:** FUELS – fuelwood (Cheek 9573).

Hypericum revolutum Vahl subsp. *revolutum*

Webbia 22: 239 (1967).

Syn.: *Hypericum lanceolatum* Lam.

Shrub or tree to 12m; leaves narrowly elliptic, 1–2(–3.5)cm long; petals yellow, 2–3cm long. Grassland, forest-grassland transition; 1940–2060m.

Distr.: tropical Africa. (Afromontane).

IUCN: LC

Kupe Village: Ryan 400 5/1996; **Mt Kupe:** Sebsebe 5091 11/1995; **Mwanenguba:** Thomas 3125 fl., fr., 2/1984; **Nyasoso:** Cable 119 fl., 9/1992; 2913 fl., fr., 6/1996; Cheek 7336 2/1995; Etuge 1676 1/1996.

Local name: no local name. **Uses:** no uses (Cheek 7336).

Hypericum roeperianum Schimp. ex A.Rich.

Webbia 22: 248 (1967).

Shrub to 3m; leaves elliptic, ovate or lanceolate, 3–8cm long; petals yellow 2.5–3.5cm long. Forest edge; 1250–2000m.

Distr.: Cameroon, Ethiopia to Tanzania. (Afromontane).

IUCN: LC

Kodmin: Ghogue 42 fl., fr., 1/1998; **Mwanenguba:** Bamps 1556 fl., fr., 12/1967; Cheek 7237 2/1995; Etuge 4381 fl., fr., 10/1998; Leeuwenberg 8600 fl., 10/1971; 8787 fl., fr., 11/1971; Sanford 5495 fl., fr., 11/1968; Thomas 3124 fl., fr., 2/1984; **Nyasoso:** Cheek 7337 fl., fr., 2/1995; Etuge 1677 fl., fr., 1/1996.

Local name: Mboro Mbore. **Uses:** MATERIALS – fruits resemble another species used near town in gumming (Cheek 7337).

Mammea cf. africana Sabine

Tree 18m, resembling *Pentadesma grandifolia*, but leaves to 25 × 11cm, apex rounded; petiole 2–5cm; flowers with sepals 2, ovate (buds only). Forest; 1100m.

Distr.: Sierra Leone to Congo (K). (Guineo-Congolian).

Nyale: Etuge 4469 11/1998.

Note: material incomplete; may be an aberrant *Pentadesma grandifolia*.

Pentadesma grandifolia Baker f.

Bull. Jard. Bot. Nat. Belg. 35: 424 (1965).

Tree 10–35m; buttresses to 2m; leaves elliptic-oblong or oblanceolate-oblong, 20 × 7cm, shortly acuminate, obtuse to rounded base, c. 50–70 pairs of secondary nerves, resin canals not seen, lower surface densely covered in black spots; petiole c. 2cm; inflorescence c. 6-flowered, terminal; pedicels c. 1.5cm; flowers globular, 6cm diameter, white; anthers with free filaments. Forest; 870–1300m.

Distr.: Nigeria to Congo (K). (Lower Guinea & Congolian).

IUCN: LC

Edib: Etuge 4094 fl., fr., 1/1998; **Kupe Village:** Cable 2514 fr., 5/1996; 3816 7/1996; 839 1/1995; Cheek 7683 11/1995; Etuge 1436 fl., 11/1995; 2611 fr., 7/1996; **Mbule:** Leeuwenberg 8819 fl., 12/1971; **Nyasoso:** Cable 3298 6/1996; Groves 99 fl., fr., 2/1995.

Note: distinguished from the better known *P. butyracea* Sabine e.g. by the presence of black dots on lower surface of the leaf blade.

Psorospermum tenuifolium Hook.f.

Shrub or small tree 5–6m, glabrous; bark peeling; leaves elliptic, to 9.5 × 4.5cm, acute, apex twisting when pressed, base acute, brown dots showing in transmitted light; petiole 4–6mm; inflorescence terminal, dense, 3–4cm diameter; peduncle 1–2.5cm; flowers white, 5mm diameter; petals hairy; berries 5mm diameter. Grassland edge; 800–1540m.

Distr.: Nigeria to Congo (K). (Lower Guinea & Congolian).

IUCN: LC

Kodmin: Satabie 1125 fr., 2/1998; **Ndoungue:** Leeuwenberg 9453 fl., 3/1972.

Symphonia globulifera L.f.

Tree to 30m, glabrous; leaves narrowly elliptic to oblong-elliptic, c. 10.5 × 2.8cm, long-acuminate, acute; lateral nerves c. 30 pairs, resin canals inconspicuous; petiole 0.7cm; inflorescence terminal, c. 10-flowered; pedicels 1.5cm; flowers red, globose, c. 1cm diameter; styles ascending, aculeate, 5; fruit ovoid, 2.5cm, lenticellate; styles persistent. Forest; 250–1050m.

Distr.: tropical America, Africa & Madagascar. (Amphi-Atlantic).

IUCN: LC

Kodmin: Plot B31 1/1998; **Kupe Village:** Cable 2511 fl., 5/1996; Etuge 1970 fl., fr., 5/1996; **Loum F.R.:** Cheek 10253 12/1999; **Nyasoso:** Sidwell 402 10/1995.

HALORAGACEAE

M. Cheek

Laurembergia tetrandra (Schott) Kanitz subsp. *brachypoda* var. (Welw. ex Hiern) A.Raynal

Webbia 19: 694 (1965).

Red, prostrate herb, rooting at nodes, glabrous; leaves opposite, linear, 10 × 1mm, sessile; stipules absent; flowers several per node, axillary, sessile, pink, 1mm, 4-merous; fruit with 8 lines of 4 tubercles. *Sphagnum* mat; 1200–1250m.

Distr.: Sierra Leone, Cameroon, Congo (K), Angola, Mozambique, Madagascar. (Tropical Africa & Madagascar).

IUCN: LC

Lake Edib: Cheek 9132 2/1998; Gosline 189 11/1998.

Note: a possible new record for Cameroon.

Myriophyllum spicatum L.

Sp. Pl.: 992 (1753).

Submerged, bottom-rooting, aquatic herb; stems up to 1m long; leaves 4 per whorl, soft, ovate in outline, 2–3cm, pinnately divided into 15–30 capillary segments; inflorescences not seen. Lakes; 1975–2000m.

Distr.: Canada, Eurasia, Sudan, Ethiopia, Malawi, Cameroon. (Circumboreal & Afromontane).

IUCN: LC

Mwanenguba: Cheek 7245 2/1995; Cheek 9408 10/1998.

HERNANDIACEAE

M. Cheek

Illigera madagascariensis H.Perrier
Arch. Bot., Caen, Bull. 1: 70 (1927).
Climber resembling *I. pentaphylla*, but has 3 leaflets; lateral leaflets unequal at base, secondary nerves evenly spread, acumen longer, apex acute; petiolules 0.5–1cm. Forest; 300m.
Distr.: Cameroon, Tanzania (E Arc Mountains) & Madagascar. (Tropical Africa & Madagascar).
IUCN: LC
Mungo River F.R.: Cheek 10123 11/1999.
Note: if this identification is correct (more material desirable), this is a remarkable new expansion of the range for this species, previously unknown in Cameroon.

Illigera pentaphylla Welw.
Climber, 5m; leaves alternate, shortly-pubescent; leaflets 5, elliptic, to 9 × 6cm, subacuminate, sub–5-nerved, nervation scalariform; petioles 12cm, twisting; petiolules 2.5cm; inflorescences axillary, 30cm, 1–5-branched; flowers 7mm; corolla white and purple; fruit 3 × 9cm, 1-seeded, 2-winged. Fallow forest, *Hypselodelphys* thicket; 750–1400m.
Distr.: Ivory Coast to Uganda. (Guineo-Congolian).
IUCN: LC
Kodmin: Plot B274 1/1998; **Manehas F.R.:** Cheek 9400 10/1998; **Ndum:** Cable 897 1/1995; **Nyasoso:** Cable 1228 2/1995; Cheek 5646 12/1993.
Note: easily confused with *Dioscorea* when sterile.

HUACEAE

M. Cheek

Afrostyrax lepidophyllus Mildbr.
Tree to 30m; bole dull-white, smelling strongly of garlic, glabrous; leaves lanceolate-elliptic, c. 21 × 7cm, acumen 2cm, base rounded or obtuse, lower surface completely obscured by peltate scales, appearing pale brown or dull-white; lateral nerves c. 6 pairs; petiole 1cm; flowers axillary, 5mm; fruit ellipsoid, 2cm, fleshy. Forest; 600–1300m.
Distr.: Ghana to Congo (K). (Guineo-Congolian).
IUCN: LC
Edib: Etuge 4102 1/1998; **Ngomboaku:** Cheek 10293 12/1999; Ghogue 443 12/1999; **Nyale:** Cheek 9649 11/1998; Etuge 4236 2/1998.
Note: the similar *A. kamerunensis* lacks peltate scales but has sparse red-stellate hairs; it is unknown in Bakossi. Assessed as VU by Hawthorne (1997) in ignorance of its large range in Congo (K) where it is common and unthreatened in Equateur and Orientale provinces. It is here reassessed as LC.
Local name: Bush Onion (Pidgin) (Etuge 4236 & Ghogue 443). **Uses:** FOOD – fruits edible, used in soup (Etuge 4102); FOOD ADDITIVES – the bark and fruit are widespread in markets for food flavouring (Ghogue 443)

HYDROPHYLLACEAE

I. Darbyshire

Hydrolea palustris (Aubl.) Raeusch.
Rhodora 90: 200 (1988).
Syn.: *Hydrolea guineensis* Choisy
Syn.: *Hydrolea glabra* Schum. & Thonn.
Herb to 50cm; stems glabrous; leaves papery, lanceolate-elliptic, 4.5–8 × 0.8–2cm, base and apex cuneate, margin subentire, glabrous; petiole to 1cm; flowers in axillary fascicles of 1–4; pedicels 3–4mm; calyx lobes lanceolate, 6mm; corolla lobes ovate, 5mm, violet; ovary globose; styles 2, divergent; capsule subglobose, 0.5cm; styles persistent. Riverbank, moist open areas; 50m.
Distr.: Sierra Leone to N Congo (K). (Upper & Lower Guinea).
IUCN: LC
Mungo River F.R.: Leeuwenberg 8691 fl., 11/1971.

ICACINACEAE

G. Gosline & I. Darbyshire

Fl. Cameroun 15 (1973).

Chlamydocarya thomsoniana Baill.
Woody climber; branchlets pubescent; leaves elliptic to obovate-elliptic, 10–28 × 4–12cm, margin denticulate, pubescent; dioecious; male flowers in catkins 3–6cm long; infructescence a spiked-sphere comprising fruits 1.5cm diameter, topped by an accrescent orange corolla, elongated into a 4–8cm tube, covered with bristly hairs. Forest; 910m.
Distr.: Sierra Leone to Congo (K). (Guineo-Congolian).
IUCN: LC
Nyasoso: Zapfack 675 fr., 6/1996; 705 fr., 6/1996.

Desmostachys tenuifolius Oliv. var. *tenuifolius*
Tree or shrub to 10m; leaf elliptic to oblanceolate, 15–24 × 5–8cm, base attenuate to cuneiform, apex broadly acuminate, 8–10 pairs lateral nerves arcing, anastomosing 2–3mm from edge; inflorescence a long pubescent rachis, solitary or paired, 15–20cm; flowers arranged in two ranks, 4-merous, 4–6mm, white; sepals glabrous, c. 1mm; petals free, 5.5 × 2mm, exterior glabrous, interior pubescent; pistil 5mm; fruit an ellipsoid drupe, 2.5 × 2cm. Forest, forest gaps; 200–350m.
Distr.: SE Nigeria to Congo (B). (Lower Guinea).
IUCN: LC
Bakolle Bakossi: Etuge 152 fl., 5/1986; **Konye:** Thomas 5176 fr., 11/1985; **Kurume:** Thomas 5457 1/1986; **Mile 12 Mamfe Road:** Nemba 312 10/1986; **Southern Bakossi F.R.:** Olorunfemi in FHI 30596 fl., 5/1951.

Icacina mannii Oliv.
Woody climber or scandent shrub with a large tuber; leaves elliptic to oblanceolate, 6–25 × 3–14cm, 4–6 pairs lateral nerves, glabrous; flowers highly fragrant, in lax axillary clusters; pentamerous, 8 × 6mm; petals shortly-pubescent outside, bearded inside; anthers extending beyond corolla. Forest; 920–1000m.
Distr.: Sierra Leone to Congo (K). (Guineo-Congolian).
IUCN: LC
Kupe Village: Lane 443 fl., 1/1995; **Mwambong:** Etuge 4124 fl., 2/1998; Onana 540 fl., 2/1998; **Nyasoso:** Groves 88 fl., 2/1995.

Iodes africana Welw. ex Oliv.
Woody climber, by means of extra-axillary tendrils to upper canopy; leaves opposite to subopposite, broadly ovate, 8–16 × 3–10cm; flowers small, yellow, in irregularly-forked

ICACINACEAE

cymose panicles; dioecious; calyx absent; petals 4–5, reflexed; fruit obovate, c. 1cm diameter, reticulate. Forest; 750–1300m.
Distr.: Nigeria to Congo (K). (Lower Guinea & Congolian).
IUCN: LC
Kupe Village: Cable 785 fl., 1/1995; 822 fl., 1/1995; Etuge 1717 2/1996; **Mwambong:** Mackinder 148 fl., 1/1998; **Nyasoso:** Cheek 7304 fl., 2/1995.

Lasianthera africana P.Beauv.
Tree or shrub to 5m; branchlets terete; leaves elliptic, 9–20 × 3–10cm, acumen long-caudate, 6–8 pairs lateral nerves; cymes leaf-opposed; flowers in small partial heads; petals 3mm; anthers densely villose, visible as white tufts; fruit flattened, oblong-elliptic, beaked, c. 1cm long, with longitudinal ridges. Forest, forest gaps; 210–1200m.
Distr.: Nigeria to Congo (K). (Lower Guinea & Congolian).
IUCN: LC
Bakolle Bakossi: Etuge 131 5/1986; **Enyandong:** Cheek 10968 10/2001; Etuge 4449r 11/2001; Gosline 329 10/2001; **Etam:** Etuge 222 fl., 8/1986; **Kupe Village:** Cable 2492 fl., 5/1996; Cheek 7064 fl., 1/1995; Etuge 1419 fl., 11/1995; Gosline 249 fl., 12/1999; Lane 393 fl., fr., 1/1995; 430 fl., 1/1995; **Mbule:** Leeuwenberg 9285 fl., 1/1972; **Mile 11, Kumba - Bakossi road.:** Dundas in FHI 15231 fl., fr., 6/1946; **Mungo River F.R.:** Etuge 4317 fl., fr., 10/1998; **Ngomboaku:** Ghogue 514 fl., 12/1999; Muasya 2028 fl., 12/1999; **Nyale:** Etuge 4240 2/1998; **Nyandong:** Ghogue 1433 fl., 3/2003; Tadjouteu 557 3/2003; **Nyasoso:** Cable 1138 fl., 2/1995; 3309 fl., 6/1996; Etuge 2089 fl., 6/1996; Lane 153 fl., 10/1994; 93 fl., 6/1994.

Lavigeria macrocarpa (Oliv.) Pierre
Woody climber to 4cm diameter; branchlets and leaves somewhat stellate-pubescent; leaves in canopy, oblong-elliptic 15–25 × 7–11cm, shortly-pointed, 6–11 pairs lateral nerves; cauliflorous in panicles on the main stem far from the leaves; flowers pentamerous; sepals hairy; petals glabrous; stamens 5; ovary bristly; fruit ovoid, c. 7cm, corrugated. Forest; 250–1200m.
Distr.: SE Nigeria to Congo (K). (Lower Guinea & Congolian).
IUCN: LC
Ekona Mombo: Etuge 412 12/1986; **Kupe Village:** Cable 2460 5/1996; 2461 5/1996; 2686 fl., 5/1996; Cheek 7135 fl., 1/1995; Etuge 1421 fl., 11/1995; Lane 399 fr., 1/1995; **Loum F.R.:** Cheek 10241 fl., 12/1999; Onana 995 fl., 12/1999; **Nyandong:** Darbyshire 141 fr., 3/2003; **Nyasoso:** Cable 2766 6/1996; Lane 127 fl., 6/1994.
Local name: Bush carrot. **Uses:** FOOD – fruits edible (Darbyshire 141).

Leptaulus daphnoides Benth.
Tree or shrub to 15m; leaves elliptic, 6.5–16 × 2.5–6.5cm, base attenuate, apex obtusely acuminate, 6–7 pairs lateral nerves ascendant, anastomosing, glabrous; inflorescence a short cyme, joined to the leaf; flowers 5-merous, white, 11 × 1.5mm; corolla tubular, linear, lobes 1mm long; anther filaments joined to the petals; style glabrous; fruit an ellipsoid drupe, 12 × 8 × 5mm, papillose. Forest; 250–600m.
Distr.: Sierra Leone to Sudan & Tanzania. (Guineo-Congolian).
IUCN: LC
Bolo: Thomas 5894 fl., 3/1986; 5898 fl., 3/1986; **Ikiliwindi:** Etuge 497 3/1987; **Mombo:** Olorunfemi in FHI 30558 fr., 5/1951; **Ngusi:** Etuge 47 fl., 4/1986; **Nlog:** Etuge 5 fl., 4/1986.

Leptaulus holstii Engl.
Fl. Cameroun 15: 62 (1973).
Shrub 1–3m; stems glabrescent, minutely puberulent when young; leaves elliptic to obovate, 6–13 × 2.5–5.5cm, acumen to 1.5cm, obtuse, prominent, base acute; lateral nerves 6–8 pairs, anastomosing, glabrous; petiole 3mm; inflorescence a few-flowered axillary cyme, c. 6mm; pedicels 1–2mm; flowers 3–4mm, yellow; drupe ovoid-ellipsoid, c. 1 × 0.7cm, smooth, shining-red. Forest; 300m.
Distr.: Cameroon to Tanzania. (Lower Guinea & Congolian).
IUCN: LC
Konye: Thomas 5190 11/1985.

Polycephalium lobatum (Pierre) Pierre ex Engl.
Fl. Cameroun 15: 96 (1973).
Woody climber; stems, leaves, petioles, flowers and fruit covered with erect red hairs; leaves highly variable, palmately nerved, often 3-lobed, 6–17 × 8–22cm; petiole 5–10cm; dioecious; male flowers in panicles of pedunculate heads 3–5mm diameter; female flowers in solitary axillary heads 1cm diameter; male flowers reduced; calyx tubular, 2mm; stamens 3; female calyx tubular, c. 3 × 2mm; fruits forming a hairy stellate ball; fruits 4 × 2cm, enveloped by the accrescent calyx. Forest; 850–1000m.
Distr.: Sierra Leone to Congo (K). (Guineo-Congolian).
IUCN: LC
Mwambong: Cheek 9116 fr., 2/1998; **Ngomboaku:** Cheek 10279 fl., 12/1999.

Pyrenacantha cordicula Villiers
Fl. Cameroun, 15: 86 (1973).
Liana, felty-pubescent throughout; leaves elliptic, c. 11.5 × 5cm, apex acuminate, base rounded; lateral nerves 5 pairs, nerves yellow below; petiole to 1.5cm; female flowers in dense sessile axillary glomerules, c. 0.7cm; male flowers in axillary fascicles of racemes 4–14cm; ovary truncate at apex; fruit ellipsoid, 1.7 × 1.2cm, green, felty, subrostrate. Forest; 760m.
Distr.: Ivory Coast, Ghana, Cameroon, Equatorial Guinea. (Upper & Lower Guinea).
IUCN: EN
Ngombombeng: Etuge 28 fl., fr., 4/1986.

Rhaphiostylis beninensis (Hook.f. ex Planch.) Planch. ex Benth.
Woody climber to 10–12m; leaves elliptic to lanceolate-elliptic, 6–15 × 2–7cm, 5–7 pairs lateral nerves; flowers rather numerous in axillary fascicles; pentamerous; petals free, 7 × 1mm, drying black; stamens 5, white; pedicels 7–8mm; fruits red, reniform, 1.3 × 1–2 × 1cm, reticulate. Forest; 650–1100m.
Distr.: Senegal to Tanzania. (Guineo-Congolian).
IUCN: LC
Ngusi: Etuge 1582 fl., 1/1996; **Nyasoso:** Lane 517 fl., 2/1995.

Stachyanthus zenkeri Engl.
Syn.: *Neostachyanthus zenkeri* (Engl.) Exell & Mendonça
Woody climber, slender, twining to 8m; young branches and petioles bristly-pilose; leaves elliptic to oblanceolate, 10–23 × 5–10cm, appressed-pubescent beneath, 6–9 lateral nerves; dioecious; flowers on long spikes in fascicles on the stem below the leaves; male spikes to 25cm; flowers 6-merous, c. 6mm; fruit a drupe, ellipsoid with 4 flat surfaces, 3 × 1.8 × 1.2cm, in neat ranks along rachis. Forest; 350–1500m.
Distr.: Nigeria to Congo (K). (Lower Guinea & Congolian).
IUCN: LC

Kodmin: <u>Barlow 8</u> fr., 1/1998; <u>Cheek 8877</u> fr., 1/1998; <u>9597</u> 11/1998; <u>Etuge 4159</u> fr., 2/1998; <u>Gosline 151</u> fr., 11/1998; <u>Onana 573</u> fr., 2/1998; **Kupe Village:** <u>Cable 2529</u> fr., 5/1996; <u>Etuge 2916</u> 7/1996; <u>Kenfack 296</u> fr., 7/1996; **Mejelet-Ehumseh:** <u>Etuge 2976</u> fr., 8/1996; **Mungo River F.R.:** <u>Tadjouteu 315</u> fr., 12/1999.

IRVINGIACEAE

D.J. Harris & M. Cheek

Desbordesia glaucescens (Engl.) Tiegh.
Tree to 50m; basal trunk cylinder sometimes rotting in mature trees to be supported only by buttresses, glabrous; leaves elliptic, c. 13 × 6cm, acumen 1cm, obtuse, glossy above, off-white, matt below, c.10 lateral nerve pairs; petiole 1cm; panicle 2–7cm, terminal; flowers 3mm; fruit papery, winged, flattened, leaf-like, oblong-elliptic, 11 × 3cm, 1-seeded. Forest; 450m.
Distr.: Nigeria to Congo (K). (Lower Guinea).
IUCN: LC
Nyandong: <u>Cheek 11373</u> 3/2003.

Irvingia gabonensis (Aubry-Lecomte ex O'Rorke) Baill.
Tree, 10–40m, glabrous; apical bud 1cm, of a single sheathing stipule; leaves elliptic, 7–17 × 3–6cm, acumen 1cm, acute; lateral nerves 8 pairs; petiole 1cm; panicles axillary, 7cm; flowers 5mm; fruit hardly laterally flattened drupes, 5cm, endocarp wall 1–2mm thick, 1-seeded. Forest; 150–400m.
Distr.: Nigeria to Congo (K). (Lower Guinea).
IUCN: NT
Bakossi F.R.: <u>Etuge 4303</u> 10/1998; **Etam:** <u>Etuge 78</u> 3/1986; **Nlonako:** <u>Etuge 1772</u> 3/1996.
Note: listed as LR/nt by WCMC 1997 (www.redlist.org). Not threatened in Cameroon.
Local name: Bush mango (Etuge 1772 & 4303). **Uses:** FOOD – fruit flesh sometimes edible; seed extracted for soups; ENVIRONMENTAL USES – a suitable species for genetic improvement and agroforestry; widely traded; MEDICINES (Etuge 1772).

Irvingia grandifolia (Engl.) Engl.
Tree to 40m, resembling *I. gabonensis*, but easily identified by the large leaves, often red below the tree; fruit larger than broad, at least 4 × 2.5cm. Forest; 300m.
Distr.: Nigeria to Angola. (Lower Guinea & Congolian).
IUCN: LC
Eboné: <u>Leeuwenberg 8379a</u> 9/1971.
Note: *I. robur* is to be looked for in Bakossi.

Klainedoxa gabonensis Pierre ex Engl.
Bull. Jard. Bot. Nat. Belg. 65: 153 (1996).
Syn.: *Klainedoxa gabonensis* Pierre ex Engl. var. *oblongifolia* Engl.
Tree to 50m; trunk spiny when young, glabrous, apical bud 5cm or 10cm in sterile growth; leaves elliptic, c. 11 × 5cm, subacuminate, acute, with 2–3 streaks parallel to midrib; fruit dark green, woody, wider than long, often 5-lobed, 3–4 × 5cm. Forest; 350m.
Distr.: Guinea (B) to Zambia. (Guineo-Congolian).
IUCN: LC
Kupe Village: <u>Etuge 2050</u> 5/1996; **Mungo River F.R.:** <u>Cheek 10143</u> 11/1999.

Klainedoxa trillesii Pierre ex Tiegh.
Bull. Jard. Bot. Nat. Belg. 65: 161 (1996).
Tree to 30m, as *K. gabonensis*, but lacking spines when young; nervation reticulate; areolae 0.5mm diameter (0.2–0.3mm in *K. gabonensis*); fruit orange, globose or longer than wide. Forest; 150m.
Distr.: Sierra Leone to Congo (K). (Guineo-Congolian).
IUCN: LC
Mungo River F.R.: <u>Cheek 10236</u> 12/1999.

LABIATAE

B.J. Pollard

Note: here follows the treatment of Labiatae *sensu lato*, adding *Clerodendrum* L., *Premna* L., and *Vitex* L., which were previously included in Verbenaceae. See Harley, R.M., Atkins, S., Budantsev, A., Cantino, P.D., Conn, B., Grayer, R.J., Harley, M.M., De Kok, R., Krestovskaja, T., Morales, A., Paton, A.J., Ryding, O. and Upson., T., (2004). Labiatae. *In*: Kadereit, J.W. (*ed.*). The Families and Genera of Vascular Plants, **vi**: (Lamiales), Springer, Berlin.

Achyrospermum africanum Hook.f. ex Baker
Erect herbaceous undershrub to about 3m; leaves ovate, c. 15 × 8cm; terminal inflorescences 8–12 × 1.5cm, the laterals smaller; bracts broadly ovate, c. 0.8cm; calyx teeth twice as long as broad; corolla exceeding calyx, purplish pink with white throat; stamens ascending. Forest; 860–1100m.
Distr.: Guinea (C) to Cameroon. (Upper & Lower Guinea (montane)).
IUCN: LC
Kupe Village: <u>Lane 348</u> fl., 1/1995; **Mt Kupe:** <u>Schoenenberger 12</u> 11/1995; **Ndum:** <u>Williams 139</u> fl., 1/1995; **Nyasoso:** <u>Cheek 5627</u> fl., 12/1993; <u>7486</u> fl., 10/1995; <u>Etuge 1304</u> fl., 10/1995.

Achyrospermum oblongifolium Baker
Erect herbaceous little-branched, undershrub, 30–70cm; stems little-branched, tomentose; leaves 10–18 × to 8cm, narrowly obovate-elliptic, acuminate; inflorescence terminal, 3–5(–10) × 2cm; bracts broadly ovate, ciliate; calyx teeth as broad as long, margin ciliate; corolla greenish white; stamens ascending. Forest; 700–1100m.
Distr.: Guinea (C) to Cabinda. (Upper & Lower Guinea (montane)).
IUCN: LC
Enyandong: <u>Pollard 729</u> fl., fr., 10/2001; **Kupe Village:** <u>Cable 3678</u> 7/1996; <u>Cheek 7128</u> fr., 1/1995; **Mt Kupe:** <u>Schoenenberger 30</u> 11/1995; **Ngomboaku:** <u>Cheek 10375</u> 12/1999; <u>Etuge 4656</u> 12/1999; **Nyasoso:** <u>Cheek 6014</u> fr., 1/1995; <u>7477</u> fr., 10/1995; <u>Lane 187</u> fl., 11/1994; <u>Pollard 87</u> 10/1998; <u>89</u> 10/1998.

Clerodendrum anomalum Letouzey
Adansonia (sér. 2) 14: 225 (1974).
Scandent shrub or climber; stem solid; leaves ovate, 12–14(–15) × 6–7(–8)cm, margin slightly undulate, long-acuminate, cuneate; inflorescences axillary pseudoracemes of opposite fascicles, 5–25cm; flowers strongly zygomorphic, green, lower lip white; fruit globose, a strongly-crested pyrene. Degraded lowland to submontane forest formations, overgrown edges to forest paths; 880m.
Distr.: Cameroon & Gabon. (Lower Guinea).
IUCN: EN

LABIATAE

Eboné: Leeuwenberg 9540 fl., 3/1972; **Kupe Village:** Cable 2502 fr., 5/1996; Cheek 7084 fl., 1/1995.

Clerodendrum bipindense Gürke
Slender cauliflorous liane to 8m; leaves elliptic, 10–20 × 4–8cm, acuminate or cuspidate; inflorescence from leafless part of stem near ground, much-branched, lax; corolla white or cream; tube c. 1.5 × 0.1cm; calyx teeth rounded, c. 1mm. Forest; 420m.
Distr.: Nigeria, Bioko, Cameroon, Gabon, Congo (K). (Lower Guinea).
IUCN: NT
Mungo River F.R.: Onana 978 12/1999.

Clerodendrum buettneri Gürke
Woody climber; leaves; branches and calyx brown-hispid, hairs 2–5mm; leaves ovate, 6–10 × 2.5–6cm; calyx 12–15mm; flowers white or yellowish, with a red centre. Forest; 250m.
Distr.: Nigeria to Gabon. (Lower Guinea).
IUCN: LC
Mile 15, Kumba-Mamfe Road: Nemba 131 fr., 7/1986.

Clerodendrum capitatum (Willd.) Schum. & Thonn. var. *capitatum*
Erect or scrambling shrub, or spiny climber, with long petiole-derived spines; internodes hollow; leaves ovate or elliptic, 7–20 × 4–10cm, with 3–6 pairs of lateral nerves; inflorescence terminal or occasionally lateral on young branches, conspicuously bracteate; bracts ovate-acuminate or ovate-lanceolate; corolla white; tube 4–8cm; calyx fimbriate on margins. Forest; 750–890m.
Distr.: tropical and subtropical Africa. (Afromontane).
IUCN: LC
Baseng: Cheek 10416 12/1999; **Nyasoso:** Cheek 7467 10/1995.

Clerodendrum chinense (Osbeck) Mabb.
F.T.E.A. Verbenaceae: 94 (1992).
Syn.: *Clerodendrum japonicum sensu* Huber, Hepper & Meikle in FWTA 2: 443
Subshrub or ± herbaceous, 0.9–1.8m; stems ± quadrangular, shortly pubescent; leaves broadly ovate, 6–29 × 5–28cm, acute; petiole 2–25cm; inflorescences terminal, compact, many-flowered cymes, 3–6 × 3.5–9cm; bracts lanceolate, leafy, scattered throughout the cymes; calyx usually red or purple, campanulate, 1–1.5cm; flowers strongly scented, white. Villages; 600–700m.
Distr.: introduced to tropical Africa, probably native to China or S tropical Asia. (Pantropical).
IUCN: LC
Kupe Village: Cable 2540 5/1996; Lane 327 fl., 1/1995.

Clerodendrum formicarum Gürke
Woody climber, or climbing shrub; internodes hollow; leaves ternate, 4–10 × 2–5cm, glabrous, acuminate, shortly cuneate, elliptic; inflorescence terminal; axis short; branches long and conspicuously horizontal; flowers small; calyx 2–3mm; corolla tube very short, 4–5mm, yellowish white. Forest; 400m.
Distr.: tropical Africa.
IUCN: LC
Banyu: Thomas 8151 fl., 6/1988.

Clerodendrum globuliflorum B.Thomas
Shrub or climber to 3m; stems hollow; leaves large, elliptic, 15–25 × 6–12cm; inflorescence globose, dense, on older

wood; calyx to 2cm long, margin ± fimbriate; corolla less than 12cm, ± glandular. Forest; 870–1500m.
Distr.: Nigeria, Bioko, Cameroon. (W Cameroon Uplands).
IUCN: LC
Enyandong: Gosline 335 10/2001; **Kodmin:** Gosline 78 1/1998; Onana 578 2/1998; **Nyasoso:** Cable 2746 6/1996; 2847 6/1996; 3534 7/1996; Etuge 2385 6/1996; 2387 6/1996; Zapfack 692 6/1996; **Nzee Mbeng:** Gosline 106 2/1998.

Clerodendrum grandifolium Gürke
Bot. Jahrb. 18: 173 (1894).
Shrub to 3m, cauliflorous; stems hollow; leaves long-petiolate, obovate, acuminate, 15–30 × 7–15cm; inflorescence globose, dense, many-flowered; calyx to 3cm, ± glabrous; corolla to 15cm or more, ± glabrous. Forest; 1000–1100m.
Distr.: Cameroon, Congo (K). (Lower Guinea & Congolian).
IUCN: NT
Nyasoso: Cable 3452 7/1996; Lane 201 fr., 11/1994; Sidwell 361 10/1995.
Note: it is the author's contention that the above two taxa may well be conspecific. The supposed differences relate to size of bracts, degree of pubescence and corolla size. However, with an accrescent calyx, and caducous hairs at maturity, it is probable that these two taxa have been descibed based on specimens at different stages of maturity when the collections were made. Further investigation is required.

Clerodendrum inaequipetiolatum Good
Scrambling shrub to 3m; stem usually with dense indumentum, in our area seemingly glabrous; leaves 12–18 × 7–12cm, broadly ovate; inflorescence capitate, globose, many-flowered; pedicels less than 5mm; calyx green; corolla white. Forest; 1550–2065m.
Distr.: Cameroon, Angola. (Lower Guinea & Congolian).
IUCN: NT
Mt Kupe: Cheek 7582 10/1995; Sebsebe 5070 10/1995; **Nyasoso:** Balding 56 fl., 8/1993; Cable 109 fl., 9/1992; Lane 234 fr., 11/1994.

Clerodendrum melanocrater Gürke
Climbing shrub to 8m, conspicuously drying black or dark purplish black; leaves ovate, 5–15cm, cordate; petiole 3cm or more; inflorescence a dense corymb of small flowers; calyx white; corolla tube glabrous, c. 1 × 0.1cm, brownish yellow. Forest; 250–1400m.
Distr.: Bioko & W Cameroon to Tanzania. (Lower Guinea & Congolian).
IUCN: LC
Kupe Village: Etuge 2692 7/1996; 2863 7/1996; Kenfack 212 7/1996; Schoenenberger 54 fr., 11/1995; **Mile 15, Kumba-Mamfe Road:** Nemba 129 fl., 7/1986; **Nyasoso:** Cable 3470 7/1996; 3603 7/1996.

Clerodendrum silvanum Henriq. var. *buchholzii* (Gürke) Verdc.
F.T.E.A. Verbenaceae: 110 (1992).
Syn.: *Clerodendrum buchholzii* Gürke
Syn.: *Clerodendrum thonneri* Gürke
Woody climber to 10m; stems with petiolar spines; leaves elliptic or ovate, glabrous, 8–20 × 3–10cm, often confined to the canopy; inflorescence an elongate, leafless panicle, frequently cauliflorous; rachis 5–30cm long; calyx tube enlarged, 8–10mm; corolla white, fragrant; tube (1.5–)1.7–2.5cm; fruits red. Forest; 700–1350m.
Distr.: tropical and subtropical Africa.
IUCN: LC

Baduma: Nemba 178 fl., 8/1986; **Bangem:** Manning 470 fr., 10/1986; **Enyandong:** Ghogue 1257 10/2001; **Kodmin:** Gosline 64 1/1998; **Kupe Village:** Cable 3674 7/1996; Cheek 6075 fr., 1/1995; Ensermu 3529 11/1995; Lane 374 fr., 1/1995; Zapfack 935 7/1996; **Mwanenguba:** Sanford 5449 fl., 11/1968; **Ndum:** Cable 905 fl., 1/1995; **Nyasoso:** Balding 20 fl., 7/1993; Cable 3220 6/1996; Cheek 5632 fr., 12/1993; Etuge 1340 10/1995; 2415 6/1996; Lane 207 fr., 11/1994.
Note: Nemba 178, Manning 470, and Sanford 5449 were all determined at MO as *C. thonneri*, a synonym of this taxon.
Local name: Mbong Mbom. **Uses:** MEDICINES (Daniel Epie in Cheek 6075).

Clerodendrum silvanum Henriq. var. *nuxioides* (S.Moore) Verdc.
F.T.E.A. Verbenaceae: 111 (1992).
Syn.: *Clerodendrum laxicymosum* De Wild.
Syn.: *Clerodendrum validipes* S.Moore
As var. *buchholzii*, except: calyx tube ± 5mm; corolla tube 0.6–1cm, with lobes mostly at smaller end of range. Forest; 250–1400m.
Mejelet-Ehumseh: Etuge 172 fl., 6/1986; **Mile 12 Mamfe Road:** Nemba 654 10/1987.
Distr.: Cameroon, Congo (K), E Africa. (Afromontane).
IUCN: LC
Note: Nemba 654 was determined at MO as *C. laxicymosum*; and Etuge 172 as *C. validipes*, which Verdcourt treats as being synonymous with this taxon. The author has not yet seen these two specimens, but includes them here, although they may belong to var. *buchholzii*.

Clerodendrum splendens G.Don
Climbing shrub, 2–4m; young shoots quadrangular; leaves opposite, subsessile, elliptic or ovate, 6–20 × 4–10cm; petioles 3–20mm; inflorescences corymbose on young leafy shoots, terminal or axillary; calyx teeth 2–4mm; corolla deep red; tube c. 2.5cm long. Forest; 1000m.
Distr.: Senegal to Congo (K) & CAR. (Guineo-Congolian).
IUCN: LC
Mwambong: Etuge 4144 2/1998; Gosline 113 2/1998; Onana 533 2/1998.

Clerodendrum umbellatum Poir.
Woody climber to 7m; leaves opposite, the lower petiolate, 6–20 × 3.5–11cm, the upper bracteate; inflorescence terminal; calyx teeth 5mm or more; corolla white, tube puberulous, sometimes with a pink throat. Forest, farmbush; 250–1050m.
Distr.: Senegal to Congo (K), CAR, Sudan, Ethiopia, E Africa. (Afromontane).
IUCN: LC
Baduma: Etuge 396 11/1986; **Kupe Village:** Cable 789A fl., 1/1995; **Mt Kupe:** Leeuwenberg 9253 fl., 1/1972; **Ndum:** Cable 917 fr., 2/1995; Williams 151 1/1995.

Clerodendrum violaceum Gürke
Straggling or climbing shrub; young shoots quadrangular; leaves distinctly petiolate, thinly membranaceous, ovate, elliptic or oblong, 5–12 × 3–10cm; inflorescence paniculate; calyx lobes obtuse; flowers about 2.5cm, a conspicuous violet, violet and white or greenish. Forest; 1200–2000m.
Distr.: Guinea (C) to Cameroon, Congo (K), Zimbabwe. (Tropical Africa).
IUCN: LC
Edib: Cheek 9126 2/1998; **Mwanenguba:** Leeuwenberg 9952 fr., 6/1972.

Haumaniastrum caeruleum (Oliv.) J.K.Morton
Perennial herb 0.2–1m; stems one to several, sparsely pubescent; leaves almost glabrous to densely pubescent, leaf base sometimes clasping stem; petioles 0–2mm long; upper leaves subtending heads, 5–25mm long, caudate or apiculate, white or bluish, apex green, often inrolled; corolla white, pink, blue or purple; stamens declinate. Grassland; c. 1600m.
Distr.: tropical Africa.
IUCN: LC
Mwanenguba: Sanford 5547 fl., 11/1968.

Hoslundia opposita Vahl
Erect or semi-climbing aromatic shrub to 3–5m; leaves opposite or ternate, ovate-lanceolate, c. 10 × 4cm, gradually acuminate; inflorescences copious, of many-flowered terminal panicles; flowers greenish cream; calyx very small in flower, campanulate, enlarging to form orange-yellow succulent berry-like fruits. Forest; 750–1200m.
Distr.: tropical & subtropical Africa. (Afromontane).
IUCN: LC
Essossong: Leeuwenberg 9231 fl., fr., 1/1972; **Kupe Village:** Etuge 2638 7/1996; Kenfack 219 7/1996; 261 7/1996; **Mt Kupe:** Nemba 910 fl., 3/1988; **Ndum:** Cable 903 fl., 1/1995; Williams 132 fl., 1/1995; **Nyasoso:** Cable 2869A 6/1996; 3512 7/1996; Etuge 1515 12/1995; 2589 7/1996; Sebsebe 4994 fl., 10/1995; Thomas 5041 fl., fr., 11/1985.

Hyptis lanceolata Poir.
Erect aromatic herb; leaves lanceolate, to 11 × 3.5cm, punctate beneath, shortly petiolate; inflorescence very dense, axillary, globose, many-flowered, c. 1.5cm across; mature calyx 4–5mm, with subulate teeth; corolla white with pale mauve markings on the lip, very small. Grassland; c. 1500m.
Distr.: tropical Africa & America. (Amphi-Atlantic).
IUCN: LC
Kodmin: Cheek 8929 1/1998.

Isodon ramosissimus (Hook.f.) Codd
Bothalia 15: 8 (1984).
Syn.: *Homalocheilos ramosissimus* (Hook.f.) J.K.Morton
An erect or straggling herb to 4m; stems hollow, strongly quadrangular, pilose; leaves ovate, up to 7 × 4cm; inflorescence an axillary panicle of many-flowered dichotomous cymes; calyx tube declinate, ventricose, teeth subequal; corolla 5mm long, white, speckled purple in throat; upper lip very small, recurved. Forest margins; 1150–2000m.
Distr.: Sierra Leone to Bioko, Cameroon, Sudan, Uganda, Zimbabwe. (Afromontane).
IUCN: LC
Kodmin: Cheek 9055 1/1998; **Mwambong:** Pollard 114 fl., 10/1998; **Mwanenguba:** Pollard 149 10/1998; Sanford 5450 fl., 11/1968; **Nyasoso:** Etuge 1521 12/1995.
Local names: Popobe (Cheek 9055); Mebon mebon (Epie Ngome in Pollard 114). **Uses:** MATERIALS – fishing – host's caterpillars used for fish bait (Cheek 9055); musical instruments – used for making whistles, stem hollow (Cheek 9055); slice the stem longitudinally with a sharp blade to produce a musical pipe; SOCIAL USES – ritual – leaves used as a constituent part of a lucky charm (Pollard 114).

Leonotis nepetifolia (L.) R.Br. var. *nepetifolia*
Robust herb to 2.5m; stems tomentellous; leaves long-petiolate, ovate-triangular, broadly cuneate, c. 12 × 6cm, serrate; flowers in large globose whorls at the upper nodes; corolla orange-brown. Farmbush, cultivated around villages.
Distr.: pantropical.
IUCN: LC
Ebonji: Olorunfemi in FHI 30592 fl., 5/1951.

LABIATAE

Leucas deflexa Hook.f.
A straggling or semi-erect aromatic herb to 2m; leaves lanceolate, cuneate at base, serrate with an entire, acute tip, petiolate; inflorescence a densely globose axillary whorl with numerous linear-subulate bracteoles; corolla white; stamens ascending; anthers often conspicuously hairy, orange. Forest, forest margins, savanna; 850–2000m.
Distr.: Ghana, Bioko, Cameroon, Angola. (Guineo-Congolian (montane)).
IUCN: LC
Enyandong: Pollard 823 10/2001; **Ngomboaku:** Cheek 10281 12/1999; **Nyasoso:** Etuge 1672 1/1996.

Ocimum gratissimum L. subsp. gratissimum var. gratissimum
A branched erect pubescent shrub to ± 3m, emanating an aroma similar to that of cloves (*Syzygium aromaticum*); leaves ovate to obovate, 6–12 × 3cm, cuneate, acutely acuminate; inflorescence of several dense spikes, > 1cm; calyx dull, densely lanate, horizontal or slightly downward pointing in fruit; corolla small, greenish white; stamens declinate. Submontane forest, woodland, savanna; 1000–1170m.
Distr.: widespread in the tropics from India to W Africa, S to Namibia & S Africa; naturalised in tropical S America.
IUCN: LC
Edib village: Cheek 9171 2/1998; **Kupe Village:** Etuge 1619 1/1996.
Local names: Messipe (Etuge 1619); Aseb (Cheek 9171).
Uses: MEDICINES (Etuge 1619 & Cheek 9171).

Ocimum lamiifolium Hochst. ex Benth.
Perennial woody herb or shrub to 3m; stems much-branched, woody; leaves ovate, 15–65 × 6–40mm, serrate, sparsely pubescent; petiole 3–14(–20)mm; calyx 3–5mm at anthesis, 8–11mm in fruit; corolla white, white marked pink, or dull mauve, 9–11mm; stamens exserted 3–4mm, declinate. Scrubby forest, clearings, wasteland, secondary bush, forest edges; 1320m.
Distr.: W Cameroon to E Africa, Zambia, Malawi (Afromontane).
IUCN: LC
Nsoung: Darbyshire 72 fl., fr., 3/2003.

Platostoma africanum P.Beauv.
A slender weedy annual herb to ± 60cm, often less; leaves long-petiolate, ovate, c. 5 × 2.5cm; inflorescences racemose, numerous, axillary, somewhat lax and many-flowered; bracts persistent; pedicels short; corolla white, speckled mauve. Forest, farmbush; 400–1300m.
Distr.: tropical Africa, India. (Palaeotropics).
IUCN: LC
Banyu: Thomas 8164 fl., 6/1988; **Kupe Village:** Etuge 1289 fl., 11/1995; 2725 7/1996; **Mwanenguba:** Leeuwenberg 8200 8/1971; **Mungo River F.R.:** Manning 560 fl., 10/1986; **Nyasoso:** Etuge 1606 1/1996; Pollard 90 10/1998; 92 10/1998; **Tombel:** Etuge 376 fl., 11/1986.
Uses: none known (Max Ebong in Pollard 92).

Platostoma denticulatum Robyns
Bull. Jard. Bot. Nat. Belg. 17: 22 (1943).
A perennial herb very similar to *P. africanum*, except the lower calyx lip has a denticulate apical margin. Roadside verges, other damp places; 1450–1520m.
Distr.: Cameroon, Congo (K), Rwanda, Burundi, E Africa, Angola. (Afromontane).
IUCN: LC

Kodmin: Cheek 8930 1/1998; 9017 1/1998; Ghogue 9 1/1998; Pollard 177 11/1998.
Local name: Ntore (Njume Edmondo in Pollard 177).

Platostoma rotundifolium (Briq.) A.J.Paton
Kew Bull. 52: 287 (1997).
Syn.: *Geniosporum rotundifolium* Briq.
A stout woody perennial to ± 2m; stems grooved, densely ferrugineous-pubescent; leaves subrotund to broadly lanceolate, 2–5 × 1–3cm, crenulate; inflorescences several, dense, cylindrical, 2.5–10cm long; bracts broadly ovate, conspicuously white or mauve-tinged; calyx tubular, c. 4mm long at maturity, 4-toothed; corolla twice as long as calyx. Forest edge, grassland; 1500–2000m.
Distr.: Sierra Leone to Cameroon, Congo (K), E Africa, Angola. (Afromontane).
IUCN: LC
Kodmin: Cheek 8871 1/1998; **Mwanenguba:** Bamps 1591 fl., 12/1967; Cheek 7278 fl., 2/1995; Leeuwenberg 8185 fl., 8/1971; 8262 fl., 9/1971; Pollard 146 10/1998; Sanford 5448 fl., 11/1968; Thomas 3101 fl., 2/1984.

Plectranthus cataractarum B.J.Pollard
Kew Bull. 56: 976 (2001).
Decumbent or rarely erect rheophytic herb to 0.6m; leaves rhombic to obtrullate, c. 20–45 × 5–20cm, attenuate, with basal half of margin entire, distally bluntly serrate; inflorescence terminal, to 35 × 4cm, of 5–14, 6–8-flowered verticillasters; calyx 4–6mm; pedicel 8–10(–14)mm; corolla mauve or rarely white. Waterfalls, fast-flowing watercourses; 1360–1450m.
Distr.: Cameroon, Bioko. (W Cameroon Uplands).
IUCN: VU
Kodmin: Pollard 207 11/1998; 208 11/1998; Satabie 1109 fl., fr., 1/1998.

Plectranthus decumbens Hook.f.
Bot. J. Linn. Soc. 7: 210 (1864).
Syn.: *Solenostemon decumbens* (Hook.f.) Baker
Epilithic or terrestrial (at Mt. Cameroon) herb, 0.2–0.8m; stems usually annual, arising from a perennial tuber; leaves petiolate at maturity; petiole 5–32mm; lamina broadly ovate-triangular to triangular, (5–)20–40(–55) × (4–)15–35(–45)mm, acute, broadly cuneate, deeply crenate at maturity; inflorescences lax, to 44 × 35mm, with 10–20 verticillasters; fruiting calyx 3–4mm. Grassland, on rocks in grassland; 1900m.
Distr.: SE Nigeria, Cameroon: Mt Cameroon, Mt Kupe, Bakossi Mts, Bamenda Highlands, Mt Oku. (W Cameroon Uplands).
IUCN: NT
Mt Kupe: Cheek 7574 fl., 10/1995.

Plectranthus decurrens (Gürke) J.K.Morton
Understorey herb to 2m; stems subwoody; leaves at least twice as long as broad, ovate-lanceolate, to 16 × 7cm, decurrent on a long petiole, crenate-serrate; inflorescence racemose, elongate, lax; mature calyx 9–10mm, teeth long-acuminate; corolla 10–15mm, red and yellow; stamens declinate. Wet forest understorey; 300–1210m.
Distr.: SE Nigeria, Bioko, Cameroon, Gabon & Tanzania. (Afromontane).
IUCN: NT
Enyandong: Gosline 362 11/2001; **Konye:** Thomas 5175 fl., 11/1985; **Mwambong:** Pollard 118 10/1998; **Nyasoso:** Cable 3262 6/1996; 3531 7/1996; Lane 209 fl., 11/1994; Pollard 100 10/1998.

Plectranthus epilithicus B.J.Pollard in press.

Syn.: *Solenostemon repens* (Gürke) J.K.Morton
A small epilithic herb, 5–15cm; stems sometimes creeping, pubescent; leaves long-petiolate, 2–3 × 2–3cm, thin, coarsely crenate; inflorescence 3–4cm, a terminal umbellate cyme of about 4–6 short racemes; calyx 6mm, median teeth deltoid; corolla greenish white. On damp rocks, often by streams, in shade in forest understorey; 760–1200m.
Distr.: Guinea (C) to Bioko, SW Cameroon. (Upper & Lower Guinea).
IUCN: LC
Enyandong: Pollard 824 10/2001; **Kupe Village:** Cable 3694 7/1996; Etuge 1446 fr., 11/1995; 2660 7/1996; **Nyasoso:** Cable 3292 6/1996; Lane 172 fr., 11/1994; Pollard 83 10/1998.
Note: the genus *Solenostemon* is no longer recognised (see Pollard & Paton 2001).

Plectranthus glandulosus Hook.f.

A coarse scrambling to erect, often robust, glandular and strongly aromatic herb to ± 3.5m; leaves to 15cm long, glandular-punctate, margin with very uneven, rather small, double or treble crenations; inflorescence of copious loose panicles to ± 65cm long; mature calyx 9mm long; corolla violet; stamens declinate. Forest; 1200–1800m.
Distr.: Mali to Bioko, Cameroon. (Upper & Lower Guinea).
IUCN: LC
Enyandong: Pollard 754 fl., fr., 10/2001; **Mejelet-Ehumseh:** Etuge 305 fl., 10/1986; **Mwanenguba:** Leeuwenberg 8784 fl., 11/1971; Pollard 126 10/1998; Sanford 5418 fl., 11/1968.
Uses: BEE PLANTS (*fide* Pollard).

Plectranthus insignis Hook.f.

A large, soft-wooded, monocarpic mass-flowering undershrub from 3–5m at maturity, usually leafless at the time of flowering; inflorescence a very lax, ample racemose panicle; mature calyx 2cm long, teeth very unequal; corolla 2cm long, yellow suffused with purple; stamens declinate. Forest; 1300m.
Distr.: W Cameroon. (Cameroon endemic).
IUCN: NT
Nsoung: Darbyshire 68 3/2003.

Plectranthus kamerunensis Gürke

Straggling densely woolly herb to 1m; leaves ovate, to 11 × 9cm, acutely acuminate, cordate, coarsely crenate; inflorescences little-branched; mature calyx 8mm long, with a long white pubescence; lower teeth lanceolate acuminate; corolla violet; stamens declinate. Forest, forest margins; 1200–1975m.
Distr.: SE Nigeria, W Cameroon, E Africa. (Afromontane).
IUCN: LC
Enyandong: Pollard 839 11/2001; **Mwambong:** Pollard 119 10/1998; **Mwanenguba:** Cheek 7283a fr., 2/1995; Leeuwenberg 8555 fl., 10/1971.
Local name: Mebon Mebon (Epie Ngome in Pollard 119).

Plectranthus luteus Gürke

A branched woody herb to ± 1m, producing conspicuous fusiform, densely brown-villose bulbils in the axils of the inflorescence and branches; leaves elliptic-lanceolate, long-acuminate, cuneate, petiolate; inflorescences axillary and terminal; flowers in sessile fascicles; pedicels c. 7mm; mature calyx glandular, c. 7mm; corolla yellow, c. 1cm; stamens declinate. Forest; 1500–1530m.
Distr.: Liberia, Cameroon, Congo (K), E Africa. (Afromontane).

IUCN: LC
Kodmin: Ghogue 20 1/1998; Onana 564 2/1998; Plot B90 1/1998.

Plectranthus monostachyus (P.Beauv.) B.J.Pollard subsp. *monostachyus*

Kew Bull. 56: 980 (2001).
Syn.: *Solenostemon monostachyus* (P.Beauv.) Briq. subsp. *monostachyus*
Annual herb to about 60cm; stems quadrangular, fleshy; leaves often with a central purple blotch, broadly ovate, base broadly cuneate narrowing abruptly to become decurrent on the petiole, coarsely crenate; inflorescence 5–40 × 1.5–1.7cm in fruit; mature calyx 4–5mm, pubescent; corolla mauve about 5mm; stamens declinate. Damp places, forest understorey; 400–1300m.
Distr.: widespread in tropical Africa. (Afromontane).
IUCN: LC
Edib village: Cheek 9174 2/1998; **Kupe Village:** Etuge 1698 2/1996; 2016 5/1996; **Loum F.R.:** Pollard 93 10/1998; **Nlonako:** Etuge 1774 3/1996; **Nyasoso:** Pollard 91 10/1998.
Local name: Echume-chum (Cheek 9174). **Uses:** MEDICINES (Cheek 9174); none known (Epie Ngome in Pollard 93).

Plectranthus occidentalis B.J.Pollard in press.

Syn.: *Solenostemon mannii* (Hook.f.) Baker
A herbaceous or somewhat woody perennial herb or shrub to c. 1m; stems climbing or erect; leaves ovate, 4–15cm, acutely acuminate, crenate, long-petiolate, often purplish tinged; inflorescence a copiously flowered, dense raceme, up to 25 × 3–4cm or more in fruit; calyx 4–5mm; corolla rich bluish purple. Forest, woodland; 1000–1550m.
Distr.: Sierra Leone to Bioko & W Cameroon. (Upper & Lower Guinea).
IUCN: LC
Enyandong: Etuge 4512r fl., 11/2001; **Kodmin:** Etuge 4019 1/1998; Pollard 191 11/1998; **Kupe Village:** Sidwell 439 11/1995; **Nyasoso:** Lane 180 fl., 11/1994; 239 fl., 11/1994.
Note: the genus *Solenostemon* is no longer recognised (see Pollard & Paton 2001).

Plectranthus tenuicaulis (Hook.f.) J.K.Morton

Syn.: *Plectranthus peulhorum* (A.Chev.) J.K.Morton
Slender branched annual herb to ± 1m; stems pubescent; leaves shortly petiolate, lanceolate, acute, 0.5–6cm long, a third as broad, crenate, pubescent; inflorescence a panicle with lateral racemose branches; mature calyx 3–5mm; corolla 8–10mm, pale blue; stamens declinate. Forest, forest margins; 1300–1600m.
Distr.: tropical Africa.
IUCN: LC
Enyandong: Pollard 849 11/2001; **Mwanenguba:** Pollard 127 10/1998; Sanford 5513 fl., 11/1968.

Plectranthus sp. nov.

Shrub to c. 1.5m; flowering stems leafless; leaves ± orbicular, serrate, about as long as wide, to c. 10 × 10cm; inflorescence dense, with no obvious central axis, but with numerous whip-like patent side axes, to c. 15cm; mature calyx to c. 1cm; corolla yellow only, to 1.2cm. Forest; 450–710m.
Nyale: Etuge 4192 fl., fr., 2/1998; Cheek 11337 fl., fr., 3/2003.
Note: this is a species new to science, probably most closely related to *P. insignis*.

Uses: MEDICINES (Etuge 4192).

LABIATAE

Premna angolensis Gürke

Tree to 12m, less often shrubby; bole often crooked, usually hollow, sometimes fluted; branches ± horizontal; leaves usually in whorls of 4, rarely paired, ovate to suborbicular, 4–21 × 3–15cm, acuminate; inflorescence a thyrsoid panicle, 12–30cm; flowers small, white, sometimes tinged green; calyx 1.5–2mm; corolla tube 2–3mm. Forest; 200–1650m.
Distr.: Senegal to Angola, Ethiopia to E Africa. (Afromontane).
IUCN: LC
Ikiliwindi: Nemba 540 fl., 6/1987; **Kodmin:** Cheek 9588 11/1998; **Ngomboaku:** Cheek 10345 12/1999.
Local name: Ehchunge. **Uses:** none known (Max Ebong in Cheek 9588).

Pycnostachys batesii Baker

F.T.A. 5: 386 (1900).
A scrambling aromatic undershrub to 4m; branches slender; leaves ovate, 5–10 × 3–6cm, acute, cuneate to truncate or even ± cordate, very loosely-crenate, long-petiolate; inflorescence a terminal spike, very dense, c. 2–3cm long; calyx-tube campanulate with conspicuously rigid spiny lanceolate-subulate teeth; corolla purple; infructescence ovoid. Forest; 850–1450m.
Distr.: Cameroon, Sudan, Uganda. (Afromontane).
IUCN: LC
Kodmin: Cheek 8946 1/1998; **Kupe Village:** Cheek 7094 fr., 1/1995; Ensermu 3558 fl., fr., 11/1995; **Ndum:** Cable 896 fl., 1/1995; **Nyasoso:** Etuge 2538 7/1996.

Pycnostachys eminii Gürke

An aromatic shrub, 2–4m; leaves about 10 × 4cm, distinctly crenate-serrate, shortly pubescent; inflorescence a cylindrical spike, c. 2.5–5.0 × 1.5cm; calyx teeth spiny, about 2.5mm long, sharp to the touch, densely tomentose at maturity; corolla pale blue. Grassland, woodland; 1200–1975m.
Distr.: W Cameroon, Congo (K), E Africa. (Afromontane).
IUCN: LC
Mwanenguba: Cheek 7277 fr., 2/1995; Leeuwenberg 8842 fl., 12/1971; Sanford 5451 fl., 11/1968.

Pycnostachys meyeri Gürke

Similar to *P. eminii*, except spike stouter, about 2.5cm or more broad; mature calyx shortly and rather thinly pubescent; corolla pale blue to purplish blue, occasionally even pinkish. Grassland, woodland, savanna; 2000m.
Distr.: tropical Africa. (Afromontane).
IUCN: LC
Mwanenguba: Pollard 147 10/1998.

Satureja punctata (Benth.) Briq.

A robust narrow erect or spreading heath-like woody herb to ± 60cm; stems wiry, branched, pubescent; leaves ovate, 4–10mm long, conspicuously glandular punctate beneath, subsessile, entire; flowers few together, in axillary whorls; calyx tubular, 4mm; corolla rich pink. Grassland, forest edges; 1560–2000m.
Distr.: Cameroon to E & S Africa. (Afromontane).
IUCN: LC
Kodmin: Onana 599 2/1998; **Mwanenguba:** Bamps 1602 fl., 12/1967; Cheek 7239 fl., 2/1995; Pollard 153 10/1998; 156 10/1998; 870 fl., fr., 11/2001; Thomas 3104 fl., 2/1984.

Satureja robusta (Hook.f.) Brenan

An erect robust, strongly aromatic, perennial, 0.8–1.4m; stems branched; leaves ovate-rotund, crenate; inflorescence broad, of many dense terminal sessile spikes; mature calyx

4–5mm; corolla white with mauve marks on the lip; stamens ascending. Grassland, woodland, forest edges; c. 1600m.
Distr.: W Cameroon. (Cameroon endemic).
IUCN: LC
Mwanenguba: Sanford 5600 fl., 11/1968.
Note: although this taxon is of limited distribution, it is quite common within its range, and so does not warrant assignation of an IUCN category of threat.

Stachys aculeolata Hook.f. var. *aculeolata*

A perennial prickly herb; stems straggly, scrambling, rather slender with stiff prickly bristles and short glandular hairs; leaves ovate-triangular, to 5.5 × 4.5cm, cordate; petiole as long as the lamina; inflorescence of distant, several-flowered verticillasters; calyx 7mm at maturity, campanulate; corolla c. 1cm, white or pale pink; lower lip 3-lobed, central lobe larger, emarginate. Damp grassland, woods, swamps, forest margins; 1450–1500m.
Distr.: Bioko, Nigeria, W Cameroon, Burundi, Rwanda, Congo (K), Ethiopia, E Africa. (Afromontane).
IUCN: LC
Kodmin: Cheek 9050 1/1998; Pollard 250 11/1998.

Vitex grandifolia Gürke

Tree, 10–30m; leaves compound, opposite, 5-foliolate; leaflets abruptly short- to long-acuminate, the middle ones 13–40 × 6–20cm, attenuated into a short petiolule, c. 1cm long; inflorescence cymose, at base of leaves; peduncle 0.5–5.0cm; flowers pale yellowish to yellowish brown. Forest; 200–1200m.
Distr.: Sierra Leone to Gabon. (Upper & Lower Guinea).
IUCN: LC
Etam: Etuge 209 fr., 8/1986; **Ikiliwindi:** Nemba 524 6/1987; **Kupe Village:** Elad 78 1/1995; Etuge 1396 11/1995; 1920 5/1996; 2724 7/1996; 2727 7/1996; Kenfack 218 7/1996; 240 7/1996; Lane 343 fr., 1/1995; **Mungo River F.R.:** Cheek 9348 10/1998; **Ngusi:** Etuge 55 4/1986.
Uses: MATERIALS – timber (Etuge 2724).

Vitex lehmbachii Gürke

Tree to 12m; branches glabrous; leaves compound, opposite; middle leaflets 10–25 × 4–10cm, with 4–8 pairs of lateral nerves, obovate, abruptly acuminate, cuneate; petiole drying black; inflorescence of long-pedunculate axillary cymes; peduncle 10–20cm; calyx 3–5mm; corolla less than 1cm, white with pink lower lip; fruits 1–1.5cm. Forest; 700–1050m.
Distr.: Mt Cameroon, Kupe-Bakossi, Bamenda Highlands. (W Cameroon Uplands).
IUCN: EN
Kupe Village: Cable 3729 7/1996; **Mwambong:** Etuge 4328 10/1998; **Ngomboaku:** Cheek 10377 12/1999.

Vitex rivularis Gürke

Tree to 15m; leaves compound, opposite, 5–7-foliolate; leaflets elliptic, elliptic-oblong or broadly lanceolate, acuminate, narrowly cuneate, the middle ones 8–18 × 3–6cm, with 12–20 pairs of lateral nerves; inflorescence cymose; peduncle 10–20cm; calyx 1–1.5mm; flowers very small, white tinged purple. Forest; 250m.
Distr.: Liberia to Congo (K). (Guineo-Congolian).
IUCN: LC
Ikiliwindi: Etuge 514 3/1987.
Note: the author has not seen this MO specimen, but it is quite reasonable to suppose this species occurs in our area.

Vitex thyrsiflora Baker

Undershrub or small tree, sometimes climbing to 7m; stems quadrangular with conspicuous 'ant-holes', harbouring biting ants; leaves compound, opposite, 5-foliolate, densely dotted with yellow glands; central leaflets 5–18 × 3.5–9.0cm; inflorescence of cymes arranged usually in terminal or terminal and axillary panicles; calyx truncate, with obsolete teeth; corolla white, with a right-angle bend, about twice as long as calyx; fruit globose, up to 5mm across. Forest; 800m.
Distr.: Guinea (C) to Cameroon, Congo (K). (Guineo-Congolian).
IUCN: LC
Kupe Village: <u>Kenfack 267</u> 7/1996.

Vitex yaundensis Gürke

Bot. Jahrb. 33: 296 (1904).
Tree 10–15m tall; bole pale grey-brown; outer slash wet dark brown; inner pale yellow; leaves compound, opposite, 5-foliolate, with extremely long petioles to 30cm; the secondary nerves conspicuously prominent beneath, close together and almost parallel; petiolules 1–2cm; inflorescence long-pedunculate axillary cymes; bracts linear; corolla tube scarcely exceeding the calyx; fruits c. 2cm long. Forest; 150m.
Distr.: Bakossi & Yaoundé. (Cameroon Endemic).
IUCN: CR
Mungo River F.R.: <u>Cheek 10139</u> 11/1999.
Note: only the second collection of this species.

LAURACEAE

M.V. Norup

Fl. Cameroun 18 (1974).

Note: species delimitation in the genus *Beilschmiedia* is poorly understood, and there is currently no active specialist working on this genus. In light of this fact, it is not considered suitable to apply conservation assessments to the taxa presented below; several are known only from the type collection and may therefore prove to be critically endangered, though this cannot be confirmed until a full revision of the genus has been carried out.

Beilschmiedia acuta Kosterm.

Fl. Cameroun 18: 15 (1974).
Tree 12m; leaves aromatic when broken, ± coriaceous, concolourous, olive green, elliptic-lanceolate, 15–20 × 4cm; petiole 1–2cm; inflorescence highly branched, 6 × 3cm; flowers red, c. 3mm wide; tepals ± densely hairy on outside, hairy within. Forest; 900m.
Distr.: Cameroon & CAR. (Lower Guinea).
Kupe Village: <u>Kenfack 287</u> 7/1996.
Note: in addition to the specimen cited, currently known only from Tscappe Pass, Adamawa Prov., Yokadouma-Moloundou, E Prov. and Boukoko, CAR (Fl. Cameroun 18: 16).

Beilschmiedia barensis (Engl. & K.Krause) Robyns & Wilczek

Fl. Cameroun 18: 48 (1974).
Tree 1–5m; leaves concolorous, drying brown, or slightly discolourous, 17–20 × 9–10cm, shortly acuminate, base rounded, lateral nerves 6–7 pairs; inflorescence 10cm, lax; branches 8–10; flowers yellow, tepals densely velvety on both sides; staminal whorls 1 and 2 sessile, back pubescent;

whorl 3 with filament hairy, glands rounded, staminodes 0.5mm, hairy on both surfaces; ovary ovoid, style as long as ovary. Forest, plantation; 1000m.
Distr.: Cameroon Endemic.
Baré: <u>Ledermann 1231</u> fl., 11/1908; **Ndoungue:** <u>Ledermann 6235</u> fl., 9/1909.
Note: specimens not seen by us. Known from only 3 collections, the third being near Douala, SW Province. This taxon, and the others species which we know only from the literature (*B. crassipes*, *B. ndungensis*, *B. robynsiana*), may be represented among the imperfect or incompletely analysed material treated as *Beilschmiedia sp. 1–4* listed below.

Beilschmiedia crassipes (Engl. & K.Krause) Robyns & Wilczek

Fl. Cameroun 18: 48 (1974).
Tree 4–5m; leaves concolorous, drying brown, oblong or oblong-lanceolate, 25–30 × 8–12cm, shortly acuminate, lateral nerves 8–10 pairs; petiole 15–20mm, very stout (3–4mm diameter); inflorescence 10–15cm, lax; tepals densely velvety on exterior, puberulent inside, receptacle tomentellous; staminal whorls 1 and 2 sessile, back pubescent, apex reflexed to interior; whorl 3 with filament hairy, glands oval, inserted on a fine pedicel against a filament; ovary spherical, style twice as long as ovary. Open forest; alt. unknown.
Distr.: W Cameroon. (Narrow Endemic).
Loum: <u>Ledermann 6460</u> fl., 9/1909.
Note: specimen not seen by us; taxon currently known only from the type collection.

Beilschmiedia cf. cuspidata 1 (K.Krause) Robyns & Wilczek

Fl. Cameroun 18: 15 (1974).
Shrub 8m; leaves concolourous, ovate-elliptic, 4–6 × 15–22cm, glabrous beneath; inflorescence higly branched; flowers yellow, tepals short-haired on the outside, hairy within. Forest; 250–300m.
Distr.: Cameroon Endemic.
Loum F.R.: <u>Cheek 10249</u> 12/1999; **Mungo River F.R.:** <u>Cheek 10121</u> 11/1999.
Note: keys out as *B. cuspidata* using Fl. Cameroun but det. uncertain since no material at K; taxon known only from the type, E of Kribi at 200m alt.

Beilschmiedia cf. cuspidata 2 (K.Krause) Robyns & Wilczek

Shrub 1.5–5m; leaves concolorous, ovate-elliptic, 4–6 × 15–22cm, glabrous beneath; inflorescence highly-branched; flowers white, tepals short-haired on the outside, hairy within. Forest; 1350–1400m.
Kodmin: <u>Etuge 4414</u> 11/1998; <u>Gosline 60</u> 1/1998; <u>Plot 189</u> 1/1998.
Note: very close to *B. cf. cuspidata 1*, differing in the smaller habit and in habitat; perhaps a montane form of this taxon.

Beilschmiedia ndongensis (Engl. & K.Krause) Robyns & Wilczek

Fl. Cameroun 18: 66 (1974).
Tree 8–10 m; leaves elliptic, 15–20 × 5–7.5cm, acuminate, discoloured, brown below, nerves 7–10 pairs, tertiary reticulum raised above and below; petiole 8–14mm; inflorescence 6–10cm; flowers green, tepals glabrous apart from puberulent apex; whorl 1 and 2 staminal filaments flat, hairy on both sides, anthers broad; whorl 3 filaments straight, hairy on both surfaces, twice the anther length, glands at

base; staminodes cordiform, stipitate; ovary elongate, style capitate. Forest.
Distr.: W Cameroon. (Narrow Endemic).
Ndoungue: Ledermann 6165 fl., 11/1909.
Note: currently known only from the type specimen, not seen at K.

Beilschmiedia preussioides Fouilloy & N.Hallé
Fl. Gabon 10: 46 (1965).
Tree 9m; leaves discolorous, subcoriaceous, elliptic-oblong, 12–15 × 5cm; inflorescence highly-branched; flowers greenish pink, not opening fully at anthesis, tepals rounded apically; staminodes heart-shaped or triangular, glands on anthers of whorl 3 stamens shorter than length of the corresponding filaments, receptacle glabrous inside, ovary glabrous. Forest; 900m.
Distr.: Cameroon & Gabon. (Lower Guinea).
Nyasoso: Etuge 2431 6/1996.
Note: previously known only from the type collection, Mongoumou, Gabon (Fl. Gabon 10: 48).

Beilschmiedia robynsiana Kosterm.
Fl. Cameroun 18: 49 (1974).
Tree 8–12m; leaves olive above, brown below, oblong, 18–26 × 6–9cm, acuminate, lateral nerves 7–8 pairs, glabrous; petiole black, 18–25mm; inflorescence lax, 12–25cm, branched 3–4 times; tepals hairy inside, receptacle tomentellous; whorls 1 and 2 sessile, straight, dorsal surface hairy; whorl 3 with cells lozenge-shaped, hairy, glands spherical, inserted at base of filament; staminodes filiform, 1mm, entirely hairy; ovary ovoid, attenuate. Forest; 860m.
Distr.: Cameroon. (W Cameroon Uplands).
Baré: Ledermann 1284 fl., 11/1909.
Note: known only from three specimens, not seen at K; in addition to the specimen cited, recorded at Banga nr. Kumba, and near Nkongsamba (Fl. Cameroun 18: 50).

Beilschmiedia sp. 1 (subgenus *Synanthoradenia*)
Shrub or tree 2.5–20m; leaves papyraceous, concolourous, olive green, 10–15 × 3–5cm; inflorescence highly-branched; calyx ± densely hairy outside. Forest; 800–1400m.
Kodmin: Etuge 4405 11/1998; **Kupe Village:** Etuge 1985 5/1996.

Beilschmiedia sp. 2
Large tree; leaves discolorous, olive-green above, brown below, ovate-elliptic, 12–18 × 5–7cm; inflorescence highly-branched. Forest; 850m.
Nyasoso: Sidwell 317 10/1995.

Beilschmiedia sp. 3
Tree (specimen from fallen branch); leaves drying reddish brown, highy coriaceous; inflorescence in bud, inadequate for determination. Forest; 1100m.
Nyasoso: Cable 2849 6/1996.

Beilschmiedia sp. 4
Tree 2.5m; leaves papyraceous, ± discolorous, 15–18cm long; petioles 1–2cm, blackish; inflorescence highly-branched; fruits c. 5 × 2mm, mucronate, with ± persistent calyx. Forest near stream; 810m.
Ngomboaku: Ghogue 456 12/1999.

Beilschmiedia sp. 5
Leaning climber to 3m; leaves discolorous, olive green above, brown below; inflorescence highly-branched; flowers light yellow; tepals hairy on the inside. Forest; 900m.
Ngomboaku: Gosline 262 12/1999.

Cinnamomum verum J.S. Presl
F.T.E.A. Lauraceae: 2 (1996).
Syn.: *Cinnamomum zeylanicum* Blume
Tree 8m; leaves opposite, ovate-elliptic-lanceolate, triplinerved or penninerved, glabrous above; inflorescence terminal, dichotomously branched; perianth bell-shaped, tepals 3.5–5mm, hairy on the outside; anthers 9, with hairy filaments, outer anther whorl opposite tepals, anthers of whorl 3 with glands at the base of filament; staminodes 3 in whorl 4, heart-arrow shaped at apex; ovary superior, encompassed in cupule formed by tepals, ovule 1, apically attached. Farms; 350m.
Distr.: native of SE Asia, cultivated widely in the tropics. (Pantropical).
IUCN: LC
Kupe Village: Etuge 2037 5/1996.
Note: occurs sporadically, but not spontaneously.
Uses: FOOD ADDITIVES – bark harvested as a spice (*fide* Cheek).

Hypodaphnis zenkeri (Engl.) Stapf
Tree 15–25m; leaves alternate, elliptic, c. 12 × 7cm, shortly acuminate, acute, lateral nerves c. 4 pairs, tertiary nervation subscalariform; petiole 2cm; inflorescences several from near stem apex; peduncle c. 8cm, dichasial cyme c. 2 × 4cm, branching 5 times; flowers dull yellow, 4mm diameter; staminal whorls 1 and 2 introrse, glands on whorl 2, whorl 3 extrorse, staminodes absent; ovary inferior; fruit fleshy, inferior, obovoid, c. 5 × 4cm, with a single pendulous seed. Forest; 150–350m.
Distr.: Nigeria to Gabon. (Lower Guinea).
IUCN: NT
Bakolle Bakossi: Etuge 150 5/1986; **Etam:** Etuge 79 3/1986; **Mt Kupe:** Etuge 150 fr., 5/1986; 79 fl., 3/1986.

Persea americana Mill.
Tree 8m; leaves drying cinnamon-brown with lighter coloured nervation, elliptic, 16–20 × 7–8cm; fruits oblong, 3–4cm, fleshy, on slightly swollen receptacle, 0.5–0.9mm long. Farms; 650–1170m.
Distr.: native of S America, cultivated widely in the tropics. (Pantropical).
IUCN: LC
Edib village: Cheek (sight record) 2/1998; **Ngusi:** Etuge 1547 1/1996.
Local names: 'Pear' or 'Avocado'. **Uses:** FOOD – commonly planted for its edible fruits (*fide* Pollard).

LECYTHIDACEAE

I. Darbyshire

Crateranthus talbotii Baker f.
Tree to 20m, 30cm d.b.h.; bark red-brown, smooth; leafy branchlets foliaceously winged, to 1cm width; leaves oblong-elliptic to oblanceolate, 26–33.5 × 8–11cm, grey-green above, acuminate, base unequally cordate, lateral nerves 10–14 pairs; petiole 0.5cm, black; flowers axillary, subsessile; bracts leathery 1cm; calyx 3-lobed, lobes obovate, 2.5cm; corolla 3.5cm, pink to magenta, margin pale. Forest, particularly swampy areas; 700–1000m.
Distr.: S Nigeria to Cameroon. (Lower Guinea).
IUCN: VU

Baseng: <u>Cheek 10414</u> 12/1999; **Nyale:** <u>Cheek 9637</u> 11/1998; <u>Etuge 4241</u> 2/1998; <u>4243</u> 2/1998; **Nyandong:** <u>Cheek 11308</u> 3/2003.

Napoleonaea

Note: species delimitation in the genus *Napoleonaea* appears unclear; our specimens, with the exception of *N. egertonii*, proved difficult to key out using the treatments in FWTA and Liben: Bull. Jard. Bot. Nat. Belg. 41: 363–382 (1971).

Napoleonaea egertonii Baker f.

Tree to 20m; leaves oblong-elliptic, 18–36.5 × 8–13cm, apex shortly acuminate, base acute, lateral nerves 9–11 pairs; petiole to 1.2cm; panicles cauliflorous, 12 × 12cm; peduncle brown-puberulent; pedicel 0.3–1cm; calyx 2.4cm diameter, 5-lobed, densely warted and puberulous; corolla c. 4.5cm diameter, white with purple veins, outer corona segments 0.45cm, puberulent, inner corona glabrous; fruit subglobose, 14 × 16.5 × 14cm, echinate, spines to 1.5cm; seeds to 5cm. Forest; 450–920m.
Distr.: S Nigeria to Gabon. (Lower Guinea).
IUCN: VU
Kupe Village: <u>Lane 438</u> fl., 1/1995; **Manehas F.R.:** <u>Pollard 125</u> 10/1998; **Nyandong:** <u>Ghogue 1463</u> fl., 3/2003.

Napoleonaea cf. talbotii Baker f.

Shrub or tree to 3m; stems subangular; leaves oblong-oblanceolate, c. 21 × 6cm, acumen 1–2cm, lanceolate to spatulate, base obtuse, margin entire, paired glands at blade base prominent, lower surface green-brown, upper dark green, lateral nerves 9–12 pairs; petiole 0.5cm; flowers cauliflorous or axillary, solitary or in few-flowered fascicles; calyx 0.8–1.1cm diameter, lobes 3-glandular, glabrous; corolla c.3.5cm diameter, white to yellow with purple streaking; fruit not seen. Forest; 300–400m.
Distr.: Nigeria to Gabon. (Lower Guinea).
Bakossi F.R.: <u>Etuge 4283</u> 10/1998; <u>Pollard 102</u> 10/1998; **Bolo Moboka:** <u>Nemba 241</u> fl., 9/1986.
Note: *N. talbotii* usually has flowers in fascicles of up to 10. The glands on the calyx lobes of Pollard 102 were much-visited by ants (Pollard, *pers. obs.*).
Local name: Ngabube (Epie Ngome in Pollard 102).

Napoleonaea cf. vogelii Hook. & Planch.

Shrub or tree to 2m; stems whitish, subangular towards apex; leaves (oblong-) elliptic, 15–22.3 × 4.7–8cm, bright green, acumen spatulate, c. 2 × 0.15cm, base acute, margin obscurely toothed, lateral nerves c. 7 pairs, paired glands at base of blade inconspicuous; petiole 0.8cm; flowers cauliflorous on leafy stems or axillary, solitary; calyx 1.2cm diameter, lobes 3-glandular, glabrous; corolla fused, 3.6cm diameter, margin dentate, pale yellow with purple streaking, outer corona slightly shorter than inner corona; fruit oblate, 3 × 5 × 5cm, shallowly c. 10-lobed, depressed at poles; seeds 2 × 1.5cm. Forest; 820m.
Distr.: Guinea (C) to Congo (K). (Guineo-Congolian).
Kupe Village: <u>Kenfack 300</u> 7/1996; <u>Lane 412</u> fl., fr., 1/1995.
Note: *N. vogelii* usually has smaller leaves and a puberulent or verrucate calyx.

Napoleonaea sp. 1

Liana; stems subangular; leaves oblanceolate, 21 × 7.3cm, vivid green below, darker above, acumen 1.6cm, base obtuse, paired glands near base of blade conspicuous, lateral nerves 8 pairs; petiole 0.5cm; inflorescence not seen; flowers cauliflorous, white (label). Forest; 420m.
Mungo River F.R.: <u>Onana 977</u> 12/1999.

Note: the leaves of this specimen match the description in Liben (1971) of the poorly known lianescent *N. reptans* Baker f. from Nigeria; more material is required.

Petersianthus macrocarpus (P.Beauv.) Liben
Bull. Jard. Bot. Nat. Belg. 38: 207 (1968).
Syn.: *Combretodendron macrocarpum* (P.Beauv.) Keay
Syn.: *Combretodendron africanum* (Welw. ex Benth. & Hook.f.) Exell
Syn.: *Petersia africana* Welw. ex Benth. & Hook. f.
Tree to 25m; leaves obovate, to 12 × 5.3cm, apex shortly acuminate, base cuneate, margin obscurely toothed, lateral nerves c. 9 pairs, domatia pitted; petiole 1cm, winged; racemes 1–3cm, on peduncles 2.5cm, puberulent; fruiting pedicels c. 3cm, articulated mid-way; fruit membranously 4-winged, 3.5 × 4cm, apex emarginate, wings parallel-nerved. Forest, secondary forest; 350–520m.
Distr.: Guinea (C) to Angola. (Guineo-Congolian).
IUCN: LC
Bakolle Bakossi: <u>Etuge 135</u> 5/1986; **Eboné:** <u>Leeuwenberg 8295</u> 9/1971.

LEEACEAE

B.J. Pollard

Fl. Cameroun 13 (1972).

Leea guineensis G.Don

Erect or suberect soft-wooded shrub to 7m; leaves bipinnate; leaflets opposite, imparipinnate, oblong-elliptic to 18cm long; flowers bright yellow, orange or red; fruits brilliant red, turning black. Forest, forest gaps; 700–1240m.
Distr.: tropical Africa.
IUCN: LC
Kupe Village: <u>Biye 53</u> 11/1999; <u>Cheek 6034</u> fl., fr., 1/1995; **Ndum:** <u>Groves 24</u> fr., 1/1995; **Ngomboaku:** <u>Ghogue 473</u> 12/1999; **Nsoung:** <u>Darbyshire 48</u> 3/2003; **Nyasoso:** <u>Balding 22</u> fr., 7/1993; <u>92</u> fl., fr., 8/1993; <u>Cable 2807</u> 6/1996; <u>Sidwell 414</u> 10/1995; <u>Sunderland 1476</u> fl., 7/1992.

LEGUMINOSAE-CAESALPINIOIDEAE

B.A. Mackinder

Fl. Cameroun 9: (1970).

Afzelia bella Harms

Shrub or small tree to 12m; leaves pinnate; petiolules twisted; leaflets 3–5 pairs, glabrous above and below, up to 15 × 7cm, midvein subcentral; inflorescence a panicle; flowers with single well-developed petal, white marked red, up to 4.8cm, apex bi-lobed; pod compressed, woody, elastically dehiscent, ± kidney shaped, up to 15cm long, brown; seeds black with bright orange-red aril. Primary or secondary forest; 250–800m.
Distr.: Guinea (C) to Congo (K). (Guineo-Congolian).
IUCN: LC
Mile 15, Kumba-Mamfe Road: <u>Nemba 145</u> fl., 7/1986 (n.v.); **Mwanenguba:** <u>Leeuwenberg 9457</u> fl., 3/1972 (n.v.).

Afzelia pachyloba Harms

Tree to 45m; leaves pinnate; leaflets 5–10 pairs, glabrous above, sparsely to moderately hairy below, up to 7.8 × 3.2cm; petiolules twisted; inflorescence a raceme often

compounded into a panicle; flowers with single well-developed petal, white marked red, up to 4cm, apex bi-lobed; pod compressed, woody, elastically dehiscent, ± kidney-shaped, 15–20cm long, pale brown; seeds black with yellow aril. Forest; 420m.
Distr.: Nigeria to Congo (K). (Lower Guinea & Congolian).
IUCN: VU
Loum F.R.: Cheek 9302 10/1998.

Anthonotha lamprophylla (Harms) J.Léonard
Tree to 10m; leaves pinnate; leaflets 4–7 pairs, glabrous above, densely silky-hairy below giving a silvery to golden sheen, up to 28 × 7cm; inflorescence a raceme, compounded into a panicle, cauliflorous; flowers with single well-developed white petal, up to 8mm, apex bi-lobed; pod compressed, woody, green mottled-brown, golden-hairy. Forest, secondary forest; 750–800m.
Distr.: Nigeria, Cameroon & Gabon. (Lower Guinea).
IUCN: NT
Kupe Village: Etuge 1987 5/1996; **Nyasoso:** Etuge 2594 7/1996.

Anthonotha macrophylla P.Beauv.
Tree to 10m; leaves pinnate; leaflets 2–5 pairs, sparsely hairy or glabrous above, somewhat glossy above, densely but minutely hairy below, terminal pair longest, up to 21 × 7.8cm, lower pairs shorter and relatively broader; inflorescence a raceme, compounded into a panicle, axillary, terminal or cauliflorous; flowers with single conspicuous white petal, up to 9mm, apex bi-lobed, other petals inconspicuous; pod compressed, woody, golden-hairy becoming glabrous, up to 26cm; seeds irregular shaped, closely packed, edges flattened at adjoining surfaces. Forest; 180–900m.
Distr.: Guinea (C) to Congo (K). (Guineo-Congolian).
IUCN: LC
Kupe Village: Groves 14 fl., 1/1995; **Loum F.R.:** Mackinder 222 11/1999; 223 11/1999; 268 12/1999; **Mungo River F.R.:** Cheek 9339 10/1998; Mackinder 229 11/1999; 235 11/1999; **Nyasoso:** Etuge 2458 6/1996; Sidwell 310 10/1995.

Baikiaea insignis Benth.
Tree to 26m; leaves imparipinnate; leaflets 3–8, glabrous, alternate, variable, often elliptic or oblong-elliptic, 9–38 × 3.5–15cm, leathery; inflorescence an axillary or (more commonly) a terminal cluster; flowers white, large, up to 20cm long; pod compressed, woody, brown, velvety, oblong, up to 30cm long, elastically dehiscent, occasionally remaining attached to the tree for some time after dehiscing. Forest; 400m.
Distr.: Sierra Leone to Tanzania. (Guineo-Congolian).
IUCN: LC
Banyu: Thomas 8227 fl., 4/1988 (n.v.).

Berlinia bracteosa Benth.
Shrub or tree (2–)8–35m; leaves paripinnate; leaflets in (3–)4–5 pairs, glabrous above, appearing glabrous below at × 10 magnification, puberulous at higher magnifications, up to 43 × 12.5cm; inflorescence a raceme; flowers subtended by conspicuous bracteoles, 4.5–10.5cm long; petals white, subequal in length, posterior petal (broadest) 5–7 × 5.5–9cm, apex bilobed; pod oblong, compressed, woody up to 30cm long, elastically dehiscent, upper suture thickened and ridged, 12mm wide; seeds 3–6, discoid 3–4.5 × 2.2–2.8cm, glossy mid-brown. Lowland forest, forest edges and clearings, often along streams; 190–310m.

Distr.: SE Nigeria to Congo (K). (Lower Guinea & Congolian).
IUCN: LC
S. Bakossi: Mackinder 225 11/1999; **Bolo:** Thomas 5715 fl., 3/1986; **Mungo River F.R.:** Mackinder 206 11/1999; 252 12/1999.

Chamaecrista kirkii (Oliv.) Standley
Fl. Cameroun 9: 58 (1970).
Erect herb to 1.2m; leaflets 20–40 pairs, oblong to linear-oblong, 6–17 × 1–3mm; leaf rachis channeled but not crested along upper side; petiole with an apical, sessile gland; flowers yellow, 2–5, axillary; fruit linear, compressed, dehiscent. Grassland; 2000m.
Distr.: tropical Africa.
IUCN: LC
Mwanenguba: de Gironcourt 514 1911–1912 (n.v.); Hédin s.n. date unknown (n.v.).
Note: specimens cited in Fl. Cameroun 9: 58 (1970).

Chamaecrista mimosoides (L.) Greene
Lock, Legumes of Africa: 32 (1989).
Prostrate or more commonly erect herb or subshrub to 1.5m; leaves paripinnate; leaflets 20–70 pairs, linear to linear oblong, 3–8 ×1–1.5mm; flowers yellow, 4–13mm; pod linear to linear oblong, up to 8cm. Grassland; 2000m.
Distr.: widespread in the palaeotropics.
IUCN: LC
Mwambong: Mackinder 160 1/1998 (field det.); **Mwanenguba:** Cheek 9435 10/1998.

Chamaecrista sp.
Shrub to 1.5m; leaves paripinnate; leaflets 28–46 pairs, narrowly-oblong, 5–8 × 1–1.5cm; flowers solitary or in pairs, orange, up to 8mm; pod linear oblong, up to 7.2cm. Edge of lake; 1800m.
Distr.: Cameroon. (W Cameroon Uplands).
Mwanenguba: Etuge 4375 10/1998.
Note: this specimen is an intermediate between C. kirkii Standley and C. mimosoides (L.) Greene

Copaifera mildbraedii Harms
Tree to 50m; leaves imparipinnate; leaflets 18–30, alternate or subopposite, oblong, 1–3.5 × 0.8–2.1cm, translucent gland dots prominent, apex emarginate; inflorescence a raceme, compounded into an axillary or terminal panicle; flowers white, inconspicuous, up to 2.2cm; petals absent; pod ellipsoid or subspherical, up to 4.5cm, black, glabrous; single seed with an orange-red aril. Gallery forest; 350m.
Distr.: Nigeria to Congo (K) & CAR. (Lower Guinea & Congolian).
IUCN: LC
Mungo River F.R.: Cheek 10200 11/1999.

Cynometra mannii Oliv.
Tree to 20m; leaves pinnate; leaflets in 3 pairs, obovate elliptic, terminal pair largest, 3.8–7 × 1.2–2.5cm; flowers white; fruit ovoid or obovoid, up to 4cm, torulose. Forest; 700m.
Distr.: SE Nigeria to Congo (K). (Lower Guinea & Congolian).
IUCN: LC
Nyandong: Thomas 7259 fl., 2/1987 (n.v.).

Daniellia klainei A.Chev.
Fl. Cameroun 9: 167 (1970).
Tree to 35m; leaves pinnate; leaflets 4–6 pairs, oblong-lanceolate, 12–30 × 4–10cm, translucent gland dots present;

flowers lilac, 3.5–4.5cm long (to stamen apices); sepals and petals with translucent gland dots; fruit compressed, slightly falcate up to 12cm long; single seed remains attached to one persistent valve. Forest; 600–710m.
Distr.: Cameroon, Gabon & Congo (K). (Lower Guinea & Congolian).
IUCN: NT
Nyale: Cheek 9663 fl., 11/1998; Etuge 4246 2/1998.
Note: conservation notes at www.redlist.org

Daniellia oliveri (Rolfe) Hutch. & Dalziel
Tree to 25m; leaves paripinnate; leaflets 4–10 pairs, ovate, 6–15 × 3.5–8cm, translucent gland dots near margins; inflorescence a terminal panicle; panicle branches horizontal; flowers creamy-white, fragrant, single petal up to 15mm; pod compressed, leathery, partially or fully elastically dehiscent, if the latter, then single seed remains attached to the valve eventually falling together like a helicopter. Forest, forest margins, savanna; 800m.
Distr.: tropical Africa.
IUCN: LC
Kupe Village: Mackinder 276 12/1999; 277 12/1999.

Delonix regia (Boj. ex Hook.) Raf.
Tree; leaves bipinnate, to 45cm; pinnae c, 15 pairs; leaflets numerous, oblong, 10 × 3mm, base asymmetric, apex mucronate; racemes terminal, lax, glabrous; petals clawed, 3–5cm long, red with ornage claw,except posterior petal, variegated red and white; fruit oblong, 30–60 × 5cm, valves woody, glabrous; seeds to 2.5 × 0.6cm. Planted; 840m.
Distr.: native of Madagascar, widely introduced. (Pantropical).
IUCN: LC
Nyasoso: Pollard (sight record), by school, 10/1998.

Dialium guineense Willd.
Tree to 20m; leaves imparipinnate; leaflets 4–5, alternate or subopposite, upper leaflets larger, 3.8–10.2 × 2.5–4.8cm, puberulous on lower surface; inflorescence a panicle; flowers whitish or greyish, inconspicuous, c. 3mm long; fruit drupaceous ovoid to globose, up to 2.2cm diameter, black-velvety, 1(–2) seeds embedded in edible pulp. Forest, forest margins; 750–950m.
Distr.: Senegal to Cameroon. (Upper & Lower Guinea).
IUCN: LC
Ngomboaku: Mackinder 326 12/1999; 328 12/1999.

Dialium pachyphyllum Harms
Tree 10m; leaves with 4–5 alternate leaflets, coriaceous, ovate-elliptic, 10–11.5 × 4–5.5cm, acuminate, base obtuse to rounded, lateral nerves 8–10 pairs, tertiary venation closely reticulate; rachis c. 5cm; petiole 2cm; petiolules 4mm; panicles lax, rachis to 23cm; branches racemose, to 10cm, puberulent; flowers 5mm; calyx grey-puberulent outside; petals white to yellow; stamens with bent filaments; fruits obovoid, 2–3 × 1.5–2cm, brown-black, velvety. Forest; 200–300m.
Distr.: Nigeria to Congo (K). (Lower Guinea & Congolian).
IUCN: LC
Ikiliwindi: Nemba 559 fr., 6/1987 (n.v.).

Dialium polyanthum Harms
Lock, Legumes of Africa: 35 (1989).
Tree to 15m; leaves imparipinnate; leaflets 3–4, alternate, ovate or broadly elliptic, 9–17 × 4.6–9cm, glabrous; inflorescence a panicle; flowers inconspicuous, yellowish, up to 2.5mm long; fruit drupaceous, globose, up to 2.5cm

diameter, brownish black velvety; 1(–2) seeds embedded in pulp. Forest; 310m.
Distr.: Cameroon to Congo (K). (Lower Guinea & Congolian).
IUCN: LC
Loum F.R.: Mackinder 219 11/1999.

Distemonanthus benthamianus Baill.
Tree to 40m; bark reddish; leaves pinnate; leaflets alternate, ovate, lanceolate, oblong-elliptic 3.5–10 × 2–3.8cm; flowers red (sepals) and white (petals), 1.8–2.5cm; pod compressed, papery, elliptic, up to 10cm. Forest; 310–350m.
Distr.: Sierra Leone to Gabon. (Upper & Lower Guinea).
IUCN: LC
Loum F.R.: Mackinder 217 11/1999; **Mungo River F.R.:** Cheek 10212 11/1999.

Duparquetia orchidacea Baill.
Sarmentose shrub or liane to 5m; leaves imparipinnate; leaflets, 2–4 pairs, ovate, elliptic or oblanceolate, 6–18 × 4–7cm; flowers "orchid-like", pink and white, red-veined; 3 upper petals 11–15mm; pod compressed, lanceolate, up to 10cm, 2 prominent nerves on face of valves. Forest, streambanks; 170–600m.
Distr.: Liberia to Congo (K). (Guineo-Congolian).
IUCN: LC
Etam: Etuge 237 fl., 8/1986 (n.v.); **Mungo River F.R.:** Mackinder 230 11/1999; 260 12/1999.

Erythrophleum ivorense A.Chev.
Tree to 35m; leaves bipinnate; leaflets 9–13 per pinna, alternate, ovate or elliptic 4.5–8.5 × 3–5cm long, often drying blackish; inflorescence a panicle; flowers small, reddish, up to 3mm; pods oblong, compressed, up to 10cm, drying black. Forest; 180m.
Distr.: Sierra Leone to Gabon. (Upper & Lower Guinea).
IUCN: LC
Mungo River F.R.: Mackinder 256 12/1999.

Eurypetalum unijugum Harms
Tree to 30m; leaves with a single pair of opposite leaflets; leaflets oblong-lanceolate, slightly falcate, 12–15 × 4.5–6.2cm, translucent gland dots present; inflorescence a raceme compounded into an "axillary panicle"; flowers small, single well developed petal; blade auriculate, 6 × 10mm; claw absent; pod compressed, oblong, up to 10cm, glabrous. Forest.
Distr.: Cameroon Endemic.
IUCN: VU
Baduma: Letouzey 14612 4/1976 (n.v.).

Gilbertiodendron sp. aff. splendidum (A.Chev. ex Hutch. & Dalziel) J.Léonard
Tree; leaves paripinnate; leaflets 2–4 pairs, oblong-elliptic, 11–34 × 5–10cm, apex long acuminate; stipules not seen; Pod compressed, oblong, woody, up to 27 × 8cm with 3 longitudinal nerves. River valley in forest; 500m.
Distr.: Cameroon Endemic.
Boubaji: Letouzey 14349 fr., 8/1975; 14350 8/1975.
Note: inflorescence not present. *G. splendidum* listed as VU A1c; B1+2c at www.redlist.org

Gilbertiodendron sp. aff. zenkeri (Harms) J.Léonard
Leaves paripinnate; leaflets 3–5 pairs, elliptic to oblong-elliptic, 11.5–26.5 × 6.8–13.2cm, leathery, stipules persistent, free, reniform, longest dimension 2.5–4cm. Forest; 200–310m.

LEGUMINOSAE-CAESALPINIOIDEAE

Distr.: Cameroon Endemic.
Mungo River F.R.: <u>Cheek 10190</u> 11/1999; <u>Mackinder 207</u> 11/1999.

Gossweilerodendron joveri Normand ex Aubrév.
Lock, Legumes of Africa: 53 (1989).
Tree to 35m; leaves imparipinnate; leaflets 3–5, alternate, elliptic, 6.3–10 × 3–4.2cm; blade with translucent gland dots; inflorescence a panicle; flower minute; sepals 4–5, up to 2mm long; petals absent; fruit narrowly ovate, curved, broadly winged at basal end, 11–13 × 2.5–3.4cm. Forest.
Distr.: Cameroon, Rio Muni & Gabon. (Lower Guinea).
IUCN: VU
Boubaji: <u>Letouzey 14339</u> 8/1975 (n.v.).
Note: treated as *Prioria joveri* (Normand *ex* Aubrév.) Breteler, by Breteler (1999).

Griffonia physocarpa Baill.
Woody climber; leaves unifoliolate, narrowly ovate or elliptic, 5–11 × 2.2–4.6cm; inflorescence a raceme; flowers with orange-brown calyx and green petals; petals up to 1.5cm; pod cylindrical, deep brown, shiny, leathery, up to 7cm. Forest, secondary forest; 400m.
Distr.: Cameroon to Congo (K). (Lower Guinea & Congolian).
IUCN: LC
Banyu: <u>Thomas 8161</u> fr., 6/1988 (n.v.).

Griffonia cf. simplicifolia (Vahl ex DC.) Baill.
Woody climber resembling *G. physocarpa*, but leaves ovate, 3-nerved at base; inflorescence tomentose (not glabrous); calyx tube shorter, tomentose outside (not glabrous). Forest; 420m.
Mungo River F.R.: <u>Onana 973</u> 12/1999.
Note: specimen mislaid, further id. not possible.

Guibourtia demeusei (Harms) J.Léonard
Tree to 40m; leaves 2-foliolate; leaflets ovate, falcate, 6.5–20 × 1.5–3cm, translucent glands present; flowers 1–1.2cm (to stamen apices); sepals 4, puberulous outside; petals absent; fruit broadly elliptic, up to 4cm; seeds lacking an aril. Forest; 350m.
Distr.: Cameroon to Congo (K). (Lower Guinea & Congolian).
IUCN: LC
Mungo River F.R.: <u>Cheek 10211</u> 11/1999.

Guibourtia tessmannii (Harms) J.Léonard
Tree to 40m; leaves 2-foliolate; leaflets ovate, falcate, 7–15 × 3–6cm, translucent glands absent; flowers c. 1cm (to stamen apices); sepals 4, glabrous outside except on margins; petals absent; fruit obliquely elliptic, up to 3.5cm; seeds enclosed in a red aril. Forest; 180–420m.
Distr.: Cameroon, Rio Muni & Gabon. (Lower Guinea).
IUCN: LC
Mungo River F.R.: <u>Mackinder 255</u> 12/1999; <u>Onana 982</u> 12/1999; **Nyandong:** <u>Tadjouteu 563</u> 3/2003 (n.v.).

Hylodendron gabunense Taub.
Tree to 30m; trunk and branches armed with stout spines; leaves imparipinnate; leaflets alternate, oblong-elliptic, 6–13.5 × 1.5–5.4cm; flowers greenish white, 3–4mm (to stamen apices); sepals 4; petals 0; fruit papery, oblong-elliptic, up to 12cm. Forest; 300–400m.
Distr.: Nigeria to Congo (K). (Guineo-Congolian).
IUCN: LC
Boubaji: <u>Letouzey 14338</u> fr., 8/1975 (n.v.); **Etam:** <u>Dundas in FHI 15225</u> fr., 5/1946 (n.v.); **Loum F.R.:** <u>Mackinder 269</u>

12/1999; **Mungo River F.R.:** <u>Cheek 10204</u> 11/1999; <u>Mackinder 214</u> 11/1999; <u>241</u> 11/1999.

Hymenostegia afzelii (Oliv.) Harms
Tree to 15m; leaves pinnate, leaf rachis winged; leaflets in 2 pairs, elliptic to ovate, upper pair larger, to 10 × 3.5cm; bracteoles 2; white or pink; sepals 4; petals 3, yellowy-green becoming pink, 6–7mm; fruit compressed, oblong-elliptic, up to 8cm. Forest; 800m.
Distr.: Guinea (C) to Cameroon. (Upper & Lower Guinea).
IUCN: LC
Kupe Village: <u>Mackinder 273</u> 12/1999.

Hymenostegia cf. floribunda (Benth.) Harms
Lock, Legumes of Africa: 55 (1989).
Tree to 15m. leaves paripinnate; leaflets in 7–9 pairs, subsessile, rhombic-oblong, middle pair(s) largest, 2.8–3.6 × 0.9–1.2cm, glabrous, paler below (dried material), mid-vein diagonal, base asymmetric, leaf rachis tomentose; pod compressed, obliquely triangular, widest at apex, 3.9–7.1 × 2.2–3.1cm, coriaceous, thickened along upper suture. Lowland deciduous forest; 300m.
Distr.: Cameroon to Congo (K). (Lower Guinea & Congolian).
Loum F.R.: <u>Mackinder 266</u> fr., 12/1999.
Note: no nodules seen.

Hymenostegia sp. 1
Tree 10m; bole sloping, buttresses absent, d.b.h. 10cm; root nodules absent; branchlets pale, puberulent when young; leaves 7–9-jugate; leaflets grey-green above, pale green below, subsessile, asymmetrically oblong, to 5 × 1.5cm, basal pair 1.4 × 0.5cm, apex subacuminate, midrib yellow-brown, prominent below, lateral nerves inconspicuous; rachis to 10cm, puberulent. Forest; 310m.
Mungo River F.R.: <u>Mackinder 208</u> 11/1999.
Note: currently awaiting naming by specialist at WAG.

Hymenostegia sp. 2
Pods woody, red, compressed oblong, 3.7–5 × 2.2–3cm, with raised veins, surfaces finely puberulent. Forest; 1000m.
Kupe Village: <u>Cheek 7170</u> fr., 1/1995.
Note: fallen fruits only, more material required for det.

Julbernardia seretii (De Wild.) Troupin
Tree 12–20(–40)m; leaves 1–2(–3)-jugate; leaflets elliptic to obovate, upper pair 17.5 × 7.5cm, lower pair 10.5 × 5.5cm, subacuminate, base obtuse to rounded, glabrous, translucent gland dots numerous, rachis to 5cm; petiole 2cm; petiolules twisted, 4mm; inflorescence golden-brown tomentose; pedicels 3–5mm; sepals free, subequal; petals white to pink, one obovate, to 2cm long, the others minute; pods woody, compressed, oblong-elliptic, c. 15 × 5cm, glabrous, drying dark brown. Forest; 300m.
Distr.: SE Nigeria to Congo (K). (Lower Guinea & Congolian).
IUCN: LC
Bolo: <u>Thomas 5713</u> fl., 3/1986 (n.v.); **Kupe Village:** <u>Mackinder 271</u> 12/1999.

Lebruniodendron leptanthum (Harms) J.Léonard
Bull. Jard. Bot. Nat. Belg. 21: 421 (1951).
Treelet to 2m, glabrous; leaves (1–)2-jugate; leaflets obovate-elliptic, 7–9.5 × 3.5–4cm, apex abruptly caudate-acuminate, 0.5–2.5cm, base asymmetrically cuneate, rachis 2.5cm; petiole 1.5cm; petiolules 4–5mm; racemes axillary, few-flowered, to 2cm; peduncles fine, with imbricate bracts at the

base; flowers c. 4mm; sepals reflexed; petals white; pedicels 5–8mm, fine; pods compressed, c. 10cm. Forest; 950–1000m.
Distr.: Cameroon & Congo (K). (Lower Guinea & Congolian).
IUCN: LC
Ngomboaku: Cheek 10297 12/1999; 10385 12/1999.
Note: our material is sterile.

Leonardoxa africana (Baill.) Aubrév. subsp. *africana*
Adansonia (sér. 2) 8: 178 (1968).
Syn.: *Schotia africana* (Baill.) Keay
Shrub or small tree to 15m; leaves pinnate; leaflets in (2–)3(–4) pairs, 10–20 × 4–7.5cm, elliptic, slightly falciform, discolorous, acuminate at apex, acuminate to rounded at base; petioles twisted; petiole and internodes sometimes swollen to house ants; inflorescence a short densely flowered raceme, usually axillary, sometimes on the old wood; flowers up to 2.5cm; petals 5, subequal, purple; stamens purple; fruit woody, compressed obovate-oblong, surface finely ridged, 10.5 × 4cm; 1–2-seeded. Forest, often along or near rivers; 700–1000m.
Distr.: SE Nigeria to Gabon. (Lower Guinea).
IUCN: NT
Bambe: McKey s.n. 3/1983 (n.v.); **Ngomboaku:** Cheek 10301 12/1999; 10311 12/1999; 10373 12/1999; **Nyale:** Etuge 4173 2/1998; 4468 11/1998.

Loesenera talbotii Baker f.
Tree to 15m; leaves pinnate; leaflets in 3(–4) pairs, elliptic or oblong-elliptic 7.2–12.8 × 2.5–5.1cm; flowers white; sepals 4; petals 5, upper 3 well-developed, lower 2 poorly-developed; fruit compressed, broadly-oblong, up to 15cm, golden-brown tomentose. Forest; 180–1050m.
Distr.: SE Nigeria & SW Cameroon. (Lower Guinea).
IUCN: VU
Bakossi F.R.: Cheek 9310 10/1998 (n.v.); **Enyandong:** Cheek 10951 10/2001 (n.v.); **Kupe Village:** Cheek 7721 11/1995; Mackinder 203 1/1998; 239 11/1999; **Manehas F.R.:** Etuge 4352 10/1998 (n.v.); **Mbule:** Letouzey 14680 fl., 4/1976 (n.v.); **Mt Kupe:** Schoenenberger 19 11/1995; 33 11/1995; **Mungo River F.R.:** Mackinder 205 11/1999; 213 11/1999; 242 11/1999; 254 12/1999; **Ngomboaku:** Cheek 10302 12/1999; Mackinder 313 12/1999; **Nyale:** Cheek 9672 11/1998; 9677 11/1998; **Nyandong:** Cheek 11294 3/2003.
Uses: MATERIALS – clothing – sandals (Cheek 7721).

Microberlinia bisulcata A.Chev.
Tree to 40m; leaves pinnate, leaf rachis slightly winged; leaflets in 12–18 pairs, oblong, 1.5–3.5 × 0.8–1m; inflorescence a terminal panicle; flowers crowded; petals 5; pod woody, compressed, oblong, up to 15cm long. Forest; 310m.
Distr.: Cameroon. (Lower Guinea).
IUCN: CR
Loum F.R.: Mackinder 215 11/1999.

Oxystigma gilbertii J.Léonard
Bull. Seanc. Inst. Roy. Col. Belge 21: 748 (1950).
Tree to 25m; buttresses absent or small, to 1.5m; outer slash red, inner straw-coloured; leaflets subcoriaceous, 1–4(–5), alternate to subopposite, elliptic, 13–18.5 × 6.5–8.5cm, glabrous, ± glaucous, lateral nerves often inconspicuous, rachis 5–7cm; petiolules swollen, 5mm; inflorescence a raceme, 5–13cm; fruit glossy green, asymmetrically ellipsoid, 5.5 × 3cm, subcompressed, conspicuously longitudinally-veined. Forest; 150–650m.
Distr.: Nigeria to Congo (K). (Lower Guinea & Congolian).

IUCN: NT
Kupe Village: Cable 2688 5/1996; Mackinder 270 12/1999; **Mungo River F.R.:** Cheek 10238 12/1999; Mackinder 232 11/1999; 258 12/1999.
Note: only 12 sites are known (fide Breteler), who treats as *Prioria gilbertii*, a genus not universally accepted (Wag. Ag. Un. Papers 99(3), (1999)).

Senna alata (L.) Roxb.
Lock, Legumes of Africa: 36 (1989).
Syn.: *Cassia alata* L.
Shrub to 5m; leaves pinnate; leaflets in 5–7 pairs elliptic to oblong-elliptic, 5–18.2 × 3.2–11.7cm; inflorescence a raceme; flowers bright yellow, 1.5–2.2cm; pod up to 16cm, winged along the middle of each valve. Villages, farmbush; 740–800m.
Distr.: pantropical.
IUCN: LC
Baseng: Bruneau, 1076 10/1995; **Tombel:** Etuge 371 fl., 11/1986.
Note: probably an escape from cultivation.

Senna hirsuta (L.) H.S.Irwin & Barneby
Lock, Legumes of Africa: 37 (1989).
Herb or shrub to 2.5m; leaves paripinnate; leaflets in 3–5 pairs; elliptic to ovate-elliptic, 14–22 × 7–11cm, densely pilose above and below; flowers yellow, 1–1.7cm; pod linear, slightly falcate, up to 15cm. Farmbush; 650–900m.
Distr.: native to the neotropics, now naturalised in the palaeotropics.
IUCN: LC
Mwanenguba: Leeuwenberg 8521 fl., 9/1971 (n.v.); **Ngusi:** Etuge 1549 1/1996.

Senna obtusifolia (L.) H.S.Irwin & Barneby
Lock, Legumes of Africa: 38 (1989).
Herb or subshrub to 2m; leaves paripinnate; leaflets in 3 pairs, obovate, 1.5–5 × 1–3cm; flowers yellow, 0.9–1.9cm; pod linear, sometimes slightly falcate, up to 23cm. Grassland and a weed of cultivation; 1200m.
Distr.: pantropical.
IUCN: LC
Mwanenguba: Leeuwenberg 8495 fl., 9/1971 (n.v.); **Tombel:** Mackinder 237 11/1999 (field det.).

Senna septemtrionalis (Viv.) H.S.Irwin & Barneby
Lock, Legumes of Africa: 39 (1989).
Syn.: *Cassia laevigata* Willd.
Shrub or small tree to 3m; leaves paripinnate; leaflets 3–4 pairs, lanceolate to ovate, 4–11 × 2–4cm; flowers yellow, 1–1.5cm; pod subterete, up to 10cm. Farmbush, villages; 500–1975m.
Distr.: pantropical.
IUCN: LC
Edib: Etuge 4104 1/1998; **Edib village:** Cheek 9172 2/1998; **Kupe Village:** Cable 3737 7/1996; **Mwanenguba:** Cheek 7257 fl., fr., 2/1995 (n.v.); **Mungo River F.R.:** Manning 531 fr., 10/1986 (n.v.).

Stemonocoleus micranthus Harms
Tree to 50m; leaves imparipinnate; leaflets alternate, elliptic, 4.5–10 × 2.2–5cm; flowers whitish green, 3–4mm, fragrant; sepals 4; petals absent; stamens 4; pod samaroid, oblong-elliptic, papery, twisted at base. Forest; 310m.
Distr.: Ivory Coast to Gabon. (Upper & Lower Guinea).
IUCN: LC
Loum F.R.: Mackinder 221 11/1999.

Talbotiella bakossiensis sp. nov.

Evergreen tree c. 20m; leaves paripinnate; stipules peltate, the lower lobes minute, triangular, upper lobe very narrowly triangular, 8–17mm; leaflets in 21–26 pairs, 12–24 × 3–4.5mm; inflorescence a panicle; flower white, 5.5–7mm; fruit and seed unknown. Submontane forest; 850–1350m.
Distr.: Cameroon. (Narrow Endemic).
Ngomboaku: Cheek 10326 12/1999; 10382 12/1999; **Nyale:** Cheek 9678 11/1998; 9688 11/1998.

Tessmannia anomala (Micheli) Harms

Fl. Congo, Rw. & Bur. 3: 293 (1952).
Tree to 10m; leaves paripinnate; leaflets 7–15 pairs, subopposite, oblong-elliptic, 1.0–3.5 × 0.3–1.2cm, translucent gland dots absent, midvein diagonal; flowers white or pink, 1.3–1.5cm; sepals 4; petals 5; pod suborbicular, compressed, up to 4cm 'diameter', armed with tiny spiny protuberances when young. Forest; 850–1000m.
Distr.: Cameroon to Congo (K). (Lower Guinea & Congolian).
IUCN: LC
Ngomboaku: Cheek 10312 12/1999; 10381 12/1999; **Ngusi:** Letouzey 14655 4/1976 (n.v.).

Zenkerella citrina Taub.

Tree or shrub to 20m; leaves 1-foliolate; leaflet narrowly-elliptic or elliptic, 8–12 × 4.5–6cm; flowers white and pink, 5–7mm; pod broadly oblong, compressed, somewhat asymmetric, up to 7cm, leathery. Forest; 300–2050m.
Distr.: SE Nigeria to Gabon. (Lower Guinea).
IUCN: NT
Enyandong: Cheek 10950 10/2001 (n.v.); **Kodmin:** Cheek 9072 1/1998; Mackinder 192 1/1998; 193 1/1998; **Kupe Village:** Cable 3886 7/1996; Cheek 7760 11/1995; 7793 fr., 11/1995 (n.v.); 7803 fl., 11/1995 (n.v.); Etuge 2613 7/1996; Mackinder 238 11/1999; Ryan 341 5/1996; **Mt Kupe:** Cheek 7610 11/1995 (n.v.); 7614 11/1995 (n.v.); **Ngomboaku:** Etuge 4684 12/1999; **Nyale:** Etuge 4451 11/1998.

LEGUMINOSAE-MIMOSOIDEAE

B.A. Mackinder

Acacia kamerunensis Gand.

Lock, Legumes of Africa: 69 (1989).
Scandent shrub or climber; stems terete, armed with prickles; leaves bipinnate, gland usually present at the junction of the apical 3–10 pinnae pairs; leaflets numerous, tiny, up to 4 × 1mm; inflorescences capitate; flowers yellowish white; pods compressed, subcoriaceous, dehiscent, pale to medium brown, up to 14 × 3cm. Forest; 700–950m.
Distr.: Liberia to Congo (K). (Guineo-Congolian).
IUCN: LC
Kupe Village: Cheek 6090 1/1995; 7850 11/1995 (n.v.).

Acacia pentagona (Schum.) Hook.f.

F.T.E.A. Mimosoideae: 100 (1959).
Syn.: *Acacia pennata sensu* Keay in FWTA 1: 500
Climber; stems somewhat angled, armed with prickles; leaves bipinnate, a gland usually present at the junction of the apical 1–4 pinnae pairs; leaflets numerous, up to 7 × 2mm; inflorescences capitate; flowers yellowish white; pods compressed, coriaceous, indehiscent, dark brown, up to 16 × 3.5cm. Forest; 1350–1530m.
Distr.: tropical Africa.

IUCN: LC
Kodmin: Cheek 9008 fr., 1/1998; Mackinder 143 1/1998.

Adenopodia scelerata (A.Chev.) Brenan

Fl. Gabon 31: 83 (1989).
Syn.: *Entada scelerata* A.Chev.
Climber with thorned stems, petioles and leaf rachises; stems subangular, sparsely pubescent, thorns to 2.5mm; leaves bipinnate, 3–6 pairs of pinnae, each pinna to 6cm (lower pinnae shorter), with 7–10 leaflet pairs; leaflets oblong, to 1.4 × 0.5cm, reducing towards the base, terminal leaflet pair obovate; petiole to 10cm, principal rachis to 6.5cm. Forest; 420m.
Distr.: Liberia to Congo (K). (Guineo-Congolian).
IUCN: LC
Mungo River F.R.: Onana 967 12/1999.
Note: our material sterile.

Albizia adianthifolia (Schum.) W.Wight

Tree to 35m; crown flat; leaves bipinnate; leaflets numerous, sometimes hairy above, usually hairy below, occasionally hairs confined below to the mid-rib and margins, not auriculate at base, up to 2.0 × 1.1cm; inflorescence capitate; calyx and corolla inconspicuous; stamens numerous, showy, up to 2.5cm, greenish becoming red towards apex, fused into a tube, the free ends extending a further 5–7mm; pod compressed, coriaceous, finely hairy, not glossy, up to 18 × 2.6cm. Forest, farmbush; 650–1250m.
Distr.: tropical & subtropical Africa.
IUCN: LC
Kodmin: Nwaga 3 1/1998; **Kupe Village:** Cheek 7152 fr., 1/1995; **Mwambong:** Mackinder 169 1/1998; **Ngusi:** Etuge 1538 1/1996.
Note: closely resembles *Albizia gummifera* (J.F.Gmel.) C.A.Sm.
Local name: Nsangah (*fide* Etuge ?). **Uses:** MATERIALS – timber (*fide* Elad in Cheek 7152 & Etuge 1538); a good shade tree (*fide* Elad in Cheek 7152).

Albizia gummifera (J.F.Gmel.) C.A.Sm.

Syn.: *Albizia gummifera* (J.F.Gmel.) C.A.Sm. var. *ealaensis* (De Wild.) Brenan
Tree to 30m; crown flat; leaves bipinnate; leaflets numerous, usually glabrous, occasionally with hairs on the mid-rib and margins, not auriculate at base, up to 2 × 1.1cm; inflorescence capitate; calyx and corolla inconspicuous; stamens numerous, showy, up to 2.5cm long, greenish becoming red towards apex, fused into a tube, the free ends extending a further 5–7mm; pod compressed, coriaceous, glabrescent, becoming glossy, up to 18 × 3.2cm wide. Forest; 800m.
Distr.: Nigeria to Congo (K). (Lower Guinea & Congolian).
IUCN: LC
Ngomboaku: Mackinder 323 12/1999.
Note: closely resembles *Albizia adiantifolia* (Schum.) W.F.Wight.
Local name: Asang. **Uses:** MATERIALS – timber (*fide* Etuge).

Albizia zygia (DC.) J.F.Macbr.

Tree to 25m; leaves bipinnate; leaflets relatively few, asymmetrical, sessile, (1–)2–3(–4) pinnae pairs, with 3–4 pairs of leaflets per pinna, apical pair largest, up to 9 × 5.2cm, lower pairs progressively smaller; inflorescence capitate; calyx and corolla inconspicuous; stamens numerous, showy, red, fused into a tube up to 13mm, the free ends extending a further 4–5mm; pod compressed, coriaceous,

glabrous, glossy, up to 20 × 3.8cm. Forest, farmbush; 700–950m.
Distr.: tropical Africa.
IUCN: LC
Kupe Village: <u>Cheek 7852</u> 11/1995 (n.v.); **Mt Kupe:** <u>Schoenenberger 32</u> 11/1995 (n.v.); **Nyandong:** <u>Cheek 11386</u> 3/2003 (n.v.); **Nyasoso:** <u>Etuge 1792</u> 3/1996.

Aubrevillea platycarpa Pellegr.

Tree to 25m; leaves bipinnate; pinnae 4–7 pairs, widely spaced; internodes more than 2cm long; leaflets oblong-elliptic, somewhat asymmetric, 1.4–3 × 0.6–1.4cm, secondary veins prominent on lower surface; inflorescence a panicle; flowers small, less than 3mm, sessile; pod compressed oblong, narrowing towards base, up to 20cm long, papery. Forest.
Distr.: Sierra Leone to Congo (K). (Guineo-Congolian).
IUCN: LC
Baduma: <u>Letouzey 14616</u> 4/1976 (n.v.).

Calliandra houstoniana (Miller) Standley var. *calothyrsus* (Meisn.) Barneby

Mem. New York Bot. Gard. 74(3): 180 (1998).
Tree to 4m; leaves bipinnate, 12–16 pinnae pairs; leaflets tiny, numerous; inflorescence an axillary or terminal panicle; flowers with red corolla and stamens; pod compressed, narrowly oblong, up to 13.5cm, upper and lower sutures thickened. Planted; c. 1000m.
Distr.: introduced in Africa, native of the neotropics.
IUCN: LC
Mwambong: <u>Mackinder 154</u> 1/1998.
Uses: ENVIRONMENTAL USES – soil improvers – planted by Mt Kupe Forest Project to provide 'green manure' (Mackinder 154).

Calpocalyx dinklagei Harms

Tree to 15m; leaves bipinnate, single pinna pair, pinnules 5–6 pairs, each pair increasing in size from base to apex of the pinna, the apical pair 6–12cm × 3.5–6cm; inflorescence a short spike arranged in terminal clusters or panicles, covered in a dense brown tomentum; pod woody, compressed, asymmetrically kidney-shaped, up to 16cm, valves with single longitudinal ridge, rolling up after dehiscence. Forest; 300m.
Distr.: Nigeria to Congo (K). (Lower Guinea & Congolian).
IUCN: LC
Mungo River F.R.: <u>Cheek 10117</u> 11/1999; **Supé:** <u>Letouzey 14365</u> fr., 8/1975 (n.v.).

Cylicodiscus gabunensis Harms

Tree to 50m; trunk armed with spines when young; leaves bipinnate; leaflets few, glabrous, 1–2 pinnae pairs, each pinna with 5–10 alternate leaflets, up to 10.2 × 5.1cm; inflorescence a dense raceme, either solitary, in clusters, or compounded into panicles, covered in stellate hairs when young; flowers inconspicuous, yellowish, up to 7mm; pod compressed, black, up to 1m × 5cm, splitting along the upper margin only. Forest; 400–450m.
Distr.: Ivory Coast to Gabon. (Upper & Lower Guinea).
IUCN: LC
Mungo River F.R.: <u>Mackinder 244</u> 11/1999; <u>248</u> 11/1999; **Nyandong:** <u>Cheek 11342</u> 3/2003 (n.v.).

Dichrostachys cinerea (L.) Wight & Arn.

Lock, Legumes of Africa: 89 (1989).
Shrub or tree to 8m; lateral spreading stems often terminating in a spine; leaves bipinnate, stalked gland between pinnae pairs; leaflets numerous, up to 5 × 1.5mm; inflorescence a

spike on long peduncle to 9cm; apical flowers hermaphrodite, yellow, up to 4mm; lower flowers neuter, mauve to very pale mauve, up to 17mm; pod compressed, coriaceous, undulating, glabrous, up to 10 × 2cm. Thicket; 1200m.
Distr.: Caribbean & N America, introduced in Africa.
IUCN: LC
Mwanenguba: <u>Leeuwenberg 8547</u> fr., 10/1971 (n.v.).

Leucaena leucocephala (Lam.) De Wit subsp. *leucocephala*

Opera Botanica 68: 38 (1983).
Tree to 7m; much-branched; leaves bipinnate; pinnae pairs 4–7, occasionally with gland between lower pair; leaflets 10–17 pairs per pinna, sparsely hairy above and below, up to 14 × 6mm; inflorescence capitate; flowers white; anthers glandular; fruits produced in abundance, compressed, subcoriaceous, up to 18 × 2cm; seeds 10–25, set slightly diagonally across the pod. Farmbush; 950m.
Distr.: native of Central America; widely introduced in tropical Africa.
IUCN: LC
Mt Kupe: <u>Etuge 1783</u> 3/1996; **Mwanenguba:** <u>Leeuwenberg 9859</u> fr., 5/1972 (n.v.).

Mimosa invisa Colla

Scandent or prostrate shrub; stems to 3m, armed with prickles; leaves bipinnate, closing up when touched; petioles prickly, longer or shorter than the leaf rachis; 3–10 pinnae pairs, prickles not present at the junction of each pinna pair, sometimes unarmed prickles present elsewhere on the rachis; leaflets numerous; inflorescence subcapitate or capitate; flowers pink; pods compressed, bristly, breaking up into single-seeded segments, margins persistent. Pathside, villages; 800m.
Distr.: introduced in the Old World, native of neotropics. (Pantropical).
IUCN: LC
Tombel: <u>Etuge 364</u> fl., 11/1986 (n.v.).

Mimosa pigra L.

Erect or scandent shrub to 3m; stems armed with prickles; leaves bipinnate, closing up when touched; 6–15 pinnae pairs, slender prickle at the junction of each pinna pair, occasionally stouter prickles present elsewhere on the rachis; leaflets numerous; petiole prickly, shorter than the rachis; inflorescence subcapitate; flowers pink or mauve; pods compressed, densely bristly, breaking up into single-seeded segments, margins persistent. Freshwater swamp and along streams in drier areas; 150–1250m.
Distr.: native throughout tropical Africa and the neotropics; introduced (probably) in Asia and Australasia. (Pantropical).
IUCN: LC
Mwambong: <u>Mackinder 176</u> 1/1998; **Mungo River F.R.:** <u>Cheek 10136</u> 11/1999; **Nyasoso:** <u>Etuge 1849</u> 4/1996.
Local names: Ekum-nehchuerch; 'touch me, I die' (Pidgin) (Etuge 1849).

Mimosa pudica L.

Herb to 1m; stems sparsely armed with prickles; leaves bipinnate, closing up when touched, unarmed; (1–)2 pinnae pairs; petiole longer than the rachis; inflorescence subcapitate; flowers pink or mauve; pods compressed, bristly on margins only, breaking up into single-seeded segments, margins persistent. Farmbush, roadsides; 150m.
Distr.: introduced in the Old World, native of the neotropics. (Pantropical).
IUCN: LC

Mungo River F.R.: Cheek 10137 11/1999 (n.v.); **Nyasoso:** Etuge 1847 4/1996; 1851 4/1996.
Local name: Touch me I die (Pidgin) (Etuge 1851).

Newtonia duncanthomasii Mackinder & Cheek
Kew Bull. 58: 447 (2003).
Gregarious tree 8–25(–40)m; bole c. 19m, 70cm d.b.h.; outer slash red-brown, exuding yellow drops, inner pale yellow; flowering branches with dense raised orange lenticels; leaves bipinnate, with one pair of pinnae; pinna rachis c. 2cm with (1–)2(–3) pairs of sessile leaflets; leaflets elliptic, 6.5–9.5 × 3.3–4.6cm, asymmetric, apex acute to acuminate, base acute, glabrous, a flat-topped conical gland protruding abaxially between the apical leaflet pair; petiole c. 1cm; panicles or spikes terminal, to 15cm long, axis densely brown-puberulent, to 7.8cm long; peduncle 4–7mm; corolla lobes oblong, 2.5 × 0.8cm, densely appressed-hairy outside; pod 9–15 × 2–2.8cm, brown, glabrous, nerves prominent, mucro to 2mm. Forest on ridges and hilltops; 1200–1800m.
Distr: Cameroon. (W Cameroon Uplands).
IUCN: NT
Kodmin: Mackinder 135 1/1998; 188 1/1998; 194 1/1998; **Mwendolengo:** Etuge 4543r 11/2001; **Ngomboaku:** Mackinder 315 12/1999; **Nyale:** Cheek 9691 11/1998.
Note: see reference for rationale of the IUCN assessment.

Parkia bicolor A.Chev.
Tree to 30m; leaves bipinnate; pinnae pairs opposite or subopposite; leaflets numerous, asymmetric; inflorescence pendant, claviform, apical part globose, then much narrower, up to 7cm broad; flowers yellow, orange, orangish red or bluish red; pods compresssed, up to 40 × 3.3cm; seeds clearly visible through valves, lying diagonally. Forest; 800–1000m.
Distr.: Guinea (C) to Congo (K). (Guineo-Congolian).
IUCN: LC
Kupe Village: Cheek 8408 5/1996; **Mbule:** Etuge 4397 10/1998; **Ngomboaku:** Mackinder 314 12/1999; **Nyale:** Mackinder 198 1/1998 (field det.); 199 1/1998 (field det.).

Pentaclethra macrophylla Benth.
Tree to 25m; leaves bipinnate, rusty hairy along the axes when young, 12–15 pairs per pinna, middle pinnae longest, up to 12.7cm, others up to 2 × 1.1cm; leaflets sessile, asymmetric, glabrous or nearly so; inflorescence a spike; flowers creamy-yellow or pinkish white, fragrant; pods compressed, woody, elastically dehiscent, up to 50 × 11cm. Forest; 400–750m.
Distr.: Senegal to Congo (K). (Guineo-Congolian).
IUCN: LC
Bakossi F.R.: Cheek 9325 10/1998 (n.v.); **Mungo River F.R.:** Mackinder 240 11/1999 (field det.); 243 11/1999; 245 11/1999; **Ngomboaku:** Mackinder 329 12/1999.

Piptadeniastrum africanum (Hook.f.) Brenan
Tree to 45m; crown flat; leaves bipinnate, 10–12- pairs of alternate (rarely opposite) pinnae; leaflets numerous, up to 7 × 2mm, glabrous; inflorescence a spike, compounded into panicles; flowers yellowish white; pods compressed, coriaceous, splitting along one margin, up to 36 × 3.2cm; seeds surrounded by a papery brown wing, up to 7.8 × 2.6cm (inc. wing). Forest; 310–1050m.
Distr.: Senegal to Uganda. (Guineo-Congolian).
IUCN: LC
Enyandong: Tanwani, 1 10/2001 (n.v.); **Kupe Village:** Cheek 7767 fr., 11/1995 (n.v.); Elad 76 1/1995; **Loum F.R.:** Mackinder 216 11/1999 (field det.); **Mungo River F.R.:** Mackinder 212 11/1999; 253 12/1999 (field det.); **Ndum:**

Elad 108 fr., 2/1995; **Ngomboaku:** Mackinder 310 12/1999; 330 12/1999; **Nyale:** Etuge 4245 2/1998; **Nyasoso:** Cheek 5670 12/1993 (n.v.).
Note: seeds resemble those of *Newtonia* spp. but point of attachment is central (apical in *Newtonia*).
Local name: Asang, of which there are two types. **Uses:** FUELS – fuelwood (Cheek 7767).

Tetrapleura tetraptera (Schum. & Thonn.) Taub.
Tree to 25m; slash red; leaves bipinnate, glabrous or nearly so, 5–9 pairs of pinnae; leaflets alternate, up to 2.6 × 1.3cm; inflorescence a raceme, usually arranged in pairs; flowers pink or creamy; pods persistent, indehiscent, dark brown and glossy, up to 26cm, strongly 4-winged, wings up to 5cm broad. Forest; 180–750m.
Distr.: tropical Africa.
IUCN: LC
Bakossi F.R.: Thomas 5388 fl., 1/1986 (n.v.); **Ikiliwindi:** Nemba 534 fl., 6/1987 (n.v.); **Konye:** Nemba 442 fl., 1/1987 (n.v.); **Mungo River:** Mackinder 257 12/1999; **Nyandong:** Cheek 11273 3/2003 (n.v.); **Nyasoso:** Etuge 1512 12/1995; 2598 7/1996.
Local name: Esisang. **Uses:** FOOD – infructescences – fruits eaten in soup (Cheek 11273); FOOD ADDITIVES – fruits – spice for preparing 'Achu' sauce or soup (*fide* Etuge).

LEGUMINOSAE-PAPILIONOIDEAE
B.A. Mackinder

Adenocarpus mannii (Hook.f.) Hook.f.
Shrub to 5m; leaves 3-foliolate; leaflets very variable in shape, 5–8 × 1.5–3.5cm; flowers yellow, 9–14mm; pod oblong, up to 2.5cm, viscose-glandular indumentum. Grassland, forest-grassland transition; 1980–2170m.
Distr.: Bioko, Cameroon, Congo (K) & E Africa. (Afromontane).
IUCN: LC
Mwanenguba: Leeuwenberg 8919 fl., fr., 12/1971 (n.v.); Sanford 5501 fl., 11/1968 (n.v.); 5542 fl., 11/1968 (n.v.); **Nyasoso:** Cable 1192 fr., 2/1995; Cheek 7338 fr., 2/1995.
Local name: none known. **Uses:** none known (Cheek 7338).

Andira inermis (Sw.) Kunth
Tree to 35m; leaves imparipinnate; leaflets 2–9 pairs plus a terminal leaflet, narrowly elliptic, elliptic to oblanceolate, 4.2–12.5 × 1.9–5.4cm, apex acuminate, acumen up to 12mm; inflorescence a panicle, usually terminal, rarely axillary; flowers pale pink to purple, up to 1.3cm long; fruit subglobose to ellipsoid, 3.5–5cm longest dimension, drying dark brown to black. Forest and wooded grassland; 1000m.
Distr.: Gambia to Cameroon & Bioko, widespread in the neotropics (Amphi-Atlantic).
IUCN: LC
Kupe Village: Etuge 1438 11/1995.
Note: Polhill (1969) recognised 2 subspecies of *Andira inermis* in Africa, *A. inermis* subsp. *inermis* which he restricted to wet forest of Cameroon and Nigeria and *A. inermis* subsp. *grandiflora* (Guill., Perr. & A.Rich) Gillett ex Polhill, found in wooded grassland of Upper Guinea. However Pennington (2003), in his monograph of *Andira*, after taking into account morphological variation within *A. inermis* subsp. *inermis* across its full geographical range, placed subsp. *grandiflora* in synonymy with the typical subspecies.

Angylocalyx oligophyllus (Baker) Baker f.

Understorey shrub to 5m; leaves imparipinnate; leaflets (3–)
5–7, alternate or subopposite, ovate, 8–25 × 3.8–8.2cm, long-
acuminate; inflorescence cauliflorous; flowers white with
green and red or purple markings, standard petal not reflexed;
pod torulose, up to 15cm, beak up to 3cm. Forest, forest
gaps; 500–840m.
Distr.: Liberia to Congo (K). (Guineo-Congolian).
IUCN: LC
Kupe Village: Cheek 7127 fr., 1/1995; 7708 11/1995 (n.v.);
Loum F.R.: Etuge 4269 10/1998; **Mombo:** Olorunfemi in
FHI 30553 fr., 5/1951 (n.v.); **Mungo River F.R.:** Mackinder
228 11/1999 (field det.); 233 11/1999 (field det.); 261
12/1999 (field det.).

Angylocalyx talbotii Baker f.

Understorey shrub to 3m; leaves imparipinnate; leaflets
imparipinnate, 4–5 alternate or subopposite, ovate or oblong-
elliptic, 14–21 × 5.5–9.9cm, long acuminate; inflorescence
cauliflorus; flowers white; standard petal blushed pink with a
green blotch at the base, strongly reflexed; pod torulose, to
24.5cm, beak up 2.4cm. Forest gaps and margins; 170–180m.
Distr.: SE Nigeria & SW Cameroon. (Lower Guinea).
IUCN: VU
Mungo River F.R.: Mackinder 227 11/1999; 259 12/1999.

Antopetitia abyssinica A.Rich.

Slender herb to 75cm; leaves imparipinnate; leaflets 3–4
pairs, narrowly oblanceolate, 1cm; umbels 2–4-flowered;
fruits torulose. Grassland, disturbed ground; 2000m.
Distr.: Cameroon to E Africa. (Afromontane).
IUCN: LC
Mwanenguba: Cheek 9425 10/1998; Sanford 5439 fr.,
11/1968 (n.v.).

Arachis hypogaea L.

Erect or scrambling herb; stems pilose when young; leaves 4-
foliolate; leaflets obovate, 3–5 × 1–3cm, apex rounded to
emarginate, mucronate, base obtuse; rachis flat above; petiole
2–5cm; flowers axillary, solitary; corolla yellow with red
veins, c. 1cm long; pods oblong, 2–6 × 1–1.5cm, geocarpic;
seeds ovoid, 1–2cm long. Cultivated.
Distr.: pantropical, originating from S America.
IUCN: LC
Konye: Cheek (sight record) 3/2003.
Local name: groundnut. **Uses:** FOOD – infructescences –
seeds edible (*fide* Pollard).

Baphia leptobotrys Harms subsp. *leptobotrys*

Scrambling or erect shrub; leaves 1-foliolate; leaflet broadly-
ovate or oblong-elliptic, 8–24 × 3.5–10cm; inflorescence a
raceme or panicle; calyx spathaceous, 8–15mm long; flowers
11–18mm, white, standard with a yellow blotch at base; pod
narrowly-oblong compressed, up to 2.5cm. Humid forest
especially along rivers, secondary forest, farmbush; 200–
310m.
Distr.: SE Nigeria & Cameroon. (Lower Guinea).
IUCN: NT
Loum F.R.: Mackinder 220 11/1999; Onana 931 11/1999.
Note: only 11 sites are recorded (Soladoye, Kew Bull. 40:
347, 1984).

Baphia maxima Baker

Scrambling shrub or tree to 14m; leaves 1-foliolate; leaflet
oblong-ovate or suborbicular, 8–20 × 4.5–11.5cm; flowers
clustered in the axils; calyx spathaceous 14–17mm long;
flowers 16–18mm, white with a yellow blotch at base; pod

narrowly-oblong, compressed, up to 20cm. Humid forest,
especially along rivers, farmbush.
Distr.: E Nigeria to Congo (K). (Lower Guinea &
Congolian).
IUCN: NT
Mungo River F.R.: Mackinder 247 11/1999.
Note: only 14 sites are known (Kew Bull. 40: 348, (1984)).

Baphia nitida Lodd.

Shrub or small tree to 10m; leaves 1-foliolate; leaflets
oblong-elliptic to ovate, 7–12 × 4–5cm, glabrous except for
nerves on underside; flowers white with central yellow
blotch, 1.5–2.1cm, fragrant; pod oblanceolate, compressed,
up to 15cm. Forest; 500–600m.
Distr.: Senegal to Gabon. (Upper & Lower Guinea).
IUCN: LC
Enyandong: Zapfack 1773 10/2001 (n.v.); **Etam:** Etuge 244
fr., 8/1986 (n.v.); **Kupe Village:** Cheek 7742 11/1995;
Zapfack 903 7/1996 (n.v.); **Mungo River F.R.:** Mackinder
249 11/1999; **Nyasoso:** Etuge 1 fr., 4/1986 (n.v.).

Baphiopsis parviflora Benth. ex Baker

Understorey tree to 15m; leaves 1-foliolate, broadly elliptic
to elliptic, 0.9–2.5 × 4–11.2cm; flowers white, 7–10mm;
petals 6, subequal; pod oblong-elliptic, compressed, up to
9cm. Forest; 650–1200m.
Distr.: Cameroon to Tanzania. (Lower Guinea & Congolian).
IUCN: LC
Kupe Village: Cable 2533 5/1996; 2691 5/1996; 3881
7/1996; Etuge 1897 5/1996; 2627 7/1996; 2685 7/1996;
Gosline 246 12/1999; **Manehas F.R.:** Etuge 4348 10/1998;
Ngomboaku: Mackinder 302 12/1999; 320 12/1999.

Calopogonium mucunoides Desv.

Prostrate or scrambling herb; leaves 3-foliolate; leaflets
broadly elliptic-rhombic, 2.5–10.5 × 4.6–8.5cm; flowers
blue, 8–11mm; pod narrowly oblong, 31–40 × 5–6mm,
golden brown pilose. Forest; 600–1250m.
Distr.: tropical America, cultivated in Africa. (Pantropical).
IUCN: LC
Kupe Village: Etuge 1373 11/1995; Groves 10 fl., fr.,
1/1995 (n.v.); **Mwambong:** Mackinder 178 1/1998;
Mwanenguba: Leeuwenberg 8672 fr., 11/1971 (n.v.);
Ngomboaku: Mackinder 338 12/1999 (field det.); **Nyasoso:**
Etuge 1646 1/1996.
Uses: ANIMAL FOOD – unspecified parts (Etuge 1646).

Centrosema pubescens Benth.

Scrambling and/or climbing herb; leaves 3-foliolate; leaflets
elliptic, 4–8.5 × 1.8–4.5cm; flowers pink and pale purple or
white and purple, 1.8–3.2cm; pod linear, up to 12.5cm,
margins thickened. Grassland, roadsides, weed of cultivation;
350–500m.
Distr.: native of neotropics, widely naturalised in tropical
Africa. (Pantropical).
IUCN: LC
Kupe Village: Cable 2697 5/1996; Etuge 2038 fl., fr.,
5/1996.

Crotalaria incana L. subsp. *incana*

Erect or spreading herb to 2m tall; leaves 3-foliolate; leaflets
elliptic-obovate to obovate, 2.5–5 × 1.5–4.0cm; flowers
yellow, veined reddish brown or purple, 8–12mm; pod
oblong, inflated, up to 4.5cm. Grassland; 1400m.
Distr.: widespread in tropics, probably native to neotropics.
(Pantropical).
IUCN: LC
Kodmin: Etuge 3984 1/1998.

Crotalaria incana L. subsp. *purpurascens* (Lam.) Milne-Redh.

F.T.E.A. Papilionoideae: 870 (1971).
Herb 0.6–1.2m; stems coarsely hairy; leaves obovate, 3–4cm long; flowers numerous, yellow with purple veins; fruit coarsely hairy. Forest edge; 1975–2000m.
Distr.: tropical Africa.
IUCN: LC
Mwanenguba: Cheek 7270 fl., fr., 2/1995 (n.v.); Cheek 9437 10/1998.

Crotalaria ledermannii Baker f.

J.M.Lock, Legumes of Africa: 187 (1989).
Erect well-branched annual or short-lived perennial 20–70cm; stems appressed puberulous; leaves 3-foliolate; leaflets oblanceolate, 7–20 × 1–5mm, apex rounded or truncate, apiculate; inflorescence 1.5–3cm long; flowers yellow, reddish veined, brown puberulous outside, 5.5–6.5mm long; pod ovoid, inflated. Montane grassland; 1200m.
Distr.: W Cameroon & N Nigeria. (Lower Guinea).
IUCN: VU
Mwanenguba: Leeuwenberg 8666 fl., 11/1971 (n.v.).
Note: *C. ledermanii* is the only species of *Crotalaria* in which the large anthers (as well as the small) are spinulose.

Crotalaria retusa L. var. *retusa*

Erect herb to 1.5m; leaves 1-foliolate; leaflet oblanceolate to oblong-obovate, 3.5–10 × 1.8–3.8cm; flowers yellow, reddish purple-veined, 1.2–1.5cm; pod oblong, inflated, up to 2cm. Roadside; 800m.
Distr.: introduced in Africa, native of Asia. (Palaeotropics).
IUCN: LC
Mt Kupe: Etuge 1781 3/1996.

Crotalaria spartea Baker

J.M.Lock, Legumes of Africa: 204 (1989).
Erect or straggling herb to 1.3m; leaves 1-foliolate; leaflet linear-lanceolate to lanceolate, 6.5–12.5 × 0.5–0.8cm; flowers bright yellow, veined reddish brown; pod cylindrical, up to 2.5cm. Montane grassland or in damp places at lower altitude; 2000m.
Distr.: Lower Guinea, SE & E Africa. (Tropical Africa).
IUCN: LC
Mwanenguba: Leeuwenberg 8477 fl., 9/1971 (n.v.).

Crotalaria subcapitata De Wild. subsp. *oreadum* (Baker f.) Polhill

Polhill, *Crotalaria* of Africa: 197 (1982).
Syn.: *Crotalaria acervata* sensu Hepper in FWTA 1: 550
Annual or perennial erect or straggling herb 0.5–1.3m; leaves 3-foliolate; leaflets very variable; inflorescence a raceme; peduncle shorter than rachis; flowers yellow, darkly-veined, 0.5–1cm. Grassland; 1250–2000m.
Distr.: tropical Africa. (Afromontane).
IUCN: LC
Mwanenguba: Cheek 9432 fl., 10/1998; Leeuwenberg 8603 fl., 10/1971 (n.v.).

Dalbergia heudelotii Stapf

Shrub (sometimes lianescent) or tree to 6m; leaves imparipinnate; leaflets 3–7, alternate, elliptic or ovate or oblong-elliptic, 10.2–15 × 5–7.5cm; flowers white, 5–7mm; pod oblong, up to 3.4cm, one margin concave and valves surface smooth when immature, becoming wrinkled. Marshes, river and streambanks, secondary forest margins; 1200m.

Distr.: Senegal to Cameroon & Congo (K). (Guineo-Congolian).
IUCN: LC
Mwanenguba: Leeuwenberg 8549 fl., 10/1971 (n.v.).

Dalbergia hostilis Benth.

Climbing or erect shrub to 2.5m; leaves imparipinnate, axillary spines often present; leaflets 11–19, alternate, oblong to obovate, 1.5–3 × 0.9–1.6cm; flowers white, 3–5mm; pod oblong, compressed, papery, up to 3.5cm. Dry deciduous forest; 800–1200m.
Distr.: W & WC Africa, Angola & Tanzania. (Guineo-Congolian).
IUCN: LC
Mwanenguba: Leeuwenberg 9267 fl., 1/1972 (n.v.); 9455 fr., 3/1972 (n.v.).

Dalbergia lactea Vatke

Scrambling shrub to 3m; leaves imparipinnate; leaflets 9–13, alternate, oblong-elliptic or obovate, 3–7 × 1.5–2.5cm; flowers white or blue, 7–10mm; pod oblong, compressed, up to 15cm. Montane forest; 1100–1800m.
Distr.: SE Nigeria to Congo (K) & E Africa. (Afromontane).
IUCN: LC
Ehumseh - Mejelet: Thomas 452 fl., 1/1987 (n.v.); **Kodmin:** Cheek 9568 fl., 11/1998; Gosline 163 fl., 11/1998; Mackinder 191 1/1998; **Mwanenguba:** Leeuwenberg 9969 fl., 6/1972 (n.v.).

Dalbergia oligophylla Baker ex Hutch. & Dalziel

Liana to 30m; leaves imparipinnate; leaflets 3–7, alternate, narrowly elliptic to elliptic, 3.5–6 × 1.4–2.8cm, terminal often larger, underside sparse-moderate pubescent, somewhat appressed (more dense and clearly appressed in *D. heudelotii*), secondary venation very fine, many pairs (unlike very similar *D. heudelotii* with fewer more prominent secondary veins); flowers white, 8–10mm; fruit papery, winged, narrowly elliptic-oblong; single central seed, dull (*D. heudelotii* fruit suborbicular). Grassland, forest edge; 900–2060m.
Distr.: Cameroon . (W Cameroon Uplands)
IUCN: EN
Kupe Village: Cheek 7740 11/1995; 7762 11/1995; **Mt Kupe:** Cheek 7567 10/1995; Sebsebe 5085 fl., 11/1995; **Mwanenguba:** Etuge 4363 fl., 10/1998; **Nyasoso:** Cable 107 fl., 9/1992; Cheek 7348 fr., 2/1995.
Local name: none known (Cheek 7348). **Uses:** none known (Cheek 7348).

Dalbergia saxatilis Hook.f.

Liana, 3–4m long; leaves imparipinnate; leaflets 9–21, alternate, oblong-elliptic, 2.4–3.8 × 1–1.8cm; inflorescence a panicle; flowers white, 4–6mm; pod oblong, compressed, papery, up to 8.8cm. Forest; 800–1500m.
Distr.: Guinea (B) to Congo (K). (Guineo-Congolian).
IUCN: LC
Kodmin: Cheek 8872 1/1998; **Mwambong:** Etuge 4132 2/1998; **Ngomboaku:** Etuge 4677 12/1999.

Desmodium adscendens (Sw.) DC. var. *adscendens*

Erect or straggling, sometimes prostrate herb up to 80cm; leaves 3-foliolate; leaflets broadly obovate or broadly elliptic, 1.6–4.5 × 1.2–3.2cm; flowers white, pale purple or pink, 4–6mm; pod narrowly oblong, up to 3.5cm, upper margin straight, lower margin indented to c. ½ the pod's width. Forest clearings, grassland, farmbush; 660–2000m.
Distr.: tropical & subtropical Africa.

IUCN: LC
Kupe Village: Cable 2445 5/1996; 3671 7/1996; Etuge 1961 5/1996; **Nyasoso:** Cable 2825 6/1996; Etuge 1691 1/1996 (n.v.).
Uses: MEDICINES (Etuge 1691).

Desmodium repandum (Vahl) DC.
Erect herb to 1.3m; leaves 3-foliolate; leaflets rhombic-elliptic, 4.2–9.5 × 2.8–7.5cm; flowers orange-red or red, 8–11mm; pod strongly indented along the upper margin, up to 2.5cm. Forest, forest-grassland transition; 860–2000m.
Distr.: palaeotropical. (Montane).
IUCN: LC
Kodmin: Satabie 1120 2/1998; **Kupe Village:** Ryan 410 5/1996; **Mwambong:** Mackinder 162 1/1998;
Mwanenguba: Cheek 9438 10/1998; Leeuwenberg 8400 fl., 9/1971; **Ndum:** Williams 154 fl., 1/1995 (n.v.); **Nyasoso:** Cable 2776 6/1996; Cheek 5659 fl., fr., 12/1993; 7472 fl., 10/1995 (n.v.); Etuge 1665 1/1996; Zapfack 703 6/1996.
Uses: MEDICINES (Etuge 1665).

Desmodium salicifolium (Poir.) DC.
Erect herb to 1.5m; leaves 3-foliolate; leaflets narrowly elliptic, 4–13.2 × 1.2–5.5cm; flowers green, tinged purple, 6–9mm; pods slightly indented along both margins, up to 4cm long. Swamps, marshy ground and riverbanks; 1975m.
Distr.: tropical & subtropical Africa.
IUCN: LC
Mwanenguba: Cheek 7271 fl., fr., 2/1995; Leeuwenberg 8551 fl., 10/1971 (n.v.).

Desmodium setigerum (E.Mey.) Benth. ex Harv.
Prostrate or scrambling herb; leaves 3-foliolate; leaflets obovate or broadly obovate, 1.5–3.7 × 1.0–2.5cm; flowers blue, pale purple or pink, 4–6mm; pod indented along both margins, more so along upper margin, up to 1.8cm. Riverine grassland, swampy forest margins; 1450–1600m.
Distr.: widespread in tropical Africa.
IUCN: LC
Kodmin: Barlow 14 1/1998; Cheek 9074 1/1998; 9603 11/1998; Gosline 162 11/1998; **Mwanenguba:** Leeuwenberg 8676 fl., fr., 11/1971 (n.v.).

Desmodium uncinatum (Jacq.) DC.
Lock, Legumes of Africa: 248 (1989).
Erect or scrambling herb to 2m; stems with hooked hairs; leaves 3-foliolate with stipules and stipels; leaflets ovate or elliptic, 2.2–9 × 0.7–4.8cm, pubescent below; flowers pink, turing pale purple or blue, up to 1.5cm; fruits indented along both margins, up to 3cm, articles up to 3mm wide covered with hooked hairs rendering fruit 'sticky' - readily attaching to clothes. Naturalised roadside weed; 600–2000m.
Distr.: native to S America, introduced elsewhere. (Pantropical).
IUCN: LC
Mwambong: Mackinder 147 1/1998; 167 1/1998;
Mwanenguba: Cheek 9436 10/1998; **Nyasoso:** Etuge 1643 1/1996; 2529 7/1996.

Eriosema glomeratum (Guill. & Perr.) Hook.f.
Herb or shrub to 2m; leaves 3-foliolate; leaflets narrowly elliptic, 3.2–11.5 × 1.0–2.1cm; flowers yellow, 9–11mm; pods elliptic, up to 1.5cm. Seasonally flooded forest, secondary forest, marshes, farmbush near rivers; 200–1250m.
Distr.: widespread in tropical Africa.
IUCN: LC
Mwanenguba: Leeuwenberg 8515 fl., 9/1971 (n.v.); **Nyale:** Cheek 9715 11/1998.

Eriosema parviflorum E.Mey. subsp. *parviflorum*
Erect or straggling herb from woody base, up to 3m; leaves 3-foliolate; leaflets ovate or ovate-elliptic, 2.2–7 × 1.6–3.5cm; flowers reddish yellow or yellow, 6–9mm; pod elliptic, up to 1.8cm, long-pilose. Montane grassland, wooded grassland; 950–1250m.
Distr.: Cameroon to Zimbabwe, & E Africa. (Tropical Africa).
IUCN: LC
Mwambong: Mackinder 177 fl., 1/1998; **Mwanenguba:** Leeuwenberg 8675 fl., fr., 11/1971.

Erythrina excelsa Baker
Tree to 15m; bole armed with stout prickles, axes of young shoots also armed with prickles; leaves 3-foliolate; leaflets with prickles along mid-vein and secondary veins when young, broadly elliptic to rhombic, laterals oblique, 9.5–15 × 4.8–10.5cm; flowers red, 3–3.5cm; pod moniliform (like a beaded necklace). Forest margins, wooded grassland, old farmland; 650–1300m.
Distr.: SW Cameroon to Tanzania. (Lower Guinea & Congolian).
IUCN: LC
Kupe Village: Lane 291 fl., 1/1995; **Mwambong:** Mackinder 145 1/1998; **Ngusi:** Etuge 1585 1/1996; **Nyasoso:** Bruneau, 1069 10/1995; Cable 1211 fl., 2/1995; Cheek 6011 fl., 1/1995.

Erythrina vogelii Hook.f.
Tree 8–15m; bole armed with stout prickles; leaves 3-foliolate; leaflets oblong, elliptic, ovate to broadly-ovate, 9.5–15 × 4.8–10.5cm; inflorescence a pseudoraceme; flowers bright red, up to 5cm; pod moniliform (like a beaded necklace), up to 13cm. Forest margins, wooded grassland, edge of cultivation; 920m.
Distr.: Ivory Coast to Cameroon. (Upper & Lower Guinea).
IUCN: LC
Nyasoso: Bruneau, 1070 10/1995.

Glycine max (L.) Merr.
Lock, Legumes of Africa: 414 (1989).
Herb to 1m; leaves 3-foliolate; leaflets ovate or elliptic, 4–10.5 × 2.5–5.8cm; flowers pink, blue or purple, 8–12mm; pod oblong, slightly falcate, up to 5.5cm, hispid. Commonly cultivated; 650m.
Distr.: native to China, widespread in cultivation throughout temperate and tropical regions. (Cosmopolitan).
IUCN: LC
Ngusi: Etuge 1539 1/1996.
Local name: Soya bean (English). **Uses:** FOOD ADDITIVES – seeds used to make edible oil, always available for sale in shops. Commonly cultivated around Ngusi village, Mt Kupe. (Etuge 1539).

Indigofera atriceps Hook.f. subsp. *atriceps*
Syn.: *Indigofera atriceps* Hook.f. subsp. *alboglandulosa* (Engl.) J.B.Gillett
Herb to 80cm; leaves imparipinnate; leaflets 2–7 pairs plus a terminal leaflet, 8–12 × 3–5mm; inflorescence an axillary raceme; flowers deep red, 5–7mm; pod narrowly oblong, up to 12mm, covered with glandular-tipped hairs. Montane grassland; 1500m.
Distr.: tropical Africa. (Afromontane).
IUCN: LC
Kodmin: Onana 568 2/1998.

Indigofera mimosoides Baker var. *mimosoides*

Shrub or scrambling herb to 2m; somewhat woody, sparingly glandular; leaves imparipinnate; leaflets elliptic, 6–14 × 3–8mm; flowers red, 4–7mm; pod linear, to 1.6cm. Upland grassland, streambanks, forest margins; 1560–2000m.

Distr.: Cameroon to E & SE Africa & Angola. (Tropical Africa).

IUCN: LC

Kodmin: Mackinder 190 fr., 1/1998; **Mwanenguba:** Cheek 9431 10/1998; Leeuwenberg 8510 fl., 9/1971 (n.v.); Sanford 5441 fl., 11/1968 (n.v.).

Kotschya strigosa (Benth.) Dewit & Duvign.

Robust herb to 2m; leaves paripinnate, 5–12 pairs; leaflets linear, slightly falcate, 3–10 × 1–2mm; flowers bright blue, small yellow blotch at base of standard, 6–9mm; pod resembling a 'caterpillar', up to 8mm. Montane grassland; 1000–2000m.

Distr.: Cameroon to Mozambique & to Angola, Indian Ocean. (Palaeotropics).

IUCN: LC

Kodmin: Etuge 4020 1/1998; Satabie 1107 fr., 1/1998; **Mwambong:** Gosline 112 2/1998; **Mwanenguba:** Cheek 9433 10/1998; Sanford 5409 fl., 11/1968 (n.v.).

Uses: FOOD – infructescences – fruits edible, slight lemon taste. fibrous (Gosline 112).

Lablab purpureus (L.) Sweet

F.T.E.A. Papilionoideae: 696 (1971).

Syn.: *Lablab niger* L.W.Medicus

Herbaceous climber; stems glabrescent; leaves trifoliolate; leaflets ovate-triangular, 2.5–10 × 1.5–8cm, acuminate, base obtuse to truncate, glabrescent; petiole c. 8cm; petiolules c. 4mm; stipules ovate, 5mm; inflorescence falsely racemose; peduncle 5–30cm; pedicels c. 3mm; calyx glabrescent, tube c. 3.5mm, upper lobes joined to form an entire to emarginate lip; standard crimson to purple or white, c. 1.3cm, wings purple, keel pale; pods linear-oblong, c. 6 × 1.3cm, indumentum of dot-like tubercular hairs; seeds oblong, compressed, c. 10 × 5 × 4mm, aril 6–14 × 1–2mm. Forest, naturalised from cultivation; 1570m.

Distr.: tropical and subtropical Africa, often cultivated.

IUCN: LC

Kodmin: Mackinder 195 1/1998.

Leptoderris cf. congolensis (De Wild.) Dunn

Climber to 15m; imparipinnate, 3 pairs plus a terminal leaflet; leaflets elliptic to obovate-elliptic, 3.5–4.8 × 8.5–9.5cm; inflorescence an axillary or terminal spike with flowers clustered (fasciculate) along the axis; flowers in bud, colour not known. Forest; 1920m.

Kupe Village: Sebsebe 5099 11/1995.

Leptoderris fasciculata (Benth.) Dunn

Climber; imparipinnate, 3–4 pairs of leaflets plus a terminal leaflet; leaflets elliptic to obovate elliptic 5–15 × 4–8cm; basic inflorescence form a spike with flowers clustered (fasciculate) along the axis, compounded into a panicle up to 45cm; flowers purple or white; fruit flat, oblong, papery, winged, single-seeded. Forest; 1900m.

Distr.: Guinea (C) to Uganda. (Guineo-Congolian).

IUCN: LC

Mt Kupe: Cheek 7579 10/1995.

Leptoderris cf. ledermannii Harms

Liana to 15m; leaves imparipinnate, 4-jugate; leaflets oblong to oblanceolate, lowest pair 8 × 3.7cm, terminal leaflet 17 × 7.5cm, apex truncate, with a triangular acumen 5–7mm, base

rounded to obtuse, lateral nerves 13–17 pairs, prominent below, lower surface rusty-puberulent, particularly on nerves; petiole 13cm; petiolules 3mm, stipules linear, 6mm; inflorescence a large branching panicle, rachis to 17cm; peduncle to 11cm, brown-pilose; branches spiciform, dense, to 22cm; bracts c. 1 × 0.8cm, pilose; flowers subsessile, 7mm long; calyx densely pubescent; corolla white with purple and brown streaks. Forest; 900m.

Ngomboaku: Gosline 257 12/1999.

Note: not matched at Kew, possibly *sp. nov.*

Leptoderris sp. 1

Sterile woody climber; imparipinnate, at least 3 pairs of leaflets plus a terminal; leaflets elliptic, c. 8.5 × 5.1cm, dark brown patent hairs on upper and underside of leaflets, apex acute, somewhat asymmetric. Montane forest valley; 1450m.

Kodmin: Cheek 8964 1/1998.

Leptoderris sp. 2

Sterile woody climber to 15m; trifoliolate; leaflets broadly-elliptic, c. 17.5 × 12cm, apex rounded or shortly (c. 10mm) acuminate. Montane forest near stream; 1350m.

Kodmin: Cheek 8986 1/1998.

Leptoderris sp. 3

Sterile woody climber to 8m, resembling *L. cf. congolensis* but brown hirsute on the underside of the leaflets. Submontane forest; 1400m.

Kodmin: Mackinder 137 1/1998.

Note: specimen sterile; more material required.

Leucomphalos capparideus Benth. ex Planch.

Liana; leaves 1-foliolate; leaflet ovate, elliptic or oblong, 9–20.5 × 2.6–7cm; flowers white, 8–10mm; stamens 11; anthers prominent; pod ellipsoid, falcate, up to 5.5cm, white or sometimes pinkish when mature. Evergreen forest; 650–1000m.

Distr.: SE Nigeria to Gabon. (Lower Guinea).

IUCN: NT

Kupe Village: Cable 2730 5/1996; Cheek 7090 fr., 1/1995 (n.v.); Etuge 2814 7/1996; Gosline 247 12/1999 (n.v.); Lane 413 fr., 1/1995 (n.v.); Mackinder 274 12/1999; **Manehas F.R.:** Etuge 4346 10/1998.

Mildbraediodendron excelsum Harms

Tree to 40m; imparipinnate; leaflets alternate or subopposite, 7–25, narrowly-elliptic to narrowly-oblong, 2.5–8 × 1–2.5cm, translucent gland dots present; inflorescence a raceme on current growth; petals absent; stamens 12–18; fruit a 1–3 seeded drupe, 4–6cm diameter. Forest; 600–700m.

Distr.: Ghana, Cameroon, Congo (K), Uganda, Sudan. (Guineo-Congolian).

IUCN: LC

Kupe Village: Cheek 8350 5/1996; Mackinder 202 1/1998.

Millettia barteri (Benth.) Dunn *vel sp. aff.*

Climber to 5m; imparipinnate, 4 pairs plus a terminal, obovate 9–10.2 × 4.8–6cm, apex rounded with a short acumen (up to 6mm). Forest; 1350m.

Distr.: Guinea (C) to Congo (K). (Guineo-Congolian).

Kodmin: Cheek 8988 1/1998.

Millettia macrophylla Benth.

Small tree to 5m; leaves imparipinnate, 3–5 pairs; leaflets oblong to elliptic, 9.5–19 × 4.1–9.1cm, golden puberulous on veins and blade of leaflet underside, secondary venation in 9–11 pairs; flowers purple, 1.4–2.3cm; pod linear, compressed, up to 15cm, golden puberulous. Forest, streambanks; 300–885m.

Distr.: SE Nigeria, Bioko, Cameroon & Congo (B). (Lower Guinea).
IUCN: VU
Kupe Village: Cable 3667 7/1996; Kenfack 208 7/1996; Lane 296 fr., 1/1995; Mackinder 275 12/1999; Ryan 280 5/1996; **Loum F.R.:** Cheek 9297 10/1998; Mackinder 262 fr., 12/1999; **Nyasoso:** Bruneau, 1068 fl., 10/1995.

Millettia cf. macrophylla Benth.
Tree; imparipinnate, 5–6 pairs plus a terminal leaflet; leaflets elliptic, 12.5–13.5 × 4.8–5.6cm, glabrous, secondary venation in c. 7 pairs, apex acuminate, acumen 10–13mm; inflorescence unknown; pod linear, compressed, up to 13cm, glabrous. Farmbush to secondary regrowth in forest; 400m.
Kupe Village: Etuge 2020 5/1996.
Note: typical *M. macrophylla* has closer secondary nerves.

Millettia sp. aff. versicolor Welw.
Tree 7m; branches lenticellate, young branchlets rusty-pubescent; leaves imparipinnate, 8-jugate; leaflets oblong-elliptic, c. 10.5 × 4cm, acuminate, base rounded, lateral nerves 8–9 pairs, prominent and sparsely pubescent below, rachis 20cm, puberulent; petiole 6.5cm; petiolules 3.5m, puberulent; spikes to 11cm on peduncles to 5.5cm, rusty-puberulent; pedicel 2.5mm; calyx cupular, 6mm, puberulent; corolla 1.5–2.5cm, purple with yellow centre. Secondary forest; 500m.
Nyasoso: Etuge 17 fl., 4/1986.

Millettia sp. 1
Climber; imparipinnate, 3 pairs plus a terminal leaflet; leaflets elliptic, 5.5–7.5 × 2.5–4cm, apex acute to subacute. Grassland, montane forest; 1550m.
Kodmin: Cheek 9189 2/1998.

Millettia sp. 2
Climber; imparipinnate, 2 pairs plus a terminal leaflet; leaflets elliptic to narrowly-obovate, 12.5–15.2 × 5.2 × 6.7cm, apex long acuminate, acumen to 15mm. Forest; 1000m.
Kupe Village: Cheek 7172 1/1995.

Mucuna flagellipes Hook.f.
Liana; leaves 3-foliolate; leaflets ovate or elliptic, 3.6–6.5 × 8.3–15.5cm; inflorence rachis clearly "zig-zag"; flowers greenish cream, 3.5–5.5cm; pod oblong, up to 16cm, transversely ridged, covered in golden irritant hairs. Secondary forest, edge of old cultivation, along roadsides, overhanging water; 310–780m.
Distr.: Sierra Leone to Uganda. (Guineo-Congolian).
IUCN: LC
Kupe Village: Etuge 2747 7/1996; **Mungo River F.R.:** Mackinder 211 11/1999; **Ngomboaku:** Mackinder 318 12/1999; 337 12/1999; **Nyasoso:** Etuge 2597 7/1996.

Neonotonia wightii (Wight & Arn.) J.A.Lackey subsp. *wightii* var. *longicauda* (Schweinf.) J.A.Lackey
Iselya 2(1): 11 (1981).
Scrambling or climbing herb to 4m; leaves 3-foliolate; leaflets ovate or elliptic, 1.3–13 × 0.8–1cm; flowers white, sometimes with purple blotch on standard, drying orange-red, 4–8mm; pod linear-oblong, up to 3cm, pubescent. Montane forest edges, montane grassland; 800–2000m.
Distr.: widespread in tropical Africa.
IUCN: LC

Kodmin: Etuge 4045 1/1998; Nwaga 1 1/1998; 22 fr., 1/1998; **Mwanenguba:** Cheek 9434 10/1998; **Nyasoso:** Cheek 7909 11/1995; Etuge 1530 1/1996.

Neonotonia wightii (Wight & Arn.) J.A.Lackey subsp. *pseudojavanica* (Taub.) J.A.Lackey
Iselya 2(1): 11 (1981).
Scrambling or climbing herb to 4m; leaves 3-foliolate; leaflets ovate or elliptic, 1.3–12.5 × 0.7–1cm; flowers white, drying red, 4–7mm; pod linear oblong, up to 3cm, glabrous. Montane grassland, old cultivation; 1450m.
Distr.: Sierra Leone to Tanzania. (Guineo-Congolian).
IUCN: LC
Kodmin: Cheek 9586 11/1998.
Local name: Ehtoneh (Max Ebong in Cheek 9586).

Ormocarpum megalophyllum Harms
Shrub to 2m; leaves imparipinnate; leaflets oblong-elliptic, alternate, 5.5–10.2 × 2.5–6cm; flowers cream, purple-veined, 1–1.8cm; pod torulose, up to 10cm, longitudinally ridged. Forest understorey, often along paths and in clearings; 400m.
Distr.: Guinea (C) to Gabon. (Upper & Lower Guinea).
IUCN: LC
Bakossi F.R.: Etuge 4293 10/1998.

Ormocarpum sennoides (Willd.) DC. subsp. *hispidum* (Willd.) Brenan & J.Léonard
Shrub or small tree to 4m; leaves imparipinnate; leaflets alternate, oblong-elliptic, 1.4–4.5 × 0.5–2.5cm; flowers creamy-yellow, veined reddish brown, 5–12mm; pod torulose, up to 8.5cm, longitudinally ridged. Forest edges, clearings and along paths, along waterways, secondary forest; 300–1400m.
Distr.: Senegal to CAR, Angola & Zambia. (Guineo-Congolian).
IUCN: LC
Kodmin: Etuge 4080 fl., 1/1998; Mackinder 138 1/1998; **Kupe Village:** Etuge 2026 5/1996; 2755 7/1996; **Loum F.R.:** Mackinder 267 12/1999; Onana 994 12/1999.

Ormocarpum verrucosum P.Beauv.
Shrub to 3m; leaves 1-foliolate; leaflets ovate or elliptic, acuminate, 6.4–11.5 × 2.5–5.8cm; flowers cream, purple-veined, 7–10mm; pod torulose, up to 10cm, longitudinally ridged, verrucose over seeds. Seasonally inundated coastal forest, along riverbanks and sandy areas.
Distr.: Senegal to Congo (K). (Guineo-Congolian).
IUCN: LC
Nyasoso: Etuge 1810 3/1996.

Phaseolus vulgaris L.
Lock, Legumes of Africa: 421 (1989).
Climber or suberect herb to 5m; leaves 3-foliolate; leaflets ovate to rhomoid, laterals oblique, 5–15 × 2.5–6.5cm; inflorescence a few-flowered raceme or flowers sometimes solitary; flowers white, yellowish, pink or purple; pod linear, up to 18cm, 10–14 seeded. Cultivated; 650m.
Distr.: native to S America, widely cultivated elsewhere. (Pantropical).
IUCN: LC
Ngusi: Etuge 1555 fr., 1/1996.
Local name: Beans (Haricot). **Uses:** FOOD – edible, always available in Cameroon restaurants. Place an order, talk to the restaurant or hotel manager to serve you with a dish of beans and rice or with plantain or any other food; edible seeds white, with black dots (Etuge 1555).

LEGUMINOSAE-PAPILIONOIDEAE

Physosstigma venenosum Balf.
Robust climber; leaves 3-foliolate; terminal leaflet broadly ovate, laterals obliquely ovate, 7–17 × 3.5–9.5cm; inflorescence rachis somewhat 'zig-zag' with swollen nodes; flowers pink or purple, 14–21mm; pod oblong, up to 18cm, irregular transverse veins. Forest margins, secondary forest, roadside thicket; 710–1000m.
Distr.: Sierra Leone to Congo (K). (Guineo-Congolian).
IUCN: LC
Muahunzum: Cheek 9638 fr., 11/1998; **Nyale:** Etuge 4200 2/1998.
Local name: Kol Nsamble. **Uses:** MEDICINES (Nestor in Cheek 9638).

Pseudarthria hookeri Wight & Arn.
Erect herb or subshrub to 2m; leaves 3-foliolate; leaflets ovate or narrowly rhomboid to rhomboid, 5.5–13.5 × 2.5–6.5cm; flowers purple, blue, deep pink or white, sometimes pale yellow, 4–8mm; pod narrowly oblong, up to 2.5cm. Grassland, farmbush, grazed grassland; 1250–1900m.
Distr.: widespread in tropical Africa.
IUCN: LC
Mwanenguba: Etuge 4364 fl., fr., 10/1998; Leeuwenberg 8602 fl., 10/1971 (n.v.).

Pterocarpus mildbraedii Harms
Tree to 40m; leaves imparipinnate; leaflets 7–11, alternate or subopposite, ovate or elliptic, 7–10.5 × 2.5–5.5cm; inflorescence an axillary raceme; flowers yellow, 16–18mm long; fruit discoid or nearly so, 7.5–11.5cm diameter. Forest, often along streams; 200m.
Distr.: Ivory Coast to Cameroon, Bioko & Tanzania. (Guineo-Congolian).
IUCN: LC
Mt Kupe: Cheek 10171 11/1999.

Pterocarpus santalinoides L'Hér. ex DC.
Tree to 15m; leaves imparipinnate; leaflets (5–)7–9, elliptic-ovate to ovate, 5–12 × 3–4cm, alternate; inflorescence a panicle; flowers yellow; fruit (floats in water) a single seed with a surrounding circular wing, 3–4cm diameter. Riverine forest; 250m.
Distr.: Upper and Lower Guinea; also in the neotropics. (Amphi-Atlantic).
IUCN: LC
Konye: Nemba 433 fl., 1/1987.

Pterocarpus soyauxii Taub.
Tree to 35m; leaves imparipinnate; leaflets 9–13, alternate or subopposite, obovate or elliptic, 2.5–6 × 2–2.7cm; inflorescence an axillary or terminal panicle; flowers deep yellow, 10–13mm; fruit discoid, 6.5–7.5cm diameter, single central seeds surrounded by a slightly glossy papery wing. Forest; 400m.
Distr.: Nigeria to Congo (K). (Guineo-Congolian).
IUCN: LC
Banyu: Thomas 8175 fr., 6/1988 (n.v.).
Local names: Hii or Redwood (Pidgin). **Uses:** MATERIALS – timber; MEDICINES (fide Etuge).

Pueraria phaseoloides (Roxb.) Benth.
Climbing or straggling herb; leaves 3-foliolate; leaflets ovate, 2.1–12 × .7.5–11cm; flowers purple, 10–20mm; pod linear, compressed, up to 11cm. In cultivation, disturbed ground, along roadsides; 150–1100m.
Distr.: tropical Asia, in widespread cultivation in Africa. (Palaeotropics).
IUCN: LC

Kupe Village: Etuge 1921 fl., 5/1996; Lane 328 fl., fr., 1/1995; **Mungo River F.R.:** Cheek 10138 11/1999; **Nyasoso:** Cable 3329 6/1996.

Rhynchosia congensis Baker
Robust climber to 6m; leaves 3-foliolate; leaflets ovate to broadly ovate, the laterals obliquely so, 3.8–12 × 2.8–10.2cm; inflorescence a pseudoraceme or few-branched panicle; calyx lobes of similar length; flowers pale yellow with red markings drying reddish brown; pods up to 2.5cm. Grassland, farmbush; 1400m.
Distr.: Guinea (B), Nigeria, Cameroon, Angola & E Africa. (Tropical Africa).
IUCN: LC
Kodmin: Gosline 79 1/1998.

Rhynchosia mannii Baker
Climber to 8m; leaves 3-foliolate; leaflets ovate to broadly ovate, the laterals obliquely so, 8–15 × 5–13.5cm; inflorescence a pseudoraceme or few-branched panicles; lower calyx lobe much longer than the other lobes; flowers pale yellow with purple veins; pods up to 1.5cm. Forest gaps; 1200–1350m.
Distr.: Nigeria to Uganda. (Lower Guinea & Congolian).
IUCN: LC
Edib: Cheek 9122 2/1998; **Nyale:** Cheek 9708 11/1998.

Tephrosia interrupta Engl. sp. vel. aff.
Shrublet 1m, pubescent; leaves 5–9-jugate; leaflets oblong, 2–3.5 × 0.4–0.9cm; infructescence terminal; pods densely black pubescent. Grassland; 1975m.
Mwanenguba: Cheek 7272 fr., 2/1995.

Tephrosia paniculata Baker
Lock, Legumes of Africa: 380 (1989).
Syn.: *Tephrosia preussii* Taub.
Erect herb to 2m; leaves imparipinnate, 1–2-jugate; leaflets elliptic-oblong or lanceolate, 5–7 × 1.5–2.5cm; flowers red, tomentose outside, 8–10mm; pod linear, slightly curved, up to 5cm. Grassland, thicket; 2000m.
Distr.: Sierra Leone to Cameroon, Angola, E & SE Africa. (Afromontane).
IUCN: LC
Mwanenguba: Cheek 9452 10/1998.

Teramnus labialis (L.f.) Spreng. subsp. *labialis* aff. var. *acutus* Verdc.
F.T.E.A. Papilionoideae: 537 (1971).
Slender twining herb to 1m; leaves 3-foliolate; leaflets elliptic, ovate or rhomboid, 3–7 × 1.7–3cm; flowers white with purple spot on standard, 4–5mm; pod linear, compressed, 4–5.8cm, beak up to 5mm. Grassland, forest margin, forest clearings; 1250m.
Distr.: Guinea (B) to Uganda. (Guineo-Congolian).
IUCN: LC
Mwanenguba: Leeuwenberg 8607 fl., 10/1971 (n.v.).

Teramnus micans (Welw. ex Baker) Baker f.
Twining herb to 3m; leaves 3-foliolate; leaflets elliptic, obovate or rhomboid, 3.3–13.5 × 1.7–9cm; flowers red, purple or blue, 4–6mm; pod linear, up to 4.5cm, beak up to 4mm. Grassland, occasionally marshes; 1250m.
Distr.: widespread in tropical Africa. (Tropical Africa).
IUCN: LC
Mwambong: Mackinder 150 1/1998; 179 1/1998.

Trifolium baccarinii Chiov.

Prostrate (occasionally ascending) herb, sometimes rooting at the nodes; leaves 3-foliolate; leaflets elliptic or obovate, finely toothed, 11–16 × 7–10mm; flowers purple or white, 3–4mm; pod broadly oblong, 2–3 × 1–1.5mm. Grazed grassland; 1000m.
Distr.: N Nigeria to Ethiopia & to Tanzania. (Afromontane).
IUCN: LC
Mwambong: Cheek 9388 10/1998.
Local name: Abat-eh. **Uses:** MEDICINES (Cheek 9388).

Trifolium simense Fresen.

Straggling perennial to 45cm; leaves 3-foliolate; leaflets linear or linear-oblanceolate, finely toothed, 15–55 × 2.4mm; inflorescence spherical up to 1.5cm diameter; calyx-nerves 17–20; flowers purple (rarely white); pod oblong c. 4 × 2mm. Montane grassland; 880m.
Distr.: Bioko & Cameroon to Zambia & E Africa. (Afromontane).
IUCN: LC
Nyasoso: Cheek 7488 fl., 10/1995 (n.v.).

Trifolium usambarense Taub.

Straggling herb to 1m; leaves 3-foliolate; leaflets oblanceolate, finely toothed, 6–13 × 3–7mm; calyx-nerves 10–12; flowers purple, occasionally white, 4–6mm; pod broadly oblong, c. 3 × 2mm. Marshy places, clearings in forest; 2000m.
Distr.: Bioko & Cameroon, Congo (B), Ethiopia to E Africa, Zambia & Rwanda. (Afromontane).
IUCN: LC
Mwanenguba: Cheek 9454 10/1998.

Vigna adenantha (G.Mey.) Maréchal, Mascherpa & Stainier

Taxon 27: 199 (1978).
Syn.: *Phaseolus adenanthus* G.Mey.
Scrambling or climbing herb to 2m; leaves 3-foliolate; leaflets broadly ovate, 7–12 × 4.2–8cm; flowers creamy-white or pale blue, 2–2.5cm; pod linear, up to 14cm. Cultivated; 650m.
Distr.: Senegal to Congo (K). (Guineo-Congolian).
IUCN: LC
Ngusi: Etuge 1536 1/1996.
Local name: Kun or Cooky beans. **Uses:** FOOD – leaves and seeds: grind seeds, mix with cocoyam, young leaves, salt, palm oil, pepper to make cooky beans. (Etuge 1536).

Vigna desmodioides R.Wilczek

Climber to 2.5m; leaves 3-foliolate; leaflets ovate, 5–12 × 4–10cm; inflorescence an axillary pseudoraceme; flowers 8–12mm, pink, purple or blue; pods linear, subcylindrical, up to 5cm, falcate, valves with raised reticulate nervation; seeds without aril. Forest; 1400m.
Distr.: Sierra Leone, Nigeria, Cameroon, Congo (K) & CAR. (Guineo-Congolian).
IUCN: LC
Kodmin: Mackinder 136 1/1998.
Note: *V. desmodioides* is very close to *V. racemosa* but the former has short inflorescences and short peduncles, shorter more reticulate pods and no aril.

Vigna gracilis (Guill. & Perr.) Hook.f. var. *gracilis*

Slender twining or semi-prostrate herb; leaves 3-foliolate; leaflets ovate, broadly elliptic or rhombic, 1–4.5 × 0.8–2.1cm; flowers pink or bluish, turning yellow; 9–16mm; pod linear, deflexed, up to 4cm. Wooded grassland, grassland, roadsides; 740–1480m.
Distr.: widespread in W & WC Africa. (Guineo-Congolian).
IUCN: LC
Baseng: Bruneau, 1078 fl., 10/1995; **Kodmin:** Nwaga 5 fl., 1/1998; **Mwambong:** Mackinder 146 fl., 1/1998; **Mwanenguba:** Leeuwenberg 8656 fl., 11/1971 (n.v.).

Vigna gracilis (Guill. & Perr.) Hook.f. var. *multiflora* (Hook.f.) Maréchal, Mascherpa & Stainer

Taxon 27: 199 (1978).
Syn.: *Vigna multiflora* Hook.f.
Slender twining herb to 2m; leaves 3-foliolate; leaflets ovate to ovate-lanceolate, 2.1–8.5 × 1.5–7.2cm; flowers pink or blue, turning yellow, blue with yellow spot on standard, 9–16mm; pods linear, deflexed, up to 4cm. Forest; 1100–1500m.
Distr.: Sierra Leone to Congo (K). (Guineo-Congolian).
IUCN: LC
Kupe Village: Cheek 7695 11/1995; **Mt Kupe:** Schoenenberger 61 11/1995; **Mwanenguba:** Sanford 5440 fl., 11/1968 (n.v.).

Vigna unguiculata (L.) Walp.

Scrambling herb to 3m; leaves 3-foliolate; leaflets very variable, 2–15.8 × 1.5–11.5cm; inflorescence a pseudoraceme; flowers white, greenish, yellow or purple; pods narrowly-oblong, up to 10cm. Open, disturbed forest; 190m.
Distr.: widespread in tropical Africa.
IUCN: LC
Ngomboaku: Mackinder 224 11/1999.

Vigna vexillata (L.) Benth.

Climbing or trailing herb to 5m; leaves 3-foliolate; leaflets 2.5–16.5 × 0.4–8.3cm, often ovate to lanceolate but very variable in shape; flowers pink or purple, sometimes yellowish, occasionally with pale spot at base of standard; pod linear, cylindrical, up to 14cm. Grassland; 950–1570m.
Distr.: pantropical.
IUCN: LC
Kodmin: Gosline 40 fl., 1/1998 (n.v.); Mackinder 196 fl., 1/1998; Nwaga 2 fl., 1/1998; **Mwambong:** Mackinder 149 fl., 1/1998; 163b fr., 1/1998; 175 fl., 1/1998; **Mwanenguba:** Leeuwenberg 8671 fl., fr., 11/1971 (n.v.).
Local name: Kunahunte. **Uses:** MEDICINES (Gosline 40).

LENTIBULARIACEAE

M. Cheek

Utricularia andongensis Welw. ex Hiern

Annual terrestrial herb to 10cm; leaves 4–5 in basal rosette, strap-shaped, 1–3 × 0.3cm; inflorescence often twining; bracts and bracteoles basifixed, ovate; calyx lobes subequal, 2–6mm; corolla c. 0.5cm, yellow, spur subulate, larger than lower lip. Rock-faces; 1050m.
Distr.: Guinea (C) to Zambia. (Guineo-Congolian).
IUCN: LC
Nyale: Cheek 9666 11/1998.

Utricularia appendiculata E.A.Bruce

Taylor, The Genus *Utricularia*: 492 (1989).

Perennial terrestrial herb to 30cm; leaves stoloniferous, strap-shaped, to 3 × 0.2cm; inflorescence twining to the left; bracts and bracteoles basifixed, narrowly deltoid; upper calyx lobe trullate, 4mm, lower half as large; corolla 0.8cm, yellowish white; capsule with dorsal and ventral slits. Sphagnum mats, lake-side; 1200–1250m.

Distr.: Cameroon to Zimbabwe, Madagascar. (Tropical Africa & Madagascar).

IUCN: LC

Lake Edib: Cheek 9134 2/1998; Gosline 187 11/1998.

Note: not recorded in FWTA.

Utricularia gibba L.

Syn.: *Utricularia gibba* L. subsp. *exoleta* (R.Br.) P.Tayl.

Submerged, non-affixed aquatic; stems c. 1mm diameter, to c. 20cm, internodes c. 1cm; leaves filiform, paired, rarely branched, 0.5–1.5cm, bearing 1–3 traps, each 1–2mm. Lakes; 1200m.

Distr.: pantropical.

IUCN: LC

Lake Edib: Gosline 190 11/1998.

Note: sterile material.

Utricularia mannii Oliv.

Epiphytic perennial herb 2–9cm, with translucent ellipsoid tubers to 1cm; leaves stoloniferous, strap-shaped to obovate, to 5 × 0.5cm, petiolate; inflorescence not twining; bracts ovate, 2mm; upper calyx lobe ovate, 4–6mm, lower one third as large; corolla 1.2–2.5cm, bright yellow; capsule with short ventral slit. Forest; 1000–2000m.

Distr.: Nigeria, Bioko & Cameroon. (W Cameroon Uplands).

IUCN: NT

Kupe Village: Cheek 8392 5/1996; Ryan 379 5/1996; **Nyale:** Cheek 9709 11/1998; **Nyasoso:** Balding 41 fl., 8/1993; Cable 2909 6/1996; 3379 6/1996; 3546 7/1996; Lane 16 fl., 7/1992; Wheatley 437 fl., 7/1992; Zapfack 722 6/1996.

Note: despite its restricted range, the habitat of this short-lived herb remains almost unthreatened.

Utricularia cf. stellaris L.f.

Taylor, The Genus *Utricularia*: 632 (1989).

Submerged, non-affixed aquatic; stems 1mm diameter, 0.5m long, internodes c. 0.5cm; leaves numerous, semicircular in outline, 1–6cm long, divided from the base into numerous filiform segments; traps numerous, 1–3mm. Lakes; 1200–1250m.

Distr.: palaeotropics.

Lake Edib: Cheek 9129 2/1998; Gosline 191 11/1998.

Note: sterile material; flowers needed. May prove to be *U. inflexa.*

Utricularia striatula Sm.

Epiphytic annual herb 1–20cm, stoloniferous; leaves petiolate, blade transversely elliptic, to 1–5mm wide; inflorescence non-twining; bracts medifixed, oblong, to 1.5mm; upper calyx obcordate c. 2mm, lower one fifth as large; corolla 0.5–1cm, white, lower lobe flat, 5-lobed, upper inconspicuous; capsule with short ventral slit. Forest; 870–1450m.

Distr.: Guinea (C) to Tanzania, India to New Guinea. (Palaeotropics).

IUCN: LC

Enyandong: Pollard 778 10/2001; **Kodmin:** Cheek 9000 1/1998; 9617 11/1998; **Nyasoso:** Cable 63 fl., 8/1992; Cheek 7449 10/1995; Lane 81 fl., 9/1992.

LEPIDOBOTRYACEAE

M. Cheek

Fl. Cameroun 14 (1972).

Lepidobotrys staudtii Engl.

Tree 7m, glabrous; leaves ovate-elliptic, to 17 × 7.5cm, acuminate, obtuse, c. 5 pairs lateral nerves, margin thickened; petiole 1.2cm, joined 0.5cm from the base, the distal 0.4cm swollen; inflorescences axillary, buds resembling a pine-cone, 3mm; fruit ovoid, 3cm, brown tomentose, 2-valved, 1-seeded. Forest; 250m.

Distr.: Nigeria to Congo (K). (Lower Guinea & Congolian).

IUCN: LC

Ikiliwindi: Etuge 511 3/1987; **Ngomboaku:** Cheek 10290 fr., 12/1999.

LINACEAE

M. Cheek

Fl. Cameroun 14 (1972).

Hugonia macrophylla Oliv.

Woody climber to 10m; stems orange-brown scurfy, climbing by opposite woody hooks, usually on short spur-shoots; leaves papery, alternate on main axis or clustered at ends of spur shoots, oblanceolate, c. 20 × 7cm, obtuse, cuneate, entire or subserrate, lateral nerves c. 16 pairs, densely orange-brown pubescent; petiole 0–5mm; stipules 1–2cm, deeply pectinate-laciniate; inflorescences few-flowered, subsessile; flowers yellow, 3–4cm, petals 5, free, obovate 2cm; stamens 10; ovary 1. Forest; 650m.

Distr.: SE Nigeria, Cameroon & Gabon. (Lower Guinea).

IUCN: VU

Kupe Village: Lane 400 fl., 1/1995.

Note: *H. obtusifolia* and *H. planchonii* are known from forest immediately adjacent (Wone and Bakaka) to our area and are likely to be found within it in future.

Hugonia micans Engl.

Fl. Cameroun 14: 36 (1972).

Climber resembling *H. macrophylla*; leaves leathery, elliptic, rarely oblanceolate, c. 12 × 6cm, acumen abrupt, c. 0.5cm, rounded, serrate, lateral nerves c. 16 pairs, minutely sparsely puberulent; petiole 1cm; stipules ligulate, with basal laciniae. Forest; 500m.

Distr.: Cameroon & Gabon. (Lower Guinea).

IUCN: VU

Wone: Nemba 588 7/1987.

Note: specimen not seen; if verified, a range extension.

Hugonia platysepala Welw. ex Oliv.

Climber resembling *H. macrophylla*, but leaves drying dark green or black, acuminate, serrate, lateral nerves c. 22 pairs, indumentum white or grey; petiole 1–2cm. Forest; 1350m.

Distr.: Guinea (C) to Bioko, Congo (K), Angola & Uganda. (Guineo-Congolian).

IUCN: LC

Kodmin: Cheek 8981 1/1998.

Hugonia spicata Oliv. var. *spicata*

Climber resembling *H. macrophylla*, but leaves leathery, elliptic or ovate-elliptic, to 17 × 8.5cm, subacuminate, rounded to truncate, serrate-crenate, lateral nerves c. 12 pairs,

lower surface densely pale grey felty-hairy; petiole 1cm; stipules digitately laciniate. Forest; 400m.
Distr.: Bioko & Cameroon. (Lower Guinea).
IUCN: NT
Banyu: Thomas 8157 fl., 6/1988.
Note: this variety appears otherwise restricted to Bioko. An identification check is required before a Red Data assessment is made.

Hugonia sp. 1
Climber resembling *H. macrophylla*; leaves c.2 8 × 8cm, acumen rounded, lateral nerves c. 22 pairs, weakly serate-crenate; petiole 1cm; stipules pectinately laciniate. Forest; 880m.
Kupe Village: Cheek 7080 1/1995.
Note: possibly a variant of *H. macrophylla*, fertile material needed.

Linum volkensii Engl.
Fl. Cameroun 14: 40 (1972).
Erect annual herb 30cm; leaves alternate, sessile, lanceolate, 2 × 0.4cm; inflorescence paniculate, terminal; flowers yellow, 2mm, sepals 5, free, margins glandular-toothed; petals 5, free; capsule globose, 3mm. Montane grassland; 2000m.
Distr.: Cameroon to Mozambique. (Afromontane).
IUCN: LC
Mwanenguba: Cheek 9426 10/1998.
Note: otherwise known in Cameroon only from 2 collections from the Bamboutos Mts. First record for FWTA area.

Radiola linoides Roth
Erect annual herb 3–5cm; leaves sessile, yellowish green, elliptic, 2–5mm; inflorescence cymose, 1–3 × 1–4cm; flowers white, 2mm. Montane grassland; 1900–2000m.
Distr.: Europe & Africa in temperate areas. (Montane).
IUCN: LC
Mwanenguba: Cheek 9428 10/1998; Leeuwenberg 8462 fl., 9/1971; Pollard 882 fl., fr., 11/2001.

LOGANIACEAE
M. Cheek & A.J.M. Leeuwenberg

Fl. Cameroun 12 (1972).

Anthocleista microphylla Wernham
Epiphytic shrub, becoming scandent and then a strangler tree; leaves narrowly elliptic to oblanceolate, c. 17 × 6.5cm, acuminate, acute, nerves c. 5 pairs; petiole 2cm; inflorescence c. 25-flowered; flowers white or bluish white, outer sepals c. 0.4 × 0.4cm. Forest; 800–1200m.
Distr.: Ghana, Nigeria, Bioko, São Tomé, Cameroon. (Upper & Lower Guinea).
IUCN: VU
Kupe Village: Cable 2663 5/1996; 3775 7/1996; Etuge 1451 11/1995; 2671 7/1996; Kenfack 249 7/1996; **Lake Edib:** Gosline 194 11/1998.
Note: the absence of *A. obanensis* is remarkable.

Anthocleista scandens Hook.f.
Epiphytic climbing shrub 7–17m; branchlets square; leaves oblong-elliptic, 6–20 × 2.5–11cm; flowers white; corolla tube c. 3cm long. Forest, forest-grassland transition; 1300–2000m.
Distr.: Bioko, São Tomé & W Cameroon. (W Cameroon Uplands).
IUCN: VU

Kupe Village: Cable 3858 7/1996; Cheek 7105 1/1995; Ryan 397 5/1996; **Mt Kupe:** Sebsebe 5071 10/1995; Thomas 3171 2/1984; 5060 11/1985; **Nyasoso:** Cable 1232 fr., 2/1995; 3396 6/1996; Etuge 1656 1/1996.
Local name: Ngim (Etuge 1656).

Anthocleista schweinfurthii Gilg
Tree 10–25m, sparsely branched; leaves below inflorescence sessile, oblanceolate, c. 28 × 10cm, apex rounded, base cuneate-decurrent, nerves 10–12 pairs; flowers pale brown, fragrant, slender and pointed in bud, to 3 × 1cm; outer sepals 1 × 1.2cm. Farmbush; 700–900m.
Distr.: Nigeria to Uganda. (Lower Guinea & Congolian).
IUCN: LC
Kupe Village: Cable 2532 5/1996; Cheek 7037 fr., 1/1995; **Ngomboaku:** Etuge 4679 12/1999.

Anthocleista vogelii Planch.
Tree to 25m, similar to *A. schweinfurthii* but stem and trunk always spiny; leaves petiolate; flower buds blunt and rounded at apex. Farmbush; 650m.
Distr.: Sierra Leone to Uganda. (Guineo-Congolian).
IUCN: LC
Ngusi: Etuge 1580 1/1996.

Mostuea batesii Baker
Shrub 0.6–1.2m, densely long patent hairy; leaves ovate-elliptic, c. 4 × 2cm, obtuse, both surfaces long-appressed-hairy; petiole 0.2cm; inflorescence terminal, 1cm; corolla white, yellow at base, tubular, curved, zygomorphic, 9mm; fruits 2-valved, obovate, flat, bilobed, hairy. Secondary forest; 200–250m.
Distr.: Cameroon to Congo (K). (Lower Guinea & Congolian).
IUCN: LC
Ikiliwindi: Etuge 513 3/1987; **Loum F.R.:** Leeuwenberg 9479 fl., 3/1972; **Nyasoso:** Schlechter 12936 date not known.

Mostuea brunonis Didr. var. brunonis
Shrub 0.3m, appressed pubescent on stem and midrib; leaves obovate-elliptic, 7 × 3.2cm, acumen 0.5cm, mucronate, obliquely obtuse-acute; nerves 3–5 pairs; petiole 1mm; inflorescence sessile, axillary; flowers white with yellow centre; sepals linear, 6mm; fruit as *M. batesii*, 7mm, brown, sparsely puberulent. Forest; 1450m.
Distr.: Ghana to Madagascar. (Tropical Africa & Madagascar).
IUCN: LC
Kodmin: Cheek 8940 1/1998.
Note: Cheek 8940 differs from all other material of this taxon in SW Province due to the filiform calyx lobes.

Strychnos aculeata Solered.
Climber, glabrous; stems grooved, often prickly; leaves elliptic or oblong, 8–18 × 3.5–8cm, apiculate or acuminate, base rounded or cuneate, palmately 3-nerved; petiole 2–12mm; tendrils in 1–3 pairs; subumbels axillary, 1.5–7 × 0.8–3cm, congested; peduncle 0.7–5cm; flowers greenish white, 3–5mm; fruit subglobose, hard, laterally compressed, c. 10cm. Forest.
Distr.: Sierra Leone to Congo (K). (Guineo-Congolian).
IUCN: LC
Southern Bakossi F.R.: Onochie in FHI 29666 4/1954.

Strychnos campicola Gilg ex Leeuwenb.
Belmontia 10: 74 (1969).
Climber, drying dark brown; stems terete, rusty-brown pubescent; leaves elliptic or ovate, 7–11 × 2.5–5cm,

LOGANIACEAE

acuminate, base rounded or cuneate, palmately 3-nerved; petiole 3–9mm, rusty pubescent; tendrils solitary; flowers 2.5–3.3mm, yellow; fruit ellipsoid, 16–21 × 13–17mm, 1-seeded, orange. Forest; 200m.
Distr.: Guinea (C) to Congo (K). (Guineo-Congolian).
IUCN: NT
Mile 15, Kumba-Mamfe Road: Nemba 615 7/1987.
Note: only 25 collections known, not yet known from Gabon or Nigeria.

Strychnos camptoneura Gilg & Busse
Climber; stems terete; leaves drying dark brown, glabrous, ovate-elliptic 24 × 10.5cm, acuminate, obtuse-decurrent, not palmate, 3-nerved, lateral nerves 5 pairs; petiole 0.7cm; tendrils 1–2 pairs; inflorescence axillary 4.5–6.5cm; flowers white or yellow; corolla 1cm; fruit globose to pear-shaped, 6–20cm, yellow, 10–100 winged seeds. Forest; 200m.
Distr.: Liberia to Congo (K). (Guineo-Congolian).
IUCN: LC
Mungo River F.R.: Leeuwenberg 6848 10/1965.

Strychnos congolana Gilg
Climber; stems terete; leaves drying dark grey green, papery-membranous, long-ovate, c. 12 × 5.5cm, long slender acuminate, rounded-truncate, palmately 5-nerved, tertiary nerves scalariform; petiole 0.5cm; tendrils paired; inflorescence terminal 6 × 6cm; flowers white, 2mm; fruit globose, 7–10cm, yellow-green, many-seeded. Forest; 400m.
Distr.: Guinea (C) to Uganda. (Guineo-Congolian).
IUCN: LC
Wone: Mambo 29 fl., 5/1986.

Strychnos dolichothyrsa Gilg ex Onochie & Hepper
Climber, smelling of cloves; stems terete; leaves thickly leathery, glossy above, subsessile below, narrowly elliptic, 9.5 × 3cm, including 1cm acumen, cuneate, palmately 3-nerved, tertiary nerves subscalariform; petiole 0.5cm; tendrils axillary, single; inflorescence terminal; flowers 3mm; corolla urceolate, greenish white; fruit globose 1.5cm. Forest; 760m.
Distr.: Cameroon to Congo (B). (Lower Guinea).
IUCN: NT
Ngombombeng: Etuge 23 fr., 4/1986.
Note: only 15 sites are known for this taxon.

Strychnos elaeocarpa Gilg ex Leeuwenb.
Meded. Land. Wag. 69(1): 114 (1969).
Tree 6m; stems terete; leaves thickly leathery, glossy above, elliptic, 11.5 × 6cm, acumen broad and short, 0.7cm; cuneate, palmately 3-nerved, marginal nerve, tertiaries open-scalariform; petiole 0.7cm; inflorescence axillary, congested, 1–1.5cm; flowers 4.5mm, white; fruit obliquely ellipsoid c. 1.3 × 2cm, dark brown, soft. Forest; 200m.
Distr.: Cameroon Endemic.
IUCN: VU
Mile 15, Kumba-Mamfe Road: Nemba 137 7/1986; **Mungo River:** Leeuwenberg 6845 fr., 10/1965.

Strychnos johnsonii Hutch. & M.B.Moss
Climber; stems square, hollow, green; leaves subglossy above, papery-leathery, elliptic, 10 × 5cm, shortly acuminate, acumen 0.5cm, obtuse, palmately 3-nerved, inconspicuously so above; petiole 0.5cm; tendrils paired; inflorescence axillary 2.5cm; flowers yellowish white, c. 4mm; fruit globose, c. 1.5cm, stipitate, single-seeded. Forest edge; 1000–1550m.
Distr.: Guinea (C) to Uganda. (Guineo-Congolian).
IUCN: LC

Kodmin: Cheek 9185 2/1998; **Mwanenguba:** Leeuwenberg 9491 fl., fr., 3/1972.

Strychnos memecyloides S.Moore
Climber; stems terete; leaves leathery, drying matt blackish green, narrowly elliptic, 12 × 4cm, gradually acute-acuminate, obtuse, palmately 3-nerved, 0.3cm; petiole 2–3mm; tendrils paired; inflorescence axillary, unbranched, 10cm; flowers pale green, 3–4mm; fruit ellipsoid, 2cm, 1-seeded. Forest; 300m.
Distr.: S Nigeria to Congo (K). (Lower Guinea & Congolian).
IUCN: LC
Mombo: Olorunfemi in FHI 30563 fr., 5/1951.

Strychnos cf. nigritiana Baker
Climber; stems terete, densely lenticillate; leaves leathery, subglossy green above, elliptic, 3 8 × 4cm, acumen oblique, slender, 1cm, obtuse, strongly palmately 3-nerved, tertiary nerves faint; petiole 1cm; tendrils not seen; inflorescence terminal; flowers not seen; fruit globose, 4cm, wall 5mm; seeds 10–20, each c. 1.5cm, plano-convex. Forest; 1350m.
Nyale: Cheek 9700 11/1998.
Note: more collections needed.

Strychnos phaeotricha Gilg
Climber; stems and midribs black-pilose; leaves papery, drying black above, pale green below, oblong-elliptic, 23 × 6cm, acumen slender, 2cm, truncate-subcordate, weakly palmately 3-nerved, c. 10 pairs secondary nerves; inflorescence terminal, lax, 5–8cm; petiole 0.3mm; tendrils paired; inflorescence terminal on branchlets, 3cm; fruit ovoid, 1cm; flowers yellowish white, 4–5mm; fruit ellipsoid 1.8–3cm, orange, soft, 3–7-seeded. Forest; 250–1100m.
Distr.: Ghana to Congo (K). (Guineo-Congolian).
IUCN: LC
Enyandong: Cheek 10941 10/2001; **Kupe Village:** Etuge 2828 7/1996; **Mile 15, Kumba-Mamfe Road:** Nemba 133 7/1986; **Nyasoso:** Cable 3321 6/1996.

Strychnos staudtii Gilg
Tree or shrub 2–20m; stems whitish brown, terete; leaves thinly leathery, pale green above, elliptic, c. 12 × 5cm, acumen to 1cm, obtuse, strongly palmately 3-nerved, scalariform; petiole 0.3cm; inflorescence axillary, subsessile, 1cm; flowers white; 3mm; fruit ellipsoid or globose, pale brown, 3.5cm. Forest; 200–1200m.
Distr.: Cameroon & Gabon. (Lower Guinea).
IUCN: VU
Ebonji: Olorunfemi in FHI 30598 fr., 5/1951; **Konye:** Thomas 7016 /1987; **Kupe Village:** Cable 731 fl., 1/1995; 798 fl., 1/1995; 836 fl., 1/1995; Cheek 6032 fl., 1/1995; 7056 fl., 1/1995; 7138 fl., 1/1995; 7722 11/1995; 7806 11/1995; Lane 445 fl., 1/1995; **Mbule:** Leeuwenberg 9309 fl., 1/1972; **Mungo River:** Breteler 2575 fl., 2/1962; Breyne 5060 fl., 1/1986; **Mungo River F.R.:** Leeuwenberg 6849 10/1965; **Ndibise:** Etuge 520 6/1987; **Nyasoso:** Cable 1188 fl., 2/1995; 3297 6/1996; Etuge 2127 6/1996; Lane 130 fr., 6/1994; Sebsebe 5029 10/1995.
Note: habit photo at Mungo Bridge, Loum-Kumba Road in Belmontia 10: 254 (1969) – Breteler *et al.* 2575.

Strychnos talbotiae S.Moore
Climber; stems slender, terete; leaves elliptic, often narrowly, c. 10–4cm, acuminate, obtuse, palmately 3-nerved; petiole 1cm; tendrils paired; inflorescence terminal, 2–3cm; flowers pale green 3cm; fruit ellipsoid, 3cm; several-seeded. Forest; 150–400m.

Distr.: Nigeria to Congo (K). (Lower Guinea & Congolian).
IUCN: LC
Banyu: Thomas 8219 fl., 4/1988; **Etam:** Etuge 183 7/1986.
Note: determined by D.Thomas (MO), specimen not seen at K; description from specimen ex Eseka.

Strychnos tricalysioides Hutch. & M.B.Moss
Fl. Cameroun 12: 119 (1972).
Climber; stems terete; leaves papery, drying dark green above, brown-green below, ovate-elliptic, 14–28 × 7–11cm, acumen 1.5cm, base rounded-obtuse, palmately 3-nerved, marginal nerves, tertiary nerves 5–10, prominent; petiole 0.5–1cm; tendrils paired; inflorescence axillary, dense, 1–1.5cm; flowers white, 6mm; fruit globose-ellipsoid, 2–4cm orange, hard, 1-seeded; seed with deep pit on one side. Forest; 200–800m.
Distr.: Nigeria to Congo (B). (Lower Guinea).
IUCN: NT
Bangem: Thomas 5246 1/1986; **Mungo River F.R.:** Leeuwenberg 6847 10/1965; **Nyale:** Etuge 4228 2/1998.
Note: known from only 12 sites.

Strychnos urceolata Leeuwenb.
Fl. Cameroun 12: 121 (1972).
Climber closely resembling *S. dolichothyrsa*, but not smelling of cloves; blades slightly shiny or matt below (not always matt); barbate in axils of second pair of secondary nerves. Forest; 400m.
Distr.: Nigeria to Uganda. (Guineo-Congolian).
IUCN: LC
Wone: Mambo 38 fl., 5/1986.

Strychnos sp. A
Climber; stem subsquare, solid, whitish brown, scabrous; leaves papery, drying black above, mid brown below, elliptic, c. 10 × 4.5cm, acumen 1.5cm, obtuse-decurrent, strongly palmately 3-nerved, secondary nerves c. 8 pairs inconspicuous; petiole 0.5–1.5cm; tendrils not seen; fertile material unknown. Forest; 1400m.
Kodmin: Plot B223 1/1998.
Note: does not match any of our other *Strychnos*. Particularly distinct in stems and lower leaf surface. More material needed.

Strychnos sp. B
Climber; stems glossy pale brown, striate and warty, leafy branches black; leaves resembling those of *S. cf. nigritiana* but acumen shorter and broader, matt below; fruit ellipsoid, 3 × 2cm, blue-green, wall brittle; seed single, deeply pitted on one side, golden sericeous. Forest; 1000m.
Nyasoso: Cheek 9516 10/1998.
Note: not keyed out in Leeuwenberg's revision. The hard blue ellipsoid fruits with single seeds, deeply pitted on one side seem to be unusual in the genus. More collections needed.
Local name: Ngele. **Uses:** related to another species which is eaten (Cheek 9516).

LORANTHACEAE
R.M. Polhill & B.J. Pollard

Fl. Cameroun 23 (1982).

Agelanthus brunneus (Engl.) Balle & N.Hallé
Parasitic shrub to 1m or more; hairs simple or absent; leaves ovate-lanceolate, 2.5–15 × 1.5–8cm, apex blunt, 3(–5)-nerved from base; flowers few–numerous, crowded at axils or older nodes, banded red, orange and white; corolla 3.2–4(–4.5)cm, 5-petaled, white or yellow over vents, conspicuously swollen basally from inception; petals erect. Forest, farmbush; 1000–1800m.
Distr.: Senegal to W Kenya, S to Angola. (Guineo-Congolian (montane)).
IUCN: LC
Edib: Etuge 4476 11/1998; **Kodmin:** Cheek 8860 1/1998; **Kupe Village:** Etuge 1456 fl., 11/1995; **Mt Kupe:** Sunderland 1527 fl., 7/1992; **Mwanenguba:** Etuge 4371 10/1998; **Ndum:** Cable 901 fl., 1/1995.

Agelanthus djurensis (Engl.) Polhill & Wiens
Polhill & Wiens, Mistletoes of Afr.: 157 (1998).
Parasitic shrub, 0.5–2 m, glabrous; leaves broadly ovate to elliptic, 5–14 × 2.5–7cm, ± oblique, triplinerved, acuminate; inflorescence of several umbels in axils and at older nodes below, 3–6-flowered; corolla 4–5cm, deep crimson or wine red (or paler yellow-green when young), paler over vents, basal swelling slight; petals reflexing. Forest; 150m.
Distr.: Cameroon to Uganda & Angola. (Guineo-Congolian).
IUCN: LC
Mungo River: Gosline 231 11/1999.

Globimetula braunii (Engl.) Tiegh.
Parasitic shrub; leaves elliptic, elliptic-oblong or ovate-elliptic; 4–10 × 1.5–7cm; apex rounded; 2nd and 3rd nerves strongly curved-ascending, not reaching tip; inflorescence of 3–6 umbels, mostly at older leafless nodes; peduncle 0.3–2.0cm; corolla red or pink, apical swelling darkening when ripe; petals conspicuously rolling down at anthesis. Forest, plantations; 700–2000m.
Distr.: Ivory Coast to W Kenya. (Guineo-Congolian).
IUCN: LC
Kodmin: Pollard 196 11/1998; **Kupe Village:** Ensermu 3524 fl., 11/1995; Lane 313 fl., fr., 1/1995; **Mwanenguba:** Etuge 4379 10/1998; Sanford 5416 11/1968; **Nyandong:** Thomas 6700 2/1987; **Nyasoso:** Cheek 7518 10/1995.
Local name: Jessi-Nkam: a generic name for all Loranthaceae (Cheek 7518).

Globimetula dinklagei (Engl.) Danser
Parasitic shrub; leaves ovate to lanceolate or oblong-elliptic, acuminate, cuneate, 4–12 × 1–5cm, 2–3 pairs of strongly ascending nerves; umbels 1–9, in axils and at leafless nodes below; peduncle 0.4–1.2cm; pedicels 4–7, 4–6(–8 in fruit) mm; basal ⅔ of corolla green or white, red above; petals conspicuously rolled down at anthesis. Forest, plantations; 450–1650m.
Distr.: Cameroon, Gabon & Congo (B). (Lower Guinea).
IUCN: LC
Bangem: Thomas 5318 1/1986; **Kodmin:** Etuge 4036 1/1998; Gosline 100 2/1998; **Kupe Village:** Cheek 7096 fl., 1/1995; Etuge 1461 11/1995; **Menyum:** Doumenge 514 5/1987; **Nyale:** Cheek 9719 11/1998.

Globimetula oreophila (Oliv.) Tiegh.
Parasitic shrub; twigs compressed; leaves lanceolate-ovate, 8–13 × 2.5–6cm with 6–12 pairs of well-spaced curved-ascending nerves; umbels 1–4, in axils, 8–21-flowered; peduncle 0.5–3.5cm; corolla red or red-purple, darkening apically as bud ripens, or red with a green or cream top, 2.5–3.5cm; basal swelling 5-shouldered. Forest-grassland transition; 1900–1975m.
Distr.: SE Nigeria & Cameroon. (W Cameroon Uplands).
IUCN: NT

LYTHRACEAE

Mwanenguba: Cheek 7260 fl., 2/1995; **Nyasoso:** Cable 1197 fl., 2/1995.
Note: this species is restricted to SW and NW Provinces and the Bamboutos Mts of W Province, Cameroon, and neighbouring areas of Nigeria. More than 35 collections are known at K, which suggests that although it has a limited distribution, it is quite common within its range, though it is possibly threatened by forest loss in the Bamenda Highlands.

Helixanthera mannii (Oliv.) Danser
Parasitic shrub, ± 1m; leaves elliptic-oblong to elliptic, ovate or oblong-lanceolate, 6–10(–14) × 1–4cm with 6–10(–12) pairs of fine lateral nerves; racemes terminal and in many axils, 2–10(–15)cm, 20–50-flowered; flowers white to pinkish; petals free, caducous, linear, (4–)8–18(–24)mm. Forest, often along rivers; 400–950m.
Distr.: SE Nigeria to Uganda. (Lower Guinea & Congolian).
IUCN: LC
Kupe Village: Cable 2695 5/1996; Etuge 1917 5/1996; 2728 7/1996; Kenfack 211 7/1996; **Nyasoso:** Cable 2944 6/1996; Etuge 2493 7/1996; Zapfack 628 6/1996.

Phragmanthera batangae (Engl.) Balle
Fl. Cameroun 23: 33 (1982).
Parasitic shrub to 1–1.5m; branchlets scurfy with pale scales and ± with reddish brown dendritic hairs to 0.5mm, glabrescent to varying degrees; leaves thinly coriaceous, paler beneath, oblong-lanceolate to oblong-elliptic, acuminate, 4.5–10 × 1–3.5cm, with 3–8 pairs of looped nerves; umbels in axils, (1–)2–4-flowered; peduncle 0.5–2mm; corolla 2.2–3.8(–4.7)cm, ciliolate, yellow, sometimes pale green above, tipped red. Forest.
Distr.: Cameroon, Gabon, CAR, Congo (K), Angola. (Lower Guinea & Congolian).
IUCN: LC
Nyasoso: Letouzey 14694 fl., 4/1976.
Note: specimen cited in Fl. Cameroun 23: 34 (1982) and Polhill & Wiens: Mistletoes of Africa (1998).

Phragmanthera capitata (Spreng.) Balle
Fl. Cameroun 23: 29 (1982).
Syn.: *Phragmanthera incana* (Schum.) Balle
Syn.: *Phragmanthera lapathifolia* (Engl. & K.Krause) Balle
Parasitic shrub to 2 m; branchlets and lower leaf surface with reddish dendritic hairs to 1mm; leaves glossy above with hairs beneath, ovate-lanceolate to ovate, elliptic or round, acuminate, 6–17 × 3–14cm, with 4–8 pairs of looped nerves; umbels 2–4-flowered; peduncle 0–3mm; corolla 4–5.5(–6)cm, sparsely lepidote to subglabrous, yellow to orange with red tips; petals erect. Forest; 800–1700m.
Distr.: Guinea (C) to Congo (K), Angola. (Guineo-Congolian (montane)).
IUCN: LC
Kodmin: Gosline 89 2/1998; Pollard 197 11/1998; **Mbulé:** Letouzey 14682 fl., 4/1976; **Mwanenguba:** Etuge 4384 10/1998; **Ndum:** Cable 927 fl., fr., 2/1995; **Nyasoso:** Cable 3437 7/1996; Sidwell 345 fl., 10/1995; Sunderland 1526 fl., 7/1992.

Phragmanthera nigritana (Hook.f. ex Benth.) Balle
Parasitic shrub to 2m; branchlets tomentose with reddish brown dendritic hairs; leaves oblong-lanceolate to ovate, blunt, 4–7.5 × 1.5–4cm, brown-pubescent abaxially, with 6–8 pairs of inconspicuous nerves; umbels several, 2–4-flowered; corolla 3.5–4cm, densely hirsute; petals reflexing at anthesis. Forest, riverine forest; 500–1500m.

Distr.: Ghana, Nigeria, Cameroon, Bioko, Angola. (Upper & Lower Guinea).
IUCN: NT
Kodmin: Cheek 8869 1/1998; **Kupe Village:** Cable 2704 5/1996; Ryan 265 5/1996; **Nyandong:** Thomas 6701 2/1987.
Note: only 15 sites are shown by Polhill & Wiens: Mistletoes of Afr. (1998).

Tapinanthus globiferus (A.Rich.) Tiegh.
Parasitic shrub to ± 1m, mostly glabrous; leaves linear-lanceolate to oblong-elliptic, ovate or somewhat oblanceolate, 4–8(–17) × 0.5–4(–12)cm, with 4–6 pairs of curved-ascending nerves; umbels mostly in axils, (4–)6–8-flowered; corolla-tube (2–)2.5–3.5(–4)cm, pink to red, spotted; basal swelling conspicuous, petals reflexing at anthesis; stigma capitate. Forest; 400–900m.
Distr.: tropical & subtropical Africa, Arabia.
IUCN: LC
Etam: Etuge 236 8/1986; **Kupe Village:** Cheek 8375 5/1996; Etuge 1916 5/1996; Kenfack 209 7/1996; 229 7/1996; **Manyemen:** Letouzey 13752 6/1975; **Nyasoso:** Cable 31 fl., 8/1992.

Tapinanthus ogowensis (Engl.) Danser
Polhill & Wiens, Mistletoes of Afr.: 187 (1998).
Parasitic shrub; stem robust, arching and pendant to 2 m; leaves mostly opposite; petiole 1–2cm; lamina 7–21 × 4–14cm, ovate or ovate-elliptic, acuminate, cordate; umbels many, crowded at older nodes, mostly 4-flowered; corolla-tube 4.5–5.2cm, white or greenish white, becoming pink-flushed; head blackish; petals reflexed at anthesis; stigma ovoid-ellipsoid; berry orange-red, 0.8 × 0.6cm. Forests, on various hosts including plantation crops; 400–900m.
Distr.: Cameroon to Congo (K). (Lower Guinea & Congolian).
IUCN: LC
Kupe Village: Cable 3802 7/1996; Etuge 1918 5/1996; Kenfack 228 7/1996; **Tombel:** Cheek 9332 10/1998.
Local name: Deokid. **Uses:** MATERIALS – traps/snares – producing fruit gum 'Nkam', used to catch dragonflies 'Nsisa' for food by children aged 4 to 8. Roasted first; gum squeezed onto hand, then threaded onto head of coconut leaf fibre; this wand then waved to hit and catch the flying dragonflies (Cheek 9332).

Tapinanthus preussii (Engl.) Tiegh.
Parasitic shrub, glabrous; leaves ovate to broadly elliptic, acuminate, 8–14 × 6–10cm with 6–10 pairs of nerves; umbels 4(–8)-flowered; peduncle 3–4mm; corolla tube 5.5–6.2cm, greyish, flushed or lined reddish or purplish; bud-heads 4–5 × 4–5mm, with conspicuous sutural wings, 1mm across. Forest; 900m.
Distr.: SE Nigeria, Cameroon, Gabon, Cabinda. (Lower Guinea & Congolian).
IUCN: VU
Nyasoso: Sunderland 1480 fl., 7/1992.

LYTHRACEAE

M. Cheek

Cuphea carthagenensis (Jacq.) J.F.Macbr.
Publ. Field Mus. Nat. Hist. Chicago, Bot. Ser. 8: 124 (1930).
Herb or subshrub 0.15–0.3m; stem glandular-hairy; leaves opposite or subopposite, elliptic to obovate, to 3.5 × 1.5cm, acute, cuneate, rounded-serrate, subsessile; inflorescence of

single, axillary, subsessile flowers; flowers pink, 4mm diameter; petals 5, obovate; calyx campanulate, longitudinally striped, apex toothed. Roadsides; 600–2000m.
Distr.: S America, now a pantropical weed.
IUCN: LC
Mwambong: Cheek 9559 fl., 11/1998; **Mwanenguba:** Cheek 9500 fl., 10/1998; **Nsoung:** Darbyshire 83 fl., 3/2003; **Nyasoso:** Etuge 1650 fl., 1/1996.
Note: newly recorded here for Africa.

MALPIGHIACEAE
C.C. Davis & B.J. Pollard

Fl. Cameroun 14 (1972).

Acridocarpus macrocalyx Engl.
Liana; stems densely brown-puberulent; leaves obovate, 9–25 × 5–13cm, emarginate-apiculate, base obtuse to subcordate, pubescent to glabrous, with 2–4 gland pairs at base; petiole 4–15mm, pubescent; stipules absent; racemes terminal; bracts up to 1cm; pedicels 1–1.5cm, pubescent; calyx with 2–3 glands; petals clawed, yellow, 10–15mm; anthers elliptic, 5–7mm, apex with 2 prongs; styles 2, curved; samaras 2, paired, 5–7cm. Forest.
Distr.: Guinea (C), Togo to Congo (K). (Guineo-Congolian).
IUCN: NT
Mungo River: Davis 9912 12/1999.
Note: only 4 collections at three locations cited in Fl. Cameroun.

Flabellaria paniculata Cav.
Climbing shrub; stems highly-lenticellate; branchlets appressed-pubescent; leaves opposite, ovate or ovate-orbicular, 10–14 × c. 10cm, glabrous above, appressed silky-tomentose beneath; lateral nerves c. 4–5 pairs; panicles terminal, lax, many-flowered; pedicels 0.8–1.0cm; petals entire, glabrous; wings of fruit highly conspicuous, suborbicular, c. 3–6cm across, papery, white. Forest; 1450–1780m.
Distr.: tropical Africa.
IUCN: LC.
Kodmin: Cheek 9087 fl., fr., 1/1998; Etuge 3963 fl., fr., 1/1998.

MALVACEAE
M. Cheek

Abelmoschus esculentus (L.) Moench
Blumea 14: 100 (1966).
Syn.: *Hibiscus esculentus* L.
Robust erect annual herb, 1–2m; stems green, often tinged red; leaves alternate; stipules narrow, caducous; petiole hispid, 15–35cm; lamina broadly cordate, 10–25 × 10–35cm, palmately 3–7-lobed, hirsute, serrate; flowers solitary, axillary; peduncle to c. 2cm; epicalyx of up to 10 narrow hairy bracteoles, to 1.5cm; calyx falling with corolla after anthesis, 2–3cm; petals 5, yellow, with crimson spot on claw, obovate, 5–7cm; staminal column united to base of petals, 2–3cm; fruit a pyramidal-oblong, beaked capsule, 10–30 × 2–3cm, dehiscing longitudinally when ripe. Farmbush.
Distr.: originating in tropical Africa, now widely spread throughout the tropics.
IUCN: LC
Nyasoso: Cheek sight record.
Local name: Okra. **Uses:** FOOD – fruits have a high mucilage content and are used in soups, usually accompanied by fufu (*fide* Pollard).

Hibiscus cf. noldeae Baker f.
Scrambling spiny herb 1–1.5m, sparsely long-hairy; resembling *H. surattensis*, but leaves lobed by less than half radius, or ovate, entire; stipules caducous; pedicels 2(–5)cm; corolla orange, with brown centre, 5cm. Forest edge and grassland; 860–1400m.
Distr.: Kupe-Bakossi. (W Cameroon Uplands).
Kodmin: Etuge 4009 1/1998; 4067 1/1998; **Nyasoso:** Etuge 1291 10/1995.
Note: possibly a new species, intermediate between *H. noldeae*, with which it shares 2cm pedicels and similar indumentum, and *H. rostellatus*, which has similar leaves. More investigation is needed.

Hibiscus rosa-sinensis L.
Sp. Pl.: 696 (1753).
Shrub 1.5m, sparingly hairy; leaves ovate-lanceolate, c. 8 × 4cm, acute, rounded; petiole 1cm; flowers single, axillary; pedicel 3cm; epicalyx 1.5cm; bracts 10, ligulate, exceeding calyx; corolla red, 6.5cm. Cultivated; 950m.
Distr.: pantropical (cultivated); origin unknown.
IUCN: LC
Mwanenguba: Leeuwenberg 9987 fl., 5/1972.

Hibiscus surattensis L.
Spiny shrub 1m; stem with long, fine hairs and prickles; leaves orbicular in outline, 3–5-lobed, divided by half to three quarters, c. 6cm, serrate; petiole 4cm; stipules ovate, foliaceous, 1.5cm; flowers single, axillary; pedicel c. 6cm; epicalyx 10, each c. 2cm, bifurcate, foliaceous, shorter than calyx; corolla 2.5cm, orange-yellow, with a red centre. Secondary forest; 700–1000m.
Distr.: palaeotropics.
IUCN: LC
Kupe Village: Etuge 1625 1/1996; Lane 326 fl., 1/1995.
Uses: MATERIALS – abrasives – leaves used to clean dishes (Etuge 1625).

Kosteletzkya adoensis (Hochst. ex A.Rich.) Mast.
Shrub 0.3–0.9m, or straggling in trees to 6m; subscabrid; leaves ovate or shallowly 3-lobed, to 5.5 × 3cm, acute, cordate, crenate; petiole 1cm; pedicel 1cm; epicalyx of c. 8 filiform bracts, 4mm; calyx 4mm; corolla pink, centred purple, 1.5cm; fruit strongly 5-ridged-winged. Forest edge; 1400–2000m.
Distr.: tropical Africa. (Afromontane).
IUCN: LC
Kodmin: Etuge 4039 1/1998; **Mwanenguba:** Cheek 9439 10/1998; Leeuwenberg 8449 fl., 9/1971; Sanford 5422 fl., fr., 11/1968; 5550 fl., fr., 11/1968.

Pavonia urens Cav. var. *glabrescens* (Ulbr.) Brenan
Shrub 3–5m, glabrescent; leaves orbicular in outline, shallowly 3–5-lobed or entire, to 8cm; panicles to 30cm; pedicel 2mm; epicalyx of 10 ligulate bracts, divided to base, slightly exceeding calyx; corolla 3cm diameter, white, centre purple; mericarps 5, awns 5mm, barbed. Forest edge; 1000–1100m.
Distr.: Guinea (C) to Kenya. (Tropical Africa).
IUCN: LC

Nyasoso: Cable 702 fl., 1/1995; Cheek 5656 fl., 12/1993; 7324 fl., fr., 2/1995.
Note: probably worthy of specific recognition.
Local name: Musum-Mwitchum, which means 'old man's stick', because the thorns on the fruits resemble such. **Uses:** SOCIAL USES – unspecified social uses (Epie Ngome in Cheek 7324).

Pavonia urens Cav. var. *urens*

Subshrub to 2m, densely persistent-pubescent on stems and leaves; leaves circular in outline, c. 15cm, ± 5-lobed; fascicles axillary, 5–10-flowered; corolla pink, 1cm; mericarps 5, awns long-exserted with retrorse spines. Forest edge; 1445–2000m.
Distr.: Guinea (C) to Madagascar. (Tropical Africa & Madagascar).
IUCN: LC
Kodmin: Biye 1 1/1998; **Mwanenguba:** Cheek 7267 fr., 2/1995; Cheek 9447 10/1998.

Sida cf. acuta Burm.f.

Subshrub, prostrate; subglabrous, but with sparse, long, simple hairs; leaves drying green, lanceolate, to 3.5 × 0.8cm; fascicles subsessile, 3–4-flowered; corolla orange-yellow; mericarps not seen. Farms; 1050m.
Distr.: Cameroon. (Lower Guinea).
Ndum: Cable 916 fl., 2/1995.
Note: possibly an aberrant individual of *S. garckeana*; *S. acuta* is normally completely glabrous.

Sida garckeana Pell.

Bol. Soc. Brot. (sér. 2) 54: 104 (1980).
Syn.: *Sida corymbosa* R.E.Fr.
Subshrub 0.2–1.5m, all parts with simple, white, patent hairs, 1mm; leaves lanceolate, to 8 × 3cm, acute, obtuse, biserrate; petiole to 1cm; flowers 1–3, axillary; pedicels 1–2mm; corolla yellow-orange, open at 11am. Roadsides, waste places; 200–900m.
Distr.: Sierra Leone to Cameroon, originally S America.
IUCN: LC
Kupe Village: Cheek 7097 fl., 1/1995; 7918a 11/1995; Etuge 2729 7/1996; Lane 325 fl., 1/1995; **Mwanenguba:** Leeuwenberg 8523 fl., fr., 9/1971; **Mungo River F.R.:** Cheek 10188 11/1999; **Nyasoso:** Etuge 2485 7/1996; **Tombel:** Etuge 373 fl., fr., 11/1986.

Sida javensis Cav.

Blumea 14: 184 (1966).
Syn.: *Sida veronicifolia* Lam.
Syn.: *Sida pilosa* Retz.
Prostrate herb, rooting at nodes; puberulent; leaves ovate, 4cm, acute, cordate, crenate; petiole 3cm; pedicel 1cm; corolla orange, c. 0.7cm. Forest edge; 700–1200m.
Distr.: pantropical.
IUCN: LC
Bangem: Thomas 5349 1/1986; **Enyandong:** Etuge 4479r 11/2001; **Kodmin:** Etuge 4074 1/1998; **Nyale:** Etuge 4184 2/1998.

Sida rhombifolia L. var. *α*

Subshrub 0.5–1m, stellate-puberulent; leaves lanceolate-rhombic, to 6.5 × 2cm, rounded, acute, serrate, lower surface completely obscured by minute, white, stellate hairs; flowers axillary, single; pedicels extending to 2.5cm in fruit; corolla yellow; mericarps c. 8, sides rugose-reticulate. Roadsides; 200–650m.
Distr.: pantropical.
IUCN: LC

Mungo River F.R.: Cheek 10189 11/1999; **Ngusi:** Etuge 1552 1/1996.

Sida rhombifolia L. var. *β*

Erect or prostrate shrub, 0.2–0.5(–2.5)m, resembling var. *α*, but leaves mostly elliptic, upper surface sometimes with some simple hairs; pedicels less than 1cm; mericarps smooth. Montane grassland; 1200–2000m.
Distr.: pantropical.
IUCN: LC
Kodmin: Ghogue 60 1/1998; **Kupe Village:** Etuge 1832 3/1996; **Mwanenguba:** Cheek 9466 10/1998; **Leeuwenberg** 8598 fl., 10/1971; **Nyasoso:** Cheek 7339 fr., 2/1995.
Local name: Sim-sim. **Uses:** MEDICINES – veterinary – boil leaves in water and give to a non-barking dog (Cheek 7339).

Sida rhombifolia L. var. *γ*

Subshrub 1–1.5m; resembling var. *β*, but upper surface of leaves mainly with long simple hairs, and stellate hairs absent. Grassland; 1850–1940m.
Distr.: Cameroon uplands & Ethiopia. (Afromontane).
IUCN: LC
Kupe Village: Ryan 413 5/1996; **Mwanenguba:** Sanford 5382 fl., 11/1968.
Note: perhaps not worthy of formal distinction from var. *β*, or distinction in FWTA needs adjustment.

Sida sp. aff. rhombifolia L.

Subshrub 0.3–2m; resembling *S. rhombifolia* var. *α*, but the lower surface of the leaf blade subglabrous, black veinlets clearly visible; pedicels to 3cm; corolla white, 1.8cm diameter, closed at 10:00am, open at 14:00pm; mericarps smooth, very shortly awned. Farms, forest edge; 300–600m.
Distr.: Cameroon Endemic.
Bakossi F.R.: Cheek 9308 10/1998; **Kupe Village:** Cheek 7917a fl., 11/1995.
Note: not matched at K, possibly a new variety - further work is needed.
Local names: Semme/Sem. **Uses:** MATERIALS – personal items – a favoured toothbrush; tools – formerly important for brooms; MEDICINES (Ngolle Abraham in Cheek 7917).

Urena lobata L.

Subshrub 0.6m, stellate-hairy; leaves elliptic, c. 6cm, slightly 3-lobed or entire, base rounded, teeth glandular, densely grey-hairy below; petiole 2cm; flowers subsessile; epicalyx 5-lobed in upper half, 7mm; corolla pink, centred purple, 1cm; mericarps 5, spines with grapnel ends. Farmbush; 200–1650m.
Distr.: pantropical.
IUCN: LC
Baduma: Etuge 402 11/1986; **Kodmin:** Cheek 9078 1/1998; **Mejelet-Ehumseh:** Manning 439 10/1986; **Mile 12 Mamfe Road:** Nemba 286 fl., fr., 10/1986; **Mwanenguba:** Leeuwenberg 8677 fr., 11/1971.

MEDUSANDRACEAE

M. Cheek

Note: *Soyauxia*, also in Bakossi, included in *Medusandraceae* in FWTA, has recently been found to belong to *Peridiscaceae*, where we have transferred it.

Medusandra mpomiana Letouzey & Satabié

Adansonia (sér 2) 14: 65 (1974).

Tree 10m, glabrous; leaves alternate, elliptic, c. 24 × 10cm acuminate, obtuse, 3-nerved; petiole 8cm, swollen at base and apex; inflorescence an axillary spike, 1.5cm; flowers white, 2mm; fruit globose, 2cm diameter, 3-valved, 1-seeded; sepals 5, accrescent, reflexed, 0.7 × 0.5cm. Forest; 420–1200m.

Distr.: Cameroon. (W Cameroon Uplands).
IUCN: NT
Enyandong: Cheek 10911 10/2001; Tadjouteu 443 fr., 10/2001; **Kodmin:** Cheek 9201 2/1998; Etuge 4053 fl., fr., 1/1998; **Nyandong:** Cheek 11282 3/2003.
Local name: none (Cheek 11282).

Medusandra richardsiana Brenan

Tree to 18m, puberulous; leaves oblong-elliptic, 10–30 × 5–14cm, acutely acuminate, cuneate-rounded, entire; lateral nerves 8 pairs; petiole 3–9cm, bipulvinate; inflorescence spikes pendulous, 3–15cm, axillary; petals 2mm; fruit hard, 3-valved, 1.5cm; seed 1. Forest; 200–600m.

Distr.: Cameroon. (Cameroon Endemic).
IUCN: VU
Bolo: Thomas 5712 fl., 3/1986; **Etam:** Etuge 238 8/1986; **Ikiliwindi:** Nemba 536 fr., 6/1987.

MELASTOMATACEAE

M. Cheek & E. Woodgyer, *Memecylonoideae* by R.D. Stone & M. Cheek

Fl. Cameroun 24 (1983).

Amphiblemma amoenum Jacq.-Fél.

Fl. Cameroun 24: 104 (1983).

Shrubby herb; stems ascending, 25cm; leaves membranous, ovate, 8 × 9cm, acute, cordate, dentate; petiole 1–5cm; inflorescence terminal, 6cm, long-pedunculate, 3–8-flowered; flowers pink; hypanthium 3 × 3mm, shallowly 5-lobed, glandular-hairy. Forest.

Distr.: Cameroon. (W Cameroon Uplands).
IUCN: EN
Boubaji: Letouzey 14342 8/1975.
Note: record taken from Fl. Cameroun, known only from 4 collections from the Mamfe-Bakossi area.

Amphiblemma mildbraedii Gilg ex Engl.

Erect, robust terrestrial herb, 1–2m; stems square, to 1cm; leaves ovate, to 29 × 18.5cm, subacuminate, cordate, 7-nerved, subscabrid above; petiole to 10cm; stipules persistent, c. 1 × 1cm; panicle terminal, c. 18 × 15cm; peduncle 7cm; flowers 3cm, 5-petalled, purple; fruit dry, 5-angled, 0.8cm. Forest edge; 450–1950m.

Distr.: SE Nigeria, Bioko, Cameroon. (W Cameroon Uplands).
IUCN: NT
Baseng: Cheek 10411 12/1999; **Kodmin:** Cheek 8926 1/1998; Etuge 4078 1/1998; Ghogue 39 fl., 1/1998; **Kupe Village:** Etuge 2674 7/1996; **Mt Kupe:** Cheek 7566 10/1995; Sebsebe 5082 11/1995; Thomas 5051 11/1985; **Mwambong:** Etuge 4329 10/1998; **Nyasoso:** Cable 2912 6/1996; **Tombel:** Cheek 9333 10/1998.

Amphiblemma molle Hook.f.

Fl. Cameroun 24: 97 (1983).

Terrestrial herb to 45cm, densely appressed-setose; leaves ovate, c. 9 × 4.5cm, acuminate, rounded, margin ciliate, softly hairy above and below; petiole to 5cm; inflorescence an erect, terminal, 1-sided dense spike, 2–8cm; fruits pubescent, dry, 7mm. Forest; 760–1500m.

Distr.: Cameroon, Gabon, Cabinda. (Lower Guinea).
IUCN: NT
Kodmin: Etuge 4006 1/1998; **Ngomboaku:** Cheek 10399 12/1999; Ghogue 513 12/1999; Mackinder 311 12/1999.
Note: only 9 other sites known in Cameroon; Fl. Cameroun 24: 90.

Amphiblemma sp. 1

Prostrate herb; stems rooting, to 60cm or more long; leaves appressed to ground, lanceolate, anisophyllous, smaller leaf of pair to 1.8 × 1cm, larger 12 × 5cm, acuminate, rounded-abruptly cordate; petiole to 3cm; inflorescence terminal, 9cm, 3–8-flowered, 1-sided; hypanthia c. 5 × 3mm; calyx lobes filiform, 6mm, glandular, hairy; petals pink. Forest: wet rock-faces and near streams; 750–1500m.

Distr.: Cameroon. (W Cameroon Uplands).
Baseng: Mackinder 331 12/1999; **Kodmin:** Etuge 4015 1/1998; **Nyale:** Cheek 9665 11/1998.
Note: Thomas 5324 (1/1986, Bangem, 800m) probably belongs here on the basis of the field description, although named as *A. monticola* at MO, presumably in error (name not found).

Antherotoma naudinii Hook.f.

Erect, annual herb, c. 15cm tall; stems hirsute; leaves ovate to oblong-lanceolate; blade c. 2 × 1cm; inflorescence capitulate; flowers 4-merous; petals pink; anthers truncate; fruit a capsule. Short grassland, usually in damp places; 1500–2000m.

Distr.: Guinea (C) to Cameroon, Ethiopia to Angola & S Africa; Madagascar. (Afromontane & Madagascar).
IUCN: LC
Mwanenguba: Cheek 9416 10/1998; Sanford 5442 fl., fr., 11/1968.

Calvoa hirsuta Hook.f.

Epiphytic, hemiepiphytic, rarely terrestrial herb, erect, 7–20cm, hirsute; leaves purplish green, ovate-elliptic, to 5.5 × 3cm, finely serrate; inflorescence terminal, dichasial branching; flowers solitary (–3), pale pink or white, 1cm; stamens 6, equal, lacking appendages; fruit top-shaped, 0.5cm diameter, 5-angled, apex with plates. Forest; 760–1350m.

Distr.: Ghana, Bioko & Cameroon. (Upper & Lower Guinea).
IUCN: NT
Kodmin: Cheek 9005 1/1998; Etuge 4147 2/1998; Gosline 62 1/1998; **Kupe Village:** Cable 781 fl., 1/1995; 805 fl., 1/1995; Etuge 2805 7/1996; **Mbule:** Cable 3357 6/1996; **Mwambong:** Cheek 9372 10/1998; Onana 537 2/1998; 547 2/1998; **Ngomboaku:** Cheek 10361 12/1999; Mackinder 283 12/1999; 295 12/1999.
Note: 12 sites known, apart from those cited.

Dicellandra barteri Hook.f. var. *barteri*

Epiphyte-climber to 4 m; stems appressed to tree bark, adventitious roots numerous, densely red-brown-hairy; hairs patent, apex curled, 1mm; leaves ovate-lanceolate, 11 × 5.5cm, subacuminate, rounded-abruptly cordate; petiole to 1.5cm; Forest; 710m.

Distr.: Liberia to Bioko & Cameroon. (Upper & Lower Guinea).
IUCN: NT

Nyale: Etuge 4221 2/1998.
Note: although sterile, the distinctive climbing habit of this specimen strongly suggests this species.

Dichaetanthera africana (Hook.f.) Jacq.-Fél.
Syn.: *Sakersia africana* Hook.f.
Shrub to small tree, c. 6–12m tall; leaves elliptic, strigose, apex acuminate; petiole c. 2cm long; inflorescence terminal panicle; flowers 4-merous; hypanthium and calyx lobes glabrous; buds c. 0.5cm long; petals pink; fruit a capsule. Secondary bush, lowland rainforest; 800–1000m.
Distr.: Sierra Leone to Angola. (Upper & Lower Guinea).
IUCN: LC
Eboné: Bamps 1613 fl., 12/1967; **Enyandong:** Tadjouteu 437 fr., 10/2001; **Kupe Village:** Cheek 7033 fl., 1/1995; 7068 fl., 1/1995; Etuge 1980 5/1996.
Uses: MEDICINES (Tadjouteu 437).

Dinophora spenneroides Benth.
Shrubby herb 1.5m, glabrous; leaves ovate-oblong, 5–14 × 3–6cm, membranous, acuminate, cordate, 5–7-nerved at base; inflorescence of lax, terminal panicles; calyx dentate, white; petals 1cm, pink; berries white. Forest edge; 200–250m.
Distr.: Guinea (C) to Congo (K) & Angola. (Guineo-Congolian).
IUCN: LC
Mile 15, Kumba-Mamfe Road: Nemba 168 7/1986; 606 7/1987.
Note: specimen not seen.

Dissotis brazzae Cogn.
Erect herb c. 1.5m tall; stems 4-winged; leaves c. 7.5 × 3cm, 9-nerved, ovate, apex acuminate; petiole c. 4mm long; inflorescence a terminal panicle; flowers 5-merous; petals pink to violet; fruit a capsule. Grassland, roadsides; 1400–1600m.
Distr.: Guinea (C) to Ethiopia & to Zambia. (Tropical Africa).
IUCN: LC
Kodmin: Cheek 9075 1/1998; Ghogue 41 fl., 1/1998; **Mwanenguba:** Sanford 5435 fl., 11/1968; 5540 fl., 11/1968.

Dissotis buettneriana (Cogn. ex Büttner) Jacq.-Fél.
Adansonia (sér. 2) 11: 547 (1971).
Syn.: *Heterotis buettneriana* (Cogn. ex Büttner) Jacq.-Fél.
Forest edge; 1050m.
Distr.: Cameroon to W Congo (K). (Lower Guinea).
IUCN: NT
Ehumseh - Mejelet: Etuge 349 10/1986.
Note: herb resembling *D. rotundifolia*, but sepals caudulate at apex; stamens homomorphic 6mm. Specimen not seen. Description from Fl. Cameroun.

Dissotis perkinsiae Gilg
Fl. Cameroun 24: 30 (1983).
Shrub 1–1.8m; leaves ovate to ovate-lanceolate, 5–8cm long; flowers 11cm diameter; petals 3cm long, reddish violet or deep mauve-purple. Grassland; 1250m.
Distr.: Togo to Uganda. (Guineo-Congolian).
IUCN: LC
Mwanenguba: Leeuwenberg 8604 fl., 10/1971.

Dissotis prostrata (Thonn.) Hook.f.
Fl. Nigrit: 349 (1849).

Herb resembling *D. rotundifolia*, but leaves elliptic, 2–4 × 3–7cm; hypanthial emergences sparsely and slightly setulose. Roadside, paths; 250m.
Distr.: Ghana to Zanzibar. (Tropical Africa).
IUCN: LC
Bakolle Bakossi: Etuge 404 11/1986.
Note: specimen not seen. Description from Fl. Cameroun, perhaps not distinct from *D. rotundifolia*.

Dissotis rotundifolia (Sm.) Triana
Scrambling herb 30–150cm long; stems pinkish, slightly 4-angled, pale hairs (much longer at nodes); leaves ovate to elliptic (sometimes widely so); blade green above, pale grey-green beneath, 0.9–5.4 × 0.7–2.6cm, with 3–5 nerves, apex acute to acuminate, margin ciliate and sometimes slightly crenulate; petiole purple, 0.5–2cm long; flowers solitary or in pairs, 5-merous; petals purple; hypanthium with emergences; stamens strongly dimorphic; fruit a capsule. Paths, roadsides; 500–900m.
Distr.: tropical and subtropical Africa.
IUCN: LC
Kupe Village: Cable 2702 5/1996; Lane 379 fl., 1/1995; **Nyasoso:** Etuge 2443 6/1996.

Dissotis rotundifolia vel. D. prostrata
210–1200m.
Kupe Village: Etuge 1840 3/1996; **Mungo River F.R.:** Etuge 4323 10/1998.
Note: the specimens cited are intermediate between the two closely related taxa (q.v.).

Dissotis thollonii Cogn. ex Büttner var. *elliotii* (Gilg) Jacq.-Fél.
Fl. Cameroun 24: 28 (1983).
Syn.: *Dissotis elliotii* Gilg var. *elliotii*
Syn.: *Dissotis elliotii* Gilg
Erect herb to c. 3m tall; stems; petioles and main inflorescence axis glabrescent, long hairs at nodes; leaves lanceolate, 17 × 3cm, 5–7 nerves; petiole c. 4cm long; inflorescence a panicle; buds enclosed by bracteoles; flowers 5-merous; petals purple, c. 1.5cm long; pedicel 1cm long; anthers c. 9mm long; fruit a capsule. Montane grassland; 1500–2000m.
Distr.: Sierra Leone to Cameroon. (Upper & Lower Guinea).
IUCN: LC
Bangem: Cheek 7290a fl., 2/1995; **Kodmin:** Cheek 8897 1/1998; **Mwanenguba:** Bamps 1540 fr., 12/1967; Cheek 9490 10/1998; Sanford 5378 fl., fr., 11/1968.

Dissotis tubulosa (Sm.) Triana
Trans. Linn. Soc. 28: 58 (1871).
Syn.: *Osbeckia tubulosa* Sm.
Herb 15–100cm tall; stems red, 4-angled, hairs appressed; leaves ovate; blade 2.4–6.5 × 0.9–3.3cm, with 5–7 pairs of nerves, setose, apex acuminate, margin crenate; petiole 0.5–1cm long; inflorescence a terminal cyme; flowers 5-merous; petals lilac; hypanthium with emergences, apex extending into a persistent tube in fruit; fruit a capsule. Roadside 950–1800m.
Distr.: Guinea (C) to CAR. (Upper & Lower Guinea).
IUCN: LC
Bangem: Manning 437 10/1986; **Kupe Village:** Cable 3752 7/1996; Cheek 7692 11/1995; **Mt Kupe:** Thomas 5090 11/1985; **Mwanenguba:** Leeuwenberg 8583 fl., fr., 10/1971.

Dissotis sp. 1

Herb 75cm; leaves ovate, c. 3 × 1.5cm, softly pubescent, sessile; inflorescence terminal; 1–3-flowered, or with single flowers in upper axil, pale pink, 4cm; hypanthium c. 1 × 0.6cm, long white-hairy, epidermis visible between hairs, emergences filiform, in a single ring beneath sepals; sepals caducous, oblong, 5mm, margin ciliate, apical hair cluster; petals 5; outer stamens with long pedoconnective and laminate, 4-lobed appendage; inner stamens lacking appendages. Grassland; 1900m.
Distr.: Mt Kupe. (W Cameroon Uplands).
Mt Kupe: Sidwell 428 11/1995.
Note: possibly a new species; more material required.

Dissotis sp. nr. hensii

Herb 75cm; leaves ovate, c. 3 × 1.5cm, softly pubescent.
Distr.: Cameroon, Congo (Kinshasa), Angola. (Lower Guinea & Congolian).
Mwanenguba: Sanford 5387 11/1968.

Guyonia ciliata Hook.f.

Prostrate herb, sparsely ciliate on stems; leaves and fruit; leaves ovate, 1.5cm, finely serrate; petiole to 2cm; flowers single, pink, 1cm diameter; stamens equal; fruit 5mm with leafy calyx. Forest edge; 800–1250m.
Distr.: Sierra Leone to Tanzania. (Guineo-Congolian (montane)).
IUCN: LC
Lake Edib: Cheek 9152b fl., 2/1998; **Ngomboaku:** Cheek 10358 12/1999; **Nyale:** Cheek 9675 11/1998.

Medinilla mannii Hook.f.

Epiphytic shrub to 1m; stems whitish brown, with raised black lenticels; leaves lanceolate, to 16 × 4cm, acute, base abruptly and shortly cordate, 3-nerved; petiole 0.6cm; inflorescence axillary, 1-several-flowered; petals 4; berries globose, 0.5cm; peduncle 1cm. Forest; 1000m.
Distr.: Liberia to Bioko, Cameroon, Congo (K) and Uganda. (Guineo-Congolian (submontane)).
IUCN: NT
Nyasoso: Cable 3570 7/1996.
Note: known from c. 12 localities.

Medinilla mirabilis (Gilg) Jacq.-Fél.

Fl. Cameroun 24: 115 (1983).
Syn.: *Myrianthemum mirabile* Gilg
Stem-twining liane; leaves in whorls of three, oblanceolate or elliptic, to 22 × 9cm, subacuminate, cuneate-obtuse, 3-nerved; petiole to 3cm; inflorescence cauliflorous, near ground, fascicles of c. 40 flowers; peduncles to 2cm, much-branched; pedicels 1.5cm; flowers 1cm, blue; berries 1cm. Forest; 700–900m.
Distr.: SE Nigeria, Bioko, Cameroon & Gabon. (Lower Guinea).
IUCN: NT
Kupe Village: Cable 3732 7/1996; **Ngomboaku:** Cheek 10371 12/1999.
Note: known from about 10 other localities in Cameroon.

Memecylon afzelii G.Don var. afzelii

Shrub 1–2m; stems obscurely 4-angular; leaves subcoriaceous, narrowly elliptic-lanceolate, 8–9 × 2.5–3cm, acumen 1–1.5cm, cuneate, nervation obscure, with 8–10 pairs of oblique transverse nerves; cymes axillary, 1–2, lax, to 2.5cm, 3–7-flowered; peduncle 1cm; pedicels 1.5–2mm; hypanthium 3 × 2mm, 4-denticulate; petals white, 2.5mm; fruit white, oblong-ellipsoid, 1.5 × 9mm. Forest.
Distr.: Guinea (C) to Cameroon. (Upper & Lower Guinea).

IUCN: LC
Mombo: Olorunfemi in FHI 30567 fl., 5/1951.

Memecylon afzelii G.Don var. amoenum Jacq.-Fél.

Adansonia (sér 2) 18: 426 (1979).
Shrub 1.5–3m; resembling var. *afzelii*, but leaves c. 11 × 4cm, acumen 1.2cm, gradually tapering, base acute; lateral nerves c. 7 pairs, inconspicuous; petiole 2–4mm; inflorescence umbellate; peduncle 1cm; pedicels 0.5cm; flowers not seen; fruit ellipsoid, 1cm. Forest; 800–1000m.
Distr.: Liberia to Cameroon. (Upper & Lower Guinea).
IUCN: LC
Bangem: Thomas 5267 1/1986; **Mwambong:** Etuge 4131 2/1998.
Note: newly recorded here for Lower Guinea.

Memecylon dasyanthum Gilg & Ledermann ex Engl.

Tree 6m; stems terete; leaves coriaceous, elliptic, 5–8 × 3–3.5cm, acumen acute, 0.5cm, acute-decurrent; lateral nerves c. 6 pairs, inconspicuous; petiole 3mm; compound umbels axillary, 2.5cm; peduncle 1.2cm, c. 20-flowered; hypanthium 0.5 × 1mm, entire; petals and fruit not seen. Forest; 1300m.
Distr.: Cameroon. (W Cameroon Uplands).
IUCN: VU
Kupe Village: Cable 3785 7/1996.

Memecylon griseo-violaceum Gilg & Ledermann ex Engl.

Fl. Cameroun 24: 144 (1983).
Tree 4–5m; leaves with a rounded base; transverse nerves 5–8, strongly compressed on the upper surface, conspicuous on the lower surface. Forest; 300m.
Distr.: Cameroon Endemic.
IUCN: DD
Loum F.R.: Ledermann s.n. 11/1909.
Note: specimen at B, presumed destroyed; treated in Fl. Cameroun as 'espèce mal connue' and as possibly a synonym of *M. virescens* Hook.f. or *M. viride* Hutch. & Dalziel. The location of the specimen is given as Lome, interpreted in Fl. Cameroun as Loum. If confirmed as a good species then IUCN status of CR will be merited.

Memecylon cf. lateriflorum (G.Don) Brem.

Shrub or small tree closely resembling *M. occultum*, but stems terete. Forest; 550–700m.
Distr.: Guinea (C) to Nigeria, ?Cameroon. (Upper & Lower Guinea).
Nyandong: Stone 2455 3/2003; 2477 3/2003.
Note: otherwise known only from Upper Guinea.

Memecylon occultum Jacq.-Fél.

Fl. Cameroun 24: 126 (1983).
Shrub or small tree, 1–2.5m; leaves opposite, elliptic. Forest; 800m.
Distr.: Upper Guinea, Cameroon & Gabon. (Upper & Lower Guinea).
IUCN: NT
Nyandong: Stone 2505 3/2003.
Note: determination tentative pending collection of fertile material. Our material is sterile.

Memecylon viride Hutch. & Dalziel

Shrub; stems terete; leaves oblong to elliptic, 10–14 × 3–5cm, acumen linear, to 2cm; inflorescence contracted, 1 × 1cm; pedicels slender, 2.5–3mm; calyx 4-sinuate; petals white or violet, 2 × 2mm; fruit globose. Forest; 250m.

Distr.: Liberia to Gabon. (Upper & Lower Guinea).
IUCN: NT
Mile 15, Kumba-Mamfe Road: Nemba 141 fr., 7/1986; 174 7/1986.
Note: only 3 collections listed in Fl. Cameroun, 1 in FWTA. "Peu recoltée, elle existe cependant Liberia au Gabon".

Memecylon zenkeri Gilg

Shrub (1–)3(–5)m; stems 4-angled-winged; leaves elliptic, 9–13 × 3.5–4.5cm, acumen to 2cm, obtuse; lateral nerves highly conspicuous, 6 pairs, united by a nerve 3–5mm from margin; petiole 3mm; fascicles 2–8-flowered, 1cm; peduncle 1–3mm, bracteose; hypanthium 2 × 3mm, entire; petals purple-white; fruit pale blue, ellipsoid, 1.7cm. Forest; 250–1250m.
Distr.: SE Nigeria & Cameroon. (Lower Guinea).
Konye: Thomas 5925 fl., 3/1986; **Kupe Village:** Cable 2689 5/1996; Cheek 6025 fr., 1/1995; Etuge 1375 11/1995; Ryan 312 5/1996; **Mbule:** Etuge 4394 10/1998; **Ngomboaku:** Cheek 10296 12/1999; **Ngusi:** Etuge 49 fl., 4/1986; **Nyale:** Cheek 9657 11/1998; **Nyandong:** Stone 2445 3/2003; 2476 3/2003; 2502 3/2003; **Nyasoso:** Cable 2765 6/1996; 3211 6/1996; Etuge 1807 3/1996; 2409 6/1996; Zapfack 664 6/1996.
Note: common and widespread in Cameroonian forest.

Memecylon Sect. *Mouririoides sp. nov. 1*

Shrub 1.5–3m; leaves elliptic, to 18 × 7.5cm, acumen acute, 1cm, base obtuse; secondary nerves fairly prominent below, c.10 pairs, linking nerves 3–4mm from margin; petiole 5mm; compound umbel axillary, c.3cm; hypanthium c. 2 × 3mm; petals pale blue; stamens numerous; fruit white, ripening black, ellipsoid, c. 1.5cm. Forest; 1100–1200m.
Distr.: Mt Kupe & Bakossi Mts. (Narrow Endemic).
Kupe Village: Cable 818 fr., 1/1995; **Ndum:** Lane 464 fl., 2/1995; **Nyasoso:** Lane 121 fl., 6/1994; 515 fl., 2/1995.
Note: known only from the specimens cited. Most likely to be confused with *M. englerianum*, which is unknown from Bakossi.

Memecylon Sect. *Mouriroides sp. nov. 2*

Shrub 1–4m; stems terete to obscurely angular; leaves thinly-coriaceous, elliptic, 9–12.5 × 4.5–7cm, acuminate, acumen 1–1.5cm; petiole 4–5mm; cymes axillary, 2.5–4cm, 10–20-flowered, solitary or paired; peduncle 13–20mm; pedicels 2–3mm; calyx shallowly sinuate; petals purple-blue, 2.5mm; fruit ellipsoid, 10–13mm long. Forest; 350–500m.
Distr.: W Bakossi. (Narrow Endemic).
Nyandong: Stone 2449 3/2003; 2475 3/2003; 2497 3/2003; 2499 3/2003; Thomas 7010 5/1987.

Preussiella kamerunensis Gilg

Syn.: *Preussiella chevalieri* Jacq.-Fél.
Epiphytic shrub to 1.5m; stems terete, smooth; leaves ovate, c. 7 × 5cm, acuminate, base rounded; nerves 3–5 pairs, red; tertiaries scalariform; petiole 2cm; inflorescence a loose terminal cyme, 8 × 8cm; flowers pink, 1.5cm; fruits elliptic, 1.5cm, dry, papery, the walls disintegrating, leaving the net-like veins. Forest; 500–1550m.
Distr.: Guinea (C) to Gabon. (Upper & Lower Guinea).
IUCN: NT
Kupe Village: Cable 2413 5/1996; 2613 5/1996; 3735 7/1996; 3842 7/1996; Cheek 7747 11/1995; 7781 11/1995; Etuge 1945 5/1996; 2662 7/1996; 2870 7/1996; Zapfack 904 7/1996; 917 7/1996; **Ngomboaku:** Cheek 10303 12/1999; **Nyale:** Cheek 9693 11/1998; **Nyandong:** Stone 2446 3/2003; **Nyasoso:** Cable 14 fl., 7/1992; 3543 7/1996; 3554 7/1996; 3576 7/1996; 3627 7/1996; Etuge 2185 6/1996; 2492 7/1996;

2563 7/1996; Lane 49 fl., 8/1992; Sidwell 405 10/1995; Wheatley 442 fl., fr., 7/1992.
Note: this species, common in Bakossi (21 specimens), is rare elsewhere. Only 20 specimens are known at K for the species throughout its range.

Tristemma demeusei De Wild.

Fl. Cameroun 24: 73 (1983).
Syn.: *Tristemma radicans* Gilg ex Engl.
Decumbent herb 10–30cm; stems rooting along ground, ascending, c. 2mm, densely soft-hairy; leaves membranous, c. 5 × 2.5cm, acute, puberulent; petiole 1–1.5cm; 1-flowered; bracts 2 pairs; hypanthium with 1 ring of hairs; flowers 1.5cm diameter; Forest edge; 1050–1700m.
Distr.: Bioko, Cameroon & Congo (K). (Lower Guinea & Congolian).
IUCN: LC
Edib: Cheek 9120b 2/1998; **Ehumseh - Mejelet:** Etuge 347 10/1986; **Kodmin:** Ghogue 40 fr., 1/1998; **Mejelet-Ehumseh:** Manning 416 10/1986.
Note: new record for FWTA area; usually collected from near sea-level.

Tristemma hirtum P.Beauv.

Erect shrub resembling *T. mauritianum*, but stems 4-angled, not winged; leaves c. 8 × 4.5cm, lower surface with dense long, spreading hairs. Forest edge; 700m.
Distr.: Senegal to Congo (K). (Guineo-Congolian).
IUCN: LC
Nyandong: Thomas 6697 2/1987.
Note: specimen not seen.

Tristemma littorale Benth. subsp. *biafranum* Jacq.-Fél.

Fl. Cameroun 24: 62 (1983).
Erect weedy shrub, 1–2m; stems 4-angled, subscabrid, hairs stout, most appressed, bases red; leaves to 11 × 5.5cm, pink-nerved below; inflorescence as T. mauritianum but hypanthium with a single ring of hair. Forest edge, swamp; 700–1750m.
Distr.: Ghana to Congo (K). (Guineo-Congolian).
IUCN: LC
Ehumseh - Mejelet: Etuge 320 10/1986; **Kodmin:** Cheek 10268 12/1999; 9051 1/1998; 9100 2/1998; 9591 11/1998; **Kupe Village:** Cable 2722 5/1996; 3720 7/1996; 3821 7/1996; Ensermu 3551 11/1995; Etuge 1905 5/1996; 2618 7/1996; 2722 7/1996; **Lake Edib:** Cheek 9140 2/1998; **Mwambong:** Etuge 4332 10/1998; **Mwanenguba:** Sanford 5436 fl., fr., 11/1968; **Nyasoso:** Cable 1126 fl., 2/1995; Etuge 1319 10/1995.
Local name: Nu-tunduhl (Max Ebong in Cheek 9591). **Uses:** FOOD – infructescences (Etuge 2722 & Max Ebong in Cheek 9591).

Tristemma mauritianum A.Juss.

Fl. Cameroun 24: 64 (1983).
Syn.: *Tristemma incompletum sensu* Keay in FWTA 1: 250
Erect weedy shrub, 0.5–2m; stems 4-winged, c. 1cm, hairs fine, soft, grey; leaves ovate, c. 17 × 8cm, puberulent below; inflorescence capitate, sessile, c. 6-flowered; involucral bracts numerous; hypanthium with 2 rings of hairs; flowers pink, 2cm; stamens yellow, equal, without connective; berries red. Farms, farmbush; 250–1700m.
Distr.: Nigeria to Congo (K). (Lower Guinea & Congolian).
IUCN: LC
Ehumseh: Manning 493 10/1986; **Kodmin:** Cheek 9043 1/1998; **Kupe Village:** Cable 2507 5/1996; 3682 7/1996; 3851 7/1996; Cheek 6065 fl., fr., 1/1995; 7077 1/1995; Etuge

2620 7/1996; Kenfack 248 7/1996; Lane 294 fl., fr., 1/1995; Zapfack 958 7/1996; **Mile 15, Kumba-Mamfe Road:** Nemba 166 7/1986; **Mt Kupe:** Schoenenberger 34 fl., fr., 11/1995; **Mwanenguba:** Leeuwenberg 8263 fl., 9/1971; **Nyandong:** Cheek 11300 3/2003; **Nyasoso:** Etuge 2442 6/1996.
Local names: Akon Achukud (Cheek 9043); Ntundon (Cheek 11330). **Uses:** FOOD – infructescences – fruits edible (Cheek 6065 & 9043); MEDICINES (Cheek 7077);

Tristemma oreophilum Gilg
Fl. Cameroun 24: 75 (1983).
Erect shrub resembling *T. littorale*; stems densely white-appressed-hairy; leaves c. 4.5 × 2.5cm; inflorescence often pedunculate, 2–3-flowered; bracts c. 8mm broad, densely hairy; hypanthium with 4 rings of hairs. Forest gaps; 900m.
Distr.: Cameroon, Gabon & Congo (K). (Upper & Lower Guinea).
IUCN: NT
Kupe Village: Etuge 2879 7/1996.
Note: only 5 other collections known in Cameroon.

Warneckea cinnamomoides (G.Don) Jacq.-Fél.
Fl. Cameroun 24: 164 (1983).
Syn.: *Memecylon cinnamomoides* G.Don
Small tree 5m; stems 4-angled; leaves resembling *W. membranifolia*, but fruits on old wood; pedicel 8mm; fruits ellipsoid, c. 6 × 4.5mm. Forest; 1000m.
Distr.: Guinea (C) to Congo (K). (Guineo-Congolian).
IUCN: LC
Ndibise: Etuge 528 fr., 6/1987.
Note: reported as very common in our area.

Warneckea membranifolia (Hook.f) Jacq.-Fél.
Fl. Cameroun 24: 172 (1983).
Syn.: *Memecylon membranifolium* Hook.f.
Shrub 1.5–3m; stems 4-angled/winged; leaves papery, elliptic to 11 × 4cm, acumen 1–2cm, acute, 3-nerved, with c. 15 transverse nerves; petiole 3mm; fascicles few-flowered, axillary; flowers subsessile, not seen; fruit ellipsoid, 6 × 4mm, blue. Forest; 500–950m.
Distr.: Ivory Coast to Congo (K). (Guineo-Congolian).
IUCN: LC
Ngomboaku: Cheek 10395 12/1999; **Nyandong:** Darbyshire 19 3/2003; Ghogue 1445 fr., 3/2003; Stone 2448 3/2003; 2450 3/2003.

Warneckea cf. memecyloides (Benth.) Jacq.-Fél.
Fl. Cameroun 24: 163 (1983).
Shrub or tree similar to *W. pulcherrima*, but leaves 11–17 × 9–11cm; cymes on leafless nodes; calyx lobes truncate; petals semi-oval, 3 × 3mm; fruit ellipsoid-oblong, 11 × 6mm. Forest; 200–300m.
Distr.: Ivory Coast to Cameroon. (Upper & Lower Guinea).
Bolo: Nemba 1 fl., 4/1986; **Ikiliwindi:** Nemba 108 fr., 6/1986; 538 6/1987.
Note: description from Fl. Cameroun account, specimens not seen by MC; taxon poorly known in Cameroon, listed as VU (www.redlist.org).

Warneckea pulcherrima (Gilg) Jacq.-Fél.
Fl. Cameroun 24: 163 (1983).
Shrub or tree 2.5–6(–12)m; stems terete, pale brown; leaves elliptic, 22–27 × 9–13cm, subacuminate, rounded and abruptly cordate, strongly 3-nerved, transverse nerves oblique, sessile; cymes 5–10 at leafless nodes, 3cm; peduncle 1.6cm, c. 20-flowered; petals pink-purple; hypanthium c. 2 × 4mm. Forest; 700–1275m.

Distr.: Cameroon to W Congo (K). (Lower Guinea).
IUCN: NT
Kupe Village: Cheek 10224 12/1999; Etuge 2680 7/1996; **Nsoung:** Stone 2471 3/2003; **Nyasoso:** Cable 1156 fl., 2/1995; Etuge 2434 6/1996; 2556 7/1996; Sebsebe 4998 10/1995; Wheatley 402 fr., 7/1992.
Note: only 4 other collections in Cameroon (Fl. Cameroun).

Warneckea reygaertii (De Wild.) Jacq.-Fél.
Fl. Cameroun 24: 170 (1983).
Shrub 1.5m; stems subangular-terete; leaves broadly elliptic, 10–17 × 5–8cm, acuminate; petiole 5mm; cymes axillary, branched, compact; bracts caducous; pedicels 2–4mm; fruit globose, 10–15mm. Forest; 200m.
Distr.: Cameroon to Cabinda. (Lower Guinea).
IUCN: NT
Ikiliwindi: Nemba 543 6/1987.
Note: specimen diseased, determined by Gereau (MO), but doubted by Stone. Only 2 other collections in Cameroon (Fl. Cameroun).

MELIACEAE

M. Cheek

Carapa
Note: *Carapa* needs revision, after which several additional taxa, further to the two accepted in FWTA, might be revealed.

Carapa grandiflora Sprague
Tree 6–20m, glabrous; leaves to 1.2m, paripinnate, 4–7-jugate; petiole c. 15cm; leaflets oblong to oblong-obovate, c. 18 × 7cm, rounded, acute; petiolules c. 1cm; inflorescence a terminal panicle, c. 30cm; flowers white, c. 8mm; sepals and petals greenish; staminal tube white; disc orange; stigma white; fruit 5-valved, subglobose, c. 10cm, warty; seeds c. 3cm. Forest; 900–2000m.
Distr.: Nigeria to Uganda. (Lower Guinea & Congolian (montane)).
IUCN: NT
Kodmin: Cheek 9068 1/1998; Plot B35 10/1998; **Mt Kupe:** Sebsebe 5090 fr., 11/1995; Thomas 5075 11/1985; 5489 2/1986; **Nyasoso:** Cable 3410 fr., 6/1996; Etuge 1682 fl., fr., 1/1996; 2140 fr., 6/1996.

Carapa procera DC.
Tree c. 15m, resembling C. grandiflora; leaves variable, to 2m, 6–21-jugate; leaflets to 50 × 15cm; panicles to 80cm; flowers 8mm. Forest; 250–500m.
Distr.: S America & Senegal to Congo (K). (Amphi-Atlantic).
IUCN: LC
Ikiliwindi: Nemba 109 6/1986; **Nyasoso:** Elad 122 2/1995; 140 2/1995; **Wone:** Nemba 592 7/1987.
Note: specimens cited not seen.
Local name: Red Mahogany. **Uses:** MATERIALS – timber (*fide* Etuge).

Carapa cf. procera DC.
Shrub 2m, glabrous, resembling C. *grandiflora*, but leaves 3-jugate; petiole 27cm; panicle 12cm. Forest; 810m.
Kupe Village: Kenfack 278 fl., 7/1996.
Note: the specimen cited does not match several aspects of C. *procera*, the only lowland species accepted in FWTA.

Entandrophragma angolense (Welw.) C.DC.

Forest tree to 55m; 3m girth; bole long and straight; crown open; buttresses blunt, broad, low; bark smooth, pale grey-brown to orange-brown with papery scales, scales flaking high up tree; slash dark red and pink; leaves paripinnate, clustered at ends of branches, 7–10-jugate; rachis 25–45cm; petioles 12–16cm, not winged, glabrous or puberulous; leaflets oblong-elliptic, 7–28 × 3–10.5cm, rounded and often mucronate, base rounded to obtuse; lateral nerves pubescent, 9–12 pairs; flowers yellowish; fruit 14–22 × 3.5–5cm; valves 2.5–3cm wide, 2.5–4mm thick; seeds winged. Forest and farmbush; 1100m.
Distr.: Guinea (C) to Uganda & Sudan. (Guineo-Congolian).
IUCN: VU
Bakossi F.R.: Lobe in FHI 8398 4/1947; **Kupe Village:** Cable 783 fr., 1/1995; Etuge 1484 fr., 11/1995; **Nyasoso:** Cable 1219 fr., 2/1995; Groves 86 fr., 2/1995.
Local names: Njobwele (Lobe in FHI 8398); mahogany (English). **Uses:** MATERIALS – furniture (Etuge 1484); timber (*fide* Etuge).

Entandrophragma cylindricum (Sprague) Sprague

Tree c. 40m, resembling *E. angolense*, but; leaves 5–9-jugate; petioles 5–10cm, slightly winged, tomentellous; leaflets oblong-laneolate 8–13 × 2.5–5cm, obtusely acuminate, base rounded to cuneate, asymmetrical; fruit 6.5–10 × 2.7–3.5cm; valves 1.4–2.2cm wide. Forest.
Distr.: Sierra Leone to Uganda. (Guineo-Congolian).
IUCN: VU
Ngab: Forteh in FHI 8394 2/1947, and in FHI 8395 fl., 2/1947.
Local name: Njobwele (Forteh in FHI 8394 & 8395).

Entandrophragma utile (Dawe & Sprague) Sprague

Tree c. 25m, resembling *E. angolense*, but: leaves 8–16-jugate; rachis 10–30cm; petioles 5–14cm, slightly-winged, puberulous; leaflets oblong-lanceolate, 3.5–15 × 1.5–5cm, subacuminate, base rounded to subcordate, asymmetric; lateral nerves 10–16, with tufts of hairs in axils; fruit 18–28 × 5–7cm; valves 3–4cm wide, 7–12mm thick. Forest.
Distr.: Sierra Leone to Congo (K). (Guineo-Congolian).
IUCN: VU
Tombel: Forteh in FHI 8396 3/1947.
Local name: Njobwele (Forteh in FHI 8396).

Guarea cf. cedrata (A.Chev.) Pellegr.

Tree c. 35m; slash yellow; leaves imparipinnate, 3–7-jugate; rachis 2–18cm; petiole 2–4.5cm, wings to 6mm wide; leaflets oblong-lanceolate, 8–32 × 2.5–10cm, obtuse to acuminate, base acute to rounded; lateral nerves 16–22 pairs; fruits resembling those of *Turraeanthus africanus*. Forest; 870m.
Distr.: Sierra Leone to Uganda. (Guineo-Congolian).
Nyasoso: Cable 2760 fr., 6/1996.
Note: description partly taken from FWTA; specimen cited incomplete.
Uses: MATERIALS – used for timber (*fide* Cheek).

Guarea glomerulata Harms

Shrub or small tree 1.8–4(–6)m; leaves c. 5-jugate; rachis 30cm; petioles 15cm, terete, puberulent; leaflets elliptic or oblanceolate-elliptic, c. 24 × 8cm, long acuminate, acute; lateral nerves c. 10 pairs; inflorescence a pendulous spike, c. 0.5 m; flowers pink, scattered on distal part; fruit subglobose, rostrate, 1.5cm, densely brown-hairy. Forest; 200–1750m.
Distr.: Nigeria to Congo (K). (Lower Guinea & Congolian)

IUCN: LC
Ehumseh - Mejelet: Etuge 338 fr., 10/1986; **Enyandong:** Ghogue 1250 fr., 10/2001; **Kodmin:** Biye 26 fl., fr., 1/1998; Cheek 8920 fr., 1/1998; Ghogue 49 fr., 1/1998; **Konye:** Thomas 5154 11/1985; **Kurume:** Thomas 5427 1/1986; 5467 1/1986; **Mejelet-Ehumseh:** Etuge 2929 8/1996; **Mekom:** Thomas 5237 fl., 1/1986; **Mile 12 Mamfe Road:** Nemba 299 fr., 10/1986; **Mwambong:** Onana 554 fr., 2/1998; **Ngomboaku:** Cheek 10315 fr., 12/1999; Ghogue 455 fr., 12/1999; **Nyandong:** Ghogue 1443 fr., 3/2003; Tchiengue 1731 fr., 3/2003; **Nzee Mbeng:** Gosline 107 2/1998.

Guarea mayombensis Pellegr.

F.T.E.A. Meliaceae: 42 (1991).
Syn.: *Leplaea mayombensis* (Pellegr.) Staner
Tree to 25m, closely resembling *G. cedrata*, but fruits 8–20cm diameter, dehiscing on ground and then only weakly and irregularly. Forest; 820–1000m.
Distr.: Cameroon to Uganda. (Lower Guinea & Congolian (montane).
IUCN: LC
Enyandong: Cheek 10969 10/2001; **Kupe Village:** Etuge 2678 fr., 7/1996; **Manehas F.R.:** Pollard 162 fr., 10/1998; **Ngomboaku:** Mackinder 319 fr., 12/1999; **Nyasoso:** Cheek 7499 10/1995; 9284 fr., 10/1998.
Uses: used for timber (*fide* Cheek).

Guarea thompsonii Sprague & Hutch.

Tree 15m; white exudate; leaves 3–8-jugate; rachis 8–30cm; petiole 7–14cm, barely winged; leaflets elliptic, 9–24 × 3–9cm, obtuse to acuminate, base acute to obtuse; lateral nerves 9–16 pairs; fruit resembling *G. cedrata*, but glabrous. Forest; 250–300m.
Distr.: Liberia to Ghana, Nigeria to Congo (K). (Upper & Lower Guinea).
IUCN: LC
Bolo: Nemba 3 fr., 4/1986; **Mile 12 Mamfe Road:** Nemba 653 10/1987.
Note: the first specimen cited is 5km outside our checklist area; only a poor duplicate has been seen. The second has not been seen.

Heckeldora staudtii (Harms) Staner

Shrub or small tree, 3–6(–10)m, sparsely puberulous; leaves c. 6-jugate; rachis c. 30cm; petiole 4cm, terete; leaflets oblong-elliptic, to 20 × 6cm, acuminate, acute; lateral nerves c. 12 pairs; inflorescences pendulous spikes, 0.3–0.6 m; flowers white, 6mm; fruit globose to lobed-ellipsoid, to 10 × 3cm, pale yellow, 1–6-seeded, indehiscent, fleshy. Forest; 200–1750m.
Distr.: Liberia, Nigeria to Congo (K). (Guineo-Congolian).
IUCN: LC
Bakolle Bakossi: Etuge 160 5/1986; **Etam:** Etuge 227 8/1986; 245 fr., 8/1986; **Kupe Village:** Cable 2630 fl., 5/1996; 2656 fr., 5/1996; 3787 fr., 7/1996; 3856 fr., 7/1996; 3882 fr., 7/1996; 748 fr., 1/1995; Cheek 7176 fr., 1/1995; 7672 11/1995; Lane 352 fl., 1/1995; 439 fr., 1/1995; Ryan 266 fl., 5/1996; 275 fl., 5/1996; 425 fl., 5/1996; **Kurume:** Thomas 5411 fl., 1/1986; **Manehas F.R.:** Pollard 124 fl., 10/1998; **Mbule:** Leeuwenberg 9281 fr., 1/1972; **Mile 15, Kumba-Mamfe Road:** Nemba 612 fr., 7/1987; **Mt Kupe:** Thomas 5103 fl., 12/1985; **Mungo River F.R.:** Cheek 9357 10/1998; **Ndum:** Lane 477 fr., 2/1995; **Ngomboaku:** Cheek 10328 12/1999; **Nyasoso:** Cable 2836 fr., 6/1996; 2861 fl., fr., 6/1996; 3499 fr., 7/1996; Etuge 1298 fl., 10/1995; 1804 fr., 3/1996; 2073 fr., 6/1996; 2398 fl., 6/1996; Groves 82 fr.,

2/1995; <u>Sebsebe 5014</u> fl., 10/1995; <u>5044</u> 10/1995; <u>Sidwell 393</u> fl., 10/1995; <u>Zapfack 684</u> fr., 6/1996; <u>702</u> fl., 6/1996.

Khaya ivorensis A.Chev.

Tree 30–45m; buttresses blunt to 3–4m from ground; slash red; bark dark, scaly; leaves paripinnate, 5–6-jugate; rachis c. 10cm; petiole 3cm; leaflets oblong, c. 10 × 5cm, bluntly acuminate, base rounded; lateral nerves 5 pairs; fruit globose, 4–7cm, 5-valved. Forest; 400m.

Distr.: Ivory Coast & Ghana, Nigeria to Gabon. (Upper & Lower Guinea).

IUCN: VU

Kupe Village: <u>Etuge 1486</u> fr., 11/1995.

Lovoa trichilioides Harms

Tree 20m; bark dark, lenticellate; leaves paripinnate, 6–jugate; leaflets elliptic-oblong to 20 × 9cm, acuminate, base obtuse to rounded; inflorescence subcorymbose, 10–15cm; flowers white, c. 4mm; fruit sharply cylindrical, c. 4–7cm, thinly-woody; seeds winged. Forest; 840–1450m.

Distr.: Sierra Leone to Ghana, Nigeria to Congo (K). (Guineo-Congolian).

IUCN: VU

Kodmin: <u>Cheek 9103</u> fr., 2/1998; **Kupe Village:** <u>Elad 89</u> fr., 1/1995.

Uses: MATERIALS – used for timber (*fide* Cheek).

Trichilia gilgiana Harms

Tree resembling *T. monadelpha* but usually taller (c. 20–30m); leaves with dots and dashes in transmitted light; flowers pink within. Forest; 350m.

Distr.: Nigeria to Congo (K). (Lower Guinea & Congolian).

IUCN: LC

Nyandong: <u>Thomas 6998</u> 5/1987.

Trichilia monadelpha (Thonn.) J.J.de Wilde

Meded. Land. Wag. 68(2): 108 (1968).

Syn.: *Trichilia heudelotii* Planch. ex Oliv.

Tree to 15–18m, densely puberulous; leaves imparipinnate, c. 5-jugate; rachis 14cm; petiole 10cm; leaflets without dots and dashes, oblanceolate to oblong, c. 13 × 3.5cm, acuminate, acute; lateral nerves c. 12 pairs; fruits globose, 2cm, yellow-brown-puberulous; stipe c. 3mm. Forest; 300–1100m.

Distr.: Guinea (B) to Congo (K). (Guineo-Congolian).

IUCN: LC

Bolo: <u>Thomas 5896</u> 3/1986; **Kupe Village:** <u>Etuge 1922</u> fr., 5/1996; <u>2750</u> fr., 7/1996; **Mejelet-Ehumseh:** <u>Etuge 490</u> 2/1987.

Trichilia prieureana A.Juss. subsp. *vermoesenii* J.J.de Wilde

Meded. Land. Wag. 68(2): 139 (1968).

Tree 3–30m, glabrous; leaves 3-jugate; rachis c. 12cm; petiole 10cm; leaflets elliptic, c. 14 × 5.5cm, subacuminate, acute; lateral nerves c. 9 pairs; infructescences c. 6cm long; fruits c. 1.5cm, glabrous, drying dark brown. Forest; 150–1000m.

Distr.: Sierra Leone to Uganda & Sudan. (Guineo-Congolian).

IUCN: LC

Etam: <u>Etuge 91</u> fl., 3/1986; **Mile 12 Mamfe Road:** <u>Nemba 288</u> 10/1986; **Nyasoso:** <u>Cable 3241</u> fr., 6/1996; <u>Etuge 1763</u> fr., 3/1996; <u>Zapfack 632</u> fr., 6/1996.

Uses: MATERIALS – wood (Etuge 1763).

Trichilia rubescens Oliv.

Tree (4–)5–12(–18)m; twigs densely lenticellate, hollow; leaves c. 5-jugate; rachis c. 18cm; petiole 5cm; leaflets

drying red-brown below, oblong-elliptic, c. 16 × 5cm, acuminate, acute; lateral nerves c. 10 pairs; inflorescences terminal, c. 25cm; flowers yellow-white, c. 4mm wide; fruit 2.5cm, glabrous, glossy dark brown. Forest; 250–1100m.

Distr.: Nigeria to Tanzania. (Lower Guinea & Congolian).

IUCN: LC

Bangem: <u>Thomas 5914</u> 3/1986; **Kodmin:** <u>Etuge 4158</u> fl., fr., 2/1998; **Kupe Village:** <u>Cheek 7070</u> fl., 1/1995; <u>Kenfack 286</u> fr., 7/1996; **Kurume:** <u>Thomas 5462</u> 1/1986; **Mt Kupe:** <u>Nemba 909</u> 3/1988; **Mwambong:** <u>Cheek 9112</u> fl., 2/1998; **Ndum:** <u>Elad 104</u> fl., 2/1995; **Nyasoso:** <u>Cable 1129</u> fl., 2/1995; <u>2875</u> fr., 6/1996; <u>3458</u> fr., 7/1996; <u>Etuge 1766</u> 3/1996; <u>Thomas 3047</u> fr., 2/1984; <u>Zapfack 665</u> fr., 6/1996.

Trichilia tessmannii Harms

Meded. Land. Wag. 68(2): 171 (1968).

Syn.: *Trichilia lanata* A.Chev.

Tree 12m, densely yellow-brown patent-tomentose; leaves 8-jugate; rachis 18cm; petiole 10cm; leaflets narrowly oblong or oblanceolate, subsessile, 12 × 3.5cm, acuminate, obtuse; lateral nerves c. 19 pairs, highly-prominent below, impressed above; fruit subglobose 3–3.5cm, sparsely long-hairy. Forest; 280m.

Distr.: Sierra Leone to Ghana, Nigeria to Congo (K). (Guineo-Congolian).

IUCN: LC

Karume: <u>Nemba 74</u> fr., 5/1986.

Turraea vogelii Hook.f. ex Benth.

Woody climber to 12m, rarely a shrub to 2m, glabrous; leaves simple, elliptic, c. 11 × 5cm, subacuminate, obtuse; petiole 0.5cm; inflorescence axillary; peduncle 3.5cm, subumbellate; flowers white, 3–10; pedicels 1cm; staminal tube 1.5cm; fruit globose, 2.5cm. Forest; 600–1000m.

Distr.: Ghana to Uganda. (Guineo-Congolian).

IUCN: LC

Kupe Village: <u>Etuge 1989</u> fl., 5/1996; <u>2734</u> fr., 7/1996; **Manehas F.R.:** <u>Pollard 123</u> fr., 10/1998; **Mwambong:** <u>Mackinder 152</u> fl., 1/1998; **Mwanenguba:** <u>Leeuwenberg 9768</u> fl., 4/1972; **Ngombombeng:** <u>Etuge 119</u> 5/1986; **Nyasoso:** <u>Cable 1131</u> fl., 2/1995; <u>2823</u> fl., fr., 6/1996; <u>3228</u> fr., 6/1996; <u>3529</u> fl., 7/1996; <u>Etuge 2523</u> fl., fr., 7/1996; <u>Sunderland 1484</u> fl., 7/1992; <u>Wheatley 458</u> fl., 7/1992; <u>Zapfack 640</u> fl., 6/1996.

Turraeanthus africanus (Welw. ex C.DC.) Pellegr.

Tree c. 20–35m; 60cm d.b.h.; slash yellow; leaves imparipinnate, c. 12-jugate; rachis 4–60cm, petioles 5–10cm, terete, densely chocolate-hairy when young; leaflets oblong, c. 16 × 4.5cm, rounded submucronate, base acute; lateral nerves c. 20 pairs; infructescence c. 19cm, stout; partial peduncles c. 1cm, 1–5-fruited; fruits leathery, globose c. 5cm, dehiscent; seeds 4, orange-segment shaped, coat yellow, pulpy. Forest; 250–1500m.

Distr.: Sierra Leone to Uganda. (Guineo-Congolian).

IUCN: LC

Baduma: <u>Nemba 669</u> 11/1987; **Kodmin:** <u>Cheek 9046</u> fr., 1/1998; <u>Etuge 3962</u> fr., 1/1998; **Mekom:** <u>Thomas 5406</u> fl., 1/1986; **Nyasoso:** <u>Cheek 6018</u> 1/1995; <u>Elad 131</u> fr., 2/1995; <u>Etuge 2170</u> fr., 6/1996.

Local names: Avodiré or African Walnut of trade. **Uses:** timber (*fide* Cheek & Etuge).

MELIANTHACEAE

M. Cheek

Bersama abyssinica Fresen.

Syn.: *Bersama maxima* Baker
Syn.: *Bersama acutidens* Welw. ex Hiern
Tree or shrub 2–8m, glabrous; leaves alternate, c. 30cm, variable, imparipinnate; leaflets 5–6 pairs, densely pubescent or glabrous, glossy, oblong-elliptic, c. 15 × 5.5cm, apex acute, base obliquely obtuse; lateral nerves c. 10 pairs, sometimes serrate in upper half; petiolule 0.5cm; rachis ± winged; petiole c. 12cm; stipules c. 1cm, intrapetiolar; inflorescence a terminal raceme to c. 40cm; rachis c. 7cm; pedicels 1cm; flowers white, 1cm; fruit magenta-red, dehiscent, ovoid, 2cm; seeds 1cm, arillate. Forest; 250–2000m.
Distr.: tropical Africa. (Afromontane).
IUCN: LC
Baduma: Nemba 671 11/1987; **Bakolle Bakossi**: Etuge 136 fl., 5/1986; **Edib**: Etuge 4503 11/1998; **Kodmin**: Etuge 4149 2/1998; 4433 11/1998; **Kupe Village**: Groves 9 fr., 1/1995; Lane 375 fr., 1/1995; **Mwambong**: Pollard 115 10/1998; **Mwanenguba**: Thomas 3133 2/1984; **Nyasoso**: Cable 1191 2/1995; Cheek 6024 fl., 1/1995; 7506 10/1995; Etuge 2099 6/1996; 2599 7/1996; Sebsebe 5058 10/1995.
Note: at least three varieties might be recognised from our material, but we follow Verdcourt in treating all the African members of the genus as one variable species.
Local name: Abi. **Uses**: none known (Epie Ngome in Pollard 115); FOOD – endocarps harvested for oily seeds (*fide* Cheek).

MENISPERMACEAE

B.J. Pollard, *Penianthus* by M. Cheek

Cissampelos owariensis P.Beauv. ex DC.

Twining suffrutescent herb to c. 10m; all or some part of plant pubescent; leaves suborbicular, peltate, 5–11cm diameter, sometimes obscurely 3-lobed, apex acuminate, truncate to rounded at base; inflorescence bracteate; bracts greenish white; ♂ and ♀ flowers large; ♂ flower with cupuliform connate petals; ♀ with large membranaceous bracts; fruits red. Forest; 700–1000m.
Distr.: tropical Africa.
IUCN: LC
Kupe Village: Cheek 7168 1/1995; **Ngomboaku**: Cheek 10379 fl., 12/1999.

Dioscoreophyllum cumminsii (Stapf) Diels var. *cumminsii*

Climber; dark rusty-brown bristly hairs usually present; leaves chartaceous, long-petiolate, entire, or irregularly dentate to strongly 3–5-lobed, base usually cordate; inflorescence long-pedunculate, racemose; flowers crowded distally; pedicellate, green; petals absent; stamens connate in an hemispherical synandrium; synandrium ± sessile, or slightly stipitate; stipe thick, to c. 1mm; fruits red when mature. Forest; 850–1000m.
Distr.: Guinea (C) to Congo (K), CAR, Sudan. (Guineo-Congolian).
IUCN: LC

Kupe Village: Cheek 8409 5/1996; Lane 356 fr., 1/1995; **Nyasoso**: Cable 2877 fl., 6/1996; 3533 fl., 7/1996; Etuge 1791 fl., 3/1996; 2540 fr., 7/1996; Zapfack 680 fl., 6/1996.
Note: *D. volkensii* differs in having scattered hairs, pale not bristly, and male flowers with a slender stipe to c. 2mm.

Jateorhiza macrantha (Hook.f.) Exell & Mendonça

Woody climber to 10m; stems, petioles and leaf-margins with stiff brown hairs; leaves to 20cm diameter, 3–5-lobed, each lobe acutely acuminate, chartaceous; venation reticulate, the nerves setose; petiole to 20cm; ♂ flowers in lax panicles, to 30cm long, sessile on lateral branchlets; ♀ in simple racemes, with conspicuous, laciniate, deciduous bracts; fruit ovoid, c. 2.5cm long, conspicuously echinate. Forest; 250–1200m.
Distr.: S Nigeria to Congo (K). (Lower Guinea & Congolian).
IUCN: LC
Edib: Cheek 9125 2/1998; **Kupe Village**: Cheek 8353 fr., 5/1996; **Loum F.R.**: Onana 991 fr., 12/1999; **Mile 15, Kumba-Mamfe Road**: Nemba 167 7/1986; **Nyandong**: Cheek 11295 3/2003.
Local name: Yoh yoh (Cheek 11295).

Kolobopetalum auriculatum Engl.

Woody climber to c. 20m; outer layer of stems pale brown-yellow, sloughing off; fruiting stems c. 5mm diameter; leaves drying ± greyish beneath, elliptic to ovate, acutely or obtusely auriculate, to widely cordate, caudate-acuminate, often remotely repand-dentate, 5–15 × 3.5–9cm; petioles twining, 2–9cm; ♂ inflorescence to 60cm; lateral branches short; ♀ flowers green; fruiting on woody parts. Forest; 760–1800m.
Distr.: Ghana, Nigeria, Cameroon, Gabon. (Guineo-Congolian).
IUCN: LC
Kodmin: Etuge 4049 fl., 1/1998; **Kupe Village**: Cable 2457 fr., 5/1996; 2458 fr., 5/1996; 3703 fr., 7/1996; Cheek 7735 fr., 11/1995; Lane 449 fl., 1/1995; **Nyasoso**: Cable 2898a fr., 6/1996; Etuge 2144 fr., 6/1996.

Penianthus camerounensis A.Dekker

Bull. Jard. Bot. Nat. Belg. 53: 43 (1983).
Shrub or treelet to 8m; leaves (22–)25–55(?) × (7–)10–27cm, elliptic to oblong-elliptic, cuneate, gradually acuminate; midrib slightly, gradually prominent above; nerves impressed above; petiole (5–)8–30cm; flowers remaining almost closed; stamens protruding at maturity, 3–5mm; fruit 3.5–4 × 1.7–2cm. Forest; 400–1100m.
Distr.: Cameroon Endemic.
IUCN: NT
Bakossi F.R.: Etuge 4305 fr., 10/1998; **Kupe Village**: Cable 825 fr., 1/1995; 2518 fr., 5/1996; Cheek 7081 fl., 1/1995; **Mbule**: Leeuwenberg 8814 fr., 12/ 1971; **Mungo River F.R.**: Onana 945 fl., fr., 11/1999; **Nyasoso**: Etuge 1335 fl., 10/1995; 2438 fr., 6/1996; Sebsebe 5011 fl., 10/1995.
Note: widespread in Cameroon.

Penianthus longifolius Miers

Shrub to 3(–4.5)m; leaves obovate to elliptic, 15–34(–40) × 6–15(–18)cm, rounded to cuneate; midrib slightly, gradually prominent above; nerves plane or slightly-raised above; petiole 5–18cm; flowers remaining almost closed; stamens not protruding at maturity, to 2.8mm; fruit (1.5–)1.8–3.1(–3.4) × 1.1–1.8cm. Forest; 220–950m.
Distr.: Nigeria to Congo (K). (Lower Guinea & Congolian).
IUCN: LC

Bakolle Bakossi: Etuge 287 9/1986; **Bakossi F.R.:** Cheek 9320 10/1998; **Bangem:** Thomas 5377 1/1986; **Bolo Moboka:** Nemba 243 9/1986; **Enyandong:** Gosline 320 10/2001; **Kupe Village:** Ryan 311 fr., 5/1996; **Mwendolengo:** Etuge 4528r 11/2001; **Mwanenguba:** Pollard 895 11/2001; **Ngomboaku:** Ghogue 444 12/1999; **Nyasoso:** Cable 1117 fr., 2/1995; Lane 92 fr.,` 6/1994; Thomas 3042 2/1984.

Local name: Elen. **Uses:** ANIMAL FOOD – infructescences – 'Bush Babies' eat fruits; MATERIALS – hunting (Epie Ngome in Cheek 9320).

Penianthus zenkeri (Engl.) Diels

Shrubbery or small treelet, 1.5–6m; leaves drying bright-green, elliptic to oblanceolate, 24–41 × 8–15cm, cuneate to shortly rounded; midrib very abruptly prominent above; perianth spreading to reflexed (not ± closed); fruits 2–4 ×1.1–1.8(–2.0)cm. Forest; 300–700m.

Distr.: SE Nigeria, Cameroon, E Congo (K). (Lower Guinea & Congolian).

IUCN: NT

Banyu: Olorunfemi in FHI 30621 fr., 6/1951; **Kupe Village:** Etuge 2741 fr., 7/1996; **Loum F.R.:** Biye 68 fl., 12/1999; **Nyale:** Cheek 9646 11/1998; **Nyasoso:** Etuge 1799 fr., 3/1996.

Perichasma laetificata Miers var. *laetificata*

Adansonia (sér. 2) 17: 226 (1977).

Syn.: *Stephania laetificata* (Miers) Benth.

Climber to 15m; stem striate/ribbed, hispid; petiole ribbed, hispid, 9–15cm, < lamina, inserted 4–6cm from base of lamina, base pulvinate and twisted, drying black; lamina peltate, to 27 × 22cm, usually palmately 9-nerved, chartaceous; margin ciliate; ♂ inflorescence a very large panicle, to 50cm, axillary, with shortly-pedicellate lateral racemose branches, dimishing in size from base to apex. Forest; 200–800m.

Distr.: Nigeria to Congo (K) & Angola. (Lower Guinea & Congolian).

IUCN: LC

Kupe Village: Kenfack 274 7/1996; **Mungo River F.R.:** Cheek 9338 10/1998.

Platytinospora buchholzii (Engl.) Diels var. *macrophylla* Diels

Engl. Pflanzenr. 46: 170 (1910).

Twining herb to 8–10m; stems and branches spindly; leaves peltate, 10–13 × 5.5–7.5(–9.5)cm, submembranaceous, ovate to ovate-oblong, rounded to very shallowly cordate, acuminate (acumen linear, 6–8mm), entire or very rarely denticulate towards base; petiole 4–5(–9)cm, geniculate at base, ± twisted; ♂ inflorescences in axillary or supra-axillary racemes, 4–6cm; pedicels c. 1.5mm; bracteoles c. 1mm; ♀ inflorescences and flowers unknown; infructescence racemose, 10–20cm; drupes ovoid, c. 2 × 1.5cm. Forest, riverine forest, farmbush; 1000m.

Distr: Cameroon. (Cameroon Endemic).

IUCN: CR

Nyasoso: Cable 2898b 6/1996; Cheek 7535 10/1995.

Note: our variety differs from the typical variety in having leaves 10–13 × 5.5–7.5cm (not 6–9 × 4.5–6cm), being submembranaceous (not coriaceous). The varietal variation could be only a simple ecological adaptation to different conditions of luminosity.

Rhigiocarya racemifera Miers

Woody climber to c. 20m, glabrous; stems drying blackish, striate; leaves broadly ovate, broadly cordate, abruptly acutely acuminate, entire, distinctly membranaceous, 7–18(–28) × 6–16cm; 7-nerved at base, or with 2 nerves arising centrally, slightly above base; petiole to 9cm, twisted basally; ♂ inflorescence axillary, to 18cm; fruits ellipsoid, c. 1.5 × 1cm. Forest; 350m.

Distr.: Sierra Leone to Angola (Cabinda). (Upper & Lower Guinea).

IUCN: LC

Kupe Village: Etuge 2051 fr., 5/1996.

Note: our specimen differs from all other K material in having a truncate leaf base (not cordate).

Stephania abyssinica (Quart.-Dill. & A.Rich.) Walp. var. *abyssinica*

Slender glabrous liana, to 10m; leaves ovate to orbicular-ovate, 5–10 × 4–13cm, entire, dark green above, glaucous beneath; petiole 4–12cm; inflorescences 4–7cm diameter, to 40cm long; pseudo-umbel on a single peduncle to 10cm; rachis fleshy, red; flowers green or purple. Forest; 1700–2000m.

Distr.: tropical Africa. (Afromontane).

IUCN: LC

Kodmin: Cheek 8922 fr., 1/1998; **Kupe Village:** Cable 2562A fr., 5/1996; **Mwanenguba:** Cheek 9473 10/1998.

Stephania dinklagei (Engl.) Diels

Woody climber to c. 10m, glabrous; leaves ovate to suborbicular, 7–15cm diameter, margin undulate; petiole 6–12cm; inflorescence to 50cm, the main axis bearing numerous, lateral, long-pedunculate, racemes or pseudo-umbels; flowers greenish. Forest; 900m.

Distr.: Guinea (C) to Cameroon, Congo (B), Angola. (Upper & Lower Guinea).

IUCN: LC

Nyasoso: Etuge 2370 fl., 6/1996.

Syntriandrium preussii Engl.

Climber to c. 10m; stems glabrous, striate, ± herbaceous; leaves coriaceous, usually conspicuously 3-foliolate, sometimes lobed or entire, often obliquely auriculate at base, cordate, drying pale green; nerves yellow; ♂ inflorescence a very long and slender panicle to 150cm; flowers greenish yellow, minute, early caducous; ♀ inflorescence racemose, c. 10–20cm, much < than ♂; fruits c. 1cm diameter. Forest; 600–1200m.

Distr.: S Nigeria to Congo (K). (Lower Guinea & Congolian).

IUCN: LC

Kodmin: Etuge 4058 fl., 1/1998; **Kupe Village:** Cable 2666 fr., 5/1996; Cheek 8334 fl., 5/1996; 8335 fr., 5/1996; Etuge 1422 fl., 11/1995; 1432 fr., 11/1995; 2615 fr., 7/1996; **Mt Kupe:** Letouzey 14683 fr., 4/1976; **Nyasoso:** Cable 2817 fr., 6/1996; 3305 fr., 6/1996; Etuge 1633 fl., 1/1996; 2460 fl., 6/1996; 2579 fl., 7/1996.

Tiliacora lehmbachii Engl.

Robust, rambling, woody climber, to c. 20m; branches thick; leaves drying grey-green above, coriaceous, ovate to ovate-oblong, obtuse or rounded, gradually acuminate, 8–13 × 4–6cm; petiole slender, rigid, 2.5–3.5cm, with a laminal pulvinus that dries black; inflorescences cauliflorous, c. 10–15cm; ♂ flowers c. 4mm, with 6 stamens, united at least to the middle; fruit obovoid, c. 7mm. Forest; 960m.

Distr.: Mt Cameroon, Mt Kupe, Congo (K). (Lower Guinea & Congolian).

IUCN: EN

Kupe Village: Cheek 7809 fl., 11/1995.

MONIMIACEAE

Triclisia dictyophylla Diels
Syn.: *Triclisia gilletii* (De Wild.) Staner
Giant liana to 20m; stems smooth, drying dark brown, with
inconspicuous striations; leaves massive, to 36 × 36cm,
broadly ovate, deeply and widely cordate; petiole 5–15(–
25)cm; petiole and nerves very shortly puberulous to
glabrous beneath; ♂ inflorescence laxly cymose; ♂ flowers
with 3 petals, densely golden-brown pubescent;
infructescence very large; fruits to 4.5 × 3.5cm. Forest; 400–
1000m.
Distr.: Ivory Coast to Congo (K). (Guineo-Congolian).
IUCN: LC
Kupe Village: Cheek 7126 fl., 1/1995; Etuge 2031 5/1996;
Nyasoso: Cheek 7528 fl., 10/1995; Sidwell 396 fl., 10/1995.

Triclisia lanceolata Troupin
Bull. Jard. Bot. Nat. Belg. 19: 413 (1949).
Woody liana; stem and branches twisting, striate,
ferrugineous-pubescent, becoming glabrous; leaves elliptic-
lanceolate, rounded, acuminate, 12–18 × 4.5–7cm,
coriaceous, 3–5-nerved from base; ♂ inflorescence
unknwown; ♀ as numerous cymes; fruits c. 1 × 0.8cm,
densely short brown-pubescent, stipitate; pedicel inserted
obliquely. Forest; 1170m.
Distr.: Cameroon, Congo (K). (Lower Guinea & Congolian).
IUCN: EN
Edib village: Cheek 9169 fr., 2/1998.
Note: this collection is of fallen fruits only, which match well
those at K.

Triclisia macrophylla Oliv.
Woody climber to 10m; stem smooth, dark brown; leaves
20–25 × 10–18cm; petiole, midrib and main nerves beneath
densely pubescent; inflorescence arising from woody stem; ♂
flowers green. Forest; 1000–1500m.
Distr.: Sierra Leone, Bioko, SW Cameroon. (Upper & Lower
Guinea).
IUCN: CR
Kodmin: Plot B96 10/1998; **Kupe Village:** Cable 2739 fl.,
5/1996.

MONIMIACEAE

N.A. Brummitt & M. Cheek

Fl. Cameroun 18 (1974).

Glossocalyx brevipes Benth.
Gregarious treelet 2–4(–8)m, aromatic when crushed; stems
hairy, main axis with branches opposite, branch bases
triangular; leaves appearing alternate, oblong-elliptic, c. 20 ×
5cm, slightly serrate, base unequal; flowers inconspicuous,
yellow-green, 0.5–1cm. Forest; 200–1200m.
Distr.: SE Nigeria to Gabon. (Lower Guinea).
IUCN: NT
Bolo: Nemba 62 3/1986; **Etam:** Etuge 230 8/1986; **Kupe
Village:** Cable 2715 5/1996; 722 fl., 1/1995; Cheek 10949
10/2001; 7041 fl., 1/1995; Etuge 1416 11/1995; Lane 351 fr.,
1/1995; **Mile 12 Mamfe Road:** Nemba 296 10/1986; **Mungo
River F.R.:** Cheek 9344 10/1998; Onana 935 11/1999;
Nyale: Cheek 9651 11/1998; **Nyandong:** Cheek 11320
3/2003; Nana 23 3/2003; **Nyasoso:** Balding 86 fl., 8/1993;
Cable 3315 6/1996; Cheek 5669 fl., fr., 12/1993; 6005 fl.,
1/1995; Etuge 1588 1/1996; Wheatley 479 fl., 7/1992.

Note: it is likely that the related *G. longicuspis* with
pendulous, long, white flowers, will be found in the Mungo
River F.R.

Xymalos monospora (Harv.) Baill. ex Warb.
Shrub or small tree 3–8(–25)m; leaves opposite, leathery,
elliptic c. 10 × 4cm, acute, serrate; inflorescences c. 4cm,
below leaves; fruit elliptic, 1cm with apical knob. Forest,
forest-grassland transition; 800–2000m.
Distr.: SE Nigeria to E & S Africa. (Afromontane).
IUCN: LC
Bangem: Thomas 5345 1/1986; **Ehumseh - Mejelet:** Etuge
334 fr., 10/1986; **Kodmin:** Cheek 8868 1/1998; Onana 603
fr., 2/1998; **Mt Kupe:** Cheek 7578 10/1995; Thomas 5076
11/1985; 5473 2/1986; **Mwambong:** Mackinder 164 1/1998;
Mwanenguba: Etuge 4374 10/1998; **Nyasoso:** Cable 2916
6/1996; Elad 124 2/1995.
Uses: FOOD ADDITIVES – leaves – young leaves are used
widely as cooking spice (Elad 124).

MORACEAE

M. Cheek, *Dorstenia* by B.J. Pollard

Fl. Cameroun 28 (1985).

Artocarpus altilis (C.E.Parkinson) Fosb.
Fl. Cameroun 28: 24 (1985).
Tree, sparingly branched, 10m, glabrous; leaves rhombic in
outline, c. 45 × 30cm, pinnately 3–5 lobed on each side, the
lobes reaching 2/3 the distance to the midrib; petiole c. 5cm;
male inflorescence cylindrical, c. 9 × 2.5cm; peduncle 3cm;
fruit spherical, c. 8cm diam, fleshy, glabrous, many-seeded.
Farmbush and cultivated; 450m.
Distr.: Polynesia, introduced in Africa. (Pantropical).
IUCN: LC
Nyandong: Cheek 11322 3/2003.
Note: formerly cultivated, now persisting in secondary forest
and farmbush, but rare.

Dorstenia africana (Baill.) C.C.Berg
Fl. Gabon 26: 30 (1984).
Syn.: *Craterogyne africana* (Baill.) Lanj.
Shrub to 2.5m; branches spreading or arching; leaves
distichous, oblong to lanceolate, (3–)7–30 × (1.5–)2.5–11cm,
margin entire to faintly dentate, midrib prominent below;
petiole 0.2–0.6cm; stipules 0.3–0.8cm; inflorescences usually
2 or more together, pendulous, bright yellow or orange;
receptacle 0.3–0.7cm across, turbinate; peduncle 0.5–2.7(–
3.5)cm. Forest, often along streams; 150–350(–850)m.
Distr.: SE Nigeria to Gabon. (Lower Guinea).
IUCN: LC
Mekom: Thomas 5408 1/1986; **Mungo River:** Gosline 225
fl., 11/1999; **Mungo River F.R.:** Cheek 9350 fl., 10/1998;
Etuge 4316 fl., 10/1998; Onana 928 fl., 11/1999;
Ngomboaku: Etuge 4687 fl., 12/1999.

Dorstenia astyanactis Aké Assi
Adansonia (sér. 2) 7: 387 (1967).
Epiphytic herb; stem pendent, ascending apically, succulent,
c. 50–150 × 1cm; leaves in spirals, 3–5 crowded at stem
apices, elliptic, 5–14 × 3–9cm, margin denticulate, glandular,
puberulous; inflorescences solitary; receptacle discoid,
obovate, 2–3.5 × 1.5–2.5cm; fringe c. 1mm, with a single
terminal appendage, 2–3.5cm, borne apically. Forest; 800–
1000m.
Distr.: Ivory Coast, W Cameroon. (Upper & Lower Guinea).

IUCN: EN
Ngomboaku: <u>Cheek 10430</u> 12/1999; Etuge <u>4672</u> fl., fr., 12/1999; **Nyale:** <u>Etuge 4187</u> fl., fr., 2/1998; <u>4470</u> 11/1998; **Nyasoso:** <u>Cheek 9286</u> 10/1998.

Dorstenia barteri Bureau var. *barteri*

Herb to 90cm; stem ascending, sometimes branched; leaves in spirals or subdistichous; lamina variable, lanceolate to elliptic, broadest above, narrowed below, (3–)6–22 × (1.5–)3–8.5cm, usually entire, but sometimes with 1–2(–5) blunt teeth in the upper third; receptacle with c. 5–6(–15) primary appendages, to 3cm, secondary appendages to 0.7cm; flowering surface (sub)orbicular; fringe 1–5(–10)mm. Forest; 1000–2000m.
Distr.: SE Nigeria, Bioko, W Cameroon & Congo (K). (Lower Guinea & Congolian).
IUCN: LC
Edib: <u>Pollard 239</u> fl., 11/1998; **Kodmin:** <u>Cheek 9623</u> fl., 11/1998; <u>Etuge 4146</u> 2/1998; **Ngomboaku:** <u>Muasya 2024</u> fl., fr., 12/1999; **Nyasoso:** <u>Cable 1198</u> fl., 2/1995; <u>676</u> fl., 12/1993; <u>Cheek 7333</u> fl., 2/1995; <u>Etuge 1662</u> 1/1996.
Local name: Didum-bwe, meaning 'Eye of a dog'. **Uses:** SOCIAL USES – unspecified social uses (Cheek 7333).

Dorstenia barteri Bureau var. *multiradiata* (Engl.) Hijman & C.C.Berg

Adansonia (sér. 2) 16: 434 (1977).
Syn.: *Dorstenia multiradiata* Engl.
As var. *barteri*, but with 12–20 primary appendages, to 4cm; flowering surface angular to star-shaped; fringe 4–10mm. Forest; 1000m.
Distr.: SE Nigeria, Cameroon. (Lower Guinea).
IUCN: NT
Nyasoso: <u>Letouzey 14693</u> fl., 4/1976.
Note. cited in Fl. Cameroun, specimen not seen.

Dorstenia barteri Bureau var. *subtriangularis* (Engl.) Hijman & C.C.Berg

Adansonia (sér. 2) 16: 436 (1977).
Syn.: *Dorstenia subtriangularis* Engl.
As var. *barteri*, but only with up to 5 primary appendages, and without or with only a few secondary appendages; fringe 3–8mm. Forest; 1100m.
Distr.: SE Nigeria, Cameroon. (Lower Guinea).
IUCN: NT
Tombel: <u>Leeuwenberg 5313</u> fl., 4/1965.
Note. cited in Fl. Cameroun, specimen not seen.

Dorstenia ciliata Engl.

Herb to 1m; stem ascending; leaves lanceolate to elliptic, 10–26 × 3–11cm, margin repand, very rarely with 1–10 blunt teeth; petiole 1–3.5cm; stipules 0.1–0.3cm, caducous; inflorescences solitary or in pairs; peduncle (0.5–)1–4cm; receptacle 0.8–2.2cm, fringe to 0.5mm; primary appendages 6–15, to 3cm, basally decurrent on the receptacle below, apically swollen; secondary appendages numerous, 0.1–0.5cm. Forest; 300–1450m.
Distr.: SE Nigeria, Cameroon & Gabon. (Lower Guinea).
IUCN: NT
Kodmin: <u>Cheek 8937</u> fl., 1/1998; <u>Etuge 4160</u> fl., fr., 2/1998; <u>Satabie 1100</u> fl., 1/1998; <u>1104</u> fl., 1/1998; **Kupe Village:** <u>Cable 2415</u> fl., 5/1996; <u>3898</u> fl., 7/1996; <u>Cheek 6030</u> fl., 1/1995; <u>Etuge 2787</u> fl., fr., 7/1996; <u>Lane 270</u> fr., 1/1995; **Mwambong:** <u>Onana 525</u> fl., 2/1998; **Ndum:** <u>Lane 469</u> fl., 2/1995; **Ngomboaku:** <u>Mackinder 291</u> fl., 12/1999; **Ngombombeng:** <u>Etuge 120</u> fl., 5/1986; **Nyasoso:** <u>Cable 1157</u> fl., 2/1995; <u>2748</u> 6/1996; <u>2942</u> fl., fr., 6/1996; <u>3213</u> 6/1996; <u>3427</u> 7/1996; <u>3500</u> fl., 7/1996; <u>3519</u> fl., 7/1996; <u>648</u>

fl., 12/1993; <u>701a</u> fl., 1/1995; <u>Cheek 5633</u> fl., 12/1993; <u>6003</u> fl., 1/1995; <u>Elad 133</u> fl., fr., 2/1995; <u>Etuge 1616</u> fl., 1/1996; <u>Sebsebe 5050</u> fl., 10/1995; <u>Sidwell 316</u> fl., 10/1995; <u>389</u> fl., 10/1995; <u>Sunderland 1519</u> fl., 7/1992; <u>Wheatley 406</u> fl., 7/1992; <u>Zapfack 591</u> fl., 6/1996.
Note: the commonest Bakossi *Dorstenia*, but very rare outside the area. At Mt Cameroon only one collection, at S Bakundu.

Dorstenia elliptica Bureau

Subshrub to 1.5m; leaves in spirals; lamina elliptic to oblong or lanceolate, usually wider apically, (3–)7–28 × (1–)2.5–9cm; petiole (0.2–)0.5–1.2(–2)cm; stipules lateral, not semi-amplexicaul, subulate, 0.3–1.7cm; inflorescences 1–3 together, brownish to purplish; receptacle broadly turbinate to discoid, 0.5–1.1cm; fringe very narrow. Humid forest, often along streams; 300–700m.
Distr.: SW Cameroon, Bioko, Gabon, Congo (B), Angola (Cabinda). (Lower Guinea).
IUCN: LC
Bakossi F.R.: <u>Etuge 4288</u> fl., 10/1998; <u>Keay 37362</u> fl., 1/1958; **Molongo:** <u>Nemba 782</u> 2/1988; **Mungo River F.R.:** <u>Cheek 10119</u> fl., 11/1999; <u>Onana 941</u> fl., 11/1999; **Nyale:** <u>Etuge 4464</u> 11/1998.

Dorstenia mannii Hook.f. var. *mungensis* (Engl.) Hijman

Adansonia (sér. 4) 3: 314 (1984).
Syn.: *Dorstenia mungensis* Engl.
Herb 40cm, puberulent; leaves lanceolate to obovate, 3–21 × 1–10cm, acute to subacuminate, base subobtuse, toothed or with 2–4 obscure lobes near apex; lateral nerves 4–10 pairs; inflorescences solitary, receptacle shallowly angular-conical, 3–12mm diameter; peduncle 0.5–4cm, appendages 4–8, each 2–3cm, linear-triangular. Forest.
Distr.: Cameroon and Gabon. (Lower Guinea).
IUCN: LC
Mungo Ndaw: <u>Buchholz s.n.</u> 4/1874.

Dorstenia poinsettiifolia Engl. var. *etugeana* B.J.Pollard

Kew Bull. 58: 188 (2003).
Herb (20–)35–70cm; stem monopodial, swollen nodally; leaves in spirals, obovate, elliptic or oblanceolate, 3–13(–16.5) × 2.5–7.5cm, entire to bluntly dentate; inflorescences zygomorphic, solitary; receptacle discoid, convex; fringe 1.5–5.0mm, reflexed; primary appendages 1, basal, (4.5–)5.9–10.2cm. Deep shade in submontane Forest; 450–1000m.
Distr.: Bakossi Mts & Rumpi Hills. (Cameroon endemic)
IUCN: EN
Nyale: <u>Etuge 4195</u> fl., fr., 2/1998; <u>4462</u> 11/1998; **Nyandong:** <u>Etuge 4891</u> fl., 3/2003.

Dorstenia prorepens Engl.

Herb to 40cm; stem creeping to ascending; leaves elliptic to oblong or obovate, 2–9 × 1–5cm, margin repand or with 1–3 blunt, broad teeth, almost 3-lobed; petiole 0.5–1.5cm; stipules persistent, 0.1–0.4(–0.5)cm; inflorescence zygomorphic, usually solitary, purplish; receptacle naviculate, 1.2–2.0 × 0.2–0.5cm; fringe 0.5mm; appendages numerous, black-purple, 0.1–0.2cm. Forest; 700–1550m.
Distr.: Nigeria, Bioko, Cameroon. (Lower Guinea).
IUCN: VU
Kupe Village: <u>Cable 796</u> fl., 1/1995; <u>Cheek 6057</u> fl., 1/1995; **Mt Kupe:** <u>Cheek 7622</u> 11/1995.

Dorstenia psilurus Welw. var. *psilurus*

Fl. Gabon 26: 77 (1984).

Syn.: *Dorstenia tenuifolia* Engl.
Herb to 60cm; stem ascending to erect, often branched, densely puberulous; leaves crowded apically; lamina glabrous, lanceolate to elliptic, (2–)5–19 × (1–)2–8cm, often papyraceous, margin usually denticulate to coarsely dentate; inflorescences solitary or paired; receptacle vertical, naviculate; fringe almost lacking, appendages terminal, 2, filiform, the upper one 2–7(–10)cm, the lower 0.3–3.0cm. Forest, particularly along rivers; 200–800m.
Distr.: Nigeria to Sudan, Uganda, S to Angola, Mozambique, Zimbabwe. (Tropical & subtropical Africa).
IUCN: LC
Bangem: Thomas 5347 1/1986; **Mombo-Bakossi:** Olorunfemi in FHI 30573 4/1951; **Mungo River F.R.:** Cheek 9349 10/1998.

Dorstenia psilurus Welw. var. *scabra* Bureau
As for var. *psilurus*, but to 2(–3)m; stems mostly unbranched; leaves not crowded, 7–26 × 6–16cm; upper surface often scabrid. Forest; c. 1000m.
Distr.: SE Nigeria to Congo (K). (Lower Guinea & Congolian).
IUCN: NT
Ndibise: Etuge 519 fl., 6/1987.

Dorstenia turbinata Engl.
Shrub 0.5–1.2(–3)m; leaves ± distichous, ovate to oblanceolate or oblong, (4–)8–21 × (1.5–)3–6cm, acumen 5–25 × 2.5mm, margin sinuate to obscurely dentate or sometimes with several large teeth distally, white-hairy beneath; petiole 2–10mm; stipules triangular, semi-amplexicaul, 2.5–5.5 × 1.5mm; inflorescences solitary or paired; peduncle 5–22mm; receptacle turbinate, 2–10 × 4–15mm; primary appendages (2–)4–6(–9), 5–27mm; sometimes with 2 tiny secondary appendages; fringe 0.5–1mm. Forest; to 1000m.
Distr.: Sierra Leone, Liberia, Ivory Coast, Nigeria, Cameroon. (Upper & Lower Guinea).
IUCN: LC
Ndoungue: Ledermann 6229 fl., 11/1909.
Note. cited in Fl. Cameroun, specimen not seen.

Ficus ardisioides Warb. subsp. *camptoneura* (Mildbr.) C.C.Berg
Fl. Cameroun 28: 238 (1985).
Syn.: *Ficus camptoneura* Mildbr.
Epiphytic shrub to 6m, occasionally a tree to 12m; epidermis flaking, glabrous; leaves obovate, elliptic, or elliptic-oblong, 9–22 × 4–9cm, acuminate, obtuse; lateral nerves c. 4 pairs, basal-most arising 0.5cm or more above leaf base, quaternary nerves conspicuous; stipules subpersistent, 9mm; figs axillary, sessile, globose 0.7–1cm diameter, often verrucose, substrostrate, basal bracts 2, 4mm. Forest; 400–1500m.
Distr.: Ivory Coast, Nigeria to E Congo (K) & N Zambia. (Guineo-Congolian).
IUCN: LC
Ehumseh - Mejelet: Etuge 357 10/1986; **Essossong:** Cable 929 fr., 2/1995; **Kodmin:** Biye 19 fr., 1/1998; 4 fr., 1/1998; Gosline 76 fr., 1/1998; Plot B39 10/1998; **Kupe Village:** Cable 3722 fr., 7/1996; 3810 fr., 7/1996; 3891 fr., 7/1996; Cheek 7213 fr., 1/1995; Etuge 1925 fr., 5/1996; 2704 fr., 7/1996; **Mwambong:** Etuge 4134 fr., 2/1998; Onana 555 fr., 2/1998; **Mwanenguba:** Leeuwenberg 9620 fr., 4/1972; **Ndum:** Cable 873 fr., 1/1995; Elad 106 fr., 2/1995; Groves 29 fr., 2/1995; **Ngomboaku:** Gosline 263 fr., 12/1999; **Nyasoso:** Cable 3453 fr., 7/1996; Cheek 7301 fr., 2/1995; 7500 fr., 10/1995; Etuge 1308 fr., 10/1995; 2059 fr., 6/1996; 2379 fr., 6/1996; Lane 196 fr., 11/1994.

Local name: Akoum. **Uses:** MATERIALS – hunting – formerly the gum used to catch birds (Cheek 7301).

Ficus artocarpoides Warb.
Kirkia 13(2): 275 (1990).
Syn.: *Ficus elegans* (Miq.) Miq.
Epiphytic shrub, glabrous; leaves thinly chartaceous, oblong or oblong-elliptic, c. 14 × 4.5cm, acuminate, truncate or rounded, entire; lateral nerves c. 12 pairs; petiole 2cm; stipules caducous; figs clustered on short spurs below the leafy branches; peduncle 2.5cm, globose, 3.5cm, glabrous, high-wrinkled when dry; bracts caducous; stipules caducous. Forest; 300–1750m.
Distr.: Ivory Coast to Uganda. (Guineo-Congolian).
IUCN: LC
Ehumseh - Mejelet: Etuge 337 10/1986; **Kupe Village:** Etuge 2023 5/1996; 2789 fr., 7/1996; **Ndum:** Lane 489 2/1995; **Nyasoso:** Etuge 2122 fr., 6/1996.

Ficus asperifolia Miq.
Syn.: *Ficus warburgii* Winkl.
Stem-twining climber 10m, twining counter-clockwise; leaves elliptic, 8–22 × 3.5–7.5cm, long-acuminate, cuneate, sinuate-denticulate, secondary and tertiary nerves white-margined; petiole 1.75cm; stipules caducous; figs on naked stems c. 2mm diameter, 0–2m above ground, 1–8 from brachyblasts, spherical, 0.8(–2)cm, glabrous, orange; pedicel 4mm; bracts 3, alternate. Forest; 150–1000m.
Distr.: Senegal to Tanzania. (Guineo-Congolian).
IUCN: LC
Kodmin: Plot B148 10/1998; B295 10/1998; **Kupe Village:** Cheek 7130 fr., 1/1995; 8373 fr., 5/1996; Gosline 238 fr., 11/1999; Zapfack 933 fr., 7/1996; **Mungo River:** Gosline 227 fr., 11/1999; **Nyasoso:** Cable 2763 fr., 6/1996; 3444 fr., 7/1996; 3497 fr., 7/1996; Cheek 9519 fr., 10/1998; Etuge 2526 fr., 7/1996; Groves 94b fr., 2/1995.
Note: our material is all stem-twining, yet Berg (1992) gives this species as only straggling or subscandent.

Ficus chlamydocarpa Mildbr. & Burret subsp. *chlamydocarpa*
Tree c. 15 m; stems and petioles with white straggly hairs; resembling *F. lutea*, but lateral nerves c. 11 pairs; calyptra persistent, 2.5cm, hairy, enveloping the lower 2/3 of the fig, fig 2.5cm diameter, sparse, densely long pale brown-hairy. Forest; 1100–1500m.
Distr.: Bioko & Cameroon. (W Cameroon Uplands).
IUCN: VU
Kodmin: Cheek 8888a fr., 1/1998; **Nyasoso:** Etuge 2392 fr., 6/1996.
Local name: Akum. **Uses:** MEDICINES (Cheek 8888a).

Ficus conraui Warb.
Syn.: *Ficus praticola* Mildbr. & Hutch.
Epiphytic shrub 1.5–4m, or free standing tree 7m, glabrous; leaves oblong-elliptic, to 17 × 5cm, acuminate, obtuse, entire; lateral nerves 8–11 pairs, quarternary nerves conspicuous, reticulate, white; petiole c. 3cm; stipules persistent, 1.5–2.5cm; figs axillary, sessile, minutely puberulent, globose, 2cm diam; bracts 2, 3mm. Forest; 600–1300m.
Distr.: Sierra Leone to Tanzania. (Guineo-Congolian).
IUCN: LC
Edib: Etuge 4489 fr., 11/1998; **Mwambong:** Etuge 4336 fr., 10/1998; **Nlonako:** Leeuwenberg 9744 fr., 4/1972; **Nyasoso:** Cable 3522 fr., 7/1996.

Ficus cyathistipula Warb. subsp. *cyathistipula*

Fl. Gabon 26: 232 (1984).
Syn.: *Ficus rederi* Hutch.
Epiphytic shrub or tree, 20m, glabrous, resembling *F. ardisioides*, but lower nerves arising at base of blade; stipules to 2.5cm, the basal halves united, persistent; figs cauliflorous on leafless branches to 15cm; peduncles 1cm, globose, 1.5cm; bracts 2, 4mm. Secondary forest; 800–1100m.
Distr.: Ivory Coast to Angola, N Zambia, Malawi, Kenya and Tanzania. (Tropical Africa).
IUCN: LC
Ngomboaku: <u>Ghogue 452</u> fr., 12/1999; **Nlonako:** <u>Leeuwenberg 9250</u> fr., 1/1972; **Nyasoso:** <u>Cable 3237</u> fr., 6/1996; <u>Etuge 2489</u> fr., 7/1996; <u>Lane 530</u> fr., 2/1995; <u>531</u> fr., 2/1995.

Ficus cyathistipula Warb. subsp. *pringsheimiana* (Braun & K.Schum.) C.C.Berg

Kew Bull. 43: 83 (1988).
Syn.: *Ficus pringsheimiana* Braun & K.Schum.
Epiphyte; leaves spirally arranged, obovate-oblong to elliptic, 7–16(–27) × 3.5–6.5(–8)cm, glabrous; figs sessile to subsessile, solitary, 1–2cm diameter. in dry state. Forest, edges of rivers and seasonally flooded areas; 220m.
Distr.: Sierra Leone to Congo (K). (Guineo-Congolian).
IUCN: LC
Bakolle Bakossi: <u>Etuge 270</u> 9/1986.

Ficus dryepondtiana Gentil ex De Wild.

Tree to 30m, or shrub, glabrous; leaves drying a characteristic dark brown, otherwise resembling *F. ottonifolia*, further differing in the globose, larger, (3.8cm diameter), wrinkled figs; bracts 2, 4mm. Secondary forest, farmbush on cinder cone; 800m.
Distr.: Cameroon to Congo (K) & CAR. (Lower Guinea and Congolian).
IUCN: LC
Kupe Village: <u>Cable 2551</u> fr., 5/1996.

Ficus exasperata Vahl

Tree c. 15m tall, beginning as strangler; leaves elliptic (lobed and longer when juvenile), 10 × 6.5cm, obtuse at base and apex, obscurely crenate-dentate, scabrid above and below, lower 2 nerve pairs with waxy axils; petiole c. 2.5cm; stipules not amplexicaul; figs c. 1.5cm on leafy branches, single; bracts 3, scattered on pedicel; ostiole circular. Farmbush and secondary, deciduous forest; 350–990m.
Distr.: tropical Africa to S India & Sri Lanka. (Palaeotropics).
IUCN: LC
Kupe Village: <u>Etuge 1465</u> fr., 11/1995; <u>2039</u> fr., 5/1996; **Loum F.R.:** <u>Cheek 9300</u> 10/1998; **Mt Kupe:** <u>Leeuwenberg 9742</u> fr., 4/1972; **Ndum:** <u>Elad 97</u> fr., 2/1995; **Nyasoso:** <u>Cable 2793</u> fr., 6/1996; <u>Etuge 2591</u> fr., 7/1996.
Note: distinctive in the combination of non-amplexicaul stipules, leaf indumentum and fig position.
Local names: Ekol or Sandpaper (fide Etuge). **Uses:** MATERIALS – abrasives – leaves used as sand-paper to clean pots in the kitchen and wooden furnitures (Etuge 1465, 2039 & *fide* Etuge).

Ficus kamerunensis Warb. ex Mildbr. & Burret

Tree 20m, or epiphytic shrub, glabrous; leaves chartaceous, drying crinkled, lanceolate to oblanceolate, 5–11 × 1.5–3.5cm, acuminate, obtuse to rounded, entire; secondary nerves c. 11 pairs, fine; petiole to 2cm; stipule 1.5cm subpersistent; figs axillary, crowded, sessile, globose, 0.7cm, smooth; bracts 2, 2mm. Farmbush; 800–870m.

Distr.: Sierra Leone to Congo (K) & Burundi. (Guineo-Congolian).
IUCN: LC
Kupe Village: <u>Cable 2417</u> fr., 5/1996; <u>Lane 285</u> fr., 1/1995.
Note: closely related to *F. thonningii*.

Ficus cf. leonensis Hutch.

Epiphytic shrub, glabrous; leaves drying dark brown on both surfaces, matt, ovate, ovate-elliptic, c. 12 × 6.5cm, acumen displaced, base obtuse or truncate, entire; lateral nerves c. 6 pairs; petiole 6cm; figs globose, 15mm, glabrous, pale yellow; peduncle 1cm; bracts 2, 2mm. Farmbush; 870m.
Distr.: Guinea (C) to Cameroon. (Upper & Lower Guinea).
Nyasoso: <u>Cable 3261</u> fr., 6/1996.
Note: *F. leonensis* is known from very few collections. The fruits described above are atypically large and possibly of another species. The specimen lacks stem and stipules.

Ficus lingua Warb. ex De Wild. & T.Durand subsp. *lingua*

Epiphytic shrub 4m, glabrous; leaves oblanceolate, c. 3 × 1.3cm, apex rounded or unequally truncate to emarginate, base obtuse, margin entire, midrib not reaching leaf apex; petiole 0.7cm; stipules caducous; figs sessile, glabrous, axillary, globose, 0.4cm diam, smooth; bracts 2, 1mm. Forest; 850–1100m.
Distr.: Ivory Coast, Cameroon to Uganda. (Guineo-Congolian).
IUCN: LC
Kupe Village: <u>Cable 2661</u> fr., 5/1996; <u>3758</u> 7/1996; **Nyasoso:** <u>Lane 6</u> fr., 7/1992.

Ficus lutea Vahl

Kew Bull. 36: 597 (1981).
Syn.: *Ficus vogelii* (Miq.) Miq.
Epiphytic shrub, later a tree, resembling *F. ardisioides*, but leaves 13–27cm; lateral nerves 5–6 pairs; petioles exfoliating; stipules caducous; figs sessile, axillary and below leaves, dense, globose, 1cm, densely brown-hairy; bracts 2, 2mm. Farmbush; 420–1400m.
Distr.: Senegal to S Africa & Madagascar.
IUCN: LC
Essossong: <u>Cable 930</u> fr., 2/1995; **Kupe Village:** <u>Etuge 1910</u> fr., 5/1996; <u>Kenfack 245</u> fr., 7/1996; <u>Lane 390</u> fr., 1/1995; **Loum F.R.:** <u>Cheek 9291</u> fr., 10/1998; **Mejelet-Ehumseh:** <u>Etuge 165</u> fr., 6/1986.

Ficus lyrata Warb.

Tree to 15m, or epiphytic strangler on *Spathodea*; leaves lyre-shaped, coriaceous, c. 30 × 20cm, rounded, cordate, entire, tertiary nerves conspicuous; petiole 4cm; stipules 2.5 × 2cm. Semi-deciduous forest; 420m.
Distr.: Sierra Leone to Cameroon. (Upper & Lower Guinea).
IUCN: LC
Loum F.R.: <u>Cheek 9292</u> 10/1998.

Ficus mucuso Welw. ex Ficalho

Tree to 40m, trunk to 1m, smooth, grey and brown; leaves broadly ovate, c. 10 × 9cm, apex obtuse to rounded, sometimes subacuminate, base deeply cordate, scabrid above, subvelvety below; petiole 3cm, long-hairy; stipules ovate, c. 12 × 6mm, acuminate, long-hairy; figs in groups on main branches, 3–4cm diameter; bracts 3, whorled. Farmbush and secondary forest; 600–850m.
Distr.: Guinea (B) to Tanzania, Ethiopia. (Guineo-Congolian).
IUCN: LC

Ndum: Elad 110 fr., 2/1995; **Ngomboaku:** Cheek 10305 12/1999; **Nlonako:** Leeuwenberg 9750 fr., 4/1972.

Ficus natalensis Hochst. subsp. *leprieurii* (Miq.) C.C.Berg
Meded. Land. Wag. 82(3): 235 (1982).
Syn.: *Ficus leprieuri* Miq.
Epiphytic shrub c. 2m tall, less usually a shrub or tree 2–15m; leaves coriaceous, obtriangular, to 7 × 6.5cm; petiole 1cm; stipules caducous; figs axillary, single (or paired), globose, to 2cm diam, yellow, minutely puberulent; pedicel c. 4mm, apex saucer-shaped, bracts caducous. Forest and grassland; 790–1600m.
Distr.: Senegal to NW Zambia. (Guineo-Congolian).
IUCN: LC
Kodmin: Etuge 3966 fr., 1/1998; Ghogue 12 fr., 1/1998; **Kupe Village:** Cable 2524 5/1996; Etuge 2834 fr., 7/1996; Kenfack 243 fr., 7/1996; Ryan 363 fr., 5/1996; **Ngomboaku:** Etuge 4675 fr., 12/1999; **Nyasoso:** Cable 3463 fr., 7/1996; Etuge 2191 fr., 6/1996.
Note: unmistakeable in the triangular leaves.

Ficus oreodryadum Mildbr.
Fl. Cameroun 28: 200 (1985).
Tree to 30m, or epiphytic shrub, glabrous; leaves coriaceous drying dark `rown below, oblanceolate, c. 20 × 5cm, obtuse, subacuminate, base acute; secondary nerves c. 13 pairs; petiole 5.5cm, exfoliating, stout; stipules caducous; figs axillary, sparse, subglobular, 1.3cm, slightly warty, glabrous, sessile, stoutly beaked; basal bracts 3, 2mm. Forest; 1950m.
Distr.: SW Cameroon to Burundi, Rwanda & Uganda. (Lower Guinea & Congolian (montane)).
IUCN: LC
Nyasoso: Cable 2905 6/1996.

Ficus ottoniifolia (Miq.) Miq. subsp. *ottoniifolia*
Shrub or tree to 17 m, usually an epiphytic strangler at first; branchlets smooth pale; leaves elliptic-oblong, to 12 × 6cm, base obtuse or acute; secondary nerves bifurcating, c. 1.5cm from margin; petiole c. 4cm; receptacles ellipsoid, 1.3–2.3cm diameter; figs long-stalked, in pairs or clusters on older branches below leaves, globose, 1.5–2.5cm in diameter; bracts caducous; peduncle 1cm. Forest; 720–1200m.
Distr.: Sierra Leone to Uganda (Guineo-Congolian)
IUCN: LC
Kupe Village: Kenfack 318 fr., 7/1996; **Ndum:** Lane 506 fr., 2/1995; **Ngomboaku:** Etuge 4674 fr., 12/1999.
Note: similar to *F. tremula*.

Ficus ovata Vahl
Epiphytic shrub, glabrous; leaves thickly coriaceous, ovate, to 28 × 17cm, acuminate, obtuse, entire; lateral nerves c. 10 pairs; petiole c. 6cm; stipules caducous; figs axillary, single, sessile, elliptic, 3.8 × 2.5cm, rarely globular, dark yellow-puberulent; basal bracts forming a sinuate-margined puberulent cup, 1cm diameter. Farmbush; 900m.
Distr.: Senegal to Mozambique. (Tropical Africa).
IUCN: LC
Ngomboaku: Etuge 4680 fr., 12/1999; **Nyasoso:** Lane 513 fr., 2/1995.

Ficus cf. polita Vahl
Shrub, initially epiphytic, glabrous; leaves ovate, c. 12 × 10cm; apex obtuse, base cordate, sinuate (juvenile); lateral nerves 6 pairs; petiole c. 6.5cm; figs from main trunk on short spurs, 1–4 together; peduncles 0.8–2cm; globose, 2–4cm, basal bracts 2, 3–5mm. Semi-deciduous forest; 420m.
Distr.: Senegal to S Africa. (Tropical Africa).

IUCN: LC
Loum F.R.: Cheek 9299 10/1998.
Note: sterile specimen, description in part from C.C.Berg (1992).

Ficus preussii Warb.
Strangler 15m; leaves coriaceous, elliptic, c. 23 × 8cm, sharply acuminate, obtuse to rounded entire; lateral nerves c. 5 pairs, prominent, white, tertiary nerves conspicuous; petiole c. 4cm; stipules c. 35 × 10mm, persistent; figs axillary, single, sessile, globose, 4cm diameter, densely pubescent; basal bracts 2, 5mm. Forest; 990–1250m.
Distr.: Nigeria to Uganda. (Lower Guinea & Congolian).
IUCN: LC
Lake Edib: Cheek 9156 2/1998; **Ndum:** Elad 105 fr., 2/1995; Groves 26 fr., 2/1995.

Ficus saussureana DC.
Fl. Gabon 26: 113 (1984).
Syn.: *Ficus eriobotryoides* Kunth & Bouché var. *eriobotryoides*
Tree 15–20m, brown-pubescent, resembling *F. chlamydocarpa*, but secondary nerves c. 18 pairs, quaternary nervation conspicuous; calyptra absent or inconspicuous; figs 2cm diam, fairly dense, red-orange hairy. Forest; 250–1050m.
Distr.: Guinea (C) to Bioko, Cameroon & Tanzania. (Guineo-Congolian (montane)).
IUCN: LC
Mile 15, Kumba-Mamfe Road: Nemba 156 7/1986; **Ndum:** Elad 102 fr., 2/1995; **Ngomboaku:** Ghogue 474 12/1999.

Ficus sur Forssk.
Fl. Cameroun 28: 135 (1985).
Syn.: *Ficus capensis* Thunb.
Tree 5–20(–30)m, but fruiting at only 5m; stem to 60cm d.b.h.; leaves elliptic-oblong, c. 14 × 7cm, shortly acuminate, rounded to obtuse, margin with c. 5 well-marked serrations on each side, subglabrous; petiole c. 4cm; stipules caducous; figs on branches c. 15cm long on main branches or trunk apex, (lowest c. 6m from ground), c. 2cm diameter; peduncle 7mm; bracts 3, c. 1.5mm, whorled. Farmbush and secondary forest; 600–1400m.
Distr.: Senegal to S Africa. (Tropical Africa).
IUCN: LC
Essossong: Groves 45 fr., 2/1995; **Etam:** Etuge 213 8/1986; **Kupe Village:** Cheek 7102 fr., 1/1995; 8333 fr., 5/1996; Etuge 1900 fr., 5/1996; Kenfack 242 fr., 7/1996; **Manehas F.R.:** Cheek 9389 fr., 10/1998; **Ngomboaku:** Etuge 4678 fr., 12/1999; **Nyasoso:** Cable 2815 fr., 6/1996.
Local name: Ekole etu. **Uses:** none known (Cheek 9389).

Ficus tesselata Warb.
Tree to 18m; leafy stems brown, lenticels large; leaves oblong to subovate, 9–25 × 4–9.5cm, acuminate, base acute to obtuse, entire, glabrous; lateral nerves 4–7 pairs, tertiary nervation reticulate; petiole 1–3cm; epidermis flaking; stipules 1.5–2cm, mostly basally connate, minutely puberulent, persistent; figs 1–2 in axils, sessile, globose, drying 1.5–2cm, yellowish hispid, scabrous and warty, apex protruding; basal bracts 2mm, persistent. Forest; 260m.
Distr.: Sierra Leone to Rwanda. (Guineo-Congolian).
IUCN: LC
Baduma: Nemba 621 9/1987.
Note: specimen not seen.

Ficus thonningii Blume
Syn.: *Ficus dekdekena* (Miq.) A.Rich.

Syn.: *Ficus iteophylla* Miq.
Epiphytic shrub, glabrous; leaves elliptic, elliptic-oblong or oblanceolate-elliptic, 7.5–13 × 3.5–5.5cm, shortly-acuminate, obtuse, entire; lateral nerves c. 12 pairs, fine, quaternary nerves conspicuous; petiole 2.3cm; stipules caducous; figs axillary, amongst and below the leaves, dense, sessile, globose, c. 0.8cm; basal bracts 2, fused to form a bilobed, brown-puberulent, saucer-shaped structure, 0.7cm diameter. Forest; 600–1400m.
Distr.: tropical and S Africa.
IUCN: LC
Etam: Etuge 243 8/1986; **Mejelet-Ehumseh:** Etuge 162 6/1986; 457 1/1987; **Nyasoso:** Cable 2889 fr., 6/1996; 699 fr., 1/1995.

Ficus tremula Warb. subsp. *kimuenzensis* (Warb.) C.C.Berg
Kew Bull. 43: 96 (1988).
Syn.: *Ficus kimuenzensis* Warb.
Epiphytic shrub, known to grow on *Dracaena* and *Shirakiopsis*; leafy stems red-brown, glabrous; leaves ovate-lanceolate, c. 6.5 × 3cm, acuminate, cordate, entire, quaternary nerves conspicuous; petiole to c. 4mm, slender; stipules caducous; figs on old, leafless wood in clusters; pedicel c. 15mm; bracts 2–3mm, paired; fig obovoid, green, 2.5 × 1.5cm. Farmbush, forest edge; 1400–1500m.
Distr.: SE Nigeria to Angola. (Lower Guinea).
IUCN: VU
Kodmin: Etuge 3964 fr., 1/1998; Pollard 232 st.,11/1998.
Note: according to C.C.Berg (1992), this is a lowland taxon.

Ficus trichopoda Baker
Meded. Land. Wag. 82(3): 240 (1982).
Syn.: *Ficus congensis* Engl.
Tree 25m; leaves resembling *F. ovata*, but leaves elliptic, 17–28 × 10–17cm, apex rounded, subcordate; lateral nerves 5–8 pairs, papillose below; figs ellipsoid, 3 × 2.5cm, pale yellow, glabrous; peduncle 1cm; bracts 2, 4mm. Semi-deciduous forest; 500m.
Distr.: Senegal to S Africa. (Tropical Africa).
IUCN: LC
Kupe Village: Etuge 1477 fr., 11/1995.
Note: fruits atypical and possibly of another species.

Ficus vallis-choudae Delile
Tree 7–20m, densely puberulent; bole short; crown spreading; leaves broadly ovate, sometimes slightly 3-lobed, c. 20 × 16cm, obtuse, gradually acuminate, base rounded, margin slightly serrate; secondary nerves 5 pairs, the basal most pair ascending in the upper half, smooth above, almost scabrid below; petiole 5–9cm; stipules 4cm, lanceolate, subpersistent; figs axillary, globose, 1.5cm, solitary; peduncle 0.8cm; bracts triangular, 3mm. Riverine forest, lakesides, ground-water forest; 1300–1350m.
Distr.: Guinea (C) to Ethiopia, E Africa, Mozambique, Zimbabwe. (Tropical Africa).
IUCN: LC
Edib: Etuge 4481 fr., 11/1998; **Kodmin:** Cheek 9002 fr., 1/1998.

Ficus vogeliana (Miq.) Miq.
Tree 6–20m, closely resembling *F. sur*, but usually specific to riverbanks and swamp forest; leaves usually conspicuously hairy and mottled white on suckers and juvenile plants; stipules subpersistent; figs at ground level on leafless branches several metres long, with bracts c. 5mm. River edges; 250–1000m.
Distr.: Guinea (B) to Uganda. (Guineo-Congolian).

IUCN: LC
Baseng: Cheek 10408 fr., 12/1999; **Kodmin:** Plot B276 10/1998; **Mt Kupe:** Cheek 10134 11/1999; **Mwambong:** Etuge 4126 fr., 2/1998; **Nyasoso:** Cable 2891 fr., 6/1996; 3429 fr., 7/1996; Etuge 1290 fr., 10/1995.
Local name: Ekol (Etuge 4126).

Ficus sp. 1
Tree, glabrous; leaves elliptic, c. 14 × 7cm, apex acuminate, base obliquely obtuse to acute, entire; lateral nerves one unequal pair arising 0.7–1.4cm from base, ascending to apex; petiole 1.5cm; stipules 1.5cm, caducous; figs unknown. Forest; 1450m.
Distr.: Bakossi Mts. (Narrow Endemic).
Kodmin: Cheek 9599 11/1998.
Note: extremely distinctive on account of the triplinerved leaves. Placed here in the absence of figs on account of white exudate and engirdling stipules.
Uses: use unknown (Max Ebong in Cheek 9599).

Ficus sp. 2
Strangling tree 15m, glabrous; leaves papery, drying dark brown, midrib white, obovate to oblanceolate c. 6 × 2.5cm, acuminate, acute, entire; lateral nerves c. 10 pairs, with numerous translucent gland-dots; petiole 4mm; stipules 2.5mm; figs unknown. Forest; 420m.
Distr.: Bakossi Mts. (Narrow Endemic).
Loum F.R.: Cheek 9289 10/1998.
Note: placed in *Ficus* despite the absence of figs owing to the presence of white exudate, amplexicaul stipules and strangling habit. Not matching any other of our species. Unusual in having translucent gland-dots.

Milicia excelsa (Welw.) C.C.Berg
Bull. Jard. Bot. Nat. Belg. 52: 227 (1982).
Syn.: *Chlorophora excelsa* (Welw.) Benth.
Tree 40m; leaves oblong-elliptic, c. 13 × 7.5cm, apex rounded or slightly acuminate, base unequally truncate to subcordate, sinuate; lateral nerves c. 15 pairs; petiole 6cm. Semi-deciduous forest; 250–450m.
Distr.: Ivory Coast to Mozambique. (Tropical Africa).
IUCN: NT
Bolo-Meboka: Thomas 7153 7/1987; **Loum F.R.:** Cheek 9296 10/1998; **Nyandong:** Cheek 11324 3/2003.
Note: listed by IUCN (2003, www.redlist.org) as LR/nt.
Local names: Emag or Iroko. **Uses:** MATERIALS – timber (*fide* Etuge).

Scyphosyce manniana Baill.
Subshrub to 0.5m, densely short-brown-hairy; leaves usually oblanceolate, acumen c. 1cm, base cuneate, abruptly rounded to subcordate, often toothed near apex; lateral nerves 10–15 pairs; petiole 0.5–1cm; stipules persistent, lanceolate, 0.5–1.5cm; inflorescence axillary; flowers 1–3 per axil; pedicel 2–3mm; flowers green, 2–3mm; fruit fleshy, subglobose, 7mm. Forest.
Distr.: SE Nigeria to W Congo (K). (Lower Guinea).
IUCN: NT
Baduma: Letouzey 14611 4/1976.
Note: not common, but often locally abundant.

Treculia africana Decne. subsp. *africana* var. *africana*
Bull. Jard. Bot. Nat. Belg. 47: 384 (1977).
Syn.: *Treculia africana* Decne. var. *africana*
Tree 25m; 45cm d.b.h.., trunk ribbed; leaves lanceolate, to 40 × 18cm, shortly acuminate, unequally rounded-truncate at base; petiole 1cm; stipules caducous; inflorescence globose,

6cm diameter; flowers white, minute; fruit globose 45–50cm diam, glabrous. Forest; 850–1300m.
Distr.: Senegal to Mozambique. (Tropical Africa).
IUCN: LC
Kupe Village: <u>Lane 349</u> fl., 1/1995; **Ngomboaku:** <u>Cheek 10341</u> 12/1999; **Nyasoso:** <u>Cable 2837</u> 6/1996; <u>Elad 121</u> fr., 2/1995.
Uses: FOODS – harvested for edible seeds (*fide* Cheek).

Treculia obovoidea N.E.Br.
Tree 5–8m, glabrous; leaves coriaceous, narrowly oblong-elliptic, c. 16 × 5cm, long acuminate, base obtuse, entire; lateral nerves 9 pairs, drying orange and prominent, looping 3mm from margins; petiole c. 1cm; stipules caducous; male inflorescence axillary, ellipsoid or obovoid, 1cm; fruit densely spiny, 4–5cm diameter. Forest and forest gaps; 300–500m.
Distr.: SE Nigeria to Congo (K). (Lower Guinea & Congolian).
IUCN: LC
Bolo: <u>Nemba 52</u> fl., 3/1986; **Manyemen:** <u>Letouzey 13920</u> 6/1975; **Mungo River F.R.:** <u>Cheek 10213</u> 11/1999; **Nlog:** <u>Etuge 6</u> fl., 4/1986; <u>19</u> 4/1986.

Trilepisium madagascariense DC.
Bull. Jard. Bot. Nat. Belg. 47: 299 (1977).
Syn.: *Bosqueia angolensis* Ficalho
Tree 20m, glabrous; leaves elliptic c. 10 × 4.5cm, acuminate, obtuse, entire; lateral nerves c. 5 pairs, basal pair acute; petiole 1cm; stipules caducous; inflorescence axillary at leafless nodes, ellipsoid, 1.5 × 0.8cm, glabrous; peduncle 1.5cm; lowers emerging from apical aperture in inflorescence, in cluster c. 0.5 × 0.5cm. Forest; 420–800m.
Distr.: Guinea (C) to Congo (K). (Guineo-Congolian).
IUCN: LC
Ekona Mombo: <u>Etuge 408</u> fl., fr., 12/1986; **Kupe Village:** <u>Elad 93</u> fr., 1/1995; <u>Lane 293</u> fl., 1/1995; **Loum F.R.:** <u>Cheek 9293</u> 10/1998; **Ngusi:** <u>Etuge 1543</u> fl., fr., 1/1996.

MYRICACEAE

M. Cheek

Morella arborea (Hutch.) Cheek
Cheek *et al.* The Plants of Mt Oku & Ijim Ridge, Cameroon: a Conservation Checklist: 149 (2000).
Syn.: *Myrica arborea* Hutch.
Tree 5–10m, densely shortly pubescent; leaves alternate, aromatic when crushed, oblong, c. 7 × 2.5cm, acute, broadly obtuse, spiny-serrate; c. 11 pairs lateral nerves; petiole 1cm; spikes axillary, 3cm; fruiting capitula 3mm. Forest edge; 1800–2000m.
Distr.: Bioko & Cameroon. (W Cameroon Uplands).
IUCN: VU
Mwanenguba: <u>Cheek 7253</u> fr., 2/1995; <u>Etuge 4385</u> fl., 10/1998; <u>Letouzey 14391</u> fl., 8/1975; <u>Sanford 5410</u> fl., 11/1968; <u>Thomas 3097</u> fr., 2/1984; <u>5885</u> 3/1986.

MYRISTICACEAE

B.J. Pollard

Fl. Cameroun 18 (1974).

Coelocaryon botryoides Vermoesen
Fl. Cameroun 18: 101 (1974).
Tree to 30m; bole long, clear, without buttresses; red exudate when cut; leaves larger than in *C. preussii*, to 15–35 × 8–12cm; veins prominent beneath; lamina often subpanduriform, wider in the apical third; apex long-acuminate; petiole 10–20mm; inflorescence once-branched; fruits less oval than in *C. preussii*. Forest; 350m.
Distr.: Cameroon, Congo (K). (Lower Guinea & Congolian).
IUCN: LC
Kupe Village: <u>Etuge 2043</u> fl., 5/1996.
Note: this specimen most closely agrees with the characters used to distinguish *C. preussii* and *C. botryoides* in Fl. Cameroun. Better material is needed to investigate this further.
Uses: MATERIALS – wood – timber; MEDICINES (Etuge 2043).

Coelocaryon preussii Warb.
Tree to 30m; bole long, clear, without buttresses; red exudate when cut; leaves 15–25 × 6–8cm; veins prominent beneath; lamina lanceolate; apex shortly acuminate; petiole 10–12mm; inflorescence branched twice; fruits rounded at the extremities, 25–40 × 15–30mm. Forest; 200–1200m.
Distr.: Nigeria to CAR & Congo (K). (Lower Guinea & Congolian).
IUCN: LC
Bakolle Bakossi: <u>Etuge 158</u> fr., 5/1986; **Kupe Village:** <u>Cable 707</u> fr., 1/1995; <u>Cheek 7853</u> fl., 11/1995; <u>Elad 74</u> 1/1995; <u>Etuge 1423</u> fl., 11/1995; <u>2910</u> fr., 7/1996; <u>Kenfack 291</u> fr., 7/1996; **Loum F.R.:** <u>Gosline 213</u> fl., 11/1999; **Mile 15, Kumba-Mamfe Road:** <u>Nemba 599</u> 7/1987; **Nyale:** <u>Etuge 4205</u> fr., 2/1998; **Nyasoso:** <u>Etuge 1800</u> fr., 3/1996; <u>2463</u> fr., 6/1996.
Note: the above two species of *Coelocaryon* have not been easy to separate on the basis of the Fl. Cameroun key, which includes characters that do not hold up to scrutiny, such as numbers of lateral nerve pairs. Further material is needed both to clarify the distinction, and to assess whether there really are two species in our area.
Uses: MATERIALS – wood (Etuge 1800).

Pycnanthus angolensis (Welw.) Warb.
Syn.: *Pycnanthus microcephalus* (Benth.) Warb.
Forest tree to 40m; specimens drying reddish brown; bole clear for 10–15(–20)m, to 5m diameter; crown of horizontal, whorled branches; red exudate when cut; leaves with 15–60 pairs of lateral nerves; lamina often with conspicuous insect damage; indumentum of rusty-brown, branched hairs; inflorescences axillary; paniculate; male flower heads c. 5mm across; fruits oblong, 3–4cm, dehiscent; seed solitary, covered by a bright red laciniate aril. Forest; 250–1000m.
Distr.: Guinea (C) to Angola, Uganda. (Guineo-Congolian).
IUCN: LC
Mile 15, Kumba-Mamfe Road: <u>Nemba 198</u> 9/1986; **Ndum:** <u>Elad 99</u> fr., 2/1995; **Nyasoso:** <u>Cable 3448</u> fr., 7/1996; <u>Etuge 2364</u> fl., 6/1996; <u>Lane 40</u> fl., 8/1992.
Local name: Ngozame. **Uses:** MATERIALS – timber (*fide* Etuge).

MYRSINACEAE

Y.B. Harvey, *Maesa* by T.M.A. Utteridge & Y.B. Harvey

Note: descriptions are taken in part from FWTA where material was not available.

Ardisia cf. buesgenii (Gilg & Schellenb.) Taton
Bull. Jard. Bot. Nat. Belg. 49: 94 (1979).
Shrub to 0.75m; leaves with tiny gland-dots; flowers unknown; fruits in axillary clusters, pointing down, immature, light green with tiny red dots. Forest; 1275m.
Nyasoso: Wheatley 415 fr., 7/1992.
Note: description from label notes, specimen not seen.

Ardisia dolichocalyx Taton
Bull. Jard. Bot. Nat. Belg. 49: 98 (1979).
Monopodial shrub, 1–2.5m; glabrous throughout, with the exception of very young growth; leaves elliptic to obovate, 14–22 × 6–10cm, with orange pellucid gland-dots and streaks, margin subentire; petiole 15–25mm, occasionally sparsely-hairy; flowers in axillary fascicles, 4–6 flowers per fascicle; peduncles to 5mm; pedicels 11–12mm, extending to 22mm in fruit; calyx c. 2.5mm wide; flowers to 6mm, red with glandular streaks; fruit globose, 8–10mm diameter, red with purplish gland-dots. Forest; 800–1200m.
Distr.: Cameroon Endemic.
IUCN: NT
Bangem: Thomas 5278 1/1986; **Kupe Village:** Cable 2485 5/1996; 752 fr., 1/1995; Cheek 7166 fr., 1/1995; 7652 11/1995; Ryan 245 5/1996; **Ngomboaku:** Ghogue 436 12/1999; **Nyasoso:** Cable 1136 fr., 2/1995; Cheek 7543 10/1995; Etuge 2083 6/1996; 2440 6/1996; Sebsebe 5015 10/1995.
Note: known from c. 13 localities, mainly in S & SW Province, Cameroon.

Ardisia sp. cf. etindensis Taton
Bull. Jard. Bot. Nat. Belg. 49: 100 (1979).
Shrub to 0.5m, glabrous throughout, with the exception of very young growth; leaves elliptic to obovate, 8.5–15 × 3.5–4.5cm, brown gland-dots, margin subentire; petiole 5–7mm; flowers in axillary fascicles, c. 2 flowers per fascicle; peduncles to 2mm; flowers not seen; fruits globose, 7mm diameter, pinkish with gland-dots. Riverine forest; 1368m.
Distr.: Mt Kupe. (Narrow Endemic).
Kodmin: Biye 41 1/1998.
Note: the reference refers to the place of publication of *A. etindensis* itself.

Ardisia kivuensis Taton
Bull. Jard. Bot. Nat. Belg. 49: 102 (1979).
Shrub, 0.3–2m; leaves ovate-elliptic or oblong, 13–18cm; petioles pink; flowers purple or reddish, in small clusters along branches; fruits bright red. Forest; 1000m.
Distr.: Cameroon, Congo (K) to Uganda. (Lower Guinea & Congolian).
IUCN: LC
Bangem: Thomas 5892 3/1986.

Ardisia koupensis Taton
Bull. Jard. Bot. Nat. Belg. 49: 104 (1979).
Shrub or treelet 1–3(–12)m, glabrous throughout; leaves elliptic (occasionally ovate), 11–18 × 3.5–8cm, acuminate, cuneate, entire or very shallowly crenate, with pellucid gland-dots and dashes; petioles 8–15mm; inflorescence axillary, 1–4 flowers per fascicle; peduncle 3–5mm; pedicels

15–25mm (extending to 35mm in fruit); calyx c. 3.5mm wide; flowers to 5mm, sparsely covered with glandular dots, pink; fruits ellipsoid with apiculate apex, 11–18 × 7–11mm, longitudinally striate. Lowland forest, secondary forest; 650–1250m.
Distr.: Mt Kupe & Mt Nlonako. (Narrow Endemic).
IUCN: EN
Kupe Village: Biye 61 11/1999; Cable 2683 5/1996; 2734 5/1996; 3701 7/1996; 3765 7/1996; Cheek 6069 fr., 1/1995; 6070 fr., 1/1995; 7006 fr., 1/1995; Etuge 1378 11/1995; 1837 3/1996; 2887 7/1996; Lane 255 fr., 11/1994; 414 fr., 1/1995; Ryan 239 5/1996; **Mbule:** Letouzey 14669 fl., 4/1976; **Nyasoso:** Cable 3623 7/1996; 3624 7/1996; Cheek 9506 10/1998; Gosline 132 11/1998; Sebsebe 5051 10/1995.
Local name: Mesip mehe. **Uses:** MATERIALS/SOCIAL USES – branches waved above face of corpse to drive flies away (Epie Ngome in Cheek 9506).

Ardisia mayumbensis (R.Good) Taton
Bull. Jard. Bot. Nat. Belg. 49: 108 (1979).
Monopodial shrub 0.5–1(–2)m tall; glabrous except for pubescent young growth; leaves obovate, 10–26 × 8–11.5cm, with pellucid gland-dots, margin subentire to very shallowly crenate, base cuneate-attenuate; petiole 10–15mm; flowers in axillary fascicles, 4–12 flowers per fascicle; peduncle 4–6mm long; pedicels 6–7(–13mm in fruit), red, fleshy; calyx c. 2.5mm wide; flower to 6mm long, red with red gland-dots; fruits globose, 6–8mm diameter, red with darker gland-dots. Lowland, submontane and transitional forest; 800–1350m.
Distr.: Cameroon, Gabon, Congo (K) & Angola. (Guineo-Congolian).
IUCN: NT
Bangem: Thomas 5279 1/1986; **Kodmin:** Etuge 4070 1/1998; Gosline 51 1/1998; **Kupe Village:** Cheek 7082 fr., 1/1995; **Mwambong:** Onana 538 2/1998; **Ngomboaku:** Etuge 4665 12/1999; **Nyasoso:** Cable 2747 6/1996; 3475 7/1996; 685 fr., 1/1995; Etuge 1309 10/1995; Wheatley 476 fl., 7/1992.
Note: rare throughout its range; known from only 13 locations; it appears most abundant in the Kupe-Bakossi region.

Ardisia staudtii Gilg
Bull. Jard. Bot. Nat. Belg. 49: 112 (1979).
Syn.: *Afrardisia staudtii* (Gilg) Mez
Syn.: *Afrardisia cymosa* (Baker) Mez
Shrub (0.5–)1.5–4(–5)m tall, glabrous; leaves elliptic to ovate, 90–180 × 30–70mm, glandular dots present on lower surface, very shallowly crenate; petioles 7–15mm; flowers in axillary fascicles; peduncles 2–5mm; 6–12 flowers per fascicle; pedicels 6–10mm; calyx c. 2.5mm wide, fimbriate margin; flowers white or pink, to 4mm with glandular spots/streaks; fruits globose, 3–6.5mm, red with red gland-dots. Lowland and submontane forest; 950–2000m.
Distr.: Nigeria to Congo (K) & CAR. (Guineo-Congolian (montane)).
IUCN: LC
Kodmin: Etuge 4012 1/1998; Onana 584 2/1998; **Kupe Village:** Cheek 7203 1/1995; 7784 11/1995; 7796 11/1995; 7879 11/1995; Ryan 406 5/1996; **Mt Kupe:** Sebsebe 5077 11/1995; **Mwanenguba:** Thomas 3120 2/1984; **Ngomboaku:** Cheek 10292 12/1999; **Nyasoso:** Cable 1199 fl., fr., 2/1995; 3403 6/1996; 97 fr., 9/1992; Etuge 1733 2/1996; Lane 525 fr., 2/1995; Sunderland 1518 fl., 7/1992.

Ardisia cf. staudtii Gilg
Shrub 1–2m, glabrous; leaves elliptic, 11–21.5 × 4–7.5cm, glandular dots present on the lower surface, margin entire;

petioles 10–12mm; flowers in axillary fascicles, 8–12 flowers per fascicle; peduncle to 10mm; pedicels 8mm; calyx c. 2mm wide; flowers to 3mm, with glandular spots; fruits globose, to 5mm. Submontane forest; 900–1200m.

Kupe Village: Cheek 8412 5/1996; Kenfack 289 7/1996; **Nyasoso:** Etuge 2095 6/1996; Lane 113 fl., 6/1994.

Ardisia sp. near staudtii Gilg

Similar to *A. staudtii* in all characters with the exception of: immature stems with rusty pubescence (not glabrous); leaves ovate and much smaller (7.5–11 × 3–4.5cm). Montane forest; 1950m.

Distr.: Mt Kupe. (Narrow Endemic).

Nyasoso: Cable 2904 6/1996.

Ardisia zenkeri Gilg

Bull. Jard. Bot. Nat. Belg. 49: 116 (1979).

Syn.: *Afrardisia zenkeri* (Gilg) Mez

Shrub 0.75–1.5m, young growth pubescent; leaves elliptic, 9–17 × 2.2–4.5cm, with pellucid gland-dots, margin subentire; petiole 14–16mm; flowers in axillary fascicles, 2–6 flowers per fascicle; peduncles 2–3mm; pedicels 7–9mm; calyx to 3mm wide; flowers not seen, red; fruits globose, 4–8mm, red with deeper red gland-dots. Forest; 1030–1200m.

Distr.: Cameroon Endemic.

IUCN: NT

Kodmin: Etuge 4076 1/1998; **Kupe Village:** Ryan 278 5/1996; **Nyasoso:** Cable 1152 fr., 2/1995.

Note: known from c. 11 localities, most common in S Province around Kribi and Bipindi.

Embelia guineensis Baker

Syn.: *Embelia sp. A sensu* Hepper in FWTA 2: 32

Woody climber to 8–9m, climbing by spur-shoots; stems and older branches brown, slightly fissured and with lenticels; leaves obovate, 9–14.5 × 4–7cm, rounded with acuminate tip, glabrous with black gland-dots on lower surface, cuneate; petiole 10–15mm; raceme 1.5–5cm, on the branchlets; pedicels (fruiting) 4–6mm; flowers not seen; fruits red/pink, spherical, 6–10mm; style persistent; 1-seeded. Grassland, primary forest; 1400–1450m.

Distr.: Mali to Congo (K) & Angola. (Guineo-Congolian).

IUCN: LC

Kodmin: Biye 21 1/1998; Cheek 9058 1/1998; 9601 11/1998; Etuge 4423 11/1998.

Uses: use unknown (Max Ebong in Cheek 9601).

Embelia mildbraedii Gilg & Schellenb.

Bull. Jard. Bot. Nat. Belg. 50: 203 (1980).

Syn.: *Embelia sp. nr. welwitschii* (Hiern) K.Schum.

Scandent shrub climbing to c. 3m; stems woody, brown; immature branches densely covered with chestnut-coloured indumentum; bark fissured and with lenticels; leaves elliptic, 6–9 × 3–4.5cm, apex acuminate, base attenuate, glabrous when mature, with gland-dots, laxly pubescent when immature, margins loosely serrate; petiole c. 10–12mm; inflorescence a raceme 30–90mm, from leaf axils on younger branches; flowers clustered towards apex; pedicels 4–6mm; calyx lobes c. 0.6mm; corollas whitish turning pink, lobes 5, to 2mm; fruits red, spherical, 5–6mm, with longitudinal striations when dry, 1-seeded. Montane grassland; 1900m.

Distr.: SW Cameroon. (W Cameroon Uplands).

IUCN: NT

Mt Kupe: Cheek 7580 10/1995.

Note: although rare, this taxon's habitat remains largely unthreatened and it is widespread in the higher altitudes of Mt Cameroon. Listed as LR/nt in www.redlist.org.

Embelia schimperi Vatke

Woody climber straggling over shrubs to about 2m; stems woody, brown, immature growth paler; bark fissured, lenticels pale; leaves elliptic (to obovate), 4.5–10.5 × 2–4.7cm, coriaceous, obtuse, with tiny gland-dots, otherwise glabrous, entire, slightly revolute, lower surface paler than upper (almost silvery); petioles 1–2cm; raceme 2–5.5cm, on leafless part of previous year's branchlet; calyx 1.5–3.5mm in diameter; pedicels c. 3mm, hairs chestnut; flowers whitish, 4–5-lobed, 5–6.5mm diameter; fruits red, globose, 5–8mm., 1-seeded. Forest-grassland transition; 1800m.

Distr.: Cameroon to Ethiopia. (Afromontane).

IUCN: LC

Mwanenguba: Etuge 4391 10/1998.

Maesa kamerunensis Mez

Liana, 2m; stem dark brown with short hairs, and slightly paler lenticels; leaves ovate, 6–14 × 4–8cm, base rounded to truncate, glabrous; petiole to 25mm, laxly pubescent; inflorescence many-branched, 12–25cm, densely covered with minute hairs; flowers white with brown-purple streaks in middle of lobes, lobes to 1.5mm; pedicels 1.5–2mm; fruits globose, 3.5mm, dark purple-brown. Forest; 420–1050m.

Distr.: Bioko, Cameroon, São Tomé, Gabon & Congo (B). (Lower Guinea).

IUCN: LC

Bangem: Thomas 5913 3/1986; **Kupe Village:** Etuge 1895 5/1996; Kenfack 220 7/1996; 328 7/1996; **Mungo River F.R.:** Onana 976 12/1999; **Mwambong:** Etuge 4330 10/1998; **Ngusi:** Etuge 1566 1/1996.

Maesa lanceolata Forssk.

Tree or shrub 6–8m tall; stem glabrous, dark brown with paler lenticels; leaves elliptic, 9–16 × 3.5–6cm, serrulate, glabrous; petioles 2–2.5cm, glabrous; inflorescence many-branched, 5–7mm, profusely covered with minute hairs (< 0.5mm); flowers pale green, to 1.5mm, subsessile; fruits globose, 4–5mm; pedicels to 3.5mm. Forest-grassland transition, montane forest; 1200–2150m.

Distr.: Guinea (C) to Madagascar. (Afromontane & Madagascar).

IUCN: LC

Bangem: Manning 469 fr., 10/1986; **Mejelet-Ehumseh:** Manning 414 10/1986; **Mwanenguba:** Leeuwenberg 8670 fl., fr., 11/1971; Sanford 5548 fr., 11/1968; Thomas 3149 fl., fr., 2/1984.

Maesa rufescens A.DC.

Prodr. 8: 81 (1884).

Shrub or small tree, 3–10(–18)m; stem glabrous, dark brown with paler lenticels; leaves elliptic-ovate, 9–16 × 4–5cm, serrate, hairs on midrib (at base) of undersurface and occasionally on secondary veins; petioles 1–4cm, glabrous; inflorescence many-branched, 3–9cm, densely hairy (to 1mm); flowers white, to 2 × 2mm; fruits globose, 3.5–4mm; pedicels to 2mm. Forest, secondary forest; 1500–2050m.

Distr.: Guinea (C) to E & S Africa, Madagascar. (Afromontane & Madagascar).

IUCN: LC

Kodmin: Cheek 8862 1/1998; Ghogue 54 fl., 1/1998; **Mt Kupe:** Cheek 7601 11/1995; **Mwanenguba:** Cheek 7288 2/1995; **Nyasoso:** Cable 2911 6/1996; 3389 6/1996; Elad 128 2/1995.

Local name: Nguele. **Uses:** FUELS – fuelwood (Cheek 8862).

MYRTACEAE

M. Cheek

Eucalyptus sp.
Tree 10–20m; bole pink-brown; leaves narrowly lanceolate, slightly falcate, c. 12 × 2.5cm, apex long-attenuate, base unequally obtuse, midrib pale below, lateral nerves numerous, inconspicuous; inflorescences 5–7cm; partial-peduncles flattened; flowers white, c. 5mm diameter; calyx c. 6 × 3mm, including slightly narrower stipe. Cultivated.
Distr.: native to Australia.
Nyasoso: Pollard (sight record) 10/1998.
Note: many species of *Eucalyptus* have been introduced into Africa and elsewhere from E Malesia (few) and Australia. The description above is meant to be a general one to cover the various species that may occur in our area.
Uses: MATERIALS – wood for building; FUELS – fuelwood (*fide* Pollard).

Eugenia fernandopoana Engl. & Brehmer
Shrub or small tree 1.5–4m; young stems brown-puberulent; leaves elliptic-oblong, c. 12 × 5.5cm, including a 2cm acumen, obtuse, margin revolute; lateral nerves c. 8 pairs, marginal nerve well-developed; petiole 0.3cm; inflorescence axillary with 1–2 flowers per axil; pedicel 1–2mm; flowers c. 3mm diameter; petals white. Forest; 800m.
Distr.: Bioko, CAR, Cameroon, Rio Muni. (Lower Guinea).
IUCN: VU
Bangem: Thomas 5277 1/1986.
Note: specimen not seen; description made from other Cameroonian material.

Eugenia obanensis Baker f.
Shrub 2–4m; resembling *E. fernandopoana*; leaves c. 7 × 3.8cm, acumen 1cm; lateral nerves 5–7 pairs; flowers subsessile; fruit globose, c. 0.8cm. Forest; 1700m.
Distr.: Ghana to Congo (K). (Guineo-Congolian).
IUCN: LC
Bangem: Manning 417 10/1986; **Banyu:** Olorunfemi in FHI 30618 fr., 6/1951.
Note: specimens not seen.

Eugenia sp. 1
Shrub or tree 2–15m; stems glabrous, epidermis pale brown, glossy; leaves drying grey above, light green below, elliptic or ovate, c. 15 × 4.5cm, including acumen 1.5cm, acute, revolute; lateral nerves 3–5 pairs, marginal nerve bold, > 0.5cm from edge; petiole 0.8cm; inflorescences axillary, fasciculate, c. 3 per axil; pedicel 5mm, bracteoles ovate 1mm; flowers white, 0.8cm (buds red); fruit globose, 4cm, dark purple; pedicel 0.5–1cm. Forest; 800–2000m.
Distr.: Cameroon. (Cameroon Endemic).
Kodmin: Etuge 4034 fr., 1/1998; Gosline 77 1/1998; Onana 566 2/1998; **Kupe Village:** Cheek 7868 11/1995; Etuge 1459 fr., 11/1995; 2809 7/1996; Ryan 348 fr., 5/1996; **Mt Kupe:** Etuge 1361 11/1995; **Nyasoso:** Cable 2900 fl., 6/1996; 2918 6/1996; 3563 7/1996; 3632 fr., 7/1996; Elad 129 fr., 2/1995; Etuge 1663 1/1996; 1679 fl., 1/1996; 1806 3/1996; 2091 6/1996; 2101 fr., 6/1996.
Note: known only from Kupe-Bakossi, possibly *sp. nov.*

Eugenia sp. 2
Shrub 4.5m; resembling *Eugenia sp. 1*, but stems and lower surface of the midrib densely red-patent-puberulent; lateral nerves c. 8 pairs; flowers unknown; fruit red, 1.8cm,

puberulent; pedicel 0–2mm, densely red-puberulent. Forest; 920m.
Distr.: Cameroon. (Cameroon Endemic).
Nyasoso: Zapfack 661 6/1996.
Note: does not match any FWTA species. Flowering material needed.

Psidium guajava L.
Cultivated shrub or small tree, persisting in farmbush, c. 6m; bark pale brown, smooth, falling as plaques, green beneath; leaves elliptic, 7–15 × 3–5cm, rounded or acute, base rounded, densely pubescent below; lateral nerves 12–20 pairs, prominent below; flowers axillary, usually in small clusters, c. 3cm diameter, white; fruit pale yellow, globose or ellipsoid, c. 5–8 × 5cm, flesh gritty; seeds numerous, 3mm. Farmbush, farms; 850m.
Distr.: originally from S America, now widely cultivated. (Pantropical).
IUCN: LC
Enyandong: Jam 4 10/2001.
Local name: Guava (English). **Uses:** FOOD – infructescences (Jam 4).

Syzygium malaccense (L.) Merr. & Perry
Journ. Arn. Arb. 19: 215 (1938).
Cultivated fruit tree, 5–10m; leaves leathery, elliptic or lanceolate, c. 14 × 5cm, weakly acuminate, obtuse; lateral nerves c. 10 pairs; petiole c.5mm; inflorescence mostly on branches below leaves or trunk, 1–7cm, 8–25-flowered; flowers 2cm, red-pink, bracteoles absent; fruit obovoid, c. 7 × 4cm, truncate, green and red, white-fleshed; seed globose, 2cm. Cultivated in home gardens for fruits in March; 400–1400m.
Distr.: originally SE Asia, now widespread. (Pantropical).
IUCN: LC
Edib village: Pollard 248 11/1998; **Nyandong:** Cheek 11478 3/2003.
Local name: Apple (Cheek 11478). **Uses:** FOOD – edible (*fide* Darbyshire).

Syzygium staudtii (Engl.) Mildbr.
Tree 8–20m; bole white, usually with 2–3(–numerous) laterally flattened root buttresses, arising up to 60cm above ground; stems near apex red when young, 4-ridged, glabrous; leaves (fruiting stems) elliptic, 6–7 × 2–3.5cm, acute; secondary nerves numerous; petiole c. 1.2cm; juvenile leaves to 11 × 5cm, sometimes briefly-acuminate; inflorescence terminal, 10cm, 10–30-flowered; flowers white, 0.6cm; fruit subumbellate, obovoid, 1cm. Forest; 1400–2000m.
Distr.: Liberia to Cameroon. (Upper & Lower Guinea (montane)).
IUCN: NT
Kodmin: Cheek 9184 2/1998; 9621 11/1998; Plot B61 1/1998; **Mt Kupe:** Cheek 7594 11/1995; **Mwanenguba:** Thomas 3127 2/1984; 3144 fl., 2/1984; **Nyasoso:** Letouzey 14699 fl., 4/1976.
Note: unjustifiably reduced to a subspecies of *S. guineense* by White. Likely to rate as VU when taxon better delimited.
Local name: Kan-kan. **Uses:** MEDICINES (Max Ebong in Cheek 9621).

NYCTAGINACEAE

M. Cheek & I. Darbyshire

Boerhavia diffusa L.

Herb c. 1m, from perennial rootstock; stems glandular-hairy; leaves opposite, ovate, c. 5 × 3.5cm, obtuse, rounded; lateral nerves c. 4 pairs, margin brown-hairy; petiole 3.5cm; inflorescence diffuse, 30cm; flowers pink, 3–4, in sessile clusters, each c. 2mm; fruit dry, clavate, 4mm, with lines of sessile glands. Farmbush.
Distr.: pantropical.
IUCN: LC
Mt Kupe: Etuge 1782 fl., 3/1996.

Bougainvillea sp.

Sarmentose shrub, 4m high; stems puberulent when young; leaves broadly ovate, to 6.5 × 5cm, apex shortly acuminate, base truncate, margin subentire; petiole 1–1.5cm; inflorescences axillary, lax cymes with hooks at the point of articulation of secondary peduncles which persist when flowers fall off; primary peduncle to 4.5cm; 2–3 flowers enclosed within paired cordiform bracts, c. 3.5 × 3cm, orange to pink; corolla pale yellow. Cultivated in gardens; 950m.
Mwanenguba: Leeuwenberg 9975 6/1972.
Note: probably a cultivar of *B. spectabilis* Willd.
Uses: ENVIRONMENTAL USES – Ornamental (*fide* Pollard).

NYMPHAEACEAE

M. Cheek

Nymphaea lotus L.

Kew Bull. 44: 484 (1989).
Aquatic herb, rhizome bottom-rooting; petioles to c. 2m; leaves floating on surface of water, orbicular, c. 20cm diameter, margin sharply toothed, primary to tertiary nerves strongly produced below; flowers white, c. 8cm diameter, floating on surface. Lake edge; 1250m.
Distr.: SE Europe, Africa, India, Malaysia.
IUCN: LC
Edib: Etuge 4096 fl., 1/1998.

Nymphaea cf. maculata Schum. & Thonn.

Aquatic herb resembling *N. lotus*, but leaf-blade ovate-elliptic, 11 × 8cm, margin entire. Lake outlet; 1300m.
Distr.: Senegal to Cameroon. (Upper & Lower Guinea).
Edib: Etuge 4484 11/1998.
Note: flowers needed to confirm determination.

OCHNACEAE

I. Darbyshire & M.S.M. Sosef

Note: species in the genus *Campylospermum* (in our area) consistently have yellow petals and fruits consisting of 2–several usually black drupelets on an enlarged, fleshy, red receptacle (torus).

Campylospermum calanthum (Gilg) Farron

Bull. Jard. Bot. Nat. Belg. 35: 394 (1965).
Syn.: *Ouratea calantha* Gilg
Syn.: *Ouratea nigroviolacea* Gilg ex De Wild.

Shrub or tree to 6m; stipules caducous; leaves elliptic or obovate, 17–28 × 6.5–10.5cm, base acute, apex acuminate, acumen c. 1.5cm, margin subentire or denticulate towards apex; secondary nerves impressed on upper surface; panicles widely-branched; rachis c.25cm; branches to 18cm; flowers lax; pedicels c. 0.8cm, in clusters of 5–8, articulated, c. 3mm from the base; peduncle subtended by numerous lanceolate bracts, c. 1.5 × 0.1cm; calyx lobes ovate, 0.6 × 0.3cm in fruit, apex obtuse. Forest; 250m.
Distr.: S Nigeria to Congo (K). (Lower Guinea & Congolian).
IUCN: LC
Ikiliwindi: Etuge 501 3/1987.
Note: *C. flavum*, *C. sulcatum* and *C. vogelii* are known from Kumba and so are likely to be found in lowland Bakossi in the future.

Campylospermum elongatum (Oliv.) Tiegh.

Fl. Congo, Rw. & Bur. Ochnaceae: 35 (1967).
Syn.: *Ouratea elongata* (Oliv.) Engl. ex Gilg
Tree to 6m; stems sparingly-branched; stipules triangular, persistent; leaves crowded towards apex, oblanceolate, 44–50 × 9.5cm, subsessile, base subcordate or truncate, apex acute, margins serrulate, midrib raised below; flowers on pendulous racemes; peduncles flattened, 0.5cm wide, c. 25cm long; rachis c.18cm; flowers single or in clusters of 2–3; calyx lobes in fruit, 0.8 × 0.3cm, apex acute, torus fleshy, 0.8cm diameter. Forest.
Distr.: Nigeria to Congo (K). (Lower Guinea & Congolian).
IUCN: LC
Kurume: Thomas 5412 fl., fr., 1/1986.

Campylospermum laxiflorum (De Wild. & T.Durand) Tiegh.

Fl. Congo, Rw. & Bur. Ochnaceae: 45 (1967).
Syn.: *Ouratea macrobotrys* Gilg
Syn.: *Ouratea laxiflora* De Wild. & T.Durand
Shrub or tree 1.5–4m; stipules and stem-scales persistent, lanceolate, 3mm long; leaves elliptic, 12–17 × 3.5–6cm, base acute, apex acute to acuminate, margin shallowly serrulate; secondary veins raised below, ± impressed above; petiole 0.2–0.4cm; inflorescence a widely-branched lax panicle, with persistent lanceolate bracts clustered at the base of the peduncle; rachis to 20cm; branches to 13.5cm; pedicels 1.2cm, articulated at base, solitary or in clusters of 2–3; calyx lobes in fruit, 0.8 × 0.3cm, apex obtuse. Forest; (700–)1000–1550m.
Distr.: Nigeria to Congo (K). (Lower Guinea & Congolian).
IUCN: LC
Bangem: Thomas 5332 1/1986; 5917 3/1986; **Kodmin:** Cheek 10272 12/1999; 8876 1/1998; 9186 2/1998; 9594 11/1998; **Kupe Village:** Cable 735 1/1995; Cheek 7779 11/1995; **Mwanenguba:** Leeuwenberg 9548 fr., 4/1972; **Ngomboaku:** Cheek 10369 12/1999; **Nyale:** Cheek 9697 11/1998.
Note: specimens with solitary or paired pedicels on the inflorescence resemble the poorly known *Ouratea leroyana* (Tiegh.) Keay, which requires reclassification following further study (Lebrun & Stork: Trop. Af. Fl. Pl. 1 (2003)). All K material of this taxon viewed by Farron has been redetermined as *C. laxiflorum*, with which *O. leroyana* may be synonymous.
Local name: Asongeh. **Uses:** MATERIALS – used in setting traps; MEDICINES (Max Ebong in Cheek 9594).

Campylospermum sp. aff. laxiflorum (De Wild. & T.Durand) Tiegh.

Shrub or tree 2–5m; as *C. laxiflorum,* but pedicels in clusters of 3–7, articulated 2–4mm above the base, leaving prominent 'stumps' when flowers have fallen; secondary venation of the leaves less prominent, not impressed above; leaf apex longer acuminate. Forest; 400–1000(–1400)m.
Kodmin: Onana 582 2/1998; **Kupe Village:** Cable 2482 5/1996; 3890 7/1996; Etuge 2022 5/1996; 2700 7/1996; Kenfack 277 7/1996; Ryan 284 5/1996; **Mt Kupe:** Cheek 10156 11/1999.
Note: it is possible that this taxon represents a lower altitude form of *C. laxiflorum.* Onana 582, collected at 1400m, appears intermediate between this taxon and *C. laxiflorum,* having (3–)4 pedicels in each cluster, articulated to 2mm from the base.

Campylospermum letouzeyi Farron

Adansonia (sér. 2) 9: 117 (1969).
Tree or shrub to 4m; stipules and stem-scales ± persistent, triangular, c. 0.7cm; leaves elliptic or oblanceolate, 30–35 × 8.5–10cm, base rounded to subcordate, apex acute or shortly acuminate, margin serrulate; petiole 0.5–1cm; infloresence a sparingly-branched panicle, only 1 branch observed in each specimen of our material; peduncle to 16cm, numerous lanceolate bracts, 0.7 × 0.2cm, at the base; rachis 21cm; branches 7cm; flowers in clusters of 6–10, each subtended by an ovate bracteole, 0.8 × 0.2cm; fruits not observed. Forest; 250–300m.
Distr.: W Cameroon. (Cameroon Endemic).
IUCN: VU
Kurume: Thomas 5460 fl., 1/1986; **Mekom:** Thomas 5236 1/1986.

Campylospermum mannii (Oliv.) Tiegh.

Bot. Helv. 95: 68 (1985).
Syn.: *Ouratea mannii* (Oliv.) Engl.
Shrub or tree to 4.5m; leaves crowded towards apex, subsessile, oblanceolate, 30–36 × 8–11cm, base cordate, apex obtuse or shortly acuminate, margin serrulate; inflorescence a raceme; peduncle 7–11cm; rachis 4–9cm; flowers in dense clusters; pedicels to 0.8cm, several triangular bracts, 0.5 × 0.2cm, present at the base of the peduncle; calyx lobes in fruit triangular, 1.5 × 0.8cm, nervosa; torus 0.7cm diameter. Forest; 425–1200m.
Distr.: E Nigeria, Bioko & W Cameroon. (W Cameroon Uplands).
IUCN: NT
Kupe Village: Cable 2439 5/1996; Etuge 1833 3/1996; Ryan 308 5/1996; **Nyandong:** Darbyshire 119 3/2003.
Note: locally not uncommon in SW Cameroon, habitat threatened by lowland forest clearance.

Campylospermum monticolum (Gilg ex Gilg) Cheek

Cable & Cheek: The Plants of Mt Cameroon: a Conservation Checklist: 92 (1998), as *C. monticola.*
Syn.: *Ouratea monticola* Gilg ex Gilg
Tree 3–10m; stipules and stem-scales caducous, scars conspicuous, as wide as stem; leaves narrowly elliptic, 13–15 × 4–4.5cm, papery, base acute, apex short-acuminate, margin serrulate, midrib rounded and raised below; petiole c. 0.5cm; panicles shorter than leaves, c. 6.5cm, with few short branches; pedicels c. 1.2cm, in clusters of 2–3; calyx lobes ovate in fruit, 0.8 × 0.3cm. Forest; 1120m.
Distr.: Nigeria, Cameroon, CAR, Congo (K). (Lower Guinea).

IUCN: NT
Kupe Village: Ryan 232 5/1996.
Note: the leaves of our material are somewhat smaller and less conspicuously serrulate than those usually observed in this species. Species is rare over most of its range.

Campylospermum reticulatum (P.Beauv.) Farron var. *reticulatum*

Fl. Congo, Rw. & Bur. Ochnaceae: 51 (1967).
Syn.: *Ouratea reticulata* (P.Beauv.) Engl.
Shrub or tree 2–6m; stems pale; stipules and scales early-caducous, scars not conspicuous; leaves long-elliptic, 10–17.5 × 2.5–4.5cm, papery, base cuneate or attenuate, apex acuminate, margin serrulate, lamina pale green, midrib prominent yellow-green below; inflorescence a widely-branched terminal, or subterminal, lax panicle, 15–25cm long; branches to 15cm; bracts absent; pedicels 1cm, in clusters of 2–3, articulated at the base; calyx lobes in fruit ovate, 0.8 × 0.3cm, apex acute. Forest, forest margins; 870–1300m.
Distr.: Gambia to Congo (K) and Kenya. (Tropical Africa).
IUCN: LC
Kupe Village: Etuge 1704 2/1996; **Ndum:** Groves 20 fl., 1/1995; Lane 454 fl., 1/1995; **Nyasoso:** Cable 1130 fl., 2/1995; 2842 6/1996; 3239 6/1996; 3433 7/1996; 3577 7/1996; Etuge 1350 10/1995; 2378 6/1996; Lane 514 fl., 2/1995; Zapfack 673 6/1996.

Lophira alata Banks ex Gaertn.f.

Emergent tree 40–60m with long, narrow, crown; leaves flushing vivid-red, obovate-oblanceolate, 18–28 × 7.5–9cm, coriaceous, base cuneate, apex emarginate (immature) to acute (mature), midrib prominent, raised; secondary veins very numerous, parallel; inflorescence paniculate; pedicels not clustered, articulated to 1.5cm from base; corolla white, 2–3cm diameter; stamens numerous; anthers linear, c. 4mm long; fruit an ellipsoid nut with persistent style; sepal lobes expand irregularly to form asymmetric wings, the longest c. 11 × 2cm. Forest; 450m.
Distr.: Guinea (C) to Congo (K). (Guineo-Congolian).
IUCN: VU
Nyandong: Cheek 11355 3/2003.

Ochna membranacea Oliv.

Shrub or small tree; bark red-brown; paired stipules linear, to 1.1cm, fimbriate towards base, straw-coloured; leaves elliptic-obovate, c. 10.5 × 3cm, base cuneate, apex acuminate, margins serrulate; midrib and secondary nerves prominent yellow-brown below in mature leaves; petiole 1–3mm, red-brown; inflorescence not observed in our material. Forest; 1000m.
Distr.: Gambia to Congo (K). (Guineo-Congolian).
IUCN: LC
Nyasoso: Etuge 1759 3/1996.

Rhabdophyllum affine (Hook.f.) Tiegh. subsp. *nov.*

Shrub or tree, 1.5–3m; branches with persistent scales numerous towards apices, triangular, forked, 1.5–4mm long; leaves oblong-oblanceolate, 17–22 × 5.5–7.3cm, thin, base acute to cuneate, apex conspicuously acuminate, acumen 1–2.2cm, margin subserrulate; secondary veins very numerous, parallel; infloresence a terminal or subterminal raceme, 10.3–16.8cm, subtended by numerous, triangular, forked bracts; pedicels in clusters of 2–4(–6), or solitary towards apex, articulated at base or on short stumps to 2mm; corolla white to pale yellow; calyx lobes in fruit ovate, c. 6mm long, apex acute; drupelets red. Forest; 700–1000m.

OLACACEAE

Distr.: Kupe-Bakossi. (Narrow Endemic).
Enyandong: Pollard 726 10/2001; **Kupe Village:** Cable 3889 7/1996; 3897 7/1996; Elad 68 fr., 1/1995; Gosline 248 12/1999; **Manehas F.R.:** Etuge 4347 10/1998; **Manyemen:** Letouzey 13765 fl., 6/1975; **Nyasoso:** Lane 97 fr., 6/1994.
Note: this subspecies is most closely related to subsp. *myrioneurum*, also recorded in Cameroon, differing in the larger leaves (*myrioneurum*: 9–18.7 × 3.4–5.8cm) with a more pronounced acumen, the higher density of stem-scales and the pedicels being articulated, 0–2mm, (not (2–)3–11mm), above the base. The taxonomy of this species complex requires further investigation.

Rhabdophyllum calophyllum (Hook.f.) Tiegh.
Bull. Jard. Bot. Nat. Belg. 35: 392 (1965).
Syn.: *Ouratea calophylla* (Hook.f.) Engl.
Shrub or tree 4–15m; stems with numerous triangular scales towards apices, c. 1.5mm long, ± persistent, scars prominent; leaves obovate c. 15 × 4.5cm, coriaceous, base acute, apex shortly-acuminate, margin entire; secondary nerves very numerous, parallel; petiole c.0.5cm; inflorescence a raceme or sparingly-branched panicle, to 5.5cm; peduncle c. 1.5mm wide; pedicels in clusters of 3, articulated at the base; sepal lobes ovate, enlarged and curved and c. 1cm long in fruit; fruits pendulous, drupelets red. Forest; 200–300m.
Distr.: Guinea (C) to Congo (B). (Upper & Lower Guinea).
IUCN: LC
Baduma: Nemba 116 fl., fr., 6/1986; **Ikiliwindi:** Nemba 549 6/1987.

OLACACEAE

G. Gosline

Fl. Cameroun 15 (1973).

Diogoa zenkeri (Engl.) Exell & Mendonça
Tree to 20m; oblong-lanceolate, 15–40 × 4–12cm; 5–8 pairs lateral nerves; inflorescence a short congested axillary raceme; flowers 5-merous; fruits subglobose; partially imbedded in the accrescent receptacle, which makes a ridge around the fruit, 4–5cm in diameter. Forest; 150–720m.
Distr.: Nigeria to Congo (K). (Lower Guinea & Congolian).
IUCN: LC
Ebonji: Olorunfemi in FHI 30594 fr., 5/1951; **Etam:** Etuge 182 7/1986; **Ikiliwindi:** Nemba 454 fl., 1/1987; 563 6/1987; **Kupe Village:** Cheek 7729 11/1995; **Mungo River F.R.:** Cheek 9340 10/1998.
Local name: M'pun. **Uses:** MATERIALS – wood (Cheek 7729).

Heisteria parvifolia Sm.
Tree or shrub to 10m; branchlets narrowly winged; leaves oblong, 8–12 × 3–6cm; lateral nerves c. 6 pairs; inflorescence axillary with 2–3 very small flowers; fruits with enlarged star-shaped calyx, bright-red, 3–5cm diameter; fruit ellipsoid, single-seeded. Forest; 250–1450m.
Distr.: Senegal to Uganda. (Guineo-Congolian).
IUCN: LC
Kodmin: Cheek 9059 fl., 1/1998; **Kurume:** Thomas 5421 1/1986; 5464 1/1986; **Lake Edib:** Cheek 9142 fl., 2/1998; **Mekom:** Thomas 5389 1/1986; 5390 1/1986; **Mile 15, Kumba-Mamfe Road:** Nemba 219 9/1986; **Mungo River F.R.:** Cheek 9343 10/1998; **Nlog:** Etuge 9 4/1986.

Octoknema affinis Pierre
Fl. Cameroun 15: 187 (1973).

Syn.: *Octoknema winkleri* Engl.
Tree to 20m, dioecious; branchlets, petioles, flowers and fruits covered with tiny, white, stellate hairs; leaves elliptic, 12–34 × 5–12cm; flowers in axillary spike-like racemes, pentamerous; fruit ovoid, orange, 1.5–3cm long, not winged. Forest; 200–1100m.
Distr.: Nigeria, Cameroon & Gabon. (Lower Guinea).
IUCN: LC
Bakolle Bakossi: Etuge 144 5/1986; **Bakossi F.R.:** Cheek 9316 10/1998; **Ikiliwindi:** Nemba 553 6/1987; **Kupe Village:** Cable 3770 7/1996; Kenfack 288 fl., 7/1996; **Mekom:** Nemba 22 fr., 4/1986; **Nyandong:** Cheek 11351 3/2003.
Local name: Mbit. **Uses:** unknown (Max Ebong in Cheek 9316).

Octoknema sp. nov.
Small tree 4–8m, resembling *O. affinis*, but indumentum scurfy-brown stellate-hairy; leaves smaller, to 10–19cm long; fruits ellipsoid, strongly-winged. Forest; 800–1400m.
Distr.: Bakossi Mts. (Narrow Endemic).
Bangem: Thomas 5359 fl., 1/1986; **Kodmin:** Cheek 9093 2/1998; Etuge 4152 2/1998; **Muahunzum:** Etuge 4440 11/1998; **Ngomboaku:** Cheek 10289 fl., 12/1999; 10321 fl., 12/1999.
Note: known only from Bakossi. This genus is currently being revised by George Gosline and Valéry Malécot (INH).

Olax gambecola Baill.
Shrub to 3m; branches 4-winged, covered with whitish dots; leaves ovate-elliptic to lanceolate, 7–15 × 3–6cm; petiole to 2mm; inflorescence an axillary raceme, 1–2.5cm long; flowers 2.5–3mm long, white; stamens 3; staminodes 5–6; fruits red; calyx a disc at base, not accrescent and enveloping fruit. Forest, espcially by streams; 300–870m.
Distr.: Guinea (C) to Uganda. (Guineo-Congolian).
IUCN: LC
Bolo: Nemba 48 3/1986; Thomas 5901 3/1986; **Kupe Village:** Zapfack 928 fr., 7/1996; **Nyasoso:** Cable 3259 fl., fr., 6/1996; Etuge 1591 fr., 1/1996; 1632 fr., 1/1996.

Olax latifolia Engl.
Shrub to 3m; branches 4-sided; leaves variable, elliptic to oblong, 7–22 × 3–10cm; 5–12 pairs of lateral nerves; petiole 0–4mm long; flowers pink, c. 4mm long; stamens 6; staminodes 3, in axillary racemes; fruits globose, 2cm diameter, completely enveloped by accrescent calyx. Forest; 200–1200m.
Distr.: Cameroon to Congo (K). (Lower Guinea & Congolian).
IUCN: LC
Bakolle Bakossi: Etuge 161 5/1986; **Bakossi F.R.:** Etuge 4294 fl., 10/1998; **Bolo:** Nemba 47 3/1986; **Enyandong:** Gosline 317 10/2001; **Kupe Village:** Mackinder 279 fl., 12/1999; **Mungo River F.R.:** Onana 957 fl., 11/1999; Pollard 106 fl., 10/1998; **Ngomboaku:** Cheek 10295 fl., 12/1999; 10331 fr., 12/1999; Mackinder 316 fl., 12/1999; **Nyale:** Cheek 9653 fl., 11/1998.
Uses: none known (Epie Ngome in Pollard 106).

Olax mannii Oliv.
Shrub to 3m; branches 4-winged; leaves lanceolate to elliptic, 6–15 × 2–7cm; petiole 0–4mm; 5–6 pairs of lateral nerves, brownish when dried; flowers greenish white, 6–7mm long; stamens 3; staminodes 4–5; fruit orange, globose 2–3cm diameter; calyx accrescent, enveloping fruit, drying light brown. Forest; 150–850m.
Distr.: Sierra Leone to Congo (K). (Guineo-Congolian).

IUCN: LC
Bolo: Nemba 44 3/1986; **Etam:** Etuge 208 7/1986;
Ngomboaku: Cheek 10333 fl., 12/1999.

Ongokea gore (Hua) Pierre

Tree to 40m; branchlets slightly 2-winged; leaves oblong-
elliptic, 4–7 × 2.5–3.5cm; 6–10 pairs of lateral nerves;
flowers in paniculate clusters, 4-merous; anthers united in a
tube; fruit globose, 3cm diameter, enveloped by an
accrescent calyx which dehisces in 2–3 valves. Forest; 250m.
Distr.: Sierra Leone to Congo (K). (Guineo-Congolian).
IUCN: LC
Ikiliwindi: Etuge 507 3/1987; **Kupe Village:** Cheek 7746
11/1995.

Ptychopetalum petiolatum Oliv.

Tree or shrub to 10m; twigs with 2 wings; leaves lanceolate,
oblong, elliptic or oblanceolate, 7–19 × 3–7cm; lateral nerves
4–8(–10) pairs, ascendant, anastomosing 5mm or more from
edge of leaf; inflorescence few-flowered, axillary; rachis 5–
12mm; bud 4.5 × 1.5–2mm, flower 5-merous; pedicel 2–
3mm; calyx absent; petals c. 7mm long, reflexed at top; fruit
a red, ellipsoid drupe, 17–21 × 12–14mm, often with 6
longitudinal grooves; a single seed. Forest; 200m.
Distr.: SE Nigeria, Cameroon, Rio Muni & Gabon. (Lower
Guinea).
IUCN: LC
Ikiliwindi: Nemba 542 fr., 6/1987.
Note: fairly common in lowland Cameroon, less so in the rest
of its range.

Strombosia grandifolia Hook.f. ex Benth.

Tree to 13m; branchlets terete; leaves ovate-elliptic to
oblong, 7–30 × 3–16cm; 4–8 pairs of lateral nerves, tertiary
venation finely parallel, with less than 2mm between veins;
flowers with sepals distinct; petals 1–3mm long; fruits ovoid
1.7 × 1.2cm. Forest; 300–650m.
Distr.: Benin to Congo (K). (Guineo-Congolian).
IUCN: LC
Bakossi F.R.: Cheek 9318 10/1998; **Bolo:** Thomas 5900 fl.,
3/1986; **Kupe Village:** Lane 406 fl., 1/1995; 410 fl., 1/1995.
Local names: M'pun (Epie Ngome in Cheek 9318); Npon
(Cheek 11277). **Uses:** MATERIALS – trapping (Epie Ngome
in Cheek 9318); timber (*fide* Etuge).

Strombosia pustulata Oliv.

Syn.: *Strombosia glaucescens* Engl.
Tree to 20m; branchlets subterete; leaves oblong to elliptic-
oblong, 4–10 × 2–5cm; 4–6 pairs of main lateral nerves,
underside of leaf with minute pustules; petiole 1–1.5cm;
flowers in axillary clusters; pedicels with small bracteoles, 5-
merous; petals 4–5mm long. Forest; 200m.
Distr.: Senegal to Congo (K). (Guineo-Congolian).
IUCN: LC
Ikiliwindi: Nemba 463 1/1987; **Kupe Village:** Elad 83
1/1995.

Strombosia scheffleri Engl.

Tree to 35m; branchlets strongly-angled; leaves ovate-elliptic
or oblong, 6–20 × 3–13cm; 5–8 pairs of main lateral nerves,
venation distinct; petioles 1–3cm long; flowers greenish
yellow or white; petals 3–5mm; fruits obconical c. 2cm.
Forest; 700–1200m.
Distr.: SE Nigeria to Uganda & E Africa. (Tropical Africa).
IUCN: LC
Kupe Village: Elad 82 1/1995; Etuge 1949 fl., 5/1996;
Ngomboaku: Cheek 10351 fl., fr., 12/1999; **Nyasoso:** Cable
3471 fr., 7/1996; 3595 7/1996; 650 12/1993; Etuge 1528 fl.,

1/1996; 1764 fr., 3/1996; Lane 182 fl., 11/1994; Thomas
3044 fr., 2/1984.

Strombosia zenkeri Engl.

Tree to 15m; branchlets slightly ridged; leaves narrowly
elliptic, 5–15 × 1.5–5.5cm; 5–7 pairs of main lateral nerves,
tertiary venation obscure; petioles 1.5–2cm; buds ovoid;
flowers with sepals distinct, ciliate; petals 1.5–2.5mm long;
fruits ellipsoid, 2–2.5 × 1.2–1.7cm. Forest; 660–1650m.
Distr.: Nigeria (Calabar) to Gabon. (Lower Guinea).
IUCN: NT
Kodmin: Cheek 9065 fl., 1/1998; Gosline 173 fl., fr.,
11/1998; **Kupe Village:** Cable 3719 fr., 7/1996; Cheek 8342
fl., 5/1996; Etuge 1694 fr., 2/1996; 2665 fr., 7/1996.
Note: local within its range; known from c. 15 locations.

Strombosia sp. 1

Tree 6–20m tall; stems and leaves resembling *S. scheffleri*,
but leaves usually cordate; pedicels 2mm; flower buds black,
4mm; petals white, 5mm, inner surface sparsely and very
shortly puberulent; fruit 1cm, resembling *Diogoa*: turbinate,
with a calyx-derived wing around the equator. Forest; 850–
1000m.
Distr.: Nigeria (Chappal Waddi) & Cameroon (Kupe-
Bakossi, Bali-Ngemba F.R., Bamboutos Mts). (W Cameroon
Uplands).
Kupe Village: Cable 2662 fl., fr., 5/1996; **Ngomboaku:**
Etuge 4668 fl., 12/1999; Ghogue 479 fl., 12/1999; **Nyasoso:**
Gosline 138 11/1998;.
Note: probably a new and threatened species. More
investigation needed. The fruits are quite unlike any other in
the genus.

Strombosiopsis tetrandra Engl.

Tree 4–30m tall, with red exudate; leaves oblong to
lanceolate, 5–25 × 3–10cm; 6–10 pairs of lateral nerves,
tertiary nerves parallel; inflorescence a short, congested
raceme; rachis 2–5mm; flowers cream-coloured, 4-merous;
fruits ellipsoid, 2.5–3.5 × 2–3cm; calyx accrescent,
enveloping fruit, with 4 teeth at summit. Forest; 150–300m.
Distr.: SE Nigeria to Uganda. (Guineo-Congolian).
IUCN: LC
Baduma: Nemba 123 6/1986; **Etam:** Etuge 204 fr., 7/1986;
Kurume: Thomas 5413 fl., 1/1986.

OLEACEAE

P.S. Green & I. Darbyshire

Chionanthus africanus (Knobl.) Stearn

Bot. J. Linn. Soc. 80: 197 (1980).
Syn.: *Linociera africana* (Knobl.) Knobl.
Tree 6–12m; bark pale brown, lenticels concolorous; leaves
oblong-oblanceolate, 19.5–22.7 × 7–9.7cm, acumen abrupt,
0.6–1.4cm, base acute, glabrous, midrib raised and bronze
below; lateral nerves 11–12 pairs, domatia minute, tufted;
petiole 1.2–2.2cm, swollen, flaking; panicles axillary, c. 5 ×
1.5cm, appressed-pubescent; calyx lobes ovate, 1mm; corolla
lobes subulate, 0.5cm, free almost to base, white; stamens 2,
filaments short; fruit ellipsoid, 1.9 × 1.4cm, grey-brown.
Forest, forest edge; 1200–1550m.
Distr.: Sierra Leone to Tanzania. (Guineo-Congolian).
IUCN: LC
Edib: Cheek 9128 2/1998; Etuge 4106 1/1998; **Kodmin:**
Cheek 9190 2/1998.

Chionanthus mannii (Soler.) Stearn subsp. *congesta* (Baker) Stearn

Bot. J. Linn. Soc. 80: 200 (1980).

Syn.: *Linociera congesta* Baker

Shrub or small tree to 4m; leaves oblong-elliptic, 9.5–14.5 × 3.6–5.5cm, acumen abrupt, to 1.5cm, tip rounded, base obtuse, glabrous; lateral nerves 7–8 pairs, inconspicuous, domatia absent; petiole to 0.6cm, slightly swollen and flaking; flowers not seen; infructescence rachis 1.5(–2)cm, few-branched, glabrous; fruit ellipsoid, 1.4 × 1.1cm, green, drying reddish brown. Forest, forest edge; 1000–1400m.

Distr.: Cameroon to Congo (K). (Lower Guinea & Congolian).

IUCN: LC

Kodmin: Plot B138 1/1998; **Ndibise:** Etuge 517 fr., 6/1987.

Note: rare in Cameroon, Gabon and Equatorial Guinea but more numerous in Congo (K); habitat largely unthreatened.

Chionanthus mildbraedii (Gilg & Schellenb.) Stearn

Bot. J. Linn. Soc. 80: 202 (1980).

Shrub or tree to 8m; bark pale; leaves oblong-oblanceolate, 11–14.2 × 3.9–5.5cm, acumen abrupt 1–2.2cm, base obtuse; lateral nerves 6–7 pairs, inconspicuous, domatia minute, tufted in young leaves; petiole to 0.7cm, swollen and slightly flaking; inflorescence axillary, 3–6-flowered, lax; rachis to 6cm, thin, glabrous; pedicels 1.2cm, fine, single on opposite, secondary peduncles, 0–1.5cm; calyx cupular with spreading triangular lobes, glabrous; corolla lobes lanceolate, 0.45 × 0.25cm, free to near base, yellow to red-green; fruit not seen. Forest, forest edge; 1000–1300m.

Distr.: Cameroon, Congo (K), Ethiopia to Tanzania. (Tropical Africa).

IUCN: LC

Mwanenguba: Leeuwenberg 8414 fl., 9/1971; **Nyasoso:** Cable 3447 7/1996.

Jasminum bakeri Scott-Elliot

Climber, glabrous; leaves leathery, (1–)3-foliolate; central leaflets ovate-elliptic, 12.5–20 × 6–10cm, apex acuminate, base acute; lateral nerves brochidodromous, 4–5 pairs, domatia minute, tufted, lower surface ± punctate; lateral leaflets 9–11 × 4.5–5cm; petioles 3–6cm; petiolules 0.2–0.7cm. Forest; 1200m.

Distr.: Guinea (C), Nigeria to Burundi. (Guineo-Congolian).

IUCN: NT

Kupe Village: Cheek 7666 11/1995; 7848 11/1995.

Note: fertile material not seen. This taxon is uncommon over much of its range, known from less than 20 localities.

Jasminum dichotomum Vahl

Liana to 8m; stems glabrescent to minutely-puberulent; leaves elliptic, 7–9.5 × 3.5–4.7cm, apex acuminate-apiculate, base acute, glabrous; petiole c. 0.6cm, puberulent; inflorescence densely corymbose, terminal on lateral shoots; flowers numerous, fragrant; calyx 3mm long, lobes triangular; corolla tube 1.7cm, red, lobes c. 8, oblanceolate, 0.7cm long, spreading, white. Forest edge; 1200–1300m.

Distr.: Senegal to Mozambique. (Afromontane).

IUCN: LC

Mwanenguba: Leeuwenberg 8591 fl., 10/1971; Sanford 5496 fl., 11/1968.

Jasminum preussii Engl. & Knobl.

Woody climber to 5m; densely brown-pubescent on young stems, leaves, pedicels and calyx, hairs to 2mm; leaves papery, subopposite borne on lateral branches, ovate, 5.7–6.1

× 2.5–3.1cm, apex acimunate, base rounded or subcordate; petiole 0.2–1.5cm; flowers solitary, terminal on lateral branches; pedicel 1.5cm in fruit, thickened to 1.5mm at apex; calyx lobes linear, 0.6cm; corolla white; fruits globose, 0.9cm diameter, white, glabrous. Forest, farmbush; 870–1400m.

Distr.: Ghana to Cameroon. (Upper & Lower Guinea).

IUCN: NT

Enyandong: Tadjouteu 438 fr., 10/2001; **Kodmin:** Pollard 168 10/1998; **Menyum:** Doumenge 477 5/1987; 543 5/1987; **Mwanenguba:** Leeuwenberg 9574 fl., 4/1972; **Nyasoso:** Cable 3268 6/1996; Etuge 2359 6/1996.

Note: uncommon to scarce over much of its range, but fairly common in Kupe-Bakossi and on Mt Cameroon.

Olea capensis L. subsp. *macrocarpa* (C.H.Wright) I.Verd.

Kew Bull. 57: 108 (2002).

Tree to 20m, glabrous; bark with pale to concolorous lenticels; leaves coriaceous, elliptic, 9.5–11 × 3–3.5cm, apex acuminate, base attenuate, margin slightly revolute, lower surface minutely-punctate, midrib raised below, yellow to reddish; lateral nerves 6–7 pairs, inconspicuous; petiole to 2cm; cymose-panicles terminal, c. 8 × 7cm, branching opposite, patent; flowers numerous; pedicels 1–3mm; calyx cupular, lobes triangular, < 1mm; corolla lobes ovate, c. 3mm, spreading, white; stamens 2, spreading; anthers 1.5mm long, medifixed; fruit ellipsoid, c. 1.3 × 0.7cm, green, smooth: Forest; 1940–2000m.

Distr.: Guinea (C) to Somalia, S Africa, Madagascar & Comoros.

IUCN: LC

Kupe Village: Ryan 404 5/1996; **Nyasoso:** Cable 2906 6/1996; Etuge 1651 1/1996.

ONAGRACEAE

M. Cheek

Fl. Cameroun 5 (1966).

Epilobium salignum Hausskn.

Herb to 1.8 m, herbaceous or thinly woody; stems often branched above, appressed grey-puberulous; leaves lanceolate to oblong-linear, 2–8cm; flowers white or cream, later pink-tinged, 6–11.5mm; fruit 3cm, 4-valved; seeds long-hairy. Streamsides, swamps; 2160m.

Distr.: SE Nigeria to S Africa. (Afromontane).

IUCN: LC

Mwanenguba: Sanford 5521 fl., 11/1968.

Ludwigia abyssinica A.Rich.

Webbia 27: 496 (1972).

Syn.: *Jussiaea abyssinica* (A.Rich.) Dandy & Brenan

Shrubby herb 0.7–2m; stems terete, glabrous; leaves narrowly elliptic or (larger leaves) oblong-elliptic, 3–7.5 × 0.5–1.5cm, apex rounded, base cuneate-decurrent, larger leaves with 1–2 teeth; petiole to 0.5cm; flowers 1–2 per axil, sessile; inferior ovary 7mm; sepals 4, triangular, 1mm; petals 4, yellow-orange, subequal to sepals; fruit 1cm, pale brown; seeds in 1 row. Lake or river edges; 1000–1300m.

Distr.: Guinea (C) to S Africa. (Tropical & subtropical Africa).

IUCN: LC

Edib: Etuge 4103 1/1998; 4485 11/1998; **Lake Edib:** Cheek 9138 2/1998; **Mwambong:** Onana 551 2/1998.

Ludwigia africana (Brenan) Hara

J. Jap. Bot. 28: 291 (1953).

Syn.: *Jussiaea africana* Brenan

Herb 0.9m, erect; resembling *L. abyssinica*, but stems, leaves and inflorescence with thin, spreading, white hairs; leaves elliptic, to c. 6 × 3.5cm, acute; petiole c. 1.5cm; inflorescence with c. 5-flowered peduncles 1.5cm; ovary 10mm; sepals 5mm; fruit not seen. Riverbanks; 1000m.

Distr.: Guinea (C) to Congo (K). (Guineo-Congolian).

IUCN: LC

Mwambong: Onana 562 2/1998.

Note: matches closely Daramola in FHI 40473, determined in 1964 as *L. africana* by Peter Raven (MO) 'but very unusual in spreading hairs'. Perhaps worthy of varietal rank. This species is rare in Upper Guinea.

Ludwigia decurrens Walt.

Reinwardtia 6: 327 (1963).

Syn.: *Jussiaea decurrens* (Walt.) DC.

Herb 0.2–1.3m; resembling *L. abyssinica*, but stems 3-winged, glabrous; largest leaves lanceolate, to 14 × 2.5cm; sepals to 7mm; fruits 1.5cm, 4-winged. Roadside; 200–730m.

Distr.: S America, introduced to W Africa.

IUCN: LC

Bakolle Bakossi: Etuge 271 9/1986; **Bakossi F.R.:** Cheek 9305 10/1998; **Bazeng:** Sebsebe 5060 10/1995; **Mile 12 Mamfe Road:** Nemba 292 fr., 10/1986; **Mungo River:** Breteler 2574 2/1962.

Local name: Ken-Ken-eh. **Uses:** MEDICINES (Epie Ngome in Cheek 9305).

Ludwigia erecta (L.) Hara

Webbia 27: 488 (1972).

Syn.: *Jussiaea erecta* L.

Herb 0.3–0.6m; resembling L. abyssinica, but stems angular; leaves lanceolate, to 5.5 × 1cm, subsessile; flowers solitary, sessile; ovary 5mm; sepals 2mm; fruit 2cm. Roadside; 300–600m.

Distr.: tropical Africa & America. (Amphi-Atlantic).

IUCN: LC

Bakossi F.R.: Cheek 9306 10/1998; **Nyasoso:** Etuge 1648 1/1996.

Note: other species, especially the weedy *L. octovalvis* (puberulent stems), are likely to be found in our area.

Local name: Mun-den. **Uses:** MEDICINES (Epie Ngome in Cheek 9306).

OPILIACEAE

G. Gosline

Fl. Cameroun 15 (1973).

Urobotrya congolana (Baill.) Hiepko subsp. *congolana*

Bot. Jahrb. 107: 142 (1985).

Syn.: *Urobotrya minutiflora* Stapf

Syn.: *Opilia congolana* Baill.

Tree or shrub to 5m; branchlets striated; leaves yellow-green, glabrous, oblong to elliptic-lanceolate, 3.5–8 × 10–22cm, base attenuate, apex obtusely acuminate; petiole 5–10mm; inflorescence an axillary spike equal to or longer than the leaf; flowers in fascicles of 2–4; pedicel 5–11mm, 5-merous, yellow; petals reflexed, pubescent internally, c. 2mm long; anthers 6–8mm; fruit an ellipsoid drupe, yellow-red, 1.5–2 × 1.5cm, single seeded; pedicel 2cm. Forest; 150–1200m.

Distr.: SE Nigeria to Congo (B). (Lower Guinea & Congolian).

IUCN: LC

Ikiliwindi: Etuge 512 fr., 3/1987; **Kupe Village:** Cable 2677 fr., 5/1996; 3699 fr., 7/1996; 3811 fr., 7/1996; Cheek 7010 fl., 1/1995; Etuge 1836 fr., 3/1996; 2907 7/1996; Lane 397 fl., 1/1995; Ryan 231 5/1996; **Mungo River F.R.:** Cheek 10235 fl., 12/1999.

OXALIDACEAE

M. Cheek

Biophytum talbotii (Baker f.) Hutch. & Dalziel

Unbranched herb, gregarious, 10–15cm; leaves whorled, paripinnate, 7–10cm; leaflets 10–20 pairs, median leaflets obliquely rectangular, 8–10 × 3mm, acute, glabrous; flowers pink, c. 1cm. Shady riverbanks; 150–800m.

Distr.: Liberia, Nigeria to S Cameroon. (Upper & Lower Guinea).

IUCN: NT

Bangem: Thomas 5384 1/1986; **Mungo River:** Gosline 221 fl., fr., 11/1999; **Nyale:** Etuge 4201 fl., 2/1998; **Nyandong:** Cheek 11271 3/2003; Thomas 7012 fl., fr., 5/1987.

Note: despite its restricted range, being known from less than 20 localities, it is locally common e.g. along the Mungo River, Bakossi. Its habitat may be threatened in e.g. E Nigeria and parts of lowland Cameroon.

Local name: none (Cheek 11271).

Biophytum umbraculum Welw.

Brittonia 33: 451 (1981).

Syn.: *Biophytum petersianum* Klotzsch

Annual herb to c. 15cm, glabrous; leaves in terminal rosette, 1–3cm long, to 6 leaflets; leaflets ovate, c. 7mm; inflorescence 1–3-flowered; peduncle 1.5cm; flowers c. 0.5cm, orange, yellow in centre. Fields, grassland; 1250–1900m.

Distr.: palaeotropical.

IUCN: LC

Mwanenguba: Leeuwenberg 8460 fl., 9/1971; 8501 fl., 9/1971.

Oxalis barrelieri L.

Willdenowia 8: 17 (1977).

Erect-stemmed annual herb, c. 30cm; leaves subopposite, c. 9cm; leaflets ovate, 3–4 × 2–3cm, apex entire, central leaflet with petiolule 1cm, lateral leaflets sessile; inflorescences axillary, many-flowered, dichasial; peduncle 4cm; flowers 1cm, pink, centre yellow. Farms, farmbush; 200–930m.

Distr.: pantropical.

IUCN: LC

Bakolle Bakossi: Etuge 277 fl., fr., 9/1986; **Kupe Village:** Etuge 2763 fl., fr., 7/1996; Ryan 251 fl., fr., 5/1996; 334 fl., fr., 5/1996; **Loum F.R.:** Pollard 96 fl., fr., 10/1998; **Mungo River F.R.:** Cheek 10177 fl., fr., 11/1999; **Nyasoso:** Etuge 1630 fl., fr., 1/1996.

Note: not in FWTA.

Local name: Korkoremba. **Uses:** MEDICINES (Epie Ngome & George Ngole in Pollard 96).

Oxalis corniculata L.

Prostrate, rooting-stemmed herb; stems to c. 30cm, borne horizontally, densely white-hairy when young; leaflets as *O. corymbosa*, but 1cm long; inflorescence 1-flowered; flower 0.7cm; yellow. Grassland, forest edge; 650–1450m.

PANDACEAE

Distr.: cosmopolitan.
Distr.: cosmopolitan.
IUCN: LC
Kodmin: <u>Cheek 9034</u> fl., 1/1998; **Mombo:** <u>Olorunfemi in FHI 30571</u> fl., fr., 5/1951; **Nyasoso:** <u>Etuge 1599</u> fl., fr., 1/1996.
Uses: MEDICINES (Etuge 1599).

Oxalis corymbosa DC.
Perennial herb, with numerous basal corms, 15–30cm; aerial stem absent; leaves erect, trifoliolate; leaflets transversely elliptic c. 3 × 4cm, apical sinus 0.5cm, base obtuse, sessile; inflorescence equal to leaves, c. 10-flowered; flowers purple, 1.3cm. Farmbush; 800m.
Distr.: S America, introduced in tropical Africa.
IUCN: LC
Nyasoso: <u>Etuge 2483</u> fl., 7/1996.

PANDACEAE

M. Cheek

Fl. Cameroun 19 (1975).

Microdesmis haumaniana J.Léonard
Fl. Gabon 22: 26 (1973).
Tree 3m, resembling *M. puberula*, but stems more sparsely pubescent; fruits 3(–4)-locular (not 2-locular). Forest; 250m.
Distr.: Cameroon to Congo (K). (Lower Guinea & Congolian).
IUCN: NT
Baduma: <u>Nemba 234</u> fr., 9/1986.
Note: these specimens determined by Duncan Thomas.

Microdesmis cf. puberula Hook.f.
Shrub 2(–8)m, densely puberulent; apical bud clawed; leaves alternate, elliptic, c. 10 × 4cm; inflorescences axillary fascicles of c. 10 florets; flowers orange or pink, flat, c. 3mm diameter; fruit globose 4mm, orange, with tuberulate endocarp. Forest; 220–1050m.
Distr.: Nigeria to Uganda. (Lower Guinea & Congolian).
Baduma: <u>Nemba 195</u> 8/1986; **Bakolle Bakossi:** <u>Etuge 251</u> 9/1986; **Bakossi F.R.:** <u>Etuge 4285</u> 10/1998; **Etam:** <u>Etuge 211</u> 8/1986; **Konye:** <u>Thomas 5179</u> 11/1985; **Mungo River F.R.:** <u>Onana 938</u> 11/1999; **Nyasoso:** <u>Cable 3612</u> 7/1996.
Note: western Cameroonian material of this genus needs critical investigation.

Panda oleosa Pierre
Tree 15–30m; slash pink with black dots and lines; leafy stems c. 15cm, ridged, glabrous; leaves alternate, ovate-elliptic, c. 12 × 6cm, leathery, acute, serrate; lateral nerves c. 4 pairs; inflorescence spikes c. 30cm; fruit ovoid, c. 4 × 3cm; endocarp pointed, dehiscing by 3 window-like flaps. Forest; 450m.
Distr.: Liberia to Congo (K). (Guineo-Congolian).
IUCN: LC
Mungo River F.R.: <u>Cheek 10144</u> 11/1999; **Nyandong:** <u>Cheek 11335</u> 3/2003.
Note: *Centroplacus*, a related genus, can be expected in Mungo River F.R.
Local name: Pob. **Uses:** FOOD – seeds eaten raw by man (Cheek 11335).

PASSIFLORACEAE

M. Cheek

Adenia cissampeloides (Planch. ex Benth.) Harms
Climber, 12m, glabrous; leaves trilobed, broadly ovate, 5 × 5.5cm, lobes 1.5cm; pinnately nerved; petiole-gland single; peduncle 2.5cm, c. 8-flowered; flowers white, 1.5cm; fruit ellipsoid 1.5–2cm. Forest; 800m.
Distr.: Guinea (C) to Kenya. (Tropical Africa).
IUCN: LC
Nyasoso: <u>Etuge 2495</u> fl., 7/1996.
Note: the single specimen cited is atypical in the small, lobed leaves and large flowers.

Adenia cynanchifolia (Benth.) Harms
Climber, 4m; glabrous; leaves lanceolate, to 7.5 × 3.5cm, long-acuminate, obtuse to truncate, coarsely serrate; petiole-gland single, stipitate, capitate; peduncle 2cm, several-flowered; flowers white, 0.5cm. Forest; 1400m.
Distr.: Nigeria to Congo (K). (Lower Guinea & Congolian).
IUCN: LC
Kodmin: <u>Etuge 4403</u> 11/1998.

Adenia gracilis Harms subsp. *gracilis*
Climber closely resembling *A. cissampeloides*, but peduncle 0.5–1cm; fruit ellipsoid, 1cm, smooth. Forest edge; 210–1000m.
Distr.: Nigeria to Uganda. (Guineo-Congolian).
IUCN: LC
Kodmin: <u>Plot B302</u> 10/1998; **Kupe Village:** <u>Cheek 7149</u> fl., 1/1995; <u>Etuge 1892</u> fl., fr., 5/1996; <u>2735</u> fl., 7/1996; <u>Ryan 250</u> fl., fr., 5/1996; <u>333</u> fl., 5/1996; **Loum F.R.:** <u>Cheek 9303</u> fr., 10/1998; **Mungo River F.R.:** <u>Etuge 4313</u> fl., 10/1998.
Note: de Wilde notes that leaves of *A. gracilis* average 4cm wide, *A. cissampeloides* 7cm.

Adenia lobata (Jacq.) Engl.
Climber to 10m; older stems swollen, with lines of tubercules, glabrous; leaves drying black, ovate, c. 14 × 8cm, acuminate, cordate, palmately nerved; petiole-glands paired; fruiting pedicel 1cm; fruit globose, angled 2 × 2cm; stipe 2cm. Forest; 400–1100m.
Distr.: Senegal to Cameroon. (Upper & Lower Guinea).
IUCN: LC
Kupe Village: <u>Cable 2538</u> fr., 5/1996; <u>3849</u> fr., 7/1996; <u>Etuge 2028</u> fr., 5/1996.

Adenia mannii (Mast.) Engl.
Syn.: *Adenia oblongifolia* Harms
Climber to 8m, glabrous; stems neither swollen nor tuberculate; leaves drying black, oblong-elliptic, c. 14 × 4.5cm, acuminate, base rounded; c. 4 pairs lateral nerves; petiole glands paired; peduncle 10cm; inflorescence continuing as tendril; flowers light green, 1.5cm; fruit ellipsoid, 2cm. Forest edge; 150–1000m.
Distr.: Sierra Leone to Uganda. (Guineo-Congolian).
IUCN: LC
Kupe Village: <u>Cable 2444</u> fl., 5/1996; <u>2711</u> fl., 5/1996; <u>774</u> fl., 1/1995; <u>800</u> fr., 1/1995; <u>Etuge 2705</u> fr., 7/1996; <u>2739</u> fr., 7/1996; <u>Lane 431</u> fl., 1/1995; **Mungo River F.R.:** <u>Cheek 10141</u> fr., 11/1999; <u>10150</u> fl., 11/1999.

Adenia rumicifolia Engl. & Harms var. *miegei* (Aké Assi) W.J.de Wilde
Meded. Land. Wag. 71(18): 154 (1971).

Climber to 7m, resembling *A. lobata*; flowers to 2.5cm; fruits obovoid or pyriform, c. 4 × 2cm, excluding the c. 1.5cm stipe. Forest; 850–1920m.
Distr.: Mali to Kenya. (Guineo-Congolian).
IUCN: LC
Ebonji: Olorunfemi in FHI 30593 fr., 1/1951; **Kupe Village:** Etuge 2631 fr., 7/1996; **Mt Kupe:** Sebsebe 5104 fr., 11/1995; **Nyasoso:** Cable 3401 fl., 6/1996; Etuge 2531 fr., 7/1996.

Adenia sp. 1
Climber; leaves drying green, ovate in outline, to 9 × 5.5cm, 3–5-lobed, the lateral lobes smallest, to 1cm, apex acute, base obtuse to cordate, margin sinuous, 3–5 palmately-veined, the midrib with another major vein pair midway; blade glands as vein endings; petiole gland single, spatulate. Forest; 1400m.
Kodmin: Plot B9 10/1998.
Note: this sterile plot voucher was not named to species but is not referable to any of the other taxa enumerated.

Barteria fistulosa Mast.
Syn.: *Barteria nigritana* Mast. subsp. *fistulosa* (Mast.) Sleumer
Tree 9m, glabrous; lateral stems horizontal, dilated (c. 1cm), hollow, inhabited by large, fierce, black, biting-ants; leaves elliptic-obovate, simple to c. 45 × 19cm, shortly acuminate, unequally obtuse; petiole 1.5cm. Forest; 150–300m.
Distr.: Nigeria to Congo (K). (Lower Guinea & Congolian).
IUCN: LC
Bakole: Nemba 785 2/1988; **Etam:** Etuge 86 fl., 3/1986; **Kupe Village:** Etuge 2909 7/1996; **Kurume:** Thomas 8207 4/1988.

Barteria solida Breteler
Adansonia (sér. 4) 21: 316 (1999).
Tree to 15m, minutely puberulous; lateral stems solid, (c. 3mm diameter), ants absent; leaves obovate-elliptic, c. 9 × 4cm, acuminate, obtuse, margin glandular-serrate, slightly revolute, brochidodromous; petiole 5mm, short; flowers axillary, single, sessile sweetly-scented, 5cm diam, white; petals c. 5; stamens numerous; stigma 1cm diameter. Forest; 950–1200m.
Distr.: SE Nigeria to Congo (K). (Lower Guinea).
IUCN: NT
Kupe Village: Etuge 2703 fl., 7/1996; **Mwanenguba:** Leeuwenberg 9544 fl., 4/1972.
Note: known from only 13 sites.

Efulensia clematoides C.H.Wright
Blumea 22: 33 (1974).
Syn.: *Deidamia clematoides* (C.H.Wright) Harms
Climber to 15m, glabrous; leaves trifoliolate; petals c. 4.5cm, with gland near base; leaflets elliptic, c. 8 × 4.5cm, subacuminate with a peltate mucro, base obtuse; petiolule 1.5cm; peduncle c. 9cm, several-flowered; fruit subglobose, 2.5cm, yellow, 3-valved, woody. Forest edge; 700–1300m.
Distr.: Nigeria to Congo (K). (Lower Guinea & Congolian).
IUCN: LC
Kupe Village: Cable 3895 fr., 7/1996; Cheek 7071 1/1995; Etuge 1964 fr., 5/1996; Kenfack 254 fr., 7/1996; Lane 433 fr., 1/1995; **Loum F.R.:** Cheek 10247 12/1999; **Nyasoso:** Etuge 1615 fr., 1/1996.

Passiflora foetida L.
Herbaceous climber 0.1–2m, foetid, softly-hispid; leaves suborbicular, 5–10cm, 3-lobed, acute, base cordate, ciliate; tendrils axillary; flowers axillary, solitary, to 5cm diameter; bracts pinnatisect, stalked-glandular; petals white; corona striped, blue-purple; fruit ovoid, 4cm, enclosed by bracts and calyx. Forest, farmbush; 260m.
Distr.: tropical Africa, introduced from tropical America.
IUCN: LC
Baduma: Nemba 629 9/1987.

PEDALIACEAE
I. Darbyshire

Sesamum radiatum Schum. & Thonn.
Herb to 1m, densely pubescent throughout; leaves narrowly elliptic, 4.5–7.7 × 1–1.7cm, apex obtuse-apiculate, base cuneate; petiole c. 1cm; flowers axillary, 1 per axil, c. 3.8cm long; pedicel 3mm; calyx lobes lanceolate, 6mm; corolla tubular, white-pink, 5-lobed, lower lobe c. 1 × 2.5cm; capsule oblong 2.2 × 0.7cm, dehiscent, apex subrostrate for 2mm, base rounded; calyx persistent; seeds numerous, 2–3mm long, rugose. Cultivated for its seed, persisting in farmbush; 800m.
Distr.: originating in tropical Africa. (Pantropical).
IUCN: LC
Tombel: Etuge 363 fl., 11/1986.

PENTADIPLANDRACEAE
M. Cheek

Fl. Cameroun 15 (1973).

Pentadiplandra brazzeana Baill.
Tree or climber, 20m; leaves alternate, elliptic 12.5 × 4cm, acuminate, 1.5cm; petiole 1cm; inflorescence axillary, 0.7cm, few-flowered; flowers 3cm, white, mottled-purple; calyx, corolla and androecium free; fruit ovoid, c. 8 × 7cm, hard, mottled-grey, pointed, flesh-orange; seeds 2cm, white. Forest; 1400–1600m.
Distr.: Cameroon to Congo (K). (Lower Guinea & Congolian).
IUCN: LC
Kodmin: Cheek 9091 1/1998; Ghogue 25 fr., 1/1998; Gosline 161 11/1998; Onana 581 2/1998.

PERIDISCACEAE
M. Cheek

Note: *Soyauxia* was previously included in Medusandraceae (FWTA); molecular analysis led to the discovery of its placement within Peridiscaceae (*fide* Vincent Savolainen, R.B.G., Kew, Dec. 2003). A revision of the genus is to be published.

Soyauxia gabonensis Oliv.
Shrub or tree to 15m, brown-puberulent; leaves alternate, oblong-elliptic, 15 × 6cm, acumen acute, 1cm, base cuneate, midrib appressed-long-hairy; lateral nerves 12 pairs, scalariform; petiole 5mm; stipules persistent, oblong, 9 × 3mm; spikes axillary, to 10cm; flowers 8mm diameter; sepals glabrous near margins; fruit not seen. Forest; 200m.
Distr.: Nigeria, Cameroon & Gabon. (Lower Guinea).
IUCN: NT
Ikiliwindi: Nemba 547 6/1987.

PHYTOLACCACEAE

Soyauxia talbotii Baker f.
Tree 5m, shortly-pubescent; leaves oblong-elliptic, c. 15 ×
4cm, acumen 1.5cm, acute; lateral nerves 15 pairs, midrib
puberulent below; petiole 7mm; stipules aristate, caducous;
spikes 10cm, arranged in clusters of 4–5 on short axillary
branches; flowers not seen; fruit ovoid, 8mm, 3-valved.
Forest; 300m.
Distr.: S Nigeria & SW Cameroon. (Lower Guinea).
IUCN: NT
Bolo: Nemba 92 fr., 5/1986.

PHYTOLACCACEAE
M. Cheek

Hilleria latifolia (Lam.) H.Walt.
Herb, 0.3–1m, glabrous; leaves alternate, elliptic, c. 8 ×
3.5cm, acuminate, acute; petiole 3cm; stipules absent;
inflorescence a raceme, terminal, 1-sided, c. 10cm; pedicels
0.4cm; flowers pink and white, 2mm; calyx lobes persistent,
unequally united at base, outermost longest; fruit a capsule,
2mm. Forest edge; 300–1650m.
Distr.: tropical Africa, introduced from tropical America.
IUCN: LC
Konye: Thomas 5174 11/1985; **Kupe Village:** Etuge 2013
fl., fr., 5/1996; **Manehas F.R.:** Cheek 9392 fl., 10/1998;
Mejelet-Ehumseh: Manning 428 10/1986; **Molongo:**
Nemba 774 2/1988; **Ndum:** Cable 898 fl., 1/1995; Lane 491
fl., 2/1995; Williams 140 fl., 1/1995; **Ngomboaku:** Cheek
10277 fl., fr., 12/1999; Ghogue 480 fl., 12/1999; **Ngusi:**
Etuge 1544 fl., fr., 1/1996; **Nyasoso:** Cable 3523 fl., fr.,
7/1996; Cheek 7907 11/1995.
Local names: Zom Mumba (Epie Ngome in Cheek 9392);
Asmi-ngalle (Cheek 10277). **Uses:** FOOD – unspecified parts
(Cheek 10277); MEDICINES (Epie Ngome in Cheek 9392).

Phytolacca dodecandra L'Hér.
Dioecious shrub, sometimes scandent, to 4m, glabrous,
slightly fleshy; leaves alternate, ovate-elliptic, c. 8 × 3cm,
acute, base rounded; petiole c. 2cm; inflorescence racemose,
c. 20cm; pedicels 0.6cm; sepals 4; petals absent; stamens c.
15; berries red, c. 0.8cm. Forest; 2000m.
Distr.: tropical & subtropical Africa.
IUCN: LC
Nyasoso: Etuge 1681 fl., 1/1996.

PIPERACEAE
M. Cheek & G. Mathieu

Peperomia bangroana C.DC.
J. Bot. 4: 134 (1866).
Syn.: *Peperomia rotundifolia* (L.) Kunth
Mat-forming epiphytic herb c. 1cm; stems protrate,
branching, minutely-puberulent; leaves alternate, circular,
0.6cm; inflorescence erect, axillary, single, 1.5cm. *Coffea*
plantations and *Musanga* leaves; 400–1350m.
Distr.: New World, introduced to tropical Africa.
IUCN: LC
Kupe Village: Zapfack 959 7/1996; **Ndum:** Cable 938 fl.,
2/1995; **Nyale:** Cheek 9710 11/1998; **Tombel:** Thorold 16
3/1953; 75 11/1950.

Note: an introduced weed of tree crops, spreading to
Musanga.

Peperomia fernandopoiana C.DC.
Epiphytic herb c. 4m from ground; stems erect, branched,
30cm, drying black, glabrous; leaves alternate, ovate-
lanceolate, c. 7 × 3.5cm, acumen long, acute; inflorescences
terminal and axillary, 2–3 per peduncle, to 6cm. Forest; 400–
2000m.
Distr.: Sierra Leone to Kenya. (Tropical Africa).
IUCN: LC
Bangem: Thomas 5273 1/1986; **Kodmin:** Biye 11 1/1998;
Gosline 47 1/1998; Nwaga 17 1/1998; **Kupe Village:** Cable
717 fl., 1/1995; 801 1/1995; Cheek 7133 fl., 1/1995; 8384
5/1996; Gosline 239 11/1999; Ryan 340 5/1996; **Nyandong:**
Cheek 11385 3/2003; **Nyasoso:** Cable 3301 6/1996; 3416
6/1996; 3455 7/1996; 3466 7/1996; Cheek 7317 2/1995;
Etuge 1690 1/1996; 2172 6/1996; Groves 96 fl., 2/1995;
Letouzey 14696 fl., 4/1976; Sunderland 1532 7/1992;
Wheatley 450 fl., 7/1992; Zapfack 716 6/1996; **Tombel:**
Thorold 15 3/1953.
Note: a smaller-leaved form with zig-zag stems is
commonest on Mt Kupe.
Local name: Achangadite. **Uses:** SOCIAL USES –
unspecified social uses – used by herbalists, not known what
for (Gosline 47).

Peperomia kamerunana C.DC.
Epiphytic herb in tree crown, erect, 7–15cm, densely brown-
pubescent; leaves opposite, elliptic, c. 2 × 1.5cm, apex
rounded; inflorescences 1–2 per stem, axillary, 3cm. Forest;
1415–2300m.
Distr.: Cameroon. (W Cameroon Uplands).
IUCN: EN
Mwanenguba: Sanford 5546 11/1968; **Nyasoso:** Lane 56 fr.,
8/1992; Wheatley 433 fl., 7/1992.

Peperomia molleri C.DC.
Herb c. 15cm, epiphytic or on rocks, rarely terrestrial, erect,
drying black; leaves alternate, membranous, broadly ovate,
obtuse to subacuminate, to c. 3.5 × 2.75cm, base obtuse to
rounded, 3-nerved, margin hairy; inflorescence axillary, to
10cm, single. Semi-deciduous and evergreen forest; 870–
1400m.
Distr.: Liberia to Kenya. (Tropical Africa).
IUCN: LC
Kodmin: Onana 580 2/1998; **Kupe Village:** Zapfack 937
7/1996; **Mwambong:** Etuge 4327 10/1998; **Nyasoso:** Cable
3267 6/1996; 3467 7/1996.

Peperomia pellucida (L.) Kunth
Terrestrial herb, erect, 20cm, glabrous; leaves alternate,
membranous, drying pale green, broadly ovate, c. 2 × 2cm,
rounded or subacuminate, base cordate; inflorescences
axillary or terminal, single, to 5cm. Roadside; 400m.
Distr.: Senegal to S Africa. (Tropical & subtropical Africa).
IUCN: LC
Loum F.R.: Pollard 94 10/1998.
Uses: none known (Epie Ngome in Pollard 94).

Peperomia retusa (L.f.) A.Dietr. var. *mannii* (Hook.f.) Düll
Bot. Jahrb. 93: 90 (1973).
Syn.: *Peperomia mannii* Hook.f. ex C.DC.
Epiphytic herb, at first mat-forming with prostrate stems and
circular leaves, 0.5–1cm; flowering from erect stems with
leaves elliptic, slightly reflexed, to 3cm, apex rounded;
inflorescence 4(–7)cm. Forest; 1050–1100m.

Distr.: Liberia, Nigeria, Cameroon, Bioko. (Upper & Lower Guinea).
IUCN: NT
Ndum: Cable 872 fr., 1/1995; Lane 456 fr., 1/1995; 495 fr., 2/1995.

Peperomia thomeana C.DC.

Bot. Jahrb. 93: 104 (1973).
Syn.: *Peperomia vaccinifolia* C.DC.
Epiphytic herb, stoloniferous, glabrous, flowering from erect stems; leaves opposite at uppermost node, obovate or elliptic, to 3 × 2cm, apex retuse; inflorescences 1–2 in uppermost axils, to 6cm. Forest; 850–2030m.
Distr.: Bioko & W Cameroon. (Lower Guinea).
IUCN: NT
Kupe Village: Cable 2478 5/1996; 2605 5/1996; 2637 5/1996; 2652 5/1996; 3754 7/1996; 3803 7/1996; Ryan 213 5/1996; 345 5/1996; 414 5/1996; 416 5/1996; **Nyasoso:** Cable 110 fr., 9/1992; 2935 6/1996; 3569 7/1996; Etuge 2565 7/1996; Letouzey 14689 4/1976; Wheatley 409 fr., 7/1992.
Note: for details on conservation status, see Cheek *et al.* (2000).

Peperomia vulcanica Baker & C.H.Wright

Bot. Jahrb. 93: 102 (1973).
Syn.: *Peperomia hygrophila* Engl.
Mid-crown epiphytic herb, erect, fleshy, robust, glabrous, 15–25cm; leaves alternate, red below (alive), drying pale yellow, ovate-rhombic, 5–8 × 2.5–4cm, subacuminate-rounded, obtuse; inflorescence single, terminal, stout, 5–12cm. Forest; 400–1800m.
Distr.: Sierra Leone, Cameroon, Bioko. (Upper & Lower Guinea).
IUCN: NT
Kupe Village: Cable 2409 5/1996; 2568 5/1996; 2669 5/1996; Cheek 7783 11/1995; Ryan 346 5/1996; **Mt Kupe:** Thomas 5055 11/1985; **Nyasoso:** Cable 3406 6/1996; Wheatley 417 fr., 7/1992; **Tombel:** Thorold 12 3/1953; 76 11/1950.
Note: resembles the much rarer *P. laeteviridis* Engl., differs in alternate leaves.

Piper capense L.f.

Pithy shrub c. 1(–5)m; peppery when crushed; leaves opposite, broadly ovate, c. 15 × 10cm, cordate, glabrous except on nerves; inflorescences of leaf-opposed single erect white spikes c. 3 × 0.5cm. Forest; 700–2000m.
Distr.: Guinea (C) to S Africa. (Tropical & subtropical Africa).
IUCN: LC
Kodmin: Biye 39 1/1998; **Kupe Village:** Cable 2455 5/1996; 3788 7/1996; Cheek 7107 fl., 1/1995; Kenfack 230 7/1996; Ryan 349 5/1996; **Mt Kupe:** Schoenenberger 73 fl., 12/1995; Sebsebe 5081 11/1995; **Ndum:** Cable 884 fr., 1/1995; Lane 478 fr., 2/1995; **Ngomboaku:** Ghogue 477 12/1999; **Nyasoso:** Balding 1 fl., 7/1993; Cable 2749 6/1996; 2932 6/1996; 3313 6/1996; Cheek 5642 fl., 12/1993; 7300 fl., 2/1995; Etuge 1668 1/1996; 2426 6/1996; Lane 175 fl., fr., 11/1994; Sidwell 321 10/1995; Sunderland 1477 7/1992; 1522 7/1992; Wheatley 430 fl., 7/1992; Zapfack 696 6/1996.
Note: above 1400m, plants commonly to 2.5m.
Local names: 2 types: Diangwe and Micribong: 'The Peaceful Leaf' (Cheek 7300). **Uses:** FOOD – leaves used as a vegetable, mostly with fresh fish (Etuge 1668); FOOD ADDITIVES – unspecified parts – spice (Etuge 2426); MEDICINES (Cheek 7300).

Piper guineense Schum. & Thonn.

Hemiepiphyte-climber reaching 20m above ground; peppery when crushed; stem twining and rooting adventitiously; leaves ovate-elliptic, to 19 × 10cm, obliquely-obtuse at base; inflorescence single, leaf-opposed, 3cm. Forest; 600–1360m.
Distr.: Guinea (B) to Uganda. (Guineo-Congolian).
IUCN: LC
Bakossi F.R.: Etuge 4298 10/1998; **Enyandong:** Etuge 4478r 11/2001; **Etam:** Etuge 235 8/1986; **Kodmin:** Etuge 4151 2/1998; **Kupe Village:** Etuge 1403 11/1995; 2605 7/1996; 2762 7/1996; Kenfack 305 7/1996; Lane 299 fr., 1/1995; **Mungo River:** Letouzey 14843 fr., 5/1976; **Ndum:** Groves 27 fr., 2/1995; **Ngomboaku:** Etuge 4667 12/1999; Ghogue 485 12/1999; **Nyandong:** Ghogue 1457 fr., 3/2003; **Nyasoso:** Cable 3242 6/1996; Cheek 7513 10/1995; Wheatley 423 fl., 7/1992.
Local name: Yob (*fide* Etuge). **Uses:** FOOD – leaves (Etuge 2762), as vegetable (*fide* Etuge); FOOD ADDITIVES – unspecified parts (Etuge 1403 & 4478r) – fruits edible as spice (Etuge 4151); infructescences (*fide* Cheek).

Piper umbellatum L.

Pithy shrub 1–2m; peppery when crushed; leaves orbicular, c. 24 × 24cm, deeply cordate, white below; inflorescences 2–5, clustered on peduncle, c. 3.5cm. Forest edge; 220–1290m.
Distr.: Guinea (C) to Cameroon. (Upper & Lower Guinea).
IUCN: LC
Baduma: Nemba 222 9/1986; **Bakolle Bakossi:** Etuge 249 9/1986; **Enyandong:** Ghogue 1220 fr., 10/2001; **Kodmin:** Cheek 9024 1/1998; **Kupe Village:** Cable 2640 5/1996; Etuge 2840 7/1996; Lane 288 fl., 1/1995; Ryan 246 5/1996; **Nsoung:** Darbyshire 87 3/2003; **Nyasoso:** Balding 76 fl., 8/1993; Etuge 1297 10/1995; 1670 1/1996; Wheatley 429 fl., 7/1992.
Local name: Chehgweh (Cheek 9024). **Uses:** FOOD – leaves used as a vegetable in soup, good with fresh fish (Etuge 1670); leaves (Etuge 2840); FOOD ADDITIVES – unspecified parts – spice (Etuge 1297); SOCIAL USES – used for peace in the village, if you see in the village, do not pass; MEDICINES (Etuge 2840 & Cheek 9024).

PITTOSPORACEAE

M. Cheek

Pittosporum viridiflorum Sims *'mannii'* L.

Meded. Land. Wag. 82(3): 260 (1982).
Shrub or small tree to 6m; leaves glabrous, alternate, obovate or elliptic, to 13 × 6cm, acumen 0.5cm, cuneate; lateral nerves c. 10 pairs; flowers whitish yellow; fruit orange, globose, 8mm; valves 2, reflexing to show 5–6 seeds, sticky, scarlet. Forest edge; 1900–2000m.
Distr.: Cameroon. (W Cameroon Uplands).
IUCN: NT
Kupe Village: Ryan 398 5/1996; **Mwanenguba:** Cheek 7259 2/1995; Etuge 4367 10/1998; Thomas 3116 fr., 2/1984; **Nyasoso:** Cable 2908 6/1996.

PLANTAGINACEAE

PLANTAGINACEAE
M. Cheek

Plantago palmata Hook.f.
Erect herb 15cm, glabrous; aerial stem absent; leaves broadly ovate, c. 7 × 6cm, obtuse, base truncate-decurrent, margin crenate-lobed, upper surface often black-splotched; petiole 15cm; inflorescences spike-like, 20cm. Path edges; 1300–2000m.
Distr.: Cameroon to Malawi. (Afromontane).
IUCN: LC
Ehumseh - Mejelet: Etuge 359 fl., 10/1986; **Kodmin:** Biye 2 fr., 1/1998; **Mejelet-Ehumseh:** Manning 432 10/1986; **Mwanenguba:** Bamps 1560 fl., 12/1967; Cheek 9449 fl., fr., 10/1998; Leeuwenberg 8409 fl., 9/1971; 9639 fl., 4/1972; Letouzey 14403 fl., 8/1975; Sanford 5492 fl., 11/1968; Thomas 3118 fl., 2/1984; **Nsoung:** Darbyshire 73 3/2003.

PLUMBAGINACEAE
M. Cheek

Plumbago zeylanica L.
Weak-stemmed shrub, 1m; stems smooth, grey-green, longitudinally ridged; leaves alternate, ovate, c. 4 × 2cm, apex acute, base obtuse, decurrent, glabrous; petiole 1cm; inflorescence a terminal spike, c. 20cm, glandular; flowers white; corolla tube 1.5cm, lobes 5, 0.7cm; fruit cylindrical, 1cm, densely stalked-glandular. Forest; 420–720m.
Distr.: pantropical.
IUCN: LC
Baseng: Cheek 7547 fl., 10/1995; **Nyandong:** Cheek 11377 3/2003.
Local name: Ndole. **Uses:** MEDICINES (Cheek 11377).

PODOSTEMACEAE
M. Cheek

Ledermanniella letouzeyi C.Cusset
Adansonia (sér. 4) 6: 260 (1984).
Rheophyte; thallus adhering to rock surface, to 10 × 10cm or more; axial stems up to 4cm, apex with 3–6 distichous, dichotomously-branched, ribbed, flattened leaves to 15 × 1cm; flowers borne on stem, inverted in bud; pedicel c. 1cm, andrioecia 2; ovary laterally-flattened, 8-ridged; styles 2. Waterfalls; 750–1350m.
Distr.: Bakossi Mts & Rumpi Hills. (Cameroon Endemic).
IUCN: EN
Baseng: Cheek 10409 12/1999; **Mwambong:** Cheek 9119 2/1998; Cheek 9706 11/1998.

Ledermanniella onanae Cheek
Kew Bull. 58: 733 (2003).
Rheophyte; thallus less than 1 × 1cm; aerial stems 5–7cm, completely covered in stiff distichous leaves, each laterally flattened, c. 1mm apart, in side view suboblong to ligulate, 5–8 × 1–1.5mm, apex acute, base clasping the stem, forming a sheath; spathellae single, axillary, subtended by a curved bract about as long as the leaf; spathellum 4mm; pedicels 7–8mm; androecia 2; ovary laterally-flattened, 2mm; styles 2. Waterfalls, rapids; 1000–1350m.
Distr.: Bakossi Mts. (Narrow Endemic).
IUCN: EN
Kodmin: Cheek 9196 2/1998; **Mwambong:** Cheek 9120a 2/1998; Cheek 9703 11/1998.

Ledermanniella thalloidea (Engl.) C.Cusset
Fl. Cameroun 30: 80 (1987).
Rheophyte; thallus adhering to rock surface, to 10 × 10cm or more; aerial stems absent; leaves distichous, 4–5 subtending each spathellum, linear, c. 10 × 0.5mm; spathellae 0.5–1cm, scattered over thallus; flower inverted in bud; ovary flattened, 2mm; styles 2. Waterfalls, rapids; 400–800m.
Distr.: Cameroon Endemic.
IUCN: VU
Ndoungue: Ledermann 6328a 11/1909; **Nyandong:** Cheek 11475 3/2003.

Macropodiella pellucida (Engl.) C.Cusset
Fl. Cameroun 30: 64 (1987).
Rheophyte; thallus 3–5cm diameter, enveloping rock surface; stems absent; leaves bract-like, two subtending each spathellum, ovate, c. 1mm; spathellae scattered, 4–5mm, flower inverted in bud; pedicel 1–1.5cm; androecia 2; ovary laterally-flattened, 8-ridged; styles 2. Waterfalls, rapids; 1000m.
Distr.: SW Cameroon. (Cameroon Endemic).
IUCN: EN
Mwambong: Cheek 9118 2/1998.

Tristicha trifaria (Bory) Spreng.
Rheophyte; polymorphic; variant with aerial stems not seen in our area; stoloniferous variant with thallus adhering to rock surface, linear, 2–3mm wide, branching, forming rosettes 30cm diameter, or more; leaves ovate, 3 × 2mm; flower erect in bud; ovary unflattened; styles 3. Rapids; 750m.
Distr.: tropical Africa, America & India.
IUCN: LC
Baseng: Cheek 10406 12/1999.
Note: genus in less-well-oxygenated water than other Podostemaceae.

POLYGALACEAE
I. Darbyshire

Atroxima sp. aff. afzeliana (Oliv.) Stapf
Liana to 5m or shrub to 1m, glabrous; leaves elliptic, 10–14.5 × 3.8–5.6cm, acumen to 1.8cm, ligulate, base acute; lateral nerves 5 pairs, brochidodromous at c. 5mm from margin, midrib yellow beneath; petiole 0.5cm; racemes axillary, to 2.5cm, minutely-puberulent; flowers strongly zygomorphic; sepals rounded, to 4mm; petals 5, to 1cm long, white, filaments free for uppermost 1.5–2mm; fruit globose, c. 3cm diameter, smooth, orange. Forest; 650m.
Distr.: Mt Kupe. (Narrow Endemic).
Kupe Village: Cable 2675 5/1996; Lane 392 fl., 1/1995.
Note: differs from *A. afzeliana* in the prominent, looping lateral venation, the long-ligulate acumen, the incompletely fused filaments (in *A. afzeliana* max. 1mm free) and the smooth (not verrucate) fruits. More material needed.

Carpolobia alba G.Don

Syn.: *Carpolobia glabrescens* Hutch. & Dalziel
Shrub or tree to 3(–6)m; stems puberulent towards apex; leaves papery, elliptic, oblong or obovate, 7.5–11.5 × 3.2–4.2cm, apex acuminate, base acute to obtuse; lateral nerves 6–9 pairs, midrib sparsely puberulent; petiole 0.3cm, puberulent; inflorescence axillary, 2–4-flowered; rachis < 1cm, puberulent; pedicel c. 5mm; sepals ovate, largest 0.7cm, others 0.4–0.5cm; corolla white, turning yellow; petals 5, keel petal to 1.5cm, limb as long as claw, 2–3mm wide when folded; fruit ovoid, 3-lobed, 1.7–2cm, rostrate, smooth, orange, with paired white rings at articulation with pedicel. Forest; 200–1000m.
Distr.: Guinea (C) to Liberia & Nigeria to Angola. (Guineo-Congolian).
IUCN: LC
Banyu: Thomas 5510 fl., 2/1986; **Baseng:** Cheek 10421 12/1999; **Ekona Mombo:** Etuge 413 fl., fr., 12/1986; **Enyandong:** Etuge 4482r 11/2001; **Konye:** Thomas 5143 fr., 11/1985; **Kupe Village:** Cable 747 fr., 1/1995; Lane 388 fr., 1/1995; **Loum F.R.:** Gosline 214 11/1999; **Mile 12 Mamfe Road:** Nemba 301 fl., fr., 10/1986; **Mungo River F.R.:** Onana 979 12/1999; **Ngusi:** Etuge 1573 1/1996; **Nyandong:** Tchiengue 1714 3/2003.
Uses: MATERIALS – famous as a source of cow-driving sticks used by Fulani (Gosline 214).

Carpolobia cf. alba G.Don

Shrub 2.5m, as *C. alba*, but leaves narrowly oblong-elliptic, 14 × 3.7cm, long-acuminate; lateral nerves 11 pairs, paired white rings at base of fruit conspicuous. Forest; 300m.
Loum F.R.: Biye 66 12/1999.
Note: probably an aberrant form of *C. alba*; the leaves of this taxon are known to be highly variable (Meded. Land. Wag. 77(18): 24, (1977)).

Heterosamara cabrae (Chodat) Paiva

Fontqueria 50: 128 (1998).
Syn.: *Polygala cabrae* Chodat
Decumbent herb or subshrub to 60cm; stems pubescent; leaves elliptic, 3.3–11 × 1.4–5cm, apex acuminate-mucronate, base attenuate, margin obscurely undulate, upper surface and veins of lower surface setulose, especially in young leaves; petiole to 1.4cm; racemes axillary or terminal, lax; rachis to 6cm, pubescent; lateral sepals oblong, 0.9cm, purple-white; remaining sepals ovate, 3mm; petals 3, purple-white; fruit to 4mm, emarginate, narrowly-winged. Forest, forest edge; 1150–1500m.
Distr.: Cameroon to Congo (K). (Lower Guinea).
IUCN: NT
Kodmin: Biye 46 1/1998; Cheek 8908 1/1998; 9203 2/1998; Etuge 4117 2/1998; 4409 11/1998; Ghogue 35 fl., fr., 1/1998; **Nlonako:** Letouzey 14480 3/1976.
Note: highly localised, known from c. 20 sites, but habitat currently largely unthreatened.

Polygala albida Schinz subsp. *stanleyana* (Chodat) Paiva

Fontqueria 50: 191 (1998).
Syn.: *Polygala stanleyana* Chodat
Erect herb to 20(–60)cm; stems puberulent; leaves oblanceolate, 4.5–7 × 0.4–0.9cm, apex acute-apiculate, base cuneate, margin revolute; petiole to 0.3cm; racemes axillary and terminal, the latter to 6cm, the former shorter, dense; lateral sepals obovate, 4 × 3mm, glabrous except on veins, pink-purple, remaining sepals 1–2mm, green; petals white;

fruit c. 4mm, emarginate, narrowly-winged, ciliate. Grassland; 1400–2000m.
Distr.: Guinea (C) to Ethiopia & Mozambique. (Afromontane).
IUCN: LC
Mwanenguba: Cheek 9422 10/1998; Leeuwenberg 8455 fl., 9/1971; Letouzey 14382 fl., 8/1975; Pollard 854 11/2001.

Polygala myriantha Chodat

Erect much-branched herb to 30cm; stems purplish, finely-puberulent; leaves ovate-lanceolate, 1.5–2.5 × 0.4–0.7cm, subsessile; racemes terminal, to 7.5cm, slender; pedicels 1mm; lateral sepals oblanceolate, 2.5–3mm, purple, remaining sepals green, c. 1.5mm; petals 3, purple; fruit c. 2mm, narrowly winged. Grassland; 1900–2000m.
Distr.: Nigeria to Zimbabwe. (Afromontane).
IUCN: LC
Mwanenguba: Cheek 9444 10/1998; Leeuwenberg 8456 fl., 9/1971; Pollard 876 fl., 11/2001.

Securidaca welwitschii Oliv.

Shrub or liana to 6m; branchlets finely-puberulent; leaves papery, ovate, c. 10 × 4cm, apex acuminate, base obtuse, glossy above, glabrous; petiole 0.6cm, puberulent; panicles axillary, to 16cm, several-branched; pedicels 1cm, puberulent; flowers 1cm; petals 3, white; samaras c. 5cm long, main wing 4.5 × 1.5cm, obliquely oblanceolate, glabrous, green, second wing vestigial. Forest edge; 450–1290m.
Distr.: Guinea (C) to Angola & Uganda. (Guineo-Congolian).
IUCN: LC
Lake Edib: Cheek 9163 2/1998; **Nsoung:** Darbyshire 61 3/2003; **Nyandong:** Cheek 11374 3/2003.

POLYGONACEAE

I. Darbyshire

Polygonum

Note: there is a trend towards placing all the *Polygonum* species recorded in our area within the genus *Persicaria*; however we here follow Lebrun & Stork: Trop. Afr. Fl. Pl. (2003) in maintaining *Polygonum*. The equivalent combinations for *Persicaria* are listed as synonyms.

Polygonum nepalense Meisn.

Syn.: *Persicaria nepalensis* (Meisn.) H.Gross
Straggling annual herb; stems sparsely-pubescent; ocrea 0.7cm, glabrous; leaves ovate-deltoid, 3.2–4.2 × 2.4–3.8cm, apex acute, base truncate, margin subcrenulate, glabrous; petiole 0.5–1.5cm, winged; inflorescence capitate, 0.5 × 0.5cm with a subtending sessile leaf, 1 × 0.6cm; peduncle to 4cm, sparsely glandular-pubescent towards apex; bracts ovate 3mm, apex acute; perianth white; nut lenticular, c. 2mm diameter, brown. Farmbush, grassland; 600–2000m.
Distr.: tropical & S Africa, Madagascar, tropical Asia.
IUCN: LC
Kodmin: Cheek 8890 1/1998; Ghogue 70 1/1998; **Mwanenguba:** Cheek 9451 10/1998; **Nyasoso:** Etuge 1641 1/1996.

Polygonum cf. pulchrum Blume

Straggling herb; stems pubescent; ocrea 2–3cm, pubescent, apical cilia dense, to 1.4cm long; leaves subsessile, linear-lanceolate, 9.5–12 × 0.9–1.1cm, base acute to obtuse,

surfaces pubescent, densely so below; spikes dense, 3–3.8cm long, 2–3 together; peduncle pubescent; bracts c. 4mm long, pubescent and ciliate; flowers shortly pedicellate; perianth 3–4mm, white; nut lenticular, shining black. Lake margins, damp areas; 1975–2000m.
Distr.: tropical & subtropical Africa & Asia. (Palaeotropics).
IUCN: LC
Mwanenguba: <u>Cheek 7250</u> fl., 2/1995; <u>9415</u> 10/1998; <u>Thomas 3098</u> 2/1984.
Note: our specimens differ from *P. pulchrum* in the short inflorescences with white, not pink, perianths, and in the often obtuse, not acute, leaf bases; *P. pulchrum* is often more robust, with broader ovate-lanceolate leaves.

Polygonum salicifolium Brouss. ex Willd.
Syn.: *Persicaria decipiens* (R.Br.) K.L.Wilson
Aquatic herb to 1m tall; lower stems often partially submerged; ocrea to 2cm, appressed-pubescent or glabrescent, apical cilia c. 7mm; leaves subsessile, lanceolate, 9–9.5 × 1–1.6cm, base obtuse, glabrous except on midrib and margins; spikes lax and slender, to 7cm long, 2–several together; peduncles glabrous; bracts c. 3mm, ciliate; perianth 2.5–3mm long, pink; nut trigonous, c. 2mm long, shining-brown. Lake margins; 1200–2000m.
Distr.: pantropical, also S Europe.
IUCN: LC
Lake Edib: <u>Cheek 9160</u> 2/1998; <u>Gosline 196</u> 11/1998;
Mwanenguba: <u>Cheek 7251</u> fl., 2/1995; <u>9406</u> 10/1998.

Polygonum senegalense Meisn. fa. *albotomentosum* R.A.Graham
F.T.E.A. Polygonaceae: 19 (1958).
Syn.: *Persicaria senegalensis* (Meisn.) Sojak fa. *albotomentosa* (R.A.Graham) K.L.Wilson
Syn.: *Polygonum lanigerum* Meisn. var. *africanum*
Herb to 1.2m, matted white-tomentose throughout; ocrea 1.7–3cm, apical cilia absent; leaves ovate-lanceolate, 15.5–27 × 3–6.5cm, base acute; petiole c. 1.5cm; spikes dense, to 7.5cm long, 1–several together; bracts c. 4mm, ciliate; perianth white-green; nut lenticular, 3.5–4mm diameter, shining dark brown. Moist open areas; 720m.
Distr.: tropical & S Africa, Madagascar & Egypt.
IUCN: LC
Baseng: <u>Cheek 7545</u> 10/1995; **Nyasoso:** <u>Pollard 576</u> fl., 9/2001.
Local name: none known. **Uses:** none known (Cheek 7545).

Polygonum setosulum A.Rich.
Syn.: *Polygonum nyikense* Baker
Syn.: *Persicaria setosula* (A.Rich.) K.L.Wilson
Decumbent herb to 60cm tall; ocrea to 3cm, glabrescent or sparsely pubescent, apical cilia 7–8mm; leaves ovate-lanceolate, 14.5–20 × 3–4.2cm, base acute, margin entire-ciliate, midrib and secondary nerves of lower surface with short coarse hairs; petiole to 2cm; spikes dense to 9cm long, 2–several together; peduncles glabrous; bracts c. 4mm, with an apical fringe of short bristles; flowers 2.5–3mm long, shortly-pedicellate; perianth white; nut lenticular or with one side swollen, 2.5mm long, shining dark brown. Lake margins, damp disturbed areas; 1200–1450m.
Distr.: tropical & southern Africa.
IUCN: LC
Kodmin: <u>Cheek 9011</u> 1/1998; **Lake Edib:** <u>Gosline 195</u> 11/1998.
Local name: Mwimbweeh. **Uses:** MATERIALS – cleansers – removes stains from hands, e.g. *Musa* stains (Cheek 9011); see also *Rumex nepalensis*.

Rumex abyssinicus Jacq.
Stout herb to 2m; stems hollow, cylindrical, to 1.3cm diameter; leaves papery, hastate, 10.5–16 × 10–13cm, uppermost becoming linear subhastate, palmately-nerved, glabrous; petiole 3–7cm; panicles many-branched, to 30cm long, leafless, many-flowered; flowers in fascicles; pedicels 3–4cm; perianth 2mm; fruiting valves orbicular, 6–7mm diameter, base cordate, tubercles absent, reticulately nervose; nut acutely trigonous, 3.5 × 1.5cm, shining pale brown. Forest edge, montane grassland; 1800–2000m.
Distr.: tropical Africa & Madagascar. (Afromontane).
IUCN: LC
Kodmin: <u>Ghogue 59</u> 1/1998; **Mwanenguba:** <u>Cheek 7263</u> 2/1995; <u>9459</u> 10/1998.

Rumex nepalensis Spreng.
Fragm. Florist. Geobot., Suppl. 2(1): 93 (1993).
Syn.: *Rumex bequaertii* De Wild.
Erect herb to 0.6(–1.5)m; stems hollow, cylindrical; leaves papery, oblong-lanceolate, to 25 × 5cm, apex and base acute, margin subentire, glabrous; petiole to 0.7cm; spikes to 35cm, rarely-branched; flowers in dense, widely-spaced fascicles, c. 13–15-flowered, reduced leaves present in the axes of the lower fascicles; pedicels 3.5–4mm; perianth c. 2mm long; fruiting valves deltoid, 3.5–4mm long, apex acute, margins with 5–7 hooked bristles, c. 1mm; nut trigonous, 2 × 1mm, shining-brown. Forest edge, disturbed areas; 1450–2000m.
Distr.: tropical Africa, S Europe & Asia-Minor to China.
IUCN: LC
Ehumseh: <u>Manning 393</u> 10/1986; **Ehumseh - Mejelet:** <u>Etuge 322</u> 10/1986; **Kodmin:** <u>Cheek 9009</u> 1/1998; **Mwanenguba:** <u>Thomas 3147</u> 2/1984.
Local name: Mwibweed. **Uses:** MATERIALS – cleansers – removes stains from hands, e.g. Musa stains (Cheek 9009).

PORTULACACEAE

M. Cheek

Portulaca oleracea L.
Succulent, spreading, annual herb, c. 15cm, glabrous; leaves oblanceolate, 2 × 0.75cm, rounded, cuneate, sessile; stipules inconspicuous; flowers terminal; petals yellow, 0.5cm; fruit circumscissile. Villages; 700m.
Distr.: pantropical.
IUCN: LC
Nyale: <u>Cheek 9643</u> 11/1998.

Talinum triangulare (Jacq.) Willd.
Succulent, annual, erect, herb, 10–15cm, glabrous; leaves obovate, c. 7 × 3cm, retuse, cuneate; inflorescence a terminal panicle; sepals 5, pink; capsule 0.5cm, superior. Farmbush and cultivated; 400–650m.
Distr.: tropical America, naturalised in Africa.
IUCN: LC
Banyu: <u>Thomas 5519</u> 2/1986; **Ngusi:** <u>Etuge 1546</u> 1/1996.

PRIMULACEAE
M. Cheek

Anagallis djalonis A.Chev.
Erect annual herb, 3–10cm; stems 4-angled, glabrous; leaves alternate, ovate, 5 × 3mm, obtuse; petiole 2mm; flowers single, axillary; pedicel 5mm; petals 5, 2mm, actinomorphic; fruit circumscissile. Footpaths; 1200–1500m.
Distr.: Guinea (C) to Kenya. (Afromontane).
IUCN: LC
Kodmin: Cheek 8895 fr., 1/1998; **Mwambong:** Cheek 9558 fl., fr., 11/1998; **Mwanenguba:** Leeuwenberg 8195 fl., fr., 8/1971.

Ardisiandra sibthorpioides Hook.f.
Prostrate, rooting herb; stems to 30cm, succulent, red, sparsely long-hairy; leaves alternate, orbicular, 3cm, palmately-incised; petiole 3cm; flowers axillary, single; pedicel 1cm; calyx green, campanulate, 5mm. Forest; 940–1200m.
Distr.: Cameroon, Bioko & E Africa. (Afromontane).
IUCN: LC
Enyandong: Etuge 4508r 11/2001; **Mwanenguba:** Leeuwenberg 9614 fl., 4/1972.

Lysimachia ruhmeriana Vatke
Erect herb 30–70cm; stems red, unbranched, glabrous; leaves opposite, ovate, c. 3.5 × 1.75cm, acute, decurrent; petiole c.1cm; raceme terminal, 20cm; pedicels 3mm; flowers inconspicuous; calyx lobes equal; fruit globose, 3mm, circumscissile. Rocky slopes; 500–2000m.
Distr.: Cameroon to Kenya. (Afromontane).
IUCN: LC
Enyandong: Cheek 10967 10/2001; **Kodmin:** Etuge 4042 fl., fr., 1/1998; **Kupe Village:** Etuge 1468 fl., fr., 11/1995; **Mwanenguba:** Cheek 9496 fr., 10/1998; Leeuwenberg 8546 fl., 10/1971; 9648 fl., 4/1972.
Note: distinguished from *Hilleria latifolia* by the opposite leaves and symmetrical calyx.

RANUNCULACEAE
M. Cheek

Clematis grandiflora DC.
Climber, 6m, sparsely puberulent; leaves opposite to ternate; leaflets ovate-lanceolate, 11 × 6cm, acuminate, cordate, serrate, secondary nerves sparsely hairy; petioles 5cm, twining; inflorescence 3-flowered, terminal on spur shoots; flowers light green, 6cm diameter; fruit globose, 10cm diameter; achenes numerous; plumes dense, white-hairy. Secondary forest; 700–1560m.
Distr.: Guinea (C) to Uganda. (Guineo-Congolian).
IUCN: LC
Kodmin: Onana 596 fr., 2/1998; **Kupe Village:** Cable 710 fr., 1/1995; Etuge 1997 fl., fr., 5/1996; **Nyasoso:** Etuge 1510 fl., fr., 12/1995.

Clematis hirsuta Guill. & Perr. var. hirsuta
Climber; leaves opposite, pinnate; leaflets ovate, slightly 3-lobed, c. 4 × 3cm, dentate, white-velvety beneath; inflorescence a dense panicle; flowers white, 2.5cm diameter. Forest edge; 1800–2000m.

Distr.: widespread in tropical Africa, Arabia. (Afromontane & Arabia).
IUCN: LC
Mwanenguba: Cheek 7286 fr., 2/1995; Etuge 4390 fl., 10/1998; Thomas 3105 fl., 2/1984.

Clematis simensis Fresen.
Climber resembling *C. hirsute*, but leaflets glabrous below; flowers 1.5cm diameter. Forest edge, farmbush; 1250–1800m.
Distr.: Cameroon to S Africa. (Afromontane).
IUCN: LC
Kodmin: Etuge 4043 fl., 1/1998; **Mwanenguba:** Etuge 4387 fl., 10/1998; Sanford 5494 fl., fr., 11/1968; **Nyasoso:** Etuge 1524 fl., 12/1995.

Thalictrum rhynchocarpum Quart.-Dill. & A.Rich. subsp. rhynchocarpum
Lidia 4(3): 89 (1998).
Herb c. 60cm, glabrous; leaves tripinnate, c. 20cm; leaflets orbicular-elliptic, c. 1.2 × 1.2cm, entire or 3–5-lobed; petiolules capillary; inflorescence c. 30cm; flowers numerous, c. 0.4cm, green. Forest edge; 2000–2150m.
Distr.: Bioko to Tanzania & S Africa. (Afromontane).
IUCN: LC
Mwanenguba: Sanford 5524 fl., 11/1968; **Nyasoso:** Cable 3387 fl., 6/1996.

RHAMNACEAE
M. Cheek & I. Darbyshire

Fl. Cameroun 33 (1991).

Gouania longipetala Hemsl.
Climber with tendrils; leaves ovate, to 7.5 × 4.5cm, subacuminate, base rounded to truncate, crenate-serrate, lateral nerves 3–4 pairs, sparsely hairy only on nerves; petiole to 2cm; stipules 1mm; inflorescence axillary, spike-like, 10–15cm; flowers in fascicles, white, 2mm; fruits 3-winged, 6mm. Forest; 200–860m.
Distr.: Guinea (C) to Congo (K). (Guineo-Congolian).
IUCN: LC
Mile 12 Mamfe Road: Nemba 300 fl., 10/1986; **Nyasoso:** Cheek 7469 10/1995.

Gouania longispicata Engl.
Shrub or climber resembling *G. longipetala* but leaves with lateral nerves 6–7 pairs, densely red-hairy, interstices very stoutly and densely white puberulent. Grassland, forest edge; 1450–1500m.
Distr.: S Nigeria to Mozambique. (Afromontane).
IUCN: LC
Kodmin: Barlow 15 fr., 1/1998; Cheek 9616 fl., 11/1998; Ghogue 26 fl., 1/1998.
Local name: Mpuna. **Uses:** MEDICINES (Max Ebong in Cheek 9616).

Gouania sp.
Nerve number equaling *G. longispicata*, indumentum equaling *G. longipetalata*, altitude intermediate; a potential hybrid between the two species. Forest; 1150m.
Mwambong: Cheek 9360 10/1998.

Maesopsis eminii Engl.
Tree, stinking when wounded, 35m, glabrous; leaves subopposite or alternate, papery, elliptic-oblong, c. 11 × 5cm,

subacuminate, obtuse, serrate, the teeth often partly covered towards the petiole, lateral nerves c. 6; petiole 2cm; inflorescences axillary, cymose, few-flowered; flowers white c. 2mm; fruit obovoid 3cm, fleshy, orange, 1-seeded. Forest, grassland; 350–1500m.
Distr.: Liberia to Uganda. (Guineo-Congolian).
IUCN: LC
Kodmin: Cheek 9624 11/1998; **Kupe Village:** Etuge 2844 fr., 7/1996; **Mungo River F.R.:** Cheek 10208 11/1999.
Local name: Ntup. **Uses:** MATERIALS – wood, furniture (Max Ebong in Cheek 9624).

Rhamnus prinoides L'Hér.
Shrub 2–3m, glabrous; leaves glossy, dark green, alternate, elliptic, 7 × 3cm, subacuminate, acute, finely serrate, lateral nerves 4 pairs; petiole 1cm; inflorescence of single axillary flowers; pedicel c. 8mm; flowers green, 2mm; berry globose, 7mm, purplish red. Forest edge; 1800–2000m.
Distr.: Cameroon to S Africa. (Afromontane).
IUCN: LC
Mwanenguba: Cheek 7265 fl., fr., 2/1995; Etuge 4368 fr., 10/1998; Leeuwenberg 8933 fl., fr., 12/1971; Onana 1971 11/2001; Thomas 3128 fr., 2/1984; 5888 3/1986.

Ventilago diffusa (G.Don) Exell
Liana; fruits ovoid-globose, to 7mm, with an oblong wing to 5.3 × 1cm, felty golden-puberulent, apex apiculate, prominently central-veined; calyx persistent, cupular, 4.5mm diameter, felty. Forest; 1100m.
Distr.: Nigeria, Cameroon, São Tomé, Congo (K), Ethiopia to Malawi. (Tropical Africa).
IUCN: LC
Nyasoso: Cheek 7319 fr., 2/1995.
Note: the cited specimen refers to a collection of fallen fruit; leaves and inflorescence not seen. This species is rare in W Africa, separated from the more widespread *V. africana* by the puberulent (not glabrous) fruits, and the paniculate (not glomerulate) inflorescence.

RHIZOPHORACEAE

I. Darbyshire & M. Cheek

Anopyxis klaineana (Pierre) Engl.
Tree to 50m; bole straight; crown dense; leaves opposite or in whorls of 3, subcoriaceous, drying dark brown above, reddish brown below, oblong-elliptic to obovate, 6.5–11 × 3–5.5cm, apex rounded, base acute, lateral nerves 7–8 pairs, lamina glabrous, shiny; petiole 0.8–2cm; cymes axillary, subumbellate, 2cm diameter; peduncle to 2.5cm; calyx densely pale puberulent, lobes oblong, 6mm, spreading; petals linear, pubescent towards apex, green-white; staminal tube 3–4mm; fruit ellipsoid, 3 × 2cm, grey-brown felty-puberulent; calyx persistent. Forest; 760m.
Distr.: Sierra Leone to Congo (K), Ethiopia & Sudan. (Guineo-Congolian).
IUCN: NT
Ngombombeng: Etuge 24 4/1986.
Note: designated by W.Hawthorne (1997) at www.redlist.org (2001) as VU A1cd on the basis of it being a timber tree and in regeneration being poor. However, this appears unmerited in light of the wide range of this taxon and in its remaining widespread in the Lower Guinea and Congolian forest where exploitation of this species for timber is very limited; it is therefore here reassessed as NT.

Cassipourea acuminata Liben
Bull. Jard. Bot. Nat. Belg. 56: 139 (1986).
Tree 4(–15)m, to 60cm diameter; leaves opposite, elliptic, 7–10 × 3–4cm, acumen pronounced, oblong-triangular, to 1cm long, base cuneate, margins serrate, lateral nerves 7–10 pairs; petiole c. 5mm; stipules ovate, 4.5 × 2mm; fascicles axillary, 2–3-flowered; pedicels c. 3mm, pubescent; calyx lobes c. 3mm; petals to 5mm. Forest, often near rivers; 1000m.
Distr.: Cameroon, Gabon & Congo (K). (Lower Guinea & Congolian).
IUCN: EN
Menyum: Doumenge 571 5/1987.
Note: specimen not seen at K; det. by J.Floret (P), a specialist in this family.

Cassipourea malosana Alston
Kew Bull.: 258 (1925).
Tree 3m; leaves opposite, elliptic, 9 × 3.5cm, including 1.5cm acumen, base cuneate, serrate in upper 2/3, lateral nerves c. 6 pairs; petiole c. 1cm, interpetiolar stipule triangular, sericeous, 6 × 3cm, rounded; flowers 1–3, axillary-fasciculate, c. 4 fertile nodes per stem; pedicel 7mm; flower 8mm. Forest; 1950m.
Distr.: Cameroon to Kenya. (Afromontane).
IUCN: NT
Nyasoso: Cable 2914 6/1996.
Note: further research may well show that the Cameroon montane population merits specific distinction from the E African populations. Likely to be confused with *Rubiaceae* but for the toothed leaves and superior ovary.

ROSACEAE

B.J. Pollard

Alchemilla cryptantha Steud. ex A.Rich.
Herb; stems prostrate, stoloniferous throughout; leaves reniform, c. 1.5 × 2.5cm, 5-palmatifid to palmatilobed, median lobe with 7–11 teeth, dentate, sparsely hairy; flowers white, inconspicuous c. 2mm across; 2–8 carpels. Grassland; 1700–1975m.
Distr.: tropical & southern Africa. (Afromontane).
IUCN: LC
Mwanenguba: Cheek 7242 2/1995; Letouzey 14394 fl., 8/1975; Pollard 857 11/2001.

Alchemilla kiwuensis Engl.
Herb with rosettes and stolons; leaves 7-palmatilobed to palmatipartite, median lobe with 13–23 teeth; flower with 5–12 carpels. Forest; 1300–2000m.
Distr.: Nigeria to Ethiopia, E Africa & S to Zambia, Zimbabwe. (Afromontane).
IUCN: LC
Kodmin: Cheek 9014 1/1998; **Mwanenguba:** Cheek 9472 10/1998.

Prunus africana (Hook.f.) Kalkman
F.T.E.A. Rosaceae: 46 (1960).
Syn.: *Pygeum africanum* Hook.f.
Tree to c. 20m; leaves alternate, lanceolate, 3–6 × 6–15cm, serrate; petiole 2cm long, bearing 2 glands near apex, or at base of lamina; inflorescence a dense panicle; flowers white, 5mm diameter; fruit a drupe, succulent, red, c. 1cm diameter. Forest, forest edge; 1800–2050m.
Distr.: tropical & subtropical Africa. (Afromontane).

IUCN: NT
Mt Kupe: Cheek 7591 11/1995; 7600 fr., 11/1995;
Mwanenguba: Cheek 7254 2/1995; Etuge 4373 10/1998;
Kongor 69 fl., 11/2001; **Nyasoso:** Cheek 7358 2/1995; Etuge
1678 1/1996.
Note: here I follow Cheek, in Cable & Cheek (1998), by
assigning the 'Near Threatened' IUCN category.

Rosa sp.
Prickly shrub to c. 2m; leaves imparipinnate; flowers of
various colurs and sizes, with many stamens. Cultivated in
gardens in villages; 1500m.
Distr.: cosmopolitan.
IUCN: LC
Kodmin: Cheek sight record fl., 1/1998.
Uses: ENVIRONMENTAL USES – Ornamentals (*fide*
Cheek).

Rubus pinnatus Willd. var. *afrotropicus* (Engl.) Gust.
Scandent prickly shrub to 5m; leaves less than 2.5 × as long
as broad, glabrous, not glandular below; inflorescences
terminal or less often axillary, many-flowered, rachis densely
appressed with short, silver, velvety-hairs; petals
inconspicuous or caducous; infructescence with many fewer
than 100 drupelets. Forest, forest margins; 750–1300m.
Distr.: tropical & southern Africa. (Afromontane).
IUCN: LC
Kodmin: Ghogue 17 fr., 1/1998; **Kupe Village:** Cheek 7104
fr., 1/1995; Etuge 1957 fr., 5/1996; 2717 fr., 7/1996; Kenfack
255 fl., fr., 7/1996; **Nyasoso:** Cable 2805 fr., 6/1996; 2878
fl., 6/1996.
Uses: FOOD – infructescences (Etuge 2717).

Rubus rosifolius Sm.
Prickly subshrub to 1m; leaves more than 2.5 × as long as
broad, glabrous, glandular below; inflorescence axillary, few-
flowered, sparsely pilose; petals white, c. 1.5cm long, showy;
infructescence of about 100 drupelets. Forest understorey;
800–1520m.
Distr.: introduced from Asia. (Palaeotropics).
IUCN: LC
Kodmin: Cheek 8969 fl., fr., 1/1998; Pollard 198 fl., fr.,
11/1998; **Kupe Village:** Ensermu 3530 fl., fr., 11/1995;
Kenfack 221 fl., fr., 7/1996; Lane 310 fl., fr., 1/1995;
Mejelet-Ehumseh: Etuge 311 10/1986; **Mwanenguba:**
Leeuwenberg 8141 fl., fr., 8/1971; Letouzey 13930 fl., fr.,
6/1975.
Local names: Akun (Edmondo Njume in Pollard 198);
'Bush Allowance' (Pidgin, *fide* Pollard). **Uses:** FOOD –
infructescences (Edmondo Njume in Pollard 198).

RUBIACEAE

M. Cheek, S.E. Dawson, B. Sonké & D.
Bridson (herbaceous taxa determined by S.E.
Dawson, *Aulacocalyx*, *Oxyanthus* &
Rothmannia by B. Sonké)

Aidia genipiflora (DC.) Dandy
Shrub or tree to 12m, glabrous; leaves drying blackish green,
oblong to elliptic, 5–17 × 2–7cm, long-acuminate, cuneate to
obtuse, lateral nerves 4–5 pairs; inflorescences axillary, at
alternate nodes; flowers white, turning yellow; calyx 6–7mm,

teeth 1.5mm; corolla tube 16–17.5mm, lobes 11–16mm; fruit
globose, 1–1.2cm; calyx tube persistent. Forest; 300m.
Distr.: Guinea (B) to Bioko, Cameroon & Sudan. (Guineo-
Congolian).
IUCN: LC
Bolo: Thomas 5893 3/1986; **Konye:** Thomas 7036 5/1987.

Aidia rhacodosepala (K.Schum.) Petit
Fl. Gabon 17: 164 (1970).
Shrub 3–7m, glabrous; leaves elliptic or elliptic-obovate, c.
18 × 7cm, acuminate, obtuse, nerves 8–9, domatia elongate,
white-hairy; petiole 0.5–1cm; stipule 5mm, sheathing, apex
triangular; flowers pink, 3–10, in 1–2cm axillary panicles;
pedicels 1.5cm; calyx tube 3–4mm, limb 8mm, lobes 4mm;
corolla tube and lobes 1–1.5cm; fruit globular, 1.2cm.
Evergreen forest; 250–1200m.
Distr.: Cameroon Endemic.
IUCN: NT
Baduma: Nemba 120 6/1986; **Kodmin:** Etuge 4056 fl.,
1/1998; **Mile 15, Kumba-Mamfe Road:** Nemba 196 9/1986;
Nlog: Etuge 20 4/1986; **Nyasoso:** Etuge 2511 fl., fr., 7/1996;
Groves 78 fl., 2/1995.
Note: less than 20 sites are known for this taxon.

Aoranthe cladantha (K.Schum.) Somers
Bull. Jard. Bot. Nat. Belg. 58: 47 (1988).
Syn.: *Porterandia cladantha* (K.Schum.) Keay
Tree 8m, glabrescent; leaves papery, obovate, 20 × 10cm,
rounded, acute; petiole 1.5cm; stipule oblong, 22 × 7cm;
cauliflorous; corolla pinkish white, tube 20–25mm, lobes
15mm; fruit axillary, few-fasciculate, ridged, ellipsoid, 1.5 ×
0.7cm; calyx 4mm; pedicel 1–2cm. Evergreen forest; 200–
600m.
Distr.: Nigeria to Cabinda. (Lower Guinea).
IUCN: LC
Baduma: Nemba 124 6/1986; **Bolo Moboka:** Nemba 255
9/1986; **Etam:** Etuge 218 8/1986; **Ikiliwindi:** Nemba 546
6/1987; **Kurume:** Thomas 5415 1/1986; **Mungo River F.R.:**
Etuge 4320 fr., 10/1998.

Argocoffeopsis scandens (K.Schum.) Lebrun
Bull. Jard. Bot. Nat. Belg. 51: 365 (1981).
Syn.: *Coffea scandens* K.Schum.
Climber, glabrous; leaves obovate-elliptic, c. 7 × 4cm,
acuminate, obtuse, nerves 4, looped; petiole 2mm;
inflorescence sessile, axillary; flowers sessile; corolla white,
tube 2mm, lobes 5mm. Forest; 800m.
Distr.: Cameroon to Congo (K). (Lower Guinea &
Congolian).
IUCN: LC
Bangem: Thomas 5366 fl., 1/1986.

Argostemma africanum K.Schum.
Epilithic, erect, succulent, annual herb, 2–14cm, glabrous;
leaves obliquely narrowly elliptic to 6 × 2cm, erect, obtuse,
cuneate, serrate; inflorescences terminal, 1–15-flowered;
pedicels 1cm; corolla white, divided to the base or almost
entire, 5mm diameter. Wet rocks in forest; 350–1450m.
Distr.: SE Nigeria, Cameroon & Rio Muni. (Lower Guinea).
IUCN: NT
Kodmin: Cheek 8953 fl., fr., 1/1998; **Ngomboaku:** Etuge
4689 fl., fr., 12/1999; **Nyasoso:** Cable 3290 6/1996; 66 fl.,
8/1992; Lane 212 fl., fr., 11/1994.
Note: only 12 sites are known for this taxon.

Atractogyne bracteata (Wernham) Hutch. & Dalziel

Woody climber to 10m, smelling of linament, glabrous; leaves ovate, to 20 × 12cm, base obtuse; petiole 4cm; flowers green, cylindric, 10 × 3mm; fruits cylindric, c. 7 × 5cm, red, soft, many-seeded; stipe 4cm. Evergreen forest; 700–950m.
Distr.: Ivory Coast to Gabon. (Guineo-Congolian).
IUCN: LC
Kupe Village: Cable 2744 fr., 5/1996; 3731 fr., 7/1996; 3879 fr., 7/1996; Etuge 2830 fr., 7/1996; Gosline 242 fr., 11/1999; Kenfack 299 fr., 7/1996; **Ngomboaku:** Cheek 10278 fl., fr., 12/1999; **Nyasoso:** Cheek 7461 fr., 10/1995; Etuge 2528 fr., 7/1996; Sidwell 312 10/1995.

Aulacocalyx caudata (Hiern) Keay

Shrub to 4m; stems appressed yellow-hairy; leaves elliptic, c. 11 × 4cm, long-acuminate, base unequally cordate, sessile; stipules triangular-awned; flowers 1(–3); calyx tube 3–5mm, teeth 2–5mm; corolla white, tube c. 4 × 0.3cm, lobes elliptic, 2 × 0.7cm. Evergreen forest; 300–1000m.
Distr.: SE Nigeria to Gabon. (Lower Guinea).
IUCN: NT
Bakole: Thomas 7030 5/1987; **Etam:** Etuge 233 8/1986; **Kupe Village:** Etuge 1929 fr., 5/1996; 1995 fr., 5/1996; Kenfack 302 fr., 7/1996; **Nyasoso:** Cable 2803 fr., 6/1996; 3265 fr., 6/1996; 3513 fr., 7/1996; Etuge 2437 fr., 6/1996; Sunderland 1546 fl., 7/1992; Zapfack 674 fr., 6/1996.

Aulacocalyx jasminiflora Hook.f. subsp. *jasminiflora*

Shrub or small tree to 4(–12)m, resembling *A. caudata*, but leaf base cuneate; petiole 5mm; flowers 3–10; calyx golden silky-hairy. Evergreen forest; 300–950m.
Distr.: Sierra Leone to Congo (K). (Guineo-Congolian).
IUCN: LC
Mekom: Thomas 5240 1/1986; **Ngomboaku:** Cheek 10387 fl., 12/1999.

Aulacocalyx talbotii (Wernham) Keay

Tree 6–25m, resembling *A. caudata*, but stems glabrous; leaves drying grey-green, glabrous, base cuneate; petiole c. 5mm; flowers 1–3; calyx tube 5–7mm, truncate. Evergreen forest; 250–1230m.
Distr.: SE Nigeria to Gabon. (Lower Guinea).
IUCN: NT
Ikiliwindi: Etuge 504 3/1987; **Kupe Village:** Cable 3852 fl., fr., 7/1996; Etuge 1909 fr., 5/1996; Ryan 353 fr., 5/1996; **Nyasoso:** Cable 3613 fr., 7/1996; Etuge 1793 fr., 3/1996.
Note: 15 sites are known in total.

Belonophora coffeoides Hook.f. subsp. *hypoglauca* (Welw. ex Hiern) S.E.Dawson & Cheek

Kew Bull. 55: 77 (2000).
Tree resembling *B. coriacea*, but stipule triangular-aristate, 7–9 × 0.75–3mm. Forest; 870m.
Distr.: Sierra Leone to Zambia. (Guineo-Congolian).
IUCN: LC
Nyasoso: Cable 3270 6/1996.

Belonophora coriacea Hoyle

Shrub or tree 2–14m; leaf blades elliptic to 27 × 11cm, acuminate, cuneate-obtuse; petiole 1–2cm; stipules 0.8–1.4 × 0.1–0.6cm; flowers fasciculate, white; corolla tube 1.3–2cm, lobes 1.5 × 0.3–0.5cm; fruit yellow, 1 × 1cm. Evergreen forest; 250–1400m.
Distr.: SE Nigeria to Congo (K). (Lower Guinea & Congolian).
IUCN: LC
Baduma: Nemba 193 8/1986; **Karume:** Nemba 66 5/1986; **Kupe Village:** Cable 2612 fr., 5/1996; Cheek 7182 fr., 1/1995; Etuge 2856 fr., 7/1996; **Nyale:** Cheek 9684 fl., 11/1998; **Nyasoso:** Balding 9 fr., 7/1993; Etuge 2082 fr., 6/1996; Lane 108 fr., 6/1994; Sebsebe 5046 fl., 10/1995.

Belonophora ongensis S.E.Dawson & Cheek

Kew Bull. 55: 75 (2000).
Shrub or tree 3(–6)m; leaf blades ovate or elliptic, to 35 × 14cm, acuminate, base acute; petiole 2cm; stipules ovate to elliptic, 3–4.5 × 1.3–2.2cm; flowers fasciculate, white; corolla tube 4cm, lobes 15 × 6mm; fruit yellow, 1 × 1cm. Evergreen forest; 200m.
Distr.: SW Cameroon. (Cameroon Endemic).
IUCN: CR
Mungo River F.R.: Cheek 10193 fr., 11/1999.

Bertiera bracteolata Hiern

Liana 7m; leaves elliptic-oblong or oblanceolate-oblong, c. 10 × 3.5cm, acumen 0.5cm, base acute-decurrent, nerves 7, obscure, puberulent below on midrib; petiole 0.5cm; stipule base triangular, 2mm, mucron 2mm; flowers c. 20, pendulous, in slender, branched panicles, terminal on spur-shoots, white; corolla tube 7mm, lobes 1mm; fruit blue. Evergreen forest; 1700m.
Distr.: Sierra Leone to Gabon. (Upper & Lower Guinea).
IUCN: LC
Bangem: Manning 418 10/1986; 454 fl., 10/1986; **Ehumseh:** Manning 449 10/1986.

Bertiera breviflora Hiern

Shrub 2m; leaves similar to *B. bracteolata*; stipules sheathing at base, limb elliptic, 5mm, acumen 2mm, slender; flowers c. 50; panicles broader than long, with 4–6 main, spike-like branches to 8cm; fruits red. Evergreen forest; 200m.
Distr.: Sierra Leone to Congo (K). (Guineo-Congolian).
IUCN: LC
Baduma: Nemba 577 fl., fr., 7/1987; **Nyandong:** Ghogue 1504 fl., 3/2003; Nguembou 644 3/2003.

Bertiera laxa Benth.

Shrub 2m; leaves oblong or elliptic, 23 × 8cm, acumen 1.5cm, base acute; petiole 1cm; stipule lanceolate, 20(–40) × 7mm; flowers c. 20, pendulous in terminal thyrse c. 20cm; branches 5–8, 15mm, 3–4-flowered, pale green; corolla tube 20mm, lobes 2mm; fruit blue. Evergreen forest; 650–1200m.
Distr.: SE Nigeria to Gabon. (Lower Guinea).
IUCN: LC
Kupe Village: Cable 2480 fr., 5/1996; 2684 fl., 5/1996; 2717 fr., 5/1996; 3823 7/1996; 3871 fr., 7/1996; 792 fl., 1/1995; 794 fl., fr., 1/1995; Cheek 7008 fr., 1/1995; Ensermu 3544 fl., fr., 11/1995; Gosline 252 fr., 12/1999; Kenfack 282 7/1996; 283 fr., 7/1996; Lane 417 fr., 1/1995; Schoenenberger 57 fl., fr., 11/1995; **Ngomboaku:** Cheek 10320 fl., fr., 12/1999; Mackinder 317 fl., fr., 12/1999; **Nyandong:** Nguembou 589 fl., fr., 3/2003; 593 fl., 3/2003; 600 fr., 3/2003; 641 3/2003; **Nyasoso:** Balding 36 fr., 8/1993; Cable 3330 fl., 6/1996; 3591 7/1996; Etuge 2077 fr., 6/1996; 2432 fr., 6/1996; 2532 fl., 7/1996; Lane 128 fr., 6/1994; Sebsebe 5022 fl., fr., 10/1995.

Bertiera laxissima K.Schum.

Fl. Gabon 17: 33 (1970).
Shrub 2m; leaves similar to *B. bracteolata*, to 14 × 6cm; stipule triangular-ligulate, c. 10 × 2.5mm, long-acuminate;

flowers c. 120, pendulous in terminal thyrse c. 20cm; branches 8–12, 50mm, forked near base, reflexed, white; corolla tube 3mm, lobes 1mm; fruit blue. Evergreen forest; 800–1530m.
Distr.: Cameroon Endemic.
IUCN: NT
Bangem: Thomas 5290 1/1986; **Kodmin:** Cheek 8913 fl., fr., 1/1998; Etuge 4404 fl., fr., 11/1998; Ghogue 13 fl., fr., 1/1998; **Nyandong:** Nguembou 654 3/2003.
Note: 14 specimens / sites at K apart from those cited here. Not known from Mt Cameroon.

Bertiera racemosa (G.Don) K.Schum. var. *elephantina* N.Hallé

Fl. Gabon 17: 56 (1970).
Shrub resembling var. *racemosa*, but leaves usually subcordate; calyx, ovary and fruit hairy. Evergreen forest; 250–1530m.
Distr.: Sierra Leone to Congo (K). (Guineo-Congolian).
IUCN: LC
Edib: Etuge 4477 fl., 11/1998; **Kodmin:** Ghogue 28 fr., 1/1998; **Konye:** Thomas 5936 fr., 3/1986; **Southern Bakossi F.R.:** Olorunfemi in FHI 30595 fr., 5/1951.
Note: Etuge 4477 has the floral attributes of var. *racemosa*.

Bertiera racemosa (G.Don) K.Schum. var. *racemosa*

Shrub 3–6m; leaves lanceolate or elliptic, to 28 × 11cm, acumen 0.5cm, base obtuse or rounded; petiole c. 1cm; stipule foliose, c. 3.5cm; flowers c. 100, pendulous, terminal, c. 25cm; branches 10–15, 1–2cm long, 5-flowered, white; calyx glabrous; corolla tube 7mm, lobes 2mm, bud apex glabrous; fruit glabrous, green. Evergreen forest; 1000–1250m.
Distr.: Guinea (C) to Congo (K). (Guineo-Congolian).
IUCN: LC
Kodmin: Etuge 4072 fl., fr., 1/1998; 4154 fl., fr., 2/1998; **Lake Edib:** Cheek 9146 2/1998; **Nyandong:** Nguembou 649 3/2003.

Bertiera retrofracta K.Schum.

Shrub 2–3m, resembling *B. racemosa* var. *racemosa*, but leaves oblanceolate, base cuneate; calyx; corolla, ovary and fruit entirely and densely pubescent. Evergreen forest; 500–1250m.
Distr.: SE Nigeria, Bioko & SW Cameroon. (Lower Guinea).
IUCN: NT
Bakossi F.R.: Etuge 4289 fr., 10/1998; **Kupe Village:** Cable 2476 fr., 5/1996; 3651 fl., 7/1996; 3870 fl., 7/1996; 724 fr., 1/1995; 795 fr., 1/1995; Cheek 7060 fl., fr., 1/1995; 8376 fr., 5/1996; Ensermu 3512 11/1995; Etuge 1377 11/1995; 2690 fl., 7/1996; Schoenenberger 66 fr., 11/1995; **Ndum:** Lane 479 fr., 2/1995; **Nyasoso:** Cable 3553 fl., fr., 7/1996; Etuge 2090 fr., 6/1996; Sebsebe 5018 fr., 10/1995; 5036 fr., 10/1995.

Bertiera sp. nov.

Tree 10m, appressed long white-hairy; leaves narrowly oblong-elliptic, to 18 × 4.5cm, acumen 1cm, cuneate, lateral nerves 14 pairs; petiole 2mm; stipule oblong-triangular, 1.5 × 0.4cm; capitulum sessile, 20–30-flowered; flowers sessile, white; fruit globose, 0.7cm, sparsely hairy. Forest; 1400m.
Distr.: Rumpi Hills & Bakossiland. (W Cameroon Uplands).
Edib: Cheek 9178 fl., 2/1998; **Kodmin:** Davis 2717 10/2002.

Calochone acuminata Keay

Shrub or climber to 10m; stems brown pubescent; leaves papery, obovate or elliptic, 20(–30) × 11(–17)cm, acuminate, truncate to subcordate, pubescent, nerves 10; petiole 2–3cm; stipules triangular, 15mm; terminal panicle 10–20-flowered; corolla orange-pink, tube funnel-shaped, 5cm, lobes 3cm. Evergreen forest; 250–1350m.
Distr.: Cameroon & Gabon. (Lower Guinea).
IUCN: VU
Edib: Cheek 9123 2/1998; **Kodmin:** Cheek 8978 fl., 1/1998; Etuge 4054 fl., 1/1998; **Konye:** Nemba 436 1/1987; **Kupe Village:** Cable 813 fl., 1/1995; Cheek 7031 fl., 1/1995; Lane 335 fl., 1/1995; **Mwambong:** Mackinder 170 fl., fr., 1/1998; **Ndum:** Cable 890 fl., 1/1995; Groves 39 fl., 2/1995; Lane 458 fl., 1/1995; **Nyasoso:** Cable 1223 fl., 2/1995; 3456 fr., 7/1996; Cheek 7290b fl., 2/1995; Elad 136 fl., 2/1995; Etuge 2421 fr., 6/1996; **Wone:** Keay in FHI 37356 1/1958.
Local name: Ekol Mba. **Uses:** MEDICINES (Cheek 7290b).

Calycosiphonia macrochlamys (K.Schum.) Robbr.

Fl. Gabon 17: 146 (1970).
Syn.: *Coffea macrochlamys* K.Schum.
Tree 8m; bark white, with black lenticels; leaves papery, drying brown, oblanceolate-elliptic, 16 × 5cm, acuminate, acute-decurrent, nerves 7; petiole 1.2cm; stipule sheath 2mm, arista 6mm; fruit axillary, sessile, globular, 1.8cm, ripening yellow, foetid; seeds 2; embryo minute. Forest; 1200m.
Distr.: Ghana to Congo (K). (Guineo-Congolian).
IUCN: VU
Edib: Cheek 9127 fr., 2/1998.

Calycosiphonia sp. A

Shrub or tree to 8m, resembling *C. macrochlamys* but leaves drying green, oblong-elliptic; fruits ellipsoid, shortly rostrate, black. Forest; 1100–2000m.
Distr.: Cameroon Endemic.
Kupe Village: Cheek 7882 fl., fr., 17/1995; **Mt Kupe:** Thomas 3087 fl., 2/1984; **Nyasoso:** Etuge 1664 fr., 1/1996; 2381 fl., fr., 6/1996; 2401 fr., 6/1996; Sunderland 1539 fr., 7/1992.

Chassalia cristata (Hiern) Bremek.

Climber or shrub to 9m, minutely papillate-puberulent when young; internodes 10cm or more; leaves obovate, to 13 × 6cm, acumen to 1.5cm, cuneate, lateral nerves 5 pairs; petiole to 4cm; panicle terminal, 8 × 12cm; pedicels fleshy, white becoming pink; fruits green, ripening black. Forest; 400–1000m.
Distr.: Nigeria to Kenya. (Tropical Africa).
IUCN: LC
Kupe Village: Etuge 1926 5/1996; **Nyasoso:** Cheek 7495 10/1995; Etuge 2539 7/1996; Sunderland 1490 fr., 7/1992; Zapfack 691 6/1996.

Chassalia ischnophylla (K.Schum.) Hepper

Monopodial shrub, 30cm; leaves clustered in litter-gathering funnel at stem apex, oblanceolate, 15 × 7cm, acumen obtuse, 0.5cm, base cuneate-subcordate, lateral nerves 8 pairs, puberulent below; petiole 1cm; spike 10cm; peduncle 7cm, fleshy, bright red; fruit black, 10 × 6 × 6mm. Forest.
Distr.: SE Nigeria to Gabon. (Lower Guinea).
IUCN: NT
Boubaji: Letouzey 14345 8/1975.
Note: about 11 sites known.

Chassalia laikomensis Cheek

Kew Bull. 55: 884 (2000).
Shrub 2–3(–8)m; leaves narrowly elliptic, 4–12 × 1.5–4cm, acuminate, lateral nerves 7–10 pairs; stipules 4mm, with conspicuous yellow raphides; panicles terminal, 5 × 5cm, loosely branched; flowers white, 6–10mm long; fruits black, ovoid, 6–9mm long. Forest; 2100m.
Distr.: S Nigeria & W Cameroon. (W Cameroon Uplands).
IUCN: CR
Mwanenguba: Leeuwenberg 8924 12/1971.

Chassalia aff. laikomensis Cheek *sp. B*

Shrub 3m, glabrous; leaves oblong-elliptic, to 14 × 7cm, acumen 1cm, acute, lateral nerves 6–9 pairs; petioles to 2cm; stipules to 4 × 5mm; panicles terminal, to 7 × 5cm; flowers white, 8mm; pedicels 0–2mm; fruit green, ripening black; peduncle white, ripening red to purple, slightly fleshy. Forest; 1100–1500m.
Distr.: Kupe-Bakossi. (Narrow Endemic).
Kodmin: Cheek 8878 1/1998; 9592 11/1998; **Kupe Village:** Cable 3743 7/1996; Cheek 7197 fr., 1/1995; Ryan 215 5/1996; 216a 5/1996; **Nyasoso:** Etuge 2133 6/1996; Wheatley 470 fr., 7/1992.
Note: probably *sp. nov.*, known only from Bakossiland.
Local name: Nnyah. **Uses:** use unknown (Max Ebong in Cheek 9592).

Chassalia aff. laikomensis Cheek *sp. C*

Shrub resembling *C. sp. B*, but lateral nerve pairs 12–16; stipules 8 × 7mm; pedicels 5mm. Forest; 1400–1800m.
Distr.: Kupe-Bakossi. (Narrow Endemic).
Kodmin: Plot B49 1/1998; **Kupe Village:** Cable 2565 5/1996; Ryan 420 5/1996; **Nyasoso:** Balding 61 fr., 8/1993; 62 fr., 8/1993; Cable 3417 6/1996; Etuge 2187 6/1996; Letouzey 14691 4/1976; Sunderland 1531 fr., 7/1992.
Note: probably *sp. nov.*, known only from Bakossiland.

Chassalia petitiana F.Piesschaert

Syst. Bot. 24(3): 315 (1999).
Syn.: *Psychotria epiphytica* Mildbr.
Epiphytic shrub, glabrous, c. 30cm; bark corky, flaking; leaves leathery, drying pale grey-green, elliptic, to 8 × 3cm, broadly acute, cuneate, lateral nerves 4–6 pairs; petiole 1–1.5cm; stipule sheathing, 6mm; panicle 6cm; peduncle 1.5cm; flowers not seen; 20-fruited; fruits ellipsoid, 4mm, dark purple; calyx 4-lobed. Forest; 1050–1300m.
Distr.: Cameroon, Bioko, Gabon, Congo (K). (Lower Guinea & Congolian (montane)).
IUCN: VU
Kupe Village: Cable 3841 7/1996; Ryan 364 5/1996; **Nyasoso:** Cable 3631 7/1996.
Note: newly recorded for Cameroon and FWTA; the only published African obligate epiphytic Rubiaceae species.

Chassalia cf. simplex K.Krause

Shrub 2.5m, minutely papillate-puberulent; leaves drying blackish green, elliptic, to 23 × 9cm, acumen 2cm, decurrent, lateral nerves 11 pairs; petiole 3cm; spike 9cm, 7 equidistant nodes bearing fascicles; pedicel white; fruit 5mm, pink. Forest; 1000m.
Distr.: SE Nigeria to Cameroon. (W Cameroon Uplands).
Enyandong: Etuge 4381r 10/2001.
Note: more material needed to confirm det. Matches Leeuwenberg 9927, Bakaka.

Chassalia cf. zenkeri K.Schum. & K.Krause

Climber 1–2m, subglabrous; leaves elliptic, 12 × 3.5cm, including acumen 2cm, obtuse, lateral nerves c. 7 pairs; petiole 1cm; panicle 5cm, including peduncle 3cm; pedicel 2mm; corolla tube c. 1.5cm, sinuate, red, throat yellow. Forest; 220–600m.
Bakolle Bakossi: Etuge 278 9/1986; **Mile 15, Kumba-Mamfe Road:** Nemba 142 7/1986; **Ngusi:** Etuge 51 4/1986.
Note: possible *sp. nov.*, close also to *C. cupularis*.

Chazaliella domatiicola (De Wild.) Petit & Verdc.

Kew Bull. 30: 269 (1975).
Syn.: *Psychotria abrupta sensu* Hepper & Keay in FWTA 2: 201
Syn.: *Psychotria domatiicola* De Wild.
Shrub 2m, glabrous; stem bark sloughing; leaves drying green, membranous, elliptic, to 13 × 5cm, acumen 1.5cm, decurrent, lateral nerves 6 pairs, domatia arcs white-hairy, conspicuous; petiole to 3mm; panicle umbellate-capitate, 1cm; peduncle 2cm; flowers yellow; pedicels 1–2mm; fruit ellipsoid, 7mm, smooth. Forest; 220–1000m.
Distr.: Ghana to Cameroon, Rio Muni & Congo (K). (Guineo-Congolian).
IUCN: NT
Bakolle Bakossi: Etuge 267 9/1986; **Bangem:** Thomas 5912 fl., 3/1986.

Chazaliella obanensis (Wernham) Petit & Verdc.

Kew Bull. 31: 798 (1977).
Syn.: *Psychotria obanensis* Wernham
Shrub 0.6–2m, resembling *C. domatiicola*, but leaves oblanceolate, to 25 × 10cm, acumen 0.5cm, lateral nerves 10 pairs, puberulous below, domatia absent; petiole to 3.5cm; panicle sessile, loosely capitate, 2–3cm. Forest; 250–1250m.
Distr.: SE Nigeria & SW Cameroon. (Lower Guinea).
IUCN: VU
Kupe Village: Cable 3650 fl., 7/1996; 3868 fl., 7/1996; Cheek 7086 fl., fr., 1/1995; Ensermu 3538 fl., fr., 11/1995; Etuge 1379 11/1995; 2616 fr., 7/1996; Kenfack 284 fl., 7/1996; Ryan 247 fl., 5/1996; 344 fr., 5/1996; **Mile 15, Kumba-Mamfe Road:** Nemba 132 fl., 7/1986; **Ndum:** Lane 476 fr., 2/1995; 492 fr., 2/1995; **Nyasoso:** Balding 66 fl., 8/1993; Cable 3204 fl., 6/1996; 3266 fl., 6/1996; 3525 fl., fr., 7/1996; 691 fr., 1/1995; Cheek 5665 fl., fr., 12/1993; 7297 fr., 2/1995; Etuge 2525 fl., 7/1996; 2545 fr., 7/1996; Lane 206 fr., 11/1994; Sebsebe 5042 fl., 10/1995; Sidwell 372 fl., fr., 10/1995; Zapfack 681 fl., 6/1996.

Chazaliella oddonii (De Wild.) Petit & Verdc. var. *cameroonensis* Verdc.

Kew Bull. 31: 802 (1977).
Shrub 0.5–1.5m, resembling *C. domatiicola*, but leaves lacking domatia; petiole to 1.5cm. Forest; 1000–1200m.
Distr.: Nigeria & Cameroon. (Lower Guinea).
IUCN: NT
Mwambong: Etuge 4333 fr., 10/1998; **Mwanenguba:** Leeuwenberg 9843 fl., 5/1972.
Note: widespread in the forests of Cameroon, known from only c. 20 specimens.

Chazaliella sciadephora (Hiern) Petit & Verdc. var. *condensata* Verdc.

Kew Bull. 31: 793 (1977).
Shrub, resembling *var. sciadephora*, but inflorescence condensed, 1–1.5cm. Forest; 150m.
Distr.: Cameroon Endemic.
IUCN: NT
Ekona Mombo: Etuge 420 fr., 12/1986; **Etam:** Etuge 82 fl., fr., 3/1986.

Chazaliella sciadephora (Hiern) Petit & Verdc. var. *sciadephora*

Kew Bull. 31: 790 (1977).

Syn.: *Psychotria sciadephora* Hiern

Shrub 2m, resembling *C. domatiicola*, but leaves papery, drying dark brown, crumpled, lateral nerves 8 pairs, matt, lacking domatia; flowers white, drying black, with white throat hairs; infructescence with several axillary peduncles, to 5cm; fruits ripening yellow, later dark brown. Forest; 300–1100m.

Distr.: Guinea (C) to Cameroon. (Upper & Lower Guinea).

IUCN: NT

Etam: Etuge 228 fr., 8/1986; **Konye:** Thomas 5202 fl., 11/1985; 5399 fl., 1/1986; **Kupe Village:** Cable 793 fl., 1/1995; Cheek 7087 fl., 1/1995; 7120 fl., 1/1995; Ensermu 3537 fr., 11/1995; Etuge 2609 fr., 7/1996; Lane 353 fl., 1/1995; **Ngomboaku:** Ghogue 463 fr., 12/1999; **Nyasoso:** Balding 89 fr., 8/1993; Cable 1218 fl., fr., 2/1995; Lane 185 fr., 11/1994; Sidwell 338 fr., 10/1995; Zapfack 689 fr., 6/1996.

Note: although widespread in W Africa, its habitat is under increasing threat.

Coffea bakossii Cheek & Bridson

Kew Bull. 57: 676 (2002).

Shrub or tree 5–20m, glabrous; leaves elliptic to obovate, 20–32 × 7.5–17cm, acumen 1.5cm, base rounded or subcordate, secondary nerves 8–12, domatia over axillary web, pit-like; petiole 2–4mm; stipules triangular-ovate, 10–17 × 7–10mm, acumen 3–4mm, midrib conspicuous, ridge-like; flowers 3–4 per axil, 7-merous; corolla white, tube 5–6mm, lobes 15 × 5mm; fruit ellipsoid, 16–20 × 10–12mm; disc 3–5mm, large and conspicuous. Forest; 710–900m.

Distr.: Bakossi. (Narrow Endemic).

IUCN: EN

Kupe Village: Etuge 2675 7/1996; Lane 361 fr., 1/1995; **Ngomboaku:** Gosline 260 12/1999; **Nyale:** Etuge 4172 2/1998.

Coffea brevipes Hiern

Shrub resembling *C. montekupensis* but twig bark matt; leaves larger; flowers white. Forest; 300–1100m.

Distr.: Nigeria to Congo (K). (Lower Guinea & Congolian).

IUCN: LC

Konye: Thomas 5204a 12/1985; **Ndum:** Lane 494 fr., 2/1995; **Nyasoso:** Lane 518 fr., 2/1995; 523 fr., 2/1995.

Coffea cf. brevipes Hiern

Shrub resembling *C. montekupensis* and *C. brevipes* in e.g. leaves lacking domatia. Closest to *C. brevipes* in foliaceous bracteoles and broad leaves (to 10cm), but leaves even broader, often with 3–4 fruits per node and older stipules with distinct ridge; at lower altitudes. Forest; 300–1000m.

Kupe Village: Etuge 2791 7/1996; **Mungo River F.R.:** Onana 964 11/1999; 971 12/1999; **Nyasoso:** Cable 3207 6/1996; 3264 6/1996; 3496 7/1996; Etuge 2054 6/1996; Zapfack 596 6/1996; 666 6/1996.

Note: possibly only a local variant of *C. brevipes*, but more analysis and flowering material is desired.

Coffea canephora Pierre ex Froehner

Shrub or tree to 6m, glabrous; leaves broadly elliptic to obovate, 12–35 × 5–12cm, acumen 1.5cm, lateral nerves 9–14 pairs, domatia pits puberulous; petiole to 2cm; flowers 10–24 per axil; flowers 5–6-merous; corolla white, tube 9–14mm, lobes 9–14mm; fruits ellipsoid, 9–17mm. Farms or wild; 1170–1300m.

Distr.: Guinea (C) to Uganda, cultivated elsewhere. (Guineo-Congolian).

IUCN: LC

Edib village: Cheek (sight record) 2/1998; **Nyasoso:** Etuge 1611 1/1996.

Note: *C. arabica* is not farmed in Bakossi.

Local name: coffee (English). **Uses:** FOOD – a main cash crop; seeds very important, grind in a powder form and drink as any other tea or coffee (Etuge 1611).

Coffea liberica Bull. ex Hiern

Evergreen tree to 10m, glabrous; leaves broadly elliptic, 7–13 × 3.5–6(–10)cm, acuminate, base acute to cuneate, lateral nerves 8–12, domatia on nerve, pit-like; petiole 0.8–2cm; stipule 2–4.5mm, apiculate, midrib inconspicuous; flowers several per axil, subsessile; corolla tube 5–6mm, lobes 6, 15 × 6mm, white; fruit red, ellipsoid, 14–18 × 8–10mm Forest; 1100–1400m.

Distr.: Guinea (B) to Uganda. (Guineo-Congolian).

IUCN: LC

Kupe Village: Cheek 7819 11/1995; Etuge 2849 7/1996; **Nyasoso:** Etuge 2123 6/1996; 2391 6/1996.

Coffea mayombensis A.Chev.

Rev. Bot. Appliq. 19: 402 (1939).

Syn.: *Coffea sp. nr. carissoi* A.Chev. FWTA 2:155 (1963)

Shrub 1–2m; leaves obovate to oblanceolate, to 12 × 5cm, lateral nerves 6–8 pairs, domatia pits glabrous; stipules acute, apex less than 1mm; flowers 1–2 per axil; corolla tube 3.5–4mm, lobes 10mm; fruit ellipsoid, 1.4cm, matt. Forest; 200–300m.

Distr.: Nigeria to Cabinda. (Lower Guinea).

IUCN: LC

Konye: Thomas 5204 12/1985; **Mile 12 Mamfe Road:** Nemba 308 10/1986.

Coffea montekupensis Stoffelen

Kew Bull. 52: 989 (1997).

Shrub 0.6–3m, glabrous; twig bark shiny; leaves narrowly ovate to obovate, 7–18 × 2.5–5cm, acumen 7–18mm, cuneate, moderately shiny on both faces, secondary nerves 12–13 pairs, domatia absent; petiole 3–7mm; stipules triangular, 2–3 × 4–5mm, acute; inflorescence 1 per axil, 1-flowered; flowers 5-merous; corolla pink, tube 2.5–4mm, lobes 14–18 × 3–6mm; fruit ellipsoid, 17–23 × 8–17mm. Forest; 700–1500m.

Distr.: Kupe-Bakossi. (Narrow Endemic).

IUCN: NT

Bangem: Thomas 5306 1/1986; **Kodmin:** Cheek 8917 1/1998; Gosline 180 11/1998; Pollard 204 11/1998; **Kupe Village:** Cable 2498 5/1996; 2643 5/1996; 3782 7/1996; 3874 7/1996; 744 fr., 1/1995; Cheek 7022 fr., 1/1995; 7160 fr., 1/1995; 7190 fr., 1/1995; 7681 11/1995; 7777 11/1995; 8383 5/1996; Etuge 1834 3/1996; 2642 7/1996; 2654 7/1996; **Mwendolengo:** Etuge 4533r 11/2001; **Nyale:** Cheek 9686 11/1998; **Nyasoso:** Balding 30 fr., 8/1993; Cable 1134 fr., 2/1995; 1148 fr., 2/1995; 3477 7/1996; 3604 7/1996; 700 fr., 1/1995; Cheek 7327 2/1995; Etuge 1336 10/1995; 2078 6/1996; 2436 6/1996; Gosline 136 11/1998; Lane 112 fr., 6/1994; 151 fr., 10/1994; 156 fl., 10/1994; 251 fl., fr., 11/1994; Sebsebe 5016 10/1995; Sidwell 371 10/1995; Wheatley 466 fr., 7/1992; **Nzee Mbeng:** Gosline 105 2/1998.

Local names: Deh A Mbine; Dea Mbone; Eye of a deer (Cheek 7160); none known (Cheek 7327). **Uses:** none known (Cheek 7327).

Corynanthe pachyceras K.Schum.

Tree to 20m, glabrous; leaves drying brown-green, elliptic to oblanceolate, to 17 × 8cm, acumen 1cm, cuneate, lateral nerves 6–10 pairs, domatia pits elliptic, large; petiole 2cm; panicle terminal, 12cm, dense, each with hundreds of flowers; peduncle 5cm; flowers white, 4mm; corolla lobes with globular terminal appendage; style exserted; stigma spherical; fruit dry, elliptic, 7mm, 2-valved; seeds winged. Forest; 220–300m.
Distr.: Sierra Leone to Congo (K). (Guineo-Congolian).
IUCN: LC
Bakolle Bakossi: Etuge 276 9/1986; **Bolo:** Nemba 50 3/1986.
Note: closely related to *Pausinystalia*.

Craterispermum caudatum Hutch.

Tree 15m, resembling *C. cerinanthum*, but leaves oblong or oblong-elliptic; stipules oblong, 3mm, persistent not caducous; fruits with pedicel to 7mm; peduncle 1.5cm. Forest.
Distr.: Sierra Leone to Cameroon. (Guineo-Congolian).
IUCN: LC
Kurume: Thomas 5414 1/1986.

Craterispermum cerinanthum Hiern

Tree 10m; leaves chartaceous, oblanceolate-elliptic, 13 × 4cm, acuminate, acute; petiole 1.5cm; stipule sheathing, subtruncate, 4mm; inflorescences lateral; peduncle 2cm, dilated and flattened at apex, dichasial; corolla white, tube 2–3mm, lobes 2–3mm; fruit 3mm. Evergreen forest; 760–1100m.
Distr.: Ivory Coast to Congo (K). (Guineo-Congolian).
IUCN: LC
Mejelet-Ehumseh: Etuge 469 1/1987; **Ngombombeng:** Etuge 39 fl., 4/1986.

Cremaspora triflora (Thonn.) K.Schum. subsp. *triflora*

Shrub or climber 1–8m; stem shortly pubescent; leaves papery, elliptic, c. 8.5 × 4.5cm, acuminate, rounded, nerves 3–5; petiole 0.5–1cm; flowers white, subsessile, 1cm; fruits red, axillary, fasciculate, ellipsoid, 13 × 6mm; pedicel 1mm. Evergreen forest; 700–1650m.
Distr.: tropical Africa.
IUCN: LC
Edib: Onana 589 fr., 2/1998; **Kodmin:** Etuge 4032 fr., 1/1998; **Kupe Village:** Lane 318 fr., 1/1995; **Mwambong:** Onana 557a fr., 2/1998; **Mwanenguba:** Bamps 1548 fr., 12/1967; Leeuwenberg 8788 fr., 11/1971.

Cuviera longiflora Hiern

Shrub or tree to 8m, glabrous; stems with ants; leaves papery, lanceolate-oblong, c. 27 × 10cm, acuminate, base subcordate or rounded, nerves 9–10; petiole 1cm; stipule sheathing, 5mm; flowers 10–20 in axillary panicles; peduncle 3–8cm; bracts and calyx lobes leafy; corolla green and white; tube c. 1cm; lobes c. 0.2cm; fruit ellipsoid, 9 × 3cm, brown, fleshy; pyrenes 5, c. 3 × 1cm. Evergreen forest; 800–1500m.
Distr.: Cameroon to Angola. (Lower Guinea).
IUCN: NT
Kodmin: Etuge 4051 fl., 1/1998; **Kupe Village:** Cable 2550 fl., fr., 5/1996; 764 fl., 1/1995; Lane 298 fl., 1/1995; **Ndum:** Elad 101 fl., 2/1995; Lane 460 fl., fr., 1/1995; **Nyasoso:** Cable 2808 fl., fr., 6/1996; 3283 fr., 6/1996; Cheek 7305 fl., 2/1995; Etuge 2534 fr., 7/1996.
Note: 10–20 sites apart from those cited here.
Local name: none known. **Uses:** ANIMAL FOOD – sunbirds visit the flowers for nectar (Cheek 7305).

Cuviera subuliflora Benth.

Tree 4–7m, resembling *C. longiflora*, but bracteoles and calyx lobes subulate to linear, 2(–4)mm broad (not 4–8mm); flowers red and green. Forest; 220m.
Distr.: Ghana, Nigeria, Bioko & SW Cameroon. (Guineo-Congolian).
IUCN: NT
Bakolle Bakossi: Etuge 257 9/1986.

Cuviera talbotii (Wernham) Verdc.

Kew Bull. 42: 189 (1987).
Syn.: *Globulostylis talbotii* Wernham
Shrub or small tree to 5m, glabrous; stems with ants; leaves elliptic, 15–23 × 7–10cm, shortly acuminate, obtuse, nerves 5–7; petiole 1.5cm; stipules sheathing, 5mm, mucron awn-like; flowers 5–6, axillary; peduncle 1cm; bracts inconspicuous; calyx leafy; corolla tube c. 4mm, lobes c. 2mm; fruit ellipsoid, 3.5 × 3cm, brown, fleshy, pyrenes 5. Evergreen forest; 840–1300m.
Distr.: SE Nigeria to Cameroon. (Lower Guinea).
IUCN: VU
Kupe Village: Cable 2517 fr., 5/1996; 756 fr., 1/1995; Cheek 7109 fr., 1/1995; 7156 fl., 1/1995; 7661 fl., 11/1995; Etuge 1433 fl., 11/1995; 1722 fr., 2/1996; 2645 fl., fr., 7/1996; 2683 fr., 7/1996; **Mbule:** Cable 3351 6/1996; 3377 fr., 6/1996; **Mt Kupe:** Cheek 10164 fl., 11/1999; **Ngomboaku:** Cheek 10309 fl., 12/1999.

Dictyandra arborescens Welw. ex Hook.f.

Tree to 7m; bole with conical spines; leaves obovate-elliptic, 12–20 × 5–9cm, acuminate, domatia hairy; stipules broadly triangular, 5mm; cymes lax; calyx lobes elliptic, 1cm, one margin thickened, recurved; corolla 3.5cm, silky-tomentose outside; fruits fleshy; calyx foliose. Forest; 300m.
Distr.: Guinea (C) to Uganda. (Guineo-Congolian).
IUCN: LC
Bolo: Thomas 5718 3/1986.

Diodia sarmentosa Sw.

F.T.E.A. Rubiaceae: 336 (1976).
Syn.: *Diodia scandens sensu* Hepper & Keay in FWTA 2: 216
Straggling herb to 5m, scabrid; leaves ovate or ovate-lanceolate, 2.5–5 × 1–2.5cm, pubescent below, scabrid above; petioles to 7mm; stipule cupular 1mm, with 5 aristae, 3mm; axillary fascicles sessile; flowers white, 4-merous; fruit ellipsoid, 3–4mm, didymous, dry. Forest. 900m.
Distr.: pantropical.
IUCN: LC
Kupe Village: Etuge 2709 fr., 7/1996; **Nyasoso:** Etuge 2444 fl., fr., 6/1996.

Euclinia longiflora Salisb.

Deciduous tree 5–8m with *Terminalia* branching; leaves clustered, elliptic, 12 × 5cm, acuminate; stipules chaffy, persistent; flower terminal, 20 × 6cm, yellowish white, marked purple; berry green, ellipsoid, 4 × 3cm; seeds numerous. Evergreen forest; 250–1800m.
Distr.: Guinea (B) to Uganda. (Guineo-Congolian).
IUCN: LC
Kodmin: Gosline 176 fl., 11/1998; **Kupe Village:** Cable 2611 fr., 5/1996; 2641 fr., 5/1996; 750 fr., 1/1995; Cheek 7778 fl., 11/1995; 7867 fl., 11/1995; 8394 fr., 5/1996; Etuge 1431 fl., 11/1995; 1962 fl., 5/1996; 2653 fr., 7/1996; Ryan 235 fr., 5/1996; 321 fr., 5/1996; Schoenenberger 62 fl., 11/1995; **Manehas F.R.:** Cheek 9396 fr., 10/1998; **Mbule:** Cable 3368 fr., 6/1996; **Mile 15, Kumba-Mamfe Road:** Nemba 164 fl., 7/1986; **Mt Kupe:** Thomas 5108 fl., 11/1985.

Local name: Ekobe sane. **Uses:** MEDICINES (Epie Ngome in Cheek 9396).

Galium simense Fresen.

Straggling herb, clinging by minute hooked hairs; leaves 8 in a whorl, linear oblanceolate, 10–40 × 2–4mm; flowers solitary, axillary; pedicels to 1.5cm; fruits globose, 4mm, black. Forest-grassland transition; 2000m.
Distr.: Bioko, W Cameroon & Ethiopia. (Afromontane).
IUCN: LC
Mwanenguba: Cheek 9424 10/1998.

Gardenia vogelii Hook.f. ex Planch.

Shrub 2m; leaves papery, elliptic or obovate, c. 20 × 7cm, long-acuminate, acute, margin sometimes deeply-toothed; petiole 3mm; stipule sheathing; corolla tube 12–15cm, lobes 3–8.5cm; fruit cylindrical, 10 × 1cm, sessile; calyx persistent. Forest, farmbush; 870–1000m.
Distr.: Liberia to Zimbabwe. (Guineo-Congolian).
IUCN: LC
Nyasoso: Cable 3250 fr., 6/1996; Groves 106 fr., 2/1995; Sidwell 307 fr., 10/1995.

Geophila afzelii Hiern

Stoloniferous herb with short vertical shoots, hairy; leaves ovate, 3–6 × 2–5.5cm, subacute, cordate; stipules bifid; inflorescence terminal; peduncle 3cm; bracts involucral, ovate, bracteoles linear; flowers white or yellow; fruit red. Forest; 300m.
Distr.: Guinea (C) to Congo (K). (Guineo-Congolian).
IUCN: LC
Baduma: Nemba 121 6/1986; **Bolo Moboka:** Nemba 261 9/1986.

Hallea stipulosa (DC.) Leroy

Adansonia (sér. 2) 15: 65 (1975).
Syn.: *Mitragyna stipulosa* (DC.) Kuntze
Tree 35m; leaves obovate, 12–45 × 8.5–26cm, rounded, acute, nerves to 20; petiole to 5cm; stipule oblong to suborbicular, 4–8 × 2.5–5cm, pubescent; inflorescence of 3–10 terminal capitula, each 5–10mm diameter. Swamp forest; 150m.
Distr.: Gambia to Congo (K). (Guineo-Congolian).
IUCN: VU
Mungo River F.R.: Cheek 10145 11/1999.

Heinsia crinita (Afzel.) G.Taylor

Syn.: *Heinsia scandens* Mildbr. *nomen*
Shrub to 7m, appressed grey-puberulous; leaves elliptic, 9 × 3cm, acuminate, 4-nerved, tertiary nerves normal to midrib; petiole 1cm; stipule caducous; flowers terminal, 1 to few; corolla white, tube c. 2cm; fruit globose, 2cm diameter, many-seeded; sepals accrescent, green, 2cm. Evergreen forest; 250–1000m.
Distr.: Guinea (C) to Angola. (Guineo-Congolian).
IUCN: LC
Baduma: Nemba 221 fr., 9/1986; **Bangem:** Thomas 5373 1/1986; **Enyandong:** Etuge 4400r 10/2001; **Kupe Village:** Cable 2718 fr., 5/1996; **Ngombombeng:** Etuge 25 4/1986; **Nyale:** Etuge 4449 fr., 11/1998.

Hekistocarpa minutiflora Hook.f.

Gregarious herb 1.5m, puberulent; leaves membranous, narrowly elliptic, c. 13 × 3cm, acuminate, acute, 8–10-nerved; petiole 1cm; stipule leafy, triangular, 7mm; inflorescence axillary, tuning-fork shaped, the stalk 5mm; branches c. 3cm; corolla white, tube 3mm. Evergreen forest; 350–700m.

Distr.: SE Nigeria & Cameroon. (Lower Guinea).
IUCN: LC
Bakolle Bakossi: Etuge 143 5/1986; **Ngomboaku:** Cheek 10370 fl., 12/1999; **Nyandong:** Nguembou 596 fl., 3/2003; 607 fl., 3/2003; 619 3/2003; 652 3/2003.

Hymenocoleus glaber Robbr.

Bull. Jard. Bot. Nat. Belg. 47: 14 (1977).
Herb to 30cm, resembling *H. neurodictyon*; leaves acuminate, differing in the deeply bilobed stipule and the much longer, linear, involucral and calyx lobes (to 7mm long). Forest; 700–1450m.
Distr.: Cameroon Endemic.
IUCN: VU
Kodmin: Cheek 8962 fr., 1/1998; Etuge 3978 fr., 1/1998; 4168 fr., 2/1998; **Kupe Village:** Cheek 6081 fr., 1/1995; Ryan 300 fr., 5/1996; **Nyasoso:** Cable 641 fr., 12/1993; Cheek 9503 fr., 10/1998; Lane 159 fr., 10/1994; 250 fl., fr., 11/1994.

Hymenocoleus hirsutus (Benth.) Robbr.

Bull. Jard. Bot. Nat. Belg. 45: 288 (1975).
Syn.: *Geophila hirsuta* Benth.
Herb resembling *Geophila afzelii*, but drying black; leaves pubescent above, ovate to ovate-oblong, 2–5 × 1–4cm; stipules deeply divided; inflorescence without conspicuous leafy bracts, subsessile; flowers white; calyx lobes oblong, puberulent; fruit orange, inflated-pithy. Forest; 750–1000m.
Distr.: Guinea (C) to Congo (K) & Tanzania. (Guineo-Congolian).
IUCN: LC
Baseng: Cheek 10420 fr., 12/1999; **Kupe Village:** Cable 2490 5/1996; 3733 fl., 7/1996; **Mwambong:** Gosline 84 fl., fr., 2/1998; **Ngomboaku:** Cheek 10314 fr., 12/1999.

Hymenocoleus neurodictyon (K.Schum.) Robbr. var. neurodictyon

Bull. Jard. Bot. Nat. Belg. 45: 291 (1975).
Syn.: *Geophila neurodictyon* (K.Schum.) Hepper
Herb resembling *Geophila afzelii*, but leaves oblong to oblanceolate, 4–10 × 1.5–4cm, rounded at both ends, glabrous above, pubescent below on nerves; flowers without bracts, subsessile. Forest; 1200m.
Distr.: Sierra Leone to Congo (B) & CAR. (Upper & Lower Guinea).
IUCN: LC
Kupe Village: Etuge 1839 fl., 3/1996.

Hymenocoleus rotundifolius (A.Chev. ex Hepper) Robbr.

Bull. Jard. Bot. Nat. Belg. 45: 287 (1975).
Syn.: *Geophila rotundifolia* A.Chev. ex Hepper
Herb resembling *Geophila afzelii* but leaves orbicular to ovate, 4.5–6 × 3–6cm, pubescent above and below; inflorescence surrounded by a cupular involucre, margin toothed by ⅓ to ½ of its height, teeth linear or triangular; flowers white; fruit orange. Forest; 400–600m.
Distr.: Sierra Leone to Congo (K). (Guineo-Congolian).
IUCN: LC
Bakossi F.R.: Etuge 4297 fr., 10/1998; **Etam:** Etuge 239 8/1986; **Nyale:** Cheek 9660 fr., 11/1998.

Hymenocoleus subipecacuanha (K.Schum.) Robbr.

Bull. Jard. Bot. Nat. Belg. 47: 20 (1977).
Syn.: *Hymenocoleus petitianus* Robbr.

Herb resembling and easily confused with *H. rotundifolius*, but involucre divided into triangular lobes for only 1/5 to ⅓ of its height. Forest; 220–1050m.

Distr.: Nigeria, Cameroon & Congo (K). (Lower Guinea & Congolian).

IUCN: NT

Bakolle Bakossi: Etuge 268 9/1986; **Kupe Village:** Cheek 7851 fr., 11/1995; Etuge 2832 fr., 7/1996; 2914 fr., 7/1996; **Nyale:** Cheek 9662 fr., 11/1998; **Nyasoso:** Cheek 7478 fr., 10/1995; Lane 210 fl., 11/1994; Sebsebe 5023 fr., 10/1995; Sidwell 332 fr., 10/1995.

Note: although widespread, rare: only 5 specimens are listed by Robbrecht (1977). However, nearly 30 are now known from Mt Cameroon (Cable & Cheek 1998).

Hymenodictyon biafranum Hiern

Shrub or small tree, often epiphytic, to 15m, glabrous; leaves leathery, narrowly elliptic to oblanceolate, c. 8 × 2.5cm, subacuminate, acute, nerves 4–5, inconspicuous, margin revolute; petiole 1cm; infructescence terminal, spike-like, c. 10 × 2cm, with two basal white or pink bracts, blades 7cm, on stalks 3cm; seeds winged, 3 times as long as broad. Evergreen forest; 1000–1750m.

Distr.: SE Nigeria to Gabon. (Lower Guinea).

IUCN: NT

Ehumseh - Mejelet: Etuge 333 fr., 10/1986; **Kodmin:** Cheek 8967 fr., 1/1998; 9570 fr., 11/1998; **Kupe Village:** Etuge 1902 fr., 5/1996; 2687 fr., 7/1996.

Note: about 15 sites known.

Local name: unknown (Edmondo Njume in Cheek 9570).

Hymenodictyon floribundum (Steud. & Hochst.) B.L.Rob.

Shrub or small tree resembling *H. biafranum*, but with tertiary nervation plainly visible on lower surface of leaf blade, and often pubescent below, not glabrous; seeds less than twice as long as broad. Evergreen forest; 350–800m.

Distr.: Guinea (C) to Kenya. (Tropical Africa).

IUCN: LC

Mekom: Nemba 19 4/1986; **Nlonako:** Leeuwenberg 9748 fl., 4/1972.

Ixora foliosa Hiern

Tree (2–)7–12m; leaves coriaceous, elliptic-oblong, c. 10 × 5cm, shortly acuminate, acute, nerves c. 12; petiole 1cm; inflorescence with reduced leaves 4cm; peduncle erect, c. 9cm; flowers subcapitate, sessile; corolla tube red, 14mm; lobes white. Forest; 1800–2000m.

Distr.: SE Nigeria & Cameroon. (W Cameroon Uplands).

IUCN: VU

Mt Kupe: Thomas 5091 11/1985; 5471 2/1986; **Kupe Village:** Ryan 395 fr., 5/1996; **Mwanenguba:** Thomas 3159 2/1984; **Nyasoso:** Cable 1204 fl., 2/1995; 1205 fl., 2/1995; 1237 fl., 2/1995; 2922 fr., 6/1996; Elad 126 fl., 2/1995; Etuge 1653 fl., 1/1996.

Ixora guineensis Benth.

Syn.: *Ixora breviflora* Hiern
Syn.: *Ixora talbotii* Wernham

Shrub 2–3m; leaves coriaceous, oblong, elliptic or obovate-elliptic, c. 14 × 6cm, subacuminate, obtuse, nerves 8, drying yellow-brown below; petiole 1cm; inflorescence sessile, with minute spreading hairs; peduncle c. 2cm; pedicels 3mm; flowers white; corolla tube 25mm. Forest; 1200–2060m.

Distr.: Liberia, Nigeria to Congo (B). (Upper & Lower Guinea).

IUCN: LC

Bakolle Bakossi: Etuge 147 5/1986; **Ekona Mombo:** Etuge 425 12/1986; **Essossong:** Groves 49 fl., 2/1995; **Kupe Village:** Cable 2564 fr., 5/1996; 2667 fl., 5/1996; 2727 fl., 5/1996; Cheek 6028 fr., 1/1995; 7075 fr., 1/1995; 7871 fl., 11/1995; 8401 fr., 5/1996; 8410 fr., 5/1996; Etuge 1960 fl., 5/1996; Lane 272 fr., 1/1995; 391 fl., 1/1995; 428 fr., 1/1995; **Kurume:** Thomas 5418 1/1986; **Mt Kupe:** Thomas 5114 12/1985; **Mungo River F.R.:** Leeuwenberg 10632 11/1972; Onana 951 11/1999; **Ngomboaku:** Mackinder 288 fr., 12/1999; **Ngombombeng:** Etuge 129 5/1986; **Nyasoso:** Balding 12 fl., 7/1993; Cable 679 fr., 12/1993; 687 fr., 1/1995; Cheek 6006 fr., 1/1995; Etuge 1294 10/1995; 1744 fr., 2/1996; 2064 fl., fr., 6/1996; 3318 10/1996; Lane 145 fl., 6/1994; 174 fr., 11/1994; Letouzey 14688 fr., 4/1976; Sidwell 337 fr., 10/1995; 397 fl., 10/1995; Wheatley 390 fl., 7/1992; Zapfack 592 fl., 6/1996.

Ixora nematopoda K.Schum.

Shrub 2.5–3m; leaves papery, oblong-elliptic, c. 20 × 8cm, long acuminate, obtuse, nerves c. 12; petiole c. 1.5cm; inflorescence lax, with reduced leaves, c. 1cm; peduncle pendulous, c. 10cm; pedicels 18mm; corolla tube white, 7mm, lobes white. Forest; 220–1000m.

Distr.: SE Nigeria to Congo (B). (Lower Guinea).

IUCN: NT

Bakolle Bakossi: Etuge 282 9/1986; **Bolo Moboka:** Nemba 249 9/1986; **Kupe Village:** Cable 3688 fl., 7/1996; Etuge 2889 fl., fr., 7/1996.

Ixora sp. A

Shrub 3–7m, resembling *I. nematopoda*, but leaves acute-decurrent at base, drying brown; pedicels 1–2mm; corolla tube 20mm. Forest; 1000–1650m.

Distr.: known only from Bakossi. (Narrow Endemic).

Kodmin: Gosline 167 fl., fr., 11/1998; **Kupe Village:** Cheek 7184 fl., 1/1995; 7786 11/1995; Etuge 2837 fr., 7/1996.

Note: probably a new submontane forest species.

Ixora sp. B

Shrub or small tree 4–12m, resembling *I. guineensis*, but leaves elliptic, to 9 × 4cm, acuminate, drying green, not yellow-brown, below; inflorescence glabrous. Forest; 950–1200m.

Distr.: known only from Bakossi. (Narrow Endemic).

Kupe Village: Etuge 1835 fr., 3/1996; 2904 fr., 7/1996; Ryan 234 fr., 5/1996; 317 fr., 5/1996.

Note: probably a new taxon, but flowers are lacking.

Keetia acuminata (De Wild.) Bridson

Kew Bull. 41: 985 (1986).

Climber 3m, pale brown scurfy; leaves oblong-elliptic, c. 10 × 6cm, acumen short, base rounded, softly pubescent below, nerves 7–9; petiole 1.5cm; stipule ovate-acuminate, 9 × 6mm, white pubescent, midrib prominent; fruiting peduncle 3cm; partial-peduncle 3cm; fruit 1 × 1.4cm. *Hypselodelphys* scrub; 1100m.

Distr.: Cameroon to Burundi. (Lower Guinea & Congolian).

IUCN: LC

Ndum: Cable 876 fr., 1/1995.

Keetia bakossii Cheek ined.

Shrub or climber 6m; stems rounded, with ants, pubescent; leaves thinly papery, elliptic, 18 × 10cm, acuminate, rounded to cordate, thinly scurfy-pilose below, margin ciliate, nerves 8–10; petiole 5mm; stipule 18 × 5mm; fruiting peduncle c. 5mm; partial-peduncle 1.5cm; fruit 2.8 × 2cm. Evergreen forest; 1400–1500m.

Distr.: Bakossi. (Narrow Endemic).

IUCN: CR
Kodmin: Etuge 4081 1/1998; Ghogue 65 1/1998; Onana 579 2/1998.

Keetia hispida (Benth.) Bridson *'rubrinerve'*
Kew Bull. 41: 986 (1986).
Syn.: *Canthium rubrinerve* (K.Krause) Hepper
Climber 8m; stem glabrous, swollen at nodes, containing small black biting ants; leaves drying grey-green, papery, elliptic or ovate-elliptic, 14 × 8.5cm, glabrous, domatia brown-hairy, nerves 4–7, purple when live; petiole 0.5–1cm; stipule ovate, 2cm; fruiting peduncle 0.5cm; partial-peduncles 0.5cm; fruit 1.2 × 1.2cm, ridged. Evergreen forest; 650–1200m.
Distr.: Sierra Leone to Congo (K). (Guineo-Congolian).
IUCN: LC
Kupe Village: Cable 2682 fr., 5/1996; 2725 fr., 5/1996; Cheek 7723 fr., 11/1995; Etuge 1401 fl., fr., 11/1995; 1898 fr., 5/1996; 2684 fr., 7/1996; Kenfack 214 fr., 7/1996; Ryan 291 fr., 5/1996; **Ngomboaku:** Cheek 10298 fl., 12/1999.

Keetia hispida (Benth.) Bridson *'setosum'*
Kew Bull. 41: 986 (1986).
Syn.: *Canthium hispidum* Benth.
Syn.: *Canthium setosum* Hiern
Climber 3m; stems glabrous; leaves drying red below, thickly papery, obovate-elliptic, c. 17 × 8cm, acuminate, unequally truncate-cordate, thinly pilose on both surfaces, domatia inconspicuous, nerves 7–9; petiole 1cm; stipule not seen; fruiting peduncle 0.5cm; partial-peduncles 1cm; fruit 1.6 × 1.4cm, ridged. Evergreen forest; 200–1200m.
Distr.: Sierra Leone to Congo (K). (Guineo-Congolian).
IUCN: LC
Bakolle Bakossi: Etuge 274 9/1986; **Kupe Village:** Etuge 1402 11/1995; **Mile 12 Mamfe Road:** Nemba 287 10/1986; **Nyasoso:** Cable 3618 fr., 7/1996; Thomas 5043 11/1985; 5044 11/1985.

Keetia mannii (Hiern) Bridson *sensu lato*
Kew Bull. 41: 988 (1986).
Syn.: *Canthium mannii* Hiern
Climber; stems glabrous, square; leaves coriaceous, glossy, elliptic, 10 × 5.5cm, acuminate, obtuse, domatia red-hairy nerves 6–7; petiole 2cm; fruiting peduncle 1.5cm; partial-peduncle 1cm; fruit 1.5 × 2cm. Evergreen forest; 1000m.
Distr.: Sierra Leone to Cameroon, CAR & Sudan. (Upper & Lower Guinea).
IUCN: LC
Mbule: Leeuwenberg 9293 fr., 1/1972.

Keetia purseglovei Bridson
Kew Bull., 41: 972 (1986).
Climber 9m; stems rounded, glabrescent; leaves thinly papery, elliptic or elliptic-obovate, c. 9.5 × 4cm, acuminate, acute, bacterial nodes, nerves 5–6; petiole 1cm; stipule linear, 6 × 1mm; fruitng peduncle 1cm, partial-peduncle 0.5–1cm; fruit 1 × 1.2cm. Evergreen forest; 800–850m.
Distr.: Cameroon to Uganda. (Lower Guinea & Congolian).
IUCN: LC
Kupe Village: Etuge 2798 fr., 7/1996; 2829 fr., 7/1996.

Keetia sp. nov. aff. ripae Bridson
Climber 6m; stems square, appressed-hairy; leaves coriaceous, oblong, c. 5.5 × 2.2cm, acute, rounded, nerves 6–7, indistinct, domatia; petiole 1cm; stipule lanceolate, 4 × 2mm; fruiting peduncle 8mm; partial peduncle 4mm; fruit 6 × 11mm. Evergreen forest; 1450m.
Distr.: Bakossi Mts. (Narrow Endemic).

Kodmin: Cheek 9600 fr., 11/1998.

Lasianthus batangensis K.Schum.
Erect shrub 0.3–2m, pilose; leaves elliptic or slightly obovate, 15–20 × 5–9cm, acutely acuminate, cuneate, midrib pilose above and below; petiole 2–6cm; flowers axillary, several, subsessile, blue or white; corolla 3mm; fruits globose, 8mm, blue. Forest; 300–800m.
Distr.: Sierra Leone to Congo (K). (Guineo-Congolian).
IUCN: LC
Bangem: Thomas 5299 1/1986; **Bolo Moboka:** Nemba 237 9/1986.

Lasianthus sp. 1
Shrub resembling *L. batangensis*, but glabrous; stem drying slightly quadrangular; leaves elliptic, 8 × 3cm, glabrous below, lateral nerves 8–10 pairs; petiole 1.5cm; corolla white; fruit 3–4mm. Forest; 1200–1500m.
Distr.: Bakossi. (Narrow Endemic).
Kodmin: Cheek 8914 1/1998; **Kupe Village:** Cable 3857 7/1996; Cheek 7198 fl., 1/1995; Ryan 370 5/1996; **Mwanenguba:** Leeuwenberg 8318 9/1971.

Leptactina involucrata Hook.f.
Pl. Syst. Evol. 145: 114 (1984).
Syn.: *Dictyandra involucrata* (Hook.f.) Hiern
Shrub or tree 10m, glabrous; leaves papery, elliptic, 27 × 12cm, acuminate, acute, nerves 10–12; petiole 2cm; stipule folicaeous, 1cm; flowers white, 10–20 in terminal panicles, 3cm; calyx lobes 8mm; corolla white, pubescent, lobes 2mm; fruit ovoid, 2 × 0.7cm; calyx lobes 1.5cm; seeds numerous. Evergreen forest; 200–800m.
Distr.: Sierra Leone to Rio Muni. (Upper & Lower Guinea).
IUCN: LC
Ikiliwindi: Nemba 459 1/1987; **Konye:** Thomas 5938 3/1986; **Kupe Village:** Etuge 1467 fr., 11/1995; **Ngomboaku:** Cheek 10359 fl., 12/1999.

Massularia acuminata (G.Don) Bullock ex Hoyle
Shrub 4m; stems glabrous; leaves obovate-elliptic, c. 30 × 10cm, acuminate, subcordate, nerves 15–20, subsessile; stipule semi-circular, c. 4 × 1.5cm; flowers 5–10, axillary; panicle branched, 5cm; corolla pink and white, tube 12cm, lobes 8mm; fruit spherical-rostrate, 8 × 6cm, green, fleshy, few-seeded. Evergreen forest; 100–1000m.
Distr.: Guinea (C) to Congo (K). (Guineo-Congolian).
IUCN: LC
Eboné: Leeuwenberg 8234 9/1971; **Ikiliwindi:** Nemba 527 6/1987; **Konye:** Thomas 5153 11/1985; **Kupe Village:** Cable 3702 fl., fr., 7/1996; 3739 fl., 7/1996; Etuge 1429 fr., 11/1995; Lane 360 fl., 1/1995; **Mekom:** Thomas 5238 1/1986; **Mungo River F.R.:** Etuge 4311 fl., fr., 10/1998; Onana 981 fl., 12/1999; Thomas 2567 11/1983; **Nyale:** Etuge 4225 2/1998; **Nyandong:** Ghogue 1425 fl., 3/2003; Nana 16 3/2003; Nguembou 597 fl., 3/2003; 608 3/2003.
Uses: VERTEBRATE POISONS – fish (Etuge 1429); fruits used to poison fish in water (Etuge 4225).

Mitracarpus villosus (Sw.) Cham. & Schltdl.
F.T.E.A. Rubiaceae: 375 (1976).
Laxly erect herb 10–40cm; stems square, puberulent; leaves elliptic, to 5 × 2cm, acute, sessile, margins and lower surface slightly scabrid; inflorescences axillary, c. 100-flowered, clasping the stem; flowers sessile; calyx lobes needle-like; corolla white, 2mm, lobes 4. Roadsides, farmbush; 800m.
Distr.: tropical Africa.
IUCN: LC
Tombel: Etuge 366 11/1986.

Mitriostigma barteri Hook.f. ex Hiern

Shrub 0.5–1(–2)m, glabrous; branching plagiotropic; leaves narrowly elliptic, to 15 × 4cm, acumen slender, 2cm, cuneate, lateral nerves 6–10, domatia absent; petiole 0.5cm; stipules triangular-aristate, 8mm; panicles alternating at nodes, 10-flowered, 1.5cm; peduncle 3mm; corolla pink-red, 15mm; fruit fusiform, 3.5 × 0.8cm, red, fleshy. Forest; 300m.
Distr.: Bioko & SW Cameroon. (Lower Guinea).
IUCN: EN
Bolo: Thomas 5897 3/1986.

Morinda longiflora G.Don

Climber, glabrous; leaves obovate, 7 × 3cm, acumen 1cm, acute, lateral nerves 3 pairs; petiole 1.5cm; inflorescences as *M. lucida*, but terminal on short twisting spur-shoots; peduncle 1cm, 3–5-flowered; corolla 3cm. Forest; 300m.
Distr.: Guinea (C) to Congo (K). (Guineo-Congolian).
IUCN: LC
Konye: Thomas 7032 5/1987.

Morinda lucida Benth.

Tree to 20m, glabrous, drying black; leaves elliptic, to 20 × 13cm, lateral nerves 5 pairs; petiole 1cm; inflorescences leaf-opposed, in 3's, capitate; peduncle 3cm; flowers 10–20; corolla white, 1.5cm incl. lobes (4mm); fruit compound, 2cm diameter, fleshy. Forest; 150m.
Distr.: Senegal to Tanzania. (Guineo-Congolian).
IUCN: LC
Etam: Etuge 206 7/1986.

Mussaenda arcuata Lam. ex Poir.

Climber, glabrous; leaves elliptic, c. 10 × 5cm, acuminate, acute, nerves 4–6; petiole to 2cm; flowers numerous without bract-like calyx lobe; corolla yellow, tube 1.5cm, lobes 1cm, centre orange, maturing red; fruit ellipsoid, 1.5cm; seeds numerous. Evergreen forest; 800–1500m.
Distr.: Gambia to S Africa. (Tropical Africa).
IUCN: LC
Kodmin: Etuge 3991 fl., fr., 1/1998; **Kupe Village:** Biye 60 fl., 11/1999; Cable 3724 fr., 7/1996; Cheek 7121 fl., fr., 1/1995; Etuge 1455 fl., fr., 11/1995; 2810 fr., 7/1996; Ryan 260 fr., 5/1996; Schoenenberger 64 fl., 11/1995.

Mussaenda elegans Schum. & Thonn.

Climber; stem densely pubescent; leaves minutely puberulous, elliptic, to 15 × 7cm, acuminate, obtuse, nerves 8–10; petiole to 2cm; flowers 1–5, resembling M. arcuata, but orange or red; corolla tube 2.5cm, lobes 2.5cm; fruit ellipsoid-cylindric, 2.5cm. Evergreen forest; 400–950m.
Distr.: Mali to Uganda. (Guineo-Congolian).
IUCN: LC
Kupe Village: Etuge 1935 fr., 5/1996; 2018 fl., 5/1996; 2713 fl., 7/1996; Kenfack 320 7/1996.

Mussaenda erythrophylla Schum. & Thonn.

Climber, densely puberulent; leaves elliptic, to 15 × 7cm, acuminate, obtuse, lateral nerves 9–13 pairs; petiole to 2cm; flowers 10–20, the calyx with one leaf-sized, bract-like, red lobe; corolla cream to orange; tube 2cm; lobes 0.5cm. Evergreen forest. 800–1600m.
Distr.: Guinea (C) to S Africa. (Tropical Africa).
IUCN: LC
Mejelet-Ehumseh: Etuge 292 10/1986; **Nyasoso:** Etuge 1813 fl., 3/1996; 2171 fr., 6/1996; **Nzee Mbeng:** Gosline 95 fl., 2/1998.
Local name: The Ashanti Blood (Etuge 1813).

Mussaenda polita Hiern

Climber, resembling *M. arcuata*, but leaves slightly smaller; inflorescence puberulous; flowers resembling those of *M. erythrophylla*, but bract-like calyx lobe white. Evergreen forest; 880–1350m.
Distr.: SE Nigeria to Gabon. (Lower Guinea).
IUCN: LC
Kodmin: Gosline 56 fl., 1/1998; **Kupe Village:** Cheek 7734 fr., 11/1995; 8374 5/1996; **Nyasoso:** Etuge 2424 fr., 6/1996.

Mussaenda tenuiflora Benth.

Climber, pubescent, resembling *M. erythrophylla*, but bract-like calyx lobes white, not red. Evergreen forest; 1300–1800m.
Distr.: Guinea (C) to Congo (K). (Guineo-Congolian).
IUCN: LC
Edib: Etuge 4483 fl., 11/1998; **Kodmin:** Etuge 4010 fr., 1/1998; Ghogue 66 1/1998; **Kupe Village:** Cheek 7884 fr., 11/1995.

Mussaenda sp. nov.

Epiphytic shrub with stems to 60cm, glabrous; leaves elliptic, dull white below, c. 8 × 5cm; flowers several, bract-like calyx lobe yellow or white; corolla yellow, tube 5cm, lobes c. 1cm; fruit subellipsoid, with persistent linear sepals 1.5cm. Evergreen forest; 760–1500m.
Distr.: Kupe-Bakossi. (Narrow Endemic).
Kupe Village: Cable 3696 fr., 7/1996; 3840 fr., 7/1996; Cheek 7829 fr., 11/1995; Etuge 2635 fl., 7/1996; 2875 fr., 7/1996; Zapfack 919 fr., 7/1996; **Nyale:** Cheek 9711 fr., 11/1998; **Nyasoso:** Cable 3562 fr., 7/1996; 3622 fr., 7/1996; Wheatley 438 fr., 7/1992.

Nauclea diderrichii (De Wild. & T.Durand) Merrill

Tree c. 30m; stems with ants; leaves elliptic or oblong, 7–15 × 4–9.5cm, larger on young, sterile shoots, base rounded, nerves 6–8; petiole 8–13mm; stipule ovate to obovate, 10–25 × 4–12mm; inflorescence spherical, 3cm diameter. Evergreen forest; 760m.
Distr.: Sierra Leone to Congo (K). (Guineo-Congolian).
IUCN: VU
Ngombombeng: Etuge 41 fr., 4/1986.
Uses: MATERIALS – a timber tree (*fide* Cheek).

Nauclea vanderguchtii (De Wild.) Petit

Tree c. 30m; stems with ants; leaves oblong-ovate, 21–35 × 8–18cm, glabrous, base cordate, nerves 9–13; petiole 12–30mm; stipule ovate, 20–40 × 10–20mm, caducous; inflorescence spherical, 4–6cm diameter. Evergreen forest; 1300m.
Distr.: Nigeria, Cameroon to Congo (K). (Lower Guinea & Congolian).
IUCN: LC
Kupe Village: Cable 2577 fr., 5/1996.
Uses: MATERIALS – a timber tree (*fide* Cheek).

Oldenlandia corymbosa L. var. caespitosa (Benth.) Verdc.

F.T.E.A. Rubiaceae: 310 (1976).
Syn.: *Oldenlandia caespitosa* (Benth.) Hiern var. *caespitosa*
Syn.: *Oldenlandia caespitosa* (Benth.) Hiern var. *lanceolata* Bremek.
Decumbent or erect weedy herb to 10cm; leaves oblong, 5–10 × 1–2mm; flowers axillary, 1–2; peduncle nil; corolla 1–2mm; capsule globose, 1–2mm. Grassland, fields; 800m.
Distr.: Senegal to Mozambique. (Tropical Africa).

IUCN: LC
Tombel: Etuge 381 11/1986.

Oldenlandia corymbosa L. var. *corymbosa*

Erect herb to 30cm, diffusely branched, glabrous; leaves linear or linear-lanceolate, to 20 × 2–7mm; stipule divided by 1/2 into 5 aristae; flowers in pedunculate, 2(–4)-flowered umbels; calyx 1mm; corolla white or mauve, 1mm. Farmbush.
Distr.: pantropical.
IUCN: LC
Mt Kupe: Etuge 1778 fl., fr., 3/1996.

Oldenlandia goreensis (DC.) Summerh. var. *goreensis*

Prostrate herb up to 45cm long; leaves elliptic or ovate-elliptic, 1.5–2.5 × 0.5–1.3cm; flowers white with mauve or pink corolla lobes. Moist, muddy places; 1250m.
Distr.: tropical & subtropical Africa, Madagascar, Seychelles & Mascarene Islands. (Afromontane, Madagascar & Seychelles).
IUCN: LC
Lake Edib: Cheek 9152a fl., 2/1998.

Oldenlandia lancifolia (Schumach.) DC. var. *lancifolia*

Herb 30–60cm; leaves narrowly lanceolate, 2–6 × 0.2–0.7cm; flowers white, sometimes pale pink or mauve. Wet areas in forest; 200–2000m.
Distr.: tropical Africa. (Afromontane).
IUCN: LC
Kodmin: Cheek 8892 fl., 1/1998; **Kupe Village:** Cable 3845 fl., 7/1996; **Mejelet-Ehumseh:** Manning 427 10/1986; 443 10/1986; **Mile 12 Mamfe Road:** Nemba 303 fr., 10/1986; **Mwambong:** Etuge 4326 10/1998; **Mwanenguba:** Cheek 9414 fr., 10/1998.
Local name: No name known (Cheek 8892). **Uses:** MEDICINES (Cable 3845).

Oldenlandia lancifolia (Schumach.) DC. var. *scabridula* Bremek.

F.T.E.A. Rubiaceae: 293 (1976).
Herb resembling var. *lancifolia*, but young parts scabridulous; calyx tube puberulent, leaf width at large side of range. Lakes, wetlands; 650–1975m.
Distr.: Sierra Leone to S Africa. (Tropical Africa).
IUCN: LC
Kodmin: Etuge 3988 fr., 1/1998; **Lake Edib:** Cheek 9167 fl., fr., 2/1998; **Mwambong:** Cheek 9364 fl., 10/1998; **Mwanenguba:** Cheek 7279 2/1995; **Nyasoso:** Etuge 1602 fl., fr., 1/1996.
Local name: Isole. **Uses:** SOCIAL USES – used in peace agreements: each person eats a leaf, the rest is put in a mortar, water is added and thrown over the persons (Epie Ngome in Cheek 9364).

Otomeria volubilis (K.Schum.) Verdc.

Climber, 8m; stem twisting, puberulent; leaves papery, ovate, c. 7 × 3cm, acuminate, obtuse-decurrent, midrib hairy below; petiole 5mm; stipule 5mm, divided to the base into 3–5 linear lobes; flowers 5–10 in dense terminal panicle, red; corolla tube 3cm, lobes 0.5cm; fruit ellipsoid, 8 × 4mm. Forest edge; 710–1050m.
Distr.: SE Nigeria to Uganda. (Lower Guinea & Congolian).
IUCN: LC
Nyale: Cheek 9674 fl., 11/1998; Etuge 4198 fl., fr., 2/1998.

Oxyanthus formosus Hook.f. ex Planch.

Shrub or tree 3–10m; leaves coriaceous, oblong to 25 × 12cm, short-acuminate, unequally obtuse; petiole 1cm; stipule lanceolate, 2–3 × 1–1.5cm; corolla tube 12 × 0.1cm; lobes 2.5 × 0.1cm; fruit ellipsoid, 5 × 2.5cm; pedicel 2.5cm. Evergreen forest; 800–1550m.
Distr.: Mali to Uganda. (Guineo-Congolian).
IUCN: LC
Bangem: Thomas 5295 1/1986; **Kodmin:** Etuge 4156 fr., 2/1998; Ghogue 30 fr., 1/1998; **Kupe Village:** Cable 3793 fr., 7/1996; Cheek 7873 fr., 11/1995; 8397 5/1996; Ryan 369 fl., fr., 5/1996; **Nyasoso:** Cable 3464 fr., 7/1996; Etuge 2129 fr., 6/1996; 2384 fr., 6/1996; Lane 143 fr., 6/1994.

Oxyanthus gracilis Hiern

Shrub 0.6–2m; leaves elliptic, c. 15 × 6cm, acuminate, acute; petiole 1cm; stipule ovate, 6–10 × 4–6mm; corolla tube 5.2–7cm, lobes 1–1.2cm; fruit spherical, 2.5cm; pedicel 6mm. Evergreen forest; 850–880m.
Distr.: S Nigeria, Bioko to Congo (K). (Lower Guinea & Congolian).
IUCN: NT
Kupe Village: Cable 789b fr., 1/1995; Cheek 7073 fr., 1/1995; 7682 fr., 11/1995; 7733 fr., 11/1995.
Note: only 17 sites are known (Sonké, Op. Bot. Belg. 8, (1999)).

Oxyanthus laxiflorus K.Schum. ex Hutch. & Dalziel

Shrub resembling *O. gracilis*, but corolla tube 2.7–3.2cm, lobes 0.5–0.9cm. Evergreen forest; 250–1000m.
Distr.: SE Nigeria to Gabon. (Lower Guinea).
IUCN: NT
Baduma: Nemba 191 8/1986; **Karume:** Nemba 72 5/1986; **Kupe Village:** Cheek 8405 fr., 5/1996; Ryan 342 fl., 5/1996; **Mt Kupe:** Cheek 10155a fr., 11/1999.
Note: nearly restricted to Cameroon where numerous sites in lowland forest.

Oxyanthus montanus Sonké

Bull. Jard. Bot. Nat. Belg. 63: 397 (1994).
Syn.: *Oxyanthus sp. A sensu* Hepper & Keay in FWTA 1: 129
Shrub 5–9m, glabrous; leaves elliptic, 10–14 × 3–6cm, acuminate, cuneate, lateral nerves 16–21 pairs, with domatia; petiole 1cm; stipule triangular, 8mm; inflorescence 5–8-flowered; bracts persistent; pedicel 5mm; corolla tube 11cm, lobes 1.5–1.8cm; fruit ellipsoid, 3–4 × 1.5cm. Forest.
Distr.: Cameroon Endemic.
IUCN: VU
Ngombombeng: Etuge 117 /1986.

Oxyanthus pallidus Hiern

Shrub 3–6m; leaves coriaceous, elliptic, 20–25 × 10–15cm, acuminate, obtuse, decurrent; petiole 1cm; stipule ovate-orbicular, 15 × 9mm, long acuminate; corolla tube 17–18cm, lobes 1.7–1.8cm; fruit spherical, 2.5cm; pedicel 1.5cm. Evergreen forest; 850–1200m.
Distr.: Ivory Coast to Congo (K). (Guineo-Congolian).
IUCN: LC
Nyasoso: Cable 2893 fr., 6/1996; 3226 6/1996; 3304 fr., 6/1996; 3536 fr., 7/1996; 3608 fr., 7/1996; 869 fl., 1/1995; Cheek 7525 10/1995; Etuge 2076 fr., 6/1996; 2416 fr., 6/1996; 2519 fr., 7/1996; Sebsebe 5007 fr., 10/1995; Zapfack 662 fr., 6/1996.

Oxyanthus speciosus DC. subsp. *speciosus*

Syn.: *Oxyanthus speciosus* DC. subsp. *globosus* Bridson

Shrub or tree 5–12m; leaves elliptic, c. 20 × 9cm, acuminate, unequally obtuse; petiole 1.5cm; stipule oblong-acuminate; corolla tube 3.5cm, lobes 1.5cm; fruit ellipsoid, 2.5 × 1.5cm. Forest; 200–850m.
Distr.: Senegal to Zambia. (Guineo-Congolian).
IUCN: LC
Kupe Village: <u>Kenfack 213</u> fl., 7/1996; **Mile 12 Mamfe Road:** <u>Nemba 289</u> 10/1986; **Ngomboaku:** <u>Cheek 10324</u> fl., 12/1999; **Nyandong:** <u>Stone 2451</u> 3/2003.

Oxyanthus subpunctatus (Hiern) Keay

Shrub 1m; leaves elliptic, c. 20 × 11cm, acuminate, obtuse-decurrent; petiole 1cm; stipule triangular, 1.2 × 0.8cm; corolla tube 11–15cm, lobes 2–2.5cm; fruit spherical, 3cm; pedicel 5mm. Evergreen forest; 950m.
Distr.: Sierra Leone to Congo (K). (Guineo-Congolian).
IUCN: LC
Mt Kupe: <u>Cheek 10155</u> fr., 11/1999.

Oxyanthus unilocularis Hiern

Shrub or tree to 15m; leaves elliptic, c. 53 × 30cm, acuminate, truncate, hispid on veins below; petiole 1.5cm; stipule oblong, 4 × 1.5cm; corolla tube 13–15cm, lobes 2cm; fruit spherical, 2cm; pedicel 1cm. Evergreen forest; 300–1000m.
Distr.: Sierra Leone to Uganda. (Guineo-Congolian).
IUCN: LC
Kupe Village: <u>Etuge 1485</u> fr., 11/1995; <u>2052</u> fr., 5/1996; <u>2785</u> fl., fr., 7/1996; <u>Zapfack 925</u> fr., 7/1996; **Manehas F.R.:** <u>Cheek 9393</u> fr., 10/1998; **Ngomboaku:** <u>Ghogue 472</u> fr., 12/1999; **Nyandong:** <u>Ghogue 1503</u> fl., 3/2003; <u>Nguembou 592</u> 3/2003; <u>643</u> 3/2003.
Local name: Mebina. **Uses:** SOCIAL USES – religious uses – used in Juju (Epie Ngome in Cheek 9393).

Parapentas setigera (Hiern) Verdc.

Straggling herb, pubescent; leaves ovate to elliptic, 2.7–3.7 × 1–2cm, glabrescent; petiole 2–11mm; stipules with 3–6 filiform lobes to 2.5mm; fascicles axillary; corolla white, tube 5mm, lobes deltoid, 2mm; fruit 3mm, dehiscing; seeds numerous. Forest; 700–1500m.
Distr.: Guinea (C) to Congo (K). (Guineo-Congolian).
IUCN: LC
Kodmin: <u>Etuge 4062</u> fl., 1/1998; <u>Ghogue 37</u> fl., 1/1998; **Kupe Village:** <u>Ensermu 3526</u> fl., 11/1995; <u>Etuge 2706</u> fl., 7/1996; <u>Lane 260</u> fl., 11/1994; <u>Williams 230</u> fl., 11/1995; **Ngomboaku:** <u>Cheek 10362</u> 12/1999; **Nyasoso:** <u>Cable 3450</u> 7/1996; <u>Sebsebe 5033</u> fl., 10/1995.

Pauridiantha callicarpoides (Hiern) Bremek.

Fl. Gabon 12: 259 (1966).
Shrub or tree 2–4m; stem hollow, with ants, sparsely pubescent; leaves narrowly oblong, 11–25 × 3–4cm, pubescent on midrib, nerves c. 20; petiole 5mm; stipule ovate-acuminate, c. 14 × 7mm; flowers foul-smelling, c. 50, in axillary inflorescences 2cm long; corolla tube 1–2m; disc hairy. Evergreen forest; 1000–1450m.
Distr.: SE Nigeria to Congo (K). (Lower Guinea & Congolian).
IUCN: LC
Kodmin: <u>Cheek 9040</u> fl., 1/1998; <u>Gosline 110</u> fl., fr., 2/1998; **Mwambong:** <u>Onana 526a</u> fl., 2/1998; **Mwanenguba:** <u>Leeuwenberg 9549</u> fr., 4/1972.
Note: most Cameroonian material, including ours, has smaller leaves with much more restricted pubescence than the typical material of Gabon, as noted by Hallé (Fl. Gabon 12: 259 (1966)).

Pauridiantha canthiiflora Hook.f.

Shrub 2m; stems, stipules, petioles, nerves densely appressed-pubescent; leaves elliptic, c. 6 × 2.5cm, acumen 1cm, base rounded or obtuse, drying white above, nerves 3–4 pairs; petiole 5mm; stipule narrowly triangular, c. 9 × 1mm; flowers 1–3, axillary, subsessile, white. Evergreen forest; 600–900m.
Distr.: SE Nigeria to Congo (K). (Lower Guinea & Congolian).
IUCN: LC
Ngusi: <u>Etuge 50</u> 4/1986; **Nyasoso:** <u>Cable 2804</u> fr., 6/1996.

Pauridiantha divaricata (K.Schum.) Bremek.

Fl. Gabon 12: 236 (1966).
Shrub or small tree 5m; leaves elliptic, 11 × 4cm, acuminate, obtuse, nerves 5, very finely reticulate, domatia conspicuous; petiole 5mm; stipule lanceolate-acuminate, 4 × 1mm, base constricted; flowers few, axillary, yellow. Evergreen forest; 150m.
Distr.: Cameroon Endemic.
IUCN: VU
Etam: <u>Etuge 185</u> fl., 7/1986.

Pauridiantha floribunda (K.Schum. & K.Krause) Bremek.

Tree 6–12m; stem lacking ant cavities, glabrous; leaves lanceolate-oblong, c. 15–20 × 5–6cm, acuminate, base rounded or obtuse, drying pink below, nerves c. 20 pairs; petiole 1.5–2cm; stipule leafy, ovate, c. 1.5 × 0.8cm, acuminate; flowers c. 70 in erect, flat-topped inflorescences, c. 10cm, white; peduncle c. 10cm; fruit green, 8mm. Evergreen forest; 250–1450m.
Distr.: SE Nigeria to São Tomé & Gabon. (Lower Guinea).
IUCN: NT
Baduma: <u>Nemba 115</u> 6/1986; **Kodmin:** <u>Cheek 9589</u> fr., 11/1998; **Kupe Village:** <u>Cable 3819</u> fr., 7/1996; <u>Cheek 7815</u> fr., 11/1995; <u>Etuge 1953</u> fr., 5/1996; <u>2672</u> fr., 7/1996; **Mile 15, Kumba-Mamfe Road:** <u>Nemba 144</u> 7/1986; **Mungo River F.R.:** <u>Olorunfemi in FHI 30603</u> month unkown/1900.
Local name: Mwah-kan. **Uses:** unknown (Max Ebong in Cheek 9589).

Pauridiantha hirtella (Benth.) Bremek.

Shrub 5m, thinly pubescent on stems and lower blade; leaves oblanceolate, 22 × 9cm, base cuneate, nerves 18; petiole 1.5cm; stipule ovate, 8 × 5mm; flowers 10–30, axillary; inflorescence 2cm, pale green; corolla 2mm. Swampy gallery forest; 1000m.
Distr.: Guinea (B) to Cameroon. (Upper & Lower Guinea).
IUCN: LC
Bangem: <u>Thomas 5905</u> 3/1986.

Pauridiantha paucinervis (Hiern) Bremek.

Shrub 2–4m; stems puberulent; leaves elliptic-oblong, c. 10 × 3cm, base acute, nerves 8, domatia usually absent; petiole 7mm; stipule subulate, 7 × 1mm; flowers 5–10, on 1–2 peduncles to 1.5cm, white; corolla tube 3mm; fruit 5mm, red or black. Evergreen forest; 800–2000m.
Distr.: Bioko & Cameroon. (Lower Guinea).
IUCN: NT
Bangem: <u>Thomas 5253</u> 1/1986; **Kodmin:** <u>Etuge 4003</u> fr., 1/1998; <u>Ghogue 31</u> fl., 1/1998; <u>Onana 565a</u> fl., 2/1998; **Kupe Village:** <u>Cable 3784</u> fr., 7/1996; <u>Cheek 7207</u> fl., 1/1995; <u>8390</u> fr., 5/1996; <u>Ryan 360</u> fr., 5/1996; <u>422</u> fr., 5/1996; **Mejelet-Ehumseh:** <u>Etuge 167</u> 6/1986; <u>450</u> 1/1987; **Mt Kupe:** <u>Sebsebe 5067</u> fr., 10/1995; <u>Thomas 5070</u> 11/1985; <u>5477</u> 2/1986; **Nyasoso:** <u>Balding 57</u> fr., 8/1993; <u>Cable 1202</u>

fl., 2/1995; 122 fr., 9/1992; 3392 fr., 6/1996; Cheek 7331 fl., fr., 2/1995; Etuge 1659 fl., 1/1996; Wheatley 447 fr., 7/1992.
Note: 15 sites known apart from those cited here.

Pauridiantha rubens (Benth.) Bremek.
Shrub 6m, glabrous; stem hollow; leaves drying red below, elliptic, 20(–35) × 10cm, subacuminate, acute, nerves 11; petiole 4cm; stipule caducous; flowers c. 50, axillary, on several peduncles, c. 4cm, white; corolla tube 4mm; fruit green, 6mm. Edge of evergreen forest; 500–1000m.
Distr.: Cameroon, Bioko to Congo (K). (Lower Guinea & Congolian).
IUCN: LC
Kupe Village: Cable 2547 fr., 5/1996; Etuge 1896 fr., 5/1996; Kenfack 241 fr., 7/1996; **Nlog:** Etuge 4 4/1986.

Pauridiantha venusta N.Hallé
Fl. Gabon 12: 237 (1966).
Shrub 2–4m, appressed puberulent; leaves drying pink or red, elliptic, 10 × 4cm, nerves 4–5; petiole 5mm; stipule subulate, 6 × 1mm; flowers 1 per axil, white; corolla tube 4mm; fruit blue, 5mm. Evergreen forest; 400–950m.
Distr.: Cameroon & Gabon. (Lower Guinea).
IUCN: VU
Bakossi F.R.: Cheek 9317 fr., 10/1998; Etuge 4291 fr., 10/1998; **Kupe Village:** Cable 3872 fr., 7/1996; Kenfack 301 fr., 7/1996; Ryan 303 fr., 5/1996; **Mungo River F.R.:** Onana 944 11/1999; **Ngomboaku:** Cheek 10397 fr., 12/1999; Muasya 2030 fr., 12/1999.
Local name: Saga sagade (small fruits). **Uses:** MATERIALS – hunting: squish leaves and put in a dogs nose to help to scent animals, and to bark (MEDICINES – veterinary) (Epie Ngome in Cheek 9317).

Pausinystalia brachythyrsa De Wild.
Bot. J. Linn. Soc. 120: 308 (1996).
Tree resembling *Corynanthe*, but inflorescence 3cm; corolla lobes with linear appendage; fruit 2cm. Forest; 250m.
Distr.: Cameroon Endemic.
IUCN: NT
Ikiliwindi: Etuge 505 3/1987.
Note: specimen cited not seen, needs confirmation, especially since taxon otherwise restricted to Bipinde.

Pausinystalia macroceras (K.Schum.) Pierre ex Beille
Tree 20m, resembling *Corynanthe*, but leaves to 14 × 5cm; inflorescence diffuse; corolla appendages linear, 1cm; fruit 2cm. Forest.
Distr.: Nigeria to Congo (K). (Lower Guinea & Congolian).
IUCN: LC
Kurume: Thomas 5417 1/1986.

Pausinystalia talbotii Wernham
Syn.: *Corynanthe dolichocarpa* W.Brandt
Tree 12m, resembling *Corynanthe*, but leaves oblong-elliptic, to 24 × 10cm; panicles terminal and axillary; corolla appendages 5mm; fruit 2cm. Forest; 250–300m.
Distr.: Nigeria & Cameroon. (Lower Guinea).
IUCN: NT
Konye: Thomas 5903 3/1986; **Mile 15, Kumba-Mamfe Road:** Nemba 157 fr., 7/1986.

Pavetta bidentata Hiern var. *bidentata*
Shrub 2(–4)m, glabrous; floriferous twigs to 30cm; leaves thinly leathery, mostly narrowly oblong-elliptic, to 25 × 6cm, acute, cuneate, midrib drying orange, lateral nerves 15 pairs, domatia pits elongate, hairy, nodules elongate along midrib,

rare in blade, tertiary venation just visible; petiole to 3.5cm; inflorescence 4cm across, subglabrous; corolla white, tube 4–10 × 1–2mm, lobes 5–10mm; fruit globose, pink or white with green stripes, then black; seeds 1–2, concave. Forest; 150–1370m.
Distr.: SE Nigeria to Congo (K). (Guineo-Congolian).
IUCN: LC
Bangem: Thomas 5266 1/1986; **Etam:** Etuge 81 3/1986; **Kodmin:** Biye 45 1/1998; Etuge 4116 2/1998; 4153 2/1998; Gosline 185 11/1998; **Ndum:** Cable 891 fl., 1/1995; **Ngomboaku:** Cheek 10388 12/1999; **Nyasoso:** Cable 3254 6/1996; 3257 6/1996; 645 fr., 12/1993; 686 fl., 1/1995; Cheek 5664 fl., fr., 12/1993; Etuge 2425 6/1996; Lane 103a 6/1994; Sidwell 369 10/1995.

Pavetta brachycalyx Hiern
Shrub 3–6m, glabrous; floriferous twigs 20cm; leaves papery, drying green, elliptic to 20 × 9cm, acumen 0.5–1cm, acute, lateral nerves 9 pairs, tertiary nerves inconspicuous, domatia small pit-like, bacterial nodules rod-like, 2–3mm, sparse; petioles 3cm; panicles 10cm across; calyx lobes 0.5mm; corolla white, tube 6mm, lobe 6mm. Forest; 300–1200m.
Distr.: Cameroon. (W Cameroon Uplands).
IUCN: EN
Konye: Thomas 5193 11/1985; **Kupe Village:** Etuge 1417 11/1995; Schoenenberger 72 fl., 11/1995.
Note: once thought restricted to Mt Cameroon. Easily confused with *P. gabonica* (q.v.).

Pavetta camerounensis S.Manning subsp. *brevirama* S.Manning
Ann. Missouri Bot. Gard. 83: 108 (1996).
Shrub 0.6–2m, subglabrous; floriferous twigs to 4cm; leaves thickly papery, drying dark green, oblong-elliptic, to 25 × 6.5cm, acumen slender 1.5cm, acute, lateral nerves 15 pairs, minutely brown puberulent, tertiary nerves conspicuous, domatia and bacterial nodules not seen; petiole 4cm; panicles dense, 2cm across; calyx lobes 0.5mm; corolla greenish white, tube 2–4mm, lobes 3–5mm. Forest; 250–350m.
Distr.: SW Cameroon. (Cameroon Endemic).
IUCN: NT
Baduma: Nemba 185 8/1986; **Bakolle Bakossi:** Etuge 134 5/1986; **Bolo Moboka:** Nemba 245 9/1986.
Note: known from c. 15 collections apart from those cited.

Pavetta gabonica Bremek.
Ann. Missouri Bot. Gard. 83: 113 (1996).
Shrub similar to *P. brachycalyx*, but flowers yellow, smaller (C tube 2–4mm, not 3–5mm, c. lobe 3–5mm, not 4–6mm); calyx lobes 0.75mm; inflorescence 1–5.5cm across (not 1–10cm); fruit orange. Streambanks in forest; 350–1000m.
Distr.: Cameroon & Gabon. (Lower Guinea).
IUCN: NT
Bakolle Bakossi: Etuge 146 5/1986; **Bangem:** Thomas 5296 1/1986; 5354 1/1986; **Ngomboaku:** Cheek 10322 12/1999.
Note: known from about 23 collections apart from those cited.

Pavetta hookeriana Hiern var. *hookeriana*
Shrub 2–3m, subglabrous; floriferous twigs 15cm; leaves papery, elliptic to 13 × 6cm, acumen to 1cm, cuneate, lateral nerves 10 pairs, domatia arched, hairy, nodules not seen, tertiary venation inconspicuous; petiole 2cm; inflorescence to 10cm across; flowers to 100; calyx lobes rotund, 2mm; corolla white; tube 2–5mm; lobes 4–8mm. Forest-grassland transition; 1900–2000m.
Distr.: Bioko & W Cameroon. (W Cameroon Uplands).

RUBIACEAE

IUCN: VU
Mwanenguba: Cheek 7262 fl., 2/1995; Etuge 4362 10/1998; Thomas 3145 2/1984.

Pavetta kupensis S.Manning

Ann. Missouri Bot. Gard. 83: 118 (1996).

Shrub or tree 3–8m, subglabrous; floriferous twigs c. 20cm; leaves leathery, oblanceolate, 15 × 5cm, acumen 0.5–1cm, cuneate, midrib thick, orange below, lateral nerves 7–8 pairs, brochidodromous, tertiary venation inconspicuous, nodules few, domatia crypts; petiole 2cm; panicles 2–9cm across; flowers 25–75; calyx lobes 0.1–1mm; corolla greenish white, tube 6–8mm, lobes 8–12mm. Forest; 300–2000m.

Distr.: Kupe-Bakossi. (Narrow Endemic).

IUCN: CR

Konye: Thomas 5197 11/1985; **Kupe Village:** Ryan 399 5/1996; **Mt Kupe:** Etuge 1362 11/1995; Thomas 5481 2/1986; **Nyasoso:** Cable 1201 fl., 2/1995; 3388 6/1996.

Pavetta muiriana S.Manning

Ann. Missouri Bot. Gard. 83: 128 (1996).

Shrub 1–2m, densely grey appressed puberulent; floriferous twigs c. 7cm; leaves papery, narrowly elliptic, to 17 × 6cm, acumen 1cm, acute, lateral nerves c. 14 pairs, tertiary nerves linking, quaternary nerves plain, baterial nodules and domatia not seen; petiole 2cm; panicles to 4cm across, to 20-flowered, densely grey puberulent; calyx lobes 0.8mm; corolla white, drying black. Forest; 1470–1780m.

Distr.: Bakossi Mts & Banyang Mbo. (Narrow Endemic).

IUCN: EN

Bangem: Thomas 5303 1/1986; **Kodmin:** Cheek 9085 1/1998; Etuge 3994 1/1998; Ghogue 47 fl., 1/1998.

Pavetta neurocarpa Benth.

Shrub 1.5–3m, glabrous; floriferous twigs to 30cm; leaves elliptic to 32 × 13cm, acumen 1–2cm, acute, lateral nerves 7–10 pairs, drying brown below, tertiary nerves conspicuous where leaves dry green, bacterial nodules elliptic, sparse, domatia not seen; petiole 3cm; inflorescence 5cm across, 20-flowered; calyx lobes 0.5mm; corolla white; fruit white with 8 green stripes. Forest; 700–1000m.

Distr.: SE Nigeria, Bioko, S & SW Cameroon. (Lower Guinea).

IUCN: NT

Kupe Village: Cable 2470 5/1996; 2495 5/1996; 3700 7/1996; 3730 7/1996; 3873 7/1996; 734 1/1995; Cheek 7083 fl., 1/1995; Etuge 2886 7/1996; Kenfack 269 7/1996; Lane 418 fl., 1/1995; **Mbule:** Leeuwenberg 9277 1/1972; **Nyasoso:** Cable 3308 6/1996; 3557 7/1996.

Note: 15 coll. known apart from those cited.

Pavetta owariensis P.Beauv. var. *glaucescens* (Hiern) S.Manning

Ann. Missouri Bot. Gard. 83: 136 (1996).

Syn.: *Pavetta glaucescens* Hiern

Shrub resembling var. *satabiei* but to 15m, domatia present (crypts); fruit lacking persistent calyces, 8mm diameter. Forest; 250m.

Distr.: SE Nigeria to Angola. (Lower Guinea).

IUCN: LC

Mile 15, Kumba-Mamfe Road: Nemba 140 7/1986.

Pavetta owariensis P.Beauv. var. *satabiei* S.Manning

Ann. Missouri Bot. Gard. 83: 137 (1996).

Shrub 3–5m, minutely papillate-puberulent, glabrescent; floriferous twigs 30cm; leaves papery, drying pale brown below, elliptic-oblong, 12–20 × 5–8cm, acumen 1cm, cuneate, lateral nerves 8 pairs, tertiary nerves few, sparse, domatia and bacterial nodules not seen; petiole 3cm; panicle 10cm across, c. 50-flowered; clayx lobes to 1.5mm; corolla white, not seen; fruit with persistent calyx, to 10mm diameter. Forest; 800–1300m.

Distr.: Kupe-Bakossi & Rumpi Hills. (Narrow Endemic).

IUCN: EN

Kodmin: Satabie 1126 2/1998; **Mejelet-Ehumseh:** Etuge 178 6/1986; **Nyasoso:** Etuge 2124 6/1996; Lane 141 fr., 6/1994.

Pavetta rigida Hiern

Shrub 1.5–4.5m, glabrous; leaves leathery, drying grey-green above, pale green below, elliptic, 22–43 × 9–15cm, acumen 1cm, cuneate, lateral nerves 9 pairs, white, quaternary nerves usually plain, bacterial nodules and domatia absent; petiole 6cm; panicle 15cm across, c. 50-flowered; calyx lobes 0.6mm; corolla white, tube 15mm, lobes 12mm. Forest; 250–1120m.

Distr.: Nigeria, Bioko & Cameroon. (W Cameroon Uplands).

IUCN: LC

Enyandong: Etuge 4389r 10/2001; **Kupe Village:** Cable 2679 5/1996; Cheek 6067 fl., 1/1995; Etuge 1950 5/1996; 2882 7/1996; Kenfack 276 7/1996; Ryan 230 5/1996; Zapfack 936 7/1996; **Manehas F.R.:** Etuge 4339 10/1998; **Mile 15, Kumba-Mamfe Road:** Nemba 134 7/1986; **Ngomboaku:** Cheek 10344 12/1999; **Nyasoso:** Cable 3319 6/1996.

Pavetta rubentifolia S.Manning

Ann. Missouri Bot. Gard. 83: 139 (1996).

Shrub 1m, glabrous; floriferous twigs 14cm; leaves red, elliptic-oblong, 13–16 × 3.5–4.5cm, acumen to 1.5cm, cuneate, lateral nerves 8–13 pairs, domatia and nodules absent; petiole 1cm; panicles 0.5–1cm wide, subglabrous to puberulent; flowers 20–35; calyx tube to 1mm, lobes valvate, to 0.5mm, glabrous; corolla white, tube to 4mm, lobes 3–5mm, pubescent near throat. Forest; 800m.

Distr.: Bakossi Mts. (Narrow Endemic).

IUCN: CR

Bangem: Thomas 5343 1/1986.

Note: known only from the type, specimen not seen.

Pavetta staudtii Hutch. & Dalziel

Shrub 2–4m, glabrous; resembling *P. bidentata* but floriferous twigs 15–20cm; leaves membranous, drying green, narrowly elliptic, 11 × 3.5cm, acumen cuneate, 1cm, acute, lateral nerves 10 pairs, inconspicuous, bacterial nodules numerous, conspicuous, punctate, domatia pitted; petiole 2cm; panicles 2–3cm across, 20-flowered; calyx lobes quadrate, 0.5mm; corolla greenish white, tube 6mm, lobes 6mm. Forest; 1100m.

Distr.: Cameroon Endemic.

IUCN: NT

Mejelet-Ehumseh: Etuge 474 1/1987.

Note: widespread in Cameroon and known from 18 collections.

Pavetta subgenus *Pavetta sp. A*

Shrub 1.5m, glabrous; epidermis white, flaking in sheets, floriferous twigs 10cm; leaves papery, elliptic, c. 12 × 4cm, acumen 1–2cm, acute, lateral nerves 6 pairs, quaternary nerves prominent, bacterial nodules punctate, very conspicuous, domatia absent; panicles 4cm across, 10–20-flowered; calyx lobes aristate, 4mm, persistent in fruit; corolla white, tube 2cm, lobes 1cm. Forest; 300m.

Distr.: W Cameroon. (Cameroon Endemic).

Bakole: <u>Thomas 7031</u> 5/1987; **Kupe Village:** <u>Etuge 2776</u> 7/1996.

Note: a very distinct new species known only from 2 locations on the border of Bakossi and 1 site near Kumba.

Pavetta subgenus *Pavetta sp. B*

Shrub 3–5m; stem whitish brown, glabrous, floriferous twigs 30cm; leaves papery, drying black, obovate, to 21 × 9cm, acumen 1cm, lateral nerves 11 pairs, white, bacterial nodules punctate, sparse; panicles globose, 7cm (16cm including styles), c. 100-flowered; calyx lobes 6mm; corolla white, tube 15mm, lobes 7mm. Forest; 500–1100m.
Loum F.R.: <u>Etuge 4279</u> 10/1998; **Nyasoso:** <u>Cable 3227</u> 6/1996; <u>3524</u> 7/1996; <u>Etuge 2388</u> 6/1996; <u>2422</u> 6/1996.

Pavetta subgenus *Pavetta sp. C*

Shrub, glabrous; floriferous twigs 25cm; leaves pale green below, lanceolate, 24 × 10cm, acumen 1.5cm, base obtuse-decurrent, lateral nerves 10 pairs, finer nerves not seen, bacterial nodules maculate, conspicuous below; petiole 3cm; panicle 12cm across, c. 60-flowered; calyx lobes 0.7mm; corolla white, 15mm in bud. Forest; 1500m.
Kodmin: <u>Onana 575</u> 2/1998.

Note: specimen not matched at K, distinct in leaf shape and nodules most conspicuous on lower surface, as in *P. urophylla* of subg. *Baconia*.

Pavetta sp. D cf. nitidula Welw. ex Hiern

Shrub 5m, glabrous; floriferous twigs obscured; leaves leathery, elliptic, 22 × 9cm, acumen 1cm, acute, lateral nerves 8 pairs, tertiary nerves conspicuous below, bacterial nodes rod-like, 2mm, conspicuous, domatial pits hairy; petiole 4cm; panicle 6cm; fruits 1–1.5cm. Forest; 1200m.
Kupe Village: <u>Etuge 2633</u> 7/1996.

Note: more ample material needed to resolve, not matched at K.

Pavetta sp. E cf. camerounensis S.Manning

Shrub 2m, glabrous; floriferous twigs 25cm, almost leafless; leaves membranous, elliptic or oblanceolate, 21 × 7cm, acumen 1cm, acute, lateral nerves 10 pairs, domatia obscured, bacterial nodules not seen; panicles 6cm; pedicels 7mm; fruit white, 7mm. Forest; 1000m.
Nyasoso: <u>Cable 3537</u> 7/1996.

Note: more material needed to resolve, not matched at K.

Pentaloncha sp. nov.

Subshrub 0.3–0.5m; leaves membranous, elliptic, 7 × 2.5cm, subacuminate, acute, lateral nerves 7–10 pairs; petiole 1–2cm; stipule aristate, 5mm; inflorescences axillary, fascicles 5–10-flowered, sessile; corolla 2–3mm, purple; fruit globose, 2mm. Forest; 700–1300m.
Distr.: Kupe-Bakossi. (Narrow Endemic).
Edib: <u>Etuge 4108</u> 1/1998; **Kodmin:** <u>Etuge 4148</u> 2/1998; **Kupe Village:** <u>Cable 3825</u> 7/1996; <u>Cheek 7183</u> fl., 1/1995; <u>Ryan 257</u> 5/1996; **Mbule:** <u>Cable 3340</u> 6/1996; **Ngomboaku:** <u>Cheek 10374</u> 12/1999; **Nyasoso:** <u>Cable 3556</u> 7/1996; <u>3592</u> 7/1996.

Note: known only from Mt Kupe and the Bakossi Mts.

Pentas ledermannii K.Krause

Kew Bull. 31: 185 (1976).
Syn.: *Pentas pubiflora* S.Moore subsp. *bamendensis* Verdc.
Weak shrub 0.6–1m, densely pubescent; leaves elliptic to 10 × 5cm, subacuminate, acute, lateral nerves 10 pairs; petiole 1cm; stipules divided into 7–10 aristae, 4mm; corymbs terminal, 5cm wide, 10–20-flowered; peduncle 3cm; calyx

lobes 3mm; corolla white; tube 5mm; lobes 5. Grassland, forest edge; 1800–2060m.
Distr.: Mt Kupe to Mt Oku. (W Cameroon Uplands).
IUCN: VU
Mt Kupe: <u>Cheek 7573</u> fl., 10/1995; <u>Sebsebe 5086</u> fl., 11/1995; <u>Thomas 3166</u> 2/1984; **Mwanenguba:** <u>Leeuwenberg 9970</u> fl., 6/1972; **Nyasoso:** <u>Cable 108</u> fl., 9/1992.

Pentas schimperiana Vatke subsp. *occidentalis* (Hook.f.) Verdc.

Herb or subshrub 1.5m; leaves ovate-elliptic, 9 × 4cm, pubescent below; petiole 1cm; flowers yellow-white; corolla tube 17mm long. Forest edge. 1300–2060m.
Distr.: Cameroon, Bioko, São Tomé & Ituri, Congo (K). (Lower Guinea & Congolian (montane)).
IUCN: LC
Mt Kupe: <u>Sidwell 429</u> fl., 11/1995; **Mwanenguba:** <u>Cheek 7275</u> fr., 2/1995; <u>Cheek 9457</u> fl., 10/1998; <u>Etuge 4378</u> fl., 10/1998; <u>Sanford 5503</u> fl., 11/1968; <u>Thomas 3148</u> 2/1984; **Nyasoso:** <u>Cable 111</u> fl., 9/1992; <u>2907</u> fl., 6/1996; <u>Cheek 7340</u> fr., 2/1995.
Local name: none known. **Uses:** none known (Cheek 7340).

Petitiocodon parviflorum (Keay) Robbr.

Bull. Jard. Bot. Nat. Belg. 58: 117 (1988).
Syn.: *Didymosalpinx parviflora* Keay
Tree 4m; leaves leathery, oblong-elliptic, 22 × 11cm, acumen 1cm, obtuse, nerves 8–10; petiole 2cm; flowers 1–6 on supra-axillary peduncles 5mm; pedicels 6mm; corolla white, tube cylindrical-campanulate, 3cm. Forest; 400–800m.
Distr.: SE Nigeria & Cameroon. (Lower Guinea).
IUCN: NT
Konye: <u>Thomas 5940</u> fl., 3/1986; **Nyasoso:** <u>Etuge 7</u> fl., 4/1986.
Note: known from 12 sites; may qualify as VU under criterion A.

Pleiocoryne fernandense (Hiern) S.Rauschert

Taxon 31: 561 (1982).
Syn.: *Polycoryne fernandensis* (Hiern) Keay
Shrub or climber, 4m, glabrous; stem 1cm, hollow, pale grey; leaves papery, elliptic, c. 22 × 11cm, acumen short, base rounded or truncate, nerves 5; petiole 4cm; stipule rounded-oblong, 1cm; flowers white and orange, 10–20 on peduncles 5cm; calyx 3mm; corolla tube 2cm, lobes 1cm. Forest; 800m.
Distr.: Sierra Leone to Congo (K). (Guineo-Congolian).
IUCN: LC
Nyasoso: <u>Etuge 1520</u> fl., 12/1995.

Poecilocalyx schumannii Bremek.

Fl. Gabon 12: 230 (1966).
Shrub 3.5m; stems villose; leaves papery, narrowly elliptic, 15 × 4cm, acuminate, rounded-obtuse, pubescent below, nerves 8–9; petiole 5mm; stipule palmately 5-lobed, villose; flower white, 5mm diameter; fruit hemispherical, 1cm; calyx lobes leafy. Evergreen forest; 300–1650m.
Distr.: Cameroon & Rio Muni. (Lower Guinea).
IUCN: NT
Bakossi F.R.: <u>Etuge 4287</u> fr., 10/1998; **Ehumseh:** <u>Manning 491</u> 10/1986; **Konye:** <u>Thomas 5195</u> 11/1985; **Mekom:** <u>Thomas 5407</u> 1/1986.
Note: 13 sites known including those cited here. May qualify as VU under criterion A.

Pouchetia africana DC.

Tree 10–15m, 20cm d.b.h.; bole matt red-brown, outer slash pale red, inner dull yellow; leaves drying brown below, oblong to lanceolate, to 13 × 4cm, gradually acuminate, 1–

1.5cm, lateral nerves 6 pairs; petiole 1cm; stipule triangular, long-aristate, 9mm; inflorescences patent, axillary, numerous, slender, 12cm (imm.); peduncle 7cm. Forest; 850m.
Distr.: Senegal to CAR. (Upper & Lower Guinea).
IUCN: LC
Ngomboaku: Cheek 10280 12/1999.

Preussiodora sulphurea (K.Schum.) Keay
Shrub 1.5m; stem pale grey, lenticellate, glabrous; leaves papery, narrowly elliptic-obovate, c. 30 × 11cm, acuminate, unequally acute, nerves 10; petiole 6–7cm; stipule triangular, 1cm; flowers yellow, terminal, few; calyx 1.5cm; corolla tube 1.5cm, lobes 1.5 × 2cm; fruit ellipsoid, 4 × 3cm; calyx lobes persistent, 1.5cm. Semi-deciduous forest; 300–400m.
Distr.: SE Nigeria to Gabon & Bioko. (Lower Guinea).
IUCN: NT
Kupe Village: Etuge 2011 fr., 5/1996; **Loum F.R.:** Onana 996 fl., 12/1999.
Note: 15 sites known including those cited here; may qualify as VU under criterion A.

Pseudosabicea batesii (Wernham) N.Hallé
Fl. Gabon 12: 202 (1966).
Climber, appressed white arachnoid hairy; leaves appearing alternate, whitish brown tomentose below, thinly leathery, c. 22 × 11cm, subacuminate, truncate-highly oblique, nerves 14; petiole 3cm; flowers numerous in sessile capitula 1cm diameter. Evergreen forest; 700–1000m.
Distr.: Cameroon & Gabon. (Lower Guinea).
IUCN: VU
Menyum: Doumenge 597 5/1987; **Nyandong:** Thomas 6695 fl., 2/1987.

Pseudosabicea medusula (K.Schum.) N.Hallé
Fl. Gabon 12: 200 (1966).
Syn.: *Sabicea medusula* K.Schum. ex Wernham
Creeping herb, densely brown pubescent; leaves appearing alternate, brownish white below, elliptic, c. 10 × 5cm, acuminate, unequally cordate, nerves c. 10; petiole 2cm; flowers c. 5, in sessile capitula, 1.5cm diameter. Evergreen forest; 200m.
Distr.: Cameroon Endemic.
IUCN: VU
Baduma: Nemba 579 fr., 7/1987.

Pseudosabicea pedicellata (Wernham) N.Hallé
Adansonia (sér. 2) 3: 172 (1963).
Syn.: *Sabicea pedicellata* Wernham
Scandent shrub, 3m, densely grey silky-hairy; leaves brownish white below, ovate-oblong, 13 × 7cm, acuminate, truncate, nerves 14; petiole 1cm; flowers c. 30 in capitula; peduncles 5cm; pedicels 1.5cm; corolla tube green, 7mm. Evergreen forest; c. 400 m.
Distr.: SE Nigeria & Cameroon. (Lower Guinea).
IUCN: VU
Mile 45 Kumba-Mamfe road: Keay in FHI 28556 fl., 1/1951.

Psilanthus mannii Hook.f.
Shrub 4m; stems whitish grey, glabrous; leaves papery, elliptic-oblong, c. 16 × 6cm, acuminate, acute, nerves 6; petiole 1cm; stipule triangular, 1.5 × 1.5mm; flowers 1–few, axillary, subsessile, white; corolla tube 6cm, lobes 3cm; fruit top-shaped, 2 × 2cm, glossy, black, with green, strap-shaped calyx lobes 4 × 0.5cm; seeds 2. Evergreen forest; 250–1650m.
Distr.: Guinea (C) to Congo (K). (Guineo-Congolian).

IUCN: LC
Baduma: Nemba 179 8/1986; **Bangem:** Thomas 5341 1/1986; **Kodmin:** Etuge 4077 1/1998; Ghogue 52 fl., fr., 1/1998; **Kupe Village:** Ryan 298 fl., fr., 5/1996; **Mungo River F.R.:** Onana 937 fr., 11/1999; **Ngomboaku:** Mackinder 321 fr., 12/1999; **Nyasoso:** Cable 3336 fr., 6/1996; Elad 116 fr., 2/1995; Etuge 1827 3/1996; 2513 fr., 7/1996; Groves 91 fl., fr., 2/1995; Wheatley 441 fr., 7/1992.

Psilanthus sp. nov.
Syn.: *Psilanthus ebracteolatus sensu auctt. non* Hiern
Syn.: *Coffea ebracteolata sensu* Hepper & Keay *pro parte* in FWTA 2: 157
Shrub 1.5–3m, glabrous; leaves ovate or elliptic, c.8 × 3.5cm, acumen 20 × 4mm, rounded, base obtuse, nerves 3–4, basal pair subpalmate; petiole 3mm; inflorescence terminal, 1-flowered; corolla tube 2cm, lobes 5, 1.2cm, white; fruit subglobose, green, bilobed, 9mm. Forest.
Distr.: W Cameroon. (Narrow Endemic).
Ekona Mombo: Etuge 416 12/1986.

Psychotria sp. aff. alatipes Wernham
Shrub 0.5–2m, glabrous; leaves membranous, shining, bacterial nodules spider or dash-like, blade elliptic, 25 × 10cm, subacuminate, decurrent, lateral nerves 12 pairs; petiole to 6cm; stipules glabrous, 7mm, bifurcate; panicle 15cm, lowest node with 2(–3) branches; peduncle 9cm, 3-winged; flowers white, 3mm; corolla 4-lobed; fruit globose, 5mm, red. Forest; 870–1550m.
Distr.: Cameroon Endemic.
Kodmin: Biye 9 fl., fr., 1/1998; Gosline 50 fr., 1/1998; Pollard 205 fr., 11/1998; **Kupe Village:** Cable 2602 fr., 5/1996; 2627 fr., 5/1996; 3745 fr., 7/1996; 803a fl., 1/1995; Cheek 7201 fl., 1/1995; 8386 fr., 5/1996; Ryan 279 fr., 5/1996; **Mbule:** Leeuwenberg 8817 fr., 12/1971; 9305 fl., 1/1972; **Mwambong:** Gosline 124 fr., 10/1998; **Nyasoso:** Balding 27b fr., 8/1993; Cable 1247 fr., 2/1995; 2767 fr., 6/1996; 2855 fr., 6/1996; 3588 fr., 7/1996; 690 fl., 1/1995; Cheek 7312 fl., 2/1995; Etuge 1311 fr., 10/1995; 2063 fr., 6/1996; 2112 fr., 6/1996; Lane 140 fr., 6/1994; 200 fr., 11/1994; Sebsebe 4982 fr., 10/1995.
Local name: none known. **Uses:** none known (Cheek 7312).

Psychotria bifaria Hiern var. *bifaria*
Erect shrublet 0.5–1m; stem 4-angled, glabrous apart from two densely puberulent purple lines; leaves papery, drying pale grey below, bacterial nodules punctate, inconspicuous, blade elliptic, to 9 × 4cm, usually much smaller, acumen 1cm, obtuse, lateral nerves 8 pairs; petiole 0.5cm; stipule 5mm; panicle sessile, diffuse, 1.5cm, 5-flowered; fruit pendulous, fleshy, fusiform, 8mm, red. Forest; 650–1250m.
Distr.: Bioko, Cameroon & Gabon. (Lower Guinea).
IUCN: LC
Kupe Village: Cable 2676 fr., 5/1996; Cheek 6080 fr., 1/1995; 7679 fr., 11/1995; Ensermu 3535 11/1995; Etuge 1372 fr., 11/1995; 2623 fr., 7/1996; 2820 fr., 7/1996; 2905 7/1996; Ryan 310 fr., 5/1996; **Ngusi:** Letouzey 14658 fl., 4/1976.

Psychotria brevipaniculata De Wild.
Bull. Jard. Bot. Nat. Belg. 36: 177 (1966).
Monopodial shrublet 30cm, densely puberulent; leaves papery, pale pink-grey below, bacterial nodules dense, conspicuous, dots and dashes, blade broadly elliptic, to 12 × 6cm, subacuminate, acute, lateral nerves 10–12 pairs; petiole to 1.5cm; stipule ovate, 8mm, bifurcate; infructescence 4cm, diffuse; fruit globose, 5mm. Forest; 800–1100m.
Distr.: Cameroon to Uganda. (Lower Guinea & Congolian).

IUCN: LC
Kupe Village: Etuge 2922 fr., 7/1996; **Nyasoso:** Cheek 5663 fl., fr., 12/1993; Etuge 1310 10/1995; Lane 192 fr., 11/1994.

Psychotria calceata Petit
Bull. Jard. Bot. Nat. Belg. 34: 201 (1964).
Shrub resembling *P. sp. aff. alatipes*, but leaves lacking bacterial nodules, drying glossy greyish white below; stipules subcylindric, to 15mm, sheathing, biaristate; lowest inflorescence node 4-branched. Forest; 250–1400m.
Distr.: Nigeria & Cameroon. (Lower Guinea).
IUCN: NT
Baduma: Nemba 223 fr., 9/1986; **Kodmin:** Cheek 8994 fr., 1/1998; **Konye:** Thomas 5172 fr., 12/1985; **Kupe Village:** Cable 2494 fl., 5/1996; 2648 fr., 5/1996; 3727 fr., 7/1996; Cheek 7677 fr., 11/1995; Ensermu 3506a 5/1995; Etuge 1370 fr., 11/1995; 2852 fr., 7/1996; 2881 fr., 7/1996; 2894 fr., 7/1996; Lane 264b fr., 11/1994; Ryan 216b 5/1996; 295 fr., 5/1996; 322 fl., 5/1996; **Mbule:** Cable 3343 fr., 6/1996; **Mile 15, Kumba-Mamfe Road:** Nemba 199 fr., 9/1986; **Mungo River F.R.:** Onana 970 fr., 12/1999; **Nyale:** Etuge 4197 fr., 2/1998; **Nyasoso:** Cable 3552 fr., 7/1996; Etuge 2097 fr., 6/1996.
Note: likely to prove to be a threatened taxon once extra-Bakossi material is correctly assigned. Holotype at BR should be viewed since isotype at K is incomplete.

Psychotria calva Hiern
Shrub 3m; stems chalky-green; leaves papery, drying pale green below, bacterial nodules linear, along midrib, blade oblong-elliptic, 20 × 9cm, acumen 1.5cm, base obtuse-rounded, abruptly decurrent, lateral nerves 10–12 pairs; petiole to 3cm; stipule broadly ovate, 4mm, biaristate; infructescence diffuse, 15 × 9cm, including peduncle 7cm; flowers white, 3mm; fruit ellipsoid, 6mm, smooth. Forest; 250–1250m.
Distr.: Senegal to Bioko & Cameroon. (Upper & Lower Guinea).
IUCN: LC
Ekona Mombo: Etuge 427 12/1986; **Kupe Village:** Cable 2706 fl., 5/1996; Cheek 7013 fr., 1/1995; 7035 fr., 1/1995; Etuge 1368 fr., 11/1995; 1386 fr., 11/1995; 2890 fr., 7/1996; Lane 395 fr., 1/1995; **Mile 15, Kumba-Mamfe Road:** Nemba 135 7/1986.

Psychotria sp. A aff. calva Hiern
Shrub 3m; stems chalky-green; leaves papery, drying black above, grey below, bacterial nodules linear, along midrib; blade oblong-elliptic, 20 × 9cm, acumen 1.5cm, base obtuse-rounded, lateral nerves 10–12 pairs; petiole to 3cm; stipule ovate, aristate, bifid, to 8mm; inflorescence c. 4cm, 10–20-flowered; peduncle 2cm; infructescence 4–8cm; fruit subglobose, red, 6mm. Forest; 700–1400m.
Distr.: Kupe-Bakossi & Bali-Ngemba. (W Cameroon Uplands).
Kupe Village: Cable 2626 fr., 5/1996; 3726 fr., 7/1996; 3762 fr., 7/1996; 3865 fl., fr., 7/1996; Cheek 7178 fr., 1/1995; 8385 fr., 5/1996; Etuge 2651 fr., 7/1996; 2846 fr., 7/1996; **Mbule:** Cable 3346 fr., 6/1996; **Mwambong:** Onana 561a fr., 2/1998; **Ndum:** Lane 468 fr., 2/1995; **Nyasoso:** Etuge 2155 fr., 6/1996; 2410 fr., 6/1996; Lane 116 fr., 6/1994.
Note: easily confused with *P. martinetugei*, distinguished by the slender fruiting pedicels.

Psychotria camerunensis Petit
Bull. Jard. Bot. Nat. Belg. 36: 158 (1966).

Shrub resembling *P. sp. aff. camerunensis*, but densely brown puberulent; leaves obovate to oblanceolate, c. 13 × 5cm. Forest; 150–1700m.
Distr.: Cameroon. (Lower Guinea).
IUCN: VU
Ehumseh: Manning 457 10/1986; **Etam:** Etuge 192 7/1986.

Psychotria sp. aff. camerunensis Petit
Shrub 0.6–1.5m, glabrous; leaves drying bright orange-brown below, bacterial nodules dense, punctate, blade elliptic, to 14 × 5cm, acumen to 2cm, acute, lateral nerves 10 pairs, diffuse; petiole to 2.5cm; stipule sheath 3mm, arista 3mm; panicle 3cm, 5–15-flowered, diffuse; fruit ellipsoid, 7mm, orange-red, often pendulous. Forest; 850–1500m.
Distr.: Bakossi. (Narrow Endemic).
Kodmin: Cheek 8875 fl., fr., 1/1998; Etuge 4426 fr., 11/1998; Gosline 43 fr., 1/1998; **Kupe Village:** Cable 736 fl., 1/1995; Cheek 7161 fl., 1/1995; 7179 fl., 1/1995; Ryan 219 fr., 5/1996; **Mbule:** Cable 3373 fr., 6/1996; **Mt Kupe:** Leeuwenberg 9274 fl., 1/1972; **Ngomboaku:** Cheek 10330 fl., 12/1999; **Nyasoso:** Cable 1142 fl., 2/1995; 695 fl., 1/1995.

Psychotria camptopus Verdc.
Kew Bull. 30: 259 (1975).
Syn.: *Cephaelis mannii* (Hook.f.) Hiern
Tree 3–5m, glabrous; leaves leathery, obovate, 26 × 13cm, acumen obtuse, 1cm, acute, lateral nerves 14 pairs, petiole 6cm; stipules 2 × 1.5cm; peduncle 1–3m, pendulous, red; flowers white, 1cm, 10–20, enveloped in fleshy bracts; involucre 3 × 5cm, glabrous. Forest; 1100–1700m.
Distr.: SE Nigeria, Bioko, SW Cameroon & Congo (K). (Guineo-Congolian (montane)).
IUCN: NT
Essossong: Groves 46 fl., 2/1995; **Kodmin:** Etuge 4083 fl., 1/1998; **Kupe Village:** Cheek 8403 fl., 5/1996; **Mt Kupe:** Thomas 3081 fr., 2/1984; 5479 2/1986; 6141 4/1986; **Nyasoso:** Etuge 1747 fl., fr., 2/1996; Sunderland 1544 fl., 7/1992; Wheatley 443 fl., 7/1992.
Local name: Gonad Tree (Sunderland 1544).

Psychotria densinervia (K.Krause) Verdc.
Kew Bull. 30: 259 (1975).
Syn.: *Cephaelis densinervia* (K.Krause) Hepper
Tree 3–5m, resembling *P. camptopus*, but lateral nerves 15–19, involucral bracts pubescent. Forest; 1120m.
Distr.: Cameroon Endemic.
IUCN: EN
Kupe Village: Ryan 225 fr., 5/1996; **Mungo Ndaw:** Dundas in FHI 15324 fl., 10/1946.

Psychotria sp. aff. dorotheae Wernham
Shrub 3m, glabrous; leaves thinly leathery, drying pale silvery-green, shining, elliptic, 22 × 12cm, obtuse, acute-obtuse, lateral nerves 10 pairs; petiole to 10cm; stipule ovate, 1cm, entire; peduncle 1cm (fruit), umbellate; pedicels to 1cm; fruits 4–12, ovoid, 1.5cm, bright red, heavily ridged. Forest; 700–1100m.
Distr.: Mt Kupe. (Narrow Endemic).
Kupe Village: Cable 743 fr., 1/1995; Cheek 6053 fr., 1/1995; 6077 fr., 1/1995; 7061 fr., 1/1995; **Nyasoso:** Cable 3307 fr., 6/1996; 3320 fl., fr., 6/1996.
Note: flowering material needed; only known from Bakossi.

Psychotria elongato-sepala (Hiern) Petit
Canopy climber; stem, petioles and midrib with dense patent red hairs, 2mm; leaves drying brown-black, oblanceolate to elliptic, to 15 × 5cm, subacuminate, lateral nerves 15 pairs;

petiole 2cm; stipule aristate, 8mm; panicle 5 × 5cm; flowers not seen; fruit ovoid, 15mm, sepals to 4mm. Forest; 900–1050m.
Distr.: Guinea (C) to Cameroon. (Upper & Lower Guinea).
IUCN: NT
Kupe Village: Cable 3822 fr., 7/1996; **Nyasoso:** Cable 3625 fr., 7/1996; Etuge 2163 fr., 6/1996.

Psychotria sp. aff. foliosa Hiern
Monopodial shrub to 0.5m, densely puberulous; stem 10cm, scarred; leaves forming a litter-gathering funnel, each oblanceolate, to 30 × 10cm, acumen 1cm, lateral nerves 20–25 pairs; petiole 1–2mm; peduncle 4cm, capitulum 1.5cm, dense; fruit 8mm, red. Forest; 760–950m.
Kupe Village: Abu 119 10/1995; Cable 2477 fr., 5/1996; 3686 fr., 7/1996; Cheek 7668 fr., 11/1995; Ryan 313 fr., 5/1996; **Ngomboaku:** Cheek 10325 fr., 12/1999.
Note: flowering material needed.

Psychotria sp. A aff. fuscescens
Shrub 1–1.5m, sparingly branched, glabrous; leaves elliptic to 16 × 8cm, acumen 0.5cm, acute, lateral nerves 10–12 pairs, brochidodromous; stipule rhombic, 4–5mm, entire, midrib evident; peduncle 1cm, capitulum 1.5–2 × 1.5–2cm, enveloped by involucral bracts; berries 8mm. Forest; 1300m.
Distr.: Bakossi Mts. (Narrow Endemic).
Kodmin: Biye 12 fr., 1/1998; Etuge 3981 fr., 1/1998.
Note: possibly *sp. nov.*, flowering material needed to describe. Only known from Bakossi. This taxon appears close to, but distinct from, *Cephaelis fuscescens* Hiern. *Cephaelis* is treated as a synonym of *Psychotria*, but the epithet *fuscescens* is already occupied there by a Thai species.

Psychotria sp. B aff. fuscescens
Monopodial shrub 0.5(–1)m, glabrous; leaves membranous, drying blackish green, elliptic to oblanceolate, to 20 × 8cm, subacuminate, cuneate-decurrent, lateral nerves 15–19 pairs; petiole 2–3cm; stipule papery, to 2 × 1cm, bifid; peduncle 1.5cm; involucre boat shaped, 1.5 × 3.5cm; flowers white, 1.5cm; fruit blue. Forest; 700–950m.
Distr.: Bakossi Mts. (Narrow Endemic).
Kupe Village: Cable 3878 7/1996; **Nyasoso:** Balding 91 fl., fr., 8/1993; Cable 3306 fl., 6/1996; 45 fl., 8/1992; 868 fr., 1/1995; Zapfack 602 fr., 6/1996.
Note: probably *sp. nov.*, only known from Bakossi. See note to *P. sp. A aff. fuscescens*.

Psychotria gabonica Hiern
Syn.: *Psychotria rowlandii* Hutch. & Dalziel
Shrub 2–3(–5)m, glabrescent; stems brown; leaves thinly leathery, subglossy grey, drying pale pink-grey below, elliptic to obovate, 13 × 5cm, subacuminate, acute, lateral nerves 8–10 pairs, brochidodromous; petiole 2cm; stipule entire, 4mm, caducous; panicle sessile, capitate, dense, 2 × 3cm; flowers white, buds pink-orange, 4mm; fruits numerous, subdiffuse, ovoid, bright red, streaked black, smooth, 2 × 1cm; disc exserted. Forest; 250–1800m.
Distr.: Liberia to Gabon. (Guineo-Congolian (montane)).
IUCN: LC
Bangem: Thomas 5891 fl., 3/1986; **Essossong:** Groves 52 fl., fr., 2/1995; **Ikiliwindi1:** Nemba 160 7/1986; **Kodmin:** Biye 8 fl., fr., 1/1998; Gosline 181 fl., fr., 11/1998; 55 fl., fr., 1/1998; **Kupe Village:** Cable 2609 fr., 5/1996; Cheek 6063 fr., 1/1995; 7051 fl., fr., 1/1995; 7088 fl., fr., 1/1995; 7122 fl., fr., 1/1995; 7175 fl., 1/1995; 8381 fr., 5/1996; Elad 70 fl., fr., 1/1995; Kenfack 275 fr., 7/1996; Ryan 214 fr., 5/1996; 220 fr., 5/1996; **Mbule:** Leeuwenberg 9270 fl., 1/1972;

Mejelet-Ehumseh: Etuge 455 fl., fr., 1/1987; **Mt Kupe:** Thomas 5058 fr., 11/1985; 5478 2/1986; **Ndum:** Lane 475 fl., fr., 2/1995; **Nyasoso:** Balding 28 fr., 8/1993; 98 fr., 8/1993; Cable 42 fr., 8/1992; 651 fl., fr., 12/1993; 678 fr., 12/1993; 688 fl., fr., 1/1995; Cheek 7306 fl., fr., 2/1995; Etuge 2088 fr., 6/1996; Lane 107 fr., 6/1994; 110 fr., 6/1994; 114 fr., 6/1994; 164 fr., 10/1994; Sebsebe 4991 fr., 10/1995; 5024 fr., 10/1995; Sidwell 376 fr., 10/1995; Thomas 3064 2/1984; Zapfack 593 fr., 6/1996; 599 fr., 6/1996.
Local name: none known (Cheek 7306). **Uses:** ANIMAL FOOD – aerial parts: sunbird observed flying from inflorescence to inflorescence dipping beak in open flowers (Cheek 7175).

Psychotria sp. A aff. gabonica Hiern
Shrub resembling *P. gabonica*, but stems bright white; inflorescence with fleshy involucre; fruits with foliaceous calyx lobes. Forest; 960–1650m.
Distr.: Bakossi Mts. (Narrow Endemic).
Kodmin: Cheek 9099 fr., 2/1998; Gosline 166 fr., 11/1998; **Kupe Village:** Cable 2618 fr., 5/1996; Cheek 7814 fr., 11/1995; Etuge 2650 fr., 7/1996; **Nyasoso:** Balding 79 fr., 8/1993; Cable 1239 fl., fr., 2/1995; 677 fr., 12/1993; Etuge 1347 fr., 10/1995; 2136 fr., 6/1996; Lane 235 fr., 11/1994; Sebsebe 5049 10/1995.
Note: in prep. as *P. bakossiensis* Cheek & Sonké, known only from Bakossi.

Psychotria sp. B aff. gabonica Hiern
Shrub resembling *P. gabonica*, but leaves 17.5–31 × 8–12cm, drying matt dark brown above and below; stipules 18–35 × 15–20mm. Forest; 350–1600m.
Distr.: Bakossi & Mt Cameroon. (W Cameroon Uplands).
Bakolle Bakossi: Etuge 157 fl., 5/1986; **Kupe Village:** Cable 2654 5/1996; 3760 fr., 7/1996; Cheek 7177 fr., 1/1995; 7800 fl., fr., 11/1995; 8371 fr., 5/1996; Ryan 238 fr., 5/1996; 274 fr., 5/1996; **Nyale:** Cheek 9661 fl., fr., 11/1998; Etuge 4193 fr., 2/1998; **Nyasoso:** Balding 35 fl., fr., 8/1993; Cable 2846 6/1996; Etuge 2141 fr., 6/1996; 2550 fr., 7/1996; Lane 231 fl., 11/1994; Sebsebe 5027 fl., 10/1995; Sidwell 416 fl., fr., 10/1995; Sunderland 1540 fr., 7/1992; Thomas 3067 fr., 2/1984; Wheatley 426 fr., 7/1992.
Note: in preparation as *P. geophylax* Cheek & Sonké.

Psychotria globiceps K.Schum.
Shrub 1.5–2m, glabrous; leaves drying pale green, elliptic, to 20 × 10cm, acumen 1cm, obtuse; petiole 2–5cm; stipule ovate 1.5cm, bifurcate, red-hairy; peduncle red-hairy, pendulous, 4cm; capitulum globose, 2cm diameter, c. 30-flowered; involucre absent; flowers white; fruit sessile, ovoid, 0.8cm, green. Forest; 250–1200m.
Distr.: Cameroon & Congo (K). (Guineo-Congolian).
IUCN: LC
Baduma: Nemba 188 8/1986; **Kupe Village:** Cable 2512 fl., 5/1996; 2673 fr., 5/1996; 3728 fr., 7/1996; 842 fr., 1/1995; Ensermu 3568 fr., 11/1995; Etuge 2652 fr., 7/1996; Lane 262 fr., 11/1994; 289 fr., 1/1995; Ryan 223 fr., 5/1996; Sidwell 445 fl., 11/1995; **Ngombombeng:** Etuge 123 5/1986.

Psychotria globosa Hiern var. *ciliata* (Hiern) Petit
Bull. Jard. Bot. Nat. Belg. 34: 160 (1964).
Syn.: *Psychotria nigerica* Hepper
Herb closely resembling var. *globosa*, but ciliate-hairy on both surfaces of leaf blade, densely long-hairy on stem, petioles, stipules and bracts. Forest; 300–1400m.
Distr.: Nigeria to Congo (K). (Lower Guinea & Congolian).
IUCN: LC

Bangem: Thomas 5911 fl., 3/1986; **Kodmin:** Etuge 3977 fl., 1/1998; 4167 fr., 2/1998; Gosline 157 fr., 11/1998; 70 fl., 1/1998; **Kupe Village:** Cable 2414 fl., 5/1996; 2467 fl., fr., 5/1996; Cheek 7801 fr., 11/1995; Etuge 2699 fr., 7/1996; 2712 fl., fr., 7/1996; **Mekom:** Thomas 5400 fl., 1/1986; **Ngombombeng:** Etuge 32 fl., 4/1986; **Nyasoso:** Balding 74 fr., 8/1993; Cable 2764 fr., 6/1996; 3436 fr., 7/1996; Cheek 7323 fr., 2/1995; Etuge 1313 fr., 10/1995; 1762 fl., 3/1996; 2086 fr., 6/1996; 2535 fr., 7/1996; Lane 165 fr., 10/1994; Sidwell 318 fl., fr., 10/1995; Wheatley 456 fr., 7/1992.

Psychotria globosa Hiern var. *globosa*
Monopodial herb to 0.6m, sparsely puberulent, glabrescent; leaves drying black, elliptic or obovate, 13 × 6cm, subacuminate, acute, lateral nerves 8–10 pairs; petiole 1cm; stipule oblong-elliptic, 12mm; panicle capitate, dense, flat-topped, 3cm; flowers white, 9mm; fruit ovoid, 1cm, orange, smooth. Forest; 900–1250m.
Distr.: Nigeria & Cameroon. (Lower Guinea).
IUCN: NT
Kodmin: Satabie 1106 fr., /; **Kupe Village:** Cable 3652 fl., 7/1996; 3824 fl., 7/1996; Etuge 2688 fr., 7/1996; **Nyasoso:** Cable 3606 fr., 7/1996.

Psychotria humilis Hiern var. *maior* Petit
Bull. Jard. Bot. Nat. Belg. 34(2): 181 (1960).
Shrublet 45cm, minutely puberulent; leaves membranous, drying dark green above, whitish green below, bacterial nodules branched, conspicuous, dense, blade elliptic, 10 × 3cm, acumen 1.5cm, acute, lateral nerves 7 pairs; petiole 1.5cm; stipule 5mm, long-biaristate; infructescence 6cm; fruits few, globose, 7mm, smooth, red. Forest; 700m.
Distr.: Nigeria & Cameroon. (Lower Guinea).
IUCN: NT
Kupe Village: Cheek 6080b fr., 1/1995.
Note: likely to prove threatened once Cameroonian material redelimited against the original Nigerian specimens.

Psychotria lanceifolia K.Schum.
Bull. Jard. Bot. Nat. Belg. 34: 202 (1964).
Shrub, (?) rheophytic; resembling *P. calva*, but leaves more slender, to 19 × 4cm, apex and base acute; petiole 1.5cm; inflorescence 1.5cm, 10-flowered, dense; peduncle 1cm; infructescence 3cm; fruit ovoid, 7mm, smooth. Forest; 1650m.
Distr.: Cameroon Endemic.
IUCN: VU
Ehumseh: Manning 481 10/1986; **Ekona Mombo:** Etuge 414 12/1986.
Note: taxon formerly known only from Kribi-Bipinde, specimens cited not seen.

Psychotria latistipula Benth.
Shrub 0.3–2m, glabrous; leaves thinly papery, drying dark brown below, elliptic, 16 × 7cm, subacuminate, decurrent, 20-nerved; petiole 2cm; stipule ovate, 1.2cm, bifurcate to half its length, glabrous; inflorescence diffuse, 10–20cm; bracts linear, 1cm, patent; flowers white, 4mm; infructescence pendulous; fruit globose, 4mm, red, ridged; pedicel not fleshy. Forest; 250–1500m.
Distr.: Nigeria, Bioko, Cameroon & Gabon. (Lower Guinea (montane)).
IUCN: LC
Bolo Moboka: Nemba 258 fr., 9/1986; **Konye:** Nemba 441 1/1987; **Kupe Village:** Cable 2571 fl., 5/1996; 2625 fl., 5/1996; 2642 fl., 5/1996; 778 fl., 1/1995; Cheek 8388 fl., fr., 5/1996; Ensermu 3541 fl., fr., 11/1995; Etuge 1719 fl., fr., 2/1996; Kenfack 266 fl., 7/1996; **Loum F.R.:** Biye 56 fr.,

11/1999; **Mile 15, Kumba-Mamfe Road:** Nemba 130 7/1986; **Nyandong:** Cheek 11272 3/2003; Nguembou 621 3/2003; 622 3/2003; Stone 2464 3/2003; **Nyasoso:** Balding 45 fl., fr., 8/1993; 72 fl., 8/1993; Cable 1149 fr., 2/1995; 3333 fl., 6/1996; 3446 fl., fr., 7/1996; Etuge 2115 fl., 6/1996; Lane 190 fr., 11/1994; 232 fr., 11/1994; Sebsebe 5055 fr., 10/1995; Sidwell 333 fl., fr., 10/1995; 334 fr., 10/1995; Wheatley 416 fl., fr., 7/1992.

Psychotria sp. aff. *latistipula* Benth.
Shrub 1.5–2m, resembling *P. latistipulata*, but lateral nerves 10; stipules triangular, 6mm, densely long red-hairy; infructescence 6cm, long-hairy; pedicel dilated; fruit smooth. Forest; 1400m.
Distr.: Bakossi. (Narrow Endemic).
Kodmin: Etuge 4014 fr., 1/1998; 4421 fr., 11/1998.

Psychotria leptophylla Hiern
Shrub 0.5–1m, glabrous, resembling *P.* sp. aff. *alatipes*, but leaves drying bright-green, bacterial nodules mostly punctate; panicle to 6cm; stipules densely brown-hairy in lower half. Forest; 150–1700m.
Distr.: SE Nigeria, Bioko, Cameroon & Congo (K). (Lower Guinea & Congolian).
IUCN: LC
Baduma: Nemba 186 8/1986; **Bangem:** Thomas 5265 1/1986; **Konye:** Thomas 5171 /1985; **Kupe Village:** Cable 3708 fr., 7/1996; 777 fl., 1/1995; 803b fl., 1/1995; Etuge 2775 fr., 7/1996; Lane 323 fr., 1/1995; 424 fl., 1/1995; **Loum F.R.:** Cheek 10255 fr., 12/1999; **Manehas F.R.:** Etuge 4343 fl., 10/1998; **Mbule:** Cable 3363 fr., 6/1996; Leeuwenberg 9303 fl., 1/1972; **Mejelet-Ehumseh:** Etuge 301 10/1986; **Mt Kupe:** Thomas 5102 12/1985; 5487 2/1986; **Mwanenguba:** Leeuwenberg 8343 fr., 9/1971; **Mungo River F.R.:** Etuge 4324 fr., 10/1998; **Ngomboaku:** Etuge 4653 fr., 12/1999; **Nyasoso:** Balding 27a fr., 8/1993; 6 fr., 7/1993; Cable 1208 fl., 2/1995; 2798 fr., 6/1996; 3248 fr., 6/1996; 3628 fr., 7/1996; Cheek 7335 fr., 2/1995; Etuge 1736 fr., 2/1996; 1761 fr., 3/1996; 2446 fl., fr., 6/1996; 2459 fl., fr., 6/1996; 2584 fr., 7/1996; Lane 144 fr., 6/1994; 163 fr., 10/1994; Thomas 3061 2/1984.

Psychotria sp. A aff. *leptophylla* Hiern
Shrub 1.5–3m, resembling *P. leptophylla*, but stem densely puberulent; leaves drying dark brown, bacterial nodules sparse; stipules bifurcate nearly to base; peduncle terete. Forest; 700–1250m.
Distr.: Bakossi. (Narrow Endemic).
Kupe Village: Cable 2466 fr., 5/1996; Cheek 6076 fr., 1/1995; 7653 fr., 11/1995; Etuge 1371 fr., 11/1995; 1420 fr., 11/1995.

Psychotria sp. B aff. *leptophylla* Hiern
Shrub resembling *P. leptophylla*, but leaves drying brown, bacterial nodules minute, clustered; leaves to 13 × 6cm; petiole to 1cm. Possibly heterogeneous. Forest; 650–2000m.
Kupe Village: Cable 2685 fr., 5/1996; **Nyasoso:** Etuge 1675 fr., 1/1996; Lane 236 fr., 11/1994.

Psychotria sp. aff. *limba* Scott-Elliot
Shrub 1–5m, fistular, glabrous; leaves papery, drying black, bacterial nodules absent, blade elliptic, to 24 × 13cm, acumen 1cm, rounded, lateral nerves 12 pairs, reticulate; petiole 3cm; stipule not seen; panicle 6 × 8cm, lowest node 4-branched; flowers white, 4mm; fruit globose, 7mm, smooth. Forest; 900–1300m.
Distr.: Bakossi & Rumpi Hills. (W Cameroon Uplands).

Kupe Village: <u>Etuge 1725</u> fl., 2/1996; **Mejelet-Ehumseh:** <u>Etuge 303</u> fr., 10/1986; **Nyasoso:** <u>Etuge 2055</u> fr., 6/1996; <u>2357</u> fr., 6/1996.

Psychotria cf. mannii Hiern

Shrub 60cm, brown patent-puberulent; leaves drying black above, blackish grey below, bacterial nodules dash-like, conspicuous, dense, blade obovate or elliptic, to 16 × 8cm, acumen 8mm, apiculate, base cuneate, lateral nerves 10 pairs, lower surface white-ciliate; petiole 3cm; stipule 1cm, deeply biaristate; panicle 3 × 1.5cm, 20-flowered, moderately dense; peduncle 1.5cm. Forest; 350m.
Mungo River F.R.: <u>Tadjouteu 313</u> fr., 12/1999.

Psychotria martinetugei Cheek

Kew Bull. 57: 377 (2002).
Syn.: *Psychotria malchairei sensu* Hepper & Keay in FWTA 2: 201 (1963).
Shrub 0.2–4m, glabrous, fistular; leaves papery, elliptic, 12–18 × 4–6cm, acumen 0.2–1.2cm, cuneate, secondary nerves 10–12 pairs; petiole 2–4cm; stipules caducous, 4–6mm, acumen 1mm, entire or bifurcate; panicles 4–6 × 5–9cm, 30–60-flowered; peduncle 3mm; flowers white, 2.5mm, corolla 5-lobed; fruit ellipsoid, 1cm, on dilated pedicel. Forest; 250–1600m.
Distr.: SE Nigeria, Bioko, Cameroon. (Lower Guinea).
IUCN: NT
Kodmin: <u>Biye 13</u> fr., 1/1998; <u>Cheek 9007</u> fr., 1/1998; <u>Etuge 4157</u> fr., 2/1998; <u>Satabie 1099</u> fr., 1/1998; **Kupe Village:** <u>Cable 2487</u> fr., 5/1996; <u>2557</u> fr., 5/1996; <u>2714</u> fl., 5/1996; <u>745</u> fr., 1/1995; <u>Cheek 7015</u> fr., 1/1995; <u>7085</u> fr., 1/1995; <u>7169</u> fr., 1/1995; <u>Ensermu 3506b</u> fr., 11/1995; <u>Etuge 2833</u> fl., 7/1996; <u>2869</u> fr., 7/1996; <u>2885</u> fr., 7/1996; <u>Lane 265</u> fr., 11/1994; <u>274</u> fr., 1/1995; <u>284</u> fr., 1/1995; <u>Ryan 297</u> fr., 5/1996; **Loum F.R.:** <u>Cheek 10261</u> fr., 12/1999; **Mwambong:** <u>Onana 527</u> fr., 2/1998; **Nyandong:** <u>Cheek 11341</u> 3/2003; **Nyasoso:** <u>Balding 48</u> fl., 8/1993; <u>54</u> fr., 8/1993; <u>63</u> fl., 8/1993; <u>Cable 3607</u> fr., 7/1996; <u>3615</u> 7/1996; <u>64</u> fl., 8/1992; <u>Cheek 5677</u> fl., 12/1993; <u>6016</u> fr., 1/1995; <u>Sidwell 407</u> fl., fr., 10/1995.

Psychotria minimicalyx K.Schum.

Shrublet closely resembling *P. bifaria*, but stem terete, uniformly puberulent; leaves to 5 × 1.5cm. Forest; 300m.
Distr.: Cameroon Endemic.
IUCN: CR
Konye: <u>Thomas 5185</u> fr., 12/1985.

Psychotria peduncularis (Salisb.) Steyerm. var. hypsophila (K.Schum. & K.Krause) Verdc.

Kew Bull. 30: 257 (1975).
Syn.: *Cephaelis peduncularis* Salisb. var. *hypsophila* (K.Schum. & K.Krause) Hepper
Shrub 1–5m, glabrous; leaves elliptic, to 15 × 8cm, acumen 0.5cm, acute, lateral nerves 12–15 pairs; petiole 2–3cm; stipule translucent, bifurcate, 1 × 0.8cm; inflorescence capitate; peduncle 2–4cm, nodding in accrescence, glabrous; involucral bracts fleshy; flowers 10–15, white, 5mm; infructescence umbellate; bracts absent; pedicels white, 1.5cm, berries blue, 7mm. Forest; 660–2050m.
Distr.: Guinea (C) to Bioko & Cameroon. (Upper & Lower Guinea (montane)).
IUCN: LC
Ehumseh: <u>Manning 498</u> fr., 10/1986; **Ehumseh - Mejelet:** <u>Etuge 360</u> fr., 10/1986; **Kodmin:** <u>Cheek 9033</u> fr., 1/1998; <u>Gosline 153</u> fr., 11/1998; **Kupe Village:** <u>Cheek 7662</u> fr., 11/1995; <u>8329</u> fr., 5/1996; <u>8404</u> fr., 5/1996; <u>Ensermu 3545</u> fr., 11/1995; <u>Ryan 396</u> fl., fr., 5/1996; **Mt Kupe:** Sebsebe

<u>5066</u> fr., 10/1995; **Nyasoso:** <u>Balding 31</u> fl., fr., 8/1993; <u>Cable 1194</u> fr., 2/1995; <u>121</u> fl., 9/1992; <u>2910</u> fl., 6/1996; <u>3235</u> fl., 6/1996; <u>3240</u> 6/1996; <u>667</u> fr., 12/1993; <u>Cheek 7354</u> fr., 2/1995; <u>7460</u> fr., 10/1995; <u>Etuge 2094</u> fr., 6/1996; <u>2518</u> fr., 7/1996; <u>Lane 195</u> fr., 11/1994; <u>237</u> fr., 11/1994; <u>Wheatley 452</u> fl., fr., 7/1992.
Local name: Ehentene. **Uses:** SOCIAL USES – in football matches, twisted around each side of goal-posts so that opponents cannot score (Cheek 9033).

Psychotria peduncularis (Salisb.) Steyerm. var. peduncularis

Kew Bull. 30: 257 (1975).
Syn.: *Cephaelis peduncularis* Salisb. var. *A sensu* Hepper & Keay in FWTA 2: 204
Syn.: *Cephaelis peduncularis* Salisb. var. *B sensu* Hepper & Keay in FWTA 2: 204
Shrub 1–5m, glabrous; leaves elliptic, to 15 × 8cm, acumen 0.5cm, acute, lateral nerves 12–15 pairs; petiole 2–3cm; stipule translucent, bifurcate, 1 × 0.8cm; inflorescence capitate; peduncle 2–4cm, nodding in accrescence, puberulent; involucral bracts fleshy; flowers 10–15, white, 5mm; infructescence umbellate; bracts absent; pedicels white, 1.5cm, berries blue, 7mm. Forest; 300–2000m.
Distr.: tropical Africa. (Afromontane).
IUCN: LC
Bolo Moboka: <u>Nemba 247</u> 9/1986; **Mejelet-Ehumseh:** <u>Manning 423</u> 10/1986; **Mwanenguba:** <u>Thomas 3129</u> 2/1984.

Psychotria sp. aff. peduncularis (Salisb.) Steyerm.

Shrub 2m, resembling *P. peduncularis*, but fruiting peduncle absent; infructescence sessile. Forest edge; 1980m.
Nyasoso: <u>Cheek 7344</u> fr., 2/1995.
Note: possibly only an aberrant specimen: more material needed from this location to resolve.
Local name: Ninten (Cheek 7344).

Psychotria podocarpa Petit

Bull. Jard. Bot. Nat. Belg. 34: 145 (1964).
Syn.: *Psychotria latistipula sensu* Hepper & Keay in FWTA 2: 198 quoad Talbot 234
Shrub 0.3–4m, glabrescent, resembling *P. latistipulata*, but fruiting pedicel 1.2cm, fleshy, bright red; fruits black, bracteoles ovate, foliaceous, 7mm. Forest; 300–820m.
Distr.: Nigeria & Cameroon. (Lower Guinea).
IUCN: VU
Konye: <u>Thomas 5173</u> fr., 12/1985; **Ngomboaku:** <u>Ghogue 449</u> fr., 12/1999; <u>Mackinder 289</u> fr., 12/1999; **Nyale:** <u>Cheek 9644</u> fr., 11/1998.

Psychotria psychotrioides (DC.) Roberty

Shrub 2m, glabrous; leaves elliptic, to 14 × 7cm, subacuminate, cuneate, lateral nerves 12 pairs; petiole 2cm; stipule obovate, sheathing, 1cm; inflorescence sessile, capitate, 2cm diameter; flowers 10–15, sessile, white, each 8mm; fruit ellipsoid, 1.5cm; calyx foliose. Forest; 1600–2000m.
Distr.: tropical Africa.
IUCN: LC
Nyasoso: <u>Balding 55</u> fl., 8/1993; <u>58</u> fr., 8/1993; <u>Cable 3418</u> fr., 6/1996; <u>Etuge 1734</u> fr., 2/1996.

Psychotria schweinfurthii Hiern

Bull. Jard. Bot. Nat. Belg. 34: 146 (1964).
Syn.: *Psychotria obscura* Benth.
Shrub c. 1m, resembling *P. latistipulata*, but infructescence 3–4cm. Forest; 600m.

Distr.: Ivory Coast to Uganda. (Guineo-Congolian).
IUCN: LC
Etam: Etuge 219 fr., 8/1986.

Psychotria subobliqua Hiern

Shrub 0.6–1m, glabrous, resembling *P. sp. aff. subobliqua*, but leaves drying very dark green above, oblanceolate, acumen to 1.5cm; petiole to 1cm; pedicel to 2mm, not white, fleshy; fruit with a persistent calyx to 3mm. Forest; 150–870m.
Distr.: Guinea (C) to Congo (K). (Guineo-Congolian).
IUCN: LC
Baduma: Nemba 183 8/1986; 232 fr., 9/1986; **Bakolle Bakossi:** Etuge 281 fr., 9/1986; **Etam:** Etuge 84 3/1986; **Nyasoso:** Cable 3222 6/1996.

Psychotria sp. aff. subobliqua Hiern

Shrub 1–2m, glabrous; leaves drying pale green, elliptic, to 12 × 4cm, acumen 1cm, acute, lateral nerves 8–10 pairs; petiole to 1.5cm; stipule caducous, ovate, to 4mm, entire; peduncle 0.5–1cm, capitulum subglobose, 0.8cm; flowers 4mm, white; fruiting pedicels white, fleshy; fruit ellipsoid, 0.8cm, abruptly expanding (aberrant?) to globose, 1.5cm, red. Forest; 700–1400m.
Distr.: Bakossi. (Narrow Endemic).
Essossong: Groves 44 fr., 2/1995; **Kupe Village:** Cable 3880 fr., 7/1996; 753 fr., 1/1995; Cheek 7079 fr., 1/1995; 7680 fl., fr., 11/1995; Ensermu 3534 fl., fr., 11/1995; Kenfack 279 fr., 7/1996; Lane 363 fr., 1/1995; 442 fr., 1/1995; **Ndum:** Cable 880 fr., 1/1995; Lane 467 fr., 2/1995; **Nyasoso:** Balding 69 fr., 8/1993; 90 fr., 8/1993; Cable 640 fl., fr., 12/1993; 664 fr., 12/1993; Cheek 7313 fl., 2/1995; Etuge 2110 fr., 6/1996; 2137 fr., 6/1996; 2547 fr., 7/1996; Lane 111 fr., 6/1994; 183 fl., 11/1994; Sidwell 335 fr., 10/1995; 347 fr., 10/1995; 378 fr., 10/1995.
Note: probably *sp. nov.*
Local name: local name means 'dark'. **Uses:** none known (Cheek 7313).

Psychotria succulenta (Hiern) Petit

Shrub 1–3m, drying dark brown, matt; leaves leathery, elliptic-oblong, 15 × 7cm, acumen 0.5cm, acute-obtuse, lateral nerves 12 pairs; petiole 1cm; stipule broadly elliptic, 1.5cm, entire; inflorescence loosely capitate, 3 × 3cm; peduncle 7cm; flowers white, 3mm; fruit ovoid, 5mm. Forest; 1000–1550m.
Distr.: Nigeria to Zimbabwe. (Afromontane).
IUCN: LC
Bangem: Thomas 5890 fl., 3/1986; **Kodmin:** Cheek 9193 fl., 2/1998; **Mwanenguba:** Leeuwenberg 8843 fr., 12/1971.

Psychotria venosa (Hiern) Petit

Tree (2.5–)8–12m, glabrous; leaves drying dark brown, elliptic, to 18 × 10cm, acumen 0.5cm, obtuse, lateral nerves 12 pairs, finely puberulent below, domatia pits; petiole to 2.5cm, slightly winged; stipule broadly ovate, 1cm, sheathing; panicle flat-topped, 15cm wide; peduncle 8cm; fruit globose, 5mm, red, faintly ridged. Forest, farmbush; 900–1800m.
Distr.: Nigeria to Congo (K). (Guineo-Congolian (montane)).
IUCN: LC
Ehumseh: Manning 405 fr., 10/1986; **Ehumseh - Mejelet:** Etuge 323 10/1986; **Kodmin:** Cheek 9564 fr., 11/1998; Ghogue 64 fr., 1/1998; **Kurume:** Thomas 5424 fr., 1/1986; **Ngomboaku:** Etuge 4660 fr., 12/1999.
Local name: Mukuna. **Uses:** none known (Max Ebong in Cheek 9564).

Psychotria sp. 1

Shrub 2m, glabrous; leaves drying green, elliptic, to 9.5 × 4.5cm, including abrupt slender acumen, 1.5cm, base rounded, lateral nerves 5–6 pairs, inconspicuous; petioles to 5mm; stipule 3mm, biaristate to the base; infructescence 3.5cm; peduncle 2.5cm; fruit globose, 5mm, red, smooth. Forest; 250–900m.
Distr.: Bakossi. (Narrow Endemic).
Baduma: Nemba 180 fr., 8/1986; **Nyasoso:** Lane 103 fr., 6/1994.
Note: probably a new species; flowers required.

Psychotria sp. 2

Shrub 4m, densely red-brown papillate-puberulent; leaves drying black above, grey below, narrowly elliptic, to 14 × 5cm, acumen 1.5cm, base cuneate, lateral nerves 10–12, base acute; petiole 1.5cm; stipule ovate-triangular, 2mm, glabrous, entire; infloresence 5cm; peduncle 4cm, 3 distal branches each with 20 flowers; fruit globose, 5mm. Forest; 1000m.
Bangem: Thomas 5911a fl., 3/1986.
Note: possibly heterogeneous, more analysis required.

Psychotria sp. 3

Shrub 3m, densely pink-hairy; leaves oblong-elliptic, 13 × 7cm, acute to subacuminate, rounded, lateral nerves 12 pairs, midrib pink below, lower surface softly hairy; petiole 1.8cm; stipule elliptic, 2cm, including arista 1cm; panicle 3cm, 20-flowered, dense; peduncle 2cm; fruit globose, 1cm. Forest; 1330m.
Distr.: Bakossi. (Narrow Endemic).
Kodmin: Cheek 10271 fl., fr., 12/1999; **Kupe Village:** Cable 3796 7/1996.
Note: probably new to science: more material desired.

Psychotria sp. 4

Shrub 1–3m; stem densely bright-red-pubescent, extending to stipule; petiole and inflorescence; leaves oblong-elliptic, 13 × 5cm, acumen 1.5cm, acute, lateral nerves 15 pairs, minutely white puberulent; petiole to 1cm; stipule triangular, 6mm; inflorescence capitate, globose, 1–1.5cm, 10–20-flowered; peduncle 1cm; flowers white, 3mm; fruit globose, 7mm. Forest; 1200–1360m.
Distr.: Bakossi. (Narrow Endemic).
Kupe Village: Cheek 7208 fr., 1/1995; **Nyasoso:** Cable 1175 fl., 2/1995; Lane 117 fr., 6/1994; 249 fr., 11/1994.

Psychotria sp. 5

Shrub (1.5–)4m, glabrous; strongly anisophyllous; leaves papery, drying grey-green, elliptic to broadly ovate-elliptic, to 17 × 11cm, subacuminate, base obtuse then abruptly decurrent, lateral nerves 8 pairs; petiole to 5cm; stipule ovate, glabrous, 3mm, entire; panicle 7 × 4cm, moderately dense; peduncle 2cm; flowers white, 3mm; infructescence with distal part often absent, diffuse; fruit ovoid, 7mm, red, smooth, often enlarged by (?) galls. Forest; 600–1000m.
Distr.: Bakossi. (Narrow Endemic).
Enyandong: Etuge 4382r 10/2001; **Kupe Village:** Etuge 2604 fr., 7/1996; **Ngomboaku:** Cheek 10334 fr., 12/1999; **Nyale:** Cheek 9658 fr., 11/1998.

Psychotria sp. 6

Shrub 2–3m, minutely papillate, glabrescent; leaves oblanceolate-elliptic, 14–17 × 6.5–7.5cm, acumen 0.5cm, cuneate, lateral nerves 10–12 pairs; petiole 2.5cm; stipule sheathing, 1cm; panicle 5 × 4cm; peduncle 3cm, c. 100-flowered. Forest; 950m.
Kupe Village: Ryan 315 fl., fr., 5/1996.
Note: more material needed to analyse further.

Psychotria sp. 7
Shrub 1–1.5m, resembling *P. psychotrioides*, but with a peduncle 2–4cm. Perhaps a variant of that species; more material desirable. Forest; 1300m.
Kodmin: Biye 51 fr., 1/1998.

Psychotria sp. 8
Tree 4m, glabrous; leaves thickly coriaceous, drying grey-brown, elliptic-oblong, to 16 × 6cm, apex not seen, base cuneate, lateral nerves 6–8 pairs; petiole 3cm; stipule not seen; infructescence 3 × 6cm; bracts sheathing, cupular, overtopped by two axillary infructescences; immature fruits with calyx 2mm. Forest; 1000m.
Ndibise: Etuge 534 fr., 6/1987.
Note: not matched at K; material available poor, better required.

Psychotria sp. 9
Shrub 2–5m, resembling *P. sp. 4*, but minutely puberulent, glabrescent, lateral nerves 9–10 pairs; petiole to 2cm; stipule 7mm; inflorescence loosely capitate with two axillary inflorescences subequal to the main inflorescence; flower cluster to 1.5 × 2cm, to 20-flowered; peduncle to 1.5cm, glabrous; flowers white, 3mm; fruit globose, 1cm, red. Forest; 1600–1800m.
Distr.: Bakossi. (Narrow Endemic).
Kupe Village: Cable 2555 fl., 5/1996; 2566 5/1996; Ryan 424 fl., 5/1996; **Nyasoso:** Etuge 1739 fl., fr., 2/1996.

Psydrax acutiflora (Hiern) Bridson
F.T.E.A. Rubiaceae: 906 (1991).
Syn.: *Canthium acutiflorum* Hiern
Syn.: *Canthium henriquesianum* (K.Schum.) G.Taylor
Climber 5m; stems glabrous, 4-angled, lacking hooks or ants; leaves coriaceous, glossy, ovate, 9 × 4.5cm, acuminate, rounded, 4-nerved; petiole 0.5cm; stipule triangular, 4mm; flowers axillary, numerous, white; panicle contracted, 5mm; fruit compressed, spherical, 8mm; aborted carpel evident. Evergreen forest; 350–1400m.
Distr.: SE Nigeria to Uganda. (Lower Guinea & Congolian).
IUCN: LC
Mekom: Nemba 17 4/1986; **Mwanenguba:** Leeuwenberg 8403 fr., 9/1971; 9649 4/1972.

Psydrax bridsoniana Cheek & Sonké ined.
Tree 12–20m; stems above nodes dilated, ant-inhabited, glabrous; leaves coriaceous, glossy, oblong, c. 25 × 15cm, shortly acuminate, truncate-subcordate, nerves 9–10; petiole 1.5cm; stipule ovate-triangular, 18mm, midrib conspicuous, often reflexed; inflorescence bracts and bracteoles caducous; fruit numerous, axillary; peduncle 6cm, endocarp 1.25cm, slightly curved. Evergreen forest; 800–850m.
Distr.: Mt Kupe. (Narrow Endemic).
IUCN: EN
Kupe Village: Cable 2546 5/1996; **Nyasoso:** Etuge 2557 7/1996.

Psydrax subcordata (DC.) Bridson var. subcordata
Kew Bull. 40: 698 (1985).
Syn.: *Canthium subcordatum* DC.
Tree 15m, buttressed, bicycle-spoke branching, with biting ants, glabrous; leaves papery, matt, ovate-elliptic, c. 12 × 6cm, shortly acuminate, rounded or subcordate, nerves 8–9; petiole 1–2cm; stipule ovate, 3mm; flowers foul-smelling, white, numerous, axillary; peduncle 4cm; corolla tube 2mm; fruit 8mm. Edge of evergreen forest; 300–1050m.
Distr.: Gambia to Zambia. (Guineo-Congolian).

IUCN: LC
Bolo Moboka: Nemba 236 9/1986; **Kupe Village:** Cheek 7768 11/1995; **Mwanenguba:** Leeuwenberg 8678 11/1971; **Ngomboaku:** Cheek 10356 12/1999.

Rothmannia ebamutensis Sonké
Syst. Geogr. Pl., 70(1): 149 (2000).
Tree 10–15m; stems glossy, pale brown; leaves coriaceous, obovate or elliptic, c. 20 × 10cm, obtuse or shortly acuminate, c. 8 pairs nerves, glabrous; petiole 1cm; flowers single, yellow-white; pedicel 2.5cm; calyx tube 1.5cm, wrinkled limb 0.7cm, teeth 0.5cm; corolla with basal tube 12 × 1cm, upper tube 8 × 4.5cm, lobes broadly ovate, 3.5 × 3.5cm, outer surface densely pale brown-pubescent. Evergreen forest; 1300m.
Distr.: W Cameroon. (Cameroon Endemic).
IUCN: EN
Kodmin: Biye 50 fl., 1/1998.

Rothmannia hispida (K.Schum.) Fagerl.
Tree 5–20m; stems, nerves and calyx hispid; leaves papery, elliptic, c. 15 × 6cm, acuminate, drying black; flowers white; calyx tube 1.5cm, limb 3cm, teeth 2cm; corolla with basal tube 12 × 0.6cm, upper tube 3 × 2.5cm, lobes 1.5 × 1cm, outer surface grey silky-hairy. Evergreen forest; 220–950m.
Distr.: Guinea (C) to Congo (K). (Guineo-Congolian).
IUCN: LC
Baduma: Nemba 225 9/1986; **Konye:** Thomas 5157 /1985; **Loum F.R.:** Cheek 10250 fr., 12/1999; **Mungo River F.R.:** Pollard 109 fr., 10/1998; **Ngomboaku:** Cheek 10394 fl., 12/1999.
Local name: Mekire. **Uses:** MATERIALS – place some leaves in the hunting path, where the fellow hunters will follow, to ensure a successful hunt (Epie Ngome in Pollard 109).

Rothmannia lateriflora (K.Schum.) Keay
Fl. Gabon 17: 234 (1970).
Shrub or tree to 12m, resembling *R. hispida*, but stems and calyx glabrous; corolla shortly pubescent. Evergreen forest; 300–850m.
Distr.: Cameroon to Congo (B). (Lower Guinea).
IUCN: NT
Loum F.R.: Biye 65 fr., 12/1999; **Ngomboaku:** Cheek 10343 fr., 12/1999.
Note: about 20 sites are known.

Rothmannia longiflora Salisb.
Shrub or tree to 5m, glabrous; leaves papery, elliptic, c. 12 × 5cm, acuminate, drying green, nerves 4–5 pairs; flowers green, blotched purple and white; calyx tube 6mm, limb 9mm, teeth 1.5mm; corolla with basal tube 14.5 × 0.9cm, upper tube c. 3 × 3cm, lobes ovate, 2 × 2cm, outer surface densely puberulent. Evergreen forest; 150–300m.
Distr.: Guinea (B) to Uganda. (Guineo-Congolian).
IUCN: LC
Bolo Moboka: Nemba 254 9/1986; **Etam:** Etuge 195 7/1986; **Kurume:** Thomas 5423 1/1986; **Loum F.R.:** Onana 988 fr., 12/1999; **Mungo River F.R.:** Onana 927 fr., 11/1999.

Rothmannia talbotii (Wernham) Keay
Tree 5m, resembling *R. hispida* and *R. urcelliformis*, but leaves whitish below, obovate-oblanceolate, 12–24 × 4–9cm, acuminate, cuneate, lateral nerves 7–9 pairs, venation obscure below, rusty-pubescent when young; calyx densely brown-tomentellous, lobes linear, 5–6mm; corolla tube 20–25cm, densely shortly-velutinous outside, 2.5cm wide at

mouth, lobes 4.5cm; fruit ellipsoid, 5-ridged, 6 × 3.5cm.
Forest; 250m.
Distr.: Nigeria to Angola (Cabinda). (Lower Guinea).
IUCN: LC
Mile 15, Kumba-Mamfe Road: Nemba 165 7/1986.

Rothmannia urcelliformis (Hiern) Bullock ex Robyns

Shrub or tree to 20m; stems glabrescent; leaves elliptic, c. 13 × 5cm, acuminate, white-tuft domatia, puberulent below, 7–8 pairs nerves; flowers white, blotched purple; calyx tube 0.3cm, limb 2cm, teeth 1.5cm; corolla with basal tube 2.5 × 0.4cm, upper tube 5 × 4cm; lobes 3 × 1.5cm. Evergreen forest. 870–1200m.
Distr.: Guinea (C) to Mozambique. (Tropical Africa).
IUCN: LC
Kupe Village: Etuge 1842 fr., 3/1996; Zapfack 923 fr., 7/1996; **Ndum:** Elad 98 fr., 2/1995; **Nyandong:** Nguembou 642 3/2003; **Nyasoso:** Cable 2797 fr., 6/1996; 3234 fr., 6/1996; Cheek 7515 fr., 10/1995; 7526 fr., 10/1995; Etuge 1519 fl., 12/1995.

Rothmannia whitfieldii (Lindl.) Dandy

Tree 12m, closely similar to *R. talbotii*, but leaf venation conspicuous below; calyx lobes 15–66mm; corolla tube long-velutinous, 3–17cm; fruit subglobose, 7cm, 10-ridged. Forest; 250m.
Distr.: tropical Africa.
IUCN: LC
Ikiliwindi: Etuge 510 3/1987; **Nyandong:** Nguembou 633 3/2003; Tchiengue 1726 fr., 3/2003.

Rutidea decorticata Hiern

Climber, glabrous or glabrescent; epidermis of stem and petiole exfoliating; leaves coriaceous, elliptic, to 15 × 8cm, base obtuse, nerves 4–8, impressed and white above, finely reticulate below, domatia inconspicuous; petiole 1.5cm; stipule triangular, aristate; inflorescence with numerous branches; corolla white, tube 3mm; fruit orange. Evergreen forest; 1200–1570m.
Distr.: Nigeria to Congo (K). (Lower Guinea & Congolian).
IUCN: LC
Kodmin: Cheek 8912 fr., 1/1998; Gosline 42 fr., 1/1998; **Kupe Village:** Cheek 7215 fl., 1/1995; 7842 fr., 11/1995; 7878 fr., 11/1995; Etuge 2866 fr., 7/1996; Ryan 380 fr., 5/1996; **Mejelet-Ehumseh:** Etuge 302 10/1986; **Mwambong:** Mackinder 157 fl., 1/1998.
Note: easily confused with *Tarenna eketensis*. Records of additional taxa of this genus from Bakossi are *R. parviflora* and *R. syringioides*, based on Manning 494 and Thomas 5309. However, Bridson (Kew Bull. 33: 243–278, 1978), revising this genus, accepts the first as occuring only to the west of Cameroon and treats the second as a synonym of four other accepted taxa. Accordingly, these taxa are not accepted here.

Rutidea glabra Hiern

Climber, glabrous; leaves elliptic or oblanceolate, to 13 × 4.5cm, long-acuminate, acute, nerves 5, inconspicuous, domatia absent; petiole 1cm; stipule oblong, aristate, 2mm; inflorescence dense; fruits orange, c. 15mm. Evergreen forest; 950m.
Distr.: SE Nigeria to Congo (B). (Lower Guinea).
IUCN: LC
Ngomboaku: Cheek 10294 fr., 12/1999.

Rutidea hispida Hiern

Kew Bull. 33: 260 (1978).

Climber, hispid, with brown hairs; leaves papery, lanceolate or elliptic, to 17 × 7cm, base obtuse to truncate, nerves c. 7, domatia absent; petiole to 12mm; stipule awn to 14mm; inflorescence with 3 subequal branches; corolla white, tube c. 15mm; fruit orange, 6mm. Evergreen forest; 650–1300m.
Distr.: SE Nigeria to Gabon. (Lower Guinea).
IUCN: NT
Kupe Village: Cable 2488 fr., 5/1996; Cheek 7058 fl., 1/1995; Etuge 1720 fl., 2/1996; Lane 394 fr., 1/1995; **Ndum:** Lane 466 fl., 2/1995; 480 fr., 2/1995; **Nyasoso:** Cable 1182 fl., 2/1995; 3504 fr., 7/1996; Cheek 7455 fl., fr., 10/1995; Etuge 2406 fr., 6/1996; Groves 98 fl., 2/1995.
Note: 12 sites in total. May qualify as VU under criterion A.

Rutidea nigerica Bridson

Kew Bull. 33: 258 (1978).
Climber resembling *R. hispida*, but central inflorescence branch 1.5 times or more as long as lateral braches; leaves often subbullate, cordate. Evergreen forest; 250–800m.
Distr.: Benin to Cameroon. (Lower Guinea).
IUCN: VU
Baduma: Nemba 177 8/1986; **Nyasoso:** Etuge 2501 fr., 7/1996.

Rutidea olenotricha Hiern

Climber, densely and shortly brown-pubescent; leaves elliptic, oblong or oblanceolate, to 15 × 8cm, shortly acuminate, obtuse to rounded, nerves 7–8, domatia large, bright-brown-hairy, extending along secondary nerves; petiole 10mm; inflorescence with numerous branches; corolla yellow, tube 5mm; fruit yellow, 6mm. Evergreen forest; 200–300m.
Distr.: Sierra Leone to Congo (K). (Guineo-Congolian).
IUCN: LC
Bolo Moboka: Nemba 250 fr., 9/1986; **Mile 12 Mamfe Road:** Nemba 294 fl., 10/1986.

Rutidea rufipilis Hiern

Kew Bull. 33: 252 (1978).
Climber, resembling *R. hispida*, but stipule 5-awned. Evergreen forest; 1400–1500m.
Distr.: SE Nigeria & Cameroon. (Lower Guinea).
IUCN: NT
Kodmin: Ghogue 8 fr., 1/1998; Gosline 148 fr., 11/1998; Satabie 1114 fr., 1/1998; **Mejelet-Ehumseh:** Etuge 180 6/1986.
Note: 14 sites in total, may qualify as VU under criterion A.

Rutidea smithii Hiern subsp. *smithii*

Climber, grey puberulent or glabrescent; leaves papery, drying matt black-brown above, grey-brown below, elliptic or elliptic-obovate, 10–17 × 4–8cm, shortly acuminate, acute, nerves 7–9, with bright-white-hairy domatia extending to tertiary nerve junctions; petiole 1–2.5cm; stipule awn 8mm; inflorescence with numerous branches; corolla white, tube 3mm; fruit green, 6mm. Evergreen forest; 600–1200m.
Distr.: Sierra Leone to Kenya. (Guineo-Congolian).
IUCN: LC
Bangem: Thomas 5356 fl., 1/1986; **Kupe Village:** Etuge 2756 fr., 7/1996; Kenfack 216 fr., 7/1996; **Ndum:** Lane 508 fr., 2/1995; **Nyasoso:** Etuge 1301 fr., 10/1995; 1637 fl., 1/1996; 2559 fr., 7/1996.

Rytigynia sp. A

Syn.: *Rytigynia neglecta sensu* Hepper & Keay in FWTA 2: 186 non (Hiern) Robyns
Shrub 2–3m, glabrous; leaves lanceolate, 7 × 3.5cm, acumen to 2cm, rounded, lateral nerves 4 pairs, domatia white-hairy;

petiole 3mm; stipule sheathing, 6mm; inflorescences axillary, umbellate, 3–4-flowered, 3mm; pedicel 2–3mm; corolla white, tube 4–5 × 2.5mm; fruit globose, 1.2cm, 2-seeded; pedicel 8mm. Forest; 1800–2000m.
Distr.: Cameroon Endemic.
Mt Kupe: Cheek 7577 10/1995; **Mwanenguba:** Cheek 7258 2/1995; Etuge 4369 10/1998; **Nyasoso:** Cable 1234 fr., 2/1995; Cheek 7357 fr., 2/1995; Etuge 1660 1/1996.

Rytigynia sp. B
Shrub or climber to 10m, resembling *R. sp. A*, but petioles to 5mm; pedicels 6–7mm; corolla tube 6 × 1mm; fruit globose, 7–10mm, 5-seeded; pedicel 1.5–2cm in fruit. Forest; 1500–1700m.
Distr.: Bakossi. (Narrow Endemic).
Nyasoso: Etuge 1737 2/1996; 2193 6/1996; Wheatley 464 fr., 7/1992.
Note: only known from Kupe-Bakossi.

Rytigynia sp. C
Shrub 1(–3.5)m, glabrous; resembling *R. sp. A*, but leaves 15 × 8cm, lateral nerves 6, domatia absent; petiole 12mm; corolla tube 2 × 2mm; pedicel 3mm; fruit 12–14mm diameter, 2-seeded; pedicel 1cm in fruit. Forest; 800–1200m.
Distr.: Bakossi. (Narrow Endemic).
Kupe Village: Cable 2481 5/1996; **Manehas F.R.:** Etuge 4345 10/1998; **Nyasoso:** Balding 85 fr., 8/1993; Cable 3212 6/1996; 3601 7/1996; Cheek 7501 10/1995; Lane 155 fr., 10/1994; Sidwell 315 10/1995.

Rytigynia sp. D
Shrub 1.5–3m, sometimes lianescent; resembling *R. sp. A*, but leaves sparsely hairy above, lateral nerves 5–6 pairs; umbels 6-flowered; pedicels 1–2mm; corolla tube 2 × 1.5mm; fruit 5-seeded (imm. only). Forest; 800–950m.
Distr.: Bakossi. (Narrow Endemic).
Mwanenguba: Leeuwenberg 9516 3/1972; **Ndoungue:** Leeuwenberg 9452 3/1972.

Sabicea calycina Benth.
Climber, puberulous; leaves membranous, oblong, c. 9 × 4cm, acuminate, cordate, whitish green below, nerves 9; petiole to 3.5cm; flowers 10–15; peduncle c. 6cm; bracts ovate 1.2cm; calyx lobes elliptic, purple, 1.2cm; corolla white, tube 2cm. Farmbush; 700–1400m.
Distr.: Sierra Leone to Congo (K). (Guineo-Congolian).
IUCN: LC
Kupe Village: Cheek 6094 fl., 1/1995; Etuge 2918 fl., 7/1996; Lane 320 fl., 1/1995; 446 fl., 1/1995; Schoenenberger 63 fl., 11/1995; **Mejelet-Ehumseh:** Etuge 171 6/1986; **Messaka:** Ghogue 1510 fl., 3/2003; **Nyandong:** Nguembou 594 fl., 3/2003; 606 fl., 3/2003; Thomas 6696 2/1987.

Sabicea capitellata Benth.
Climber, densely puberulous; leaves papery, elliptic, c. 9 × 4.5cm, subacuminate, acute, silky brownish white below, nerves 16; petiole 2cm; flowers 10–15; bracts united, cup-like, brown-pilose; calyx lobes linear, 0.7cm; corolla white, tube 7mm. Forest edge; 800–1570m.
Distr.: SE Nigeria to Gabon. (Lower Guinea).
IUCN: LC
Bangem: Thomas 5319 1/1986; **Kodmin:** Cheek 8859 fl., 1/1998; Gosline 150 fl., 11/1998; Onana 593a fr., 2/1998; **Mwambong:** Mackinder 159 fl., fr., 1/1998; **Ndibise:** Etuge 532 6/1987.

Sabicea gabonica (Hiern) Hepper
Syn.: *Sabicea efulenensis* (Hutch.) Hepper
Climber, subglabrous; leaves leathery, elliptic, c. 11 × 6cm, subacuminate, obtuse, drying brown below, nerves 11; petiole 2cm; flowers c. 20; peduncle 1.5cm; bracts not seen; calyx lobes linear, 9mm; fruits glossy, red, elliptic. Forest; 880m.
Distr.: SE Nigeria to Gabon. (Lower Guinea).
IUCN: NT
Kupe Village: Cheek 7057 fr., 1/1995.
Note: 14 sites known.

Sabicea pilosa Hiern
Climber, but flowering from prostrate stems, glabrescent to subscabrid; leaves membranous, elliptic, c. 11 × 5cm, acuminate, acute, nerves 10–12; petiole to 2cm; flowers 2–5; peduncle c. 1cm, up to 7 at one node; bracts united at base, forming cup; calyx lobes ovate, 1cm, glossy, purple; corolla tube pubescent, white, 2cm; fruit white or pale red, sparsely long-hairy, sepals elongating. Forest; 760–1200m.
Distr.: Ivory Coast to Gabon. (Upper & Lower Guinea).
IUCN: LC
Kodmin: Etuge 4163 fl., 2/1998; **Kupe Village:** Cable 3691 fr., 7/1996; 3861 fl., 7/1996; Etuge 2917 fr., 7/1996; **Nyasoso:** Cable 3331 fr., 6/1996.

Sabicea speciosa K.Schum.
Cauliflorous climber; stems pilose, to over 1cm diameter, 4-lobed; leaves papery, subelliptic, c. 14 × 8cm, acuminate, cordate to obtuse, nerves 14; petiole 2cm, pilose; flowers sessile, in clusters of 3–7; calyx lobes linear, hairy, 2cm; corolla bright red, inside white, tube 3cm. Evergreen forest; 660m.
Distr.: Togo, SE Nigeria & Cameroon. (Lower Guinea).
IUCN: NT
Kupe Village: Cheek 8326 fl., 5/1996.
Note: 13 sites known.

Sabicea cf. *tchapensis* Krause
Bot. Jahrb. 48: 408 (1912).
Climber, puberulent; stems to 1cm, 4-lobed; leaves papery, elliptic, to 15 × 8cm, subacuminate, obtuse, nerves to 14; petiole 1.5–5.5cm; flowers 5–10, on spur-shoots; peduncle 2cm; bracts broadly ovate, 1.5 × 1.8cm; fruits red, ovoid, 2cm, appressed white-hairy; calyx lobes ligulate, 1.4cm. Evergreen forest; 800–950m.
Kupe Village: Cable 2723 fr., 5/1996; Etuge 2666 fr., 7/1996.

Sabicea venosa Benth.
Climber, appressed white-pubescent; leaves thinly papery, elliptic, to 9 × 3.5cm, acuminate, rounded to obtuse, nerves 16; petiole 1cm; flowers 5–10; peduncle 0.5cm; branches 0.5cm; bracts inconspicuous; calyx lobes 3mm; corolla white, tube 5mm; fruit 1cm, globose, white. Evergreen forest; 1000m.
Distr.: Senegal to Congo (K). (Guineo-Congolian).
IUCN: LC
Nyasoso: Cable 2880 fr., 6/1996; 3434 fr., 7/1996; Sunderland 1491 fl., fr., 7/1992.

Sabicea xanthotricha Wernham
Shrub 3m, cauliflorous, pubescent; leaves membranous, elliptic, c. 30 × 17cm, subacuminate, rounded-decurrent, nerves 20; petiole 10cm; flowers numerous, sessile; calyx lobes filiform, c. 1cm; corolla white, tube c. 1cm. Forest; 1000–1400m.
Distr.: SE Nigeria & Cameroon. (W Cameroon Uplands).

IUCN: EN
Kodmin: Cheek 9097 fl., 2/1998; Etuge 4065 fl., fr., 1/1998.

Sabicea sp. A

Climber or procumbent herb, white-brown pilose; leaves papery, elliptic, c. 8 × 4.5cm, subacuminate, cordate, nerves c. 20; petiole 2cm; flowers 1-several, sessile; calyx lobes elliptic-oblong, 7 × 3mm; corolla white, tube 2cm; fruit red, drying white, ellipsoid, 2.5cm. Evergreen forest; 800–1100m.
Distr.: Cameroon. (Narrow Endemic).
Kodmin: Etuge 4061 fl., fr., 1/1998; **Kupe Village:** Cable 2447 fr., 5/1996; 2743 fr., 5/1996; Etuge 1983 fr., 5/1996; Ryan 202 fr., 5/1996; **Nyasoso:** Cable 2890 fr., 6/1996; 3331a fl., 6/1996; 3430 fr., 7/1996; Etuge 2156 fr., 6/1996.

Sacosperma paniculatum (Benth.) G.Taylor

Climber to 3m or more, glabrous; leaves papery, elliptic or oblong, c. 9 × 4cm, subacuminate, acute, nerves 8–10, inconspicuous; petiole 1cm; inflorescence c. 30 × 15cm, diffuse with 4 pairs of branches; fruits dry, elliptic, 5mm long; seeds numerous, winged. Lake edge; 1200–1250m.
Distr.: Gambia to Congo (K). (Guineo-Congolian).
IUCN: LC
Lake Edib: Cheek 9139 fr., 2/1998; **Mwanenguba:** Leeuwenberg 8592 10/1971.

Schumanniophyton magnificum (K.Schum.) Harms

Monopodial treelet to 5m; leaf-like branches opposite, 1.5m long, bearing 3 leaflet-like leaves each c. 1.5 × 0.6m; flowers sessile in erect clusters from branch ends; corolla white, tube 9cm; fruit globose, 4cm. Forest; 700–900m.
Distr.: S Nigeria to Congo (K). (Lower Guinea & Congolian).
IUCN: LC
Enyandong: Tadjouteu 455 fr., 10/2001; **Kodmin:** Cheek 8933 1/1998; **Kupe Village:** Cable 3772 7/1996; Etuge 1992 fl., 5/1996; Lane 336 fr., 1/1995; **Nyandong:** Nguembou 627 3/2003; **Nyasoso:** Wheatley 481 fl., 7/1992.

Sericanthe sp. A

Tree 2–10m, finely pale brown appressed pubescent; leaves narrowly oblong-oblanceolate, to 14 × 3.5cm, acuminate, acute, nerves 8, tertiary veins patent to midrib, white, domatia prominent, hairy; petiole 1cm; inflorescence 1-flowered; bracts and bracteoles united, awned; calyx cup-like, truncate, cleft; fruit on stem below leaves, red, globose, 1.5cm. Forest; 700–1750m.
Distr.: W Cameroon. (Narrow Endemic).
Ehumseh - Mejelet: Etuge 328 fr., 10/1986; **Kodmin:** Cheek 8925 fr., 1/1998; **Kupe Village:** Cable 2731 fr., 5/1996; 3869 fr., 7/1996.
Note: a probable new species, apparently endemic to our area. More flowering material is required.

Sericanthe sp. B

Tree 12m, glabrous, otherwise resembling *Sericanthe sp. A*, but leaves to 6.5cm wide, nerves 12, domatia absent; fruit 1.4cm. Forest; 1500m.
Distr.: W Cameroon. (Narrow Endemic).
Kodmin: Etuge 3961 fr., 1/1998.
Note: flowering material required of this probable *sp. nov.*

Sherbournia bignoniiflora (Welw.) Hua

Woody climber to 10m; stems glabrous; leaves elliptic, c. 11 × 6cm, subacuminate, obtuse-truncate; petiole 1cm, hairy; stipule elliptic, 7mm; flowers axillary, 1–3, violet and white; calyx tube 5 × 7mm, lobes elliptic, 15mm; corolla tube 25mm, lobes orbicular, 10mm. Evergreen forest; 350–1500m.
Distr.: Sierra Leone to Zambia. (Guineo-Congolian).
IUCN: LC
Bakolle Bakossi: Etuge 151 5/1986; **Kodmin:** Ghogue 7 fr., 1/1998; Gosline 184 fr., 11/1998.

Sherbournia millenii (Wernham) Hepper

Kew Bull. 16: 459 (1963).
Climber resembling *S. bignoniflora*, but leaves 7–17 × 2.5–7.5cm; calyx purple, tube 3–4mm, lobes 6–10 × 3.5–4.5mm; corolla red, tube 24–28mm, lobes 4–7 × 5–8mm; fruits 2.5–3.2 × 1–1.2cm, ribbed. Forest; 300m.
Distr.: Nigeria & Cameroon. (Lower Guinea).
IUCN: NT
Bolo Moboka: Nemba 242 9/1986.

Sherbournia zenkeri Hua

Climber resembling *S. bignoniflora*, but leaves 9–14 × 3.3–7cm, lateral nerves 10–12 pairs, densely appressed-puberulent below; calyx purple, tube 3mm, lobes 12–15mm; corolla white, tube 26–32mm, lobes 10mm; fruits cylindric, 3.5 × 1.5cm, strongly ribbed. Forest; 300–500m.
Distr.: Nigeria to Cabinda. (Lower Guinea).
IUCN: NT
Konye: Thomas 5188 11/1985; **Nlog:** Etuge 10 4/1986.

Spermacoce intricans (Hepper) H.M.Burkill

Kew Bull. 41: 1006 (1986).
Syn.: *Borreria intricans* Hepper
Suberect or prostrate herb, branched, 0.2–0.6cm; leaves ovate, 2–5 × 1–2cm, long acuminate, decurrent, lateral nerves 5 pairs, smooth above; petioles 2cm; stipule sheaths 2–3mm, laciniae 2mm; flowers few, white; calyx lobes 1.5mm; corolla 4–15mm; fruit 2–5mm. Forest, rockfaces or roadsides, forest edge, lake margins; 800–1450m.
Distr.: Senegal to Gabon. (Upper & Lower Guinea).
IUCN: LC
Kodmin: Cheek 9057 fr., 1/1998; 9625 fl., 11/1998; **Kupe Village:** Cable 3747 fl., 7/1996; Cheek 7701 fl., 11/1995; Etuge 1975 fl., 5/1996; Ryan 210 fl., 5/1996; **Lake Edib:** Cheek 9148 fl., 2/1998; **Nyasoso:** Cable 3217 fl., 6/1996; Cheek 9510 fl., 10/1998; Sebsebe 4997 fl., 10/1995; Sidwell 356 fl., 10/1995.
Local name: Nyam-kubg, "Beef for the farm" (Epie Ngome in Cheek 9510). **Uses:** FOOD – cut the leaves finely and cook with chicken, tastes good (Cheek 9510); MEDICINES (Epie Ngome in Cheek 9510 & Cheek 9625).

Spermacoce mauritiana Gideon

F.T.E.A. Rubiaceae: 927 (1991).
Straggling herb to 20cm, resembling *Mitracarpus*, but sparsely soft-hairy; leaves to 1.5 × 0.7cm. Roadsides, villages; 1480m.
Distr.: tropical Africa.
IUCN: LC
Kodmin: Ghogue 1 fl., fr., 1/1998.

Spermacoce pusilla Wall.

F.T.E.A. Rubiaceae: 356 (1976).
Syn.: *Borreria pusilla* (Wall.) DC.
Erect annual herb 10–20cm; stems wiry, terete; leaves linear 2.5 × 0.2cm; corolla pink; lobes 4. Roadsides, villages. 1250–2000m.
Distr.: tropical Africa & Asia. (Palaeotropics).
IUCN: LC
Mwanenguba: Cheek 9421 fl., 10/1998; Leeuwenberg 8606 10/1971.

Stelechantha arcuata S.E.Dawson

Kew Bull. 57: 398 (2002).

Shrub 2.5m, glabrous; leaves coriaceous, narrowly obovate, 12–22 × 3.5–7.5cm, shortly acuminate, cuneate, nerves 7; petiole 1–2.5cm; stipule lanceolate, 0.4–1.4 × 0.2–0.4cm, arching away from the stem; cauliflorous; flowers blue, 1–16, single or in subsessile clusters; calyx tube c. 2 × 2mm, slightly lobed; corolla tube 8mm, lobes 2mm; fruit globose, 6mm. Evergreen forest; 200–950m.

Distr.: W Cameroon. (Cameroon Endemic).

IUCN: CR

Kupe Village: Etuge 2892 fl., 7/1996; **Mile 15, Kumba-Mamfe Road:** Nemba 613 fr., 7/1987.

Stipularia africana P.Beauv.

Fl. Gabon 12: 158 (1966).

Syn.: Sabicea africana (P.Beauv.) Hepper

Shrub 1–2m; leaves lanceolate to oblanceolate, 10–15 × 4–7cm, lower surface densely white-felty, margins ciliate; stipules 3 × 3cm, enclosing sessile inflorescences. Forest; 450m.

Distr.: Sierra Leone to Congo (K). (Guineo-Congolian).

IUCN: LC

Nyandong: Cheek 11328 3/2003.

Note: specimen not seen.

Tarenna baconioides Wernham var. baconioides

Climber, glabrous; leaves elliptic, 9 × 5cm, acuminate, unequally rounded-obtuse, nerves 4–5, subinvolute, pit-domatia; flowers 15–20; peduncle 2cm; pedicel 2mm; calyx appressed-hairy, lobes 1cm; fruit ovoid, 1.8 × 1.1cm. Evergreen forest; 250m.

Distr.: SE Nigeria to Gabon. (Lower Guinea).

IUCN: VU

Baduma: Nemba 159 fr., 7/1986.

Tarenna bipindensis (K.Schum.) Bremek.

Climber, glabrous; leaves drying black, elliptic-oblong or oblanceolate, c. 13 × 5.5cm, acuminate, unequally obtuse, nerves 6, domatia hairy, extending to secondary branches; petiole 1–1.5cm; flowers c. 10 in diffuse panicles; pedicel 2.5cm; calyx glabrous, lobes subulate, 2cm; fruit globose, 1.2cm, 5-angled when live. Evergreen forest; 600–850m.

Distr.: Sierra Leone to Gabon. (Upper & Lower Guinea).

IUCN: LC

Etam: Etuge 246 fr., 8/1986; **Kupe Village:** Etuge 2757 7/1996; **Nyasoso:** Cheek 9513 fr., 10/1998.

Tarenna cf. calliblepharis N.Hallé

Fl. Gabon 17: 111 (1970).

Shrub resembling T. conferta and T. lasiorachis, distinguished by the large, sublinear sepals, 2.5–4 × 0.9mm, arched, hirsute with numerous white hairs c. 1mm long. Evergreen forest; 650m.

Distr.: Cameroon & Gabon. (Lower Guinea).

Kupe Village: Cable 2678 fr., 5/1996.

Note: more material required; delimitation unsatisfactory.

Tarenna cf. conferta (Benth.) Hiern

Shrub 2m; stems, petioles and nerves below sparsely appressed brown-hairy, midrib glabrous above; leaves elliptic to 20 × 10cm, acuminate, unequally acute, nerves 7, domatia absent; petiole 1–1.5cm; flowers c. 30 in dense corymb; peduncle 3cm; pedicel 2mm; fruit globose, hairy; calyx non-persistent. Evergreen forest; 900–1000m.

Distr.: SE Nigeria to Gabon. (Lower Guinea).

Nyasoso: Cable 3472 7/1996; Etuge 2154 fr., 6/1996.

Note: sometimes a climber (fide FWTA); fruit glabrous (fide Fl. Gabon). Material poor. Forms a complex with T. lasiorachis and T. calliblepharis which needs more work to delimit properly (fide Bridson, pers. comm.).

Tarenna eketensis Wernham

Climber, densely brown-puberulent on stem, lower surface of leaves and inflorescence; leaves elliptic or obovate, 9(–14) × 3.5(–6.5)cm, acumen broad, rounded, base acute, asymmetric, nerves 5–7, domatia hairy, extending to secondary branches; petiole 7mm; flowers c. 30; peduncles 15mm; pedicels 2–3mm, white; corolla tube 5mm, lobes 3mm; fruit globose, 8mm, arachnoid hairy. Evergreen forest; 350–1400m.

Distr.: Liberia to Congo (K). (Guineo-Congolian).

IUCN: LC

Banyu: Thomas 8159 6/1988; **Kodmin:** Etuge 4419 fr., 11/1998; **Mejelet-Ehumseh:** Etuge 493 fr., 2/1987; **Nyale:** Cheek 9676 fl., 11/1998; **Nyandong:** Thomas 7011 fl., fr., 5/1987.

Note: a variable species. In Fl. Gabon three varieties are recognised; at least one new variety could be erected from the Kupe-Bakossi material.

Tarenna fusco-flava (K.Schum.) N.Hallé

Adansonia (sér. 2) 7: 506 (1967).

Syn.: Tarenna flavo-fusca (K.Schum.) S.Moore

Climber 7m, glabrous; leaves elliptic, c. 6 × 3cm, spatulate-acuminate, unequally acute, nerves 2–3, hairy pocket domatia; petiole 4mm; stipule ovate-acuminate, 4 × 2mm; flowers c. 3 in subterminal axils; pedicels 3cm; corolla tube pink, 1.5cm, lobes 0.8cm; fruit globose, 8mm. Evergreen forest; 300–1100m.

Distr.: Liberia to Uganda. (Guineo-Congolian).

IUCN: LC

Konye: Thomas 5152 fr., /1985; **Ndum:** Cable 881 fr., 1/1995; Lane 453 fr., 1/1995; **Ngusi:** Etuge 1581 fr., 1/1996; **Nyasoso:** Cable 3249 fr., 6/1996; 3527 fl., 7/1996; 864 fr., 1/1995; Cheek 7451 fr., 10/1995; 7895 fr., 11/1995; Etuge 2065 fl., 6/1996; 2521 fl., 7/1996; Groves 95 fr., 2/1995; Lane 94 fr., 6/1994; Sebsebe 4983 10/1995; Sidwell 330 fr., 10/1995; Sunderland 1545 fl., 7/1992.

Note: Gabonese material 4-nerved, partial inflorescences 6–10-flowered (fide Fl. Gabon).

Tarenna grandiflora Hiern

Shrub or tree to 8m, glabrous; leaves elliptic, to 20 × 7cm, unequally acute, nerves 6–8, domatia glabrous, inconspicuous; petiole c. 2cm; stipule triangular, 3 × 4mm; flowers c. 15; pedicels c. 3mm, ebracteolate; corolla tube yellow-green, 25mm, lobes 8mm; fruit globose, 14mm, orange. Evergreen forest; 250–1300m.

Distr.: SE Nigeria, Bioko & Cameroon. (Lower Guinea).

IUCN: NT

Ikiliwindi: Nemba 111 fr., 6/1986; **Ikiliwindi1:** Nemba 163 fr., 7/1986; **Kupe Village:** Cable 2572 fr., 5/1996; Etuge 2048 fr., 5/1996; Kenfack 270 fl., fr., 7/1996; Lane 387 fr., 1/1995; 427 fr., 1/1995; **Mile 15, Kumba-Mamfe Road:** Nemba 172 7/1986.

Note: about 18 sites known.

Tarenna cf. lasiorachis (K.Schum. & K.Krause) Bremek.

Shrub resembling T. conferta, but midrib hairy above, not glabrous; stems densely hairy, not moderately; calyx lobes 2.5mm long, not up to 1mm long. Evergreen forest; 300–520m.

Distr.: Cameroon & Gabon. (Lower Guinea).

Eboné: Leeuwenberg 8152 fr., 8/1971; **Konye:** Thomas 5178 fl., 11/1985.

Note: material too poor to be certain of identification; see notes for *T. conferta* and *T. calliblepharis*.

Tarenna cf. pallidula Hiern

Shrub or tree resembling *T. grandiflora*, but leaves more slender (to 5cm wide); inflorescence densely white puberulent; pedicel 4–7mm long, with 2–3 bracteoles; corolla tube 7–9mm. Evergreen forest; 250–1750m.

Distr.: Cameroon & Gabon. (Lower Guinea).

Bangem: Manning 460 fr., 10/1986; Thomas 5264 fr., 1/1986; **Ehumseh - Mejelet:** Etuge 326 fr., 10/1986; **Kodmin:** Etuge 4079 fl., fr., 1/1998; Nwaga 18 fr., 1/1998; **Kupe Village:** Cable 2604 fr., 5/1996; 2607 fr., 5/1996; 2638 fr., 5/1996; 3721 fr., 7/1996; 3792 7/1996; 790 fr., 1/1995; Cheek 7187 fl., fr., 1/1995; 7205 fr., 1/1995; 7889 fr., 11/1995; Etuge 1718 fl., 2/1996; 2644 fr., 7/1996; 2867 fr., 7/1996; **Mbule:** Cable 3360 fr., 6/1996; **Mile 15, Kumba-Mamfe Road:** Nemba 205 fr., 9/1986; **Mt Kupe:** Thomas 5099 fr., 12/1985; **Nyasoso:** Cable 1240 fr., 2/1995; 3334 fr., 6/1996; 3605 fr., 7/1996; 663 fr., 12/1993; Etuge 2080 fr., 6/1996; 2372 6/1996; 2524 fr., 7/1996; Lane 197 fr., 11/1994; 252 fr., 11/1994; Wheatley 435 fr., 7/1992.

Note: the Kupe-Bakossi material, with other specimens from Cameroon and Gabon, differs from *T. pallidula sensu stricto* in the longer pedicels with 2–3 bracteoles (not 1–2.5mm with 0–1 bracteoles), the inflorescence densely (not sparsely) puberulent and the corolla tube 7–9mm (not 4–6mm) long.

Tricalysia atherura N.Hallé

Fl. Gabon 17: 292 (1970).

Shrub 1–4(–8)m, glabrous; leaves elliptic, c.15 × 6cm, acumen apex rounded, base acute, nerves 4–5; petiole 6mm; flowers 1 per axil, bracteolar cup sheathing pedicel, truncate 2mm; calyx subcylindrical, truncate, 7mm; corolla tube 20–34mm, lobes 7–8, 15mm; fruit orange or red, juicy, ellipsoid, fusiform or globular-rostrate, to 4 × 2cm; pedicel accrescent. Forest; 300–1250m.

Distr.: Cameroon & Gabon. (Lower Guinea).

IUCN: VU

Kupe Village: Biye 59 fl., 11/1999; Cable 3705 fr., 7/1996; 3741 fr., 7/1996; Cheek 6072 fl., 1/1995; 6074 fr., 1/1995; 7065 fl., 1/1995; 7663 fl., 11/1995; Etuge 1392 fr., 11/1995; 2649 fr., 7/1996; 2694 fr., 7/1996; 2784 fr., 7/1996; Lane 358 fl., 1/1995; 365 fl., 1/1995; Ryan 263 fr., 5/1996; 299 fr., 5/1996; **Loum F.R.:** Biye 73 fr., 12/1999.

Tricalysia biafrana Hiern

Shrub or small tree 2.5–3m, glabrous; stems white; leaves elliptic to 12 × 5cm, acuminate, cuneate, nerves 4–6; petiole c. 1cm; inflorescence 3–10-flowered; peduncle 2mm; bracts and bracteoles cup-shaped; bracts ligulate; pedicel 4mm; calyx 1.5mm, truncate; corolla white, tube 3–7mm, lobes 5–7, 3–5mm; fruit globose, 8mm. Forest; 1000m.

Distr.: Liberia to Congo (K). (Guineo-Congolian).

IUCN: LC

Bangem: Thomas 5915 fr., 3/1986.

Tricalysia bifida De Wild.

Bull. Jard. Bot. Nat. Belg. 57: 72 (1987).

Syn.: *Tricalysia pleiomera* Hutch.

Tree 12–15m, glabrous; leaves glossy, oblong, c. 25 × 8cm, acuminate, obtuse, nerves 6–8; petiole 2cm; flowers 3–5; inflorescence sessile, rachis 5mm; bracts and bracteoles cup-like; pedicel concealed; calyx cup-shaped, 5mm, split; corolla white, tube 8mm, lobes c. 10; fruit orange, spherical, 1–2cm. Forest; 150–700m.

Distr.: SE Nigeria to Congo (K). (Lower Guinea & Congolian).

IUCN: LC

Ikiliwindi1: Nemba 161 7/1986; **Kupe Village:** Etuge 2753 fl., 7/1996; **Mile 12 Mamfe Road:** Nemba 295 10/1986; 649 10/1987; **Mungo River F.R.:** Cheek 10148 fr., 11/1999.

Tricalysia discolor Brenan

Syn.: *Tricalysia mildbraedii* Keay

Tree 8–10m, glabrous; leaves drying grey-brown, papery, elliptic or elliptic-oblong, acuminate, acute, nerves 7, tertiary nervation scalariform; petiole 1.5cm; inflorescences several, subsessile; bracts cup-shaped; flowers numerous; bracteoles alternate on pedicel, 5mm; fruit elliptic, 5mm; calyx lobes triangular. Forest; 350–750m.

Distr.: Liberia to Cameroon. (Upper & Lower Guinea).

IUCN: LC

Kupe Village: Etuge 2046 fr., 5/1996; **Nyale:** Etuge 4467 fr., 11/1998.

Tricalysia gossweileri S.Moore

Bull. Jard. Bot. Nat. Belg. 49: 300 (1979).

Shrub 1.5–3m, glabrous; leaves papery, drying pale grey-green, elliptic or obovate, c. 17 × 6.5cm, acuminate, acute, nerves 4–5; petiole 7mm; inflorescences 3–5-flowered, sessile; bracts and bracteoles cup-shaped; pedicel concealed; calyx 2mm, shortly lobed; corolla white, tube 3mm; fruit ellipsoid, to 1.5 × 1cm, violet, flushed white. Forest; 200–1050m.

Distr.: Cameroon to Angola. (Lower Guinea).

IUCN: NT

Baduma: Nemba 578 7/1987; **Kupe Village:** Cable 2493 fr., 5/1996; 3725 fr., 7/1996; 3866 fr., 7/1996; Etuge 1954 fr., 5/1996; 2673 fr., 7/1996; Kenfack 281 fr., 7/1996; Ryan 302 fr., 5/1996; **Manehas F.R.:** Cheek 9402 fr., 10/1998; **Mwambong:** Etuge 4141 fr., 2/1998; Onana 531 fr., 2/1998; **Nyasoso:** Balding 14 fr., 7/1993; Cable 2745 fr., 6/1996; 3258 fr., 6/1996; 3293 fr., 6/1996; 3545 fr., 7/1996; Cheek 6013 fl., fr., 1/1995; Etuge 2060 fr., 6/1996; Lane 102 fr., 6/1994; 208 fr., 11/1994; Sebsebe 5017 fr., 10/1995; Sidwell 308 fr., 10/1995; Zapfack 590 fr., 6/1996.

Note: 22 sites known (Robbrecht op. cit. 1979: 301).

Local name: Mboge. **Uses:** MATERIALS – hunting; MEDICINES – veterinary: used in a combination of 3 herbs to treat a dog's nose, so it can smell better (Epie Ngome in Cheek 9402).

Tricalysia macrophylla K.Schum.

Tree, 7m, minutely puberulent; leaves oblong, 20 × 7cm, acuminate, obtuse-decurrent, nerves c. 9; petiole 1cm; flowers numerous in sessile inflorescences, rachis c. 1cm; bracts and bracteoles opposite, free; flower sessile; calyx lobed; fruit globose, 5mm. Forest; 200–1000m.

Distr.: Ivory Coast to Togo, Nigeria to Gabon. (Upper & Lower Guinea).

IUCN: NT

Bangem: Thomas 5374 1/1986; 5904 3/1986; **Ikiliwindi:** Nemba 466 fr., 1/1987.

Note: 18 sites known (Robbrecht op. cit. 1979: 342) apart from those listed here.

Tricalysia okelensis Hiern var. *okelensis*

Syn.: *Tricalysia pobeguinii* Hutch. & Dalziel

Tree 10m, densely patent-puberulent; leaves papery,drying dark brown, elliptic oblong, 12 × 5cm, acuminate, obtuse-decurrent; petiole 1.5cm; inflorescences several per axil, sessile, 1–3-flowered; in fruit bracts cup-shaped, awned, bracteoles cup-shaped, almost concealing pedicel; fruit

globose, red, 7mm; calyx densely white-pubescent. Forest; 900–1000m.
Distr.: Mali to Congo (K). (Guineo-Congolian).
IUCN: LC
Kupe Village: Etuge 1946 fr., 5/1996; Kenfack 294 fr., 7/1996.

Tricalysia pallens Hiern
Syn.: *Tricalysia pallens* Hiern var. *gabonica* (Hiern) N.Hallé
Shrub or tree (2–)5–6(–14)m, appressed white-puberulent; leaves obovate or elliptic, c. 10 × 3.5cm, acuminate, base acute-decurrent, nerves 5, inconspicuous, domatia pits rimmed with white hairs; petiole 3mm; inflorescences contracted, sessile, 10–20 flowers per node; fruit globose, 5mm, red; calyx; bracts and bracteoles cup-like, densely white-puberulent. Forest; 650–1500m.
Distr.: Liberia to Mozambique. (Tropical Africa).
IUCN: LC
Bangem: Thomas 5312 fl., 1/1986; **Kupe Village:** Ryan 373 fr., 5/1996; **Mejelet-Ehumseh:** Etuge 495 2/1987; **Mt Kupe:** Thomas 3088 fl., 2/1984; **Ngusi:** Etuge 1570 fl., 1/1996; **Nyasoso:** Etuge 2148 fr., 6/1996; 2530 fr., 7/1996; Zapfack 637 fr., 6/1996.

Tricalysia talbotii (Wernham) Keay
Tree 5m, densely pale brown patent-pubescent; leaves oblong-elliptic, 15 × 4.5cm, acuminate, acute, nerves 6–8; petiole 1cm; inflorescence 3–10-flowered, pubescent, bracteoles opposite, free; calyx 5-lobed; corolla tube 4mm; fruit globose, 8mm white to purple. Forest; 250–300m.
Distr.: SE Nigeria & Cameroon. (Lower Guinea).
IUCN: VU
Ikiliwindi: Etuge 498 3/1987; **Konye:** Thomas 5151 fl., 12/1985.

Tricalysia sp. A aff. coriacea (Benth.) Hiern
Shrub 3m, glabrous; leaves papery, elliptic-oblong, c. 16 × 6cm, acuminate, obtuse, nerves 6, domatia cryptic; petiole 1cm; inflorescences (?) single, 3-flowered, sessile, bracteoles cup-like, concealing pedicel; calyx cylindric, truncate, split; corolla tube 5mm, lobes 7. Forest; 1350m.
Kodmin: Cheek 9003 fl., 1/1998.
Note: known from a single specimen. More material is required to evaluate further.

Tricalysia sp. B aff. ferorum Robbr.
Tree 7–13m, densely white-appressed-hairy; leaves thinly coriaceous, oblong-acuminate, acumen abrupt, 7mm, base acute-decurrent, nerves 9; petiole 1.2cm; inflorescence on naked stem, 3-flowered, both bracts and bracteoles united, dish like; calyx 5-lobed; corolla tube 5mm; lobes 5; fruit green, globose, 1cm; disc 5mm diameter. Forest; 1550–1850m.
Distr.: Mt Kupe & Bali-Ngemba F.R. (W Cameroon Uplands).
Kupe Village: Cable 2559 fr., 5/1996; **Mt Kupe:** Cheek 7619 fl., 11/1995; **Nyasoso:** Cable 2926 fr., 6/1996; Etuge 1735 fr., 2/1996; 2176 fr., 6/1996.
Note: a new species, apparently endemic to Mt. Kupe and Bali-Ngemba F.R.

Tricalysia sp. C aff. gossweileri S.Moore
Shrub or tree 2–10m, minutely patent-puberulent; leaves oblong-elliptic, 14 × 5.5cm, acuminate, acute-decurrent, nerves 5; petiole 6mm; inflorescence 1–3-flowered; bracts united, boat-shaped; pedicel 5mm; bracteoles alternate; fruit globose, 6mm; calyx 5-lobed. Forest; 800–1300m.

Bangem: Thomas 5361 fl., fr., 1/1986; **Kodmin:** Etuge 4118 fr., 2/1998; Gosline 182 fr., 11/1998.
Note: known from three fruiting specimens.
Local name: Kofa hinte (Gosline 182).

Tricalysia sp. D cf. oligoneura K.Schum.
Tree 5m, minutely appressed-puberulent; leaves drying brown, elliptic, c. 10 × 4cm, apex acuminate, base acute, nerves 4; petiole 5mm; inflorescence single, 3-flowered; bracts and bracteoles united, awned; pedicel concealed; fruit globose, 5mm; calyx cylindrical, denticulate. Forest; 1100m.
Mejelet-Ehumseh: Etuge 473 fl., fr., 1/1987.
Note: known from a single fruiting specimen.

Tricalysia sp. E
Shrub 2.5–3m; stems slender, epidermis sloughing, minutely puberulent; leaves elliptic, 12 × 5.5cm, acuminate, obtuse, nerves 4; petiole 3mm; inflorescences 1-flowered, ramiflorous; bracts and bracteoles united, congested; calyx cylindrical, long-denticulate, split; corolla tube 3mm. Forest; 1500–1780m.
Kodmin: Cheek 9086 fl., 1/1998; Ghogue 6 fl., 1/1998.
Note: probably a new species in sect. *Rosea*. Fruits lacking.

Tricalysia sp. F
Tree 3–7m, minutely puberulous, glabrescent; leaves drying pale brown-green or fawn, elliptic or oblong, 9–13 × 2.8–4cm, acumen subspatulate, laterally displaced, base acute, nerves 4, drying white below; petiole 5–7mm; inflorescences single, 3-flowered; bracts and bracteoles cup-shaped, ± awned; calyx cup-shaped, denticulate; corolla tube 2mm; fruit orange, elliptic, 7mm, stalk to 10mm. Forest; 900–1400m.
Kodmin: Gosline 158 fl., 11/1998; **Kupe Village:** Etuge 1955 fr., 5/1996; **Nyasoso:** Balding 10 fr., 7/1993; 33 fr., 8/1993; Etuge 2423 fr., 6/1996; Lane 253 fl., 11/1994.

Tricalysia sp. G
Shrub 2.5–7m, resembling *Tricalysia sp. F*, but leaves drying dark brown-green, acumen acute, straight, shorter; bracteole cup directly under the fruit. Forest; 930–1300m.
Kupe Village: Cable 3853 fr., 7/1996; Ryan 222 fr., 5/1996; **Nyasoso:** Etuge 2058 fr., 6/1996.
Note: probably *sp.nov.*, but flowering material required.

Tricalysia sp. H
Shrub 2m, glabrous; leaves elliptic, 14 × 6cm, acumen spatulate, slender, 2cm, base acute, nerves 5, drying white above; inflorescence sessile, 3-flowered; bracts opposite, awned; bracteoles cup-like; fruit ellipsoid-ovoid, 12mm (immature); calyx cylindrical, denticulate. Forest; 1450m.
Kodmin: Cheek 8955 fr., 1/1998.
Note: a highly distinct entity, but additional specimens are needed to elucidate it.

Trichostachys aurea Hiern
Syn.: *Trichostachys zenkeri* De Wild.
Monopodial erect shrub to 0.5m, ciliate-hairy; new shoots arising from base of stem; leaves obovate, to 16 × 6cm, acute-acuminate, cuneate, sessile or petiole 2mm; stipule narrowly triangular, 7–9mm; inflorescence terminal; peduncle 2cm; flowers 30–50 in a subglobose head, 1–2cm. Forest; 200m.
Distr.: Sierra Leone to Cabinda. (Upper & Lower Guinea).
IUCN: NT
Baduma: Nemba 580 7/1987; **Mahole:** Olorunfemi in FHI 30590 5/1951.
Note: fairly rare within its range.

Trichostachys interrupta K.Schum.

Rhizomatous subshrub; aerial stems erect, monopodial, to 30cm, densely puberulent, internodes 2cm; leaves at 3–4 nodes, drying black above, obovate, to 12 × 6cm, obtuse or subacuminate, cuneate, lateral nerves 8 pairs, dirt-gathering; petiole 7mm; stipule ovate, 5 × 4mm, black, glabrescent; spike 4.5cm, erect; peduncle 0.5–1cm, appressed-hairy; rachis interrupted; flowers white, 1mm; fruit 2mm, pink, hairy. Forest; 1000m.
Distr.: SE Nigeria & SW Cameroon. (Lower Guinea).
IUCN: VU
Nyale: Etuge 4176 2/1998; 4461 11/1998.

Trichostachys sp. 1

Subshrub resembling *T. interrupta*, but leaves held to ground, 8 × 5cm, apex and base rounded, lateral nerves 10 pairs; petiole 1cm; stipule triangular; spike 2cm, not interrupted. Forest; 400m.
Bakossi F.R.: Etuge 4296 10/1998.
Note: more material required.

Uncaria africana G.Don

Climber with pairs of hooks 2cm long at nodes, puberulent; stems square; leaves elliptic-oblong, 11 × 5cm, acuminate, obtuse, nerves 5, drying pink below; petiole 1cm; inflorescence globular. Evergreen forest; 1200–1350m.
Distr.: Guinea (B) to Uganda. (Guineo-Congolian).
IUCN: LC
Kodmin: Cheek 8987 1/1998; **Ndum:** Groves 22 fl., 1/1995.

Vangueriella campylacantha (Mildbr.) Verdc.

Kew Bull. 42: 196 (1987).
Syn.: *Vangueriopsis subulata* Robyns
Climber, pilose; stem spines 1cm; leaves membranous, elliptic, c. 9 × 4cm, acumen 1cm, acute, nerves 5; petiole 4mm; inflorescences c. 4-flowered; peduncle c.3mm; pedicel 10mm; flowers green; corolla tube 5 × 5mm. Forest; 200m.
Distr.: Sierra Leone to Congo (K). (Guineo-Congolian).
IUCN: LC
Ikiliwindi: Nemba 522 6/1987.

Vangueriella laxiflora (K.Schum.) Verdc.

Kew Bull. 42: 192 (1987).
Syn.: *Vangueriopsis calycophila* (K.Schum.) Robyns
Scandent shrub to at least 4m; stem pale matt-grey; spines 3cm, initially axillary; leaves in whorls of 3, obovate, to 11 × 6cm, acumen oblong, rounded, 1cm, base rounded, lateral nerves 4 pairs, domatia pits angled; petiole 4mm. Forest; 650m.
Distr.: S Nigeria to Gabon. (Lower Guinea).
IUCN: NT
Kupe Village: Cable 2680 5/1996.
Note: possibly a red data taxon.

Virectaria angustifolia (Hiern) Bremek.

Rheophytic erect herb 10–20cm, appressed-hairy; stem wiry; leaves linear, to 3.5 × 0.5cm, sessile; stipule narrowly triangular, 2–3mm; inflorescence 3–5-flowered, terminal; corolla pink, lobes 5. Rocks in rivers; 800–1350m.
Distr.: Ghana, Nigeria to Gabon. (Upper & Lower Guinea).
IUCN: NT
Nyale: Cheek 9707 fl., fr., 11/1998; Etuge 4463 fl., 11/1998; **Nyandong:** Cheek 11281 3/2003; Nguembou 618 3/2003; 651 3/2003.
Note: 11 sites known.
Local name: none known (Cheek 11281).

Virectaria major (K.Schum.) Verdc. var. *major*

Weak-stemmed shrub 60cm, terete, puberulent; leaves ovate-elliptic, to 12 × 6cm, acumen 1cm, base obtuse, abruptly decurrent, lateral nerves 10–12 pairs, sparsely softly-hairy on both surfaces; petiole 2cm; stipule 5mm, entire; flowers in erect terminal clusters; calyx lobes linear, 1cm; corolla pale purple, 2 × 1.5cm; stamens 10, exserted 2cm. Forest, grassland edges; 1200–2000m.
Distr.: Nigeria to Zimbabwe. (Afromontane).
IUCN: LC
Kodmin: Cheek 9081 fl., 1/1998; 9574 fl., 11/1998; **Mejelet-Ehumseh:** Etuge 306 10/1986; **Mt Kupe:** Cheek 7570 fl., 10/1995; Thomas 5069 11/1985; **Mwanenguba:** Cheek 9469 fl., 10/1998; Leeuwenberg 8288 9/1971; 8841 12/1971.
Local name: Nsit. **Uses:** ANIMAL FOOD – aerial parts – birds visit for nectar, then caught by children (Edmondo Njume in Cheek 9574).

Virectaria multiflora (Sm.) Bremek

Weak-stemmed herb to 45cm, hairs patent, soft; leaves elliptic, 5 × 2cm, subacuminate, decurrent, lateral nerves 7 pairs; petiole winged, 5mm; stipule 5mm, bifid to base; flower-cluster terminal; corolla white, tube 10mm, lobes 5, 6mm. Rock outcrops; 1100m.
Distr.: Mali to Cabinda. (Upper & Lower Guinea).
IUCN: LC
Kupe Village: Cheek 7694 fl., 11/1995.

Virectaria procumbens (Sm.) Bremek

Herb 20cm, resembling *V. angustifolia*, but straggling; stems with 2 lines of pubescence; leaves ovate-oblong to subspatulate, 1–6 × 0.5–3.5cm; stipules triangular, 3mm, entire or cleft; calyx lobes spatulate; corolla white, 1cm. Forest; 910m.
Distr.: Guinea (B) to Congo (K). (Guineo-Congolian).
IUCN: LC
Ngomboaku: Ghogue 496 fr., 12/1999; Muasya 2039 fr., 12/1999.

RUTACEAE

T.M. Heller & M. Cheek

Fl. Cameroun 1 (1963).

Citropsis articulata (Spreng.) Swingle & M.Kellerm.

Monopodial shrub 1.5–3m, glabrous; axillary spines 2cm; leaves 15–30cm, 1–2-jugate; terminal leaflet largest, elliptic, to 18 × 7cm, obscurely crenate; petiole and rachis broadly-winged, to 2cm wide; panicles axillary, to 2cm; fruit globose, 1.5cm. Forest; 220–1000m.
Distr.: Sierra Leone to Gabon. (Upper & Lower Guinea).
IUCN: LC
Bakolle Bakossi: Etuge 291 fr., 9/1986; **Manehas F.R.:** Etuge 4357 10/1998; **Mungo River F.R.:** Cheek 10198 11/1999; **Nyasoso:** Cable 3273 6/1996; Groves 89 fr., 2/1995.

Citrus sp.

Purseglove: tropical Crops, Dicots: 495 (1968).
Tree c. 5m, glabrous, often spiny; leaves alternate, 1-foliolate, blades with translucent gland dots, elliptic, coriaceous, c. 8 × 4cm, glossy, acute, entire; petiole winged; inflorescences subfasciculate, axillary, few-flowered; calyx 2–3mm, shallowly lobed; petals free, oblong, 2cm, white;

stamens 20–40; fruit a 'hesperidium', oblate, yellow or orange, 3–15cm. Cultivated, persisting at abandoned settlements.
Distr.: pantropical, wild in SE Asia.
Nyandong: Cheek (sight record) 3/2003.
Note: *Citrus sinensis* (L.) Osbeck (the sweet orange or 'orange'), *C. reticulata* Blanco ('mandarine') and *C. paradisi* Macf. (grapefruit or 'grape') are the usual taxa cultivated as single trees in home gardens; they vary in degree of spininess, petiole alation etc. The description above is meant to be a general one to cover all the various species which occur in our area.
Uses: FOOD – infructescences (*fide* Cheek).

Clausena anisata (Willd.) Hook.f. ex Benth.
Shrub or tree 3–8m, non-spiny, puberulent, strongly aromatic; leaves imparipinnate, 15cm, 4–9-jugate; leaflets alternate, lanceolate-oblique, c. 6 × 2.5cm, acuminate, obtuse, lateral nerves c. 10 pairs; petiolules 1mm; panicle c. 12cm, slender; flowers white, 5mm; fruit indehiscent. Forest; 1975m.
Distr.: Guinea (C) to Malawi. (Tropical Africa).
IUCN: LC
Mwanenguba: Cheek 7255 fl., 2/1995.

Oricia lecomteana Pierre
Fl. Cameroun 1: 94 (1963).
Syn.: *Oricia sp. B sensu* Keay in FWTA 1: 688
Syn.: *Araliopsis ? sp. sensu* Keay in FWTA 1: 688
Monopodial tree 2–9m, glabrous; leaves to 60cm, digitately 5-foliolate; leaflets elliptic, c. 30 × 12cm, acumen 1cm, acute, lateral nerves c. 12; petiolule 5–10mm; petiole 30cm; panicle terminal, 18 × 10cm; flowers white, 7mm, petals 4; stamens 8; fruit ellipsoid, 2cm, flattened on one side. Forest; 700–1000m.
Distr.: Nigeria, Cameroon & Gabon. (Lower Guinea).
IUCN: VU
Kupe Village: Cheek 10222 11/1999; 7027 fl., 1/1995; **Nyale:** Etuge 4452 11/1998.
Note: paper in prep. to transfer this taxon to Vepris.

Vepris suaveolens (Engl.) Mziray
Symb. Bot. Ups. 30(1): 76 (1992).
Syn.: *Oricia suaveolens* (Engl.) Verdoorn
Tree 8m, 7cm d.b.h., glabrous; stems white, corrugated; leaves 3-foliolate, 40cm; leaflets elliptic to 23 × 9cm, subacuminate, base acute, lateral nerves 10 pairs; petiolules 1–2cm; petioles 12cm. Forest; 1400m.
Distr.: Guinea (C) to Cameroon. (Upper & Lower Guinea).
IUCN: NT
Kodmin: Plot B38 1/1998.
Note: fertile material needed to confirm identity; listed as LR/nt on www.redlist.org.

Vepris trifoliolata (Engl.) Mziray
Symb. Bot. Ups. 30(1): 76 (1992).
Syn.: *Oricia trifoliolata* I.Verd.
Tree 15m, dark brown velvety-hairy; trifoliolate, 26cm; leaflets elliptic, to 21 × 9cm, acumen 1cm, acute, lateral nerves 15 pairs, sessile; petioles 5cm; panicle dense, terminal, 23 × 20cm; flowers white, 4mm. Forest; 1600m.
Distr.: Cameroon Endemic.
IUCN: VU
Ekona Mombo: Etuge 417 12/1986; **Nyasoso:** Etuge 1814 3/1996.
Note: also similar to *V. gabonensis*, more material needed to confirm identity; altitude exceptionally high for both taxa.

Vepris verdoorniana (Exell & Mendonça) Mziray
Symb. Bot. Ups. 30(1): 76 (1992).
Syn.: *Teclea verdoorniana* Exell & Mendonça
Tree 4m; stems pale grey; lenticels sulcate, inconspicuous, glabrous; leaves (1–)3-foliolate, 20cm; leaflets drying green, elliptic, 15 × 5cm, acumen 1cm, acute, revolute, midrib drying yellow below, lateral nerves 10–15, intercalary nerves numerous; petiolules 2mm, pulvinate; petiole 2cm, yellow; panicles axillary, numerous, 2cm; flowers not seen; young fruit 7mm, rostrate. Forest; 1000m.
Distr.: Sierra Leone to Congo (K). (Guineo-Congolian).
IUCN: LC
Nyasoso: Cable 3473 7/1996.
Note: more material needed to confirm identity.

Vepris sp. 1
Tree c. 6m, minutely puberulent; leaves trifoliolate, 12cm; leaflets densely gland-dotted, narrowly elliptic, 9 × 3.5cm, acumen 0.25cm, lateral nerves 8, brochidodromous, revolute; petiolule 7mm; panicle 6cm; flowers white, 3mm; apocarpous, carpels 2; fruitlets ellipsoid, 1cm, apex rounded. Forest; 1940m.
Distr.: Cameroon. (Narrow Endemic).
Kupe Village: Ryan 407 5/1996; **Mt Kupe:** Sebsebe 5097 11/1995.
Note: probably a new species to science of the Oricioid group.

Vepris sp. 2
Shrub 3m, glabrous; leaves 3-foliolate, 60cm; leaflets papery, drying green, narrowly oblong, to 30 × 6cm, acumen 3cm, cuneate, entire, lateral nerves c. 20 pairs, gland dots sparse and small; petiolule 1.5cm; petiole 10cm. Forest; 900m.
Mbule: Etuge 4395 10/1998.
Note: remarkable for leaf morphology. Fertile material needed for further study, possibly *sp. nov.* Not matched, though venation similar to *V. glaberrima*.

Zanthoxylum dinklagei (Engl.) P.G.Waterman
Taxon 24: 363 (1975).
Syn.: *Fagara dinklagei* Engl.
Shrub 0.6m or climber to 9m, spiny, otherwise resembling *Z. gilletii*, but leaves to 40cm, 7-jugate; rachis spiny, 2mm thick; leaflets papery, drying green, oblong, 11 × 3cm, acumen 1.5cm, base obliqely rounded to cordate, lateral nerves 8 pairs, brochidodromous, margin serrate; panicles slender, 30cm; flowers white, 5mm; fruit as *Z. rubescens*. Forest; 250–1530m.
Distr.: Nigeria to Gabon. (Lower Guinea).
IUCN: LC
Bolo-Meboka: Thomas 7158 fl., 7/1987; **Kodmin:** Cheek 9098 2/1998; Etuge 4162 2/1998; 4417 11/1998; Ghogue 51 1/1998; **Mbule:** Cable 3361 6/1996; **Ngomboaku:** Cheek 10339 12/1999; **Nyasoso:** Cable 1168 fr., 2/1995; 3457 7/1996; 3567 7/1996; Etuge 2084 6/1996; 2405 6/1996; 2435 6/1996; 2560 7/1996; Lane 184 fr., 11/1994; Sidwell 366 10/1995; Wheatley 432 fl., 7/1992.

Zanthoxylum gilletii (De Wild.) P.G.Waterman
Taxon 24: 363 (1975).
Syn.: *Fagara macrophylla* Engl.
Syn.: *Fagara melanorachis* Hoyle
Syn.: *Fagara tessmannii* Engl.
Syn.: *Fagara inaequalis* Engl.
Tree c. 25m; bole with spines; leaves imparipinnate, c. 1m, rachis to 1cm thick, spiny, 16+ pairs leaflets; leaflets leathery, drying blackish brown, oblong-lanceolate, c. 15 ×

6cm, acumen 0.5cm, base rounded, lateral nerves c. 12 pairs (main), revolute, subsessile. Forest; 240–900m.
Distr.: Sierra Leone to Uganda. (Guineo-Congolian).
IUCN: LC
Kupe Village: <u>Cable 3742</u> 7/1996; <u>Elad 75</u> 1/1995;
Kurume: <u>Thomas 8212</u> 4/1988.
Local name: Soré. **Uses:** FOOD ADDITIVES – seeds as spice; MATERIALS – timber (*fide* Etuge).

Zanthoxylum leprieurii Guill. & Perr.
Bull. Jard. Bot. Nat. Belg. 30: 403 (1960).
Syn.: *Fagara leprieurii* (Guill. & Perr.) Engl.
Tree 8m, spiny; leaves c. 30cm, c. 6-jugate; rachis sparingly spiny or smooth; leaflets drying dark brown, papery, oblong, c. 10 × 3cm, acumen 1.5cm, acute, cryptically serrate, lateral nerves 12–15 pairs; petiolule 1mm; panicle 9cm. Forest; 1950m.
Distr.: Senegal to Uganda. (Guineo-Congolian).
IUCN: LC
Nyasoso: <u>Cable 2919</u> 6/1996.

Zanthoxylum rubescens Hook.f.
F.T.E.A. Rutaceae: 44 (1982).
Syn.: *Fagara rubescens* (Hook.f.) Engl.
Shrub or tree 2.5–20m, resembling *Z. gilletii*, but leaves to c. 0.6m, 4(–6)-jugate; rachis smooth; leaflets drying pale green below, oblong, to 18 × 7cm, lateral nerves c. 10; petiolule 12mm; panicle terminal, 20cm; flowers numerous, 5mm; fruit globose, pink, 1cm, dehiscent; seed glossy. Farmbush; 350–900m.
Distr.: Guinea (B) to Angola. (Guineo-Congolian).
IUCN: LC
Kupe Village: <u>Cheek 7731</u> 11/1995; <u>Etuge 2030</u> 5/1996; <u>2041</u> 5/1996; **Nyasoso:** <u>Etuge 1516</u> 12/1995; <u>Lane 511</u> fr., 2/1995.
Uses: Materials – timber; MEDICINES (Etuge 2041).

Zanthoxylum rubescens × *dinklagei*
Treelet 2m, spiny; leaves 35cm; rachis spiny, 3-jugate; leaflets drying green, obovate, 17 × 7cm, acumen 2cm, unequally cuneate, lateral nerves 7 pairs, serrate-crenate; panicle 9cm. Forest; 1350m.
Nyale: <u>Cheek 9687</u> 11/1998.
Note: specimen not matched; appears intermediate between taxa indicated.

SALICACEAE

M. Cheek

Salix ledermannii Seemen
Tree or shrub, 3m, glabrous apart from silvery pubescent axils; leaves alternate narrowly oblong-elliptic, 13 × 2cm, acute, cuneate, finely serrate; petiole 1cm; stipules dimidiate ovate, 2mm, acuminate, with 2 teeth on larger side, glabrous, dark red; inflorescences terminal on short shoots, erect, spike-like, c. 4 × 0.75cm; fruits dehiscent; seeds with abundant white hairs. Lake edge; 1975–2000m.
Distr.: Niger to Cameroon. (Upper & Lower Guinea).
IUCN: LC
Mwanenguba: <u>Cheek 7252</u> 2/1995; <u>Cheek 9409</u> 10/1998; <u>Thomas 3135</u> 2/1984.
Note: there is confusion over the delimitation and name of this taxon with *S. subserrata* Willd.

SAPINDACEAE

M. Cheek & M. Etuge

Fl. Cameroun 16 (1973).

Allophylus africanus P.Beauv. var. *africanus*
Tree 10–15m; leaves trifoliolate, drying brownish above, pale green below; leaflets elliptic c. 15 × 7cm, long acuminate, cuneate, margin serrate, secondary nerves 10–13 pairs, conspicuous white-tufted domatia along the midrib and secondary nerves; petiole c. 7cm; inflorescence in leaf-axils, 6–12cm; branches 6–12 in the upper half, up to 10cm long; flowers white, buds hairy brown. Forest; 600–1550m.
Distr.: Gambia to Cameroon. (Upper & Lower Guinea).
IUCN: LC
Kodmin: <u>Cheek 9188</u> 2/1998; **Kupe Village:** <u>Lane 330</u> fl., fr., 1/1995; <u>331</u> fl., 1/1995; **Ndibise:** <u>Etuge 529</u> 6/1987; **Ngombombeng:** <u>Etuge 124</u> 5/1986.

Allophylus bullatus Radlk.
Tree 15–18m; leaves trifoliolate, drying blackish green above, brown below, secondary nerves 10–12 pairs, domatia conspicuous, white tufted, along midrib and secondary nerves; leaflets elliptic c. 19 × 8cm, long acuminate, cuneate, margin serrate; petiole c. 8cm; inflorescence in the leaf axils 10–21cm; branches 6–12 in the upper half to 10cm long; flowers white, 2mm. Forest; 1600m.
Distr.: Nigeria, Cameroon, Principe & São Tomé. (W Cameroon Uplands).
IUCN: VU
Nyasoso: <u>Etuge 1797</u> 3/1996; <u>2174</u> 6/1996.
Note: the Kupe material lacks the bullate character seen in the type specimen from Mt. Cameroon.

Allophylus grandifolius (Baker) Radlk.
Scrambling shrub 3–7m, glabrescent; leaves drying pale green below, trifoliolate; leaflets elliptic or ovate-elliptic, c. 19 × 9cm, subacuminate, obtuse to truncate, margin sinuate, secondary nerves 10–12 pairs, ending in minute teeth; petiolule 1cm; petiole c. 11cm; inflorescence terminal, 18–25cm, c. 3 branches, longest 10cm; flowers white, 2mm; fruit obovoid, 5mm (imm.). Forest; 400–1000m.
Distr.: Nigeria, Cameroon, Principe & São Tomé. (Lower Guinea).
IUCN: NT
Kupe Village: <u>Etuge 2027</u> 5/1996; **Nyasoso:** <u>Sunderland 1487</u> fr., 7/1992.
Note: potentially a Red Data species, being known from only seven localities (FWTA, Fl. Cameroun) apart from those cited here.

Allophylus hirtellus (Hook.f.) Radlk.
Shrub, 1.5–3m; stems and petioles white patent-pubescent; leaves 1-foliolate, papery, drying dark brown, elliptic-oblong, 13–20 × 5.5–8cm, acumen slender, 1.5cm, cuneate-abruptly rounded, lateral nerves c. 15 pairs, serrate-undulate; petiole 5cm; inflorescences axillary, 2cm, sessile, unbranched. Forest.
Distr.: SE Nigeria, Bioko & SW Cameroon. (Lower Guinea).
IUCN: NT
Ekona Mombo: <u>Etuge 428</u> 12/1986.
Note: specimen not seen.

Allophylus longicuneatus Vermoesen ex Hauman
Fl. Cameroun 16: 44 (1973).
Shrub or tree to 10m, glabrous; leaves trifoliolate, papery, drying bright brown below, dark above; leaflets elliptic, 16 ×

7cm, acumen 1.5cm, cuneate, lateral nerves c. 9 pairs, entire; petiolule 2mm; petiole 6cm; inflorescence axillary, c. 15cm; peduncle 4–6cm, basal branches 2, ascending, c. 7cm. Forest; 200m.
Distr.: Cameroon, Gabon & Congo (K). (Guineo-Congolian).
IUCN: NT
Ikiliwindi: Nemba 518 6/1987.
Note: specimen not seen.

Allophylus cf. schweinfurthii Gilg
Shrub or scrambling climber to 10m; stems glaucous-grey; leaves trifoliolate; leaflets papery, drying green, elliptic to 10 × 4cm, acumen 1.5m, entire, lateral nerves 4 pairs, brown, domatia absent; inflorescences in axils of spur shoots, 12–14cm, basal branches 2, patent, to 10cm; fruit ellipsoid, 5mm, glabrous. Forest; 300–400m.
Distr.: Mt Kupe. (Narrow Endemic).
Kupe Village: Etuge 1934 5/1996; 2783 7/1996.
Note: differs from *A. schweinfurthii* in the obtuse-acute based, smaller leaflets and the geog. range.

Allophylus sp. 1
Shrub 0.8–1.5m, densely long white-hairy, hairs to 3mm, patent; leaves trifoliolate, central leaflet obovate to oblanceolate, c. 15 × 5cm, long-acuminate, cuneate-decurrent, deeply serrate; petiolules c. 2mm or absent; petiole 4cm; inflorescence c. 3cm, unbranched, cymules with peduncles 1–2mm; flowers white, 1.5mm; fruit obovoid, 8mm. Forest; 1000–1100m.
Distr.: Kupe-Bakossi & Banyang Mbo. (Narrow Endemic).
Menyum: Doumenge 592 5/1987; **Nyasoso:** Cable 3327 6/1996; 3559 7/1996; Gosline 137 11/1998.
Note: probably a species new to science, close to *A. conraui* Radlk., restricted to Kupe-Bakossi.
Local name: Nyong Nyong. **Uses:** MEDICINES (Epie Ngome in Gosline 137).

Blighia unijugata Baker
Tree 8m, glabrous; leaves unifoliolate or 1–2-jugate; leaflets elliptic or obovate, c. 10 × 6cm, shortly acuminate, cuneate, lateral nerves 10 pairs, raised above; racemes axillary, 6cm; pedicels 5mm; flowers white, 4mm. Forest; 200–850m.
Distr.: Sierra Leone to Uganda. (Tropical Africa).
IUCN: LC
Ikiliwindi: Nemba 533 6/1987; **Nyasoso:** Thomas 3071 fl., 2/1984.

Cardiospermum grandiflorum Sw.
Herbaceous climber 2.5m; leaves biternate, c. 13 × 13cm, terminal leaflets ovate, 6 × 3.5cm, upper 2/3 laciniate-serrate, domatia white pubescent; inflorescence axillary, 20cm, base with tendril pair; peduncle 13cm; flowers white, 1.2cm; fruit inflated, papery, 4cm; seed black, 5mm, hilum white, 2mm. Farmbush; 600–800m.
Distr.: pantropical.
IUCN: LC
Bangem: Thomas 5300 1/1986; **Kupe Village:** Lane 386 fl., fr., 1/1995.

Cardiospermum halicacabum L.
Herbaceous climber resembling *C. grandiflorum*, but leaves lacking domatia; flowers c. 3mm; seed with hilum c. 4mm. Farmbush.
Distr.: pantropical.
IUCN: LC
Mt Kupe: Etuge 1777 3/1996.
Uses: MEDICINES (Etuge 1777).

Chytranthus atroviolaceus Baker f. ex Hutch. & Dalziel
Syn.: *Chytranthus brunneo-tomentosus* Gilg ex Radlk.
Monopodial treelet c. 3m; leaves 3-jugate, 60cm, paripinnate; leaflets leathery, elliptic to 30 × 10cm, acumen 1.5cm, obtuse, lateral nerves c. 15cm; petiole 20cm; inflorescences numerous from ground to 1m up trunk, spicate, c. 15cm; flowers 4mm, densely purple pubescent. Forest; 800m.
Distr.: Sierra Leone to Congo (B). (Upper & Lower Guinea).
IUCN: NT
Nyasoso: Cheek 6010 fl., 1/1995.
Note: known from about 15 sites.

Chytranthus edulis Pierre
Fl. Cameroun 16: 88 (1973).
Shrub resembling *C. atroviolaceus*, finely pubescent; leaves drying red, 4–7-jugate, to 30 × 10cm, acumen caudate, obtuse, lateral nerves 10–20 pairs; inflorescences 3–6cm, red or dark pink; flower buds 5mm. Forest; 200–1000m.
Distr.: Cameroon (?) & Gabon. (Lower Guinea).
Menyum: Doumenge 568 5/1987; **Mile 15, Kumba-Mamfe Road:** Nemba 600 7/1987.
Note: specimen not seen. Identification doubtful since native to Gabon; no conservation assessment is made in view of this uncertainty.

Chytranthus setosus Radlk.
Cauliflorous monopodial shrub 3m; resembling *C. atroviolaceus*, but leaves setose on petiole, rhaJides etc., c. 70cm; leaflets oblong-elliptic, c. 40 × 12cm; inflorescences c. 8cm; flowers 4mm, white. Forest; 1200m.
Distr.: Benin to Congo (K). (Guineo-Congolian).
IUCN: NT
Kupe Village: Etuge 1425 11/1995.
Note: about 12 sites known.

Chytranthus talbotii (Baker f.) Keay
Tree 12m, resembling *C. atroviolaceus*, but glabrous; leaves to 1m, 8–11-jugate; leaflets subequal, oblong, 11–24 × 4.2–6.8cm; petioles 16–22cm; racemes to 60cm; calyx 1–1.2cm, petals white. Forest; 250m.
Distr.: SE Nigeria to Gabon. (Lower Guinea).
IUCN: NT
Bolo-Meboka: Thomas 7150 7/1987; **Mombo:** Olorunfemi in FHI 30564 fr., 5/1951.
Note: about 14 sites known.

Deinbollia calophylla Gilg ex Radlk.
Treelet, 1m. Forest; 350m.
Mekom: Nemba 12 4/1986.
Note: determined by Assi (MO), specimen not seen. May be a distinct taxon from other *Deinbollia*. *D. calophylla* is known only from Upper Guinea, however. No conservation assessment is made in light of this uncertainty.

Deinbollia cuneifolia Baker
Treelet to 2m; flower buds cream; fruit orange. Forest; 760–1200m.
Mt Kupe: Thomas 5109 12/1985; **Ngombombeng:** Etuge 34 4/1986.
Note: determined by Assi (MO), specimens not seen. May be a distinct taxon from other *Deinbollia*. *D. cuneifolia* is known only from Upper Guinea, however. No conservation assessment is made in light of this uncertainty.

Deinbollia insignis Hook.f.
Fl. Cameroun 16: 62 (1973).

Tree c. 8m; leaves c. 1m, c. 10-jugate; leaflets green, elliptic, 35 ×15cm, rachis to 1.1cm thick, hollow; panicles c. 1m, terminal; fruit 2–3cm. Forest; 400m.
Distr.: SE Nigeria to Cameroon. (Lower Guinea).
IUCN: VU
Kupe Village: Etuge 2015 5/1996; **Mt Kupe:** Etuge 1841 3/1996.

Deinbollia maxima Gilg
Tree or shrub to 6m, glabrous; leaves 0.8m, 4–6-jugate, rachis terete, glossy; leaflets drying black, oblong-lanceolate, to 21 × 9cm, acumen 1.5cm, unequally obtuse-decurrent, lateral nerves c. 10 pairs; petioles 1cm; inflorescence axillary or on stems below leaves, 16–30 × 10cm; flowers 2–3mm. Forest; 900m.
Distr.: Sierra Leone, Nigeria, Cameroon & Gabon. (Guineo-Congolian).
IUCN: VU
Mt Kupe: Thomas 5474 2/1986.

Deinbollia sp. aff. maxima Gilg
Monopodial shrub 4m; lenticels brown; leaves c. 90cm, 6–7-jugate; leaflets brown, c. 25 × 10cm; petiole c. 18cm; panicles axillary, below leaves; flowers not seen; fruit 2–3cm. Forest; 1600m.
Nyasoso: Sunderland 1541 fr., 7/1992.
Note: flowers needed to complete determination. May be an aberrant *D. sp. 1*.

Deinbollia sp. 1
Monopodial shrub 0.8–3(–5)m; stems brown with bright white raised lenticels; leaves 25–63cm, (2–)3–4(–5)-jugate; leaflets pale green, nerves yellow, oblong-elliptic, c. 15–24 × 5.5–9cm, acuminate; petiole 9–16.5cm; panicle terminal, 8–20 × 5–20cm; flowers white, 3–4mm, glabrous; fruit orange, all but 1 carpel aborting, globose, 3cm diameter. Forest; 650–2050m.
Distr.: Cameroon & SE Nigeria. (W Cameroon Uplands).
Bangem: Thomas 5286 1/1986; **Edib:** Etuge 4482 11/1998; **Kodmin:** Cheek 10263 12/1999; 8993 1/1998; 9001 1/1998; 9580 fl., 11/1998; Etuge 4004 1/1998; Pollard 234 11/1998; **Kupe Village:** Cable 3786 7/1996; 829 fl., 1/1995; 855 fl., 1/1995; Cheek 7091 fl., 1/1995; 7189 fl., 1/1995; 7887 11/1995; Ryan 378 5/1996; **Mbule:** Cable 3352 6/1996; **Mt Kupe:** Cheek 7602 11/1995; **Nyasoso:** Cable 1181 fl., 2/1995; 3538 7/1996; 3555 7/1996; 48 fr., 8/1992; Etuge 1589 1/1996; 1658 1/1996; 1743 2/1996; 2186 6/1996; 2414 6/1996; 2478 6/1996; Groves 90 fr., 2/1995; Lane 222 fl., 11/1994.
Note: a new species in the process of being described.

Deinbollia sp. 2
Tree 4m; leaves 60cm, 10-jugate; leaflets narrowly elliptic, c. 14.5cm, acumen 1cm, lateral nerves 16 pairs, brochidodromous; petiole 20cm, terete, white, upper edge black scurfy puberulent; panicle 18cm, terminal, densely red pubescent; flowers 4mm. Forest; 1400–2000m.
Distr.: Mt Kupe & Bali-Ngemba F.R. (W Cameroon Uplands).
Kodmin: Plot B55 1/1998; **Nyasoso:** Cable 3386 6/1996.
Note: specimen needs more investigation. Not matched.

Eriocoelum macrocarpum Gilg ex Engl.
Tree 20m, glabrescent; leaves punctate, 2–3-jugate, paripinnate, 30cm, uppermost leaflets largest, obviate, to 30 × 14.5cm, basal leaflets 1cm, resembling stipules, shortly acuminate, soon glabrous below; inflorescence spike-like, 10–20cm; flowers c. 3mm; fruit hard, woody, orange,

smooth, depressed globose, to 3 × 4cm, valves 3, hairy inside. Forest; 300–1000m.
Distr.: SE Nigeria to Congo (K). (Lower Guinea & Congolian).
IUCN: LC
Bolo Moboka: Nemba 257 9/1986; **Kupe Village:** Cheek 7802 11/1995; Etuge 2752 7/1996; Lane 258 fr., 11/1994.

Eriocoelum petiolare Radlk.
Fl. Cameroun 16: 180 (1973).
Tree 8–20m resembling *E. macrocarpum*, but leaflets not punctate, 14–18 × 5–6cm, lacking stipuloid leaflets; petiole 3–4cm; inflorescence 12–15cm, branched 3–4 times. Forest; 1000m.
Distr.: Cameroon & Gabon. (Lower Guinea).
IUCN: NT
Menyum: Doumenge 517 5/1987.
Note: specimen not seen.

Eriocoelum racemosum Baker
Tree 8m; stems pubescent, resembling *E. macrocarpum*, but leaves not punctate, uppermost leaflets elliptic, c. 15 × 5cm, acumen 1cm, acute, midrib with sparse long hairs below, stipuloid leaflets 4cm; inflorescence pendulous, c. 30cm, often with a basal branch, brown-hairy. Forest; 1000m.
Distr.: Sierra Leone to Gabon. (Upper & Lower Guinea).
IUCN: NT
Nyale: Etuge 4447 11/1998.
Note: possibly a new record for Cameroon.

Laccodiscus ferrugineus (Baker) Radlk.
Monopodial treelet, 2–3m, pubescent; leaves paripinnate, c. 6-jugate, c. 40cm; leaflets elliptic or oblong, to 17 × 6cm, acumen 1cm, broad, base acute, lateral nerves c. 15 pairs, serrate, basal leaflets smaller, basal-most leaflets inserted at petiole base, resembling stipules, ovate c. 4cm; inflorescence red tomentose, terminal, 15–20cm, 5–8 branched; flowers white, 3mm; capsule red, 3-angular, 1.5cm, long-hairy; seeds 5mm, villose. Forest; 250–1300m.
Distr.: Nigeria to Gabon. (Lower Guinea).
IUCN: LC
Bangem: Thomas 5907 3/1986; **Kupe Village:** Cable 2484 5/1996; 2515 5/1996; 2672 5/1996; 746 fl., 1/1995; Cheek 7014 fl., 1/1995; 7062 fl., 1/1995; Etuge 1721 2/1996; **Kurume:** Thomas 5458 1/1986; **Mekom:** Thomas 5393 1/1986; **Ngomboaku:** Cheek 10390 12/1999; **Nyandong:** Tchiengue 1727 fr., 3/2003; **Nyasoso:** Cable 1132 fl., 2/1995; Cheek 7534 10/1995; Groves 105 fl., fr., 2/1995.

Lychnodiscus grandifolius Radlk.
Fl. Cameroun 16: 165 (1973).
Tree 7m; stems stout, densely pubescent; leaves glossy, leathery, c. 1m, c. 5-jugate, paripinnate; leaflets elliptic, c. 30 × 12cm, subacuminate, cuneate, lateral nerves c. 10 pairs; petiolule 5mm; petiole 30cm; inflorescence 30cm or more, dark brown puberulent, few-branched; flowers 5mm, white; capsules 3-valved, 3 × 2.5cm, sepals persistent, reflexed; seeds brilliant orange-red. Forest; 2000m.
Distr.: Cameroon & Gabon. (Lower Guinea).
IUCN: NT
Mwanenguba: Thomas 3110 2/1984.
Note: specimen not seen.

Majidea fosteri (Sprague) Radlk.
Tree c. 15m; slash scented of toothpaste, glabrous; leaves c. 7-jugate, c.18cm, imparipinnate; leaflets oblong, subacuminate, base unequal, entire; petiolule 1mm; petiole c. 3cm; panicles terminal, c. 10cm; flowers 5mm, white;

capsule 3-lobed, c. 3 × 5cm, leathery, inner surface vivid pink; seeds pink, downy, 6mm. Forest.
Distr.: Ivory Coast to Uganda. (Guineo-Congolian).
IUCN: LC
Tombel: Forteh in FHI 9288 3/1947.
Note: specimen not seen.

Pancovia pedicellaris Radlk. & Gilg

Shrub or tree to 10m; leaves 3–4-jugate, lanceolate to elliptic, 4–20 × 1.5–8cm, acuminate, obtuse to cuneate; petioles 4–9cm; inflorescences from amongst the leaves, 4–13cm; pedicels 3–5mm. Forest.
Distr.: Cameroon to Gabon. (Lower Guinea).
IUCN: LC
Ekona Mombo: Etuge 415 12/1986.
Note: specimen not seen.

Pancovia sp. 1

Cauliflorous monopodial treelet 2–5m; stems glabrous, zig-zagging slightly; leaves 3–5-jugate, 1m plus, paripinnate; leaflets elliptic to 35 × 10cm, lateral nerves 15–20 pairs; spikes 20–30cm; pedicels 3–5mm; calyx 6mm, petals white; fruit orange, fleshy, 3-lobed, rostrate, indehiscent, 3 × 3.5cm. Forest; 700–1230m.
Distr.: Kupe-Bakossi. (Narrow Endemic).
Kupe Village: Cable 2519 5/1996; 2742 5/1996; 3791 7/1996; 3808 7/1996; 3855 7/1996; Cheek 6045 fr., 1/1995; 7685 11/1995; Ryan 324 5/1996; 359 5/1996; **Ngomboaku:** Ghogue 445 12/1999; Mackinder 336 12/1999.
Note: probably a new species endemic to Bakossi, close to *P. polyantha*. Several other species of the genus, such as *P. bijuga* and *P. turbinata*, occur close by and may yet be found in Bakossi.

Paullinia pinnata L.

Woody liana to 25m; leaves c. 12cm, 2-jugate, rachis winged; leaflets elliptic, to 11 × 6cm, obscurely toothed, apex rounded; petiole c. 10cm; inflorescence tendriliform, as long as leaves, spicate; flowers white, 3mm; fruit red, 3-lobed, obovoid, 4 × 1cm, stipitate; seed white and red, 0.5cm. Forest, farmbush; 300–1450m.
Distr.: tropical Africa & America. (Amphi-Atlantic).
IUCN: LC
Kodmin: Cheek 9025 1/1998; Ghogue 74 fr., 2/1998; **Kupe Village:** Cable 712 fl., 1/1995; 755 fl., fr., 1/1995; Cheek 7093 1/1995; Kenfack 217 7/1996; 311 7/1996; Lane 300 fl., fr., 1/1995; Ryan 205 5/1996; Zapfack 938 7/1996; **Loum F.R.:** Biye 71 12/1999; **Mwanenguba:** Bamps 1547 fl., 12/1967; **Ndum:** Elad 114 fl., fr., 2/1995; Lane 484 fl., 2/1995; **Nyasoso:** Balding 24 fr., 7/1993; Cable 2883 6/1996; Etuge 1642 1/1996; Zapfack 639 6/1996; **Nzee Mbeng:** Gosline 97 2/1998.
Local name: Abah-Nyoh. **Uses:** not known (Cheek 9025).

Placodiscus caudatus Pierre ex Radlk.

Fl. Cameroun 16: 128 (1973).
Cauliflorous tree 2–8m, glabrous; leaves c. 40cm, 3–4-jugate; leaflets glossy, oblong, c. 18 × 6cm, acumen 1.5cm, lateral nerves c. 10; petiolule 5mm; spikes 15–25cm; bracts nil; pedicel 4mm, patent, persistent; calyx 3mm; fruit yellow, 3.5 × 2.5cm, felty, apex retuse, 1–3 lobed, style persistent, long. Forest; 300m.
Distr.: Cameroon, CAR & Gabon. (Lower Guinea).
IUCN: EN
Bolo: Nemba 8 4/1986.
Note: specimen not seen; identification requires confirmation.

Placodiscus opacus Radlk.

Fl. Cameroun 16: 133 (1973).
Shrub, resembling *P. caudatus*, but leaves 4–6-jugate; leaflets 10–20(–30) × 6–8cm, nerves 7–8 pairs; spikes 8–20cm; pedicel 2mm; buds 2mm. Forest; 200–1000m.
Distr.: Cameroon & Gabon. (Lower Guinea).
IUCN: VU
Ikiliwindi: Nemba 529 6/1987; **Menyum:** Doumenge 569 5/1987.

Placodiscus sp. 1

Cauliflorous tree to 14m, trunk angular, 20cm d.b.h.; leaves c. 70cm, 4–5-jugate, paripinnate; leaflets subequal, elliptic-oblong, to 40 × 15cm, acumen abrupt, 1cm, base acute, lateral nerves c. 30 pairs, brochidodromous; petiolule 1.5cm; petiole 25cm; inflorescence 2–10m from ground, c. 30cm, rarely branched; flowers 3–4mm; pedicels 3–4mm; fruit unknown. Forest; 1300m.
Distr.: Bakossi Mts & Mt Cameroon. (Narrow Endemic).
Edib village: Cheek 9176 2/1998.
Note: possibly a new species; more specimens needed. Several other members of the genus, e.g. *P. glandulosus*, *P. boya* and *P. turbinatus*, are known nearby and likely to occur in Bakossi.

Sapindaceae sp. 1

Shrub 4m; leaves 40cm, 2-jugate; leaflets alternate, black-brown above, yellow below, narrowly oblong, 25 × 6cm, acumen 2cm, base cuneate, each margin with 2–3 black glandular inconspicuous teeth, lateral nerves 14 pairs, marginal nerves prominent, tertiary nerves incompletely narrowly scalariform; petiolule pulvinate, 4mm. Forest; 710m.
Nyale: Etuge 4237 2/1998.
Note: fertile material needed. Not matched with any other taxon. Vegetatively very distinct.

SAPOTACEAE

Y.B. Harvey

Fl. Cameroun 2 (1964).

Baillonella toxisperma Pierre

Tree to 12m tall, c. 80cm d.b.h.; bole nearly black, ridged, hard; slash yellow, white exudate; leaves clustered at ends of branches; stipules lanceolate, blades 20–30 × 6–10cm, rounded with acuminate apex, cuneate at base, young leaves with chestnut pubescence, subglabrous when mature although hairs persistent on midrib; inflorescence of dense flowering fascicles at the branch tips; pedicels 2–3cm, pubescent; calyx c. 1cm long, with 8 lobes, 4 inner and 4 outer, pubescent on exterior surface, corolla with 8 lobes, each with 2 dorsal appendages longer than the lobes (5.5mm), tube 2.5mm long, lobes c. 4mm long; fruits large, spherical, c. 6.5cm diameter, grey-green; 1–2-seeded in a yellowish white pulp; seeds ellipsoid, c. 4.2 × 2.5 × 2cm, ventral scar nearly the entire length of the convex shaped ventral face. Evergreen forest, secondary forest growth; 200–450m.
Distr.: Nigeria to W Congo (K). (Lower Guinea).
IUCN: VU
Ikiliwindi: Nemba 541 6/1987; **Nyandong:** Cheek 11354 3/2003.

Chrysophyllum africanum A.DC.
F.T.A. 3: 500 (1877).
Syn.: *Gambeya africana* (Baker) Pierre
Syn.: *Chrysophyllum delevoyi* De Wild.
Tree to 20–40m; branching at 10–15m; young growth covered in chestnut-brown hairs; slash with white exudate; leaves obovate, 20–26.5 × 8–9cm, 14–16 pairs of secondary veins, apex rounded with acuminate tip, base cuneate, upper surface glabrous, lower surface pubescent with appressed chestnut coloured hairs, denser on midribs and secondary veins, lightening in colour with maturity, petioles 15–25mm long, with chestnut coloured hairs; flowers cauliflorus, in fascicles, from the axils of the leaves, 8 or more flowers per fascicle; pedicels 3–5mm long, with dense chestnut coloured appressed hairs; calyx 5-lobed, lobes c. 4mm long, with dense chestnut coloured appressed hairs, corolla c. 5–6mm long, tubular, lobes 5, c. 2.5mm long, densely covered with chestnut coloured hairs on the outer surface, with cream-coloured hairs at the margins of the lobes; fruits ovoid, yellow-orange, 60 × 50mm; seeds 5, brown, c. 3 × 1.5cm. Secondary forest; 220–1000m.
Distr.: Sierra Leone to Angola and Uganda (Guineo-Congolian).
IUCN: LC
Bakolle Bakossi: Etuge 256 fr., 9/1986; **Kupe Village:** Elad 77 fl., fr., 1/1995; **Nyasoso:** Groves 87 fl., 2/1995.

cf. Chrysophyllum albidum G.Don
Shrub 1.5m to tree 30m; young growth sparsely covered with pale chestnut-coloured hairs; leaves narrowly elliptic to obovate, 8.5–25 × 7–18cm, 10–14 pairs of secondary veins, apex long acuminate (to 4.5cm), base cuneate, upper surface green, glabrous, lower surface irridescent fawn-white with small appressed pale ferruginous hairs (particularly on midrib and veins); petioles 10–15mm long, with pale chestnut coloured hairs. Forest; 900–1150m.
Ikiliwindi: Nemba 114 fr., 6/1986 (n.v.); **Kupe Village:** Cable 2671 5/1996; **Mt Kupe:** Cheek 10161 11/1999; **Nyasoso:** Cable 2841 6/1996.
Note: specimens sterile.

Chrysophyllum lacourtianum De Wild.
Govaerts *et al.*, World Checkl. & Bibl. Sapotaceae: 65 (2001).
Syn.: *Gambeya lacourtiana* (De Wild.) Aubrév. & Pellegr.
Tree 20–35m tall, to c. 70cm d.b.h.; trunk with shallow buttresses up to 2m from ground; bark pale brown, flaking off in small pieces; outer slash dark red, oozing white exudate, oxidising orange-brown; *Terminalia*-style branching; leaves obovate, 15–26 × 4.5–8.5cm, 8–12 pairs of secondary veins, apex acuminate, base cuneate, with a network of tertiary veins perpendicular to the secondary veins, glabrous, or with scattered hairs on midrib, drying dark reddish brown; petiole 1.5–3cm; flowers in fascicles; pedicels c. 3mm long; calyx 3.5–4mm long, pubescent; corolla to 3.5mm long, lobes 1mm, margins ciliate; tube 2.5mm; anthers not exceeding corolla lobes; ovary 5-locular; fruits ovoid, subglobose, c. 10 × 7cm, red or orange; seeds 5, 2–3.5 × 1.5–1.8 × 1cm, with a prominent oblong scar, borne high in the crown from the leafy branches. Undisturbed evergreen forest; 800–850m.
Distr.: Cameroon to Congo (K). (Lower Guinea & Congolian).
IUCN: NT
Kupe Village: Cheek 7863 11/1995; **Ngomboaku:** Cheek 10317 12/1999.
Note: known from c. 15 locations.

Chrysophyllum welwitschii Engl.
Syn.: *Donella welwitschii* (Engl.) Pierre ex Aubrév. & Pellegr.
Liana or shrub; young branches covered with a chestnut pubescence, older branches glabrous; leaves variable, usually oblong or oblong-elliptic, sometimes narrowly oblong or elliptic, 5–15 × 2.5–5cm, apex acuminate (sometimes emarginate at the tip), base rounded, young leaves with chestnut pubescence, becoming glabrous; small flowers in fascicles; pedicels c. 2mm long; calyx glabrous or with some chestnut hairs; corolla 1mm long with ciliate lobes; fruits ovoid, c. 3.5 × 2.5cm, with apiculate apex, yellow, sessile. Secondary growth, old forest; 200m.
Distr.: Sierra Leone to N Zambia. (Guineo-Congolian).
IUCN: LC
Ikiliwindi: Nemba 520 fl. 6/1987.

cf. Chrysophyllum sp. A
Fruit ellipsoid, c. 4cm diameter, yellow, white inside with white latex, glabrous exterior, 5-seeded. Forest; 1200m.
Nyasoso: Cable 1166 fr., 2/1995.
Note: specimen a single fruit; more material required.

cf. Chrysophyllum sp. B
Tree 30m tall; immature branches subglabrous, appearing reddish brown; bark slightly fissured, pale brown; pedicel to 2mm long; calyx 1.5–2mm long, 5-lobed, lobes ovate, apices rounded; fruit green, immature; c. 2-seeded; seeds 21 × 11 × 4mm, reddish brown. Forest; 1000m.
Nyasoso: Cable 3502 7/1996.
Note: specimen lacking leaves and flowers; more material required.

cf. Chrysophyllum sp. C
Tree c. 10m tall, 10cm d.b.h.; underbark red, outer slash orange, inner yellowish, fibrous, with white exudate; leaves clustered at the ends of branches, obovate, 11–12.7 × 3.6–4.5cm, apex slightly acuminate, base cuneate, upper surface green, lower surface appearing white, almost irridescent; petioles (1.5–)2.5–3mm long, glabrous. Forest; 1650m.
Kodmin: Cheek 9064 1/1998.
Note: sterile specimen only; more material required.

cf. Chrysophyllum sp. D
Tree to 15m tall, 7cm d.b.h. with white exudate; *Terminalia*-style branching; young growth chestnut brown coloured; leaves obovate, 15.5–20.5 × 3.8–6cm, apex acuminate, base cuneate, glabrous, appearing olive-green, veining much yellower, especially apparent below; petioles 8–10mm long, glabrous, chestnut coloured, swollen. Submontane grassland, forest.
Kodmin: Cheek 9631 11/1998.
Note: sterile specimen only; more material required.

Englerophytum stelechanthum K.Krause
Bull. Jard. Bot. Nat. Belg. 59: 161 (1989).
Syn.: *Englerophytum hallei* Aubrév. & Pellegr.
Small tree, 6–8(–10)m tall; bark whitish; bole with brachyblasts from ground to c. 4m up; Terminalia-style branching; slash with thin white latex; leaves (narrowly) obovate with many parallel secondary veins, apex acuminate, base acute, 10–23.5 × 3–5.5cm, upper surface glabrous, green, lower surface with short, appressed chestnut-brown hairs, appearing white below; petiole 6–15mm long, glabrous; inflorescence in cauliflorous fascicles of c. 5 flowers nearer the base of the trunk/branches; pedicels c. 6mm long, with fine hairs and sessile glands; calyx with 5 lobes; corolla 7–8mm long; fruit ellipsoid, c. 2 × 1cm,

densely covered with chestnut hairs; seeds red. Forest with some farmbush; 800–1125m.

Distr.: Nigeria, Cameroon & Gabon. (Lower Guinea).

IUCN: NT

Bangem: Thomas 5304 1/1986; **Kupe Village:** Cheek 7684 11/1995; **Menyum:** Doumenge 455 5/1987; **Mt Kupe:** Cheek 10159 11/1999; 10160 11/1999; **Ndibise:** Etuge 535 6/1987; **Ngomboaku:** Cheek 10391 12/1999; **Nyasoso:** Cheek 9509 10/1998; Wheatley 394 fr., 7/1992.

Note: fairly widespread in S & SW Provinces, Cameroon, uncommon elsewhere. Lower altitude sites in Cameroon are threatened by continued clearance for agriculture and logging.

Englerophytum cf. stelechanthum K.Krause

Small tree 2–8m tall, cauliflorous, with white exudate; *Terminalia*-style branching; leaves ovate to obovate with many parallel secondary veins, 15–23.5 × 5.5–6.3cm, upper surface glabrous, green, lower surface with short, appressed hairs, appearing white below; young fruits and old flowers in fascicles of c. 15 flowers from burrs on bole beginning 1.5m above ground, extending to lateral branches; pedicels c. 1.5–2cm long; calyx 5-lobed, lobes 3.5mm long; fruits orange, turbinate. Submontane grassland, forest; 1450–1770m.

Kodmin: Cheek 9073 1/1998; 9630 11/1998.

Note: Cheek 9630 most closely resembles *E. stelechanthum*; however, it is a considerably shorter tree (2–3m as opposed to 6–8m tall).

Englerophytum sp. [E. kennedyi]

Tree 8–15m tall; trunk to 15cm d.b.h.; bole red-brown, cylindrical; slash with white exudates; *Terminalia*-style ; branching; young branches chestnut coloured; leaves oblanceolate, 34–57 × 11–15.5cm, with numerous curved secondary veins, apex rounded, with apiculate tip, base attenuate, glabrous, petioles 1.5–4.5cm long, chestnut coloured; stipules 1–2cm long, chestnut coloured; inflorescence in dense cauliflorous fascicles from 10–60cm above the ground; pedicels 2.5–3cm long, chestnut coloured; calyx lobes 5, to 5mm long; corolla tubular, tube 3mm long; lobes to 4mm long; fruits red, smooth, cauliflorous on trunk and branches, c. 35 × 35mm. Evergreen forest; 200–1100m.

Distr.: Cameroon endemic.

Baduma: Nemba 571 7/1987; **Mungo River F.R.:** Cheek 10192 11/1999; **Nyale:** Cheek 9683 11/1998; Etuge 4233 2/1998.

Note: fruits said to be edible by sucking (Manning 1720). The name *E. kennedyi* has never been validly published. Kupe specimens = Letouzey 14333; Manning 1720; Thomas 2248; Gentry *et al.* 52694 from other Cameroon locations. Likely to warrant a conservation rating of VU under criterion B once the taxon is fully delimited.

Englerophytum near E. sp. [E. kennedyi]

Tree 18m tall; slash red outside with white latex, pale brown inside; leaves obovate, 12.5–18 × 6–7.2cm, apex rounded, tip apiculate, base attenuate, upper surface glabrous, lower surface loosely covered with appressed chestnut hairs, appearing white; petiole 8–11mm long; flower buds green, cauliflorous. Forest; 1050m.

Nyasoso: Cable 3596 7/1996.

Note: specimen cited is sterile; leaves far too small for *E. sp. [E. kennedyi]*.

Manilkara sp.

Tree, 12m tall with white latex; branches dark reddish brown; leaves alternate, loosely clustered at the ends of branches, obovate, 12.5–19.5 × 5–7.5cm, apex rounded with acuminate

tip, base obtuse, glabrous, upper surface olive-green, lower surface pale chestnut, secondary veins faintly noticeable on upper and lower surface, numerous, parallel; petiole 7–24mm long, glabrous, dark reddish brown. Forest; 1000m.

Nyale: Etuge 4457 11/1998.

Note: near *M. multinervis* but leaves too long and clearly obovate. Fertile material required.

Omphalocarpum elatum Miers

Large tree c. 20–25m tall; trunk c. 70cm diameter, bark brown-red; wood pinkish brown; latex white; leaves narrowly to broadly oblanceolate, (7–)12–25 × (3–)4–8cm, apex obtuse, base attenuate, glabrous; petiole 0–5mm long; inflorescence cauliflorous, 4–6 flowers per fascicle; pedicels (4–)8–10mm; bracts at base of pedicel to 2mm long, sepals 5, (5–)7–12mm long, ovate; corollas white, 9–17mm long, lobes 5–6, elliptic, 6–7 or 10–13mm long; tube (2–3 or) 7–9mm long, filaments c. 9mm long; anthers 3.5mm long, staminodes c. 6mm long; fruits subglobose, depressed at both apex and base, 10–15.5 × 6.5–9cm, yellow-green maturing to dull brown; up to 30 seeds; seeds 3.7–5.7cm long, brown. Evergreen forest; 950m.

Distr.: Sierra Leone to Gabon. (Upper & Lower Guinea).

IUCN: NT

Ngomboaku: Cheek 10300 12/1999; 10396 12/1999.

Note: although widespread; seed germination in this species is often, though not exclusively, assisted by elephants, thus the decline in forest elephant populations throughout the Guinea region due to hunting is likely to threaten populations of this taxon.

Omphalocarpum pachysteloides Mildbr. ex Hutch. & Dalziel

Pennington, Gen. of Sapot.: 261 (1991).

Syn.: *Ituridendron bequaertii* De Wild.

Tree to 20m tall; bark grey; wood pale brown; slash red with white latex darkening slightly and slowly; leaves (broadly) obovate, 16–26 × 6–9cm, apex rounded, tip acuminate, base cuneate to angustate, upper surface subglabrous, lower surface with scattered appressed hairs, dark green above, grey-green to silvery below; petioles swollen, 11–15mm long, laxly pubescent; inflorescence of flowers in fascicles, cauliflorous on main branches, 8–20 flowers per fascicle; bracts at base of pedicels, 2–3mm long, ovate, outer surface densely tomentose, pale chestnut coloured; pedicels 3–4mm long, tomentose with chestnut coloured hairs; calyx 7–10mm long, tomentose (fawn coloured), 5-lobed, lobes ovate, 5–6mm; corolla subtending calyx, 8–11mm long, 5-lobed, lobes ovate, laxly covered with chestnut-coloured hairs, style subtending corolla by c. 2mm, glabrous in upper half, tomentose towards base; fruits subglobose, 3.5–5.5 × 3.5–4cm, with dimpled apex, pale orange in colour; seeds 20–25 × 12–15 × 0.6mm. Evergreen rainforest; 700–1200m.

Distr.: Sierra Leone to Congo (K). (Guineo-Congolian).

IUCN: NT

Kupe Village: Cheek 7212 fl., 1/1995; Lane 337 fl., 1/1995.

Note: see note to *O. elatum*.

Pouteria pierrei (A.Chev.) Baehni

Candollea 9: 292 (1942).

Syn.: *Aningeria robusta* (A.Chev.) Aubrév. & Pellegr.

Syn.: *Pouteria aningeri* Baehni

Tree (4–)10–15m tall without *Terminalia*-style branching; trunk to 25cm diameter; bark pale grey-brown with darker longitudinal lines, outer slash very pale pink, inner white; very slowing running white latex; branches pubescent, appearing floccose; leaves oblong to oblanceolate, 11.5–26 × 5.5–11cm, glabrous upper surface, covered with chestnut

pubescence on lower, especially on midrib and veins, apex rounded with acuminate tip, base rounded to narrowly cuneate, occasionally truncate; olive green above, paler below; petioles 1–2.5cm long, densely covered with chestnut hairs when immature, subglabrous when mature; inflorescence of fascicles produced on the previous season's branches; flowers with velutinous pedicels c. 3mm long, sepals 5, ovate, imbricate, tomentose outer, glabrous within; corolla to 4.5mm long, 5-lobed; fruits globose with acuminate apex, 15–20mm diameter, red, tomentose; 1 seed per fruit, dark brown, c. 14 × 10 × 8mm. Evergreen forest; 750–1000m.
Distr.: Sierra Leone to CAR. (Upper & Lower Guinea).
IUCN: NT
Manehas F.R.: Cheek 9398 10/1998; **Mt Kupe:** Cheek 10163 11/1999; **Nyasoso:** Cheek 9521 10/1998.
Note: this taxon is said to reach a height of 40m (Fl. Cameroun 2: 1964); it is more common in Upper Guinea where it is threatened by overexploitation, and may qualify as VU under criterion A.
Local name: Poga (fide Etuge). **Uses:** MATERIALS – timber, traded under the name *Aningeria robusta* (fide Cheek & Etuge).

Synsepalum cf. brevipes (Baker) T.D.Penn.
Pennington, Gen. of Sapot.: 249 (1991).
Treelet 4m tall; bark reddish brown, slightly fissured; *Terminalia*-style branching; leaves alternate throughout branch, oblanceolate, 9.5–19 × 3.8–6.2cm, glabrous, apex obtuse, but with acuminate tip, base cuneate, upper surface dull olive green in dried state, lower surface paler; petioles 14–20mm long; stipules absent. Evergreen forest; 1330m.
Kodmin: Cheek 10270 12/1999.
Note: only sterile material seen.

Synsepalum sp. near cerasiferum (Welw.) T.D.Penn.
Tree with white latex; bark brown with striations, lenticelled; without conspicuous *Terminalia*-style branching; leaves not clustered at branch tips; blades ovate to obovate, 14.5–19.5 × 5–6.2cm, apex obtuse, base shortly attenuate, glabrous, in dry state olive-green, margin entire, wavy along edge; petioles 1–1.7cm long; fruits borne on older wood; pedicels to 10mm, sepals 2 × 2mm, oblong, apex obtuse; fruits ellipsoid, 2.5 × 1.7cm, with a single seed. Forest; 1350m.
Nyale: Cheek 9694 11/1998.
Note: *S. cerasiferum* fruits have accrescent calyces, whilst those of Cheek 9694 are distinctly not accrescent.

Synsepalum msolo (Engl.) T.D.Penn.
Pennington, Gen. of Sapot.: 249 (1991).
Syn.: *Pachystela msolo* (Engl.) Engl.
Tree 5–25m tall, 15–30cm d.b.h.; bole smooth, grey and green, slightly twisted with pairs of smooth knobs at intervals of 30–60cm; slash pink-red and white to pale brown inside, with white latex; *Terminalia*-style branching; leaves in whorls at the ends of branches, oblanceolate, 14.5–51 × 5–14cm, glabrous, apex rounded with acuminate tip, towards base narrowly cuneate (base is truncate), upper surface pale green, lower surface pale chestnut and irridescent, secondary nerves prominent below, 14–22 pairs of nerves, petiole 9–25mm long, inflated, to 5mm wide and fissured in dried state; stipules triangular, 3–5mm long; inflorescence in dense fascicles on branches; pedicels 4–6mm long; sepals 5, 6.5–7mm long; corollas with 5 lobes, 7.5mm long, tube 3–4mm long, irregularly sized staminodes; fruit ovoid, orange-brown, 3 × 2.5cm in diameter; single seeded, seeds ovate. Evergreen forest; 420–1050m.

Distr.: Ghana to Tanzania. (Guineo-Congolian).
IUCN: LC
Kupe Village: Cheek 10223 12/1999; **Mungo River F.R.:** Onana 946 11/1999; **Nyasoso:** Cable 3598 7/1996; Cheek 9504 10/1998.
Local name: Mbit. **Uses:** MEDICINES (Epie Ngome in Cheek 9504).

Synsepalum revolutum (Baker) T.D.Penn.
Pennington, Gen. of Sapot.: 249 (1991).
Syn.: *Vincentella revoluta* (Baker) Pierre
Small tree to 10m tall, with white exudate; leaves c. 15 × 4.5cm, oblong-oblanceolate, apex attenuate, base cuneate, glabrous; petioles pubescent, glabrous when mature; stipules narrowly triangular; inflorescence of many flowered fascicles on at least 1–2 year old branches; pedicels c. 10mm long, pubescent; sepals lanceolate; corolla white, tubular but split almost to base, petals c. 2.2mm long, staminodes alternate with the petals; fruit ellipsoid, c. 15mm long. Forest; 1100m.
Distr.: Ivory Coast to Congo (K). (Guineo-Congolian).
IUCN: LC
Mejelet-Ehumseh: Etuge 459 fl., 1/1987.

Synsepalum cf. revolutum (Baker) T.D.Penn.
Shrub 2–3m tall; bark mid brown, slightly fissured; with white exudate; *Terminalia*-style branching; leaves in whorls at the ends of branches/branch nodes, obovate, 11.5–19 × 4–7.2cm, apex obtuse, but with acuminate tip, base attenuate, in dry state upper surface pale olive-grey, lower surface pale chestnut coloured, 12 or less secondary nerves per leaf; petioles 15–30mm long; stipules narrowly triangular, 6–8mm long. Submontane grassland, forest; 1450m.
Kodmin: Cheek 9632 11/1998.
Note: specimen sterile; more material required.

Synsepalum sp.
Tree c. 20m tall; trunk 60cm d.b.h. with c. 12 twisted, deep flutes bearing spiny roots 1–10cm long; bark brown, outer slash pinkish white, inner white; with white latex; *Terminalia*-style branching; leaves in lax whorls at the ends of branches, 12–17 × 3.7–5.7cm, obovate, apex obtuse but with acuminate tip, base attenuate, in dry state upper surface pale olive-grey, lower surface appressed pubescent; petiole 7–10mm long, thick, ridged, appressed pubescence; stipules to 3mm long. Forest; 1650m.
Kodmin: Cheek 9066 1/1998.
Note: specimen sterile; more material required.

Tridesmostemon omphalocarpoides Engl.
Tree c. 25–30m tall; stems with numerous prominent buds in old leaf scars; outer slash thin, white, streaked red, inner yellowish white; *Terminalia*-style branching; leaves 9.5–19(–25) × 3–6(–8.5)cm, obovate-oblong, apex acuminate, base obtuse, glabrous; petiole 1.5–3cm, glabrous; inflorescence either solitary or paired at the base of petioles; sepals red, to 6mm long, petals oblong, joined at the base, truncate apices, staminodes 6.5–7mm long, 3-toothed; fruits held below on branches, subspherical, 10–12cm diameter. at maturity, 10-seeded; seeds 4 × 1.8 × 1cm. Forest; 1050m.
Distr.: Cameroon to Congo (K). (Lower Guinea & Congolian).
IUCN: LC
Nyale: Cheek 9673 11/1998.

SCROPHULARIACEAE

M. Cheek

Note: *Bartsia*, *Hedbergia*, *Sopubia mannii* and *S. ramosa*, are all likely to be found at Mwanenguba in future surveys.

Alectra sessiliflora (Vahl) Kuntze var. *monticola* (Engl.) Melch.

Erect scabrid herb 30–90cm; slightly hispid or glabrous; leaves petiolate, lanceolate to elliptic, serrate, 10–50 × 10–20mm; flowers pale yellow, 5mm. Grassland; 2000m.
Distr.: tropical Africa & Asia. (Palaeotropics).
IUCN: LC
Enyandong: Ghogue 1213 fl., 10/2001; **Mwanenguba:** Cheek 9471 10/1998; Ghogue 1294 fl., 11/2001; Leeuwenberg 8585 fl., 10/1971; Sanford 5444 11/1968.

Alectra sessiliflora (Vahl) Kuntze var. *senegalensis* (Benth.) Hepper

Slender, erect, roughly pilose herb c. 30cm, usually drying black; leaves opposite, alternate within inflorescence, sessile, ovate, serrate, 10–50 × 10–20mm; flowers yellow, 5mm. Grassland; 800m.
Distr.: Senegal to Mozambique. (Tropical Africa).
IUCN: LC
Nyasoso: Etuge 2496 7/1996.

Artanema longifolium (L.) Vatke

Erect herb 1m, glabrous; stem 4-angular; leaves opposite, narrow elliptic, c. 15 × 4cm, gradually long-acute, base cuneate-decurrent, serrate; petiole 1cm, winged; raceme terminal, 15cm; pedicel 0.5cm; corolla 3cm, purple; fruit globose, 1cm. Swampy roadside in forest.
Distr.: Liberia to Uganda & tropical Asia. (Palaeotropics).
IUCN: LC
Mombo: Olorunfemi in FHI 30568 fl., 5/1951.

Buchnera capitata Benth.

Erect herb 60cm, sparsely long-pubescent; basal leaves elliptic, c. 5 × 3cm, 1–2-toothed per side, cauline leaves opposite, ligulate, c. 5 × 0.5cm, internodes 5cm; spike cylindrical, c. 5 × 1cm; flowers numerous, dense, 1cm, 5-parted, blue-purple. Grassland; 1850m.
Distr.: Ivory Coast to Tanzania, Madagascar. (Tropical Africa & Madagascar).
IUCN: LC
Mwanenguba: Pollard 861 fl., 11/2001; Sanford 5417 fl., 11/1968; 5428 fl., 11/1968.

Lindernia diffusa (L.) Wettst.

Annual herb, c. 3cm, resembling *Torenia dinklagei*, but stems densely white-puberulent; leaves purple below; flowers few, sessile, white; fruit narrowly ovoid, far exceeding calyx. Roadside; 1000m.
Distr.: Mali to Tanzania. (Tropical Africa).
IUCN: LC
Mwanenguba: Sanford 5453 fl., 11/1968; **Nyasoso:** Cable 3480 7/1996.

Lindernia nummulariifolia (D.Don) Wettst.

Erect herb, resembling *Torenia dinklagei*, but pedicel c. 1cm; fruit narrowly ovoid, far exceeding calyx. Roadside; 1200m.
Distr.: Sierra Leone to Zambia, tropical Asia. (Palaeotropics).
IUCN: LC
Mwambong: Cheek 9560b 11/1998; 9563 11/1998.

Lindernia senegalensis (Benth.) Skan

Annual herb; stems mostly prostrate, sometimes 30cm long, resembling *Torenia dinklagei*, but glabrous; leaves to 2cm; petiole 2mm; pedicel 3mm; flowers 7mm, white, with 2 yellow palate bumps; fruit cylindrical, 2cm, far exceeding calyx. Roadside; 1200–1530m.
Distr.: Senegal to Cameroon. (Upper & Lower Guinea).
IUCN: LC
Kodmin: Cheek 8909 1/1998; Ghogue 14 1/1998; **Muahunzum:** Etuge 4436 11/1998; **Mwambong:** Cheek 9562 11/1998.
Note: newly recorded here for Cameroon.

Rhabdotosperma densifolia (Hook.f.) Hartl

Beitr. Biol. Pfl. 53(1): 58 (1977).
Syn.: *Celsia densifolia* Hook.f.
Robust erect herb to 90cm; stems pithy, woody at base, tomentose; leaves tomentose beneath, lanceolate, closely serrate, 2–7 × 0.7–2cm; inflorescence a terminal raceme; pedicels 1.5cm in fruit; flowers yellow, c. 2cm diameter; fruits 6–8mm. Grassland; 2000m.
Distr.: Bioko, W Cameroon. (W Cameroon Uplands).
IUCN: NT
Mwanenguba: Leeuwenberg 10696 fl., 11/1972.

Rhabdotosperma ledermannii (Murb.) Hartl

Beitr. Biol. Pfl. 53(1): 58 (1977).
Syn.: *Celsia ledermannii* Schltr. ex Murb.
Stout erect herb 1.2–1.6m; stems sparsely tomentose, woody at base; leaves oblong-lanceolate, sparsely tomentose on nerves beneath, up to 16cm; pedicels 1.5–3cm in fruit; flowers yellow, c. 2.5cm diameter; fruits 8–10mm. Grassland; 2100m.
Distr.: Cameroon Endemic.
IUCN: VU
Mwanenguba: Sanford 5522 fl., fr., 11/1968.

Scoparia dulcis L.

Erect herb to 0.5m, puberulent; leaves opposite, elliptic, c. 1.5cm, decurrent, deeply serrate; petiole to 1cm; flowers 4mm, axillary, white, symmetrical; pedicels 3mm; fruit globose, 2mm. Roadside weed; 350–600m.
Distr.: pantropical.
IUCN: LC
Mungo River F.R.: Cheek 10221 11/1999; **Nlonako:** Etuge 1780 3/1996; **Nyasoso:** Etuge 1649 1/1996.
Uses: FOOD – leaves edible, taste sweet, gives energy; MEDICINAL (Etuge 1780).

Stemodia verticillata (Mill.) Boldingh

Sprawling herb, 10cm, densely glandular pubescent; leaves opposite, ovate, 6mm, slightly acute, serrate; petiole 6mm; flowers 1–2 per axil, subsessile, pinkish blue, 3mm; fruit globose, 2mm, enveloped by ligulate free sepals. Roadside.
Distr.: W Africa; introduced from S America.
IUCN: LC
Ebonji: Olorunfemi in FHI 30586 5/1951.

Torenia dinklagei Engl.

Annual herb, sparsely pubescent; stems square, erect or protrate; leaves opposite, broadly ovate, 1cm, serrate, sessile; flowers single, axillary; pedicel 1.5cm, tubular, two-lipped, white, purple above, throat yellow, 4mm; fruit enveloped in winged calyx, 7mm. Roadside; 1200m.
Distr.: Sierra Leone to Congo (K). (Guineo-Congolian).
IUCN: LC
Kodmin: Cheek 9560a 11/1998; **Mwanenguba:** Leeuwenberg 8593 fl., 10/1971.

Torenia thouarsii (Cham. & Schltdl.) Kuntze

Annual herb resembling *T. dinklagei*, but subglabrous; leaves longer, ovate-lanceolate; petiole to 1cm; flowers white, rimmed with purple, throat white-hairy. Roadside; 1200m.
Distr.: pantropical.
IUCN: LC
Mwambong: Cheek 9561 11/1998.

Veronica abyssinica Fresen.

Prostrate creeping herb, usually drying dark brown; stem branched from the base, pilose; leaves opposite, petiolate, ovate, serrate except towards base, 2–4 × 1–2cm; inflorescence a slender axillary peduncle; flowers blue or pinkish, paired or a few together, 8–10mm diameter; fruit bilobed, pubescent. Grassland, forest-grassland transition, roadsides; 800–2000m.
Distr.: Nigeria to Zimbabwe. (Afromontane).
IUCN: LC
Bangem: Thomas 5370 1/1986; **Ehumseh - Mejelet:** Etuge 324 fl., 10/1986; **Mwanenguba:** Cheek 7276 2/1995; Leeuwenberg 8472 fl., 9/1971; Sanford 5443 fl., fr., 11/1968; Thomas 3103 fl., 2/1984.

SCYTOPETALACEAE

M. Cheek

Fl. Cameroun 20 (1978).

Oubanguia cf. africana Baill.

Bull. Soc. Linn. Paris. 2: 869 (1890).
Tree 12m; stems c. 1mm diameter, with a groove on the upper surface, glabrous; leaves leathery, elliptic 9–11 × 4–5cm, acumen ligulate, 1cm, base unequally obtuse, lateral nerves c. 6 pairs, weakly brochidodromous, entire, glabrous, subsessile. Forest, along rivers; 400m.
Distr.: Cameroon to Gabon. (Lower Guinea).
Bakossi F.R.: Etuge 4302 10/1998.
Note: fertile material needed to confirm identification.

Oubanguia alata Baker f.

Tree c. 20m; bole straight, red-brown; leafy stems with 3–4 green wings 2–3mm deep; leaves leathery, oblong-elliptic, c. 18 × 7cm, long-acuminate, base acute, c. 5 pairs of nerves, brochidodromous; petiole 2mm; thyrse terminal, c. 15 × 10cm; flowers white, c. 1cm diameter, petals caducous; stamens numerous; berry 2cm. Forest; 460m.
Distr.: SE Nigeria & Cameroon. (Lower Guinea).
IUCN: NT
Nyandong: Cheek 11404 3/2003.
Note: specimen not yet at Kew, description from other Cameroon material.

Rhaptopetalum cf. coriaceum Oliv.

Tree 6–9m; resembling *R. geophylax*, but leaves smaller, 13–15(–22) × 5–7cm, drying dark brown below; fruits lacking cylindrical receptacle (flat) and lacking corky ridges (smooth). Forest; 800–1000m.
Distr.: SE Nigeria to Gabon. (Lower Guinea).
Kupe Village: Cable 3717 7/1996; Etuge 2822 7/1996; 2919 7/1996; Kenfack 298 7/1996; Ryan 237 5/1996.

Rhaptopetalum geophylax Cheek & Gosline

Kew Bull. 57: 662 (2002).
Tree 6–10m; stems terete, glabrous; leaves obovate-oblong (14–)21–28 × 11–15cm, acumen 1cm, base unequally rounded, glabrous; inflorescence ramiflorous, subfasciculate; pedicel 3–6mm, articulated; flowers pink, buds 9–15mm, ovary superior; fruit orange, ovoid, 27–38mm, receptacle covered in sinuous corky ridges. Forest; 900–1530m.
Distr.: Cameroon Endemic.
IUCN: NT
Edib village: Cheek 9175 2/1998; **Kodmin:** Etuge 3969 1/1998; Gosline 48 1/1998; Mackinder 139 1/1998; **Kupe Village:** Cable 2574 5/1996; 766 fl., fr., 1/1995; 830 fr., 1/1995; Cheek 7196 fl., fr., 1/1995; 7757 11/1995; 7792 11/1995; 7816 fr., 11/1995; 7872 fr., 11/1995; 8378 5/1996; Etuge 1453 fr., 11/1995; 1693 2/1996; **Mbule:** Cable 3345 6/1996; **Mwambong:** Gosline 81 2/1998; **Ngomboaku:** Gosline 261 12/1999; 265 12/1999; **Nyasoso:** Cable 3614 7/1996; Etuge 2106 6/1996; 2162 6/1996; 2429 6/1996; Sidwell 370 10/1995.
Note: only known from Bakossi and Rumpi Hills.
Local name: Aton (Chief Abweh D'Epie of Kodmin in Cheek 9175); Cheylo (Gosline 48). **Uses:** FOOD ADDITIVES – preservatives: Bakossi women use the leaves to cover baskets of freshly caught tadpoles, to stop them rotting (Chief Abweh D'Epie of Kodmin in Cheek 9175); used to cover fish after they have been caught (Gosline 48).

Rhaptopetalum sp. nov.

Tree 5–15m; resembling *R. geophylax* but stems and leaves densely puberulent; flowers white; fruits with smooth receptacle. Forest; 760m.
Distr.: Cameroon endemic.
Ngombombeng: Etuge 35 fr., 4/1986.

SIMAROUBACEAE

M. Cheek

Brucea guineensis G.Don

Tree 2–4m, glabrescent; leaves alternate, pinnate, 40cm, 4-jugate; leaflets elliptic, c. 12 × 5cm, attenuate, obtuse to rounded, secondary nerves each with a gland near margin; petiolule c. 0.5cm; inflorescences racemose, pendulous, 20cm; fruit red, fleshy, ovoid 1.2 × 0.8cm, style oblique. Forest; 750–1200m.
Distr.: Sierra Leone to Cameroon & Rio Muni. (Upper & Lower Guinea).
IUCN: LC
Kupe Village: Cable 3709 fr., 7/1996; Groves 6 fl., fr., 1/1995; Kenfack 239 fr., 7/1996; **Ndum:** Cable 926 fr., 2/1995; Groves 32 fr., 2/1995; **Nyasoso:** Cable 2795 fr., 6/1996; 2888 fr., 6/1996; Cheek 7292 fr., 2/1995; Etuge 2098 fr., 6/1996; Lane 101 fr., 6/1994; Sunderland 1473 fr., 7/1992.
Local name: Esonjo. **Uses:** MEDICINES (Cheek 7292).

Quassia sanguinea Cheek & Jongkind ined.

Syn.: *Hannoa ferruginea* Engl.
Tree (1.5–)2–4(–6)m, glabrous; resembling *Q. silvestris* but 3(–4)-jugate, rachis and midrib violet; leaflets usually drying green below, c. 13 × 5cm, apex subacuminate. Forest; 800–1750m.
Distr.: Nigeria & Cameroon. (W Cameroon Uplands).
IUCN: VU
Bangem: Thomas 5301 fl., fr., 1/1986; 5338 1/1986; **Ehumseh:** Manning 465 10/1986; **Ehumseh - Mejelet:** Etuge 329 10/1986; **Kodmin:** Cheek 8873 fr., 1/1998; Etuge 4401 fl., 11/1998; Onana 563 fr., 2/1998; **Kupe Village:** Cable 2567 fr., 5/1996; 3781 fl., 7/1996; 3854 fr., 7/1996;

802 fr., 1/1995; Cheek 7217 fl., fr., 1/1995; 7775 fl., fr., 11/1995; 8396 fr., 5/1996; **Nyasoso:** Etuge 2121 fl., 6/1996; 2383 fl., fr., 6/1996; Lane 228 fl., 11/1994; Sebsebe 5056 fl., fr., 10/1995; Thomas 3075 fl., 2/1984; Wheatley 467 fl., 7/1992.
Uses: MATERIALS – wood used for carving tools (Etuge 2383).

Quassia silvestris Cheek & Jongkind ined.
Syn.: *Hannoa klaineana* Pierre & Engl.
Tree (10–)25–40m, glabrous; leaves alternate, pinnate, 20–30cm, (4–)5-jugate; leaflets drying brown below, obovate, c. 14 × 7cm, apex retuse-mucronate, base attenuate-acute; petiolule 0.5cm; inflorescence a panicle 15–30cm; flowers white, 4mm; fruit apocarpous, ovoid, fleshy, 1-seeded, 2.5cm. Farmbush, secondary forest; 220–500m.
Distr.: Guinea (C) to Congo (K). (Guineo-Congolian).
IUCN: LC
Bakolle Bakossi: Etuge 254 fl., 9/1986; **Kupe Village:** Etuge 1476 fr., 11/1995; **Kurume:** Thomas 8213 4/1988; **Mile 15, Kumba-Mamfe Road:** Nemba 209 fr., 9/1986.
Uses: ANIMAL FOOD – eaten by parrots (Nemba 209); MATERIALS – timber (Etuge 1476).

SOLANACEAE
B.J. Pollard

Browallia americana L.
Sp. Pl. 2: 631 (1753).
Weedy herb to c. 50cm; stems herbaceous, subwoody at base, several-branched; leaves ± ovate, obtuse, cuneate, rounded or truncate, entire, shortly ciliate on nerves, c. 3–6 × 2–3.5cm; flowers borne singly along the stems, interspersed between leaves; pedicels c. 0.5–1.0cm; flowers showy, bright bluish purple; corolla rim lobed; calyx campanulate in fruit; fruit c. 0.5 × 0.5cm. Roadside verges, path edges, farmbush; 650–1200m.
Distr: C & S America, Caribbean, naturalised in tropical Africa, Malaya and New Guinea. (Pantropical).
IUCN: LC
Kodmin: Onana 602 fl., 2/1998; **Nyasoso:** Bruneau, 1075 fl., 10/1995; Etuge 1610 fl., 1/1996; 2925 fl., fr., 7/1996; **Nzee Mbeng:** Gosline 94 fl., 2/1998.

Brugmansia × *candida* Pers.
Fl. Rwanda: Spermatophytes 3: 359 (1985).
Syn.: *Datura candida* (Pers.) Safford
Shrub or small tree to 3m; leaves ovate-acuminate, to 24 × 12cm; flowers white, fragrant, pendent, funnel-shaped, 25–30cm long; calyx spathiform, terminating in a point, finely pubescent outside, appressed to the funnel-shaped lower part of the corolla-tube; anthers free; fruit lemon-shaped. Cultivated; 1000m.
Distr: tropical and subtropical countries around the world.
IUCN: LC
Bangem: Pollard sight record fl., 10/1998.

Brugmansia suaveolens (Humb. & Bonpl. ex Willd.) Bercht. & Presl.
Fl. Rwanda: Spermatophytes 3: 359 (1985).
Syn.: *Datura suaveolens* Humb. & Bonpl. ex Willd.
Shrub or small tree, 2–5m; leaves large, up to 40 × 15cm; flowers 25–30cm, pendulous, white, fragrant; calyx 5-toothed at apex, very inflated, not appressed to the thin pipe-

like lower part of the corolla-tube, nearly glabrous; anthers coherent; fruit fusiform. Cultivated; 850–1000m.
Distr: native of Brazil, introduced throughout the tropics. (Pantropical).
IUCN: LC
Enyandong: Ghogue 1297 fl., 11/2001; **Mt Kupe:** Leeuwenberg 9235 fl., 1/1972; **Mwambong:** Onana 546A fl., 2/1998; **Nyasoso:** Thomas 3060 fl., 2/1984.

Capsicum annuum L.
Annual or biennial herb; leaves broadly lanceolate to ovate, apex acutely acuminate, 5–8 × 2–5cm; inflorescences axillary, 3–8-flowered; pedicels to c. 1.5cm, nodding at maturity, dilated distally; calyx obscurely 5-toothed, 10-ribbed; corolla rotate-campanulate, deeply 5-lobed, usually straight, white or greenish; fruits single, a ± elongated berry, 1–several cm. Cultivated in gardens and farms, sometimes escaping into forest understorey; 650–1300m.
Distr: widely dispersed throughout the tropics, cultivated and sometimes naturalised. (Pantropical).
IUCN: LC
Ngusi: Etuge 1572 fr., 1/1996; **Nyasoso:** Etuge 1532 fl., fr., 1/1996; 1612 fl., fr., 1/1996.

Capsicum frutescens L.
As for *C. annuum*, except: perennial; pedicels erect at maturity; corolla lobes slightly revolute; fruits usually 2 or more from each flower-cluster, and rarely exceeding 2cm. widely cultivated; 300–650m.
Distr: as for *C. annuum*.
IUCN: LC
Loum F.R.: Biye 57 fl., fr., 11/1999; 69 fl., fr., 12/1999; **Ngusi:** Etuge 1548 fl., fr., 1/1996.
Local name: Hot Pepper. **Uses:** FOOD ADDITIVES – infructescences – chew fruits, fruits hot in the mouth; grind fruits, put in sauce as a stimulant and make sauce hot for the mouth, also adds flavour in soup. (Etuge 1548).

Cestrum nocturnum L.
Sp. Pl. 2: 191 (1753).
Shrub or treelet to 4m, ± glabrous; leaves narrowly ovate or lanceolate, to 11 × 4cm; petioles c. 0.5–1.0cm; inflorescences many-flowered, strongly scented at night; calyx campanulate, c. 2–3mm; corolla yellow to greenish white, to 25(–28)mm, slender, tube to 21(–24)mm; fruits bright white, 7.5–10mm. widely cultivated; 460–1170m.
Distr: originally from the West Indies, Mexico, Central America and northern S America, now commonly cultivated as a garden ornamental in the Old World. (Pantropical).
IUCN: LC
Edib village: Cheek (sight record) 2/1998; **Nyandong:** Cheek 11414 fl., fr., 3/2003.
Local name: 'Queen of the Night' (*fide* Pollard).

Nicotiana tabacum L.
Robust annual to 2m; upper leaves oblong-lanceolate to elliptic, 8–15 × 1.5–6cm; inflorescence a terminal cyme; flowers viscid-glandular outside; corolla tubular, c. 4cm long, lobes acute, white, cream or pinkish. Cultivated, sometimes near houses; 1000m.
Distr: native of S America, widely cultivated in warmer parts of the world. (Pantropical).
IUCN: LC
Kupe Village: Etuge 1622 fl., fr., 1/1996.
Local names: Tobacco (English); Akwande. **Uses:** SOCIAL USES – smoking materials/drugs – dry leaves used, smoked as tobacco, sometimes cultivated in nearby houses (Etuge 1622).

Physalis angulata L.

Erect glabrous annual, to 1m; leaves coarsely sinuous-dentate, ovate, 5–10cm at maturity; flowers borne singly; pedicellate, cream-coloured; calyx acrescent, ovoid, to 3 × 2cm, eventually enveloping the fruit. Farmbush; 200m.
Distr: native of Peru, widely distributed in the tropics. (Pantropical).
IUCN: LC
Mungo River F.R.: Pollard 107 fl., fr., 10/1998.
Local name: Mbabe. **Uses:** MEDICINES (Epie Ngome in Pollard 107).

Physalis lagascae Roem. & Schult.

Rhodora 69: 220 (1967).
Syn.: *Physalis micrantha* Link
Erect or ± prostrate herb, 30–100cm; leaves subentire, ovate, 1.5–3(–5)cm at maturity, acuminate; flowers cream-coloured; calyx accrescent, subglobular in fruit, to 1.5 × 1.2cm. Roadsides, farms; 1500m.
Distr: native to S America, but widely distributed across the tropics. (Pantropical).
IUCN: LC
Kodmin: Cheek 8906 1/1998.
Local name: Soon frum. **Uses:** no use, not eaten (Epie Ngome in Cheek 8906).

Physalis peruviana L.

Erect or straggling perennial to 1m, densely hairy; from creeping rootstock; leaves rhomboid to deltoid, entire or with a few large teeth, 8–10 × 6–7.5cm; flowers yellow with purple centre, 15mm; fruiting calyx large, to 4 × 3cm, villous. Fallow, rocky grassland; 2000m.
Distr: tropical America, naturalized in West Africa. (Pantropical).
IUCN: LC
Mwanenguba: Pollard 165 fl., fr., 10/1998; 862 fl., fr., 11/2001.
Uses: MEDICINES – veterinary (Martin Mwene (CERUT biologist) in Pollard 862).

Schwenckia americana L.

Glabrous herb to c. 1m; branching freely; lower leaves petiolate (upper sessile), elliptic to ovate, to 4 × 2.5cm, entire; inflorescence a lax panicle to c. 30cm; pedicels 0.5cm; flowers white or greenish yellow; corolla narrowly tubular, 7–9mm; fruit a capsule, not berry-like, subglobose, c. 0.5cm across. Weed of cultivation, escaping into forest understorey occasionally; 400m.
Distr: native to S America, introduced to Africa.
IUCN: LC
Kupe Village: Etuge 1919 fl., fr., 5/1996.

Solanum

A.E.Gonçalves' treatment of *Solanum* spp. (ined.) for Flora Zambesiaca is followed here. This is particularly relevant with regard to the taxonomy of the difficult *Solanum anguivi* Lam. complex.

Solanum aculeastrum Dunal var. *aculeastrum*

Tree or shrub to 7m, armed; white tomentum on all parts except upper surface of leaves; spines plentiful, straight or sharply recurved, to 15mm, compressed; leaves conspicuously deeply lobed, to c. 10 × 10cm; inflorescence lateral, axillary, densely brown-scurfy; flowers pinkish white, c. 1cm across; fruits globose to c. 4 × 4cm. Farms, rocky grassland; 1600–1800m.

Distr: widely distributed in the tropics, and often cultivated as a hedge plant. (Pantropical).
IUCN: LC
Mwanenguba: Etuge 4380 fr., 10/1998; Swarbrick 2395 fl., 4/1961.

Solanum aethiopicum L.

Kew Bull. 40: 391 (1985).
Syn.: *Solanum gilo* Raddi
Herb to c. 1m; unarmed; covered with stellate hairs on stems and leaves; leaves sinuate to ± laxly dentate, felty beneath, to c. 10 × 5cm; inflorescences densely stellate-hairy; flowers white, subsolitary, to c. 4–5 together, berries green-orange. Farmbush; 1500m.
Distr: Guinea (C) to Ethiopia. (Afromontane).
IUCN: LC
Kodmin: Biye 35 fl., fr., 1/1998; **Nyasoso:** Biye 63 fl., fr., 12/1999.

Solanum americanum Mill.

Gard. Dict. ed. 8: *Solanum* 5 (1768).
Glabrescent to moderately pilose herbs to c. 60cm; stems angled; leaves ovate-lanceolate to lanceolate, c. 3–6(–11) × 2–6cm, entire to sinuate; inflorescences conspicuously arising internodally (sub- or supra-axillarially), as simple umbellate cymes, 3–6(–10)-flowered; flowers white, occasionally purple, c. 4.5mm across; berries 6–8mm diameter, spherical, black, shiny. Farmbush, forest; 860–1150m.
Distr: cosmopolitan.
IUCN: LC
Kupe Village: Biye 55 fl., fr., 11/1999; Ryan 289 fl., fr., 5/1996; **Nyasoso:** Cheek 7470 fl., fr., 10/1995.
Uses: FOOD – unspecified parts (Ryan 289).

Solanum anguivi Lam.

Fl. Rwanda: Spermatophytes 3: 376 (1985).
Syn.: *Solanum distichum* Schum. & Thonn.
Syn.: *Solanum indicum* sensu auctt. et collectt. afric. plur., non L., 1753
Syn.: *Solanum indicum* subsp. *distichum* (Schum. & Thonn.) Bitter
Syn.: *Solanum indicum* subsp. *distichum* var. *grandemunitum* Bitter
Syn.: *Solanum indicum* subsp. *distichum* var. *modicearmatum* Bitter
Coarse tomentose undershrub to 2m, spiny or not, with stellate hairs; leaves elliptic, very shortly pubescent above, subtomentose beneath, to 10–16cm; inflorescence a racemose-like cyme; flowers white, c. 5mm; fruits erect, globose, red, 1–1.5cm across. Forest; 1400–2000m.
Distr: tropical Africa, Madagascar & Mascarene Is., Arabia.
IUCN: LC
Bangem: Leeuwenberg 9526 fl., fr., 3/1972; **Essossong:** Groves 50 fl., fr., 2/1995; **Kodmin:** Biye 25 fl., fr., 1/1998; **Nyasoso:** Cable 1242 fl., fr., 2/1995; Etuge 1673 fr., 1/1996.
Local name: Ndereh. **Uses:** FOOD – infructescences – fruits edible (Etuge 1673).

Solanum betaceum Cav.

Taxon 44: 584 (1995).
Syn.: *Cyphomandra betacea* (Cav.) Sendtn.
Tree to 6m; leaves alternate, ovate, to c. 20 × 15cm, base cordate; petiole c. 10cm long; inflorescence few-flowered, axillary, pendulous; flowers campanulate, c. 1.5 cm × 0.7cm wide; petals pink; fruit ellipsoid, c. 7 × 4cm, orange, edible. Forest; 1500m.

Distr: native to Peru, but naturalised in many parts of the tropics. (Pantropical).
IUCN: LC
Kodmin: Biye 34 fr., 1/1998.
Local name: Tree Tomato. **Uses:** FOOD – fruits (*fide* Pollard).

Solanum lycopersicum L.

Annual unarmed herb 0.7–2m, erect with thick solid stems, or spreading and later becoming prostrate, coarsely hairy, glandular, with characteristic strong odour; leaves spirally arranged, uppermost simple, the lower ones imparipinnate; inflorescences terminal, 4–12-flowered; flowers pendent, hypogynous, yellow, to c. 2cm across; fruit a fleshy berry, green, maturing red, 2–15cm across. Forest; 1500m.
Distr: native to Peru and Ecaudor, but now cultivated throughout the world. (Cosmpolitan).
IUCN: LC
Nyasoso: Pollard sight record fr., 5/2002.
Note. widely cultivated in our area.
Local name: Tomato. **Uses:** FOOD – fruits (*fide* Pollard).

Solanum melongena L.

Shrub to c. 2m; stems unarmed, white-green hairy, glabrous apically; leaves lanceolate, to elliptic, or obliquely ovate, subsucculent, to c. 16 × 10cm; flowers axillary, apparently borne singly, to c. 3cm, white; fruit egg-shaped, light green, sometimes elongate or pyriform. Farmbush, disturbed forest; 200m.
Distr: temperate and tropical parts of the world.
IUCN: LC
Mungo River F.R.: Biye 58 11/1999.
Local name: this is the 'aubergine' or 'egg-plant' (American colloquialism) (*fide* Pollard).

Solanum nigrum L.

Herb, 30–60cm; stems decumbent to erect; leaves mostly glabrous, shortly ciliate; inflorescence umbellate; flowers small, white, on a common peduncle; berries 0.6–2cm diameter, broadly ovoid, dull purple to blackish or yellowish green. Farmbush: weed, sometimes cultivated; 650m.
Distr: cosmopolitan.
IUCN: LC
Ngusi: Etuge 1556 fl., 1/1996.
Uses: FOOD – edible leaves prepared as any other vegetable (Etuge 1556).

Solanum terminale Forssk.

Syn.: *Solanum terminale* Forssk. subsp. *inconstans* (C.H.Wright) Heine
Syn.: *Solanum terminale* Forssk. subsp. *sanaganum* (Bitter) Heine
Syn.: *Solanum terminale* Forssk. subsp. *terminale*
Woody climber to c. 10–15m; leaves elliptic, c. 12 × 5cm, acuminate, glabrous; petiole 1–2 cm; inflorescence terminal or lateral, paniculate or cymose, very rarely spicate, 10–20cm; flowers c. 5 × 8mm, petals purple, staminal tube yellow. Forest; 300–1300m.
Distr: tropical Africa.
IUCN: LC
Bakole: Nemba 786 2/1988; **Kupe Village:** Cable 3780 fl., 7/1996; Etuge 2754 fl., 7/1996; **Ngombombeng:** Etuge 31 fl., 4/1986; **Nyasoso:** Etuge 1324 fl., fr., 10/1995; 1811 fl., 3/1996; Lane 115 fl., 6/1994; Zapfack 682 fl., 6/1996.
Uses: FOOD – young leaves edible as a vegetable (Etuge 1324).

Solanum torvum Sw.

Shrub to 3m; stems occasionally armed, densely stellate hairy; leaves large, ± elliptic, to 10–16 × 4–12cm, subscabrid, lobate-sinuate to subentire; inflorescence of corymbose cymes, 2–5(–14)-flowered; corolla white (rarely purple), to 2.5cm; fruit c. 1cm, globose, dirty brown, occasionally drying black. A common weed in farmbush or forest; 600–1400m.
Distr: pantropical.
IUCN: LC
Enyandong: Pollard 749 fl., 10/2001; **Essossong:** Groves 51 fl., fr., 2/1995; **Kodmin:** Ghogue 67 fl., fr., 1/1998; **Kupe Village:** Cable 2552 fl., fr., 5/1996; Cheek 7059 fr., 1/1995; Kenfack 324 fl., fr., 7/1996; Ryan 336 fl., fr., 5/1996; **Mwanenguba:** Leeuwenberg 8532 fl., 9/1971; **Ndum:** Cable 907 fl., fr., 2/1995; **Nyasoso:** Cable 2780 fl., 6/1996; Cheek 7903 fr., 11/1995; Etuge 1644 fl., fr., 1/1996; Sunderland 1485 fr., 7/1992; **Southern Bakossi F.R.:** Binuyo in FHI 35550 fl., fr., 2/1956; **Tombel:** Etuge 377 11/1986.
Local name: Ndare-Nzoa (Pollard 749); Ngaka (Lyonga Daniel – Bakweri in Cheek 7903). **Uses:** MEDICINES (Pollard 749 & Lyonga Daniel in Cheek 7903).

Solanum welwitschii C.H.Wright

Bothalia 25: 49 (1995).
Syn.: *Solanum terminale* Forssk. subsp. *welwitschii* (C.H.Wright) Heine
Woody climber, to c. 15m; ± glabrous; leaves elliptic, oblong to obovate, 7–15 × 3–6cm; inflorescences spicate and terminal; pedicels c. 3–5mm in flower, to 1.5cm in fruit. flowers purple; fruits c. 5mm, globose, reddish at maturity, smooth. Forest; 700–1550m.
Distr: Guinea (C) to Congo (K), Angola. (Guineo-Congolian).
IUCN: LC
Bakossi F.R.: Olorunfemi in FHI 30589 fl., 5/1951; **Edib:** Etuge 4101 fl., 1/1998; **Kodmin:** Satabie 1119 fl., fr., 2/1998; **Kupe Village:** Cable 3860 fr., 7/1996; Cheek 7039 fl., fr., 1/1995; 7885 fl., 11/1995; Ryan 259 fl., 5/1996; **Nyasoso:** Cable 3289 fl., fr., 6/1996; 3586 fr., 7/1996; Etuge 2150 fr., 6/1996.

STERCULIACEAE

M. Cheek

Byttneria catalpifolia Jacq. subsp. africana (Mast.) Exell & Mendonça

Climber to 20m; stem twisting, unarmed; leaves papery, entire, ovate c. 18 × 13cm, acuminate, broadly cordate, entire; petiole c. 9cm; stipules caducous; cymes c. 7cm; flowers white, 3mm; fruit spiny (spines 1cm), dry, dehiscent, body 3cm, resembling a 'Euphorb', but 5-locular. Forest; 750m.
Distr.: Ghana to Uganda. (Guineo-Congolian).
IUCN: LC
Manehas F.R.: Cheek 9391 10/1998; **Nyasoso:** Etuge 2588 fl., 7/1996.

Cola altissima Engl.

Tree 15–20m; crown compact; leaves drying black, alternate, thickly leathery, elliptic-oblong, 20 × 8cm, acumen 1cm, broadly acute, lateral nerves 8 pairs, very prominent below, sinuate; petiole 0.5–7cm; stipules caducous; ramiflorous, perianth campanulate, 5 × 5cm; pedicel 2cm. Forest; 400m.

Distr.: SE Nigeria to Congo (K). (Lower Guinea & Congolian).
IUCN: NT
Molongo: Nemba 772 2/1988.
Note: specimen cited not seen.

Cola anomala K.Schum.

Tree 15–20m; crown dense; stems bright white from waxy cuticle; leaves in whorls of 3(–4), simple, entire, elliptic, to 17 × 7.5cm, subacuminate, base rounded to obtuse, lateral nerves c. 7 pairs; petiole c. 2cm; stipules caducous; panicles 3cm in leaf axils; flowers yellow without red markings, 1.5cm; fruit follicles to 12cm, 2-seeded, with knobs and ridges. Forest; 1500m.
Distr.: Cameroon & Nigeria. (W Cameroon Uplands).
IUCN: NT
Mwanenguba: Letouzey 13934 fl., 6/1975.
Note: the specimen cited here, from Manenguba, is the most southerly known.
Local name: this is the Bamilike Cola of trade (*fide* Cheek).

Cola argentea Mast.

Syn.: *Cola sp. D sensu* Keay in FWTA 1: 332
Tree 5–6(–12)m; leaves 5–7-digitately compound; petiole 3–21cm; leaflets oblanceolate, to 40 × 13cm, long acuminate, cuneate, silvery below, initially orange-brown scurfy; petiolules 0.1–0.5cm; stipules 4 × 0.3cm; inflorescences few-flowered fascicles from boss at trunk base; flowers purple brown, cylindrical, c. 2.5 × 1cm including lobes 0.8cm. Forest; 700–1100m.
Distr.: SE Nigeria, Cameroon & Gabon. (Lower Guinea).
IUCN: NT
Kupe Village: Cable 3657 fl., 7/1996; 810 fl., 1/1995; Cheek 7001 1/1995; 7145 1/1995; 7719 11/1995; Ryan 255 fl., 5/1996; **Mbule:** Letouzey 14672 4/1976; **Nyasoso:** Cable 3620 fl., 7/1996; Lane 132 fl., 6/1994; Wheatley 388 fl., 7/1992.
Local name: a 'monkey cola'. **Uses:** FOOD – fruits harvested for edible seed-coat (*fide* Cheek).

Cola cauliflora Mast.

Shrub or small tree 2–3(–6)m; leaves in clusters, simple, narrowly lanceolate, to 30 × 6.5cm, acuminate, long-cuneate, glabrous; petioles uniform, c. 1.5cm; stipules caducous; inflorescences below leaves, fasciculate; flowers orange, c. 3mm wide; fruit red, subglobose 5cm, glabrous. Forest; 150–1000m.
Distr.: SE Nigeria, Cameroon & Gabon. (Lower Guinea).
IUCN: NT
Bakossi F.R.: Cheek 9319 10/1998; **Bolo Moboka:** Nemba 260 9/1986; **Etam:** Etuge 197 fr., 7/1986; **Konye:** Thomas 5186 11/1985; **Mungo River F.R.:** Cheek 9347 10/1998; **Ngomboaku:** Etuge 4686 fr., 12/1999; **Nyale:** Etuge 4222 2/1998; **Nyasoso:** Cable 3574 7/1996; 83 fr., 8/1992; Etuge 2574 fl., 7/1996; Sebsebe 5013 10/1995; Wheatley 385 fr., 7/1992.
Local name: Ngop. **Uses:** MATERIALS – personal items & tools – used as a chewing stick and also to make racks for mortar pieces (Epie Ngome in Cheek 9319).

Cola chlamydantha K.Schum.

Syn.: *Chlamydocola chlamydantha* (K.Schum.) Bodard.
Syn.: *Sterculia mirabilis* (A.Chev.) Roberty
Tree 15–20m; leaves 7-digitately compound; petiole to 1 m; leaflets thickly leathery, oblanceolate-obovate, c. 45 × 20cm, apex rounded, base acute, glabrous; petiolule c. 2cm, pulvinate; stipules c. 9 × 1cm, folded; inflorescences sessile, scattered along trunk above 2–3 m high; flowers cup-shaped,

dark red and purple, c. 8cm wide; fruits with 8 or more orange-red glabrous follicles, each 30 × 5cm. Forest, farmbush; 500–1200m.
Distr.: Guinea (C) to Congo (K). (Guineo-Congolian).
IUCN: LC
Kupe Village: Cheek 7043 fl., 1/1995; 7211 fl., 1/1995; **Nyasoso:** Cable 1214 fl., 2/1995; 40 fr., 8/1992; Cheek 7464 10/1995; **Wone:** Nemba 584 7/1987.

Cola digitata Mast.

Slender tree c. 4.5(–10)m; leaves 7-digitately compound; petiole c. 60cm; leaflets oblanceolate, entire or pinnately incised, c. 30 × 11cm, shortly acuminate, acute-decurrent, glabrous; petiolule 2cm; stipules caducous; inflorescences axillary fascicles; fruits follicles dehiscing flat, c. 20 × 20cm. Forest; 650m.
Distr.: Liberia to Congo (K). (Guineo-Congolian).
IUCN: LC
Banyu: Olorunfemi in FHI 30620 fr., 6/1951; **Kupe Village:** Lane 407 fr., 1/1995; **Loum F.R.:** Cheek 10254 12/1999; **Nyandong:** Tadjouteu 556 3/2003.

Cola ficifolia Mast.

Monopodial shrub or small tree, 3(–6)m; leaves simple, heteromorphic on one stem, entire, oblong, c. 30 × 11cm, truncate-caudate, acute, subsilvery below or orbicular in outline, 3-lobed, by 3/4 incised; petiole variable, 2–20cm on one stem; stipules c. 2 × 0.4cm, scurfy; inflorescences sessile, on stem below leaves. Forest; 200–450m.
Distr.: SE Nigeria to Congo (K). (Lower Guinea & Congolian).
IUCN: LC
Bakossi F.R.: Cheek 9327 10/1998; **Mt Kupe:** Cheek 10131 11/1999; **Mungo River F.R.:** Cheek 9341 10/1998.
Local name: a 'monkey cola' (see *C. argentea*).

Cola flaviflora Engl. & K.Krause

Small tree 2–8m, resembling *C. ricinifolia*, but leaves all 3-lobed, green below. Forest; 300m.
Distr.: SE Nigeria & W Cameroon. (Lower Guinea).
IUCN: NT
Konye: Thomas 5239 fl., 1/1986.

Cola lateritia K.Schum. var. *lateritia*

Tree 15–30m, minutely appressed brown stellate hairy; leaves leathery, ovate or orbicular in outline, to 24 × 20cm, shallowly 3–5-lobed, or subentire, apex rounded or subacuminate, base cordate, sinus obtuse, nervation palmate, minutely and sparingly stellate hairy; petiole c. 14cm; inflorescences on leafy stems; fruit carpels globose, red, glabrous, c. 6 × 6.5cm, verrucate. Forest; 400m.
Distr.: SE Nigeria to Congo (K). (Lower Guinea and Congolian).
IUCN: LC
Kupe Village: Etuge 2032 fr., 5/1996.
Uses: FOOD – unspecified parts; MATERIALS – timber (Etuge 2032).

Cola lateritia K.Schum. var. *1*

Tree resembling var. *lateritia*, but carpels oblong, c. 9 × 3cm, including rostrum 1–2cm, stipe 1cm, with shallow, intermittent, longitudinal ridges. Forest; 940–1050m.
Distr.: Cameroon endemic.
Kupe Village: Cable 2508 fr., 5/1996; **Nyasoso:** Lane 99 fr., 6/1994.
Note: probably a new variety, or possibly a new species, but flowers needed.

Cola lepidota K.Schum.

Tree 5–20m, glabrescent; leaves trifoliolate; leaflets oblanceolate-obovate, 30 × 12cm, acumen 2cm, acute, densely silvery lepidote below; petiolule 1cm; petiole to 30cm; stipules caducous; cauliflorous, inflorescences to 10cm diameter, with numerous 1–5-flowered peduncles to 4cm; perianth red, cup-shaped, 1cm; fruit oblong, red, 10 × 3cm. Forest; 450m.
Distr.: SE Nigeria, Cameroon & Gabon. (Lower Guinea).
IUCN: LC
Mt Kupe: Cheek 10132 11/1999; **Nyandong:** Cheek 11349 3/2003.
Note: the main 'monkey cola' (see *C. argentea*).
Local name: Mbwed. **Uses:** FOOD – fruits and seeds (*fide* Etuge).

Cola mahoundensis Pellegr.

Fl. Congo, Rw. & Bur. 10: 301 (1963).
Tree to 8m, scurfy brown puberulent; leaves alternate, orbicular in outline, 25cm, 3–5-lobed by 4/5–9/10 the radius, acumen to 3cm, shallowly cordate; petiole 1.5–30cm; flowers below leaves, 10mm, yellowish brown; fruit follicles sessile, subovoid, 5.5 × 3.5cm, beak obtuse. Forest; 300m.
Distr.: Cameroon to Congo (K). (Lower Guinea & Congolian).
IUCN: NT
Bolo: Nemba 6 4/1986.
Note: specimen not seen.

Cola marsupium K.Schum.

Shrub or small tree 4–8(–10)m; densely brown patent-pubescent, hairs 2mm; leaves simple, elliptic-oblong, to 20 × 8.5cm, acuminate, cordate, hairy on both surfaces, often with basal 'leaf-pockets'; petioles heteromorphic, 1.5–6cm; stipules filiform, 1cm; inflorescences axillary, sessile, near stem apex; flowers yellow, 6mm diameter. Forest; 200–500m.
Distr.: SE Nigeria to Congo (K). (Lower Guinea & Congolian).
IUCN: LC
Ikiliwindi: Nemba 460 fl., 1/1987; **Nyasoso:** Etuge 8 fl., 4/1986.

Cola metallica Cheek

Kew Bull. 57: 409 (2002).
Shrub 1m; leaves drying metallic grey, simple, evenly scattered, oblanceolate, 13–22 × 4–7cm, acumen subspatulate, acute, entire, brochidodromous; petioles 5–9mm, not pulvinate, with short simple black hairs; stipules caducous; inflorescences axillary, fasciculate; fruitlets 4–5-fruit, obovoid, 1.5–2cm, glossy orange or red. Forest; 400m.
Distr.: Cameroon endemic.
IUCN: CR
Bakossi F.R.: Etuge 4299 fr., 10/1998.

Cola nitida (Vent.) Schott & Endl.

Tree 12–25m; leaves simple, elliptic-oblong, c. 16 × 6cm, gradually acuminate, c. 1.5cm, base obtuse to rounded, entire, glabrous; petiole 0.5–5cm, variable on one stem; inflorescence a panicle to 4cm, below the leaves, densely hairy; flower 1.5cm, white, centre purple. Farmbush and around villages; 710m.
Distr.: Sierra Leone to Cameroon, cultivated widely in tropics. (Pantropical).
IUCN: LC
Nyale: Etuge 4196 fl., 2/1998.
Uses: SOCIAL USES – planted for cola nuts (*fide* Cheek).

Cola pachycarpa K.Schum.

Tree to 6m; leaves alternate, digitate; leaflets 7, elliptic-oblong, to 23 × 11cm, acumen 1cm, obtuse-decurrent, lateral nerves 12–15 pairs; petiolules to 8cm; stipules 1.5cm; cauliflorous; flowers in fascicles; pedicel 2cm; corolla cylindrical, 1.5cm, deep pink. Forest; 650m.
Distr.: SE Nigeria to Congo (K). (Lower Guinea & Congolian).
IUCN: LC
Nyandong: Darbyshire 23 3/2003.

Cola praeacuta Brenan & Keay

Monopodial treelet 8–10m; stems dull white; leaves elliptic or obovate-elliptic, to 15.5 × 6cm, acute-caudate, rounded to obtuse, entire; petioles variable on same stem, 0.7–4cm, minutely puberulent; stipules lanceolate, 6 × 2mm, not persistent. Forest; 350m.
Distr.: Mt Cameroon & Bakossi Mts. (Cameroon Endemic).
IUCN: CR
Mungo River F.R.: Cheek 10209 11/1999.

Cola ricinifolia Engl. & K.Krause

Tree or shrub c. 3–4m; leaves orbicular in outline, to 35cm, digitately 5-lobed by 1/2 to 3/4, copper-red below; petiole to 20cm; inflorescence axillary, sessile; flowers urceolate, brown, 1cm; fruit follicles elongate. Forest; 300m.
Distr.: SE Nigeria, SW Cameroon & Rio Muni. (Lower Guinea).
IUCN: NT
Bolo: Nemba 10 4/1986; 58 3/1986.
Note: the specimens cited are c. 7km outside our boundary. (Mundame).

Cola rostrata K.Schum.

Tree c. 15m, densely scurfy grey-brown pubescent, resembling *C. chlamydantha*, but leaflets papery, apex truncate-cordate; inflorescence axillary, amongst the leaves; fruits globose c. 10cm, brown, deeply verrucate. Forest; 250–300m.
Distr.: SE Nigeria, Cameroon & Gabon. (Lower Guinea).
IUCN: NT
Baduma: Nemba 192 8/1986; **Bakossi F.R.:** Cheek 9311 fr., 10/1998; **Karume:** Nemba 73 fr., 5/1986.

Cola cf. verticillata (Thonn.) Stapf

Tree 10–20m; leaves (3–)4-whorled, simple, elliptic, c. 18 × 8cm, acuminate, base obtuse to rounded, lateral nerves c. 9–10 pairs; petiole c. 4cm; inflorescences c. 3cm; flowers white with purple veins, 0.9cm; fruit smooth (rarely warty), follicles to 12 × 6cm; seeds c. 8; cotyledons 4, red. Forest; 650–1920m.
Kodmin: Cheek 9069 1/1998; Plot B17 1/1998; **Kupe Village:** Cable 2505 fl., 5/1996; Cheek 7089 fl., fr., 1/1995; Kenfack 250 fl., 7/1996; **Mt Kupe:** Sebsebe 5098 fl., 11/1995; **Ndum:** Cable 899 fr., 1/1995; **Ngusi:** Etuge 1583 fl., 1/1996; **Nyale:** Cheek 9680 11/1998; **Nyasoso:** Cable 2831 fr., 6/1996; 3400 fr., 6/1996; Cheek 9507 fr., 10/1998; Etuge 2181 fr., 6/1996; Letouzey 14712 fr., 4/1976; Sidwell 399 fr., 10/1995.
Note: resembles *C. anomala*. Cable 3400 and Etuge 2181 have warty young fruits.
Local name: Ekone; 'Slimy Cola' (Edmondo Njume in Cheek 9680); Abi (Epie Ngome in Cheek 9507). **Uses:** no use known, 'too sticky/slimy to eat' (Cheek 7089), ANIMAL FOOD – unspecified parts (Cheek 9507).

Cola sp. nov. 'etugei'

Shrub to 3m; leaves simple, loosely clustered in groups, elliptic or ovate-elliptic, c. 16 × 6cm, spatulate-acuminate, obtuse-decurrent, entire, glabrous; petioles heteromorphic, 1–4cm on same shoot, apex pulvinate; inflorescences in fascicles on stem between leaf groups; fruit follicles often paired, cylindrical, c. 6 × 1.5cm, rostrate, stipitate, glabrous, orange. Forest; 860–1100m.
Distr.: Bakossi Mts. (Narrow Endemic).
Nyasoso: Cheek 5668 fr., 12/1993; 9287 fr., 10/1998; 9518 10/1998; Etuge 1293 fr., 10/1995; Gosline 135 fr., 11/1998; Lane 245 fr., 11/1994; Sebsebe 4989 fr., 10/1995; Wheatley 480 fr., 7/1992.
Note: known only from Kupe-Bakossi.

Cola sp. nov. 'kodminensis'

Treelet 3–6m, puberulent; leaves simple, oblanceolate, c. 18 × 5.5cm, acumen 3cm, base acute, lower surface with crater-like galls; petioles dark brown, to 1cm; stipules filiform, long-tomentose, 1–5cm, not persistent; inflorescence axillary, amongst leaves; fruit carpels ellipsoid, 5 × 1.4cm, rostrate, stipulate, bright red. Forest; 1100m.
Distr.: Bakossi Mts & Korup. (Narrow Endemic).
Nyale: Cheek 9682 fr., 11/1998.

Leptonychia multiflora K.Schum.

Small tree or shrub to 8m, glabrescent; leaves ovate-elliptic, 12–20 × 4–8cm, acumen 2–3cm, cuneate, puberulent below, lateral nerves 5–7 pairs, domatia glabrous; petiole 1cm; stipules caducous; cymes axillary, subsessile, 3-flowered; pedicel 3mm, sepals 8–10 × 1–2mm, petals 1.5mm; capsule globose, 1–1.5cm, golden-green tomentellous. Forest; 1500m.
Distr.: Cameroon to Congo (K). (Lower Guinea & Congolian).
IUCN: LC
Nyasoso: Lane 230 fl., 11/1994.
Note: specimen not seen, descr. ex Fl. Congo, Rw. & Bur. 10: 237 (1963).

Leptonychia pallida K.Schum.

Tree 4–10m, glabrous; leaves elliptic-oblong to ovate-elliptic, c. 35 × 17cm, acuminate, rounded, drying pale green above; petiole 3cm; stipules 0.5cm; inflorescences of fascicles from woody burrs on main stem; pedicel c. 5mm; flowers white, c. 1cm; fruit green, dehiscent, 2cm; seeds fleshy, bicoloured, c. 1cm. Forest; 700–1230m.
Distr.: SE Nigeria, Bioko & Cameroon. (W Cameroon Uplands).
IUCN: LC
Kupe Village: Cable 2531 fl., fr., 5/1996; 3847 fl., fr., 7/1996; 739 fr., 1/1995; 784 fr., 1/1995; Cheek 7670 11/1995; Etuge 2751 fl., 7/1996; Groves 8 fr., 1/1995; Kenfack 319 fl., fr., 7/1996; Lane 420 fr., 1/1995; 432 fr., 1/1995; Ryan 350 fl., 5/1996; **Mwambong:** Cheek 9106 2/1998; **Ndum:** Cable 925 fr., 2/1995; Lane 465 fr., 2/1995; **Nyasoso:** Cable 3571 7/1996; Cheek 6019 fr., 1/1995; Etuge 1292 10/1995; 2428 fl., 6/1996; Lane 203 fl., fr., 11/1994; Whitlock 401 10/1998.

Leptonychia sp. 1 of Kupe-Bakossi checklist

Tree (2–)3.5–9m, puberulent; leaves obovate or narrowly elliptic, c. 21 × 7cm, acumen c. 3cm, base obtuse, entire, lateral nerves c. 8 pairs, white on upper surface, scalariform; petiole 1.5cm; inflorescence axillary, rachis 1cm; flowers pale green, 1.5cm; fruit globose c. 5cm. Forest; 1000–2000m.
Distr.: Bakossi Mts. (Narrow Endemic).

Bakossi F.R.: Thomas 5918 fr., 3/1986; **Kupe Village:** Cable 2563 fl., 5/1996; Cheek 7845 fl., 11/1995; **Nyasoso:** Cable 2937 fr., 6/1996; Etuge 1348 fl., 10/1995; 1666 fr., 1/1996; 2132 fr., 6/1996; 2184 fr., 6/1996; 2473 fr., 6/1996.
Note: matches none of the FWTA taxa. Genus needs revision.

Leptonychia sp. 2 of Kupe-Bakossi checklist

Shrub resembling Leptonychia sp. 1, but glabrous; leaves elliptic c. 16 × 6cm, acumen 2cm, lateral nerves c. 5 pairs, drying brown above, subreticulate; fruit 1.5cm. Forest; 300m.
Distr.: Bakossi Mts. (Narrow Endemic).
Konye: Thomas 5156 fl., fr., 11/1985.

Melochia melissifolia Benth. var. mollis K.Schum.

Shrub 1m, long-hairy; leaves lanceolate, 8 × 2.5cm, acute, truncate, serrate; petiole 1.5cm; inflorescences dense axillary fascicles; corolla 8mm, pink with white centre. Gardens; 500m.
Distr.: Liberia to Sudan. (Tropical Africa N of Congo basin).
IUCN: LC
Kupe Village: Cheek 7913a fl., fr., 11/1995.
Uses: MEDICINES (Cheek 7913a).

Octolobus spectabilis Welw.

Fl. Gabon 2: 107 (1961).
Syn.: Octolobus angustatus Hutch.
Shrub or small tree 2–4m; stems brown puberulent, glabrescent, becoming white; leaves elliptic or oblanceolate, 16 × 6cm, acumen 1cm, cuneate, lateral nerves 6–8 pairs; petioles 1–5cm or more; stipules 3mm, caduceus; flowers axillary, single, sessile; bracts ligulate, 1cm; perianth 2.5 × 2cm, lobes 5–6, 7mm, tube conical-funnel-shaped, softly pale brown pubescent; fruit 10cm diameter, including about 20 stipitate fleshy follicles. Forest; 500m.
Distr.: Sierra Leone to Angola. (Guineo-Congolian).
IUCN: LC
Nlog: Etuge 13 4/1986.

Pterygota cf. macrocarpa K.Schum.

Tree c. 30m; leaves (seedling: 3–5-lobed) ovate, papery c. 28 × 22cm, acuminate, deeply cordate, glabrous; petiole 30cm; fruit follicles woody (c. 2cm thick), globose, c. 20cm; seeds numerous, dry, winged, c. 9cm. Forest; 450m.
Bakossi F.R.: Cheek 9324 10/1998.
Note: identification tentative since based on seedling specimen only; in the absence of flowers or fruits, the specimen cited could be Cola lateritia.

Scaphopetalum sp. 1 of Kupe-Bakossi checklist

Shrub to 3m; densely grey tomentellous; leaves elliptic, to 15 × 7.5cm, shortly acuminate, base rounded to obtuse, lateral nerves c. 10 pairs, lower surface glabrous; petiole 0.7cm; stipule 4cm; flowers axillary, single; peduncle-pedicel 1.4cm; petals yellow or orange, 1.2cm; fruit ellipsoid, 2 × 1.5cm, ridged. Forest; 300m.
Distr.: Bakossi Mts. (Narrow Endemic).
Konye: Thomas 5155 fl., 11/1985; **Manyemen:** Letouzey 13921 fl., fr., 6/1975.
Note: matches none of the FWTA taxa. Scaphopetalum needs revision.

Sterculia oblonga Mast.

Syn.: Eribroma oblonga (Mast.) Pierre ex A.Chev.
Deciduous tree 20–40m; buttresses c. 60 × 60cm; bole 60cm d.b.h., white, with rectangular plates; slash pink, inner pale yellow; leaves (flushing) oblong or obovate-oblong, c. 6 ×

2cm, subacuminate, acute, lateral nerves c. 14 pairs, midrib and petiole densely hairy, glabrescent. Semi-deciduous forest; 100–500m.
Distr.: Ivory Coast to Gabon. (Upper & Lower Guinea).
IUCN: LC
Etam: Thomas 7161 7/1987; **Kupe Village:** Etuge 1474 11/1995; **Mt Kupe:** Cheek 10126 11/1999.
Uses: MATERIALS – a timber species (*fide* Cheek).

Sterculia rhinopetala K.Schum.
Tree to 40m, buttressed, glabrous; leaves drying dark brown above, pale brown below, elliptic-oblong, to 20 × 8cm, subacuminate, rounded, lateral nerves 12 pairs; petiole 6cm; panicles to 18 × 4cm, puberulent; perianth 3–7mm, puberulent; follicles oblong, 6 × 3cm. Forest; 260m.
Distr.: Ivory Coast, Ghana, Nigeria & Cameroon. (Upper & Lower Guinea).
IUCN: LC
Baduma: Nemba 630 9/1987.

Sterculia tragacantha Lindl.
Tree 10–25m; leaves simple, oblong, oblong-obovate or elliptic, c. 19 × 11cm, rounded to subacuminate, base obtuse, stellate-velvety below; petiole c. 5cm; stipules caducous; panicles axillary, slender, 15cm; flowers pink, 0.7cm, perianth lobes adhering at apex; follicles five c. 8.5 × 2.5cm, golden-brown-hairy. Secondary forest; 250–990m.
Distr.: Mali to Mozambique. (Tropical Africa).
IUCN: LC
Etam: Etuge 247 fr., 8/1986; **Loum F.R.:** Onana 932 fr., 11/1999; **Mungo River F.R.:** Thomas 5463 fl., 1/1986; **Ndum:** Elad 107 2/1995; **Ngomboaku:** Ghogue 471 fr., 12/1999; **Nlonako:** Etuge 1787 fl., 3/1996; **Nyale:** Etuge 4202 fl., 2/1998; **Nyasoso:** Cheek 7531 10/1995; Thomas 3069 fl., 2/1984.

Theobroma cacao L.
Purseglove, tropical Crops, Dicots.: 570 (1966).
Tree 5–8m, puberulent; leaves alternate, oblong-elliptic, c. 30 × 10cm, acute; petiole swollen at base and apex; stipules caducous; inflorescence of single to few-flowered fascicles on trunk and main branches; pedicels 1cm; flowers pink and white, 7mm; fruit ovoid-ellipsoid, 15–20cm, smooth or ridged, yellow or red, indehiscent, containing many seeds in white pulp. Cultivated, persisting for many years in seemingly natural forest when abandoned; 200–600m.
Distr.: pantropical, wild in Brazil.
IUCN: LC
Loum F.R.: Pollard (sight record) 10/1998; Nyandong: Darbyshire (sight record) 3/2003.
Note: probably the main smallholder cash crop in our checklist area.
Local name: cocoa or cacao. **Uses:** FOOD – seeds – chocolate (*fide* Pollard).

Triplochiton scleroxylon K.Schum.
Tree to 60m; bole whitish brown with buttresses concave, projecting 2m; leaves orbicular in outline, c. 12 × 15cm, 5-lobed by c. 1/3 the radius, lobes subtriangular, blade base truncate; petiole c. 5cm. Semi-deciduous forest; 420m.
Distr.: Guinea (C) to Congo (K). (Guineo-Congolian).
IUCN: LC
Loum F.R.: Cheek 9301 10/1998.
Uses: MATERIALS – a timber species (*fide* Cheek).

THEACEAE
M. Cheek

Camellia sinensis (L.) Kuntze
Purseglove, tropical Crops, Dicots: 599 (1968).
Tree c. 6m, glabrous; leaves alternate, elliptic, coriaceous, c. 7 × 4cm, acute, obtuse; petiole c. 5mm; flowers axillary, single; petals white, obovate, free, 2cm; anthers numerous; fruit 3–4cm diameter. Cultivated.
Distr.: cultivated in the palaeotropics, wild in China.
IUCN: LC
Essosong: Cheek (sight record) 11/1995.
Note: occurs within the forest at the decades-long abandoned Essosong plantation. Elsewhere in Cameroon (Tole and Ndu), tea is still cultivated.
Local name: tea. **Uses:** FOOD – leaves – mixed with *hot water and drunk; the 'tea' of commerce* (*fide* Cheek).

THYMELAEACEAE
M. Cheek

Fl. Cameroun 5 (1966).

Craterosiphon scandens Engl. & Gilg
Liana to 15m; principal branches to 40cm diameter, lenticellate; leaves (sub)opposite, ovate, 5–6.5 × 2cm, acuminate, base obtuse to rounded, margin revolute, lateral nerves numerous, parallel; petiole 3–5mm; flowers in axillary clusters of 1–3(–6), subsessile; calyx tube 1.3cm, narrow, lobes oblong, 5mm, green-yellow; fruit ellipsoid, c. 2.2 × 1.4cm. Forest; 1080m.
Distr.: Guinea (C) to Ivory Coast, Nigeria, Cameroon & Uganda. (Guineo-Congolian).
IUCN: NT
Banyo: Leeuwenberg 10129 7/1972.
Note: widespread but rare.

Dicranolepis buchholzii Engl. & Gilg
Treelet to 1m; leaves alternate, obliquely elliptic, c.10.5 × 5cm, including linear acumen c.2cm, base obtuse; flowers sessile, axillary, 1–3 per node; calyx tube c.1cm, glabrous, petals white, 5, c. 5mm; fruit orange, elliptic 0.5–1cm, rostrum slender, elongate. Forest; 300m.
Distr.: Cameroon to Uganda. (Lower Guinea & Congolian).
IUCN: LC
Mekom: Thomas 5242 1/1986.
Note: distinguished from *Dicranolepis disticha* by the broader leaves. Rare in the Kupe-Bakossi area.

Dicranolepis disticha Planch.
Shrub, 1–3.5m; resembling *D. buchholzii*, but leaves less than 3cm wide, drying green not brown below, secondary nerves raised, conspicuous on lower surface. Forest; 700–1250m.
Distr.: Guinea (C) to Congo (K). (Guineo-Congolian).
IUCN: LC
Kupe Village: Cable 2712 fr., 5/1996; 3698 7/1996; Cheek 6071 fr., 1/1995; Etuge 1391 11/1995; 2920 7/1996; Lane 264 11/1994; Ryan 283 fr., 5/1996.

Dicranolepis glandulosa H.H.W.Pearson
Shrub to 1.5m. resembling *D. vestita*, but leaves c.14 × 5cm; calyx tube with large capitate glands. Inundated forest; 300–760m.

Distr.: Cameroon endemic.
IUCN: NT
Konye: <u>Thomas 5165</u> 11/1985; **Ngomboaku:** <u>Mackinder 287</u> fl., 12/1999.
Note: known only from the area between Grand Batanga (Kribi) and Mt Cameroon.

Dicranolepis grandiflora Engl.
Shrub, 3–10m; resembling *D. vestita* (and possibly not distinct from it); differing in the stamens long-exserted, not inserted at the throat; calyx tube 30–40mm, petals 15–22mm. Forest; 900m.
Distr.: Ghana to Cameroon. (Upper & Lower Guinea).
IUCN: LC
Kupe Village: <u>Cheek 7144</u> fl., 1/1995; <u>Elad 71</u> fl., 1/1995; **Mbule:** <u>Leeuwenberg 9283</u> fl., 1/1972.
Note: our material, cited here, mostly falls between *D. grandiflora* and *D. vestita* in the FWTA key, but is closer to the first.

Dicranolepis polygaloides Gilg ex H.H.W.Pearson
Shrub, 0.5m; leaves 3 × 1.5cm; calyx tube 2.5cm, sepals 0.7cm, both densely white puberulent. Forest; 870m.
Distr.: Cameroon endemic.
IUCN: VU
Mombo: <u>Olorunfemi in FHI 30562</u> 5/1951; **Nyasoso:** <u>Cable 3210</u> 6/1996.
Note: previously known only from 6 specimens cited in Fl. Cameroun.

Dicranolepis vestita Engl.
Shrub or tree (2–)3–8m; leaves obliquely oblong-elliptic, c. 8 × 2.5cm, acumen 1.5cm; flowers erect, sessile, 1–several in old leaf axils; flowers 2.5cm wide, white, fragrant; calyx tube c.2cm, densely white appressed-hairy, dilated at base; petals 10, entire; anthers sessile; stigma capitate; fruit pendulous, 2.5 × 1.5cm including a 1cm robust rostrum, densely white-hairy. Forest; 650–1800m.
Distr.: SE Nigeria, Bioko & Cameroon. (W Cameroon Uplands).
IUCN: NT
Bangem: <u>Thomas 5920</u> 3/1986; **Ekona Mombo:** <u>Etuge 407</u> fl., 12/1986; **Kodmin:** <u>Cheek 9607</u> fl., 11/1998; <u>Satabie 1116</u> fl., 1/1998; **Kupe Village:** <u>Cable 2497</u> 5/1996; <u>Cheek 7054</u> fl., fr., 1/1995; <u>7881</u> fl., 11/1995; <u>Lane 389</u> fr., 1/1995; **Mt Kupe:** <u>Thomas 5107</u> fl., 12/1985; **Nyasoso:** <u>Cable 689</u> fl., 1/1995; <u>Etuge 1590</u> 1/1996; <u>2068</u> fr., 6/1996; <u>Lane 100</u> fr., 6/1994; <u>Sunderland 1510</u> fr., 7/1992; <u>Thomas 3089</u> fl., 2/1984.
Note: variable, integrading with *D. grandiflora*.
Local name: Sinsimahanteh (Cheek 9607).

Octolepis casearia Oliv.
Shrub or tree, 1.5–3m; bark white, sloughing, glabrous; leaves alternate, to 25 × 8cm, acuminate, acute; petiole 1cm; inflorescence axillary, 1–4-flowered fascicles; pedicels 0.5cm; flowers white, tepals 4, 5mm; fruit ovoid, 1.5cm, 4-valved. Forest; 600–1000m.
Distr.: SE Nigeria to Gabon. (Lower Guinea).
IUCN: NT
Bakossi F.R.: <u>Olorunfemi in FHI 30738</u> fl., 4/1951; **Bangem:** <u>Thomas 5906</u> fr., 3/1986; **Etam:** <u>Etuge 226</u> 8/1986; **Kupe Village:** <u>Mackinder 278</u> fl., 12/1999; **Ngomboaku:** <u>Cheek 10336</u> 12/1999; **Ngombombeng:** <u>Etuge 37</u> 4/1986; **Nyale:** <u>Cheek 4190</u> 2/1998; <u>Etuge 4190</u> 2/1998; **Nyandong:** <u>Ghogue 1533</u> fr., 3/2003; <u>Stone 2461</u> 3/2003;

Southern Bakossi F.R.: <u>Olorunfemi in FHI 30597</u> fr., 5/1951.

Peddiea africana Harv.
Bull. Jard. Bot. Nat. Belg. 63: 206 (1994).
Syn.: *Peddiea parviflora* Hook.f.
Syn.: *Peddiea fischeri* Engl.
Shrub or tree 3–7m; leaves alternate, elliptic, c. 15 × 5cm, base and apex acute; petiole 0.3cm; inflorescence terminal; peduncle 0.8cm, umbellate, c. 10-flowered; pedicel 0.5cm; perianth tubular, pale green, 1.8cm; lobes 4, 0.2cm; fruit ovoid, 1 × 0.7cm, bilobed, apex white-hairy. Forest; 800–2000m.
Distr.: Guinea (C) to Zambia. (Afromontane).
IUCN: LC
Bangem: <u>Thomas 5310</u> 1/1986; **Kodmin:** <u>Etuge 4150</u> 2/1998; **Kupe Village:** <u>Cable 3771</u> 7/1996; <u>Ryan 227</u> fr., 5/1996; <u>371</u> fr., 5/1996; <u>Sidwell 442</u> fr., 11/1995; **Nyasoso:** <u>Cable 2845</u> fr., 6/1996; <u>3460</u> fr., 7/1996; <u>Etuge 1661</u> fr., 1/1996; <u>2178</u> fr., 6/1996.

TILIACEAE

M. Cheek

Ancistrocarpus densispinosus Oliv.
Climber, puberulent; leaves elliptic, c. 15 × 7cm, acuminate, rounded, finely serrate, nerves thinly puberulent below; stipule entire, 4mm; inflorescence leaf-opposed, c. 10-flowered; flowers yellow, 2.5cm diameter; stamens in bundles; fruit globose, 4.5cm including dense, glabrous, hooked spines c. 1cm. Forest; 250–950m.
Distr.: Nigeria to Congo (K). (Lower Guinea & Congolian).
IUCN: LC
Enyandong: <u>Tadjouteu 445</u> fr., 10/2001; **Mt Kupe:** <u>Cheek 10128</u> 11/1999; **Ngomboaku:** <u>Gosline 268</u> fl., 12/1999; **Nyasoso:** <u>Etuge 2418</u> fr., 6/1996; <u>Sebsebe 5012</u> 10/1995.

Clappertonia ficifolia (Willd.) Decne.
Shrub 2m, tomentose; leaves ovate, 3–5-lobed, to 13 × 7cm, serrate, tomentose below; stipules lanceolate, 1cm; inflorescence terminal, few-flowered; flowers blue, 2–3cm; fruit cylindric-oblong, 4–6cm, spiny. Riverside; 1400m.
Distr.: tropical Africa.
IUCN: LC
Kodmin: <u>Gosline 80</u> 1/1998.

Corchorus olitorius L.
Herb 1m, glabrous; leaves ovate-elliptic, c. 6 × 2.5cm, acute, rounded, serrate, the basal pair with filaments c. 7mm; inflorescence leaf-opposed, subsessile; flowers yellow, 0.5cm diameter; fruit cylindrical 6 × 0.5cm, indehiscent. Weed in villages; 700m.
Distr.: tropical Africa.
IUCN: LC
Enyandong: <u>Cheek 10925</u> 10/2001; **Nyale:** <u>Cheek 9641</u> fl., 11/1998.
Local name: Keneh keneh (Cheek 9641); Nkene nkene (Cheek 10925). **Uses:** FOOD – leaves used as a vegetable (Cheek 9641 & 10925).

Desplatsia chrysochlamys (Mildbr. & Burret) Mildbr. & Burret
Tree 7m, densely and thickly golden-brown tomentose; leaves obovate-oblong c. 27 × 12cm, acuminate, rounded to cordate, shallowly dentate-denticulate, thinly pubescent

below; stipules linear, bifid, hairy, 15mm; inflorescence axillary, subsessile, 2.5cm; flowers white, 1.3cm diameter; fruit c. 10 × 8–9cm. Forest; 800–900m.
Distr.: Sierra Leone to Uganda. (Guineo-Congolian).
IUCN: LC
Kupe Village: <u>Kenfack 292</u> fl., 7/1996; **Nyasoso:** <u>Cheek 6004</u> 1/1995.

Desplatsia dewevrei (De Wild. & T.Durand) Burret

Tree 10m, trunk 30cm d.b.h.; stems minutely puberulent; leaves oblong-elliptic, to c. 24 × 10cm, long acuminate, base unequally truncate, deeply serrate, drying brown below, glabrous; stipules triangular, 2mm, puberulent; fruit oblong-ellipsoid, to 12 × 12cm. Forest; 200m.
Distr.: Ivory Coast to Congo (K). (Guineo-Congolian).
IUCN: LC
Mungo River: <u>Leeuwenberg 6844</u> fr., 10/1965.

Desplatsia subericarpa Bocq.

Shrub 3–6m, slender, dark brown pubescent, resembling *D. chrysochlamys* but leaves c. 17 × 6cm, obscurely serrate, glabrous below; stipule 5mm, digitately divided; flowers pink, 1cm; fruit c. 5 × 5cm. Forest; 500–1260m.
Distr.: Sierra Leone to Congo (K). (Guineo-Congolian).
IUCN: LC
Kupe Village: <u>Cheek 7824</u> fl., fr., 11/1995; <u>Etuge 1478</u> fr., 11/1995; <u>Lane 357</u> fl., fr., 1/1995; **Nyasoso:** <u>Etuge 2537</u> fr., 7/1996.

Duboscia macrocarpa Bocq.

Syn.: *Duboscia viridiflora* (K.Schum.) Mildbr.
Tree 30m, 40–65cm d.b.h.; bole crooked, slight buttresses; slash fibrous, oxidizing in 4 minutes; stems softly stellate brown pubescent; leaves obovate-oblong, to 18 × 7cm, long acuminate, rounded or cordate, serrate-dentate, greyish white below, densely, softly tomentose; inflorescence leaf-opposed, subumbellate, many-flowered; flowers greenish white, 1cm diameter; fruit obovoid, 3.5 × 2.5cm, woody, 6–8 ribbed. Forest; 500–900m.
Distr.: Ivory Coast to Congo (K). (Guineo-Congolian).
IUCN: LC
Kupe Village: <u>Etuge 2718</u> fr., 7/1996; **Mombo:** <u>Olorunfemi in FHI 30569</u> fr., 5/1951; **Wone:** <u>Nemba 594</u> 7/1987.

Glyphaea brevis (Spreng.) Monach.

Shrub 7m; stems brown with scattered stellate hairs; leaves obovate or oblong, c. 24 × 9cm, long acuminate, obtuse to rounded, serrate finely hairy below; petiole 3.5cm, bipulvinate; inflorescence leaf-opposed, few-flowered; flowers 3cm, yellow; fruit subcylindrical, c. 7 × 1cm, ridged, rostrate, indehiscent, green. Farmbush; 150–1200m.
Distr.: tropical Africa.
IUCN: LC
Bangem: <u>Thomas 5380</u> 1/1986; **Etam:** <u>Etuge 203</u> fr., 7/1986; **Manehas F.R.:** <u>Whitlock 404</u> 10/1998; **Mejelet-Ehumseh:** <u>Etuge 300</u> fr., 10/1986; **Mwanenguba:** <u>Leeuwenberg 8177</u> fl., fr., 8/1971; **Mungo River F.R.:** <u>Etuge 4314</u> fr., 10/1998; **Nyasoso:** <u>Etuge 16</u> fl., fr., 4/1986; <u>2596</u> 7/1996; <u>Groves 104</u> fl., 2/1995.

Microcos barombiensis (K.Schum.) Cheek *comb. nov.*

Basionym: *Grewia barombiensis* K. Schum. Bot. Jahrb. 15: 124 (1892).
Climber; stems brown puberulent; leaves ovate-elliptic, c. 17 × 9cm, acuminate, rounded or cordate, entire, subglossy above, glabrous below apart from nerves; stipules c. 6mm,

bifurcate; inflorescence terminal, 10 × 7cm; main bracts digitately divided to base; flowers in umbels of 2–3, subtended by 5–6 epicalycular bracteoles; flowers white, 13mm diameter; fruit spindle-shaped, 3 × 1.5cm, glossy red, with white thinly scattered hairs. Forest; 200–1000m.
Distr.: Ivory Coast to Angola. (Upper & Lower Guinea).
IUCN: NT
Baduma: <u>Nemba 572</u> 7/1987; **Konye:** <u>Thomas 5160</u> 11/1985; **Kupe Village:** <u>Etuge 2042</u> fl., 5/1996; <u>2737</u> fl., 7/1996; **Loum F.R.:** <u>Tadjouteu 318</u> fl., 12/1999; **Mile 15, Kumba-Mamfe Road:** <u>Nemba 128</u> fl., 7/1986; **Nyasoso:** <u>Cable 3276</u> fl., 6/1996; <u>3530</u> fl., 7/1996; <u>Etuge 1517</u> fr., 12/1995; <u>2487</u> fl., 7/1996; <u>2595</u> fl., fr., 7/1996; <u>Zapfack 670</u> fl., 6/1996.
Note: *Microcos* is not always maintained as distinct from *Grewia* in Africa despite Burret's excellent revisionary work of 1926 (Notizbl. Bot. Gart. Berlin 9).

Microcos coriacea (Mast.) Burret

Notizbl. Bot. Gart. Berlin 9: 759 (1926).
Syn.: *Grewia coriacea* Mast.
Evergreen tree c. 20m; leaves ovate-elliptic, 30 × 9–14cm, acuminate, obtuse to rounded, entire, glabrous below; petiole 1cm; inflorescence a terminal panicle; fruit obovoid, glossy, 2.5 × 2cm, red, fleshy, 1-seeded. Evergreen forest; 200–420m.
Distr.: S Nigeria to Angola. (Lower Guinea & Congolian).
IUCN: LC
Baduma: <u>Nemba 574</u> 7/1987; **Bolo:** <u>Nemba 97</u> fr., 5/1986; **Kupe Village:** <u>Etuge 2047</u> fr., 5/1996; **Mungo River F.R.:** <u>Onana 950</u> fr., 11/1999.

Microcos malacocarpa (Mast.) Burret

Notizbl. Bot. Gart. Berlin 9: 760 (1926).
Syn.: *Grewia malacocarpa* Mast.
Climber; stem obscurely puberulent; leaves elliptic-oblong 7.2 × 2.4cm, shortly acuminate, obtuse, grey-white below with short, dense stellate hairs; inflorescence terminal; flowers numerous; fruit obovoid, 2.4 × 1.2cm, densely, shortly yellow-stellate hairy. Forest; 300–850m.
Distr.: Sierra Leone to Cameroon. (Upper & Lower Guinea).
Konye: <u>Thomas 5184</u> fl., 11/1985; **Ngomboaku:** <u>Cheek 10338</u> 12/1999.

Microcos sp. A

Evergreen tree c. 20m; similar to *M. coriacea*, but stem, lower leaf surface and fruit shortly pale brown stellate-hairy; fruit warty, pink-brown when mature. Evergreen forest; 900–1000m.
Distr.: Mt Kupe. (Narrow Endemic).
Kupe Village: <u>Etuge 2686</u> fr., 7/1996; **Nyasoso:** <u>Cable 2806</u> fr., 6/1996; <u>Elad 118</u> 2/1995.
Note: a very distinct new species apparently endemic to Mt Kupe. Flowers required.

Triumfetta annua L.

F.T.E.A. Tiliaceae: 79 (2001).
Herb 0.5m, glabrous apart from a line of hairs on stem and petiole; basal leaves ovate c. 7 × 4cm, acuminate, obtuse, serrate; inflorescences at uppermost nodes, c. 1cm long; flowers yellow, 3mm; fruit indehiscent, glabrous, globose, 8mm diameter, including hooked bristles 3mm. Grassland; 2000m.
Distr.: Cameroon to China. (Palaeotropics).
IUCN: LC
Mwanenguba: <u>Cheek 9461</u> fr., 10/1998.

Triumfetta cordifolia A.Rich. var. *cordifolia*

Shrubby herb 1.5–2.5m, puberulent; basal leaves drying black, ovate, c. 13 × 10cm, long-acuminate, base cordate to obtuse, coarsely dentate, glabrescent below; inflorescence as *T. rhomboidea*; flowers 8mm; fruit drying black, globose, 10mm including hooked bristles to 4mm, very sparsely to densely white-hairy. Forest edge; 600–1100m.
Distr.: Cape Verde to Congo (K). (Guineo-Congolian & Cape Verde).
IUCN: LC
Enyandong: Ghogue 1223 10/2001; **Kodmin:** Biye 5 fr., 1/1998; **Kupe Village:** Cable 2542 fl., 5/1996; 3888 fl., 7/1996; 714 fr., 1/1995; **Manehas F.R.:** Whitlock 405 10/1998; **Mwambong:** Cheek 9361 10/1998; **Ndum:** Cable 904 fl., fr., 1/1995; **Ngomboaku:** Etuge 4663 fl., 12/1999; Ghogue 448 fl., 12/1999; **Nyasoso:** Cheek 5645 fl., 12/1993; 7308 fr., 2/1995; Etuge 1635 fr., 1/1996; 2465 fl., 6/1996.
Local name: Nkon (Epie Bakossi in Cheek 7308). **Uses:** ANIMAL FOOD – antelopes eat the leaves, stems contain mucilage (Cheek 7308); MATERIALS – cleansers – leaves used to clean dishes and pots (Etuge 1635).

Triumfetta cordifolia A.Rich. var. *tomentosa*
Sprague
F.T.E.A. Tiliaceae: 87 (2001).
Shrubby herb to 3m, resembling *T. cordifolia* var. *cordifolia* but leaves slightly 3-lobed, lower surface completely obscured by dense but short, brownish white hairs. Roadside.
Distr.: Nigeria to Zimbabwe. (Tropical Africa).
IUCN: LC
Ekona Mombo: Etuge 422 fl., 12/1986.

Triumfetta rhomboidea Jacq.

Herb 1–1.5m, puberulent; basal leaves usually 3-lobed to c. ¼ the radius, c. 10 × 8cm, serrate, the lower 2–3 teeth pairs glandular, black; inflorescence raceme-like, sparingly branched; flowers yellow, 4mm long; fruit indehiscent, spherical, 5mm diameter, densely coated in white hairs from which emerge glabrous bristles, 1.5mm with a terminal hook. Farmbush, secondary scrub, roadsides; 750–2000m.
Distr.: tropical Africa, introduced elsewhere.
IUCN: LC
Bazeng: Sebsebe 5062 fl., 10/1995; **Essossong:** Whitlock 402 10/1998; **Kodmin:** Cheek 8880 fl., fr., 1/1998; Etuge 4046 fl., 1/1998; **Mwanenguba:** Cheek 7287 fl., fr., 2/1995; Cheek 9446 fl., 10/1998; Leeuwenberg 8576 fl., fr., 10/1971; **Nyasoso:** Cheek 5652 fl., 12/1993; Etuge 1796 fl., fr., 3/1996; 2499 fl., fr., 7/1996.
Uses: MATERIALS – abrasives – leaves used to clean dishes and pots (Etuge 1796).

ULMACEAE

I. Darbyshire

Fl. Cameroun 8 (1968).

Celtis adolfi-friderici Engl.

Tree to 30m; mature leaves asymmetric oblong-elliptic, 10.5–17 × 6.5–7cm, acumen obtuse, 1.2cm, base acute, margin entire, paired basal lateral nerves extend almost to the apex, subsequent 2 lateral nerve pairs in upper half of leaf, alternate, tertiary nerves reticulate, not prominent, lower surface slightly scabridulous, upper surface glossy, glabrous; petiole stout, 1.5cm; fruit globose, 1.7cm diameter, smooth to

foveolate, style bases persistent. Semi-deciduous forest; 250–300m.
Distr.: Ivory Coast to Sudan & Congo (K). (Guineo-Congolian).
IUCN: LC
Bolo-Meboka: Thomas 7157 7/1987; **Loum F.R.:** Onana 998 12/1999.

Celtis gomphophylla Baker
F.T.E.A. Ulmaceae: 6 (1966).
Syn.: *Celtis durandii* Engl.
Small tree to 7m; stems glabrous; leaves ovate-oblong, 9.5–19 × 3.5–8cm, acumen 1.8cm, acute, base obtuse, ± asymmetrical, margin serrate in upper third (imm.) to entire (mature), lateral nerves 5(–6) pairs, alternate after basal (or rarely second) pair, basal pair not extending into upper half of leaf, veins yellow and prominent below, surfaces scabridulous when imm., soon glabrous; fruit in axillary fascicles of 2–3; pedicels 0.4–1cm, globose or ovoid, 0.6–0.7cm diameter; calyx persistent, lobes triangular, 1mm, paired styles persistent, c. 1mm, simple. Wet (and semi-deciduous) forest; 800–1100m.
Distr.: Ivory Coast to Congo (K), Zimbabwe & Madagascar. (Tropical & subtropical Africa & Madagascar).
IUCN: LC
Nyasoso: Cable 2874 6/1996; Etuge 2486 7/1996.

Celtis philippensis Blanco
Fl. Cameroun 8: 26 (1968).
Syn.: *Celtis brownii* Rendle
Tree to 15m; mature leaves subsymmetrical, oblong-elliptic, 12.5–15 × 6–7cm, acumen 1.2cm, obtuse-rounded, base rounded or obtuse, margin coarsely toothed in upper half, basal paired lateral nerves extend almost to apex, subsequent 2–3 lateral nerve pairs in uppermost third of leaf, alternate, tertiary nerves reticulate, glabrous; petiole c. 1cm; fruit ellipsoid, c. 1cm long, smooth, style bases persistent. Semi-deciduous forest; 300m.
Distr.: tropical Africa, Asia & Australia. (Palaeotropics).
IUCN: LC
Loum F.R.: Onana 999 12/1999.

Celtis zenkeri Engl.
Tree to 15m; crown conical, compact; immature stems puberulent; mature leaves subsymmetrical, elliptic, 10.5–12 × 4.2–5cm, acumen to 1.7cm, acute, base acute, margin entire, lateral nerves 4–5 pairs, alternate beyond basal pair, yellowish below, tertiary nerves markedly parallel, lower surface softly puberulent, upper surface glabrous; petiole c. 6mm, puberulent; fruits axillary, paired or single, globose, 6mm, rugulose, paired styles ± persistent, divaricating; peduncle c. 1cm; pedicels 3mm. Semi-deciduous forest; 250–600m.
Distr.: Guinea (C) to Angola & Tanzania. (Tropical Africa).
IUCN: LC
Bolo-Meboka: Thomas 7156 7/1987; **Etam:** Etuge 214 8/1986.

Trema orientalis (L.) Blume
Syn.: *Trema guineensis* (Schum. & Thonn.) Ficalho
Tree to 8m; young stems densely pubescent; leaves variable, distichous, ovate-lanceolate, 6.5–13.5 × 2.8–5.3cm, apex acuminate, base truncate, margin serrulate, lateral nerves 4(–6) pairs, alternate above basal pair, upper surface scabrid, lower surface scabrid or sparsely pubescent to densely pubescent; cymes axillary, c. 10–20-flowered; peduncle 0–0.5cm; flowers white, c. 2mm, sepals broadly elliptic, obtuse,

puberulent; fruit globose, 2–3mm diameter, green, styles and sepals persistent. Forest, farmbush; 150–1980m.

Distr.: widespread in tropical Africa & Asia. (Palaeotropics).

IUCN: LC

Etam: Etuge 188 7/1986; **Kodmin:** Cheek 8973 1/1998; Ghogue 16 fr., 1/1998; **Kupe Village:** Etuge 1463 11/1995; Ryan 405 5/1996; **Mt Kupe:** Cheek 7590 11/1995; **Mwanenguba:** Cheek 7264 2/1995; Etuge 4370 10/1998; **Nyandong:** Ghogue 1531 fr., 3/2003; **Nyasoso:** Cable 2902 6/1996; Etuge 1639 1/1996.

UMBELLIFERAE

I. Darbyshire

Fl. Cameroun 10 (1970).

Agrocharis melanantha Hochst.

Fl. Cameroun 10: 52 (1970).

Syn.: *Caucalis melanantha* (Hochst.) Hiern

Suberect perennial herb to 60cm; rootstock woody; robust stems rounded, finely ridged; petioles sheathing, membranous at node, densely pubescent; leaves bi-pinnate on rachis 7–15cm, c. 6 pairs of pinnae, each divided into c. 3 pinnules, serrate to approx. half their width, blades finely pubescent; umbel dense, globular, terminal c. 1.5–2cm diameter; peduncle to 20cm, pubescent; involucral bracts numerous, lanceolate, ciliate, 3–4mm; c. 12 subsessile flowers per umbel, each with a subtending involucel of lanceolate bracts; corolla white; fruits green, ellipsoid, c. 5 × 3mm, ridged, with reflexed barbed bristles along ridges, ciliate hairs between ridges. Montane grassland; 1450–2100m.

Distr.: Cameroon, Bioko, Congo (K), E Africa. (Afromontane).

IUCN: LC

Kodmin: Cheek 9041 1/1998; **Mwanenguba:** Cheek 9470 10/1998; Leeuwenberg 8791 fr., 11/1971; 8927 fl., fr., 12/1971; Sanford 5408 fl., 11/1968; Thomas 3132 fl., fr., 2/1984.

Note: recognised by some authors (e.g. Lee: Israel J. Pl. Sc. 50(3): 211 (2002)) as a synonym of *A. gracilis* Hook.f.

Centella asiatica (L.) Urb.

Creeping perennial herb; long glabrous internodal stolons to 10cm, nodal rooting; petioles 5–20cm, pubescent particularly when young, sheathing, with subtending leafy lanceolate stipule c. 5mm long; blade reniform with regular crenate margin, non-lobed, glabrous, c. 2–3cm diameter; umbel 3–5mm; peduncle 1–1.5cm, pubescent; 1–5 umbels per node, subtended by leafy bract; 3–4 subsessile flowers per umbel, petals pink-purple, subtended by 2 pubescent bracts 1–2mm; fruits ellipsoid, truncate at apex, 2(–4) × 1(–3)mm, with reticulate sculpturing. Damp grassland; 800–2000m.

Distr.: pantropical. (Montane).

IUCN: LC

Kupe Village: Etuge 1845 3/1996; **Mwanenguba:** Cheek 7244 2/1995; Cheek 9453 10/1998; **Nyasoso:** Thomas 5024 fr., 11/1985.

Uses: FOOD – leaves edible (Etuge 1845).

Cryptotaenia africana (Hook.f.) Drude

Upright rhizomatous herb 0.3–1m tall; stems circular, ridged, glabrous, max. 2mm diameter; leaves concentrated towards base, sheathing; petioles 6–10cm glabrous; compound, bi-ternate or with ternate terminal pinna and lobed lateral pinnae; petiolules to 2.5cm; leaflets c. 3.5 × 2cm, ovate, irregularly dentate, pilose especially on adaxial surface; cauline leaflets sublanceolate, 3 × 0.5cm; umbels sparse, florets borne on fine peduncles / pedicels; primary peduncles to 2cm; pedicels to 1cm, involucre absent; corolla white; fruits 2 × 1mm, ellipsoid to ovoid, glabrous, green, with persistent reflexed styles. Forest; 1450m.

Distr.: Cameroon, E Africa. (Afromontane).

IUCN: LC

Kodmin: Cheek 9583 11/1998.

Eryngium foetidum L.

Rankly smelling upright perennial herb to 60cm; rootstock fleshy; stems robust, green, glabrous, furrowed; basal rosette of oblanceolate leaves c. 17 × 3cm, finely dentate, spiny, glabrous, on membranous, sheathing petioles; cauline leaves sessile, ± lobed, spiny; 2-bifurcating cymes above cauline leaves, each inflorescence a dense, rounded, green spike c. 0.5–1 × 0.5cm, subtended by involucre of 4–6 spiny lanceolate bracts c. 2–3 × 0.3–0.5cm, with spiny teeth. Streambanks in forest; 650m.

Distr.: originating in the Americas, introduced in W Africa.

IUCN: LC

Mwanenguba: Sanford 5434 fl., 11/1968; **Ngusi:** Etuge 1568 1/1996.

USES: MEDICINES (Etuge 1568).

Hydrocotyle hirta R.Br. ex A.Rich.

Fl. Cameroun 10: 34 (1970).

Syn.: *Hydrocotyle mannii* Hook.f.

Creeping herb; stems ± glabrous with nodal rooting; petioles 8–12cm, pubescent particularly towards leaf bases; leaves reniform to suborbicular, c. 2.7 × 3cm, 7-lobed indented to 2mm, with further sublobing, distinctive pilose hairs to 1mm on both surfaces; umbel axillary, c. 10 subsessile flowers on peduncle c. 1cm, pubescent; umbel axillary, c.10 subsessile flowers on peduncle c. 1cm, pubescent. Open, damp montane habitats; 1000–1500m.

Distr.: tropical Africa, Madagascar, Mascarenes, Australia. (Palaeotropics).

IUCN: LC

Kodmin: Cheek 9013 1/1998; **Mwanenguba:** Leeuwenberg 8348 fl., fr., 9/1971; 8420 fl., fr., 9/1971; Sanford 5426 fl., 11/1968.

Hydrocotyle sibthorpioides Lam.

Creeping, mat-forming herb; resembling *H. hirta*, but leaves 5–7mm diameter, glabrous; petioles c. 1cm; peduncle glabrous, c. 1–5mm; flowers c. 5. Damp montane habitats including grassland; 1975m.

Distr.: tropical and southern Africa, Madagascar, Mascarenes.

IUCN: LC

Mwanenguba: Cheek 7280 2/1995.

Peucedanum kupense I.Darbysh. & Cheek

Kew Bull. 59: 133 (2004).

Erect annual (?) herb to 2m; stems robust, circular, hollow, 1cm diameter at base, shallowly ridged, glabrous; cauline leaves bipinnately compound, biternate or with lateral pinnae of two leaflets; petioles sheathing, membranous for c. 6cm; petiolule between lateral and terminal pinnae to 6.8cm, pinnae leaflets sessile, ovate, 7–9 × 2.5–3cm, glabrous, margin acutely serrate or bidentate, to 7mm depth, spinulate; compound umbel axillary or terminal; florets rarely subtended by 1–2 involucral bracts, lanceolate, to 3.5mm; pedicels subtended by involucel of shorter lanceolate bracts; flowers minute, yellow; mericarps elliptic, 4.5–6 × 2.5–

3.5mm, flattened dorso-ventrally, with 0.5mm wings, slightly thickened towards base, where notched; stylopodium c. 0.7mm length, style persists, reflexed. Montane grassland; 1900–1980m.
Distr.: Mt Kupe. (Narrow Endemic).
IUCN: VU
Mt Kupe: Cheek 7586 fl., fr., 10/1995; **Nyasoso:** Cheek 7356 fr., 2/1995.
Note: the only Umbellifer known from the summit of Mt Kupe. Flowers mistakenly reported as white in Kew Bull.; rediscovered field notes for Cheek 7586 record the flowers as "minute, golden". The related P. zenkeri and P. angustisectum have purple and cream flowers respectively.

Sanicula elata Buch.-Ham.
Upright herb, 0.5(–1)m; short stolon at base; upright stems c. 2mm diameter, glabrous, ridged; basal rosette of 2–4 leaves on petioles to 15cm, 3(–5)-lobed almost to base, with irregular sublobing and dentate-mucronate margins, c. 6 × 8cm, glabrous; cauline leaves smaller on petioles to 5cm; inflorescence cymose, bifurcating 3–4 times with a central cyme on peduncle 1–1.5cm long; lateral primary peduncles 5–6cm, secondary peduncles c. 1–2cm; each floret of 2–3 sessile flowers; florets and peduncles subtended by lanceolate bracts 6–12mm; flowers 1–2mm; corolla white; fruits ellipsoid, 3 × 2mm, covered in hooked bristles, green. Forest including stream edges; 1000–2000m.
Distr.: tropical & S Africa, Madagascar, Comores, temperate Asia. (Montane).
IUCN: LC
Kodmin: Etuge 4071 1/1998; Ghogue 61 1/1998; **Mbule:** Leeuwenberg 9291 fr., 1/1972; **Mwanenguba:** Leeuwenberg 8445 fr., 9/1971; Sanford 5429 fl., 11/1968; **Nsoung:** Darbyshire 37 3/2003; **Nyasoso:** Cheek 7315 fr., 2/1995; Etuge 1669 1/1996; Lane 226 fl., 11/1994; Sidwell 404 10/1995; Wheatley 455 fl., fr., 7/1992.
Local name: none known. **Uses:** none known (Epie Ngome in Cheek 7315).

URTICACEAE

C.M. Wilmot-Dear & I. Darbyshire

Fl. Cameroun 8 (1968).

Boehmeria macrophylla Hornem.
F.T.E.A. Urticaceae: 44 (1989).
Syn.: *Boehmeria platyphylla* D.Don
Shrub to 2(–3)m; branches glabrous except when young; leaves opposite, anisophyllous, ovate, 10–13.5 × 5.5–9cm, acuminate, base acute to rounded, margin serrate, basal lateral nerves prominent, upper surface sparsely pubescent, cystoliths punctiform, lower surface glabrescent; petiole to 6cm; spikes axillary, 7–50cm, whip-like, with glomerules of flowers spaced 1–10mm apart, male glomerules 1–2mm, female 2–3mm. Forest, forest edge; 600–1420m.
Distr.: tropical Africa & Madagascar, tropical Asia to SW China. (Palaeotropics).
IUCN: LC
Bakossi F.R.: Olorunfemi in FHI 30588 5/1951; **Kupe Village:** Cable 2554 5/1996; 3658 7/1996; Cheek 8389 5/1996; Etuge 1981 5/1996; 2857 7/1996; Ryan 357 5/1996; **Mejelet-Ehumseh:** Etuge 308 10/1986; **Menyum:** Doumenge 499 5/1987; **Nyasoso:** Balding 59 fr., 8/1993; Cable 2820 6/1996; 3252 6/1996; Etuge 1808 3/1996; 2079

6/1996; Sidwell 374 10/1995; Wheatley 428 fl., 7/1992; Zapfack 683 6/1996.

Elatostema mannii Wedd.
Fleshy decumbent unbranched herb; stipules lanceolate-caudate, 4–5mm; leaves clustered towards apex, reduced below, alternate, obliquely oblanceolate, 5.5–8 × 1.8–2.8cm, acuminate, base cuneate on both sides, margin serrate, 9–11 teeth on each side, cystoliths numerous on upper surface, rod-like, blade dark green above, paler below, subsessile; male inflorescence on a puberulent peduncle 1.5–2cm, capitate, 5–7mm, with an involucre of ovate bracts, 3mm. Rocks by streams in forest; 250–870m.
Distr.: SE Nigeria, Bioko & SW Cameroon. (Lower Guinea).
IUCN: LC
Baduma: Letouzey 14610 4/1976; **Kupe Village:** Cable 2451 5/1996; Ryan 301 5/1996; **Loum F.R.:** Cheek 10245 12/1999; **Nyandong:** Darbyshire 105 3/2003.

Elatostema monticola Hook.f.
Herb resembling *E. mannii*, but leaves smaller, 3(–6) × 1.5cm, apex acute to obtuse, base rounded distally, cuneate proximally, marginal teeth 9–10 distally, c. 6 proximally, blade sparsely pubescent; stipules 2–4mm; inflorescence sessile, 7mm wide; bracts oblong, 2.5mm, ciliate. Rocks and forest floor; 760–1650m.
Distr.: Bioko & Cameroon to E Africa & Zimbabwe. (Afromontane).
IUCN: LC
Bangem: Thomas 5323 1/1986; **Ehumseh:** Manning 486 10/1986; **Kupe Village:** Cable 3677 7/1996.

Elatostema paivaeanum Wedd.
Herb to 50cm, rarely branched, resembling *E. mannii*, but stipules conspicuous, lanceolate, 7–10mm; leaves drying green-black, large, 7.5–16 × 3–5.5cm, highly asymmetric with distal base subcordate, proximal base cuneate, marginal teeth 12–18 distally, 9–14 proximally, cystoliths dense, conspicuous; inflorescence sessile, c. 13cm wide; bracts broadly ovate (male) to lanceolate (female), c. 5mm, ciliate; bracteoles pilose, clearly so in female inflorescence. Forest; 840–1700m.
Distr.: Guinea (C) to E Africa & Malawi. (Tropical Africa).
IUCN: LC
Kodmin: Plot B289 1/1998; **Kupe Village:** Cable 2608 5/1996; 2644 5/1996; 3797 7/1996; Cheek 7129 fr., 1/1995; 7142 1/1995; 7222 fl., 1/1995; Ensermu 3561 11/1995; Ryan 276 5/1996; **Nyasoso:** Cable 2757 6/1996; 3214 6/1996; 3303 6/1996; 3402 6/1996; 3428 7/1996; Etuge 1731 2/1996; 2069 6/1996; 2118 6/1996; Lane 91 fl., 6/1994; Letouzey 14711 4/1976; Sunderland 1521 fl., 7/1992; 1538 fl., 7/1992; Wheatley 408 fl., fr., 7/1992; Zapfack 594 6/1996.

Elatostema cf. *paivaeanum* Wedd.
Herb to 30cm, as *E. paivaeanum*, but several-branched; leaves to only 6.5 × 2.7cm; stipules to only 5.5mm; inflorescence only 7mm, densely pilose; bracts 2mm. Rocks by forest stream; 1100m.
Nyasoso: Sidwell 363a 10/1995.
Note: probably a variant of *E. paivaeanum*.

Elatostema welwitschii Engl.
Herb to 30cm, occasionally branched; resembling *E. paivaeanum*, but leaves drying fresh green, marginal teeth finer, 22–29(–40) distally, 16–24 proximally, sometimes biserrate, tips ciliate, sparsely pubescent in midrib below, distal base rounded; stipules to 7mm; inflorescence 0.4–

0.8cm wide, dense, pale green. Rocks by forest streams; 750–1100m.
Distr.: Nigeria to Tanzania & Malawi. (Afromontane).
IUCN: LC
Bangem: <u>Thomas 5372</u> 1/1986; **Enyandong:** <u>Ghogue 1210</u> fl., 10/2001; **Kupe Village:** <u>Cable 780</u> 1/1995; **Ngomboaku:** <u>Mackinder 334</u> 12/1999; **Nyasoso:** <u>Cheek 7311</u> fl., 2/1995; <u>Sidwell 363b</u> 10/1995.

Girardinia diversifolia (Link) Friis
Kew Bull. 36: 145 (1981).
Syn.: *Girardinia condensata* (Steud.) Wedd.
Erect herb 0.5(–2)m, few-branched, densely setose throughout, bristles to 5mm, brown; leaves alternate, c. 12 × 10cm, deeply 5–7-lobed, lobing subpalmate, lobes acuminate, margin coarsely serrate, blade pilose and setose; petiole 3.5cm; stipules oblong-lanceolate, pilose, 1.5cm, caducous; male inflorescence a few-branched panicle to 10cm, female a dense many-branched panicle to 3cm, densely hairy, extending in fruit. Clearings in forest; 2000m.
Distr.: tropical & S Africa, Madagascar, Yemen, India to Indonesia. (Palaeotropics).
IUCN: LC
Mwanenguba: <u>Cheek 9412</u> 10/1998.

Laportea aestuans (L.) Chew
Fl. Cameroun 8: 121 (1968).
Syn.: *Fleurya aestuans* (L.) Gaudich.
Herb to 60cm; stems glabrescent; stipules lanceolate; leaves ovate, 4–8 × 3.5–6cm, acuminate, base truncate, margin serrate, surfaces stinging-setulose, lateral nerves 5–8 pairs; petioles 5.5(–8.5)cm, setulose with some glandular hairs; panicles many-branched on peduncles 4–10cm, glandular-hairy; flowers in clusters, 5mm, female flower pedicels 0.5mm, subwinged; achene flattened ovoid, centre warted. Farmbush, forest; 200–800m.
Distr.: pantropical.
IUCN: LC
Bakolle Bakossi: <u>Etuge 266</u> 9/1986; **Loum F.R.:** <u>Meurillon 935</u> /1967; **Mungo River F.R.:** <u>Cheek 10179</u> 11/1999; <u>10181</u> 11/1999.

Laportea alatipes Hook.f.
Erect herb to 1m, stinging-setulose throughout; stipules lanceolate, to 1.2cm; leaves alternate, ovate, 8–18 × 4–10cm, acuminate, base rounded, margin serrate, cystoliths punctiform, lateral nerves c. 6 pairs; petiole to 6.5cm; female panicles towards apex of shoots, to 13cm with c. 8 branches, on peduncles to 11cm, puberulent and stinging-hairy, male panicles shorter; peduncle to 1cm, found lower on the shoots; pedicels 1–2mm, membranously 2-winged in fruit; achene flattened-globose, 1.5–2mm, rugose in pale brown centre, margins smooth, dark brown. Forest; 1650–1950m.
Distr.: Cameroon to E & S Africa. (Afromontane).
IUCN: LC
Mejelet-Ehumseh: <u>Manning 422</u> 10/1986; **Mt Kupe:** <u>Sebsebe 5083</u> 11/1995.

Laportea ovalifolia (Schumach.) Chew
Fl. Cameroun 8: 131 (1968).
Syn.: *Fleurya ovalifolia* (Schumach.) Dandy
Stoloniferous herb to 45cm, erect to prostrate except for male inflorescence; stems stinging-hairy; stipules 7mm, lanceolate; leaves as *L. alatipes* but 3–9 × 2–5cm, base obtuse; male inflorescence axillary or from stolons, erect, paniculate, c. 13cm on peduncle to 22cm; branches short, flower clusters 0.5–1cm; female inflorescence geocarpic or rarely in axils of upper leaves, densely racemose, 1.5(–5)cm; peduncles to

14cm, densely stinging-hairy; achene flattened ovoid with a membranous margin and warted centre surrounded by a ridge. Roadside, forest; 650–1700m.
Distr.: Sierra Leone to Zimbabwe. (Tropical Africa).
IUCN: LC
Kupe Village: <u>Ryan 204</u> 5/1996; **Mwambong:** <u>Gosline 114</u> 2/1998; **Nyasoso:** <u>Etuge 1609</u> 1/1996; <u>1746</u> 2/1996.
Local name: Tulembuog (Gosline 114); Tolebuog (Etuge 1609). **Uses:** FOOD ADDITIVES – unspecified parts – used in herbal mixes for cooking meats (Gosline 114); MEDICINES (Etuge 1609).

Parietaria debilis G.Forst.
Syn.: *Parietaria laxiflora* Engl.
Decumbent or erect non-stinging herb to 50cm tall; stems puberulent; stipules absent; leaves alternate, ovate, 3.5–6 × 1.4–2.7cm, long acuminate, base obliquely subcordate, margin entire, ciliate; petiole to 3cm; cymes axillary, dense, 0.4–1cm diameter, with mixed male and female flowers; bracts pubescent. Farmbush, forest edge; 2000m.
Distr.: pantropical. (Montane).
IUCN: LC
Mwanenguba: <u>Cheek 9442</u> 10/1998.

Pilea angolensis (Hiern) Rendle subsp. *angolensis*
Succulent herb to 20cm, few-branched; stems glabrous; stipule < 1mm, inconspicuous; leaves opposite, ovate, 2.5–5 × 2–3.5cm, acuminate, base truncate or rounded, margin deeply serrate-apiculate, surfaces sparsely pubescent or glabrescent, cystoliths punctiform; petiole 1.5–5cm; panicles axillary, few- or unbranched, 0.4–1.5cm; flowers in cymose clusters, 3mm diameter. Rocks in forest; 870–1120m.
Distr.: Guinea (C) to Sudan & Tanzania. (Guineo-Congolian).
IUCN: LC
Kupe Village: <u>Cable 3746</u> 7/1996; <u>Ryan 212</u> 5/1996; **Nyasoso:** <u>Cable 3201</u> 6/1996; <u>Wheatley 477</u> fl., 7/1992.

Pilea rivularis Wedd.
Fl. Cameroun 8: 163 (1968).
Syn.: *Pilea ceratomera* Wedd.
Erect herb to 60cm, resembling *P. angolensis*, but stems with linear cystoliths; stipules prominent, oblong, 7mm; leaves to 7.5 × 5cm, base rounded, margin serrate-crenate, cystoliths linear; inflorescence a dense axillary cluster to 1.5cm diameter. Forest, including rocky stream margins; 1300–1950m.
Distr.: tropical & subtropical Africa. (Afromontane).
IUCN: LC
Kodmin: <u>Cheek 9038</u> 1/1998; <u>9627</u> 11/1998; <u>Gosline 155</u> 11/1998; <u>Satabie 1108</u> 1/1998; **Mejelet-Ehumseh:** <u>Manning 415</u> 10/1986; **Nyasoso:** <u>Cable 2934</u> 6/1996; <u>Lane 244</u> fl., 11/1994.
Uses: MEDICINES (Cheek 9627).

Pilea sublucens Wedd.
Syn.: *Pilea chevalieri* R.Schnell
Succulent herb 0.15–1m; stems with linear cystoliths; leaf pairs anisophyllous, variable, ovate to elliptic, 2–10.5 × 1.5–4cm, acuminate, base acute to rounded, margin convexly serrate-apiculate, basal lateral nerves prominent, extending to near apex, cystoliths linear, hydathodes prominent on lower surface; petiole 1.5–4cm; panicles lax, 0.7–4cm on peduncle 0.3–3cm; flowers in cymose clusters. Rocks and streams in forest; 800–1950m.
Distr.: Liberia, Ivory Coast, Nigeria to Uganda. (Guineo-Congolian (montane)).

IUCN: LC
Kodmin: Cheek 8999 1/1998; 9626 11/1998; Etuge 4031 1/1998; Gosline 154 11/1998; **Kupe Village:** Etuge 2817 7/1996; Ryan 431 5/1996; **Mt Kupe:** Sebsebe 5093 11/1995; **Mwambong:** Etuge 4335 10/1998; Gosline 127 10/1998; **Mwanenguba:** Leeuwenberg 9617 4/1972; **Nyasoso:** Letouzey 14697 4/1976; Sidwell 362 10/1995; Wheatley 448 fl., 7/1992.
Uses: MEDICINES (Cheek 9626).

Pilea ? sublucens Wedd.
Herb to 10cm, resembling *P. sublucens*, but leaves ovate-lanceolate, to 2 × 0.9cm, marginal teeth not convex; petiole to 0.7cm; inflorescence dense, to only 0.6cm diameter; peduncle only 1.5mm. Waterfall edge; 750–1350m.
Baseng: Cheek 10404 12/1999; **Kodmin:** Cheek 8998 1/1998.
Note: possibly a dwarf form of *P. sublucens*, requires further investigation.

Pilea tetraphylla (Steud.) Blume
Herb to 20cm; internodes decreasing and leaf size increasing towards apex; leaf pairs decussate, blade as *P. angolensis*, but to 3 × 2.2cm, apex scarcely acuminate, cystoliths linear, sparsely long-hairy, uppermost leaves subsessile; inflorescence appearing terminal, a corymb of 4 uneven parts, 2–3cm broad, dense. Forest, villages; 1650m.
Distr.: tropical Africa. (Afromontane).
IUCN: LC
Mejelet-Ehumseh: Manning 442 10/1986.

Pouzolzia denudata De Wild. & T.Durand
Fl. Cameroun 8: 190 (1968).
Perennial herb to 1m; stems pubesecent; leaves alternate, ovate, c. 10 × 4.5cm, acumen c. 2cm, base obtuse, margin entire, ciliate, upper surface and lower midrib sparsely pubescent; glomerules axillary, c. 7mm diameter, several-flowered; bracts puberulent; female perianth golden, gleaming, semi-translucent, glabrous; stigma c. 2mm; achene 1.5mm, shiny whitish brown. Farmbush, forest; 870–1000m.
Distr.: Ivory Coast to Uganda & Angola. (Guineo-Congolian).
IUCN: LC
Nyasoso: Cable 3230 6/1996; Sunderland 1492 fr., 7/1992.

Pouzolzia guineensis Benth.
Herb to 1m, resembling *P. denudata*, but leaves to only 6.5 × 2cm, base acute; petiole to 2.5cm, more densely pubescent; inflorescence c. 0.5cm, few-flowered; bracts and flowers with prominent hooked hairs; stigma tightly curloed, only up to 0.5mm; achene 1.5mm, pale brown, often enclosed in persistent green perianth. Open forest, roadsides; 50–800m.
Distr.: Senegal to Ethiopia & Angola. (Tropical Africa).
IUCN: LC
Kupe Village: Etuge 2733 7/1996; **Loum F.R.:** Etuge 4281 10/1998; **Mungo River:** Leeuwenberg 10627 11/1972; **Mungo River F.R.:** Cheek 10176 11/1999; **Tombel:** Etuge 379 11/1986.

Pouzolzia parasitica (Forssk.) Schweinf.
Erect or scandent herb, pubescent throughout; leaves ovate, 5.5–8 × 3–4.5cm, acuminate, base subcordate or rounded, margin serrate, cystoliths punctiform; petiole 0.5–3.5cm; glomerules axillary, c. 5–7mm diameter, dense; achene 1.5mm, whitish brown. Grassland, stream margins; 1250–1500m.
Distr.: tropical & S Africa, Yemen, C & W tropical S America. (Amphi-Atlantic (montane)).

IUCN: LC
Edib: Etuge 4098 1/1998; Onana 591 2/1998; **Kodmin:** Cheek 8907 1/1998.
Local name: Toola-boh. **Uses:** MATERIALS – traps/snares – stems used in trapping (Cheek 8907).

Procris crenata C.B.Rob.
Unbranched epiphytic herb to 40cm; stems succulent, 0.5cm diameter. (dried); leaves only in upper section of stems, alternate, narrowly elliptic, 11–16 × 2.5–3cm, acuminate, base cuneate, margin shallowly serrate-crenate, glabrous; petiole 5mm; inflorescences on leafless lower stems, 4–5 together (female) or single (male); peduncle 5mm, succulent, clusters capitate, 1–2mm diameter. Forest; 800–1750m.
Distr.: tropical Africa, Madagascar, India to Philippines. (Palaeotropics).
IUCN: LC
Bangem: Thomas 5271 1/1986; **Ehumseh:** Manning 495 10/1986; **Ehumseh - Mejelet:** Etuge 336 10/1986; **Kodmin:** Etuge 4055 1/1998; Nwaga 12 1/1998; Plot B143 1/1998; **Mbule:** Letouzey 14676 4/1976; **Mwambong:** Cheek 9384 10/1998; Onana 534 2/1998; **Nsoung:** Darbyshire 52 3/2003; **Nyasoso:** Cable 29 fl., 7/1992; 3425 7/1996; 3540 7/1996; Wheatley 407 fl., 7/1992; Zapfack 717 6/1996.

Urera cordifolia Engl.
Robust climber or scrambler to 8m; stems to 1cm diameter, with numerous epidermal appendages; leaves alternate, ovate-orbicular, 16–27 × 11.5–21cm, apex short-acuminate, base shallowly cordate, margin shallowly crenate, cystoliths linear, glabrous; petiole 7–23cm; panicles axillary, many-branched, 10–17cm; flowers numerous, < 1mm with short stinging hairs; pedicel 1mm. Forest; 260–1000m.
Distr.: Nigeria, Cameroon, Gabon & CAR. (Lower Guinea).
IUCN: LC
Baduma: Nemba 618 9/1987; **Ikiliwindi:** Letouzey 14609 4/1976; **Kupe Village:** Cable 2528 5/1996; Etuge 2773 7/1996; **Nyasoso:** Cheek 5666 12/1993; Etuge 2366 6/1996.

Urera flamigniana Lambinon
Bull. Soc. Roy. Bot. Belg. 91: 199 (1959).
Robust climber to 7m, resembling *U. cordifolia*, but stem appendages absent, plant generally more pubescent, densely so in some Cameroon collections; leaves narrower, to 14.5 × 8cm, base rounded to subcordate, margin serrate-apiculate; panicles corymbiform, to 12 × 13cm, on peduncle to 8.5cm; flowers subsessile. Forest, farmbush, roadsides; 800–1000m.
Distr.: S Nigeria, Cameroon, Congo (K). (Lower Guinea & Congolian).
IUCN: NT
Kupe Village: Etuge 2815 7/1996; **Nyasoso:** Lane 204 fr., 11/1994.
Note: currently known from only 12 collections, a potential red data candidate but the genus requires further study before a full assessment can be made.

Urera repens (Wedd.) Rendle
Scrambling or creeping herb; stems glabrous; leaves ovate, 8.5–16.5 × 6.5–13cm, short-acuminate, base deeply cordate, rarely rounded, margin serrate or subcrenate, cystoliths punctiform; petiole 4–10cm; male panicles corymbiform, lax, 5–9cm diameter; peduncle to 10cm; flowers 2–3mm; female panicles denser, to 5cm; peduncle to 4cm, puberulent and with stinging hairs. Forest, farmbush; 600–1600m.
Distr.: Liberia to Congo (K). (Guineo-Congolian).
IUCN: LC

Kupe Village: <u>Cable 2468</u> 5/1996; <u>2692</u> 5/1996; <u>Etuge 2759</u> 7/1996; <u>2873</u> 7/1996; **Nyasoso:** <u>Cable 3413</u> 6/1996; <u>3414</u> 6/1996; <u>3510</u> 7/1996; <u>Etuge 2451</u> 6/1996.

Urera trinervis (Hochst.) Friis & Immelman
F.T.E.A. Urticaceae: 6 (1989).
Syn.: *Urera cameroonensis* Wedd.
Robust climber, glabrous; stems cylindrical; leaves ovate-elliptic, 10.5–14 × 5.5–7.5cm, acumen to 1.7cm, base acute to rounded, margin entire, basal lateral nerve pair prominent, cystoliths inconspicuous; petiole 2–4.5cm; male panicles to 6cm; peduncles 1cm; pedicels 1.5mm; female panicles denser, c. 4cm, with clusters of stinging hairs; flowers sessile. Forest; 800–1530m.
Distr.: Ghana to E & S Africa, Madagascar. (Tropical Africa & Madagascar).
IUCN: LC
Kodmin: <u>Cheek 8888b</u> 1/1998; <u>Ghogue 23</u> 1/1998; <u>Nwaga 9</u> 1/1998; **Lake Edib:** <u>Cheek 9149</u> 2/1998; **Ndum:** <u>Lane 493</u> fr., 2/1995; **Nyasoso:** <u>Cable 2876</u> 6/1996; <u>3277</u> 6/1996; <u>Cheek 5637</u> fl., 12/1993; <u>7309</u> fr., 2/1995; <u>7902</u> 11/1995; <u>Etuge 1531</u> 1/1996; <u>Zapfack 629</u> 6/1996.
Local name: Toti (Lyonga Daniel - Bakweri - Cheek 7902); none known (Cheek 7309). **Uses:** MATERIALS – fibres used for making local rope for carrying baskets (*fide* Etuge); SOCIAL USES – stem fibre used to tie children to rid them of 'extra eyes' (i.e. not of this world) (Lyonga Daniel in Cheek 7902); none known (Cheek 7309).

Urera sp. 1
Robust climber, resembling *U. cordifolia*, but stems heavily gnarled, lacking epidermal appendages; leaves narrower, 10.5–19 × 6–12.5cm, base rounded to subcordate; petiole 5–10cm; male panicles to 10cm; pedicels persistent, 1.5mm. Forest, farmbush; 500–700m.
Kupe Village: <u>Cable 2705</u> 5/1996; <u>Cheek 8320</u> 5/1996.
Note: species delimitation in *Urera* is not currently fully worked out; the specimens cited likely fall into one of the existing taxa but further investigation required.
Local name: Nn dap (Cheek 8320).

VERBENACEAE

B.J. Pollard

Note: here follows the treatment of Verbenaceae *sensu stricto* without *Clerodendrum* L., *Premna* L., and *Vitex* L., which are now treated under Labiatae *sensu lato*. See reference under that account.

Duranta erecta L.
F.T.E.A. Verbenaceae: 48 (1992).
Shrub or small tree, 1.8–6m; leaves opposite or verticillate, ovate, obovate, elliptic or ovate-lanceolate, 1.5–8 × 1–4.5cm, entire, serrate or crenate; racemes ± lax, 3–16cm; pedicels 1–5mm; calyx 3–4(–6)mm; corolla lilac, lavender-blue or white, tube 7mm, limb spreading, 0.7–1.3cm; fruit orange-yellow, shining, globose, 0.5–1.1cm, fleshy. Farmbush; 460m.
Distr: Aregntina to Mexico & Florida; now widely cultivated and naturalized throughout the tropics and subtropics, even in Europe, Cyprus etc. (Pantropical and Temperate).
IUCN: LC
Enyandong: <u>Cheek 11413</u> fl., fr., 3/2003.
Note: not mentioned in FWTA.

Uses: ENVIRONMENTAL USES – ornamentals (*fide* Cheek).

Lantana camara L.
Shrub; stems erect or spreading, much-branched, quadrangular, usually armed with short recurved prickles; leaves; petioles and peduncles pilose or strigose; flowers in convex heads; corolla much longer than subtending bract, white, yellow, red, orange or pink. Introduced weed of roadsides, 1030–1170m.
Distr: pantropical.
IUCN: LC
Edib village: <u>Cheek (sight record)</u> 2/1998; **Enyandong:** <u>Pollard 775</u> fl 10/2001.
Note: infraspecific ranks not recognised here.

Lantana ukambensis (Vatke) Verdc.
F.T.E.A. Verbenaceae: 43 (1992).
Aromatic subshrub to 1.5m; stems quadrangular, often ± unbranched; leaves paired or ternate, ovate to ovate-lanceolate, 1.5–10 × 0.5–5cm, subsessile or petiole 1–5(–8)mm; inflorescence axillary, up to 6 per node, shortly pedunculate, 0.3–3.5(–5) × 1–1.5cm; bracts very conspicuous; corolla deep purple, crimson or magenta, scarcely exceeding the bracts. Grassland, woodland; 1400–1450m.
Distr: Guinea (B) to Ethiopia, S to Congo (K) & Zimbabwe. (Afromontane).
IUCN: LC
Kodmin: <u>Cheek 9571</u> fl fr 11/1998; <u>Etuge 4416</u> 11/1998.

Stachytarpheta cayennensis (Rich.) Vahl
Shrubby herb to 2m; stems glabrous; leaves ovate or elliptic, 1.8–8 × 0.5–4cm, attenuate into petiole, 1–1.5cm; inflorescence a slender spike, up to 20–25(–34)cm, with some pubescence; bracts linear to triangular-subulate, 4–5mm, acuminate; calyx 4–5mm, with 4 equal teeth; corolla white or mostly pale blue; tube 4–5mm, scarcely exceeding the calyx. Roadsides, fallow; 200–900m.
Distr: Sierra Leone to Mozambique, widespread in tropical America, naturalised throughout the tropics. (Pantropical).
IUCN: LC
Enyandong: <u>Pollard 719</u> 10/2001; **Mungo River F.R.:** <u>Cheek 10186</u> fl fr 11/1999.

VIOLACEAE

G. Achoundong & I. Darbyshire

Hybanthus enneaspermus (L.) F.Muell.
Syn.: *Hybanthus thesiifolius* (Juss. ex Poir.) Hutch. & Dalziel.
Erect subshrub to 70cm; stems shallowly ridged, puberulent towards apex; stipules lanceolate, 1.5mm; leaves papery, narrowly elliptic, 7 × 1.2cm, apex acute, base cuneate, margin serrulate, glabrous; petiole 3mm; flowers axillary, solitary; pedicel 1.2cm, sparsely puberulent; sepals lanceolate, 2–3mm, setulose, lower petal c. 5mm, lilac; capsule ellipsoid, 6mm long; seeds subglobose with numerous parallel ridges. Riverside; 750m.
Distr.: tropical & S Africa, Madagascar, tropical Asia & Australia. (Palaeotropics).
IUCN: LC
Baseng: <u>Cheek 10410</u> fr., 12/1999.

Rinorea ? batesii Chipp
Kew Bull.: 297 (1923).
Shrub 0.6–3.5m, resembling *R. preussii*, but leaves smaller, to 17 × 6cm, lateral nerves 9–11 pairs; petiole to 2cm; inflorescence less elongate, 5.5–10cm; calyx densely puberulent-ciliate, petals reflexed at tips. Forest; 150m.
Distr.: Cameroon endemic.
Etam: Etuge 88 3/1986.
Note: specimen not seen by G. Achoundong (YA); requires confirmation.

Rinorea dentata (P.Beauv.) Kuntze
Shrub or small tree to 5m; stems puberulent; leaves papery, elliptic, 15.5–19.5 × 6–8cm, acumen 1.7cm, base acute, margin denticulate, lateral nerves 11–13 pairs, midrib puberulous below, laminae glabrous except when young, eglandular; petiole 0.6–1.2cm; panicles terminal, c. 3cm, puberulent, c. 8–10-flowered; peduncle 3cm; sepals triangular, 1.5mm, petals lanceolate, 4mm, yellow. Forest; 150–900m.
Distr.: Liberia to Uganda. (Guineo-Congolian).
IUCN: LC
Etam: Etuge 90 3/1986; **Kupe Village:** Elad 73 fl., 1/1995; Etuge 2788 7/1996; **Mungo River F.R.:** Tadjouteu 314 fl., 12/1999.

Rinorea fausteana Achoundong
Kew Bull. 58: 958 (2003).
Shrub to 2.5m; stems glabrescent; leaves subcoriaceous, (oblong-)elliptic, 13.5–19 × 5.5–9cm, acumen to 1.8cm, base cuneate to obtuse, margin crenulate, lateral nerves 9–10 pairs, glabrous, lower surface glandular; petiole 1–2cm; inflorescence terminal, thyrsiform, to 5cm, finely puberulent; sepals triangular, 1.5mm, petals ovate-lanceolate, 4mm, yellow; fruit minutely tuberculate. Forest; 1100–1400m.
Distr.: SW Cameroon. (Cameroon Endemic).
IUCN: EN
Kodmin: Etuge 4408 fl., 11/1998; **Muahunzum:** Etuge 4438 fl., 11/1998; **Nyale:** Etuge 4466 fr., 11/1998.

Rinorea gabunensis Engl.
Bot. Jahrb. 33: 140 (1902).
Treelet to 5m; stems glabrous; leaves chartaceous, oblong-elliptic to oblanceolate, 20–27 × 8–13cm, acumen 1.5cm, base obtuse, margin (sub)serrulate, midrib drying yellow, lateral nerves prominent, 9–11 pairs, glabrous, eglandular; petiole 3–6.5cm; panicles terminal, to 7.5cm, minutely puberulent, many-flowered; peduncle to 2cm; sepals broadly ovate, 2mm; corolla ovate, yellow. Forest; 350–400m.
Distr.: Cameroon, Gabon & Congo (K). (Lower Guinea & Congolian).
IUCN: LC
Mungo River F.R.: Cheek 10207 fl., 11/1999; **Ngomboaku:** Etuge 4688 fl., 12/1999.

Rinorea oblongifolia (C.H.Wright) Marquand ex Chipp
Tree 4–15m; stems glabrous; leaves chartaceous, (oblong-)elliptic, 22–34 × 9–12.5cm, acumen 1.7cm, base acute-cuneate, margin serrulate, lateral nerves 9–12 pairs, glabrous, eglandular; petiole 1.5–5.5cm; panicles terminal, 6–10cm, puberulent, many-flowered; sepals broadly ovate, 2mm, petals triangular, 3–4mm, yellow, densely puberulent outside. Forest; 150–1300m.
Distr.: Sierra Leone to Uganda. (Guineo-Congolian).
IUCN: LC
Etam: Etuge 184 fr., 7/1986; **Kupe Village:** Cable 758 fl., 1/1995; Etuge 1724 fl., 2/1996; 2827 7/1996; **Kurume:**

Thomas 5422 fl., 1/1986; **Mile 15, Kumba-Mamfe Road:** Nemba 153 fr., 7/1986; **Nyandong:** Tchiengue 1734 fl., 3/2003; Thomas 6698 fl., 2/1987; **Nyasoso:** Etuge 2580 7/1996.

Rinorea preussii Engl.
Tree 2–12m; stems glabrous, pale; leaves (oblong-)elliptic, 14–21 × 6–11cm, acumen 1.5cm, base acute to obtuse, margin serrulate, lateral nerves prominent, 11–12 pairs, glabrous, eglandular; petioles 2.5–5cm; panicles terminal, elongate, to 20cm, somewhat lax, finely puberulent, many-flowered; sepals rounded, 2.5mm, petals ovate, 4.5mm, white to pale yellow; fruit glabrous. Forest and forest edge; 150–1700m.
Distr.: Liberia, SW Cameroon. (Upper & Lower Guinea).
IUCN: NT
Etam: Etuge 89 fl., 3/1986; **Kodmin:** Etuge 4033 fl., 1/1998; **Ndum:** Cable 921 fl., 2/1995; **Nyasoso:** Etuge 1631 fl., 1/1996; Groves 100 fr., 2/1995.
Note: specimens from Congo (K) previously identified as this sp. are in fact *R. mildbraedii*.

Rinorea subsessilis Brandt
Fl. Congo, Rw. & Bur. Violaceae: 63 (1969).
Shrub 1–3m, resembling *R. fausteana*, but leaves to 23.5 × 8cm, acumen to 2.3cm, margin subentire, lateral nerves 11–12 pairs; inflorescence (sub)corymbiform, 1.5–5cm long; fruit smooth. Forest; 750–850m.
Distr.: Cameroon & Congo (K). (Lower Guinea & Congolian).
IUCN: LC
Baseng: Cheek 10418 fl., 12/1999; **Ngomboaku:** Cheek 10316 fl., 12/1999.

Rinorea thomasii Achoundong
Kew Bull. 58: 960 (2003).
Shrub 3m; stems glabrous; leaves subcoriaceous, (ovate-)elliptic, 11–21 × 5–8.5cm, acumen to 2.5cm, base cuneate to obtuse, margin subentire, lateral nerves 7–9 pairs, glabrous, eglandular; petiole 1.5–2.5cm; panicles terminal, to 5cm, glabrous, lateral branches 2–3-flowered; sepals triangular, 1.5–3cm, petals ovate-elliptic, c. 4mm, white; fruit fusiform, conspicuously nerved. Forest near rivers; 750m.
Distr.: Cameroon endemic.
IUCN: VU
Baseng: Cheek 10413 12/1999.

Rinorea sp. nov. Achoundong in prep.
Shrub 1.5–6m, glabrous; leaves subpapyraceous, elliptic to obovate-oblong, 20–25 × 1–3.5cm, base acute, margin conspicuously serrulate, lateral nerves c. 10 pairs, eglandular; petiole 0.5–5cm; inflorescence terminal, thyrsoid, up to 2.5cm, glabrescent, c. 30-flowered; sepals ovate, 1.5mm, ribbed, petals oblong, 3.5mm, reflexed at tips, pale yellow. Forest; 800m.
Distr.: Cameroon endemic.
Bangem: Thomas 5381 1/1986.
Note: known from 19 collections in lowland Cameroon.

Viola abyssinica Steud. ex Oliv.
Prostrate herb; stems to 30cm long, purple, pubescent; stipules 4mm, deeply divided into lanceolate segments; leaves ovate, 1.4 × 1.3cm, apex acute, base cordate, margin crenate, pubescent mainly along veins; petiole 6mm; flowers axillary, solitary; pedicel 1.5–2.7cm, pubescent; sepals lanceolate, 4mm, pubescent; corolla zygomorphic, 0.6–1cm long, white edged with purple; stamens orange; capsule 4.5mm; seeds black. Grassland; 1500–1850m.

Distr.: tropical and subtropical Africa. (Afromontane).
IUCN: LC
Kodmin: Cheek 9180 fl., 2/1998; **Mejelet-Ehumseh:** Manning 426 10/1986; **Mwanenguba:** Sanford 5509 fl., 11/1968.

VISCACEAE

R.M. Polhill & B.J. Pollard

Viscum congolense De Wild.

Globose dioecious parasitic shrub, < 0.5m; nodes often dilated; leaves highly variable, elliptic-ovate to oblong, 4–6(–10) × 2–4cm, conspicuously triplinerved above and below; base cuneate; margin sometimes crisped; ♂ and ♀ flowers occurring in triads (occasionally up to 6), < 3mm; berries subsessile, 6–9mm, greenish white, translucent. Forest; 220–1700m.
Distr: Ivory Coast to Congo (K), Rwanda, Burundi, Ethiopia, Angola. (Guineo-Congolian).
IUCN: LC
Bakolle Bakossi: Etuge 263 fr., 9/1986; **Eboné:** Leeuwenberg 8158 8/1971; **Kodmin:** Gosline 90 2/1998; **Kupe Village:** Cheek 8391 5/1996; Etuge 2044 5/1996; **Mwanenguba:** Leeuwenberg 8662 11/1971; 8663 11/1971; **Nyasoso:** Cable 3412 6/1996; Wheatley 440 fl 7/1992.

VITACEAE

I. Darbyshire

Fl. Cameroun 13 (1972).

Ampelocissus bombycina (Baker) Planch.

Robust climber > 10m; stems matted reddish tomentose; leaves ovate, to 5-lobed, 16 × 14cm, apex short-acuminate-mucronate, base deeply cordate, margin serrate and irregularly lobed, lower surface densely soft red-brown tomentose; petiole to 16cm; compound cymes many-flowered; flowers in dense clusters; peduncle 9cm, tomentose; fruit subglobose, 1–1.5cm diameter; seeds flattened ellipsoid, 0.6 × 0.4 × 0.25cm, several-ridged. Thicket; 1200m.
Distr.: Guinea (C) to Ethiopia & Tanzania. (Afromontane).
IUCN: LC
Mwanenguba: Leeuwenberg 8308 fr., 9/1971.

Ampelocissus macrocirrha Gilg & Brandt

Robust climber; stems glabrous; leaves ovate, to 22 × 20.5cm, apex subacuminate, base deeply cordate, margin finely dentate, glabrous except principal veins puberulent beneath, petiole 14cm, glabrescent; tendrils borne on the peduncle, bifid; inflorescence of compound cymes, to 5 × 5cm, dense; primary peduncle to 13cm, sparsely pubescent; bracts triangular, 1mm; flowers c. 2mm long; calyx copular; corolla buds rounded at apex, pedicels 1mm, glabrescent. Farmbush; 490m.
Distr.: Bioko, Cameroon, Rio Muni. (Lower Guinea).
IUCN: LC
Nyandong: Tchiengue 1752 fl., 3/2003.
Note: both FWTA 1: 682 and Enum. Pl. Afr. Trop. 2: 191 state that *A. macrocirrha* is doubtfully distinct from *A. abyssinica* (Hochst. ex A.Rich.) Planch., of which *A. cavicaulis* (Baker) Planch. of Fl. Cameroun 13: 14 is now

considered synonymous. However, Fl. Cameroun omits *A. macrocirrha* in its treatment of the genus; it is therefore here maintained as a distinct taxon, awaiting further analysis.

Ampelocissus sp. aff. africana (Lour.) Merr.

Large climber; stems succulent, 1–1.5cm width, glabrous; leaves not seen in our material; tendrils borne on the peduncle, bifid; inflorescence a many-flowered compound cyme, c. 10 × 9cm; flowers in dense clusters; primary peduncle c. 21cm, secondary peduncles puberulent; bracts triangular, 1–2mm; flowers resembling *A. macrocirrha*, but pedicels puberulent. Forest; 700m.
Kupe Village: Cheek 7011 1/1995.
Note: the inflorescence closely resembles *A. sp. aff. africana* of Mt Cameroon; leaves are required to confirm identification.

Cayratia gracilis (Guill. & Perr.) Suess.

Fl. Cameroun 13: 20 (1972).
Syn.: *Cissus gracilis* Guill. & Perr.
Herbaceous climber to 6m; stems cylindrical, fine, glabrous; tendrils bifid; leaves papery, pedately 5-foliolate; leaflets ovate, central 5.2–8.5 × 2.1–4.2cm, apex acute or acuminate, base obtuse-rounded, margin serrate, sparsely pubescent or glabrous below; petiole 4–6cm; compound cyme c. 5cm long; flowers 2mm long; calyx cupular; corolla buds truncate; fruit globose, 0.8cm diameter, glabrous; seeds cordiform, 0.4cm long with 2 depressions on the ventral side and a dorsal furrow. Forest, thicket; 1400–1450m.
Distr.: widespread in (sub)tropical Africa, also Yemen (Afromontane).
IUCN: LC
Kodmin: Cheek 9609 fr., 11/1998; Etuge 4428 11/1998.

Cissus amoena Gilg & Brandt

Climber; stems cylindrical, densely pubescent when young; leaves ovate, 6–8 × 5–6cm, apex acuminate, base cordate, margin finely denticulate, lower surface densely puberulent, nerves with medifixed V-shaped hairs; petiole 2–5cm, pubescent; panicles with numerous subumbellate clusters, peduncle to 6cm, partial peduncles pubescent; calyx cupular; flower buds cylindric, to 3mm long, puberulent; pedicels 2–3mm, puberulent; fruit ellipsoid, pointed, c. 8mm long; seeds flattened ellipsoid, highly sculptured. Forest.
Distr.: Cameroon & Gabon (Lower Guinea).
IUCN: NT
Mwanenguba: De Wit 287 fr., date unknown.
Note: specimen cited in Fl. Cameroun 13: 110 (1972).

Cissus aralioides (Welw. ex Baker) Planch.

Climber to 15m; stems terete, succulent, glabrous, drying yellow-green; tendrils simple or bifid; leaves palmately (3–) 5-foliolate, 11.5–14 × 10–12cm, central leaflet obovate, c. 9 × 4.8cm, apex acuminate, base acute, margin finely toothed, lateral leaflets asymmetric, ovate-elliptic, glabrous; pedicel 7–9.5cm; inflorescence a compound, many-flowered cyme to 25 × 9.5cm; peduncle glabrescent; flowers 0.35cm; calyx cupular, puberulent; corolla buds rounded at apex; pedicels 0.7cm, minutely puberulent; fruit ellipsoid, 2.5 × 1.5cm, glabrous, green-red; seeds flattened ellipsoid, 1.8 × 0.9cm, smooth. Forest, farmbush; 990–1500m.
Distr.: Senegal to Mozambique. (Tropical Africa).
IUCN: LC
Edib: Etuge 4495 11/1998; **Kodmin:** Cheek 8886 1/1998; Ghogue 50 fr., 1/1998; Gosline 183 11/1998; **Kupe Village:** Etuge 1427 11/1995; **Mwambong:** Gosline 123 10/1998; **Ndum:** Cable 909 fr., 2/1995; Elad 112 2/1995.

Local name: Ichin gul (Cheek 8886); Wale wole (Gosline 183). **Uses:** no use known (Cheek 8886).

Cissus barbeyana De Wild. & T.Durand

Closely resembling *C. leonardii*, of which it is sometimes treated as a synonym, but with cylindrical mature stems; leaves less undulate and with finer teeth, flowering pedicels glabrous; fruits larger, 8–9 × 5–7mm. Secondary forest; 200m.
Distr.: Nigeria to Angola. (Lower Guinea & Congolian).
IUCN: LC
Ikiliwindi: Nemba 523 6/1987.

Cissus barteri (Baker) Planch.

Herbaceous climber; stems cylindrical, glabrous; tendrils simple; leaves ovate, 10.5–15 × 4.5–7cm, apex acuminate, base obtuse-truncate or subcordate, margin finely dentate, glabrous, drying pale green below, darker green above; petiole to 3cm; inflorescence a compound cyme c. 10 × 10cm; flowers in sparse umbellate clusters, glabrous throughout; flowers c. 3mm; calyx cupular; corolla buds acute at apex; pedicel 0.4–0.6cm; fruits subglobose or obovoid, c. 0.8cm, glabrous, green; seeds c. 6mm long, flattened ovoid, subreniform laterally, smooth with a narrow dorsal ridge. Forest; 850–1200m.
Distr.: Nigeria to Congo (K). (Lower Guinea & Congolian).
IUCN: LC
Kupe Village: Cable 763 fr., 1/1995; Etuge 1409 11/1995; **Ngomboaku:** Muasya 2025 12/1999; **Nyasoso:** Etuge 1768 3/1996; Lane 202 fl., fr., 11/1994; Sebsebe 5052 10/1995.

Cissus diffusiflora (Baker) Planch.

Herbaceous climber to 7m; stems cylindrical, pubescent; tendrils few, bifid; leaves papery, ovate-lanceolate, 9–16 × 3.3–6.8cm, apex attenuate-acuminate, base truncate to shallowly cordate, margin finely dentate, principal veins of lower surface pilose; petiole 0.4–4.5cm, pubescent; inflorescence to 1.7cm long; flowers in umbellate clusters; peduncles 1–5mm, pubescent; flowers to 1.5mm long; calyx cupular, buds rounded at apex; petiole to 2.5mm, glabrous; fruit pyriform or globose, 0.6–0.8 × 0.5cm, glabrous, red-black; seeds c. 0.45cm long, flattened ovoid with a prominent dorsal ridge. Forest and thicket; 750–1400m.
Distr.: Guinea (B) to Uganda. (Guineo-Congolian).
IUCN: LC
Baseng: Cheek 10419 12/1999; **Ehumseh - Mejelet:** Etuge 350 10/1986; **Manehas F.R.:** Cheek 9401 10/1998; **Mejelet-Ehumseh:** Etuge 173 fl., 6/1986; **Mwambong:** Cheek 9383 10/1998; **Nyasoso:** Etuge 2522 7/1996; Lane 176 fr., 11/1994; Sidwell 314 10/1995; Zapfack 690 6/1996.

Cissus leonardii Dewit

Fl. Cameroun 13: 116 (1972).
Herbaceous climber to 6m; mature stems subquadrangular, glabrous; leaves ovate-oblong, 6–11 × 2–6.5cm, acumen to 1.2cm, base truncate to subcordate, margin shallowly undulate-dentate, glabrous, drying green-brown below, brown above; inflorescence to 6.5cm long with several umbellate cymes; peduncle 1.5–2cm, puberulent; flowers 1.5–2mm; calyx shallowly round-lobed; corolla buds rounded; pedicel c. 6mm, puberulent; fruit obovoid, 0.6–0.7 × 0.5cm, black when mature; seeds c. 0.5cm long, subreniform, several-ridged dorsally. Forest, forest edges; 150–1050m.
Distr.: Cameroon to Congo (K). (Lower Guinea & Congolian).
IUCN: LC

Kupe Village: Cable 2665 5/1996; 2716 5/1996; 3820 7/1996; Cheek 7040 fr., 1/1995; **Mungo River:** Gosline 230 11/1999; **Nyasoso:** Etuge 2164 6/1996.
Note: *C. dinklagei* Gilg & Brandt, with entire leaf margins and prominent domatia at the axes of *the midrib and lateral nerves, is also likely in the checklist area.*

Cissus oreophila Gilg & Brandt

Climber to 6m; stems cylindrical, red-brown-pubescent when young, becoming glabrescent; leaves broadly ovate, 7.5–11 × 6.5–8.5cm, apex acuminate, base deeply cordate, margin denticulate, undersurface finely white-puberulent throughout and with reddish medifixed V-shaped hairs mainly along the veins; petiole 3–5cm; panicles with numerous umbellate clusters; peduncle to 8.5cm, partial peduncles and pedicels pubescent; calyx cupular; buds conical, c. 1.5mm long, minutely puberulent; petals white; fruit subglobose, c. 5mm; seeds subreniform, highly sculptured. Forest, forest edges, 1240m.
Distr.: Liberia to Gabon. (Upper & Lower Guinea).
IUCN: LC
Nsoung: Etuge 4955 fl., 3/2003.

Cissus petiolata Hook.f.

Climber to 6(–15)m; stems square, ridged when mature, glabrous; leaves ovate to pentagonal, 10.5–16 × 7–13cm, apex acuminate, base deeply cordate, margin finely dentate, surfaces glabrescent; petiole to 14cm; inflorescence of c. 4 umbellate cymes; peduncle 3.5cm; flowers not seen in our material; fruit subglobose, 0.8cm diameter, tip apiculate, glabrous, green; seeds 0.7cm long, flattened ovoid, subreniform laterally, smooth. Forest; 400m.
Distr.: Guinea (C) to Zimbabwe. (Tropical Africa).
IUCN: LC
Kupe Village: Etuge 1933 5/1996.

Cissus producta Afzel.

Herbaceous climber to 20m; stems robust, glabrous; tendrils simple; leaves variable, ovate to oblong, 5–11.7(–13) × 2.2–6(–10.2)cm, acumen 0.8cm, base truncate to shallowly cordate, margins toothed, prominent when young, glabrous, usually drying brown-green; inflorescence to 3cm with 3–4 dense umbellate cymes; peduncle c. 1cm; flowers c. 3mm; calyx cupular; corolla buds acute at apex; pedicels 0.5–1cm, puberulent; fruit obovoid, 1(–1.6) × 0.8(–1.2)cm, glabrous, black when ripe; seeds 0.8–1.4cm long, ovoid, ventral side with 2 cavities, dorsal ridge c. 1mm broad. Farmbush and open forest; 50–1200m.
Distr.: Senegal to Zambia. (Tropical Africa).
IUCN: LC
Kodmin: Biye 3 1/1998; **Kupe Village:** Etuge 2765 7/1996; Kenfack 215 7/1996; 312 7/1996; Zapfack 924 7/1996; **Mejelet-Ehumseh:** Etuge 293 fr., 10/1986; **Mile 15, Kumba-Mamfe Road:** Nemba 207 fr., 9/1986; **Mungo River:** Gosline 229 11/1999; **Mungo River F.R.:** Etuge 4312 10/1998; Leeuwenberg 8697 fl., 11/1971; **Nyasoso:** Cable 3243 6/1996; Sidwell 391 10/1995.
Note: Gosline 229 is a large-leaved and large-fruited specimen, similar to several specimens from Mt. Cameroon; the leaves are more coriaceous than usual for this species.

Cissus smithiana (Baker) Planch.

Fl. Cameroun 13: 94 (1972).
Climber to 5m; stems cylindrical, glabrous; tendrils few; leaves ovate, 8–11 × 7–9.3cm, apex short-acuminate, base deeply cordate, margin dentate, teeth apices revolute, glabrescent; petiole 5–6.5cm; infructescence c. 9 × 8cm, of several subumbellate cymes, many-fruited; peduncle 3.5cm;

pedicels 0.6cm; fruits obovoid, 5.5 × 3cm, glabrous, green; seed subreniform, 5.5 × 3 × 2.5cm, ridged ventrally, irregularly sculptured. Forest; 900m.

Distr.: Cameroon to Congo (K). (Lower Guinea & Congolian).

IUCN: LC

Enyandong: Etuge 4440r fr., 10/2001.

Note: separation of this species from *C. glaucophylla* Hook.f., of which *C. smithiana* was treated as a synonym in FWTA, is unclear. Fl. Cameroun, which we follow here, does not recognize *C. glaucophylla* in Cameroon. Our material, however, lacks the prominent acumen usual in *C. smithiana*.

Cyphostemma mannii (Baker) Desc.

Fl. Cameroun 13: 68 (1972).

Syn.: *Cissus mannii* (Baker) Planch.

Herbaceous climber; stems cylindrical, pubescent; tendrils bifid; leaves palmately 5-foliolate, c. 11 × 12cm; leaflets obovate-elliptic, central 8–9.5 × 3.8–4.6cm, apex shortly acuminate, base acute, margin dentate, puberulent along veins; petiole 3.5–4.5cm, pubescent; compound cyme c. 10 × 14cm, subcorymbiform, puberulent throughout; flowers 0.4cm; calyx cupular; corolla pinched in centre, apex truncate; fruit globose, 6mm diameter, glabrous; seeds ellipsoid, 4mm, ridged. Forest and thicket; 2000m.

Distr.: Bioko & Cameroon. (W Cameroon Uplands).

IUCN: LC

Mwanenguba: Pollard 166 10/1998.

Cyphostemma rubrosetosum (Gilg & Brandt) Desc.

Fl. Cameroun 13: 56 (1972).

Syn.: *Cissus rubrosetosa* Gilg & Brandt

Herbaceous climber; stems cylindrical, densely glandular-hairy, hairs to 5mm, red; tendrils bifid; leaves papery, palmately 5-foliolate, 12.5 × 14cm; leaflets obovate, central 12 × 4.8cm, apex shortly acuminate, base cuneate, margin crenate-dentate, sparsely pilose along veins; petiole 7–10cm, pilose; compound cyme, c. 11 × 17cm; peduncles pubescent; flower 0.4cm; calyx cupular; corolla pinched at the centre, apex truncate, glabrous; pedicels 0.3cm, pubescent; fruit subglobose, 0.5cm diameter, glabrous; seeds subglobose, c. 4mm, striate. Farmbush, open forest; 800–1000m.

Distr.: Guinea (C) to CAR. (Upper & Lower Guinea).

IUCN: LC

Kupe Village: Kenfack 223 7/1996; **Nyasoso:** Cable 3454 7/1996; Etuge 2464 6/1996.

FAMILY INDET.

B.J. Pollard

Tree to 20m; leaves simple, alternate, to c. 22 × 8cm, elliptic, oblong-elliptic or slightly obovate, entire, brochidodromous, exstipulate; petioles conspicuously 'exfoliating', woody, pulvinate; inflorescences of axillary brachyblasts, with many bracteoles; flowers pentamerous; sepals connate forming a short tube; petals 'clawed' as in Malpighiaceae; stamens 10; ovary unilocular with 1 ovule; style terminal, unbranched; fruits unknown.

Nyandong: Cheek 11384 fl., 3/2003.

Note: we have not been able to establish which family this taxon belongs to, despite having abundant fertile material. It could prove to represent an undescribed taxon, but at which rank we are not yet sure, as its affinities are unclear. Fruiting material is desirable.

MONOCOTYLEDONAE

AMARYLLIDACEAE

I. Nordal & I. Darbyshire

Fl. Cameroun 30 (1987).

Crinum jagus (Thomps.) Dandy

Bulbous herb, 0.6–1m; leaves erect, often petiolate, c. 50 × 7cm, c. 30-nerved; margin often undulate; inflorescence above leaves, 3-flowered, with 2 involucral bracts; flowers irregular, infundibuliform, sweetly scented, 7 × 3.5cm, white with green markings; stamens black. Forest, savanna, plantations, usually riverine; 800–1450m.
Distr.: Guinea (C) to Angola, Sudan, Uganda. (Guineo-Congolian).
IUCN: LC
Kodmin: Satabie 1110 fl., 1/1998; **Kupe Village:** Cable 2660 5/1996; Etuge 1710 2/1996; **Nsoung:** Darbyshire 91 3/2003; **Nyasoso:** Cheek 7899 11/1995.

Crinum natans Baker

Submerged aquatic herb; bulb cylindrical, c. 3 × 2cm; leaves in a rosette of 5-9, strap-shaped, 100–180 × 3–3.5(–5)cm, apex rounded, margin strongly undulate, fleshy, glabrous, midrib prominent above; inflorescence erect, umbellate, single, 3–5(–7)-flowered; peduncle 9–55cm, bearing 2 lanceolate bracts 6.5–8 × 1.5–3cm; flowers borne above the water, sweetly scented, white; perianth tube 12–16 × 0.2–0.4cm, lobes 6, spreading, 7–8.5 × 0.75–1.5cm; stamens 6, filaments white, almost as long as petals, erect; anthers black; style slightly longer than anthers; fruit globular, c. 1.2cm diameter. Partially shaded forest rivers and streams.
Distr.: Guinea (C) to Gabon. (Upper & Lower Guinea).
IUCN: LC
Manengolé: Bamps 1584 fl., 12/1967.
Note: specimen cited in Fl. Cameroun 30: 16 (1987).

Crinum purpurascens Herb.

Bulbous herb to 1m; bulb subglobose, 3–5cm diameter; leaves 20–70(–100) × 1–4cm; scape reddish, 20–50cm; bracts subtending 2–10 subsessile flowers; flowers with perianth segments spreading, c. 7 × 7cm, white; filaments purple apically; stamens orange. Farmbush; 800m.
Distr.: Gambia to Angola. (Upper & Lower Guinea).
IUCN: LC
Nyasoso: Etuge 2491 7/1996.

Scadoxus cinnabarinus (Decne.) Friis & Nordal

Fl. Gabon 28: 25 (1986).
Syn.: *Haemanthus cinnabarinus* Decne.
Herb, 0.4–0.5m; lacking bulb, acaulous; leaves numerous; blades elliptic, c. 20 × 8cm; petiole 14cm, winged; inflorescence terminal; peduncle 20–30cm; flowers pink, in umbel 10–15cm diameter, each flower 4cm wide; fruit globose, orange. Forest; 300–2050m.
Distr.: Sierra Leone to Uganda. (Guineo-Congolian).
IUCN: LC
Essossong: Groves 48 fl., 2/1995; **Kodmin:** Gosline 63 1/1998; Satabie 1105 fl., 1/1998; **Kupe Village:** Cable 2448 5/1996; 2687 5/1996; 728 fl., 1/1995; 850 fl., 1/1995; Cheek 6039 fl., fr., 1/1995; Etuge 1387 11/1995; 1937 5/1996; 2786 7/1996; 2896 7/1996; Lane 261 fl., 11/1994; 419 fr., 1/1995; Ryan 261 5/1996; 282 5/1996; 319 5/1996; Sidwell 444 11/1995; **Mt Kupe:** Sebsebe 5068 10/1995; Thomas 5469 2/1986; **Ndum:** Lane 481 fl., 2/1995; **Nyasoso:** Cable 120

fr., 9/1992; 3274 6/1996; Etuge 1829 3/1996; 2188 6/1996; 2439 6/1996; 2474 6/1996; Lane 215 fl., 11/1994; Zapfack 695 6/1996.

Scadoxus multiflorus (Martyn) Raf.

Fl. Cameroun 30: 8 (1987).
Syn.: *Haemanthus multiflorus* Martyn
Bulbous herb 25–80cm; bulb cylindrical, c. 2 × 1.5cm; leaves expanding after flowering, ovate-lanceolate, to 25 × 8cm, base attenuate; inflorescence lateral, 7–25cm, globose, many-flowered; flowers scarlet; pedicels 1.5–3.5cm. Forest edge and semi-deciduous forest; 1500m.
Distr.: Senegal to Somalia & to S Africa, also Yemen. (Tropical Africa).
IUCN: LC
Mt Kupe: Thomas 5493 2/1986; **Nyandong:** Ghogue 1451 fr., 3/2003.
Note: specimens not seen by I.Nordal; determinations require confirmation.

Scadoxus pseudocaulus (Bjornst. & Friis) Friis & Nordal

Fl. Cameroun 30: 6 (1987).
Syn.: *Haemanthus sp. A sensu* Hepper in FWTA 3: 132
Herb to 80cm; rhizome short; sheathing leaf bases form a false stem to 40cm; leaves elliptic-lanceolate, 15–40 × 5–10cm, appearing with the flowers; scape derived from the centre of the false stem, to 65cm; inflorescence subglobose, many-flowered; flowers pale red; pedicels c. 2cm. Forest; 1100m.
Distr.: Nigeria to Gabon. (Lower Guinea).
IUCN: NT
Mejelet-Ehumseh: Etuge 476 1/1987.
Note: specimen not seen by I.Nordal; description from Fl. Cameroun. Taxon known from only c. 16 sites.

ANTHERICACEAE

A.D. Poulsen & I. Nordal

Chlorophytum alismifolium Baker

Herb 15–20cm; leaves distichous, lanceolate; blade c. 15 × 3.5cm, acute, base rounded-decurrent; petiole 6cm; inflorescence as long as or shorter than the leaves. Forest; 750–1000m.
Distr.: Sierra Leone to Congo (K). (Guineo-Congolian).
IUCN: LC
Nyasoso: Cable 3253 6/1996; 3521 7/1996; Etuge 2585 7/1996; Zapfack 693 6/1996.

Chlorophytum comosum (Thunb.) Jacq.

Bothalia 7: 698 (1962).
Herb 40–60cm; leaves in rosette, oblanceolate-ligulate, c. 35 × 3cm, lower 10cm subpetiolate; inflorescence to 60cm. Forest, farmbush; 1200–1900m.
Distr.: tropical & southern Africa. (Afromontane).
IUCN: LC
Mt Kupe: Etuge 1364 11/1995; **Nyasoso:** Etuge 1318 10/1995.

Chlorophytum orchidastrum Lindl.

Herb 40–100cm; drying black; leaves ovate, 25 × 10cm, acuminate-mucronate, obtuse; petiole c. 30cm; inflorescence about as long as petioles. Forest; 700–1300m.
Distr.: Guinea (C) to Cameroon, Congo (K) & Zambia. (Guineo-Congolian).

IUCN: LC
Kodmin: Etuge 3970a 1/1998; **Kupe Village:** Cable 3859
7/1996; 3892 7/1996; 720 fl., 1/1995; Cheek 7115 1/1995;
Etuge 2710 7/1996; 2860 7/1996; Kenfack 316 7/1996; Lane
322 fl., 1/1995; 444 fl., 1/1995; Ryan 262 5/1996;
Ngomboaku: Muasya 2035 12/1999; **Nyasoso:** Balding 32
fl., fr., 8/1993; Cable 2887 6/1996; 3281 6/1996; 3325
6/1996; 3441 7/1996; Etuge 1316 10/1995; 2551 7/1996;
Zapfack 685 6/1996.

Chlorophytum cf. *orchidastrum* Lindl.
Forest; 940–1050m.
Nyasoso: Cable 27 fl., fr., 7/1992; Lane 41 fl., 8/1992;
Wheatley 384 fl., fr., 7/1992.
Note: leaves unusually large for this species; specimens not
seen by A.Poulsen or I.Nordal.

Chlorophytum sparsiflorum Baker
Herb 25–60cm, drying light green, sometimes viviparous;
leaves oblanceolate or oblanceolate-ligulate, c. 25 × 6cm,
acute-mucronate, base tapering into a variably defined
petiole; inflorescence about as long as leaves, or longer.
Forest; 400–1850m.
Distr.: Sierra Leone to Kenya. (Afromontane).
IUCN: LC
Bakossi F.R.: Etuge 4300 10/1998; **Kodmin:** Cheek 9047
1/1998; Etuge 4029 1/1998; 4155 2/1998; Pollard 195 fl., fr.,
11/1998; **Kupe Village:** Cable 3756 7/1996; 3757 7/1996;
Etuge 2876 fl., fr., 7/1996; **Mwambong:** Onana 536 2/1998;
Nyasoso: Balding 64 fl., 8/1993; Cable 99 fl., fr., 9/1992;
Etuge 2470 6/1996; 2510 7/1996; Pollard 85 fr., 10/1998;
Sidwell 359 10/1995; Sunderland 1508 fl., fr., 7/1992.
Note: this taxon is soon to be reduced to a variety of *C.
comosum* (Thunb.) Jacq. By I.Nordal & A.Poulsen.

ARACEAE
P.C. Boyce & I. Darbyshire

Fl. Cameroun 31 (1988).

Amorphophallus calabaricus N.E.Br.
Herb; leaf like a tattered umbrella; leaflets acute, not fishtail
shaped; petiole to 1.2m, smooth; inflorescence dark purple;
peduncle to 1m; spathe base hairy inside; spadix base
swollen. Forest; 890–1100m.
Distr.: Burkina Faso, Nigeria, Cameroon, Uganda & Kenya.
(Guineo-Congolian).
IUCN: NT
Nyasoso: Cheek 7463 10/1995; Sidwell 367 10/1995; 395
10/1995.
Note: known from only 13 locations, 6 of which are pre–
1960's collections from Nigeria, where habitat loss may
threaten this species.

Amorphophallus preussii (Engl.) N.E.Br.
Herb resembling *A. calabaricus*, but spathe base hairless
inside. Forest, sometimes on rocks; 800–1600m.
Distr.: W Cameroon. (W Cameroon Uplands).
IUCN: VU
Bangem: Thomas 5270 1/1986; **Nyasoso:** Cable 1200
2/1995; 697 fl., 1/1995; 867 1/1995.

Amorphophallus staudtii (Engl.) N.E.Br.
Herb resembling *A. calabaricus*, but inflorescence pale dirty
cream-white; peduncle very short; spadix base not swollen.
Forest; 1000–1460m.

Distr.: Cameroon, Equatorial Guinea. (Lower Guinea).
IUCN: NT
Kodmin: Cheek 9061 1/1998; 9578 11/1998; **Mwambong:**
Mackinder 155 1/1998; **Nyasoso:** Sebsebe 5001 10/1995.
Note: known from only 10 sites in Cameroon, where it is
likely threatened in the Bamenda Highlands by agricultural
encroachment into existing forest patches. Also recorded
from Equatorial Guinea, but poorly documented there.
Local name: Echin-ngwek. **Uses:** MEDICINES (Max Ebong
in Cheek 9578).

Anchomanes difformis (Blume) Engl.
Syn.: *Anchomanes difformis* (Blume) Engl. var. *pallidus*
(Hook.) Hepper
Syn.: *Anchomanes welwitschii* Rendle
Herb; leaf like a tattered umbrella; leaflets fishtail-shaped;
petiole spiny (prickles); spathe green, tinged purple; styles
curved, scabrid. Forest, forest margins; 300–1200m.
Distr.: Sierra Leone to Congo (K), Angola & Sudan.
(Guineo-Congolian).
IUCN: LC
Bakole: Nemba 784 2/1988; **Mejelet-Ehumseh:** Etuge 295
10/1986; **Nyasoso:** Sunderland 1511 fr., 7/1992.
Note: *A. hookeri* was until recently treated as a synonym of
A. difformis but the two are separable on differences in the
styles. However, the distributions of these taxa have not been
fully defined as yet; no conservation assessments can be
made until this is clarified.

Anchomanes hookeri Schott
Oestr. Bot. Wochenbl. 3: 314. (1853).
Herb resembling *A. difformis* but spathe interior purple;
styles curved downwards, warty, deep purple. Forest; 930–
1300m.
Distr.: Guineo-Congolian.
Kupe Village: Cable 2501 5/1996; 2576 5/1996; 2632
5/1996; 827 fl., 1/1995; Ryan 254 5/1996; **Mwambong:**
Mackinder 166 1/1998; **Nyasoso:** Cable 2872 6/1996.
Note: see note to *A. difformis*.

Anubias barteri Schott var. *barteri*
Meded. Land. Wag. 79(14): 12 (1979).
Terrestrial herb; leaves ovate-lanceolate, 7–20 × 4–10cm,
base truncate to cordate, with many prominent lateral nerves;
fruits green, enclosed within a green, persistent spathe.
Rivers and streams in forest; 1100m.
Distr.: SE Nigeria, Bioko, Cameroon. (W Cameroon
Uplands).
IUCN: LC
Nyasoso: Cheek 7320 fr., 2/1995; Sidwell 364 10/1995.
Note: *A. barteri* var. *glabra* N.E.Br. is also likely in our area,
being recorded S of Nkongsamba (*op. cit.:* 19); it differs in
having a cuneate to obtuse leaf base and leaves more than 2.5
× as long as broad.
Local name: none known (Cheek 7320); Machang (Cheek
11302). **Uses:** none known (Cheek 7320).

Anubias barteri Schott var. *caladiifolia* Engl.
Meded. Land. Wag. 79(14): 15 (1979).
Herb resembling var. *barteri*, but leaf base sagittate, sinus
wide, main veins passing downwards into the lobes. Rivers
and streams in forest; 700–1430m.
Distr.: SE Nigeria, Bioko, Cameroon. (W Cameroon
Uplands).
IUCN: LC
Kodmin: Ghogue 38 1/1998; **Kupe Village:** Cable 767
1/1995; Cheek 6026 fl., fr., 1/1995; Lane 317 fr., 1/1995;
Menyum: Doumenge 469 5/1987; **Ngomboaku:** Ghogue

461 12/1999; Mackinder 293 12/1999; 305 12/1999; **Nyasoso:** Lane 194 fl., fr., 11/1994.

Cercestis camerunensis (Ntépé-Nyamè) Bogner
Aroideana 8(3): 73 (1986).
Syn.: *Rhektophyllum camerunense* Ntépé-Nyamè
Robust climber; stem rooting along whole length; leaves scattered along whole stem; leaf blade pinnately divided, up to 50cm long; spathe up to 12cm long. Forest; 400–900m.
Distr.: Nigeria, Cameroon & Gabon. (Lower Guinea).
IUCN: LC
Enyandong: Salazar 6306 10/2001; **Kupe Village:** Etuge 2019 5/1996; 2758 7/1996; **Ngomboaku:** Ghogue 467 12/1999; **Nyasoso:** Cheek 5671 12/1993.
Note: locally numerous in Cameroon, excluding the east of the country (Fl. Cameroun 31: 64 (1988)).

Cercestis dinklagei Engl.
Fl. Cameroun 31: 66 (1988).
Syn.: *Cercestis elliotii* Engl.
Syn.: *Cercestis ledermannii* Engl.
Syn.: *Cercestis sagittatus* Engl.
Syn.: *Cercestis stigmaticus* N.E.Br.
Slender climber; rooting at same nodes as leaf-clusters, clusters spaced intermittently along stem; leaf-blade entire, lamina to 30cm, base deeply cordate, auricles oblong; spathe to 7cm long. Forest; 700–1300m.
Distr.: Guinea (C) to Congo (B). (Upper & Lower Guinea).
IUCN: LC
Kupe Village: Cable 2513 5/1996; 3783 7/1996; 3826 7/1996; Cheek 7185 fr., 1/1995; 7664 11/1995; Etuge 2888 7/1996; **Nyale:** Etuge 4232 2/1998.

Cercestis kamerunianus (Engl.) N.E.Br.
Climber resembling *C. dinklagei*, but leaf bases less deeply cordate, auricles short, rounded, close to the petiole, lamina occasionally with translucent glandular lines; spathe to 4.5cm. Forest; 600m.
Distr.: Nigeria, Cameroon & Gabon. (Lower Guinea).
IUCN: NT
Etam: Etuge 225a 8/1986.
Note: specimen not seen; det. by T.Croat (MO). Taxon known from less than 20 localities.

Colocasia esculenta (L.) Schott
Erect herb with subglobose tubers to 10cm diameter, not staining yellow when cut; leaves peltate, lamina cordiform, 20–40 × 15–35cm, basal lobes obtuse to rounded, 3–7cm; peduncle c. 15cm; spathe 1.8cm, acuminate; spadix much shorter. Cultivated; 400–1170m.
Distr.: originating in Polynesia, widely cultivated. (Pantropical).
IUCN: LC
Edib village: Cheek (sight record) 2/1998; **Nyandong:** Cheek 11480 3/2003.
Local name: Cocoyam. **Uses:** FOOD – a common vegetable in Cameroon (*fide* Cheek).

Culcasia angolensis Welw. ex Schott
Syn.: *Culcasia barombensis* N.E.Br.
Robust climber; leaf-blade usually more than 20 × 10cm, oblong-elliptic to ovate; inflorescences several together; infructescences to 5cm long, without persistent spathe. Forest; 650–1150m.
Distr.: Guinea (C) to Congo (K). (Guineo-Congolian).
IUCN: LC
Kupe Village: Lane 411 fr., 1/1995; **Ndum:** Cable 949 fr., 2/1995; **Nyale:** Etuge 4208 2/1998; **Nyasoso:** Cable 2852

6/1996; Cheek 7511 10/1995; Etuge 2130 6/1996; 2412 6/1996; Sidwell 387 10/1995; Thomas 3048 2/1984.

Culcasia dinklagei Engl.
Fl. Cameroun 31: 80 (1988).
Herbaceous climber or held erect by stilt roots; leaf blade to 12–35 × 4–15cm, obovate to oblanceolate, with transparent lines; petiole shorter than lamina, to 9cm; infructescence solitary or paired. Forest; 150–800m.
Distr.: Guinea (C) to Congo (B). (Upper & Lower Guinea).
IUCN: LC
Bangem: Thomas 5365 1/1986; **Etam:** Etuge 93 3/1986.
Note: specimens not seen at K, determined by T.Croat (MO).

Culcasia ekongoloi Ntépé-Nyamè
Fl. Cameroun 31: 87 (1988).
Slender climber; stems minutely roughened; leaves with transparent lines; petiole as long as lamina. Streambanks in forest; 700–1500m.
Distr.: Nigeria to Congo (K) & CAR. (Guineo-Congolian).
IUCN: LC
Kodmin: Cheek 9628 11/1998; Etuge 4164 2/1998; Gosline 177 11/1998; Plot B75 1/1998; **Kupe Village:** Cheek 6041 fr., 1/1995; 8402 5/1996; Ryan 372 5/1996.
Local name: Ntonkem (Max Ebong in Cheek 9628).

Culcasia insulana N.E.Br.
Fl. Cameroun 31: 96 (1988).
Slender climber resembling *C. ekongoloi*, but stems smooth; lamina lanceolate, asymmetric, drying pale, without transparent lines. Forest; 1150–1700m.
Distr.: Bioko, Cameroon, Congo (K) & possibly CAR. (Lower Guinea & Congolian (montane)).
IUCN: NT
Kupe Village: Cheek 7202 fr., 1/1995; 7785 11/1995; **Nyasoso:** Cheek 7332 fr., 2/1995; Wheatley 449 fr., 7/1992.
Note: rare throughout its range, only c. 10 locations known, though range in Congo (K) is poorly recorded. This taxon's submontane habitat remains largely unthreatened.

Culcasia mannii (Hook.f.) Engl.
As *C. barombensis*, but a moderately robust terrestrial herb; lamina subelliptic, lacking translucent lines; with solitary infructescences. Forest; 150–1600m.
Distr.: SE Nigeria, Cameroon & Rio Muni. (Lower Guinea).
IUCN: LC
Baseng: Cheek 10422 12/1999; **Etam:** Etuge 200 7/1986; **Kupe Village:** Cable 2516 5/1996; 2647 5/1996; 2738 5/1996; 837 fr., 1/1995; Cheek 6056 fr., 1/1995; 7124 fr., 1/1995; Etuge 2807 7/1996; Lane 364 fr., 1/1995; Ryan 323 5/1996; 358 5/1996; **Ngomboaku:** Cheek 10329 12/1999; **Nyasoso:** Balding 7 fr., 7/1993; Cable 1167 fr., 2/1995; 2844 6/1996; 3323 6/1996; 656 fr., 12/1993; Etuge 1303 10/1995; 1818 3/1996; 2104 6/1996; 2157 6/1996; 2365 6/1996; Lane 149 fr., 6/1994; 162 fr., 10/1994; Sebsebe 4981 10/1995; Sunderland 1534 fr., 7/1992.

Culcasia obliquifolia Engl.
Fl. Cameroun 31: 95 (1988).
Climber resembling *C. insulana*, but lamina not asymmetric, drying dark. Forest; 1200m.
Distr.: Cameroon, Gabon & Cabinda. (Lower Guinea).
IUCN: NT
Kupe Village: Cheek 7214 fl., 1/1995.
Note: recorded from only 11 locations, mainly in S Province, Cameroon; the specimen cited is recorded at an unusually high altitude for this taxon.

ARACEAE

Culcasia parviflora N.E.Br.
Herbaceous climber; distinguished from other *Culcasia* by the cordate lamina base and the petiole being c. 1/5 the length of the lamina. Forest; 300–1200m.
Distr.: Guinea (C) to Congo (K). (Guineo-Congolian).
IUCN: LC
Kodmin: Biye 15 1/1998; **Kupe Village:** Cable 2603 5/1996; 3663 7/1996; 3850 7/1996; Cheek 7158 fr., 1/1995; 7660 11/1995; 7675 11/1995; Etuge 1907 5/1996; 1959 5/1996; 2777 7/1996; Kenfack 244 7/1996; **Nyasoso:** Cable 3573 7/1996; Cheek 7456 10/1995; Sidwell 381 10/1995.

Culcasia sapinii De Wild.
Syn.: *Culcasia seretii* De Wild.
Herbaceous climber; innovations (new shoots) copper-orange; petiole more than half the length but always shorter than lamina; leaves with translucent dots. Forest; 1000–1600m.
Distr.: W Africa to Congo (K). (Guineo-Congolian).
IUCN: LC
Kupe Village: Cheek 7186 fl., fr., 1/1995; 8393 5/1996; **Nyasoso:** Etuge 2179 6/1996.
Note: uncommon throughout much of its range; only 5 previous collections at K, but 16 locations recorded in Fl. Cameroun. Submontane forest habitat largely unthreatened. P.Boyce contends that *C. seretii* may prove to be a good species following further investigation.

Culcasia scandens P.Beauv.
Syn.: *Culcasia lancifolia* N.E.Br.
Syn.: *Culcasia saxatilis* A.Chev.
Herbaceous climber or epiphyte; stems verrucate, with numerous internodal roots; leaves asymmetrically ovate, to 15cm long, translucent lines very short, lamina not decurrent into the petiole; petiole with a spreading sheath, terminating c. 5mm below the blade; spathe to 3.5cm; fruiting spadix terminating in an appendage. Forest; 250–1450m.
Distr.: Senegal to Congo (K). (Guineo-Congolian).
IUCN: LC
Bangem: Thomas 5321 1/1986; **Etam:** Etuge 225 8/1986; **Mile 15, Kumba-Mamfe Road:** Nemba 150 7/1986; **Mt Kupe:** Thomas 5488 2/1986.
Note: specimens not seen at K; det. by T.Croat (MO).

Culcasia striolata Engl.
Terrestrial herb with stilt roots; lamina oblong, with numerous prominent short translucent lines. Forest; 300–1400m.
Distr.: Guinea (C) to Gabon. (Upper & Lower Guinea).
IUCN: LC
Kodmin: Barlow 2 1/1998; Plot B33 1/1998; **Kupe Village:** Cable 2633 5/1996; 2646 5/1996; **Molongo:** Nemba 779 2/1988; **Muahunzum:** Etuge 4439 11/1998; **Mungo River F.R.:** Cheek 10125 11/1999; **Nyasoso:** Cable 3311 6/1996.

Culcasia tenuifolia Engl.
Climbing herb, resembling *C. scandens*, but leaf lamina decurrent into the petiole; stems smooth. Forest; 300m.
Distr.: Sierra Leone to Gabon & CAR. (Guineo-Congolian).
IUCN: LC
Konye: Thomas 5168 11/1985.
Note: specimen not seen at K; det. by T.Croat (MO).

Culcasia sp. 1
Fl. Cameroun 31: 97 (1988).
Climber to 3m with stilt roots; leaves to 15 × 10cm, obovate, apex cuspidate, short translucent and black lines numerous; petiole c. ⅓ the length of the lamina. Forest; 900–1100m.

Distr.: Cameroon Endemic.
Kupe Village: Cable 726 fr., 1/1995; 828 fr., 1/1995.

Nephthytis poissonii (Engl.) N.E.Br.
Syn.: *Nephthytis gravenreuthii* (Engl.) Engl.
Syn.: *Nephthytis constricta* N.E.Br.
Syn.: *Nephthytis poissonii* (Engl.) N.E.Br. var. *constricta* (N.E.Br.) Ntépé-Nyamè
Terrestrial creeping herb; leaves triangular, posterior lobes considerably more developed than the anterior lobes; fruits orange-red subtended by a spreading, persistent green spathe. Forest; 300–1700m.
Distr.: Sierra Leone to Gabon. (Upper & Lower Guinea).
IUCN: LC
Ehumseh: Manning 450 10/1986; **Kodmin:** Biye 6 1/1998; Nwaga 11 11/1998; 13 1/1998; **Konye:** Thomas 5144 11/1985; **Kupe Village:** Cable 2631 5/1996; 2732 5/1996; Cheek 6049 fl., fr., 1/1995; 7063 fl., fr., 1/1995; 8377 5/1996; Etuge 1390 11/1995; 2782 7/1996; 2897 7/1996; Lane 269 fr., 1/1995; Ryan 221 5/1996; **Mejelet-Ehumseh:** Etuge 310 10/1986; **Mt Kupe:** Thomas 5096 12/1985; **Mwambong:** Pollard 116 10/1998; **Nyasoso:** Cable 30 fl., fr., 7/1992; 660 fl., 12/1993; 698 fl., 1/1995; NaN fl., 8/1992; Etuge 1798 3/1996; 2093 6/1996; Lane 124 fl., 6/1994; Sebsebe 5039 10/1995; 5041 10/1995; Sidwell 365 10/1995; 394 10/1995; Thomas 3077 2/1984; Wheatley 396 fr., 7/1992; Zapfack 704 6/1996.
Note: of the specimens cited, Thomas 3077 & 5096, Manning 450 & Etuge 310 not seen at K, determined at MO as *N. afzelii* Schott by T.Croat, but this species does not occur in Cameroon (Fl. Cameroun 31: 52 (1988)); further study required to confirm these as *N. poissonii*.
Uses: none known (Epie Ngome in Pollard 116).

Remusatia vivipara (Roxb.) Schott
Tuberous epiphytic herb; stolons erect, bulbil-bearing; leaves peltate, cordiform, 10–30 × 5–15cm; petiole to 15cm. Forest; 1200–1550m.
Distr.: palaeotropical. (Montane).
IUCN: LC
Mt Kupe: Cheek 7616 11/1995; **Nyasoso:** Balding 70 7/1993; Cheek 9508 10/1998.

Rhaphidophora africana N.E.Br.
Moderately robust climber, resembling *Cercestis camerunensis*, but lamina entire, torn edges with numerous silky hairs; inflorescences in pendent clusters; spathe to 10cm long, yellow. Forest; 300–1300m.
Distr.: Sierra Leone to Congo (K) & CAR. (Guineo-Congolian).
IUCN: LC
Bakole: Nemba 789 2/1988; **Kodmin:** Ghogue 71 2/1998; Plot B144 1/1998; B203 /1998; **Kupe Village:** Cheek 6073 fr., 1/1995; 7069 1/1995; Lane 448 fl., 1/1995; **Ndum:** Cable 948 fr., 2/1995; Groves 31 2/1995; **Nyasoso:** Sidwell 373 10/1995.

Rhaphidophora pusilla N.E.Br.
Kew Bull.: 286 (1897).
Very slender creeping or climbing herb; inflorescences solitary, erect; spathe up to 2cm long, white. Forest; 1520m.
Distr.: Cameroon & Gabon. (Lower Guinea).
IUCN: VU
Kodmin: Mackinder 141 1/1998.

Stylochaeton zenkeri Engl.

Terrestrial rhizomatous herb; lamina elliptic-sagittate, posterior lobes only weakly developed; fruits red, without spathe remains. Forest; 1450m.
Distr.: Sierra Leone to Congo (B). (Upper & Lower Guinea (montane)).
IUCN: LC
Kodmin: Cheek 8932 1/1998.

Xanthosoma sagittifolium (L.) Schott

Bot. J. Linn. Soc. 110: 267 (1992).
Syn.: *Xanthosoma mafaffa* Schott
Rhizomatous herb with tubers to 35 × 10cm, staining yellow when cut; leaves sagittate, 30–45(–80) × 20–35(–60)cm, cordiform to sagittate, basal lobes rounded to oblong; petiole to 60cm; spathe c. 20cm; spadix slightly longer than spathe. Cultivated; 420–1170m.
Distr.: originally from S America, now widely cultivated. (Pantropical).
IUCN: LC
Edib village: Cheek (sight record) 2/1998; **Nyandong:** Cheek 11382 3/2003.
Local names: Akwaneh or 'Cocoyam' (together with *Colocasia esculenta*). **Uses:** FOOD – grown as a vegetable (Cheek 11382)

ASPARAGACEAE

Sebsebe Demissew

Asparagus warneckei (Engl.) Hutch.

Woody climber to 8m; stems smooth, glabrous; spines short and recurved, only on the main shoots; 'cladodes' flattened, ± falcate, c. 4 × 0.2cm; inflorescence racemose, c. 5–7cm, often clustered; pedicels jointed nearly at the base, about 0.5cm in flower; flowers white or cream-coloured with a strong sickly odour. Thicket, forest, scrub; 850–1975m.
Distr.: Guinea (C), Ghana, Togo, W Cameroon. (Upper & Lower Guinea).
IUCN: NT
Mwanenguba: Cheek 7246 fr., 2/1995; Etuge 4372 10/1998; **Ndum:** Cable 893 1/1995; Lane 452 fr., 1/1995; **Nyasoso:** Etuge 2542 7/1996.
Note: uncommon over much of its range.

BROMELIACEAE

I. Darbyshire

Ananas comosus (L.) Merrill

Robust erect herb; scape stout, short; leaves rosulate, coriaceous, lanceolate, margin coarsely spinose-serrate; inflorescence strobiliform, many-flowered, dense, crowned by stiff, foliaceous, lanceolate bracts to 30cm; flowers sessile; floral bracts triangular, serrulate or entire; syncarp to 20–40cm long when mature, fleshy. Commonly cultivated, sometimes persisting in abandoned farms.
Distr.: widely cultivated in the tropics, originating from Brazil. (Pantropical).
IUCN: LC
Nyandong: Darbyshire (sight record) 3/2003.

Local names: Pineapple or Ananas. **Uses:** FOOD – an important cash crop, commonly cultivated in Bakossi (*fide* Cheek).

BURMANNIACEAE

M. Cheek

Afrothismia pachyantha Schltr.

Herb, 2–4cm; leaves and chlorophyll absent; flowers single; perianth white with purple triangles, tube sigmoid, 1cm, mouth dull purple, held horizontally; tepals triangular, equal, 2mm. Forest; 800m.
Distr.: SW Cameroon. (Narrow Endemic).
IUCN: CR
Kupe Village: Williams s.n. 11/1995.

Afrothismia saingei T.Franke

Syst. Geogr. Pl. 74: 27–33 (2004).
Herb, 3cm; leaves and chlorophyll absent; flowers single; perianth tube pendulous, obovoid, 2 × 1cm, whitish translucent with basal red veins, the mouth held horizontal to erect, the six tepals yellow, strongly uneqal, lowest pair (of six) c. 7cm long, upper pair 2–3cm. Forest; 970m.
Distr.: Mt Kupe. (Narrow Endemic).
IUCN: CR
Mbule: Sainge 1053 10/2002.

Afrothismia winkleri (Engl.) Schltr.

Herb, 2–4cm; leaves and chlorophyll absent; flowers 1–2; perianth mouth bright yellow, held vertically; tepals linear, equal, c. 1cm, yellow. Forest; 1040–1150m.
Distr.: SE Nigeria, Cameroon & Uganda. (Lower Guinea & Congolian).
IUCN: CR
Kupe Village: Cable 3806 fl., 7/1996; Cheek 8354 5/1996; **Nyasoso:** Cable 2830 fl., 6/1996.

Burmannia congesta (C.H.Wright) Jonker

Herb, 4–6cm; leaves and chlorophyll absent; flowers 3–8, congested at apex of peduncle, cylindrical, erect, c. 7mm long, white. Forest; 950–1100m.
Distr.: Liberia to Congo (K). (Guineo-Congolian).
IUCN: NT
Kupe Village: Cheek 7857 fl., 11/1995; Williams 261 11/1995; **Ngomboaku:** Ghogue 508 12/1999.

Burmannia hexaptera Schltr.

Herb c. 6cm, resembling *B. congesta*, but more robust and often with a single flower, c. 12mm long. Forest; 850m.
Distr.: Nigeria & Cameroon. (Lower Guinea).
IUCN: NT
Kupe Village: Williams 260 11/1995.

Gymnosiphon longistylus (Benth.) Hutch.

Herb 4–7cm; leaves and chlorophyll absent; flowers 1–10; perianth white, 3-lobed, 5mm diameter, lobes rounded. Forest; 900–950m.
Distr.: Sierra Leone to Gabon. (Upper & Lower Guinea).
IUCN: NT
Kupe Village: Williams 262 11/1995; **Ngomboaku:** Ghogue 507 fl., 12/1999.

CANNACEAE

Gymnosiphon sp. A
Herb resembling *G. longistylus*, but perianth lobes with long white tassles. Forest; 1275m.
Distr.: Bakossi Mts (Narrow Endemic).
Kodmin: Cheek 9635 fl., fr., 11/1998.

CANNACEAE
B.J. Pollard

Fl. Cameroun 4 (1964).

Canna indica L.
Cultivated herb to c. 1m; stems erect, glabrous; leaves ovate-elliptic, broadly acuminate, abruptly cuneate, to 40 × 25cm, glabrous; inflorescences terminal, racemose, few-flowered; pedicels very short; bracts ovate, c. 1.3cm; flowers usually scarlet, with centres orange; petals linear-lanceolate, c. 4cm; fruit a capsule; pericarp muricate. Gardens, villages, roadsides; 1170–1650m.
Distr.: native to tropical America, introduced in Africa and other tropical parts. (Pantropical).
IUCN: LC
Bangem: Ranaivojaoana 4 fl., 11/2001; **Edib village:** Cheek (sight record) 2/1998; **Mejelet-Ehumseh:** Manning 424 fl., 10/1986.
Uses: ENVIRONMENTAL USES – ornamental (*fide* Pollard).

COLCHICACEAE
B.J. Pollard

Gloriosa superba L.
Syn.: *Gloriosa simplex* L.
Herb, ± climbing; leaves sessile, in whorls of 3 or opposite or alternate, c. 8–10 × 4cm; often apically tendriliform; flowers yellow, turning red; perianth segments narrowly linear, 6–9 × c. 1cm with very crispy-waved margins; style longer than the perianth; filaments shorter; anthers c. 1–2cm long; fruits c. 9 × 2cm. Forest, farmbush; 900–1700m.
Distr.: widely distributed in tropical & S Africa; tropical Asia. (Palaeotropics).
IUCN: LC
Nsoung: Darbyshire 71 3/2003; **Nyasoso:** Cable 10 fl., 7/1992; Etuge 1749 fl., 2/1996; 2189 fl., fr., 6/1996; Pollard 103 fl., 10/1998.
Local name: Njungela. **Uses:** MATERIALS – hunting – take leaves and crush, put a small quantity up each nostril when hunting to help to smell the animals in the forest (Epie Ngome in Pollard 103); MEDICINES (*fide* Daniel Abbiw); ENVIRONMENTAL USES – ornamentals (*fide* Pollard).

COMMELINACEAE
R.B. Faden & M. Cheek

Amischotolype tenuis (C.B.Clarke) R.S.Rao
Maharashtra Vidnyan Mandir Patrika 6(3): 53 (1971).
Syn.: *Forrestia tenuis* (C.B.Clarke) Benth.
Decumbent herb 30–60cm tall; leaves variegated white and pink above in juvenile plants, up to 5 × 3cm, in adults elliptic, purple below, c. 10 × 5cm, acuminate; petiole 1–2cm; inflorescence axillary, perforating leaf-sheath, sessile, bifurcate; branches 1–3cm long; flowers purple. Lowland forest; 450–1000m.
Distr.: Nigeria to Congo (K). (Lower Guinea & Congolian (montane)).
IUCN: NT
Mbule: Etuge 4398 10/1998; **Nyale:** Etuge 4189 2/1998; **Nyandong:** Cheek 11358 3/2003.

Aneilema beniniense (P.Beauv.) Kunth
Weak erect herb to c. 1m; leaves elliptic, c. 15 × 5cm, acuminate, sessile or shortly petiolate; inflorescence terminal, dense, c. 3–4 × 3–4cm; flowers white or mauve; fruits longer than broad, apex rounded. Lowland farmbush, forest, streambanks and *Aframomum* thicket; 200–900m.
Distr.: tropical Africa.
IUCN: LC
Baduma: Etuge 390 11/1986; **Kodmin:** Barlow 10 1/1998; Biye 20 1/1998; **Kupe Village:** Cable 2449 5/1996; Cheek 6048 fl., 1/1995; Ensermu 3519 11/1995; Lane 368 fl., 1/1995; **Mile 15, Kumba-Mamfe Road:** Nemba 605 7/1987; **Mungo River F.R.:** Cheek 10175 11/1999, **Nyasoso:** Cable 3260 6/1996; Cheek 7490 10/1995; Zapfack 679 6/1996.

Aneilema dispermum Brenan
Weak erect herb to c. 1m, resembling *A. beniniense*, but capsules broader than long, locules 1-seeded (not 2–3-seeded). Montane forest edge, forest-grassland transition; 1100–2000m.
Distr.: Bioko, SW Cameroon, Malawi & Tanzania. (Afromontane).
IUCN: NT
Mejelet-Ehumseh: Etuge 449 1/1987; **Nyasoso:** Cable 1203 fl., 2/1995.

Aneilema silvaticum Brenan
Weak erect herb, 15–30cm tall; leaves ovate-lanceolate, c. 9 × 3cm, acuminate, petiolate, margin ciliate; inflorescence terminal, dense, < 2 × 2cm; flowers 10–15, white; fruit pointed. Lowland forest; 300–880m.
Distr.: Nigeria, Cameroon & Congo (K). (Lower Guinea & Congolian).
IUCN: VU
Kupe Village: Etuge 2778 7/1996; **Ngomboaku:** Ghogue 493 12/1999.

Aneilema umbrosum (Vahl) Kunth subsp. *ovato-oblongum* (P.Beauv.) J.K.Morton
Straggling herb 10–30cm high, resembling *A. umbrosum* subsp. *umbrosum*, but rarer, smaller and more slender; leaves to 8 × 3cm; sheaths lacking rusty hairs; inflorescence with 2–8 branches. Lowland forest gaps; 700–800m.
Distr.: tropical Africa & tropical America. (Amphi-Atlantic).
IUCN: LC
Kupe Village: Ensermu 3547 11/1995; 3550 11/1995; Ryan 325 5/1996; **Ngomboaku:** Ghogue 460 12/1999.

Aneilema umbrosum (Vahl) Kunth subsp. umbrosum
Straggling herb to 1m; leaves elliptic or lanceolate, up to 13 × 4cm, acuminate; petiolate; sheath with rusty hairs at apex and sometimes on surface; inflorescence terminal, lax, up to 12 × 7cm; flowers borne on 8–30 long, scattered branches, white. Lowland farmbush, forest; 650–1500m.
Distr.: Sierra Leone to Congo (K). (Guineo-Congolian (montane)).
IUCN: LC

Kodmin: <u>Cheek 8883</u> 1/1998; **Kupe Village:** <u>Cable 820</u> fl., 1/1995; **Mwambong:** <u>Cheek 9373</u> 10/1998; **Ngomboaku:** <u>Etuge 4661</u> 12/1999; <u>Ghogue 488</u> 12/1999; **Ngusi:** <u>Etuge 1578</u> 1/1996; **Nyasoso:** <u>Cable 2818</u> 6/1996; <u>Cheek 7491</u> 10/1995; <u>Etuge 2554</u> 7/1996.
Local name: Nkona ke. **Uses:** SOCIAL USES – used to ward off accidents; MEDICINES (Cheek 8883).

Buforrestia mannii C.B.Clarke
Robust herb 1(–2)m high, gregarious; leaves elliptic c. 30 × 10cm, acuminate, petiolate; inflorescences axillary, perforating leaf-sheath and concealed below the leaf-blade, patent, 10–15cm long; flowers 5–6, white, inconspicuous. Lowland and submontane forest, streambanks; 350–1200m.
Distr.: Nigeria, Bioko & Cameroon. (Lower Guinea).
IUCN: NT
Kupe Village: <u>Cable 3679</u> 7/1996; <u>749</u> fl., 1/1995; <u>Cheek 6051</u> 1/1995; <u>Etuge 2648</u> 7/1996; <u>Schoenenberger 47</u> 11/1995; **Manehas F.R.:** <u>Etuge 4356</u> 10/1998; **Ngomboaku:** <u>Etuge 4682</u> fl., fr., 12/1999; **Nyasoso:** <u>Etuge 2546</u> 7/1996; <u>Sebsebe 5008</u> 10/1995.

Coleotrype laurentii K.Schum.
Scandent or decumbent herb to 1m; leaves elliptic, c. 15 × 4cm, long-acuminate, base attenuate; inflorescences axillary, perforating the leaf-sheath, sessile, fasciculate, c. 3 × 4cm, enclosed in leafy bracts; flowers white, numerous. Lowland forest; 900m.
Distr.: Ivory Coast to Uganda. (Guineo-Congolian).
IUCN: NT
Nyasoso: <u>Etuge 2358</u> 6/1996.

Commelina africana L. var. *africana*
Prostrate herb with rooting stems to 1m long; leaves ovate-elliptic, c. 4 × 2cm, apex acute; inflorescensces subtended by spathes; flowers yellow, open 07:00 to 10:00 am. Montane grassland; 1975m.
Distr.: Guinea (B) to S Africa. (Afromontane).
IUCN: LC
Mwanenguba: <u>Cheek 7282</u> st., 2/1995.
Note: this collection is sterile so identification is uncertain.

Commelina benghalensis L. var. *hirsuta* C.B.Clarke
Erect herb c. 30(–150)cm tall; leaves ovate, to 6 × 3.5cm, subacuminate, truncate; petiolate; sheath with conspicuous rusty hairs all over outside; spathe c. 2 × 1cm; flowers bright blue, open 08:30 am to 12:00 pm. Lower montane to submontane farmbush; 1450m.
Distr.: Guinea (C) to Malawi. (Afromontane).
IUCN: LC
Kodmin: <u>Cheek 9015</u> 1/1998; **Nyasoso:** <u>Etuge 1613</u> 1/1996.
Local name: Nkole-ke (Etuge 1613); Nkolekeh (Cheek 9015). **Uses:** SOCIAL USES – ritual – put in a pocket to protect against injury from accidents (Cheek 9015); MEDICINES (Etuge 1613); ENVIRONMENTAL USES – boundaries – planted on farms to protect against intruders: they cannot have children (SOCIAL USES – magic) (Cheek 9015).

Commelina bracteosa Hassk.
Decumbent herb to 30cm tall; leaves lanceolate, 7.5 × 2cm, apex acuminate, base cuneate, strongly oblique; petiolate; spathe 1.5 × 1cm; peduncle 6–22mm; flowers pale mauve, open at 12:00 pm. Lowland to lower montane forest; 1450m.
Distr.: Senegal to Mozambique. (Tropical Africa).
IUCN: LC
Kodmin: <u>Cheek 8949</u> 1/1998.

Commelina cameroonensis J.K.Morton
Erect herb 0.5(–1)m tall, gregarious; leaves elliptic, c. 12 × 4.5cm, acumen well-defined, base rounded, strongly oblique; petiole 0.5cm; spathes 1–3, 1.5–3 × 0.5–1cm, margin brown-hairy, base drying pale yellow, subsessile; flowers white; capsules 2-loculate, 1 smooth seed per locule. Montane forest, forest-grassland transition; 950–1950m.
Distr.: SE Nigeria, Bioko & W Cameroon. (W Cameroon Uplands).
IUCN: NT
Kodmin: <u>Etuge 4402</u> 11/1998; **Mt Kupe:** <u>Sebsebe 5095</u> 11/1995; **Nyasoso:** <u>Zapfack 600</u> 6/1996.
Note: local in the W Cameroon highlands, where it is possibly threatened in E Nigeria and NW Province, Cameroon by intensified agriculture.

Commelina capitata Benth.
Clustering, robust, erect herb 0.3–1m tall; leaves oblong-elliptic, c. 12 × 4.5cm, long-acuminate, base obtuse, strongly oblique, sheaths with rusty bristles at apex; spathes 3–7, c. 1.5–2 × 1cm, margin often rusty-hirsute; peduncle 0.5–1cm; flowers pale yellow, open 08:00 to 11:00 am. Lowland to montane forest; 300–1300m.
Distr.: Senegal to Bioko, Cameroon, Kenya & Angola. (Guineo-Congolian).
IUCN: LC
Kodmin: <u>Barlow 3</u> 1/1998; <u>Plot B209</u> 1/1998; **Kupe Village:** <u>Cable 791</u> fr., 1/1995; <u>Cheek 6058</u> fr., 1/1995; <u>7078</u> fr., 1/1995; <u>7732</u> 11/1995; <u>7772</u> 11/1995; <u>Ensermu 3523</u> 11/1995; <u>3555</u> 11/1995; <u>Etuge 1702</u> 2/1996; <u>Lane 290</u> fr., 1/1995; <u>Williams 227</u> 11/1995; **Mbule:** <u>Cable 3376</u> 6/1996; **Mwambong:** <u>Cheek 9381</u> 10/1998; **Mungo River F.R.:** <u>Cheek 10124</u> 11/1999; **Nyasoso:** <u>Cable 3599</u> 7/1996; <u>Cheek 7326</u> fr., 2/1995; <u>7494</u> 10/1995; <u>Lane 158</u> fl., 10/1994; <u>Sebsebe 5045</u> 10/1995; <u>Sidwell 382</u> 10/1995; **Nzee Mbeng:** <u>Gosline 103</u> 2/1998.
Local name: Nkawloko. **Uses:** SOCIAL USES (?) – widely used by herbalists, but how, and for what is not known (Cheek 7326).

Commelina congesta C.B.Clarke
Erect herb c. 60cm tall; leaves lanceolate, c. 9 × 3cm, acuminate, base obtuse, strongly asymmetric, subsessile, sheaths glabrous, rarely weakly hairy; spathes c. 3, crowded, c. 1.5 × 3cm, sessile, margins glabrous; flowers white to pale violet, open dawn to 12:00 pm; fruit with 3, 1-seeded locules. Lowland forest, roadsides and villages.
Distr.: Guinea (C) to Congo (K). (Guineo-Congolian).
IUCN: LC
Kupe Village: <u>Ensermu 3522</u> 11/1995.

Commelina diffusa Burm.f. subsp. *diffusa*
Decumbent to straggling herb 15–60cm; leaves ovate-lanceolate c. 5 × 1.5cm, acute, base rounded, subsessile; spathes 1, 1.5–3 × 0.8cm; peduncle 1cm; flowers blue; fruit with reticulate seeds. Forest edge, plantations, farmbush; 930m.
Distr.: pantropical.
IUCN: LC
Kupe Village: <u>Ensermu 3521</u> 11/1995; <u>Ryan 252</u> 5/1996.

Commelina diffusa Burm.f. subsp. *montana* J.K.Morton
Straggling herb to c. 25cm; leaves ovate-lanceolate 3.5–6 × 1.5–2cm, acute-acuminate, base cordate to subcordate, sessile; spathes 1–2, 1–2.5cm long; peduncle c. 1cm; flowers blue; fruit with smooth seeds. Montane forest edge and fallow; 800–2000m.

Distr.: Nigeria, Bioko & W Cameroon. (W Cameroon Uplands).
IUCN: NT
Kupe Village: Etuge 2670 7/1996; **Mwanenguba:** Cheek 9443 10/1998; **Nyasoso:** Sidwell 386 10/1995.

Commelina longicapsa C.B.Clarke

Decumbent herb to c. 30cm; leaves 3–4 clustered at apex of stem, obovate, c. 13 × 6cm, shortly acuminate, base acute; petiole 3–4cm; spathes subleafy, pinkish, 2–3, crowded, c. 2 × 1.5cm; flowers white. Lowland forest, near streams; 1000m.
Distr.: Liberia to Congo (K). (Guineo-Congolian).
IUCN: NT
Nyale: Etuge 4454 11/1998.

Cyanotis barbata D.Don

Erect herb, 10–30cm; with underground rootstock; leaves linear-lanceolate, to 12 × 1cm, white-pubescent, sessile; spathes 4–5, c. 1 × 0.5cm, pedunculate; flowers blue, actinomorphic; filaments bearded. Montane grassland; 2000m.
Distr.: tropical Africa & Asia. (Palaeotropics (montane)).
IUCN: LC
Mwanenguba: Cheek 9430 10/1998; Pollard 871 fl., 11/2001.

Floscopa africana (P.Beauv.) C.B.Clarke subsp. *africana*

Straggling herb 20–60cm; stems and leaves red-purple; leaves lanceolate, to c. 8 × 1.5–2cm, acuminate; petiole 0.5cm; inflorescence paniculate, with up to 15 bracts, glandular-hairy; flowers purple; sepals 2–2.75mm; style not exserted; fruiting pedicel 0.5–1.5mm. Lowland farmbush, often in damp areas; 420–1000m.
Distr.: Gambia to Uganda. (Guineo-Congolian).
IUCN: LC
Loum F.R.: Cheek 9298 10/1998; Etuge 4270 10/1998; **Mwambong:** Onana 549 2/1998; **Ngomboaku:** Etuge 4659 12/1999; Ghogue 453 12/1999; **Nyasoso:** Sidwell 340 10/1995.

Floscopa africana (P.Beauv.) C.B.Clarke subsp. *majuscula* (C.B.Clarke) Brenan

Resembling *F. africana* subsp. *africana*, but robust, up to 1.2m; panicle 7–10cm; pedicels 1.5–2mm; sepals 3–4mm; style exserted beyond the sepals. Lowland farmbush, forest; 950m.
Distr.: Guinea (B) to Congo (K). (Guineo-Congolian).
IUCN: LC
Kupe Village: Ensermu 3520 11/1995; **Ngomboaku:** Ghogue 5506 fl., 12/1999.

Floscopa africana (P.Beauv.) C.B.Clarke subsp. *petrophila* J.K.Morton

Resembling *F. africana* subsp. *africana*, but stems low, creeping, up to 20cm high, lacking red pigment; leaves to three times as long as broad, 2.5–6cm long. Lowland farmbush, forest edge, sometimes on rocks; 800–1000m.
Distr.: Liberia to Uganda. (Guineo-Congolian).
IUCN: LC
Kupe Village: Etuge 2812 7/1996; **Ndum:** Williams 148 1/1995; **Nyasoso:** Cheek 7527 10/1995.

Floscopa glomerata (Willd. ex Schult. & Schult.f.) Hassk subsp. *lelyi* (Hutch.) Brenan

Kew Bull. 22: 387 (1968).

Stems prostrate, ascending to 30cm; leaves ligular-lanceolate, c. 10 × 1cm, acute, sessile; panicle c. 6cm long, compact, glandular-hairy; flowers violet, one flower open per panicle, covered in glandular hairs. Marshes and swamps; 1200m.
Distr.: S America & W Africa. (Amphi-Atlantic).
IUCN: LC
Lake Edib: Gosline 197 11/1998.

Floscopa mannii C.B.Clarke

Herb; stems prostrate, ascending to 5cm; leaves ovate, to c. 4 × 2cm, apex acute, base cordate, sessile; sheath white-villose; inflorescence 4–8cm long, diffuse, with 5–6 slender parent branches, each c. 2cm long, glandular-hairy; petals white. Lowland forest clearings near streams; 700m.
Distr.: SE Nigeria, Cameroon, Gabon & Rio Muni. (Lower Guinea).
IUCN: EN
Ngomboaku: Cheek 10349 12/1999.

Murdannia simplex (Vahl) Brenan

Straggling herb to 65cm; leaves linear-lanceolate, 20 × 1.5cm, acute, base sessile to amplexicaul; inflorescence 'terminal', panicle c. 15 × 5cm, with c. 6 branches in the upper ¼; flowers purple and white; petals equal, open 16:00 to 18:30 pm. Lowland farmbush; 400–1300m.
Distr.: tropical Africa & Asia. (Palaeotropics).
IUCN: LC
Kupe Village: Ensermu 3594 11/1995; Etuge 1487 11/1995; Ryan 391 5/1996; **Nyasoso:** Etuge 1618 1/1996.
Uses: SOCIAL USES – boiled leaves and stems make slow infants smart; MEDICINES (Etuge 1618).

Palisota ambigua (P.Beauv.) C.B.Clarke

Herb resembling *P. hirsuta*, but inflorescence usually one per leaf rosette, only 3–10cm, puberulous, partial peduncles to only 3mm; flowers blue to purple. Forest; 300m.
Distr.: Nigeria to Congo (K). (Guineo-Congolian).
IUCN: LC
Mekom: Thomas 5398 1/1986.
Note: specimen cited not seen.

Palisota barteri Hook.

Herb 20–80cm; lacking aerial stem; leaves forming a basal rosette, blades elliptic, c. 25–45 × 8–15cm, apex acute or acuminate, base cuneate to rounded, margin often rusty-hairy, lower surface green, sparsely appressed-pubescent; petioles 20–30cm; inflorescence with peduncle 10–50cm long, rarely bracteose; panicle subglobose to ovoid, 5–10 × 3–7cm; flowers white; fruits red. Lowland to lower montane forest; 220–1600m.
Distr.: Sierra Leone to Bioko, Cameroon & Congo (K). (Guineo-Congolian).
IUCN: LC
Bakolle Bakossi: Etuge 289 9/1986; **Kodmin:** Cheek 8947 1/1998; Etuge 3976 1/1998; **Kupe Village:** Cable 2459 5/1996; 2558 5/1996; 2699 5/1996; Cheek 8406 5/1996; Etuge 1958 5/1996; 2908 7/1996; Ryan 203 5/1996; **Ndum:** Lane 474 fr., 2/1995; **Nyasoso:** Cable 2751 fr., 6/1996; 2835 6/1996; 3326 6/1996; 3483 7/1996; Etuge 1802 3/1996; 2135 6/1996; 2180 6/1996; Sidwell 323 10/1995; Sunderland 1507 fr., 7/1992.

Palisota bracteosa C.B.Clarke

Rosette herb; leaves to 80cm, blades elliptic, to 60 × 15cm; petiole winged; inflorescences 15cm; peduncles 10cm, with 2–3 reduced leaves 3cm, rachis with large, conspicuous bracts to 1.5 × 0.5cm. Forest; 700m.

Distr.: Guinea (C) to Cameroon & São Tomé. (Upper & Lower Guinea).
IUCN: LC
Nyale: Etuge 4194 2/1998.

Palisota hirsuta (Thunb.) K.Schum.
Herb to 3m; aerial stems present; leaves clustered at apices, oblanceolate, c. 30 × 10cm, apex acuminate, base cuneate, midrib brown pubescent, sessile or petiole 2–3cm, brown-hairy; inflorescences often clustered at apex, c. 30cm, terminal, lax; peduncle to 10cm; partial peduncles to 2cm; flowers white. Lowland forest; 600–800m.
Distr.: Senegal to Congo (K). (Guineo-Congolian).
IUCN: LC
Kupe Village: Cable 2535 fr., 5/1996; Kenfack 308 7/1996; **Ngusi:** Etuge 46 4/1986.

Palisota mannii C.B.Clarke
Herb resembling *P. barteri*, but leaves often white below; inflorescence cylindrical c. 12–18 × 3.5cm; flowers white; pedicels > flowers. Lowland and submontane forest, forest-grassland transition; 800–1900m.
Distr.: S Nigeria to Uganda. (Lower Guinea & Congolian).
IUCN: LC
Kodmin: Cheek 8963 1/1998; Plot NaN 1/1998; NaN 1/1998; **Kupe Village:** Cable 2486 fr., 5/1996; 2639 fr., 5/1996; Etuge 2862 7/1996; **Nyasoso:** Cable 2754 fr., 6/1996; 2940 fl., 6/1996; Cheek 6020 fl., 1/1995; Etuge 2548 fr., 7/1996; Sebsebe 4990 10/1995; Sidwell 324 fr., 10/1995. Note: of the above specimens seen by R.B.Faden, all but Cable 2639, 2940 and Sidwell 324 lack the dense indumentum below.

Palisota sp. aff. mannii, sp. nov. ?
Herb resembling *P. mannii*, but with an aerial stem 1–2m; leaves often in pseudowhorls. Montane forest; 1450–1940m.
Distr.: W Cameroon. (W Cameroon Uplands).
Kodmin: Cheek 9060 1/1998; Etuge 4037 1/1998; **Kupe Village:** Ryan 402 5/1996; **Nyasoso:** Cable 1250 fl., 2/1995; Etuge 2183 6/1996.

Palisota preussiana K.Schum. ex C.B.Clarke
Herb c. 1m; stem erect or scrambling; leaves attenuate, not in pseudowhorls, ovate, c. 17 × 9cm, acuminate, base rounded; petioles 3–4cm; inflorescence solitary, terminal, dense, cylindrical, 7 × 3cm; peduncle 4cm; flowers mauve; fruits red. Montane forest and *Aframomum* thicket; 2000m.
Distr.: Mt Cameroon, Mt Kupe & Bioko. (W Cameroon Uplands).
IUCN: VU
Nyasoso: Cable 3383 6/1996.

Palisota schweinfurthii C.B.Clarke
Herb 0.5–1.5m; aerial stem stout; leaves ovate-lanceolate, c. 30 × 12cm, acuminate; petiole c. 10cm; inflorescences numerous, axillary, cylindrical, c. 25 × 2cm, subsessile; flowers white; pedicels oblong and equaling flower length. Lower montane forest; 1200–2000m.
Distr.: W Cameroon to Sudan, Tanzania & Zambia. (Guineo-Congolian).
Kupe Village: Etuge 2646 7/1996; **Nyasoso:** Cable 2941 6/1996; Etuge 1667 1/1996.

Pollia condensata C.B.Clarke
Erect herb with aerial stem 0.3–1m; leaves narrowly elliptic as oblanceolate, c. 20 × 6cm, long acuminate, subsessile; inflorescence terminal, dense, ovoid, c. 2 × 1.5cm; flowers white, inconspicuous; fruits iridescent-blue, hard, spherical, c. 4mm diameter. Lowland forest, streambanks; 150–1700m.
Distr.: Sierra Leone to Tanzania. (Guineo-Congolian).
IUCN: LC
Ehumseh: Manning 458 10/1986; **Etam:** Etuge 191 7/1986; **Kodmin:** Etuge 3971 1/1998; Satabie 1103 fr., 1/1998; **Kupe Village:** Cable 2635 5/1996; 849 fl., 1/1995; Cheek 6037 fr., 1/1995; Kenfack 247 7/1996; 306 7/1996; 307 7/1996; Lane 378 fr., 1/1995; Zapfack 932 fr., 7/1996; **Loum F.R.:** Cheek 10260 12/1999; **Mekom:** Nemba 23 4/1986; **Mile 12 Mamfe Road:** Nemba 647 10/1987; **Nyasoso:** Balding 26 fr., 8/1993; Cable 2819 6/1996; 3498 7/1996; Etuge 1299 10/1995; 1795 3/1996; 2153 6/1996; Wheatley 434 fr., 7/1992.

Pollia mannii C.B.Clarke
Sprawling herb, 30cm; leaves elliptic, c. 7 × 2.5cm, acuminate; petiole 1cm; inflorescence terminal, diffuse, c. 6cm, of 6–8 1-flowered branches; flowers white, actinomorphic; fruits hard, glossy, grey, 6mm. Lowland forest; 750–1300m.
Distr.: Ivory Coast to Cameroon, São Tomé, SW Ethiopia, Uganda & W Tanzania. (Guineo-Congolian).
IUCN: LC
Kupe Village: Etuge 1707 2/1996; 2721 7/1996; Kenfack 252 7/1996; **Manehas F.R.:** Cheek 9390 10/1998.

Polyspatha paniculata Benth.
Erect herb 10–30cm; leaves elliptic, to 12 × 6cm, acuminate; petiole 0.5cm; inflorescence terminal, single, elongate, 6–20cm; spathes 4–10, c. 7 × 5mm; flowers white. Forest; 700–1450m.
Distr.: Guinea (C) to Uganda. (Guineo-Congolian).
IUCN: LC
Bangem: Thomas 5371 1/1986; **Kodmin:** Cheek 8966 1/1998; Gosline 178 11/1998; Satabie 1098 fr., 1/1998; **Kupe Village:** Cheek 6052 fl., fr., 1/1995; Etuge 2808 7/1996; **Mwambong:** Cheek 9108 2/1998; **Ngomboaku:** Etuge 4657 12/1999; Ghogue 475 12/1999; 490 12/1999; **Nyale:** Etuge 4472 11/1998; **Nyasoso:** Cable 2896 6/1996; 865 fl., 1/1995; Cheek 7325 fl., 2/1995; Etuge 1322 10/1995; Lane 193 fr., 11/1994; Pollard 88 10/1998.
Local name: unknown. **Uses:** unknown (Cheek 7325).

Stanfieldiella brachycarpa (Gilg & Lederm. ex Mildbr.) Brenan var. hirsuta (Brenan) Brenan
Ascending herb 10–50cm; leaves elliptic, clustered at stem apex, to 11 × 3cm, acuminate, subsessile, base of leaf pilose; inflorescences terminal, c. 1.5cm; flowers greenish white; capsule not exceeding sepals; seeds verrucose. Submontane forest; 1100–1230m.
Distr.: Nigeria to Equatorial Guinea. (Lower Guinea).
IUCN: NT
Kupe Village: Cable 817 fl., 1/1995; Ryan 368 5/1996; 385 5/1996.

Stanfieldiella imperforata (C.B.Clarke) Brenan var. imperforata
Straggling herb, resembling *S. brachycarpa*, but leaves glabrous, not pilose; inflorescences lax, diffuse, 2–6cm long; capsules exceeding sepals; seeds smooth. Lowland to submontane forest; 250–1200m.
Distr.: Sierra Leone to Uganda. (Guineo-Congolian).
IUCN: LC
Bangem: Thomas 5368 1/1986; **Kupe Village:** Cheek 6054 1/1995; Ensermu 3559 11/1995; Kenfack 246 7/1996; Ryan 326 5/1996; Zapfack 934 7/1996; **Loum F.R.:** Etuge 4271 10/1998; **Mile 15, Kumba-Mamfe Road:** Nemba 213

9/1986; **Mwambong:** Cheek 9380 10/1998; **Ngomboaku:** Ghogue 457 12/1999; 459 12/1999.

Stanfieldiella oligantha (Mildbr.) Brenan

Straggling herb, resembling *S. brachycarpa*, but leaves glabrous; inflorescence dense, branched, c. 1.5cm long; capsules exceeding sepals; seeds verrucose. Lowland and submontane forest; 100–1350m.
Distr.: Liberia to Cameroon. (Upper & Lower Guinea).
IUCN: LC
Kodmin: Cheek 8983 1/1998; **Kupe Village:** Cable 3764 7/1996; 3817 7/1996; Cheek 7822 11/1995; Ensermu 3532 11/1995; **Loum F.R.:** Etuge 4272 10/1998; **Mungo River F.R.:** Thomas 2565 11/1983.
Local name: Nkolek-nsad (Etuge 4272).

COSTACEAE

M. Cheek

Fl. Cameroun 4 (1965).

Costus afer Ker Gawl.

Terrestrial herb, 2–3m, glabrous; ligule shortly cylindrical, truncate, c. 5mm, glabrous; inflorescence terminal on leafy culms; bracts 2-flowered, longer than calyces; flowers white, margin red, central band yellow. Forest; 1460m.
Distr.: Sierra Leone to Mozambique. (Tropical Africa).
IUCN: LC
Mwambong: Mackinder 156 1/1998.

Costus dubius (Afzel.) K.Schum.

Syn.: *Costus albus* A.Chev. ex J.Koechlin
Perennial herb to 1.5(–2)m; leaves obovate-elliptic, long-acuminate, base cuneate, appressed-pubescent; inflorescences terminal or on short lateral shoots at the base of the leafy stems; flowers numerous, white with yellow throat. Forest; 1000m.
Distr.: tropical Africa.
IUCN: LC
Kupe Village: Etuge 2864 fr., 7/1996; **Nyasoso:** Sidwell 349 fl., 10/1995.

Costus englerianus K.Schum.

Terrestrial herb to 5cm; stoloniferous; leaves leathery, glossy, held at 45 degrees, broadly elliptic, 12 × 8cm; flowers white, centre yellow. Forest; 700–850m.
Distr.: Sierra Leone to Gabon. (Upper & Lower Guinea).
IUCN: LC
Kupe Village: Cheek 7718 fl., 11/1995; **Nyale:** Etuge 4204 2/1998; **Nyandong:** Ghogue 1442 fl., 3/2003; **Nyasoso:** Etuge 2571 fl., 7/1996.

Costus letestui Pellegr.

Fl. Cameroun 4: 86 (1965).
Epiphytic herb; stems pendulous to 2(–6)m; leaves elliptic, c. 15 × 4cm; ligule cylindrical, truncate, 3cm or more; inflorescence axillary; flowers 5; bracts shorter than calcyes; flowers pale pink; labellum pink with yellow centre. Forest; 750–1550m.
Distr.: Cameroon & Gabon. (Lower Guinea).
IUCN: NT
Kodmin: Cheek 8864 fl., 1/1998; 8952 fl., 1/1998; Mackinder 144 fl., 1/1998; **Ngomboaku:** Mackinder 309 fl., 12/1999; **Nyandong:** Cheek 11288 3/2003.
Note: confirmed from only 8 locations, mainly in Cameroon. However, unconfirmed field determinations exist from several sites on Mt. Cameroon and the Rumpi Hills. This taxon may qualify as VU, if these field determinations prove incorrect.
Local names: Wananko (Cheek 8864); Achang (Cheek 11288). **Uses:** MEDICINES (Cheek 8864).

Costus lucanusianus J.Braun & K.Schum. var. *lucanusianus*

Terrestrial herb, 2–3m, resembling *C. afer*, but lower surface of leaves appressed-pubescent, with hairy rim, hairs c. 5mm; calyces much longer than bracts. Forest, farmbush, roadsides; 550–1370m.
Distr.: Sierra Leone to Bioko, Cameroon, Gabon & Uganda. (Guineo-Congolian (montane)).
IUCN: LC
Enyandong: Pollard 780 fl., 10/2001; **Kodmin:** Biye 42 1/1998; **Kupe Village:** Lane 339 fl., 1/1995; **Loum F.R.:** Tadjouteu 311 11/1999; **Nyasoso:** Cable 2750 fl., 6/1996; Sunderland 1483 fl., 7/1992.

Costus sp. A

Terrestrial herb, 2–3m; resembling *C. afer*, but culms purple, pubescent; ligule ovate; flowers yellow. Forest; 400–870m.
Distr.: Cameroon. (Narrow Endemic).
Bakossi F.R.: Etuge 4307 fr., 10/; **Kupe Village:** Cable 2471 5/1996; Cheek 7111 fl., 1/1995.
Note: probable *sp. nov.*, to be described.

CYANASTRACEAE

M. Cheek

Cyanastrum cordifolium Oliv.

Herb, c. 15cm; acaulous; tubers orange; leaves 2–4, ovate, dark green, 8 × 6cm, subacuminate, deeply cordate; petiole 10cm; scape 10cm; flowers 5–6, blue, 1.5cm; fruit orange. Forest; 200–1300m.
Distr.: Nigeria, Cameroon & Gabon. (Lower Guinea).
IUCN: LC
Kupe Village: Cable 2408 fl., 5/1996; Etuge 1714 fl., 2/1996; 2813 fr., 7/1996; Groves 15 fl., 1/1995; Schoenenberger 353 5/1996; **Loum F.R.:** Onana 989 12/1999; **Nyandong:** Cheek 11278 3/2003, **Nyasoso:** Etuge 2087 fr., 6/1996.

CYPERACEAE

K.Å. Lye & B.J. Pollard

Bulbostylis densa (Wall.) Hand.-Mazz. var. *cameroonensis* Hooper

Annual herb to ± 30cm; stem deeply grooved, 0.2–0.4mm thick; leaves canaliculate, grooved, 0.2–0.3mm broad; inflorescence usually a compact umbel, somewhat contracted, with 3–8 shortly-pedicellate spikelets, each one 2–5 × 1.5–3mm; glumes few, each standing out from its neighbour, dark brown with conspicuous pale green or grey midrib. Grassland; 1900–2000m.
Distr.: Mt Cameroon to Bali Ngemba F.R. (W Cameroon Uplands).
IUCN: VU
Mt Kupe: Cheek 7575 10/1995; **Mwanenguba:** Pollard 134 10/1998.

Bulbostylis hensii (C.B.Clarke) R.W.Haines

Haines & Lye, Sedges & Rushes of E Africa, App. I: 1 (1983).

Syn.:*Fimbristylis hispidula* (Vahl) Kunth subsp. *brachyphylla* (Cherm.) Napper

Tussocky perennial, maybe annual; culm 20–40cm × 0.3–0.6mm, ridged, densely set with 0.5mm transparent hairs; leaves similarly hairy, filiform; inflorescence a simple umbel-like anthela with 2–5 spikelets, rarely only 1; bracts brown, < 10mm, the margin with flexuose hairs, 1 –3mm; spikelets 5–15 × 2–3mm; glumes 3–3.5mm, light reddish brown below and near margin, ± black along the midrib above. Grassland; 1500m.

Distr.: Cameroon, Congo (K), Uganda, Kenya, Zimbabwe. (Afromontane).

IUCN: LC

Kodmin: Cheek 8901 1/1998.

Local name: none. **Uses:** none (Cheek 8901).

Carex echinochloë Kunze

Tufted leafy perennial to ± 1m; rhizome creeping; stems 50–100cm, distinctly triangular below, rounded-triangular above; largest leaf-blades 40–120 × 0.6–1.4cm, flat or ± plicate; inflorescence a slender, very-much-branched panicle, 20–50cm long, widening with age; spikelets 5–10 × 5mm, male above and female below; glumes 4–5mm. Grassland, forest edges; 1980m.

Distr.: W Cameroon, E & NE Africa. (Afromontane).

IUCN: LC

Nyasoso: Cheek 7349 fr., 2/1995; 7353 fr., 2/1995.

Local name: Tchum-eboom. **Uses:** MATERIALS – used for making sleeping mats (Cheek 7349).

Coleochloa abyssinica (Hochst ex A.Rich) Gilly var. *abyssinica*

Robust plant with branching, scaly, stolons; stems 40–80 × 0.1–0.4cm; leaf-blades to 30 × 0.2–0.7cm, folded; ligule 1–2mm; inflorescence a diffuse panicle, 2–6-branched from upper leaf-sheaths; spikes 5–8 × c. 3mm, of numerous densely clustered spikelets; glumes pale or dark brown or red-brown, glossy, glabrous. In rock crevices, shallow soils over rocks, grasslands, often epiphytic on trees in montane forest; 1100m.

Distr.: NE Nigeria, W Cameroon, E & NE tropical Africa, Angola. (Afromontane).

IUCN: LC

Kupe Village: Cheek 7688 fl., 11/1995.

Cyperus atrorubidus (Nelmes) Raymond var. *nov.* Lye ined.

Grassland; 2000m.

Distr.: Mwanenguba. (Narrow Endemic).

Mwanenguba: Pollard 129 10/1998.

Note: this collection represents an undescribed variety which we intend to publish in the near future.

Cyperus atroviridis C.B.Clarke

Robust perennial herb; culms 0.3–1.3m, the lower part covered by leaf-sheaths or crowded leaves; largest leaf-blades 8–40 × 0.4–1.2cm, flat, scabrid on margin and main ribs; inflorescence anthelate, 6–25 × 3–25cm, of 2–many clustered spikes on 0.5–22cm rays; involucral bracts leafy, up to 40cm long; spikes often appearing brush-like, 2–6cm; spikelets 0.6–1.2 (–2.0) × 0.1–0.2cm, linear-lanceolate, blackish with 5–12 rather distant flowers; glumes dark reddish brown. Forest margins, farmbush; 1540m.

Distr.: tropical Africa. (Afromontane).

IUCN: LC

Kodmin: Pollard 194 11/1998.

Cyperus cyperoides (L.) Kuntze subsp. *cyperoides*

Rev. Gen. Pl. 3(2): 333 (1898).

A robust perennial herb; stem-base swollen; rhizome woody; culms 20–80cm, triangular; largest leaves 10–25 × 0.3–0.9cm, flat; inflorescence a 2–6cm wide anthela of 5–18 spikes on peduncles up to 12cm, or rarely sessile, 7–20mm long, each with 25–100 spreading spikelets; major inflorescence-bracts 5–15, leafy, to 30cm; spikelets oval, 1–3-flowered, greenish yellow. Fallow, farmbush, forest paths, grassland, savanna. 220–1500m.

Distr.: palaeotropics; introduced to West Indies, S Africa. (Pantropical).

IUCN: LC

Kodmin: Cheek 8974 1/1998; **Kupe Village:** Etuge 1706 2/1996; 2628 7/1996; **Mungo River F.R.:** Pollard 108 10/1998; **Ngomboaku:** Muasya 2033 12/1999; 2036 12/1999; **Nyasoso:** Cable 3516 7/1996; 3517 7/1996; Etuge 2586 7/1996.

Cyperus cyperoides (L.) Kuntze subsp. *flavus* Lye

Nord. J. Bot. 3: 231 (1983).

Syn.:*Cyperus subumbellatus* Kük.

Syn.:*Mariscus alternifolius sensu* Hooper in FWTA 3: 296

Tufted plant to c. 80cm; rhizome woody, ± composed of swollen stem bases; culms 25–70cm × 0.8–3.0mm leaf-sheaths purple; leaves from lowest 4–12cm of culms, 10–30cm × 2–6mm; inflorescence variable, rays ± well-developed; major inflorescence bracts 6–13, leafy, the largest 8–25cm × 2–7mm; spikes of ± crowded, small greenish or reddish, 1–2-flowered spikelets. Seasonally burnt grassland, fallow, clearings in forest; 700m.

Distr.: Senegal to Cameroon, E Africa. (Tropical Africa).

IUCN: LC

Kupe Village: Lane 329 fr., 1/1995.

Cyperus densicaespitosus Mattf. & Kük.

Haines & Lye, Sedges & Rushes of E Africa: 243 (1936).

Syn.:*Kyllinga pumila* Michx.

Densely caespitose annual herb; slender root system; culms 10–45cm, the lowest part covered by dark purple bladeless sheaths; leaves flat or enrolled, margin and midrib with short spine-like hairs; inflorescence an irregular pale green to greyish leaf, 5–8 × 5–10mm, of 1 larger central spike and several smaller lateral spikes; involucral bracts 3–5, leafy, erect or spreading to 13cm; spikelets ± 2mm, 1-flowered. Sandy river- and roadside banks, drains and ditches; 400–900m.

Distr.: tropical Africa, tropical and temperate America. (Amphi-Atlantic).

IUCN: LC

Loum F.R.: Pollard 101 10/1998; **Ngomboaku:** Muasya 2037 12/1999.

Uses: none known (Epie Ngome & George Ngole in Pollard 101).

Cyperus distans L.f. subsp. *distans*

Tufted perennial herb; culms in a row or solitary, 20–60 × 0.15–0.5cm, basally covered with grey to dark purple leaf-sheaths; leaves 5–30 × 0.2–0.8cm, flat; inflorescence anthelate, 5–26 × 3–20cm; main branches 5–15, to 15cm; secondary and tertiary branches short or spikes sessile; spikes 2–4 × 1–4cm; spikelets rather loosely set, often spreading at 90 degrees; involucral bracts 10–25cm; spikelets 7–20 × 0.5–1.0mm, linear or zigzag when glumes spreading; glumes

reddish brown, rarely falling off entire. Cultivated and waste ground, damp places, grassland, streambanks in forest; 400–2000m.
Distr.: pantropical.
IUCN: LC
Loum F.R.: Pollard 97 10/1998; **Mwanenguba:** Pollard 138 10/1998; **Ngomboaku:** Muasya 2038 12/1999; **Nyasoso:** Cable 1249 fl., 2/1995; Cheek 5658 fl., 12/1993; Pollard 112 10/1998.
Local name: Nkoko. **Uses:** MEDICINES (Pollard 97).

Cyperus distans L.f. subsp. *longibracteatus* var. *longibracteatus* (Cherm.) Lye
Nord. J. Bot. 3: 231 (1983).
Syn.:*Cyperus longibracteatus* (Cherm.) Kük.
Syn.:*Mariscus longibracteatus* Cherm.
As for the typical subspecies, but differs in having denser inflorescence, longer involucral bracts (25cm +), and larger glumes, which probably always fall off entire. Swampy ground; 1000–1150m.
Distr.: tropical Africa.
IUCN: LC
Kupe Village: Etuge 1628 1/1996; **Mwambong:** Cheek 9365 10/1998.

Cyperus fertilis Boeck.
Tufted annual herb; stem 2–10cm; 5–10 basal leaves distinctly wider above than below; inflorescence with 2–8 major peduncles of unequal length, up to 40cm, each triangular, flattened, carrying 1–3 spikelets; the peduncles sometimes recurved and proliferating (viviparous); involucral bracts leafy, elliptical, 5–10 per culm, to 20cm; spikelets light brown, flattened, 5–10 × 3–5mm. Roadsides, villages, damp places in forest; 1200m.
Distr.: Liberia to Angola, Uganda. (Guineo-Congolian).
IUCN: LC
Kupe Village: Etuge 1424 fl., fr., 11/1995; 2624 7/1996.

Cyperus flavescens L. subsp. *flavescens*
Haines & Lye, Sedges & Rushes of E Africa: 281 (1983).
Syn.:*Pycreus flavescens* (L.) Rchb.
Annual (rarely stoloniferous perennial) herb; stems crowded, 6–50cm × 0.3–2.5mm; the base enclosed by reddish brown expanded leaf bases; leaf-blades to 30 × 0.4cm, 2–4 per culm; inflorescence a dense anthela of clustered spikelets; involucral bracts 2–5, to 15cm; peduncles to 5cm, with a tubular basal purple prophyll; spikelets lanceolate to linear, 5–18 × 1.2–2.5mm, pale yellowish brown. Swamp margins, streamside-silts, seasonally wet grassland, often on bare soil between taller plants; 300–880m.
Distr.: pantropical, temperate regions.
IUCN: LC
Bakossi F.R.: Cheek 9309 10/1998; **Nyasoso:** Pollard 111 10/1998.
Local name: Ntoge. **Uses:** MEDICINES – veterinary (Epie Ngome in Cheek 9309).

Cyperus fluminalis Ridl.
Haines & Lye, Sedges & Rushes of E Africa: 278 (1983).
Syn.:*Pycreus smithianus* (Ridley) C.B.Clarke
Caespitose perennial herb; culms 15–40 × 0.1–0.2cm, basally ± swollen; leaf-blades 5–20 × 0.2–0.5cm, flat, stiff, numerous at base of plant; inflorescence a compact, globose, head-like anthela of crowded spikelets; involucral bracts 2–4, leafy, to 3–12cm; spikelets linear to lanceolate, 10–25 × 2–3mm, conspicuously white or rarely greenish white. Seasonally wet grassland, swamp edges, ditches throughout grassland; 1450–2000m.

Distr.: W Cameroon, Angola, Uganda, Tanzania. (Guineo-Congolian).
IUCN: LC
Kodmin: Cheek 9181 2/1998; Pollard 192 11/1998; **Mwanenguba:** Pollard 130 10/1998.

Cyperus kyllingia Endl.
Haines & Lye, Sedges & Rushes of E Africa: 247 (1983).
Syn.:*Kyllinga nemoralis* (J.R.Forst. & G.Forst.) Dandy ex Hutch. & Dalziel
Very leafy perennial herb; rhizome flexible, branching; culms usually distant, 10–25cm × 0.8–1.5mm, ridged; leaves 10–35 × 0.2–0.5cm, usually far overtopping the culms, flaccid; inflorescence a globose or ± ovate head, usually of a single spike, 3–8 × 3–8mm; involucral bracts 3–4 (rarely with a dwarf 5th), to 20cm; spikelets 2.0–2.5mm, 1-flowered; glumes whitish, fading pale reddish brown. Shaded parts of forest floor, forest paths; 550–1650m.
Distr.: pantropical.
IUCN: LC
Kupe Village: Etuge 2736 7/1996; Kenfack 265 7/1996; Zapfack 939 7/1996; **Mejelet-Ehumseh:** Manning 434 fl., 10/1986; **Ngomboaku:** Cheek 10355 12/1999.

Cyperus laxus Lam. subsp. *buchholzii* (Boeck.) Lye
Nord. J. Bot. 3: 231 (1983).
Syn.:*Cyperus diffusus* Vahl subsp. *buchholzii* (Boeck.) Kük.
Leafy, caespitose perennial herb; culms 25–50cm × 1.5–3.0mm, basal part covered with numerous crowded leaves; leaf sheaths purple to reddish brown; inflorescence a ± open anthela, 3–7 × 5–10cm; major involucral bracts leafy, 20–40cm, much overtopping inflorescence; largest peduncles 2–5cm bearing a subumbellate spikelet cluster; sessile clusters with 3–5 spikelets; spikelets 4–6 × 1.5–2.5mm, light reddish brown. Forest and secondary vegetation, often near streams; 570–1050m.
Distr.: tropical Africa.
IUCN: LC
Kodmin: Plot B164 1/1998; **Kupe Village:** Cable 3697 7/1996; Etuge 2806 7/1996; 2901 7/1996; Kenfack 264 7/1996; Ryan 318 5/1996; 329 5/1996; **Mwambong:** Etuge 4337 10/1998; **Ngomboaku:** Muasya 2032 12/1999; **Nyasoso:** Cable 3215 6/1996; 3468 7/1996; Etuge 2517 7/1996; 2553 7/1996; Pollard 84 10/1998.

Cyperus mannii C.B.Clarke
Stout perennial to 2m; culms tufted or with a short rhizome; leaves often well-developed, to 1m or more; inflorescence anthelate, to c. 25 × 30cm, thrice-branched, bearing small clusters of spikelets; inflorescence branches grooved, each with a conspicuous basal prophyll 1–2cm; involucral bracts leafy to 50 × 2cm; spikelets red-brown. Forest, forest edges, forest paths; 1950–2060m.
Distr.: Sierra Leone to Bioko, W Cameroon. (Upper & Lower Guinea).
IUCN: LC
Nyasoso: Cable 104 9/1992; 2903 6/1996.

Cyperus microcristatus Lye in press
Perennial herb to 50cm; rhizome short, slender, with many densely set stems; culms 30–50cm × 0.5–1.0mm; leaves 3–4 per culm; blades 5–12cm × 1.5–2.5mm; inflorescence a congested greyish to pale brown head, 10 × 10mm, consisting of 2–3 crowded spikes; involucral bracts 3–4, foliaceous; central spike 6–8 × 4mm, with numerous crowded spikelets. Forest; 500m.
Distr.: Kupe-Bakossi. (Narrow Endemic).

IUCN: CR
Kupe Village: Patterson 11 11/1995.

Cyperus pectinatus Vahl
Haines & Lye, Sedges & Rushes of E Africa: 172 (1983).
Syn.:*Cyperus nudicaulis* Poir.
Tufted leafless perennial; culms 0.3–1.2m × 0.6–2.0mm; stems slender, leafless; roots light brown to bright-orange; inflorescence a solitary reddish brown head of 3–20 sessile spikelets, 1–4cm across, sometimes viviparous; floating in mats; spikelets 5–20 × 2–6mm, flattened, 15–40-flowered, produced continually. Swamps, lake-edges, along streams and rivers, sometimes floating; 1250–1300m.
Distr.: tropical Africa & Madagascar.
IUCN: LC
Lake Edib: Cheek 9135 2/1998; Etuge 4107 1/1998; 4499 11/1998; Satabie 1122 2/1998.

Cyperus pinguis (C.B.Clarke) Mattf. & Kük.
Haines & Lye, Sedges & Rushes of E Africa: 239 (1983).
Syn.:*Kyllinga elatior* Kunth
Perennial herb; rhizome to 25cm × 1.5–3.5mm, creeping; culms 10–60cm × 1.5–2.5mm, regularly spaced at first but later with basal accessory culms; 2–4 leaf-blades per culm, 2–10cm × 3.5–5.0mm; inflorescence a solitary ovate yellowish green spike, rarely with smaller lateral spikes; central spike 8–15 × 4–8mm; involucral bracts 5–6, leafy, to 20 × 0.6cm; spikelets 1–2-flowered, 2.5–3.5mm. Forest margins, streamsides, damp hollows in grazed and cultivated areas, roadside ditches, other disturbed habitats; 1400–1500m.
Distr.: tropical & subtropical Africa. (Afromontane).
IUCN: LC
Ehumseh - Mejelet: Etuge 361 fl., 10/1986; **Kodmin:** Cheek 8899 1/1998; 9018 1/1998; **Ngomboaku:** Muasya 2034 12/1999.
Local name: Nzi-nzion. **Uses:** MEDICINES (Cheek 8899).

Cyperus pulchellus R.Br.
Slender perennial herb; stem bases many, crowded; culms 6–40 × 0.4–1.0mm, glabrous or slightly scabrid; leaf-blades 2–15 × 0.5–2.0mm, flat or V-shaped, scabrid on margin and midrib; inflorescence a congested anthela; spikelets sessile only; major involucral bract leafy, 2–10cm × 1.0–1.5mm; anthela 0.7–1.5cm, of 15–60 crowded spikelets; spikelets 4–8 × 1.0–2.5mm, greyish white, tinged light pinkish brown, 10–20-flowered; glumes 1.3–1.5mm. Seasonally wet habitats, often in temporary swamps, grassland; 2000m.
Distr.: palaeotropics.
IUCN: LC
Mwanenguba: Thomas 3121 fl., 2/1984.

Cyperus renschii Boeck. var. *renschii*
Large, robust perennial to 1.5m; rhizome woody, 1.0–1.5cm thick; culms 50–150cm × 2–8mm; many large basal leaves 60–120 × 1–2.5cm; leaf-sheaths purplish near base; inflorescence a large anthela, 12–40 × 15–30cm, with 6–12 'pseudoumbels', which, in larger specimens bear further 'pseudoumbels' of the 3rd or 4th order; peduncles 0.5–30cm; involucral bracts leafy, to 90 × 4cm; spikelets 2–3 × 1–2mm, brown, 5–8-flowered. Forest swamps, along forest paths and streams; 900–1520m.
Distr.: tropical Africa.
IUCN: LC
Ehumseh - Mejelet: Etuge 362 10/1986; **Kodmin:** Cheek 8881 1/1998; Ghogue 4 1/1998; Pollard 199 11/1998; **Kupe Village:** Etuge 1914 5/1996; Lane 345 fl., 1/1995.
Local name: Achang. **Uses:** MEDICINES (Cheek 8881).

Cyperus rheophytorum Lye in press
Slender rheophytic herb; culms crowded, 5–30cm × 0.3–0.4mm; leaves from the lower 8cm only; largest blades to 10cm × 0.5–1.3mm, flat; inflorescence a single, terminal, globose, whitish, congested anthela, c. 3mm diameter; involucral bracts usually 3, foliaceous, the largest 2–6cm × 1.0–1.4mm; spikelets 1.5–1.7 × 0.7–0.9mm, 1-flowered. On rocks and stones in or beside streams and rivers, usually submerged during rains; 750–1350m.
Distr.: Kupe-Bakossi & Mt Cameroon. (Narrow Endemic).
IUCN: VU
Baseng: Cheek 10405 12/1999; **Kodmin:** Etuge 4063 1/1998; **Mwambong:** Onana 585 2/1998; **Nyale:** Cheek 9702 11/1998.

Cyperus richardii Steud.
Haines & Lye, Sedges & Rushes of E Africa: 227 (1983).
Syn.:*Kyllinga bulbosa* P.Beauv.
Small perennial herb; culms solitary from long, slender, whitish stolon, 5–30cm × 0.7–1.5mm; leaves 5–25 × 0.2–0.5cm; inflorescence a dense irregular head of several whitish spikes, 0.7–1.3 × 0.6–1.3cm; involucral bracts leafy, c. 3–5, up to 8cm; spikelets 2.5–3.5mm, white, darkening as fruits develop. Open grassland, roadsides, disturbed ground; 2000m.
Distr.: tropical Africa. (Afromontane).
IUCN: LC
Mwanenguba: Pollard 139a fl., 10/1998.

Cyperus sesquiflorus (Torr.) Mattf. & Kük. subsp. *cylindricus* (Nees) Koyama
Haines & Lye, Sedges & Rushes of E Africa: 241 (1983).
Syn.:*Kyllinga odorata* Vahl subsp. *cyllindrica* (Nees) Koyama
Tufted perennial; rhizome creeping to 4cm; culms crowded, 3–60cm, ridged; leaves 3–30cm (as long as culm or much shorter); inflorescence white, either a single cylindrical or rarely globose spike or a compound head of one larger cylindrical spike and several smaller lateral spikes; involucral bracts leafy, 2–4, to 12cm; spikelets 1.8–2.5mm, 1-flowered. Open grassland, especially in disturbed ground, heavily grazed or cut or burnt, often on sandy, or gravelly soil; 2000m.
Distr.: palaeotropics. (Montane).
IUCN: LC
Mwanenguba: Pollard 139b fl., fr., 10/1998.

Cyperus sphacelatus Rottb.
Slender annual; culms 15–50cm × 0.8–2.0mm; leaf-blades 6–15 × 0.1–0.3mm, flat or W-shaped; inflorescence anthelate, 2.5–10 × 3–10cm, of 1 sessile and 2–8 pedunculate spikes; peduncles to 2cm; involucral bracts leafy, 8–15 × 0.2–0.4cm; spikes 1–3 × 1.5–4.0cm; spikelets 3–10, spreading; glumes green with dark reddish brown patches. Open grassland, damp roadside ditches; 880m.
Distr.: tropical Africa and America, introduced elsewhere. (Pantropical).
IUCN: LC
Nyasoso: Pollard 110 fl., 10/1998.

Cyperus tenuiculmis Boeck. subsp. *mutica* B.J.Pollard & Lye in prep.
Damp roadside ditches; 880m.
Distr.: Mt Kupe. (Narrow Endemic).
IUCN: CR
Nyasoso: Pollard 113 10/1998.

Cyperus tenuis Sw.

Prod. Ind. Occ. 20 (1788).

Syn.:*Mariscus flabelliformis* Kunth var. *flabelliformis*
Perennial caespitose herb; culms crowded, 20–60cm, ridged; leaf-blades to 40 × 0.6cm; inflorescence anthelate, of 5–10 sessile or shortly pedunculate spikes; involucral bracts leafy, to 20 × 0.5cm; spikelets 1.0–1.5cm × 0.7–1.0mm, greenish. Farmbush, open areas; 700–1300m.
Distr.: Senegal to Bioko, Cameroon, tropical America. (Amphi-Atlantic).
IUCN: LC
Kupe Village: Etuge 1699 2/1996; 2767 7/1996; **Patterson** 10 11/1995; **Nyasoso:** Cable 3518 7/1996.

Cyperus tenuispica Steud.

Slender annual herb; stems few to many; root system small; culms 5–20cm × 0.5–1.5mm, triangular or 6-angular; leaf-blades to 10cm × 0.5–3.0mm; inflorescence anthelate, 3–10cm across, with 2–12 spikes; spikes usually pedunculate, 1–3 sessile; major involucral bract(-s) leafy, 2–10cm; spikelets green or reddish brown, 2–12 × 0.8–1.3mm. Seasonally wet habitats, near swamps and streams; 2000m.
Distr.: tropical Africa, India, Nepal, E to Malesia, China & Japan. (Palaeotropics).
IUCN: LC
Mwanenguba: Pollard 141 10/1998.

Eleocharis variegata (Poir.) C.Presl

Perennial stoloniferous herb; culms crowded in small tussocks, 25–90 × 0.1–0.5cm; older plants sometimes with thick, erect rhizome; stolons to 30 × 0.3cm, reddish yellow with greyish black nodal papery scales; inflorescence a cylindric spike, 1–5.5 × 0.2–0.5cm; glumes 3–5mm, purplish or brown with a broad greenish central area. Swamps and seasonal pools, shallow water at lake edges; 1200–1250m.
Distr.: pantropical.
IUCN: LC
Lake Edib: Cheek 9130 fr., 2/1998; Etuge 4111 1/1998; Pollard 246 11/1998.

Fimbristylis dichotoma (L.) Vahl

Caespitose leafy perennial herb; rhizome short, woody; culms 10–60 × 0.1cm; leaves numerous, 5–50cm × 0.8–3.0mm, usually much shorter than culms; inflorescence anthelate, with numerous ovate spikelets arranged with 2–3 orders of branching; one conspicuous leafy involucral bract, 2–10cm; spikelets 0.4–1.2 × 0.2–0.3mm; glumes 2.5–3.0mm, reddish brown with paler midrib and margin. Wet grassland, streambanks, roadside banks, swamp-margins; 450–800m.
Distr.: pantropical.
IUCN: LC
Bakossi F.R.: Cheek 9328 10/1998; **Tombel:** Etuge 365 11/1986.

Fuirena umbellata Rottb.

Stout perennial herb; rhizome horizontal, creeping, woody, about 5mm thick; culms 60–150cm × 3–6mm, basally swollen, bulb-like, 5-angled; lower leaves sheathing; upper to 12–30 × 0.8–2.5cm, flat, distinctly horizontally-ridged; inflorescence of numerous pedunculate compound corymbs on upper 10–40cm of the stem; each corymb of 2–20 clusters of spikelets; involucral bracts similar to upper leaves; spikelets 5–8mm, green when young, maturing dark brown. Wet grasslands, swamps, ditches and pools, along streams and lake edges; 1200–1250m.
Distr.: pantropical.
IUCN: LC
Lake Edib: Cheek 9137 2/1998; Pollard 242 11/1998.

Hypolytrum heteromorphum Nelmes

Stout rhizomatous perennial herb; culms slender, 25–40cm × 1–2mm; leaves densely set near base, 30–60 × 1.0–1.5cm, flat or ± plicate, margin slightly toothed; inflorescence a series of panicles, broader than tall; panicles of a few closely-set major branches arising from the axils of the main bracts; bracts greyish brown, to 3.5cm; lowest branches to 2.5cm, with 2–6 sessile spikes at the tip; spikes 1.0–1.5 × 0.1–0.2mm, brown. Wet forest; 250–1000m.
Distr.: tropical Africa. (Afromontane).
IUCN: LC
Baduma: Nemba 670 11/1987; **Konye:** Thomas 5189 11/1985; **Nyale:** Etuge 4183 2/1998; 4450 11/1998.

Hypolytrum pseudomapanioides D.A.Simpson & Lye in press

Robust tussocky perennial; culm 1, erect, central, 60–70cm × 2–3mm; leaves mostly basal, 1 cauline, to 100 × 1.3–1.6cm; inflorescence an irregular subspherical brownish white head, 2.5–3.5 × 3–3.5cm, comprising numerous, crowded, fairly distinct spikes; involucral bracts usually 4, foliaceous, very unequal, the largest to 50 × 1.6cm; spikes ovoid to lanceolate, 10–14 × 3–5mm, pale brown. Montane forest; 1470–1500m.
Distr.: Bakossi Mts. (Narrow Endemic).
IUCN: EN
Kodmin: Cheek 8919 fr., 1/1998; Ghogue 33 1/1998; Plot B68 1/1998.

Hypolytrum subcompositus Lye & D.A.Simpson in press

Robust, rhizomatous perennial; culms 1-several, erect, lateral, 40–50cm × 1.8mm; leaf-blade linear, flat, mid-green, coriaceous, 60–120 × 1.3–2cm; inflorescence a congested terminal subhemispherical corymb with up to 20 spikes and a few major branches still visible, 1.5–3 × 2.5–3.5cm, arising laterally from rhizome, capitate; flowers spicoid with 2 floral bracts and 2 stamens. Montane forest; 1500m.
Distr.: Bakossi Mts. (Cameroon Endemic).
IUCN: CR
Kodmin: Etuge 4007 fr., 1/1998.

Mapania amplivaginata K.Schum.

Robust herb; culm solitary, erect, central, 10–30cm × 1–4mm, glabrous; leaves basal, to 70cm +; blade linear to oblong, or oblong-elliptic, 15–36 × 2.4–8.5cm; apex abruptly narrowed, serrulate; base narrowed into pseudopetiole, it being 4.2–24 × 2.5–9.3mm; inflorescence terminal, globose, 1.5–2.5cm, many-spiked, chestnut-brown to dark reddish brown. Deep shade in forest, sometimes by water; 850–1000m.
Distr.: SE Nigeria, Cameroon, Gabon, Angola (Cabinda). (Lower Guinea).
IUCN: LC
Ngomboaku: Cheek 10291 fr., 12/1999; 10327 12/1999; Ghogue 498 12/1999; Muasya 2023 12/1999.

Mapania ferruginea Ridl.

Simpson, *Mapania*, a revision of the genus: 137 (1992).
Moderately robust herb; culm solitary, erect, central, 35–45cm × 2–3mm, subtriquetrous, glabrous; leaves basal, to 75cm; blade linear, 40–60 × 2.6–2.9cm; pseudopetiole absent; inflorescence terminal, globose, 2–3.5cm wide, mid reddish brown, consisting of numerous spikes. Montane forest; 1100–1400m.
Distr.: São Tomé, Principé, W Cameroon. (W Cameroon Uplands).

IUCN: VU
Kodmin: Plot B57 1/1998; **Kupe Village:** Cable 3773 7/1996; Etuge 2853 7/1996.
Note: this is the first record of this species from (mainland) continental Africa.

Mapania macrantha (Boeck) H.Pfeiffer
Robust herb; culm solitary, erect, central, 42–96cm × 4–6mm, subtriquetrous; leaves basal, to 170cm +; blade widely linear, 108–149 × 49–6.3cm; margins entire to serrulate; pseudopetiole absent; inflorescence terminal, globose, 3.8–5.4cm wide, mid- to dark brown, with numerous spikes. Forest; 800–1350m.
Distr.: E Nigeria, Cameroon & Gabon. (Lower Guinea).
IUCN: NT
Kupe Village: Cheek 7023 fl., 1/1995; 7194 fr., 1/1995; **Nyasoso:** Cable 1173 fl., 2/1995; Sebsebe 5053 10/1995; Wheatley 404 fl., 7/1992.
Note: known from only 12 sites, with recent collections only from Mt Cameroon, Mt Kupe and Cap Estérias, Gabon.

Mapania rhynchocarpa Lorougnon & J.Raynal
Robust herb; culm solitary, erect, central, 43cm × 3.5mm; leaves basal, to 105cm +; blade linear, 60–93 × 3.9–3.2cm; pseudopetiole absent; inflorescence terminal, globose, 4–6cm wide, whitish, with numerous spikes. Forest; 1000m.
Distr.: Sierra Leone, Liberia, Ivory Coast, Nigeria, Cameroon. (Upper & Lower Guinea).
IUCN: LC
Menyum: Doumenge 530 5/1987.
Note: this specimen was determined at MO, and has not been seen by the authors. Although this taxon has not previously been reported from Cameroon, it is known to occur in SE Nigeria, and it is quite possible that its distribution extends into SW Cameroon.

Rhynchospora brownii Roem. & Schult.
Haines & Lye, Sedges & Rushes of E Africa: 317 (1983).
Slender, leafy perennial; culms 0.4–1m; leaves 5–40cm × 1–3mm; inflorescence a slender panicle of terminal corymbose clusters, each of 5–15 spikelets; spikelets elliptical, 4–5mm, brown; lower 3–5 glumes sterile, 1.5–2.8mm, a few fertile ones above, 3–4mm. Marshes, near streams, grassland; 2000m.
Distr.: Cameroon to E Africa, southern Africa, Madagascar, Asia, Australia. (Palaeotropics).
IUCN: LC
Mwanenguba: Thomas 3151 fl., 2/1984.
Note: we have not seen this Thomas collection, but conclude that it is quite likely to occur in the grassland areas of Mwanenguba, and so include this MO determination here.

Rhynchospora corymbosa (L.) Britt.
Coarse, leafy, caespitose perennial; rhizome thick, creeping with numerous soft rootlets; culms 0.6–2.5 m, to about 1cm across towards the base; leaves numerous, densely crowded, the basal ones 50–100 × 0.1 –0.2cm, very tough, with minute spinose teeth on the margin and midrib; inflorescence of one terminal and several lateral corymbs; involucral bracts leafy, 10–50 × 0.5–2.0cm; spikelets 6–10 × 1.5–2.5mm, in pedunculate clusters; glumes reddish brown. Lake shores, riverbanks, in shallow pools, tolerant of light shade; 1200–2000m.
Distr.: pantropical.
IUCN: LC
Lake Edib: Cheek 9165 2/1998; Etuge 4112 1/1998; Pollard 245 11/1998; **Mwanenguba:** Thomas 3131 2/1984.

Schoenoplectus corymbosus (Roth ex Roem. & Schult.) J.Raynal var. *brachyceras* (A.Rich.) Lye
Fabregues & Lebrun, Catal. Plant. Vasc. Niger: 343 (1976).
Syn.: *Scirpus brachyceras* Hochst. ex A.Rich.
Stout, leafless perennial herb; rhizome short, thick, woody; culms numerous, 0.5–2.0m × 2–8mm, rounded, pith-filled; inflorescence anthelate, simple or compound, with clusters of spikelets on very unequal branches; main branches to 12cm, often only to 2–3cm, secondary branches to 4cm; involucral bract leaf-like, continuing in the direction of the main stem. Bogs, lake edges, along streams. 1975–2000m.
Distr.: tropical & S Africa, Madagascar.
IUCN: LC
Mwanenguba: Cheek 7249 2/1995; 9404 10/1998; Thomas 3099 2/1984.

Schoenoplectus mucronatus (L.) A.Kern.
Haines & Lye, Sedges and Rushes of E Africa: 55 (1983).
Syn.: *Scirpus mucronatus* L.
Leafless perennial herb; rhizome erect or flat; culms 3–10, clustered, 40–80 × 0.3–0.8cm, sharply triangular; inflorescence of 4–25 sessile spikelets (may appear stalked when lower scales fall off); involucral bract 1.5–2.5cm, continuing in the direction of the stem, occasionally bent back; spikelets c. 5mm in flower, but to 30 × 6mm in fruit. Streambeds, roots often submerged; 1200–1300m.
Distr.: widely distributed from S Europe through Africa to India, E to Malesia, Australia. (Palaeotropics).
IUCN: LC
Lake Edib: Etuge 4491 11/1998; Pollard 247 11/1998.

Scleria afroreflexa Lye in press
Nord. J. Bot. (2004, in press).
Delicate annual; culms 10–50cm × 0.3–0.8mm; leaves 2–4 per stem, but only 1–3 perfecting leaf-blades; sheaths densely covered by retrorse white hairs 0.2–0.4mm; blades 2–9cm × 0.8–1.8mm; inflorescence 3–9 × 1–2cm, appearing spike-like with sessile glomerules above, but in fact a narrow panicle with 1-several reflexed branches with 1–3 glomerules below; glomerules 4–5 × 3–6mm, consisting of 2–10 spreading spikelets. Grassland; 1500m.
Distr.: Bakossi Mts, Bali-Ngemba F.R., Mt Oku. (W Cameroon Uplands).
IUCN: EN
Kodmin: Muasya 2022 12/1999.

Scleria melanomphala Kunth
Robust perennial herb; stems 0.6–1.8m × 2–6mm; leaves 20–60 × 0.7–1.8cm, with recurved hooks on the angles; inflorescence consisting of one terminal and 5–9 lateral, often pedunculous panicles; panicles 2–10 × 1–3cm; peduncles to 25cm; spikelets 1.0–1.3cm; glumes straw-coloured with conspicuous dark reddish brown margins. Wet marshes and grasslands, lake edges and swamps, growing in large free-standing clumps, avoided by cattle; 1450–1560m.
Distr.: tropical & S Africa, Madagascar & S America. (Amphi-Atlantic).
IUCN: LC
Kodmin: Cheek 9604 11/1998; **Mwambong:** Mackinder 165 1/1998.
Local name: Lehlen (Max Ebong in Cheek 9604).

Scleria vogelii C.B.Clarke
Stout perennial herb to 2.5m; leaves with sharp marginal recurved hooks; inflorescence a loose terminal panicle; achene globose to subovoid, shining ivory-white to honey-grey at maturity, 2.5–3.5mm across. Swampy places in forest and savanna; 1000–1250m.

DIOSCOREACEAE

Distr.: Liberia to Gabon. (Upper & Lower Guinea).
IUCN: LC
Lake Edib: Cheek 9154 2/1998; Etuge 4109 1/1998;
Mwambong: Onana 530 2/1998.

DIOSCOREACEAE

P. Wilkin

Dioscorea alata L.

Herbaceous climber to 9m, glabrous; usually reproducing
vegetatively; stem 4-winged, square, often with aerial tubers;
leaves opposite, rarely alternate, blades usually subsaggitate
or subhastately ovate, apex acuminate; inflorescence axis
flexuous; rarely flowers in Cameroon. In cultivated areas and
around village margins; 300m.
Distr.: pantropical cultigen, probably domesticated in
Indochina.
IUCN: LC
Kupe Village: Etuge 2794 fr., 7/1996; **Nyasoso:** Cheek 7510
10/1995.

Dioscorea bulbifera L.

Herbaceous climber, glabrous, to 3–7m; stems left-twining
(sinistrorse); leaves alternate with a pair of membranous
semicircular lateral projections clasping stem at petiole base,
apex short-acuminate, not thickened; warty bulbils usually
present. Farmbush; 600–1450m.
Distr.: widespread in tropical Africa & Asia. (Palaeotropics).
IUCN: LC
Kodmin: Cheek 8934 1/1998; **Kupe Village:** Cable 2424
5/1996; 2693 5/1996; 3687 fr., 7/1996; **Nyasoso:** Cheek 7453
10/1995; Wheatley 380 fr., 7/1992; Zapfack 635 6/1996; 641
fl., 6/1996.

Dioscorea dumetorum (Kunth) Pax

Herbaceous climber to c. 3m; stems left-twining, with fleshy
to hard prickles, especially towards base; leaves alternate,
compound, usually 3-foliolate with 3–7 veins per leaflet,
softly white-tomentose abaxially; leaflets asymmetrical;
petioles often spiny; capsule 3–5 × 1.5–2.5cm. Forest; 750–
1050m.
Distr.: Senegal to Ethiopia, S to Zambia, Zimbabwe &
Mozambique. (Tropical and subtropical Africa).
IUCN: LC
Ehumseh - Mejelet: Etuge 344 10/1986; **Kupe Village:**
Cable 3711 fr., 7/1996; **Nyasoso:** Cheek 9288 fr., 10/1998;
Etuge 2494 fl., 7/1996.

Dioscorea minutiflora Engl.

Perennial climber to 10m; stems rooting at nodes to form a
network on the ground with multiple tubers; stems right-
twining (dextrorse), with few or no prickles; leaves opposite,
orbicular or suborbicular, leathery, 5–7-veined, 6–17 × 6–
17cm. Forest; 200–1600m.
Distr.: Senegal to Uganda. (Guineo-Congolian).
IUCN: LC
Kodmin: Cheek 8923 fl., 1/1998; Etuge 4035 fr., 1/1998;
Kupe Village: Cheek 7103 1/1995; Etuge 1927 fl., 5/1996;
Mungo River F.R.: Cheek 9334 10/1998; 9336 fl., 10/1998;
Ngomboaku: Mackinder 301 fl., 12/1999; **Nyasoso:** Etuge
1634 fl., 1/1996; 2368 fr., 6/1996.

Dioscorea praehensilis Benth.

Sturdy climber to 8m; single tuber per growing season; stems
right-twining, prickly, glabrous; leaves opposite, chartaceous,
not leathery, shortly cordate, acuminate to long-acuminate.
Forest, forest edge; 200–760m.
Distr.: Sierra Leone to Cameroon & E Africa, S to Zambia,
Zimbabwe, Malawi & Mozambique. (Tropical Africa).
IUCN: LC
Kupe Village: Cable 3710 fl., 7/1996; **Mungo River F.R.:**
Etuge 4315 fr., 10/1998.

Dioscorea preussii Pax subsp. *preussii*

Robust non-spiny climber to 10m; stems often 6-winged,
subglabrous, with a few ± caducous, medifixed (T-shaped)
hairs, also on the inflorescence and leaf apices; leaves
alternate, broadly ovate, obliquely acuminate, deeply cordate,
10–30 × 8–35cm, villose-tomentose beneath. Forest,
farmbush; 700–1000m.
Distr.: Senegal to Uganda, Angola, Mozambique. (Tropical
Africa).
IUCN: LC
Kupe Village: Cheek 7099 1/1995; 7100 1/1995; Etuge 1965
fr., 5/1996; Kenfack 313 fl., 7/1996; **Nyasoso:** Cable 4 fl.,
7/1992; Etuge 2454 fl., 6/1996; 2558 fl., 7/1996; Wheatley
379 fl., 7/1992.

Dioscorea sansibarensis Pax

Glabrous, non-spiny climber to 7m; stems left-twining, often
with axillary bulbils; leaf tip conspicuously long-acuminate,
thickened and with inrolled margin; petiole with a pair of
fleshy oblong, basal, projections to c. 1cm long, margins
lobed in young plants. Forest, farmbush; 900m.
Distr.: tropical & subtropical Africa & Madagascar.
IUCN: LC
Nyasoso: Cable 2788 fr., 6/1996.

Dioscorea smilacifolia De Wild.

Perennial climber to c. 4m, resembling *D. minutiflora*, but
not rooting at nodes; stems almost always prickly, right-
twining; leaves opposite, ovate, ovate-lanceolate or ovate-
oblong, 6–10 × 4–7cm, leathery, 3(–5)-nerved, veins
displaced towards margin. Forest gaps; 1000–1400m.
Distr.: Senegal to Uganda. (Guineo-Congolian).
IUCN: LC
Kodmin: Plot B156 1/1998; **Kupe Village:** Cheek 7164 fl.,
1/1995.

DRACAENACEAE

J.J. Bos, G. Mwachala & M. Cheek

Dracaena arborea Link

Tree 10–20m, trunk 30cm, with aerial roots, several-
branched; leaves in dense heads, sword-shaped, 50–120 ×
4.5–6cm, widest above the middle, apex acute, mucro to
3mm, base clasping stem for 3/4 circumference;
inflorescence pendulous, to 1.5m; perianth cream-white, c.
1.5cm; fruit to 2cm, orange-red. Submontane forest and
planted.
Distr.: Sierra Leone to Angola. (Guineo-Congolian).
IUCN: LC
Nyasoso: Etuge 5007 3/2003.
Note: a conspicuous element of the Bakossi landscape.
Local name: Peregun (*fide* Wild). **Uses:** MATERIALS /
ENVIRONMENTAL USES – formerly planted as cattle
hedge and persisting (*fide* Cheek) – boundary plants (*fide*
Pollard).

Dracaena cf. bueana Engl.

Robust herb 4m; leaves 1–2.5m × 10cm; inflorescence terminal, erect; perianth 1.5cm long. Secondary forest; 900m.
Distr.: Cameroon. (Lower Guinea).
Kupe Village: Elad 72 fl., 1/1995.
Note: a putative hybrid of *D. fragrans* and *D. arborea*, with very large leaves.

Dracaena fragrans (L.) Ker-Gawl.

Syn.: *Dracaena deisteliana* Engl.
Herb to 3m, few-stemmed; stalk 1cm diameter; leaves sword-shaped, to 70 × 9cm; inflorescence terminal, erect; flowers white with pink lines, very fragrant. Forest and cultivated; 840–1600m.
Distr.: tropical Africa. (Afromontane).
IUCN: LC
Kupe Village: Cable 834 fl., 1/1995; Cheek 7045 fl., 1/1995; 7114 fl., fr., 1/1995; 7193 fl., fr., 1/1995; Lane 350 fl., 1/1995; Ryan 271 fr., 5/1996; **Nyasoso:** Cable 1207 fl., 2/1995; Sunderland 1535 fr., 7/1992.
Note: specimens with leaves c. 2cm wide and more highly branched were formerly known as *D. deisteliana* Engl.
Uses: SOCIAL USES – peace emblems;
ENVIRONMENTAL USES – boundaries – used on farms for boundary markers (Cheek 7045).

Dracaena mildbraedii K.Krause

Bot. Jahrb. 51: 447 (1914).
Herb, sometimes tree-like, 0.5–0.8m; leaves oblanceolate-ligulate, 13 × 3cm, acute, mucro 7mm, base sessile, leaf-sheath c. 1cm; inflorescence unbranched, 10cm, nodes c. 7. Secondary forest; 1090–1400m.
Distr.: Ghana, E Nigeria to Cabinda. (Guineo-Congolian).
IUCN: NT
Essossong: Groves 43 fr., 2/1995; **Nyasoso:** Cable 1236 fr., 2/1995; Lane 526 fr., 2/1995.
Note: rare throughout its range.

Dracaena cf. phanerophlebia Baker

Herb 50cm, without aerial stem; leaf-blades 3–6, ovate-lanceolate, c. 25 × 9cm, acuminate, decurrent; petiole c. 20 × 0.5–1cm, basal 4cm dilated; inflorescence terminal, c. 6cm, erect; fruiting umbel 3cm diameter; fruit globose, 1cm, orange. Forest, secondary forest; 800–1250m.
Distr.: Cameroon. (Lower Guinea).
Kodmin: Satabie 1097 fr., 1/1998; **Kupe Village:** Lane 286 fr., 1/1995; **Nyasoso:** Cheek 7476 fr., 10/1995; Etuge 2533 7/1996; Lane 533 fr., 2/1995; Sebsebe 4986 fr., 10/1995; 5009 10/1995; Sidwell 346 fr., 10/1995.

Dracaena phrynioides Hook.

Herb to 0.7m, lacking aerial stem; leaves 4–6, blades lanceolate, 18 × 7cm, spotted yellow; petiole c. 1mm wide; inflorescence terminal; peduncle 5cm; fruit bilobed, orange, sessile, head c. 2.5cm diameter. Forest; 420–1200m.
Distr.: Liberia to Bioko, Rio Muni & Gabon. (Upper & Lower Guinea (montane)).
IUCN: LC
Enyandong: Cheek 10912 10/2001; **Kodmin:** Etuge 4073 fr., 1/1998; **Kupe Village:** Cable 2436 5/1996; 2735 fr., 5/1996; Ensermu 3507 fr., 11/1995; Etuge 1956 fr., 5/1996; Ryan 236 fr., 5/1996; 316 fr., 5/1996; **Mungo River F.R.:** Onana 962 11/1999; **Nyandong:** Ghogue 1446 fr., 3/2003; Tchiengue 1739 fr., 3/2003; **Nyasoso:** Etuge 1765 fr., 3/1996; Sebsebe 4992 10/1995.

Dracaena sp. aff. phrynioides Hook.

Herb resembling *D. phrynioides*, but to 2m, leaf-blades to 35cm long, lacking yellow spots; inflorescence peduncle 25cm, head 5cm. Forest; 700–1500m.
Kupe Village: Etuge 2839 7/1996; **Ngomboaku:** Cheek 10353 12/1999; **Nyasoso:** Wheatley 454 fl., 7/1992.

Dracaena sp. aff. praetermissa Bos

Fleshy rhizomatous herb to 50cm; leaves in basal rosette, oblong; fruiting spikes in clusters; fruit pyriform with flattened apex, green; seeds white. Forest; 850m.
Distr.: Cameroon. (Lower Guinea).
Nyasoso: Sunderland 1547 fr., 7/1992.
Note: specimen not seen by the authors; determination to be confirmed.

Dracaena viridiflora Engl. & K.Krause

Shrub to 3m; stems long, slender; leaves linear, c. 20 × 3cm, sheathing and amplexicaul at base, lateral nerves 8–10 pairs; inflorescence spikes terminal and axillary, to 10cm; flowers clustered; bracts ovate, inconspicuous; fruit globose, green. Forest; 1200m.
Distr.: SE Nigeria, Cameroon & Rio Muni. (Lower Guinea).
IUCN: VU
Kupe Summit: Thomas 3168 2/1984; **Nyasoso:** Lane 109 fr., 6/1994.

Dracaena sp. A

Herb, sometimes tree-like, 2.5–9m, resembling *D. mildbraedii*, but leafy stems with numerous leaf-scars, nodes 5–10mm; leaves lanceolate-ligulate, c. 13 × 1cm. Forest; 1600–2000m.
Kodmin: Ghogue 53 1/1998; **Nyasoso:** Etuge 1654 fr., 1/1996; 2175 fr., 6/1996.
Uses: MEDICINES; ENVIRONMENTAL USES – boundaries – planted as a hedge (Etuge 1654).

Dracaena sp. B

Tree 15–20cm; bole 30cm d.b.h., first branch at 8m, only 4–5 branches; leaves in rosettes, blades oblanceolate, 30 × 3cm, base cuneate, clasping the stem for ½ the circumference, thickened, black. Forest; 1000m.
Ngomboaku: Cheek 10380 12/1999.
Note: known only from this specimen.

Dracaena sp. C

Herb 1.5m; leaves spatulate, 'blade' elliptic, 11 × 6.5cm, acuminate; mucro 4mm, 'petiole' c. 15 × 0.6cm, ± sheathing stem for lower 2cm, amplexicaul. Forest; 1400m.
Kodmin: Plot B141 1/1998.
Note: sterile plot voucher. In the spatulate leaf, not resembling any other species known from our area; more material required.

Sanseviera trifasciata Prain

Fleshy herb to 1.2m; leaves mottled white and green or yellow and green, c. 50 × 5cm, oblanceolate, margins green when dry (not red-brown); flowers in interrupted racemes; perianth at most 3cm long; bracts small and membranous; pedicels jointed about the middle. Forest, cultivated; 1200m.
Distr.: S Nigeria, Cameroon, Congo (K). (Lower Guinea & Congolian).
IUCN: LC
Nyandong: Cheek 11378 fr., 3/2003.
Note: this species is widely cultivated around the world as 'good luck plant', 'mother-in-law's tongue', 'snake plant'.
Local name: Achang deicum (Cheek 11378).

ERIOCAULACEAE

S.M. Phillips & B.J. Pollard

Mesanthemum jaegeri Jacq.-Fél.

Robust terrestrial or epilithic perennial; leaves, scapes and basal sheaths hirsute (glabrous in *Eriocaulon*); stem often swollen, but not elongate; leaves in basal rosette, 15–30cm; inflorescence capitate, capitula 1–1.8cm diameter, flattened; involucral bracts in several series, broadly ovate, villous, dirty-white; scape 25–45cm; scape many-ribbed; flowers densely white-hairy, immersed in a cushion of black, receptacular hairs; petals of ♀ flowers connate except at apex and base (free in *Eriocaulon*); ♂ flowers with 3 white anthers (usually black in *Eriocaulon*). Damp rockfaces of inselbergs; 1400m.
Distr.: Sierra Leone, Ivory Coast, SE Nigeria, SW Cameroon. (Upper & Lower Guinea).
IUCN: NT
Mwendolengo: Etuge 4434r fl., fr., 10/2001.
Note: known from only 8 sites, 5 of which are in Sierra Leone, but its habitat is largely unthreatened. Etuge 4434r represents the first record of this species for Cameroon.

GRAMINEAE

T.A. Cope & P. Doyle

Acroceras zizanioides (Kunth) Dandy

Perennial to 1m; culms ascending, nodes glabrous; leaf-blades 8–14cm × 9–18mm, rounded, sometimes amplexicaul; sheaths glabrous to pubescent; inflorescence often lax, 2–8 branches, each 4–11cm, along central axis 15–40cm; spikelets 5.5–6mm, ovate-lanceolate, green or purplish, tips of keels of upper palea reflexed, upper glume and lower lemma with compressed tip. Not very heavily flooded soils in forests, fallow; 500–750m.
Distr.: mostly Guineo-Congolian, but also E Africa, tropical America, India. (Pantropical).
IUCN: LC
Kupe Village: Etuge 2740 7/1996; Kenfack 327 7/1996; Patterson 5 11/1995.

Arthraxon micans (Nees) Hochst

F.T.E.A. Gramineae: 742 (1982).

Annual 10-60cm; culms slender, ascending, much branched; leaf-blades 1–8 cm × 3–20 mm, lanceolate to ovate, amplexicaul at base, almost glabrous, rough at edges; sheaths swollen, pubescent; inflorescence 2–30 (8–14) racemes, 2–6cm, digitate; rachis filiform; pedicelled spikelets 2.5–4mm, narrowly lanceolate, sessile; upper glume reddish purple; awns 4–8mm. Shady humid areas of savannahs and montane forests; 2000m.
Distr.: W, C & E Africa, tropical Asia, Australia, introduced in tropical America (montane).
IUCN: LC
Nyasoso: Cheek 7898 11/1995.

Arundinella pumila (Hochst. ex A.Rich.) Steud.

Annual 20–40cm, slender; culms erect or ascending, branching, sometimes pilose; leaf-blades linear-lanceolate, glabrous, 9–15cm × 10–14mm; ligule 1mm, membranous; inflorescence a loose panicle, obovate, 12–25 × 8–10cm; branches flexible, 10cm; spikelets 1.5–2mm, yellow-green or purple-tinged, obovate, lower floret male or sterile, glumes membranous; lower lemma ovate-elliptic, upper lemma coriaceous, awned; awns 4–6mm; palea hyaline. Stony slopes of hills; 970–1450m.
Distr.: Guinea (C) to W Cameroon, NE Africa to India, SE Asia. (Palaeotropics).
IUCN: LC
Kodmin: Cheek 8956 1/1998; **Nyasoso:** Cheek 7521 10/1995.

Axonopus flexuosus (Peter) C.E.Hubb. ex Troupin

Perennial, stoloniferous, up to 40cm; nodes glabrous; leaf-blades lanceolate, 3–30cm × 7–13mm, obtuse to acute, edges scabrous; ligule membranous, slightly ciliate, 0.6mm; sheaths folded, large, c. 25–30cm × 10mm, carinate; inflorescence 2 or 3 sessile racemes, 6–15cm, two at end of rachis, third c. 3cm beneath them; spikelets tightly appressed; spikelets 3–4mm, lanceolate, glabrous to slightly pubescent, upper lemma about ⅔ length of spikelet. Marshy areas in forests, flooded banks, grassland; 2000m.
Distr.: tropical Africa.
IUCN: LC
Mwanenguba: Pollard 137 10/1998.

Bambusa vulgaris Schrad. ex H.Wendl.

Woody bamboo, in tufts diameter 3–5m at base; culms 12–15m × 10cm; internodes 30cm; sheaths coriaceous, 27 × 35cm at base, pubescent, 13 × 10cm; leaf-blades lanceolate, 2 sizes on the same stem, 10–15 or 25–30cm × 1–4.5cm; short false petiole; auricules 3–7mm, linear, scabrous; inflorescence a spreading panicle 2–3m, several spikelets at nodes; spikelets 1–2m, in clusters of 1–10 or more along branches of panicle; florets 4–12, distichous, laterally compressed; lemmas merging into glumes below, the lowermost of which bear spikelet buds in their axils; style 1, divided at the apex into 2–3 stigmas; lodicules 3. Planted around villages in dense humid forest and by watercourses; 900m.
Distr.: native to Asia, introduced throughout the tropics. (Pantropical).
IUCN: LC
Nyasoso: Cheek 7466 10/1995.

Centotheca lappacea (L.) Desv.

Perennial rhizomatous, with erect culms to 1m; leaf-blades lanceolate, asymmetric, 10–25 × 1.5–3cm, acuminate; hairs at base of leaf-blade and on sheath; inflorescence an open panicle, 15cm, lateral branches erect then spreading; spikelets lanceolate, 4–8mm, terete or slightly compressed; lemmas emarginate and mucronate at the tip, the lowest glabrous, the upper with reflexed tubercle-based bristles; short pedicels. Semi-forested land, undergrowth, fallow clearings in forest; 500m.
Distr.: pantropical, but not E Africa.
IUCN: LC
Kupe Village: Zapfack 927 7/1996; **Loum F.R.:** Etuge 4280 10/1998.

Chloris pycnothrix Trin.

Annual up to 55cm, stoloniferous; culms geniculate, ascending; leaf-blades 4–8cm × 4mm, narrowing to a subacute point; sheaths carinate, glabrous; inflorescence digitate, 3–10 racemes, each 2–8cm, purplish green; spikelets c. 2.5mm, 2 florets, with 2 awns, 10–25mm, one ¼ length of the other; glumes narrowly acuminate. Fallow, open land in humid areas; 2000m.
Distr.: Africa, Arabia, Argentina, Brazil, Paraguay. (Amphi-Atlantic).

IUCN: LC
Nyasoso: Etuge 1685 fl. 1/1996.

Coelachne africana Pilg.
Fries, Wiss. Ergebn. Schwed. Rhod.-Kongo-Exped. 1911–12: 208 (1916).
Perennial, procumbent, mat-forming; culms rooting from lower nodes, 8–10cm; leaf-blades lanceolate, 10–13 × 1.5–3mm; inflorescence an ovate panicle, 2–4cm, with branches, 5–10mm, each carrying several spikelets; spikelets ovate, 1.8mm; short internode between lemmas. Damp areas near waterfalls; 1350–1700m.
Distr.: tropical Africa, particularly E Africa.
IUCN: LC
Kodmin: Cheek 8997 1/1998; Etuge 4030 1/1998.

Coix lacryma-jobi L.
Annual 1–3m, erect; leaf-blades 10–45 × 2–7cm, lanceolate, cordate at base; ligule narrow; inflorescence spicate racemes on peduncles; male spikelets paired or in threes, one pedicelled, the other(s) sessile, in a raceme 2.5–5cm long, subtended by a hard, polished, white or bluish utricle 8–10mm long; female inflorescence reduced to a solitary spikelet enclosed within the sheath and forming with it a false fruit. Roadsides, marshy areas and shallows; 950m.
Distr.: Asiatic origin; now pantropical.
IUCN: LC
Nyasoso: Cable 2873 6/1996.

Digitaria abyssinica (Hochst. ex A.Rich.) Stapf
Perennial 20–60cm, rhizomatous, glabrous to villose; leaf-blades linear to lanceolate, 5–15cm × 3–11mm; ligule 1.5–3mm, glabrous or with hairs; inflorescence 3–10 racemes, each 3–8cm, digitate or subdigitate, along 2–8cm rachis; racemes 2–9cm, very slender, rachis triquetrous; pedicels 0.3mm; spikelets in pairs, 1.5–2.4mm, about 3 times as long as broad, ovate-elliptic, acute at tip, glabrous, rarely pubescent; lower glume a green ovate scale 0.3mm long; upper glume ⅔–¾ length of spikelet; back of upper lemma partly exposed; anthers and stigmas crimson. Grassland; 2000m.
Distr.: tropical and subtropical Africa.
IUCN: LC
Mwanenguba: Pollard 148 10/1998.

Digitaria ciliaris (Retz.) Koeler
Annual 35–60cm, ascending; nodes lightly pubescent; leaf-blades 3–16cm × 3.5–8mm, glabrous or occasionally pubescent; hairy ligule 1.25mm; inflorescence 2–11 racemes, each 5–11cm, along axis up to 4cm, verticillate; rachis scabrous, triquetrous; spikelets in pairs, one short pedicel and one long pedicel; pedicels 0.5–2mm; spikelets 2.6–4mm, mostly 2–2.3mm, lanceolate-elliptic, imbricate, overlapping by at least half their own length; upper glume usually half as long as spikelet or more; back of upper lemma partly exposed; pedicelled spikelet often with a rufous ciliate fringe at maturity. Wasteland; 2000m.
Distr.: pantropical & warm temperate regions.
IUCN: LC
Nyasoso: Etuge 1687 1/1996.

Digitaria debilis (Desf.) Willd.
Annual 20–80cm, in small tufts or spreading, rooting at nodes with culms ascending; leaf-blades 3–8cm × 2–6mm, ribbon-like; sheaths pubescent at base; ligule membranous, glabrous, 1mm; inflorescence 5–15 racemes, 3–15cm, digitate or subdigitate, very slender; rachis triquetrous, slightly alate; spikelets in pairs; pedicels serrulate; spikelet

2.5–3mm, narrowly lanceolate, obscurely puberulous; upper glume as long as lower lemma, 5- or rarely 3-nerved, concealing back of upper lemma; lower glume a membranous collar separated from rest of spikelet by a distinct internode 0.1–0.2mm. Along water-courses, shallow water, fallow; 2000m.
Distr.: Africa & Mediterranean areas.
IUCN: LC
Mwanenguba: Pollard 143 10/1998.

Digitaria velutina (Forssk.) P.Beauv.
F.T.E.A. Gramineae: 652 (1982).
Annual 20–80cm, ascending or decumbent; leaf-blades 2–15cm × 3–17mm, broadly linear to lanceolate; inflorescence (3–)7–20 racemes, 3–13cm, digitate, slender; rachis triquetrous, alate; spikelets 1.5–2.1mm, narrowly ovate-elliptic, obscurely and appressed-pubescent; upper glume up to 4/5 as long as lower lemma, 3-nerved; lower glume obscure or an ovate scale up to 0.2mm. Paths, waste ground; 650m.
Distr.: Cameroon, E Africa from Egypt & Yemen to S Africa. (Tropical & subtropical Africa, & Arabian peninsula).
IUCN: LC
Nyasoso: Etuge 1601 1/1996.
Note: apparently the first record of this species in Cameroon.

Echinochloa colona (L.) Link
Variable annual 20–80cm, erect or ascending in small tufts; leaf-blades linear, 5–30cm × 2–6mm; no ligule; inflorescence linear, 6–15cm, with racemes 2–3cm; spikelets 2–3mm, acute or cuspidate, sometimes with lower lemma drawn out into a short awn-point up to 1mm long. Forest; 300m.
Distr.: pantropical.
Bakossi F.R.: Cheek 9307 10/1998.
Local name: Nkoko mbwe. Uses: MEDICINES (Cheek 9307).

Echinochloa crus-pavonis (Kunth) Schult.
Perennial or rarely annual 1–2m; culms erect or ascending, spongy; leaf-blades 25–35cm × 10–20mm, no ligules; inflorescence large, pendent; numerous racemes along rachis, 24–40cm; fasciculate spikelets; spikelets elliptic, reddish brown to dark brown, 2.5–3mm; lower lemma usually with a short awn up to 2mm long; stamens yellow/orange; stigmas black. Wetland areas, grassland; 2000m.
Distr.: tropical Africa & America. (Amphi-Atlantic).
IUCN: LC
Mwanenguba: Cheek 9405 10/1998.

Eleusine indica (L.) Gaertn.
Annual, prostrate at base; culms erect, up to 50–60cm; leaf-blades often folded, 5–35cm × 3–6mm; sheaths carinate; inflorescence spicate, of 1–10 ascending racemes, 3–15cm × 3–7mm; spikelets in 2 rows, sessile, narrowly elliptic, 4–8mm, green or brownish green; glumes unequal, lemmas 2.7–3mm long, narrow, membranous, keeled, longer than glumes; grain concealed within floret. Nitrophile of waste ground, roadsides; 750–1450m.
Distr.: pantropical.
IUCN: LC
Kodmin: Cheek 9029 1/1998; Kupe Village: Cable 3713 7/1996; Etuge 1627 1/1996; Mwanenguba: Pollard 887 fl. 11/2001.
Local name: Nse nseg. Uses: MEDICINES (Cheek 9029).

Eragrostis tenuifolia (A.Rich.) Hochst. ex Steud.
Perennial 35–70cm, caespitose, glabrous; basal sheaths laterally compressed; leaf-blades 5–20cm × 1–4mm, ribbon-

like, attenuate; ligule hyaline with a row of white hairs, 2mm; inflorescence an open elliptic panicle, 5–15cm, with long feathery branches; spikelets on long pedicels, 4–11mm; white hairs at axils of branches; spikelets 7–15 × 1.5mm, with 4–15 florets, linear, dark green, margins conspicuously saw-toothed; lower glume about ⅔ length of upper, the latter 0.8–1.4mm long and barely reaching base of adjacent lemma; paleas persisting long after lemmas have fallen. Wasteland, fallow, grassland; 2000m.

Distr.: pantropical.

IUCN: LC

Mwanenguba: <u>Pollard 140</u> 10/1998; **Nyasoso:** <u>Etuge 1686</u> 1/1996.

Guaduella densiflora Pilg.

Syn.:*Guaduella ledermannii* Pilg.

Perennial, rhizomatous, erect, leafy culms 40–120cm; aphyllous culms 6–35cm; leaf-blades 10–35cm × 8cm, elliptic, acuminate, slightly asymmetric, attenuate to rounded at base, glabrous or pubescent, slightly shiny; short false petiole, 1–5mm; sheaths with rounded auricules; inflorescence a dense raceme, 4–8cm, generally at end of aphyllous culms, with up to 25 spikelets; spikelets 1.5–4cm, linear, light green to brown; internodes often visible between lemmas; lemma and palea subequal; stamens conspicuous, white. Dense humid forest; 720m.

Distr.: SE Nigeria to Angola (Cabinda) & Gabon. (Lower Guinea).

IUCN: LC

Kupe Village: <u>Cheek 7712</u> 11/1995.

Guaduella macrostachys (K.Schum.) Pilg.

Perennial, 30–90cm, resembling *G. densiflora*, but leaves and inflorescences borne on the same culm. Humid forests; 750–950m.

Distr.: Ghana, Nigeria, Cameroon. (Upper & Lower Guinea).

IUCN: LC

Kupe Village: <u>Cable 854</u> fl. 1/1995; **Ngomboaku:** <u>Mackinder 282</u> 12/1999.

Hyparrhenia newtonii (Hack.) Stapf

Perennial; culms 60–120cm, caespitose in dense tufts; leaf-blades linear, 30cm × 4mm, hirsute, edges scabrous; ligule 1mm; basal sheaths glabrous or tomentose; inflorescence a loose panicle, 15–30cm, straight; spathes 2.5–5cm, glabrous or villose along edges; peduncles shorter than spathes; racemes 1.5–2cm, 2–4 awns per pair, deep-crimson; raceme base produced into a scarious appendage; sessile spikelets 6–10mm, linear-oblong, glabrous, lower glume coriaceous, awn 2.2–5.5cm; pedicelled spikelets 5–10mm, glabrous, terminating in bristle 1–5mm; pedicel tooth subulate, 0.2–1.5mm. Grassland, stony soils, often on hills; 2000m.

Distr.: tropical Africa & SE Asia. (Palaeotropics).

IUCN: LC

Mwanenguba: <u>Pollard 155</u> 10/1998.

Hyparrhenia umbrosa (Hochst.) T.Anderson ex Clayton

Perennial; culms 1–2m; resembling *H. newtonii*, but leaf-blades 60cm × 12mm; ligule 2mm; spathes navicular, 1.5–2.5cm, glabrous or lightly pilose along edges, raceme base unappendaged; sessile spikelets 4mm, villose with white hairs, awn 0.7–1.3cm; pedicelled spikelets 5–6mm, villose with white hairs, muticous or mucronate. Meadows, roadsides, fallow; 1975–2000m.

Distr.: N Nigeria, W Cameroon, widely scattered in tropical and S Africa, but not common. (Tropical Africa).

IUCN: LC

Mwanenguba: <u>Cheek 7269</u> fr., 2/1995; <u>Pollard 152</u> 10/1998.

Isachne buettneri Hack.

Perennial, spreading; culms ascending, 30–50cm, rooting at nodes; leaf-blades 5–17cm × 5–20mm, linear-lanceolate, slender, glabrous or scabrous to pubescent; inflorescence an ovate panicle, 8–20cm, lax and open, reddish; branches up to 10cm, straight, extending obliquely; spikelets 0.8–1.4mm, light green; florets subrotund, indurated; lemmas similar in size, minutely pubescent; glumes slightly shorter than florets. Humid forests, often at edges of streams; 560–1100m.

Distr.: Guinea (B) Sudan & Zambia. (Afromontane).

IUCN: LC

Kupe Village: <u>Cable 3761</u> 7/1996; <u>3818</u> 7/1996; <u>Kenfack 262</u> 7/1996; **Nyasoso:** <u>Cable 3312</u> 6/1996; <u>3503</u> 7/1996; <u>Etuge 1803</u> 3/1996.

Isachne mauritiana Kunth

Perennial up to 60cm resembling *I. buettneri*, but spikelets 1.2–1.8mm; lemmas glabrous; glumes as long as or slightly longer than florets. Undergrowth of dense forest; 1940–1980m.

Distr.: Ghana, Nigeria, Cameroon, Congo (K) to E Africa & Madagascar. (Afromontane).

IUCN: LC

Kupe Village: <u>Ryan 408</u> 5/1996; **Nyasoso:** <u>Cheek 7350</u> fr. 2/1995.

Local name: Korri Korri Mba (Cheek 7350).

Leptaspis zeylanica Nees

Wag. Agric. Univ. Papers 92(1): 39 (1992).

Syn.:*Leptaspis cochleata* Thw.

Perennial up to 1m tall; stoloniferous, trailing, rooting at lower nodes, culms erect; leaf-blades oblong-oblanceolate, 10–35cm × 2.5–6cm, sharply pointed at tip, asymmetrical; sheaths longer than internodes, imbricate; inflorescence a panicle, 45–60cm; branches subverticillate, lower branches up to 20cm, 2–3 branches in a whorl, bearing short side branches with 1 male and 1 female spikelet, the male above; male spikelets 1-flowered, finely pubescent, the lemma 4mm long, the glumes half as long; stamens 6; female spikelets 4–6mm, 1–flowered, reddish, lemma inflated, conchiform, closed except for a small hole near the apex, covered in hooked hairs. Undergrowth of dense, humid forest; edges of fields and roads; 850–1350m.

Distr.: tropical Africa and Asia. (Palaeotropics).

IUCN: LC

Kupe Village: <u>Cable 2657</u> 5/1996; <u>727</u> fl. 1/1995; <u>Etuge 1700</u> 2/1996; <u>Ryan 388</u> 5/1996; **Mt Kupe:** <u>Schoenenberger 56</u> 11/1995; **Ndum:** <u>Lane 496</u> fl. 2/1995; **Nyandong:** <u>Ghogue 1440</u> fr. 3/2003; **Nyasoso:** <u>Cable 1180</u> fl. 2/1995; <u>Cheek 7503</u> 10/1995; <u>Etuge 2536</u> 7/1996; <u>Sidwell 339</u> 10/1995; <u>Wheatley 425</u> fl. 7/1992; <u>Zapfack 669</u> 6/1996; **Nzee Mbeng:** <u>Gosline 101</u> 2/1998.

Loudetia simplex (Nees) C.E.Hubb.

Syn.:*Loudetia camerunensis* (Stapf) C.E.Hubb.

Perennial; culms caespitose, erect, up to 1.5m; nodes glabrous or having a ring of hairs; basal sheaths decaying into fibres, pubescent, sometimes glabrous; leaf-blades linear, 10–30cm × 2–5mm; inflorescence a lax panicle, linear–ovate, 10–30cm; branches occasionally verticillate; spikelets 8–14mm, narrowly lanceolate, brown, glabrous; upper lemma acutely 2–lobed, the lobes 0.3–1mm long; lower glume obtuse, usually ⅓ length of spikelet; awns 2.5–5cm; stamens 2. Grassland, stony soils, marshes, swamps; 2000m.

Distr.: tropical & southern Africa. (Afromontane).
IUCN: LC
Mwanenguba: Pollard 132 10/1998.

Loudetiopsis trigemina (C.E.Hubb.) Conert
Perennial 50–90cm, densely caespitose; basal sheaths
tomentose, others pubescent; leaf-blades 30cm × 2–3mm,
linear, rigid; hairy ligule, 0.5mm; inflorescence a lax panicle,
10(–15)cm, spikelets in groups of 3; spikelets 6–7mm,
lanceolate to oblong, golden brown, variably pubescent;
lower glume ¾ or more the length of upper; glumes glabrous
or with a few tuburcle-based hairs; lower lemma 5–9-nerved;
awns 15–20mm; stamens 3. Widespread on hills; 1900m.
Distr.: N Nigeria, Cameroon. (Lower Guinea (montane)).
IUCN: LC
Mt Kupe: Cheek 7585 10/1995.

Melinis repens (Willd.) Zizka
Fl. Zamb. 10: 116 (1989).
Syn.:*Rhynchelytrum repens* (Willd.) C.E.Hubb.
Annual or perennial 0.3–1m, erect or ascending in lax tufts;
leaf-blades 5–30cm × 2–10mm, ribbon-like or convolute;
sheath and blade creased; inflorescence an ovate panicle, 5–
20cm, silver-maroon; branching from base; pedicels
capillary, pilose, discoid at top; spikelets 2.5–6mm, ovate;
lower glume about 1mm long, separated from the upper by a
short internode up to 0.3mm long; palea of lower floret
ciliate on the keels; upper glume and lower lemma bilobed or
emarginate, shortly awned, gibbous, with silky hairs below,
narrowing above into glabrous beak ¼–½ length of spikelet.
Wasteland, roadsides, fallow; 500m.
Distr.: tropical & subtropical Africa; introduced elsewhere.
IUCN: LC
Kupe Village: Cable 2698 5/1996.

Monocymbium ceresiiforme (Nees) Stapf
Perennial to 1.3m; caespitose, rarely weak and straggling;
base pubescent, slightly fibrous; leaf-blades 5–25cm × 2–
5mm, linear-lanceolate; inflorescence a narrow panicle, 2–
5cm × 2–4mm, spathes lanceolate, reddish brown; sessile
spikelets 3–5mm, elliptic, pale reddish brown; lower glume
villose; awns 6–18mm; callus longer than broad, ciliate to
bearded on base and sides; pedicellate spikelets 5–6mm,
similar but without awns. Savanna, humid areas, slopes of
hills; 1900m.
Distr.: tropical & subtropical Africa.
IUCN: LC
Mt Kupe: Cheek 7569 10/1995.

Olyra latifolia L.
Perennial 3–4m, climbing or erect, often woody, glabrous or
pubescent at nodes; leaf-blades 5–20 × 2.5–7cm, ovate,
assymetric at base with short false petioles, acuminate; short
ligule; inflorescence a panicle, 7–25cm, pyramidal; rachis
scabrous; male spikelets at base of female; male spikelets on
filiform pedicels, glumes rudimentary, lemma lanceolate,
4mm, awned; female spikelets ovate, on claviform pedicels,
purplish; glumes 8–20mm; lemma and palea 4–6mm, white,
shiny; awns 3–4mm. Edges of clearings, slopes in dense
humid forest; 900–1100m.
Distr.: tropical Africa & America. (Amphi-Atlantic).
IUCN: LC
Nyasoso: Balding 23 fl. 7/1993; Cheek 7502 10/1995; Etuge
1341 10/1995; Lane 86 7/1992.

Oplismenus burmannii (Retz.) P.Beauv.
Annual up to 50cm; branching, culms either ascending or
creeping, often with long aerial roots from the nodes; leaf-

blades 7cm × 15mm, lanceolate, asymmetric at base, pilose;
sheath ciliate; short ligule; inflorescence 2–10cm, 4–10
racemes, each 0.5–2cm; rachis glabrous to pilose; spikelets
2.5–3.5mm, pubescent; awns 8–15mm, antrorsely barbellate,
slender, flexuous; anthers white; stigmas deep-red. Open
parts of forests, savanna, edges of marshland; 500–1000m.
Distr.: pantropical.
IUCN: LC
Kupe Village: Patterson 4 11/1995; **Ngomboaku:** Ghogue
481 12/1999.

Oplismenus hirtellus (L.) P.Beauv.
Variable perennial resembling *O. burmannii*, but culms
slender, erect or climbing, rooting at nodes, 15–100cm;
inflorescence 0.5–1.5cm; spikelets glabrous to pubescent;
awns 2–4(–15)mm, viscid, smooth, stiff, the tips truncate.
Shady places in forests and along roadsides; 200–2050m.
Distr.: pantropical.
IUCN: LC
Kodmin: Etuge 4407 11/1998; Ghogue 45 fl. fr. 1/1998;
Kupe Village: Cheek 6066 fr. 1/1995; 7698 11/1995; Etuge
2780 7/1996; Patterson 3 11/1995; **Lake Edib:** Cheek 9151
2/1998; **Loum F.R.:** Etuge 4278 10/1998; **Mt Kupe:** Cheek
7608 11/1995; Sebsebe 5079 11/1995; 5080 11/1995;
Mungo River F.R.: Cheek 10180 11/1999; **Nyasoso:**
Asongani 1285 6/1996; Lane 177 fl. 11/1994; Sidwell 328
10/1995.

Panicum brevifolium L.
Annual, ascending or climbing to 80cm; culms branching,
rooting at lower nodes; leaf-blades ovate–oblanceolate, 5–
10cm × 1.5–3.5cm, with transverse veins; ligules hairy,
4mm; inflorescence a panicle, 8–18cm; branches 6–7cm,
occasionally with long hairs; second rhachilla internode
swollen, and covered by a small bulge at base of upper
glume; spikelets 1.5–2mm, elliptic, slightly oblique to
gibbous in profile, glabrous to pubescent, upper glume and
lower lemma 5-nerved, upper lemma white, shiny, smooth.
Shade in humid forests, farmbush; 350–1200m.
Distr.: pantropical.
IUCN: LC
Kupe Village: Cheek 7150 1/1995; Etuge 1430 11/1995;
2053 5/1996; 2639 7/1996; Patterson 2 11/1995; **Loum F.R.:**
Etuge 4276 10/1998; **Mwambong:** Pollard 122 10/1998.

Panicum hochstetteri Steud.
Perennial to 1m; sheaths shorter than internodes; leaf-blades
7–10cm × 5–8mm, pubescent; inflorescence an ovate panicle,
7–10cm; lower branches ascending or patent, up to 7cm;
inflorescence with or without long glistening white hairs on
axis; spikelets up to 2.5mm, ovate-oblanceolate, purplish,
glabrous or sometimes scabrid-hairy; nerves prominent and
often forming raised ribs; lower glume 3-nerved, ⅔–¾ length
of spikelet, lanceolate; lemma 1.5mm, white, tough, shiny;
anthers yellow, 0.8mm long. Forest edges; 1900–2000m.
Distr.: Cameroon, São Tomé, Congo (K), Burundi, Kenya,
Ethiopia. (Afromontane).
IUCN: LC
Mt Kupe: Cheek 7568 10/1995; Sebsebe 5087 11/1995.

Panicum sp. aff. hochstetteri Steud.
Resembling *P. hochstetteri*, but spikelets 1.6mm long (not 2–
2.3mm) fide T.A.Cope (2001). 1150m.
Distr.: W Cameroon.
Mwambong: Cheek 9363 10/1998.

Panicum maximum Jacq.

Perennial 1–3m; densely caespitose, sometimes with stolons, green to greyish; sheaths pubescent; nodes pubescent to villose; leaf-blades 40cm × 30mm, linear, glabrous or hirsute; inflorescence an oblong panicle, 10–50cm, with numerous flexible, filiform branches, lower ones usually whorled; axis and branches generally glabrous; spikelets 3–3.5mm, oblong, coriaceous, symmetrical in profile; upper glume and lower lemma entire; upper lemma transversely rugose; lower glume obtuse, about ⅓ length of spikelet; stigmas feathery, maroon. Savanna, roadsides, fallow; 650m.
Distr.: tropical & subtropical Africa, introduced elsewhere.
IUCN: LC
Ngusi: Etuge 1545 1/1996.

Panicum monticola Hook.f.

Perennial 25–80cm; culms procumbent below, ascending above; leaf-blades lanceolate, asymmetric, 2–15 × 0.3–2cm; short false petiole; inflorescence a panicle 5–15 × 10–20cm; branches ascending or patent; pedicels 1–2.5mm; spikelets 2.5–3.5mm, ovate, pale green to crimson, glabrous; lower glume ovate, ⅓–½ length of spikelet; lower lemma 5–11-nerved. Forest; 1450m.
Distr.: tropical Africa. (Afromontane).
IUCN: LC
Kodmin: Cheek 9048 1/1998.

Panicum pusillum Hook.f.

Annual, prostrate, 10–25cm, rooting at nodes; leaf-blades lanceolate, hairy, 1–4cm × 3–6mm; ligules hairy; inflorescence a panicle, 2–5cm; pedicels 1–3mm; spikelets 1.7–2mm, lanceolate to oblong, purplish, pubescent to glabrous; upper lemma 1.3mm, coriacous, smooth, white; lower glume 3-nerved, ¾ length of spikelet. Savanna, mountains and plateaux; 2000m.
Distr.: Sierra Leone to Congo (K), Ethiopia, Malawi, E Africa. (Afromontane).
IUCN: LC
Mwanenguba: Pollard 150 10/1998.

Paspalum conjugatum Berg

Perennial; stoloniferous; culms erect, 20–80cm; leaf-blades linear to narrowly lanceolate, 4–20cm × 5–12mm, slightly ciliate at edges; ligules hairy; inflorescence 2 racemes, ± joined, 5–15cm; pedicels short; spikelets orbicular, 1.5–1.7mm, yellow-green, closely appressed to the rachis of paired slender racemes; ciliate fringe from margins of upper glume, inner glume missing. Damp places in forest clearings, grassland, rough ground; 500–1550m.
Distr.: pantropical.
IUCN: LC
Kodmin: Cheek 9030 1/1998; **Kupe Village:** Cable 3714 8/1996; Patterson 1 11/1995; **Nyasoso:** Lane 227 fl. 11/1994.
Local name: Nkuku. **Uses:** SOCIAL USES – cooked before hunting or football for success (Cheek 9030).

Paspalum paniculatum L.

Perennial, rhizomatous, erect, 30–100cm; leaf-blades 10–45cm × 6–20mm, sheath carinate; ligule hyaline, with hairs 5–10mm; inflorescence numerous fastigiate racemes 4–11cm on 5–15cm rachis; long hairs on branches of racemes; spikelets orbicular to broadly elliptic, conspicuously plano-convex, dark brown, 1.3–1.4cm, pubescent, inner glume missing; anthers white, stigmas crimson. Fallow in humid forests; 400–850m.
Distr.: Liberia to Angola and Uganda, also in Polynesia, S America, New Guinea, Australia. (Pantropical).
IUCN: LC

Kupe Village: Cable 3714a fr. 7/1996; Cheek 8321 5/1996;
Loum F.R.: Pollard 98 10/1998; **Nyasoso:** Cable 26 7/1992.
Local name: Akwor kwor (Cheek 8321). **Uses:**
MEDICINES (Koge Lawrence in Cheek 8321); none known (Epie Ngome & George Ngole in Pollard 98).

Paspalum scrobiculatum L. var. scrobiculatum

Syn.:*Paspalum orbiculare* G.Forst.
Syn.:*Paspalum polystachyum* R.Br.
Perennial; erect or ascending, 10–100cm, sheaths glabrous or pubescent along edges; leaf-blades linear, 5–30cm × 2–16mm; inflorescence of 1–5(rarely –17) racemes, 2–8cm, on rachis 5cm; spikelets 1.9–2.8mm, solitary, ovate-obovate, green, becoming brownish in maturity, fairly glabrous, inner glume missing. Humid areas in savanna, forest edges; 2000m.
Distr.: palaeotropics.
IUCN: LC
Mwanenguba: Pollard 135 10/1998.

Pennisetum clandestinum Hochst. ex Chiov.

Perennial c. 4cm tall; rhizomatous and stoloniferous, forming a dense mat; lower leaf-blades distichous, sheaths flattened laterally; inflorescence within an involucre of fine hairs, each with 2–4 subsessile spikelets enclosed in the uppermost leaf sheath; spikelets 10–20mm, narrowly lanceolate; stigmas white, 4cm; stamens 3–5cm with silvery filaments conspicuously exserted from leaf–sheath when in flower; anthers light brown, 4.5mm. Meadows and savanna at high altitude; 2000m.
Distr.: E Africa; introduced into Cameroon (Afromontane).
IUCN: LC
Mwanenguba: Pollard 144 10/1998.
Uses: as fodder and to combat erosion (*fide* Pollard).

Pennisetum laxior (Clayton) Clayton

Kew Bull. 32: 580 (1978).
Syn.:*Beckeropsis laxior* Clayton
Annual up to 80cm; leaf-blades slender, 10–25cm × 6–14mm; lower part of leaf blades reduced to false petioles; inflorescence a raceme 1–4cm; peduncles and rachis scabrous; spikelets 3–4mm, acuminate, overlapping by ½ their own length; glumes subequal, the upper 0.75–1mm long; both lemmas spinulose-hispidulous above. Rocks and isolated inselbergs; 1050–1100m.
Distr.: Ghana, Nigeria, Cameron, São Tomé, Annobon, Sudan. (Guineo-Congolian).
IUCN: LC
Kupe Village: Cheek 7705 11/1995; **Nyale:** Cheek 9670 11/1998.
Note: difficult to distinguish from *P. unisetum*.

Pennisetum macrourum Trin.

Mem. Acad. Petersb. Ser. 6(3:2): 178 (1835).
Perennial; rhizomatous, up to 1(–5)m; stem often horizontal; leaf-blades 10–30cm × 3–12mm, glaucous; hairy ligule; edges of sheaths pilose; inflorescence a terminal panicle, 3–22cm, linear, sometimes lax at base; stipe short; rachis cylindrical, glabrous or hispidulous, terminal bristle twice as long as the rest, up to 18mm; spikelets 3–5mm, lower floret male, as long as spikelet; lower glume reduced or absent; lemmas acute to acuminate. Riverbanks, streambeds, in watercourses, between rocks; 950m.
Distr.: tropical & S Africa, but rare W of Cameroon; Yemen. (Tropical Africa).
IUCN: LC
Ngomboaku: Ghogue 505 fl. 12/1999.

Pennisetum monostigma Pilg.

Perennial; caespitose; culms erect or ascending, up to 1.2m, pubescent below inflorescence; leaf-blades linear, up to 30cm × 6–8mm, sheaths densely pubescent, particularly towards top; inflorescence a terminal panicle (false spike), 5–10cm, cylindrical, often dense, pubescent, with hairs 9–15mm; spikelets up to 5mm, solitary, usually sessile, lanceolate, pointed, lower floret reduced to an empty lemma as long as the spikelet; anthers yellow, anther-tips smooth. Meadows; 1000–1350m.
Distr.: Sierra Leone, Cameroon, Bioko. (Upper & Lower Guinea (montane)).
IUCN: LC
Mwambong: Etuge 4129 2/1998; Onana 586 2/1998; **Nyale:** Cheek 9705 11/1998.

Pennisetum polystachion (L.) Schult. subsp. *polystachion*

Syn.:*Pennisetum subangustum* (Schumach.) Stapf & C.E.Hubbard
Annual or perennial 0.5–2m; leaf-blades 10–40cm × 5–15mm, often pubescent; inflorescence a narrow panicle or false spike, 5–20cm × (6–)8–10(–15)mm excluding bristles; longest bristle 15–25mm, inner bristles usually more than twice length of spikelet; rachis with sharp edges below involucral scars; spikelets 3–5mm, solitary, sessile, lanceolate, with brownish/crimson bristles 10–27mm, plumose in lower half; upper lemma tough, shiny; anthers 1.5mm. Savanna, fallow, roadsides, disturbed soils; 800–1500m.
Distr.: pantropical.
IUCN: LC
Kodmin: Cheek 9080 1/1998; Etuge 3965 1/1998; **Kupe Village:** Etuge 1629 1/1996; **Nyasoso:** Etuge 2488 7/1996.

Pennisetum purpureum Schumach.

Perennial, stoloniferous, usually 1–3m, but can reach 6m; culm pubescent just below spike; leaf-blades 0.5–1m × 1–4cm, edges serrulate; ligules ciliate; sheath and leaf-blade scabrous, pubescent; inflorescence a terminal panicle, linear, 20–35cm, 15–30mm diameter; involucres subsessile; rachis densely pubescent to loosely pilose; spikelets 4–7.5mm, glabrous, very short pedicels; lower floret variable; tuft of small hairs at tip of anthers. Streambanks in forest, river valleys, fallow; 500–1300m.
Distr.: tropical Africa, widely introduced elsewhere in tropics. (Pantropical).
IUCN: LC
Edib: Etuge 4487 11/1998; **Kupe Village:** Patterson 6 11/1995; **Ndum:** Cable 894 fl. 1/1995.

Pennisetum unisetum (Nees) Benth.

F.T.E.A. Gramineae: 681 (1982).
Syn.:*Beckeropsis uniseta* (Nees) K.Schum.
Perennial 1–4m, caespitose; culms 10–15mm diam; nodes and basal sheath pubescent; apex of sheath usually bearded with a line of hairs on abaxial side; leaf-blades 60cm × 0.5–3cm, usually with slender petiole-like base; inflorescence a false panicle comprising numerous axillary racemes 3.5–4cm; spikelets 2–3mm, elliptic, acute, overlapping by ¾ their own length or more; glumes subequal up to 0.5mm long; lower lemma hispidulous, upper scaberulous; anthers orange-red. Shady humid places in woodlands and savannas; 2000m.
Distr.: tropical & southern Africa, Yemen. (Afromontane).
IUCN: LC
Mwanenguba: Pollard 154 10/1998.

Poecilostachys oplismenoides (Hack.) Clayton

Wag. Agric. Univ. Papers 92(1): 189 (1992).
Syn.:*Chloachne oplismenoides* (Hack.) Stapf ex Robyns
Perennial 60–80cm; spreading, trailing or erect, sometimes with long aerial roots from lower nodes; leaf-blades 5–13cm × 4–25mm, lanceolate, acuminate, asymmetrical at base, pubescent; slightly hairy ligules; inflorescence loose racemes, 2–5cm, along 5–15cm rachis; short, strong pedicels; spikelets 5–7mm, narrowly lanceolate, light green, covered with long rigid hairs on tubercules; lower glume abaxial, c. ⅔ length of spikelet. Undergrowth of montane forest; 1500–1780m.
Distr.: Nigeria, Bioko, Cameroon, Burundi, Sudan, Ethiopia, E Africa, Malawi, Zimbabwe. (Afromontane).
IUCN: LC
Kodmin: Cheek 9089 1/1998; Ghogue 18 1/1998; Onana 576 2/1998.

Pseudechinolaena polystachya (Kunth) Stapf

Annual 10–30cm; spreading, rooting at nodes, with erect culms; leaf-blades 1–8.5cm × 3–10mm, lanceolate, pilose, with numerous lateral nerves; sheaths striate and ciliate at edges; ligule 0.5–1.5mm, membranous; inflorescence slender, 3–8 lax racemes along rachis of 10–20cm, lower raceme 1–6cm; spikelets distant; pedicels up to 2mm; spikelets 3.5–4.5mm, those at base often reduced, obliquely ovoid, gaping; lower glume almost as long as spikelet, adaxial; upper glume shorter, gibbous, usually armed with conspicuous hooked bristles, but these sometimes short and appressed; lower lemma chartaceous, with membranous margins; anthers white. Open undergrowth of forests, humid and shady places, fallow; 650m.
Distr.: pantropical.
IUCN: LC
Ngusi: Etuge 1567 1/1996; **Nyasoso:** Asongani 1291 6/1996.

Rhytachne rottboellioides Desv.

Perennial 20–100cm; caespitose in dense tufts; leaf-blades 5–25cm, filiform, involute, glabrous; ligule 0.5mm; inflorescence in racemes 2–22cm, cylindrical, internodes 3–4mm; spikelets 3–5mm, narrowly ovate to oblong, glabrous; glumes of sessile spikelet and rudimentary pedicelled spikelet with or without awns 0.5mm (lower glume) and 2.5mm (upper glume); upper glume smooth, concealed; very variable in length of spikelets and awns, and in degree of corrugation of lower glume. Marshland and swamps, permanently flooded areas, shallow soils over ironpan; 2000m.
Distr.: tropical Africa & tropical America. (Amphi-Atlantic).
IUCN: LC
Mwanenguba: Pollard 151 10/1998.

Saccharum officinarum L.

Stoloniferous perennial, 2–4(–6)m, culms glabrous; leaf blade c. 50–90 × 2–4cm; inflorescence a large pyramidal panicle up to 1m long; axis glabrous to pubescent; spikelets to 4mm long; callus hairs fine, off-white. Cultivated on moist soils; 1170m.
Distr.: cultivated throughout the tropics.
IUCN: LC
Edib village: Cheek (sight record) 2/1998.
Local name: sugar cane. **Uses:** FOOD – particularly as chewing cane (*fide* Etuge).

Setaria longiseta P.Beauv.

Perennial 40–150cm; caespitose, ascending or erect; leaf-blades linear, flat, not pleated, 10–30cm × 4–6mm, finely hirsute; inflorescence a branched panicle, 10–20cm, lanceolate; branches up to 5cm, often crimson-tinted, glabrous or minutely scabrid, except for sparse tubercle–based hairs, 4–7mm; spikelets subtended by bristle(s), 2mm, elliptic, pale or crimson-tinged, upper lemma strongly transversely rugose, turning brown at maturity; upper glume ½–⅔ length of spikelet. Savanna near woods, forest edges, shady places on pathsides, fallow; 1300–2000m.

Distr.: Guinea (C) to E Africa. (Tropical Africa).

IUCN: LC

Edib: Etuge 4490 11/1998; **Kodmin:** Cheek 8975 1/1998; **Nyasoso:** Etuge 1692 1/1996.

Setaria megaphylla (Steud.) T.Durand & Schinz

Syn.: *Setaria chevalieri* Stapf

Perennial 1–3m, rhizomatous; culms 5–10mm diameter at base; leaf-blades linear to lanceolate, 30–70cm × 1–10cm; inflorescence a linear-lanceolate panicle, length 20–50cm; branches 4–15cm, straight, rigid; spikelets subtended by bristle(s), 2.5–3.5mm, ovate to elliptic; lower lemma equalling or very shortly exceeding upper lemma; upper lemma smooth, shiny; palea of lower floret as long as the lemma. Forested areas near roads; forest edges; 800–1150m.

Distr.: tropical Africa, tropical America. (Amphi-Atlantic).

IUCN: LC

Kodmin: Cheek 9198 2/1998; **Kupe Village:** Lane 371 1/1995; **Nyasoso:** Cable 3244 6/1996.

Setaria sphacelata (Schumach.) Stapf & C.E.Hubb. ex M.B.Moss var. *sphacelata*

Perennial 0.5–2m, densely caespitose; culms erect, sometimes ascending, glabrous; leaf-blades linear, 15–50cm × 2–15mm; ligule short, ciliate; inflorescence a spike-shaped panicle, 5–40cm × 4–8mm; spikelets in glomerules or solitary; 6–10 hairs, 4–8mm, per glomerule, pale reddish or yellowish; spikelets subtended by bristle(s), 2–2.3mm, elliptic, glabrous, ± dorsally compressed; upper glume ¼–¾ length of spikelet, exposing conspicuously rugose upper lemma;. Flood regions, marshes, swamps, montane meadowlands; 2000m.

Distr.: tropical & southern Africa.

IUCN: LC

Mwanenguba: Pollard 136 10/1998; 145 10/1998.

Note: species highly variable.

Sporobolus africanus (Poir.) Robyns & Tournay

Syn.: *Sporobolus indicus* (L.) R.Br. var. *capensis* Engl.

Perennial 60cm, densely caespitose; leaf-blades rigid, 10–35cm × 3–4mm; ligule ciliate; inflorescence a linear panicle, 6–25cm, narrow to subspiciform; branches appressed and mostly 1–2cm long; spikelets 1.8–2.1mm, not acuminate, olive-green; upper glume up to ⅔ length of spikelet. Humid montane areas, swamps; 2000m.

Distr.: tropical & subtropical Africa, Sri Lanka, Philippines, Australia, New Zealand. (Montane).

IUCN: LC

Mwanenguba: Pollard 133 10/1998.

Sporobolus molleri Hack.

Annual 6–60cm, erect, often in small tufts; leaf-blades linear, 2–25cm × 1–5mm, glabrous, slender; inflorescence a panicle, linear to pyramidal, 5–17cm; branches in racemes 5–25mm; spikelets 1.7–2mm, acuminate, yellowish; lower glume a tiny hyaline scale; upper glume up to ⅔ length of spikelet, abruptly acuminate to a short awn-point up to 0.5mm long; palea ½–⅔ length of lemma. Roadsides, fallow; 1300m.

Distr.: tropical Africa.

IUCN: LC

Edib: Etuge 4498 11/1998.

Sporobolus paniculatus (Trin.) T.Durand & Schinz

Annual 25–70cm; solitary stems or in tufts, erect or ascending; leaf-blades 2–20cm × 1–5mm, slender, ciliate at edges; inflorescence a panicle, lanceolate, verticillate, 8–18cm × 2–4(–8)cm; main branches up to 12 verticils, 15–25mm, each with 4–12 terminal spikelets; spikelets 2–2.2mm, dark red, glabrous or scaberulous; lower glume lanceolate, acute to acuminate; stamens 3; grain ellipsoid or obovoid, ± laterally flattened. Shallow soil on rocks, verges, hillsides, grassland, fallow, rocky soils, slopes, roadsides; 1500–2000m.

Distr.: tropical Africa & Mexico. (Amphi-Atlantic).

IUCN: LC

Mwanenguba: Pollard 128 10/1998; 142 10/1998; Sanford 5392 fl. 11/1968; 5396 fl. 11/1968.

Sporobolus pyramidalis P.Beauv.

Syn.: *Sporobolus indicus* (L.) R.Br. var. *pyramidalis* (P.Beauv.) Veldkamp

Perennial 30–160cm; caespitose in dense tufts, culms 2–5mm diameter. at base; leaf-blades up to 50cm × 3–10mm, ribbon–like; basal sheaths narrow; inflorescence a panicle, 10–40cm, linear to pyramidal; branches spicate, those at base 5–10cm; spikelets not clustered; spikelets 1.7–2mm, elliptic, green, occasionally crimson, not acuminate; upper glume obtuse, ¼–⅓ length of spikelet. Savanna, often degraded or over–grazed, rough tracks, roadsides, villages; 500m.

Distr.: tropical & subtropical Africa.

IUCN: LC

Kupe Village: Cable 2700 5/1996.

Trichopteryx elegantula (Hook.f.) Stapf

Annual 5–20(–30)cm; gregarious; culms erect or geniculately ascending; leaf-blades 1–2cm × 2–4mm, lanceolate, pilose; ligule a row of hairs 1–1.5mm; inflorescence a panicle, 3–10cm, obovate, lax, scarcely extending from sheath; branches filiform, ascending; spikelets solitary or in pairs; pedicels capillary, 4–8mm; spikelets 2.5–3.5mm, lanceolate, pilose or glabrous; glumes glabrous or bearing stiff tubercle-based hairs; main awn 10–13mm, caudate, base black, upper part green; lateral awns 4mm. Slopes of hills, between rocks; 1100m.

Distr.: Sierra Leone, Nigeria, Cameroon, Congo (K), Rwanda, Burundi, E Africa. (Afromontane).

IUCN: LC

Kupe Village: Cheek 7696 11/1995.

Zea mays L.

Perennial to 2.5m; culms erect; leaf blade linear-lanceolate, to 1 × 0.1m; male spikelets paired in large terminal panicles; female spikelets in sheathed axillary inflorescences, spikelets paired, sessile, in longitudinal rows on a spongy axis; styles long-exserted in a tassel. Cultivated.

Distr.: originating in the Americas, now found throughout the tropics and warm temperate areas.

IUCN: LC

Nloé: Cheek (sight record).

Local names: Corn or Maize. **Uses:** FOOD – an important food and trade crop (*fide* Pollard).

HYDROCHARITACEAE

M. Cheek

Fl. Cameroun 26 (1984).

Ottelia ulvifolia (Planch.) Walp.
Aquatic herb; bottom-rooting, c. 30cm, sometimes submerged; leaves numerous, oblanceolate, c. 30 × 10cm, acute-acuminate; scapes as long as leaves; spathe 2-winged, enveloping ovary; flowers white, outer tepals oblong, 2–4cm, patent. Lakes, mat-forming; 1150m.
Distr.: Senegal to S Africa. (Tropical Africa).
IUCN: LC
Mwambong: Cheek 9362 10/1998.

HYPOXIDACEAE

I. Darbyshire

Fl. Cameroun 30 (1987).

Hypoxis angustifolia Lam.
Herb with rhizome 1–2cm diameter, fusiform, usually without old leaf fibres; leaves in a rosette, linear, 5–35 × 0.2–0.6cm, margin and midrib villose; inflorescence corymbose, 1–3(–5)-flowered, densely appressed-pilose; scape shorter than leaves; bracts lanceolate, to 1cm; pedicel 0.5–2cm; tepals 6, lanceolate, c. 5mm long, yellow; fruit dehiscent, to 1.5cm long; seeds c. 25. Montane grassland; 2000m.
Disr.: tropical Africa. (Afromontane).
IUCN: LC
Mwanenguba: Villiers 1375 fl., 5/1982.
Note: specimen cited in Fl. Cameroun 30: 36 (1987).

IRIDACEAE

B.J. Pollard

Note: it is unusual that there are no records of Iridaceae from the Mwanenguba grasslands.

Gladiolus aequinoctialis Herb.
Goldblatt, *Gladiolus* in tropical Africa: 287 (1996).
Syn.: *Acidanthera aequinoctialis* (Herb.) Baker
Syn.: *Gladiolus aequinoctialis* Herb. var. *aequinoctialis*
Herb (40–)90–120cm; corm 2–3cm across; leaves 4–10, lower 2–5 basal, lanceolate, (0.6–)1.2–1.7cm across; spike (3–)5–8-flowered; bracts 5–8cm long; flowers white, showy, the lower 3 tepals streaked purple; perianth tube cylindric, (8.5–)12–14cm; tepals lanceolate, ± equal, 3.5–4 × 1.8–2.0cm; capsules 1.8–2.0cm long. Rocky places, often on wet ledges on steep cliffs, stony grassland; 1900–2060m.
Distr.: Sierra Leone, W Cameroon & Bioko. (Upper & Lower Guinea).
IUCN: NT
Mt Kupe: Cheek 7584 fl., 10/1995; 7598 fl., 11/1995; Sebsebe 5088 fl., 11/1995; **Nyasoso:** Cable 118 fl., 9/1992.
Note: here we follow Goldblatt (1996), who, in his monograph of the tropical African species of *Gladiolus*, does not recognise varieties under this name.

LEMNACEAE

M. Cheek

Lemna paucicostata Hegelm. ex Engelm.
Aquatic, floating herb; thallus elliptic, c. 3 × 2mm, solitary root arising towards one side, new thallus budding from near the root insertion point; fertile material not seen. Outlet stream of lake; 1250m.
Distr.: pantropical & warm temperate areas.
IUCN: LC
Lake Edib: Cheek 9157 2/1998.
Note: the specimen cited was determined in the field, it has not yet been studied by a family specialist.

MARANTACEAE

B.J. Pollard, H. Kennedy

Fl. Cameroun 4 (1965).

Halopegia azurea (K.Schum.) K.Schum.
Erect herb to c. 3m, forming dense clumps; leaves several, to 50 × 15cm, linear-oblong, the sides ± parallel; apex long-acuminate; petiole to 1m, the calloused portion 2–5cm; inflorescence of 1–3 racemes, to 20–25cm long; axes pubescent; bracts to 2.5–3.5cm long, each bearing two cymes; flowers appearing 1 or 2 at a time, a decorative purple with blue and yellow staminodes exserted, 2.5cm; ovary pubescent; fruits cylindrical, 11–12 × 3–4mm. Forest, by streams; 400–1000m.
Distr.: Sierra Leone to Congo (K). (Guineo-Congolian).
IUCN: LC
Kupe Village: Cheek 6043 fl., 1/1995; **Molongo:** Nemba 776 2/1988; **Mwambong:** Etuge 4135 2/1998; **Nyale:** Etuge 4239 2/1998.

Hypselodelphys poggeana (K.Schum.) Milne-Redh.
Lianescent herb to several metres; leaves linear-oblong, 8–15 × 3–9cm, abruptly acuminate, subtruncate; calloused portion of petiole above point of articulation 1–2cm; inflorescence loose, little-branched spikes, nearly straight, 5–9cm; internodes c. 5mm; bracts 2–3.5cm; flowers violet and white; fruit muricate, 3-lobed, 4.5–5.0 × 3.0–3.5cm; tubercles short and dense, not curved, less than 2mm. Lowland forest; 150–210m.
Distr.: Sierra Leone to Congo (K). (Guineo-Congolian).
IUCN: LC
Mungo River: Gosline 220 fl., fr., 11/1999; **Mungo River F.R.:** Etuge 4322 10/1998.

Hypselodelphys scandens Louis & Mullend.
Lianescent herb to several metres; branched; leaves elliptic or oblong-linear, 12–35 × 5–17cm, shortly acuminate, subtruncate; calloused portion of petiole above point of articulation to 3cm, conspicuously beaked at junction with midrib adaxially; inflorescence pendulous with a number of bifurcations; branches usually zig-zag, c. 20cm long; internodes c. 1cm; bracts 3.5–4.5cm; flowers pale violet, white and brown; fruit muricate, 3-lobed, c. 5 × 2–3cm; tubercles long, often curved, to 5mm. Forest; 700–1300m.
Distr.: Ivory Coast to Congo (K). (Guineo-Congolian).
IUCN: LC

Kodmin: <u>Satabie 1124</u> fl., 2/1998; **Kupe Village:** <u>Cable 2416</u> 5/1996; <u>Cheek 6046</u> fl., 1/1995; <u>8318</u> 5/1996; <u>Etuge 1708</u> 2/1996; **Mwambong:** <u>Onana 542</u> 2/1998; **Ndum:** <u>Williams 131</u> 1/1995; **Nyasoso:** <u>Balding 93</u> fr., 8/1993; <u>Cable 2787</u> 6/1996; <u>Etuge 2376</u> 6/1996; <u>Sidwell 329</u> 10/1995; <u>Sunderland 1475</u> fr., 7/1992; <u>Zapfack 633</u> 6/1996.
Local name: Mbesel (*fide* Etuge). **Uses:** MATERIALS – leaves used to tie 'Miondo'; stems for basket making (Koge Lawrence in Cheek 8318 & *fide* Etuge).

Hypselodelphys violacea (Ridl.) Milne-Redh.

Climbing herb; leaves often with an entirely calloused petiole, 1–2cm, or callous 1–3cm; lamina ovate to linear-oblong; inflorescence as simple spikes or dichotomously branched; abaxial bracts 2.5–3cm, adaxial bracts bicarinate, 10–15mm; flowers white, c. 2cm. Forest; 250–300m.
Distr.: Sierra Leone to Congo (K). (Guineo-Congolian).
IUCN: LC
Bakole: <u>Nemba 787</u> 2/1988; **Mile 15, Kumba-Mamfe Road:** <u>Nemba 202</u> 9/1986.

Marantochloa congensis (K.Schum.) J.Léonard & Mullend. var. *congensis*

Slender herb, thicket-forming, to c. 1m; leaves antitropic, strongly asymmetric, 7–15(–25) × 4–5cm; apex progressively acuminate; petiole with calloused portion 3–6mm, bulging, pubescent above; inflorescence lax and delicate, simple or with few branches; bracts 1.2–2.0cm; flowers white or yellowish; fruits spherical, smooth (before drying), c. 5mm diameter. Forest; 1300m.
Distr.: Sierra Leone to Congo (K) & Burundi. (Guineo-Congolian).
IUCN: LC
Kupe Village: <u>Cable 3779</u> 7/1996.

Marantochloa cordifolia (K.Schum.) Koechlin

Fl. Gabon 9: 129 (1964).
Branched herb; leaves homotropic; blades glabrous, of variable size, to 30 × 15cm, often tinged violet below; petiole callus ± 1cm, non-calloused portion c. 4cm; inflorescence densely racemose, simple or sometimes bifurcate at base, 5–15cm; bracts pinkish violet, 30–35 × 8–10mm. Forest; 400m.
Distr.: Nigeria to Congo (K). (Lower Guinea & Congolian).
IUCN: LC
Molongo: <u>Nemba 783</u> 2/1988.
Note: this record is based upon a Missouri specimen which we have not seen, and cannot be sure of this determination. This taxon is quite likely to occur in our area, well within its known range.

Marantochloa filipes (Benth.) Hutch.

A ± climbing herb to 2.5m; leaves homotropic, ± asymmetric, 7–15(–20) × 3–7cm; lamina often pruinous abaxially, apex abruptly acuminate; petiole with calloused portion c. 5mm long; inflorescence paniculate, very spindly and delicate; rachis simple or once-branched; bracts 1.5–3.5cm; flowers white or pinkish; fruits smooth (before drying), spherical with persistent perianth, c. 7mm diameter, yellow, ripening red. Forest and streambanks; 300–1200m.
Distr.: Guinea (C) to Congo (K). (Guineo-Congolian).
IUCN: LC
Kupe Village: <u>Cheek 6035</u> fr., 1/1995; <u>Etuge 1405</u> 11/1995; <u>1470</u> 11/1995; <u>1932</u> 5/1996; <u>2774</u> 7/1996; <u>2906</u> 7/1996; <u>Lane 347</u> fl., fr., 1/1995; <u>377</u> fr., 1/1995; **Loum F.R.:** <u>Biye 67</u> 12/1999; <u>Etuge 4274</u> 10/1998; **Ngomboaku:** <u>Mackinder 308</u> 12/1999; **Nyale:** <u>Etuge 4199</u> 2/1998; **Nyandong:** <u>Tadjouteu 540</u> 3/2003.
Local name: Tomblele-Nbu (Etuge 4273).

Marantochloa leucantha (K.Schum.) Milne-Redh.

Erect or climbing herb to 5m; leaves homotropic, 15–20(–40) × 7–12(–25)cm, abruptly long-acuminate, not pruinose; petiole with calloused portion c. 1cm; inflorescence of long loose panicles, much-branched, pendent, 30–40cm; flowers whitish; fruits smooth (before drying), spherical, 1cm diameter, reddish becoming yellowish on drying; perianth not persistent. Forest; 220–1200m.
Distr.: tropical Africa. (Afromontane).
IUCN: LC
Baduma: <u>Nemba 227</u> 9/1986; **Bakolle Bakossi:** <u>Etuge 280</u> 9/1986; **Kodmin:** <u>Etuge 4075</u> 1/1998; **Kupe Village:** <u>Cable 2442</u> 5/1996; <u>2713</u> 5/1996; <u>3759</u> 7/1996; <u>Cheek 6036</u> fr., 1/1995; <u>6038</u> fr., 1/1995; <u>7118</u> fr., 1/1995; <u>Etuge 1435</u> 11/1995; <u>Kenfack 314</u> 7/1996; <u>Lane 278</u> fr., 1/1995; <u>Ryan 218</u> 5/1996; **Loum F.R.:** <u>Cheek 10243</u> 12/1999; <u>Pollard 99</u> 10/1998; **Manehas F.R.:** <u>Cheek 9397</u> 10/1998; **Mt Kupe:** <u>Schoenenberger 29</u> fr., 11/1995; **Ndum:** <u>Groves 30</u> fr., 2/1995; <u>Williams 149</u> fr., 1/1995; **Nyale:** <u>Etuge 4182</u> 2/1998; **Nyasoso:** <u>Balding 87</u> fr., 8/1993; <u>Cable 2810</u> 6/1996; <u>3476</u> 7/1996; <u>657</u> fr., 12/1993; <u>Cheek 5635</u> fl., fr., 12/1993; <u>7537</u> 10/1995; <u>Etuge 2081</u> 6/1996; <u>Groves 101</u> fr., 2/1995; <u>Lane 161</u> fr., 10/1994; <u>Sidwell 311</u> 10/1995; <u>Sunderland 1478</u> fr., 7/1992; <u>Zapfack 598</u> 6/1996.
Local name: Tombel (Cable 2713); (Cheek 7118 & 9397).
Uses: MATERIALS – stems used for basket weaving (Cheek 7118 & 9397); leaves used to wrap and carry food such as rice or fish (Daniel Abbiw in Pollard 99).

Marantochloa mildbraedii Loes. ex Koechlin.

Fl. Gabon 9: 102 (1964).
Erect herb to c. 1m tall, branched; leaves homotropic to c. 30 × 15cm; apex abruptly narrowly acuminate, often suffused purple beneath; petiole with calloused portion 1–4cm long; inflorescence a dense raceme, c. 1–2cm across, simple or basally bifurcate, rachis 5–15cm long; bracts persistent, 30–35 × 8–10mm, imbricate; fruit spherical, finely and densely pubescent, c. 7mm diameter. Forest; 210m.
Distr.: Cameroon & Gabon. (Lower Guinea).
IUCN: EN
Mungo River F.R.: <u>Etuge 4319</u> fl., 10/1998.
Note: similar to *M. cordifolia*, but differing in the fine pubescence of the bract margins, leaf-sheaths and flowering pedicels.

Marantochloa purpurea (Ridl.) Milne-Redh.

Erect herb to 2.5m, branched; leaves 10–48 × 5–18cm; petiole with calloused portion 1.5cm; inflorescence loose, branched; bracts pale pink, subtending two, 2-flowered cymes; peduncle 3–3.5cm; internodes of rachis pubescent; flowers deep-purple; perianth persistent; ovary pubescent; fruits red, c. 7mm diameter. Wet areas in forest; gallery forest in savanna areas; 250m.
Distr.: Sierra Leone to Tanzania. (Guineo-Congolian)
IUCN: LC
Loum F.R.: <u>Cheek 10259</u> fr., 12/1999.

Megaphrynium macrostachyum (Benth.) Milne-Redh.

Erect herb to 2–3m, thicket-forming; stem simple bearing a single terminal leaf; leaves elliptic, c. 30–60 × 12–30cm; petiole with calloused portion to c. 15cm long; inflorescence lateral, much-branched, each raceme with up to 30 internodes of ± 5mm; flowers whitish with red or purplish calyx; fruits glossy red, smooth, with 3 conspicuous sutures; seed black, arillate. Forest; 250–800m.

Distr.: Sierra Leone to Sudan & Uganda. (Guineo-Congolian).
IUCN: LC
Bakole: <u>Nemba 788</u> 2/1988; **Loum F.R.:** <u>Cheek 10252</u> 12/1999; **Manehas F.R.:** <u>Etuge 4360</u> 10/1998; **Nyandong:** <u>Ghogue 1490</u> fr., 3/2003.

Sarcophrynium brachystachyum (Benth.) K.Schum. var. *brachystachyum*
Erect herb to 2m; culms several, each bearing a single terminal leaf; leaves ovate-elliptic, 20–35 × 10–15cm, with many closely parallel secondary nerves; calloused portion of petiole 4–6cm; inflorescence sessile, branched from base, to 7cm long, each cymule with 2 fleshy bracteoles; bracts persistent, overlapping, 1.5–2cm; flowers white; fruits subglobose, red, c. 1.5cm diameter. Forest; 350–800m.
Distr.: Senegal to Congo (B). (Upper & Lower Guinea).
IUCN: LC
Manehas F.R.: <u>Etuge 4361</u> 10/1998; **Mungo River F.R.:** <u>Cheek 10202</u> 11/1999.

Sarcophrynium schweinfurthianum (Kuntze) Milne-Redh.
Bull. Soc. Roy. Bot. Belg. 83: 30 (1950).
Erect herb to ± 2m; culms in large clumps, simple, bearing a single leaf; leaves variable in size, c. 50 × 20cm; callous up to 12cm; inflorescence lateral, lightly-branched, to 20cm long; rachis internodes c. 2cm; flowers white, ± tinted red in the throat, c. 6mm long; fruit globular or pyriform, indehiscent, dull-red, juicy, c. 1.5cm diameter; seeds 2, in a sticky mucilage. Forest; 250–900m.
Distr.: Cameroon, Gabon, Congo (K), CAR, Sudan, Uganda. (Guineo-Congolian).
IUCN: LC
Kupe Village: <u>Cable 2441</u> 5/1996; <u>3662</u> 7/1996; <u>Cheek 6042</u> fr., 1/1995; <u>6044</u> fr., 1/1995; <u>6068</u> fr., 1/1995; <u>Lane 267</u> fr., 1/1995; **Loum F.R.:** <u>Cheek 10257</u> 12/1999; **Ngomboaku:** <u>Mackinder 298</u> 12/1999; **Nyasoso:** <u>Cable 3310</u> 6/1996.

Sarcophrynium villosum (Benth.) K.Schum.
Engl. Pflanzenreich, Marant.: 38 (1902).
Syn.: *Phrynium villosum* Benth.
Erect herb to 2m; leaves oblong-lanceolate, to 27 × 6(–10)cm, shortly acuminate, glabrous; leaf-sheaths densely villose; peduncle, rachis and floral pedicels very pubescent; bracts pubescent, 2–3cm long, bearing a single 2-flowered cyme; sepals 3mm long; corolla 5mm long; pinkish red; fruits red. Forest; 1250–1500m.
Distr.: Bakossi Mts & Gabon. (Lower Guinea).
IUCN: EN
Edib: <u>Etuge 4479</u> fl., fr., 11/1998; **Ehumseh:** <u>Etuge</u> 170, fr., 6/1986; **Kodmin:** <u>Biye 36</u> fr., 1/1998; <u>Cheek 8936</u> fr., 1/1998; <u>Etuge 4120</u> fl., fr., 2/1998.

Thaumatococcus daniellii (Benn.) Benth.
Erect herb to ± 3m; culms forming large clumps; leaves very large, to 60 × 40cm or more; calloused portion of petiole ± 10cm; inflorescence emerging at ground level, from the lowest node of the leaf, simple or forked, racemose, to c. 12cm; flowers white-lilac or violet; fruits 3-winged, crimson, indehiscent; seeds hard, black, rugose. Forest; 500–1200m.
Distr.: Sierra Leone to Congo (K). (Guineo-Congolian).
IUCN: LC
Enyandong: <u>Etuge 4420r</u> 10/2001; **Kupe Village:** <u>Etuge 1426</u> 11/1995; <u>Lane 359</u> fr., 1/1995; <u>Ryan 226</u> 5/1996; **Loum F.R.:** <u>Etuge 4275</u> 10/1998.
Local name: Megum megum. **Uses:** MATERIALS – for making mats (Etuge 4275).

MUSACEAE
M. Cheek

Fl. Cameroun 4 (1965).

Ensete gilletii (De Wild.) E.E.Cheesman
Robust, tree-like herb, 3m; single-stemmed; lacking suckers, lacking a leafless stem at maturity; leaves sessile, oblong, to 1.5m, at base of stem; inflorescence terminal, nodding slightly; bracts 5–9 × 17–25cm; flowers in lines of 8 in axils, 4cm long, females at base of inflorescence, males at apex; fruit subcylindric, 5–6cm. Forest edge, grassland; 1900m.
Distr.: Guinea (C) to Malawi. (Guineo-Congolian & S Central Africa).
IUCN: LC
Mwanenguba: Etuge (sight record) 2000.

Musa spp.
Robust, tree-like herbs to 3–7m; single-stemmed; with basal suckers; stem leafless at maturity; leaves oblong, to 2m, petiole distinct, at least 30cm; inflorescence pendulous; bracts purple, 10–15 × 10–15cm; flowers in lines in axils, females at base of inflorescence, males at apex; fruits yellow, cylindric-angular, 10–30cm. Cultivated for fruit, sometimes persisting for a few years in abandoned farms; 450–1170m.
Distr.: SE Asia, cultivated throughout tropics. (Pantropical).
IUCN: LC
Edib village: Cheek (sight record) 2/1998; **Nyandong:** Darbyshire (sight record) 3/2003.
Note: the description above is meant to be a general one, to cover all the various *Musa* spp. and cultivars occurring in our area.
Uses: FOOD – infructescences – several cultivars are grown in Bakossi, including the banana and plantain (cooking banana) (*fide* Pollard).

ORCHIDACEAE
P.J. Cribb & B.J. Pollard

Fl. Cameroun 34 (1998), 35 (2001) & 36 (2001).

Aërangis gravenreuthii (Kraenzl.) Schltr.
Epiphyte; stem woody to 7 × 0.5cm; leaves distichous, oblanceolate, strongly falcate, to 15 × 1.5–3cm, narrowing basally; inflorescences pendent, axillary, racemose, 10–20cm, 2–5-flowered; flowers white, sometimes with an orange flush; dorsal sepal 20–32 × 5–7mm; spur reddish, 4–8(–12)cm. Forest, woodland; 1450m.
Distr.: Bioko, Cameroon, Tanzania. (Lower Guinea & Congolian (montane)).
IUCN: NT
Mwanenguba: <u>Sanford 5473</u> fl., 11/1968; **Nyasoso:** <u>Balding 82</u> fl., 8/1993.

Aërangis luteoalba (Kraenzl.) Schltr. var. *rhodosticta* (Kraenzl.) Stewart
Fl. Cameroun 36: 838 (2001).
Epiphyte; stem 1–3cm, pendent; leaves 2–8, to 15 × 0.6–1.5cm, linear-ligulate to linear-lanceolate or linear-oblanceolate, ± falcate; inflorescence lax, to 35cm, several–25-flowered, arcuate or pendent; rachis ± zigzag bent; flowers particularly ornamental, resupinate, white; dorsal sepal 10–15 × 3–7mm; labellum 15–20 × 7–15mm, obovate

to rhombiform; column red (white in var. *luteoalba*); spur 23–40mm. Forest, riverine forest; 760–925m.
Distr.: Cameroon, CAR, Congo (K), Ethiopia, E Africa. (Afromontane).
IUCN: LC
Enyandong: Cheek 10916 fl., 10/2001; Simo 15 fl., 10/2001; 26 fl., 10/2001; Zapfack 1760 fl., 10/2001.

Ancistrochilus rothschildianus O'Brien
Epiphytic herb; pseudobulbs clustered, to 5.5cm across, 1–2-leaved; leaves caducous, lanceolate or oblanceolate, acuminate, 10–30 × 3–7cm, prominently rib-veined; inflorescences 1–2, to 20cm, 2–5-flowered, pubescent; flowers showy, 5–8cm across, pale to dark pink; labellum pink to bright purple. Forest, riverine forest; 780–970m.
Distr.: Guinea (C) to Uganda. (Guineo-Congolian).
IUCN: LC
Enyandong: Salazar 6316 fl., 10/2001; Simo 33 fl., 10/2001; **Kupe Village:** Schoenenberger 17 fl., 11/1995; **Nyasoso:** Cheek 7523 10/1995; Lane 168 fl., 11/1994; Williams 199 11/1995.

Ancistrorhynchus capitatus (Lindl.) Summerh.
Epiphyte; stem 1–10cm; leaves 3–8, coriaceous, linear, 16–32 × 1.4–2.1cm, unequally bilobed at apex, each lobe 2–3-toothed, leaf bases 2–2.7cm; inflorescence cauliflorous, capitate, densely many-flowered, to 1.5 × 2.5(–4)cm; flowers small, white or pale rose; labellum 4.5–5.7 × 4–5mm; spur 7–8.5mm. Forest; 750–900m.
Distr.: Sierra Leone to Uganda. (Guineo-Congolian).
IUCN: LC
Kupe Village: Cable 2418 5/1996; Cheek 7749 11/1995; Etuge 2668 7/1996; 2819 7/1996; 2900 7/1996; Kenfack 226 7/1996; Zapfack 911 7/1996; 912 7/1996.

Ancistrorhynchus clandestinus (Lindl.) Schltr.
Epiphyte; stem short, 7(–18)cm; leaves 5–10, in a fan, linear or tapering, acuminate at unequally bilobed apex, 15–180 × 0.7–0.9cm; inflorescence cauliflorous, to 2cm, dense, multiflorous, capitate; flowers white; labellum 4.5–6.5 × 3.4–6.6mm; spur 3.2–6.2mm, ± sigmoid. Forest; 1000m.
Distr.: Sierra Leone to Uganda. (Guineo-Congolian).
IUCN: LC
Nyasoso: Gosline 140 11/1998.

Ancistrorhynchus metteniae (Kraenzl.) Summerh.
Epiphyte; stem 2–7(–20)cm, pendent; leaves to 14, linear, 5–25 × 0.7–2.4cm, unequally bilobed at apex, each lobe 2-toothed; leaf-sheaths 1–1.5cm; inflorescence cauliflorous, dense, capitate, 1–3cm, to 20-flowered; flowers white; labellum 3.5–4 × 4.3–4.7mm; spur 3–4.3mm. Forest, gallery forest; 910–960m.
Distr.: Sierra Leone, Nigeria to Congo (K), Ethiopia, Uganda, Tanzania. (Afromontane).
IUCN: LC
Kupe Village: Zapfack 941 7/1996; 962 7/1996; **Nyasoso:** Zapfack 646 6/1996; 723 6/1996.

Ancistrorhynchus serratus Summerh.
Epiphyte; stem 5–10cm; leaves 5–9, 5–18 × 0.7–1.2cm, with parallel sides; apex irregularly serrate; inflorescence < 1cm, several–multi-flowered; flowers white; labellum 2.5–3 × 2.5–4.5mm; spur 3.5–4.5mm. Forest; 2000m.
Distr.: SE Nigeria, W Cameroon, Bioko, Rio Muni, São Tomé. (Lower Guinea).
IUCN: NT
Nyasoso: Cable 3409 6/1996.

Ancistrorhynchus straussii (Schltr.) Schltr.
Epiphyte; stem to 2.5cm; leaves ensiform, slightly falcate, 5–20 × 0.3–1.3cm, acute at the unequally bilobed apex, the 2 lobes divergent; inflorescence cauliflorous, dense, capitate, c. 0.5cm, 5–14-flowered; flowers white; labellum 2.7–3mm, weakly trilobed; spur 1.5–3.5mm. Forest; 910m.
Distr.: Ivory Coast, Nigeria, Cameroon, Congo (K), Uganda. (Guineo-Congolian).
IUCN: LC
Nyasoso: Zapfack 645 6/1996.

Angraecopsis elliptica Summerh.
Epiphyte; leaves elliptic-oblong or oblanceolate, 4–10 × 1–2.5cm; inflorescence 3.5–18cm; flowers pale green, sometimes with orange tinge on labellum; spur ± same length as labellum, not much if at all swollen apically, 2.7–5.2mm; side lobes of labellum distinctly shorter than middle. Forest, streamsides in forest; 1450–1500m.
Distr.: Liberia, Ivory Coast, Nigeria, Cameroon, Congo (K). (Guineo-Congolian).
IUCN: LC
Kodmin: Gosline 108A fr., 2/1998; **Mwanenguba:** Sanford 5483 fr., 11/1968.

Angraecopsis ischnopus (Schltr.) Schltr.
Epiphyte; stem to 5cm; leaves elliptic-oblong, elliptic-ligulate or ligulate, 2–11 × 0.4–1.3cm; inflorescence lax, 3–9cm, to 10-flowered; flowers small, resupinate, dull greenish white, fading brown; labellum to 8.5mm; spur 1.2–3.8cm, very slightly or not at all swollen apically. Forest, plantations; 1250–1500m.
Distr.: Guinea (C), Sierra Leone, Nigeria, Cameroon. (Upper & Lower Guinea).
IUCN: LC
Enyandong: Zapfack 1783 fl., 10/2001; **Kodmin:** Pollard 201 11/1998; **Kupe Village:** Etuge 1367 11/1995.

Angraecopsis parviflora (Thouars) Schltr.
Epiphyte; stem 1–3(–10)cm; leaves 4–6(–9), falcate, linear-lanceolate, 6–16(–22) × 0.4–1.2cm; inflorescences 1–several, 5–17cm, densely 2–8-flowered; peduncle slender, wiry, 4–9cm; flowers white or greenish white; spur longer than the lip, 0.6–0.9cm, very slightly or not at all swollen apically. Forest; 800–1400m.
Distr.: Cameroon, Tanzania, Malawi, Mozambique, Zimbabwe, Madagascar, Mascarene Is. (Afromontane).
IUCN: LC
Kupe Village: Cable 3836 7/1996; **Nyasoso:** Balding 29 fl., 8/1993; Cable 124 fl., 9/1992; Etuge 2504 7/1996.

Angraecopsis tridens (Lindl.) Schltr.
Epiphyte; stem 1–2(–5)cm; leaves linear, narrowly ligulate or oblanceolate, ± falcate, 2–9.5 × 0.4–1.5cm; inflorescence 4–8cm, lax, few–several-flowered; flowers very pale green or yellow-green; labellum trilobed towards the middle; median lobe 1.5 × 0.5mm; lateral lobes 2.3 × 0.6mm; spur distinctly swollen apically, 0.2–0.3cm. Farmbush, forest; 1170–1500m.
Distr.: Bioko, Cameroon. (W Cameroon Uplands).
IUCN: VU
Enyandong: Simo 38 fr., 10/2001; **Kodmin:** Gosline 108B fr., 2/1998; Pollard 233 11/1998; **Mwanenguba:** Sanford 5482 fl., fr., 11/1968.

Angraecum affine Schltr.
Bot. Jahrb. 38: 19 (1905).
Epiphyte; stem 15–25 × 0.3cm, pendent or ascending; leaves several, 4.5–16 × 1–2.5cm, twisted at base to sit in the same plane, ligulate-oblanceolate to oblong-obovate; inflorescence

3–7cm, 3–4-flowered; flowers white; dorsal sepal 9–14 × 2–4.7mm; labellum 8–10.5 × 4–5.7mm; spur 18–24mm, slightly dilated at the base and at summit. Forest; 1300m.
Distr.: Cameroon, Bioko, Congo (B) & (K), Uganda. (Lower Guinea & Congolian).
IUCN: LC
Muandelengoh: Pollard 892 fl., 11/2001.

Angraecum angustipetalum Rendle
Epiphyte; stem to 60cm, internodes 1.1–1.7 × 0.5–0.6cm, erect or pendent; leaves twisted at base, distichous, 3–10.5 × 1.0–2.2cm, linear-lanceolate to oblong-ovate, oblique; inflorescence 1-flowered, 1–2cm, shorter than the leaves; labellum 12–17 × 4–7mm; spur 12–16mm, inflated apically. Forest.
Distr.: Ghana, Nigeria to Congo (K). (Guineo-Congolian).
IUCN: LC
Ngusi: Letouzey 14650 fl., 4/1976.

Angraecum aporoides Summerh.
Epiphyte; stem to 40cm, pendent or ascending; leaves numerous, 11–25 × 5–10mm, distichous, falcately oblong-elliptic, groove on upper margin broad, extending nearly to apex; inflorescence 1–4-flowered; flowers small, white; labellum distinctly 3-lobed, 2.5–4.5mm; spur 6–7mm, straight, cylindrical. Forest; 1000m.
Distr.: Nigeria, Bioko, Cameroon, Congo (K). (Lower Guinea & Congolian).
IUCN: LC
Nyasoso: Gosline 141 11/1998.

Angraecum birrimense Rolfe
Epiphyte or epilith; stem to 100cm; leaves distichous, flat, ± unequally bilobed at the apex, 7–14 × 1.5–3.5cm; flowers solitary on long peduncles or peduncles several-flowered; sepals and petals pale green, the sepals 3.5–5cm long; labellum white, centred green, ± orbicular, 4.0 × 2.5–3.5cm; spur 3.5–4.5cm. Forest, gallery forest, swampy places or terrestrial and climbing over mossy rocks; 800–1400m.
Distr.: Sierra Leone to Cameroon. (Upper & Lower Guinea).
IUCN: LC
Kodmin: Etuge 4041 1/1998; **Kupe Village:** Etuge 1915 fl., 5/1996; Kenfack 236 7/1996; Zapfack 947 7/1996; **Mt Kupe:** Cable 23 fl., 7/1992; **Nyasoso:** Cable 2777 6/1996; 3280 6/1996.

Angraecum distichum Lindl.
Epiphyte, often forming considerable clumps; stems arcuate, seldom branching, 3–25cm long; leaves without sharp points, closely imbricate at base, falcately oblong-elliptic, obtuse, 5–11 × 3–7mm; flowers white, solitary on short peduncles (6mm or less); sepals 7mm long or less; labellum indistinctly 3-lobed; spur 5–7mm. Humid forest; 1500m.
Distr.: Guinea (C) to Cameroon, E to Uganda, S to Angola. (Guineo-Congolian).
IUCN: LC
Kodmin: Pollard 217 11/1998.

Angraecum eichlerianum Kraenzl.
Epiphyte; stem stout, to 500 × 0.8cm; internodes 3–6 × 0.3–0.5cm; leaves numerous, distichous, lanceolate or elliptic, 7–16 × 1.5–4.5cm; inflorescence 4–20cm, lax, 1–2(–4)-flowered; peduncle 5–10cm; flowers large, resupinate, greenish white or white; dorsal sepal 2.7–5.5 × 0.6–1.0cm; labellum 2.5–5.0 × 2.5–5.0cm, ± cordate; spur 2.3–4.0cm, to 1.5cm across. Riverine forest; 150–250m.
Distr.: SE Nigeria to Congo (K), Angola. (Lower Guinea & Congolian).

IUCN: LC
Etam: Etuge 194 fl., 7/1986; **Mile 15, Kumba-Mamfe Road:** Nemba 152 fl., 7/1986.

Angraecum multinominatum Rendle
Epiphyte; stem internodes 1.2–1.5cm; leaves several, 3–10 × 1–2cm, markedly unequally bilobed; inflorescence lax, 4–5cm, 2-flowered; flowers non-resupinate, green/yellow-green, often orange-tinged; dorsal sepal 9–12 × 3–5.6mm; labellum 7–8.5 × 4–6mm; spur 12.5mm, widest just beyond middle, tapering in both directions. Forest; 1500m.
Distr.: Guinea (C), Sierra Leone, Ghana, Togo, S Nigeria, W Cameroon, Gabon. (Upper & Lower Guinea).
IUCN: LC
Kodmin: Etuge 3997 1/1998.

Angraecum pungens Schltr.
Epiphyte; stem pendent, to 50 × 0.2–0.3cm; leaves with sharp points, not closely imbricate at base, oblong-lanceolate, flattened, fleshy, 20–40 × 3–6mm; flowers white; sepals 6–7mm long; labellum much broader than long; spur 4.5mm long. Forest, riverine forest, swamp forest; 870–1720m.
Distr.: SE Nigeria, Bioko, Cameroon, Congo (K). (Lower Guinea & Congolian).
IUCN: LC
Kodmin: Pollard 211 11/1998; **Kupe Village:** Cable 2434 5/1996; 3777 7/1996; 3837 7/1996; Etuge 2872 7/1996; Ryan 356 5/1996; **Mwendolengo:** Etuge 4541r fr., 11/2001; **Nyasoso:** Gosline 142 11/1998.

Angraecum pyriforme Summerh.
Epiphyte; leaves oblong, 2.5–11 × 0.7–2.2cm, unequally bilobed; flowers green and white; sepals and petals 6–11mm; labellum 5.5–7.5mm; spur apically swollen, 10–14mm. Forest; 1450m.
Distr.: Ivory Coast, S Nigeria, W Cameroon. (Upper & Lower Guinea).
IUCN: VU
Nyasoso: Balding 81 fl., 8/1993.

Angraecum sanfordii P.J.Cribb & B.J.Pollard
Kew Bull. 57: 653 (2002).
Epiphyte; stem leafy, 8–25cm; leaves oblong to slightly obovate, 6.5–17 × 1.3–2.7cm, unequally bilobed; inflorescences several, 1–2-flowered, 4–5(–7)cm; flowers white, fragrant by day; sepals linear-lanceolate, 2–3 × 0.3–0.35cm, acuminate; labellum porrect, lanceolate, 1.9–2.7 × 0.3–0.4cm; spur erect-ascending, cylindrical, 3.5–5.3cm, slightly inflated apically. Forest; 1100–2000m.
Distr.: Mt Cameroon & Mt Kupe. (W Cameroon Uplands).
IUCN: EN
Kupe Village: Cable 3778 7/1996; **Nyasoso:** Cable 3398 6/1996; Etuge 2389 fl., 6/1996.

Auxopus macranthus Summerh.
Terrestrial heteromycotrophic (saprophytic) herb; stem 15–30cm with 1–several cataphylls; ± leafless; inflorescence with scape to 33cm, arising at apex of tubers; peduncle slender to 3cm; rachis up to 8cm, closely or loosely several-many-flowered; flowers orange, yellowish brown or brown; tepals connate into an obscurely 2-lipped tube. Understorey leaf-mould in dense forest; 850m.
Distr.: Liberia, Ivory Coast, Ghana, Cameroon, Congo (K), Uganda. (Guineo-Congolian).
IUCN: LC
Ngomboaku: Etuge 4673 12/1999.

Bolusiella iridifolia (Rolfe) Schltr. subsp. *iridifolia*

Epiphyte or epilith; stem 0.5–2cm; leaves 4–6(–10), pronouncedly sulcate above, ± V-shaped in cross-section; ensiform, 1–6 × 0.15–0.6cm; inflorescence dense, 2–2.6cm, many-flowered; flowers small, translucent, white; labellum 2–2.5(–5)mm; spur 1.7–2mm. Forest, savanna, rocks in rivers, plantations; 860m.

Distr.: Ivory Coast, Ghana, Cameroon, Rio Muni, Congo (K), Rwanda, Burundi, Ethiopia, E Africa, Angola, Comoros Is. (Tropical Africa).

IUCN: LC

Enyandong: Simo 1 fl., 10/2001.

Bolusiella talbotii (Rendle) Summerh.

Dwarf epiphyte; stem 0.3–1.5cm; leaves 3–6(–12), arranged in a fan, ensiform, (1.5–)2.5–6(–8) × 0.5–0.8(–1.2)cm; inflorescence lax, 3–11cm, 10–24-flowered; flowers white, centrally pink; labellum 1.7–2.5mm; spur 2mm, often green. Forest, coffee plantations; 860–1500m.

Distr.: Sierra Leone to Cameroon, Bioko, Rio Muni, Annobon, Tanzania. (Afromontane).

IUCN: LC

Enyandong: Salazar 6333 fr., 11/2001; **Kodmin:** Pollard 186 11/1998; **Nyasoso:** Zapfack 650 fr., 6/1996.

Brachycorythis kalbreyeri Rchb.f.

Ornamental epiphytic or terrestrial herb, 15–40cm; stem delicate; leaves up to 15, lanceolate to broadly-lanceolate, acute, to 11 × 2.5cm; inflorescence a spike, laxly up to 22-flowered, usually much less, to 17cm; flowers whitish, tinged mauve or lilac; labellum 2–3cm. Riverine forest, forest, plantations; 660m.

Distr.: Guinea (C), Sierra Leone, Liberia, Cameroon, Congo (B) & (K), E Africa. (Afromontane).

IUCN: LC

Kupe Village: Cheek 8343 5/1996; **Mt Kupe:** Letouzey 14706 fl., 4/1976.

Brachycorythis macrantha (Lindl.) Summerh.

Terrestrial or epilithic herb to 70cm; leaves membranous, broadly lanceolate or elliptic-lanceolate, to 13 × 5cm; inflorescence lax, 6.5–16cm, up to 20-flowered; flowers (except lower ones) considerably longer than the bracts, green and mauve; lamina of labellum broadly obcordate, 1.5–2 × 1.5–2cm; spur 7–9 × 3.5–5mm. Submontane forest, rocks in grassland, gallery forest or along watercourses. 1100m.

Distr.: Guinea (C), Sierra Leone, Liberia, Nigeria, Cameroon, CAR, Gabon. (Upper & Lower Guinea).

IUCN: LC

Kupe Village: Cable 3750 7/1996.

Brachycorythis sceptrum Schltr.

Terrestrial herb 0.7–1m; largest leaves in middle or upper part of stem; up to 35 cauline, 4 sheathing basally, 4–9 × 3cm; inflorescence dense 10–18(–25) × 3–4cm, many-flowered; flowers white and purple; labellum 10–11 × 11–14mm; spur rounded, to 2mm. Savanna, grassland.

Distr.: Senegal, Guinea, Ghana, Nigeria, Cameroon, CAR. (Upper & Lower Guinea).

IUCN: LC

Bangem: Walker s.n. 1.

Bulbophyllum bidenticulatum J.J.Verm. subsp. *bidenticulatum*

Bull. Soc. Roy. Bot. Belg. 54: 144 (1984).

Epiphyte or epilith; pseudobulbs bifoliate, 0.7–2.0cm apart; petiole 0.5–2.0mm; leaf blade (1.6–)3–11 × 0.4–1.0cm; inflorescence 5–12cm, 6–22-flowered; rachis 4-angled, ± zig-zag, 2–6.5cm × 1.8–5.0mm; floral bracts straw-coloured to orange-red; sepals pale purplish to whitish. Submontane forest; 880–1400m.

Distr.: Guinea (C), Sierra Leone, Liberia, Ivory Coast, W Cameroon. (Upper & Lower Guinea).

IUCN: LC

Kodmin: Pollard 254 fl., 11/1998; **Kupe Village:** Cheek 7042 fl., 1/1995.

Bulbophyllum bifarium Hook.f.

Epiphyte; pseudobulbs bifoliate, 1.5–3.5cm apart, 1.2–3.5 × 0.8–1.2cm; leaves lanceolate to linear-lanceolate, 3.2–10 × 0.7–1cm; rachis 4-angled, 6.5–9cm × 3.5mm; inflorescence 16–19cm, 16–23-flowered; floral bracts straw yellow; tepals whitish, bluish or pinkish. Submontane forest; 800–1800m.

Distr.: W Cameroon. (W Cameroon Uplands).

IUCN: VU

Kodmin: Cheek 8968 fr., 1/1998; Pollard 214 fl., 11/1998; 215 fl., 11/1998; **Mt Kupe:** Thomas 5062 11/1985; **Nyasoso:** Schlechter 12896 fl., 4/1899.

Bulbophyllum calyptratum Kraenzl.

Syn.: *Bulbophyllum calyptratum* Kraenzl. var. *calyptratum* sensu J.J.Verm.

Epiphyte; pseudobulbs bifoliate, 1.3–5cm apart; petiole 5–25mm; leaves linear-lanceolate to linear, 7.2–26 × 0.5–1.6cm; inflorescence 22–47cm, 8–50-flowered; peduncle 11–21cm; rachis greenish to ± white, terete or ± flattened; flowers white or greenish, ± suffused purple; labellum 1–2 × 0.4–0.8mm. Forest, secondary growth, plantations; 1050–1500m.

Distr.: Guinea to Cameroon, Gabon, Rio Muni, Congo (B & K). (Guineo-Congolian).

IUCN: LC

Kodmin: Pollard 253 11/1998; **Kupe Village:** Zapfack 976 7/1996; **Tombel:** Thorold CM18 fl., 3/1953.

Note: we do not follow Vermeulen (1987: 129), in reducing this species to varietal rank.

Bulbophyllum cochleatum Lindl. var. *cochleatum*

Orchid Monographs 2: 41 (1987).

Epiphyte; pseudobulbs bifoliate, 0.8–7cm apart, 1.5–11 × 0.4–1.3cm; leaves 2.8–23 × 0.3–1.8cm; inflorescence 8–55cm; peduncle 5.2–43cm; rachis not thickened, 2.8–12cm, 14–64(–84)-flowered; sepals and petals green, often stained purple-red or entirely so, with a green base; labellum dark purple red, occasionally with a yellow centre, with marginal hairs ≥labellum width. Forest, woodland; 1300–1800m.

Distr.: Guinea to Bioko, São Tomé, Cameroon, Gabon, Sudan to E and S Africa. (Afromontane).

IUCN: LC

Ehumseh - Mejelet: Etuge 355 fl., 10/1986; **Kodmin:** Cheek 9531 10/1998; Pollard 226 11/1998; **Mt Kupe:** Thomas 5050 11/1985; **Mwanenguba:** Sanford 5457 fl., 11/1968; **Nyasoso:** Zapfack 726 6/1996.

Bulbophyllum cochleatum Lindl. var. *tenuicaule* (Lindl.) J.J.Verm.

Bull. Jard. Bot. Nat. Belg. 56: 230 (1986).

Syn.: *Bulbophyllum tenuicaule* Lindl.

Epiphytic herb; pseudobulbs narrowly conical or cylindrical, 1–10 × 0.4–1.2cm, bifoliate, 1–9cm apart; leaves 1.8–16 × 0.4–1.5cm; inflorescence 6.5–22cm; peduncle 3–14.5cm; rachis slightly thickened, or not, 2–12cm, 8–60-flowered;

flowers red-purple, whitish at base; labellum dark brown-red or purple-red, with marginal hairs ≥labellum width. Forest, woodland; 1550m.
Distr.: SW Nigeria, W Cameroon, Bioko, São Tomé, Congo (K), Rwanda, Uganda, Kenya. (Afromontane).
IUCN: LC
Mt Kupe: <u>Cheek 7618</u> 11/1995.

Bulbophyllum cocoinum Bateman ex Lindl.
Epiphyte; pseudobulbs unifoliate, 1–3cm apart, 2–5 × 1–2.7cm, sharply 3–4-angled; petiole 0.7–2.5cm; lamina lanceolate or broadly linear, 9–27 × 1.2–3.5cm; inflorescence 9–38cm, 15–150-flowered; rachis 4–29cm, terete; sepals white, often pink tinged towards tip; petals white; lip white or cream. Lowland to submontane forest; 860–1500m.
Distr.: Sierra Leone to Uganda. (Guineo-Congolian).
IUCN: LC
Enyandong: <u>Simo 5</u> fl., 10/2001; **Kodmin:** <u>Cheek 9530</u> 10/1998; <u>Pollard 209</u> 11/1998.

Bulbophyllum colubrinum (Rchb.f.) Rchb.f.
Epiphyte; pseudobulbs unifoliate, ellipsoid, 2.7–6cm apart, 2.5–6.5 × 1–1.9cm; petiole 0.4–1.0cm; lamina lanceolate, linear lanceolate to oblanceolate, 11–22 × 1.4–3.6cm; inflorescence 14.5–51cm, 20–100-flowered; rachis 6.5–38 × 0.4–2.4cm, widened and flattened (blade-like); tepals yellowish, purple or spotted purple; labellum yellowish, stained purple. Lowland forest and lower montane forest; 300–1250m.
Distr.: Sierra Leone to Congo (K), Angola. (Guineo-Congolian).
IUCN: LC
Bolo: <u>Nemba 9A</u> 4/1986; **Mwambong:** <u>Mackinder 183</u> 1/1998.

Bulbophyllum comatum Lindl. var. *comatum*
Orchid Monographs 2: 86 (1987).
Epiphyte; pseudobulbs unifoliate, 1–3.5cm apart, (0.6–)1.5–4 × (0.5–)1–2.2cm; leaves with maximum width ± above or below the middle, falcate, (2.8–)9–27 × 1.2–5cm, acuminate; inflorescence 4.8–14cm, 25–65-flowered; peduncle 6–9cm; rachis pendulous, distinctly swollen, 26 × 1.8cm, often with fine woolly hairs; flowers reddish; ovary and sepals hirsute; labellum 1.6–2.5 × 0.6–1mm. Submontane to montane forest.
Distr.: SE Nigeria, W Cameroon, Bioko. (W Cameroon Uplands).
IUCN: EN
Loum F.R.: <u>Zapfack 944</u> 7/1996.

Bulbophyllum encephalodes Summerh.
Bot. Mus. Leafl. Harv. Un. 14: 228 (1951).
Epiphyte; pseudobulbs unifoliate, (1.5–)3–8cm apart, 1.2–3.5 × 1–2.2cm, c. sharply 4-angled; petiole 4–6mm; lamina lanceolate, 3–14.5 × 1.2–3.2cm; inflorescence 13–43cm, 14–36-flowered; rachis 4-angled in section, 2–11 × 0.2–0.35cm, reddish brown; sepals greenish, suffused, spotted, striated or entirely purple; petals pale greenish or purple; lip very dark purple or white, marked purple. Forest, plantations; 1300m.
Distr.: W Cameroon, Congo (K), Burundi, E Africa, Zambia, Malawi, Zimbabwe. (Afromontane).
IUCN: LC
Kupe Village: <u>Etuge 2859</u> 7/1996.

Bulbophyllum falcatum (Lindl.) Rchb.f. var. *bufo* (Lindl.) J.J.Verm.
Bull. Jard. Bot. Nat. Belg. 56: 235 (1986).
Syn.: *Bulbophyllum bufo* (Lindl.) Rchb.f.
Syn.: *Bulbophyllum longibulbum* Schltr.

Epiphyte; pseudobulbs bifoliate, 3.5–8cm tall, 0.4–5cm apart; leaves broadly lanceolate to linear or oblanceolate, 8.5–21 × 1.3–4.5cm; inflorescence 11–40cm, 8–60-flowered, each one 8–30mm apart; peduncle 3–12cm; rachis flattened, blade-like, 7–28 × 0.2–1.8cm; flowers whitish, purple-spotted; ratio of length of median sepal to petal: 2.4–5; top part of petals not thickened, acute to acuminate, or subacute. Lowland and submontane forest; 750–990m.
Distr.: Guinea (C) to Cameroon, Congo (K). (Guineo-Congolian).
IUCN: LC
Enyandong: <u>Salazar 6318</u> fl., 10/2001; <u>6336</u> fl., 11/2001; <u>6337</u> fl., 11/2001; <u>Simo 6</u> fl., 10/2001; **Kupe Village:** <u>Etuge 1444</u> 11/1995; <u>1499</u> fl., 11/1995; <u>1500</u> 11/1995; <u>Schoenenberger 40</u> fl., 11/1995; <u>Williams 235</u> 11/1995; **Ngomboaku:** <u>Ghogue 476</u> 12/1999; **Nyasoso:** <u>Schlechter 12898</u> fl., 1/1900.

Bulbophyllum falcatum (Lindl.) Rchb.f. var. *falcatum*
Orchid Monographs 2: 124 (1987).
Epiphyte; psudobulbs bifoliate, 1.8–6cm, 0.4–5cm apart; leaves as for var. *bufo*, except 2–14 × 0.5–3cm; inflorescence 7–40cm, 8–60-flowered, each one 5–18(–32)mm apart; peduncle 4–17cm; rachis flattened, blade-like, 3–23 × 0.4–1.3cm; flowers yellowish, often spotted dark red; labellum 1.5–2 × 1–1.8mm, top part of petals ± thickened, rounded to subacute. Lowland or submontane forest, relic forest in savanna; 1030m.
Distr.: Guinea (C) to Bioko, SW Cameroon, Congo (K), Uganda. (Guineo-Congolian).
IUCN: LC
Enyandong: <u>Simo 18</u> fl., 10/2001.

Bulbophyllum falcatum (Lindl.) Rchb.f. var. *velutinum* (Lindl.) J.J.Verm.
Bull. Jard. Bot. Nat. Belg. 56: 235 (1986).
Epiphyte; pseudobulbs bifoliate, 0.9–4.2cm tall, 0.4–5cm apart; leaves broadly lanceolate to linear, 2–15.5 × 0.5–2cm; inflorescence 3–28cm; 8–60-flowered; peduncle 1–7(–10)cm; rachis 2–20 × 0.1–1xm; flowers 2–15(–30)mm apart; flowers yellowish often suffused red; ratio of length of median sepal to petal: 1.3–2.3; top part of petals not thickened, acute to acuminate, or subacute. Lowland to submontane forest, coffee plantations; 600–1050m.
Distr.: Sierra Leone to Congo (K). (Guineo-Congolian).
IUCN: LC
Kodmin: <u>Pollard 228</u> 11/1998; <u>230</u> 11/1998; **Nyale:** <u>Cheek 9655</u> fl., 11/1998; **Nyasoso:** <u>Lane 85</u> fl., 9/1992.

Bulbophyllum fuscum Lindl. var. *fuscum*
Orchid Monographs 2: 147 (1987).
Epiphyte; pseudobulbs bifoliate, (0.8–)2.5–13cm apart, 1–5 × 0.4–1.4cm; petiole 0.1–0.9cm; lamina broadly elliptic to linear-lanceolate, 1.2–5.5(–9)cm; inflorescence 4–12cm; rachis 1.6–7cm, 6–20-flowered, slightly swollen and flattened; flowers yellow, greenish or red; labellum with distinct, usually denticulate lateral lobes near its base. Forest; 700–1450m.
Distr.: Guinea (C) to Congo (K), Angola. (Guineo-Congolian).
IUCN: LC
Kodmin: <u>Gosline 73</u> 1/1998; **Nyale:** <u>Etuge 4203</u> 2/1998.

ORCHIDACEAE

Bulbophyllum fuscum Lind. var. melinostachyum (Schltr.) J.J.Verm.

Syn.: *Bulbophyllum melinostachyum* Schltr.
Bull. Jard. Bot. Nat. Belg. 56: 240 (1986).
Epiphyte; pseudobulbs 1 or 2-leaved; leaves 1.8–11cm; inflorescence 4.5–23cm; peduncle 2.5–10cm; rachis 2.8–17cm, 6–34(–64)-flowered, slightly swollen and flattened; sepals yellowish, greenish or ochrish, often suffused or entirely dark red or reddish brown; labellum without lateral lobes. Lowland to montane forest; 960–1760m.
Distr.: Sierra Leone to Mozambique. (Tropical and subtropical Africa).
IUCN: LC
Enyandong: Pollard 912 fl., 11/2001; **Kodmin:** Cheek 9529 10/1998; Pollard 185 11/1998; 213 11/1998; **Kupe Village:** Cheek 7780 11/1995; **Nyasoso:** Cable 652 fl., 12/1993; Sebsebe 4996 10/1995.

Bulbophyllum graminifolium Summerh.

Syn.: *Bulbophyllum calyptratum* Kraenzl. var. *graminifolium* (Summerh.) J.J.Verm.
Epiphyte; pseudobulbs bifoliate, 1.3–5cm apart; leaves narrowly linear, 11–26 × 0.3–1.2cm; inflorescence 8–28cm, 11–40-flowered; rachis ± terete or slightly flattened, not blade-like; petals white; labellum white suffused purple or pink; labellum 1.8–2.5 × 0.9–2mm. Lowland forest.
Distr.: Guinea, Sierra Leone, Liberia, Ivory Coast, Ghana, Cameroon, Congo (K). (Guineo-Congolian)
IUCN: LC
Loum F.R.: Zapfack 945 7/1996.
Note: we do not follow Vermeulen (1986: 235), in reducing this species to varietal rank.

Bulbophyllum gravidum Lindl.

Epiphytic herb; pseudobulbs 0.7–4.5 × 0.4–1.3cm, bifoliate, 0.6–10cm apart; leaves 2.5–12 × 0.3–1.8cm; inflorescence 7–23cm; peduncle 6.2–19cm; rachis not thickened, 1.5–9.5cm, 10–40-flowered; tepals yellowish or greenish, often suffused with purple-red or entirely dark purple-red; labellum dark purple-red, 3.5–5.8 × 0.7–1.5mm, marginal hairs ≥labellum width. Forest; woodland; 1500m.
Distr.: Cameroon, Bioko, Congo (K), Tanzania, Zambia, Malawi. (Afromontane).
IUCN: LC
Kodmin: Pollard 227 11/1998.
Note: we do not recognise Vermeulen's (1986: 230) treatment of this taxon at varietal rank.

Bulbophyllum imbricatum Lindl.

Syn.: *Bulbophyllum leucorachis* (Rolfe) Schltr.
Epiphyte; pseudobulbs 1–2(–3)-leaved, (0.7–)1.5–6cm apart, ± ovoid; petiole 0.2–1.3cm; lamina narrowly lanceolate to linear, (4.5–)6–25 × (0.6–)1–3.7cm; inflorescence (5–)17–60(–70)cm; (8–)26–120-flowered; peduncle (3–)9–51cm × 2–5.5mm; rachis moderately widened and flattened, (2–)4.5–26cm × 5–13mm × 1.5–5mm (dumb-bell shaped in section); tepals cream, orange-yellow, often suffused or entirely purple. Lowland to submontane Forest; 1000m.
Distr.: Sierra Leone to Congo (K). (Guineo-Congolian).
IUCN: LC
Nyasoso: Cable 3501 7/1996.

Bulbophyllum intertextum Lindl.

Diminutive epiphyte; pseudobulbs unifoliate, 0.2–2.5(–4)cm apart, 0.4–1.0 × 0.3–0.7cm; leaves elliptic to linear lanceolate, 0.7–10 × 0.3–1.1cm; inflorescence 2–30cm, 2–14(–20)-flowered; rachis arching, nodding, terete, usually zigzag bent, 0.2–19cm; floral bracts 1.5–4 × 1–2mm; flowers

very pale yellowish or greenish, often suffused red. Lowland to submontane forest, forest patches in grassland. 1450–1520m.
Distr.: tropical & subtropical Africa.
IUCN: LC
Kodmin: Pollard 178 11/1998; **Mwanenguba:** Sanford 5485 fl., 11/1968.
Local name: Ajeh-bwel (Edmondo Njume in Pollard 178).

Bulbophyllum jaapii Szlatch. & Olszweski

Fl. Cameroun 35(2): 387 (2001).
Epiphyte; pseudobulbs 6.5–9.5 × 0.2–0.4cm, violet, bifoliate; leaves sessile, 5.5–8 × 1cm, lanceolate, midrib and margins violet; inflorescence quite dense, 14–22cm, 22–50-flowered; flowers small, glabrous; lateral sepals and petals yellowish green; dorsal sepal 5.5 × 2mm, violet-tinged; labellum 4 × 0.6mm, linear and shortly ciliate. Forest; 1800m.
Distr.: Mt Kupe. (Narrow Endemic).
IUCN: VU
Mt Kupe: Thomas 5049 fl., 11/1985.

Bulbophyllum josephi (Kuntze) Summerh. var. josephi

Orchid Monographs 2: 68 (1987).
Epiphytic or epilithic herb; pseudobulbs unifoliate, 0.7–3cm apart, 1.5–4 × 0.6–2.4cm; petiole 2–35mm; lamina lanceolate, 4.5–28 × 0.9–3.2cm, slightly emarginate; inflorescence 8.5–40cm, 7–80-flowered; rachis arching to pendulous; flowers cream, tinged pink, pale green or yellowish; column without teeth along the adaxial margins. Forest; 990–1500m.
Distr.: Cameroon, Congo (K), Rwanda, Burundi, Ethiopia, Kenya, Tanzania, Malawi, Mozambique. (Afromontane).
IUCN: LC
Ehumseh - Mejelet: Etuge 353 fl., 10/1986; **Kodmin:** Pollard 202 11/1998; **Kupe Village:** Etuge 1494 11/1995; **Nyasoso:** Lane 167 fl., 10/1994.

Bulbophyllum josephi (Kuntze) Summerh. var. mahonii (Rolfe) J.J.Verm.

Orchid Monographs 2: 69 (1987).
Epiphyte; same as var. *josephi*, except: column with rather weak, deltoid, obtuse teeth along the adaxial margins, just below the stelidia. Lowland to montane forest; 1150m.
Distr.: Guinea (C) to Bioko, W Cameroon, Congo (K), Zambia, Malawi. (Afromontane).
IUCN: LC
Nyasoso: Lane 69 fl., 8/1992.

Bulbophyllum kupense P.J.Cribb & B.J.Pollard

Kew Bull. 59: 137 (2004).
Pendent epiphyte; pseudobulbs at intervals of c. 3–4cm, cigar-shaped, somewhat oblique, 4–6.5cm × 7–9mm, unifoliate at apex; leaf linear, terete, acute, 20–28cm × 3–4mm; inflorescences somewhat incurved-sinuous, subdensely many-flowered; sterile bracts up to 1cm long; flowers fleshy, distichous, purple and pink; lip very fleshy, 3-lobed.
Phorophyte: mango tree; 830m.
Distr.: Mt Kupe. (Narrow Endemic).
IUCN: CR
Nyasoso: Pollard 574 fl., 12/2000.

Bulbophyllum lucifugum Summerh.

Syn.: *Bulbophyllum calyptratum* Kraenzl. var. *lucifugum* (Summerh.) J.J.Verm.
Epiphyte; closely related to *B. calyptratum*, but easily identifiable by its conspicuous rachis, flattened and widened,

456

with an unusual unevenly crenulated margin. Lowland and submontane forest; 830–1030m.

Distr.: Sierra Leone, Liberia, Ivory Coast, W Cameroon. (Upper & Lower Guinea).

IUCN: LC

Enyandong: Salazar 6320 fl., 10/2001; Simo 19 fl., 10/2001. Note: Vermeulen (1987: 131) treats this as a variety of *B. calyptratum*, a view not concurrent with ours; our specimens differ from the other material collected further W in Africa, by having leaves 7–7.5 × 1–1.5cm (not 19–24 × 0.2–0.8), a patent inflorescence (not pendent), and petals 2–2.5mm (not 1–1.5).

Bulbophyllum lupulinum Lindl.

Epiphytic or epilithic herb; pseudobulbs 1–2-leaved, 3–13cm apart, 2.7–7.5 × 1.2–2.5cm; petiole 2–15mm; lamina linear-lanceolate, maximum width slightly above middle, 8–23 × 1.2–5cm; inflorescence 15–38cm, 28–68-flowered; peduncle rather sturdy, to 0.7mm across; flowers distichous, not much open, yellow spotted red. Lowland to montane forest, semi-deciduous forest, frequently epilithic; 1000–1250m.

Distr.: Guinea (C) to Cameroon, Congo (K), Ethiopia, Zambia. (Afromontane).

IUCN: LC

Mwambong: Mackinder 173 1/1998; **Ndum:** Cable 912 fl., fr., 2/1995.

Bulbophyllum maximum (Lindl.) Rchb.f.

Syn.: *Bulbophyllum oxypterum* (Lindl.) Rchb.f.
Epiphyte; pseudobulbs 2(–3)-leaved, 2–10cm apart, 3.5–10 × 1–3cm; leaves oblong to linear-lanceolate, maximum width usually just above middle, 3.8–20 × 1.3–5. cm; inflorescence 15–90cm; 16–120-flowered; rachis bladelike, 6–56 × 8–50mm; floral bracts spreading to reflexed, 2.5–7 × 2–4mm; flowers yellowish or greenish, spotted purple. Lowland primary and secondary forest, montane forest, savanna woodland; 400–1250m.

Distr.: tropical & subtropical Africa.

IUCN: LC

Kupe Village: Etuge 2009 5/1996; **Mwambong:** Mackinder 168 1/1998; 174 1/1998; 184A 1/1998; 187 1/1998; **Nyasoso:** Zapfack 619 6/1996.

Bulbophyllum nigericum Summerh.

Epiphytic or epilithic herb; pseudobulbs bifoliate, ovoid, 0.8–2.5cm apart, 1.3–2.7 × 0.7–1.5cm, obtusely 4-angled, drying bright yellow; petiole 1–2mm; lamina lanceolate to linear-lanceolate, 3–7 × 0.5–1.2cm, ± emarginate; midrib prominent abaxially; inflorescence 8–23cm, 7–30-flowered; peduncle erect, 4–9.5cm × 3.5mm, elliptic in section, with 9–14 tubular scales, the longest 7–12mm; rachis 4-angled in section, ± zigzag bent, 3.5–14 × 0.4cm; floral bracts conspicuous, spreading, concave, ovate, 7–12 × 4–6.5mm, fibrous; flowers distichous, many open simultaneously; sepals very pale green or yellow, occasionally stained purple; petals purplish, spotted deep purple or yellowish; labellum bright yellow, suffused purplish brown basally, or orange. Forest; 830m.

Distr.: S Nigeria, W Cameroon. (W Cameroon Uplands).

IUCN: VU

Enyandong: Salazar 6322 10/2001; **Mt Kupe:** Letouzey 408 11/ year unknown.

Bulbophyllum nigritianum Rendle

Epiphyte; pseudobulbs unifoliate, 0.3–2.5cm apart; 1.4–4 × 0.6–1.5cm; petiole 0.3–3.0cm; lamina narrowly linear-lanceolate to linear, 7–21 × 0.5–1.5cm; inflorescence a dense raceme, 7–22cm, 35–150-flowered; flowers white or cream; labellum 0.6–1.2 × 0.4–0.7mm. Forest; 800–1200m.

Distr.: Sierra Leone to Congo (K). (Guineo-Congolian).

IUCN: LC

Kupe Village: Williams 219 11/1995; **Nyasoso:** Etuge 1332 10/1995.

Bulbophyllum oreonastes Rchb.f.

Syn.: *Bulbophyllum zenkerianum* Kraenzl
Epiphytic or epilithic herb; pseudobulbs bifoliate, 0.7–4.0cm apart, 0.4–3.5 × 0.4–1.2cm; leaves elliptic to linear-lanceolate, 0.6–8.2 × 0.4–2.0cm; inflorescence 1.5–17.5cm, 5–36-flowered; floral bracts yellow, often suffused red; tepals yellow, orange to dark red-purple, with conspicuous dark longitudinal stripes; labellum 1.4–2.5 × 0.6–1.2mm. Forest; 830–1200m.

Distr.: tropical and subtropical Africa. (Afromontane).

IUCN: LC

Enyandong: Salazar 6319 fl., 10/2001; **Ngusi:** Letouzey 14661 4/1976; **Nyasoso:** Lane 166 fl., 10/1994. Note: this taxon was included in FWTA, under the name *B. zenkeranum*.

Bulbophyllum pandanetorum Summerh.

Kew Bull. 8: 580 (1954).
Epiphyte; pseudobulbs bifoliate, 2.5–4cm apart, 2.2–4 × 0.8–1.3cm; leaves linear-lanceolate, 6–15.5 × 0.8–1cm; inflorescence 20–31cm, 8–16-flowered; peduncle 12–23.5cm × 1.8–2mm; rachis 4-angled, 6–8cm × c. 4mm; floral bracts recurved to spreading, distinctly concave, c. 14 × 12.5mm; flowers distichous, red; labellum c. 4 × 2–2.4mm. Lowland or submontane forest; 950m.

Distr.: Gabon, W Cameroon. (Lower Guinea).

IUCN: EN

Nyasoso: Zapfack 607 6/1996.

Bulbophyllum pumilum (Sw.) Lindl.

Gen. and Sp. Orch. Pl.: 54 (1830).
Epiphyte; pseudobulbs unifoliate, 0.2–2.8cm apart, 0.3–4 × 0.3–1.8cm; leaves c. orbicular to broadly linear, somewhat falcate, wider above the middle, acuminate, 1.1–23 × 0.6–2(–4)cm; petiole 1–20mm; inflorescence 2–31cm, 3–65-flowered; peduncle erect to pendent, 1.4–20cm; rachis terete or ± zigzag bent, 0.6–20cm; flowers white, cream or greenish, suffused red; labellum recurved, 0.7–2.3 × 0.5–1.2mm. Lowland forest, relict savanna forest, montane forest; 910m.

Distr.: Guinea (C) to Cameroon, Congo (K). (Guineo-Congolian).

IUCN: LC

Kupe Village: Zapfack 968 7/1996; **Nyasoso:** Zapfack 643 6/1996.

Bulbophyllum renkinianum (Laurent) De Wild.

Kew Bull. 12: 122 (1957).
Epiphyte; pseudobulbs bifoliate, ± cylindrical, 5.5–10.5cm apart, 5.5–10 × 0.9–2cm; leaves ± elliptic, 9.5–20 × 3–6cm; inflorescence 28–48-flowered; peduncle 14–22cm; rachis dark green with dark purple stains, conspicuously flattened, bladelike, 9–19 × 2.7–5cm, much broader than any other species in our area; sepals yellow suffused purple; labellum greenish violet. Lowland Forest; 1250m.

Distr.: W Cameroon, Gabon, Congo (K). (Lower Guinea & Congolian).

IUCN: NT

Kupe Village: Cable 2653 5/1996; Etuge 2005 fl., 5/1996.

Bulbophyllum resupinatum Ridl. var. *filiforme* (Kraenzl.) J.J.Verm.

Orchid Monographs 2: 120 (1987).
Syn.: *Bulbophyllum filiforme* Kraenzl.
Epiphyte; pseudobulbs bifoliate, narrowly ovoid to ellipsoid, 0.5–3.5cm apart, 0.8–4 × 0.5–1.4cm, obtusely to rather sharply 2–4-angled; petiole 1–3m; lamina oblong to linear-lanceolate, 1.3–5(–8.5) × 0.5–1.6cm, tip obliquely emarginate; inflorescence 3.5–40cm, 8–60-flowered; peduncle 1.5–12cm × 0.8–1.5mm; rachis 2–28cm; sepals yellow, spotted or entirely brownish purple; dorsal sepal with fine dark hairs; labellum purple, with a whitish base. Lowland to montane forest; 1300m.
Distr.: Sierra Leone, Liberia, Ivory Coast, Cameroon, Gabon, Congo (K). (Upper & Lower Guinea).
IUCN: LC
Menyum: Thomas 7137 5/1987.

Bulbophyllum saltatorium Lindl. var. *saltatorium*

Orchid Monographs 2: 33 (1987).
Epiphyte; pseudobulbs unifoliate, 1.4–3cm apart, 1.2–2.7cm, distinctly flattened; leaves narrowly elliptic to linear-lanceolate, 3–11.5 × 1–2.3cm; inflorescence 3.5–8cm, 15–30-flowered; peduncle 1.5–4cm; rachis 1.5–4cm; floral bracts 4.5–6.5 × 1.8–2.7mm; flowers purple; labellum with marginal hairs ≥labellum width, dark purple-red with purple or brownish purple hairs. Primary forest; 1250m.
Distr.: Sierra Leone to Rio Muni. (Upper & Lower Guinea).
IUCN: LC
Mwambong: Mackinder 181 1/1998; 184B 1/1998.

Bulbophyllum sandersonii (Hook.f.) Rchb.f. subsp. *sandersonii*

F.T.E.A. Orchidaceae: 320 (1984).
Syn.: *Bulbophyllum tentaculigerum* Rchb.f.
Epiphyte; pseudobulbs bifoliate, 1.2–6.5cm apart, narrowly ovoid; leaves lanceolate to linear, 3.5–26 × 0.5–2.5cm, oblique; inflorescence 5.5–30cm; rachis 1.5–9 × 0.3–1.1cm; floral bracts distinctly narrower than the fully developed part of the rachis; flowers usually placed along an excentric line on the rachis, 3–10mm apart; flowers yellowish or greenish suffused purple. Lowland to montane forest; 850–1700m.
Distr.: Cameroon to Mozambique, S Africa. (Tropical & subtropical Africa).
IUCN: LC
Kodmin: Pollard 220 11/1998; **Kupe Village:** Cable 2569 5/1996; Etuge 1418 11/1995; Williams 229 11/1995; Zapfack 920 7/1996; 942 7/1996; **Nyasoso:** Cable 76 fr., 8/1992; Zapfack 608 6/1996; 651 6/1996; 728 6/1996.

Bulbophyllum sandersonii (Hook.f.) Rchb.f. subsp. *stenopetalum* (Kraenzl.) J.J.Verm.

Bull. Jard. Bot. Nat. Belg. 56: 234 (1986).
Epiphyte; pseudobulbs bifoliate, 1.2–6.5cm apart, narrowly ellipsoid to narrowly ovoid; leaves narrowly lanceolate to linear-lanceolate, 3.5–26 × 0.5–2.5cm, oblique; inflorescence 5.5–30cm; rachis 6–14.5 × 0.2–0.5cm; floral bracts as wide as the fully developed part of the rachis; flowers usually placed along the median line of the rachis, 5.5–11mm apart. flowers yellowish or greenish suffused purple; labellum yellowish with purple dots. Lowland to lower montane forest, occasionally in secondary forest; 400–930m.
Distr.: Liberia to Congo (K). (Guineo-Congolian).
IUCN: LC
Mbule: Letouzey 14678 fl., 4/1976; **Ngusi:** Letouzey 14652 fl., 4/1976; **Tombel:** Thorold CM16 fl., 3/1953.

Bulbophyllum scaberulum (Rolfe) Bolus var. *fuerstenbergianum* (De Wild.) J.J.Verm.

Orchid Monographs 2: 118 (1987).
Syn.: *Bulbophyllum fuerstenbergianum* De Wild.
Epiphyte; pseudobulbs bifoliate, 2.3–13cm apart; 1.8–9 × 0.8–3.6cm; leaves elliptic to linear, 3–28 × 0.7–6cm; rachis bladelike, 3–30 × 0.2–1.4cm; flowers yellowish or greenish suffused purple, median sepal narrowly oblong to broadly lanceolate; petals oblong to broadly lanceolate. Lowland to montane forest; 800–1500m.
Distr.: SE Nigeria, W Cameroon, Bioko, Congo (K). (Lower Guinea & Congolian).
IUCN: VU
Kodmin: Pollard 252 11/1998; **Kupe Village:** Gosline 253 12/1999.

Bulbophyllum scaberulum (Rolfe) Bolus var. *scaberulum*

F.T.E.A. Orchidaceae: 320 (1984).
Epiphyte; pseudobulbs bifoliate, 2.3–13cm apart; 1.8–9 × 0.8–3.6cm; leaves elliptic to linear, 3–28 × 0.7–6cm; rachis bladelike, 3–30 × 0.2–1.4cm; flowers yellow, greenish or crimson; median sepal ovate-lanceolate to broadly linear-lanceolate; petals broadly ovate-lanceolate to linear-lanceolate. Lowland to montane forest; 1000–1500m.
Distr.: Liberia to Uganda. (Guineo-Congolian (montane)).
IUCN: LC
Kodmin: Pollard 176 11/1998; **Kupe Village:** Lane 259 fl., 25/1994; **Nyasoso:** Etuge 2472 6/1996.

Bulbophyllum scariosum Summerh.

Epiphyte; pseudobulbs bifoliate, narrowly ovoid, 2–4.6cm apart, (1.2–)2.5–4 × 0.6–1.7cm, ± flatteneed, 4(–5)-angled; petiole 0.5–3mm; lamina narrowly lanceolate to linear-lanceolate, 3–9 × 0.6–1.3cm, rather thick; inflorescence racemose, lax, 5.5–20cm, 2–8-flowered; peduncle 4.3–15.5cm, with 10–19 scales; rachis 4-angled, zigzag bent apically, 1.2–5.5cm; flowers distichous, white or orange; labellum 2–4.9 × 1.4–2.5mm. Submontane forest; 1500m.
Distr.: Guinea (C), Sierra Leone, Liberia, Ivory Coast, Bioko, W Cameroon. (Upper & Lower Guinea).
IUCN: LC
Kodmin: Pollard 180 fl., 11/1998.

Bulbophyllum schimperianum Kraenzl.

Epiphyte; pseudobulbs tightly clustered, unifoliate, ovoid or conical, 1.2–2.5 × 0.8–1.7cm, yellow; leaf coriaceous, oblong or oblong-lanceolate, slightly falcate, 4.5–13.5 × 1.2–2.5cm, minutely apiculate; inflorescence 10–33cm, densely many-flowered; rachis terete; flowers white or cream, ± flushed pink. Forest; 830–860m.
Distr.: Liberia, Nigeria to Uganda. (Guineo-Congolian).
IUCN: LC
Enyandong: Simo 7 fr., 10/2001; **Nyasoso:** Thorold TN36 fr., 11/1950.

Bulbophyllum schinzianum Kraenzl. var. *schinzianum*

Epiphyte; pseudobulbs unifoliate, orbicular to narrowly ovoid, 0.6–7cm apart, 1–6 × 0.8–3cm, 2–4-angled; petiole 8–40mm; leaf oblong to linear-lanceolate, 7.5–32 × 1.4–6.5cm; inflorescence 28–102cm, 11–102-flowered; peduncle 25–77cm × 1.8–6mm, with 7–12 scales; floral bracts recurved, 16–30 × 5–13mm; flowers purplish; labellum oblong to broadly linear-lanceolate, top part ≥basal part, with hairy margins, hairs ≥labellum. Forest, farmbush; 850–930m.
Distr.: Liberia to Congo (K). (Guineo-Congolian).

IUCN: LC
Ndum: Cable 931 fl., 2/1995; **Nyasoso:** Etuge 2512 7/1996.

Bulbophyllum teretifolium Schltr.

Epiphyte; pseudobulbs unifoliate; leaves erect to erect-patent, terete; inflorescences unknown. Forest; 800m.
Distr.: Mt Cameroon, Bakossi Mts. (Cameroon Endemic).
IUCN: CR
Bangem: Thomas 5281 1/1986; 5950 1/1986.
Note: it is not possible to assign these two Thomas collections to this name with any certainty, as the species was described from a sterile Schlechter collection, destroyed at Berlin during World War II. It is possible that they will fall within the circumscription of *B. kupense*, described by the authors, but are here treated under *B. teretifolium*, as the data is from Missouri Botanic Gardens' Tropicos database and we have not yet seen the specimens on which these determinations were based.

Bulbophyllum unifoliatum De Wild. subsp. *unifoliatum*

Orchid Monographs 2: 154 (1987).
Epiphyte; pseudobulb unifoliate; 5.5–9cm apart, 1.4–4cm, narrowly ellipsoid; leaves linear-lanceolate to linear, 6–17 × 0.6–1.4cm; rachis glabrous; flowers yellowish or brownish orange; sepals abaxially glabrous or at most very finely papillose; labellum orange-red, basally purple, 1.8–3 × 1–1.6mm. Lowland to montane forest; 1200m.
Distr.: Cameroon, Congo (K), Rwanda, Tanzania, Angola, Zambia (Afromontane).
IUCN: LC
Kupe Village: Cable 2622 5/1996.

Calyptrochilum emarginatum (Sw.) Schltr.

Epiphyte; stem elongate, pendent, to 1(–3)m; leaves numerous, distichous, lanceolate-elliptic or oblong, 6–17 × 2.5–5cm; inflorescence many-flowered, very dense, up to 5cm; flowers white; labellum yellow or yellow-green, very fragrant, 7–10 × 7–10mm. Forest, gallery forest, plantations. 300m.
Distr.: Guinea (C) to Cameroon, Bioko, Gabon, Congo (K), CAR, Angola. (Guineo-Congolian).
IUCN: LC
Kupe Village: Etuge 2779 7/1996; Zapfack 961 7/1996.

Chamaeangis ichneumonea (Lindl.) Schltr.

Epiphyte; stem covered in persistent sheathing leaf bases; leaves arranged in a fan, oblanceolate, 17–43 × 2.5–5cm; inflorescence 20–50cm, many-flowered; flowers white, cream or greenish brown, fragrant - especially at night; spur 13–18mm. Forest; 1000m.
Distr.: Sierra Leone, Liberia, Ghana, S Nigeria, W Cameroon, Gabon. (Upper & Lower Guinea).
IUCN: LC
Mwambong: Etuge 4137 2/1998.

Chamaeangis vesicata (Lindl.) Schltr.

Epiphyte, pendent; leaves 3–10, 20–40 × 0.7–1.6cm, linear, falcate, margins revolute, distichous; inflorescence dense, 7–30cm, to 100-flowered; flowers single in lower part of inflorescence, in twos or threes in upper part, lime-green, yellowish or orange, fragrant at night; spur 7–12mm. Forest, woodland; 860m.
Distr.: Guinea (C) to Cameroon, Congo (K), Rwanda, E Africa. (Afromontane).
IUCN: LC
Enyandong: Salazar 6338 fl., 11/2001.

Cheirostylis lepida (Rchb.f.) Rolfe

Terrestrial herb 15–30cm; stem decumbent, rooting basally; leaves membranous, petiolate, ovate, shortly acuminate, rounded or ± cordate, 1–5 × 0.8–2.5cm; inflorescence a dense terminal raceme, to 1.3cm, up to 20-flowered; flowers white; labellum to 5.6mm. Dense shade in leaf mould on forest floor; 800–1250m.
Distr.: SE Nigeria, SW Cameroon, São Tomé, Congo (K), Rwanda, E Africa. (Afromontane).
IUCN: LC
Bangem: Thomas 5255 1/1986; **Kodmin:** Cheek 8874 1/1998; Etuge 4087 1/1998.

Corymborkis corymbis Thouars

Bot. Tidsskr. 71: 161 (1977).
Syn.: *Corymborkis corymbosa* Thouars
Robust terrestrial herb; stems subwoody, 0.5–2m; leaves large, elliptic-lanceolate, acuminate, with narrow sheathing base, shortly petiolate, 11–35 × 3–10cm, with several conspicuous prominent yellowish veins; inflorescence a panicle, ± secund, 10–20-flowered; flowers white, turning creamy-white, to 7cm. Forest understorey, or streambanks in forest, in dense shade; 700–1200m.
Distr.: widespread in tropical Africa to S Africa, Madagascar & Mascarene Is. (Tropical and subtropical Africa).
IUCN: LC
Enyandong: Salazar 6304 10/2001; 6328 11/2001; **Nyasoso:** Etuge 2151 6/1996; Lane 119 fl., 6/1994.

Cribbia brachyceras (Summerh.) Senghas

Kew Bull. 52: 744 (1997).
Syn.: *Rangaeris brachyceras* (Summerh.) Summerh.
Epiphyte; stems erect or ascending, to 20cm; leaves distichous, straight or ± falcate, linear-oblong, 8–13 × 0.7–1.3cm; inflorescence 1–several, suberect, laxly 7–15-flowered, 4–17cm; flowers pale yellow, tinged with orange or green, ± diaphanous; spur 5–9.5mm. Forest; 870–1500m.
Distr.: Liberia to Congo (K), Uganda, Kenya, S to Zambia, Malawi. (Tropical and subtropical Africa)..
IUCN: LC
Enyandong: Pollard 831 10/2001; **Kodmin:** Pollard 174 11/1998; **Kupe Village:** Cable 2433 5/1996; **Mbule:** Cable 3356 6/1996; **Nyasoso:** Lane 136 fl., 6/1994.

Cribbia confusa P.J.Cribb

Kew Bull. 51: 359 (1996).
Epiphyte; stem to 4cm, leafy; leaves distichous, in one plane, linear-oblanceolate, 5–7 × 0.8–1.3cm; inflorescences pendent, several, very slender, rather laxly to 12-flowered; flowers diaphanous, pale yellow-green, sometimes orange-tinged; spur 5–5.5mm. Forest; 940m.
Distr.: Liberia, Ivory Coast, SW Cameroon, São Tomé. (Upper & Lower Guinea).
IUCN: LC
Nyasoso: Cheek 7514 10/1995.

Cynorkis anacamptoides Kraenzl. var. *anacamptoides*

Fl. Zamb. Orchidaceae 1: 51 (1995).
Terrestrial herb, 10–60cm; leaves 2–6, lanceolate or lanceolate-elliptic, acuminate, lower ones petiolate, to 2–13 × 0.8–1.8cm; inflorescence densely many-flowered, 2.5–20cm; flowers pink, mauve or purple; spur present. Grassland, swamp, streamsides, forest edges; 1250–1500m.
Distr.: Bioko, W Cameroon, Congo (K), Ethiopia, E Africa, S to Angola, Zambia, Malawi, Zimbabwe. (Afromontane).
IUCN: LC

ORCHIDACEAE

Edib: Etuge 4090 1/1998; Onana 587 2/1998; **Kodmin:** Pollard 188 11/1998.

Cynorkis debilis (Hook.f.) Summerh.
Terrestrial tuberous herb to 20cm, occasionally epiphytic; leaves 1–2, basal, lanceolate, oblanceolate or oblong, to 10.5 × 3cm; inflorescence a dense spike, to 8cm, (2–)10–20-flowered; flowers small, white; labellum with claret spots. Farmbush, forest and forest-grassland transition; 1050–1500m.
Distr.: Nigeria to Tanzania, S to Zimbabwe. (Afromontane).
IUCN: LC
Kodmin: Pollard 235 11/1998; **Nyasoso:** Cable 61 fl., 8/1992.

Cyrtorchis aschersonii (Kraenzl.) Schltr.
Epiphyte to 50cm; stem erect, to 25cm, woody, clothed in leaf-sheaths; leaves fleshy, long and narrow, 8–22 × 0.2–1.5cm, with ± parallel sides; ± equally bilobed apically; inflorescences several, 4–15cm, 6–8-flowered, leaf-opposed; flowers white, fragrant; labellum 9–15(–20) × 6mm; spur greenish or brownish, (15–)20–40mm, recurved or ± sigmoid. Dense forest.
Distr.: Sierra Leone, Ghana, Nigeria, Cameroon, Congo (K). (Upper & Lower Guinea).
IUCN: LC
Kupe Village: Zapfack 975 7/1996.

Cyrtorchis brownii (Rolfe) Schltr.
Epiphyte to 8–20(–40)cm; stem robust, branched; leaves 4–9, 3–12 × 0.7–1.5cm, linear or oblong-elliptic, unequally bilobed at the apex; inflorescence very dense, 3–9cm, 8–20-flowered; flowers small, non-resupinate, white, fragrant; labellum 5.5–10.5 × 2.5–3mm; spur 2–3cm. Evergreen forest; 1300m.
Distr.: Sierra Leone, Cameroon, Gabon, CAR, Congo (B) & (K), Uganda, Tanzania, Malawi. (Tropical Africa)
IUCN: LC
Kupe Village: Cheek 8379 fl., 5/1996.

Cyrtorchis chailluana (Hook.f.) Schltr.
Epiphyte; stems to 70cm, becoming pendent; leaves 11–25 × 1.9–5.0cm, linear-oblanceolate; inflorescence 13–25cm, 7–10-flowered; flowers waxy white or cream, turning apricot; spur 9–16cm. Forest; 800m.
Distr.: Sierra Leone, Nigeria to Uganda. (Guineo-Congolian).
IUCN: LC
Kupe Village: Kenfack 234 7/1996; **Nyasoso:** Cable 33 fl., 8/1992.

Cyrtorchis monteiroae (Rchb.f.) Schltr.
Epiphyte; aerial roots to 200cm; stem 20–80(–200)cm, pendent; leaves many, oblanceolate or narrowly elliptic, 8–21 × (2.5–)3–5cm, unequally bilobed at apex; inflorescence lax, pendent, 18–32cm, 10–25-flowered; flowers resupinate, white or cream with a greenish or orange-tinged spur, becoming entirely brown-orange at full maturity; spur 3.5–4.5cm. Lakeside and riverine forest, gallery forest, swamp forest; 600–850m.
Distr.: Sierra Leone to Uganda. (Guineo-Congolian).
IUCN: LC
Etam: Etuge 216 fl., 8/1986; **Kupe Village:** Zapfack 974 7/1996; **Nyasoso:** Cable 101 fl., 9/1992.

Cyrtorchis ringens (Rchb.f.) Summerh.
Epiphyte; stem arcuate, woody, to 30cm in old plants; leaves linear, usually 6–7, thick and leathery, 7.5–12.5 × 1.2–2.5cm;

inflorescence 6–7(–16)cm, to 16-flowered; flowers closely placed, creamy white, sweetly scented; spur to 3cm. Forest; 900–2000m.
Distr.: Senegal, Sierra Leone to Cameroon, São Tomé, Congo (K), Burundi, Uganda, Tanzania, Zambia, Malawi, Zimbabwe. (Afromontane).
IUCN: LC
Kodmin: Pollard 218 11/1998; **Kupe Village:** Cable 2620 fl., 5/1996; 2621 5/1996; Etuge 2630 7/1996; Ryan 272 fl., 5/1996; 412 5/1996; **Mwanenguba:** Sanford 5553 fr., 11/1968; **Nyasoso:** Cable 3407 6/1996; Etuge 2158 6/1996; 2448 6/1996; 2468 6/1996; Lane 139 fl., 6/1994; Wheatley 399 fl., 7/1992; Zapfack 616 6/1996; 715 6/1996; 729 6/1996; 735 6/1996.

Diaphananthe bidens (Sw.) Schltr.
Epiphyte; stem pendent to 50cm; leaves up to 22, distichous, oblong-elliptic to ovate-elliptic, sometimes ± oblique, 6.5–17 × 1.1–4.5cm; inflorescence 4–22cm, up to 25-flowered; flowers salmon-pink, yellowish pink, flesh-coloured or whitish; labellum 3.5–5mm. Forest; 700–1100m.
Distr.: Guinea (B) to Cameroon, E to Uganda, S to Angola. (Guineo-Congolian).
IUCN: LC
Enyandong: Zapfack 1780 fr., 10/2001; **Kupe Village:** Zapfack 943 7/1996; **Nyale:** Etuge 4188 2/1998; **Nyasoso:** Etuge 2117 6/1996.

Diaphananthe kamerunensis (Schltr.) Schltr.
Large pendent epiphyte; stem stout, 6–8 × 1.1–1.8cm; leaves 3–7, falcate, oblanceolate, 20–35 × 3–5.5cm; inflorescences pendent, 1–several, 10–12cm, laxly 8–10-flowered; peduncle 3.5–4.5cm; flowers pale translucent green. Forest; 1050–1100m.
Distr.: Cameroon, Congo (K), Uganda, Zambia. (Lower Guinea & Congolian).
IUCN: LC
Nyasoso: Lane 1 fl., 7/1992; Sunderland 1501 fl., 7/1992.

Diaphananthe pellucida (Lindl.) Schltr.
Robust epiphyte; stem short, up to 12 × 1cm; leaves distichous, oblanceolate, 18–70 × 2–9cm; inflorescences 1–7, pendent, 15–55cm, to 50–or more-flowered; flowers white, translucent creamy yellow, pale green or pinkish; labellum 0.8–1.1cm, shortly fimbriate. Forest; 750–1300m.
Distr.: Guinea (C) to Uganda. (Guineo-Congolian).
IUCN: LC
Enyandong: Simo 10 fl., 10/2001; 17 10/2001; 31 10/2001; **Kodmin:** Pollard 212 11/1998; **Kupe Village:** Cheek 7748 11/1995; 7788 11/1995; **Nyasoso:** Zapfack 636 6/1996.

Diaphananthe polyantha (Kraenzl.) Rasm.
Norw. J. Bot. 21: 229 (1974).
Syn.: *Sarcorhynchus polyanthus* (Kraenzl.) Schltr.
Epiphyte; stem to 15cm; leaves linear-oblong, distichous, ± oblique, 3–9 × 0.5–1.3cm; inflorescence 4–22cm, many-flowered; flowers green in bud, opening greenish white, white or very pale yellow, vanilla-scented; labellum 2.7–3.6mm. On scattered trees in montane grassland or pastures; 1650–2000m.
Distr.: Bioko, W Cameroon. (W Cameroon Uplands).
IUCN: NT
Mt Kupe: Cheek 7576 10/1995; Etuge 1365 11/1995; **Mwanenguba:** Sanford 5479 fl., fr., 11/1968; 5554 fr., 11/1968.
Note: known from only 14 locations, half of which are within the protected Kilum-Ijim area, NW Province, Cameroon.

Diaphananthe polydactyla (Kraenzl.) Summerh.
Epiphyte; stem very short, c. 1cm, pendent; leaves in a bunch at its apex, linear-lanceolate, 6–10 × 0.7–1cm; inflorescence 6–18cm, 5–16-flowered; flowers ornamental, white or greenish white; labellum c. 1cm, 3-lobed, lateral lobes long-pectinate. Forest, woodland; 1400–1850m.
Distr.: W Cameroon. (W Cameroon Uplands).
IUCN: VU
Bangem: Manning 440 fl., 10/1986; **Ehumseh - Mejelet:** Etuge 354 fl., 10/1986; **Enyandong:** Pollard 905 fl., 11/2001; **Kodmin:** Pollard 172 11/1998; **Mwanenguba:** Sanford 5478 fl., 11/1968; 5486 fl., 11/1968; **Nyasoso:** Cheek 7617 11/1995.

Diaphananthe rohrii (Rchb.f.) Summerh.
Kew Bull. 14: 140 (1960).
Syn.: *Diaphananthe quintasii* (Rolfe) Schltr.
Epiphyte; leaves linear to obovate, 2–7, ± falcate, 5–15 × 0.7–2.2cm; inflorescence 6–28cm, 6–20-flowered; flowers greenish white or greenish yellow; sepals 3.5–5.5mm. Montane forest; 1300m.
Distr.: Liberia to Cameroon, São Tomé, Bioko, Congo (K), Burundi, Ethiopia, E Africa. (Afromontane).
IUCN: LC
Kodmin: Cheek 9528 fl., 10/1998.

Diaphananthe rutila (Rchb.f.) Summerh.
Epiphyte, usually in canopy of tall forest; stems 3–40cm, occasionally branched; leaves linear to obovate or narrowly elliptic, 3–12 × 0.5–2.2cm; inflorescences 1–many, pendent, 5–19cm, densely up to 25-flowered or more; flowers translucent, pale green, pale yellow, often tinged rose, brown or purple. Riverine and montane forest, occasionally in old coffee plantations; 760–925m.
Distr.: Guinea (C) to E Africa, S to Angola, Malawi, Mozambique, Zimbabwe. (Tropical and subtropical Africa).
IUCN: LC
Enyandong: Cheek 10917 fl., 10/2001; Simo 2 fl., 10/2001; 27 fl., 10/2001; 3 10/2001; Zapfack 1779 10/2001; **Kupe Village:** Zapfack 967 7/1996.
Note: *Diaphananthe* is in the process of being split, and this species is likely to be known as *Rhipidoglossum rutilum* (Rchb.f.) Schltr. at some point in the future.

Diceratostele gabonensis Summerh.
Terrestrial herb; stem 36–60cm, erect; leaves 5–10, ± resembling young palm fronds; petiole 3.5–8cm, robust, canaliculate; blade 8–18 × 2–6cm, elliptic, folded, trinerved; inflorescences 1–4 per plant, to 4cm, 10–15-flowered; rachis pubescent; floral bracts to 1.5cm; flowers not easily visible, tubular, resupinate, white to greenish. Swampy forest or riverine Forest; 760m.
Distr.: Liberia, Ivory Coast, Cameroon, Gabon, Congo (K). (Guineo-Congolian).
IUCN: LC
Ngomboaku: Mackinder 297 fl., 12/1999.
Note: only the second known collection from Cameroon.

Dinklageella liberica Mansf.
Epiphyte; roots very long, much branched, often in opposite pairs along the stem; stems very rarely branched, creeping; leaves alternate, narrowly elliptic, 1.7–3.5 × 0.8–1.5cm, bilobed; inflorescence 5–8cm, c. 3–6-flowered; flowers white, tinged orange; spur 2–3cm, swollen distally; capsules 2–3(–5.5) × 1(–2)cm. Forest, plantations; 1250m.
Distr.: Liberia, Ivory Coast, S Nigeria, Cameroon, Rio Muni, Congo (B) & (K). (Guineo-Congolian).

IUCN: LC
Lake Edib: Cheek 9144 2/1998.

Disa erubescens Rendle subsp. *erubescens*
A terrestrial herb, 0.3–1m; leaves 9–16, linear or linear-lanceolate, to 37 × 0.5–2cm; inflorescence to 19 × 4–7.5cm, 2–12-flowered; flowers suberect, deep orange, scarlet or deep red, apices of petals orange or yellow, spotted red. Wet montane grassland and swamp; 1850–2000m.
Distr.: N Nigeria, W Cameroon, Congo (K), Rwanda, Burundi, Sudan, E Africa, Zambia, Zimbabwe, Malawi. (Afromontane).
IUCN: LC
Mwanenguba: Gregory 301 fl., 6/1948; Leeuwenberg 9951 fl., 6/1972.

Disperis mildbraedii Schltr. ex Summerh.
Terrestrial herb, 10–25cm; stem slender; leaves 2, opposite, broadly ovate, acute, cordate at base, to 6 × 4.5cm; inflorescence a short, bracteate raceme, to 2.5cm; flowers 1–4, white. Forest, moist rocks in moderate shade in undisturbed forest, woodland; 2000m.
Distr.: SE Nigeria, W Cameroon, Bioko. (W Cameroon Uplands).
IUCN: VU
Kupe Village: Cable 117 fl., 9/1992; **Nyasoso:** Cable 3378 6/1996.

Disperis nitida Summerh.
An epiphytic or terrestrial herb 12–30(–50)cm; leaves 2, opposite, broadly ovate, acute, cordate at base; inflorescence a short raceme, lax, 2–3(–5)-flowered, 2–2.5cm; flowers 1–4, white. Montane forest, in deep shade; 2100m.
Distr.: W Cameroon. (W Cameroon Uplands).
IUCN: EN
Mwanenguba: Letouzey 14404 fl., 8/1975.

Disperis thomensis Summerh.
Terrestrial or epiphytic or epilithic herb, 5–25cm; leaves 2, opposite, purplish beneath, ovate, acute, cordate at base, c. 3 × 2cm; petiole 7mm; flowers 1–6, in a short raceme, white with faint purplish markings. Forest, riverine forest, in humus in dense shade, on rocks; 900–1200m.
Distr.: Guinea (C), Sierra Leone, Liberia, Ghana, Cameroon, Bioko, São Tomé, Angola, Tanzania. (Afromontane).
IUCN: LC
Nyasoso: Balding 73 fl., 8/1993; Cable 54 fl., 8/1992.

Epipogium roseum (D.Don) Lindl.
Terrestrial achlorophyllose heteromycotrophic (saprophytic) herb; leafless; flowering scape to 60cm, hollow, fleshy, brownish at base, almost white apically; inflorescence to 20cm, many-flowered; flowers creamy-white to dull dirty pink, with small dark purple or crimson spots on labellum. Forest floor; 650–1650m.
Distr.: Ghana to Malawi, Indo-Malaysia to Australia, Vanuatu. (Palaeotropical).
IUCN: LC
Ngusi: Etuge 1560 1/1996; **Nyasoso:** Williams 180 fl., 2/1995.

Eulophia horsfallii (Bateman) Summerh.
Robust terrestrial herb, 1–3m; leaves 3–5, lanceolate to oblanceolate, 30–200 × 5–15cm, ribbed; scapes to 3m; inflorescence a long raceme, laxly 5–50-flowered; flowers large, fleshy, purple with maroon sepals and petals; labellum 2–4 × 1.4–4cm. Forest edge, swamp, fallow; 900–1500m.
Distr.: tropical and subtropical Africa.

ORCHIDACEAE

IUCN: LC
Edib: <u>Etuge 4113</u> 1/1998; **Kodmin:** <u>Pollard 223</u> 11/1998;
Nyasoso: <u>Etuge 3269</u> 9/1996.

Eulophia odontoglossa Rchb.f.
Linnea 19: 373 (1846).
Syn.: *Eulophia shupangae* (Rchb.f.) Kraenzl.
Terrestrial herb, 0.6–1.0m; leaves 5–6, erect, plicate,
oblanceolate, acuminate, 40–70 × 1–2.1cm, basal 3
sheathing; inflorescence densely many-flowered; flowers
yellow; labellum with yellow, orange or red papillae.
Grassland, bushland, rocky areas; 1850m.
Distr.: tropical & subtropical Africa. (Afromontane).
IUCN: LC
Mwanenguba: <u>Gregory 304</u> fl., 6/1948.

Eurychone rothschildiana (O'Brien) Schltr.
Epiphyte; stem 2–8cm, with persistent leaf-sheaths; leaves 3–
8, 6–21 × 1.5–7cm, unequally and subacutely bilobed,
undulate; inflorescence laxly 2–6–12-flowered, deflexed or
pendent, 3–10cm; flowers large, pleasantly scented; tepals
white or greenish, 18–25 × 6.5–10mm; labellum white,
centrally green, with a purple-brown marking at throat, 20–
27 × 20–25mm. Forest; 1000m.
Distr.: Guinea (C) to Nigeria, Bioko, Cameroon, Congo (K),
Uganda. (Guineo-Congolian).
IUCN: LC
Kupe Village: <u>Zapfack 970</u> 7/1996; **Ndibise:** <u>Etuge 522</u>
6/1987.

Genyorchis micropetala (Lindl.) Schltr.
Epiphyte; pseudobulbs 1.2cm × 4.5mm, 1.5–3cm apart,
ovoid, bifoliate; leaves 1–3 × 0.2–0.3cm, linear-lanceolate;
inflorescence lax, 2–6cm, to 10-flowered; peduncle with 4
bracts; rachis zigzag bent; flowers small, non-resupinate,
violet-pink; dorsal sepal 1.7–2.2 × 0.6–0.9mm; labellum 1–
1.5 × 1.5–1.6mm. Forest.
Distr.: Bioko, W Cameroon. (W Cameroon Uplands).
IUCN: EN
Mt Kupe: <u>Letouzey 477</u> 11/1970.

Graphorkis lurida (Sw.) Kuntze
Robust epiphyte; pseudobulbs cylindrical-fusiform or conical
ovoid, 3–9 × 1–3cm, 4–6-leaved, yellowish and ribbed;
inflorescence appearing before the leaves, erect, paniculate,
15–50cm; flowers yellowish with brown sepals and petals;
sepals about 5–6mm; spur sharply bent forward, ± as long as
labellum. Forest; 1250m.
Distr.: Senegal to Bioko, Cameroon, Gabon, Congo (K),
Burundi, Uganda, Tanzania. (Guineo-Congolian).
IUCN: LC
Mwambong: <u>Mackinder 180</u> 1/1998.

Habenaria

Note: we do not follow the generic delimitations of
Szlachetko & Olszewski (1998) with regard to *Habenaria*
sens. lat., and so do not recognise their generic names
Kryptostoma, *Podandriella*, *Pseudoperistylus*, *Roeperocharis*
or *Veyretella*.

Habenaria attenuata Hook.f.
Syn.: *Pseudoperistylus attenuatus* (Hook.f.) Szlach. &
Olszewski
Terrestrial herb or epiphyte on mossy tree-trunks, 15–50cm;
leaves 4–7, lower ones lanceolate-oblong, upper ones much
smaller; inflorescence to 2cm, laxly up to 25-flowered; lower
bracts often longer than flowers; flowers c. 1cm across, green

or yellowish green; spur pendent, 10–16mm, swollen distally.
Upland grassland; 1900m.
Distr.: Cameroon, Bioko, Congo (K), Ethiopia, Uganda,
Kenya. (Afromontane).
IUCN: LC
Mt Kupe: <u>Schoenenberger 9</u> fl., fr., 11/1995.

Habenaria batesii la Croix
Kew Bull. 48: 371 (as *H. praetermissa*, a later homonym of
Seidenfaden's name for an Asian species) (1993); Kew Bull.
51: 364 (1996).
Syn.: *Habenaria praetermissa* la Croix
Syn.: *Podandriella batesii* (la Croix) Szlach. & Olszewski
Robust terrestrial herb, 30–40cm; 3–4 leaves in a basal tuft;
petiole to 5cm, blade c. 15 × 5cm, obliquely ovate to
obliquely obovate, acute; stem leaves several, sheathing;
bracteate; inflorescence c. 10 × 8cm, (10–)20–30-flowered;
flowers white, sepals with green veins; labellum porrect,
trilobed almost to base, lobes linear, 15–18 × 0.5–1.0mm;
spur 4–4.5cm, recurved upwards, white, distally green.
Forest; 900m.
Distr.: Bakossi Mts & Efulen. (Cameroon Endemic).
IUCN: EN
Enyandong: <u>Pollard 731</u> fl., 10/2001.
Note: this is the second ever known collection.

Habenaria bracteosa A.Rich.
Syn.: *Pseudoperistylus bracteosus* (A.Rich.) Szlach. &
Olszewski
An erect, leafy terrestrial herb 0.15–1.0m; leaves 5–10,
lowest 2 sheathing, the middle 3–5 lanceolate or oblong-
lanceolate, acute, to 30 × 3cm, uppermost bracteate;
inflorescence a densely many-flowered spike, 5–50cm;
flowers suberect, green or yellowish green; spur > 15mm.
Grassy glades in forest, especially by streams, damp places;
1250m.
Distr.: W Cameroon, Bioko, Ethiopia, Sudan, E Africa.
(Afromontane).
IUCN: LC
Edib: <u>Etuge 4095</u> 1/1998.

Habenaria cornuta Lindl.
Terrestrial herb, 0.2–0.8m; leaves 9–15, lowermost
sheathing, largest 2–10 × 0.7–4.5cm; inflorescence a rather
dense raceme, 5–19 × 3–6cm, 4–many-flowered; flowers
pale green or yellow-green; lip tripartite, 0.8–1.8cm; spur
15–25mm. Open, often badly drained, grassland, or in
woodland; 2000m.
Distr.: N Nigeria, W Cameroon to Ethiopia, E Africa, S to
Malawi, Zambia, Zimbabwe, S Africa. (Afromontane).
IUCN: LC
Mwanenguba: <u>Cheek 9495</u> 10/1998.

Habenaria gabonensis Rchb.f. var. *gabonensis*
Syn.: *Habenaria procera* (Sw.) Lindl. var. *gabonensis*
(Rchb.f.) Geerinck
Fl. Congo, Rw. & Bur. Orchidaceae: 89 (1984).
Terrestrial (or epiphytic) herb to 90cm; leaves narrowly
lanceolate or lanceolate, 1–4cm across; inflorescence lax,
almost always with < 12 flowers; floral bracts 2cm; pedicel
2–3.5cm; flowers white; spur 5–12cm, distally greenish.
Humus, grassy places amongst rocks, damp places; 950–
1250m.
Distr.: Sierra Leone to Congo (K). (Guineo-Congolian).
IUCN: LC
Edib: <u>Etuge 4088</u> 1/1998; **Kodmin:** <u>Pollard 236</u> 11/1998;
Mwambong: <u>Pollard 117</u> 10/1998; **Nyasoso:** <u>Cable 51</u> fl.,
8/1992; <u>Sunderland 1502</u> fl., 7/1992.

Habenaria malacophylla Rchb.f. var. malacophylla

A slender terrestrial herb, 0.3–1m; stem leafy in centre part, bare below; inflorescence a long loose raceme, 8–34 × 2.5–3.5cm; bracts lanceolate, 0.9–2.0cm; flowers numerous, small, green; petals bipartite near the base, anterior lobe 5–9.5 × 0.3–0.8mm, posterior lobe 4–7.3 × 0.4–1.5mm; labellum trilobed, median lobe 4.5–8 × 0.8–1mm; spur 9–18mm. Woodland and grassland; 900m.
Distr.: Sierra Leone, Nigeria, Cameroon, Ethiopia, E Africa, Malawi, Zambia, Zimbabwe, S Africa, Oman. (Afromontane & Arabia (montane)).
IUCN: LC
Ngomboaku: Ghogue 489 12/1999.

Habenaria mannii Hook.f.

Syn.: *Kryptostoma mannii* (Hook.f) Szlach.
Terrestrial herb to 70cm; stem leafy; leaves linear-lanceolate, 5–15 × 0.5–2.0cm; inflorescence a dense raceme, 6–22 × 3–4.5cm, of few–25 flowers; flowers pale green; labellum tripartite, easily recognised by outside of lateral lobes being conspicuously pectinate; spur 11–14(–28)mm. Grassland; 1300–2300m.
Distr.: Nigeria, Bioko, Cameroon. (W Cameroon Uplands).
IUCN: NT
Mt Kupe: Schoenenberger 8 fl., fr., 11/1995; **Mwanenguba:** Cheek 9493 10/1998; de Gironcourt 524 6/ year unknown; Gregory 300 fl., 6/1948; Leeuwenberg 8410 fl., 9/1971; Pollard 880 fl., fr., 11/2001; Salazar 6314 fl., 10/2001; Sanford 5520 fl., 11/1968; 5570 11/1968.

Habenaria nigrescens Summerh.

Syn.: *Roeperocharis nigrescens* (Summerh.) Szlach. & Olszewski
Terrestrial herb, 15–70cm; leaves 9, arranged spirally, sheathing basally, linear to linear-lanceolate, 7–13 × 0.5–1.0cm; inflorescence spicate, lax to dense, 5–13cm, 4–17-flowered; floral bracts 10–15mm; flowers pale green or yellow-green; labellum deeply trilobed, median lobe 4.3–4.5 × 1.2mm; spur 6–8.5mm. Montane grassland; 2250m.
Distr.: SE Nigeria, W Cameroon. (W Cameroon Uplands).
IUCN: VU
Mwanenguba: Leeuwenberg 8473 fl., 9/1971.

Habenaria peristyloides A.Rich.

Syn.: *Roeperocharis peristyloides* (A.Rich.) Szlach. & Olszewski
Terrestrial herb to 0.8m; stem leafy; leaves ± erect, 4–9, lower sheath-like, otherwise ± lanceolate, acute, up to 6–24 × 0.5–2.5cm; inflorescence a dense spike, 5–30cm; floral bracts to 19mm; flowers green or yellowish green; labellum markedly trilobed; median lobe 5–9.5 × 1.2–3mm; spur 1.5–3.5mm. Short upland grassland, marshes and open scrub; 2000m.
Distr.: Nigeria, Cameroon to E Africa. (Afromontane).
IUCN: LC
Mwanenguba: Cheek 9494 10/1998; de Gironcourt 506 date unknown.

Habenaria procera (Afzel. ex Sw.) Lindl.

Fl. Congo, Rw. & Bur. Orchidaceae: 89 (1984).
Syn.: *Habenaria ochyrae* Szlach. & Olszewski
Epiphyte (usually) to 60cm; leaves lanceolate to ovate, up to 30 × 27cm; inflorescence up to 15cm, rather dense, usually 10–30-flowered; floral bract to 3.5cm; pedicel 8–15mm; flowers white; spur 6–10cm. Epiphyte in lowland rain-forest in the humus of fissured bark and axils of branches of several tree species; 600–950m.

Distr.: Sierra Leone to W Cameroon, Gabon, Uganda. (Guineo-Congolian).
IUCN: LC
Kupe Village: Cable 3734 7/1996; Zapfack 916 7/1996; 948 7/1996; **Ngomboaku:** Cheek 10400 12/1999; **Nyale:** Pollard 222 11/1998; **Nyasoso:** Lane 3 fl., 7/1992; 78 fl., 9/1992.

Habenaria thomana Rchb.f.

Kew Bull. 51: 353 (1996).
Terrestrial herb to 50cm; leaves (3–)5–10(–12), the lower radical, long-petiolate; petiole sheathing; lamina lanceolate or ovate-lanceolate, to 6–20 × 3–6cm, acuminate, the upper bracteate; inflorescence to 40cm, laxly 10–25-flowered; flowers very pale green; labellum trilobed; median lobe 5.5–7 × 0.8–1.5mm; spur 13–15mm. Submontane forest, secondary forest; 800–1200m.
Distr.: SW Cameroon, Bioko, São Tomé. (W Cameroon Uplands).
IUCN: VU
Kupe Village: Cheek 7807 11/1995; 7864 11/1995; **Nyasoso:** Balding 15 fl., 7/1993; Cable 28 fl., 7/1992; 52 fl., 8/1992; 3421 7/1996; 3610 7/1996.

Habenaria weileriana Schltr.

Epilithic, occasionally epiphytic, or rarely terrestrial herb to 40cm, often occurring in large colonies; leaves linear or narrowly lanceolate, 8–10 cauline, 6.5–11.5 × 0.7–1cm; inflorescence lax, 2.7–6cm, to (2–)3(–4)-flowered; flowers white; labellum trilobed; median lobe 8–9 × 2–2.5mm; spur 4–5.5cm. Forest and streambanks, rocks in rivers and streams; 750–1250m.
Distr.: SE Nigeria, W Cameroon, Gabon. (Lower Guinea).
IUCN: NT
Baseng: Cheek 10412 12/1999; **Enyandong:** Salazar 6310 fl., 10/2001; **Lake Edib:** Cheek 9133 2/1998.
Note: rare throughout its range and known from only 12 locations.

Hetaeria occidentalis Summerh.

Terrestrial herb, 30–60cm; leaves obliquely lanceolate or elliptic-lanceolate, 6–16 × 2–5cm; inflorescence dense, 7–17cm, 20–40-flowered; flowers non-resupinate; bracts and sepals white, tinged pink; labellum white with yellow markings, 2-lobed, lobes spreading, 2.5–4mm; spur short, 2-lobed. Forest, damp areas, lowland wooded areas; 150–760m.
Distr.: Guinea (B) to Cameroon, CAR, Congo (K). (Guineo-Congolian).
IUCN: LC
Loum F.R.: Cheek 10256 fl., 12/1999; **Mungo River F.R.:** Cheek 10142 11/1999; **Ngomboaku:** Mackinder 286 12/1999.

Hetaeria tetraptera (Rchb.f.) Summerh.

Terrestrial herb to 70cm; leaves obliquely ovate or elliptic-ovate, 3–10 × 1.5–4.5cm; inflorescence lax below, dense above, 5–8cm; flowers non-resupinate; bracts pale green, basally ± white; labellum white with pale yellow-green markings, 6.5mm, lateral lobes 4–6mm. Forest and streambanks; 1300–1368m.
Distr.: Nigeria to Congo (K). (Lower Guinea & Congolian).
IUCN: LC
Kodmin: Biye 44 1/1998; Etuge 3972 1/1998; **Kupe Village:** Cheek 6055A fr., 1/1995.

Liparis ascendens P.J.Cribb

Kew Bull. 51(2): 357 (1996).

Creeping terrestrial herb with an ascending rhizome, to 9cm long; pseudobulbs conical, 3-noded, 2–3-leaved; leaves suberect, ovate-lanceolate, acute, 5–9 × 1.5–2.5cm; petiole conduplicate, to 4cm; inflorescence erect, subcorymbose, many-flowered, 7–15cm; flowers green, marked purple; pedicel and ovary 10–14mm; lip porrect, conduplicate, subcircular, apiculate, 4.5–5 × 4–4.5mm. In soil and moss on rocks in forest; 900–1550m.
Distr.: Cameroon, Gabon, Congo (K). (Lower Guinea).
IUCN: NT
Bangem: Villiers 1410 fl., 6/1982.

Liparis deistelii Schltr.
Terrestrial or epiphytic herb to 15cm; rhizomatous; pseudobulbs fleshy, to 4 × 0.4cm, tapering to 1mm across apically; with 2–3 small sheathing leaves at base, and 3 larger ones, ovate to oblong-lanceolate, 5–8(–12) × 1.5–3cm, acute to acuminate; inflorescence terminal, to 12cm, 2–6(–10)-flowered; flowers cream or yellowish green to light and dark reddish purple. In moss and leaf-mould on rocks and fallen trees in shade near rivers; 1500–1550m.
Distr.: Cameroon, Bioko, Gabon, Congo (K), Ethiopia, E Africa, Malawi. (Afromontane).
IUCN: LC
Mwanenguba: Sanford 5500 fr., 11/1968; **Nyasoso:** Cable 123 fl., 9/1992; Wheatley 436 fl., 7/1992.

Liparis epiphytica Schltr.
Epiphyte; pseudobulbs 1.2 × 0.2–0.4cm, ovoid; leaves usually 4, 5–11 × 0.6–0.9cm, oblong-lanceolate to oblong-elliptic, 1 or 2 sheathing, 2 or 3 above regular; inflorescence 2.8–10cm, lax, 2–20-flowered; peduncle with several cauline bracts; rachis winged; flowers ± resupinate, light green-yellow. Forest; 780–930m.
Distr.: Ivory Coast, Nigeria to Uganda. (Guineo-Congolian).
IUCN: LC
Enyandong: Salazar 6303 10/2001; Zapfack 1781 fr., 10/2001.

Liparis nervosa (Thunb.) Lindl. var. *nervosa*
Gen. & Sp. Orch.: 26 (1830); Orchid Flowers, 2: 77 (1971).
Syn.: *Liparis guineensis* Lindl.
Syn.: *Liparis rufina* (Ridl.) Rchb.f. ex Rolfe
Terrestrial, epilithic or rarely epiphytic herb to 70cm; stem basally swollen; leaves 2–5, petiolate, sheathing, lanceolate, to 35 × 7.5cm; peduncle to 55cm; rachis many-flowered, to 15cm; flowers green or yellow to reddish or purplish brown. var. khasiana occurs from Bhutan to Assam. Forest, marshy grassland, or stony grassland and on wet rock outcrops; 900–930m.
Distr.: tropics and subtropics of the World, including Africa, India to Japan, Philippines, Costa Rica, W Indies, S America. (Pantropical).
IUCN: LC
Enyandong: Salazar 6301 fr., 10/2001; **Kupe Village:** Etuge 2878 7/1996.

Liparis platyglossa Schltr.
Epiphyte or rarely an epilith, 7–15cm; leaves shortly petiolate, ovate, rarely elliptic, acuminate, 6–9 × 1.5–4cm; inflorescence 5–9cm, densely 4–10-flowered towards the apex; flowers large, greenish yellow; labellum conspicuously lined with purple veins, 12 × 15mm. Forest; 875–1150m.
Distr.: Ivory Coast, Nigeria, Bioko, Cameroon, Uganda. (Guineo-Congolian).
IUCN: LC
Nyasoso: Etuge 2071 6/1996; Lane 73 fl., 9/1992; 106 fl., 6/1994; Wheatley 382 fl., 7/1992; Zapfack 713 fl., 6/1996.

Liparis tridens Kraenzl.
Epiphyte, 5–12cm, usually found in clumps; stem basally swollen, forming a small ovoid pseudobulb; leaves ± 5(–7.5) × 1(–1.4)cm, oblong-lanceolate, margin conspicuously undulate; inflorescence many-flowered; peduncle 5–6(–10)cm; flowers greenish yellow. Epiphytic on trees in wet places; 700–1500m.
Distr.: Ivory Coast, S Nigeria, Cameroon, Bioko, Uganda, Tanzania. (Afromontane)
IUCN: LC
Enyandong: Salazar 6309 10/2001; 6327 fr., 11/2001; **Kodmin:** Pollard 225 11/1998; **Kupe Village:** Etuge 2715 7/1996; **Ngomboaku:** Cheek 10352 12/1999; **Nyasoso:** Cable 3422 7/1996; Etuge 2505 7/1996; Lane 28 fl., 7/1992.

Malaxis maclaudii (Finet) Summerh.
Terrestrial (or rarely epilithic) herb, erect to 30cm; stems 3–6cm, 3–4-leaved; leaves ovate or ovate-lanceolate, 4–9 × 2.5–6cm; inflorescence apex with flowers close together, forming a false umbel, to 15cm, to 40-flowered; peduncle with 3–4 cauline bracts; flowers flesh to deep rose-coloured; labellum 3.5–4.6 × 3.8–5.6mm. Forest understorey or on rocks covered with humic layers; 1100m.
Distr.: Guinea (C) to Cameroon, Congo (K), CAR, Sudan, Zambia. (Afromontane).
IUCN: LC
Nyasoso: Cheek 7316 fr., 2/1995.

Malaxis prorepens (Kraenzl.) Summerh.
Terrestrial herb, 15 to ± 35cm; pseudobulbs to 15 × 0.2cm; leaves 3, rarely more, ovate, acute or subacute, to 6 × 4cm; inflorescence ± dense, 5–16cm, to 50-flowered; peduncle with several sterile cauline bracts; flowers resupinate, very small, flat, purple or green; labellum 1.3–2 × 1.7–2mm. Forest; 1000m.
Distr.: Guinea (C), Sierra Leone, Nigeria, Cameroon, CAR, Ethiopia, Uganda, Tanzania, Mozambique. (Afromontane).
IUCN: LC
Bakossi F.R.: Doumenge 461 fl., 5/1987.

Malaxis weberbaueriana (Kraenzl.) Summerh.
Terrestrial herb to 25cm; rhizome creeping, slender; pseudobulbs 2.5–8 × 0.1–0.2cm, fusiform; leaves thin-textured, ovate or elliptic-ovate, to 5.5 × 3cm; inflorescence lax, 5–20-flowered, 4–17cm; flowers small, flat, purple or green; labellum 2–2.6 × 2–3mm. Deciduous woodland, forest, plantations; 860–1760m.
Distr.: Cameroon, Bioko, Congo (K), Kenya, Tanzania, Zambia, Malawi, Zimbabwe. (Afromontane).
IUCN: LC
Enyandong: Pollard 904 fr., 11/2001; Salazar 6326 fr., 11/2001; **Nyasoso:** Cable 2755 6/1996; 3200 6/1996; Zapfack 611 6/1996.

Manniella cypripedioides Salazar, T.Franke, Zapfack & Beenken
Lindleyana 17(4): 240 (2002).
Terrestrial herb to 30–65cm; rhizome elongate, creeping; leaves several, obliquely elliptic or elliptic-ovate; margins ± wavy, homogeneously deep-green, 4.5–10.5 × 2.5–7.5cm; petiole 4–9cm; inflorescence glabrous, racemose, erect, 20–45 × 0.2–0.4cm, pinkish green; raceme 9.5–20cm, loosely many-flowered; floral bracts glabrous with entire margins; flowers resupinate, distinctly honey-scented during the day; dorsal sepal pinkish red; lateral sepals pinkish white, spreading, concave; petals white; labellum white. Forest understorey and along streams; 1300m.
Distr.: SW Cameroon. (Cameroon Endemic).

IUCN: EN
Kodmin: <u>Muasya 2021</u> fl., 12/1999; **Mwendolengo:** <u>Etuge 4515r</u> fl., 11/2001.

Manniella gustavi Rchb.f.
Terrestrial herb, 50–90cm with stout fleshy roots; stem very short; leaves radical, membranous, 4–7; petiole to 15cm; blades obliquely ovate, to 16 × 7cm, usually white-spotted; inflorescence glandular-pubescent; scape to 45cm, raceme to 40cm; floral bracts pubescent-papillose with ciliate margins; flowers brownish pink; lateral sepals revolute, convex; labellum 2–3 × 3mm. Forest; 700–1350m.
Distr.: Sierra Leone to Bioko, São Tomé, Cameroon, Congo (B) & (K), Uganda, Tanzania. (Afromontane).
IUCN: LC
Bakossi F.R.: <u>Thomas 5275</u> fl., 1/1986; **Enyandong:** <u>Salazar 6300</u> 10/2001; **Kodmin:** <u>Cheek 9006</u> 1/1998; **Kupe Village:** <u>Cheek 6055B</u> fl., fr., 1/1995; <u>7167</u> fl., fr., 1/1995; <u>7770</u> 11/1995; **Nyasoso:** <u>Cable 646</u> fl., 12/1993.

Microcoelia caespitosa (Rolfe) Summerh.
Epiphyte; stem 1.0–3.1 × 0.25–5.0cm, almost leafless; roots very long, to 90cm; scale-leaves acuminate, 6–7-nerved, to 0.5–0.7cm; inflorescences up to 10 simultaneously, erect, spreading, to 3(–6.5)cm, each to 22-flowered; flowers to 1.8cm, whitish or greenish white. Forest, often near rivers; 400m.
Distr.: Sierra Leone to Uganda. (Guineo-Congolian).
IUCN: LC
Bakossi F.R.: <u>Schlechter 12779</u> 9/1899; **Mwanenguba:** <u>Schlechter 12892</u> fl., 1/1900.

Nervilia adolphi Schltr. var. *adolphi*
Terrestrial herb, 5–8cm; petiole sulcate, 3cm; blade uniformly green, reniform, very cordate at base, 3–3.5 × 4.3–6cm, appearing after and separately from inflorescence; scape 2–13cm, single-flowered; flower small, horizontal, greenish yellow or yellowish; labellum 9.5–20mm. Forest, grasslands, edge of riverine forests; 1000m.
Distr.: Nigeria, Cameroon, Tanzania, Malawi, Zambia, Zimbabwe, S Africa. (Afromontane).
IUCN: LC
Mbule: <u>Etuge 4396</u> 10/1998.

Nervilia subintegra Summerh.
Fl. Cameroun 34: 300 (1998).
Terrestrial herb, 5–13cm; leaf solitary, appearing after flowering, horizontal (not prostrate); petiole (2–)3–9cm; lamina star-shaped, (7–)9(–11)-lobed, (2.5–)4–9(–13)cm across; flower small, ± horizontal, 2.5–4.0cm across; labellum 10–17 × 3.5–7mm. Dense and humid lowland and submontane forest, forest edge, gallery forest, sometimes secondary forest and tree savanna; 850–960m.
Distr.: Guinea (C) to Cameroon, CAR, Congo (K), Sudan (Guineo-Congolian).
IUCN: LC
Enyandong: <u>Etuge 4412r</u> 10/2001; **Nyasoso:** <u>Cheek 9502</u> 10/1998.
Local name: Abati (Cheek 9502).

Oberonia disticha (Lam.) Schltr.
Fl. Cameroun 35: 378 (2001).
Dwarf epiphyte; stems clustered, leafy, 2–15cm; leaves distichous, fleshy, bilaterally compressed, 2–5 × 0.6–1cm, narrowing apically; inflorescence terminal, 4–10cm; flowers minute, yellow to orange. Forest; 1050m.

Distr.: Cameroon, Congo (K), E Africa to S Africa, Comores, Madagascar, Mascarene Is. (Tropical and subtropical Africa)..
IUCN: LC
Nyasoso: <u>Lane 39</u> fl., fr., 8/1992.

Oeceoclades maculata (Lindl.) Lindl.
Gen. & Sp. Orch.: 237 (1833).
Terrestrial herb; pseudobulbs 2–4 × 3cm, unifoliate; leaves 10–40 × 2.5–6cm, elliptic-oblong to oblanceolate-oblong, with transverse stripes and dark marbling; inflorescence lax, to 20cm, usually simple, but sometimes with 1 or 2 short branches, 6–10(–20)-flowered; flowers small; tepals translucent, greenish white, pinkish to brownish; labellum 7–10 × 7–10mm, greenish white or cream with pink nerves; spur 4–5mm, greenish white or brownish. Forest; 300m.
Distr.: tropical & subtropical Africa, naturalised in Florida, Panama, S America, W Indies. (Amphi-Atlantic).
IUCN: LC
Kupe Village: <u>Etuge 2781</u> fl., 7/1996.

Ossiculum aurantiacum P.J.Cribb & Laan
Kew Bull. 41: 824 (1986).
Epiphyte to 16cm; stem ± flattened in the plane of the leaves, to 15 × 0.5–0.6cm; leaves distichous, fleshy, 2–4.5 × 0.6–1cm; V-shaped in cross-section; inflorescences axillary, densely 9–15-flowered, to 1.8cm; flowers non-resupinate; tepals bright orange-red; lip yellow, entire, 4.5–5 × 3mm, oblong-ovate, apiculate; spur claviform, 1.5–2 × 1.5mm, ± sigmoid. Primary forest; 150m.
Distr.: W Cameroon. (Narrow Endemic).
IUCN: CR
Mungo River F.R.: <u>Beentje 1460A</u> fl., fr., 12/1980.
Note: this is the only known collection for the genus.

Platylepis glandulosa (Lindl.) Rchb.f.
Terrestrial herb, 15–50cm; stem decumbent; leaves 6–8, basal; petiole sheathing, 2–9cm; lamina obliquely ovate, 4–16 × 2–5.5cm; inflorescence racemose, densely 30–60-flowered, 5–12cm; flowers white; labellum 5.5–8.5 × 2.2–2.5mm. In shady, marshy places, on riverbanks, rocks in forests; 600m.
Distr.: Guinea (C) to Cameroon, CAR, Congo (K), Sudan, Angola, Zambia, Mozambique, S Africa. (Tropical & subtropical Africa).
IUCN: LC
Nyale: <u>Cheek 9652</u> 11/1998.

Plectrelminthus caudatus (Lindl.) Summerh.
Robust epiphyte; stem leafy, to 15cm; leaves oblong or elliptic-oblong, 10–30 × 1.5–3.5cm; inflorescence 25–60cm, 4–10-flowered; tepals yellow-green, sometimes flushed bronzen; labellum white; spur 17–25cm, twisted spirally. Forest; 850m.
Distr.: Guinea (C) to Congo (K). (Guineo-Congolian).
IUCN: LC
Nyasoso: <u>Cable 32</u> fl., 8/1992; <u>Wheatley 482</u> fl., 7/1992; <u>Zapfack 718</u> 6/1996.

Podangis dactyloceras (Rchb.f.) Schltr.
Epiphyte; stem leafy, 1–8(–11)cm; leaves 4–8, linear-ligulate, often falcate, 4–10(–16) × 0.5–1.2cm; inflorescence to 6cm, borne laterally in leaf axil, few–20-flowered; peduncle and rachis 1–4cm; flowers glistening white, semi-transparent; labellum 4–6 × 5mm; spur 9–11 × 2–3mm, inflated into 2 apical lobules. Forest; 800–1400m.
Distr.: Guinea (C) to Cameroon, Congo (K), Uganda, Tanzania, Angola, Madagascar. (Afromontane).

IUCN: LC
Bangem: Thomas 5387 1/1986; **Kodmin:** Etuge 4023
1/1998; **Nyasoso:** Zapfack 730 6/1996.

Polystachya adansoniae Rchb.f. var. *adansoniae*
Epiphyte or epilith to 20cm; pseudobulbs nearly cylindric-
conical, 2.5–9cm, 2–4-leaved; leaves linear, narrowly oblong
or linear-ligulate, 8–19 × 0.6–1.3cm; inflorescence racemose,
5–12(–20)cm, many-flowered; flowers non-resupinate, white
to greenish yellow; petals and labellum purple; labellum 1.8–
2.8 × 1.1–2.8mm; spur 1.5–2.5mm, sacciform. On trees, or
rocks, in forest; 400–1300m.
Distr.: Guinea (C) to Congo (K), E Africa, Zambia, Malawi,
Angola, Zimbabwe. (Tropical & subtropical Africa).
IUCN: LC
Kupe Village: Cable 2425 fl., 5/1996; Etuge 1488 11/1995;
1931 5/1996; Kenfack 237 7/1996; Ryan 241 5/1996; 258
5/1996; 288 5/1996; Zapfack 966 7/1996; **Nyasoso:** Cable
3544 fl., 7/1996; Etuge 2450 6/1996; Lane 125 fl., 6/1994;
Wheatley 461 fl., 7/1992; Zapfack 614 6/1996; 644 6/1996;
Tombel: Thorold CM13 fr., 3/1953.

Polystachya affinis Lindl.
Epiphyte to 50cm; pseudobulbs several, subspherical, 1–
4.8cm across, 2–3-leaved; leaves petiolate, oblanceolate,
oblong-lanceolate or oblong-elliptic, 9–28 × 2.6–6cm, shortly
acuminate; inflorescence lax, 8–60-flowered, 6–40cm;
flowers non-resupinate, white, yellow or mustard-yellow,
marked red or brown, fragrant; labellum 6.3–8 × 4.7–5.7mm;
spur to 6mm, sacciform. Forest, riverine forest; 930m.
Distr.: Guinea (C) to Uganda. (Guineo-Congolian).
IUCN: LC
Ndum: Cable 937 fl., 2/1995.

Polystachya albescens Ridl. subsp. *albescens*
Epiphyte or epilith, 15–25(–40)cm; pseudobulbs absent;
leaves linear-lanceolate, 10–13 × 1.1–1.3cm; inflorescence
racemose or weakly once-branched, 2–5cm; flowers non-
resupinate, greenish or whitish, sometimes with reddish
veins; labellum tinged or veined red, indistinctly trilobed, 6–
6.25 × 5mm; spur to 8mm, sacciform-conical. Forest,
riverine forest, sometimes on rocks; 600m.
Distr.: SE Nigeria, W Cameroon, Bioko, São Tomé,
Principé, Annobon. (W Cameroon Uplands).
IUCN: NT
Kupe Village: Cable 3862 7/1996; **Nyasoso:** Etuge 2160 fl.,
6/1996.

Polystachya albescens Ridl. subsp. *manengouba*
Sanford
Fl. Cameroun 35(2): 550 (2001).
Epiphyte; as for subsp. *albescens*, except: inflorescence 6cm
with numerous dense, short branches, multi-flowered;
labellum 9 × 8.5mm, conspicuously trilobed. Woodland; c.
2000m.
Distr.: SW Province (Mwanenguba). (Narrow Endemic).
IUCN: CR
Mwanenguba: Sanford 5557 fl., 11/1968.

Polystachya alpina Lindl.
Epiphyte to 20cm; stems swollen at base; leaves 2.5 × 0.3–
1.2cm; inflorescence lax, 1.5–6(–9)cm, 7–20-flowered;
peduncle and rachis densely pubescent; flowers small, non-
resupinate; tepals a sparkling 'crystalline-white', tinged pink;
labellum 7–9 × 4mm; spur 6mm, cylindrical-sacciform.
Forest or woodland; 1500–2300m.
Distr.: SE Nigeria, W Cameroon, Bioko. (W Cameroon
Uplands).

IUCN: NT
Kodmin: Pollard 184 11/1998; 190 11/1998; **Mwanenguba:**
Leeuwenberg 9953 fl., 6/1972; Salazar 6312 10/2001;
Sanford 5488 fr., 11/1968; 5489 fr., 11/1968; **Nyasoso:**
Cable 3380 6/1996.

Polystachya bicalcarata Kraenzl.
A small densely tufted epiphyte to 20cm; pseudobulbs 1.5–
7.5 × 0.05–0.2cm, unifoliate; leaves linear, fleshy, 4–16 ×
0.2–0.4cm, articulated 2 to 3mm above the apex of the
pseudobulb; inflorescence paniculate, 2.5–12cm, 3(–11)-
flowered, very slender, axis shortly branching; flowers non-
resupinate, borne on a tight head on each branch, white,
purple and green; mentum bifid; labellum 5–6 × 3–4mm.
Forest; 1250–1450m.
Distr.: Nigeria, W Cameroon, Bioko. (W Cameroon
Uplands).
IUCN: VU
Mwambong: Mackinder 185 1/1998; **Nyasoso:** Balding 84
fl., 8/1993.

Polystachya bifida Lindl.
Epiphyte 20–60cm; stems often pendent, 20–40cm, fusiform,
4–12-leaved; leaves linear-lanceolate, 4–21 × 0.4–1.5cm;
inflorescence lax, unbranched, 4–20cm; 3–25-flowered;
peduncle and rachis glabrous; flowers non-resupinate, white;
dorsal sepal purple; labellum 7–8 × 3.5–4mm, ± rhombiform;
spur to 7mm, conical-sacciform. Forest, marshy forest; 1100–
1560m.
Distr.: SE Nigeria, Cameroon, Bioko, São Tomé, Gabon,
Congo (K), Rwanda. (Lower Guinea & Congolian).
IUCN: LC
Kodmin: Cheek 9527 10/1998; Etuge 4008 1/1998; 4400
11/1998; Gosline 71 1/1998; Nwaga 20 fr., 1/1998; Onana
597 2/1998; Pollard 179 11/1998; **Kupe Village:** Cheek 7877
11/1995; **Nyasoso:** Etuge 2116 6/1996; 2469 6/1996.

Polystachya calluniflora Kraenzl.
Epiphyte, 2–30cm; stems arching; pseudobulbs superposed,
fusiform, (2.5–)4–7(–11.5) × 0.3–0.5cm; bifoliate; leaves
linear, grass-like, 4.5–20 × 0.5–0.8cm; inflorescence erect,
racemose, 4–10cm, 5–30-flowered; flowers white, non-
resupinate; labellum and dorsal sepal purple, but whole
flower drying orange; labellum 2.5–4 × 1.8–2mm; spur 0.5–
1mm, sacciform. Forest; 400–1700m.
Distr.: Nigeria, Cameroon, Bioko, Rwanda, Uganda. (Lower
Guinea & Congolian).
IUCN: LC
Kodmin: Etuge 4166 2/1998; **Kupe Village:** Etuge 1490
11/1995; **Menyum:** Doumenge 511 5/1987; **Mwanenguba:**
Sanford 5466 fr., 11/1968; **Nyasoso:** Cable 3408 6/1996; 57
fl., 8/1992; Etuge 1333 10/1995; 2466 6/1996; Lane 20 fr.,
7/1992; 70 fl., 8/1992; 77 fl., 9/1992; Sunderland 1528 fl., fr.,
7/1992; Wheatley 421 fl., 7/1992; Zapfack 727 6/1996.

Polystachya caloglossa Rchb.f.
Epiphyte 10–70cm; without pseudobulbs; leaves 3–8,
obovate or oblanceolate, undulate, 5–18 × 1.3–6.5cm;
inflorescence terminal, simply racemose or with 1–5
branches, 4–12cm; rachises markedly zigzag with bracts
arranged distichously at the angles; flowers non-resupinate,
yellow or orange; labellum 9–11 × 10–12mm; spur to 8mm;
capsule ± 3cm. Forest, gallery forest; 700–1650m.
Distr.: Cameroon, Bioko, Gabon, Congo (K), Rwanda,
Burundi, Uganda. (Lower Guinea & Congolian).
IUCN: LC
Enyandong: Simo 32 10/2001; **Kodmin:** Pollard 229
11/1998; **Kupe Village:** Cable 2619 5/1996; Etuge 1440

11/1995; <u>1969</u> 5/1996; <u>2818</u> 7/1996; <u>Kenfack 235</u> 7/1996; <u>332</u> 7/1996; **Mwambong:** <u>Gosline 87</u> 2/1998; **Mwanenguba:** <u>Sanford 5458</u> fl., 11/1968; **Nyasoso:** <u>Cable 2778</u> 6/1996; <u>2943</u> 6/1996; <u>60</u> fl., fr., 8/1992; <u>Etuge 2109</u> 6/1996; <u>2445</u> 6/1996; <u>2590</u> 7/1996; <u>Zapfack 613</u> 6/1996; <u>615</u> 6/1996.

Polystachya cooperi Summerh.
Epiphyte to 20cm; without pseudobulbs; leaves 5 or 6, elliptic or lanceolate-elliptic, 4–15 × 2–3cm; inflorescence subdense, few-flowered, 4.5–5cm, glabrous; flowers large, to 2cm across, greenish white with purple striations; labellum to 1.6 × 1.2cm; spur 11mm, conical, obtuse. Forest and woodland; 1400–1650m.
Distr.: SE Nigeria, W Cameroon. (W Cameroon Uplands).
IUCN: EN
Mwanenguba: <u>Sanford 5464</u> fl., 11/1968; **Nyasoso:** <u>Balding 43</u> fl., 8/1993.

Polystachya cultriformis (Thouars) Spreng.
Epiphyte 6–40cm; pseudobulbs unifoliate, narrowly cylindric to conical, 1.4–18 × 0.2–1.2cm; leaf obovate, ovate or elliptic, auriculate at base, 3.2–36 × 1.2–5.5cm, ± undulate; inflorescence terminal, branched, 4.4–29cm, bearing up to 60 flowers successively; flowers non-resupinate, very variable in size and colour; labellum 4–7.8 × 3–6mm; spur 4–7mm, sacciform. Forest; 720–1600m.
Distr.: Ivory Coast, Cameroon to Congo (K), Burundi, E Africa to S Africa, Madagascar, Mascarene Is., Seychelles. (Tropical & subtropical Africa).
IUCN: LC
Kodmin: <u>Biye 43</u> 1/1998; <u>Etuge 4038</u> 1/1998; <u>Gosline 74</u> 1/1998; <u>Nwaga 14</u> fr., 1/1998; <u>Satabie 1127</u> fr., 2/1998; **Kupe Village:** <u>Etuge 1943</u> 5/1996; <u>Kenfack 333</u> 7/1996; **Mejelet-Ehumseh:** <u>Etuge 297</u> 10/1986; **Nyasoso:** <u>Etuge 2196</u> 6/1996; <u>Lane 74</u> fl., 9/1992; <u>Pollard 575</u> fl., 9/2001.

Polystachya elegans Rchb.f.
Epiphyte 20–30(–40)cm; pseudobulbs superposed, 7–15 × 0.5–0.7cm, 3–5-leaved; leaves strap-shaped or lanceolate-oblong, 5–22 × 0.5–1.6(–2.5)cm, borne apically; inflorescence 6–15cm, branched basally, many-flowered; flowers non-resupinate, greenish white or yellowish white, tinged purple; labellum 4.6–5.9 × 2.5mm; spur 2.8mm, clylindric-sacciform. Forest; 830–1550m.
Distr.: Nigeria, W Cameroon, Bioko, Congo (K). (Lower Guinea & Congolian).
IUCN: LC
Enyandong: <u>Salazar 6317</u> fl., fr., 10/2001; **Kodmin:** <u>Pollard 181</u> 11/1998; **Nyasoso:** <u>Balding 83</u> fl., 8/1993; <u>Cable 126</u> fl., 9/1992; <u>56</u> fl., 8/1992.

Polystachya farinosa Kraenzl.
Fl. Cameroun 35: 536 (2001).
Epiphyte to 60cm; without pseudobulbs; leaves 7–11, to 19 × 0.9cm, linear to linear-lanceolate; inflorescence simple, dense, to 16cm, 15–30-flowered; flowers non-resupinate, white; labellum deeply trilobed, 8–8.5 × 3.2–3.5mm; spur 7–8.2mm. Submontane forest; 1650m.
Distr.: Cameroon, São Tomé. (W Cameroon Uplands).
IUCN: EN
Mwanenguba: <u>Sanford 5463</u> fl., 11/1968.
Note: Summerhayes (1968) and Geerinck (1992) consider this taxon to be conspecific with *P. bifida*, but here we follow Szlachetko & Olszewski (2001) in separating the two. They differ in *P. farinosa* having a labellum deeply lobed in the upper half (not ± rhombiform), the spur 7–8.2mm (not 7mm).

Polystachya fractiflexa Summerh.
Epiphyte, 10–60cm; stem (5–)18–30(–45) × 0.3–0.5cm, without pseudobulbs; leaves 3–5, oblanceolate or elliptic-oblanceolate, 6–12 × 0.8–2.3cm, margins not undulate; inflorescence terminal, 3–11cm, densely branched; branches 5–8, to 2cm, each 8–14-flowered; rachis of branches markedly zigzag with bracts arranged distichously at the angles; flowers small, non-resupinate, yellowish green tinged purple-red, cream or orange-brown; labellum 5–8 × 6–8mm; capsules to 4cm. Forest; 350–1500m.
Distr.: Ivory Coast to Congo (K). (Guineo-Congolian).
IUCN: LC
Kodmin: <u>Pollard 183</u> fl., 11/1998; **Kupe Village:** <u>Cable 2422</u> 5/1996; <u>Zapfack 965</u> fr., 7/1996; <u>969</u> fr., 7/1996; **Mungo River F.R.:** <u>Cheek 10215</u> 11/1999; **Nyasoso:** <u>Cable 58</u> fl., fr., 8/1992; **Tombel:** <u>Thorold TN37</u> fr., 11/1950.

Polystachya fulvilabia Schltr.
Fl. Cameroun 35: 517 (2001).
Epiphyte 14–38cm; pseudobulbs unifoliate, loosely clustered from a short creeping rhizome, 5.2–9.5 × 0.2–1.0cm; leaf linear-lanceolate, base cuneate and articulated, 7.2–20 × 0.6–1.2cm; inflorescence a simple or rarely-branched lax raceme, 5.5–14.5cm, pubescent, to 20-flowered; peduncle and rachis densely pubescent; flowers non-resupinate, to 1cm across, yellow, yellowish green or greenish white, marbled violet; labellum 7.5–10 × 3.6–5mm; spur 6.5–8mm, conical. Montane forest, woodland; 660–1650m.
Distr.: Nigeria to Uganda. (Lower Guinea & Congolian).
IUCN: LC
Enyandong: <u>Simo 24</u> fr., 10/2001; **Kupe Village:** <u>Cable 2421</u> 5/1996; <u>Cheek 8340</u> 5/1996; <u>Etuge 1445</u> 11/1995; <u>2679</u> 7/1996; <u>Kenfack 225</u> 7/1996; <u>Zapfack 971</u> 7/1996; **Mwanenguba:** <u>Sanford 5465</u> fl., 11/1968; **Ndum:** <u>Cable 964</u> 2/1995; **Nyasoso:** <u>Etuge 2452</u> 6/1996; <u>2602</u> 7/1996; <u>Zapfack 618</u> 6/1996.

Polystachya fusiformis (Thouars) Lindl.
Suberect or pendent epiphyte or epilith to ± 60cm; stems (pseudobulbs) cylindrical to fusiform, superposed, longitudinally ridged, drying yellow, to 22 × 0.3–0.4cm; leaves 3–7, oblong-lanceolate, lanceolate to oblanceolate, 5–16 × 0.6–1.6(–3.2)cm, largest borne apically; inflorescence terminal, paniculate, 3–8(–15)cm, densely 20–80-flowered; peduncle pubescent; flowers miniscule, non-resupinate, persistent on developed ovary, cream, yellow-green tinged or entirely purple or mauve; pedicel and ovary 4mm, glabrous; labellum 2–2.5 × 2.5mm; spur 1mm, sacciform. Forest; 770–1960m.
Distr.: Ghana, Cameroon, Bioko, Congo (K), Rwanda, Burundi, E Africa, Zambia, Malawi, Zimbabwe, S Africa, Mascarene Is. (Afromontane).
IUCN: LC
Kodmin: <u>Onana 595</u> 2/1998; <u>Pollard 231</u> 11/1998; **Kupe Village:** <u>Ryan 376</u> 29/1996; <u>Schoenenberger 39</u> fl., 11/1995; **Mwambong:** <u>Mackinder 182</u> 1/1998; **Nyasoso:** <u>Zapfack 731</u> 6/1996.

Polystachya geniculata Summerh.
Terrestrial herb; stem 27–53cm; pseudobulb c. 3 × 0.6–0.7cm; leaves 2 or 3, 14–16 × 0.6–0.8cm, linear; inflorescence quite dense, 5–10cm, many-flowered, unbranched, silky; flowers purpurescent or mauve and cream or yellow; labellum 7–8.5 × 1.9–2.7mm; spur 3.7–4.7mm, cylindrical. Terrestrial in swampy grasslands or on granitic rocks; 1850m.
Distr.: SW Cameroon. (Cameroon Endemic).

ORCHIDACEAE

IUCN: EN
Mwwanenguba: Gregory 302 fl., 6/1948.

Polystachya golungensis Rchb.f.
Epiphyte or epilith, 10–47cm; stems densely clustered, obscurely pseudobulbous; pseudobulbs 2–6 × 0.4–0.7cm leaves 2–4, ligulate, conduplicate, 5–17(–28) × 0.3–2.5cm; inflorescence paniculate, 6–40cm, much > leaves; 50–200-flowered, with up to 8 secund branches, each to 5cm; peduncle and rachis pubescent; flowers non-resupinate, yellow, yellow-green or brownish cream; labellum 2.2–3.0 × 1.7–2.5mm; spur 2mm, sacciform-conical. On trees or rocks in riverine forest, rainforest or drier scrub; 910m.
Distr.: Mali, Liberia to Cameroon, Congo (K), E Africa, S to Angola, Zimbabwe, Mozambique. (Tropical & subtropical Africa).
IUCN: LC
Nyasoso: Zapfack 653 6/1996.

Polystachya kupensis P.J.Cribb & B.J.Pollard
Kew Bull. 57: 656 (2002).
Pendent epiphyte; stem very short, leafy; leaves 5, linear-lanceolate to oblanceolate, to 20 × 1.4cm; midvein prominent when dried; inflorescence to 15.5cm, racemose, once-branched, lax, to 12-flowered; peduncle pubescent; flowers non-resupinate, purple, fleshy; pedicel and ovary 6–8mm; labellum 3-lobed, 4 × 2mm, densely pubescent in basal part. Submontane forest; 1050m.
Distr.: W Cameroon (Mt Kupe). (Narrow Endemic).
IUCN: CR
Kupe Village: Cable 2521 5/1996.
Note: known only from this collection.

Polystachya laxiflora Lindl.
Epiphytic herb; stems (6–)15–30(–50) × 0.3–0.6cm, without pseudobulbs; leaves 2–6, (8–)10–26 × (2–)3–4(–6)cm, elliptic to oblanceolate; inflorescences (4–)8–30cm, ± branched, each branch dense, to 2–5(–12)cm, to 13-flowered; flowers white, yellow or orange-yellow, sometimes with red markings on the sides of the labellum; labellum 7–10 × 6.5–10mm, ± pubescent on the central vein. Forest, woodland, plantations; 200m.
Distr.: Guinea (C) to Congo (K), Zambia. (Guineo-Congolian).
IUCN: LC
Ikiliwindi: Nemba 566 fl., 6/1987.

Polystachya mukandaensis De Wild.
Epiphyte, 25–60cm; stems obscurely pseudobulbous, pseudobulbs 1.5–2.5 × 0.5–1.0cm, 2–6-leaved; leaves oblanceolate or linear-lanceolate, 12–33 × 1–3.5cm, acute; veins prominent beneath; inflorescence dense, 50–200-flowered, paniculate, 10–45cm, simple or to 4-branched, each branch to 4cm, secund; peduncle and rachis subdensely pubescent; flowers small, non-resupinate, brownish green, ± marked purple; labellum pale greenish yellow, 5.4–7 × 4.3–5mm; spur 7mm, conical-sacciform. On trunks and larger branches in riverine forest, in rain-forest elsewhere; 50m.
Distr.: Ivory Coast, Ghana, Nigeria, W Cameroon, Congo (K), Uganda, Angola. (Guineo-Congolian).
IUCN: LC
Mungo River F.R.: Leeuwenberg 8695 fr., 11/1971.

Polystachya odorata Lindl. var. *odorata*
Epiphyte or rarely epilith, 20–60cm; pseudobulbs 2–4.5 × 0.6–1.5cm; leaves 4–8, oblanceolate to oblong-elliptic, 8–26 × 2.8–5.5cm, prominently veined beneath; inflorescence paniculate, 10–30cm, conspicuously branched; branches 6–15, puberulous, many-flowered, to 8cm; flowers non-resupinate, white, yellow, dull red-brown or pale green; labellum 6–7(–8) × 5–7.5mm; spur 4–5mm, sacciform. Forest; 350–400m.
Distr.: Ivory Coast, Burkina Faso, Ghana, Nigeria to Tanzania, Angola. (Guineo-Congolian).
IUCN: LC
Kupe Village: Etuge 2008 5/1996; 2035 5/1996.

Polystachya paniculata (Sw.) Rolfe
Epiphyte 22–45cm; pseudobulbs clustered, 4–18 × 1–2.2cm; leaves 3–7, distichous, lanceolate, oblong-elliptic or ligulate, 8–33 × 2–4.5cm; inflorescence 6–21cm, a many-flowered raceme, to 15-branched, each to 6cm; peduncle and rachis glabrous; flowers non-resupinate, flame red or a wondrous bright orange; labellum 2.8–4 × 1.5–2mm; spur 1.4mm, sacciform. Forest; 350–1100m.
Distr.: Guinea (C) to Uganda. (Guineo-Congolian).
IUCN: LC
Kupe Village: Kenfack 233 7/1996; Zapfack 973 7/1996; **Nyasoso:** Cable 13 fl., 7/1992; Etuge 2572 7/1996; Sunderland 1503 fl., fr., 7/1992; Zapfack 719 6/1996; **Tombel:** Thorold 17 fr., 3/1953.

Polystachya polychaete Kraenzl.
Stout epiphyte, 15–50cm; pseudobulbs 8–10 × 1cm, 3–7-foliate; leaves ligulate, (6–)12–18(–30) × 0.8–2.5cm, apex conspicuously bilobed; inflorescence a dense unbranched terminal raceme, 10–26cm, densely many-flowered; rachis ciliate; flowers miniscule, non-resupinate, yellow, greenish yellow, white, greenish white or cream; labellum 1.2–2.2 × 1.5–2.2mm; spur 1.3–2mm, sacciform-conical. Forest; 770–1500m.
Distr.: Sierra Leone to Rwanda, Uganda, E Africa. (Guineo-Congolian).
IUCN: LC
Enyandong: Simo 8 fl., 10/2001; **Kodmin:** Pollard 210 11/1998; **Kupe Village:** Cable 2617 5/1996; Etuge 1501 11/1995; 2629 7/1996; 2865 7/1996; Ryan 273 5/1996; Schoenenberger 38 fl., 11/1995; Williams 226 11/1995; **Ngomboaku:** Cheek 10337 12/1999; **Nyasoso:** Cable 3423 7/1996; Etuge 1331 10/1995; 2400 6/1996; Lane 17 fl., 7/1992; Wheatley 420A fl., fr., 7/1992.

Polystachya ramulosa Lindl.
Epiphyte, 8–34cm; pseudobulbs 2.5 × 0.3–0.5cm, ovoid, 3–5-foliate; leaves 5.5–16 × 2.6cm, elliptic, oblanceolate or oblong-lanceolate; inflorescence dense, 5–31cm, 25–100-flowered, to 10-branched, each to 5cm; flowers non-resupinate, yellow, orange, greenish yellow; labellum 2.6–2.8 × 1.6–1.7mm, clear pink; spur 1.6–2.5mm. Forest, swampy Forest; 860m.
Distr.: Sierra Leone to Tanzania. (Guineo-Congolian).
IUCN: LC
Enyandong: Simo 12 fr., 10/2001.

Polystachya rhodoptera Rchb.f.
Epiphyte to 50cm; without pseudobulbs; leaves 3–9, 5–20 × 0.3–2.5cm, lanceolate to oblong, acute; inflorescence dense, to 10.5cm, 8–40-flowered, simple or 1–2-branched; branches very short; peduncle and rachis densely pubescent; flowers non-resupinate, white or yellow, often pink-tinged; labellum 5–6.6 × 4–5.5mm; spur 4–5mm, conical. Forest, shaded branches above water; 700–1200m.
Distr.: Sierra Leone to Congo (K). (Guineo-Congolian).
IUCN: LC
Enyandong: Simo 29 fl., 10/2001; **Kupe Village:** Etuge 1457 fl., 11/1995.

Polystachya stauroglossa Kraenzl.

F.T.E.A. Orchidaceae 2: 359 (1984).
Epiphyte to 50cm; stems clustered, 7–50 × 0.1–0.4cm, the apical two-thirds 4–7(–12)-leaved; leaves 10–25 × 0.5–2.5cm, linear to linear-lanceolate; inflorescence a dense terminal panicle, 8–20cm, to 100-flowered, with 1–5 simple branches, which may branch again, to 9 × 1cm; peduncle and rachis glabrous; flowers small, non-resupinate, white or yellow-green to marbled violet, or a waxy pink; labellum 2–2.8 × 2.5–3.3mm; spur 2.6mm, subglobular. Forest; 1500m.
Distr.: W Cameroon, E Congo (K), Uganda, Tanzania (Afromontane).
IUCN: NT
Kodmin: Pollard 175 fl., 11/1998.
Note: this is the only record from West-Central Africa.

Polystachya supfiana Schltr.

Epiphyte, 10–20cm, drying ± black; pseudobulbs very narrowly clylindrical, 1–3.5(–10) × 0.1–0.3cm, unifoliate; leaves linear-oblong to oblanceolate, 5–16 × 0.6–1.5cm; inflorescence terminal, rarely exceeding lamina, loose, c. 1cm, 2–4-flowered; flowers non-resupinate, yellow, marked brown; labellum 8 × 3mm; spur c. 10mm. Forest; 710–1250m.
Distr.: SE Nigeria, Cameroon, Gabon. (Lower Guinea).
IUCN: NT
Kupe Village: Cable 3655 fr., 7/1996; Etuge 2667 fr., 7/1996; **Mwambong:** Mackinder 186 fl., 1/1998; **Nyale:** Etuge 4174 fr., 2/1998.

Polystachya tenuissima Kraenzl.

Syn.: *Polystachya inconspicua* Rendle
Epiphyte to 40cm; pseudobulbs 0.7–10cm × 0.6–3mm, closely grouped, unifoliate; leaf 2–14.5(–33) × 0.2–0.7cm, linear; inflorescence dense, 1.7–30cm, few–40-flowered, 1–6-branched; peduncle and rachis glabrous; flowers small, non-resupinate, mauve, yellow or green-yellow; mentum conspicuously 2-pronged; labellum 2.5–6.5 × 1.7–4mm, the lateral lobes stained brownish purple or brownish red; spur 2(–5)mm, cylindrical or cylindric-claviform. Forest, cocoa plantations; 1000–2300m.
Distr.: Ivory Coast, Ghana, Cameroon, Congo (K), E Africa. (Afromontane).
IUCN: LC
Menyum: Doumenge 500 fl., 5/1987; **Mwanenguba:** Salazar 6311 fl., fr., 10/2001.

Polystachya tessellata Lindl.

Syn.: *Polystachya concreta* (Jacq.) Garay & H.R.Sweet
Epiphyte, (10–)20–60cm; pseudobulbs 15 × 0.5–0.7cm, 3–5-foliate; leaves oblanceolate or elliptic, (3–)10–30 × 0.8–6.0cm; inflorescence paniculate, 10–50cm; branches secund, distant, densely 20–200-flowered; rachis and peduncle covered in sheaths; flowers small, non-resupinate, cream, yellow, clear green or red-purple. Forest, savanna or woodland; 400–1450m.
Distr.: tropical & subtropical Africa.
IUCN: LC
Enyandong: Simo 21 10/2001; **Kodmin:** Gosline 72 1/1998; **Kupe Village:** Cable 2423 5/1996; 3669 7/1996; Etuge 1489 11/1995; 2871 7/1996; Ryan 286 5/1996; Zapfack 963 7/1996; 964 7/1996; **Mwanenguba:** Sanford 5474 fl., 11/1968; **Nyasoso:** Cable 3443 7/1996; 47 fl., 8/1992; 55 8/1992; Etuge 1330 10/1995; Zapfack 617 6/1996; 647 6/1996.

Rangaeris muscicola (Rchb.f.) Summerh.

Epiphyte or epilith; stem very short, 1–6cm; leaves arranged in a fan, conduplicate, 5–11, linear, 6.5–20 × 0.6–1.3cm; inflorescence 5–22(–42)cm, 5–16-flowered; bracts amplexicaul; flowers fragrant, white; labellum 6.7–8.5 × 4–7mm; spur pinkish olive, 5.5–7cm. Forest; 910–1500m.
Distr.: Guinea (C) to Cameroon, Congo (K), E Africa to S Africa. (Tropical & subtropical Africa).
IUCN: LC
Kodmin: Pollard 189 fr., 11/1998; **Kupe Village:** Zapfack 946 7/1996; **Nyasoso:** Zapfack 649 6/1996.

Rangaeris rhipsalisocia (Rchb.f.) Summerh.

Epiphyte; stem covered with persistent leaf bases, 5–25cm; leaves iridiform, coriaceous, distichous, conduplicate, acute and entire at apex, to 8–20 × 0.3–0.8cm; inflorescence basal, lax, 2.5–20cm, 6–15-flowered; flowers small, tubular, resupinate, fragrant at night, white or cream; labellum 7–8 × 4–4.5mm; spur greenish, 12–15mm. Forest, forestry exploitations, coffee plantations, very tolerant of dessicant conditions; 910m.
Distr.: Senegal to Cameroon, Congo (K), Angola. (Guineo-Congolian).
IUCN: LC
Nyasoso: Zapfack 648 6/1996.

Satyrium crassicaule Rendle

Terrestrial herb, 0.3–1.2m; stem slender to robust, leafy; leaves 8–13, lower ones broadly lanceolate to ligulate, to 8–48 × 2.5(–7.5)cm, the upper 3–5, smaller, appressed to stem; inflorescence 5–37 × 2–3.5cm, densely many-flowered; flowers non-resupinate, pink to mauve, rarely white; labellum 5–7.5 × 6–7.5mm; spurs 2, 8–13mm. Damp grassland or swamps, especially by streams, sometimes in running water; grassland at edge of or in glades in montane forest; 1200–2050m.
Distr.: Nigeria, Cameroon, Congo (K), Rwanda, Burundi, Ethiopia, E Africa, S to Zambia, Malawi, Zimbabwe. (Afromontane).
IUCN: LC
Lake Edib: Cheek 9136 2/1998; Pollard 240 11/1998; **Mwanenguba:** Sanford 5515 fl., 11/1968.

Satyrium volkensii Schltr.

Terrestrial herb; sterile stems to 5cm, with 3–4 leaves, 5–18 × 2–6cm, lanceolate to oblong-elliptic; fertile stem 20–110cm, entirely covered with imbricate cauline bracts which are 7–15, 2.5–11 × 1–3cm; inflorescence lax, 5–42cm, several-flowered; flowers non-resupinate, green to yellowish green; spurs 2, 11–23mm. Grassland; c. 2000m.
Distr.: Nigeria, Cameroon, Congo (K), Kenya, Tanzania, Malawi, Zambia, Zimbabwe. (Afromontane).
IUCN: LC
Mwanenguba: Nditapah 380 fl., 4/year unknown.

Solenangis clavata (Rolfe) Schltr.

Epiphyte; roots long, produced laterally along stem; stem to 100cm, climbing; internodes 1.2–2 × 0.3cm; leaves numerous, 2.5–5 × 0.7–2cm, oblong-elliptic to oblong-ovate; inflorescence ± lax, 0.5–1.5(–2)cm, 6–10(–14)-flowered; flowers small, resupinate, whitish or greenish; labellum white, 2 × 2mm; spur 5–10 × 1.2–2mm, pendent, oblong-claviform. Forest, swamp forest, coffee and avocado plantations; c. 500m.
Distr.: Liberia to Congo (K), Rwanda. (Guineo-Congolian).
IUCN: LC
Tombel: Olorunfemi FHI 30602 fl., 5/1951.

ORCHIDACEAE

Solenangis scandens (Schltr.) Schltr.

Epiphyte; roots all along stem; stem climbing, internodes 1.2–2 × 0.3cm; leaves numerous, elliptic-lanceolate, 3–7.5 × 1.2–2.5cm; inflorescence axillary, lax, 2–10cm, 3–15-flowered; flowers small, resupinate, white, greenish white, greenish yellow or pinkish; sepals and petals 0.5–0.7cm; labellum 6–8 × 2.5–4mm, entire; spur 2–2.5cm. Forest, swamp-forest, plantations, always at the extremities of branches, tolerant of light, most often above a humid soil; c. 900m.

Distr.: Sierra Leone to Congo (K). (Guineo-Congolian).

IUCN: LC

Kupe Village: Zapfack 972 7/1996.

Stolzia peperomioides (Kraenzl.) Summerh.

Kew Bull. 8: 142 (1953).

Creeping dwarf epiphyte to 10cm high (when generative), or 4cm (vegetative); pseudobulbs 1.7cm apart, 0.5 × 0.3cm, more swollen below the leaf; leaves solitary; petiole as long as blade; lamina elliptic, 3.7 × 1.5cm; inflorescence with a long several-flowered peduncle; flowers brown. Forest; 850m.

Distr.: W Cameroon, E Congo (K). (Lower Guinea & Congolian (montane)).

IUCN: NT

Nyasoso: Etuge 2582 7/1996.

Stolzia repens (Rolfe) Summerh. var. repens

Creeping dwarf epiphytic herb to 1cm high; pseudobulbs prostrate except at apex, 1.3–4cm apart, to 3 × 0.3cm, with 2 leaves near insertion of next pseudobulb; leaves subsessile, 0.5–1.4 × 0.3–0.8cm; peduncle with a single flower; flowers yellow, brown or reddish, striped red or brown. Forest, woodland; 1000–1300m.

Distr.: Ghana, S Nigeria, W Cameroon, Congo (K), Ethiopia, E Africa, Malawi, Zambia, Zimbabwe. (Afromontane).

IUCN: LC

Kupe Village: Cable 3839 fl., 7/1996; **Menyum:** Doumenge 503 5/1987; Thomas 7130 fl., 5/1987; **Nyasoso:** Lane 126 fl., 6/1994.

Stolzia sp. nov.

Dwarf epiphyte; rhizome creeping; leaves arising in pairs from the nodes, c. 6 × 5mm, ± orbicular, apex minutely emarginate-mucronulate, semi-translucent, with c. 5 conspicuous arching veins; inflorescence appearing at same nodes as leaves, c. 11mm, 2-flowered; bracts c. 2mm; complete flowers unknown; labellum papillose; capsule c. 5 × 2mm. Forest; 1760m.

Distr.: Bakossi Mts. (Narrow Endemic).

Enyandong: Pollard 906 fr., 11/2001.

Note: even without flowers, we are sure that this collection does not match any of the taxa in Cribb's revision (1978), but complete flowering material is needed in order to describe it fully.

Tridactyle anthomaniaca (Rchb.f.) Summerh.

Epiphyte; stems semi-pendent to 2m × 0.5–0.6cm; leaves numerous, fleshy-coriaceous, linear to narrowly elliptic-oblong, 3.5–11 × 0.6–1.9cm, shiny above, matt below; inflorescence very short, to 1cm, 2–4-flowered; flowers small, resupinate, green, pale green, yellow or white; labellum 3–6 × 1–2mm; spur filiform, (6–)11–16mm. Riverine, swamp and lower montane forest, plantations (coffee, cocoa, guava), often above water, or in sunny positions; 600–1500m.

Distr.: Sierra Leone to Congo (K), CAR, E Africa, S to Mozambique, Malawi, Zambia, Zimbabwe. (Afromontane).

IUCN: LC

Enyandong: Pollard 914 11/2001; **Kodmin:** Pollard 182 11/1998; **Mwanenguba:** Sanford 5468 fl., 11/1968; 5491 fl., 11/1968; **Nyale:** Pollard 221 11/1998.

Tridactyle bicaudata (Lindl.) Schltr.

Epiphyte; stems pendent to 0.1–0.8m × 0.35–0.6cm, usually several-branched; leaves numerous, linear, distichous, 9–17.5 × 0.5–1.3cm; inflorescence axillary, lax, 3.5–13cm, 8–25-flowered; flowers small to medium, resupinate, white, pink or yellow, often tinged green; labellum 3–6.5 × 8–11.5mm, trilobed, lateral lobes to 7mm, much longer than mid-lobe, apically divided; spur 11–17mm. Forest, woodland, riverine forest, plantations, secondary forest; 1340–1950m.

Distr.: Sierra Leone to Cameroon, to Ethiopia, E Africa, S to Zimbabwe, S Africa. (Tropical and subtropical Africa)..

IUCN: LC

Kodmin: Etuge 4022 1/1998; Pollard 206 11/1998; **Mt Kupe:** Sebsebe 5074 11/1995.

Tridactyle tridactylites (Rolfe) Schltr.

Epiphyte or epilith; stems pendent, 0.4 to 1.6m, robust; leaves numerous, 6–21 × 0.6–1.3cm, linear or linear-lanceolate; inflorescence lax, 1.7–10cm, to 18-flowered; flowers small, resupinate, yellow, orange or brownish orange, sometimes tinged green, fragrant; labellum 4–5 × 6.5mm, side lobes ± equal in length to mid-lobe, entire or rarely slightly bifid; spur 6–11mm. Forest; 1000–1500m.

Distr.: Sierra Leone to Congo (K), E Africa, S to Mozambique, Malawi, Zambia, Angola, Zimbabwe. (Afromontane).

IUCN: LC

Kodmin: Etuge 4114 2/1998; Pollard 219 11/1998; **Ndum:** Cable 911 fl., 2/1995; Groves 56 fl., 2/1995.

Tridactyle tridentata (Harv.) Schltr.

Epiphyte or epilith; stem 10–50 × 0.2–0.5cm, pendent or arcuate, usually branched; leaves numerous, 6–10.5 × 0.1–0.35cm; inflorescence lax, (0.5–)1.8–2.7cm, 4–5-flowered; flowers small, whitish, pale ochre-yellow or salmon-pink; labellum 2.1–5 × 1.4–4mm, auriculate at base, weakly 3-lobed; spur 6–18mm, filiform, straight or incurved. Forest, secondary forest; branches of exposed trees; 760m.

Distr.: Cameroon, CAR, Congo (B) & (K), Uganda, Tanzania, Mozambique, Malawi, Zimbabwe, S Africa. (Afromontane).

IUCN: LC

Ngombombeng: Etuge 26 4/1986.

Vanilla ramosa Rolfe

Climbing or epiphytic herb; internodes 5.5–13 × 1.5–7cm; leaves shortly petiolate; petiole 1–1.5cm; lamina 8–17.5 × 1.5–6.5cm, oblong-elliptic, oblong-ovate to linear-lanceolate; inflorescence dense, usually axillary and branched basally, 2.5–8cm, 12–40-flowered; tepals ± 4 × longer than wide; labellum 14–25 × 22–27mm; median lobe oblong-elliptic. Forest; c. 850m.

Distr.: Ghana to Tanzania. (Guineo-Congolian).

IUCN: LC

Enyandong: Salazar 6334 11/2001; 6335 11/2001.

Zeuxine elongata Rolfe

Terrestrial herb, 25–45cm; stem very delicate; leaves 4–7, aggregated basally; petiole and sheath 5–15mm; lamina membranous, ovate-lanceolate, acute, to 2.5–5.5(–7) × 1–2(–3)cm, often drying pinkish; inflorescence a slender raceme, 6–13cm, 13–25-flowered; flowers small, semi-campanulate, resupinate, greenish white, sometimes tinged pink, hardly opening; tepals to 2.5mm; labellum with involute margins. In dense shade, in humus and on rocks of rain-forest floor, by rivers and waterfalls; 300–1000m.

Distr.: Sierra Leone to Cameroon, Congo (K), CAR, E Africa, Zambia, Angola. (Afromontane).

IUCN: LC

Kupe Village: Cable 713 fl., 1/1995; **Mekom:** Thomas 5404 fl., 1/1986; **Ngusi:** Etuge 1559 1/1996; **Nyasoso:** Cable 1124 fl., 2/1995; 1209 fl., 2/1995; 870 fl., 1/1995.

Zeuxine gilgiana Kraenzl. & Schltr.

Fl. Cameroun 34: 258 (1998).

Terrestrial herb; stem to 26cm, erect, delicate; leaves 3–5, grouped basally; petiole and sheath 9–15mm; lamina 3.3–3.5 × 1.2–2.1cm, obliquely ovate-lanceolate; bracts cauline, 2–3, to 14mm; inflorescence 8–10cm, 20–30-flowered, lax below, dense above; flowers small, resupinate; perianth green and pink; labellum orange to reddish orange, with involute margins. Forest, gallery forest; 850m.

Distr.: Cameroon to Congo (K). (Lower Guinea & Congolian).

IUCN: LC

Nyasoso: Cable 1251 fr., 2/1995.

Zeuxine heterosepala (Rchb.f.) Geerinck

Bull. Jard. Bot. Nat. Belg. 50: 120 (1980).

Syn.: *Hetaeria heterosepala* (Rchb.f.) Summerh.

Slender terrestrial herb to 20cm; leaves 5–6(–10), aggregated in middle of stem, lanceolate, acute, subcordate or rounded at base, to 3.5–5.5 × 1.2–1.8cm, petiolate, sheathing the stem; inflorescence racemose, very dense, 1.5–2(–5)cm, 7–10(–20)-flowered; flowers small, resupinate, green and white, sometimes flushed purple; labellum with 2 conspicuous large lateral lobes. Forest; 860–1100m.

Distr.: Liberia, Ivory Coast, Cameroon, São Tomé, Congo (K), Tanzania. (Guineo-Congolian (montane)).

IUCN: LC

Enyandong: Salazar 6324 fl., 11/2001; **Kupe Village:** Cheek 7813 11/1995; 7856 11/1995; Williams 228 11/1995; **Nyasoso:** Cheek 9520 10/1998.

Local name: Mimo-koo ≡'The eyes of a chicken'. **Uses:** MEDICINES – used in a combination of 3 leaves (Epie Ngome in Cheek 9520).

PALMAE

T.C.H. Sunderland, W.J. Baker, J. Dransfield & M. Cheek

Elaeis guineensis Jacq.

Single-stemmed tree to 20m; leaves crowded, pinnately-compound, to 5m, arching, basal leaflets modified as spines; inflorescences partially hidden at leaf bases; fruits oblong-globose, angular by mutual compression, c. 4 × 3cm, ripening orange-red, marked brown. Plantations, farmbush, forest; 1170–1400m.

Distr.: tropical Africa, but cultivated throughout the tropics.

IUCN: LC

Edib village: Cheek (sight record) 2/1998; **Kodmin:** Plot B220 1/1998.

Local names: Oil Palm (English) (*fide* Cheek); Dii or Palm Tree (*fide* Etuge). **Uses:** FOOD ADDITIVES – cultivated for fruit oil, but also native (*fide* Cheek); harvest palm nuts, make red palm oil, kernels for country oil called 'Mayanga'; SOCIAL USES – drugs – tapped for palm wine (*fide* Etuge).

Eremospatha cuspidata (G.Mann & H.Wendl.) H.Wendl.

Kerch. Palm.: 244 (1878).

Climber, clustering, to 20m; leaf-sheaths unarmed, with truncate ligule-like extension; leaves with only the basal pair of leaflets weakly reflexed or not reflexed; main leaflets lanceolate, c. 15 × 3cm, apex apiculate. Forest; 750–1300m.

Distr.: Cameroon to Congo (K) & Zambia. (Lower Guinea & Congolian).

IUCN: NT

Baseng: Cheek 10423 12/1999; **Kodmin:** Etuge 3982 1/1998; **Kupe Village:** Etuge 1397 11/1995; **Nyasoso:** Bruneau, 1074 10/1995.

Note: This species is local, being most common in the coastal forests of Lower Guinea (T.C.H. Sunderland, Ph.D. thesis).

Local names: Ndumme (Etuge 1397); Ratan (Etuge 3982); Ndum (Cheek 10423). **Uses:** MATERIALS – weaving, building etc. (Etuge 1397); stems used in building houses (Etuge 3982); much used in basket weaving (Cheek 10423).

Eremospatha macrocarpa (G.Mann & H.Wendl.) H.Wendl.

Climber, clustering, to c. 15m; leaf-sheaths unarmed, with truncate ligule-like extension, often characterized by a linear "wrinkle" on the upper portion; leaves with numerous basal, ovate, spiny, strongly-relexed leaflets, when adult (not seen); main leaflets lanceolate, c. 18 × 4cm; juvenile leaves fish tailed (bifid). Forest; 200–950m.

Distr.: Sierra Leone to Angola and the Congo Basin. (Guineo-Congolian).

IUCN: LC

Kupe Village: Cheek 7005 1/1995; **Mungo River F.R.:** Cheek 9346 10/1998; **Nyasoso:** Etuge 2072 6/1996.

Uses: MATERIALS – used in weaving baskets, tying houses: 1) over-shell removed, 2) central core split (Cheek 7005).

Laccosperma acutiflorum (Becc.) J.Dransf.

Kew Bull., 37: 456 (1982).

Robust climber, clustering; leaf-sheaths spiny, with ragged ligule-like extension at mouth of leaf sheath; leaflets 50+ pairs, irregularly lanceolate-sigmoid, held horizontally from rachis, c. 40 × 2.5cm. Forest; 200m.

Distr.: Sierra Leone, Ghana, Nigeria to Rio Muni, Congo (K). (Guineo-Congolian).

IUCN: LC

Mungo River F.R.: Cheek 9345 10/1998.

Note: determination requires confirmation by T.C.H.Sunderland.

Laccosperma opacum (G.Mann & H.Wendl.) Drude

Kew Bull. 37: 456 (1982).

Syn.: *Ancistrophyllum opacum* (G.Mann & H.Wendl.) Drude

Slender climber, clustering, to c. 20m, often branching in canopy; leaf-sheath spiny, with ragged ligule-like extension at mouth of leaf sheath; leaflets 10–15 pairs, irregularly lanceolate-sigmoid, held horizontally from rachis, 20–30 × 1–1.5cm. Forest; 750–1400m.

Distr.: Ghana, Nigeria, Bioko, Cameroon & Gabon. (Upper & Lower Guinea).

IUCN: LC

Baseng: Cheek 10424 12/1999; **Kodmin:** Plot B176 1/1998; **Nyasoso:** Cheek 9505 10/1998.

Local names: Ndom (Cheek 9505); Ekat (Cheek 10424).

Uses: not widely used, apart from MATERIALS – basket

uprights; not good for weaving, the stems will not split (Cheek 9505).

Laccosperma secundiflorum (P.Beauv.) Kuntze
Kew Bull. 37: 456 (1982).
Syn.: *Ancistrophyllum secundiflorum* (P.Beauv.) Wendl.
Robust climber, clustering; leaf-sheaths spiny, with ragged, ligule-like extension at mouth of leaf sheath; leaflets linear-lanceolate, pendulous from rachis, 50 × 1–2cm. Forest; 1200m.
Distr.: Guinea (B) to Congo (K). (Guineo-Congolian).
IUCN: LC
Nzee Mbeng: Gosline 102 2/1998.
Note: specimen cited has not been seen by T.C.H.Sunderland, and may refer to *L. robustum* (Burr.) J.Dransf..

Oncocalamus tuleyi Sunderl.
J. Bamboo and Rattan 1(4): 365 (2002).
Slender to robust climber, clustering, to 30m; leaf sheaths with scattered, easily detached black spines, with truncate ligule-like extension; leaflets 30–50 on each side of the rachis, c. 36 × 3cm; inflorescences long-pedunculate, to 30cm. Forest; 415m.
Distr.: SE Nigeria & W Cameroon. (Lower Guinea).
IUCN: NT
Wone: Dransfield 7646 fr., 11/1997.
Note: known from only 11 sites but well protected in some of these, including the Korup N.P., Takamanda F.R. (Cameroon), and Cross River N.P. (Nigeria). Therefore, although provisionally listed as vulnerable by Sunderland (Unasylva 52(205): 18–24), it is here reassessed as near-threatened. Future losses of populations in, for example, the threatened S Bakundu F.R., may result in a need to upgrade this species to vulnerable.

Phoenix reclinata Jacq.
Tree or shrub; stems clustered; fruiting when 1–10m tall; leaves pinnately compound, basal leaflets modified as spines, leaflet apices often spine-like; fruits ellipsoid, fleshy, 2 × 1cm, orange, ripening brown. Forest, farms, farmbush; 1200m.
Distr.: tropical and subtropical Africa.
IUCN: LC
Bangem: Cheek 7291a fr., 2/1995.
Note: this species is extremely variable throughout Africa. Several forms have in the past been described as separate species but these have proved to be impossible to maintain.
Uses: SOCIAL USES – drugs – cultivated for Bakossi palm wine. Leaves and stem not used (Cheek 7291a).

Raphia cf. africana Otedoh
J. Niger. Inst. Oil Palm Res. 6: 156 (1982).
Tree; stems clustered; trunks to 6 × 2 m; sheath fibres subcylindrical; inflorescences pendulous; fruits ellipsoid, 9 × 5.5cm, lacking beak, pale yellow. Swamp forest; 700m.
Distr.: SE Nigeria & Cameroon. (Lower Guinea).
Ngomboaku: Cheek 10347 fr., 12/1999.
Note: if the determination is confirmed, this is likely to be a Red Data taxon.
Local name: Mboh. **Uses:** MATERIALS – used for building (thatches and spars); SOCIAL USES – drugs – tapped for wine (Cheek 10347).

Raphia hookeri G.Mann & H.Wendl.
Tree; stems clustered; trunks 8 × 0.35m; sheath fibres ribbon-like, curling, numerous; inflorescences pendulous; fruit narrowly obovoid, with 1cm beak, 10 × 3cm, pale yellow. Forest edge and cultivated; 200–900m.
Distr.: Guinea (B) to Gabon. (Upper & Lower Guinea).
IUCN: LC
Mungo River F.R.: Cheek 10191 fl., fr., 11/1999; **Nyasoso:** Cheek 9285 10/1998.
Note: widespread and common in our area.
Local names: Ehtoo (Cheek 10191); Etud, Mbu (*fide* Etuge).
Uses: MATERIALS – leaves for making thatches, building traditional houses (*fide* Etuge); SOCIAL USES – drugs – semi-cultivated for wine tapping (Cheek 10191 & *fide* Etuge).

Raphia regalis Becc.
Tree; stems single; trunkless; inflorescences huge, erect; fruit narrowly obovoid, 7 × 3cm, dark red-brown; beak 0.3cm. Forest slopes; 850m.
Distr.: SE Nigeria to Cabinda. (Lower Guinea).
IUCN: VU
Ngomboaku: Cheek 10310 12/1999.
Note: the largest leaves ever recorded in the plant kingdom (30m) belong to this species (*fide* Hallé 1970).
Local name: Ateen. **Uses:** MATERIALS – fronds used in building, not tapped for wine, little used (Cheek 10310).

PANDANACEAE
M. Cheek

Pandanus sp.
Palm-like herb 6m; stems sparingly branched; leaves strap-shaped, c. 90 × 5cm, margins toothed; inflorescence pineapple-like, c. 20 × 10cm. Open forest; 400m.
Distr.: Cameroon. (Lower Guinea).
Nyandong: Cheek & Gosline (sight record) 11/1998; no specimen collected.
Note: FWTA contains no inland records of *Pandanus* species in Cameroon; collection of a specimen is required to elucidate the identity of this taxon.

POTAMOGETONACEAE
J.J. Symoens, Z. Kaplan & M. Cheek

Fl. Cameroun 26 (1984).

Potamogeton nodosus Poiret
Fl. Cameroun 26: 58 (1984).
Aquatic, bottom-rooting herb; leaves floating on surface, lanceolate, 6.5 × 2cm, obtuse, base rounded; petiole c. 12cm. Lakes; 1200–1250m.
Distr.: pantropical & Europe.
IUCN: LC
Lake Edib: Cheek 9158 2/1998; Gosline 186 11/1998.
Note: recent work by the first two authors (Kaplan & Symoens in press) has elucidated that the material referred to above should be placed in *P. richardsii* Solms, an Afromontane species otherwise unknown from W & C Africa apart from a specimen near Mbi crater in NW Province. We intend to confirm this.

SMILACACEAE

M. Cheek

Smilax anceps Willd.
Meded. Land. Wag. 82(3): 219 (1982).
Syn.: *Smilax kraussiana* Meisn.
Climber to 7m; stem spiny; leaves coriaceous, alternate, elliptic, c. 14 × 8cm, mucro 0.5cm, base obtuse, nerves palmate, 3–5; petiole c. 2cm; inflorescence terminal, umbellate, 5cm diameter. Forest; 750–1100m.
Distr.: Senegal to S Africa. (Tropical & subtropical Africa).
IUCN: LC
Kupe Village: <u>Etuge 1899</u> fr., 5/1996; **Nyasoso:** <u>Balding 65</u> fr., 8/1993; <u>Etuge 2456</u> 6/1996; <u>2587</u> fr., 7/1996.

TRIURIDACEAE

M. Cheek

Kupea martinetugei Cheek & S.Williams
Kew Bull. 58: 225 (2003).
Dioecious herb, 3–8cm; leaves and green tissue absent; inflorescence erect, subfleshy, dull pinkish orange; flowers sessile, c. 3mm diameter; carpels bilobed; male flowers with 4 tepals and 4 anthers. Forest; 700–1200m.
Distr.: Mt Kupe. (Narrow Endemic).
IUCN: CR
Kupe Village: <u>Cheek 10225</u> 12/1999; <u>7720</u> fl., 11/1995; <u>Etuge 1428</u> fl., 11/1995; <u>1623</u> 1/1996.

Sciaphila ledermannii Engl.
Monoecious myco-heterotrophic herb, 4–10cm; leaves and green tissue absent; inflorescence wiry, glossy purple-red; flowers pedicellate, c. 3mm diameter, female below, male above. Forest; 750–1000m.
Distr.: SE Nigeria & Cameroon. (Lower Guinea).
IUCN: NT
Kupe Village: <u>Cheek 7763</u> 11/1995; <u>Williams 159</u> fl., fr., 2/1995; **Nyasoso:** <u>Cable 3426</u> 7/1996.

ZINGIBERACEAE

Aframomum by D.J. Harris, H.J. Atkins & I. Darbyshire; *Renealmia* by J.M. Lock

Fl. Cameroun 4 (1965).

Note: both *Aframomum* and *Renealmia* are genera requiring revision in Africa; D.J.Harris has begun to work out species delimitation in *Aframomum* but this work is not yet complete. The determinations for *Renealmia* are provisional. Many specimens of *Aframomum* and some of *Renealmia* from our checklist area remain unassigned to species at present. It is inappropriate to assess the conservation status of taxa within these genera until they have been fully revised.

Formal descriptions for several *Aframomum* taxa were not available, therefore the brief descriptions here are compiled largely from field data, aided where possible by the Flore du Cameroun account

Uses: there have been many uses recorded by us with regard particularly to *Aframomum*, but do not include most of them here until the taxonomy of the genus is sorted out.

Aframomum arundinaceum (Oliv. & Hanb.) K.Schum.
Fl. Cameroun 4: 63 (1965).
Herb 1.5–6m tall with creeping rhizomes; leaves linear-lanceolate, c. 30 × 5cm, acuminate, base attenuate, glabrous except along the midrib below; ligule 4mm, scarious; inflorescence single-flowered (?); peduncle 5–7cm; bracts slightly puberulent, mucronate; corolla lobes 7cm, mauve, labellum broadly elliptic, to 7.5cm diameter, throat yellow; fruit 7 × 4cm. Forest; 700m.
Distr.: Guinea (C), Ivory Coast, Bioko, Cameroon, Gabon. (Upper & Lower Guinea).
Nyandong: <u>Thomas 6691</u> 2/1987.

Aframomum citratum (Pereira) K.Schum.
Rhizomatous herb to 4m; leaves linear-oblong, acuminate, base abruptly rounded, asymmetric; ligule membranous, bifid, 2.5cm, false petiole to 3cm; inflorescence at stem base, subsessile, globose, 7cm diameter, dense; bracts reddish, broadly elliptic, margins curled; corolla lobes to 7cm, mauve, labellum obovate, margin curled, 7 × 5cm; fruit 3cm diameter; calyx persistent; seeds sweet. Forest; 700m.
Distr.: Guinea (C), SE Nigeria, Cameroon & Gabon. (Upper & Lower Guinea).
Nyandong: <u>Thomas 6690</u> 2/1987.

Aframomum flavum Lock
Bull. Jard. Bot. Nat. Belg. 48: 393 (1978).
Syn.: *Aframomum hanburyi sensu* Koechlin
Herb to 4m; leaves narrowly elliptic, c. 45 × 8–11cm, acuminate, base cuneate, glabrous; ligule 5–8mm, sparsely pilose, suborbicular; inflorescence 4–6-flowered; peduncle 4–6(–20)cm; bracts broadly ovate, coriaceous, puberulent, 4.5 × 3.5cm; flowers yellow; fruit smooth, red; calyx persistent. Forest, clearings and thicket; 420–1500m.
Distr.: Cameroon & Rio Muni. (Lower Guinea).
Kodmin: <u>Cheek 8945</u> 1/1998; <u>Etuge 4021</u> 1/1998; **Harris** <u>5765</u> 1/1998; **Kupe Village:** <u>Cheek 6096</u> fr., 1/1995; **Mwambong:** <u>Mackinder 171</u> 1/1998; <u>Onana 553</u> 2/1998; **Mungo River F.R.:** <u>Onana 963</u> 11/1999; **Ndum:** <u>Cable 900</u> fl., fr., 1/1995; <u>Lane 503</u> fl., 2/1995; **Ngomboaku:** <u>Ghogue 483</u> 12/1999; **Nyandong:** <u>Thomas 6693</u> 2/1987; **Nyasoso:** <u>Sunderland 1489</u> fl., 7/1992.
Note: this species can only be reliably separated from the similar *A. daniellii*, not recorded within our checklist area to date, when it is in flower.
Uses: FOOD – fruits edible, taste soft, soon strong (Etuge 4021).

Aframomum leptolepis (K.Schum.) K.Schum.
Syn.: *Aframomum sp. A sensu* Hepper in FWTA 3: 76
Syn.: *Aframomum dalzielii* Hutch. *ined.*
Herb to 4m tall with long rhizomes; leaves with a puberulent midrib below; ligule short and coriaceous; inflorescence arising from rhizomes away from stems and at leafy stem bases, 4–5-flowered; flowers purple; fruits smooth to subridged, red. Forest; 710–1400m.
Distr.: Bioko & Cameroon. (Lower Guinea).
Kodmin: <u>Cheek 9083</u> 1/1998; <u>Etuge 4052</u> 1/1998; **Nyale:** <u>Etuge 4213</u> 2/1998; **Nyasoso:** <u>Harris 5777</u> 1/1998.

ZINGIBERACEAE

Aframomum limbatum (Oliv. & Hanb.) K.Schum.

Herb to 3–5m with creeping rhizomes; leaves oblong-lanceolate, c. 40 × 10cm, caudate-acuminate, base asymmetrically attenuate, pubescent on the lower midrib and margins; ligule rounded, 3mm; inflorescence 2–3-flowered on short spikes from the rhizome; bracts 2.5cm, submembranous; flowers purple, lobes c. 2.5cm; labellum obovate, undulate, 4.5–5cm diameter; fruit globose, smooth, pale brown, (?) underground. Forest; 700–1120m.
Distr.: Nigeria to Gabon & Uganda. (Lower Guinea).
Kupe Village: Ryan 229 5/1996; **Nyandong:** Thomas 6694 2/1987.

Aframomum melegueta K.Schum.

Herb to 2m; leaves (sub)sessile, lanceolate, 18–22 × 2cm, acuminate, base attenuate, glabrous; ligule <1mm, scarious; inflorescence at base of stems, a 1-flowered spike, 8cm; bracts c. 5cm, obtuse, mucronate; flowers pink to purple, tube 5cm, lobes 4.5cm, labellum obovate, to 10cm wide, throat yellow; fruit 5cm long; calyx persistent, smooth, red; seeds tuberculate. Cultivated and forest.
Distr.: Cameroon to Congo (K). (Lower Guinea & Congolian).
Nyasoso: Etuge 1511 12/1995.
Local names: 'Alacata pepper' (Etuge 1511); N'Lowo (Etuge Martin in Cheek 7904). **Uses:** FOOD ADDITIVES – seeds used in seasoning; MEDICINES (Etuge 1511).

Aframomum pilosum (Oliv. & Hanb.) K.Schum.

Rhizomatous herb to 3m; leaves oblong-lanceolate, 25–30 × 6cm, caudate-acuminate, base asymmetrically attenuate to obtuse, surfaces and petiole with long stiff hairs; ligule inconspicuous; inflorescence at stem base, 1-flowered spikes in clusters of up to 20; flower yellow; corolla lobes to 2cm; labellum obovate, 1.5cm wide; fruit to 2cm long, smooth, pinkish. Forest; 600–1400m.
Distr.: Nigeria, Bioko & Cameroon. (Lower Guinea).
Kodmin: Etuge 4057 1/1998; Harris 5770 1/1998; **Kupe Village:** Cable 2694 5/1996; **Manehas F.R.:** Etuge 4354 10/1998; **Nyandong:** Thomas 6689 2/1987.
Note: easily recognised even when sterile by the long stiff hairs on its leaves and petioles.

Aframomum subsericeum (Oliv. & Hanb.) K.Schum. subsp. *glaucophyllum* (K.Schum.) Lock

Kew Bull. 35: 306 (1980).
Herb to 4m with long rhizomes; leaves oblong-lanceolate, c. 35 × 6.5cm, caudate-acuminate, base attenuate, hairs on the lower leaf surface strongly-appressed, spreading parallel to the veins; leaves whitish below when fresh; ligule coriaceous, 5mm; inflorescence a 2–3-flowered spike; peduncle to 10cm; bracts coriaceous, emarginate, 3cm; corolla pink, lobes 3.5–7cm, labellum purple with yellow throat, obovate, 6cm diameter, margin undulate; fruit 6 × 3cm, smooth, red. Forest; 1200–1400m.
Distr.: Nigeria & Cameroon. (W Cameroon Uplands).
Edib: Etuge 4091 1/1998; **Kodmin:** Biye 37 fl., 1/1998; Harris 5759 1/1998; **Nyasoso:** Harris 5781 1/1998.
Local name: Etuin. **Uses:** MEDICINES (Biye 37).

Aframomum zambesiacum (Baker) K.Schum.

Syn.: *Aframomum chlamydanthum* Loes. & Mildbr.
Herb to 3m; leaves narrowly elliptic, base cuneate, margins and midrib of lower leaf hairy; ligule short, rounded; inflorescence capitate, > 20-flowered; flowers whitish,

labellum with purple centre; fruits deeply grooved, thick-walled; seeds shiny, dark brown, rough. Forest; 1940–2000m.
Distr.: Nigeria, Cameroon, Congo (K) & E Africa to Malawi. (Afromontane).
Kupe Village: Ryan 409 5/1996; **Mt Kupe:** Harris 5797 1/1998; 5805 1/1998; **Nyasoso:** Cable 1235 fr., 2/1995; 2921 6/1996; Etuge 1683 1/1996.

Aframomum sp. 1 of Kupe-Bakossi

Herb to 2m; flower yellowish white; fruit smooth, red. Forest; 1000–1600m.
Distr.: Cameroon Endemic.
Mt Kupe: Harris 6763 1/1999; **Nyasoso:** Cheek 7540 10/1995; Etuge 1334 10/1995; 1817 3/1996.

Aframomum sp. 2 of Kupe-Bakossi

Herb to 3m; leaf base clasping stem; flower purple; fruit smooth, red. Forest; 700m.
Distr.: Cameroon Endemic.
Mt Kupe: Harris 5814 1/1998; **Nyandong:** Thomas 6692 2/1987.

Aframomum sp. 3 of Kupe-Bakossi

Herb to 1.5m; flower purple; fruit smooth, red. Forest.
Distr.: Cameroon Endemic.
Mt Kupe: Harris 5820 1/1998.

Aframomum sp. 4 of Kupe-Bakossi

Herb to 2m; flower purple; fruit smooth, red. Forest-grassland edge.
Distr.: Cameroon Endemic.
Mt Kupe: Harris 5760 1/1998.

Aframomum sp. 4 sensu Harris, D.J. 1997

Herb c. 2–3m; leaves auriculate at base, apex cirrhose, leafy stalk waxy. Forest; 880m.
Distr.: Cameroon Endemic.
Kupe Village: Cheek 7053 1/1995.

Hedychium sp.

Ornamental herb to 1m tall. Cultivated; 1170m.
Edib village: Cheek (sight record) 2/1998.
Note: the cited specimen was not collected, therefore no further determination or description can be made.

Renealmia africana (K.Schum.) Benth.

Herb from short thick rhizome; leaves with a distinct false petiole, blades elliptic, to 30 × 8cm or more, apex acuminate, base narrowly cuneate, veins prominent on both surfaces; inflorescence arising from the rhizome, near the leaves; flowers small, delicate, whitish translucent, lateral inflorescence branches spreading to upright; fruits spherical to ellipsoid, c. 8mm diameter, reddish, becoming black. Forest; 350–1950m.
Distr.: Nigeria to Congo (K). (Lower Guinea & Congolian (montane)).
Kodmin: Cheek 9070 fl., 1/1998; **Kupe Village:** Cheek 7076 fl., 1/1995; Ryan 427 fl., 5/1996; **Mejelet-Ehumseh:** Etuge 169 6/1986; **Mekom:** Nemba 25 4/1986; **Mt Kupe:** Sebsebe 5096 fl., fr., 11/1995; **Mwambong:** Cheek 9385 10/1998; **Ndum:** Groves 33 fl., 2/1995; Williams 147 fl., 1/1995; **Nyasoso:** Cable 1196 fr., 2/1995; 2832 fl., 6/1996; Cheek 7542 fl., 10/1995; Etuge 1312 fr., 10/1995; Sunderland 1509 fr., 7/1992; Wheatley 453 fr., 7/1992.

Renealmia sp. aff. africana (K.Schum.) Benth.

Herb resembling *R. africana*, but distinct in the long, thin rhizomes and small stature. Forest; 1150m.

Kupe Village: <u>Cheek 7789</u> fl., 11/1995.

Renealmia polyantha K.Schum.

Fl. Cameroun 4: 32 (1965).

Herb resembling *R. africana*, but leaf lamina usually longer and narrower, often glaucous, veins prominent below only; inflorescences like those of *R. africana*, but longer, lateral inflorescence branches longer, conspicuously bracteate, usually reflexed; fruits ellipsoid, c. 1 × 5cm. Forest; 700–1100m.

Distr.: Cameroon & Gabon. (Lower Guinea).

Ngomboaku: <u>Mackinder 307a</u> fl., 12/1999; **Nyale:** <u>Etuge 4175</u> fl., fr., 2/1998; **Nyasoso:** <u>Etuge 2139</u> fr., 6/1996.

Renealmia polypus Gagnep.

Fl. Cameroun 4: 36 (1965).

Herb resembling *R. africana*, but smaller; each inflorescence branch with more and smaller fruits. Forest; 1350m.

Distr.: Cameroon & Gabon. (Lower Guinea).

Kupe Village: <u>Cheek 8398</u> fr., 5/1996.

Renealmia stenostachys K.Schum.

Fl. Cameroun 4: 41 (1965).

Herb differing from other *Renealmia* spp. in the simple, spiciform inflorescence with single flowers in each bract axil. Forest.

Distr.: Cameroon. (Lower Guinea).

Nyasoso: <u>Letouzey 481</u> fl., month unknown/1970.

Note: J.Koechlin in Fl. Cameroon 4: 41 (1965) states that this specimen displays the spiciform inflorescence characteristic of this poorly known taxon but that it differs in several minor characters from the species description, including the inflorescence length.

GYMNOSPERMAE
PINOPSIDA

CUPRESSACEAE

B.J. Pollard

Cupressus lusitanica Mill.
Dallimore & Jackson, Handbook of *Coniferae*: 206 (1923).
Tree to 30–35m, evergreen, monoecious; trunk monopodial, large trees buttressed, up to 2 m dbh; leaves scale-like; seed cones solitary or in groups near the upper ends of lateral branches, terminal on short leafy branchlets, maturing in 2 growing seasons, persistent. From cultivation; 460m.
Distr.: pantropical and temperate regions.
IUCN: LC
Nyandong: Cheek 11416 3/2003.
Uses: MATERIALS (*fide* Pollard).

PODOCARPACEAE

I. Darbyshire

Podocarpus mannii Hook.f.
Dioecious tree resembling *P. milanjianus*, but leaves narrower, 6–16 × 0.4–1.1cm, stomata on both surfaces; receptacle not large and fleshy; seeds solitary, pyriform, 2.5–3.5cm long. Planted in gardens.
Distr.: São Tomé, introduced in Cameroon & Nigeria. (Lower Guinea).
IUCN: NT
Kupe village: Cable 2561 st., 05/1996.
Note: uncommon in its native São Tomé but apparently widely planted in Cameroon, less so in Nigeria.

Podocarpus milanjianus Rendle
Dioecious shrub or tree to 35m; bark exfoliating in papery flakes; slash pale brown; stems much-branched, sympodial; leaves spreading, alternate, linear-lanceolate, 5–10–15 × 0.5–1.5cm, stomata on lower side only, midrib prominent and raised below; male cones solitary or paired, flesh-pink, c. 3cm; female cones solitary; fruit green, obovoid to subglobose, c. 1cm long; receptacle well-developed, obconical to subglobose, fleshy, red; seeds 1–2, subglobose, 8–9mm. Forest; 900–2000m.
Distr.: Cameroon, Congo (K), Angola, Sudan to Zimbabwe. (Afromontane).
IUCN: LC
Kupe Village: Cable 2560 st., 5/1996; Cheek 7204 st., 1/1995; **Mt Kupe:** Sebsebe 5072 st., 10/1995; Thomas 3161 2/1984; 5475 2/1986; **Mwanenguba:** Letouzey 13933 st., 06/1975; **Nyasoso:** Cable 2920 6/1996; 2931 6/1996; Elad 123 2/1995.

GNETOPSIDA

GNETACEAE

I. Darbyshire

Gnetum africanum Welw.
Liana; leaves opposite, ovate-oblong to elliptic-oblong, 10–13 × 3.5–5cm, acuminate, base attenuate; dioecious, axillary catkin-like spikes jointed with flowers in whorls; male spikes with internodes of equal diameter throughout their length; staminal column exserted from the mouth of a tubular envelope, formed by 2 connate scales. Secondary forest, farms; 600m.
Distr.: Cameroon to Congo (K). (Lower Guinea & Congolian).
IUCN: LC
Ngusi: Etuge 54 4/1986.
Local name: Eru. **Uses:** FOOD – leaves eaten as a green vegetable after shredding and steaming or boiling (Etuge 54).

PTERIDOPHYTA
LYCOPSIDA

P.J. Edwards & J.-M. Onana

Fl. Cameroun 3 (1964).

LYCOPODIACEAE

Huperzia ophioglossoides (Lam.) Rothm.
Acta Botanica Barcinonensia 31: 8 (1978).
Syn.: *Lycopodium ophioglossoides* Lam.
1950m.
Nyasoso: Cable 2936 6/1996; **Tombel:** Thorold TN10 3/1953.

Huperzia verticillata (L.f.) Trevis.
Biologiske Skrifter 34: 22 (1989).
Syn.: *Lycopodium verticillatum* L.f.
1500m.
Kodmin: Pollard 224 11/1998.

Lycopodiella cernua (L.) Pic.Serm.
Acta Botanica Barcinonensia 31: 11 (1978).
Syn.: *Lycopodium cernuum* L.
1200m.
Lake Edib: Pollard 241 11/1998.

Pseudolycopodiella affinis (Bory) J.Holub
Folia Geobot. Phytotax. 20(1): 79 (1985).
Syn.: *Lycopodium affine* Bory
1200m.
Lake Edib: Gosline 193 11/1998.

SELAGINELLACEAE

Selaginella abyssinica Spring
Mwanenguba: <u>Nicklès 51</u> 1946–48.
Note: specimen cited in Fl. Cameroun 3: 32 (1964).

Selaginella kraussiana (Kunze) A.Braun
1050–2000m.
Mwanenguba: <u>Cheek 9440</u> 10/1998; **Nyale:** <u>Cheek 9669</u> 11/1998.

Selaginella myosurus (Sw.) Alston
Mwanenguba: <u>de Gironcourt 494</u> 1911–12.
Note: specimen cited in Fl. Cameroun 3: 24 (1964).

Selaginella soyauxii Hieron.
Mwanenguba: <u>de Gironcourt 457</u> 1911–12.
Note: specimen cited in Fl. Cameroun 3: 34 (1964).

Selaginella versicolor Spring
750–1450m.
Kodmin: <u>Biye 14</u> 1/1998; <u>Cheek 8944</u> 1/1998; <u>Plot B160</u> 1/1998; **Kupe Village:** <u>Cheek 7101</u> 1/1995; <u>Etuge 2707</u> 7/1996; <u>Lane 301</u> 1/1995; <u>Zapfack 913</u> 7/1996; **Mwanenguba:** <u>de Gironcourt 426</u> 1911–12; **Nyasoso:** <u>Cable 3442</u> 7/1996; <u>Wheatley 395</u> 7/1992.

Selaginella vogelii Spring
900–1050m.
Ngomboaku: <u>Ghogue 487</u> 12/1999; **Nyale:** <u>Cheek 9668</u> 11/1998; **Nyasoso:** <u>Cheek 5641</u> 12/1993.

FILICOPSIDA

P.J. Edwards & J.-M. Onana

ADIANTACEAE

Adiantum philippense L.
700m.
Kupe Village: <u>Etuge 2766</u> 7/1996.

Adiantum poiretii Wikstr.
2000m.
Mwanenguba: <u>Cheek 9476</u> 10/1998; <u>Nicklès 1</u> 1946–48.

Coniogramme africana Hieron.
760–1450m.
Kodmin: <u>Cheek 9023</u> 1/1998; **Kupe Village:** <u>Cable 3685</u> 7/1996; <u>Etuge 2823</u> 7/1996.
Local name: Mpoo mwakum. **Uses:** Traditional use unexplained (Cheek 9023).

Pellaea doniana J.Sm. ex Hook.
870–900m.
Nyasoso: <u>Balding 94</u> 8/1993; <u>Cable 2770</u> 6/1996; <u>3203</u> 6/1996.

Pityrogramma calomelanos (L.) Link
600–760m.
Kupe Village: <u>Cable 3684</u> 7/1996; <u>Etuge 2768</u> 7/1996; **Nyandong:** <u>Fay 4745</u> 3/2003; **Nyasoso:** <u>Etuge 1640</u> 1/1996.

ASPLENIACEAE

Asplenium aethiopicum (Burm.f.) Bech.
1200m.
Nyasoso: <u>Etuge 2482</u> 6/1996.

Asplenium africanum Desv.
Mwanenguba: <u>de Gironcourt 448</u> 1911–12.
Note: specimen cited in Fl. Cameroun 3: 180 (1964).

Asplenium barteri Hook.
700–1400m.
Kodmin: <u>Plot B213</u> 1/1998; <u>B218</u> 1/1998; <u>B253</u> 1/1998; <u>B299</u> 1/1998; **Kupe Village:** <u>Cable 3659</u> 7/1996; <u>3885</u> 7/1996; <u>Kenfack 315</u> 7/1996; <u>Zapfack 949</u> 7/1996; **Mwanenguba:** <u>de Gironcourt 443, 446</u> 1911–12; **Nyasoso:** <u>Cable 95</u> 9/1992; <u>Zapfack 658</u> 6/1996.

Asplenium biafranum Alston & Ballard
1570m.
Mwambong: <u>Mackinder 158</u> 1/1998.

Asplenium cancellatum Alston
870–1250m.
Kupe Village: <u>Cable 2400</u> 5/1996; <u>2741</u> 5/1996; <u>Etuge 2877</u> 7/1996; <u>Zapfack 955</u> 7/1996; **Lake Edib:** <u>Cheek 9155</u> 2/1998; **Nyasoso:** <u>Cable 3302</u> 6/1996; <u>Thorold 13</u> /1900; <u>Zapfack 654</u> 6/1996.

Asplenium cf. *cancellatum* Alston
1100–1450m.
Nyasoso: <u>Cable 49</u> 8/1992; <u>Lane 15</u> 7/1992.

Asplenium currori Hook.
Tombel: <u>Thorold 11</u> date unknown.
Note: specimen cited in Fl. Cameroun 3: 181 (1964).

Asplenium dregeanum Kunze
870–1950m.
Kodmin: <u>Barlow 7</u> 1/1998; <u>Etuge 4000</u> 1/1998; <u>Mackinder 140</u> 1/1998; <u>Plot B18</u> 1/1998; <u>B23</u> 1/1998; **Kupe Village:** <u>Cable 2504</u> 5/1996; <u>Cheek 7810</u> 11/1995; <u>Ryan 417</u> 5/1996; **Mt Kupe:** <u>Sebsebe 5094</u> 11/1995; **Nyasoso:** <u>Balding 42</u> 8/1993; <u>Cable 3218</u> 6/1996; <u>3585</u> 7/1996; <u>68</u> 8/1992; <u>Cheek 7355</u> 2/1995; <u>Lane 26</u> 7/1992; <u>Sidwell 388</u> 10/1995; <u>410</u> 10/1995; <u>Wheatley 389</u> 7/1992; <u>Zapfack 606</u> 6/1996.

Asplenium erectum Bory ex Willd. var. *usambarense* (Hieron.) Schelpe
Fl. Zamb. Pteridophyta: 176 (1970).
Syn.: *Asplenium quintasii* Gand.
1190m.
Mt Kupe: <u>Letouzey 423</u> 11/1954; **Mwanenguba:** <u>Nicklès 13</u> 1946–48; **Nyasoso:** <u>Etuge 1357</u> 10/1995.

Asplenium gemmascens Alston
1125–1500m.
Kodmin: <u>Etuge 4005</u> 1/1998; **Nyasoso:** <u>Wheatley 393</u> 7/1992.

Asplenium gemmiferum Schrad.
950–1550m.
Nyasoso: <u>Cable 105</u> 9/1992; <u>Zapfack 620</u> 6/1996.

BLECHNACEAE

Asplenium geppii Carruth.
Mt Kupe: Letouzey 416 11/1954.

Asplenium hemitomum Hieron.
870–1200m.
Mwambong: Etuge 4136 2/1998; **Nyasoso:** Cable 3236
6/1996; 85 8/1992; Etuge 2066 6/1996; Zapfack 687 6/1996.

Asplenium hypomelas Kuhn
1500m.
Kodmin: Etuge 3993 1/1998.

Asplenium mannii Hook.
920–1000m.
Kupe Village: Cheek 8413 5/1996; **Mwanenguba:** Nicklès
5 1946–48; **Nyasoso:** Zapfack 712 6/1996.

Asplenium paucijugum Ballard
Bull. Jard. Bot. Nat. Belg. 55: 147 (1985).
Syn.: *Asplenium variabile* Hook. var. *paucijugum* (Ballard)
Alston
760–1900m.
Kodmin: Etuge 4085 1/1998; Plot B66 1/1998; **Kupe
Village:** Cable 3798 7/1996; Ryan 384 5/1996; **Mbule:**
Cable 3366 6/1996; **Mt Kupe:** Etuge 1363 11/1995;
Ngomboaku: Mackinder 304 12/1999; **Nyasoso:** Cheek
9523 10/1998; Etuge 2085 6/1996; 2152 6/1996; 2479
6/1996.
Local name: Apak-kwor-le. **Uses:** SOCIAL USES – magic –
used in disturbing witchcraft (Epie Ngome in Cheek 9523).

Asplenium preussii Hieron.
870–2000m.
Kupe Village: Cable 3767 7/1996; **Nyasoso:** Cable 2769
6/1996; 3382 6/1996; Etuge 1323 10/1995; 1745 2/1996;
Lane 59 8/1992.

Asplenium protensum Schrad.
1000–1050m.
Mwanenguba: Nicklès 12 1946–48; **Nyasoso:** Cable 3440
7/1996; Lane 24 7/1992.

Asplenium sandersonii Hook.
Fl. Zamb. Pteridophyta: 183 (1970).
Syn.: *Asplenium sandersonii* Hook. var. *vagans* (Baker)
C.Chr.
Syn.: *Asplenium vagans* Baker
850m.
Kupe Village: Zapfack 950 7/1996; **Nyasoso:** Etuge 2575
7/1996.

Asplenium sp. aff. subaequilaterale (Baker)
Hieron.
1050m.
Kupe Village: Zapfack 954 7/1996; **Nyasoso:** Lane 23
7/1992.

Asplenium subintegrum C.Chr.
Loum: Nicklès 87 1946–48.
Note: specimen cited in Fl. Cameroun 3: 182 (1964).

Asplenium theciferum (Kunth) Mett. var.
cornutum (Alston) Benl
Acta Botanica Barcinonensia 40: 32 (1991).
Syn.: *Asplenium cornutum* Alston
950–1960m.

Kupe Village: Ryan 339 5/1996; **Mwanenguba:** Nicklès 18
1946–48; **Nyasoso:** Lane 10 7/1992; 4 7/1992; Sidwell 409
10/1995; Zapfack 622 6/1996; 736 6/1996.

Asplenium unilaterale Lam. var. **unilaterale**
800–1650m.
Kodmin: Cheek 8960 1/1998; **Kupe Village:** Cheek 7220
1/1995; 7808 11/1995; Etuge 2803 7/1996; **Nyasoso:**
Balding 77 8/1993; Cable 15. 7/1992; 3582 7/1996; 53
8/1992; Etuge 1306 10/1995; 2360 6/1996; Gosline 143
11/1998; Lane 58 8/1992; Sunderland 1516 7/1992;
Wheatley 386 7/1992.

Asplenium warneckei Hieron.
870–1050m.
Kupe Village: Cable 2443 5/1996; **Nyasoso:** Cable 3621
7/1996.

Asplenium sp. 2
850m.
Nyasoso: Cable 861 1/1995.
Note: *Asplenium sp. 2–6* are simple-fronded species currently
being worked on by R.J.Johns.

Asplenium sp. 4
900m.
Nyasoso: Cable 69 8/1992.

Asplenium sp. 5
1050m.
Nyasoso: Lane 31 7/1992.

Asplenium sp. 6
900m.
Kupe Village: Groves 13 1/1995.

BLECHNACEAE

Blechnum attenuatum (Sw.) Mett. var.
attenuatum
1200–1500m.
Kodmin: Etuge 4002 1/1998; **Lake Edib:** Pollard 244
11/1998; **Mwambong:** Cheek 9376 10/1998.

CYATHEACEAE

Cyathea camerooniana Hook. var. **zenkeri**
(Diels) Tardieu
Fl. Cameroun 3: 67 (1964).
Syn.: *Alsophila camerooniana* (Hook.) R.M.Tryon var.
zenkeri (Diels) Benl
940–1000m.
Kupe Village: Cable 2509 5/1996; **Nyasoso:** Cable 3479
7/1996.
Local name: Esuk. **Uses:** MATERIALS – stem poles used
for building round traditional Bakossi houses (*fide* Etuge).

Cyathea manniana Hook.
Syn.: *Alsophila manniana* (Hook.) R.M.Tryon
1200m.
Nyasoso: de Gironcourt s.n. 1911–12; Sunderland 1523
7/1992.

Cyathea obtusiloba (Hook.) Domin
Pteridophyta: 263 (1929).
1400–1450m.
Kodmin: Cheek 8951 1/1998; **Kupe Village:** Etuge 2845
7/1996.

Cyathea sp. nov. ?
1100-1400m.
Kodmin: Plot B69 1/1998; **Kupe Village:** Cable 823
1/1995; **Mbule:** Cable 3358 6/1996; **Nyasoso:** Etuge 2145
6/1996.
Note: has characteristics of *Cyathea camerooniana* and *C.
obtusiloba*; conceivably a hybrid (but spores even and good)
or *sp. nov.*

DAVALLIACEAE

Davallia chaerophylloides (Poir.) Steud.
400m.
Kupe Village: Etuge 1936 5/1996.

DENNSTAEDTIACEAE

Blotiella currori (Hook.) R.M.Tryon
Fl. Zamb. Pteridophyta: 84 (1970).
Syn.: *Lonchitis currorii* (Hook.) Mett. ex Kuhn
1100–1300m.
Kodmin: Plot B13 1/1998; **Kupe Village:** Cable 3801
7/1996; **Mwambong:** Cheek 9386 10/1998; **Mwanenguba:**
de Gironcourt 423, 445 1911–12; **Nyasoso:** Cable 3328
6/1996.

Blotiella mannii (Baker) Pic.Serm.
Acta Botanica Barcinonensia 38: 30 (1988).
Syn.: *Lonchitis mannii* (Baker) Alston
1100–1200m.
Kupe Village: Cable 2628 5/1996; **Mbule:** Cable 3353
6/1996; **Mt Kupe:** Letouzey 407 11/1954.

Microlepia speluncae (L.) T.Moore var.
speluncae
700m.
Kupe Village: Etuge 2745 7/1996; **Mwanenguba:** Nicklès
s.n. 1946–48.

Pteridium aquilinum (L.) Kuhn subsp. *aquilinum*
1450m.
Kodmin: Cheek 9207 2/1998.

DRYOPTERIDACEAE

Didymochlaena truncatula (Sw.) J.Sm.
1100–1650m.
Kupe Village: Ryan 390 5/1996; **Nyasoso:** Balding 50
8/1993; Etuge 2402 6/1996; Lane 62 8/1992; Sidwell 411
10/1995.

Dryopteris inaequalis (Schltdl.) Kuntze
Fl. Zamb. Pteridophyta: 222 (1970).
870–1100m.

Nyasoso: Cable 3219 6/1996; 70 8/1992.

Dryopteris manniana (Hook.) C.Chr.
1190–2050m.
Kodmin: Etuge 4406 11/1998; Plot B16 1/1998; **Kupe
Village:** Ryan 418 5/1996; **Mwanenguba:** Nicklès 32 1946–
48; **Nyasoso:** Balding 60 8/1993; Etuge 1356 10/1995; 1358
10/1995.

Lastreopsis barteriana (Hook.) Tardieu
Fl. Cameroun 3: 279 (1964).
Syn.: *Ctenitis barteriana* (Hook.) Alston
900–1600m.
Kodmin: Etuge 4050 1/1998; **Kupe Village:** Cable 3800
7/1996; Etuge 2843 7/1996; 2898 7/1996; Ryan 389 5/1996;
Nyasoso: Cable 3584 7/1996.

Lastreopsis nigritiana (Baker) Tindale
Fl. Cameroun 3: 280 (1964).
Syn.: *Ctenitis pubigera* Alston
1200m.
Nyasoso: Etuge 2096 6/1996.

Lastreopsis subsimilis (Hook.) Tindale
Fl. Cameroun 3: 278 (1964).
Syn.: *Ctenitis subsimilis* (Hook.) Tardieu
800–1000m.
Kupe Village: Cable 2726 5/1996; Etuge 2610 7/1996;
Manehas F.R.: Etuge 4353 10/1998; **Nyasoso:** Etuge 2520
7/1996.

Tectaria barteri (J.Sm.) C.Chr.
400–950m.
Bakossi F.R.: Etuge 4292 10/1998; **Kupe Village:** Etuge
2606 7/1996; **Manehas F.R.:** Etuge 4350 10/1998; **Nyasoso:**
Cable 3288 6/1996; Etuge 1587 1/1996; 2408 6/1996;
Zapfack 604 6/1996.

Tectaria camerooniana (Hook.) Alston
400–1700m.
Kodmin: Cheek 8995 1/1998; **Kupe Village:** Cable 3664
7/1996; 3712 7/1996; 3799 7/1996; 852 1/1995; Cheek 6095
1/1995; Etuge 2012 5/1996; 2821 7/1996; **Mbule:** Cable
3370 6/1996; **Mwambong:** Etuge 4128 2/1998; **Nyasoso:**
Cable 2774 6/1996; 3251a 6/1996; Etuge 2142 6/1996; Lane
60 8/1992; Sidwell 392 10/1995; Viane 2424 2/1983; 2431
2/1983; 2444 2/1983; 2457 2/1983; Zapfack 605 6/1996.

Tectaria fernandensis (Baker) C.Chr.
700–2000m.
Kodmin: Biye 48 1/1998; **Kupe Village:** Cable 3768
7/1996; Etuge 2880 7/1996; Ryan 328 5/1996; **Nyasoso:**
Cable 3251B 6/1996; 3381 6/1996; Lane 34 8/1992; 46
8/1992; Zapfack 711 6/1996.

Tectaria cf. gemmifera (Fée) Alston
J. Bot. 77: 288 (1939).
1200m.
Nyasoso: Viane 2421 2/1983.

Triplophyllum jenseniae (C.Chr.) Holttum
1050–1230m.
Kupe Village: Ryan 366 5/1996; **Nyasoso:** Cable 3583
7/1996.

Triplophyllum protensum (Sw.) Holttum
Kew Bull. 41: 247 (1986).
Syn.: *Ctenitis protensa* (Afzel. ex Sw.) Ching
710m.
Nyale: Etuge 4206 2/1998.

Triplophyllum securidiforme (Hook.) Holttum var. *nanum* (Bonap.) Holttum
Kew Bull. 41: 243 (1986).
Syn.: *Ctenitis securidiformis* (Hook.) Copel. var. *nana* (Bonap.) Tardieu
750m.
Baseng: Cheek 10402 12/1999.

Triplophyllum securidiforme (Hook.) Holttum var. *securidiforme*
Kew Bull. 41: 242 (1986).
Syn.: *Ctenitis securidiformis* (Hook.) Copel. var. *securidiformis*
760–1450m.
Kodmin: Cheek 8961 1/1998; Etuge 3975 1/1998; **Kupe Village:** Cable 3689 7/1996; Cheek 7817 11/1995; Ryan 296 5/1996; **Ngomboaku:** Cheek 10288 12/1999; **Nyasoso:** Cable 67 8/1992.

GLEICHENIACEAE

Dicranopteris linearis (Burm.f.) Underw. var. *linearis*
Fl. Zamb. Pteridophyta: 50 (1970).
Syn.: *Gleichenia linearis* (Burm.f.) C.B.Clarke
850m.
Ngomboaku: Cheek 10332 12/1999.

GRAMMITIDACEAE

Lellingeria oosora (Baker) A.R.Smith & R.C.Moran
Amer. Fern J. 81(3): 85 (1991).
Syn.: *Xiphopteris oosora* (Baker) Alston var. *oosora*
1050–1760m.
Kodmin: Pollard 216 11/1998; **Nyasoso:** Cable 77 8/1992; Lane 57 8/1992; 82 9/1992.

Zygophlebia villosissima (Hook.) L.E.Bishop
Amer. Fern J. 79(3): 117 (1989).
Syn.: *Ctenopteris villosissima* (Hook.) W.J.Harley
Syn.: *Xiphopteris villosissima* (Hook.) Alston var. *villosissima*
1150m.
Nyasoso: Lane 72 8/1992.

HYMENOPHYLLACEAE

Hymenophyllum kuhnii C.Chr.
Syn.: *Hymenophyllum polyanthos* Sw. var. *kuhnii* (C.Chr.) Schelpe
2050m.
Mt Kupe: Cheek 7609 11/1995.

Hymenophyllum splendidum Bosch
1200–1800m.
Kodmin: Cheek 9076 1/1998; 9094 2/1998; **Kupe Village:** Cheek 7209 1/1995; **Nyasoso:** Zapfack 732 6/1996.

Hymenophyllum triangulare Baker
1450m.
Kupe Village: Ryan 375 5/1996.

Trichomanes africanum Christ
750m.
Baseng: Cheek 10403 12/1999.

Trichomanes chevalieri Christ
910m.
Kupe Village: Zapfack 953 7/1996; **Nyasoso:** Zapfack 660 6/1996.

Trichomanes clarenceanum F.Ballard
870–1200m.
Kupe Village: Cheek 7667 11/1995; **Mwambong:** Cheek 9378 10/1998.

Trichomanes erosum Willd. var. *erosum*
Syn.: *Trichomanes erosum* Willd. var. *aerugineum* (Bosch) Bonap.
Syn.: *Trichomanes aerugineum* Bosch
Syn.: *Trichomanes chamaedrys* Taton
Syn.: *Microgonium chamaedrys* (Taton) Pic.Serm.
810–2050m.
Mt Kupe: Cheek 7609A 11/1995; **Ngomboaku:** Ghogue 462A 12/1999.

Trichomanes guineense Afzel. ex Sw.
450m.
Bakossi F.R.: Cheek 9323 10/1998.
Local name: Mboh. **Uses:** FOOD – larger ones edible (Cheek 9323).

Trichomanes mannii Hook.
1500–1600m.
Kupe Village: Cheek 8400 5/1996; **Nyasoso:** Zapfack 725 6/1996.

Trichomanes mettenii C.Chr.
950m.
Nyasoso: Zapfack 627 6/1996.

Trichomanes pyxidiferum L. var. *melanotrichum* (Schltdl.) Schelpe
Acta Botanica Barcinonensia 32: 24 (1980).
Syn.: *Trichomanes melanotrichum* Schltdl.
950m.
Nyasoso: Zapfack 624 6/1996.

Trichomanes rigidum Sw.
Fl. Cameroun 3: 90 (1964).
900–1100m.
Kodmin: Etuge 4066 1/1998; **Kupe Village:** Cheek 7221 1/1995; **Nyale:** Etuge 4475 11/1998.

LOMARIOPSIDACEAE

Bolbitis acrostichoides (Afzel. ex Sw.) Ching
1000–1100m.
Kupe Village: Sidwell 440 11/1995; **Nyasoso:** Cable 1224
2/1995; 662 12/1993.

Bolbitis auriculata (Lam.) Alston
860–1400m.
Kodmin: Plot B158 1/1998; **Mwambong:** Etuge 4130
2/1998; **Mwambong:** Cheek 9107 2/1998; **Nyasoso:** Etuge
1527 1/1996.

Bolbitis fluviatilis (Hook.) Ching
900–1200m.
Kodmin: Etuge 4059 1/1998; **Kupe Village:** Cheek 7216
1/1995; Etuge 2902 7/1996; **Nyasoso:** Etuge 2371 6/1996.

Elaphoglossum barteri (Baker) C.Chr.
1400m.
Kodmin: Etuge 4434 11/1998; **Nyasoso:** Cable 86 8/1992.

Elaphoglossum sp. aff. barteri (Baker) C.Chr.
1200m.
Kupe Village: Cable 2623 5/1996.

Elaphoglossum cinnamomeum (Baker) Diels
1050–1200m.
Kupe Village: Etuge 2658 7/1996; **Nyasoso:** Lane 83
9/1992.

Elaphoglossum isabelense Brause
1275m.
Nyasoso: Wheatley 400 7/1992.

Elaphoglossum kuhnii Hieron.
1100–2000m.
Kupe Village: Cable 2570 5/1996; **Nyasoso:** Etuge 1680
1/1996; Lane 12 7/1992; 68A 8/1992.

Elaphoglossum salicifolium (Willd. ex Kaulf.)
Alston
1100–1650m.
Kupe Village: Etuge 2868 7/1996; **Nyasoso:** Cable 79
8/1992; Etuge 2397 6/1996; Zapfack 733 6/1996.

Elaphoglossum sp. nov.
1150m.
Nyasoso: Lane 71 8/1992.

Lomariopsis guineensis (Underw.) Alston
250–1500m.
Kodmin: Etuge 3990 1/1998; **Kupe Village:** Cable 826
1/1995; **Mungo River F.R.:** Cheek 9355 10/1998; **Nyasoso:**
Cable 3322 6/1996.

Lomariopsis muriculata Holttum
1100m.
Nyasoso: Cable 3335 6/1996.

MARATTIACEAE

Marattia fraxinea J.Sm. var. *fraxinea*
870–1450m.

Kodmin: Cheek 8939 1/1998; Plot B215 1/1998; **Kupe
Village:** Cable 2655 5/1996; Etuge 2850 7/1996; Ryan 386
5/1996; **Ndum:** Lane 505 2/1995; **Nyasoso:** Cable 3206
6/1996; Cheek 6021 1/1995; Lane 120 6/1994; Zapfack 603
6/1996.

OLEANDRACEAE

Arthropteris cameroonensis Alston
Ngol: Nicklès 25 1946–48.
Note: specimen cited in Fl. Cameroon 3: 116 (1964).

Arthropteris monocarpa (Cordem.) C.Chr.
850–1960m.
Kodmin: Plot B288 1/1998; **Kupe Village:** Etuge 1942
5/1996; 2634 7/1996; Kenfack 256 7/1996; Zapfack 918
7/1996; **Nyasoso:** Cable 2927 6/1996; 3415 6/1996; Lane 84
9/1992; Thorold 15 date unknown; Wheatley 439 7/1992;
Zapfack 721 6/1996.

Arthropteris orientalis (J.F.Gmel.) Posth.
900–1200m.
Kupe Village: Etuge 2636 7/1996; Zapfack 957 7/1996;
Nyasoso: Cable 21 7/1992; 3300 6/1996; 3438 7/1996;
Gosline 145 11/1998; Lane 11 7/1992; Thorold 19 date
unknown.

Arthropteris palisoti (Desv.) Alston
850–1000m.
Kupe Village: Etuge 1951 5/1996; **Mwanenguba:** de
Gironcourt 432 1911–12; **Ngomboaku:** Cheek 10306
12/1999; **Nyasoso:** Cable 3462 7/1996.

Nephrolepis biserrata (Sw.) Schott
660–850m.
Kupe Village: Cheek 8325 5/1996; Etuge 2748 7/1996;
Nyandong: Fay 4751 3/2003; **Nyasoso:** Etuge 2561 7/1996.

Nephrolepis undulata (Afzel. ex Sw.) J.Sm. var.
undulata
750–2000m.
Kupe Village: Cable 2410 5/1996; 3846 7/1996; Etuge 1395
11/1995; 2619 7/1996; 2816 7/1996; Zapfack 914 7/1996;
956 7/1996; **Mwanenguba:** Cheek 9417 10/1998; de
Gironcourt 489 1911–12; **Nyasoso:** Balding 51 8/1993;
Cable 25 7/1992; 2792 6/1996; 3229 6/1996; Etuge 1315
10/1995; 2395 6/1996; 2527 7/1996; Lane 5 7/1992; Sidwell
379 10/1995; Zapfack 625 6/1996.

Oleandra annetii Tardieu
Fl. Cameroun 3: 107 (1964).
1000–1400m.
Kupe Village: Cable 2740 5/1996; **Nyasoso:** Cable 88
8/1992.

Oleandra distenta Kunze var. *distenta*
850–1340m.
Kupe Village: Cable 2670 5/1996; 3753 7/1996; **Nyasoso:**
Cable 3581 7/1996; 87 8/1992; Etuge 2475 6/1996; Lane 18
7/1992; 66 8/1992; Wheatley 422 7/1992; Zapfack 663
6/1996.

OSMUNDACEAE

Osmunda regalis L. var. ***regalis***
1975m.
Mwanenguba: Cheek 7247 2/1995.

POLYPODIACEAE

Anapeltis lycopodioides (L.) J.Sm. var.
lycopodioides
Cat. Cult. Ferns 6 (1857).
2000m.
Mwanenguba: Pollard 159 10/1998.

Anapeltis lycopodioides (L.) J.Sm. var.
owariensis (Desv.) Benl
Acta Botanica Barcinonensia 33: 18 (1982).
Syn.: *Microgramma owariensis* (Desv.) Alston
Syn.: *Anapeltis owariensis* (Desv.) J.Sm.
850–1100m.
Kupe Village: Etuge 1944 5/1996; **Nyasoso:** Cable 59
8/1992; Lane 29 7/1992.

Belvisia spicata (L.f.) Mirb.
1275m.
Nyasoso: Thorold 18 date unknown.
Note: specimen cited in Fl. Cameroon 3: 342 (1964).

Drynaria laurentii (Christ ex De Wild. &
Durand) Hieron.
1000–1200m.
Kupe Village: Etuge 1963 5/1996; **Nyasoso:** Etuge 2105
6/1996; Lane 48 8/1992.

Drynaria volkensii Hieron.
1300m.
Kupe Village: Cheek 8382 5/1996.

Lepisorus excavatus (Bory ex Willd.) Ching
Zink M., Systematics of *Lepisorus*: 37 (1993).
Syn.: *Pleopeltis excavata* (Bory ex Willd.) Sledge
870–1960m.
Kodmin: Pollard 203 11/1998; **Kupe Village:** Cable 2403
5/1996; Ryan 285 5/1996; **Nyasoso:** Cable 2790 6/1996;
2928 6/1996; Sidwell 375 10/1995; Zapfack 655 6/1996; 656
6/1996; 720 6/1996.

Loxogramme abyssinica (Baker) M.G.Price
Amer. Fern. J. 74(2): 61 (1984).
Syn.: *Loxogramme lanceolata* (Sw.) C.Presl
800–2000m.
Kupe Village: Cable 2406 5/1996; 2407 5/1996; 2624
5/1996; 3660 7/1996; 3666 7/1996; Ryan 411 5/1996;
Mwanenguba: Pollard 160 10/1998; **Nyasoso:** Cable 2791
6/1996; 2929 6/1996; 3279 6/1996; 3579 7/1996; 94 9/1992;
Cheek 7509 10/1995; Etuge 2194 6/1996; 2508 7/1996; Lane
13 7/1992; 25 7/1992; 9 7/1992; Sunderland 1517 7/1992;
Wheatley 387 7/1992.

Loxogramme latifolia Bonap.
660–1450m.
Kodmin: Cheek 9593 11/1998; **Kupe Village:** Cable 2404
5/1996; 2405 5/1996; 3665 7/1996; Cheek 8338 5/1996;
Etuge 2697 7/1996; Kenfack 322 7/1996; Zapfack 951

7/1996; **Nyasoso:** Cable 24 7/1992; 3221 6/1996; 3580
7/1996; Etuge 2576 7/1996; Gosline 144 11/1998; Zapfack
621 6/1996.

Microsorum punctatum (L.) Copel.
200–1450m.
Kodmin: Cheek 8935 1/1998; Plot B219 1/1998; B271
1/1998; **Kupe Village:** Cable 2411 5/1996; Etuge 1705
2/1996; **Mungo River F.R.:** Cheek 10183 11/1999;
Nyandong: Fay 4740 3/2003; **Nyasoso:** Cable 2894 6/1996;
3278 6/1996; 3451 7/1996; Etuge 2577 7/1996; Lane 19
7/1992; 90 6/1994.

Microsorum scolopendria (Burm.f.) Copel.
Univ. Calif. Publ. Bot. 16: 112 (1929).
Syn.: *Phymatodes scolopendria* (Burm.f.) Ching
Syn.: *Phymatosorus scolopendria* (Burm.f) Pic.Serm.
660–1100m.
Kupe Village: Cable 2402 5/1996; 831 1/1995; Cheek 8336
5/1996; Etuge 1450 11/1995; **Nyasoso:** Cable 100 9/1992;
2800 6/1996; Etuge 2433 6/1996; Lane 21 7/1992; Sidwell
380 10/1995; Zapfack 623 6/1996.

Platycerium stemaria (P.Beauv.) Desv.
1000m.
Kupe Village: Etuge 1913 5/1996; **Mungo River F.R.:**
Cheek 10216 11/1999.

Pleopeltis macrocarpa (Bory ex Willd.) Kaulf.
var. ***macrocarpa***
Fl. Zamb. Pteridophyta: 152 (1970).
Syn.: *Pleopeltis lanceolata* (L.) Kaulf.
1050–1700m.
Kodmin: Pollard 203A 11/1998; **Nyasoso:** Lane 30 7/1992;
Sunderland 1543 7/1992.

Pyrrosia schimperiana (Mett.) Alston var.
schimperiana
Fl. Zamb. Pteridophyta: 147 (1970).
Syn.: *Pyrrosia mechowii* (Hieron.) Alston
350–1150m.
Kupe Village: Etuge 2040 5/1996; Zapfack 930 7/1996;
Nyasoso: Cable 3435 7/1996; Etuge 2457 6/1996; Lane 67
8/1992.

PTERIDACEAE

Pteris atrovirens Willd. fa. ***atrovirens***
1400m.
Kodmin: Plot B255 1/1998.

Pteris catoptera Kunze
F.T.E.A. Pteridaceae: 24 (2002).
650–2000m.
Kupe Village: Ryan 383 5/1996; **Ngusi:** Etuge 1535 1/1996;
Nyasoso: Cable 3385 6/1996.

Pteris hamulosa (Christ) Christ
Fl. Zamb. Pteridophyta: 120 (1970).
700–800m.
Kupe Village: Etuge 2743 7/1996; **Nyasoso:** Etuge 2506
7/1996.

Pteris intricata C.H.Wright
700m.
Nyale: Etuge 4185 2/1998.

Pteris linearis Poir.
700m.
Kupe Village: <u>Cable 832</u> 1/1995; **Mwanenguba:** <u>de Gironcourt 488</u> 1911–12.

Pteris manniana Mett. ex Kuhn
Acta Botanica Barcinonensia 38: 6 (1988).
Syn.: *Pteris camerooniana* Kühn
800–1200m.
Kupe Village: <u>Etuge 1404</u> 11/1995; <u>1991</u> 5/1996; **Nyasoso:** <u>Balding 68</u> 8/1993; <u>Cable 3314</u> 6/1996; <u>658</u> 12/1993; <u>Cheek 5648</u> 12/1993; <u>7475</u> 10/1995.

Pteris preussii Hieron.
Syn.: *Pteris prolifera* Hieron.
1450m.
Kodmin: <u>Cheek 9037</u> 1/1998.

Pteris togoensis Hieron.
Mwanenguba: <u>de Gironcourt 486</u> 1911–12.
Note: specimen cited in Fl. Cameroun 3: 162 (1964).

Pteris cf. togoensis Hieron.
1100m.
Nyasoso: <u>Cable 666</u> 12/1993.

THELYPTERIDACEAE

Amauropelta bergiana (Schltdl.) Holttum
J. S. Afr. Bot. 40: 133 (1974).
Syn.: *Thelypteris bergiana* (Schltr.) Ching
Mwanenguba: <u>Nicklès 28</u> 1946–48.
Note: specimen cited in Fl. Cameroun 3: 242 (1964).

Christella dentata (Forssk.) Brownsey & Jermy
Acta Botanica Barcinonensia 38: 53 (1988).
Syn.: *Cyclosorus dentatus* (Forssk.) Ching
Syn.: *Thelypteris dentata* (Forssk.) E.P.St.John
700m.
Kupe Village: <u>Etuge 2746</u> 7/1996.

Cyclosorus striatus (Schumach.) Ching
Syn.: *Thelypteris striata* (Schumach.) Schelpe
1200–1300m.
Edib: <u>Etuge 4486</u> 11/1998; <u>Onana 590</u> 2/1998; **Kupe Village:** <u>Kenfack 331</u> 7/1996.

Menisorus pauciflorus Alston
750–1350m.
Baseng: <u>Cheek 10407</u> 12/1999; **Kodmin:** <u>Etuge 4064</u> 1/1998; **Nyale:** <u>Cheek 9704</u> 11/1998.

Pneumatopteris afra (Christ) Holttum
Bull. Jard. Bot. Nat. Belg. 53: 283 (1983).
Syn.: *Cyclosorus afer* (Christ) Ching
200–1090m.
Kupe Village: <u>Cable 3740</u> 7/1996; <u>754</u> 1/1995; <u>Etuge 2742</u> 7/1996; <u>Lane 302</u> 1/1995; **Mwambong:** <u>Cheek 9358</u> 10/1998; **Mungo River F.R.:** <u>Cheek 10172</u> 11/1999; **Nyasoso:** <u>Balding 80</u> 8/1993; <u>Cable 1230</u> 2/1995; <u>Cheek 7912B</u> 11/1995; <u>Lane 42</u> 8/1992.

Pneumatopteris blastophora (Alston) Holttum
J. S. Afr. Bot. 40: 156 (1974).
900–1400m.
Kodmin: <u>Plot B235</u> 1/1998; **Nyasoso:** <u>Etuge 2411</u> 6/1996.

Pseudocyclosorus pulcher (Bory ex Willd.) Holttum
J. S. Afr. Bot. 40: 138 (1974).
Syn.: *Thelypteris zambesiaca* (Baker) Tardieu
Mwanenguba: <u>Nicklès 22</u> 1946–48.
Note: specimen cited in Fl. Cameroun 3: 243 (1964).

Thelypteris confluens C.V.Morton
Contrib. U.S. Nation. Herb. 38: 71 (1967).
2000m.
Mwanenguba: <u>Pollard 164</u> 10/1998.

VITTARIACEAE

Antrophyum annetii (Jeanp.) Tard.
Fl. Cameroun 3: 123 (1964).
810m.
Ngomboaku: <u>Ghogue 462</u> 12/1999.

Antrophyum mannianum Hook.
870–1650m.
Enyandong: <u>Cheek 10953</u> 10/2001; <u>Onana 1974</u> 11/2001; **Kupe Village:** <u>Cable 2463</u> 5/1996; **Mwambong:** <u>Cheek 9375</u> 10/1998; <u>Etuge 4138</u> 2/1998; **Nyasoso:** <u>Cable 2752</u> 6/1996; <u>Etuge 2067</u> 6/1996; <u>Lane 51</u> 8/1992.

Vittaria guineensis Desv. var. *guineensis*
1500–1650m.
Nyasoso: <u>Cable 127</u> 9/1992; <u>Lane 52</u> 8/1992; <u>Wheatley 465</u> 7/1992.

Vittaria schaeferi Hieron.
Fl. Cameroun 3: 126 (1964).
Mwanenguba: <u>Schäfer 77</u> date unknown.

WOODSIACEAE

Athyrium ammifolium (Mett.) C.Chr.
1250–1300m.
Edib: <u>Etuge 4480</u> 11/1998; **Lake Edib:** <u>Cheek 9150</u> 2/1998.

Diplazium proliferum (Lam.) Kaulf.
Syn.: *Callipteris prolifera* (Lam.) Bory
800m.
Kupe Village: <u>Cable 2553</u> 5/1996; <u>Etuge 2824</u> 7/1996.

Diplazium velaminosum (Diels) Pic.Serm.
Webbia 27: 443 (1973).
Syn.: *Diplazium zanzibaricum sensu auct.*
1650m.
Nyasoso: <u>Lane 61</u> 8/1992.

Lunathyrium boryanum (Willd.) H.Ohba
Yokosuka City Mus. 11: 53 (1965).
Syn.: *Athyrium glabratum* (Mett.) Alston
Syn.: *Dryoathyrium boryanum* (Willd.) Ching
Syn.: *Deparia boryanum* (Willd.) M.Kato
700–1600m.
Kupe Village: <u>Etuge 2723</u> 7/1996; **Nyasoso:** <u>Etuge 1816</u> 3/1996; <u>2471</u> 6/1996.
Uses: FOOD – edible, slice fresh young leaves like any other vegetable, then fry (Etuge 1816); leaves eaten as a vegetable (Etuge 2723).

INDEX TO VASCULAR PLANT CHECKLIST

Note: Generic names are left-aligned, in 10pt.
Species names are indented, and in 9pt.
Accepted epithets are in roman, and their page numbers in bold.
Synonyms are in italics (except for synonyms at generic rank), and their page numbers in roman.
Infraspecific names are indented further.

486

491

496

498

502